Practical Handbook of Photovoltaics

Fundamentals and Applications

Second Edition

Edited by

AUGUSTIN McEVOY
TOM MARKVART
LUIS CASTAÑER

Amsterdam • Boston • Heidelberg • London
New York • Oxford • Paris • San Diego
San Francisco • Singapore • Sydney • Tokyo

Academic Press is an imprint of Elsevier

Academic Press is an imprint of Elsevier
225 Wyman Street, Waltham, MA 02451, USA
The Boulevard, Langford Lane, Kidlington, Oxford, OX5 1GB, UK

Copyright © 2012 Published by Elsevier Ltd. All rights reserved

No part of this publication may be reproduced or transmitted in any form or by any means, electronic or mechanical, including photocopying, recording, or any information storage and retrieval system, without permission in writing from the publisher. Details on how to seek permission, further information about the Publisher's permissions policies and our arrangements with organizations such as the Copyright Clearance Center and the Copyright Licensing Agency, can be found at our website: www.elsevier.com/permissions This book and the individual contributions contained in it are protected under copyright by the Publisher (other than as may be noted herein).

Notices

Knowledge and best practice in this field are constantly changing. As new research and experience broaden our understanding, changes in research methods, professional practices, or medical treatment may become necessary.

Practitioners and researchers must always rely on their own experience and knowledge in evaluating and using any information, methods, compounds, or experiments described herein. In using such information or methods they should be mindful of their own safety and the safety of others, including parties for whom they have a professional responsibility.

To the fullest extent of the law, neither the Publisher nor the authors, contributors, or editors, assume any liability for any injury and/or damage to persons or property as a matter of products liability, negligence or otherwise, or from any use or operation of any methods, products, instructions, or ideas contained in the material herein.

Library of Congress Cataloging-in-Publication Data
Practical handbook of photovoltaics : fundamentals and applications / edited by Augustin McEvoy, Tom Markvart, and Luis Castañer. – 2nd ed.
  p. cm.
Includes bibliographical references and index.
ISBN 978-0-12-385934-1 (hardback)
1. Photovoltaic cells–Handbooks, manuals, etc. 2. Photovoltaic power generation–Handbooks, manuals, etc. 3. Photovoltaic power systems–Handbooks, manuals, etc. I. McEvoy, A. J. (Augustin Joseph) II. Markvart, T. III. Castañer, Luis.
TK8322.P73 2011
621.3815'42–dc23

011027611

British Library Cataloguing-in-Publication Data
A catalogue record for this book is available from the British Library.
ISBN: 978-0-12-385934-1

For information on all Academic Press publications
visit our website at: www.elsevierdirect.com

Printed in the United States of America
11 12 13 14 15   8 7 6 5 4 3 2 1

Working together to grow
libraries in developing countries

www.elsevier.com | www.bookaid.org | www.sabre.org

ELSEVIER   BOOK AID International   Sabre Foundation

Contents

TK
8322
.P73
2012

Preface to the Second Edition   xv
Preface to the First Edition   xix
List of Contributors   xx

Introduction   1

Part IA: Solar Cells

IA-1. Principles of Solar Cell Operation   7
T. Markvart and L. Castañer
1. Introduction   7
2. Electrical Characteristics   10
3. Optical Properties   16
4. Typical Solar Cell Structures   19

IA-2. Semiconductor Materials and Modelling   33
T. Markvart and L. Castañer
1. Introduction   33
2. Semiconductor Band Structure   34
3. Carrier Statistics in Semiconductors   38
4. The Transport Equations   40
5. Carrier Mobility   42
6. Carrier Generation by Optical Absorption   44
7. Recombination   48
8. Radiation Damage   53
9. Heavy Doping Effects   55
10. Properties of Hydrogenated Amorphous Silicon   57
Acknowledgements   59

IA-3. Ideal Efficiencies   63
P.T. Landsberg and T. Markvart
1. Introduction   63
2. Thermodynamic Efficiencies   64
3. Efficiencies in Terms of Energies   65
4. Efficiencies Using the Shockley Solar Cell Equation   67
5. General Comments on Efficiencies   72

Part IB: Crystalline Silicon Solar Cells

IB-1. Crystalline Silicon: Manufacture and Properties   79
F. Ferrazza
1. Introduction   79
2. Characteristics of Silicon Wafers for Use in PV Manufacturing   80
3. Feedstock Silicon   86
4. Crystal-Preparation Methods   87
5. Shaping and Wafering   94

IB-2. High-Efficiency Silicon Solar Cell Concepts   99
M.A. Green
1. Introduction   100
2. High-Efficiency Laboratory Cells   100
3. Screen-Printed Cells   111
4. Laser-Processed Cells   116
5. HIT Cell   120
6. Rear-Contacted Cells   121
7. Conclusions   124
Acknowledgements   125

IB-3. Low-Cost Industrial Technologies for Crystalline Silicon Solar Cells   129
J. Szlufcik, S. Sivoththaman, J. Nijs, R.P. Mertens and R. Van Overstraeten
1. Introduction   130
2. Cell Processing   131
3. Industrial Solar Cell Technologies   145
4. Cost of Commercial Photovoltaic Modules   152

IB-4. Thin Silicon Solar Cells   161
M. Mauk, P. Sims, J. Rand and A. Barnett
1. Introduction, Background, and Scope of Review   162
2. Light Trapping in Thin Silicon Solar Cells   165
3. Voltage Enhancements in Thin Silicon Solar Cells   174
4. Silicon Deposition and Crystal Growth for Thin Solar Cells   179
5. Thin Silicon Solar Cells Based on Substrate Thinning   188
6. Summary of Device Results   190

Part IC: Thin Film Technologies

IC-1. Thin-Film Silicon Solar Cells — 209
A. Shah
1. Introduction — 210
2. Hydrogenated Amorphous Silicon (a-Si:H) Layers — 215
3. Hydrogenated Microcrystalline Silicon (μc-Si:H) Layers — 225
4. Functioning of Thin-Film Silicon Solar Cells with p–i–n and n–i–p Structures — 234
5. Tandem and Multijunction Solar Cells — 252
6. Module Production and Performance — 259
7. Conclusions — 273

IC-2. CdTe Thin-Film PV Modules — 283
D. Bonnet
1. Introduction — 284
2. Steps for Making Thin-Film CdTe Solar Cells — 285
3. Making of Integrated Modules — 303
4. Production of CdTe Thin-Film Modules — 306
5. The Product and Its Application — 316
6. The Future — 320

IC-3. Cu(In,Ga)Se$_2$ Thin-Film Solar Cells — 323
U. Ran and H.W. Schock
1. Introduction — 324
2. Material Properties — 325
3. Cell and Module Technology — 331
4. Device Physics — 346
5. Wide-Gap Chalcopyrites — 354
6. Conclusions — 361
Acknowledgements — 362

IC-4. Progress in Chalcopyrite Compound Semiconductor Research for Photovoltaic Applications and Transfer of Results into Actual Solar Cell Production — 373
A. Jäger-Waldau
1. Introduction — 374
2. Research Directions — 375
3. Industrialisation — 379
4. Conclusions and Outlook — 390

Part ID: Space and Concentrator Cells

ID-1. GaAs and High-Efficiency Space Cells — 399
V.M. Andreev

1. Historical Review of III–V Solar Cells — 399
2. Single-Junction III–V Space Solar Cells — 403
3. Multijunction Space Solar Cells — 407

Acknowledgements — 412

ID-2. High-Efficiency III–V Multijunction Solar Cells — 417
S.P. Philipps, F. Dimroth and A.W. Bett

1. Introduction — 417
2. Special Aspects III–V Multijunction Solar Cells — 419
3. III–V Solar Cell Concepts — 427
4. Conclusions — 440

Acknowledgements — 440

ID-3. High-Efficiency Back-Contact Silicon Solar Cells for One-Sun and Concentrator Applications — 449
P.J. Verlinden

1. Introduction — 449
2. Concentrator Applications of IBC Solar Cells — 450
3. Back-Contact Silicon Solar Cells — 453
4. Modelling of Back-Contact Solar Cells — 458
5. Perimeter and Edge Recombination — 462
6. Manufacturing Process for Back-Contact Solar Cells — 464
7. Stability of Back-Contact Cells — 464
8. Toward 30% Efficiency Silicon Cells — 467
9. How to Improve the Efficiency of Back-Contact Solar Cells — 468
10. Conclusions — 472

Acknowledgements — 473

Part IE: Dye-Sensitized and Organic Solar Cells

IE-1. Dye-Sensitized Photoelectrochemical Cells — 479
A. Hagfeldt, U.B. Cappel, G. Boschloo, L. Sun, L. Kloo, H. Pettersson and E. A. Gibson

1. Introduction — 480
2. Photoelectrochemical Cells — 483
3. Dye-Sensitized Solar Cells — 496
4. Future Outlook — 536

IE-2. Organic Solar Cells 543
C. Dyer-Smith and J. Nelson
1. Introduction 544
2. Organic Electronic Materials 544
3. Principles of Device Operation 548
4. Optimising Solar Cell Performance 552
5. Production Issues 559
6. Conclusions 563

Part IIA: Photovoltaic Systems

IIA-1. The Role of Solar-Radiation Climatology in the Design of Photovoltaic Systems 573
J. Page
1. Introduction 575
2. Key Features of the Radiation Climatology in Various Parts of the World 577
3. Quantitative Processing of Solar Radiation for Photovoltaic Design 605
4. The Stochastic Generation of Solar-Radiation Data 610
5. Computing the Solar Geometry 618
6. The Estimation of Hourly Global and Diffuse Horizontal Irradiation 627
7. The Estimation of the All Sky Irradiation on Inclined Planes from Hourly Time Series of Horizontal Irradiation 632
8. Conclusion 637
Acknowledgements 638
Appendix Solar Energy Data for Selected Sites 642

IIA-2. Energy Production by a PV Array 645
L. Castañer, S. Bermejo, T. Markvart and K. Fragaki
1. Annual Energy Production 645
2. Peak Solar Hours: Concept, Definition, and Illustration 646
3. Nominal Array Power 648
4. Temperature Dependence of Array Power Output 650
5. Module Orientation 651
6. Statistical Analysis of the Energy Production 653
7. Mismatch Losses and Blocking/Bypass Diodes 655
Acknowledgement 658

IIA-3. Energy Balance in Stand-Alone Systems 659
L. Castañer, S. Bermejo, T. Markvart and K. Fragaki
1. Introduction 659

2.	Load Description	663
3.	Seasonal Energy Balance	668

IIA-4. Review of System Design and Sizing Tools — 673
S. Silvestre

1.	Introduction	673
2.	Stand-Alone PV Systems Sizing	674
3.	Grid-Connected PV Systems	681
4.	PV System Design and Sizing Tools	684

Part IIB: Balance-of-system Components

IIB-1. System Electronics — 697
J.N. Ross

1.	Introduction	697
2.	DC to DC Power Conversion	698
3.	DC to AC Power Conversion (Inversion)	703
4.	Stand-Alone PV Systems	708
5.	PV Systems Connected to the Local Electricity Utility	714
6.	Available Products and Practical Considerations	715
7.	Electromagnetic Compatibility	718

IIB-2. Batteries in PV Systems — 721
D. Spiers

1.	Introduction	722
2.	What Is a Battery?	723
3.	Why Use a Battery in PV Systems?	724
4.	Battery Duty Cycle in PV Systems	726
5.	The Battery as a 'Black Box'	727
6.	The Battery as a Complex Electrochemical System	737
7.	Types of Battery Used in PV Systems	740
8.	Lead—Acid Batteries	741
9.	Nickel—Cadmium Batteries	754
10.	How Long Will the Battery Last in a PV System?	756
11.	Selecting the Best Battery for a PV Application	765
12.	Calculating Battery Size for a PV System	767
13.	Looking After the Battery Properly	770
14.	Summary and Conclusions	774
	Acknowledgements	775

Part IIC: Grid-Connected Systems

IIC-1. Grid-Connection of PV Generators: Technical and Regulatory Issues — 779
J. Thornycroft and T. Markvart
1. Introduction — 780
2. Principal Integration Issues — 783
3. Inverter Structure and Operating Principles — 784
4. Islanding — 785
5. Regulatory Issues — 789
Acknowledgements — 803

IIC-2. Installation Guidelines: Construction — 805
B. Cross
1. Roofs — 806
2. Facades — 811
3. Ground-Mounted Systems — 817

IIC-3. Installation Guidelines: Electrical — 819
M. Cotterell
1. Introduction — 820
2. Codes and Regulations — 820
3. DC Ratings (Array Voltage and Current Maxima) — 821
4. Device Ratings and Component Selection — 823
5. Array Fault Protection — 824
6. Earthing Arrangements — 830
7. Protection by Design — 832
8. Labelling — 834

Part IID: Space and Concentrator Systems

IID-1. Concentrator Systems — 837
G. Sala
1. Objectives of PV Concentration — 837
2. Physical Principles of PV Concentration — 839
3. Description of a Typical Concentrator: Components and Operation — 843
4. Classification of Concentrator Systems — 845
5. Tracking-Control Strategies — 851
6. Applications of C Systems — 854

7. Rating and Specification of PV Systems	854
8. Energy Produced by a C System	859
9. The Future of Concentrators	860

IID-2. Operation of Solar Cells in a Space Environment — 863
S. Bailey and R. Rafaelle

1. Introduction — 863
2. Space Missions and their Environments — 865
3. Space Solar Cells — 872
4. Small Power Systems — 875
5. Large Power Systems — 877

IID-3. Calibration, Testing and Monitoring of Space Solar Cells — 881
E. Fernandez Lisbona

1. Introduction — 882
2. Calibration of Solar Cells — 883
3. Testing of Space Solar Cells and Arrays — 886
4. Monitoring of Space Solar Cells and Arrays — 895
   Acknowledgements — 908

Part IIE: Case Studies

IIE-1. Architectural Integration of Solar Cells — 917
R. Serra i Florensa and R. Leal Cueva

1. Introduction — 918
2. Architectural Possibilities for PV Technology — 919
3. Building-Integrated Photovoltaics (BIPVs) — 923
4. Aesthetics in PV Technology — 926
5. Built Examples — 934

IIE-2. Solar Parks and Solar Farms — 943
P.R. Wolfe

1. What Is a Solar Park? — 944
2. Design Issues for Solar Parks — 946
3. Solar Park Project Development Issues — 953
4. Regulatory Issues for Solar Parks — 959
5. The End Game — 960
   Acknowledgements — 962

IIE-3. Performance, Reliability, and User Experience 963
U. Jahn

1. Operational Performance Results 964
2. Trends in Long-Term Performance and Reliability 972
3. User Experience 978

Acknowledgements 982

Appendix. Specifications of Performance Database of IEA PVPS 983

IIE-4. Solar-Powered Products 987
P.R. Wolfe

1. The Genesis of Solar-Powered Products 988
2. Stand-Alone Consumer Products 989
3. Solar Products for Grid Connection 1001
4. Nonconsumer Products 1003
5. Designing PV for Products 1004
6. Solar Products of the Future 1006

Acknowledgements 1007

Part III: Testing, Monitoring, and Calibration

III-1. Characterization and Diagnosis of Silicon Wafers, Ingots, and Solar Cells 1011
A. Cuevas, D. Macdonald, R.A. Sinton

1. Introduction 1012
2. Factors Affecting Carrier Recombination 1012
3. Measurement of the Minority-Carrier Lifetime 1015
4. Relationship Between Device Voltage and Carrier Lifetime 1030
5. Applications to Process Monitoring and Control of Silicon Solar Cells 1031
6. Conclusions 1040

Acknowledgements 1040

III-2. Standards, Calibration, and Testing of PV Modules and Solar Cells 1045
C.R. Osterwald

1. PV Performance Measurements 1046
2. Diagnostic Measurements 1054
3. Commercial Equipment 1056
4. Module Reliability and Qualification Testing 1057
5. Module Degradation Case Study 1061

Acknowledgments 1063

III-3. PV System Monitoring — 1071
B. Cross
1. Introduction — 1071
2. Equipment — 1073
3. Calibration and Recalibration — 1075
4. Data Storage and Transmission — 1076
5. Monitoring Regimes — 1076

Part IV: Environment and Health

IV-1. Overview of Potential Hazards — 1083
V.M. Fthenakis
1. Introduction — 1083
2. Overview of Hazards in PV Manufacture — 1084
3. Crystalline Silicon (x-Si) Solar Cells — 1084
4. Amorphous Silicon (a-Si) Solar Cells — 1088
5. Cadmium Telluride (CdTe) Solar Cells — 1089
6. Copper Indium Diselenide (CIS) Solar Cells — 1090
7. Gallium Arsenide (GaAs) High-Efficiency Solar Cells — 1092
8. Operation of PV Modules — 1093
9. Photovoltaic Module Decommissioning — 1094
10. Conclusion — 1095

IV-2. Energy Payback Time and CO_2 Emissions of PV Systems — 1097
E. Alsema
1. Introduction — 1097
2. Energy Analysis Methodology — 1099
3. Energy Requirements of PV Systems — 1100
4. Energy Balance of PV Systems — 1105
5. Outlook for Future PV Systems — 1107
6. CO_2 Emissions — 1112
7. Conclusions — 1114

APPENDICES
Appendix A Constants, Physical Quantities and Conversion Factors — 1121
Appendix B List of Principal Symbols — 1123
Appendix C Abbreviations and Acronyms — 1131
Appendix D The Photovoltaic Market — 1137
Appendix E The Photovoltaic Industrty — 1153

Appendix F	Useful Web Sites and Journals	1173
Appendix G	International Standards With Relevance to Photovoltaics	1177
Appendix H	Books About Solar Cells, Photovoltaic Systems, and Applications	1185
Index		**1189**

PREFACE TO THE SECOND EDITION

"And there was light." Seeing our Earth for the first time as an emerald sphere against the cold of space, the Apollo 8 crew in December 1968 associated a founding text of human culture with its highest technical expression as they orbited the moon. It was a time of dynamic scientific and engineering confidence, almost hubris: the prospect of nuclear power "too cheap to meter," supersonic passenger aircraft, semiconductor devices, telecommunications, new materials, and of course the emergence and earliest applications of silicon solar cells. The image of our blue Earth, the planet of light and water, became an icon of technical achievement, but even more, communicated a sense of its fragility, its isolation in an otherwise barren universe. Then there was an overwhelming confidence that with competence, application and resources anything is possible for our burgeoning technology. Now, in contrast, and perhaps to some extent due to this new vision of our planet, there is a pessimism, a sense that nature is beyond our control, and maybe even contemptuous of our efforts. As at Fukushima with an earthquake and tsunami, or in New Orleans flooded by a hurricane, it can overwhelm our best engineering achievements. In the context of the energy crisis, and with an overall "systems" vision of our world, it appears that an unavoidable consequence of our technology is a perturbation of the equilibria which make our blue planet so comfortably habitable. The image of earth rising over the barren moon has returned, this time as an icon of the environmental movement and a challenge to render ourselves, our technology, economy and society permanently geocompatible. Exploitation of light, the solar energy resource, through photovoltaics, photothermal systems, perhaps genetically engineered photoorganisms, and by every aspect of human ingenuity which can be brought to bear, will certainly provide a key contribution to that objective. It is a measure of the challenge that in 2007 over 80% of the primary commercial energy demand was still supplied by combustible fossil fuels. Aside from hydroelectric and nuclear sources, other energy technologies which are genuinely geocompatible, particularly by eliminating emissions capable of inducing a climatic disequilibrium, make only a marginal contribution at this time. To change that situation decisively will require a level of commitment comparable with the Apollo programme. It is a daunting task.

At the same time, the potential is there, and solar photovoltaics are just one option to provide the sustainable energy systems required. Solar irradiation of earth represents a continuous power input of some 173 petawatts, or 173×10^{15} W; 1 PW-hr represents more than the global *annual* energy requirement. However, it is a dilute and intermittent source at any given location. Large-area, low-cost systems must be made available for the capture and conversion of incident solar energy; ultimately also, when the solar electrical resource becomes a significant fraction of total installed generating capacity, the issue of storage will become critical.

A revision of the *Practical Handbook of Photovoltaics* is an opportunity to contribute to the understanding and promotion of photovoltaic technology. As Editor of this edition, I would first thank Tom Markvart and Luis Castañer for their pioneering work in editing the first edition. Equally, thanks are due to the authors, especially those responsible for new or extensively revised chapters. It has been a real pleasure to interact with these specialists, some veterans who contributed to the bases of photovoltaic science and engineering, some prominent established figures in the photovoltaic world, many current practitioners of the art, contributing by their research efforts to the future. I would mention two whose work appears in this Handbook, who are sadly no longer with us: Peter Landsberg, and Roger Van Overstraaeten, both remembered with respect and whose work is still relevant. I should thank also the team at Elsevier without whom this work could never have seen the light of day, particularly Jill Leonard, Tiffany Gasbarrini, and Meredith Benson.

The present commercial status of photovoltaics is not presented as a chapter since the situation is evolving so rapidly, given that the market and the total installed PV capacity continue to grow exponentially, by some 40% per year since 2000. It does, however, appear together with a survey of the industry, as appendices. Here the work of Dr. Arnulf Jäger-Waldau of the European Commission Joint Research Centre, Ispra, Italy is particularly acknowledged as he made available the very latest statistics. PV is now the fastest-growing energy technology, with installed capacity increasing from 100 MW in 1992 to 14 GW in 2008 (International Energy Agency figures, Oct. 2010, <*http://www.iea.org/papers/2010/pv_roadmap.pdf*>). To maintain and if possible accelerate this process, to keep our earth as it was when seen from space 40 years ago, is a scientific, engineering, economic, and social engagement of historic proportions and significance, an adventure in which photovoltaic technology is called upon to take a significant role.

A.J. McEvoy,
Lausanne, Switzerland
2011

PREFACE TO THE FIRST EDITION

Photovoltaics is about to celebrate 50 years of its modern era. During this time, the industry has grown from small satellite power supplies to utility-scale systems that are now routinely installed in many countries of the world. Solar cells capable of producing power in excess of 500 MW were manufactured in 2002, providing electricity to a variety of applications ranging from small consumer products, power systems for isolated dwellings and remote industrial equipment to building-integrated solar arrays and megawatt-size power stations.

This *Practical Handbook of Photovoltaics* addresses the need for a book that summarises the current status of know-how in this field. It represents a detailed source of information across the breadth of solar photovoltaics and is contributed to by top-level specialists from all over the world. Over 1,000 references, bibliographies and Web sites guide the reader to further details, be it specific information for industrial production and research or a broad overview for policy makers. Thirty-seven chapters in the handbook cover topics from fundamentals of solar cell operation to industrial production processes, from molecular photovoltaics to system modelling, from a detailed overview of solar radiation to guidelines for installers and power engineers, and from architectural integration of solar cells to energy payback, CO_2 emissions, and photovoltaic markets. Appendices include extensive bibliography and lists of standards, journals, and other sources of information which can be found in a printed or electronic form.

The main credit for this handbook must go to the chapter authors who have produced a unique compilation of the contemporary knowledge in photovoltaic science and technology.

Our thanks go to our families for their patience and support without which this book would have never seen the light of day.

Luis Castañer
Barcelona

Tom Markvart
Southampton

LIST OF CONTRIBUTORS

Erik Alsema
Department of Science, Technology and Society, Copernicus Institute for Sustainable Development and Innovation, Utrecht University, Padualaan 14, NL-3584 CH Utrecht, The Netherlands
email: e.a.alsema@uu.nl

Vyacheslav M. Andreev
Ioffe Physico Technical Institute, 26 Polytekhnichcskaya str., St. Petersburg 194021, Russia
email: vmandreev@mail.ioffe.ru

Sheila Bailey
Photovoltaic and Space Environments Branch 5410, NASA Glenn Research Center, MS 301-3, 21000 Brookpark Rd., Cleveland, OH 44135, USA
email: Sheila.Bailey@grc.nasa.gov

Allen Barnett
Electrical and Computer Engineering, University of Delaware Newark, DE 19716, USA
email: abarnett@udel.edu

Sandra Bermejo
GDS, Modulo C4 Campus Nord, Universidad Politecnica de Catalunya, Calle Jordi Girona 1, 08034 Barcelona, Spain
email: bermejo@eel.upc.es

Andreas W. Bett
Fraunhofer Institute for Solar Energy Systems ISE, Freiburg, Germany
Email: andreas.bett@ise.fraunhofer.de

Dieter Bonnet
Breslauer Ring 9a, D-61381 Friedrichsdorf, Germany / CTF Solar GmbH
email: DieterBonnet@aol.com

Gerrit Boschloo
Uppsala University, Box 259, SE-75105 Uppsala, Sweden

Ute B. Cappel
Uppsala University, Box 259, SE-75105 Uppsala, Sweden

Luis Castañer
GDS, Modulo C4 Campus Nord, Universidad Politecnica de Catalunya, Calle Jordi Girona 1,08034 Barcelona, Spain
email: castaner @eel.upc.es

Martin Cotterell
SunDog Energy Ltd, Matterdale End, Penrith, CA11 0LF, UK
email: martin@sundog-energy.co.uk

Bruce Cross
PV Systems / EETS Ltd, Unit 2, Glan-y-Llyn Industrial Estate, Taffs Well, CF15 7JD, UK
email: bcross@eets.co.uk

Andres Cuevas
Department of Engineering, The Australian National University, Canberra, Australia 0200
email: Andres.Cuevas@anu.edu.au

Frank Dimroth
Fraunhofer Institute for Solar Energy Systems ISE, Freiburg, Germany
email: Frank.dimroth@ise.fraunhofer.de

Clare Dyer-Smith
Imperial College London, Blackett Laboratory, Prince Consort Road, London SW7 2AZ, UK
email: c.dyer-smith07@imperial.ac.uk

Emilio Fernandez Lisbona
ESA-Estec, Noordwijk, The Netherlands

Francesca Ferrazza
Eurosolare S.p.A. Via Augusto D'Andrea 6, 00048 Nettuno, Italy
email: francesca.ferrazza@eurosolare.agip.it

Katerina Fragaki
Durham University, Mountjoy Research Centre, Block 2 Stockton Road, Durham DH1 3UP, UK
email: katerina.fragaki@durham.ac.uk

Vasilis M. Fthenakis
National PV EHS Assistance Center, Department of Environmental Sciences, Brookhaven National Laboratory, Upton, New York, USA

Elizabeth A. Gibson
University of Nottingham, University Park, Nottingham NG7 2RD, England

Martin A. Green
Centre of Excellence for Advanced Silicon Photovoltaics and Photonics, University of New South Wales, Sydney NSW 2052, Australia
email: m.green@unsw.edu.au

Anders Hagfeldt
Uppsala University, Dept. of Physical and Analytical Chemistry, P.O. Box 259 SE-751 05 Uppsala, Sweden
email: anders.hagfeldt@fki.uu.se

Arnulf Jäger-Waldau
European Commission, DGJRC, Institute for Energy /Renewable Energies Unit, JRC Ispra, 21027, Italy
email: arnulf-jaeger-waldau@ec.europa.eu

Ulrike Jahn
TÜV Rheinland Immissionsschutz und Energiesysteme GmbH, Am Grauen Stein, D-51105 Köln, Germany
email: ulrike.jahn@de.tuv.com

Lars Kloo
KTH—Royal Institute of Technology, Teknikringen 30, SE-10044 Stockholm, Sweden

Peter T. Landsberg
Faculty of Mathematical Studies, University of Southampton, Southampton SO17 1BJ, UK
✠ 14 Feb. 2010

Rogelio Leal Cueva
Av. Gomez Morín 402, Colonia Villa de Aragón CP66273, San Pedro Garza García, Nuevo León, México
email: info@soleco.com.mx

Daniel Macdonald
School of Engineering, The Australian National University, Canberra, ACT 0200, Australia
email : daniel.macdonald@anu.edu.au

Tom Markvart
School of Engineering Sciences, University of Southampton, Southampton SO17 1BJ, UK
email: t.markvart@soton.ac.uk

Michael G. Mauk
Drexel University, Goodwin College of Prof Studies, 3001 Market St, Philadelphia, PA 19104, USA
email: michael.g.mauk@drexel.edu

Augustin McEvoy
Dyesol Ltd, Queanbayan, Australia
email: mcevoy@physics.org

Robert P. Mertens
Interuniversity Microelectronic Center (IMEC), Kapeldreef 75, 3001 Leuven, Belgium
email: robert.mertens@imec.be

Jenny Nelson
Imperial College London, Blackett Laboratory, Prince Consort Road, London SW7 2AZ, UK
Email: Jenny.nelson@imperial.ac.uk

Johan F. Nijs
Photovoltech, rue de l'Industrie 52/Nijverheidsstraat, 1040 Brussel, Belgium
email: johan.nijs@photovoltech.be

Carl R. Osterwald
National Renewable Energy Laboratory, Golden, CO 80401, USA
email : carl.osterwald@nrel.gov

List of Contributors

Roger van Overstraeten (honorary, by request)
Interuniversity Microelectronic Center (IMEC), Kapeldreef 75, Leuven B-3001, Belgium
✠ 25 april 1999

John Page
5 Brincliffe Gardens, Sheffield, S11 9BG, UK / Emeritus, Univ. Sheffield
email: johnpage@univshef.freeserve.co.uk

Henrik Pettersson
Swerea IVF AB, Box 104, SE-431 22 Mölndal, Sweden

Simon P. Philipps
Fraunhofer Institute for Solar Energy Systems ISE, Freiburg, Germany
email: simon.philips@ise.fraunhofer.de

Ryne P. Raffaelle
National Center for Photovoltaics, National Renewable Energy Laboratory (NREL),
1617 Cole Blvd., Golden, CO 80401-3305, USA
email: ryne_raffaelle@nrel.gov

James Rand
AstroPower, Inc., now GE Power, 300 Executive Drive, Newark, DE 19702-3316, USA
email: jimrand@AstroPower.com

U Rau
IPE, Universität Stuttgart, Germany

J. Neil Ross
Department of Electronics and Computer Science, University of Southampton,
Southampton SO17 1BJ, UK
email: jnr@ecs.soton.ac.uk

Gabriel Sala
Instituto de Energia Solar, Universidad Politecnica de Madrid, Ciudad Universitaria,
28040 Madrid, Spain
email: sala@ies-def.upm.es

H.W. Schock
IPE, Universität Stuttgart, Germany

Rafael Serra I Florensa
Universitat Politècnica de Catalunya, Barcelona, Spain

Arvind Shah
Chemin des Pommiers 37, 2022 Bevaix, Switzerland /Emeritus, IMT, Université de
Neuchâtel, Switzerland
email: arvind.shah@unine.ch

Santiago Silvestre
GDS, Modulo C4 Campus Nord, Universidad Politecnica de Catalunya, Calle Jordi
Girona 1, 08034 Barcelona, Spain
email: santi@ecl.upc.es

Paul Sims
AstroPower, Inc., now GE Power, 300 Executive Drive, Newark, DE 19702, USA

email: pesims@AstroPower.com

Ronald A. Sinton
Sinton Instruments, 4720 Walnut St., Boulder, CO 80301, USA
email: ron @sintonconsulting.com

Siva Sivoththaman
University of Waterloo, Faculty of Electrical and Computer Engineering, 200 University Ave, Waterloo, Ontario, Canada N2L 3G1
email: sivoththman@uwaterloo.ca

David Spiers
Naps Systems, PO Box 83, Abingdon, Oxfordshire OX14 2TB, UK
email: david.spiers@napssystems.com

Licheng Sun
KTH—Royal Institute of Technology, Teknikringen 30, SE-10044 Stockholm, Sweden/ State Key Laboratory of Fine Chemicals, Dalian University of Technology (DUT), Dalian 116012, China

Jozef Szlufcik
Photovoltech, rue de l'Industrie 52 Nijverrheidsstraat, 1040 Brussel, Belgium
email: Jozef.Szlufcik@photovoltech.com

Jim Thornycroft
Consultant, Burderop Park, Swindon SN4 0QD, UK
email: jim@jamesthornycroft.co.uk

Pierre Verlinden
Amrock Pty Ltd., Adelaide, Australia
email: pjverlinden@ieee.org

Philip Wolfe
WolfeWare, Rose Cottage, Dunsomer Hill, North Moreton, OX11 9AR, UK
email: philip@wolfeware.com

Introduction

This *Practical Handbook of Photovoltaics* aims to give a detailed overview of all aspects of solar photovoltaics in a way that can be easily accessed by the expert and non-specialist alike. It reflects the current status of this modern power-generating technology which, despite its mature status, continues to explore new directions to improve performance and reduce costs. A focus on practical aspects, however, docs not imply the neglect of fundamental research and theory which are both covered in depth, as are the environmental impacts, commercial aspects, and policy views.

Solar cell manufacturing technologies are discussed in Part II. An overview of the principal issues, including the physics of solar cell operation, materials and modelling, and the fundamental theoretical framework form an introduction to the device aspects of photovoltaics given in Part IIa. Part IIb gives a detailed account of crystalline silicon technology, from the manufacture and properties of silicon (Chapter IIb-1) to industrial and high-efficiency solar cells based on wafer silicon (Chapters IIb-2 and -5), and thin silicon cells (Chapter IIb-3). Part IIc examines all aspects of thin-film solar cells, including amorphous silicon, cadmium telluride, and copper—indium diselenide and its derivatives (Chapters IIc-1, -3 and -4). Chapter IIc-2 describes a novel amorphous/microcrystalline silicon cell which is rapidly gaining ground. Part IId focuses on high-efficiency cells for space and concentrator use. Part IIe deals with devices based on molecular structures.

The testing and calibration of both terrestrial and space solar cells is discussed in Part IV. An overview of material characterisation methods for silicon wafers and devices will be found in Chapter IIb-4.

Solar radiation has been called the fuel of photovoltaics, and its characteristics form the basis of system design, from array construction to the reliability of electricity supply by stand-alone photovoltaic systems. The understanding of solar radiation forms arguably the most ancient part of physical science but it is only recently that the statistical nature of solar energy has been understood in some detail. A number of sophisticated computer models are now available and are described in detail in Part I, which also summarises the relevant aspects of solar radiation as an energy source and examines the principal attributes and limitations of the available computer tools.

System engineering is discussed in Part III. Part IIIa provides an introduction to this field by giving an overview of the generic aspects of photovoltaic system design, including a review of the relevant modelling and simulation tools. The balance of system components is discussed in Part IIIb, providing an in-depth analysis of battery operation in photovoltaic systems (Chapter IIIb-2) and an overview of the electronic control and power conditioning equipment (Chapter IIIb-1).

The rapidly growing area of grid-connected systems is considered from several viewpoints in Part IIIc which examines the technical and regulatory issues of the grid connection, and the building and electrical installation of domestic systems. The International Energy Agency data about user experience and performance indicators are analysed in Chapter IIIe-2; the allied subject of system monitoring is reviewed in Chapter IV-2.

Solar cells as a source of power for consumer products are reviewed in Chapter IIIe-3. Arguably the most exciting aspect of photovoltaics—the visual impact of the new solar architecture—is discussed in Chapter IIIe-1, which also gives numerous breathtaking examples of the new trends.

An in-depth analysis of the world photovoltaic markets is given in Part V alongside an overview of support mechanisms. Part V also reviews the potential hazards in solar cell manufacture, and examines broader environmental issues, including CO_2 emissions and the energy payback times.

The space application of solar cells has always been considered a unique and a special area of photovoltaics. The device aspects, including radiation damage, are discussed in Chapters IId-1 and IIId-2, with a brief introduction to the material aspects in Chapter IIa-2. Chapter IIId-2 gives a thorough review of the operation of solar cells in the space environment together with its history and space mission requirements.

The use of high-efficiency cells is not confined to space but is finding an increasing application in concentrator systems which are examined Chapter IIId-1. The corresponding solar cells are discussed in Part IId.

The handbook contains a number of chapters that contain a strong research element. The ultimate efficiencies that can be reached by a solar cell are discussed in Chapter IIa-3. High-efficiency concepts in crystalline silicon photovoltaics, which have driven the progress in this field over several decades, are reviewed in Chapter IIb-5. The dye-sensitised and organic/plastic solar cells that are attracting a large research investment are examined in Part IIe. Part IId and Chapter IIc-2 cover wide areas at the boundary between industrial production and research, as do many other chapters on thin-film solar cells in Part IIc.

Each chapter gives a self-contained overview of the relevant aspect of photovoltaic science and technology. They can be read on their own although ample cross-referencing provides links that can be followed to build a knowledge base for any particular purpose at hand. For the non-specialist, the introductory chapters of Parts II and III can serve as a starting point before proceeding to explore other parts of the handbook. The supplementary chapters and appendices, including bibliography, solar radiation data for selected sites, and the lists of standards, web sites and news sheets, act as pointers to more specific details and textbooks if a more didactic approach is required.

PART IA

Solar Cells

CHAPTER IA-1

Principles of Solar Cell Operation

Tom Markvart[a] and Luis Castañer[b]
[a]School of Engineering Sciences, University of Southampton, UK
[b]Universidad Politecnica de Catalunya, Barcelona, Spain

1. Introduction   7
2. Electrical Characteristics   10
    2.1 The Ideal Solar Cell   10
    2.2 Solar Cell Characteristics in Practice   13
    2.3 The Quantum Efficiency and Spectral Response   15
3. Optical Properties   16
    3.1 The Antireflection Coating   16
    3.2 Light Trapping   17
4. Typical Solar Cell Structures   19
    4.1 The p–n Junction Solar Cell   19
       4.1.1 The p–n Junction   *19*
       4.1.2 Uniform Emitter and Base   *23*
       4.1.3 Diffused Emitter   *23*
    4.2 Heterojunction Cells   25
    4.3 The p–i–n Structure   27
    4.4 Series Resistance   29
References   30

1. INTRODUCTION

Photovoltaic energy conversion in solar cells consists of two essential steps. First, absorption of light generates an electron–hole pair. The electron and hole are then separated by the structure of the device—electrons to the negative terminal and holes to the positive terminal—thus generating electrical power.

This process is illustrated in Figure 1, which shows the principal features of the typical solar cells in use today. Each cell is depicted in two ways. One diagram shows the physical structure of the device and the dominant electron-transport processes that contribute to the energy-conversion process.

Practical Handbook of Photovoltaics.
© 2012 Elsevier Ltd. All rights reserved.

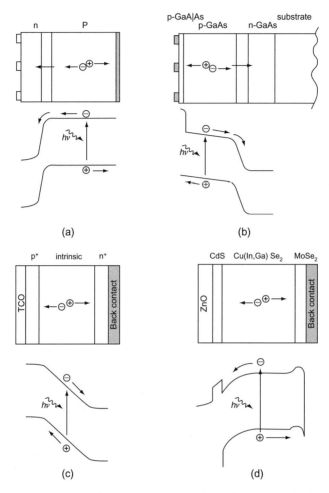

Figure 1 (a) The structure of crystalline silicon solar cell, the typical solar cell in use today. The bulk of the cell is formed by a thick p-type base in which most of the incident light is absorbed and most power is generated. After light absorption, the minority carriers (electrons) diffuse to the junction where they are swept across by the strong built-in electric field. The electrical power is collected by metal contacts to the front and back of the cell (Chapters Ib-2 and Ib-3). (b) The typical gallium—arsenide solar cell has what is sometimes called a heteroface structure, by virtue of the thin passivating GaAlAs layer that covers the top surface. The GaAlAs 'window' layer prevents minority carriers from the emitter (electrons) to reach the surface and recombine but transmits most of the incident light into the emitter layer where most of the power is generated. The operation of this p—n junction solar cell is similar in many respects to the operation of the crystalline silicon solar cell in (a), but the substantial difference in thickness should be noted. (Chapters Id-1 and Id-2). (c) The structure of a typical single-junction amorphous silicon solar cells. Based on p—i—n junction, this

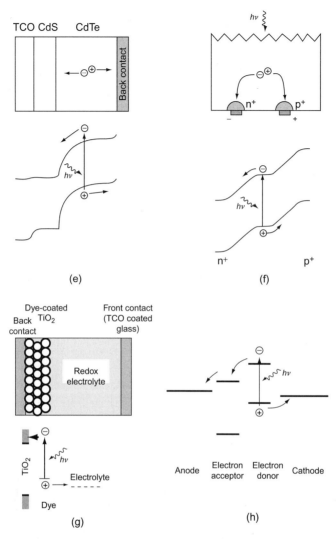

Figure 1 *(Continued)* cell contains a layer of intrinsic semiconductor that separates two heavily doped p and n regions near the contacts. Generation of electrons and holes occurs principally within the space-charge region, with the advantage that charge separation can be assisted by the built-in electric field, thus enhancing the collection efficiency. The contacts are usually formed by a *transparent conducting oxide* (TCO) at the top of the cell and a metal contact at the back. Light-trapping features in TCO can help reduce the thickness and reduce degradation. The thickness of a-Si solar cells ranges typically from a fraction of a micrometer to several micrometers. (Chapter Ic-1). (d), (e) The typical structures of solar cells based on compound semiconductors copper indium–gallium diselenide (d) and cadmium telluride (e). The front part of the junction is formed by a wide-band-gap material (CdS 'window') that

The same processes are shown on the band diagram of the semiconductor, or energy levels in the molecular devices.

The diagrams in Figure 1 are schematic in nature, and a word of warning is in place regarding the differences in scale: whilst the thickness of crystalline silicon cells (shown in Figures 1(a) and 1(f)) is of the order of 100 micrometres or more, the thickness of the various devices in Figures 1(b)−1(e) (thin-film and GaAs-based cells) might be several micrometres or less. The top surface of the semiconductor structures shown in Figure 1 would normally be covered with antireflection coating. The figure caption can also be used to locate the specific chapter in this book where full details for each type of device can be found.

2. ELECTRICAL CHARACTERISTICS

2.1 The Ideal Solar Cell

An ideal solar cell can be represented by a current source connected in parallel with a rectifying diode, as shown in the equivalent circuit of Figure 2. The corresponding I−V characteristic is described by the Shockley solar cell equation

$$I = I_{ph} - I_o \left(e^{\frac{qV}{k_B T}} - 1 \right) \qquad (1)$$

Figure 1 *(Continued)* transmits most of the incident light to the absorber layer (Cu(In, Ga)Se$_2$ or CdTe) where virtually all electron−hole pairs are produced. The top contact is formed by a transparent conducting oxide. These solar cells are typically a few micrometers thick (Chapters Ic-2 and Ic-3). (f) Contacts can be arranged on the same side of the solar cell, as in this point contact solar cell. The electron hole pairs are generated in the bulk of this crystalline silicon cell, which is near intrinsic, usually slightly n-type. Because this cell is slightly thinner than the usual crystalline silicon solar cell, efficient light absorption is aided here by light trapping: a textured top surface and a reflecting back surface (Chapter Ib-3). (g), (h) The most recent types of solar cell are based on molecular materials. In these cells, light is absorbed by a dye molecule, transferring an electron from the ground state to an excited state rather than from the valence band to the conduction band as in the semiconductor cells. The electron is subsequently removed to an electron acceptor and the electron deficiency (hole) in the ground state is replenished from an electron donor. A number of choices exist for the electron acceptor and donor. In the dye-sensitised cell (g, Chapter Ie-1), the electron donor is a redox electrolyte and the role of electron acceptor is the conduction band of titanium dioxide. In plastic solar cells (h, Chapter Ie-2), both electron donor and electron acceptor are molecular materials.

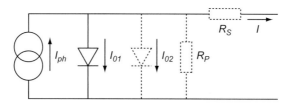

Figure 2 The equivalent circuit of an ideal solar cell (full lines). Nonideal components are shown by the dotted line.

where k_B is the Boltzmann constant, T is the absolute temperature, q (>0) is the electron charge, and V is the voltage at the terminals of the cell. I_o is well known to electronic device engineers as the diode saturation current (see, for example, [1]), serving as a reminder that a solar cell in the dark is simply a semiconductor current rectifier, or diode. The photogenerated current I_{ph} is closely related to the photon flux incident on the cell, and its dependence on the wavelength of light is frequently discussed in terms of the quantum efficiency or spectral response (see Section 2.3). The photogenerated current is usually independent of the applied voltage with possible exceptions in the case of a-Si and some other thin-film materials [2–4].

Figure 3(a) shows the I–V characteristic (Equation (1)). In the ideal case, the short-circuit current I_{sc} is equal to the photogenerated current I_{ph}, and the open-circuit voltage V_{oc} is given by

$$V_{oc} = \frac{k_B T}{q} \ln\left(1 + \frac{I_{ph}}{I_0}\right) \qquad (2)$$

The maximum theoretically achievable values of the short-circuit current density J_{ph} and of the open-circuit voltage for different materials are discussed and compared with the best measured values in Chapter Ia-3.

The power $P = IV$ produced by the cell is shown in Figure 3(b). The cell generates the maximum power P_{max} at a voltage V_m and current I_m, and it is convenient to define the fill factor FF by

$$FF = \frac{I_m V_m}{I_{sc} V_{oc}} = \frac{P_{max}}{I_{sc} V_{oc}} \qquad (3)$$

The fill factor FF of a solar cell with the ideal characteristic (1) will be furnished by the subscript 0. It cannot be determined analytically, but it

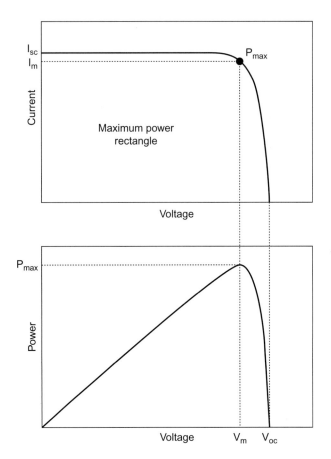

Figure 3 The I–V characteristic of an ideal solar cell (a) and the power produced by the cell (b). The power generated at the maximum power point is equal to the shaded rectangle in (a).

can be shown that FF_0 depends only on the ratio $v_{oc} = V_{oc}/k_BT$. FF_0 is determined, to an excellent accuracy, by the approximate expression [5]

$$FF_0 = \frac{v_{oc} - \ln(v_{oc} + 0.72)}{v_{oc} + 1}$$

The I–V characteristics of an ideal solar cell complies with the *superposition principle*: the functional dependence (1) can be obtained from the corresponding characteristic of a diode in the dark by shifting the diode characteristic along the current axis by I_{ph} (Figure 4).

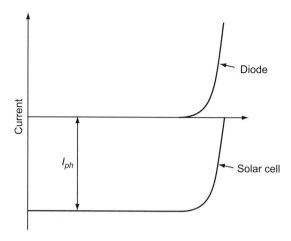

Figure 4 The superposition principle for solar cells.

2.2 Solar Cell Characteristics in Practice

The I—V characteristic of a solar cell in practice usually differs to some extent from the ideal characteristic (1). A two-diode model is often used to fit an observed curve, with the second diode containing an 'ideality factor' of 2 in the denominator of the argument of the exponential term. The solar cell (or circuit) may also contain series (R_s) and parallel (or shunt, R_p) resistances, leading to a characteristic of the form

$$I = I_{ph} - I_{o1}\left\{\exp\left(\frac{V + IR_s}{k_B T}\right)\right\} - I_{o2}\left\{\exp\left(\frac{V + IR_s}{2k_B T}\right) - 1\right\} - \frac{V + IR_S}{R_p} \quad (4)$$

where the light-generated current I_{ph} may, in some instances, depend on the voltage, as we have already noted. These features are shown in the equivalent circuit of Figure 2 by the dotted lines. The effect of the second diode, and of the series and parallel resistances, on the I—V characteristic of the solar cell is shown in Figures 5 and 6, respectively; further information about these parameters can be obtained from the dark characteristic (Figure 7). The effect of the series resistance on the fill factor can be allowed for by writing where $r_s = R_s I_{sc}/V_{oc}$. An analogous expression exists also for the parallel resistance. Instead of the two-diode equation (4), an empirical nonideality factor n_{id} can be introduced in the single-diode equation (1) that usually lies between 1

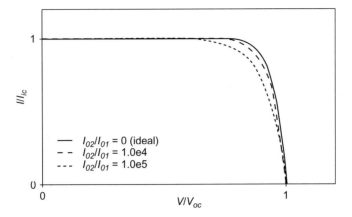

Figure 5 The I–V characteristic of the solar cell in the two-diode model for three values of the ratio I_{02}/I_{01}.

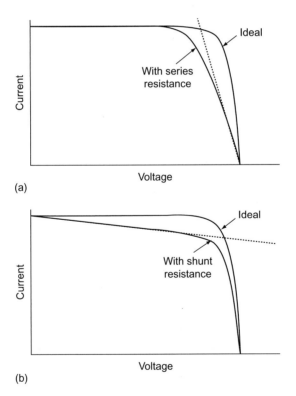

Figure 6 The effect of series (a) and parallel (b) resistance on the I–V characteristic of the solar cell.

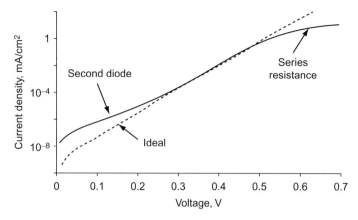

Figure 7 The dark I–V characteristic of a solar cell for the two-diode model including the series resistance. The shunt resistance has a similar effect to the second diode.

and 2. Two among a number of possible sources of nonideal behaviour—recombination in the depletion region and the series resistance—are discussed in Sections 4.1.1 and 4.4.

$$FF = FF_0(1 - r_s) \qquad (5)$$

2.3 The Quantum Efficiency and Spectral Response

The quantum efficiency of a solar cell is defined as the ratio of the number of electrons in the external circuit produced by an incident photon of a given wavelength. Thus, one can define external and internal quantum efficiencies (denoted by $EQE(\lambda)$ and $IQE(\lambda)$, respectively). They differ in the treatment of photons reflected from the cell: all photons impinging on the cell surface are taken into account in the value of the EQE, but only photons that are not reflected are considered in the value of IQE.

If the internal quantum efficiency is known, the total photogenerated current is given by

$$I_{ph} = q \int_{(\lambda)} \Phi(\lambda)\{1 - R(\lambda)\} IQE(\lambda) d\lambda \qquad (6)$$

where $\Phi(\lambda)$ is the photon flux incident on the cell at wavelength λ, $R(\lambda)$ is the reflection coefficient from the top surface (see Section 3.1), and the integration is carried out over all wavelength λ of light absorbed by the solar cell. The values of the internal and external quantum efficiency are routinely measured to assess the performance of a solar cell by using interference filters or monochromators.

The *spectral response* (denoted by $SR(\lambda)$, with the units A/W) is defined as the ratio of the photocurrent generated by a solar cell under monochromatic illumination of a given wavelength to the value of the spectral irradiance at the same wavelength. Since the number of photons and irradiance are related, the spectral response can be written in terms of the quantum efficiency as (see, for instance, [6])

$$SR(\lambda) = \frac{q\lambda}{hc} QE(\lambda) = 0.808 \cdot \lambda \cdot QE(\lambda) \qquad (7)$$

where λ is in micrometres. Spectral response in (7) can be either internal or external, depending on which value is used for the quantum efficiency.

3. OPTICAL PROPERTIES
3.1 The Antireflection Coating

Most solar cells rely on a thin layer of a dielectric (an antireflection coating) to reduce the reflection of light from the front surface of the cell. This section gives a brief description of the reflection of light from a bare semiconductor, and from a semiconductor with a single-layer antireflection coating. The discussion is confined to the case of normal incidence of light onto a smooth planar surface.

The reflection coefficient from bare silicon for light incident from air is given by

$$R = \frac{(\mathbf{n}-1)^2 + \kappa^2}{(\mathbf{n}+1)^2 + \kappa^2} \qquad (8)$$

where $\mathbf{n}$ and κ are the refractive index and the extinction coefficient of the semiconductor, both in general functions of the wavelength λ of light in vacuum. The extinction coefficient is related to the absorption coefficient α by

$$\kappa = \frac{\alpha\lambda}{4\pi\mathbf{n}} \qquad (9)$$

For single-layer antireflection coating of refractive index $\mathbf{n}_{ar}$ between a top medium of refractive index $\mathbf{n}_0$ (for example, glass or air) and a semiconductor, the reflection coefficient becomes, neglecting light absorption in the semiconductor,

$$R = \frac{r_0^2 + r_{sc}^2 + 2r_0 r_{sc} \cos 2\beta}{1 + r_0^2 + 2r_0 r_{sc} \cos 2\beta} \qquad (10)$$

where

$$r_0 = \frac{\mathbf{n}_{ar} - \mathbf{n}_0}{\mathbf{n}_{ar} + \mathbf{n}_0}; \quad r_{sc} = \frac{\mathbf{n}_{sc} - \mathbf{n}_{ar}}{\mathbf{n}_{sc} + \mathbf{n}_{ar}} \qquad \beta = \frac{2\pi}{\lambda} \mathbf{n}_{ar} d$$

and d denotes the thickness of the coating. The transmission coefficient is, in both cases, simply

$$T = 1 - R \qquad (11)$$

In most cases of interest, both r_{sc} and r_0 are positive and R vanishes when

$$d = \frac{\lambda}{4\mathbf{n}_{ar}}; \frac{3\lambda}{4\mathbf{n}_{ar}}; \frac{5\lambda}{4\mathbf{n}_{ar}}, \ldots \qquad (12)$$

and

$$\mathbf{n}_{ar} = \sqrt{\mathbf{n}_0 \mathbf{n}_{sc}} \qquad (13)$$

The first value of d in (12) is often used in practice under the name of *quarter-wavelength rule* since $\lambda/\mathbf{n}_{ar}$ is the wavelength of light in the antireflection coating.

Reflection from the top surface can be reduced further by the use of a multilayer coating. The details of such coatings as well as a general theory for an oblique incidence of light can be found, for example, in [7]. Figure 8 compares the reflection coefficients for a smooth bare silicon surface, a smooth surface covered with antireflection coating, and a textured surface with antireflection coating.

3.2 Light Trapping

In solar cells with a simple geometry, light rays enter the cell through the front surface and, if not absorbed, leave through the rear surface of the cell. More sophisticated arrangements exist that extend the path of light inside the cell, and they are usually referred to as *optical confinement* or *light trapping*. In crystalline or amorphous silicon solar cells, light trapping is used to reduce the thickness of the cell without lowering the light absorption within the cell. Light trapping can also be used to enhance the open-circuit voltage [8,9].

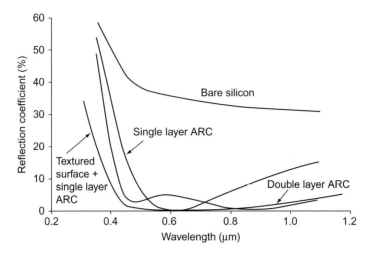

Figure 8 The reflection coefficient from polished bare silicon and a polished silicon surface covered with a single- and double-layer antireflection coating (after [31,33]). The reflection coefficient for a textured surface is also shown.

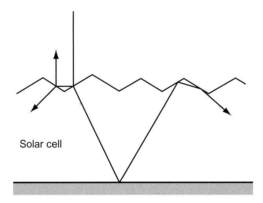

Figure 9 The textured top surface reduces reflection from the solar cell and, when combined with a reflecting back surface, helps to confine or 'trap' light within the cell.

The most common light-trapping features include a textured top surface combined with an optically reflecting back surface (Figure 9). In the ideal case, Yablonovich [10,11] (see also [12]) has shown that a randomly textured (so-called Lambertian) top surface in combination with a perfect back-surface reflector produces a light-trapping scheme that enhances the light intensity inside the cell by a factor of $\mathbf{n}_{sc}^2$ where, as in Section 3.1, $\mathbf{n}_{sc}$ is the refractive index of the solar cell material. This arrangement also increases the average path length of light rays inside the cell from 2 W, in the case of single pass through the cell, to $4\mathbf{n}_{sc}^{2}\ W$ in the case of complete light trapping, where W the cell thickness. Schemes have been developed to enhance the operation of practical devices including crystalline, polycrystalline, and amorphous silicon cells (discussed in Chapters Ib-2, -3, and -4, and in Chapter Ic-1). With application to the latter cells, Schropp and Zeman [13] consider the trapping and scattering of light at rough interfaces in some detail. In gallium—arsenide cells, multilayer Bragg reflectors (in place of the back-surface reflector) have been used with success (see Chapter Id-1).

4. TYPICAL SOLAR CELL STRUCTURES

4.1 The p—n Junction Solar Cell

The planar p—n junction solar cell under low injection is usually singled out for special analysis since realistic approximations exist that allow analytic solutions to be developed and used successfully for the description of practical devices. The success of this model is due, to a large extent, to the clear way the cell can be divided into three regions—emitter, junction region, and base—that serve a different purpose in solar cell operation.

The emitter and the base—which remain largely neutral during the cell operation—absorb the main part of the incident light and transport the photogenerated minority carriers to the junction. The p—n junction—which contains a strong electric field and a fixed space charge—separates the minority carriers that are collected from the emitter and the base. The junction is effectively devoid of mobile charge carriers and is sometimes called the *depletion region*.

4.1.1 The p—n Junction

Figure 10 shows the principal parameters of a p—n junction in equilibrium along the spatial coordinate perpendicular to the junction. In operation, the Fermi-level E_F splits into two quasi-Fermi levels E_{Fn} and E_{Fp}, one each for the electrons and the holes, with the corresponding potentials $\phi_n = -q/E_{Fn}$ and $\phi_p = -q/E_{Fp}$. Near the open circuit, the quasi-Fermi levels are parallel in the junction, their gradients are small, and their splitting is equal to the observed voltage at the junction (Figure 11). The charge carrier statistics in terms of the quasi-Fermi levels is discussed in Section 3 of Chapter Ia-2.

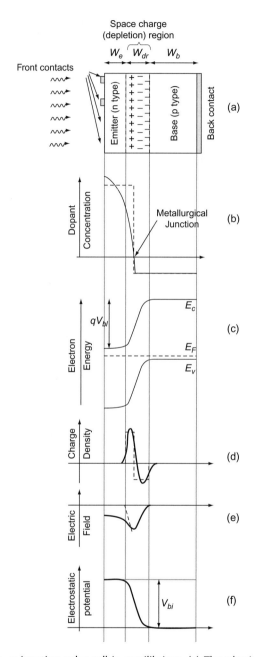

Figure 10 The p–n junction solar cell in equilibrium. (a) The physical layout (not to scale); (b) the difference of dopant concentrations $N_D - N_A$; (c) the band diagram; (d) charge density; (e) electric field; (f) electrostatic potential. The quantities shown by the dashed line correspond to an idealised abrupt junction with constant dopant concentrations in the base and in the emitter; the full line corresponds to a typical industrial solar cell with a diffused emitter.

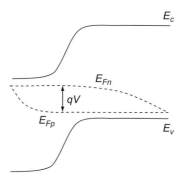

Figure 11 The p–n junction at open circuit.

Under illumination or under applied bias in the dark, the electrostatic potential difference $\Delta\Psi$ between the two sides of the junction is a difference of two terms: the equilibrium built-in voltage V_{bi} and the voltage V at the junction edges,

$$\Delta\psi = V_{bi} - V \tag{14}$$

$$qV_{bi} = k_B T \ln\left(\frac{N_D N_A}{N_i^2}\right) \tag{15}$$

where N_A and N_D are the acceptor and the donor concentrations on the p- and n-sides of the junction, respectively. In the absence of resistive losses, V is equal to the voltage measured at the terminals of the cell. The junction width W_j is given by

$$W_j = L_D \sqrt{\frac{2q\Delta\psi}{k_B T}} \tag{16}$$

Here, L_D is the Debye length,

$$L_D = \frac{\sqrt{\varepsilon k_B T}}{q^2 N_B} \tag{17}$$

where ε is the static dielectric constant, and

$$N_B = \frac{N_A N_D}{N_A + N_D}$$

In an ideal p–n junction solar cell, the junction (or depletion) region serves as a lossless mechanism for extracting and separating the minority carriers from the quasi-neutral regions—the base and the emitter. The function of the junction can then be summarised in the form of boundary conditions that link the majority carrier concentration on one side of the junction with the minority carrier concentration on the other. For an n-type emitter and a p-type base, for example, the following relations hold:

$$n(base) = n_0(base)e^{qV/k_BT} = n_0(emitter)e^{q(V-V_{bi})/k_BT} \qquad (18)$$

Equation (18) relates the electron concentration n(base) at the edge of the depletion region of the base to its equilibrium value n_0(base) and to the equilibrium electron concentration n_0(emitter) at junction edge of the emitter. A similar relationship exists for the hole concentration at the junction edge of the emitter and the base:

$$p(emitter) = p_0(emitter)e^{qV/k_BT} = p_0(base)e^{q(V-V_{bi})/k_BT} \qquad (19)$$

Equations (18) and (19) are the boundary conditions for an analytical solution of the transport equations (discussed in Chapter Ia-2, Section 4) in the quasi-neutral regions. A rigorous discussion of this *depletion approximation*, which forms the basis for the analytical treatment, can be found in reference [14].

The photogenerated and dark saturation currents for the cell are obtained by adding the relevant quantities for the base and the emitter:

$$\begin{aligned} I_{ph} &= I_{phb} + I_{phe} \\ I_0 &= I_{0b} + I_{0e} \end{aligned} \qquad (20)$$

A similar results holds also for the quantum efficiency. It is sometimes convenient to define the collection efficiency ϑ_i for a region i (where i stands for the base, the emitter, or the depletion region) as the probability that an electron−hole pair generated in this region reaches the junction

$$EQE_i(\lambda) = a_1(\lambda)\vartheta_1(\lambda) \qquad (21)$$

where $a_i(\lambda)$ is the (fractional) number of electron−hole pairs generated by each photon of incident light in region i.

No recombination occurs in an ideal p−n junction, but the (small) light-generated current produced here can be added to the first Equation (20). Recombination is included in more realistic analytical theories: the original treatment by Sah et al. [15] uses the Shockley−Read−Hall model of recombination via defects (see Chapter Ia-2, Section 7) with the principal result that the current in (I) is reduced by a term of the form

$$I_{02}\left(e^{\frac{qV}{2k_BT}} - 1\right) \tag{22}$$

In other words, recombination in the depletion region gives rise to an additional dark current corresponding to the second diode in the I–V characteristic (4), as already discussed in Section 2.2.

4.1.2 Uniform Emitter and Base

Analytical expressions for the photogenerated and dark saturation current densities for the emitter or base can be obtained if the dopant concentration and all other parameters are assumed constant. To this end, we define

$$\zeta = \frac{SL}{D} \tag{23}$$

$$\gamma_+ = (\zeta + 1)e^{W/L} + (\zeta - 1)e^{-W/L} \tag{24}$$

$$\gamma_- = (\zeta + 1)e^{W/L} - (\zeta - 1)e^{-W/L} \tag{25}$$

where S is the surface recombination velocity at external surface (front surface in the case of emitter and rear surface in the case of base); W is the width of the relevant region (W_e for the emitter and W_b for the base); and $L = \sqrt{D\tau}$ is the minority-carrier diffusion length, where τ is the minority-carrier lifetime and D is the minority-carrier diffusion constant. The photogenerated and dark saturation currents for each region are then given by

$$J_0 = \frac{qD}{L}\frac{n_i^2}{N_{dop}}\frac{\gamma_+}{\gamma_-} \tag{26}$$

where N_{dop} is the dopant concentration N_A or N_D appropriate for the relevant region. The internal quantum efficiency for each region is given in Table 1.

4.1.3 Diffused Emitter

In practical silicon solar cells, the emitter is generally fabricated by diffusion of impurities into the semiconductor wafer. This creates a thin layer where the impurity gradient is very high and the approximation of constant doping concentration does not hold. Simultaneously, the continuity and current equations do not combine into a second-order differential

Table 1 The internal quantum efficiency (IQE) for the emitter and the base in the uniform doping model. The subscripts e or b of γ, ζ, L, and W refers to the emitter or base, respectively. In the case of base, IQE is understood per unit photon entering from the junction.

	IQE(λ)
Base	$\dfrac{\alpha L_b}{\gamma_{b-}} \left\{ \dfrac{\zeta_b + 1}{1 + \alpha L_b}(e^{W_b/L_b} - e^{-\alpha W_b}) + \dfrac{\zeta_b - 1}{1 - \alpha L_b}(e^{-W_b/L_b} - e^{-\alpha W_b}) \right\}$ $\rightarrow \dfrac{\alpha L_b}{1 + \alpha L_b}$ for an infinite base $(W_b \rightarrow \infty)$
Emitter	$\dfrac{\alpha L_b e^{-\alpha W_e}}{\gamma_{e-}} \left\{ \dfrac{\zeta_e + 1}{1 + \alpha L_e}(e^{-W_e/L_e} - e^{+\alpha W_e}) + \dfrac{\zeta_e - 1}{1 - \alpha L_e}(e^{+W_e/L_e} - e^{+\alpha W_e}) \right\}$

equation with constant coefficients, and a simple analytical solution cannot be found. Several approaches have been followed besides the numerical integration of the equations [16] to reach a reasonably simple analytical or truncated series solutions. Analytical solutions were reviewed in reference [17], where the errors have been estimated for the transparent emitter [18] and the quasi-transparent emitter [19] solutions. An emitter is considered transparent when the recombination inside the emitter bulk is negligible and quasi-transparent when this recombination can be considered as a perturbation to the transparent solution. Solutions based on an infinite series that can be truncated to provide different order approximations were proposed in reference [20] and were extended as a succession of asymptotic expansions in [21]. One of the simplest yet accurate solutions is given in [22] based on the superposition of zero-input and zero-state solutions of the continuity equation with a boundary condition at the surface given by a surface recombination velocity S as follows:

$$J_0 = \dfrac{qn_i^2}{\int_0^{W_e} \dfrac{N_{\mathit{eff}}}{D} dx + \dfrac{N_{\mathit{eff}}(W_e)}{S}} + qn_i^2 \int_0^{W_e} \dfrac{dx}{N_{\mathit{eff}}\tau} \qquad (27)$$

where $N_{\mathit{eff}}(x)$ is the effective doping concentration at depth x taking into account the effect of band-gap narrowing. A systematic and general formulation of the several approximations is given in [23]. An elegant formalism to deal with inhomogeneously doped emitters can be found in [24].

When the emitter in illuminated, the problem can be solved using the same approaches as used in the dark, computing the emitter collection efficiency ϑ_{em} equal, as in (21), to the ratio of the photogenerated current at the emitter boundary of the space charge region divided by the integrated carrier generation in the emitter. Bisschop et al. [25] extended Park's solution in the dark [20] to illuminated emitters. Cuevas et al. [23] provided a formulation in terms of a series expansion and Alcubilla et al. extended the dark superposition model [22]. The first-order result for the photocurrent is given by (see, for instance, [23])

$$J_{ph} = \frac{q \int_0^{W_e} g(x) dx}{1 + \frac{S}{N_{eff}(W_e)} \int_0^{W_e} \frac{N_{eff}}{D} dx + \frac{N_{eff}(W_e)}{S}} \tag{28}$$

where $g(x)$ is the generation rate (see Section 6.1, Chapter Ia-2).

4.2 Heterojunction Cells

Heterostructures represent an opportunity to manufacture efficient solar cells from highly absorbing thin-film materials without substantial losses through electron–hole recombination at the front surface. This is illustrated by the structures of the CdS/CdTe and CdS/CIGS solar cells where a wide-band-gap semiconductor (here, CdS) serves as a 'window' partner to a lower-band-gap 'absorber' where most of the power is generated.

An important consideration in the heterojunction design includes the band-gap lineups at the interface between the two semiconductors. Figure 12 shows the equilibrium band diagrams of typical heterojunctions between a wide-gap window A and an absorber B. The band diagram

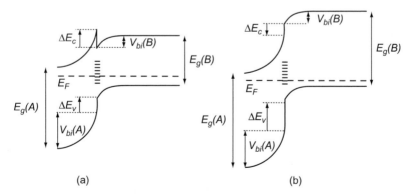

Figure 12 The band diagrams of typical heterojunction solar cells consisting of a window layer of wide-gap n-type semiconductor A, and an absorber of p-type semiconductor B. The energy levels of an interface defect layer and the band-bending potential $V_{bi}(A)$ and $V_{bi}(B)$ are also shown. A spike in the conduction band occurs for positive ΔE_c, as shown in the structure (a).

corresponds the usual situation encountered in CdTe and CIGS solar cells where an n-type wide-gap window and a p-type emitter are the most common arrangements. Similarly to the p–n junction, the built-in potentials $V_{bi}(A)$ and $V_{bi}(B)$ on the two sides of the junction can be determined by solution of the Poisson equation (see Equation (7) in Chapter 1a-2). The band-gap discontinuities ΔE_c and ΔE_v have been subject to much discussion over the years, and a number of theories have evolved that provide an understanding in terms of electron affinities and the electron dipole moments at the interface. The discontinuity in the conduction band edge, for example, can be written in the form

$$\Delta E_c = \chi_B - \chi_A + \text{interface dipole terms}$$

where χ_A and χ_B are the electron affinities of semiconductors A and B [26]. The classical Shockley–Anderson model [27] neglects the interface dipole terms. Its limited validity has been discussed extensively (see, for example, [28]), although it does seem to provide a reasonable description for some heterojunctions (Figure 13). In the application to solar cells, a full understanding of the problem is hindered further by the polycrystalline nature of the materials and frequently the presence of more than two layers that need to be considered in the analysis.

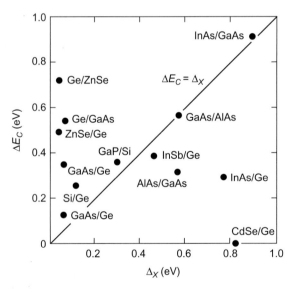

Figure 13 The conduction band discontinuity ΔE_c and the difference of electron affinities $\Delta \chi$ for a number of heterojunctions. *(reference[26], p. 391).* © *Elsevier, Reprinted with permission.*

On account of the wide band gap and the weak generation in the window material, both the dark and photogenerated currents from the emitter are significantly smaller that the corresponding quantities from the base. In addition, dark current may contain a component due to recombination at the interface defect states; less frequently, these states also reduce the collection probability and the photogenerated current.

Because of the short minority-carrier diffusion lengths, it is desirable to ensure that minority carriers are generated predominantly in a region where electric field assists collection through drift rather than diffusion. As in the homojunction cell, this can be achieved by employing sufficiently low doping concentrations in the absorber to obtain a wide depletion region; a similar philosophy is also employed in amorphous silicon solar cells, as discussed in Section 4.3 and Chapter Ic-1. More detailed description of heterojunctions in application to practical solar cells will be found in Chapters Ic-2 and Ic-3.

Somewhat similar to heterojunction is the *heteroface* solar cell, a common structure in GaAs solar cells where a thin layer of wide-gap GaAlAs is deposited to reduce recombination at the top surface. It is more convenient, however, to describe this structure as a homojunction cell with surface passivation that can be treated by the methods described in Section 4.1.

4.3 The p—i—n Structure

The analysis of p—i—n junction solar cells is of considerable importance for the understanding of operation of amorphous silicon solar cells. Furthermore, similar principles have been invoked in the description of other thin-film solar cells where the carrier diffusion is ineffective and the electric field is used to enhance carrier transport and collection. Despite this importance, however, the theoretical understanding of these structures is limited, hampered by the fundamental complexity of the problem. Indeed, the less-than-complete knowledge of the parameters of amorphous or polycrystalline material is compounded by mathematical difficulties arising principally from the need to solve the nonlinear transport equations. Although a detailed description is possible only with the use of numerical computational techniques, a broad understanding can be gained through judicious approximations based on a physical insight [29].

A schematic band diagram of a p—i—n structure is shown in Figure 14. Noting that the carrier transport dominant chiefly by drift in the electric field of the junction rather than by diffusion, carrier collection will be described by the drift lengths l_n and l_p rather than by the diffusion lengths

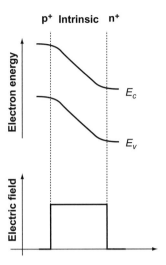

Figure 14 An idealised model of a p–i–n junction amorphous silicon solar cell with a constant electric field in the intrinsic region.

L_n and L_p (see Chapter Ia-2, Section 7.3). The recombination can be conveniently approximated with the use of minority-carrier lifetimes τ_n and τ_p on the two sides of the junction where electrons and holes are minority carriers, respectively. The use of constant electric field ε is obviously an approximation, but it is usually a good one if carrier injection is not too high. A reasonably simple analysis is then possible that, in the limit of weak absorption, results in an analytical expression in the form

$$J = qg\ell_c(1 - e^{-W_i/\ell_c}) \qquad (29)$$

for the current density produced by illumination. In Equation (29), W_i is the width of the intrinsic region, and

$$\ell_c = \ell_n + \ell_p \qquad (30)$$

is the collection length, d is the width of the i layer, and g is the generation function, which is assumed here to be constant. Equations similar to (29) have been used with success to interpret various characteristics of p–i–n solar cells (see, for example, [4]).

An extension of this theory was later proposed that allows for the three charge states of the dangling bonds in amorphous silicon rather than the two charge states usually considered in the Shockley–Read–Hall theory [30].

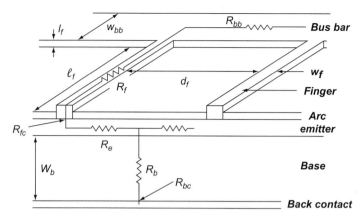

Figure 15 Components of the series resistance in a p–n junction solar cell (after [32] and [33]).

Table 2 Expressions for the various components of the series resistance; the bus-bar resistance assumes that connection is made at one end. Here, R_{sp} is the sheet resistance of the emitter layer (in ohm/square); ρ_{cf} and ρ_{cr} are the contact resistances (in ohm/cm^2) of the front and rear contact, respectively; ρ_b is the base resistivity; and ρ_m is the resistivity of the front metallisation. The geometrical dimensions are defined in Figure 15

Component of resistance	Notation	Expression
Emitter resistance	R_e	$R_e = \frac{R_{sp} d_f}{7 \ell_f}$
Resistance of the base	R_b	$R_b = A W_b \rho_b$
Contact resistance: front contact	R_{fc}	$R_{fc} = \frac{\sqrt{R_{sp} \rho d}}{\ell_f} \coth\left(W_f \sqrt{\frac{R_{sp}}{\rho_{cf}}}\right)$
Contact resistance: rear contact	R_{bc}	$R_{bc} = A \rho_{cr}$
Resistance of the finger contact	R_f	$R_f = \frac{\ell_f \rho_m}{3 t_f W_f}$
Resistance of the collecting bus bar (per unit length)	R_{bb}	$R_{bb} = \frac{\rho_m}{3 t_f W_{bb}}$

4.4 Series Resistance

Considerations regarding series resistance form an important part of the solar cell design. The main components of the series resistance of a typical crystalline silicon solar cell are shown in Figure 15 and expressions are given in Table 2.

REFERENCES

[1] S.M. Sze, Physics of Semiconductor Devices, second ed., John Wiley & Sons, New York, 1981.
[2] Y. Hishikawa, Y. Imura, T. Oshiro, Irradiance dependence and translation of the I–V characteristics of crystalline silicon solar cells, Proc. 28th IEEE Photovoltaic Specialists Conf., Anchorage, 2000, pp. 1464–1467.
[3] J.E. Philips, J. Titus, D. Hofmann, Determining the voltage dependence of the light generated current in CuInSe$_2$-based solar cells using I-V measurements made at different light intensities, Proc. 26th IEEE Photovoltaic Specialist Conf., Anaheim, 1997, pp. 463–466.
[4] S.S. Hegedus, Current-voltage analysis of a-Si and a-SiGe solar cells including voltage-dependent photocurrent collection, Prog. Photovolt: Res. Appl. 5 (1997) 151–168.
[5] M.A. Green, Silicon Solar Cells: Advanced Principles and Practice, Centre for Photovoltaic Devices and Systems, University of New South Wales, 1995.
[6] L. Castañer, S. Silvestre, Modelling Photovoltaic Systems Using Pspice, John Wiley & Sons, Chichester, 2002.
[7] M. Born, E. Wolf, Principles of Optics, seventh ed., Cambridge University Press, Cambridge, 1999, Section 1.6.
[8] R. Brendel, H.J. Queisser, On the thickness dependence of open circuit voltages of p–n junction solar cells, Sol. Energy Mater. Sol. Cells 29 (1993) 397.
[9] T. Markvart, Light harvesting for quantum solar energy conversion, Prog. Quantum Electron. 24 (2000) 107.
[10] E. Yablonovich, Statistical ray optics, J. Opt. Soc. Am. 72 (1982) 899.
[11] E. Yablonovich, G.C. Cody, Intensity enhancement in textured optical sheets for solar cells, IEEE Trans. Electron Devices ED-29 (1982) 300.
[12] J.C. Miñano, Optical confinement in photovoltaics, in: A. Luque, G.L. Araujo (Eds.), Physical Limitations to Photovoltaic Energy Conversion, Adam Hilger, Bristol, 1990, p. 50.
[13] R. Schropp, M. Zeman, Amorphous and Microcrystalline Silicon Solar Cells: Modelling, Materials and Device Technology, Kluwer, Boston, 1998.
[14] S. Selberherr, Analysis and Simulation of Semiconductors Devices, Springer, Vienna, New York, 1984, pp. 141–146.
[15] C.T. Sah, R.N. Noyce, W. Shockley, Carrier generation and recombination in p–n junctions and p–n junction characteristics, Proc. IRE 45 (1957) 1228.
[16] D.T. Rover, P.A. Basore, G.M. Thorson, Proc. 18th IEEE Photovoltaic Specialist Conf., Las Vegas, 1985, pp. 703–709.
[17] A. Cuevas, M. Balbuena, Review of analytical models for the study of highly doped regions of silicon devices, IEEE Trans. Electron Devices ED-31 (1989) 553–560.
[18] M.A. Shibib, F.A. Lindholm, F. Therez, IEEE Trans. Electron Devices ED-26 (1978) 958.
[19] J.A. del Alamo, R.M. Swanson, Proc. 17th IEEE Photovoltaic Specialist Conf., Orlando, 1984, pp. 1303–1308.
[20] J.S. Park, A. Neugroschel, F.A. Lindholm, IEEE Trans. Electron Devices ED-33 (1986) 240.
[21] N. Rinaldi, Modelling of minority carrier transport in non-uniformly doped silicon regions with asymptotic expansions, IEEE Trans. Electron Devices ED-40 (1993) 2307–2317.
[22] R. Alcubilla, J. Pons, L. Castañer, Superposition solution for minority carrier current in the emitter of bipolar devices, Solid State Electron. 35 (1992) 529–533.

[23] A. Cuevas, R. Merchan, J.C. Ramos, On the systematic analytical solutions for minority carrier transport in non-uniform doped semiconductors: application to solar cells, IEEE Trans. Electron Devices ED-40 (1993) 1181–1183.
[24] J.A. del Alamo, R.M. Swanson, The physics and modelling of heavily doped emitters, IEEE Trans. Electron Devices ED-31 (1984) 1878.
[25] F.J. Bisschop, L.A. Verhoef, W.C. Sinke, An analytical solution for the collection efficiency of solar cell emitters with arbitrary doping profile, IEEE Trans. Electron Devices ED-37 (1990) 358–364.
[26] L.J. Brillson, Surfaces and interfaces: atomic-scale structure, band bending and band offsets, in: P.T. Landsberg (Ed.), Handbook of Semiconductors, vol. 1, Elsevier, 1992, pp. 281–417.
[27] R.L. Anderson, Experiments on Ge–As heterojunctions, Solid State Electron. 5 (1962) 341–351.
[28] H. Kroemer, Heterostructure devices: a device physicist looks at interfaces, Surf. Sci. 132 (1983) 543–576.
[29] R.S. Crandall, Modelling of thin film solar cells: uniform field approximation, J. Appl. Phys. 54 (1983) 7176.
[30] J. Hubin, A.V. Shah, Effect of the recombination function on the collection in a p–i–n solar cell, Phil. Mag. B72 (1995) 589.
[31] K. Zweibel, P. Hersch, Basic Photovoltaic Principles and Methods, Van Nostrand Reinhold, New York, 1984.
[32] R. van Overstraeten, R.P. Mertens, Physics, Technology and Use of Photovoltaics, Adam Hilger, Bristol, 1986.
[33] A. Goetzberger, J. Knobloch, B. Voss, Crystalline Silicon Solar Cells, John Wiley & Sons, Chichester, 1998.

CHAPTER IA-2

Semiconductor Materials and Modelling

Tom Markvart[a] and Luis Castañer[b]
[a]School of Engineering Sciences, University of Southampton, UK
[b]Universidad Politecnica de Catalunya, Barcelona, Spain

1. Introduction	33
2. Semiconductor Band Structure	34
3. Carrier Statistics in Semiconductors	38
4. The Transport Equations	40
5. Carrier Mobility	42
6. Carrier Generation by Optical Absorption	44
6.1 Band-to-Band Transitions	44
6.2 Free-Carrier Absorption	46
7. Recombination	48
7.1 Bulk Recombination Processes	48
7.2 Surface Recombination	50
7.3 Minority-Carrier Lifetime	51
8. Radiation Damage	53
9. Heavy Doping Effects	55
10. Properties of Hydrogenated Amorphous Silicon	57
Acknowledgements	59
References	59

1. INTRODUCTION

Solar cell modelling is fundamental to a detailed understanding of the device operation, and a comprehensive model requires a detailed knowledge of the material parameters. A brief overview of the semiconductor properties relevant to solar cell operation is given in this chapter, including the semiconductor band structure and carrier statistics, transport and optical properties, recombination processes, material aspects of radiation damage to solar cells in space and effects observed under heavy doping, with special attention given to the properties of hydrogenated amorphous silicon. The principal semiconductor parameters encountered in photovoltaic applications are summarised in Tables 1 and 2. The refractive indices of materials used for antireflection coating can be found in Table 3.

Numerous computer programs that use the material parameters to model solar cell operation have been developed over the years, and several are now available commercially:

- PC1D, developed by P.A. Basore and colleagues at the University of New South Wales, Australia, is the standard one-dimensional simulator used by the PV community.
- ATLAS, a device simulation software by SILVACO International, uses physical models in two and three dimensions. It includes the Luminous tool, which computes ray tracing and response of solar cells. It allows the use of monochromatic or multispectral sources of light.
- MEDICI by Technology Modelling Associates models the two-dimensional distribution of potential and carrier concentration in a semiconductor device. It also includes an optical device advanced application module in which photogeneration can be computed for multispectral sources.

Some programs are available free of charge. These include, for example, SimWindows which can be downloaded from www-ocs.colorado.edu/SimWindows/simwin.html.

A discussion of the main principles of the numerical techniques can be found in specialised texts that deal with the modelling of solar cells (see, for example, the review [1]) or with the modelling of semiconductor devices in general [2].

2. SEMICONDUCTOR BAND STRUCTURE

The energy gap (or band gap) E_g and its structure as a function of the wave vector are key characteristics of the semiconductor material and of fundamental importance to the operation of the solar cell (see Figure 1). The principal features of interest are the temperature variation of the bandgap energy E_g and the magnitude of wave vector associated with low-energy transitions.

Table 1 Properties of the principal semiconductors with photovoltaic applications (all at 300 K). Bandgap: d = direct; i = indirect. Crystal structure: dia = diamond, zb = zinc blende, ch = chalcopyrite. The refractive index is given at the wavelength 590 nm (2.1 eV) unless otherwise stated. Principal sources of data: c-Si [20]; GaAs [3]; InP [4,41]; a-Si [5] and Section 10; CdTe [6]; CIS [7]; Al$_x$Ga$_{1-x}$As [9]. Details of the absorption coefficient and refractive index, including the wavelength dependence, can be found in references [27] and [28]. X is the electron affinity and TEC stands for thermal expansion coefficient

	E_g(eV)	Crystal structure	ε	n	X(eV)	Lattice const. (Å)	Density (g/cm^{-3})	TEC (10^{-6} K^{-1})	Melting point (K)	Comments
c-Si	1.12(i)	dia	11.9	3.97	4.05	5.431	2.328	2.6	1687	Cell material (Part Ib)
GaAs	1.424(d)	zb	13.18	3.90	4.07	5.653	5.32	6.03	1510	Cell material (Ch. Id-2)
InP	1.35(d)	zb	12.56	3.60	4.38	5.869	4.787	4.55	1340	Cell material (Ch. IId-1)
a-Si	~1.8(d)	–	~11	3.32						Cell material, invariably Si: H alloy; sometimes also alloyed with germanium or carbon (Ch. Ic-1)
CdTe	1.45 – 1.5 (d)	zb	10.2	2.89★	4.28	6.477	6.2	4.9	1365	Cell material (Ch. Ic-2)

(continued)

Table 1 (continued)

	E_g(eV)	Crystal structure	ε	n	χ(eV)	Lattice const. (Å)	Density (g/cm^{-3})	TEC (10^{-6} K^{-1})	Melting point (K)	Comments
CuInSe$_2$ (CIS)	0.96 – 1.04 (d)	ch			4.58			6.6	~1600	Cell material, often alloyed with gallium (Ch. Ic-3)
Al$_x$Ga$_{1-x}$As ($0 \leq x \leq 0.45$)	1.424 + 1.247x (d)	zb	13.18 – 3.12x		4.07 – 1.1x	5.653 + 0.0078x	5.36 – 1.6x	6.4 – 1.2x		Window layer for GaAs solar cells
($0.45 < x \leq 1$)	1.9 + 0.125x + 0.143x^2 (i)				3.64 – 0.14x					

*At 600 nm.

Table 2 The energy gap E_g, refractive index n and the electron affinity X of transparent conducting semiconductors used as window layers in thin-film solar cells

Material	E_g(eV)	n	X
CdS	2.42	2.5	4.5
ZnS	3.58	2.4	3.9
$Zn_{0.3}Cd_{0.7}S$	2.8		4.3
ZnO	3.3	2.02	4.35
In_2O_3:Sn	3.7–4.4		4.5
SnO_2:F	3.9–4.6		4.8

Table 3 The refractive index at 590 nm (2.1 eV) of the common materials used for antireflection coating. The full-wavelength dependence of most of these substances can be found in references [27] and [28]

Material	n
MgF_2	1.38
SiO_2	1.46
Al_2O_3	1.76
Si_3N_4	2.05
Ta_2O_5	2.2
ZnS	2.36
SiO_x	1.8–1.9
TiO_2	2.62

	E_{L1}	E_{L2}	E_X	E_L	E_{SO}	n_s
Si	3.4	4.2	1.12	1.9	0.035	6
GaAs	1.42		1.90	1.71	0.34	1
InP	1.34		2.19	1.93	0.11	1

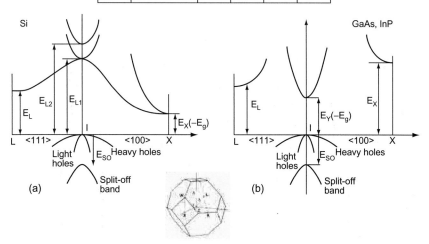

Figure 1 The energy gaps in Si (a), and GaAs and InP (b) as functions of the wave vector **k**. The inset shows the Brillouin zone of the corresponding face-centred cubic crystal lattice.

Table 4 The parameters E_g0, α, and β in Equation (1). Sources of data: (a) Thurmond [11]; (b) Varshni [10]

	$E_g(T = 0$ K), eV	$\alpha \times 10^{-4}$, eV/K^2	β, K
Si (a)	1.17	4.730	636
GaAs(a)	1.52	5.405	204
InP(b)	1.42	4.906	327

The variation of the band gap with temperature can be described by an expression originally suggested by Varshni [10]

$$E_g(T) = E_{g0} - \frac{\alpha T^2}{T + \beta} \qquad (1)$$

where T is the absolute temperature and the parameters α and β are given in Table 4.

The current produced by solar cells is generated by optical transitions across the band gap. Two types of such transitions can be distinguished: (1) direct transitions where the momentum of the resulting electron−hole pair is very close to zero, and (2) indirect transitions where the resulting electron−hole pair has a finite momentum. The latter transitions require the assistance of a phonon (quantum of lattice vibration). Thus, there are two types of semiconductors:

- **direct gap semiconductors** where the top of the valence band and the bottom of the conduction band lie at the Γ point of the first Brillouin zone (i.e., at zero wave vector $\mathbf{k} = 0$); and
- **indirect gap semiconductors** where the minima of the conduction band (in general, more than one) lie at a another point of the first Brillouin zone, with a different value of the wave vector $\mathbf{k}$.

The optical absorption in indirect-gap semiconductors is considerably weaker than in direct-gap semiconductors, as shown in Figure 2 on the example of silicon and gallium arsenide.

3. CARRIER STATISTICS IN SEMICONDUCTORS

In thermal equilibrium, the temperature and the electrochemical potential (the Fermi level, denoted by E_F) are constant throughout the

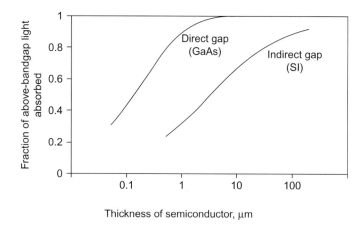

Figure 2 A comparison of the difference in strength of the optical absorption in direct and indirect semiconductors, illustrated on the examples of silicon and gallium arsenide.

Table 5 The densities of states in the conduction and valence band (N_c and N_v, respectively), the intrinsic carrier concentration n_1 and the density-of-states effective masses m_e and m_h (all at 300 K) in Si, GaAs and InP. m_0 denotes the free-electron mass. Data from references [20,35]; intrinsic carrier concentration in Si from reference [12]

	N_c(cm^{-3})	N_v(cm^{-3})	n_1(cm^{-3})	m_e/m_0	m_h/m_0
Si	2.8×10^{19}	1.04×10^{19}	1.00×10^{10}	1.08	0.55
GaAs	4.7×10^{17}	7.0×10^{18}	1.79×10^{6}	0.063	0.53
InP	5.7×10^{17}	1.1×10^{19}	1.3×10^{7}	0.08	0.6

device. The product of the electron concentration n and the hole concentration p is then independent of doping and obeys the mass action law

$$np = n_i^2 = N_c N_v \exp\left(-\frac{E_g}{K_B T}\right) \quad (2)$$

where k_B is the Boltzmann constant, n_i is the electron (or the hole) concentration in the intrinsic semiconductor, and the effective densities of states N_c and N_v are given by

$$N_c = 2\left(\frac{2\pi m_e k_B T}{h^2}\right)^{3/2} \quad N_v = 2\left(\frac{2\pi m_h k_B T}{h^2}\right)^{3/2} \quad (3)$$

Here, h is the Planck constant and m_e and m_h are the electron and hole density-of-states effective masses (see Table 5). In Equation (3), the

Table 6 Carrier concentration in degenerate and nondegenerate semiconductors. $F_{1/2}$ denotes the integral $F_{1/2}(z) = \frac{2}{\sqrt{\pi}} \int_0^\infty \frac{x^{1/2}}{1+\exp(x-z)} dx$

	Nondegenerate semiconductors		
	In terms of the band parameters	n terms of the parameters of intrinsic semiconductor	General expressions
n	$N_c \exp\left(\frac{E_{Fn}-E_c}{k_B T}\right)$	$n_i \exp\left\{\frac{q(\psi-\varphi_n)}{k_B T}\right\}$	$N_c F_{1/2}\left(\frac{E_{Fn}-E_c}{k_B T}\right)$
p	$N_v \exp\left(\frac{E_v-E_{Fp}}{k_B T}\right)$	$n_i \exp\left\{\frac{q(\varphi_p-\psi)}{k_B T}\right\}$	$N_v F_{1/2}\left(\frac{E_v-E_{Fp}}{k_B T}\right)$

effective mass m_e includes a factor that allows for several equivalent minima of the conduction band in indirect-gap semiconductors (see Figure 1).

Equation (2) does not hold for a solar cell in operation. Current flow as well as electron transitions between different bands or other quantum states are induced by differences and gradients of the electrochemical potentials, and temperature gradients may also exist. It is then usual to assign a quasi-Fermi level to each band that describes the appropriate type of carriers. Thus, electrons in the conduction band will be described by the quasi-Fermi level E_{Fn}, and holes by E_{Fp}. It is convenient to define also the appropriate potentials Φ_n and Φ_p by

$$\begin{aligned} E_{Fn} &= -q\phi_n \\ E_{Fp} &= -q\phi_p \end{aligned} \quad (4)$$

The use of quasi-Fermi levels to describe solar cell operation was already mentioned in Section 4.1 of Chapter Ia-1. The formalism leads to expressions for electron and hole concentrations, which are summarised in Table 6.

4. THE TRANSPORT EQUATIONS

The electron and the hole current densities J_n and J_p are governed transport by the transport equations

$$\begin{aligned} J_n &= q\mu_n n\varepsilon + qD_n \nabla n \\ J_p &= q\mu_p p\varepsilon - qD_p \nabla p \end{aligned} \quad (5)$$

where n and p are the electron and the hole; the concentrations μ_n and μ_p are the electron and the hole mobilities; D_n and D_p are the electron and the hole diffusion constants; and ε is the electric field. The first term in each equation is due to drift in the electric field ε, and the second term corresponds to carrier diffusion. With the help of the quasi-Fermi levels, the equations in (5) can be written as

$$\begin{aligned} J_n &= -q\mu_n n \nabla \phi_n \\ J_p &= -q\mu_p n \nabla \phi_p \end{aligned} \qquad (6)$$

The equations in (6) are valid for a semiconductor with a homogeneous composition and when position-dependent dopant concentration is included. A generalisation to semiconductors with position-dependent band gaps can be found, for example, in [13].

In a region where space charge exists (for example, in the junction), the *Poisson equation* is needed to link the electrostatic potential ψ with the charge density ρ:

$$\nabla \cdot \varepsilon = \frac{\rho}{\varepsilon} \qquad (7)$$

with

$$\rho = q(p - n + N_D - N_A) \qquad (8)$$

$$\varepsilon = -\nabla \cdot \psi \qquad (9)$$

where ε is the static dielectric constant of the semiconductor (see Table 1).

In nondegenerate semiconductors, the diffusion constants are related to mobilities by the *Einstein relations*

$$D_n = \frac{kT}{q}\mu_n; \; D_p = \frac{kT}{q}\mu_p \qquad (10)$$

The generalisation of the Einstein relations to degenerate semiconductors is discussed, for example, in [14].

The conservation of electrons and holes is expressed by the *continuity equations*

$$\begin{aligned} \frac{\partial n}{\partial t} &= G - U + \frac{1}{q}\nabla \cdot J_n \\ \frac{\partial p}{\partial t} &= G - U + \frac{1}{q}\nabla \cdot J_p \end{aligned} \qquad (11)$$

where G and U are the generation and recombination rates, which may be different for electrons and holes if there are transitions into or from localised states.

5. CARRIER MOBILITY

In weak fields, the drift mobilities in Equations (5) represent the ratio between the mean carrier velocity and the electric field. The mobilities—which are generally different for majority and minority carriers—depend on the concentration of charged impurities and on the temperature. For silicon, these empirical dependencies are generally expressed in the Caughey–Thomas form [15]:

$$\mu = \mu_{min} + \frac{\mu_0}{\left(\dfrac{N}{N_{ref}}\right)^\alpha} \tag{12}$$

The values of the various constants for majority and minority carriers are given in Tables 7 and 8. A full model that includes the effects of lattice scattering, impurity scattering, carrier–carrier scattering, and impurity clustering effects at high concentration is described in [16], where the reader can find further details.

In strong fields, the mobility of carriers accelerated in an electric field parallel to the current flow is reduced since the carrier velocity saturates. The field dependence of the mean velocity v (only quoted reliably for majority carriers) is given by

$$v = \frac{\mu_{min}}{1 + \left(1 + \left(\dfrac{\mu f \varepsilon}{V_{sat}}\right)^\beta\right)^{1/\beta}} \tag{13}$$

Table 7 The values of parameters in Equation (12) for majority-carrier mobility in silicon (from reference [17]; $T_n = T/300$)

	$\mu_{min} = AT_n^{-\beta}$		$\mu_0 = BT_n^{-\beta_2}$		$N_{ref} = CT_n^{\beta_3}$		$\alpha = DT_n^{-\beta_4}$	
	A	β_1	B	β_2	C	β_3	D	β_4
Electrons	88		7.4×10^8		1.26×10^{17}	2.4	0.88	0.146
Holes	54.3	0.57	1.36×10^8	2.33	2.35×10^{17}			

Table 8 The values of parameters in Equation (12) for minority-carrier mobility in silicon

	μ_{min}	μ_0	N_{ref}	α
Electrons	232	1180	8×10^{16}	0.9
Holes	130	370	8×10^{17}	1.25

(from references [18,19])

where μ_{lf} is the appropriate low field value of the mobility (Equation (12)), the parameter $\beta = 2$ for electrons and $\beta = 1$ for holes, and v_{sat} is the saturation velocity, identical for electrons and holes [20]:

$$v_{sat} = \frac{2.4 \times 10^7}{1 + 0.8e^T/600} \quad (14)$$

In gallium arsenide, the empirical fitting of the mobility data is of a more complex nature, and the reader is referred to the reference [21] for more information regarding the majority-carrier mobility. The temperature and dopant concentration dependences of the minority-carrier mobility are shown in Figures 3 and 4.

The electric field dependence of carrier mobility in GaAs is different from silicon as the velocity has an 'overshoot' that is normally modelled by the following equation [24]:

$$\mu_n = \frac{\mu_{lf}\varepsilon + v_{sat}(\varepsilon/\varepsilon_0)^\beta}{1 + (\varepsilon/\varepsilon_0)^\beta} \quad (15)$$

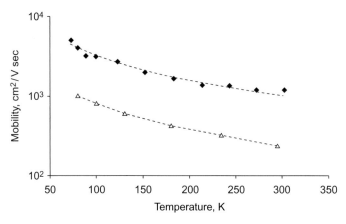

Figure 3 The temperature dependence of the minority-carrier mobility in GaAs. Full symbols: electron mobility in p-type GaAs ($N_A = 4 \times 10^{18}$ cm^{-3}); data from [22]. Empty symbols: hole mobility in n-type GaAs ($N_D = 1.8 \times 10^{18}$ cm^{-3}) [23]. The dashed lines show fits to the data with expressions

$$\mu_n = \frac{337{,}100}{T} - 116; \mu_p = \frac{85{,}980}{T} - 49.84;$$

where T is the temperature in K.

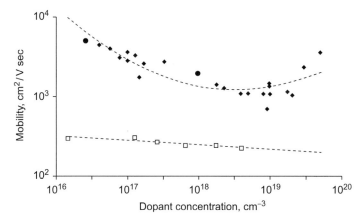

Figure 4 The dopant concentration dependence of the room-temperature minority-carrier mobility in GaAs. Full symbols: electron mobility in p-type GaAs; data from [21]. Empty symbols: hole mobility in n-type GaAs [23]. The dashed lines show fits to the data with the expressions

$$\log \mu_n = 0.16(\log N_A)^2 - 5.93 \log N_A + 58$$
$$\log \mu_p = -0.0575 \log N_D + 3.416$$

where N_D and N_A are the donor and acceptor concentrations in cm^{-3}.

where μ_{lf} is the appropriate low-field mobility, $\varepsilon_0 = 4 \times 10^3$ V/cm, β equals 4 for electrons and 1 for holes, and v_{sat} is given by [25]

$$v_{sat} = 11.3 \times 10^6 - 1.2 \times 10^4 T \tag{16}$$

where T is the temperature in K.

The majority-carrier mobility in indium phosphide reported in reference [26] is shown in Figure 5.

6. CARRIER GENERATION BY OPTICAL ABSORPTION

6.1 Band-to-Band Transitions

The principal means of carrier generation in solar cells is the absorption of light. For a planar slab (Figure 6), a photon that enters the semiconductor generates $g(x)\ \delta x$ electron–hole pairs in a thin layer at depth $x \rightarrow x + \delta x$. The generation function $g(x)$ is given by

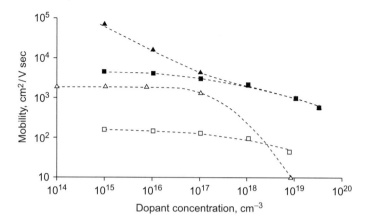

Figure 5 The majority-carrier mobilities in InP [26]. Full symbols correspond to electron mobility, empty symbols to hole mobility. Room temperature (300 K) data are shown by squares; data at 77 K by triangles. The room-temperature data were fitted with a Caughey–Thomas-type expression with parameters shown in Table 9.

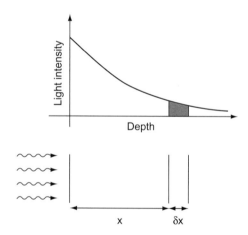

Figure 6 The geometry used to discuss the light absorption in semiconductors.

$$g(x) = \alpha(\lambda)exp\{-\alpha(\lambda)x\} \qquad (17)$$

where $\alpha(\lambda)$ is the absorption coefficient, shown in Figure 7 for a number of semiconductors with photovoltaic applications. The generation rate G per unit volume that appears in Equation (11) is related to the generation function g in Equation (17) by $G = g/A$, where A is the illuminated area of the sample.

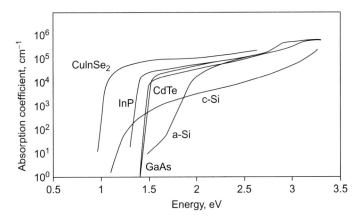

Figure 7 The absorption coefficients of the principal semiconductors used in the manufacture of solar cells. Full details of these and other semiconductors can be found in references [27] and [28]. Further details of silicon absorption are given in the text.

A useful formula for the absorption coefficient of silicon is provided by the expression of Rajkanan et al. [29]:

$$\alpha(T) = \sum_{\substack{I-1,2 \\ J-1,2}} C_i A_j \left[\frac{\{h\nu - E_{gj}(T) + E_{pi}\}^2}{\{exp(E_{pi}kT) - 1\}} + \frac{\{h\nu - E_{gj}(T) - E_{pi}\}^2}{\{1 - exp(-E_{pi}kT)\}} \right]$$
$$+ A_d \{h\nu - E_{gd}(t)\}^{1/2} \qquad (18)$$

where $h\nu$ is the photon energy, $E_{g1}(0) = 1.1557$ eV, $E_{g2}(0) = 2.5$ eV, and $E_{gd}(0) = 3.2$ eV are the two lowest indirect and the lowest direct band gap, respectively (used here as parameters to obtain a fit to the spectrum); $E_{p1} = 1.827 \times 10^{-2}$ eV and $E_{p2} = 5.773 \times 10^{-2}$ eV are the Debye frequencies of the transverse optical and transverse acoustic phonons, respectively; and $C_1 = 5.5$, $C_2 = 4.0$, $A_1 = 3.231 \times 10^2$ cm^{-1} eV^{-2}, $A_2 = 7.237 \times 10^3$ cm^{-1} eV^{-2}, and $A_1 = 1.052 \times 10^6$ cm^{-1} eV^{-2}. The temperature variation of the band gaps is described by Equation (1) where the original Varshni coefficients $\alpha = 7.021 \times 10^{-4}$, eV/K^2 and $\beta = 1108$ K are used for all three band gaps E_{g1}, E_{g2}, and E_{gd}.

6.2 Free-Carrier Absorption

In regions with high-carrier concentration (due to doping or strong illumination, for example), photon absorption can also occur by electron transitions with initial and final states inside the same band. This free-carrier absorption does not generate electron–hole pairs and competes with the band-to-band transitions that produce the photogenerated current, which were discussed above. Free-carrier absorption might be significant in the case of photon energies near the band gap and should not be included in the absorption coefficient α in front of the exponential in Equation (17). Figure 8 illustrates the different phenomena that occur near the band edge for high doping concentrations.

Based on experimental data of [30] and [31], the PC1D model, for example, uses the following expression for the absorption coefficient due to free-carrier absorption:

$$\alpha FC = K_1 n \lambda^a + K_2 p \lambda^b \qquad (19)$$

where λ is in nanometres and the empirically determined constants K_1, K_2, a, and b are given in Table 10.

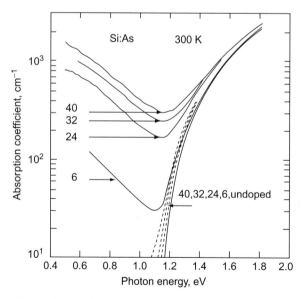

Figure 8 The effects observed near the absorption edge at high doping concentrations, illustrated here on the example of n-type silicon *(from reference [30] as adapted by M.A. Green, Silicon Solar Cells—see Bibliography)*.

Table 9 Parameters in Equation (12) that were used to fit the room-temperature electron and hole mobilities in InP

	μ_{min}	μ_0	N_{ref}	α
Electrons	0	4990	4.02×10^{17}	0.433
Holes		168	1.24×10^{18}	

Table 10 The constants for the free-carrier absorption coefficient in Equation (19)

	K_1	α	K_2	b
Si	2.6×10^{-27}	3	2.7×10^{-24}	2
GaAs	4×10^{-29}	3	—	—
InP	5×10^{-27}	2.5	—	—

7. RECOMBINATION

Recombination processes can be classified in a number of ways. Most texts distinguish between bulk and surface recombination, and between band-to-band recombination as opposed to transitions with the participation of defect levels within the band gap. Recombination processes can also be classified according to the medium that absorbs the energy of the recombining electron–hole pair: radiative recombination (associated with photon emission) or the two principal nonradiative mechanisms by Auger and multiphonon transitions, where the recombination energy is absorbed by a free charge carrier or by lattice vibrations, respectively. An opposite process to Auger recombination (in which an electron–hole pair is generated rather than consumed) is called *impact ionisation*.

The following sections give a brief summary of the main recombination mechanisms with relevance to solar cell operation, which are summarised schematically in Figure 9.

7.1 Bulk Recombination Processes

A detailed discussion of the variety of the bulk recombination mechanisms can be found in [32]. Here, we confine ourselves to a brief overview of the radiative, Auger and defect-assisted recombination processes that are most frequently encountered in practical operation of solar cells. These processes are depicted schematically in Figure 9, which also shows the notation used to describe the relevant parameters.

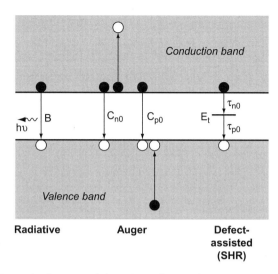

Figure 9 A schematic diagram of the principal recombination processes in semiconductors and the notation for the rate constant adopted in this book. The direction of arrows indicates electron transitions.

Table 11 The coefficients B of the radiative recombination rate (Equation (20)) and C_{n0} and C_{p0} of the Auger recombination rate (Equation (21)). Sources of data: (a) [34]; (b) [32]; (c) [35]; (d) [26]

	$B.cm^3 s^{-1}$	$C_{n0}\, cm^6\, s^{-1}$	$C_{p0}\, cm^6\, s^{-1}$
Si	1.8×10^{-15} (a)	2.8×10^{-31} (b)	0.99×10^{-31} (b)
GaAs	7.2×10^{-10} (a)	$\sim 10^{-30}$ (c)	
InP	6.25×10^{-10} (d)	$\sim 9 \times 10^{-31}$ (c)	

The rate of *band-to-band radiative recombination* can be written in the form

$$U_{rad} = B(np - n_i^2) \qquad (20)$$

where the coefficient B is sometimes written as R/n_i^2. Radiative transitions between a free electron and a localised state within the band gap may also be important in certain situations—for example, in novel concepts such as the impurity photovoltaic effect. The values of B for silicon, gallium arsenide, and indium phosphide can be found in Table 11. In other situations, the radiative recombination coefficient B can be determined from optical absorption using the detailed balance argument due to van Roosbroeck and Shockley [33].

The rate of band-to-band Auger recombination can be written as

$$U_{Auger} = (C_{p0}p + C_{n0})(np - n_i^2) \tag{21}$$

where the first bracket gives the two, usually most important, Auger terms. Table 11 gives values of the coefficients C_{po} and C_{no} for Si, GaAs and InP.

The *recombination rate via defects* of concentration N_t with a level at energy E_t within the band gap is described by the Shockley–Read–Hall formula [36,37]:

$$U_{SHR} = \frac{np - n_i^2}{\tau_p(n + n_1) + \tau_n(p + p_1)} \tag{22}$$

where

$$n_1 = n_i exp\left(\frac{E_t - E_i}{k_B T}\right) p_1 = p_i exp\left(\frac{E_t - E_i}{k_B T}\right) \tag{23}$$

and τ_n, τ_p are parameters, proportional to the defect concentration N_t, that are characteristic for the particular defect and energy level. At low injection, τ_n and τ_p assume the meaning of minority-carrier lifetimes. With an appropriate dependence on the doping concentration and temperature, τ_n and τ_p are also used extensively in material and device modelling (see reference [20], Section 1.5.3 and Section 7.3 for a fuller discussion).

7.2 Surface Recombination

Surface recombination velocity is an important parameter that affects the dark saturation current and the quantum efficiency of solar cells. Similarly to dislocations and planar defects such as grain boundaries, surfaces (and interfaces in general) introduce band of electronic states in the band gap that can be ascribed to broken (or strained) bonds and impurities. A complete characterisation of surface recombination must also take into account the surface charge, which may give rise to band bending. To achieve optimal operation, surface recombination is reduced by a passivating or window layer that prevents minority carriers from reaching the surface. Passivation of silicon surface by an oxide layer or the deposition of a thin 'window' layer of GaAlAs on GaAs solar cells are just two examples of such practice.

For an oxidised silicon surface, surface recombination velocity is strongly dependent on the surface roughness, contamination, and ambient gases used during oxidation and the annealing conditions. Under identical process parameters, however, one can identify trends in the dependence of the surface recombination velocity on the surface doping concentration. Cuevas et al. [38] proposed the following analytical relationship between surface recombination velocity and doping concentration:

$$S = 70 \text{ cm/s for } N < 7 \times 10^{17} \text{cm}^{-3}$$
$$S = 70 \left(\frac{N}{7 \times 10^{17}} \right) \text{ for } N > 7 \times 10^{17} \text{cm}^{-3} \quad (24)$$

Equation (24) models several experimental results such as those reported in [39].

In gallium arsenide, the surface recombination velocity is very high (of the order of 10^6 cm/s). The deposition of a thin layer of GaAlAs, however, reduces the recombination velocity at the interface to $10-10^3$ cm/s (see reference [40], p. 41). Coutts and Yamaguchi [41] quote the values of $10^3 - 2 \times 10^4$ cm/s and 1.5×10^5 cm/s for the surface recombination velocity in n- and p-type InP, respectively.

7.3 Minority-Carrier Lifetime

Under low injection—a regime of particular importance for solar cell operation—the majority-carrier concentration can be assumed excitation independent, and the effect of recombination can be discussed in terms of minority-carrier lifetime. In p-type material, for example, the recombination rate can be written as

$$U = \frac{1}{\tau_n}(n - n_0) \quad (25)$$

where τ_n is the minority-carrier (electron) lifetime. An analogous equation can be written for the hole lifetime τ_p in n-type material. The inverse of the lifetime—the rate constant—is a sum of the different contributions to the lifetime:

$$\frac{1}{\tau} = \frac{1}{\tau_{rad}} + \frac{1}{\tau_{Auger}} + \frac{1}{\tau_{SRH}} \quad (26)$$

where τ stands for τ_n or τ_p, as appropriate. This additive nature of the recombination rate constant is also useful when discussing the radiation damage (see Section 8).

The effect of lifetime on transport properties by carrier diffusion can be discussed in terms of the diffusion length defined by

$$L = \sqrt{D\tau} \qquad (27)$$

where D is the diffusion constant for the minority carriers in question. If, however, drift in electric field ε is the dominant transport mechanism, it is appropriate to define the drift length as

$$\ell_n = \varepsilon \tau_n \mu_n; \ell_p = \varepsilon \tau_p \mu_p \qquad (28)$$

for electrons and holes as minority carriers. This parameter plays an important role in the analysis of p–i–n junction solar cells (see Section 4.3 in Chapter Ia-1).

The wealth of available data for crystalline silicon have made it possible to arrive at a consensus as to the magnitude as well as the temperature and doping-concentration dependence of the contributions (26) to minority-carrier lifetime [42]. The contribution to lifetime due to defects, when combined with recombination in intrinsic material, has been empirically observed to follow the equations

$$\frac{1}{\tau_{n \cdot SRH}} = \left(\frac{1}{2.5 \times 10^{-3}} + 3 \times 10^{-13} N_D \right) \left(\frac{300}{T} \right)^{1.77}$$

$$\frac{1}{\tau_{n \cdot SRH}} = \left(\frac{1}{2.5 \times 10^{-3}} + 11.76 \times 10^{-13} N_A \right) \left(\frac{300}{T} \right)^{0.57} \qquad (29)$$

where the first term in the brackets applies for recombination in intrinsic semiconductor. Similarly, the contribution to Equation (26) by Auger recombination can be described by the expressions

$$\frac{1}{\tau_{n \cdot Auger}} = 1.83 \times 10^{-31} p^2 \left(\frac{T}{300} \right)^{1.18}$$

$$\frac{1}{\tau_{p \cdot Auger}} = 2.78 \times 10^{-31} n^2 \left(\frac{T}{300} \right)^{0.72} \qquad (30)$$

Although the concept of minority-carrier lifetime is most commonly applied to bulk recombination, a similar notion can be relevant for surface processes. For example, the effective lifetime observed of minority carriers with uniform concentration in a wafer can be written as

$$\frac{1}{\tau_{eff}} = \frac{1}{\tau_{SRH}} + \frac{2W}{A} S \qquad (31)$$

where S is the value of the recombination velocity, W is the wafer thickness, and A is the area of the sample.

8. RADIATION DAMAGE

Solar cells that operate on board a satellite in an orbit that passes through the Van Allen belts are subjected to fluxes of energetic electrons and protons trapped in the magnetic field of Earth and by fluxes of particles associated with high solar activity [43,44] (see Chapter IId-2). When slowed down in matter, most of the energy of the incident proton or electron is dissipated by interaction with the electron cloud. A relatively small fraction of this energy is dissipated in collisions with the nuclei, resulting in the formation of a lattice defect when energy in excess of a minimum threshold value is transferred to the nucleus [45,46]. A proton with energy in excess of this threshold causes the most damage (typically, of the order of 10–100 atomic displacements) near the end of its range in the crystal. At high energy, on the other hand, a proton creates simple point defects, and the displacement rate decreases with increasing proton energy. In contrast, the displacement rate by electrons increases rapidly with energy and approaches a constant value at higher energies. Electron damage can usually be assumed to be uniform throughout the crystal.

Some defects that are thus created act as recombination centres and reduce the minority-carrier lifetime τ. Thus, it is the lifetime τ (or, equivalently, the diffusion length L) that is the principal quantity of concern when the cell is subjected to the particle radiation in space. Under low injection, the reduction of the diffusion length L is described by the Messenger–Spratt equation (see, for example, [47], p. 151):

$$\frac{1}{L^2} = \frac{1}{L_0^2} + K_L \phi \qquad (32)$$

where L_0 is the diffusion length in the unirradiated cell, ϕ is the particle fluence (integrated flux), and K_L is a (dimensionless) diffusion-length damage constant characteristic for the material and the type of irradiation.

The damage constant K_L generally depends on the type of dopant, and even in single-junction cells, a different damage constant should therefore be introduced for each region of the cell. Moreover, care should be exercised when dealing with low-energy protons (with energy of the order of 0.1–1 MeV), which are stopped near the surface and may create nonuniform damage near the junction. Figures 10 and 11 show the doping concentration and energy dependence of the damage constant K_L for electrons and protons in p-type silicon. Details of the damage constants for other materials can be found in references [43], [44], [48], and [49].

Damage constants are usually quoted for 1 MeV electrons and converted to other energies and particles (such as protons) by using the concept of damage equivalent. Conversion tables are available in the Solar Radiation Handbooks that will suit most circumstances. Radiation damage equivalence works well when applied to uniform damage throughout the cell, but care should be exercised in the case of nonuniform damage—for example, for low-frequency protons.

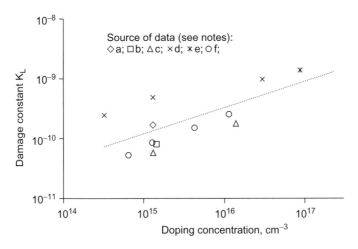

Figure 10 The radiation damage constant K_L for 1 MeV electrons in p-type silicon as a function of the dopant concentration. Source of data: (a) [50]: (b) [51]; (c) [52]; (d) [53]; (e) [54]; (f) [55]. The dashed line shows a fit to the data with the formula $K_L = 3.43 \times 10^{-17} N_A^{0.436}$, where N_A is the acceptor concentration in cm^{-3}.

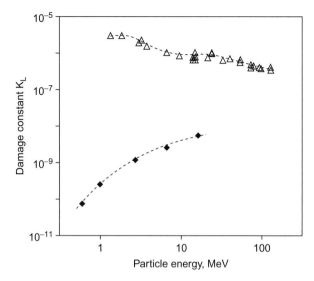

Figure 11 The energy dependence of the damage constant K_L in 1 Ω-cm p-type silicon by electron (full symbols, [55]) and protons (empty symbols, [56]).

The reduction of the diffusion length describes usually the most significant part of the damage, but changes in other parameters on irradiation also sometimes need to be considered. The dark diode current I_0 may increase as a result of compensation by the radiation-induced defects. This, however, occurs usually at a fluence orders of magnitude higher than that which degrades τ or L. Other source of degradation may be defects created in the depletion region, at the interfaces between different regions of the cell, or at the external surfaces.

9. HEAVY DOPING EFFECTS

At high doping densities, the picture of independent electrons interacting with isolated impurities becomes insufficient to describe the electron properties of semiconductors. As doping concentration increases, an impurity band is formed from the separate Coulomb wave functions located at individual doping impurities. This band gradually merges with the parent band and eventually gives rise to a tail of localised states. Electron—electron interaction becomes important, and the exchange and correlation energy terms have to be taken into account for a satisfactory description of the optical parameters and semiconductor device operation. An early overview of this multifaceted physical problem can be found in [57].

In practical situations, this complex phenomenon is usually described in terms of band-gap narrowing, with a possible correction to effective masses and band anisotropy. Different manifestations of this effect are normally found by optical measurements (including absorption and low-temperature luminescence) and from data that pertain to device operation. Parameters that have been inferred from the latter are usually referred to as *apparent band-gap narrowing,* denoted by ΔE_g. For n-type material, for example,

$$\Delta E_g = kT \ln \left(\frac{p_0 N_D}{n_i^2} \right) \quad (33)$$

where p_0 is the minority-carrier (hole) concentration and N_D is the dopant (donor) concentration.

The experimental data obtained by various methods have been reviewed in [58]. where the reader can find references to much of the earlier work. Jain et al. [59,60] obtain a fit for band-gap narrowing in various materials using a relatively simple and physically transparent expression that, however, is less easy to apply to device modelling. Values for silicon that are now frequently used in semiconductor devices modelling are based on the work of Klaassen et al. [61]. After reviewing the existing experimental data and correcting for the new mobility models and a new value for the intrinsic concentration, Klaassen et al. show that the apparent band-gap narrowing of n- and p-type silicon can be accurately modelled by a single expression:

$$\Delta E_g(meV) = 6.92 \left[\ln \left(\frac{N_{dop}}{1.3 \times 10^{17}} \right) + \sqrt{\ln \left(\frac{N_{dop}}{1.3 \times 10^{17}} \right)^2 + 0.5} \right] \quad (34)$$

where N_{dop} is the dopant concentration.

For gallium arsenide, Lundstrom et al. [62] recommend the following formula based on fitting empirical data:

$$\Delta E_g = A N_{dop}^{1/3} + k_B T \ln \{E_F\} - E_F$$

where E_F is the Fermi energy, the function $F_{1/2}$ is defined in the caption to Table 5, and

$$A = \begin{cases} 2.55 \times 10^{-8} eV & (p - GaAS) \\ 3.23 \times 10^{-8} eV & (p - GaAS) \end{cases}$$

10. PROPERTIES OF HYDROGENATED AMORPHOUS SILICON[1]

In hydrogenated amorphous silicon (a-Si:H), the effective band gap between the conduction and valence band edges is around 1.8 eV, but a thermal shrinking of the band gap with temperature has been reported [63]:

$$E_g = E_{go} - (T - T_o)\gamma \approx 5k_B \qquad (35)$$

For statistical calculations, the corresponding effective densities of states can be approximated by $N_c \approx N_v \approx 4 \times 10^{19}$ cm^{-3}. In contrast to crystalline silicon, the conduction and valence bands show evidence of exponential tails of localised states within the band gap [64]:

$$D_{Ct} = D_{CO} exp\left(\frac{E - E_C}{kT_C}\right) \quad D_{CO} \approx 0.8 \times 10^{21} cm^{-3} \quad kT_C \approx 30\, meV$$
$$D_{Vt} = D_{VO} exp\left(\frac{E_V - E}{kT_V}\right) \quad D_{VO} \approx 1.1 \times 10^{21} cm^{-3} \quad kT_V \approx 50\, meV \qquad (36)$$

In addition, there is a Gaussian distribution of dangling bond states (states corresponding to nonsaturated silicon bonds) around midgap (Figure 12):

$$D_{DB} = \frac{N_{DB}}{\sqrt{2\pi w^2}} exp\left(\frac{(E - E_{DB})}{2w^2}\right) \quad w \approx 100\, meV \quad E_{DB} = \frac{E_C + E_V}{2} \qquad (37)$$

In device-quality intrinsic a-Si:H films, the density of dangling bonds N_{DB} ranges from 10^{15} to 10^{16} cm^{-3}.

The carrier mobility in the localised states (band tails and dangling bonds) is negligible. The accepted values for extended states in the valence and

---

[1] A rigorous description of the band structure and charge carrier transport in amorphous silicon is a complex matter well beyond the scope of this handbook. This section gives a simplified picture in terms of effective parameters that has been used in success in semiconductor device modelling.

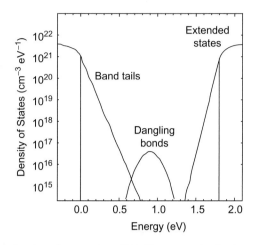

Figure 12 Typical density of states in a-Si:H. The states in the band tails behave as carrier traps whereas dangling bonds are recombination centres. Only those carriers in the extended states contribute to the electron transport [65].

conduction band are $\mu_p = 0.5$ cm^2/V-s and $\mu_n = 10$ cm^2/V-s, respectively. As in other semiconductors, the electron transport is given by the drift-diffusion equations (5) where the Einstein relations (10) apply.

The optical absorption coefficient shows three different zones (Figure 13):

- Band-to-band transitions (Tauc plot) for photon energies in excess of the band gap (>1.8 eV).
- Transitions involving tail states (Urbach's front) for photon energies in the range 1.5–1.8 eV.
- Transitions involving dangling bonds for photon energies below 1.5 eV.

Once the optical absorption coefficient is known, the carrier generation can be easily calculated according to Equation (17), but it is important to note that only photons with energies in excess of the band gap yield useful electron–hole pairs for photovoltaic conversion.

Finally, the dominant recombination mechanism in intrinsic a-Si:H is given by the modified Shockley–Read–Hall equation, as applied to the amphoteric distribution of dangling bonds:

$$U = v_{th}\left(n\sigma_n^o + p\sigma_p^o\right) \frac{N_{DB}}{1 + \frac{p}{n}\frac{\sigma_p^o}{\sigma_n^o} + \frac{n}{p}\frac{\sigma_n^o}{\sigma_p^-}} \qquad (38)$$

where the parameters are given in Table 12 [66].

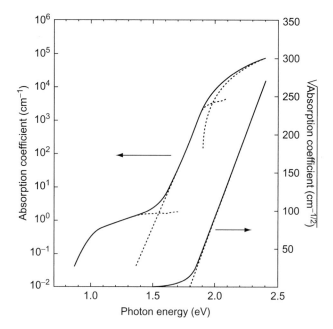

Figure 13 Typical optical absorption coefficient in a-Si:H. Three different regions can be observed corresponding to band-to-band absorption, transitions involving tail states, and defect-associated absorption [65]

Table 12 Recombination parameters in amorphous silicon.

$v_{th}(cm \cdot s^{-1})$	$\sigma_p^o(cm^2)$	$\sigma_n^o(cm^2)$	$\sigma_n^+(cm^2)$	$\sigma_p^-(cm^2)$
10^7	10^{-15}	$\sigma_n^o/3$	$50\sigma_n^o$	$50\sigma_p^o$

ACKNOWLEDGEMENTS

We are grateful to J. Puigdollers and C. Voz for providing graphs of the absorption coefficient and the density of states in amorphous silicon.

REFERENCES

[1] R.J. Schwartz, The application of numerical techniques to the analysis and design of solar cells, in: T.J. Coutts, J.D. Meakin (Eds.), Current Topics in Photovoltaics, vol. 4, 1990, p. 25.
[2] S. Selberherr, Analysis and Simulation of Semiconductor Devices, Springer, Vienna, New York, 1984.
[3] M.R. Brozel, G.E. Stillman (Eds.), Properties of Gallium Arsenide, third ed., IEE/INSPEC, Institution of Electrical Engineers, London, 1996.
[4] T.P. Pearsall (Ed.), Properties, Processing and Applications of Indium Phosphide, INSPEC/IEE, London, 2000.

[5] M. Shur, Physics of Semiconductor Devices, Prentice Hall, Englewood Cliffs, NJ, 1990.
[6] T.L. Chu, Cadmium telluride solar cells, in: T.J. Coutts, J.D. Meakin (Eds.), Current Topics in Photovoltaics, vol. 3, 1988, p. 235.
[7] L.L. Kazmerski, S. Wagner, Cu-ternary chalcopyrite solar cells, in: T.J. Coutts, J.D. Meakin (Eds.), Current Topics in Photovoltaics, vol. 1, 1985, p.41.
[8] Handbook of Physics and Chemistry of Solids (82nd edition), CRC Press, Boca Raton, 2001.
[9] S. Adachi, GaAs, AlAs and $Al_xGa_{1-x}As$: material parameters for use in research and device applications, J. Appl. Phys. 58 (1985) R 1.
[10] Y.P. Varshni, Temperature dependence of the energy gap in semiconductors, Physica 34 (1967) 149.
[11] C.D. Thurmond, The standard thermodynamic functions for the formation of electrons and holes in Ge, Si, GaAs and GaP, J. Electrochem. Soc. 122 (1975) 1133.
[12] A.B. Sproul, M.A. Green, Intrinsic carrier concentration and minority-carrier mobility of silicon from 77-K to 300-K, J. Appl. Phys. 73 (1993) 1214–1225.
[13] A.H. Marshak, K.M. van Vliet, Electrical currents in solids with position dependent band structure, Solid-State Electron. 21 (1978) 417.
[14] R.R. Smith, Semiconductors, second ed., Cambridge University Press, Cambridge, 1978.
[15] D.M. Caughey, R.E. Thomas, Carrier mobilities in silicon empirically related to doping and field, Proc. IEEE 55 (1967) 2192.
[16] D.B.M. Klaassen, A unified mobility model for device simulation-l. Model equations and concentration dependence, Solid-State Electron. 35 (1992) 953–959.
[17] N.D. Arora, T.R. Hauser, D.J. Roulston, Electron and hole mobilities in silicon as a function of concentration and temperature, IEEE Trans. Electron Dev. ED-29 (1982) 292.
[18] S.E. Swirhun, Y.-H. Kwark, R.M. Swanson, Measurement of electron lifetime, electron mobility and band-dap narrowing in heavily doped p-type silicon, IEDM'86 (1986) 24–27.
[19] J. del Alamo, S. Swirhun, R.M. Swanson, Simultaneous measurement of hole lifetime, hole mobility and band-gap narrowing in heavily doped n-type silicon, IEDM'85 (1985) 290–293.
[20] S.M. Sze, Physics of Semiconductor Devices, second ed., John Wiley & Sons, New York, 1981.
[21] D. Lancefield, Electron mobility in GaAs: overview, in: M.R. Brozel, G.E. Stillman (Eds.), Properties of Gallium Arsenide, 3rd edition, IEE/INSPEC, Institution of Electrical Engineers, London, 1996, p. 41.
[22] E.S. Harmon, M.L. Lovejoy, M.S. Lundstrom, M.R. Melloch, Minority electron mobility in doped GaAs, in: M.R. Brozel, G.E. Stillman (Eds.), Properties of Gallium Arsenide, third ed., IEE/INSPEC, Institution of Electrical Engineers, London, 1996, p. 81.
[23] M.L. Lovejoy, M.R. Melloch, M.S. Lundstrom, Minority hole mobility in GaAs, in: M.R. Brozel, G.E. Stillman (Eds.), Properties of Gallium Arsenide, third ed., IEE/INSPEC, Institution of Electrical Engineers, London, 1996, p. 123.
[24] J.J. Barnes, R.J. Lomax, G.I. Haddad, Finite element simulation of GaAs MESFET's with lateral doping profiles and submicron gates, IEEE Trans. Electron Dev. ED-23 (1976) 1042.
[25] M.A. Littlejohn, J.R. Hauser, T.H. Glisson, Velocity-field characteristics of GaAs with $\Gamma_6^c - L_6^c - X_6^c$ ordering, J. Appl. Phys. 48 (1977) 4587.
[26] R.K. Ahrenkiel, Minority carrier lifetime in InP, in: T.P. Pearsall (Ed.), Properties, Processing and Applications of Indium Phosphide, INSPEC/IEE, London, 2000.

[27] E.D. Palik (Ed.), Handbook of Optical Constants of Solids, Academic Press Handbook Series, Orlando, 1985.
[28] E.D. Palik (Ed.), Handbook of Optical Constants of Solids II, Academic Press, San Diego, 1991.
[29] K. Rajkanan, R. Singh, J. Shewchun, Absorption coefficient of silicon for solar cell calculations, Solid-State Electron. 22 (1979) 793.
[30] P.E. Schmid, Optical absorption in heavily doped silicon, Phys. Rev. B23 (1981) 5531.
[31] H.Y. Fan, in: R.K. Willardson, A.C. Beer (Eds.), Semiconductors and Semimetals, vol. 3, Academic Press, 1967, p. 409.
[32] P.T. Landsberg, Recombination in Semiconductors, Cambridge University Press, 1991.
[33] W. van Roosbroeck, W. Shockley, Photon-radiative recombination of electrons and holes in germanium, Phys. Rev. 94 (1954) 1558.
[34] M.H. Pilkuhn, Light emitting diodes, in: T.S. Moss (Ed.), vol. 4, Handbook of Semiconductors, North Holland, p. 539.
[35] M. Levinstein, S. Rumyantsev, M. Shur (Eds.), Handbook Series on Semiconductor Parameters, vols. 1 and 2, World Scientific, London, 1996, 1999 (See also http://www.ioffe.rssi.ru/SVA/NSM/Semicond/).
[36] W. Shockley, W.T. Read, Statistics of the recombination of holes and electrons, Phys. Rev. 87 (1952) 835.
[37] R.N. Hall, Electron hole recombination in germanium, Phys. Rev. 87 (1952) 387.
[38] A. Cuevas, G. Giroult-Matlakowski, P.A. Basore, C. du Bois, R. King, Extraction of the surface recombination velocity of passivated phosphorus doped emitters, Proc. 1st World Conference on Photovoltaic Energy Conversion, Hawaii, 1994, pp. 1446–1449.
[39] R.R. King, R.A. Sinton, R.M. Swanson, Studies of diffused emitters: saturation current, surface recombination velocity and quantum efficiency, IEEE Trans. Electron Dev. ED-37 (1990) 365.
[40] V.M. Andreev, V.A. Grilikhes, V.D. Rumyantsev, Photovoltaic Conversion of Concentrated Sunlight, Wiley, Chichester, 1997.
[41] T.J. Coutts, M. Yamaguchi, Indium phosphide based solar cells: a critical review of their fabrication, performance and operation, in: T.J. Coutts, J.D. Meakin (Eds.), Current Topics in Photovoltaics, vol. 3, 1988, p. 79.
[42] D.B.M. Klaassen, A unified mobility model for device simulation—II. Temperature dependence of carrier mobility and lifetime, Solid-State Electron. 35 (1992) 961.
[43] H.Y. Tada, J.R. Carter, B.E. Anspaugh, R.G. Downing, Solar-Cell Radiation Handbook, Jet Propulsion Laboratory, California Institute of Technology, Pasadena, CA, 1982JPL Publication 82–69
[44] B.E. Anspaugh, GaAs Solar Cell Radiation Handbook, Jet Propulsion Laboratory, California Institute of Technology, Pasadena, CA, 1996JPL Publication 96–9.
[45] G.H. Kinchin, R.S. Pease, The displacement of atoms in solids by radiation, Rep. Prog. Phys. 18 (1955) 1.
[46] F. Seitz, J.S. Koehler, Displacement of atoms during irradiation, Solid State Phy. 2 (1956) 307.
[47] H.J. Hovel, Semiconductor solar cells, in: R.K. Willardson, A.C. Beer (Eds.), Semiconductors and Semimetals, vol. 11, Academic Press, New York, 1975.
[48] T. Markvart, Radiation damage in solar cells, J. Mater. Sci.: Mater. Electron. 1 (1990) 1.
[49] M. Yamaguchi, K. Ando, Mechanism for radiation resistance of InP solar cells, J. Appl. Phys. 63 (1988) 5555.
[50] F.M. Smits, IEEE Trans. Nucl. Sci. NS-10 (1963) 88.

[51] A. Meulenberg, F.C. Treble, Damage in silicon solar cells from 2 to 155 MeV protons, Proc. 10th IEEE Photovoltaic Specialists Conf., Palo Alto, 1973, p. 359.
[52] W. Rosenzweig, Diffusion length measurement by means of ionizing radiation, Bell. Syst. Tech. J. 41 (1962) 1573—1588.
[53] N.D. Wilsey, Proc. 9th IEEE Photovoltaic Specialists Conf., Silver Springs, 1972, p. 338.
[54] J.A. Minahan, M.J. Green, Proc. 18th IEEE Photovoltaic Specialists Conf., 1985, p. 350.
[55] R.G Downing, J.R. Carter Jr., J.M. Denney, The energy dependence of electron damage in silicon, Proc. 4th IEEE Photovoltaic Specialists Conf. 1 (1964) A-5—1.
[56] W. Rosenzweig, F.M. Smits, W.L. Brown, Energy dependence of proton irradiation damage in silicon, J. Appl. Phys. 35 (1964) 2707.
[57] R.W. Keyes, The energy gap of impure silicon, Comm. Solid State Phys. 7(6) (1977) 149.
[58] J. Wagner, J.A. delAlamo, Band-gap narrowing in heavily doped silicon: a comparison of optical and electrical data, J. Appl. Phys. 63 (1988) 425.
[59] S.C. Jain, D.J. Roulston, A simple expression for bandgap narrowing in heavily doped Si. Ge, GaAs and Ge_xSi_{1-x} strained layers, Solid-State Electron. 34 (1991) 453.
[60] S.C. Jain, J.M. McGregor, D.J. Roulston, P. Balk, Modified simple expression for bandgap narrowing in n-type GaAs, Solid-State Electron. 35 (1992) 639.
[61] D.B.M. Klaassen, J.W Slotboom, H.C. de Graaf, Unified apparent bandgap narrowing in n and p-type silicon, Solid-State Electron. 35 (1992) 125.
[62] M.S. Lundstrom, E.S. Harmon, M.R. Melloch, Effective bandgap narrowing in doped GaAs, in: M.R. Brozel, G.E. Stillman (Eds.), Properties of Gallium Arsenide, third ed., IEE/INSPEC, Institution of Electrical Engineers, London, 1996, p. 186.
[63] C. Tsang, R.A. Street, Phys. Rev. B19 (1979) 3027.
[64] H. Fritzsche (Ed.), Amorphous Silicon and Related Materials, University of Chicago, 1989.
[65] J. Puigdollers, C. Voz, personal communication.
[66] R.A. Street, Hydrogenated Amorphous Silicon, Cambridge University Press, 1991.

CHAPTER IA-3

Ideal Efficiencies

Peter T. Landsberg[a] and Tom Markvart[b]
[a]Faculty of Mathematical Studies, University of Southampton, UK
[b]School of Engineering Sciences, University of Southampton, UK

1. Introduction 63
2. Thermodynamic Efficiencies 64
3. Efficiencies in Terms of Energies 65
4. Efficiencies Using the Shockley Solar Cell Equation 67
5. General Comments on Efficiencies 72
References 74

1. INTRODUCTION

In this chapter we deal with the simplest ideas that have been used in the past to attain an understanding of solar cell efficiencies from a theoretical point of view. The first and most obvious attack on this problem is to use thermodynamics, and we offer four such estimates in Section 2. Only the first of these is the famous Carnot efficiency. The other three demonstrate that one has more possibilities even within the framework of thermodynamics. To make progress, however, one has to introduce at least one solid-state characteristic, and the obvious one is the energy gap, E_g. That this represents an advance in the direction of a more realistic model is obvious, but it is also indicated by the fact that the efficiency now calculated is lower than the (unrealistically high) thermodynamic efficiencies (Section 3). In order to get closer to reality, we introduce in Section 4 the fact that the radiation is effectively reduced from the normal blackbody value (Equation (6)) owing to the finite size of the solar disc. This still leaves important special design features such as the number of series-connected tandem cells and higher-order impact ionisation, and these are noted in Section 5.

Practical Handbook of Photovoltaics.
© 2012 Elsevier Ltd. All rights reserved.

2. THERMODYNAMIC EFFICIENCIES

The formulae for ideal efficiencies of solar cells are simplest when based on purely thermodynamic arguments. We here offer four of these: they involve only (absolute) temperatures:
- T_a, temperature of the surroundings (or the ambient),
- T_s, temperature of the pump (i.e., the sun),
- T_c, temperature of the actual cell that converts the incoming radiation into electricity.

From these temperatures, we form the following efficiencies [1]:

$$\eta_C \equiv 1 - T_a/T_s, \text{ the Carnot efficiency} \tag{1}$$

$$\eta_{CA} \equiv 1 - (T_a/T_s)^{\frac{1}{2}}, \text{ the Curzon–Ahlborn efficiency} \tag{2}$$

$$\eta_L \equiv 1 - (4/3)(T_a/T_s) + (1/3)(T_a/T_s)^4, \text{ the Landsberg efficiency} \tag{3}$$

$$\eta_{PT} = \left[1 - (T_c/T_s)^4\right]\left[1 - T_a/T_c\right], \text{ the photo–thermal efficiency due to Müser} \tag{4}$$

In the latter efficiency, the cell temperature is determined by the quintic equation

$$4T_C^5 - 3T_a T_c^4 - T_a T_s^4 = 0 \tag{5}$$

The names associated with these efficiencies are not historically strictly correct: for example, in Equations (2) and (3) other authors have played a significant part.

Figure 1 [1] shows curves of the four efficiencies, which all start at unity when $T_a/T_s \equiv 0$, and they all end at zero when $T_a = T_s$. No efficiency ever beats the Carnot efficiency, of course, in accordance with the rules of thermodynamics. Values near $T_s = 5760-5770$ K seem to give the best agreement with the observed solar spectrum and the total energy received on Earth, but a less accurate but more convenient value of $T_s = 6000$ K is also commonly used. Using the latter value of T_s and $T_a = 300$ K as the temperature for Earth, one finds

$$\eta_C = 95\%, \eta_{CA} = 77.6\%, \eta_L = 93.3\%, \eta_{PT} = 85\%$$

If $T_s = T_a = T_c$ one has in effect an equilibrium situation, so that the theoretical efficiencies are expected to vanish.

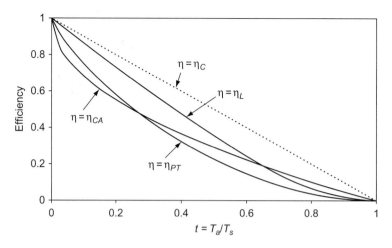

Figure 1 The efficiencies (1)–(4) as functions of T_a/T_s.

The above thermodynamic efficiencies utilise merely temperatures, and they lie well above experimental results. One needs an energy gap (E_g) as well to take us from pure thermodynamics to solid-state physics. Incident photons can excite electrons across this gap, thus enabling the solar cell to produce an electric current as the electrons drop back again. The thermodynamic results presented earlier, on the other hand, are obtained simply by considering energy and entropy fluxes.

3. EFFICIENCIES IN TERMS OF ENERGIES

In order to proceed, we need next an expression for the number of photons in blackbody radiation with an energy in excess of the energy gap, E_g say, so that they can excite electrons across the gap. At blackbody temperature T_s the number of photons incident on unit area in unit time is given by standard theory as an integral over the photon energy [2]:

$$\Phi(E_g, T_s) = \frac{2\pi k^3}{h^3 c^2} T_s^3 \int_{E_g/kT_s}^{\infty} \frac{x^2 dx}{e^x - 1} \qquad (6)$$

Now suppose that each of these photons contributes only an energy equal to the energy gap to the output of the device, i.e., a quantity proportional to

$$x_g \int_{x_g}^{\infty} \frac{x^2 dx}{e^x - 1} \quad (x_g \equiv E_g/kT_s) \tag{7}$$

To obtain the efficiency η of energy conversion, we must divide this quantity by the whole energy that is, in principle, available from the radiation:

$$\eta = x_g \int_{x_G}^{\infty} \frac{x^2 dx}{e^x - 1} \bigg/ \int_{0}^{\infty} \frac{x^2 dx}{e^x - 1} \tag{8}$$

Equation (8) gives the first of the Shockley–Queisser estimates for the limiting efficiency of a solar cell, the *ultimate efficiency* (see Figure 5). The argument neglects recombination in the semiconductor device, even radiative recombination, which is always present (a substance that absorbs radiation can always emit it!). It is also based on the blackbody photon flux (Equation (6)) rather than on a more realistic spectrum incident on Earth.

We shall return to these points in Section 4, but first a brief discussion of Equation (8) is in order. There is a maximum value of η for some energy gap that may be seen by noting that $\eta = 0$ for both $x_g = 0$ and for x_g very large. So there is a maximum efficiency between these values. Differentiating η with respect to x_g and equating to zero, the condition for a maximum is

$$x_g = x_{gopt} = 2.17$$

corresponding to $\eta = 44\%$.

This is still higher than most experimental efficiencies, but the beauty of it is that it is a rather general result that assumes merely properties of blackbody radiation.

Let $f(x)$ be a generalised photon distribution function; then a generalised efficiency can be defined by

$$\eta = \frac{x_g \int_{x_g}^{\infty} f(x)dx}{\int_{0}^{\infty} xf(x)dx} \tag{9}$$

The maximum efficiency with respect to x_g is then given by

$$x_{gopt} f(x_{gopt}) = \int_{x_{gopt}}^{\infty} f(x)dx \tag{10}$$

This is rather general and will serve also when the photon distribution departs from the blackbody forms and even for radiation in different numbers of dimensions.

4. EFFICIENCIES USING THE SHOCKLEY SOLAR CELL EQUATION

A further step in finding the appropriate efficiency limits for single-junction solar cells can be made by estimating the relevant terms in the Shockley ideal solar cell equation (Equation (1) in Chapter Ia-1). To this end, further remarks must be made about the solar spectrum and solar energy incident on Earth's surface. The ultimate efficiency, discussed in Section 3, was based on the blackbody photon flux (Equation (6)), a rigorous thermodynamic quantity but not a very good estimate of the solar spectrum as seen on Earth. By virtue of the large distance between the Sun and Earth, the radiative energy incident on Earth's surface is less than that of Equation (6) by a factor f_ω, which describes the size of the solar disk (of solid angle ω_s) as perceived from Earth:

$$f\omega = \left(\frac{R_s}{R_{SE}}\right)^2 = \frac{\omega_s}{\pi} \qquad (11)$$

where R_s is the radius of the Sun (696×10^3 km), and R_{SE} is the mean distance between the Sun and Earth (149.6×10^6 km), giving $\omega_s = 6.85 \times 10^{-5}$ sterad and $f_\omega = 2.18 \times 10^{-5}$. The resulting spectrum is shown in Figure 2 alongside the standard terrestrial AM1.5 spectrum (a further discussion of the

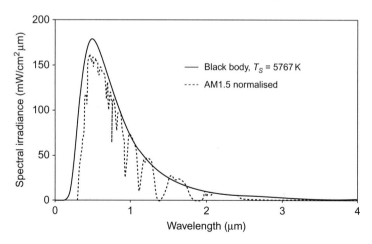

Figure 2 The blackbody spectrum of solar radiation and the AM1.5 spectrum, normalised to total irradiance 1 kW/m², which is used for the calibration of terrestrial cells and modules.

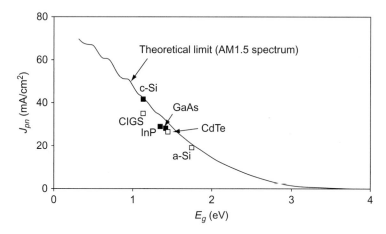

Figure 3 The theoretical limit on photogenerated current, compared with the best measured values. The curve is obtained by replacing the product $f_\omega \Phi(E_g, T_s)$ in Equation (12) by the appropriate AM1.5 photon flux. Full symbols correspond to crystalline materials, open symbols to thin films.

spectra that are used for solar cell measurements in practice can be found in Chapter III-2, which also shows the extraterrestrial spectrum AMO).

The maximum value of the photogenerated current I_{ph} now follows if we assume that one absorbed photon contributes exactly one electron to the current in the external circuit:

$$I_{ph} = A q f \omega \Phi(E_g, T_s) \qquad (12)$$

where A is the illuminated area of the solar cell and q is the electron charge. The maximum photogenerated current density $J_{ph} = I_{ph}/A$ that can be produced by a solar cell with band gap E_g is shown in Figure 3. To allow a comparison with photocurrents measured in actual devices, Figure 3 is plotted for the AM1.5 solar spectrum, which is used for calibration of terrestrial solar cells, rather than for the blackbody spectrum used in Section 3.

The open-circuit voltage V_{oc} can now be obtained using the photogenerated current I_{ph} (Equation (12)) and the (dark) saturation current I_0 that appears in the ideal solar cell equation:

$$V_{oc} = \frac{kT}{q} \ln\left(1 + \frac{I_{ph}}{I_o}\right) \qquad (13)$$

The current I_0 can be obtained by a similar argument as the photogenerated current I_{ph}, since, as argued by Shockley and Queisser, it can be equated to the blackbody photon flux at the cell temperature T_c (in what follows, the cell temperature T_c will be assumed to be equal to the ambient temperature T_a):

$$I_0 = A q f_0 \Phi(E_g, T_a) \quad (14)$$

where the coefficient f_0 has been inserted to describe correctly the total area $f_0 A$ exposed to the ambient photon flux. Various values of f_0 (some dependent on the refractive index **n** of the cell material) can be found, appropriate for different device structures and geometries. The usual value is $f_0 = 2$, as suggested by Shockley and Queisser [2], since this radiation is incident through the two (front and rear) surfaces of the cell. A similar argument for a spherical solar cell yields an effective value $f_0 = 4$ [3]. Henry [4] gives $f_0 = 1 + \mathbf{n}^2$ for a cell grown on a semiconductor substrate, but the value $f_0 = 1$ is also sometimes used (see, for example, [5]). Green [6] gives a semi-empirical expression for the dark saturation current density $J_0 = I_o/A$:

$$J_0 (\text{in Amps}/cm^2) = 1.5 \times 10^5 \exp\left(-\frac{E_g}{kT_a}\right) \quad (15)$$

An approximate analytical method for estimating V_{oc} can also be useful, particularly as it stresses the thermodynamic origin of V_{oc}. Indeed, it can be shown [7] that, near the open circuit, the solar cell behaves as an ideal thermodynamic engine with Carnot efficiency $(1 - T_c/T_s)$. Ruppel and Würfel [3] and Araùjo [8] show that V_{oc} can be approximated to a reasonable accuracy by the expression

$$V_{oc} \frac{E_g}{q}\left(1 - \frac{T_c}{T_s}\right) + \frac{kT}{q}\ln\frac{f_\omega}{f_0} + \frac{kT_c}{q}\ln\frac{T_s}{T_c} \quad (16)$$

which depicts the dependence of V_{oc} on the band gap E_g and on the cell temperature T_c. Figure 4 compares this theoretical values for the open-circuit voltage with data for the best solar cells to date from different materials.

Using now an expression for the fill factor (defined by Equation (3) in Chapter Ia-1), one readily obtains a theoretical estimate for the efficiency. Slightly different results may be encountered, principally by virtue of the different ways one can estimate the current and the voltage. Figure 5 shows the best-known result, the celebrated Shockley–Queisser ideal efficiency limit [2]. Shockley and Queisser call this limit the *nominal*

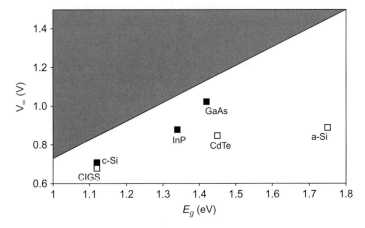

Figure 4 The theoretical Shockley–Queisser limit on open-circuit voltage: values exceeding this limit lie in the shaded area of the graph. Line corresponding to Equation (16) appears as identical to within the accuracy of this graph. Full symbols correspond to crystalline materials, open symbols to thin films.

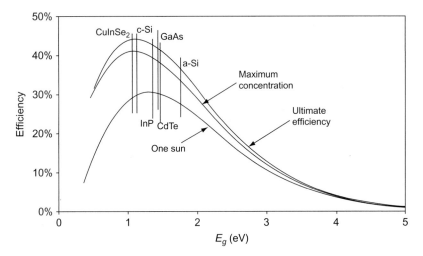

Figure 5 The 'ultimate' and two 'nominal' Shockley–Queisser efficiencies. Note that the blackbody radiation with temperature $T_s = 6000$ K has been used here, in keeping with the Shockley–Queisser work [2].

efficiency, to be compared with the *ultimate efficiency*, which is discussed in Section 3. Figure 5 shows two such curves: one labelled 'one-sun' corresponds to the AMO solar intensity, as observed outside Earth's atmosphere. A second curve, labelled 'maximum concentration' corresponds

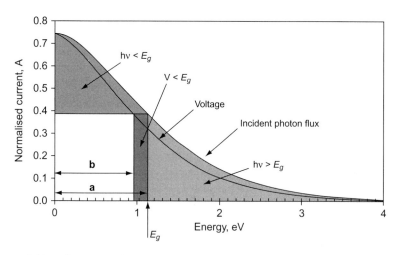

Figure 6 Henry's construction.

to light focused on the cell, by a mirror or a lens, at the maximum concentration ratio of $1/f_\omega = 45,872$ [9].

The various unavoidable losses in photovoltaic energy conversion by single-junction solar cells can be depicted in a graph constructed by Henry [4] and analogous to Figure 6. There are two curves in this graph:

- The photogenerated current density J_{ph} from Equation (12) as a function of photon energy. J_{ph} is divided here by the total irradiance, making the area under this curve equal to unity by construction.
- The maximum voltage that can be extracted from the cell at the maximum power point. This curve is drawn in such a way that the ratio of lengths of the two arrows b/a is equal to V_m/E_g.

The three shaded areas then depict the three fundamental losses in a single junction solar cell (shown here for silicon with band gap E_g equal to 1.12 eV):

- Shaded area marked $h\nu < E_g$ is equal to the loss of current due to the inability of the semiconductor to absorb below-band-gap light.
- Shaded area marked $h\nu > E_g$ represents energy losses due to the thermalization of electron–hole pairs to the band-gap energy.
- Hatched area marked $V < E_g$ corresponds to the combined thermodynamic losses due to V_{oc} being less than E_g and losses represented by the fill factor FF.

The area of the blank rectangle then represents the maximum efficiency that can be obtained for a single junction cell made from a semiconductor with band gap E_g. The graph is drawn here for light with maximum possible concentration. A different 'voltage curve' would result if light with one-sun intensity were used.

5. GENERAL COMMENTS ON EFFICIENCIES

The ideal solar cell efficiencies discussed above refer to single-junction semiconductor devices. The limitations considered in the ultimate efficiency of Section 3 are due to the fact that the simplest semiconductor (i.e., one whose defects and impurities can be ignored) cannot absorb below-band-gap photons. Furthermore, it is also due to the fact that the part of the energy of the absorbed photons in excess of the band gap is lost as heat. Radiative recombination at the necessary fundamental level was taken into account in the treatment of Section 4. It is sometimes argued that there are other 'unavoidable' losses due to electronic energy transfer to other electrons by the Auger effect (electron–electron collisions) [10–12]. There is also the effect of band-gap shrinkage, discussed in Chapter Ia-2, and light trapping may also play a part [11]. None of these effects are discussed here, and the reader is referred to the relevant literature.

It is clear that it is most beneficial if one can improve the effect of a typical photon on the electron and hole density. This can be achieved, for example, if the photon is energetic enough to produce two or more electron–hole pairs. This is called *impact ionisation* and has been studied quite extensively. A very energetic photon can also project an electron high enough in to the conduction band so that it can, by collision, excite a second electron from the valence band. This also improves the performance of the cell. On the other hand, an electron can combine with a hole, and the energy release can be used to excite a conduction-band electron higher into the band. In this case, energy is uselessly dissipated with a loss of useful carriers and hence of conversion efficiency. This is one type of Auger effect. For a survey of these and related effects, see [12]. These phenomena suggest a number of interesting design problems. For example, is there a way of limiting the deleterious results of Auger recombination [13]? One way is to try to 'tune' the split-off and the fundamental band gaps appropriately. If one is dealing with parabolic bands,

Table 1 The maximum efficiencies of tandem cells as a function of the number of cells in the stack for different concentration ratios [17]. Note that de Vos [17] uses a slightly smaller value of f_ω than Shockley and Queisser, resulting in a marginally different maximum concentration ratio than used in Figure 5

Concentration ratio	Number of cells in the stack	Maximum efficiency (%)
1	1	31.0
	2	49.9
	3	49.3
	...	
	∞	68.2
46.300	1	40.8
	2	55.7
	3	63.9
	...	
	∞	86.8

then the obvious way is to examine the threshold energies that an electron has to have in order to jump across the gap and make these large so as to make this jump difficult.

Then there is the possibility of placing impurities on the energy band scale in such a way as to help better use to be made of low-energy photons, so that they can now increase the density of electrons in the system. This is sometimes referred to as the *impurity photovoltaic effect*. So one can make use of it [14].

One can also utilise excitons to improve the efficiencies of solar cells. There may be as many as 10^{17} cm^{-3} excitons in silicon at room temperature. If they are split up in the field of a p–n junction, this will increase the concentration of current carriers and so increase the light generated current, which is of course beneficial.

We have here indicated some useful ideas for improving solar cells, There are of course many others, some of which are discussed in Chapters Ib-2 and Id-2 as well as elsewhere [15]. Note, in particular, the idea of developing tandem cells in which photons hit a large band-gap material first and then proceed gradually to smaller band-gap materials. Tandem cells are now available with three of more stages. Solar cells with efficiency of order 20% arc predicted to be produced on a large scale in the near future [16]. Table 2 shows the best laboratory efficiencies at the present time for different materials.

Table 2 The best reported efficiencies, at time of writing, for different types of solar cells [18]

	Efficiency (%)	J_{se} (mA/cm^2)	V_{oc} (V)	FF (%)
Crystalline: single junction				
c-Si	24.7	42.2	0.706	82.8
GaAs	25.1	28.2	1.022	87.1
InP	21.9	29.3	0.878	85.4
Crystalline: multijunction				
GaInP/GaAs/Ge tandem	31.0	14.11	2.548	86.2
Thin-film: single junction				
CdTe	16.5	25.9	0.845	75.5
CIGS	18.9	34.8	0.696	78.0
Thin-film: multijunction				
a-Si/a-SiGe tandem	13.5	7.72	2.375	74.4
Photoelectrochemical				
Dye-sensitised TiO$_2$	11.0	19.4	0.795	71.0

REFERENCES

[1] P.T. Landsberg, V. Badescu, Solar energy conversion: list of efficiencies and some theoretical considerations, Prog. Quantum Electron. 22 (1998) 211 and 231.
[2] W. Shockley, H.J. Queisser, Detailed balance limit of efficiency of pn junction solar cells, J. Appl. Phys. 32 (1961) 510.
[3] W. Ruppel, P. Würfel, Upper limit for the conversion of solar energy, IEEE Trans. Electron Dev. ED-27 (1980) 877.
[4] C.H. Henry, Limiting efficiencies of ideal single and multiple energy gap terrestrial solar cells, J. Appl. Phys. 51 (1980) 4494.
[5] H Kiess, W. Rehwald, On the ultimate efficiency of solar cells, Solar Energy Mater Solar Cells 38(55) (1995) 45.
[6] M.A. Green, Solar Cells, Prentice Hall, New York, 1982.
[7] P. Baruch, J.E. Parrott, A thermodynamic cycle for photovoltaic energy conversion, J. Phys. D: Appl. Phys. 23 (1990) 739.
[8] G.L. Araùjo, Limits to efficiency of single and multiple band gap solar cells, in: A. Luque, G.L. Araùjo (Eds.), Physical Limitations to Photovoltaic Energy Conversion, Adam Hilger, Bristol, 1990, p. 106.
[9] W.T. Welford, R. Winston, The Physics of Non-imaging Concentrators, Academic Press, New York, 1978 (Chapter 1).
[10] M.A. Green, Limits on the open-circuit voltage and efficiency of silicon solar cells imposed by intrinsic Auger process, IEEE Trans Electron Dev. ED-31 (1984) 671.
[11] T. Tiedje, E. Yablonovich, G.C. Cody, B.C. Brooks, Limiting efficiency of silicon solar cells, IEEE Trans Electron Dev. ED-31 (1984) 711.
[12] P.T. Landsbcrg, The band-band Auger effect in semiconductors, Solid-State Electron. 30 (1987) 1107.

[13] C.R. Pidgeon, C.M. Ciesla, B.N. Murdin, Suppression of nonradiative processes in semiconductor mid-infrared emitters and detectors, Prog. Quantum Electron. 21 (1997) 361.
[14] H. Kasai, H. Matsumura, Study for improvement of solar cell efficiency by impurity photovoltaic effect, Solar Energy Mater. Solar Cells 48 (1997) 93.
[15] M.A. Green, Third generation photovoltaics: Ultra high conversion efficiency at low cost, Prog. Photovoltaics Res. Appl. 9 (2001) 123–135.
[16] G.P. Wileke, The Frauenhoffer ISE roadmap for crystalline silicon solar cell technology, Proc. 29th IEEE Photovoltaic Specialists Conf., New Orleans, 2002.
[17] A. deVos, Detailed balance limit of the efficiency of tandem solar cells, J. Phys. D: Appl. Phys. 13 (1980) 839. (See also A. deVos, Endoreversible Thermodynamics of Solar Energy Conversion, Oxford University Press, 1992.).
[18] M.S. Green, K.I Emery, D.I King, S. Igari, W. Warta, Solar cell efficiency tables (version 20), Prog. Photovoltaics Res. Appl. 10 (2002) 355–360.

PART IB

Crystalline Silicon Solar Cells

CHAPTER IB-1

Crystalline Silicon: Manufacture and Properties

Francesca Ferrazza
Eurosolare S.p.A, Nettuno, Italy

1. Introduction	79
2. Characteristics of Silicon Wafers for Use in PV Manufacturing	80
2.1 Geometrical Specifications	80
2.2 Physical Specifications	81
2.3 Physical Specifications	82
3. Feedstock Silicon	86
4. Crystal-Preparation Methods	87
4.1 Czochrahki Silicon	87
4.2 Multicrystalline Silicon	88
4.2.1 Charge Preparation	90
4.2.2 Crucibles	90
4.3 Electromagnetic Continuous Casting	90
4.4 Float-Zone Silicon	91
4.5 Nonwafer Technologies	92
5. Shaping and Wafering	94
5.1 Shaping	94
5.2 Wafering	94
References	96

1. INTRODUCTION

The majority of silicon wafers used for solar cells are Czochralski (CZ) single crystalline and directional solidification, or cast, multicrystalline (mc) material. The split between the two types of wafer is presently about 55% mc-Si and 45% CZ-Si. Until 1995, CZ wafers represented 60% of the substrates used by industry and mc-Si wafers around 25%. The fast scale-up of commercially available multicrystalline wafers changed the picture rapidly. The remainder of the silicon substrates used by the industry

Practical Handbook of Photovoltaics.
© 2012 Elsevier Ltd. All rights reserved.

are nonwafered sheets or ribbons that are of different types and have recently gained significant production figures following long development phases. Nonwafer silicon accounted for about 4% of the market in 2001, up from 1—2% in the mid-1990s [1,2].

2. CHARACTERISTICS OF SILICON WAFERS FOR USE IN PV MANUFACTURING

2.1 Geometrical Specifications

Most of the wafer substrates used in production facilities have dimensions relating to the diameters of monocrystalline silicon cylinders for the semiconductor industry (essentially 5 and 6 inches) that, in turn, have influenced standards for wafer carriers, automation, packaging, etc. However, in order to maximise the power density of the modules, wafers are square, or pseudo-square in the case of monocrystalline silicon—that is, cylinders are shaped as squares with rounded-off corners. This reduces the surface area of the wafers by between 2% and 5% compared with a full square of same dimensions.

In the case of mc-Si, ingot sizes are designed to be compatible with multiple numbers of each of the standard wafer dimensions in order to maximise geometrical yield. Yield considerations limit the possible wafer sizes achievable for any given ingot dimension, as much as expensive wafer cassettes, automation, and packaging do later in the process. Table 1 reports the different sizes for commercially available wafers, including typical tolerances.

A SEMI™ (Semiconductor Equipment and Materials International) standard, M61000, was developed with the purpose of covering the requirements for silicon wafers for use in solar cell manufacturing [3], including dimensional specifications, defects, and electronic properties. Most commercial suppliers sell their wafer products using specifications

Table 1 Commercially available wafer sizes

Nominal size	Dimension (mm)	Diagonal (mulli) (mm ± 1)	Diameter (mono) (inch)
103	103±0.5	146	5
125	125±0.5	177	6
150	150±0.5	212	—

Table 2 Other dimensional specifications for typical PV wafers

Parameter	Value
Thickness of a batch	530 ± 40 μm
Total Thickness Variation (TTV) of a wafer	50 μm
Cracks	<1 mm
Saw marks	<10 μm
Bow	<50 μm

that are close to those described by M61000. However, smaller wafers are usually 103 mm rather than 100 mm as specified, and there are some notable exceptions to the specifications—e.g., dimensions of wafers produced in-house by some of the early players who have developed their own standard and do not usually buy wafers on the market. Another obvious exception is provided by nonwafer substrates, the dimensions of which are in general determined by the growth equipment and the technique. Some manufacturers use rectangular wafers. Other requirements, besides the geometrical definitions of the wafers, are thickness uniformity and reduced levels of cracks and saw marks that could adversely affect later processing.

Typical specifications for commercially available wafers are described in Table 2.

The absolute value of the wafer thickness has dropped by about 100 μm in the last decade, as a consistent cost-reduction measure [4,5], and is expected to decrease further in the next years [2] as automation and cell processing become more sophisticated and can allow effective handling of thin wafers. Some wafer and cell producers already have less than 300 μm wafers in their production lines, although in general wafer sizes in such cases are limited to the 100 cm^2 range in order to maintain mechanical yields in the high 90s. Similarly to the case of the area dimensions, thickness in nonwafer substrates is determined by the process and is in general less homogeneous, providing one of the major differences between wafer and nonwafer cell technologies. A great deal of effort was put in the last decade in developing automated thickness measurement tools for manufacturing plants to inspect wafer thickness variations in lots, an extremely difficult task in manually inspected wafer-fabrication sequences.

2.2 Physical Specifications

Wafers are generally classified in terms of resistivity, type, and oxygen and carbon content. These data are generally present in all commercial

Table 3 Physical specifications of commercial silicon wafers

Parameter	Value
Type	P—boron doped
Resistivity	0.5–3 ohm cm
Oxygen (mc-Si)	$<8 \times 10^{17}$ at/cm^3
Oxygen (CZ-Si)	$<1 \times 10^{18}$ at/cm^3
Carbon (mc-Si)	$<1 \times 10^{18}$ at/cm^3
Carbon (CZ-Si)	$<2 \times 10^{17}$ at/cm^3

specifications related to single and multicrystalline wafers, and they refer to ASTM or equivalent standards. However, the PV community has had to face the unavoidable departure from standard test conditions of all parameters when measuring the properties of the inherently inhomogeneous nature of multicrystalline wafers, which led to agreement on relatively broad ranges for resistivity or upper thresholds for oxygen and carbon contents. Early concerns, for instance, of the influence of grain boundaries on the determination of resistivity using the four-point probe method are now somewhat more relaxed after significant statistical feedback has provided comfort in the values proposed. Still, in strict terms, standards related to the measurement of resistivity in multicrystalline wafers do not exist, which is true of course for nonwafer silicon technologies as well. This is also true for other kinds of measurements, and the effort to develop meaningful characterisation tools for lower or inhomogeneous quality materials as compared to the semiconductor industry is a clear indication of such a need (see Chapter III-1). Furthermore, the increasing volumes of wafers in the growing PV market has forced a second, big departure from semiconductor wafer characterisation standards, imposing fast, nondestructive test methods to optimise costs and yields. In most cases, for instance, resistivity and type are measured at block rather than at wafer level. Table 3 shows the typical values for physical parameters of commercially available wafers for industrial processing. These are either multicrystalline or monocrystalline Czochralski. Float-zone (FZ) wafers for PV may become commercial products and will be discussed in a later paragraph.

2.3 Physical Specifications

Minority-carrier lifetime characterisation of commercial silicon is worth a paragraph on its own. This is by far the most complicated parameter to measure and to effectively relate to subsequent processing quality and yield. It is also most influenced by the inhomogeneity of multicrystalline silicon, as well as by thermal treatments. It became immediately evident to all PV manufacturers at the very beginning of the expansion of the multicrystalline silicon market that the identification of an appropriate tool for analysing and understanding the properties of mc-Si would have been one of the keys for the commercial success of the material. A generous number of attempts were made to adapt the existing lifetime measurements—rigorously valid for high-quality polished single crystalline wafers—to provide meaningful values for mc-Si and even CZ-Si for the PV community. Also, as mentioned before, any acceptable test would need to be fast, cost-effective, and obviously nondestructive as the number of samples to inspect was bound to be large. This focussed effort led to the development of a number of automated lifetime analysers, which at the end of the development process had relatively low resemblance with the semiconductor industry counterparts. Probably the most successful commercial methods are the microwave photoconductance decay method (μ-PCD), performed directly on silicon blocks, and the photoconductance decay or quasi-steady state method developed by Ron Sinton, which is discussed in Chapter III-1.

For the purposes of the present chapter, we focus on the μ-PCD characterisation of mc-Si blocks. This measurement technique is commercially available and widely used, although strong debates on the validity of the results occurred for many years. The measurement is based on the detection of the amplitude of the microwave field reflected by the sample surface. This amplitude variation depends on the conductivity, and thus also on the number of minority carriers generated by a short laser pulse [6]. The time in which the system recovers the initial state is associated with the quality of the semiconductor material and with the recombination mechanisms in the bulk and at the surface. It is generally rather complicated to separate different contributions, and PV silicon has peculiar characteristics that enhance difficulties, such as relatively high doping, rough surfaces, and short diffusion lengths for minority carriers. Furthermore, as mentioned before, the industry requirements are for fast nondestructive techniques that enable prediction of later behaviour of the material in the processing line in order to minimise the costs of processing low-quality material as early as possible. For this reason, the industry has pushed towards the use of fast, noninvasive block scanners since the early 1990s, despite the inability of the measurement systems to conform to any of the existing standards. A certain effort has been directed until relatively recently toward developing uniform measurement systems and procedures, which has proved once more the difficulty of the problem [7] and finally led to generally accepted principles subject to bilateral confirmation in the case of commercial relationships between wafer vendors and cell producers.

The main problems with block scanners, besides separation of bulk and surface components, lie in the fact that the measurement is actually performed on a very thin portion of the block. This is due to the absorption of the laser pulse in silicon (usually in the near infrared range) and to the high reflectivity of microwaves, which only allow the field to penetrate a skin depth of the sample under examination. Other difficulties lie in the unpredictable behaviour of the material in three dimensions, lateral distribution of carriers, trapping effects, unknown injection levels, and the macroscopic saw damage affecting the control of the distance of the measurement head from the sample. Microwave block scanners are unable to handle these problems, which are better taken into account by the technique described in Chapter III-1. The experience of crystal growers and the feedback from cell processing lines, however, led to the establishment of a method for the analysis of silicon blocks that is able to reject low-quality material at block level and that allows the identification of the correct cropping position for the rejection of tops and tails.

A typical map, performed with a commercially available automated system [8], of a standard block of mc-Si grown by the directional solidification method is shown in Figure 1. The silicon is boron doped to a resistivity corresponding to 1×10^{16} at/cm^3, and the measurement is performed with a microwave field in the GHz range coupled by an antenna for a sample irradiated by a laser diode pulse at 904 nm.

The absolute value of the minority-carrier lifetime is surface limited, as the sample is measured 'as cut.' Normally, in fact, no impractical etching or passivation treatments are applied to the surface, which therefore has a high recombination velocity. It is assumed that the surface is always in the same conditions, so any change in the relaxation time is associated with bulk properties. A map such as the one in Figure 1 takes a few minutes to be realised with a modern lifetime scanner such as the one in [8]. Early systems could take several hours to perform measurements with the same resolution. The low-lifetime regions (in red) have different physical origins, and this is where the extensive material-to-cell correlation work performed over the years has been essential in comfortably introducing these instruments in the production environment [4,5,9]. The red zones

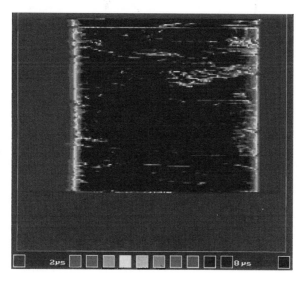

Figure 1 Lifetime map.

at the top, in fact, are determined by the segregation of metals due to the refining process during solidification, and the wafers cannot effectively be used in cell processing, so they are rejected and possibly remelted. The red zones at the bottom of the block are composed of a highly defected area—the initial crystal growth, highly dislocated and unusable for solar cells within about 1 cm from the start—and the 2–3 cm region of oxygen rich material that, despite its low initial lifetime, recovers after the thermal treatments used in cell processing and normally produces good-quality cells. Care must be taken, therefore, to exactly determine the cut off (by any means possible!) between good- and low-quality material at the bottom of the block, and this is probably the most difficult part of the quality-control procedure at block level.

Figure 2 shows a typical correlation between the lifetime at block level and cell performance, being evidence of a good performance of initially low-quality material as detected by the block scanner [9]. The central part of the block is instead relatively uniform and produces good-quality cells, in the range of 12–15%, depending on the particular process used, the higher value provided by silicon nitride–based sequences.

Typical values in the central part of the blocks are around 5–10 μs, depending on the specific measurement system used. A fundamental assumption of this method is a relative uniformity of the material in any

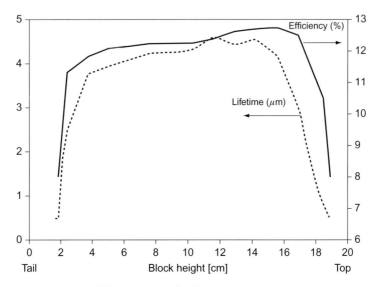

Figure 2 Correlation of lifetime with cell efficiency.

given region of the block (i.e., all central regions behave similarly in same conditions), which is also a result of extensive correlation work [5,9].

μ-PCD testers are also used to inspect incoming wafers in production lines, and in this case as well there has been important correlation work to be able to confidently accept material for subsequent processing. This applies to CZ-Si wafers, which are inspected for uniformity as well as for the acceptance threshold value (which varies from case to case).

3. FEEDSTOCK SILICON

The commercial success of PV is driven critically by its cost. Silicon wafers account for about 50% of the total production cost of a module, a figure that has increased over the years, from the 33% of about 10 years ago, thanks to the constant improvements in technology that have identified the wafer as the ultimate cost-limiting factor [10].

There is no source of silicon feedstock unique to the PV industry, so the issue of a possible feedstock shortage has been largely debated and is still not concluded. About 10–15% of the silicon used by the microelectronics industry is available in various forms for PV use. This is in the range of

1800−2500 tons per year of higher quality scrap and an extra 1500−2000 tons of lower-quality material (e.g., pot scrap). Based on an effective usage rate of 10−15 tons per MWp produced, the amounts considered cannot feed the fast-growing PV market for long. Whilst extensive research programmes have been conducted for many years to upgrade cheap metallurgical silicon to be an independent low-cost silicon source for PV, none of the techniques proposed has reached commercial maturity [5]. The scare of a silicon shortage as early as the middle of the present decade has instead favoured several proposals for processes similar to those used for the production of polysilicon but with looser specifications [11,12].

From a practical point of view, there is, in general, no constraint related to the geometrical specification of the starting material, so for the moment PV can enjoy low-cost scrap such as silicon chips from the cutting processes of semiconductor manufacturing, popcorn silicon rods, tops and tails from crystal growth processes, etc. [13].

However, different crystallisation methods require different specifications. In general, monocrystalline and nonwafer technologies require high-quality starting material, while the multicrystalline technology can allow a looser specification if some care is taken, due to its purifying characteristics—another point in its favour.

4. CRYSTAL-PREPARATION METHODS

A number of techniques are available for the production of silicon wafers for the PV industry: CZ-Si and multicrystalline silicon (which have already been mentioned), magnetically confined multicrystalline silicon, float-zone silicon and the nonwafer technologies (also already mentioned). In this paragraph we briefly introduce the main features of each of them. The reader is encouraged to consult specific references for further details, as we focus on the most relevant recent developments of the technologies under discussion.

4.1 Czochrahki Silicon

The most common method for the growth of single crystalline ingots consists of pulling an oriented seed slowly out of the molten silicon contained in a pure quartz crucible (see Figure 3). The method is well known and extensively described in literature [10,14].

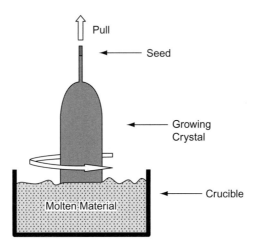

Figure 3 Schematic of CZ growth principle.

What we will mention here is that a number of actions have been taken in the last 10 years to reduce the cost of CZ material and regain competitiveness against multicrystalline silicon.

For instance, crystal growers now quite commonly use some kind of scrap silicon from the semiconductor industry as well as virgin poly as feedstock. Lower energy consumption, from the standard 100 kWh/kg figure to a promising 40 kWh/kg, was recently reported due to improved furnace design, including heaters and gas-distribution systems [15]. A crystallisation yield up to 70% from the standard 50% was also reported in the same study.

4.2 Multicrystalline Silicon

The realisation of multicrystalline silicon ingots is a relative simple process and is based on controlling the extraction of heat from the melt in a quartz crucible in such a way that the interface between the growing solid and the ingot is as flat as possible. In this way, silicon grows in large *columns* of a few centimetres in section and as tall as 25 cm, and most detrimental impurities are segregated toward the top of the ingot. The critical steps to ensure a high-quality and high-yield process are in the design of the furnace for appropriate heat control and in the quality of the quartz crucibles. A schematic of the general method is given in Figure 4.

Modern mc-Si furnaces are designed to minimise inhomogeneity and maximise productivity [16,17], and in the last few years a great deal of

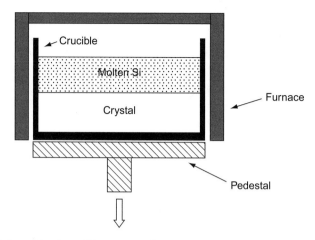

Figure 4 Schematic of mc-Si ingot growth.

Table 4 Typical features of DS mc-Si

Parameter	Typical value
Energy consumption	10 kWh/kg
Crystallisation yield	70–80%
Growth rate	5–10 mm/h
Ingot size	100–300 kg
Ingot base	Square. 66 cm × 66 cm
Ingot height	20–25 cm

Table 5 Impurity levels of a typical mc-Si ingot

Impurity	Typical value (ppma)
Fe	<0.1
Al	0.5–2
Cu, Mn, Cr, Mg, Sr	<0.1

effort has been put in the study of appropriate models for growth control and optimisation [18]. Some of the distinctive features of mc-Si ingot growing are summarised in Tables 4 and 5.

As discussed in Section 2.3, the central part of the ingot enjoys a relatively uniform quality, and the purification ability of the process is witnessed by the following typical values for impurities other than oxygen and carbon.

4.2.1 Charge Preparation

Ingots are normally doped at the level of about 1×10^{16} at/cm^3 of boron. This can be achieved with different mixtures of starting feedstock, and it is a common practice in the PV industry, which has to use feedstock form different sources and of different nature. For the range of resistivity values used in PV, no special requirements are needed for doping the ingot. If the feedstock is virgin poly or lowly doped silicon, then highly doped silicon powder, available commercially with specifications of the B content, can be added to the charge. It is easy to control the final resistivity of an ingot given the starting characteristics of the material. A simple set of equations determines the amounts of each kind of feedstock to be added to the mix. The constraints are the weight of the ingot and the doping level, the latter given by the difference between donor and acceptor concentrations, assuming all impurities are ionised at room temperature. The relationship between resistivity and doping level is known from the literature [14], and a simple spreadsheet can be used to do the conversion.

4.2.2 Crucibles

Crucibles are one of the critical points of mc-Si technology. They are made of slip-cast silica, a technique known since the medieval age, which consists of letting a plaster mould slowly absorb the quartz present in a water suspension. A layer of up to about 2 cm thick can be realised in this way, and the mould can have a double jacket to improve thickness uniformity. The crucible is then baked for mechanical resistance.

The crucibles currently used have been developed to withstand the high temperatures of a heavy silicon ingot growth process in order to avoid unwanted failures in the presence of liquid silicon. Crucibles are lined with a Si_3N_4-based coating to prevent liquid silicon sticking to the walls and subsequent cracking of the ingot due to the strong stress during solidification and cooling. However, only a limited number of companies manufacture crucibles worldwide, and the maximum size of a 'safe' crucible has probably already been reached (68 × 68 cm). This imposes a boundary condition on the design of future mc-Si furnaces [17].

4.3 Electromagnetic Continuous Casting

Electromagnetic continuous casting (EMC) uses an RF coil to induce currents in an appropriately designed circuit able to push the melt away from the walls, therefore making it unnecessary to use crucibles. A

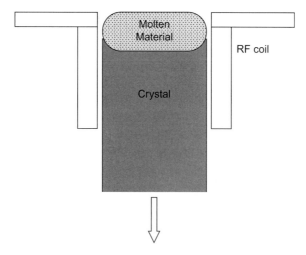

Figure 5 Schematic of EMC furnace.

schematic of the furnace is shown in Figure 5 [19]. The process is carried out in an argon ambient at slight overpressure. The top end is open for the ingot to be pulled down while new feed material is added. The resulting ingot is a long bar of about 240 kg in weight. The idea behind this growth method is to completely avoid the use of any physical crucible by confining the charge electromagnetically. This gets rid, at one time, of two major issues of the DS techniques that have previously been described: expensive crucibles and related contamination. However, inhomogeneous nucleation occurs, and grain size is also rather small, resulting in a low starting quality of the material [20], although a low oxygen content is reported. This kind of material is not commercially available yet, but there are announcements that it could be shortly.

4.4 Float-Zone Silicon

As float-zone silicon typically is used for power electronic components and detectors, the advantage of using large-diameter substrates has been limited, and even today the majority of all float-zone crystals are only 100–125 mm in diameter, which is advantageous for the PV industry needs. Most product and process optimisation activities have been focused on achieving highly predictable yields when running many small series of different types, constantly varying with respect to crystalline orientation (or), diameter (25–150 mm), dopant type (n- or p-type), and resistivity range (0.01–100,000 Ω cm).

During the float-zone growth method, a molten zone is passed along the silicon rod, melting the raw polycrystalline silicon material and leaving behind a purified monocrystal, shown in Figure 6. Modern FZ machines are now capable of accepting feedrods up to 2 m long with a weight between 60 and 100 kg. The bottom end of the feedrod is coned by a grinding operation; during the process, the surface of this cone is heated to the melting temperature of silicon. This results in a thin layer of molten silicon continuously running down the feedrod bottom tip and through the centre hole of the induction coil. The feedrod is heated by a skin current induced by an electromagnetic field. As the feedrod, the molten silicon and the finished crystal are freely suspended in the growth chamber; there is never direct physical contact between silicon and the surroundings except for the ambient gas, typically argon. Contamination is therefore very low, and the process also allows purification of impurities that segregate in the melt.

As the feed rates of the feedrod and finished crystal can be controlled independently, there are no constraints on the diameter of the monocrystal and the diameter of the feedrod. Typical growth rates of the monocrystal are between 2 and 3 mm/min.

Beside the physical dimension of the monocrystal, the fundamental parameter that must be controlled is the shape of the phase boundaries—i.e., the free surface of the melt and the melt–solid interface where the crystallisation takes place. Factors affecting the phase boundaries are the induction coil current, the pull velocities of the feedrod and monocrystal, the rotation rates, the eccentricity of the rod and monocrystal with respect to the coil centre, and additional heat sources.

The monocrystalline perfection of the finished crystal is very high, as volume defects (precipitates or voids), planar defects (twins grain boundaries or stacking faults), or line defects dislocations) are present. The purity of the finished monocrystal is very high, with oxygen and carbon as the two impurities of highest concentrations (upper limit at 1.0 and 2.0×10^{16} cm^{-3}, respectively). Also due to the purity of the feedrod material, the concentration of other impurities is very low, and the total concentration of all metal atoms typically lies below 10^{13} at/cm^3 [21,22].

4.5 Nonwafer Technologies

The idea of lowering the wafer manufacturing costs by avoiding the wafer-cutting step with its silicon loss was the main motivation for the development of a number of silicon ribbon-growth technologies. The common feature to all of them is the principle of a continuous production of a thin

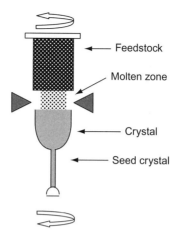

Figure 6 Schematic of FZ growth.

Table 6 Comparison of different ribbon technologies

Material type	Pull rate (cm/min)	Throughput (m^2/h)	Furnaces per 100 MWp
EFG	1.7	1	100
SR	1–2	0.03–0.1	1200
RGS	600	45	2–3

foil or sheet directly from the silicon melt, using different techniques to confine or stabilise the edges. From these different technologies, developed in R&D programmes such as the JPL flat-plate solar module project [23], only a few are used in commercial wafer production today. The most relevant to date are the edge-defined film fed growth (EFG) [24] (by far the most advanced in terms of industrial performance), the string ribbon (SR) [25], and the dendritic web [26] technology. A great deal of improvement was reported recently in all technologies, and in all cases industrial facilities are described or anticipated. In all cases, tailored cell processing is needed to improve the starting quality of the material, which is generally low. Also, in all cases the technologies appear to be capable of producing very thin sheets, in the 100 μm range. A promising technique from the point of view of productivity is the ribbon-growth on substrate (RGS), originally developed by Bayer with further development at ECN [27].

Other silicon ribbon technologies with the potential for high production rates by decoupling ribbon production from crystal growth (such as the low-angle silicon sheet [LASS] or the supporting web [S-Web]) are not yet developed to industrial production. Table 6 compares production speed and capacity of different silicon ribbon production technologies. The last column shows the number of furnaces for a 100-MWp production line [27].

5. SHAPING AND WAFERING

5.1 Shaping

The large, square-based mc ingots are cut into smaller blocks using large blade or band machines. Blade machines are in general more robust and easy to use and maintain, but they have the disadvantage of producing a relatively high kerf loss, up to 3—4 mm. Band saws, on the other hand, suffer from frequent band breakage and may produce waviness in the blocks, which will then need rectifying. However, modern band saws seem to have greatly improved from this point of view.

Monocrystalline silicon ingots instead are treated as the semiconductor counterpart for removing heads and tails and are shaped to pseudo-square by removing parts of the rounded edges, a process that does not present particular problems, thanks to the relatively small dimension of the ingots.

In the case of mc-Si blocks, it is after the shaping step that blocks are inspected for minority-carrier lifetime and resistivity, as described earlier, so finally tops and tails can be removed to leave the material ready to be wafered.

5.2 Wafering

Wafering of Si ingots for the PV industry is probably one of the only examples of technology successfully transferred to the semiconductor industry, which was originally developed for the PV industry. Cost constraints in PV in fact imposed the development of a slicing technique able to reduce kerf loss and increase productivity, as an alternative to slow, large kerf-loss blade-cutting techniques used until about 10 years ago [15]. On the other hand, the specifications for semiconductor-grade wafers up to 300 mm in diameter needed a totally new concept of machines for the control of taper, thickness variation, and surface smoothness, so at the end both industries enjoy the development of wire-saw technology.

Modern slicing technology is based on wire sawing, where a thin wire (160 μm diameter) web pushes an abrasive-based slurry into the silicon to

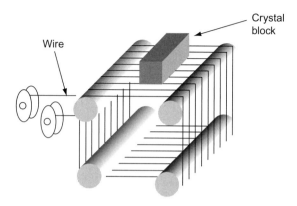

Figure 7 Principle of wire sawing.

be cut. In this way, several wafers are cut at the same time, with high mechanical precision and in a highly automated process.

The principle of wire sawing is shown in Figure 7. A spool of single bronze-coated stainless steel wire up to several hundreds of kilometres long (!) is fed on high-precision grooved wire guides. The feeding system, not detailed in the figure, is designed to allow high-precision control of the wire tension, one of the critical parameters of the process, throughout the cut. The silicon block, or blocks, is glued on a low-cost glass support, which in turn is mounted on a motor-driven table that translates downward through the web. The abrasive slurry is fed to the wire web through a nozzle and allows the silicon to be cut. The abrasive is fine-mesh silicon carbide powder. The process is completed when the wires reach the glass, thus allowing the wafers to be separated without any damage as they are attached to the support through a very thin layer of glue. Wire speed and tension, table speed, slurry viscosity and temperature, and abrasive characteristics are the main parameters to control in order to produce wafers that match the specifications indicated in an earlier paragraph. As the conditions tend to change during the process—e.g., because of the increase of the temperature of the slurry or because of the contamination of the slurry by the silicon dust produced by the process—it is critical to be able to adjust the parameters to avoid waves or thickness variations. The slurry can be based on oil or on other fluids that are water washable and can improve subsequent cleaning and handling steps, although the waste-treatment system tends to be more sophisticated than with oil-based procedures, needing for example an oxygen demand control device.

Table 7 Typical watering conditions Tor 350 μm thick, 125 mm square mc-Si wafers

Parameter	Value
Wire speed (m/s)	5–10
Table speed (μm/min)	300–400
Wire diameter (μm)	160–180
Wire guide pitch (μm)	550–570
SiC mesh	500–600
Slurry temperature (°C)	25 ± 5
Slurry mixture	1:1 (glycol based)
	5:8 (oil based)
Viscosity (g/l)	1600
Wire tension (N)	23–25

Typical parameters for cutting 125-mm square wafers over a length of about 30 cm are given in Table 7.

Large wafer manufacturers have developed automated washing and handling equipment for the steps following the cutting one.

REFERENCES

[1] Photon International, March 2002.
[2] T. Bruton, General trends about photovoltaics based on crystalline silicon. Proc. E-MRS 2001 Spring Meeting, Symposium E on Crystalline Silicon Solar Cells, Sol. Energy Mater. Sol. Cells, vol. 72, 2002, pp. 3–10.
[3] SEMI™ Standard M61000—SEMI™; International Standards www.semi.org.
[4] F. Ferrazza, New developments and industrial perspectives of crystalline silicon technologies for PV, Proc. 13th European Photovoltaic Solar Energy Conf., Nice, 1995, p. 3.
[5] F. Ferrazza, et al., The status of crystalline silicon modules, World Renewable Energy Conf., Florence, 1998.
[6] H. Kunst, G. Beck, The study of charge carrier kinetics in semiconductors by microwave conductivity measurements, J. Appl. Phys. 60(10) (1986) 3558.
[7] A. Schonecker, et al., Results of five solar wafer minority carrier lifetime round robins organised by the SEMI M6 Solar Silicon Standardisation Task Force, Proc. 14th European Photovoltaic Solar Energy Conf., Barcelona, 1997, p. 666.
[8] Semilab homepage www.semilab.hu.
[9] F. Ferrazza, et al., Cost effective solar silicon technology, Proc. 2nd World Conference on Photovoltaic Solar Energy Conversion, Vienna, 1998, p. 1220.
[10] A. Endroes, Mono- and tri-crystalline Si for PV application, Proc. E-MRS 2001 Spring Meeting, Symposium E on Crystalline Silicon Solar Cells. Sol. Energy Mater. Sol. Cells, vol. 72, 2002, pp. 109–124.
[11] J. Maurits, Policrystalline Silicon-World Demand and Supply, Eighth NREL Workshop on Crystalline Silicon Solar Cell Materials and Processes, Colorado, 1998.
[12] F. Woditsch, W. Koch, Solar grade silicon feedstock supply for PV industry, Proc. E-MRS 2001 Spring Meeting, Symposium E on Crystalline Silicon Solar Cells, Sol. Energy Mater. Sol. Cells, vol. 72, 2002, pp. 11–26.

[13] H. Aulich, F. Schulze, Crystalline silicon feedstock for solar cells, Prog. Photovolt: Res. Appl. 10 (2002) 141–147.
[14] W. O'Mara, R. Herring, L. Hunt (Eds.), Handbook of Semiconductor Silicon Technology, Noyes Publication, 1990.
[15] T. Jester, Crystalline silicon manufacturing progress, Prog. Photovolt: Res. Appl. 10 (2002) 99–106.
[16] F. Ferrazza, Growth and Post growth Processes of multicrystalline silicon for photovoltaic use, in: S. Pizzini, H. P. Strunk and J. H. Werner (Eds.), Polycrystalline Semiconductors IV—Physics, Chemistry and Technology, in Solid State Phenomena, vols. 51–52, Transtec, Switzerland, 1995, pp. 449–460.
[17] F. Ferrazza, Large size multicrystalline silicon ingots. Proc. E-MRS 2001 Spring Meeting, Symposium E on Crystalline Silicon Solar Cells. Sol. Energy Mater. Sol. Cells, vol. 72, 2002, pp. 77–81.
[18] D. Franke, et al., Silicon ingot casting: process development by numerical simulations, Proc. E-MRS 2001 Spring Meeting, Symposium E on Crystalline Silicon Solar Cells, Sol. Energy Mater. Sol. Cells, vol. 72, 2002, pp. 83–92.
[19] F. Durand, Electromagnetic continuous pulling process compared to current casting processes with respect to solidification characteristics, Proc. of the E-MRS 2001 Spring Meeting, Symposium E on Crystalline Silicon Solar Cells, Sol. Energy Mater. Sol. Cells, vol. 72, 2002, pp. 125–132.
[20] I. Perichaud, S. Martinuzzi, F. Durand, Multicrystalline silicon prepared by electromagnetic continuous pulling: recent results and comparison to directional solidification material, Sol. Energy Mater. Sol. Cells 72 (2002) 101–107.
[21] W. Dietze, W. Keller, A. Muhlbauer, Float-zone grown silicon, in Crystals, Growth, Properties, and Applications, vol. 5, Silicon, Springer-Verlag, 1981, p. 1.
[22] A. Luedge, H. Riemann, B. Hallmann, H. Wawra, L. Jensen, T.L. Larsen, A. Nielsen, High-speed growth of FZ silicon for photovoltaics, Proc. High Purity Silicon VII, Electrochemical. Society, Philadelphia, 2002.
[23] Flat plate solar array project: vol. III, Silicon sheet: wafers and ribbons, Report DOE/JPL-1012–125, 1986.
[24] J. Kaleis, Silicon ribbons and foils—state of the art, Proc. E-MRS 2001 Spring Meeting, Symposium E on Crystalline Silicon Solar Cells, Sol. Energy Mater. Sol. Cells, vol. 72, 2002, pp. 139–153.
[25] J. Hanoka, PVM at contribution towards Evergreen Solar's new factory, Proc. 29th IEEE Photovoltaic Specialists Conference, New Orleans, 2002, p.66.
[26] D.L. Meyer, et al., Production of thin (70–100 mm) crystalline silicon cells for conformable modules, Proc. 29th IEEE Photovoltaic Specialists Conference, New Orleans, 2002, p. 110.
[27] A. Schonecker, et al., Ribbon growth-on-substrate: progress in high speed crystalline silicon wafer manufacturing, Proc. 29th IEEE Photovoltaic Specialists Conf., New Orleans, 2002, p. 316.

CHAPTER IB-2

High-Efficiency Silicon Solar Cell Concepts

Martin A. Green
School of Photovoltaic and Renewable Energy Engineering, University of New South Wales, Sydney, Australia, 2052

1. Introduction	100
2. High-Efficiency Laboratory Cells	100
2.1 Silicon Space Cell Development	100
2.2 High-Efficiency Terrestrial Cells	104
2.3 Rear Passivated Cells	108
2.4 PERL Cell Design Features	110
3. Screen-Printed Cells	111
3.1 Structure	111
3.2 Typical Screen-Printed Cell Performance	113
3.3 Improved Screen-Printing Technology	114
3.3.1 Improved Pastes	*114*
3.3.2 Selective Emitter and Double Printing	*114*
3.3.3 Hot-Melt and Stencil Printing	*115*
3.3.4 Plated Seed Layers	*115*
3.4 Screen-Printing Limitations	115
4. Laser-Processed Cells	116
4.1 Buried-Contact Cells	116
4.2 Semiconductor Finger Solar Cell	118
4.3 Laser-Doped, Selective-Emitter Solar Cells	119
5. HIT Cell	120
6. Rear-Contacted Cells	121
6.1 Rear-Junction Solar Cells	122
6.2 Emitter Wrap-Through (EWT) Cells	123
6.3 Metal Wrap-Through (MWT) Cells	124
7. Conclusions	124
Acknowledgements	125
References	125

Practical Handbook of Photovoltaics.
© 2012 Elsevier Ltd. All rights reserved.

1. INTRODUCTION

The vast majority of solar cells sold up to the present use crystalline or multicrystalline silicon wafers in combination with a relatively simple screen-printing approach to applying the metal contacts. This approach has the advantage of being widely used and well established with the ready availability of appropriate and ever-improving equipment, such as screen printers and furnaces for drying and firing the screened metal patterns. This limits the capital requirements and risks involved in setting up cell manufacture and provides access to a skill and resource base far broader than that of any single manufacturer.

However, there are disadvantages arising from the simplicity of the screen-printing approach. One is the limited cell performance that results. Commercial solar cell modules based on this approach are limited to efficiencies in the 12–15% range [1], corresponding to cell efficiencies of 15–19%. As the photovoltaic industry matures, approaches that offer higher performance than is possible with screen printing will most likely be required.

With this as the basic rationale, this chapter explores approaches that offer this higher efficiency potential. First, the history of high-efficiency laboratory cell development will be outlined, highlighting features responsible for each successive jump in performance. Features that limit screen-printed cells to the relatively modest performance levels previously mentioned will then be discussed. Several commercial high-efficiency cell designs that overcome some of these limitations will then be described, followed by an exploration of other approaches that show potential.

2. HIGH-EFFICIENCY LABORATORY CELLS
2.1 Silicon Space Cell Development

The evolution of record silicon laboratory cell efficiency is shown in Figure 1, reflecting several stages in the evolution of cell design. After an initial period of rapid evolution in the 1950s, design stabilised for more than a decade on the "conventional" space cell of Figure 2(a). Key features include the use of 10-Ωcm p-type substrates to maximize high-energy radiation resistance and the use of a nominally 40-Ω/square, 0.5-μm deep phosphorus diffusion to form the emitter region of the cell. Although it was known that

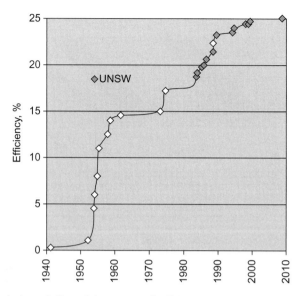

Figure 1 Evolution of silicon laboratory cell efficiency.

lighter diffusions gave better blue response, this value was chosen because it was found to be more resistant to shunting by the top-contact metallisation during processing [2]. These cells remained the standard for space use for more than a decade and, until recently, were still specified for some space missions. Energy conversion efficiency was 10–11% under space radiation and 10–20% higher on a relative basis under terrestrial test conditions.

Toward the end of the 1960s, the benefits of a rear Al treatment became apparent, particularly for cells that were thinner than normal [3,4]. The corresponding increase in space cell efficiency to 12.4% was attributed to the gettering action of the Al treatment [5].

More detailed work showed that it was the presence of a heavily doped region beneath the rear contact that gave rise to these beneficial effects [6]. These benefits were postulated to arise from spillage of majority carriers from the rear doped region into the bulk region of the cell, thus increasing the effective bulk concentration and hence the open-circuit voltage [6]. Although the correct explanation (in terms of a reduction in the effective rear-surface recombination velocity) was soon forthcoming [7], the effect is still rather inappropriately, but almost universally, referred to as the *back-surface-field* (BSF) *effect* [6].

The conventional space cells had a relatively poor response to wavelengths shorter than 0.5 μm, due to the relatively heavy top junction

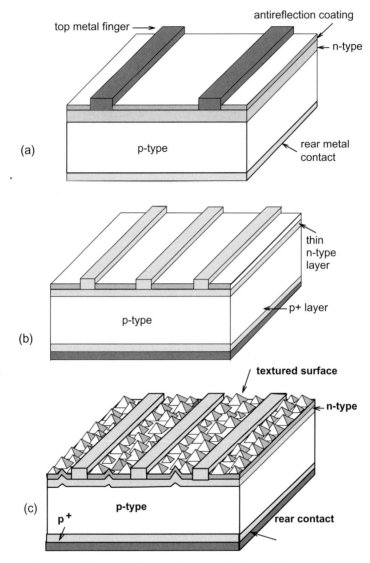

Figure 2 (a) Space silicon cell design developed in the early 1960s, which became a standard design for more than a decade; (b) shallow junction "violet" cell; (c) chemically textured nonreflecting "black" cell.

diffusion, as previously noted, and the use of a silicon monoxide (SiO) AR coating that absorbs light below this wavelength. A marked improvement in performance was demonstrated in the early 1970s by replacing such junctions with much shallower (0.25 μm) and higher sheet resistivity junctions and by redesigning the entire cell to accommodate this change as shown in Figure 2(b).

To accommodate the increased sheet resistivity of the diffused layer, much finer metal finger patterns were also defined using photolithography. As a consequence, the cell resistance was actually lower than in earlier cells. Improved antireflection coatings, based upon TiO_2 and Ta_2O_5, were also incorporated. These dielectrics were less absorbing than SiO and provided a better optical match between the cells and the cover glass to which space cells are normally bonded. The thickness of these dielectric coatings was also selected so they would be more effective at shorter wavelengths than are traditional coatings, giving the cells their characteristic violet appearance. (A subsequent development, made possible by the higher refractive index of the new AR coating materials, was the use of double-layer AR coatings).

The final design change in these "violet" cells was the use of lower-resistivity 2-Ω cm substrates. Since the improved output of the cells at blue wavelengths was quite tolerant to radiation exposure, the overall radiation tolerance remained at least equal to that of the conventional devices, while giving improved voltage output. The combination of improved open-circuit voltage (due to the change in substrate resistivity), improved current output (due to the removal of "dead layers," better antireflection coatings, lower top-contact coverage) and improved fill factor (due to the improved open-circuit voltage and decreased cell resistance) resulted in a massive increase of 30% in performance as compared to an average space cell of conventional design. Efficiencies of 13.5% were obtained under space radiation, with terrestrial efficiencies close to 16% being demonstrated [8].

Not long after the superior performance of the violet cell had been established, a further performance boost was obtained in 1974 by texturing the top surface of the cell [9]. The idea of mechanically forming pyramids on the top surface of the cell to reduce reflection had been suggested some time earlier [10]. In the "black" cell, a simpler chemical approach was used that relied upon the random nucleation of selective etches to expose (111) crystallographic planes in a substrate originally of (100) surface orientation. The intersecting (111) planes so exposed formed small, square-based pyramids of random size distributed randomly over the cell surface.

This has two advantages for cell performance. One is that light striking the sides of the pyramids is reflected downward and hence has at least one more chance of being coupled into the cell. A second advantage is that light coupled into the cell enters obliquely. Most light will be coupled in at the first point of incidence on the pyramids. This light is refracted in at an angle of about 48° to the original surface, resulting in an increase in the path length of weakly absorbed light by a factor of 1.35 compared to a nontextured cell. The effect is similar to that of an increase in the silicon absorption coefficient or in the bulk minority-carrier diffusion length by the same factor. A third feature of the texturing approach is the high degree of trapping of light within the cell that is possible.

For terrestrial cells, this is an advantage since it improves the long-wavelength response of cells. However, for space cells, it is a disadvantage since light trapping increases the absorption of sub-band-gap photons in the rear contact of the cell. This causes the cells to operate at a higher temperature in the space environment, largely negating the previous advantages. Combined with a greater potential for mechanical damage during array assembly by knocking peaks from pyramids meant that textured cells were not as widely used in space as their apparent performance advantage would seem to warrant.

These "black" cells gave an energy conversion efficiency of 15.5% under AM0 radiation, corresponding to an energy conversion efficiency of about 17.2% under the current terrestrial standard (Global AM1.5, 100 mW/cm^2, 25°C). Such were the strengths of the texturing concept, when combined with the technological improvements incorporated into the violet cell, that it was almost a decade before any further significant improvement in cell performance was demonstrated. These improvements ultimately resulted from an increased open-circuit voltage due to the development of improved surface and contact passivation approaches.

The "black cell" of Figure 2(c) represents an interesting stage in cell development in that in both its efficiency (circa 17–18%) and design features it closely resembles a modern screen-printed monocrystalline cell (Figure 3). In effect, the strengths of the screen-printing approach have locked cell design into best practice in the mid-1970s when the screen-printed cell was first developed [11]. In future, cell design is likely to evolve to access the 40–50% gain in relative performance demonstrated since then (Figure 1).

2.2 High-Efficiency Terrestrial Cells

The simplicity of the surface passivation afforded by its thermal oxide is one of the key features of silicon technology, which explains why it rose to dominance in microelectronics. Unfortunately for photovoltaics, the refractive index of silicon dioxide is too low for use as an effective antireflection coating in high-performance cells. In fact, if present in any reasonable thickness (greater than 20 nm) on the top surface of the cell, it will limit the ability to

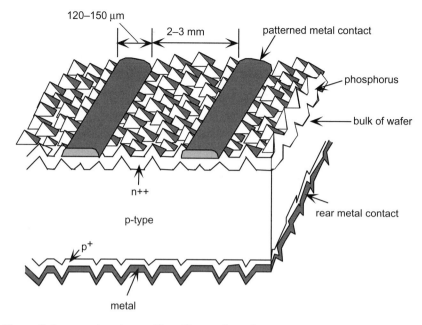

Figure 3 Screen-printed crystalline silicon solar cell (not to scale).

reduce reflection by the subsequent deposition of any number of compensating layers [12]. Hence, if oxide passivation is to be used on the cell surface exposed to light, the oxide layer has to be very thin.

The potential of such oxide passivation became clear around 1978 [13,14]. Subsequent high-efficiency silicon cells have taken advantage of the passivation provided by thin thermal oxide layers to maximise their open-circuit voltage and short-wavelength response.

Contacts made to the surface are generally regions of high recombination velocity. Best cell performance will be obtained when the electronic activity at such contacts is "passivated." The earliest approach was to passivate by isolating the contact from minority carriers by interposing a heavily doped region. As already discussed, rear-contact passivation via the back-surface-field effect resulted in significant gains in cell performance in this way. Heavily doped regions, localised to those areas where the top contact is made to the cell, are used in most recent high-efficiency cell designs.

A second approach to reducing contact effects is to minimise the contact area [15,16]. The benefits were demonstrated by increased open-circuit voltages on low-resistivity substrates [17]. Most modern high-efficiency cells employ this low-contact area approach. A third approach is to

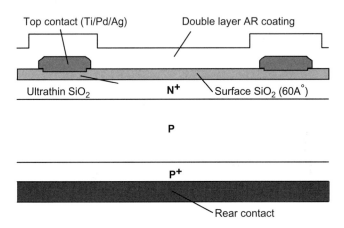

Figure 4 Metal–insulator–NP (MINP) junction solar cell.

employ a contacting scheme in which the contact itself has an inherently low-recombination velocity. The MINP (metal–insulator–NP junction) cell of Figure 4 was the first successfully to exploit this approach [18]. The thin-surface passifying oxide is continued under the metal, thus reducing its effective recombination velocity. Polysilicon [19] and semi-insulating polysilicon (SIPOS) [20] contact passivation have also been demonstrated. It appears that a thin interfacial oxide layer may play an important role in both of these schemes [20,21]. Subsequently, excellent surface passivation was demonstrated by a combination of a very thin layer of lightly doped amorphous silicon followed by a layer of doped amorphous silicon to form the HIT (heterojunction with intrinsic layer) cell [22], a somewhat related approach.

As seen in Figure 1, the performance levels established by "violet" and "black" cells in the mid-1970s remained unchallenged for close to a decade. Cells successfully incorporating oxide and contact passivation approaches along the top surface were the first to exceed these levels.

The MINP cell (Figure 4), the first silicon cell to demonstrate 18% efficiency in 1983, employs top-contact passivation by the use of a thin oxide layer underlying this contact, as well as top-surface passivation by a slightly thicker oxide layer. This difference in thickness complicates processing but was found necessary for maximum performance. The top-contact metallisation is a Ti/Pd/Ag multilayer. The use of a low-work function metal such as Ti as the contact layer is essential with this approach. This is to produce an electrostatically induced accumulation layer in the underlying silicon, an important factor in reducing contact recombination. The cells used alloyed aluminium to give a heavily doped region near the rear contact

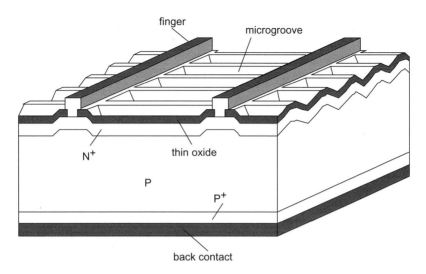

Figure 5 The microgrooved PESC cell, the first silicon cell to exceed 20% efficiency in 1985.

and were fabricated on polished (100)-oriented 0.2-Ω-cm substrates. Surface passivation is easier to achieve on polished rather than textured or "as-lapped" surfaces. To minimise reflection losses, a double-layer antireflection coating was used, consisting of approximately a quarter-wavelength of ZnS on top of the thin oxide, followed by a quarter-wavelength of MgF_2.

The passivated emitter solar cell (PESC) structure, shown in Figure 5, further improved cell efficiency. It is similar to the MINP cell structure, except that electrical contact is made directly through narrow slots in the thin oxide. In this case, contact passivation is obtained by minimising the contact area.

By combining surface texture with the strengths of the PESC approach, the first nonconcentrating silicon cells with a confirmed efficiency above 20% were fabricated in 1985 [23]. Rather than pyramidal texturing, the cells used microgrooving for the same purpose. The microgrooves were defined by using selective etches to expose crystal planes. Photolithographically designed oxide patterns were used to protect the surface against etching where this was not required and so determine the final pattern of microgrooves. This approach was found to be easier to combine with fine-line photolithography than was the normal pyramidal texturing.

The key characteristics of the PESC sequence are oxide surface passivation, self-aligned contacts through this oxide, high sheet resistivity topjunction diffusion, alloyed aluminium rear-surface passivation and antireflection control by texturing or double-layer antireflection coating.

The PESC sequence proved to be very rugged and repeatable. Within one year of the initial 20% results, two groups had reported results approaching this figure (when referred to present calibrations) using almost identical structures [24,25]. The sequence has since been reproduced in many laboratories, with commercial product available at different points in time for space and concentrator cells or for high-value-added applications such as solar car racing.

2.3 Rear Passivated Cells

The next major advance in cell performance highlights an important demarcation point in the history of cell evolution. The improvement came as a result of applying surface and contact passivation approaches to both top and rear surfaces. The rear-point contact solar cell of Figure 6 achieved this landmark in cell design. Since both contacts are on the rear of the cell, the design places enormous pressure upon the quality of surface passivation along both top and rear surfaces and upon post-processing carrier lifetimes.

Although originally developed for concentrator cells [26], the device design was modified for one-sun use by adding a phosphorus diffusion

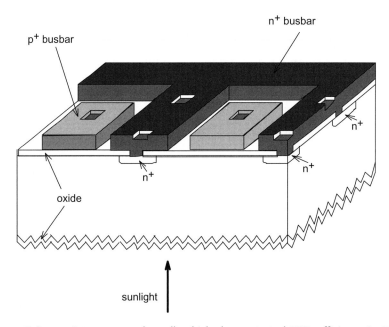

Figure 6 Rear-point contact solar cell, which demonstrated 22% efficiency in 1988 (cell rear shown uppermost).

along the illuminated surface [27]. This produced the first one-sun silicon cells of efficiency above 22%.

Looking at what this history so far means for commercial cells, improvements up to this stage (white square in Figure 1) represent what may ultimately be achieved in a commercial sequence that retains a standard rear-contacting approach but removes deficiencies in the standard screen-printing approach that are associated with the top surface. An efficiency in the 21—22% range is feasible following this strategy.

Combining the earlier PESC sequence with similar double-sided surface passivation and chlorine-based processing [28] produced an improved device, the PERL cell (passivated emitter, rear locally diffused cell) shown in Figure 7. This took silicon cell efficiency to 23% by the end of the 1980s, an enormous improvement over the figure of 17%, the highest value only 7 years earlier. The PERL cell shares many features in common with the rear-point contact cell, including almost complete enshroudment in a passivating oxide layer and small area contacts passified by local heavy diffusions. However, it is inherently a more robust design, being more tolerant of poor surface passivation and poor bulk lifetimes.

Since then, further improvements in PERL cells have taken their efficiency to 25% [29], which remains the highest to date for a silicon cell. Major improvements include the growth of much thinner oxide for top-surface passivation which allows the direct application of a double-layer antireflection coating to increase short-circuit current [12], the use of an Al-based "alnealing" sequence for this top oxide and localised top-contact points to increase open-circuit voltage, improved rear-surface passivation and reduced metallisation resistance to improve fill factor.

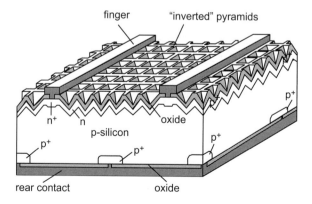

Figure 7 The PERL cell, which took efficiency above 24% in the early 1990s.

2.4 PERL Cell Design Features

To maximise performance, as much light as possible of useful wavelengths should be coupled into and absorbed by the cell. Modern cell designs such as the PERL cell (Figure 7) incorporate several primarily optical features to achieve this result.

The inverted pyramids along the top surface serve primarily in such an optical role. Most light incident on this structure will hit one of the sloping side walls of the pyramids at the first point of incidence with the majority of this light coupled into the cell. That reflected will be reflected downward, ensuring that it has at least a second chance of entering into the cell. More than 40% of the light such as incident near the bottom and corners of the pyramids has three such chances. The pyramids are covered by an oxide layer of appropriate thickness to act as a quarter-wavelength antireflection coating. Alternatively, this oxide can be grown thin and a double-layer antireflection coating applied [12].

Light coupled into the cell moves obliquely across the cell toward the rear surface with most absorbed on the way. Weakly absorbed light reaching the rear is reflected by the very efficient reflector formed by the combination of the rear oxide layer covered by an aluminium layer [30]. The reflectance from this combination depends upon the angle of incidence of the light and the thickness of the oxide layer but is typically above 95% for angles of incidence close to the normal, decreasing to below 90% as the incidence angle approaches that for total internal reflection at the silicon/oxide interface (24.7°), and increasing to close to 100% once this angle is exceeded.

Light reflected from the rear then moves toward the top surface. Some reaching this surface strikes a face of a pyramid of opposite orientation to that which coupled it into the cell. Most of this immediately escapes from the cell. Light striking any other pyramid face is totally internally reflected. This results in about half the light striking the top surface internally at this stage being reflected back across the cell toward the rear contact. The amount of light escaping after the first double pass depends on the precise geometry involved. It can be reduced by destroying some of the symmetries involved—for example, by using tilted inverted pyramids or by using a "tiler's pattern" [30]. The latter approach is currently used in current PERL cell designs.

The combination of the inverted pyramids and the rear reflector therefore forms a very efficient light-trapping scheme, increasing the path length of weakly absorbed light within the cell. Effective path-length enhancement factors [30] above 40 compared to a single pass across the cell are measured. This light trapping boosts the infrared response of the cell. The external responsivity (amps per watt of incident light) of PERL cells peaks at longer wavelengths at higher values than previous silicon cells with values of 0.75 A/W measured at 1.02 μm wavelength. Energy-conversion efficiency under monochromatic light peaks at the same wavelength with values above 45% measured [31]. Further improvements could push this figure to above 50% at 1.06 μm.

Other optical losses are due to reflection from, and absorption in, the top metal fingers of the cell. This can be minimised by making these lines as fine as possible with, ideally, as large an aspect ratio (height to width ratio) as possible. Alternatively, optical approaches can be used to steer incoming light away from these lines or to ensure that light reflected from them eventually finds its way to the cell surface [32,33].

Present PERL cells lose about 5% of incoming light due to absorption or reflection loss associated with these metal fingers when combined with reflection from the unmetallised top surface of the cell. They also lose 1—2% in performance from the use of a less than optimal light-trapping scheme and from less than 100% reflection of light from the rear surface of the cell. There is therefore some scope for small to moderate gains in performance by further improving the optical properties of these cells.

Although such advanced cell designs have been used for spacecraft and high-value terrestrial applications such as solar car racing, the multiple photolithographic steps required in their fabrication make them too expensive for low-cost terrestrial applications [34,35]. They do, however, provide a reference point for the discussion of the compromises involved in lower-cost designs, as discussed in the following sections.

3. SCREEN-PRINTED CELLS

3.1 Structure

The structure of a standard crystalline screen-printed cell is shown in Figure 3. The normal cell-processing sequence would consist of [36] saw-damage removal from the starting wafer by etching; chemical texturing of the top surface; top-surface diffusion to about 40—60 ohms/square; etching to remove diffusion oxides and the junction at the rear of the cell; deposition of a silicon nitride antireflection coating to the top cell surface;

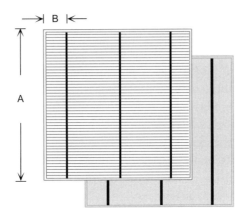

Figure 8 Standard 156 mm × 156 mm solar cell "H" metallisation pattern. Screen-printed silver regions are shown coloured black. The pattern consists of six unit cells of size A × B.

screening and drying of three metal layers sequentially, including a localised Ag rich region on the rear of the cell to allow soldering; an Al rich paste over the rest of the rear and a Ag rich paste on the front side to form the fingers and busbars; simultaneous firing of these layers; and cell testing and sorting.

The typical metallisation pattern used for such cells is shown in Figure 8. Using the normal boron-doped Czochralski silicon wafers of 0.5–5-Ω-cm resistivity, the resulting cell efficiency is typically 16–18%, while for multicrystalline wafers of similar resistivity efficiency is in the 15–17% range.

The major disadvantages of the screen-printing approach relate to the cost of the metal pastes used in the process [37] and the relatively low cell efficiency that results. The low efficiency is due most fundamentally to the restricted line width possible by screen printing. The relatively high contact resistance between the paste and the silicon is another constraint. The low aspect ratio (height/width) of the final lines due to paste thickness shrinkage during firing is another problem compounded by the low conductivity of the fired paste (up to 3 times lower than that of pure silver).

There have been several investigations of the feasibility of screening pastes other than those based on silver. Nickel and copper have been investigated without success to date [38,39]. The contact resistance between the paste and silicon can be a sensitive function of the precise firing environment and temperature, as well as of the doping of the contacted region. The glass frit (dispersed glass particles used in the paste as a binder) forms an oxide precipitate along the interface between the screened paste and the silicon surface. This contributes to the high contact resistance, although often phosphorus is added to the paste to decrease the contact resistance to n-type material.

The rear-surface contact resistance is generally less of a problem. Even though contact is being made to more lightly doped material, a much larger contact area is available. Generally, as previously mentioned and shown in Figure 8, a small region of the rear is printed with a silver-rich paste to allow soldering of the cell interconnect to these regions. The remainder of the rear is printed with an aluminium paste used to increase the doping level in the surface region by alloying. After firing, a significant back-surface-field effect can be obtained by the use of such aluminium.

3.2 Typical Screen-Printed Cell Performance

Typically, the screen-printing approach will produce cell open-circuit voltages in the 600- to 630-mV range, depending on substrate resistivity, short-circuit current densities in the 30- to 36-mA/cm^2 range and fill factors for large area cells in the 72—78% range. For a large area cell, typically 10% of the top surface of the cell will be shaded by the screened metallisation and busbars. The top-surface metallisation as indicated in Figure 8 typically consists of metal fingers of about 100—150 microns width, spaced about 2—3 mm apart. Due to the low conductivity of the paste, an interconnect busbar or H-pattern design is mandatory for large area cells. Although contributing to the large shading loss of this approach, this design also has the advantage of improving the cell tolerance to cracks. When scaling to large area cells, the number of busbar interconnections is increased.

To maintain reasonable contact resistance, quite low sheet resistivities for the top-surface diffusion are required. The 50- to 60-Ω/square typically used results in a significant loss in blue response of the cells due to dead layers along the surface. Higher sheet resistivities will improve the blue response but at the expense of cell fill factor. The heavy diffusion also limits the open-circuit voltage output of the cells. Oxide surface passivation is not of much benefit in improving performance due to this limitation. Using improved quality substrates such as float-zone silicon also generally do not result in any substantial performance improvement for screen-printed cells again due to this limitation.

3.3 Improved Screen-Printing Technology

3.3.1 Improved Pastes

The limitations of the screen-printing approach may not be fundamental but may be able to be overcome by ongoing development. Paste manufacturers have quite ambitious plans for improving the performance of their pastes. The aim is to improve the ability of pastes to better contact much lighter top-surface diffusions, with the ability to contact 100-Ω/square emitters targetted by the end of the decade [40].

3.3.2 Selective Emitter and Double Printing

In addition to the improvements in screen-printing pastes over the last decade, there have been substantial improvements in screen-printing equipment. In particular, it is now possible to align printed features very accurately.

This improved alignment accuracy allows the implementation of higher efficiency structures. In particular, the selective-emitter structure of Figure 9(a) now can be implemented in a number of different ways [41]. The end result is a heavily doped emitter region underlying the contact region for the top-surface metallisation fingers. This ensures good contact resistance while allowing other regions of the emitter to be more lightly doped, improving blue response. A gain of 1% absolute (e.g., from 18% to 19% energy conversion efficiency) results from the additional processing involved.

Another possibility made possible by improved alignment accuracy is "double printing," which is illustrated in Figure 9(b). By printing the top metallisation fingers twice, the aspect ratio (height/width) of these fingers

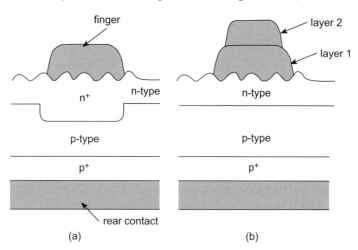

Figure 9 The selective-emitter approach.

can be improved, decreasing the resistance loss associated with these fingers. Additionally, the underlying layer can be optimised for low-contact resistance to silicon while the second layer can be optimised for good lateral conduction. Gains are more modest compared to the use of selective emitters, with about 0.2–0.3% absolute gain in efficiency.

3.3.3 Hot-Melt and Stencil Printing

In hot-melt printing, similar pastes to conventional printing are used but these are solid at room temperature. The paste is heated to 60–75°C during the screen-printing step using heated screen-printing equipment and screens; the paste solidifies as it cools after printing without needing the normal drying step. The key advantage is a higher aspect ratio with an efficiency gain of up to 1% absolute reported [42]. Several pastes manufacturers now sell suitable pastes.

Stencil printing offers both higher aspect ratio and finer line width. Instead of using a metal mesh mask covered with a patterned emulsion as the screen-printing mask, a patterned metal mask is used allowing a wider range of mask properties. An improvement in both current output and paste utilisation has been claimed [43].

3.3.4 Plated Seed Layers

Another approach to achieving high aspect ratio and fine line width involves printing a seed layer followed by plating. Since high aspect ratio of the seed layer is not important in this approach, this layer can be printed with quite narrow line width. Alternatively, more adventurous approaches such as aerosol or ink-jet printing [44] can be used to print this layer. After deposition of the seed layer by any of these techniques, it can be plated to a much larger cross-section either with Ag or with less-expensive metals such as Cu, with a thin Sn or Ag overcoating to temper the reactivity of copper [45]. The use of such approaches may become mandatory to reduce sensitivity to volatile Ag prices [36].

3.4 Screen-Printing Limitations

Despite the potential for improvement outlined above, it is doubtful that an extremely high-efficiency cell can ever be produced using the screen-printed approach, with the best cells combining all available options expected to give only 19–20% efficiency. Even if both the line width and the contact-resistance problems are successfully solved in production, there are still challenges due to conductivity and small aspect ratio.

Perhaps even more important, there is a looming issue with Ag supply. The 2010 production of more than 20 GW of silicon photovoltaics corresponded to the use of close to 2000 tonnes of Ag in this product, from a total of 28,600 tonnes of Ag supply [36]. If module production continues to double every year or two, considerable pressure will be placed on Ag supply in the coming decade with a continuation or even an acceleration of the rapid escalation in Ag price seen over recent years. A major change in the way most cells have been made for more than 30 years seems essential over the coming decade.

4. LASER-PROCESSED CELLS
4.1 Buried-Contact Cells

The buried-contact solar cell of Figure 10 was developed to overcome the efficiency limitations of the screen-printed cell approach. The most distinctive feature of this approach is the use of laser-defined grooves in the top surface to locate the cell metallisation. Although originally investigated using screen-printed metallisation sequences (where the metal was forced into the groove during the screening operation), the most successful designs have used electrolessly plated metal contacts [46].

Cell processing bears some resemblance to the screen-printing sequence previously described. After saw-damage removal and texturing,

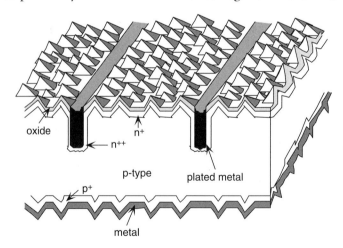

Figure 10 Buried-contact solar cell.

the surface is lightly diffused and an oxide is grown over the entire surface and a nitride antireflection coating is deposited. This dielectric layer serves multiple purposes during cell processing and is the key to the relative simplicity of the processing. Grooves are then cut into the top cell surface using a laser scribing machine, although mechanical cutting wheels [47] or other mechanical or chemical approaches could be used.

After cleaning of the grooves by chemical etching, these are subject to a second phosphorus diffusion, much heavier than the first. This produces selective doping in the contact areas. Aluminium is then deposited on the rear surface by evaporation, sputtering, screen printing, or plasma deposition.

After firing of the aluminium and etching to remove oxides, cell metallisation is then completed by electroless plating of a nickel—copper—silver layer. A thin layer of nickel is first deposited. This is then fired, and a substantial thickness of copper is then deposited. Finally, the thin layer of silver is formed on the top surface by displacement plating. All these processes can be electroless, meaning that canisters containing the wafers are simply immersed in the plating solution.

The first commercial use of buried-contact cell technology was when a high-efficiency array was fabricated by Telefunken [48] for the Swiss car "Spirit of Biel," which convincingly won the 1990 World Solar Challenge, the solar car race from Darwin to Adelaide. Array efficiency was 17%, then the highest ever for silicon. The array gave 25% more power than that of the second-place car, which used enhanced screen-printing technology [49].

BP Solar has reported on both manufacturing yields and process economics [50]. Using the same "solar grade" CZ substrates as in its screen-printing process, BP reported substantial efficiency improvement for the technology (circa 30%) and cell efficiencies of 17.5—18%. Economic analysis shows that the approach, as developed by BP with nitride antireflection coating, is well suited for polycrystalline material [50]. In 1994, BP Solar launched the Saturn 585 module, an 85-W module based on 123-mm-square buried-contact solar cells as its "top-of-the-line" commercial product [51], with module size subsequently doubled to 170 W. Production capacity at BP Solar peaked in 2003 when 50 MW/year of buried-contact cells were manufactured, about 3% of total world production.

The consensus of production experience is that the buried-contact process can give a substantial efficiency margin over screen printing. Although more processing steps are involved than in the simplest screen-printing approach, expensive silver pastes are eliminated so that processing costs per unit area are not greatly different, with costs per watt of product lower [34,35]. Marketing experience has shown that higher-selling prices are feasible for this product due to the lower balance of systems costs in installed systems and the perceived higher quality due to the superior performance.

4.2 Semiconductor Finger Solar Cell

The semiconductor finger solar cell shown in Figure 11 incorporates the advantages of either the buried-contact solar cell of the previous section or the laser-doped solar cell of the following section while retaining a standard screen-printed contacting approach.

In the buried-contact variant, the initial steps prior to metallisation are similar to those in the previous section. Metallisation however is applied by screen printing with the printed Ag top contact running perpendicular to the grooves as shown in Figure 11. Paste composition in this case would be selected so that the paste did not fire through the dielectric, reducing contact area. It would, however, contact the exposed, heavily doped regions in the grooves. Cell efficiency in the 18–19% range has been reported in large-scale pilot production using the approach.

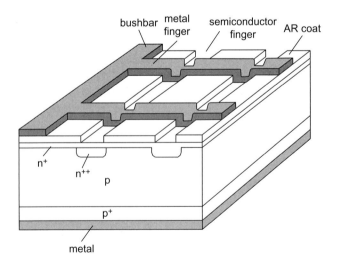

Figure 11 Semiconductor finger solar cell. Heavily doped semiconductor regions provide an additional layer of fingers contacted by the traditional metal finger layer.

A simpler version uses the laser-doped, selective-emitter sequence of the following section to form the "semiconductor" fingers with similarly good outcomes.

4.3 Laser-Doped, Selective-Emitter Solar Cells

The laser-doped, selective-emitter (LDSE) cell [52] uses a laser scan to both dope the emitter selectively in regions scanned as well as to remove the overlying dielectric to expose the silicon surface, allowing selective contacting to this exposed surface region as shown in Figure 12.

The processing starts as for a selective-emitter, screen-printed or buried-contact cell, with the initial top-surface diffusion quite light, to maintain good blue response in the completed cell. After deposition of the top-surface dielectric, the top surface is scanned by a laser to produce the selective emitter and to remove the dielectric in the scanned area. The dopant source could be from within this dielectric, from a doped layer deposited on top of the dielectric, or even in liquid form mixed in the water jet sometimes used to guide the laser in such applications.

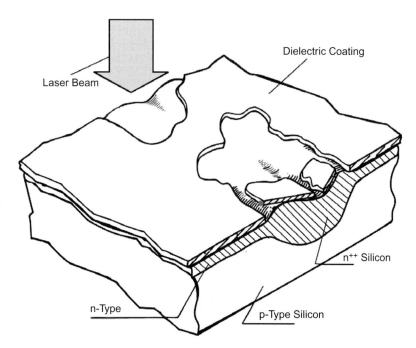

Figure 12 Laser-doped selective-emitter (LDSE) solar cell [18]. Metal is selectively plated to the heavily doped regions exposed during the laser-doping step.

Electroless or electrolytic plating can be used to produce Ni–Cu–Ag or related layers confined to the regions scanned by the laser, as described in connection with the buried-contact cell of Section 4.1.

The combination of selective-emitter, narrow, conductive top-finger metallisation and the good surface passivation possible with this approach results in excellent final performance with efficiencies more than 19% confirmed for cells manufactured on a commercial production line apart from the laser-processing steps. Efficiencies above 20% are anticipated using "double-sided" versions of the technology.

5. HIT CELL

The HIT cell combines both crystalline and amorphous silicon cell technology to produce high-conversion-efficiency cells in production and also some of the highest-efficiency large-area laboratory devices ever reported. The basic device structure is shown in Figure 13.

The starting substrate is an n-type silicon wafer, the opposite polarity from most previous commercial product. This is a fortunate choice since such substrates are free from the light-induced degradation effects that limit the performance of cells made on p-type, boron-doped substrates attributed to the formation of boron–oxygen complexes under illumination [53].

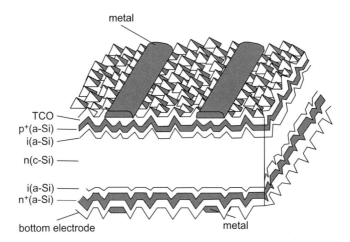

Figure 13 HIT cell structure using a textured n-type silicon wafer with amorphous silicon heterojunctions on both front and rear surfaces.

The n-type substrates are also more tolerant to Fe, a common substrate impurity. After texturing, very thin layers of intrinsic hydrogenated and p-type amorphous silicon (a-Si) are deposited on the top surface and intrinsic and n-type a-Si on the rear surface. As is usual in amorphous silicon technology, these layers are contacted by transparent conducting oxide (TCO) layers, in turn contacted by screen-printed metal contacts.

The low sheet resistivities possible with these TCO layers relax the constraints previously described on the screen-printed metallisation parameters. However, the TCO layers are more absorbing than Si_3N_4, and the underlying doped layers of amorphous silicon are inactive for photocurrent collection, resulting in poor blue response of the cell. About 10% of available current is lost by absorption in these layers. However, the band gap of amorphous silicon is very much higher than in silicon, and the quality of the interface between the amorphous and crystalline material is excellent. This has produced some of the highest open-circuit voltages seen in silicon cells (up to 745 mV).

This approach has produced record large-area laboratory cell performances of up to 23.0% [54]. The differences between these laboratory devices and the less highly performing commercial product are not clear. Commercial modules are offered in the 16.1–19.0% total area efficiency range [1]. Sanyo produced an estimated 300 MW of HIT cell product during 2010 [55].

6. REAR-CONTACTED CELLS

The first solar cell of the modern photovoltaic era had both contacts on the rear [30]. The design subsequently evolved to bifacial contacts (Figure 2) to reduce resistance losses. There has, however, long been an interest in reverting to designs with both contacts on the rear since this offers scope for more densely packing cells and for simplifying cell interconnection.

The initial rear-contact cell can be described as an "emitter wrap-around" cell, part of a more general class of emitter wrap-through (EWT) cells that have no metal on the top surface. An alternative approach is the metal wrap-through (MWT) cell where both front and rear surfaces of the cell are metallised but the top metallisation connects through holes in the cell to the appropriate polarity rear contact. However, the most successful rear contact cell to date has been a more sophisticated design initially suggested by Schwartz [56] with both polarity junctions on the rear surface.

6.1 Rear-Junction Solar Cells

As previously mentioned, the first silicon cell to demonstrate efficiency above 22% was the rear-point contact cell of Figure 6 developed at Stanford University. With some simplifications to the rear-contacting scheme [57], this approach has now been commercialised as the mainstream product offered by a major silicon cell and module manufacture, SunPower, with its cell structure shown in Figure 14.

Using this cell, SunPower offers modules of 18.4–19.7% total area efficiency [1], where the area includes the module frames. These are presently the highest-efficiency commercial modules available by a clear margin. Recently, a large-area cell (225 cm^2) with some improved features was measured at 24.2% efficiency, the highest ever for a cell of this size [58].

The rear-junction approach to rear contact cell design is quite challenging. The silicon wafer has to be both thin and of good quality, and both front and rear surfaces must have good surface properties. Most carriers in a silicon cell are generated within the first 2 microns of entering the cell. The wafer and surface quality needs to be good so these carriers can find their way to the rear contact without recombining.

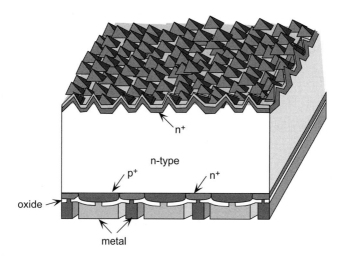

Figure 14 SunPower rear-junction solar cell.

Minority-carrier diffusion lengths have to be at least 5 times the wafer thickness to keep the loss of these migrating carriers to less than 2%. As such, the approach is not suited to multicrystalline or lower-solar-grade materials but results in excellent cells on good-quality materials.

6.2 Emitter Wrap-Through (EWT) Cells

As previously noted, the silicon cells that introduced the modern photovoltaic era are related to modern EWT designs.

For EWT cells, there is no metal on the top surface of the cell with the top-surface dopant providing the contact path to the rear as shown in Figure 15(a). The advantages are zero shading by metal on the top surface plus closer cell packing and simplified cell interconnection. The disadvantage is the high density of holes required due to the relatively poor conductivity of the top layer. The MWT cell of the following section is closely related but does not require as many through-wafer contacts.

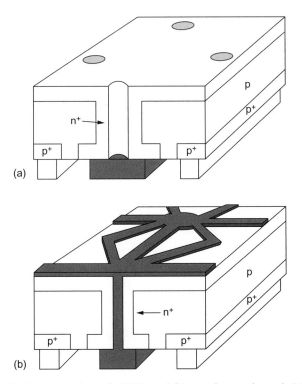

Figure 15 (a) Emitter wrap-through (EWT) and (b) metal wrap-through (MWT) cells.

6.3 Metal Wrap-Through (MWT) Cells

The metal wrap-around cell has been investigated from the 1960s for space use, given the clear advantages of having both cell contacts on the rear. Of more interest for terrestrial use has been the metal wrap-through (MWT) cell with metal on both surfaces connected by holes through the wafer (Figure 15(b)).

Advantages are reduced shading of the front surface of the cell since thick busbars are eliminated, closer cell packing, and simplified module assembly. Disadvantages are the extra processing required in drilling holes through the cell by laser.

Unlike the rear-junction cells of Section 6.1, MWT cells are well suited to low-quality substrates. In fact, the first solar modules of more than 17% aperture area efficiency using multicrystalline wafers were fabricated using MWT cells [59].

7. CONCLUSIONS

With the costs of present wafer-based silicon approaches dominated by material costs, particularly those of the wafers, encapsulants, low-iron tempered glass superstrates, and metal pastes, increasing cell efficiency is an effective, if often counterintuitive, approach to reducing the cost of the final product.

Recent years have seen a diversification of manufacturing into higher-efficiency approaches. Quite revolutionary departures from the standard approach have been made by BP Solar, with its former top-of-the-line Saturn processes, based on laser-grooved buried contacts; by Sanyo, with its crystalline—amorphous silicon hybrid HIT cells; SunPower, with the company's very-high-performance rear-junction cells; and, more recently, Suntech with its Pluto cell [60].

In mainstream production, there is still a large potential to capture more of the improvements in silicon cell performance that have been demonstrated in the laboratory. The escalating price of Ag over recent years is likely to require a response from across the photovoltaic manufacturing industry, which may finally induce the industry to depart from the screen-printing process that has been its backbone for more than 30 years.

To reduce Ag consumption, a transition to Cu seems mandatory over the coming decade, whether by plating Cu to layers applied themselves by plating approaches or by screen printing or related physical deposition processes.

The laser-processing approaches, such as demonstrated by the buried-contact solar cell and, more recently, the laser-doped, selective-emitter cell, seem the natural choice for such plating sequences, particularly given their demonstrated high-efficiency potential.

The limiting efficiency of cells relying on traditional Al rear contacts lies in the 21—22% range, as demonstrated by the best laboratory devices using this approach. To progress closer to the 25% efficiency value that may ultimately be achievable in production by a silicon device, a more sophisticated rear-contacting approach is required. Double-sided laser-processed approaches are again a promising option in this area.

ACKNOWLEDGEMENTS

The author gratefully acknowledges support by an Australian Government Federation Fellowship in the initial stages of this work and thanks the School of Photovoltaic and Renewable Energy Engineering for permission to reproduce material and figures from the author's text *Silicon Solar Cells: Advanced Principles and Practice*.

REFERENCES

[1] C. Haase, C. Podewils, More of everything: market survey on solar modules 2011, Photon Int. February (2011) 174—221.
[2] K.D. Smith, H.K. Gummel, J.D. Bode, D.B. Cuttriss, R.J. Nielson, W. Rosenzweig, The solar cells and their mounting, Bell Sys. Tech. J. 41 (1963) 1765—1816.
[3] P.A. Iles, Increased output from silicon solar cells, Conference Record, Eigth IEEE Photovoltaic Specialists Conference, Seattle, 1970, pp. 345—352.
[4] R. Gereth, H. Fischer, E. Link, S. Mattes, W. Pschunder, Silicon solar technology of the seventies, Conference Record, Eigth IEEE Photovoltaic Specialists Conference, Seattle, 1970, p. 353.
[5] R. Gereth, H. Fischer, E. Link, S. Mattes, W. Pschunder, Contribution to the solar cell technology, Energy Conversion 12 (1972) 103—107.
[6] J. Mandelkorn, J.H. Lamneck, A new electric field effect in silicon solar cells, J. Appl. Phys. 44 (1973) 4785.
[7] M.P. Godlewski, C.R. Baraona, H.W. Brandhorst, Low-high junction theory applied to solar cells, Tenth IEEE Photovoltaic Specialists Conference, Palo Alto, 1973, pp. 40—49.
[8] J. Lindmayer, J. Allison, The violet cell: an improved silicon solar cell, COMSAT Tech. Rev. 3 (1973) 1—22.
[9] J. Haynos, J. Allison, R. Arndt, A. Meulenberg, The comsat non-reflective silicon solar cell: a second generation improved cell, International Conference on Photovoltaic Power Generation, Hamburg, September, 1974, p. 487.
[10] H.G. Rudenberg, B. Dale, Radiant energy transducer, US Patent 3,150,999 (filed 17 February, 1961).
[11] E.L. Ralph, Recent advancements in low cost solar cell processing, Conference Proceedings, Eleventh IEEE Photovoltaic Specialists Conference, Scottsdale, 1975, pp. 315—316.

[12] J. Zhao, M.A. Green, Optimized antireflection coatings for high efficiency silicon solar cells, IEEE Trans. Electron Devices 38 (1991) 1925–1934.
[13] J.G. Fossum, E.L. Burgess, High efficiency p^+nn^+ Back-surface-field silicon solar cells, Appl. Phys. Lett. 33 (1978) 238–240.
[14] R.B. Godfrey, M.A. Green, 655 mV open circuit voltage, 17.6% efficient silicon MIS solar cells, Appl. Phys. Lett. 34 (1979) 790–793.
[15] M.A. Green, Enhancement of schottky solar cell efficiency above its semiempirical limit, Appl. Phys. Lett. 28 (1975) 268–287.
[16] J. Lindmayer, J.F. Allison, Dotted contact fine geometry solar cell, U.S. Patent 3,982,964, September 1976.
[17] R.A. Arndt, A. Meulenberg, J.F. Allison, Advances in high output voltage silicon solar cells, Conference Record, Fifteenth IEEE Photovoltaic Specialists Conference, Orlando, 1981, pp. 92–96.
[18] M.A. Green, A.W. Blakers, J. Shi, E.M. Keller, S.R. Wenham, High-efficiency silicon solar cells, IEEE Trans. on Electron Devices ED-31 (1984) 671–678.
[19] F.A. Lindholm, A. Neugroschel, M. Arienze, P.A. Iles, Heavily doped polysilicon–contact solar cells, Electron Device Lett. EDL-6 (1985) 363–365.
[20] E. Yablonovitch, T. Gmitter, R.M. Swanson, Y.H. Kwark, A 7209 mV open circuit voltage, SiO_X:c−Si:SiO_X double heterostructure solar cell, Appl. Phys. Lett. 47 (1985) 1211–1213.
[21] P. Van Halen, D.L. Pulfrey, High-gain bipolar transistors with polysilicon tunnel junction emitter contacts, IEEE Trans. on Electron Devices 32 (1985) 1307.
[22] M. Tanaka, M. Taguchi, T. Takahama, T. Sawada, S. Kuroda, T. Matsuyama, et al., Development of a new heterojunction structure (ACJ-HIT) and its application to polycrystalline silicon solar cells, Prog. Photovolt. 1 (1993) 85–92.
[23] A.W. Blakers, M.A. Green, 20% Efficiency silicon solar cells, Appl. Phys. Lett. 48 (1986) 215–217.
[24] T. Saitoh, T. Uematsu, T. Kida, K. Matsukuma, K. Morita, Design and fabrication of 20% efficiency, medium-resistivity silicon solar cells, Conference Record, Ninteenth IEEE Photovoltaic Specialists Conference, New Orleans, 1987, pp. 1518–1519.
[25] W.T. Callaghan, Evening presentation on Jet Propulsion Lab. Photovoltaic activities, Seventh E.C. Photovoltaic Solar Energy Conference, Sevilla, Oct., 1986.
[26] R.A. Sinton, Y. Kwark, J.Y. Gan, R.M. Swanson, 27.5% Si concentrator solar cells, Electron Device Lett. EDL-7 (1986) 567.
[27] R.R. King, R.A. Sinton, R.M. Swanson, Front and back surface fields for point-contact solar cells, Conference Record, twentieth IEEE Photovoltaic Specialists Conference, Las Vegas, September, 1988, pp. 538–544.
[28] A.W. Blakers, A. Wang, A.M. Milne, J. Zhao, X. Dai, M.A. Green, 22.6% Efficient silicon solar cells, Proceedings, Fourth International Photovoltaic Science and Engineering Conference, Sydney, February, 1989, pp. 801–806.
[29] M.A. Green, The path to 25% silicon solar cell efficiency: history of silicon cell evolution, Prog. Photovolt: Res. Appl. 17(3) (2009) 183–189.
[30] M.A. Green, Silicon Solar Cells: Advanced Principles and Practice, Bridge Printery, Sydney, 1995.
[31] M.A. Green, J. Zhao, A. Wang, S.R. Wenham, 45% Efficient silicon photovoltaic cell under monochromatic light, IEEE Electron Device Lett. 13 (1992) 317–318.
[32] M.A. Green, J. Zhao, A.W. Blakers, M. Taouk, S. Narayanan, 25-percent efficient low-resistivity silicon concentrator solar cells, IEEE Electron Device Lett. EDL-7 (1986) 583–585.
[33] A. Cuevas, R.A. Sinton and R.M. Swanson, Point- and planar-junction p-i-n silicon solar cells for concentration applications, fabrication, performance and stability,

Conference Record, Twenty-first IEEE Photovoltaic Specialists Conference, Kissimmee, May, 1990, pp. 327–332.
[34] T. Bruton, G. Luthardt, K.-D. Rasch, K. Roy, I.A. Dorrity, B. Garrard, et al., A study of the manufacture at 500 mwp p.a. of crystalline silicon photovoltaic modules, Conference Record, Fourteenth European Photovoltaic Solar Energy Conference, Barcelona, June–July, 1997, pp. 11–26.
[35] T.M. Bruton, MUSIC FM five years on: fantasy or reality? PV in Europe, Rome, October 2002.
[36] D.-H. Neuhaus, A. Münzer, Industrial siliconwafer solar cells, Adv. OptoElectronics (2007) Article ID 24521, 1–15.
[37] M.A.Green, Ag requirements for silicon wafer-based solar cells, Prog. Photovolt., accepted for publication.
[38] K. Firor, S. Hogan, Effects of processing parameters on thick film inks used for solar cell front metallization, Solar Cells 5 (1981–1982) 87–100.
[39] Final Report, Flat Plate Solar Array Project, Vol. V, Jet Propulsion Laboratory, Publication 86–31, October, 1986.
[40] Crystalline Silicon PV Technology and Manufacturing Group, International Technology Roadmap for Photovoltaics: Results 2010, Second ed., SEMI PV Group Europe, Berlin, March 2011.
[41] S.K. Chunduri, New choices for 'Selective' shoppers, Photon Int. December (2010) 158–172.
[42] H. Kerp, K. McVicker, B. Cruz, P. Van Eijk, D. Arola, E. Graddy, et al., Leading high aspect ratio front contact metallization techniques for crystalline silicon solar cells, Proceedings of the Twenty-fifth European Photovoltaic Solar Energy Conference and Exhibition, 6–10 September 2010, Valencia, Spain, pp. 2460–2463.
[43] B. Heurtault, J. Hoornstra, Towards industrial application of stencil printing for crystalline silicon solar cells, Proceedings of the Twenty-fifth European Photovoltaic Solar Energy Conference and Exhibition, 6–10 September 2010, Valencia, Spain, pp. 1912–1916.
[44] J. Ebong, B. Renshaw, I. B. Rounsaville, K. Cooper A. Tate, S. Rohatgi, et al., Ink jetted seed and plated grid solar cells with homogeneous high sheet resistance emitters, Proceedings of the Twenty-fifth European Photovoltaic Solar Energy Conference and Exhibition, 6–10 September 2010, Valencia, Spain, pp. 2390–2394.
[45] J. Bartsch, A. Mondon, C. Schetter, M. Hörteis, S. W. Glunz, Copper as conducting layer in the front side metallization of crystalline silicon solar cells—challenges, processes and characterization, Proceedings of the Second Workshop on Metallization, April 14th & 15th, 2010, Constance, Germany, pp. 32–37.
[46] S.R. Wenham, Buried-contact silicon solar cells, Prog. Photovolt. 1 (1993) 3–10.
[47] J. Wohlgemuth, S. Narayanan, Buried contact concentrator solar cells, conference record, twenty-second IEEE Photovoltaic Specialists Conference, Las Vegas, pp. 273–277, 1991.
[48] H.-W. Boller, W. Ebner, Transfer of the BCSC-Concepts into an industrial production line, Proceedings of the Ninth E.C. Photovoltaic Solar Energy Conference, Freiburg, September, pp. 411–413, 1989.
[49] C. Kyle, Racing with the sun: The 1990 world solar challenge, Engineering Society for Advancing Mobility: Land, Sea, Air and Space, SAE Order No. R-111, 1991.
[50] T.M. Bruton, A. Mitchell, L. Teale, Maximizing minority carrier lifetime in high efficiency screen printed silicon bsf cells, Conference Proceedings, Tenth E.C. Photovoltaic Solar Energy Conference, Lisbon, April, 1991, pp. 667–669.
[51] BP Saturn product sheet, April, 1994.

[52] S.R. Wenham, M.A. Green, Self aligning method for forming a selective emitter and metallization in a solar cell, United States Patent US6,429,037 B1, 6 August 2002.
[53] J. Schmidt, K. Bothe, Structure and transformation of the metastable boron- and oxygen-related defect center in crystalline silicon, Phys. Rev. B 69 (2004) 0241071−02410718.
[54] E. Maruyama, A. Terakawa, M. Taguchi, Y. Yoshimine, D. Ide, T. Baba, et al., Sanyo's challenges to the development of high-efficiency HIT solar cells and the expansion of HIT business, Proceedings of the Fourth World Conference on Photovoltaic Energy Conversion (WCEP-4), Hawaii, May 2006.
[55] G. Hering, Year of the tiger: PV cell output roared in 2010 to over 27 gw—beating 2006 through 2009 combined—but can the year of the rabbit bring more multiples?, Photon Int. March (2011) 186−218.
[56] R.J. Schwartz, Review of silicon solar cells for high concentrations, Solar Cells 6 (1982) 17−38.
[57] W.P. Mulligan, M.J. Cudzinovic, T. Pass, D. Smith, N. Kaminar, K. McIntosh, et al., Solar cell and method of manufacture, United States Patent US7,897,867 B1, 1 March 2011.
[58] P.J. Cousins, D.D. Smith, H.C. Luan, J. Manning, T.D. Dennis, A. Waldhauer, et al., Gen III: Improved Performance at Lower Cost, Proceedings of the Thirty-fifth IEEE PVSC, Honolulu, June 2010.
[59] M. W. P. E. Lamers, C. Tjengdrawira, M. Koppes, I. J. Bennett, E. E. Bende, T. P. Visser, et al., 17.9% Metal-wrap-through mc-si cells resulting in module efficiency of 17.0%, Prog. Photovolt., article first published online: 23 March 2011 | DOI: 10.1002/pip.1110.
[60] Z. Shi, S.R. Wenham, J.J. Ji, mass production of the innovative PLUTO solar cell technology, Proceedings of the Thirty-fourth IEEE Photovoltaic Specialists Conference, Philadelphia, 7−12 June 2009.

CHAPTER IB-3

Low-Cost Industrial Technologies for Crystalline Silicon Cells[1]

Jozef Szlufcik[a], S. Sivoththaman[b], Johan F. Nijs[a], Robert P. Mertens, and Roger Van Overstraeten[c]

Interuniversity Microelectronics Centre, Leuven, Belgium
[a]Now at Photovoltech, Brussels, Belgium
[b]Now at University of Waterloo, Ontario, Canada
[c]Died 25 April 1999

Contents

1. Introduction	130
2. Cell Processing	131
2.1 Substrates	131
2.2 Etching, Texturing, and Optical Confinement	132
2.3 Cleaning	134
2.4 Junction Formation	136
2.5 Front Surface Passivation and Antireflection Coating	137
2.6 Back-Surface-Field (BSF) and Back Side Passivation	138
2.7 Front Contact Formation	139
2.8 Substrate Material Improvement	141
2.8.1 Gettering by Phosphorus Diffusion	*141*
2.8.2 Gettering by Aluminium Treatment	*142*
2.9 'Fast Processing' Techniques	143
3. Industrial Solar Cell Technologies	145
3.1 Screen Printing Solar Cells	145
3.2 Buried Contact Solar Cells (BCSC)	147
3.3 Metal−Insulator−Semiconductor Inversion Layer (MIS-IL) Solar Cells	148
3.4 Solar Cells on EFG Silicon Sheets	150
3.5 Commercial Thin Film Crystalline Silicon Solar Cells	151
4. Cost of Commercial Photovoltaic Modules	152
References	154

[1] Portions reprinted, with permission, from Proceedings of the IEEE. Vol. 85. No. 5. May 1997. © 1997 IEEE.

1. INTRODUCTION

Although efficiency is important, the principal requirement for industry is low cost. Processing techniques and materials are selected for the maximal cost reduction while maintaining a relatively good efficiency. Industrial solar cells are fabricated in large volumes mainly on large area (≥ 100 cm^2) Czochralski monocrystalline or multicrystalline silicon substrates.

Analysis indicates that the market price of the commercial PV modules lies in the range 3.5–4.5 \$/Wp. 40–50% of the PV module cost is due to ingot growth (including the polysilicon feedstock material), single crystal ingot formation and wafering. The tendency here is to develop a cheap, good-quality solar-grade polysilicon feedstock material, to increase the substrate size, to reduce the kerf loss in slicing and to decrease the thickness of the substrates below 200 μm. Cell fabrication and module assembly are each responsible for 25% to 30% of the final module cost.

Typical efficiency of commercially produced crystalline silicon solar cells lies in the range 13–16%. Because the efficiency of the cell influences the production cost at all production stages, substantial effort is directed towards efficiency improvement. The required future efficiency goals for industrial cells are 18–20% for monocrystalline and 16–18% for multicrystalline silicon. Based on laboratory scale achievements one can consider that production type cells able to fulfil the efficiency goal should will possess most of the following features (providing that they can be introduced in a cost effective way):

- front surface texturing;
- optimised emitter surface concentration and doping profile;
- effective front surface passivation;
- fine-line front electrode;
- front electrode passivation:
  - point contact,
  - deep and highly doped emitter under the contact,
  - MIS contact;
- thin base, i.e., much smaller than the minority carrier diffusion length;
- back surface passivation:
  - oxide and/or nitride passivation + local BSF(PERL),
  - floating junction structure,

- or back surface field;
- back electrode passivation:
  * point contact,
  * deep back-surface diffusion under the contact;
- back reflector;
- back surface texture; and
- antireflection coating optimised for encapsulation.

The current status in the development of industrial-type processing steps leading to an improved cell efficiency will be described in detail in the sections below.

2. CELL PROCESSING

2.1 Substrates

Although the standard substrate size is still 10×10 cm^2, there is a clear tendency to larger sizes. Many solar cell manufacturers base their production lines on 12.5×12.5 cm^2 wafers. Efficient cells of 15×15 cm^2 and even 20×20 cm^2 [1] have been reported. The driving force towards these larger cell sizes results from the fact that the cell manufacturing and the module assembly costs show little area dependence and, therefore, the cost per Wp decreases with increasing cell size. The optimum cell size, however, is limited by series resistance and by a limitation on the module size due to handling, wind loads, module transportation and system assembly. Therefore, cell sizes larger than 20×20 cm^2 appear to be excluded. Due to the successful developments in multiwire sawing, wafer thickness of 150 μm or thinner are becoming feasible, corresponding to a final cell thickness of 120 μm or lower [2]. This allows an important material saving and, at the same time, thinner cells correspond to the optimum thickness if efficient light trapping and surface passivation are possible. In spite of significant progress in slicing techniques, around 200 μm of high-quality silicon per wafer is still lost in kerf waste.

Kerf loss can be completely avoided in the ribbon and sheet silicon technologies. Although many technologies have been tried on the laboratory or pilot scale [3–6] only the Edge-defined-Film-fed Growth (EFG) polysilicon sheets have been introduced into high volume production [5]. An individual crystal is grown in the form of hollow octagonal tube, with eight 10-cm-wide faces and an average tube wall thickness of 300 μm. A total tube length is usually 4.6 m. The faces are then separated and cut to lengths appropriate for cell processing.

2.2 Etching, Texturing, and Optical Confinement

Silicon substrates used in commercial solar cell processes contain a near-surface saw-damaged layer which has to be removed at the beginning of the process. Thickness of the damage depends on the technique used in wafering of the ingot. A layer with thickness of 20 to 30 μm has to be etched from both sides of wafers cut by an inner-diameter (ID) blade saw, while only 10 to 20 μm is enough when a wire saw is used. The damage removal etch is typically based on 20–30 wt.% aqueous solution of NaOH or KOH heated to 80–90°C. The etching process has to be slightly modified when applied to multicrystalline substrates. Too fast or prolonged etching can produce steps at grain boundaries. This can lead to problems with interruptions of metal contacts. This problem can be avoided by an isotropic etching based on a mixture of nitric, acetic, and hydrofluoric acids. However, a strong exothermic reaction makes this etching process difficult to control and toxicity of the solution creates safety and waste disposal problems.

The silicon surface after saw damage etching is shiny and reflects more than 35% of incident light. The reflection losses in commercial solar cells are reduced mainly by random chemical texturing [7,8]. Surface texturing reduces the optical reflection from the single crystalline silicon surface to less than 10% by allowing the reflected ray to be recoupled into the cell. Monocrystalline silicon substrates with a surface orientation <100> can be textured by anisotropic etching at temperature of 70–80°C in a weak, usually 2 wt.%, solution of NaOH or KOH with addition of isopropanol. This etch produces randomly distributed upside pyramids [8]. However, this process brings often production problems of repeatability, lack of pyramid size control and the presence of untextured regions [7]. The important parameters are adequate surface preparation, temperature control, mixing rate and isopropanol concentration [8]. The solution to this problem requires the use of appropriate additives which enhance the pyramid nucleation process [7]. When the process is under control, uniformly distributed pyramids with height of 3–5 μm are optimal for low reflection losses and later metallization process. Figure 1 shows the SEM micrograph of a randomly textured <100> oriented silicon surface.

The random texturization process is not effective on multicrystalline substrates due to its anisotropic nature. Isotropic texturing methods

Figure 1 SEM micrograph of a randomly textured <100> oriented silicon surface.

based on photolithography and wet etching are not cost-effective. Many techniques such as defect etching, reactive ion etching or laser scribing have been tried by many groups [9–14]. The best results, from the optical point of view, have been obtained by mechanical texturing and by reactive ion etching [13]. Mechanical grooving is a method of forming V-grooves in Si wafer by mechanical abrasion, using a conventional dicing saw and bevelled blades [14]. The optical quality of the mechanically textured surface depends on the blade tip angle, groove depth, and damage layer etching. An average reflection of 6.6% in the range 500–1000 nm was obtained with a minimum reflection of 5.6% at 950 nm on multi-Si grooved with a blade having tip angle of 35° [14]. Figure 2 shows the influence of the groove depth on the reflectance. In addition to the reduced reflection, an improvement in internal quantum efficiency in the range 750–1000 nm has been observed in multicrystalline cells, indicating the effect of light trapping [15]. The light trapping effect is very important when thin silicon substrates (<200 μm) are used for material saving. Optimised grooving can bring as much as 0.5–1% absolute improvement in cell performance [16]. The efficiency of 17.2%, the highest ever reported for 10×10 cm^2 multicrystalline cells, has been achieved by Sharp with mechanical grooving and screen printing [17]. The best results are obtained when a single-blade, bevelled saw is used. The main disadvantage of this approach is the low throughput

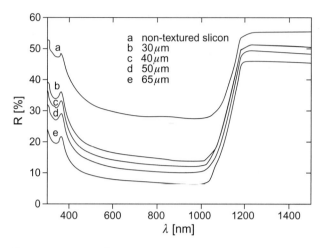

Figure 2 Reflectance curves of mechanically V-grooved multicrystalline silicon substrates. Influence of the groove depth is presented.

of one wafer every two hours. A multiblade system can reduce the grooving process to a few seconds but the quality of the grooves are poorer than obtained with single blade grooving [18]. Lately a new technique of the mechanical surface structuring, wire grooving has been introduced [19]. This process has a much higher throughput than V-grooving with a bevelled blade and a dicing saw. A homogeneous web of stainless steel wires of about 180 μm in diameter and at a certain distance are guided by four grooved rollers as in the standard wafer cutting technique.

A very elegant technique of isotropic surface texturing of multicrystalline surface has been developed [20]. Here the proprietary acid solution gives isotropic surface structuring which, in combination with a TiOx antireflection coating, decreases the surface reflection to the value of monocrystalline randomly textured wafers.

2.3 Cleaning

In an industrial high-efficiency silicon solar cell process, wafers are typically cleaned after texturing and before surface passivation (oxide growth). Traditionally RCA clean [21], originally developed for use in microelectronics, is the most widespread cleaning recipe in solar cell processing. Although not often discussed in the solar cell technical literature, cleaning is very important for solar cell performance. Long diffusion lengths of minority carriers, necessary for high efficiency cells, require low levels of metal contamination at the silicon surface before a high temperature treatment. Moreover, a growing concern in the photovoltaic community is the chemical waste produced during cell processing. The conventional RCA cleaning consists of two steps normally referred to as SC1 and SC2. The SC1 step consisting of a $NH_4OH/H_2O_2/H_2O$ mixture, aims at organic particle removal, whereas the SC2 step (an $HCl/H_2O_2/H_2O$ mixture) is used to remove metal contaminants. A more detailed analysis reveals that, because of their large feature size and the absence of photo-lithographic processes, organic particles are not an important issue and therefore the SC1 step is not essential for industrial solar cells. On the contrary, it is known that the metal contamination resulting from the SC1 step may be high, and a one-to-one correlation between the metal concentration of the SC1 bath and the metal contamination of the silicon surface was found. This requires the use of sub ppb metal contamination specification of chemicals.

Recently, a new cleaning concept has been introduced [22] as a potential replacement for the standard RCA clean. This new TMEC-clean, which reveals a perfect removal of metallic particles, usually consists of H_2SO_4/H_2O_2 step, followed by a 1% diluted HF step. There are, however, processes developed already for surface cleaning which use ozonated mixtures as H_2SO_4/O_3, H_2O/O_3 and $UV-O_3$. Table 1 presents the metal contamination removal for RCA-clean and IMEC-clean.

If removal of metallic contamination is the only issue, a single cleaning step in 1% diluted HCl yields excellent results. This is especially important for the cleaning step performed after texturization. A further advantage is the low consumption of chemicals. This results not only in important cost savings but also in a considerable reduction of chemical waste products.

Table 1 Metal contamination removal (in 10^{10} at/cm^2 after IMEC vs. RCA + HF clean) [22]

Cleaning treatment	Ca	Fe	Cu	Zn
Starting level	15	1	2	1
IMEC	0.7	0.3	<0.1	0.7
RCA + HF	2.5	0.8	0.5	4.9
5 × (IMEC) + HF	<0.1	<0.1	<0.1	0.1
5 × (RCA) + HF	4.5	0.3	<0.1	<0.1

2.4 Junction Formation

It has been proved by many workers [23–26] that the optimum emitter doping profile should be relatively deep and moderately doped, or a shallow emitter should be formed with a high surface concentration. Both profiles combined with surface passivation by high quality thermal oxides show reduced surface recombination losses and increased emitter collection efficiency. However, both industrial techniques used for front contact fabrication—i.e., screen printing of silver pastes and electroless plating of Ni—require a highly doped P surface concentration (above 10^{20} cm^{-3}) and a deep junction to obtain acceptable contact resistance and to avoid metallic impurity penetration towards the junction region. Typical emitter sheet resistance used in a screen printing metallization process is between 30 and 50 ohm/sq. This can be achieved by diffusion from liquid POCl$_3$ or solid P$_2$O$_5$ sources in open tube furnaces or, more industrially, from screen printed, sprayed- or spin-on P-sources followed by a conveyor belt furnace diffusion. Deep emitter and poor surface passivation lead to voltage loss and collection losses in the short light wavelengths. The perfect solution brings a selective emitter structure shown in Figure 3.

The heavy and deep diffusion under the contact fingers not only assures low contact resistance, giving good fill factor, but also reduces the

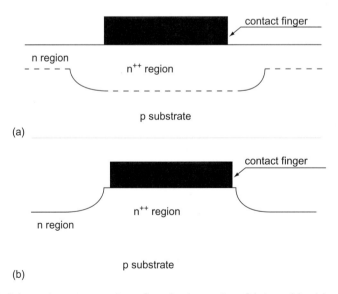

Figure 3 Schematic representation of a selective emitter fabricated by (a) two diffusions, (b) one deep diffusion and selective etch-back.

contact contribution to front surface recombination losses. The emitter between fingers is optimised for high spectral response and high Voc thanks to relatively low surface concentration. The selective emitter structure is realised in practice by double diffusion [27] or single diffusion and etching back the active emitter region to the desired sheet resistance [28]. The high thermal budget of the emitter diffusion process can be significantly reduced by adapting rapid thermal processing RTP [29,30] to the industrial cell process. Although this technology is at an early stage of development, the first trials bring promising results.

An entirely different approach is presented by MIS inversion layer solar cell technology. Here, the emitter diffusion process is completely eliminated by inducing an n^+ inversion layer in the p-type silicon substrate. The inversion layer is induced by positive charge in the top silicon nitride antireflection coating layer. The high positive charge density is achieved by the incorporation of cesium [31].

2.5 Front Surface Passivation and Antireflection Coating

Surface recombination can be effectively decreased by many techniques. The most common one for the laboratory cells is to grow a thin thermal oxide and deposit a double antireflection coating by evaporating ZnS and MgF_2 layers [32], or by growing a thick thermal oxide up to 110 nm which serves, at the same time, as a passivating and anti-reflection coating (ARC) layer [33]. This gives a decreased surface recombination velocity at the $Si-SiO_2$ interface depending on the P-surface concentration of the emitter and density of interface states. The interface quality is further improved by a low temperature anneal in forming gas. Such surface passivation is very effective especially with low emitter doping concentration, and regularly leads to an enhancement in the photogenerated current and open circuit voltage. Both approaches are excluded in industrial processes due to the high cost and low throughput of vacuum processes and long time needed to grow a thick thermal oxide The thickness of the passivating oxide in case of industrial cells is in the range 6–15 nm which is thin enough not to disturb the optical system in combination with antireflection coating and thick enough to ensure an effective surface passivation.

Industrial solar cells today widely use TiO_2 anti-reflection coating deposited mainly by atmospheric pressure chemical vapour deposition (APCVD) in a conveyor belt furnace, spin-on or spray-on techniques. An investigation revealed that perfect passivation for mono- as well as multi-crystalline cells is obtained by a combination of a thin dry oxide and a plasma enhanced chemical vapour deposition (PECVD) silicon nitride [13,34]. The thermal treatment of the PECVD SiNx layer during the contact firing through the SiNx layer is a crucial process to get good surface and bulk passivation. This process is particularly attractive for multicrystalline cells since it can eliminate lengthy Si-bulk passivation processes in atomic hydrogen ambient [34,35].

2.6 Back-Surface-Field (BSF) and Back Side Passivation

Since the trend in industrial solar cell manufacturing is towards producing cells on thinner wafers, the role of back side recombination becomes important also in industrial mono- and multicrystalline solar cells. Currently, in most of the industrial cell structures, back side passivation is performed by alloying a screen printed aluminium paste with silicon. Aluminium forms an eutectic alloy with silicon at a temperature of 577°C. During the firing process, a liquid Al—Si phase is formed according to the Al—Si phase diagram. The molten Al—Si region acts as a sink for many impurities, giving a perfect gettering effect. During cooling down, the silicon recrystallises and is doped with aluminium at its solubility limit of Al at given temperature, creating a p^+ Back Surface Field (BSF) layer (Figure 4(a)) [36,37]. A sufficient thickness of Al is required to achieve a significant contribution of Si in the formation of the liquid phase. A very low back surface recombination velocity down to 200 cm/s can be achieved when a thick aluminium layer above 20 μm is printed and fired at a temperature above 800°C for a duration of 1 to 5 minutes [37,38]. This process, however, causes significant wafer warping which can create a problem of low mechanical yield if thin substrates are used. The p^+ BSF can be created by evaporating an Al layer and sintering for a few hours at high temperature [39]. The wafer warping problem is significantly reduced, but this process, however, is not able to achieve as low a surface recombination velocity as screen-printed Al BSF. A rear surface recombination velocity above 1000 cm/s at the p-p^+ interface has been measured [40,41].

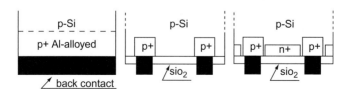

Figure 4 Schematic representations of different backside structures: (a) Al-alloyed BSF, (b) locally diffused BSF, and (c) floating junction.

Another method which proved to be very effective is the so-called local BSF process (Figure 4(b)). In this case most of the back side surface is passivated with thermally grown oxide while the gridded back side electrode covers between 1% and 4% of the total back surface area [42]. An emerging and very promising process for back side passivation is the floating junction approach [43,44] (Figure 4(c)). Here, a back side n^+ junction which is created during $POCl_3$ diffusion of the emitter, is at a floating and forward-biased potential. In theory, this should give a very effective back side passivation not strongly dependent on the back surface recombination velocity. The back contact grid can be made by screen printing Ag/Al pastes [20,45], mechanical of laser grooving, boron diffusion and plating [44].

2.7 Front Contact Formation

The process of front contact formation is one of the most important solar cell processing steps. The applied metallization technique determines the shadowing and series resistance losses, determines the emitter diffusion profile and surface doping concentration, and dictates the choice of certain surface passivation techniques. High-efficiency, large-area solar cells require front electrodes with low series resistance and a low area coverage. In order to meet these requirements, two basic techniques are implemented in solar cell mass production: laser-grooved buried contact metallization and advanced screen-printing processes.

In the first method, plated metal contacts are formed in deep grooves cut by laser or mechanically into a lightly diffused and protected (by nitride or oxide) front surface. After etching and cleaning, the grooves are subjected to a second very heavy diffusion. The metallization is then obtained by a self-aligned plating process of nickel, copper, and thin layer of silver [46]. The advantages of these cells have been described in the literature [40] and include the very large height-to-width ratio of the finger metallization and the fine line width (typically 20–25 μm, see Figure 5(a)). Based on this cell structure and using simplified processing, a cost effective production technology is now in operation, yielding average efficiencies between 16–17% and occasionally up to 18% on 100-cm^2 CZ pseudo-square cells [47].

The main problems related to this technology are the environmental issues. The external costs of meeting environmental specifications of developed countries with processes producing an enormous amount of rinse water containing nickel and copper must be taken into account

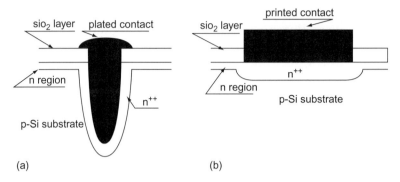

Figure 5 Cross-section of a laser-grooved buried plated contact (a) in comparison with a fine-line screen-printed contact (b).

[48,49]. It can be a major issue for planning large-volume plants above 100 MW.

In the screen-printing method, a stainless steel or polyester mesh screen stretched on a metal frame is covered by a photo-emulsion layer. Then openings, which define the front contact pattern, are photolithographically formed in the screen. Highly conductive silver paste is pushed by a squeegee through openings in the screen onto substrates with a well-defined adjustable pressure.

Screen printing is a traditional industrial solar cell technology, existing since the beginning of the 1970s. The advantages of screen-printing solar cells are the fact that the technology is well established and can be improved step by step without requiring large capital investment, the robustness of the production equipment and the low amount of chemical wastes. The first generation of photovoltaic devices made with this technology suffered from severe limitations:

- Screen printed contacts were typically 150–200 μm wide, giving rise to large shading losses.
- Fill factors were very low (below 75%) because of the high contact resistance and low metal conductivity of screen-printed contacts.
- Effective emitter surface passivation was difficult because a high emitter surface concentration and deep emitter are needed to limit the contact resistance and avoid junction shunting. This also resulted in a poor response to short wavelength light.

Due to all these factors the efficiency of screen-printed cells was typically about 25% to 35% lower than the efficiency of buried contact cells.

Since the early stages of this technology, an important effort has been made by several research groups and companies, mainly in Japan and Europe, to improve the screen printing process [13,17,35,50]. It appears possible to reduce the finger width to values between 50 and 100 μm with a high aspect ratio. By optimising pastes and firing temperatures, fill factors between 78% and 79% for 100-cm^2 screen-printed cells can be obtained [34]. Screen printed metallization can also be combined with effective surface passivation in selective emitter structures. Figure 5 depicts the cross-section of buried and screen printed contacts drawn in the same scale.

Among other techniques for front contact formation, there are reports of direct pen writing, offset printing [51] and roller printing [18] The direct pen writing with a very high aspect ratio has been applied for a front contact metallization of EFG silicon sheets solar cells [52].

2.8 Substrate Material Improvement

Maintaining a high-enough bulk minority-carrier lifetime is essential to improving the performance of crystalline Si solar cells. Unlike the high quality and expensive FZ Si, the 'solar-grade quality' CZ-Si and multicrystalline Si (mc-Si) substrates generally exhibit moderate bulk lifetimes depending on various factors that introduce generation-recombination (GR) centres within the band gap. The major sources of these centres include the oxygen and carbon content, metallic impurities, high densities of crystallographic defects, and the presence of grain and sub-grain boundaries. Many of these factors are not completely avoidable given the low-cost substrate requirement for silicon photovoltaics. Gettering is a technique which either reduces or eliminates metallic impurities in the substrates by localising and blocking them away from the device active regions, or by completely removing them from the substrate. The former is referred to as intrinsic gettering and the latter as extrinsic gettering. Gettering techniques are well-established in Si IC technology for various purposes, for example, to reduce the leakage currents induced by GR centres in CCD image sensors, CMOS devices, etc. In solar cell fabrication, it is desirable that the gettering treatment remains part of the cell processing and does not considerably increase the production cost. Phosphorus diffusion and aluminium diffusion are the most efficient gettering schemes used in Si solar cell processing.

2.8.1 Gettering by Phosphorus Diffusion

Impurity migration towards gettering sites takes place as a consequence of a strong emission of Si interstitials due to the formation of SiP particles by heavy P-diffusion [53]. Enhanced solubility of metallic impurities in such heavily P-diffused regions, and impurity segregation at Si_3P_4 precipitates lead to efficient gettering [54]. In solar cell fabrication, P-diffusion can be performed prior to cell fabrication followed by the removal of the heavily diffused layers (pre-gettering) or as part of emitter formation, depending on the optimal gettering conditions required for the material and the costs involved. Certain selective emitter processes that involve two P-diffusions (see Section 2.4) benefit both from gettering and from the removal of heavily diffused layers with accumulated impurities. The gettering process is more complicated in mc-Si since, at high temperatures, the dissolution of impurities that are precipitated near bulk crystal defects is also initiated. Therefore, in mc-Si, high temperature gettering is not desirable due to the competition between the gettering rate and dissolution rate of the precipitated impurities [54]. Gettering efficiency depends also on certain material properties, for example, the interstitial oxygen $\{O_i\}$ content [55]. Some mc-Si wafers with lower $\{O_i\}$ reportedly showed a better improvement after pre-gettering compared to those with higher $\{O_i\}$ [56]. Diffusion length exceeding substrate thickness (200 μm) was obtained even on low-resistivity mc-Si wafers after pre-gettering at low (<900°C) temperatures [57]. However, the incorporation of an additional pre-gettering step in a production line adds to the production cost. An additional cost increment of 0.3 $/wafer due to the pre-gettering step, based on a 5% relative efficiency improvement, has been reported [58]. Based on the same study, this additional cost can be halved by keeping the heavy P-diffused layer and etching back to the desired sheet resistance. Selective emitter structures, which can additionally lead to a better blue response and surface passivation, can therefore provide an acceptable solution.

2.8.2 Gettering by Aluminium Treatment

Formation of a p^+/p high/low junction at the rear side of the cell by regrowth from a 'fast-alloyed' Al—Si melt is the most commonly used process for creating BSF in industrial silicon cells (see Section 2.6). This treatment has an additional advantage of bulk gettering by prolonged firing or by a thermal anneal after the initial firing. At alloying times >1 minute with screen-printed Al, evidence of bulk impurity gettering has been noted at the grain boundaries of (100)/(111) CZ bi-crystals [59]. The effective grain boundary passivation obtained after Al diffusion has also been attributed to a much higher diffusion coefficient of Al along the grain boundaries [60]. Al-gettering effects have also been noted in single crystalline CZ-Si materials. From the relatively low temperatures generally used in Al gettering, it can be said that the gettered species are the fast moving interstitial impurities such as Cu, Fe, etc. Pilot line processes involving fast-alloying of screen printed Al-paste by firing, followed by removal of excess aluminium and a subsequent thermal anneal for up to 1 h, resulted in considerable enhancement in bulk diffusion length in large area mc-Si wafers [61]. This process of fast firing also prevents impurities contained in the Al source from diffusing into Si. High-performance, screen-printed CZ-Si cells with Al-gettering have also been reported [50]. The thickness of the Al-layer deposited on the rear side of the wafers is also of great importance in getting ideal pp$^+$ junctions for low-enough surface recombination velocity S_{eff} [36,37,62−65]. In ribbon materials such as EFG, efficiency enhancements of 1.4% absolute along with an increment of >60 μm in diffusion length have been reported after an 850°C gettering treatment with evaporated aluminium [66]. Due to the large thickness requirement for the Al-source and for the cost effectiveness of the process, it appears that screen printing is the most efficient way for depositing the Al-source.

2.9 'Fast Processing' Techniques

In recent years, there has been an increased interest in the application of Rapid Thermal Processing (RTP) in solar cell fabrication. RTP is widely used in IC processing because of the low thermal budget for which it is essential to have a strict control over junction depth variation, lateral diffusion of dopants, etc. In RTP, which is based on annealing by radiation from incoherent light sources, the wafers are rapidly heated (50−150°C/s) and rapidly cooled. The process time is generally <1 minute (Figure 6). A few examples of such RTP applications are dopant activation after implantation, implantation damage anneal, silicidation, sintering, and oxide anneal.

A more recent application of RTP is the development of Si solar cells. RTP has been used in various cell processing steps such as junction formation, growth of high-quality thin passivating oxides, annealing of Si/dielectric interface, and Si/metal contact firing. RTP differs from conventional furnace processing in that the wafer is optically heated by incoherent lamp radiation (tungsten−halogen or UV lamps) with much higher colour temperatures compared to conventional heating elements. Due to the availability of highly energetic, short-wavelength photons the

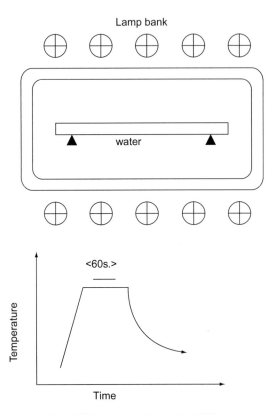

Figure 6 Cross-section of an RTP furnace and a typical RTP temperature cycle.

electronically excited states influence and enhance diffusion, oxidation, and gettering phenomena [67].

In solar cell application, RTP looks attractive because of its reduced process time, reduced cross contamination (reactor parts are almost transparent to lamp spectrum), and the possibility to diffuse front and rear junctions together. Furthermore, extensive wafer cleaning steps can by avoided by, for example, UV cleaning. For junction formation, diffusion sources such as spin-on or spray-on P or B sources or APCVD-deposited $SiO_2:P_2O_5$ or $SiO_2:B_2O_3$ sources and evaporated or screen-printed Al are used. In most cases, the emitter and the BSF are formed simultaneously in a single cycle [68,69]. The emitter junction depths are generally of the order of 0.15–0.25 μm. Although such shallow junctions yield an excellent blue response, it had been somewhat difficult to apply a cost-effective metallization scheme such as screen printing. However, successful application of screen-printed front metallization to RTP emitters has recently been reported [29]. For surface passivation, either high quality, thin RT-grown oxides [68] or PECVD SiO_2/Si_3N_4 [69] followed by RT-anneal have been efficiently applied. Solar cells fabricated without involving any conventional furnace processes exhibit encouraging conversion efficiencies of 17% in 1 cm^2 FZ-Si [69] and 15.4% in 100 cm^2 CZ-Si [70]. MIS solar cells have also been fabricated with high quality, ultra-thin RT-oxides and oxynitrides [71]. Nevertheless, for the RTP technique to become a competitive alternative to existing conventional processing, the wafer yield of the process should be drastically increased. This can be achieved by building either continuous process systems with spectrum-engineered lamps or batch-process systems capable of handling a large number of wafers. RTP system manufacturers and the PV industry are presently working on this.

3. INDUSTRIAL SOLAR CELL TECHNOLOGIES

Most of the commercially fabricated crystalline solar cells are still based on screen printing. The industrial production of more advanced solar cells technologies such as laser-grooved buried contact, MIS, and EFG solar cells contribute less than 10% to the total commercial production. The way in which the high-efficiency solar cell features described previously can be incorporated in the commercial processing sequence is outlined below.

3.1 Screen Printing Solar Cells

Screen printing is a well established, simple, robust, continuous and easily adaptable process. Fully automated screen-printing solar cell production lines are offered by many companies. From all high-efficiency features described in Section 2 (see also Chapter Ib-2), usually only the front surface texturing and Al-alloyed back-surface field are included in the screen printing solar cell process. This explains why the efficiency of some industrial crystalline solar cells still ranges from 13% to 15%. An important effort has been made by several research groups and companies, mainly in Japan and Europe, to improve the screen-printing process [13,17,35,50]. The main progress has been made in the paste formulation, fabrication of new type of fine-line screens, and the development of modern screen printers.

The new type of silver pastes contains additives which permit a selective dissolving of silicon dioxide and ARC layers of TiOx or SiNx during the firing process while preventing deep penetration into bulk silicon. This gives the possibility of using a very simplified process with firing through passivating oxide and ARC. The improved screen-printability of new pastes and new types of metal screens [50] give the possibility to print narrower lines down to 50–60 μm with an improved aspect ratio. Therefore, spacing of the screen-printed contact lines can be reduced, leading to more lightly doped emitters with improved short wavelength response. The problem of non-optimal surface passivation is tackled by the use of a selective emitter structure.

One applied process sequence for high efficiency screen printed solar cells is shown in Table 2 [72]. Figure 7 presents a cross section of such a processed cell.

Alignment of front contacts on top of a highly doped n^{++} region is done automatically by a new type screen printer equipped with digital cameras. The movement of the table and the squeegee is driven by linear motors controlled by a central computer.

The high-efficiency features of these new screen-printed solar cells are front-surface texturing, deep diffusion under front metallization contacts, and reduced total front-surface shadowing losses down to 6%, shallow emitter with optimised profile and front-surface passivation, Back Surface Field (BSF).

Pilot line efficiencies of 16.6% for homogeneous emitter CZ large area solar cells have been obtained and independently measured [73]. The selective emitter process gives efficiencies close to 17.3% for CZ-Si [50]

Table 2 Processing sequence of screen printing solar cell based on selective emitter and Bring-through passivating oxide and ARC [72]

Step no.	Process description
1	Saw damage etching
2	Deep n^{++} diffusion over the whole front surface (15 ohm/sq.)
3	Selective texture etching of diffused layer between contact fingers
4	Second light n^+ diffusion (80–120 ohm/sq.)
5	Growing of passivating dry thermal oxide
6	Edge junction Isolation
7	ARC deposition (APCVD TiOx or PECVD SiNx)
8	Top and back contacts printing and drying
9	Co-firing of both contacts and firing through passivating oxide and ARC layer

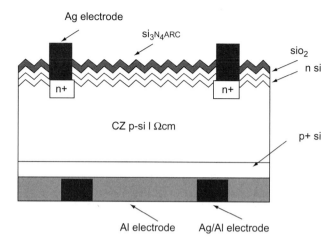

Figure 7 Cross-section of an advanced screen-printed solar cell with selective emitter and Al-alloyed BSF[72].

and 15.9% for large-area multicrystalline cells [73]. All processing steps can easily be transferred to large-volume production lines. A process that combines screen printing and grooving has demonstrated efficiencies above 17% on 100 cm^2 multicrystalline solar cells [17]. This is a record efficiency for a large-area silicon multicrystalline solar cell fabricated by any type of technology. The screen-printing process is also well suited for larger cell sizes: 15- × 15-cm^2 multicrystalline cells with efficiencies of 15.6% has also been demonstrated [74].

3.2 Buried Contact Solar Cells (BCSC)

The buried contact solar cell (BCSC) process has been developed at the University of New South Wales [46,75]. Many aspects of the BCSC structure and its processing have been extensively described in the technical literature [39,42,44,75,76–79] (see also Chapter Ib-2). A laboratory efficiency as high as 21.3% on small area FZ material has been reported [80]. Figure 10 in Chapter Ib-2 shows the structure of the buried contact solar cell. A conventional commercial BCSC processing sequence licensed to many solar cell manufactures is presented in Table 3.

The buried contact solar cell structure embodies almost all characteristic features of high efficiency laboratory cells described in Section 2: shallow emitter diffusion with a very good surface passivation by a thick thermal oxide, very fine metallization line width, and front contact passivation by heavy diffusion in the contact area, BSF. One of the important

Table 3 Conventional commercial process sequence of buried contact solar cell [78]

Step no.	Process description
1	Saw damage etching and random texturing
2	Light n^+ diffusion over the whole surface
3	Growing of thick thermal oxide
4	Mechanical or laser groove formation
5	Groove damage etching and cleaning
6	Second heavy diffusion in grooved areas only
7	Aluminium evaporation on a back side
8	High-temperature Al alloying
9	Electroless plating of nickel, sintering and etching
10	Electroless plating of copper and silver
11	Edge junction isolation by laser scribing

processing steps is growing a very thick thermal oxide on the top surface which simultaneously acts as a diffusion mask, plating mask, and surface passivation layer. In some processing sequences, the thick oxide is replaced by silicon nitride which reduces the front surface reflection.

This process, however, has its disadvantages when commercial applications are considered: a large number of lengthy processing steps at high temperature (above 950°C for a total time of up to 16 hours), expensive equipment, and many careful pre-cleaning steps that make the process complex and labour intensive [81]. Although the BCSC process has been licensed to many leading solar cell manufacturers, only one has succeeded introducing it into large-volume production by simplifying many processing steps [82]. Efficiencies close to 17% are rudimentarily obtained.

A simplified buried contact solar cell process has been proposed in reference [79]. The aim of this process simplification is to suit infrastructure and equipment existing already in many solar cell production plants based on screen-printed contacts. The number of high-temperature processing steps has been reduced to one, front surface passivation is achieved by retention of the diffusion oxide, and a sprayed-on TiO_2 layer acts as an ARC and plating mask. There are no data published about the cell performance achieved with this process.

3.3 Metal−Insulator−Semiconductor Inversion Layer (MIS-IL) Solar Cells

The commercial version of MIS-IL cells was developed in the 1980s [83]. Industrial production of 10- × 10-cm^2 cells was introduced in the early

1990s [84]. MIS-IL solar cells received much attention because of the elimination of high temperature steps, such as junction diffusion or thermal oxide passivation, from the cell processing sequence. The processing sequence of MIS is presented in Table 4. The diffused n-type front emitter is replaced by an inversion n-conducting layer at the front surface which is induced by a high-density fixed positive charge present in the transparent insulator film. Evaporated Al front contacts are separated from the silicon interface by a thin tunnel oxide that reduces surface recombination. An efficiency of 15.3% on 100-cm^2 CZ silicon has been measured and confirmed [73]. It appears that further improvement of the MIS inversion layer requires the PECVD nitride deposition at 400°C prior to the MIS contact formation. This leads to several modifications of the first-generation cell, for example, truncated pyramid or surface-grooved abrased-ridge-top MIS-IL cells [31]. The MIS-IL solar cells are especially suitable for a high-quality material like FZ silicon since low-temperature processing preserves a high initial lifetime. For industrial application, however, where low-quality materials are used, an impurity gettering which accompanies the phosphorus diffusion or aluminium alloying are prerequisites for obtaining high efficiency.

The combination of p−n junction and MIS solar cells has been demonstrated in a MIS-contacted p−n junction solar cell (see Figure 8). Evaporated aluminium front contacts are separated from a phosphorus diffused p-n junction by a thin oxide layer. This process brings the advantages of phosphorus gettering, front contact passivation by tunnel oxide and front surface passivation by PECVD silicon nitride. An efficiency of 16.5% has been reported on a 98-cm^2 CZ-silicon substrate [73].

Table 4 Processing sequence of MIS solar cells [84]

Step no.	Process description
1	Saw damage etching and surface texturing
2	Chemical texturing
3	Evaporation of 4 μm Al on the back side
4	Sintering of contacts at the back at 500°C and growth of 1.5-nm tunnel oxide on the front side
5	Front 6 μm Al grid evaporation through metal mask on the top of the tunnel oxide
6	Etch excess Al along the grid lines
7	Dip in CsCl solution to increased fixed positive charge density
8	Deposition of PECVD silicon nitride onto the entire front surface at 250°C for 5 min

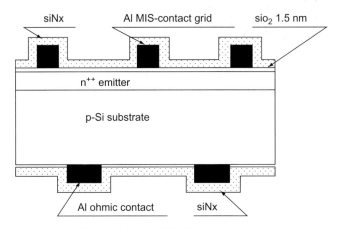

Figure 8 Cross-section of an MIS solar cell [31].

3.4 Solar Cells on EFG Silicon Sheets

EFG silicon sheets offer a significant cost advantage over traditional crystalline silicon technology like CZ pulling or casting multicrystalline blocks. The cost saving arises from elimination of the slicing process which is a significant cost contributor to CZ and multicrystalline wafers. However, the EFG material has high crystal defect density such as grain boundaries, twins, and dislocations [5].

High mechanical stress and the uneven surface of EFG sheets make application of standard screen-printing processes difficult because of high breakage. Several patented processing steps have been developed to passivate the highly defected EFG bulk material and to tackle the problem of contacting the uneven surface. A detailed description of the EFG solar cell process is not published but some information can be found in [85]. The process comprises several patented processing steps: spray-on of liquid P-source and diffusion in an IR-belt furnace, PECVD of silicon nitride antireflection coating preceded by ammonia plasma treatment in order to produce hydrogen implantation, 'pad printing' of solderable silver contacts on the back and drying, 'pad printing' of aluminium paste on the back and drying, direct writing of silver paste on the front, and drying and co-firing of all contacts in an IR-furnace.

During the firing process the front contact paste penetrates through PECVD silicon nitride layer and forms a good ohmic contact. The aluminium paste creates a deep p^+ alloyed BSF region on the back, and at the same time the heating process tends to release hydrogen from silicon nitride and drives it deeper into the substrate providing good bulk passivation. A record average efficiency of 14.3% on large-area EFG substrates has been reported [86].

3.5 Commercial Thin Film Crystalline Silicon Solar Cells

There is a growing interest in thin film silicon solar cells consisting of a thin (20–50 μm) silicon film deposited on potentially cheap substrates, as reviewed in detail Chapter Ib-4. Such thin structures offer the opportunity for silicon cells to use much less high-purity silicon than conventional ingot-type solar cells. When an effective light trapping scheme is utilised, efficiencies higher than for standard-type thick cells are possible [87]. Potential substrates include metallurgical-grade silicon, stainless steel, graphite, ceramics, or even glass. Among the many deposition methods, the most common are chemical vapour deposition (CVD) or liquid phase epitaxial growth (LPE).

Although most of the development in this field is still in an early stage, the formation of continuous sheets of polycrystalline silicon on conductive ceramics, the so called Silicon-Film™ process has already been put in test production [88]. Since these sheets are fabricated at the desired thickness, ingot sawing is avoided, leading to a significant cost reduction. The cross section of a so-called product II Si-Film™ solar cell is shown in Figure 9.

The Silicon-Film™ is a proprietary process and only a very general process sequence has been published. The generic process consists of ceramic formation, metallurgical barrier formation, polycrystalline layer deposition, emitter diffusion and contact fabrication. The conductive ceramic substrate is fabricated from low-cost materials. The metallurgical barrier prevents substrate impurities from entering and contaminating the active thin silicon layer. The randomly textured and highly reflective metallurgical barrier layer improves the light trapping. A suitable p-type doped 30–100 μm active layer is deposited from a liquid solution. Phosphorus and aluminium impurity gettering are used for bulk quality improvement. The rest of the cell process is similar to a standard screen-printing solar-cell manufacturing sequence. Cells with large areas of 240, 300, and 700 cm^2 have been developed. A cell with an area of 675 cm^2 has demonstrated the record efficiency of 11.6% [89]. PV modules with Silicon Film™ cells are now in test production [6,90]. A further discussion of this field can be found in Chapter Ib-4.

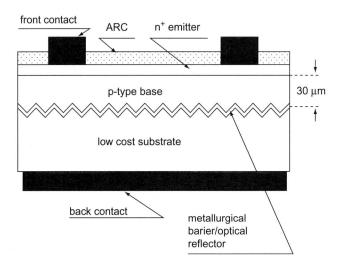

Figure 9 Cross-section of a Silicon Film™ solar cell [88].

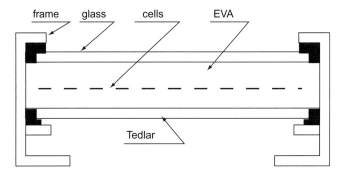

Figure 10 Cross-section of a photovoltaic module

4. COST OF COMMERCIAL PHOTOVOLTAIC MODULES

A typical crystalline Si solar cell produces a voltage of 0.5 V around the maximum power point. In a module, the individual cells are usually connected in series to produce a voltage useful for practical application. Most industrial modules today comprise 36 series-connected cells in order to charge a 12 V battery. A cross-section of a typical PV module is shown on Figure 10. The cells are placed in a sandwich structure glass/encapsulant sheet/cells/encapsulant sheet/back substrate. The cover glass normally has a low iron content to minimise light absorption and is generally either chemically or thermally tempered to increase the mechanical strength, especially against hailstorms. The encapsulant sheet is a transparent polymer that can be laminated onto glass. The most commonly used encapsulants are polyvinyl butyral (PVB) or ethylene vinylacetate (EVA). For the back substrate, several materials are used, including anodised aluminium, glass or polymers such as Mylar or Tedlar.

The price of commercially available photovoltaic modules is 3.5–4.5 \$/Wp. The average total cost breakdown from five cell manufacturers is shown in Table 5. Of the total module cost, the silicon wafer is responsible for 46%, cell fabrication 24%, and module fabrication 30%. Sensitivity analysis of different cost elements, performed for a base-line screen-printing process, shows that the cell efficiency is the most significant factor in the cost reduction [58,91]. Cost reduction is a driving force for introducing efficiency enhancement techniques into commercial cell production. However, care must be taken that novel techniques do not bring significant cost from increased capital investment and lower yield associated with the higher complexity process [91]. The cost effectiveness of several efficiency enhancement techniques has been analysed [58] and has revealed that all techniques described in Section 2 are commercially viable. Processing sequences such as evaporated front contacts and selective emitter involving photolithography are not commercially applicable in their present form.

A study in the framework of the European project APAS RENA CT94 0008 [93] has demonstrated that practical implementation of new industrial processing schemes and increasing the market size towards 500 MWp/year will lead to drastic crystalline silicon photovoltaic module price reductions to 0.95–1.73 \$/Wp depending on the wafer type and cell process. The different commercial technologies of mono- and multicrystalline solar cells have been studied include screen printing, laser-grooved buried contact, MIS-contacted diffused emitter, and a proprietary cell process on EFG silicon sheets. The study assumptions and cost breakdown are presented in Table 6. These results demonstrate that cell fabrication based on EFG

Table 5 Average values for five manufacturers of total cost breakdown of crystalline silicon photovoltaic modules

Total cost (\$/Wp)	Si feedstock	Crystallisation	Wafering	Cell fab.	Module fab.
4	0.38	0.73	0.73	0.96	1.2

Table 6 Manufacturing cost in $/Wp for different crystalline manufacturing scenarios at 500 MWp/year production. Case 1: EFG Si sheets, printed contacts = 15%: Case 2: direct solidification multi-Si, screen-printed contacts, $\eta = 15\%$; Case 3: CZ mono-Si, LGBC, $\eta = 18\%$; Case 4: CZ mono-Si, screen-printed contacts, $\eta = 16\%$; Case 5: mono-Si, MIS contacted, $\eta = 17\%$. Wafer size in all cases is 125 mm × 125 mm and thickness 250 μm [93]

Fabrication step	Case 1	Case 2	Case 3	Case 4	Case 5
Ingot growing	0.37	0.37	0.73	0.82	0.78
Wafering	0.00	0.29	0.24	0.28	0.28
Solar cell fab.	0.15	0.15	0.19	0.16	0.24
Module fab.	0.43	0.40	0.37	0.41	0.43
Factory cost	0.95	1.21	1.53	1.67	1.73

silicon sheets gives the lowest module cost mainly due to the avoidance of the costly wafering step. The well-known and established screen-printed cells on multicrystalline wafers fabricated by directional solidification shows the next lowest cost of 1.21 $/Wp. It is worth mentioning that efficiencies between 16% and 17%, higher than the 15.5% assumed by the model, have already been achieved on large-area screen-printed multicrystalline cells. All the processes based on CZ wafers show higher costs because of the costly mono-Si wafers.

REFERENCES

[1] H. Watanabe, K. Shirasawa, K. Masuri, K. Okada, M. Takayama, K. Fukui, et al., Technical progress on large area multicrystalline silicon solar cells and its applications, Optoelectronics – Devices Technol. 5 (1990) 223–238.
[2] T. De Villers, G.R. Smekens, Monochess II – Final Report, 1995, JOU 2-CT92–0140.
[3] A. Eyer, N. Schilinger, A. Rauber, J.G. Grabmaier, Continuous processing of silicon sheet material by SSP-Method. Proceedings of the 9th European Photovoltaic Solar Energy Conference, Freiburg, 1989, pp.17–18.
[4] T. Vermeulen, O. Evrard, W. Laureys, J. Poortmans, M. Caymax, J. Nijs, et al., Realisation of thin film solar cells in epitaxial layers grown on highly doped RGS-ribbons. Proceedings of the 13th European Photovoltaic Solar Energy Conference, Nice, 1995, pp.1501–1504.
[5] M. Kardauskas, Processing of large-area silicon substrates with high defect densities into higher efficiency solar cells. Proceedings of the 6th Workshop on the Role of Impurities and Defects in Silicon Device Processing, 1996, pp.172–176.
[6] A.E. Ingram, A.M. Barnett, J.E. Cotter, D.H. Ford, R.B. Hall, J.A. Rand, et al., 13% Silicon film™ solar cells on low cost barrier-coated substrates. Proceedings of the 25th IEEE Photovoltaic Specialists Conference, Washington DC, 1996, pp. 477–480.
[7] J. Gee, S. Wenham, Advanced processing of silicon solar cells. Tutorial Notebook of Proceedings of the 23rd IEEE Photovoltaic Specialists Conference, Louisville, 1993.

[8] D.L. King, M.E. Buck, Experimental optimization of an anisotropic etching process for random texturization of silicon solar Cells. Proceedings of the 22nd IEEE Photovoltaic Specialists Conference, Las Vegas, 1991, pp. 303–308.
[9] M. Grauvogl, A. Aberle, R. Hezel, 17.1% Efficient truncated-pyramid inversion-layer silicon solar cells. Proceedings of the 25th IEEE Photovoltaic Specialists Conference, Washington DC, 1996, pp. 433–436.
[10] U. Kaiser, M. Kaiser, R. Schindler, Texture etching of multicrystalline silicon. Proceedings of the 10th European Photovoltaic Solar Energy Conference, Lisbon, 1990, pp. 293–294.
[11] S. Narayanan, J. Zolper, F. Yung, S. Wenham, A. Sproul, C. Chong, et al., 18% Efficient polycrystalline silicon solar cells. Proceedings of the 21st IEEE Photovoltaic Specialists Conference, Orlando, 1990, pp. 678–680.
[12] L. Pirozzi, M. Garozzo, E. Salza, G. Ginocchietti, D. Margadona, The laser texturization in a full screen printing fabrication process of large area poy silicon solar cells. Proceedings of the 12th European Photovoltaic Solar Energy Conference, Amsterdam, 1994, pp. 1025–1028.
[13] K. Shirasawa, H. Takahasashi, Y. Inomata, K. Fukui, K. Okada, M. Takayaka, et al., Large area high efficiency multicrystalline silicon solar cells. Proceedings of the 12th European Photovoltaic Solar Energy Conference, Amsterdam, 1994, pp. 757–760.
[14] G. Willeke, H. Nussbaumer, H. Bender, E. Bucher, A simple and effective light trapping technique for polycrystalline silicon solar cells, Sol. Energy Mat. and Sol. Cells 26 (1992) 345–356.
[15] H. Bender, J. Szlufcik, H. Nussbaumer, J. Nijs, R. Mertens, G. Willeke, et al., Polycrystalline silicon solar cells with a mechanically formed texturization. Appl. Phys. Lett. 62, 2941–2943.
[16] J. Szlufcik, P. Fath, J. Nijs, R. Mertens, G. Willeke, E. Bucher, Screen printed multicrystalline silicon solar cells with a mechanically prepared v-groove front texturization. Proceedings of the 12th European Photovoltaic Solar Energy Conference, Amsterdam, 1994, pp. 769–772.
[17] H. Nakaya, M. Nishida, Y. Takeda, S. Moriuchi, T. Tonegawa, T. Machida, et al., Polycrystalline silicon solar cells with V-grooved surface, Sol. Energy Mat. Sol. Cells 34 (1994) 219–225.
[18] P. Fath, C. Marckmann, E. Bucher, G. Willeke, Multicrystalline silicon solar cells using a new high throughput mechanical texturization technology and a roller printing metallization technique. Proceedings of the 13th European Photovoltaic Solar Energy Conference, Nice, 1995, pp. 29–32.
[19] R. Hezel, A novel approach to cost effective high efficiency solar cells. Proceedings of the 13th European Photovoltaic Solar Energy Conference, Nice, 1995, pp. 115–118.
[20] D. Sarti, Q.N. Le, S. Bastide, G. Goaer, D. Ferry, Thin industrial multicrystaline solar cells and improved optical absorption. Proceedings of the 13th European Photovoltaic Solar Energy Conference, Nice, 1995, pp. 25–28.
[21] W. Kern, D.A. Poutinen, Cleaning solution based on hydrogen peroxide for use in silicon semiconductor technology, RCA Rev. (1970) 187–206.
[22] M. Meuris, P.W. Mertens, A. Opdebeeck, H.F. Schmidt, M. Depas, G. Vereecke, et al., The IMEC clean: a new concept for particle and metal removal on Si surfaces, Solid St. Technol. July (1995) 109–113.
[23] R.R. King, R.A. Sinton, R.M. Swanson, Studies of diffused phosphorus emitters: saturation current, surface recombination velocity, and quantum efficiency, IEEE Trans. Electron. Dev. ED-37 (1990) 365–371.
[24] A. Morales-Acevado, Optimization of surface impurity concentration of passivated emitter solar cells, J. Appl. Phys. 60 (1986) 815–819.

[25] M. Wolf, The influence of heavy doping effects on silicon solar cells performance — I, Solar Cells 17 (1986) 53—63.
[26] A. Morales-Acevado, Theoretical study of thin and thick emitter silicon solar cells, J. Appl. Phys. Vol.70 (1991) 3345—3347.
[27] S. Wenham, Buried-contact solar cells, Prog. Photovolt. 1 (1993) 3—10.
[28] J. Szlufcik, H. Elgamel, M. Ghannam, J. Nijs, R. Mertens, Simple integral screen printing process for selective emitter polycrystaline silicon solar cells, Appl. Phys. Lett. 59 (1991) 1583—1584.
[29] P. Doshi, J. Mejia, K. Tate, S. Kamra, A. Rohatgi, S. Narayanan, et al., High-efficiency silicon solar cells by low-cost rapid thermal processing, screen printing and plasma-enhanced chemical vapour deposition. Proceedings of the 25th IEEE Photovoltaic Specialists Conference, Washington DC, 1996, pp. 421—424.
[30] S. Sivoththaman, W. Laureys, P. De Schepper, J. Nijs, R. Mertens, Rapid thermal processing of conventionally and electromagnetically cast 100 cm^2 multicrystalline silicon. Proceedings of the 25th IEEE Photovoltaic Specialists Conference, Washington DC, 1996, pp. 621—624.
[31] R. Hezel, A review of recent advances in MIS solar cells. Proceedings of the 6th Workshop on the Role of Impurities and Defects in Silicon Device Processing, 1996, pp. 139—153.
[32] J. Zhao, M. Green, Optimized antireflection coatings for high-efficiency silicon solar cells, IEEE Trans. Electron. Dev. 38 (1991) 1925—1934.
[33] A. Aberle, S. Glunz, W. Warta, J. Knopp, J. Knobloch, SiO_2-passivated high efficiency silicon solar cells: process dependence of $Si-SiO_2$ interface recombination. Proceedings of the 10th European Photovoltaic Solar Energy Conference, Lisbon, 1991, pp. 631—635.
[34] J. Szlufcik, K. De Clercq, P. De Schepper, J. Poortmans, A. Buczkowski, J. Nijs, et al., Improvement in multicrystalline silicon solar cells after thermal treatment of PECVD silicon nitride AR Coating. Proceedings of the 12th European Photovoltaic Solar Energy Conference, Amsterdam, 1994, pp.1018—1021.
[35] M. Takayama, H. Yamashita, K. Fukui, K. Masuri, K. Shirasawa, H. Watanabe, Large area high efficiency multicrystalline silicon solar cells. Technical Digest International PVSEC-5, 1990, pp. 319—322.
[36] G.C. Cheeck, R.P. Mertens, P. Van Overstraeten, L. Frisson, Thick film metallization for silicon solar cells, IEEE Trans. Electron Dev. ED-31 (1984) 602—609.
[37] P. Lolgen, C. Leguit, J.A. Eikelboom, R.A. Steeman, W.C. Sinke, L.A. Verhoef, et al., Aluminium back-surface field doping profiles with surface recombination velocities below 200 cm/sec. Proceedings of the 23rd IEEE Photovoltaic Specialists Conference, Louisville, 1993, p. 231.
[38] J.A. Amick, F.J. Battari, J.I. Hanoka, The effect of aluminium thickness on solar cell performance, J. Electroch. Soc. 141 (1994) 1577—1585.
[39] C.M. Chong, S.R. Wenham, M.A. Green, High-efficiency, laser grooved buried contact silicon solar cells, Appl. Phys. Lett. 52 (1988) 407—409.
[40] S. Wenham, Buried-contact silicon solar cells, Prog. Photovolt. 1 (1993) 3—10.
[41] S. Narasimha, S. Kamra, A. Rohatgi, C.P. Khattak, D. Ruby, The optimization and fabrication of high efficiency hem multicrystalline silicon solar cells. Proceedings of the 25th IEEE Photovoltaic Specialists Conference, Washington DC, 1996, pp. 449—452.
[42] S.R. Wenham, C.B. Honsberg, M.A. Green, Buried contact solar cells, Sol. Energy Mat. Sol. Cells 34 (1994) 101—110.
[43] M. Ghannam, E. Demesmaeker, J. Nijs, R. Mertens, R. Van Overstraeten, Two dimensional study of alternative back surface passivation methods for high efficiency silicon solar cells. Proceedings of the 11th European Photovoltaic Solar Energy Conference, Montreux, 1992, pp. 45—48.

[44] C.B. Honsberg, F. Yun, A. Ebong, M. Tauk, S.R. Wenham, M.A. Green, 685 mV Open-circuit voltage laser grooved silicon solar cell, Sol. Energy Mat. Sol. Cells 34 (1994) 117−123.
[45] R. Gutierrez, J.C. Jimeno, F. Hernanado, F. Recart, G. Bueno, Evaluation of standard screen printed solar cells. Proceedings of the 13th European Photovoltaic Solar Energy Conference, Nice, 1995, pp. 1508−1511.
[46] S.R. Wenham, M.A. Green, Buried Contact Solar Cell. US Patent 4,726,850, 1988.
[47] N.B. Mason, D. Jordan, J.G. Summers, A high efficiency silicon solar cell production technology. Proceedings of the 10th European Photovoltaic Solar Energy Conference, Lisbon, 1991, pp. 280−283.
[48] A. Munzer, MONOCEPT Firs Progress Report EC project JOR 3-CT-95−0011, 1996.
[49] F. Wald, EFG Crystal growth technology for low cost terrestrial photovoltaics: review and outlook, Sol. Energy Mat. Sol. Cells 23 (1991) 175−182.
[50] J. Nijs, E. Demesmaeker, J. Szlufcik, J. Poortmans, L. Frisson, K. De Clercq, et al., Latest efficiency results with the screenprinting technology and comparison with the buried contact structure. Proceedings of the First World Conference on Photovoltaic Energy Conversion, Hawaii, 1994, pp. 1242−1249.
[51] A. Dziedzic, J. Nijs, J. Szlufcik, Thick-film fine-line fabrication techniques − application to front metallisation of solar cells, Hybrid Circ. No. 30 (1993) 18−22.
[52] J.I. Hanoka, S.E. Danielson, Method for Forming Contacts. US Patent 5,151,377, 1992.
[53] A. Ourmazd, W. Schroter, Phosphorus gettering and intrinsic gettering of nickel in silicon, Appl. Phys. Lett. 45 (1984) 781.
[54] B. Sopori, L. Jastrzebski, T. Tan, A comparison of gettering in single and multicrystalline silicon for solar cells. Proceedings of the 25th IEEE Photovoltaic Specialists Conference, Washington DC, 1996, pp. 625−628.
[55] J. Gee, Phosphorus diffusions for gettering-induced improvement of lifetime in various silicon materials. Proceedings of the 22nd IEEE Photovoltaic Specialists Conference, Las Vegas, 1991, pp. 118−123.
[56] I. Perichaud, F. Floret, S. Martinuzzi, Limiting factors in phosphorus external gettering efficiency in multicrystalline silicon. Proceedings of the 23rd IEEE Photovoltaic Specialists Conference, Louisville, 1993, pp. 243−247.
[57] S. Sivoththaman, M. Rodot, L. Nam, D. Sarti, M. Ghannam, J. Nijs, Spectral response and dark I−V characterization of polycrystalline silicon solar cells with conventional and selective emitters. Proceedings of the 23rd IEEE Photovoltaic Specialists Conference, Louisville, 1993, pp. 335−339.
[58] S. Narayanan, J. Wohlgemuth, Cost-benefit analysis of high efficiency cast polycrystalline silicon solar cell sequences, Prog. Photovolt. 2 (1994) 121−128.
[59] W. Orr, M. Arienzo, Investigation of polycrystalline silicon BSF solar cells, IEEE Trans. Electron Dev. ED-29 (1982) 1151.
[60] L. Kazmerski, Polycrystalline silicon: Impurity incorporation and passivation. Proceedings of the 6th European Photovoltaic Solar Energy Conference, London, 1985, pp. 83−89.
[61] L. Verhoef, S. Roorda, R. Van Zolingen, W. Sinke, Improved bulk and emitter quality by backside aluminum doping and annealing of polycrystalline silicon solar cells. Proceedings of the 20th IEEE Photovoltaic Specialists Conference, Las Vegas, 1988, pp.1551−1556.
[62] J. Mandelkorn, J.H. Lamneck, Simplified fabrication of back surface electric field silicon cells and novel characteristics of such cells, Solar Cells (1990) 121−130.
[63] J. Del Alamo, J. Eguren, A. Luque, Operating limits of al-alloyed high-low junctions for BSF solar cells, Solid St. Electron. 24 (1981) 415−420.

[64] P. Lolgen, W.C. Sinke, L.A. Verhoef, Bulk and surface contribution to enhanced solar-cell performance induced by aluminium alloying. Technical Digest International PVSEC-5, 1990, pp. 239–243.
[65] J.A. Amick, F.J. Battari, J.I. Hanoka, The effect of aluminium thickness on solar cell performance, J. Electrochem. Soc 141 (1994) 1577–1585.
[66] P. Sana, A. Rohatgi, J. Kalejs, R. Bell, Gettering and hydrogen passivation of EFG multicrystalline silicon solar cells by aluminum diffusion and forming gas anneal, Appl. Phys. Lett. 64 (1994) 97–99.
[67] R. Singh, F. Radpour, P. Chou, Comparative study of dielectric formation by furnace and rapid isothermal processing, J. Vac. Sci. Technol A7 (1989) 1456–1460.
[68] S. Sivoththaman, W. Laureys, J. Nijs, R. Mertens, Fabrication of large area silicon solar cells by rapid thermal processing, Appl. Phys. Lett. 67 (1995) 2335–2337.
[69] A. Rohatgi, Z. Chen, P. Doshi, T. Pham, D. Ruby, High efficiency solar cells by rapid thermal processing, Appl. Phys. Lett. 65 (1994) 2087–2089.
[70] S. Sivoththaman, W. Laureys, P. De Schepper, J. Nijs, R. Mertens, Large area silicon solar cells fabricated by rapid thermal processing. Proceedings of the 13th European Photovoltaic Solar Energy Conference, Nice, 1995, pp. 1574–1577.
[71] A. Beyer, G. Ebest, R. Reich, MIS solar cells with silicon oxynitride tunnel insulators by using rapid thermal processing, Appl. Phys. Lett. 68 (1996) 508–510.
[72] J. Szlufcik, F. Duerinckx, E. Van Kerschaver, R. Einhaus, A. Ziebakowski, E. Vazsonyi, et al., Simplified industrial type processes for high efficiency crystalline silicon solar cells. Proceedings of the 14th European Photovoltaic Solar Energy Conference, Barcelona, 1997.
[73] ISE PV Chart, Fraunhofer Institute FhG-ISE, Freiburg, Germany, November 1996.
[74] T. Saitoh, R. Shimokawa, Y. Hayashi, Recent improvements of crystalline silicon solar cells in Japan. Proceedings of the 22nd IEEE Photovoltaic Specialists Conference, Las Vegas, 1991, pp. 1026–1029.
[75] S.R. Wenham, M.A. Green, Laser Grooved Solar Cell, in US Patent 4,748,130.
[76] M.A. Green, S.R. Wenham, Present status of buried contact solar cells. Proceedings of the 22nd IEEE Photovoltaic Specialists Conference, Las Vegas, 1991, pp. 46–49.
[77] M.A. Green, S.R. Wenham, C.B. Honsberg, D. Hogg, Transfer of buried contact cell laboratory sequences into commercial production, Sol. Energy Mat. Sol. Cells 34 (1994) 83–89.
[78] C.B. Honsberg, S.R. Wenham, New insights gained through pilot production of high-efficiency silicon solar cells, Prog. Photovolt. 3 (1995) 79–87.
[79] S.R. Wenham, M.A. Green, Silicon solar cells, Prog. Photovolt. 4 (1996) 3–33.
[80] M.A. Green, S.R. Wenham, J. Zhao, Progress in high efficiency silicon solar cells and module research. Proceedings of the 23rd IEEE Photovoltaic Specialists Conference, Louisville, 1993, pp. 8–13.
[81] S. Narayanan, J.H. Wohlgemuth, J.B. Creager, S.P. Roncin, J.M. Perry, Buried contact solar cells. Proceedings of the 23rd IEEE Photovoltaic Specialists Conference, Louisville,1993, pp. 277–280.
[82] D. Jordan, J.P. Nagle, New generation of high-efficiency solar cells: development, processing and marketing, Prog. Photovolt. 2 (1994) 171–176.
[83] R. Hezel, R. Schroner, Plasma Si Nitride – A promising dielectric to achieve high-quality silicon MIS/IL solar cells, J. Appl. Phys 27 (1981) 3076.
[84] R. Hezel, W. Hoffmann, K. Jager, Recent advances in silicon inversion layer solar cells and their transfer to industrial pilot production. Proceedings of the 10th European Photovoltaic Solar Energy Conference, Lisbon, 1991, p. 511.
[85] J. Amick, F.J. Bottari, J.I. Hanoka, Solar cell and method of making same. US Patent 5,320,684, 1994.

[86] R.O. Bell, M. Prince, F.V. Wald, W. Schmidt, K.D. Rasch, A comparison of the behavior of solar silicon material in different production processes, Sol. Energy Mat. Sol. Cells 33 (1996) 71–86.
[87] P. Campbell, M.A. Green, The limiting efficiency of silicon solar cells under concentrated sunlight, IEEE Trans. Electron Dev. ED-33 (1986) 234–239.
[88] A.M. Barnett, S.R. Collins, J.E. Cotter, D.H. Ford, R.B. Hall, J.A. Rand, Polycrystalline silicon film™ solar cells: present and future, Prog. Photovolt. 2 (1994) 163–170.
[89] D.H. Ford, A.M. Barnett, J.C. Checchi, S.R. Collins, R.B. Hall, C.L. Kendall, et al., 675-cm2 silicon-film™ solar cells. Technical Digest Ninth International PVSEC, 1996, pp. 247–248.
[90] D.H. Ford, A.M. Barnett, R.B. Hall, C.L. Kerndall, J.A. Rand, High power, commercial silicon film™ solar cells. Proceedings of the 25th IEEE Photovoltaic Specialists Conference, Washington DC, 1996, pp. 601–604.
[91] S. Hogan, D. Darkazalli, R. Wolfson, An analysis of high efficiency si cells processing. Proceedings of the 10th European Photovoltaic Solar Energy Conference, Lisbon, 1991, pp. 276–279.
[92] J. Nijs, Photovoltaic cells and modules: technical and economic outlook towards the year 2000, Int. J. Solar Energy 15 (1994) 91–122.
[93] Final Report of the EC project 'Multi-Megawatt Upscaling of Silicon and Thin Film Solar Cell and Module Manufacturing 'Music FM' – APAS RENA CT94', in press.

CHAPTER IB-4

Thin Silicon Solar Cells

Michael Mauk, Paul Sims, James Rand, and Allen Barnett[1]
AstroPower Inc., Solar Park, Newark, Delaware, USA

Contents

1. Introduction, Background, and Scope of Review	162
2. Light Trapping in Thin Silicon Solar Cells	165
2.1 Methods of Implementing Light Trapping	167
2.1.1 Random Texturing	*168*
2.1.2 Geometrical or Regular Structuring	*168*
2.1.3 External Optical Elements	*169*
2.2 Assessment of Light-Trapping Effects	170
2.2.1 Short-Circuit Current Analysis of Light Trapping	*170*
2.2.2 'Sub-Band-Gap' Reflection Analysis of Light Trapping	*172*
2.2.3 Extended Spectral-Response Analysis of Light Trapping	*173*
3. Voltage Enhancements in Thin Silicon Solar Cells	174
3.1 Minority-Carrier Recombination Issues in Thin Silicon Solar Cells	176
4. Silicon Deposition and Crystal Growth for Thin Solar Cells	179
4.1 Substrate Considerations	179
4.2 High-Temperature Silicon-Deposition Methods	181
4.2.1 Melt-Growth Techniques	*181*
4.2.2 Recrystallisation of Silicon	*181*
4.2.3 High-Temperature Silicon Chemical Vapour Deposition	*182*
4.2.4 Low-Temperature Chemical Vapour Deposition	*186*
5. Thin Silicon Solar Cells Based on Substrate Thinning	188
6. Summary of Device Results	190
References	196

[1] For current address, see List of Contributors on page xx

Practical Handbook of Photovoltaics.
© 2012 Elsevier Ltd. All rights reserved.

1. INTRODUCTION, BACKGROUND, AND SCOPE OF REVIEW

Thin silicon solar cells are an important class of photovoltaics that are currently the subject of intense research, development, and commercialisation efforts. The potential cost reductions realised by manufacturing solar cells in a thin device configuration are highly compelling and have long been appreciated. However, most work on thin-film approaches to solar cells has centred on materials other than crystalline silicon because it was believed that the optical properties of crystalline silicon would limit its usefulness as a thin-film solar cell, and for various reasons crystalline silicon did not readily lend itself to the common thin-film deposition technologies. Several early experimental efforts on making thin-film crystalline silicon solar cells seemed to confirm at least some of these perceived difficulties. Still, there were proponents of thin crystalline silicon solar cells including Redfield [1], Spitzer et al. [2], and Barnett [3], who advocated their potential advantages and articulated design principles of light trapping and high-open-circuit voltage needed to achieve high efficiencies. Starting in the late 1980s and gaining momentum in the 1990s, thin-film crystalline silicon solar cells emerged (or perhaps more aptly 'reemerged') as a promising approach. This was partly due to several device design and materials processing innovations proposed to overcome the difficulties and limitations in using crystalline silicon as a thin-film solar cell material and also partly due to the stubborn lack of progress in many competing photovoltaic technologies. Progress on thin crystalline silicon solar cells has now reached a level where they are positioned to capture a significant share of the photovoltaics market, and a thin silicon solar cell may well become the successor to the conventional (thick) silicon wafer solar cells that are presently the mainstay of the photovoltaics industry. This chapter reviews the issues and technical achievements that motivate the current interest in thin silicon solar cells and surveys developments and technology options.

Thin silicon solar cells is actually an umbrella term describing a wide variety of silicon photovoltaic device structures utilising various forms of silicon (monocrystalline, multicrystalline, polycrystalline, microcrystalline, amorphous, and porous) and made with an almost incredibly diverse selection of deposition or crystal growth processes and fabrication techniques. Thin silicon solar cells are distinguished from traditional silicon solar cells that are comprised of ~0.3-mm-thick wafers or sheets of silicon. The common defining feature of a thin silicon solar cell is a relatively thin (<0.1-mm) 'active' layer or film of silicon formed on, or attached to, a passive supporting substrate. Nevertheless, even this very general description may not subsume all the different variations of 'thin silicon solar cell' designs currently under investigation. One purpose of a review such as this is to provide some perspective and objective criteria with which to assess the merits and prospects of different approaches. However, such comparisons must be tempered with the realisation that the technology is still in a state of flux and relatively early development, and there are many disparate solar cell applications, each with a different emphasis on cost and performance. Thus, it is possible—if not probable—that no single solar cell technology will satisfy or otherwise be the best choice for every present or future application. For example, a thin-film solar cell design for large-scale (~megawatt) utility grid-connected power applications may not be the best choice for a small (~1 watt) minimodule battery charger for walk lights. Further, the projected economics of solar cells is based on complicated and sometimes speculative assumptions of materials costs, manufacturing throughput and scale-up issues, and balance of systems constraints and costs, and it would be imprudent at this stage to 'downselect' the most promising technical path for the solar industry. From this vantage, it is fortunate that there is such a diverse choice of technology options under development that should lead to thin silicon solar cells with a wide range of cost and performance characteristics.

Such considerations not withstanding, Bergmann [4] has categorised thin silicon solar cells into three groups, and this delineation serves as rational and useful framework to discuss and review the subject.

1. Thin-film solar cells based on small-grained (<1 micron) nanocrystalline or microcrystalline silicon films (2–3 microns in thickness) deposited on glass substrates typically using technologies adapted from thin-film amorphous silicon solar cells. These types of cells are usually made as p–i–n structures, sometimes in combination with amorphous silicon layers in heterojunction or tandem-cell designs.
2. Thick (~30 micron) silicon layers deposited on substrates that are compatible with recrystallisation of silicon to obtain millimeter size grains. A diffused p–n junction is the preferred device design. Also

included in this category are epitaxial silicon solar cells on upgraded metallurgical silicon substrates.
3. Transfer techniques for films of silicon wherein an epitaxial monocrystalline silicon film is separated from the silicon substrate upon which it seeded and bonded to a glass superstrate. The anticipated cost reductions in this technology are based on reuse of the seeding substrate.

In this chapter, we concentrate on the second and third types of solar cells listed above. The first type is more properly considered as an extension or outgrowth of amorphous silicon technology, although much of the technology and design considerations are also relevant to other kinds of thin silicon solar cells. We will briefly survey device designs deposition technologies for making nanocrystalline and microcrystalline thin-film silicon solar cells, but we will not discuss materials and device physics of this type of solar cell. Amorphous silicon and related thin-film solar cells are considered in Chapter Ic-1 of this book. Bergman [4] has provided several penetrating reviews of thin-film microcrystalline and nanocrystalline silicon solar cells. Our emphasis is on polycrystalline silicon solar cells formed on ceramics, because most of our experience is with this type of solar cell. The third type of solar cell listed above is of general interest as it provides the shortest route to making thin silicon devices in which device issues can be analysed independently of material properties issues. For instance, light trapping and surface passivation effects can be studied in thin device structures with and without grain boundaries.

The plan of the chapter is as follows. We first provide a general discussion of effects and features that are common to most thin silicon solar cells such as light-trapping designs, modelling, and methods of analysis. We next discuss voltage-enhancement issues including surface passivation, followed by a review of grain-boundary effects in thin silicon solar cells. A survey of technologies for producing thin silicon solar cells is then presented. This will include descriptions of various silicon-deposition techniques such as chemical vapour deposition, melt and solution growth, and substrate issues and postdeposition recrystallisation to enhance grain size and texture. Low-temperature chemical vapour deposition methods for microcrystalline silicon solar cells is also covered, as are techniques for wafer thinning and transfer of thin layers to surrogate substrates. Finally, we end with a tabulation summarising some of the prominent experimental results for thin silicon solar cells.

2. LIGHT TRAPPING IN THIN SILICON SOLAR CELLS

Thin silicon solar cells can greatly benefit from light-trapping effects, which can offset the relatively weak absorption near-band-gap energy photons by increasing the optical path length of light within the solar cell structure. The basic idea of light trapping in a thin solar cell is shown in Figure 1. The back side of the solar cell with a grooved, blazed, textured, or otherwise roughened surface is made reflective by, for example, a change in refractive index or by coating with a reflective material such as a metal. In a free-standing solar cell, this surface is the backside of the solar cell itself, but in a thin solar cell formed on a substrate, the reflective backside 'mirror' is situated at the interface between the silicon film and the substrate. This latter design for a solar cell made on a substrate is obviously more difficult to implement than with a solar cell formed as a free-standing wafer or sheet. As suggested by the schematic ray tracing of Figure 1, it appears in this case the light makes at least several passes of the silicon layers. Thus, the effective optical thickness is several times the actual silicon physical thickness, in which case a thin silicon solar cell with light trapping could reap the same absorption and carrier generation as a conventional solar cell (without light trapping) of much greater thickness.

Further, the generation of minority carriers would be relatively close to the p–n junction formed near the top surface of the solar cell, thus providing a high collection efficiency. To effect light trapping, it is necessary that at least one interface (front or back) deviates from planarity. If both front and back surfaces are smooth and parallel, a simple analysis shows that the first internal reflection at the front side will result in a large loss of the light due to transmission. This is especially true since the front surface of the solar cell will use an optical coating to reduce reflection (maximise transmission) of incident light. Since the transmission

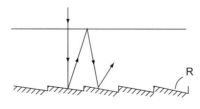

Figure 1 Illustration of the light-trapping concept [5].

characteristics are symmetrical, the internal reflection from the front surface would be small. Instead, if the back surface is grooved, textured, or roughened as indicated in Figure 1, then light reflected from the back surface will in general be obliquely incident to the front surface. If the angle of incidence exceeds a critical angle of about 16 degrees, then the light will be totally internally reflected as shown.

Nevertheless, in a real device with imperfect interfaces and diffuse components of light, a certain fraction of internally trapped light will eventually fall within the near-normal incidence angle for transmission, so some light leakage is inevitable and perfect confinement is impossible. An alternative approach is to texture or groove both the top and front surface, in which case, external reflection at the top surface is also reduced [5]. In theory, this is the best approach. In such cases, Yablonovitch and Cody [6] have predicted that for weakly absorbed light, the effective optical thickness a silicon solar cell with both surfaces textured to form Lambertian diffuse reflectors can be about 50 times greater than its actual thickness.

Enhancing the performance of thin crystalline silicon layers with light trapping has been actively discussed in the literature since the 1970s [1,2,7], In general, texturing one or both surfaces and maximising the reflection at the back surface obtains optical path lengths greater than the thickness of the device. Texturing results in oblique paths for internally confined light and maximises total internal reflection at the illuminated device surface.

To quantify the effects of light trapping on device performance and optimisation, we define a parameter Z that indicates the ratio of effective optical thickness to actual thickness for weakly absorbed light. Z can thus be interpreted as the number of passes trapped light makes in the solar cell. Z can vary from 1 (no light trapping) to about 50 and is regarded as an adjustable parameter in the optimisation. Although this is an overly simplistic way to describe light trapping in a solar cell, it does not appear that the main conclusions of such and similar modelling depend on the details of the optical absorption and carrier generation due to light-trapping phenomena. The main effect on solar cell efficiency is the total level of enhanced absorption and generation in the thin silicon active layers rather than the microscopic details. However, such detailed modelling is necessary to optimise light-trapping designs.

In Figure 2, an example result of the determination of optimal solar cell thickness as a function of light trapping (Z) is illustrated. In this

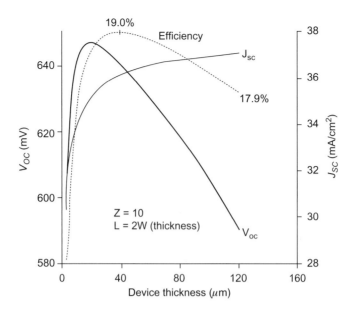

Figure 2 Thin silicon device performance predictions for the case where the optical thickness due to light trapping (Z) is 10 times the device thickness and the diffusion length is twice the actual device thickness. The optimum silicon thickness using these assumptions is in the range of 30–40 microns [8].

analysis, the doping level was adjusted to yield a minority-carrier diffusion length twice the layer thickness, thus ensuring good collection efficiency. (We regarded this as a general design principle of thin silicon solar cells.) As a conservative estimate, the diffusion length for a given doping concentration was degraded by a factor of 5 from typical single-crystal values to account for the relatively inferior material quality generally expected for silicon deposited on a substrate. Modest surface passivation corresponding to front and back minority-carrier surface recombination velocities of 1000 cm/s, a front-surface solar-spectrum averaged reflection of 5%, and a series resistance of 0.1 ohm-cm were also assumed. The results of this analysis are shown in Figure 2 for a Z factor of 10. For this set of assumptions, the optimum efficiency occurs at silicon thicknesses between 30 and 40 microns.

2.1 Methods of Implementing Light Trapping

As might be imagined, there is ample opportunity for creative designs to effect light trapping in silicon solar cells, and light-trapping structures have been realised by many different methods. For purposes of review, three types of reflective surfaces can be distinguished—random texture, geometric or regular structuring, and the use of optical elements external to the silicon solar cell structure.

2.1.1 Random Texturing

Random texturing holds promise for two reasons: modelling predicts that such random texturing can provide very reflective light trapping and the perceived ease at which random texture can be experimentally realised. For Lambertian diffuse reflectors made by random surface or interface texturing, and where the angular distribution of reflected light follows a cosine law, Goetzberger [5] has shown that the fraction of light reflected by total internal reflection at the front surface is equal to $1-1/n^2$, where n is the refractive index of silicon (approximately 3.4). In this case, when weakly absorbed light is reflected from a diffuse back reflector, 92% will be internally reflected from the front surface. The 50-fold increase in optical path length suggested by Yablonovitch and Cody [6] is based on front and back-surface texturing to effect Lambertian reflection. Unfortunately, the experimental realisation of a Lambertian reflector [9] has proved to be more difficult.

Random textures of varying effectiveness have been experimentally realised by the following methods:
- reactive ion etching [10–12],
- sand blasting [13],
- photolithography [14],
- natural lithography [15],
- porous etching [16],
- rapid thermal processing of an aluminium—silicon interface [17], and
- random-textured ancillary dielectric layers such as ZnO or SnO_2 [18–21].

2.1.2 Geometrical or Regular Structuring

Surface structuring is relatively easy to realise in single-crystal silicon wafer surfaces by taking advantage of the anisotropic (crystal orientation—dependent) etch rates of alkaline solutions. This approach has been used to fabricate the majority of light-trapping structures demonstrated to date, a few examples of which are shown in Figure 3. Common structures include 54-degree pyramids, inverted pyramids, slats, and perpendicular slats. Regular patterns are achieved with photolithography in combination with anisotropic etching [22–24]. Random pyramids are formed with the use of anisotropic etchants on nonpatterned surfaces [25,26], A similar

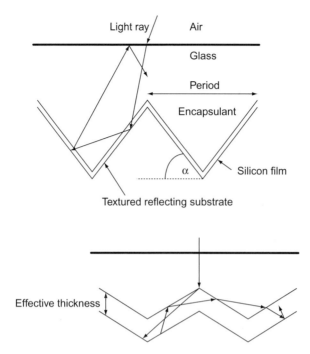

Figure 3 Examples of structured surfaces to effect light trapping in silicon solar cells [34].

result can be achieved in multicrystalline substrates (with much more effort) with the use of mechanical scribing, abrading or grinding, or laser ablation [27–32]. An option for very thin layers is conformal growth on a textured substrate [33] (Figure 3).

Modelling of geometrical textures is often carried out by computer-aided ray-tracing techniques [35–38]. Ray-tracing analysis makes no assumptions about the distribution of light within the absorber layer. Instead, the path of representative incident rays of light are plotted using geometric optics to follow the light path through as it is reflected and refracted at surfaces and interfaces. Attenuation of light due to optical absorption in bulk silicon is also incorporated in the model. This approach is useful for the analysis of regularly structured surfaces [39] with complex geometries as well as unusual solar cell shapes such as the Spheral Solar™ Cell [40].

2.1.3 External Optical Elements
The solar cell can be overlayed with refractive optics elements such as prismatic cover slips (Figure 4) to effect a degree of light trapping. These

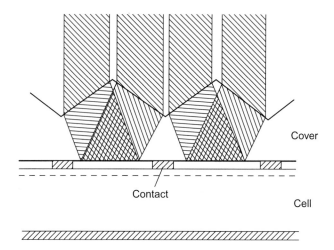

Figure 4 A prismatic cover slip can be used to effect light trapping in solar cells [42].

optical elements were originally developed to ameliorate shading by the front contact grid metallisation, but they can also be used to redirect light back into the solar cell after it has escaped. External reflectors and optical cavities are two other examples of the use of external optical elements to implement light trapping. Miñiano et al. [41] have analysed light-confining cavities for concentrator solar cells.

2.2 Assessment of Light-Trapping Effects

Light trapping has been incorporated in structures with thickness ranging from less than 1 micron to 400 microns with varying degrees of success. The short-circuit current is the ultimate figure of merit for comparisons, as the objective of light trapping is to enhance absorption and contribute to minority-carrier generation. Using short-current current to assess the effectiveness of light-trapping schemes is complicated by other losses such as bulk recombination and front-surface reflection. In laboratory settings, spectral-response and reflection data can be analysed to extract detailed information about light trapping; however, these techniques require assumptions that limit their usefulness in predicting final cell performance. Each analysis method is reviewed in the following sections.

2.2.1 Short-Circuit Current Analysis of Light Trapping

Modelling the current generated by a known thickness of silicon is straightforward when all absorbed photons contribute a charge carrier. A 'no light-trapping' scenario can be generated that has perfect antireflection properties (no front-surface reflection) and only one pass of light through the device thickness (back surface is 100% absorbing with no contribution to current). To the extent an experimental result exceeds this model, light-trapping features are indicated. In real devices shading, imperfect AR, and parasitic absorption make the 'no light-trapping' model impossible to achieve [43]. Calibration of the light source is another source of error in this method.

Figure 5 shows modelling results and experimental data for short-circuit current as a function of thickness. The effect of light trapping becomes more pronounced as the device thickness is reduced and the light not absorbed on the first pass through the device grows to a significant level. Theoretical analysis of Green [44] and Tiedje [45] are included in Figure 5 and serve as the 'best case' limit for perfect light trapping. The analysis of Green appears to predict perfect collection ($J_{SC} = 44$ mA/cm^2)

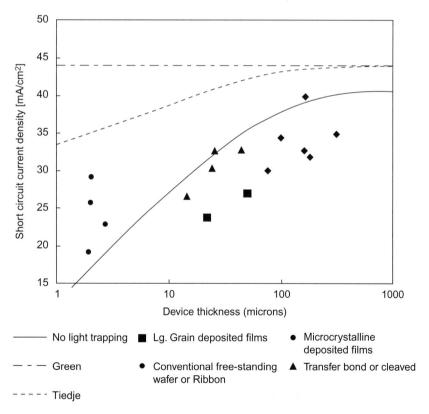

Figure 5 Summary of the last 5 years of published data on silicon-layer thickness and short-circuit current. Theoretical analyses are shown as lines. Laboratory results are shown as solid symbols [47–68].

independently of thickness. The method of Tiedje assumes random texture, Lambertian-type reflection, and more realistic modelling using a loss-cone analysis for front-surface internal reflections. The baseline case with no light trapping was computed with PC-1D [46].

The experimental results shown in Figure 5 indicate that most laboratory and commercial devices have not produced a level of photogenerated current exceeding that which would be possible in a device of comparable thickness without light trapping. Thus, based solely on total current, it cannot be concluded that in these particular solar cells light trapping makes a significant contribution to solar cell performance. These devices may include some light-trapping enhancement of current, but additional losses due to imperfect antireflection coatings, bulk and surface recombination, possibly exacerbated by adding light-trapping features to the solar cell, may have offset any gains from light trapping. This has prompted researchers to investigate other methods based on reflection and spectral-response measurement to analyse light-trapping effects in ways that are not obscured by various losses [69].

2.2.2 'Sub-Band-Gap' Reflection Analysis of Light Trapping

The effectiveness of backside reflectors can be evaluated at long wavelengths ('sub-band-gap' energy photons)—i.e., longer than about 1100 nm. The silicon layer is approximately transparent to light in this wavelength range. As shown in the inset of Figure 6, the measured total reflection for nonabsorbed light at these wavelengths of a thin solar cell with a backside mirror is due to contributions of multiple internal reflections from the front and back surface. If absorption is negligible, then the ratio between the measured reflection and the reflection expected of an infinitely thick slab of silicon gives an estimate of the effective light trapping in a device. This technique does not measure light trapping *per se* but instead provides an indicator of the effectiveness of the backside mirror and the level of optical confinement for near-band-gap, weakly absorbed photons. This analysis can be especially decisive if test structures, both *with* and *without* backside reflectors, are compared. In thin silicon structures, the effect of the buried reflector is shown as an enhancement of sub-band-gap (>1050-nm-wavelength) reflection, which is attributed to unabsorbed light reflected at the backside mirror and transmitted through the front surface. This escaping light boosts the measured front-surface reflection (Figure 6), and the increased reflectance permits an estimate of the internal reflections in the silicon layer.

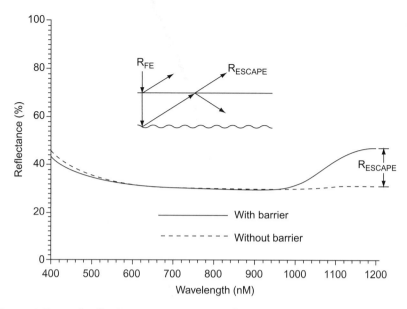

Figure 6 External reflection measurements in thin silicon structures with reflecting barrier layer between silicon layer and substrate [70].

2.2.3 Extended Spectral-Response Analysis of Light Trapping

The extended spectral-response method was developed by Basore [71] as a means to estimate an effective optical path length for near-band-gap light in a thin silicon solar cell, and it is based on an analysis of the internal quantum efficiency of a solar cell as a function of wavelength.

The method utilises a plot of $1/\text{IQE}$ versus $1/\alpha$, where α is the optical absorption coefficient for silicon and IQE is the internal quantum efficiency. Two representative examples of such plots are shown in Figure 7, for a thick silicon wafer-based solar cell and a thin (4 micron) silicon solar cell, both of which have textured surfaces to effect light trapping. The $1/\text{IQE}(\lambda)$ versus $1/\alpha$ curves show two linear regions, which appears to be a general feature of solar cells with light trapping.

We will not discuss the details of the somewhat involved analysis in which Basore shows that the slopes and intercepts of the linearised parts of these curves can be used to estimate the collection efficiency, the back reflectance, the back-surface recombination velocity, and the effective minority-carrier diffusion length. This extended spectral-response analysis, although more complicated than other techniques, appears to be the most useful method for assessing light trapping in solar cells.

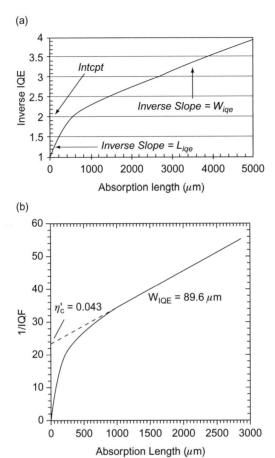

Figure 7 (a) Plot of $1/\alpha$ vs. 1/IQE for a textured (thick) wafer-based silicon solar cell [71] and (b) textured 4-micron-thick silicon solar cell [72].

3. VOLTAGE ENHANCEMENTS IN THIN SILICON SOLAR CELLS

The factors that influence the open-circuit voltage of a silicon solar cell are the same whether the device is thin or thick. These include doping levels, the various bulk recombination mechanisms (defect-mediated Shockley–Read–Hall, radiative, and Auger), and surface recombination. The open-circuit voltage V_{OC} depends logarithmically on the dark diode current density J_O. The diode is analysed in terms of current components from the silicon layer (base) and a thin emitter layer formed on the surface of the base, usually by impurity diffusion. J_O, can vary by orders of magnitude, depending on device parameters and silicon properties. The dark current can be calculated as

$$J_O = \frac{q \cdot n_i^2}{N_D} \frac{D}{L} \left[\frac{(SL/D)\cosh(W_b/L) + \sinh(W_b/L)}{\cosh(W_b/L) + (SL/D)\sinh(W_b/L)} \right]$$

where q is the electronic charge, n_i is the intrinsic carrier concentration of silicon; N_D is the base doping concentration; S is the back-surface recombination velocity that characterises the extent of minority recombination at the silicon−substrate interface; L and D are the minority-carrier diffusion length and diffusivity in the base, respectively, and are sensitive to doping levels and material quality; and W_b is the thickness of the base. Actually, the above equation represents only the base layer contribution to J_O. There is an analogous equation for emitter contribution to J_O, but in well-designed solar cells, normally the base component is the dominant contribution to J_O. At any rate, our purpose here is to simply highlight the factors that contribute to J_O and indicate the design principles to reduce J_O. In a thin silicon solar cell, the diffusion length will normally be longer than the layer thickness—i.e., $W_b/L < 1$, in which case the above equation simplifies to

$$J_O = \frac{q \cdot n_i^2 \cdot S}{N_D}$$

In such cases, the dark current does not depend on diffusion length but is directly proportional to the surface recombination velocity, thus underscoring the importance of surface passivation in thin silicon devices. Higher doping concentrations N_D will decrease J_O, so long as the doping does not degrade L such that $L < W$ and bulk recombination becomes significant. Another consideration of high doping that is not evident from the above equations are band-gap narrowing effects. As doping levels exceed about 10^{19} atoms/cm^3, the effective band gap of silicon is reduced, leading to increased intrinsic carrier concentrations n_i, and correspondingly increased J_O. Thus, increasing doping to increase V_{OC} becomes self-defeating after a certain point.

3.1 Minority-Carrier Recombination Issues in Thin Silicon Solar Cells

As already pointed out, an important potential, although not completely experimentally verified, advantage of thin silicon solar cells is their decreased sensitivity to minority-carrier recombination. This permits higher doping levels to enhance open-circuit voltage and leads to better tolerance of impurities and defects. A sensitivity analysis of solar cell efficiency to device thickness and minority-carrier lifetime is shown in Figure 8. Minority-carrier lifetime τ and diffusion length L are related as

$$L = \sqrt{D \cdot \tau}$$

where D is the minority-carrier diffusivity, which is proportional to the minority-carrier mobility. While D is not highly sensitive to impurity and defects, τ can easily vary by more than an order of magnitude in silicon of various purity and quality. The analysis summarised in Figure 11 shows the interesting result that for a given minority-carrier lifetime there is an optimum thickness, and that for material with relative low minority-carrier lifetimes the optimum thickness is less than 50 microns.

Films of silicon deposited on substrates will usually be polycrystalline, although not all thin silicon solar cells are necessarily polycrystalline; epitaxial films removed from a monocrystalline silicon substrate and

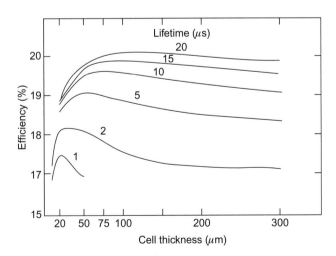

Figure 8 Sensitivity of solar cell efficiency to device thickness and minority-carrier lifetime [74].

bonded to a superstrate, as well as solar cells made by thinning monocrystalline silicon wafers, need not have grain boundaries. In fact, these types of silicon solar cells provide an interesting control for exploring the effects of grain boundaries. Nevertheless, many low-cost approaches to thin silicon solar cells will produce material with varying grain sizes and textures. The effects of grain boundaries are complex and depend on the microstructure, film thickness, grain size distribution, junction depth, doping, etc. Diffused junctions can not only spike down grain boundaries, which may improve collection efficiency but also make the solar cell more prone to shunting effects (Figure 9). These issues are not unique to thin silicon solar cells, and, in fact, polycrystalline cast silicon solar cells have been an established line of solar cells for many years. Grain boundaries act as surfaces for minority-carrier recombination and can be depleted or accumulated and can exhibit space charge regions much as junctions and free surfaces. For single-crystal silicon solar cells, space-charge recombination is usually so small that it can be neglected, and the performance of the device, particularly open-circuit voltage, is controlled by bulk, surface, and shunt losses, but this may not be the case with multicrystalline solar cells. It was initially thought that polycrystalline silicon solar cells would suffer major short-circuit current and open-circuit voltage losses from grain-boundary recombination, which would severely restrict the maximum light-generated current [73]. However, it has been shown that when the grain diameter is several times larger than the intragrain (bulk) minority-carrier diffusion length, the short-circuit current is controlled not by grain-boundary recombination, but by the intragrain diffusion length.

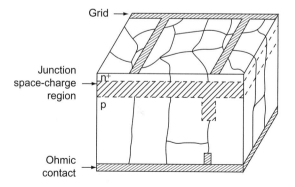

Figure 9 Geometry of grain boundaries in silicon solar cells [75].

Other factors that may play a substantial role in determining the electrical performance of polycrystalline silicon solar cells are the presence of inclusions and tunnel junctions, both of which act as resistive shunts and degrade the open-circuit voltage and fill factor locally. Although their impact on performance is fairly straightforward, it is not clear that either of these two possible defects is intrinsic to any polycrystalline silicon solar cell material or process. Accordingly, the analysis of the thin polycrystalline silicon solar cell is based on the relaxation of single crystal material and device properties due to the polycrystalline characteristics of the semiconductor.

As expected, increasing grain size results in better solar cell performance (Figure 10), but note the discontinuity between trends for p—i—n microcrystalline silicon solar cells and p—n multicrystalline silicon solar cells. The electric field in thin p—i—n cells is probably aiding collection efficiency and mediating the effect of grain-boundary recombination. The grain structure in these thin microcrystalline silicon solar cells may also have a texture resulting in grain boundaries with less electrical activity (e.g., minority-carrier recombination). Grain-boundary effects are one of the most active areas for research in silicon solar cells, and passivation techniques can be very effective in all types of multicrystalline silicon solar cells (Figure 11).

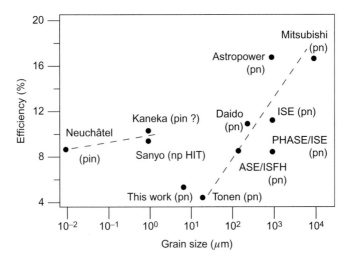

Figure 10 Effect of grain size on solar cell efficiency for p—i—n microcrystalline silicon devices and p—n multicrystalline thin silicon devices [76].

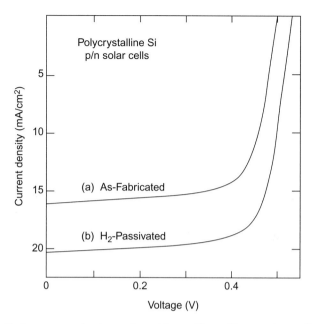

Figure 11 Hydrogen passivation of multicrystalline silicon solar cells showing improvement in current and voltage [77].

4. SILICON DEPOSITION AND CRYSTAL GROWTH FOR THIN SOLAR CELLS

In this section, we review some of the more important technologies used to realise thin silicon solar cells on supporting substrates (Figure 12). There is a wide range of methods used to deposit semiconductor materials, virtually all of which have been applied to some extent or degree to the production of thin silicon solar cells. This section is offered as a survey of the diverse approaches; space limitations do not permit an in-depth review. Where possible, common issues and criteria and unifying design principles are noted.

4.1 Substrate Considerations

One of the key technological challenges to achieving a commercially viable thin-layer polycrystalline silicon solar cell technology is the development of a low-cost supporting substrate. The requirements for the substrate material are severe. Mechanical strength and thermal coefficient of expansion (TCE) matching are needed to prevent the film from breaking or

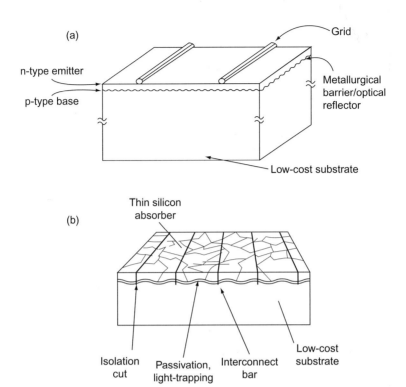

Figure 12 Generic polycrystalline thin-film-silicon device structures: (a) the electrically active part of the device consists of a thin silicon layer on top of a passive mechanically supporting substrate [78]; (b) similar structure with interconnection achieved monolithically, a benefit unavailable to conventional wafer-based devices [79].

deforming during handling and high-temperature processing. There are several good candidates for thermal expansion matched substrates, including mullite, a compound of alumina and silica. The substrate must also provide good wetting and nucleation during the film-growth process without contaminating the film. The substrate can be conducting or insulating, depending on device requirements. Finally, the substrate–silicon interface must provide a high degree of diffuse reflectivity and surface passivation.

The following substrates have been utilised to fabricate thin silicon solar cells:
- glass [80–82],
- ceramics [83–90],
- steel [91–93],
- graphite [94–99], and
- upgraded metallurgical silicon sheet or wafers [100–103].

4.2 High-Temperature Silicon-Deposition Methods

It is useful to distinguish silicon-deposition methods that employ high temperatures (>1000°C) from low-temperature deposition techniques. The high temperatures impose significant constraints on and limit the choice of substrates, especially if a postdeposition recrystallisation step is desired. High-temperature deposition methods are probably the only way to achieve high silicon-deposition rates (e.g., 1—20 microns/min), and therefore, if silicon layers of 10—50-microns thickness are required, such high-temperature steps may be the only viable option. High-temperature deposition methods include

- Melt-growth or melt-coating techniques where elemental silicon is melted and then deposited as a film or layer on a substrate.
- Chemical vapour deposition (CVD) where a silicon-containing gaseous precursor is thermally decomposed on a substrate.
- Liquid-phase epitaxy (LPE) where silicon is precipitated from a molten metal solution.

4.2.1 Melt-Growth Techniques

Melt-growth processes characteristically have both high growth rates and good material quality [83,84,104]. An example melt-coating process is shown in Figure 13. In these examples, a substrate is contacted with molten silicon, which wets the substrate and then solidifies as a silicon layer. In many cases, the substrate is drawn through a bath of molten silicon. Generally, such processes cannot produce layers much less than 100 microns in thickness.

4.2.2 Recrystallisation of Silicon

Related to melt growth are various recrystallisation techniques. These are generally not deposition processes per se but instead are used to melt already-deposited silicon layers and recrystallise them in order to achieve a more favourable grain structure. In this case, the grain structure of the as-deposited silicon layer is not critical, and the deposition process can be optimised for high-growth rates, large areas, and purity specifications. For instance, a plasma-enhanced CVD process such as shown in Figure 14 can be used to plate a silicon layer of desired thickness on a suitable substrate, such as a ceramic, which is compatible with a recrystallisation step.

In the preferred techniques of recrystallisation, the deposited silicon layer is not usually simultaneously melted in its entirety. Instead, a zone-melting recrystallisation (ZMR) process is effected by localised heating to

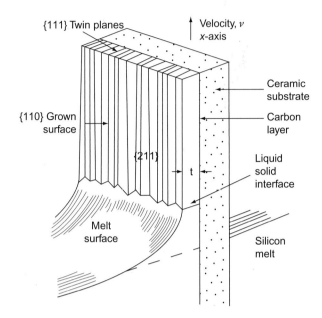

Figure 13 Honeywell silicon-on-ceramic dip-coating process [83].

create a melted zone that moves or scans across the deposited silicon layer, melting the silicon at the leading edge and resolidifying a silicon layer at the trailing edge. Such ZMR techniques can yield millimetre- to centimetre-sized grain structures. There are several ways to induce localised or zone melting of layers, including moving point- or line-focused infrared lamps, travelling resistively heated strip heaters, and laser and electron beams. Figure 15 shows several of these ZMR techniques commonly used for silicon solar cell applications.

Much work has been done on optimising the quality of silicon layers produced by ZMR. For example, Figure 16 shows the effects of silicon layer thickness and zone-melting scan speed on the defect density of recrystallised silicon layers.

4.2.3 High-Temperature Silicon Chemical Vapour Deposition

Chemical vapour deposition or CVD is defined as the formation of a solid film on a substrate by reacting vapour-phase chemicals, or 'precursors,' that contain the desired constituents [110]. For example, substrates can be coated with silicon layers by decomposition of gaseous silane (SiH_4) or trichlorosilane ($SiHCl_3$). In fact, many precursors are possible for silicon CVD, and silane or the chlorosilanes are probably the most commonly

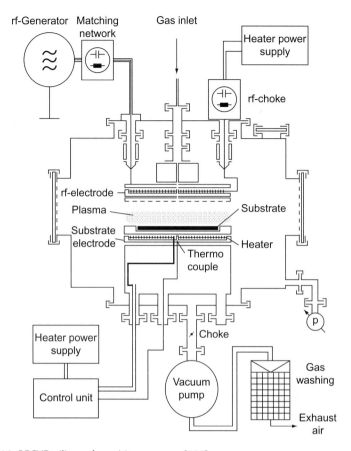

Figure 14 PECVD silicon-deposition process [105].

used—although, for example, iodine and bromine compounds are also sometimes considered as silicon precursors. In general, silicon CVD is a well-developed technology commonly used in integrated circuit fabrication, in which case it is often used for epitaxial growth of silicon layers on monocrystalline silicon substrates. For solar cell applications where CVD is used to deposit a 10- to 50-micron-thick silicon layer for subsequent, post-deposition ZMR, the CVD is optimised for high-precursor utilisation (i.e., the fraction of precursor converted to silicon), deposition efficiency (i.e., the fraction of deposited silicon that ends up on the substrate rather than the walls of the reactor chamber or the substrate susceptor), the deposition rate, the purity of the deposited silicon, areal uniformity, the potential to recover unreacted precursors or reaction product, and various safety and environmental issues.

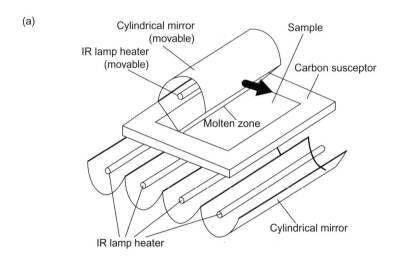

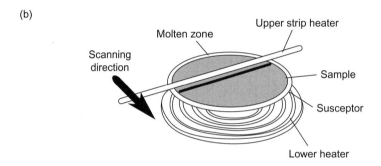

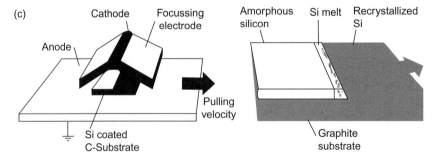

Process parameter
- Pulling velocity: v = 5...45 mm/s
- Power/length: p = 80...320 W/cm
- Haltwidth: FWHM = 0.8...1.5 mm
- Acceleration voltage: U_H = 12...16 kV
- Substrate temperature: T_{sub} = 650...850 °C
- SI layer thickness: d = 3...10 µm

Figure 15 (a) Halogen IR lamp heating for ZMR [106]. (b) Travelling strip heaters for ZMR [107]. (c) Electron beam heating [108].

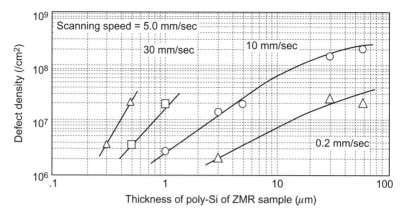

Figure 16 Relation between deposited-silicon-layer thickness, scan speed, and silicon-defect density in ZMR [109].

For solar cell applications, three types of CVD are most used:
- atmospheric pressure (APCVD)
- rapid thermal (RTCVD)
- low-pressure (LPCVD)

In all these types of CVD, a gaseous silicon precursor (e.g., SiH_4 or $SiHCl_3$), generally mixed with a dilution carrier gas such as hydrogen or nitrogen, is delivered into a reaction chamber. These gases move from the inlet to the outlet in a continuous stream to form the main gas-flow region that brings the precursor into close proximity of the heated substrate. Some further description and details of the three main types of CVD processes used for thin silicon solar cells follows.

Atmospheric Pressure CVD

APCVD systems were historically the first used for applications in the microelectronics industry [111,112]. These systems are simple in design and are generally composed of three subsystems: a gas delivery system, a reactor, and an effluent-abatement system. For the most part, APCVD is carried out at relatively high temperatures for silicon-containing precursors (1100–1250°C). This allows for high deposition rates to be achieved. At these high temperatures, APCVD is in the mass-transport limited regime. This requires that a very uniform gas flow be achieved within the reactor to ensure that all areas of the heated substrate are exposed to equal amounts of precursor. This requirement is the primary concern during APCVD reactor geometry design and, to date, has limited this process to batch-type processes with respect to CVD silicon.

Rapid Thermal CVD

RTCVD is a variation of APCVD. It is based on the energy transfer between a radiant heat source and an object with very short processing times—seconds or minutes. This is typically from an optical heating system such as tungsten halogen lamps. The obvious benefit of this process is the fast cycle times for heating substrates to their required deposition temperature. A detailed explanation of this process along with its advantages and disadvantages can be found in Faller et al. [113].

Low-pressure CVD

LPCVD systems are inherently more complex than APCVD systems since they require robust vacuum systems that are capable of handling the often toxic corrosive precursor effluents. With respect to silicon deposition, LPCVD is generally conducted using vacuum pressures of 0.25–2.0 torr and temperatures of 550–700°C. At these pressures and temperatures LPCVD is in the surface rate-limited regime. It is important to note that at reduced pressures the diffusivity of the precursor is greatly enhanced. This allows for multiple wafers to be stacked very closely together, on the order of a few millimetres, and still achieve a highly uniform deposition. However, in order to ensure this, very precise temperature control is necessary across the entire reactor, within 0.5–1°C is not uncommon. Since LPCVD is in the surface rate-limited regime it has the constraint of very low deposition rates. These low growth rates have limited its application in silicon solar cell fabrication.

A summary of silicon CVD growth rates for various precursors and deposition temperatures is given in Table 1.

4.2.4 Low-Temperature Chemical Vapour Deposition

Some low-temperature chemical vapour techniques can be distinguished from the high-temperature (>1000°C) CVD processes discussed previously. These methods are employed with substrates such as glass that are

Table 1 Common silicon precursors

Silicon precursor	Deposition temperature (°C)	Growth rates (μm/min)
Silicon tetrachloride ($SiCl_4$)	1150–1250	0.4–1.5
Trichlorosilane ($SiCl_3H$)	1000–1150	0.4–4.0
Dichlorosilane (SiH_2Cl_2)	1020–1120	0.4–3.0
Silane (SiH_4)	650–900	0.2–0.3
Disilane (Si_2H_6)	400–600	<0.1

not compatible with either a high-temperature deposition step or a post-deposition recrystallisation step. The relatively slow growth rates inherent in a low-temperature deposition process necessitates thin device structures on the order of several microns thickness or less. The as-deposited silicon layers have average grain sizes of 1 micron or less and are characterised as microcrystalline. Hydrogenated microcrystalline silicon solar cells using a p—i—n structure, which is similar to the amorphous silicon solar cell structure, can achieve very respectable conversion efficiencies in excess of 10%. The benefits of such microcrystalline silicon solar cells over amorphous silicon solar cells are a greater stability to light-induced degradation processes. The deposition processes for microcrystalline silicon solar cells are typically adaptations of those used for amorphous silicon solar cells.

Liquid-phase Epitaxy (LPE)

Liquid-phase epitaxy is a metallic solution growth technique that can be used to grow semiconductor layers on substrates. Silicon can be precipitated from solutions of a number of molten metals in the temperature range 600—1200°C. This method has been used to grow thin silicon solar cells on low-cost metallurgical grade (MG) silicon substrates. In this case, the MG silicon substrate is too impure for direct use in photovoltaics. Instead, the substrate is used as a substrate for the growth of high-purity layers of silicon by either LPE or CVD. The MG silicon provides a thermal-expansion matched substrate for the thin silicon solar cell. The grain size of MG substrates is relatively large (several millimetres to centimetres in lateral dimension); as the epitaxial layer will replicate the grain structure of the silicon substrate, this approach will yield thin silicon solar cells with large grain sizes. One issue with using a MG silicon substrate is contamination of the solar cell device by outdiffusing substrate impurities. Other favourable features of LPE are the high mobility of adatoms in the liquid phase (as compared to surface diffusion upon which vapour-phase techniques depend) and the near-equilibrium growth conditions that reduce point defects and dislocations originating from the substrate. A conventional slideboat LPE system, similar to that used for making compound semiconductor optoelectronics devices and suitable for R&D of LPE thin silicon solar cells, is shown in Figure 17. This type of LPE system employs a programmed transient cooling mode and is essentially a batch process. Steady-state LPE processes (Figure 18) using an imposed temperature difference across the growth solution in conjunction with a solid silicon source to replenish the solution have been proposed and developed for high-throughput production.

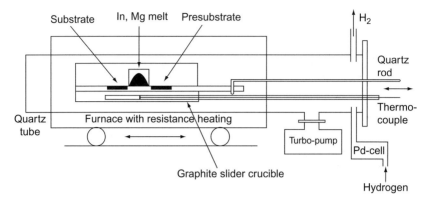

Figure 17 Small-scale LPE system [114].

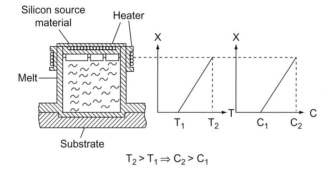

Figure 18 Principle of the temperature-difference method [115].

5. THIN SILICON SOLAR CELLS BASED ON SUBSTRATE THINNING

Figure 19 shows a solar cell made by thinning and grooving the backside of a silicon wafer. Such solar cells obviously have little cost advantage in that they utilise a high-quality silicon wafer and add considerable processing complexity. Because of the decreased sensitivity of performance to lifetime in such thin solar cells, they have application to space solar cells due to their potential radiation hardness. Further, these cells provide a means of studying basic effects such as light trapping and surface passivation in thin solar cell structures without the complicating issues of material quality and grain boundaries. A similar type of thin solar cell is shown in Figure 20. This

Thin Silicon Solar Cells

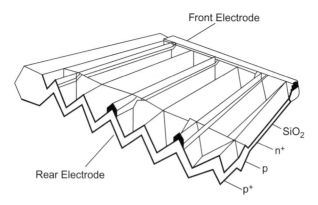

Figure 19 Thinned and grooved wafer-based silicon solar cell [116].

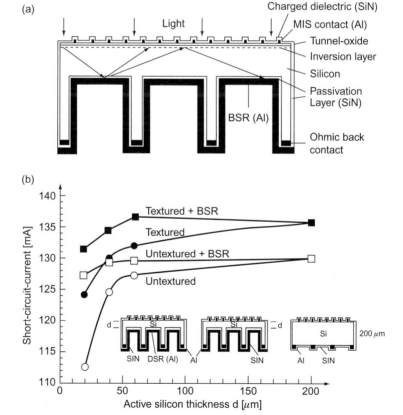

Figure 20 (a) Cross-section of ultra-thin, self-supporting MIS solar cell made by structuring a silicon wafer [117]. (b) Short-circuit current as a function of active-layer silicon thickness for solar cell structures shown in inset [117].

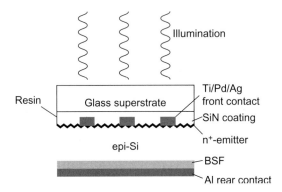

Figure 21 Monocrystalline silicon substrate made by layer transfer and bonding to a glass superstrate [119].

device structure permits a planar back mirror to be effected close to the front surface (Figure 20(a)). The dependence of short-circuit current on effective device thickness can be readily studied with this type of thin solar cell (Figure 20(b)).

Even thinner silicon solar cells can be made with silicon wafers using epitaxial growth, provided a superstrate is used for mechanical support. For instance, Figure 21 shows a thin silicon solar cell made by layer transfer and wafer-bonding techniques. An epitaxial silicon solar cell structure is grown on a monocrystalline silicon substrate. The structure is then bonded to a glass superstrate. The silicon substrate is then removed. Most simply, the removal of the silicon substrate can be effected by controlled etching, in which case the substrate is dissolved away. Various schemes have been proposed to separate the substrate from the epitaxial layer after bonding the solar cell structure to a superstrate. One method of achieving this shown in detail in Figure 22. Some creative variations on this approach have been reported; for example, see Figure 23.

6. SUMMARY OF DEVICE RESULTS

We end this review by summarising results for thin silicon solar cells. Table 2 lists reports for thin silicon layers made by high-temperature growth methods, often including a postdeposition recrystallisation (e.g.,

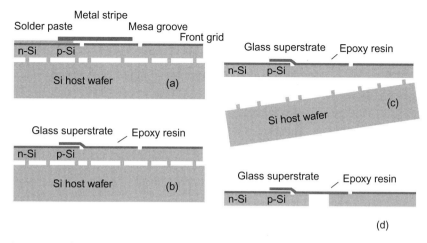

Figure 22 Schematic representation of series connection of thin-film Si-transfer solar cells. (a) Two epitaxial thin-film silicon solar cells connected to the host wafer with the separation layer (columns). A metal stripe (Ag) is soldered to the front-side grid of the left solar cell. Mesa grooves provide electrically isolation of the emitter in the interconnection area. (b) Epoxy resin fixes the superstrate glass to the surface of the cell. (c) Mechanical force removes the host wafer from the cell. (d) A groove structured via chip dicing sawing separates the two solar cells. Oblique deposition of aluminium creates the back contact of the solar cells and electrically connects the metal stripe that is in contact with the front side grid of the left solar cell with the back-side contact of the right solar cell [120].

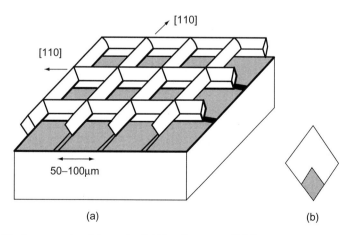

Figure 23 An example of the epitaxial lift-off process [121].

Table 2 High-temperature growth on foreign substrates (after Catchpole et al. [122])

Institution	Substrate	Deposition method	Grain size (μm)	Electrical	Reference
PHASE, ECN and CNRS-LMPM	Alumina	RTCVD	10	$\tau = 0.3\ \mu s$, $L = 10\ \mu m$	[123,124]
PHASE, LPM and GEMPPM	Alumina	LPE with RTCVD seed	10		[125]
IMEC and PHASE	Alumina	RTCVD	5–10	$\tau = 0.5\ \mu s$	[126,127]
ETL	Alumina	ECR–PCVD with EB-ZMR, diff. Barrier	10 × 200		[128]
ETL	Alumina	CVD with laser recryst., diff. Barrier		$\eta = 6.5\%$	[129]
TU-Berlin	Alumina	CVD	10	$L = 8\ \mu m$	[130]
PHASE and IMEC	Mullite	RTCVD	10–15	$\tau = 0.5\ \mu s$	[131]
MPI-F, MPI-M and LSG	Glassy carbon	LPE with RF plasma seed	10		[132]
MPI-F, Siemens and	Graphite	PECVD or sputtering + ZMR + CVD	100 × 1000		[133]
TU-Hamburg Fraunhofer ISE	Graphite	LPCVD, diff, Barrier, ZMR	mm × cm	$L = 30\ \mu m$, $\eta = 11.0\%$	[134]
Daido Hoxan	Graphite	LPE (temp diff) with diff. Barrier	100–300	$L = 30\ \mu m$	[155,136]

Institution	Substrate	Method			Ref
PHASE, IMEC and Fraunhofer	Graphite	RTCVD with diff. Barrier	0.1–6	$L = 3\ \mu m$	[137]
ASE GmbH	Graphite	CVD with ZMR, diff. Barrier	100	$L = 10\text{–}15\ \mu m$	[138]
TU-Hamburg	Graphite	LPCVD with EB-ZMR	$100\ \mu m \times$ cm		[139]
ECN	Si/SiAlON	LPE with plasma-spray	10–100		[140]
UNSW	High-temperature glass	LPE with a-Si seed	50		[141]
UNSW	High-temperature glass	LPE	100		[142]
MPI-F, U Erlangen and U Stuttgart NTT	High-temperature glass	CVD with LPCVD and SPC seed	2	$L = 2\ \mu m$, $\eta = 2\%$	[143,144]
AstroPower	Tape cast ceramic	APCVD + ZMR	mm $\times$ cm	9.18%, 545 mV	[145]

Table 3 Summary of thin silicon-on-silicon device results (after McCann et al. [146])

Institution	Substrate	Deposition method	Electrical	Reference
Single-crystal substrates				
ANU	p-type sc-Si	LPE and substrate thinning	18%, 666 mV	[147]
UNSW	p^+ sc-S	CVD	17.6%, 664 mV	[148]
Fraunhofer ISE	p^+ sc-Si	RTCVD	17.6%	[149]
MPI-F	sc-Si	CVD	17.3%, 661 mV	[150]
ANU	p^+ sc-Si	LPE and substrate thinning	17%, 651 mV	[151]
UNSW	p^+ sc-Si	LPE	16.4%, 645 mV	[152]
ASE GmbH	p^+ sc-Si	CVD	15.4%, 623 mV	[153]
Imec	p^+ sc-Si	CVD	14.9%, 635 mV	[154]
MPI-F	p^+ sc-Si	LPE	14.7%, 659 mV	[155]
Beijing SERI	p^+ sc-Si	RTCVD	12.1%, 626 mV	[156]
Mc-Si, ribbon and MG–Si substrates				
ANU	p-type mc-Si	LPE	15.4%, 639 mV	[157]
ANU	p^+ mc-Si	LPE	15.2%, 639 mV	[157]

Institution	Substrate	Process	Results	Ref.
IMEC and KU	p⁺ mc-Si	APCVD	13.3%, 615 mV	[158]
Fraunhofer ISE	p⁺ mc-Si	RTCVD	13.2%, 614 mV	[149]
IMEC	p⁺ mc-Si	CVD and industrial cell Process	12.1%	[159]
IMEC, KU Leuven and Bayer	RGS-ribbons	CVD	10.4%, 558 mV	[160]
Fraunhofer ISE	SSP ribbons	RTCVD	8.0%, 553 mV	[161]
IMEC and KU	SSP pre-ribbon	CVD	7.6%	[162]
Leuven 1 Kristallzuchtung	mc-Si	Temp. diff. LPE	$\tau = 5$–$10\ \mu m$	[163]
NREL	MG-Si	LPE	$L = 42\ \mu m$	[164,165]
Substrates with diffusion barriers				
Mitsubishi Electric	SiO$_2$ on Si	ZMR and CVD	16.4%, 608 mV	[166]
Fraunhofer ISE	SiO$_2$ (perforated) on SSP-Si	RTCVD and large area heating	11.5%, 562 mV	[167]
IMEC and Fraunhofer ISE	SiO$_2$ on Si	ZMR and CVD	9.3%, 529 mV	[168]
Fraunhofer ISE	SiO$_2$ on Si	LPCVD and ZMR	6.1%	[169]
Delft UT	SiO$_2$ on Si	CVD	Grain size 1–2 μm	[170]

ZMR) step, on nonsilicon substrates such as ceramics, graphite, and high-temperature glasses. Samples are characterised either by diffusion length or minority-carrier lifetime or, in cases where a solar cell was made, by efficiency. Table 3 shows a similar summary of results for thin silicon solar cell structures on silicon-based substrates, which include oxidised silicon, MG silicon, and various types of silicon sheet.

REFERENCES

[1] D. Redfield, Enhanced photovoltaic performance of thin silicon films by multiple light passes. Proceedings of the 11th IEEE Photovoltaic Specialists Conference, Scottsdale, 1975, pp. 431–432.
[2] M. Spitzer, J. Shewchun, E.S.Vera, J.J. Loferski, ultra high efficiency thin silicon p−n junction solar cells using reflecting surfaces. Proceedings of the 14th IEEE Photovoltaic Specialists Conference, San Diego, 1980, pp. 375–380.
[3] A.M. Barnett, thin film solar cell comparison methodology. Proceedings of the 14th IEEE Photovoltaic Specialists Conference, San Diego, 1980, pp. 273–280.
[4] J.H. Werner, R.B. Bergmann, Crystalline silicon thin film solar cells. Technical Digest International PVSEC-12, Jeju, 2001, pp. 69–72.
[5] A. Goetzberger, Optical confinement in thin si-solar cells by diffuse back reflectors. Proceedings of the 15th IEEE Photovoltaic Specialists Conference Orlando, 1981, pp. 867–870.
[6] E. Yablonovitch, G.D. Cody, Intensity enhancement in textured optical sheets for solar cells, IEEE Trans. Electron Devices ED-29 (1982) 300–305.
[7] D. Redfield, Thin-film silicon solar cell multiple-pass. Appl. Phys. Lett. 25(11), 647–648.
[8] A.M. Barnett, J.A. Rand, F.A. Domian, D.H. Ford, C.L. Kendall, M.L. Rock, et al., Efficient thin silicon-film solar cells on low cost substrates. Proceedings of the 8th European Photovoltaic Solar Energy Conference, Florence, Italy, 1988, pp. 149–155.
[9] Green, M.A., P. Campbell, Light trapping properties of pyramidally textured and grooved surfaces. Proceedings of the 9th IEEE Photovoltaic Specialists Conference, New Orleans, 1987, pp. 912–917.
[10] D.S. Ruby, P. Yang, S. Zaidi, et al., Improved performance of self-aligned, selective-emitter silicon solar cells. Proceedings of the 2nd World Conference on Photovoltaic Solar Energy Conversion, Vienna, 1988, pp. 1460–1463.
[11] M. Schnell, R. Lüdemann, S. Schaefer, Plasma surface texturization for multicrystal-line silicon solar cells. Proceedings of 28th IEEE Photovoltaic Specialists Conference, Anchorage, 2000, pp. 367–370.
[12] T. Wells, M.M. El-Gomati, J. Wood, Low temperature reactive ion etching of silicon with SF_6/O_2 plasmas, J. Vac. Sci. Technol. B 15 (1997) 397.
[13] H.W. Deckman, C.B. Roxlo, C.R. Wronski, E. Yablonovitch, Optical enhancement of solar cells. Proceedings of the 17th IEEE Photovoltaic Specialists Conference, 1984, pp. 955–960.
[14] M.J. Stocks, A.J. Carr, A.W. Blakers, Texturing of polycrystalline silicon. Proceedings of the 24th IEEE Photovoltaic Specialists Conference, Hawaii, 1994, pp. 1551–1554.
[15] H.W. Deckman, C.R. Wronski, H Witzke, E. Yablonovitch, Optically enhanced amorphous solar cells, Appl. Phys. Lett. 42(11) (1983) 968–970.
[16] Y.S. Tsuo, Y. Xiao, M.J. Heben, X. Wu, F.J. Pern, S.K. Deb, potential applications of porous silicon in photovoltaics. Proceedings of the 23rd IEEE Photovoltaic Specialists Conference, Louisville, 1993, pp. 287–293.

[17] M. Cudzinovic, B. Sopori, Control of back surface reflectance from aluminum alloyed contacts on silicon solar cells. Proceedings of the 25th IEEE Photovoltaic Specialists Conference, Washington DC, 1996, pp. 501–503.
[18] A. Rothwarf, Enhanced solar cell performance by front surface light scattering. Proceedings of the 18th IEEE Photovoltaic Specialists Conference, Las Vegas, 1985, pp. 809–812.
[19] J.M. Gee, R.R. King, K.W. Mitchell, High-efficiency cell structures and processes applied to photovoltaic-grade czochralski silicon. Proceedings of the 25th IEEE Photovoltaic Specialists Conference, Washington DC, 1996, pp. 409–412.
[20] J.M. Gee, R. Gordon, H. Liang, Optimization of textured-dielectric coatings for crystalline-silicon solar cells. Proceedings of the 25th IEEE Photovoltaic Specialists Conference, Washington DC, 1996, pp. 733–736.
[21] S.S. Hegedus, X. Deng, Analysis of optical enhancement in a-si n–i–p solar cells using a detachable back reflector. Proceedings of the 25th IEEE Photovoltaic Specialists Conference, Washington DC, 1996, pp. 1061–1064.
[22] S.P. Tobin, C.J. Keavney, L.M. Geoffroy, M.M. Sanfacon, Experimental comparison of light-trapping structures for silicon solar cells. Proceedings of the 20th IEEE Photovoltaic Specialists Conference, Las Vegas, 1988, pp. 545–548.
[23] J.A. Rand, R.B. Hall, A.M. Barnett, light trapping in thin crystalline silicon solar cells. Proceedings of the 21st IEEE Photovoltaic Specialists Conference, Orlando, 1990, pp. 263–268.
[24] F Restrepo, C.E. Backus, IEEE Trans. Electron Devices 23 (1976) 1195–1197.
[25] P. Campbell, S.R.Wenham, M.A. Green, Light-trapping and reflection control with tilted Pyramids and Grooves. Proceedings of the 20th IEEE Photovoltaic Specialists Conference, Las Vegas, 1988, pp. 713–716.
[26] D.L. King, M.E. Buck, Experimental optimization of an anisotropic etching process for random texturization of silicon solar cells. Proceedings of the 22nd IEEE Photovoltaic Specialists Conference, Las Vegas, 1991, pp. 303–308.
[27] B. Terheiden, P. Fath, E. Bucher, The MECOR (mechanically corrugated) silicon solar cell concept. Proceedings of the 28th IEEE Photovoltaic Specialists Conference, Anchorage, 2000, pp. 399–402.
[28] P. Fath. et al., Multicrystalline silicon solar cells using a new high-throughput mechanical texturization technology and a roller printing metallization technique. Proceedings of the 13th European Photovoltaic Solar Energy Conference, Nice, 1995, pp. 29–32.
[29] S. Narayanan, S.R.Wenham, M.A. Green, Technical Digest International PVSEC-4, Sydney, 1989, p. 1111.
[30] U. Kaiser, M. Kaiser, R. Schindler, Texture etching of multicrystalline silicon. Proceedings of the 10th European Photovoltaic Solar Energy Conference, Lisbon, 1991, pp. 293–294.
[31] G. Willeke, H. Nussbaumer, H. Bender, E. Bucher, Mechanical texturization of multicrystalline silicon using a conventional dicing saw and bevelled blades. Proceedings of the 11th European Photovoltaic Solar Energy Conference, Montreux, 1992, pp. 480–483.
[32] M.J. Stocks, J.J. Carr, A.W. Blakers, Texturing of polycrystalline silicon. Proceedings of the First World Conference on Photovoltaic Energy Conversion, Hawaii, 1994, pp. 1551–1554.
[33] P. Campbell, M. Keevers, Light trapping and reflection control for silicon thin films deposited on glass substrates textured by embossing. Proceedings of the 28th IEEE Photovoltaic Specialists Conference, Anchorage, 2000, pp. 355–358.
[34] D. Thorp, P. Campbell, S.R. Wenham, Absorption enhancement in conformally textured thin-film silicon solar cells. Proceedings of the 25th IEEE Photovoltaic Specialists Conference, Washington DC, 1996, pp. 705–708.

[35] A.W. Smith, A. Rohatgi, S.C. Neel, Texture: A ray-tracing program for the photovoltaic community. Proceedings of the 21st IEEE Photovoltaic Specialists Conference, Orlando, 1990, pp. 426−431.

[36] B.L. Sopori, T. Marshall, Optical confinement in thin silicon films: a comprehensive ray optical theory. Proceedings of the 23rd IEEE Photovoltaic Specialists Conference, Louisville, 1993, pp. 127−132.

[37] U. Rau, T. Meyer, M. Goldbach, R. Brendel, J.H. Werner, Numerical simulation of innovative device structures for silicon thin-film solar cells. Proceedings of the 25th IEEE Photovoltaic Specialists Conference, Washington DC, 1996, pp. 469−472.

[38] A.A. Abouelsaood, M.Y. Ghannam, J. Poortmans, R.P. Mertens, Accurate modeling of light trapping in thin film silicon solar cells. Proceedings of the Twenty-sixth IEEE Photovoltaic Specialists Conference, 1997, pp. 183−186.

[39] P. Campbell, M.A. Green, light trapping properties of pyramidally textured surfaces, J. Appl. Physics July (1987) 243−249.

[40] R. Bisconti, H.A. Ossenbrink, Light trapping in spheral solar™ cells. Proceedings of the 13th European Photovoltaic Solar Energy Conference, Nice, 1995, pp. 386−389.

[41] J.C. Minano, A Luque, I. Tobias, Light-confining cavities for photovoltaic applications based on the angular-spatial limitation of the escaping beam, Appl. Opt. 31(16) (1992) 3114−3122.

[42] J. Zhao, A. Wang, A.W. Blakers, M.A. Green, High efficiency prismatic cover silicon concentrator solar cells. Proceedings of the 20th IEEE Photovoltaic Specialists Conference, Las Vegas, 1988, pp. 529−531.

[43] J.M. Gee, The effect of parasitic absorption losses on light trapping in thin silicon solar cells. Proceedings of the 20th IEEE Photovoltaic Specialists Conference, Las Vegas, 1988, pp. 549−554.

[44] M.A. Green. et al., Enhanced light-trapping in 21.5% efficient thin silicon solar cells. Proceedings of the 13th European Photovoltaic Solar Energy Conference, Nice, 1995, pp. 13−16.

[45] T. Tiedje, E. Yablonovitch, G.D. Cody, B.G. Brooks, Limiting efficiency of silicon solar cells, IEEE Trans. Electron Devices ED-31 (1984) 711−716.

[46] P.A. Basore, Numerical modeling of textured silicon solar cells using PC-ID, ZEEE Trans. Electron Devices ED-37 (1990) 337.

[47] W. Zimmerman, A. Eyer, Coarse-grained crystalline silicon thin film solar cells on laser perforated SIO_2 Barrier Layers. Proceedings of the 28th IEEE Photovoltaic Specialists Conference, Anchorage, 2000, pp. 233−236.

[48] T.M. Bruton, S. Roberts, K.C. Heasman, R. Russell, Prospects for high efficiency silicon solar cells in thin czochralski wafers using industrial processes. Proceedings of the 28th IEEE Photovoltaic Specialists Conference, Anchorage, 2000, pp. 180−183.

[49] C. Berge, R.B. Bergmann, T.J. Rinke, J.H. Werner, Monocrystalline silicon thin film solar cells by layer transfer. Seventeenth European Photovoltaic Solar Energy Conference, Munich, 2001, pp. 1277−1281.

[50] W. Zimmerman, S. Bau, A. Eyer, F. Haas, D. Oßwald, Crystalline silicon thin film solar cells on low quality silicon substrates with and without SiO_2 intermediate layer. Proceedings of the 16th European Photovoltaic Solar Energy Conference, Glasgow, 2000, pp. 1144−1147.

[51] B. Finck von Finckenstein, H. Horst, M. Spiegel, P. Fath, E. Bucher, Thin MC SI low cost solar cells with 15% efficiency. Proceedings of the 28th IEEE Photovoltaic Specialists Conference, Anchorage, 2000, pp. 198−200.

[52] C. Zahedi, F. Ferrazza, A. Eyer, W. Warta, H. Riemann, N.V. Abrosimov, et al., Thin film silicon solar cells on low-cost metallurgical silicon substrates by liquid phase epitaxy. Proceedings of the 16th European Photovoltaic Solar Energy Conference, Glasgow, 2000, pp. 1381−1384.

[53] M. Tanda, T. Wada, H. Yamamoto, M. Isomura, M. Kondo, A. Matsuda, Key technology for μc-Si thin-film solar cells prepared at a high deposition rate. Technical Digest International PVSEC-11. Sapporo, 1999, pp. 237–238.
[54] J.I. Hanoka, An overview of silicon ribbon-growth technology. Technical Digest International PVSEC-11, Sapporo, 1999, pp. 533–534.
[55] R. Brendel, R. Auer, K. Feldrapp, D. Scholten, M. Steinof, R. Hezel, et al., Crystalline thin-film si cells from layer transfer using Porous Si (PSI-Process). Proceedings of the 29th IEEE Photovoltaic Specialists Conference, New Orleans, 2002, pp. 86–89.
[56] D.L. Meier, J.A. Jessup, P. Hacke Jr., S.J. Granata, production of thin (70–100 μm) crystalline silicon cells for conformable modules. Proceedings of the 29th IEEE Photovoltaic Specialists Conference, New Orleans, 2002, pp. 110–113.
[57] M.J. Cudzinovic, K.R. McIntosh, Process simplifications to the pegasus solar cell—sunpower's high-efficiency bifacial silicon solar cell. Proceedings of the 29th IEEE Photovoltaic Specialists Conference, New Orleans, 2002, pp. 70–73.
[58] J. Schmidt, L. Oberbeck, T.J. Rinke, C. Berge, R.B. Bergmann, Application of plasma silicon nitride to crystalline thin-film silicon solar cells. Proceedings of the 17th European Photovoltaic Solar Energy Conference, Munich, 2001, pp. 1351–1354.
[59] K.A. Münzer, K.H. Eisenrith, R.E. Schlosser, M.G. Winstel, 18% PEBSCO—Silicon solar cells for manufacturing. Proceedings of the 17th European Photovoltaic Solar Energy Conference, Munich, 2001, pp. 1363–1366.
[60] H. Tayanaka, A. Nagasawa, N. Hiroshimaya, K. Sato, Y. Haraguchi, T. Matsushita, Effects of crystal defects in single-crystalline silicon thin-film solar cell. Proceedings of the 17th European Photovoltaic Solar Energy Conference, Munich, 2001, pp. 1400–1403.
[61] B. Terheiden, B. Fischer, P. Fath, E. Bucher, Highly efficient mechanically V-textured silicon solar cells applying a novel shallow angle contacting scheme. Proceedings of the 17th European Photovoltaic Solar Energy Conference, Munich, 2001, pp. 1331–1334.
[62] E. Schneiderlöchner, R. Preu, R. Lüodemann, S.W. Glunz, G. Willeke, Laser-fired contacts. Proceedings of the 17th European Photovoltaic Solar Energy Conference, Munich, 2001, pp. 1303–1306.
[63] S.W. Glunz, J. Dicker, D. Kray, J.Y. Lee, R. Preu, S. Rein, et al., High efficiency cell structures for medium-quality silicon. Proceedings of the 17th European Photovoltaic Solar Energy Conference, Munich, 2001, pp. 1286–1292.
[64] K. Yamamoto, M. Yoshimi, T. Suzuki, Y. Okamoto, Y. Tawada, A. Nakajima, Thin film poly-si solar cell with star structure on glass substrate fabricated at low temperature. Proceedings of the Twenty-sixth IEEE Photovoltaic Specialists Conference, Anaheim, 1997, pp. 575–580.
[65] K. Yamamoto, M. Yoshimi, Y. Tawada, Y. Okamoto, A. Nakajima, Cost effective and high performance thin film si solar cell towards the twenty-first century. Technical Digest International PVSEC-11, Sapporo, 1999, pp. 225–228.
[66] S. Golay, J. Meier, S. Dubail, S. Fay, U.Kroll, A. Shah, 2000. First pin/pin micromorph modules by laser patterning. Proceedings of the 28th IEEE Photovoltaic Specialists Conference, Anchorage, 2000, pp. 1456–1459.
[67] J. Meier, E. Vallat-Sauvain, S. Dubail, U. Kroll, J. Dubail, S. Golay, et al., Microcrystalline silicon thin-film solar cells by the VHF-GD technique. Technical Digest International PVSEC-11, Sapporo, 1999, pp. 221–223.
[68] A. Shah, J. Meier, P. Torres, U. Kroll, D. Fischer, N. Beck, et al., Recent progress on microcrystalline solar cells. Proceedings of the 26th IEEE Photovoltaic Specialists Conference, Anaheim, 1997, pp. 569–574.

[69] J.A. Rand, P.A. Basore, Light-trapping silicon solar cells experimental results and analysis. Proceedings of the 22nd IEEE Photovoltaic Specialists Conference, Las Vegas, 1991, pp. 192−197.

[70] J.A. Rand, D.H. Ford, C. Bacon, A.E. Ingram, T.R. Ruffins, R.B. Hall, et al., Silicon-film product II: initial light trapping results. Proceedings of the 10th European Photovoltaic Solar Energy Conference, Lisbon, 1991, pp. 306−309.

[71] P.A. Basore, Extended spectral analysis of internal quantum efficiency. Proceedings of the 23rd IEEE Photovoltaic Specialists Conference, Louisville, 1993, pp. 147−152.

[72] K. Yamamoto, T. Suzuki, M. Yoshimi, A. Nakajima, Low temperature fabrication of thin film polycrystalline si solar cell on the glass substrate and its application to the a-Si:h/polycrystalline Si tandem solar cell. Proceedings of the 25th IEEE Photovoltaic Specialists Conference, Washington DC, 1996, pp. 661−664.

[73] A. Rothwarf, Crystallite size considerations in polycrystalline solar cells. Proceedings of the 12th IEEE Photovoltaic Specialists Conference, Baton Rouge, 1976, pp. 488−495.

[74] A.R. Mokashi, T. Daud, A.H. Kachare, Simulation analysis of a novel high efficiency silicon solar cell. Proceedings of the 18th IEEE Photovoltaic Specialists Conference, Las Vegas, 1985, pp. 573−577.

[75] J.B. Milstein, Y.S. Tsuo, R.W. Hardy, T. Surek, The influence of grain boundaries on solar cell performance. Proceedings of the 15th IEEE Photovoltaic Specialists Conference, Orlando, 1981, pp. 1399−1404.

[76] G. Beaucarne, S. Bourdais, A. Slaoui, J. Poortmans, Carrier collection in fine-grained p−n junction polysilicon solar cells. Proceedings of the 28th IEEE Photovoltaic Specialists Conference, Anchorage, 2000, pp. 128−133.

[77] L.L. Kazmerski, Silicon grain boundaries: correlated chemical and electro-optical characterization. Proceedings of the 17th IEEE Photovoltaic Specialists Conference, Orlando, 1984, pp. 379−385.

[78] A.M. Barnett, J.A. Rand, F.A. Domian, D.H. Ford, C.L. Kendall, M.L. Rock, et al., Efficient thin silicon-film solar cells on low-cost substrate. Proceedings of the 8th European Photovoltaic Energy Conference, Florence, 1988, pp. 149−155.

[79] D.H. Ford, J.A. Rand, A.M. Barnett, E.J. DelleDonne, A.E. Ingram, R.B. Hall, Development of light-trapped, interconnected, silicon-film modules. Proceedings of the 26th IEEE Photovoltaic Specialists Conference, Anaheim, 1997, pp. 631−634.

[80] M. Silier, A. Konuma, E. Gutjahr, F. Bauser, C. Banhart, V. Zizler, et al., High-quality polycrystalline silicon layers grown on dissimilar substrates from metallic solution, I. Proceedings of the 25th IEEE Photovoltaic Specialists Conference, Washington, DC, 1996, pp. 681−684.

[81] R. Bergmann, J. Kühnle, J.H. Werner, S. Oelting, M. Albrecht, H.P. Strunk, et al., Polycrystalline silicon for thin film solar cells. Proceedings of the 24th IEEE Photovoltaic Specialists Conference, Hawaii, 1994, pp. 1398−1401.

[82] G. Andrä, J. Bergmann, E. Ose, M. Schmidt, N.D. Sinh, F. Falk, Multicrystalline LLC-Silicon thin film cells on glass. Proceedings of the 29th IEEE Photovoltaic Specialists Conference, New Orleans, 2002, pp. 1306−1309.

[83] J.D. Zook, S.B. Shuldt, R.B. Maciolek, J.D. Heaps, Growth, evaluation and modeling of silicon-on-ceramic solar cells. Proceedings of the 13th IEEE Photovoltaic Specialists Conference, Washington DC, 1978, pp. 472−478.

[84] J.D. Heaps, S.B. Schuldt, B.L. Grung, J.D. Zook, C.D. Butter, Continuous coating of silicon-on-ceramic. Proceedings of the 14th IEEE Photovoltaic Specialists Conference, San Diego, 1980, pp. 39−48.

[85] S. Minagawa, T. Saitoh, T. Warabisako, N. Nakamura, H. Itoh, T. Tokuyama, Fabrication and characterization of solar cells using dendritic silicon thin films grown on alumina ceramic. Proceedings of the 12th IEEE Photovoltaic Specialists Conference, Baton Rouge, 1976, pp. 77−81.

[86] A.M. Barnett, D.A. Fardig, R.B. Hall, J.A. Rand, D.H. Ford, Development of thin silicon-film solar cells on low-cost substrates. Proceedings of the 9th IEEE Photovoltaics Specialists Conference, New Orleans, 1987, pp. 1266−1270.

[87] J.A.M. van Roosmalen, C.J.J. Tool, R.C. Huiberts, R.J.G. Beenen, J.P.P. Huijsmans, W.C. Sinke, Ceramic substrates for thin-film crystalline silicon solar cells. Proceedings of the 25th IEEE Photovoltaic Specialists Conference, Washington DC, 1996, pp. 657−660.

[88] S.B. Shuldt, J.D. Heaps, F.M. Schmit, J.D. Zook, B.L. Grung, Large area silicon-on-ceramic substrates for low cost solar cells. Proceedings of the 15th IEEE Photovoltaic Specialists Conference, Orlando, 1981, pp. 934−940.

[89] A. Slaoui, M. Rusu, A. Fosca, R. Torrecillas, E. Alvarez, A. Gutjar, Investigation of barrier layers on ceramics for silicon thin film solar cells. Proceedings of the 29th IEEE Photovoltaic Specialists Conference, New Orleans, 2002, pp. 90−93.

[90] E. DelleDonne, A. Ingram, R. Jonczyk, J. Yaskoff, P. Sims, J. Rand, et al., Thin silicon-on-ceramic solar cells. Proceedings of the 29th IEEE Photovoltaic Specialists Conference, New Orleans, 2002, pp. 82−85.

[91] A.M. Barnett, M.G. Mauk, J.C. Zolper, R.B. Hall, J.B. McNeely, Thin-film silicon and gaAs solar cells on metal and glass substrates. Technical Digest International PVSEC-1, Kobe, 1984, pp. 241−244.

[92] A.M. Barnett, M.G. Mauk, R.B. Hall, D.A. Fardig, J.B. McNeely, Design and development of efficient thin-film crystalline silicon solar cells on steel substrates. Proceedings of the 6th European Photovoltaic Solar Energy Conference, London, 1985, pp. 866−870.

[93] A.M. Barnett, R.B. Hall, D.A. Fardig, J.S. Culik, Silicon-film solar cells on steel substrates. Proceedings of the 18th IEEE Photovoltaics Specialists Conference, Las Vegas, 1985, pp. 1094−1099.

[94] T. Kunze, S. Hauttmann, J. Seekamp, J. Müller, Recrystallized and epitaxially thickened poly-silicon layers on graphite substrates. Proceedings of the 26th IEEE Photovoltaic Specialists Conference, Anaheim, 1997, pp. 735−738.

[95] T.L. Chu, H.C. Mollenkopf, K.N. Singh, S.S. Chu, I.C. Wu, Polycrystalline silicon solar cells for terrestrial applications. Proceedings of the 11th IEEE Photovoltaic Specialists Conference, Scottsdale, 1975, pp. 303−305.

[96] M. Merber, M. Bettini, E. Gornik, Large grain polycrystalline silicon films on graphite for solar cell applications. Proceedings of the 17th IEEE Photovoltaic Specialists Conference, Orlando, 1984, pp. 275−280.

[97] M. Pauli, T. Reindl, W. Krühler, F. Homberg, J. Müller, A new fabrication method for multicrystalline silicon layers on graphite substrates suited for low-cost thin film solar Cells. Proceedings of the 24th IEEE Photovoltaic Specialists Conference, Hawaii, 1994, pp. 1387−1390.

[98] A.Z. Lin, Z.Q. Fan, H.Y. Sheng, X.W. Zhao, Thin-film polycrystalline silicon solar cell. Proceedings of the 16th IEEE Photovoltaic Specialists Conference, San Diego, 1982, pp. 140−145.

[99] R. Lüdemann, S. Schaefer, C. Schüle, C. Hebling, Dry processing of mc-silicon thin-film solar cells on foreign substrates leading to 11% efficiency. Proceedings of the 26th IEEE Photovoltaic Specialists Conference, Anaheim, 1997, p. 159

[100] T.L. Chu, S.S. Chu, K.Y. Duh, H.I. Yoo, Silicon solar cells on metallurgical silicon substrates. Proceedings of the 12th IEEE Photovoltaic Specialists Conference Baton Rouge, 1976, pp. 74−78.

[101] T.L. Chu, S.S. Chu, E.D. Stokes, C.L. Lin, R. Abderrassoul, Thin film polycrystalline silicon solar cells. Proceedings of the 13th IEEE Photovoltaic Specialists Conference, Washington DC, 1978, pp. 1106−1110.

[102] J. Hötzel, K. Peter, R. Kopecek, P. Fath, E. Bucher, C. Zahedi, Characterization of LPE Thin film silicon on low cost silicon substrates. Proceedings of the 28th IEEE Photovoltaic Specialists Conference, Anchorage, 2000, p. 225.

[103] T.L. Chu, E.D. Stokes, S.S. Chu, R. Abderrassoul, Chemical and structural defects in thin film polycrystalline silicon solar cells. Proceedings of the 14th IEEE Photovoltaic Specialists Conference, San Diego, 1980, pp. 224—227.

[104] C. Belouet, C. Hervo, M. Mautref, C. Pages, J. Hervo, Achievement and properties of self-supporting polysilicon solar cells made from RAD Ribbons. Proceedings of the 16th IEEE Photovoltaic Specialists Conference, San Diego, 1982, pp. 80—85.

[105] J. Heemeier, M. Rostalsky, F. Gromball, N. Linke, J. Müller, Thin film technology for electron beam crystallized silicon solar cells on low cost substrates. Proceedings of the 29th IEEE Photovoltaic Specialists Conference, New Orleans, 2002, pp. 1310—1313.

[106] M. Deguchi, H. Morikawa, T. Itagaki, T. Ishihara, H. Namizaki, Large grain thin film polycrystalline silicon solar cells using zone melting recrystallization. Proceedings of the 22nd IEEE Photovoltaic Specialists Conference, Las Vegas, 1991, pp. 986—991.

[107] A. Takami, S. Arimoto, H. Naomoto, S. Hamamoto, T. Ishihara, H. Kumabe, et al., Thickness dependence of defect density in thin film polycrystalline silicon formed on insulator by zone-melting recrystallization. Proceedings of the 24th IEEE Photovoltaic Specialists Conference, Hawaii, 1994, pp. 1394—1397.

[108] T. Reindl, W. Krühler, M. Pauli, J. Müller, Electrical and structural properties of the Si/C interface in Poly-Si thin films on graphite substrates. Proceedings of the 24th IEEE Photovoltaic Specialists Conference, Hawaii, 1994, pp. 1406—1409.

[109] Y. Kawama, A. Takami, H. Naomoto, S. Hamamoto, T. Ishihara, In-situ control in zone-melting recrystallization process for formation of high-quality thin film polycrystalline Si. Proceedings of the 25th IEEE Photovoltaic Specialists Conference, Washington DC, 1996, pp. 481—484.

[110] R.N. Tauber, S. Wolf, Silicon Processing for the VLSI Era, Volume 1—Process Technology, second ed., Lattice Press, Sunset Beach, 2000.

[111] W. Kern, V. Ban, chemical vapor deposition of inorganic thin films, in: J.L. Vossen, W. Kern (Eds.), Thin Film Processes, Academic Press, New York, 1978, pp. 257—331.

[112] M. Hammond, Introduction to chemical vapor deposition, Solid State Tech. December (1979) 61.

[113] F.R. Faller, V. Henninger, A. Hurrle, N. Schillinger, Optimization of the CVD Process for low cost crystalline silicon thin film solar cells. Proceedings of the 2nd World Conference on Photovoltaic Solar Energy Conversion, Vienna, 1998, pp. 1278—1283.

[114] B.F. Wagner, Ch. Schetter, O.V. Sulima, A. Bett, 15.9% Efficiency for Si thin film concentrator solar cell grown by LPE. Proceedings of the 23rd IEEE Photovoltaic Specialists Conference, Louisville, 1993, pp. 356—359.

[115] B. Thomas, G. Müller, P.-M. Wilde, H. Wawra, Properties of silicon thin films grown by the temperature difference method (TDM). Proceedings of the 26th IEEE Photovoltaic Specialists Conference, Anaheim, 1997, pp. 771—774.

[116] M. Ida, K. Hane, T. Uematsu, T.Saitoh, Y. Hayashi, A novel design for very-thin, high efficiency silicon solar cells with a new light trapping structure. Technical Digest PVSEC-4, Sydney, 1989, pp. 827—831.

[117] R. Hezel, R. Ziegler, Ultrathin self-supporting crystalline silicon solar cells with light trapping. Proceedings of the 23rd IEEE Photovoltaic Specialists Conference, Louisville, 1993, pp. 260—264.

[118] T. Markvart, Solar Electricity, second ed., John Wiley & Sons, Chichester, 2000.

[119] J. Schmidt, L. Oberbeck, T.J. Rinke, C.Berge, R.B. Bergmann, Application of plasma silicon nitride to crystalline thin-film silicon solar cells. Proceedings of the 17th European Photovoltaic Solar Energy Conference, Munich, 2001, p. 1351.
[120] T.J. Rinke, G. Hanna, K. Orgassa, H.W. Schock, J.H. Werner, Novel self-aligning series-interconnection technology for thin film solar modules. Proceedings of the 17th European Solar Energy Conference, Munich, 2001, p. 474.
[121] K.J. Weber, K. Catchpole, M.Stocks, A.W. Blakers, Lift-off of silicon epitaxial layers for solar cell applications. Proceedings of the 26th IEEE Photovoltaic Specialists Conference, Anaheim, 1997, p. 474.
[122] K.R. Catchpole, M.J. McCann, K.J. Weber, A.W. Blakers, A review of thin-film crystalline silicon for solar cell applications. Part 2: Foreign substrates, Sol. Energy Mater. Sol. Cells 68 (2001) 173–215.
[123] D. Angermeier, R. Monna, A. Slaoui, J.C. Muller, C.J. Tool, J.A. Roosmalen, et al., Analysis of silicon thin films on dissimilar substrates deposited by RTCVD for photovoltaic application. Proceedings of the 14th European Photovoltaic Solar Energy Conference, Barcelona, 1997, p. 1452.
[124] A. Slaoui, R. Monna, D. Angermeier, S.Bourdias, J.C. Muller, Polycrystalline silicon films on foreign substrates by a rapid thermal-CVD technique. Proceedings of the 26th IEEE Photovoltaic Specialist Conference, Anaheim, 1997, p. 627.
[125] S. Bourdais, R. Monna, D. Angermeier, A. Slaoui, N. Rauf, A. Laugier, et al., Combination of RT-CVD and LPE for thin silicon-film formation on alumina substrates. Proceedings of the 2nd World Conference on Photovoltaic Solar Energy Conversion, Vienna, 1998, pp. 1774–1777.
[126] G. Beaucarne, C. Hebling, R. Scheer, J. Poortmans, Thin silicon solar cells based on re-crystallized layers on insulating substrates. Proceedings of the 2nd World Conference on Photovoltaic Solar Energy Conversion, Vienna, 1998, p. 1794.
[127] G. Beaucarne, J. Poortmans, M. Caymax, J.Nijs, R. Mertens, CVD-Growth of crystalline Si on amorphous or microcrystalline substrates. Proceedings of the 14th European Photovoltaic Solar Energy Conference, Barcelona, 1997, p. 1007.
[128] T. Takahashi, R. Shimokawa, Y. Matsumoto, K. Ishii, T. Sekigawa, Sol. Energy Mater. Sol. Cells 48 (1997) 327.
[129] R. Shimokawa, K. Ishii, H. Nishikawa, T. Takahashi, Y. Hayashi, I. Saito, et al., Sol. Energy Mater. Sol. Cells 34 (1994) 277.
[130] M.E. Nell, A. Braun, B. von Ehrenwell, C.Schmidt, L. Elstner, Solar cells from thin silicon layers on Al_2O_3. Technical Digest International PVSEC-11, Sapporo, 1999, p. 749–750.
[131] D. Angermeier, R. Monna, S. Bourdais, A. Slaoui, J.C. Muller, G. Beaucarne, et al., Thin polysilicon films on mullite substrates for photovoltaic cell application. Proceedings of the 2nd World Conference on Photovoltaic Solar Energy Conversion, Vienna, 1998, p. 1778.
[132] A. Gutjahr, I. Silier, G. Cristiani, M. Konuma, F. Banhart, V. Schöllkopf, et al., Silicon solar cell structure grown by liquid epitaxy on glass carbon. Proceedings of the 14th European Photovoltaic Solar Energy Conference, Barcelona, 1997, p. 1460.
[133] M. Pauli, T. Reindl, W. Krühler, F. Homberg, J. Müller, Sol. Energy Mater. Sol. Cells 41/42 (1996) 119.
[134] R. Ludemann, S. Schaefer, C. Schule, C. Hebling, Dry processing of mc-Silicon thin-film solar cells on foreign substrates leading to 11% efficiency. Proceedings of the 26th IEEE Photovoltaic Specialists Conference, Anaheim, 1997, p. 159.
[135] T. Mishima, Y. Kitagawa, S. Ito, T. Yokoyama, Polycrystalline silicon films for solar cells by liquid phase epitaxy. Proceedings of the 2nd World Conference on Photovoltaic Solar Energy Conversion. Vienna, 1998, p. 1724.

[136] S. Ito, Y. Kitagawa, T.Mishima, T. Yokoyama, Direct-grown polycrystalline Si film on carbon substrate by LPE. Technical Digest International PVSEC-11, Sapporo, 1999, pp. 539–540.

[137] R. Monna, D. Angermeier, A. Slaoui, J.C. Muller, G. Beaucarne, J. Poortmans, et al., Poly-Si thin films on graphite substrates by rapid thermal chemical vapor deposition for photovoltaic application. Proceedings of the 14th European Photovoltaic Solar Energy Conference, Barcelona, 1997, p. 1456.

[138] H.V. Campe, D. Nikl, W. Schmidt, F. Schomann, Crystalline silicon thin film solar cells. Proceedings of the 13th European Photovoltaic Solar Energy Conference, Nice, 1995, p. 1489.

[139] T. Kunze, S. Hauttmann, S. Kramp, J. Muller, Thin recrystallized silicon seed layers on graphite substrates. Proceedings of the 14th European Photovoltaic Solar Energy Conference, Barcelona, 1997, p. 1407.

[140] S.E. Schiermeier, C.J. Tool, J.A. van Roosmalen, L.J. Laas, A.von Keitz, W.C. Sinke, 1998. LPE-growth of crystalline silicon layers on ceramic substrates. Proceedings of the 2nd World Conference on Photovoltaic Solar Energy Conversion, Vienna, 1998, p. 1673.

[141] A. Shi, T.L. Young, G.F. Zheng, M.A. Green, Sol. Energy Mater. Sol. Cells 31 (1993) 51.

[142] Z. Shi, T.L. Young, M.A. Green, Solution growth of polycrystalline silicon on glass at low temperatures. Proceedings of the First World Conference on Photovoltaic Energy Conversion, Hawaii, 1994, p. 1579.

[143] R.B. Bergmann, B. Brendel, M. Wolf, P. Lölgen, J.H. Werner, High rate, low temperature deposition of crystalline silicon film solar cells on glass. Proceedings of the 2nd World Conference on Photovoltaic Solar Energy Conversion, Vienna, 1998, pp. 1260–1265.

[144] R. Brendel, R.B. Bergmann, B. Fischer, J. Krinke, R. Plieninger, U. Rau, et al., Transport analysis for polycrystalline silicon solar cells on glass substrates. Proceedings of the 26th Photovoltaic Solar Conference, Anaheim, 1997, p. 635.

[145] E. DelleDonne, A. Ingram, R. Jonczyk, J. Yaskoff, P. Sims, J. Rand, et al., Thin silicon-on-ceramic solar cells. Proceedings of the 29th IEEE Photovoltaic Specialists Conference, New Orleans, 2002, p. 82.

[146] M.J. McCann, K.R. Catchpole, K.J. Weber, A.W. Blakers, A review of thin-film crystalline silicon for solar cell applications. Part 1: native substrates, Sol. Energy Mater. Sol. Cells 68 (2001) 135–171.

[147] A.W. Blakers, K.J. Weber, M.F. Stuckings, S. Armand, G. Matlakowski, M.J. Stocks, et al., 18% Efficient thin silicon solar cell by liquid phase epitaxy. Proceedings of the 13th European Photovoltaic Solar Energy Conference, Nice, 1995, p. 33.

[148] G.F. Zheng, S.R. Wenham, M.A. Green, Prog. Photovoltaics 4 (1996) 369.

[149] F.R. Faller, V. Henninger, A. Hurrle, N. Schillinger, Optimization of the CVD Process for low-cost crystalline-silicon thin-film solar cells. Proceedings of the 2nd World Conference on Photovoltaic Solar Energy Conversion, Vienna, 1998, p. 1278.

[150] J.H. Werner, J.K. Arch, R. Brendel, G. Langguth, M. Konuma, E. Bauser, et al., Crystalline thin film silicon solar cells. Proceedings of the 12th European Photovoltaic Solar Energy Conference, Amsterdam, 1994, pp. 1823–1826.

[151] A.W. Blakers, K.J. Weber, M.F. Stuckings, S. Armand, G. Matlakowski, A.J. Carr, et al., Prog. Photovoltaics Vol. 3 (1995) 193.

[152] G.F. Zheng, W. Zhang, Z. Shi, M. Gross, A.B. Sproul, S.R. Wenham, et al., Sol. Energy Mater. Sol. Cells 40 (1996) 231.

[153] H.V. Campe, D. Nikl, W.Schmidt, F. Schomann, Crystalline silicon thin film solar cells. Proceedings of the 13th European Photovoltaic Solar Energy Conference, Nice, 1995, p. 1489.
[154] O. Evrard, E. Demesmaeker, T. Vermeulen, M. Zagrebnov, M. Caymax, W. Laureys, et al., The analysis of the limiting recombination mechanisms on high efficiency thin film cells grown with CVD epitaxy. Proceedings of the 13th European Photovoltaic Solar Energy Conference, Nice, 1995, p. 440.
[155] J.H. Werner, S. Kolodinski, U. Rau, J.K. Arch, E. Bauser, Appl. Phys. Lett. 62 (1993) 2998.
[156] W. Wang, Y. Zhao, Y. Xu, X. Luo, M. Yu, Y. Yu, The polycrystalline silicon thin film solar cells deposited on SiO_2 and Si_3N_4 by RTCVD. Proceedings of the 2nd World Conference on Photovoltaic Solar Energy Conversion, Vienna, 1998, p. 1740.
[157] G. Ballhorn, K.J. Weber, S. Armand, M.J. Stocks, A.W. Blakers, High efficiency thin multicrystalline silicon solar cells by liquid phase epitaxy. Proceedings of the 14th European Photovoltaic Solar Energy Conference, Barcelona, 1997, p. 1011.
[158] T. Vermeulen, J. Poortmans, M. Caymax, J. Nijs, R. Mertens, C. Vinckier, 1997. The role of hydrogen passivation in 20 μm thin-film solar cells on p^+ Multicrystalline-Si substrates. Proceedings of the 14th European Photovoltaic Solar Energy Conference, Barcelona, 1997, p. 728.
[159] T. Vermeulen, F. Deurinckx, K. DeClercq, J. Szlufcik, J. Poortmans, P. Laermans, et al., Cost-effective thin film solar cell processing on multicrystalline silicon. Proceedings of the 26th IEEE Photovoltaic Specialists Conference, Anaheim, 1997, p. 267.
[160] T. Vermeulen, O. Evrard, W. Laureys, J. Poortmans, M. Caymax, J. Nijs, et al., Realization of thin film solar cells in epitaxial layers grown on highly doped RGS-Ribbons. Proceedings of the 13th European Photovoltaic Solar Energy Conference, Nice, 1995, p. 1501.
[161] F.R. Faller, N. Schillinger, A. Hurrle, C. Schetter, Improvement and characterization of Mi-Si thin-film solar cells on low cost SSP Ribbons. Proceedings of the 14th European Photovoltaic Solar Energy Conference, Barcelona, 1997, p. 784.
[162] T. Vermeulen, J. Poortmans, K. Said, O. Evrard, W. Laureys, M. Caymax, et al., Interaction between bulk and surface passivation mechanisms in thin film solar cells on defected silicon substrates. Proceedings of the 25th IEEE Photovoltaic Specialists Conference, Washington, DC, 1996, p. 653.
[163] B. Thomas, G. Muller, P.Heidborn, H. Wartra, Growth of polycrystalline silicon thin films using the temperature difference method. Proceedings of the 14th European Photovoltaic Solar Energy Conference, Barcelona, 1997, p. 1483.
[164] T.H. Wang, T.F. Ciszek, C.R. Schwerdtfeger, H. Moutinho, R. Matson, Sol. Energy Mater. Sol. Cells 41–42 (1996) 19.
[165] T.F. Ciszek, J.M. Gee, Crystalline silicon R&D at the US national center for photovoltaics. Proceedings of the 14th European Photovoltaic Solar Energy Conference, Barcelona, 1997, p. 53.
[166] T. Ishihara, S. Arimoto, H. Kumabe, T. Murotani, Progr. Photovoltaics 3 (1995) 105.
[167] W. Zimmerman, S. Bau, F. Haas, K. Schmidt, A. Eyer, Silicon sheets from powder as low cost substrates for crystalline silicon thin film solar cells. Proceedings of the 2nd World Conference on Photovoltaic Solar Energy Conversion. Vienna, 1998, p. 1790.
[168] G. Beaucarne, C. Hebling, R. Scheer, J. Poortmans, Thin silicon solar cells based on recrystallized layers on insulating substrates. Proceedings of the 2nd World Conference on Photovoltaic Solar Energy Conversion, Vienna, 1998, p. 1794.

[169] C. Hebling, R. Gaffke, P. Lanyi, H. Lautenschlager, C. Schetter, B. Wagner, et al., Recrystallized silicon on SiO_2-Layers for thin-film solar cells. Proceedings of the 25th IEEE Photovoltaic Specialists Conference, Washington, DC, 1996, p. 649.

[170] A.J. van Zutphen, M. Zeman, F.D. Tichelaar, J.W. Metselaar, Deposition of thin film silicon by thermal CVD Processes for. 1997.

PART IC

Thin Film Technologies

CHAPTER IC-1

Thin-Film Silicon Solar Cells[1]

Arvind Shah
with the collaboration of Horst Schade

Contents

1. Introduction — 210
   1.1 Tandem and Multijunction Solar Cells — 214
2. Hydrogenated Amorphous Silicon (a-Si:H) Layers — 215
   2.1 Structure of Amorphous Silicon — 215
   2.2 Gap States in Amorphous Silicon: Mobility Gap and Optical Gap — 220
   2.3 Conductivity and Doping of Amorphous Silicon — 223
     2.3.1 Conductivities — 223
     2.3.2 Doping — 224
3. Hydrogenated Microcrystalline Silicon (μc-Si:H) Layers — 225
   3.1 Structure of Microcrystalline Silicon — 225
   3.2 Optical Absorption, Gap States, and Defects in Microcrystalline Silicon — 228
   3.3 Conductivities, Doping, Impurities, and Ageing in Microcrystalline Silicon — 232
     3.3.1 Conductivities — 232
     3.3.2 Doping — 233
     3.3.3 Impurities — 233
     3.3.4 Ageing — 233
4. Functioning of Thin-Film Silicon Solar Cells with p–i–n and n–i–p Structures — 234
   4.1 Role of the Internal Electric Field — 234
     4.1.1 Formation of the Internal Electric Field in the i Layer — 238
     4.1.2 Reduction and Deformation of the Internal Electric Field in the i Layer — 241
   4.2 Recombination and Collection — 241
   4.3 Shunts — 244
   4.4 Series Resistance Problems — 248
   4.5 Light Trapping — 249
5. Tandem and Multijunction Solar Cells — 252
   5.1 General Principles — 252
   5.2 a-Si:H/a-Si:H Tandems — 254
   5.3 Triple-Junction Amorphous Cells with Silicon–Germanium Alloys — 255
   5.4 Microcrystalline–Amorphous or "Micromorph" Tandems — 257
6. Module Production and Performance — 259

---

[1] The present chapter is partly an excerpt from the book *Thin-Film Silicon Solar Cells*, edited by Arvind Shah and published in 2010 by the EPFL Press, Lausanne [1], with contributions by Horst Schade and Friedhelm Finger. For further specialized study and for details, the reader is referred to this book.

6.1	Deposition of the Thin-Film Silicon Layers	259
6.2	Substrate Materials and Transparent Contacts	263
6.3	Laser Scribing and Cell Interconnection	267
6.4	Module Encapsulation	269
6.5	Module Performance	270
6.6	Field Experience	272
7. Conclusions		273
References		274

1. INTRODUCTION

Silicon thin films for solar cells are at present predominantly deposited by plasma-enhanced chemical vapour deposition (PECVD) either from silane (SiH_4) or preferably from a mixture of silane and hydrogen. They are either amorphous or microcrystalline. They contain about 5% to 15% of hydrogen atoms. The hydrogen atoms are essential, as they passivate a large part of the inherent defects in these semiconductor films.

Amorphous silicon thin films were first deposited by PECVD by R.C. Chittick et al. [2]; this work was continued in a systematic manner by Walter Spear and Peter Le Comber and their research group at the University of Dundee in the 1970s. In a landmark paper published in 1975 [3] (see also [4]), they demonstrated that amorphous silicon layers deposited from silane by PECVD could be doped by adding to the plasma discharge either phosphine (PH_3) to form n-type layers or diborane (B_2H_6) to form p-type layers: They showed that the conductivity of these thin amorphous silicon layers (which contain about 10% to 15% hydrogen) could be increased by several orders of magnitude. Their pioneering work made it possible to use *hydrogenated amorphous silicon* (a-Si:H) to fabricate diodes and thin-film transistors, which can be used for the active addressing matrix in liquid crystal displays. It was Dave Carlson and Chris Wronski who fabricated the first amorphous silicon solar cells at the RCA Laboratories; the first publication in 1976 described cells with an efficiency of 2% [5], this value being increased to 5% within the same year [6]. A year later, Staebler and Wronski reported [7] on a reversible photodegradation process that occurs within amorphous silicon solar cells when the latter are exposed to light during long periods (tens to hundreds of hours). This effect is called the *Staebler–Wronski effect* (SWE), and it is a major limitation of amorphous silicon for solar cell technology. It is due to an increase of midgap defects, which act as recombination centres. It is a reversible effect: the initial, nondegraded state can be restored by annealing at 150°C for several hours. By affecting the quality of the photoactive layer within the cell, the SWE causes the efficiency of amorphous silicon solar cells to decrease during the first months of operation. After about a thousand hours of operation, the efficiency more or less stabilizes at a lower value. This is why it is important to always specify stabilized efficiencies for amorphous silicon solar cells. In the initial phase of amorphous silicon solar cell development, it was hoped to overcome this degradation effect. So far, nobody has succeeded in fabricating amorphous silicon layers that do not show any photodegradation. However, by adding hydrogen to silane during the plasma deposition of the layers, and by increasing the deposition temperature, the photodegradation can be somewhat reduced. Furthermore, by keeping the solar cells very thin (i-layer thickness below 300 nm), one can reduce the *impact* of the Staebler–Wronski effect on the cell's efficiency. An important feature of amorphous silicon solar cells, introduced also by Carlson and Wronski, is that one does not use the classical structure of a p–n diode, as in almost all other solar cells, but one uses a *p–i–n diode*, keeping the doped layers (p- and n-type layers) very thin and employing the i layer (i.e., an intrinsic or undoped layer) as the photogeneration layer, where the light is mainly absorbed and its energy transferred to the charge carriers (holes and electrons). There are two reasons for this: (1) the electronic quality of doped amorphous layers is very poor; they have a very high density of midgap defects or recombination centres, so that practically all carriers, which are photogenerated within the doped layers are lost through recombination; (2) within the whole i layer of a p–i–n diode an internal electric field is created that separates the photogenerated electrons and holes and helps in collecting them in the n and p layers, respectively. The internal electric field is absolutely essential for the functioning of an amorphous silicon solar cell—without this field most of the photogenerated carriers would not be collected, and, thus, the cell's performance would be totally unsatisfactory. The theory of p–i–n diodes has not been studied to the same extent as that of classical p–n diodes, and further work is clearly called for.

Amorphous silicon solar cells at first found only "niche" applications, especially as the power source for electronic calculators. For 15 years or so, they have been increasingly used for electricity generation: they seem particularly well suited for wide applications in building-integrated photovoltaics (BIPV). One of their main advantages is that they are available in the form of monolithically integrated large-area modules (and even as flexible modules based on stainless steel or polymer substrates). Another significant advantage is that their temperature coefficient is only −0.2%/°C—i.e., less than half of that prevailing in wafer-based crystalline silicon solar cells. At present, single-junction amorphous silicon solar cells attain in the laboratory stabilized efficiencies of more than 10% [8], whereas single-junction commercial modules have stabilized total-area efficiencies between 6% and 7%.

Microcrystalline silicon thin films containing hydrogen (μc-Si:H films) were first described in detail by S. Veprek and co-workers [9], who used a chemical transport technique to fabricate them. The first report of depositing μc-Si:H films with PECVD, from a plasma of silane strongly diluted with hydrogen, was published by Usui and Kikuchi in 1979 [10]. The plasma-deposition techniques for microcrystalline silicon layers were extensively investigated during the following years—in all cases, one obtains μc-Si:H instead of a-Si:H by increasing the hydrogen-to-silane ratio in the gas fed into the plasma. The first solar cells using μc-Si:H films as photogeneration layers (i layers) were reported in the early 1990s [11–13]. In 1994, the Neuchâtel group published solar cell results with efficiencies of more than 4% and showed that these cells had virtually no photodegradation at all [14]. By reducing the oxygen contamination in the intrinsic μc-Si:H layers, the Neuchâtel group was able to enhance the efficiency of small-area laboratory cells to more than 7% in 1996 [15–16]. After that, many other research groups started optimizing microcrystalline silicon solar cells. Plasma-deposited μc-Si:H solar cells generally also use the p−i−n configuration, just like a-Si:H solar cells, although doped microcrystalline silicon layers (p- and n-type layers) have much better electronic quality (and much higher conductivities) than doped amorphous silicon layers. Such doped microcrystalline silicon layers could basically be used as photogeneration layers. However, the use of the p−i−n configuration is still necessary in order to reduce recombination and collect the charge carriers with the help of the internal electric field within the i layer, which here again plays a key role. Because of the low optical absorption coefficient of microcrystalline silicon, the i layer of μc-Si:H solar cells has to be kept relatively thick (1 to 2 μm). At present, the best single-junction microcrystalline silicon solar cells attain stabilized efficiencies in the laboratory around 10% [17].

So far, single-junction microcrystalline silicon solar cells are not used within commercial modules. Microcrystalline silicon solar cells are, however, used as "bottom cells" within tandem cells—i.e., within microcrystalline–amorphous (so-called micromorph) tandem cells (see Section 5.4). Indeed, microcrystalline silicon, like wafer-based crystalline silicon, has a band gap around 1.1 eV and can absorb light in the near infrared range and is therefore complementary to hydrogenated amorphous silicon with its band gap around 1.75 eV, which limits light absorption to the visible range of sunlight.

Hydrogenated microcrystalline silicon (μc-Si:H), as deposited by PECVD, is not a uniform, standard material; rather, it is a mixture of crystallites, amorphous regions, and what are often referred to as "voids" or "cracks" (and which are in reality low-density regions). As we increase the hydrogen dilution in the deposition plasma, we obtain layers that are more and more crystalline and have less amorphous volume fraction and an increasing fraction of voids or cracks. The solar cells with the highest open-circuit voltage V_{oc}, and also those with the highest conversion efficiency η, are fabricated with microcrystalline intrinsic layers having approximately 50% amorphous volume fraction and 50% crystalline volume fraction. These layers have a low density of cracks or voids and contain around 6% hydrogen.

When studying μc-Si:H layers, one faces the following peculiarities and difficulties: (1) growth is strongly substrate dependent; (2) if the hydrogen-to-silane dilution ratio is kept constant, the layers start growing with a relatively high amorphous volume fraction but become more and more crystalline as they become thicker (for this reason the i layer of a μc-Si:H solar cell has to be grown with a variable hydrogen-to-silane dilution ratio—e.g., see [18]); (3) μc-Si:H layers and cells are much more sensitive to oxygen and other impurities than a-Si:H layers and cells; and (4) individual μc-Si:H layers (especially those with a high crystalline volume fraction) often show, during storage, degradation effects even in the dark [19], these being probably due to adsorption of oxygen and to oxidation; nitrogen possibly also plays a role in this ageing process. These degradation effects are less pronounced in cells and can be avoided by storing the layers in vacuum or in an inert gas.

Finally, we may mention here that instead of using the term *microcrystalline*, many scientists and engineers use the term *nanocrystalline* to describe very the same layers and cells. The reason is the following: within μc-Si:H, the smallest features—i.e., the individual crystallites (grains)—have indeed *nano*metric dimensions (around 10 to 100 nm), but they are packaged together into "conglomerates" of columnar shape with dimensions often extending for more than 1 µm. One generally assumes that it is the conglomerate boundaries and not the grain boundaries that limit transport in µc-Si:H layers and collection in µc-Si:H solar cells. In state-of-the-art µc-Si:H solar cells, the columnar conglomerates will extend through the whole i layer, right from the p layer up to the n layer, and carriers can be collected without having to cross any conglomerate boundaries. Furthermore, at the conglomerate boundaries themselves, most of the defects are passivated by the amorphous regions present there. This explains why hydrogenated microcrystalline silicon solar cells generally have excellent collection properties and allow for almost perfect collection at i-layer thicknesses up to a few µm, even though the crystallites or grains themselves are indeed very small. This is an essential difference between microcrystalline silicon solar cells and classical polycrystalline (multicrystalline) silicon solar cells; the latter only function properly for grain sizes at the mm level.

1.1 Tandem and Multijunction Solar Cells

Researchers and industries working in the field of thin-film silicon solar cells have made extensive use of the tandem and multijunction concept. Various designs have been studied and commercialized; the main designs used are the following:

1. In a simple a-Si:H/a-Si:H tandem [20], both subcells of the tandem have approximately the same band gap. The advantage of the tandem concept is that the i layers of the subcells can be made thinner for the same light absorption compared to a thicker i layer in a single-junction cell. A tandem will therefore be basically less prone to light-induced degradation (i.e., to the Staebler–Wronski effect). Stabilized module efficiencies (total area) of as much as 7.1% have been obtained with this concept by the firm SCHOTT Solar Thin Film GmbH.

2. Triple-junction cells use an amorphous silicon top subcell and middle and bottom subcells based on amorphous silicon–germanium alloys: a-Si:H/a-Si,Ge:H/a-Si,Ge:H. Here the band gaps of the individual subcells are varied (through alloying with germanium) in such a way that the solar spectrum is well covered. With a corresponding laboratory cell, a record stabilized efficiency of 13.0% was achieved [21]. The firm United Solar Ovonic sells commercial modules based on this design, with stabilized total area module efficiencies in the 6% to

7% range. The advantage of these modules is that they are flexible, because the substrate material is stainless steel.
3. Micromorph (μc-Si:H/a-Si:H) tandem cells use a microcrystalline silicon bottom cell and an amorphous silicon top cell. Here the solar spectrum is ideally shared between the two subcells [22]; furthermore, the bottom μc-Si:H subcell is not subject to light-induced degradation (SWE). Commercial modules with stabilized total-area efficiencies in the 8% to 9% range are sold at present by several companies. Moreover, various industrial research laboratories have just recently announced having obtained 10% stabilized efficiency for large-area modules (see, e.g., [23]).

On the research front, many other designs for triple-junction cells are being studied. Results have been obtained for the following designs:
1. a-Si:H/a-Si:H/μc-Si:H
2. a-Si:H/a-Si,Ge:H/μc-Si:H
3. a-Si:H/μc-Si:H/μc-Si:H

At this stage, it is not clear which of these designs will be the most successful; it may well be a completely different design using a novel microcrystalline alloy for one of the subcells. Such alloys are presently being developed (e.g., see [24]).

2. HYDROGENATED AMORPHOUS SILICON (A-SI:H) LAYERS

2.1 Structure of Amorphous Silicon

Crystalline solids, such as monocrystalline silicon wafers, have a fully regular and periodic structure; they possess what is called both *short-* and *long-range order*. In such a crystalline silicon network (or crystalline silicon *matrix* as it is also called), each silicon atom is bonded to four neighbouring silicon atoms. The *bond angle*—i.e., the angle between two adjacent bonds—is fixed at a value of 109° 28′ and remains the same throughout the whole crystalline network. Figure 1 schematically shows this situation. The *bond length*, or distance between two neighbouring silicon atoms within such a network, is also fixed and remains constant throughout the whole network at a value of approximately 0.235 nm.

In amorphous silicon thin films, both the bond angles and the bond lengths vary in a random fashion: there is a whole distribution of values. For instance, the bond angles have a random distribution centred around

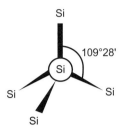

Figure 1 Atomic model for a silicon atom within a crystalline silicon network, indicating the bond angle formed between two adjacent bonds. In amorphous silicon, this angle has a distribution of values. *(Reproduced from [1] with permission of the EPFL Press.)*

109° 28′ and a standard deviation of 6° to 9°. If the amorphous silicon layer has just a low "amount of disorder," then the distributions for bond angles and bond lengths will be very narrow. In this case, we will obtain "device quality" amorphous layers with satisfactory electronic properties. If the amorphous layer has a "high amount of disorder," we will obtain broad distributions and unsatisfactory electronic properties. The disorder will directly affect the band tail states: the band tails will be more pronounced for strongly disordered layers. Amorphous silicon layers possess some amount of short-range order, the nearest atomic neighbours being in almost the same positions as would be the case for crystalline silicon. However there is no long-range order at all. Furthermore, due to the disorder prevailing in the network, about one in every 10^4 silicon atoms is unable to have four regular bonds with neighbouring silicon atoms; it has a broken or "dangling" bond as it is called (Figure 2(a)). These *dangling bonds* give rise to "midgap states" that act as recombination centres. In hydrogenated amorphous silicon, most (but not all) dangling bonds are "passivated" by a hydrogen atom (Figure 2(b)); in this case, they no longer contribute to the midgap states and do not at all act as recombination centres. The density of remaining, unpassivated dangling bonds in device-quality hydrogenated amorphous silicon (a-Si:H) is somewhere between 10^{14} and 10^{17} cm^{-3}. (The value 10^{14} dangling bonds per cm^3 refers to the bulk of the very best a-Si:H layers in the as-deposited or annealed state—i.e., before light-induced degradation or after its removal by annealing; the value 10^{17} dangling bonds per cm^3 refers to a-Si:H layers after light-induced degradation.) Under the influence of light shining on the amorphous silicon layer, a degradation effect takes place that is characterized by an increase of unpassivated dangling bonds; this is the

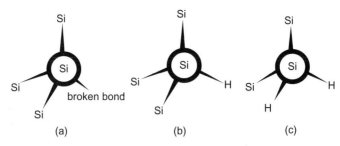

Figure 2 Atomic models showing (a) silicon atom with "broken bond" or "dangling bond," (b) silicon atom with hydrogen atom passivating what was originally a broken bond, and (c) silicon atom with two hydrogen atoms, where one or both of the H atoms can easily be separated from the Si atom under the influence of light. Such an atomic configuration contributes to a pronounced light-induced degradation effect (Staebler–Wronski effect, or SWE). *(Reproduced from [1] with permission of the EPFL Press.)*

Staebler–Wronski effect. After about 1000 hours of light exposure (with light intensity equivalent to full sunlight), the dangling bond density tends to saturate at a higher value. If there are many silicon atoms with two hydrogen atoms passivating a broken bond (so-called SiH_2 configuration, Figure 2(c)); it will be relatively easy to break the Si-H bonds and the SWE will be more pronounced. This is seen in porous layers, having a relatively high density of microvoids; a typical microvoid is schematically represented in Figure 3.

In general, one can consider that in amorphous silicon layers the microstructure is an important structural property, which often may not have much effect on the initial layer properties but may strongly influence the *light-induced degradation effect* (SWE). This is a particularly disturbing situation, because one may have amorphous silicon layers and cells with reasonably good initial properties, which show their deficiencies only after several hundred hours of light-induced degradation. During the 1980s and 1990s, a huge amount of work was undertaken to clarify, understand, and reduce the SWE (for a summary, see [25]). In spite of this tremendous effort, there is to date still no complete understanding of the SWE. Experimentally, one has made the following observations:

1. The SWE can be reduced by the use of hydrogen dilution during the plasma deposition of a-Si:H and also by increasing the deposition temperatures.
2. Layers with a high density of microvoids tend to have an enhanced SWE. (Such layers can be identified with the help of the so-called

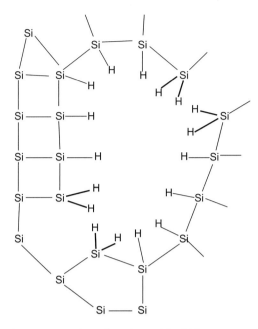

Figure 3 Schematic representation of a microvoid within an amorphous silicon layer, containing here four SiH$_2$-configurations—i.e., four silicon atoms, each bonded to two hydrogen atoms. Such SiH$_2$-configurations give very easily way to the formation of new broken bonds and thus lead to enhanced light-induced degradation. *(Reproduced from [1] with permission of the EPFL Press.)*

microstructure factor, which is evaluated from Fourier transform infrared thermography (FTIR); see [26,27].

3. Layers with a high density of certain impurities, such as oxygen and possibly nitrogen (e.g., oxygen in excess of 2×10^{19} atoms per cm^3), have an enhanced SWE.
4. Layers deposited at high deposition rates tend to exhibit enhanced SWE. For economic reasons (to obtain the highest production throughput possible), production facilities for amorphous silicon solar modules are always operated at the highest possible deposition rates; the limitation is then given by the SWE. For conventional PECVD with an excitation frequency of 13.56 MHz, deposition rates are limited to about 0.1 nm/s. Through the use of modified deposition techniques such as very high frequency (VHF) plasma deposition, with plasma excitation frequencies of 60 MHz and more, deposition rates of more than 1 nm/s have been obtained without any noteworthy increase of the SWE [28].

5. The magnitude of the SWE depends on light intensity and on the temperature of the layer during exposure to light. The higher the light intensity and the lower the temperature, the more pronounced the SWE will be.
6. Under constant illumination conditions, the SWE tends to saturate after the initial degradation phase. This tendency to saturate is more pronounced in complete silicon solar cells than in individual layers. The efficiency of these cells then stabilizes at a lower value.

In amorphous silicon solar cells, the p—i—n-configuration is used as already stated. Here the light enters into the cell generally through the p layer. For these amorphous silicon p layers, it is customary to employ *amorphous alloys of silicon and carbon* (a-Si,C:H). Such alloyed p layers have a higher band gap than do unalloyed a-Si:H layers [29]. They are used as so-called window layers: they absorb less light than unalloyed a-Si:H layers. Now all doped amorphous layers have a poor electronic quality and a very high density of recombination centres so that the light absorbed in the p layer is lost and does not contribute to the collected photocurrent. Therefore, it is of advantage for the solar cell if the p layer absorbs less light. If one increases the carbon content in the a-Si,C:H layer too much (more than approximately 40%), the gap does increase further (more than 2.1 eV), and the unwanted absorption further decreases, but the electrical conductivity of the layer decreases to values below 10^{-6} S/cm, and the layers are no longer suitable for solar cells [30].

For the photoactive i layers of tandem and multijunction solar cells, the use of *amorphous alloys of silicon and germanium* (a-Si,Ge:H) has been extensively studied. Such alloys have lower band gaps than unalloyed a-Si:H layers and allow the tuning of the spectrum absorbed in each subcell of the multijunction device to complementary parts of the solar spectrum [31–34]. However, if one increases the germanium content in the a-Si,Ge:H layer over a certain threshold (about 40% Ge content) one obtains (up to now) layers that have high-defect densities, especially in the "degraded" or "stabilized" state—i.e., after light-induced degradation. This is one reason why multijunction cells containing a-Si,Ge:H alloys have so far not led to substantial improvements in solar module efficiencies. Another disadvantage of using a-Si,Ge:H alloys is that germane is much more expensive than silane; this leads to higher overall costs for source gases, even though the utilisation ratio of germane is higher than that for silane. Furthermore, the global availability of germanium as a raw material is at present a reason for concern.

2.2 Gap States in Amorphous Silicon: Mobility Gap and Optical Gap

Classical crystalline semiconductors, such as wafer-based crystalline silicon, have a well-defined energy gap between the valence band and the conduction band. Within this energy gap, practically no electronic states can be seen (except for those due to impurities and crystal defects). In amorphous semiconductors, such as hydrogenated amorphous silicon (a-Si:H) and its alloys, there is a continuous band of states throughout and no actual band gap (Figure 4), though valence and conduction bands can still be identified. These bands have delocalized electronic states; this means that electrons in the conduction band and holes in the valence band can move about, albeit with much lower values of mobility than in the corresponding bands of (mono)-crystalline semiconductors. The highest energy level in the valence band is now given by the mobility edge

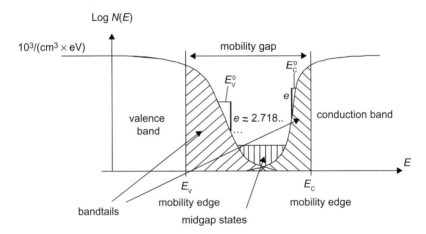

Figure 4 Density $N(E)$ of electronic states for a-Si:H layers. The states within the mobility gap—i.e., between E_V and E_C—are localized states. The states in the valence and conduction band are delocalized or extended states, which are occupied by free holes and free electrons, respectively. The states in the valence and conduction band tails act as traps. Holes trapped within the valence band tail are in constant exchange with the free holes in the valence band; for every free hole in the valence band there are 100 to 1000 trapped holes in the valence band tail. A similar behaviour (but less pronounced) applies to electrons trapped in the conduction band tail; here there are about 10 trapped electrons for each free electron. Midgap states are associated with dangling bonds; they act as recombination centres. Their density increases by two to three orders of magnitude under the influence of the Staebler–Wronski effect. *(Reproduced from [1] with permission of the EPFL Press.)*

E_V; for energies $E > E_V$, the states are localized, which means that the charge carriers are "trapped" and cannot move about freely.

For energies above E_V, we first have the *valence band tail*, where the density of states $N(E)$ decreases exponentially. Thereby, the energy constant E_V^0 denotes the energy needed for the exponential function to fall off by a factor of e ($= 2.718...$). In the best device-quality a-Si:H layers, E_V^0 is about 45 to 50 meV. E_V^0 is a measure for the disorder in the amorphous network; the higher the value of E_V^0, the higher the disorder. E_V^0 can be evaluated by measuring the optical absorption coefficient as a function of photon energy (see Figure 10). Due to the capture and recombination kinetics, states in the valence band tail do not play the role of recombination centres; rather, they act as traps for the free holes during solar cell operation. One can assume that for every free hole in the valence band (i.e., for every free hole contributing to the photocurrent in the solar cell), there are 100 to 1000 "trapped" holes in the valence band tail. These trapped holes constitute a positive charge that can deform and reduce the internal electric field within the i layer of p–i–n-type solar cells.

Now, if the light enters into a p–i–n-type solar cell from the p side, the majority of the photogenerated holes have less far to travel than if the light enters from the n side: in the first case, they are mostly generated near the p–i interface; in the second case, they are mostly generated near the n–i interface. Thus, the field deformation through trapped charge is less pronounced in a-Si:H p–i–n-type solar cells, if the light enters from the p side: this is what is generally done.

Note that the band tails are not modified by light-induced degradation (SWE).

Above the valence band tail (for higher energies—i.e., to the right in Figure 4) we have the *midgap states*, given by the dangling bonds. These states act as recombination centres and are therefore detrimental to the functioning of solar cells as they directly limit the collection of the photogenerated carriers. The density of midgap states is increased by a factor of 10^2 to 10^3 by light-induced degradation (SWE).

At higher energies we can see the *conduction band tail*, where the density of states $N(E)$ also follows an exponential law, but with an energy constant E_C^0 that is about half the value of E_V^0. As the conduction band tail is much less pronounced than the valence band tail, it does not play a great role in solar cells—the electrons trapped within the conduction band tail do not noticeably deform the electric field within the i layer of p–i–n-type solar cells. (The conduction band tail does play an important role in n-channel thin-film transistors, where the electrons are the dominant charge carriers, and the density of holes is very low and of no importance at all.)

A mobility edge E_C separates now the localized states in the conduction band tail from the states in the conduction band with its delocalized electronic states. Instead of a "true" band gap as in (mono)-crystalline semiconductors, we now have, in amorphous silicon, a *mobility gap* ($E_C - E_V$), where there are the localized gap states. It is not straightforward to determine the mobility gap. It is easier to determine the so-called optical gap, a quantity that is extrapolated from measurements of the optical absorption coefficient. The optical gap is found to be about 100 meV smaller than the mobility gap [35–36].

There are different methods used for determining the *optical gap*. The most common method is the one proposed by Tauc et al. ([37], see also [38]). The method consists of measuring the absorption coefficient α as a function of photon energy $E = h\nu = hc/e\lambda \approx 1.240 \, [eV]/(\lambda[\mu m])$, for photon energies above the band gap energy. Here ν and λ are the frequency and the wavelength of light; h is Planck's constant, c the velocity of light, e the charge of an electron (unit charge). One then plots $(\alpha(E)E)^{1/2}$ as a function of E; this plot gives us more or less a straight line (Figure 5). The intersection of this straight line with the abscissa

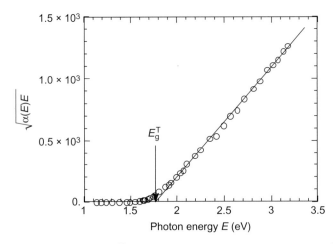

Figure 5 Typical plot of $\{\alpha(E)E\}^{1/2}$ versus photon energy E (with α in cm^{-1} and E in eV) as used for the determination of the Tauc optical gap E_g^T in amorphous silicon layers. (Reproduced from [1] with permission of the EPFL Press.)

$((\alpha(E)E)^{1/2} = 0)$ gives us the value of the *Tauc gap* E_g^T, generally used as estimate for the optical gap.

The values measured for the optical gap in a-Si:H layers are significantly higher than are the band-gap values for crystalline silicon (c-Si); they are in the range 1.6 eV to 1.85 eV, compared to 1.1 eV for c-Si. Furthermore, the band-gap values of a-Si:H layers vary according to the deposition conditions: layers deposited at higher temperatures have lower band-gap values; layers deposited with high values of hydrogen dilution have higher band-gap values, as long as they remain amorphous and do not become microcrystalline (with a substantial crystalline volume fraction—more than 20%). In fact, if one uses high hydrogen dilution values but just avoids crossing the transition from amorphous to microcrystalline silicon, one can obtain so-called protocrystalline [39] or polymorphous [40] silicon layers that have band gaps around 1.9 to 2.0 eV, more short-range and medium-range order than do standard a-Si:H layers and a very small fraction of tiny crystallites. These layers constitute a promising topic for future research. Protocrystalline p layers are apparently used as window layers in certain amorphous silicon solar cells.

2.3 Conductivity and Doping of Amorphous Silicon
2.3.1 Conductivities

Because of its relatively high band gap, the conductivity of undoped amorphous silicon layers in the dark (without illumination) σ_{dark} is very low, between 10^{-8} and 10^{-12} ($\Omega^{-1}\text{cm}^{-1}$). The value of 10^{-12} ($\Omega^{-1}\text{cm}^{-1}$) corresponds to pure a-Si:H layers with a very low density of oxygen atoms and other impurities. Under the influence of white light of an intensity of 100 mW/cm^2 (corresponding to full sunlight—i.e., to an intensity of "1 sun") the conductivity (now called *photoconductivity* σ_{photo}) increases considerably, and attains, for device quality, as-deposited (or annealed) layers values around 10^{-4} to 10^{-5} ($\Omega^{-1}\text{cm}^{-1}$). The lower the density of midgap defects (or dangling bonds), the higher will be the photoconductivity σ_{photo}. On the other hand, impurities, such as oxygen, will also, to a certain extent, increase the photoconductivity σ_{photo}. Thus, the photosensitivity ratio ($\sigma_{photo}/\sigma_{dark}$) is a measure of layer quality; it should be higher than 10^5 for device quality layers, even in the degraded, stabilized state.

The dark conductivity σ_{dark} of a-Si:H layers is strongly dependent on the measurement temperature. If we plot σ_{dark} in a logarithmic scale as a function of (1/T)—i.e., as a function of the inverse of the absolute temperature T—we obtain more or less a straight line. The slope of this line is E_{act}/k, where k is the Boltzmann constant and E_{act} is called the "activation energy of the dark conductivity"; it is a measure of the distance between the Fermi level E_F of the layer and the nearest band edge or mobility edge. A high value of E_{act} means that the E_F is near the middle of the mobility gap, whereas a low value means that E_F is near the band edge and that the layer is strongly doped.

2.3.2 Doping

If phosphine (PH_3) is fed into the PECVD deposition chamber (along with the other gases such as silane and hydrogen and possibly also methane and germane) we will obtain n-doped layers with higher dark conductivities. At the same time, the dangling bond density will also increase [41]. Similarly, if we add diborane (B_2H_6) or trimethylboron ($B(CH_3)_3$) to the deposition gas mixture, we will obtain p-doped layers, also with higher dark conductivities and increased dangling bond densities. Figure 6 (adapted from [42]) shows how the Fermi level E_F is shifted and the conductivity is increased by doping. Thereby, the position of E_F has been evaluated by measuring the activation energy of the dark conductivity and correcting it according to the "statistical shift." The correction due to the "statistical shift" is based on [43], Section 8.1.1, assuming a constant density for the deep states (or midgap states) of $10^{16}/cm^3 eV$ and an exponential band tail, as shown by a full line in Figure 8.3 of [43] ($E_C^0 \approx 25$ meV).

Note that it is not possible to dope a-Si:H layers in such a way that the Fermi level E_F approaches the mobility edges. There remains, even for strong dopant concentrations, a distance of about 400 meV; this is caused by the effect of midgap states and band tails. Due to this difficulty in doping, the open-circuit voltage in a-Si:H solar cells is always much lower than the "theoretical" limit value it should have based on its bandgap value.

Note also that, at least in the original data published in [3] (on which Figure 6 is based), layers produced with pure silane (and without any dopant gases) have a slightly n-type character. It took here a slight p-type doping (with a gas-phase doping ratio $N_{B2H6}/N_{SiH4} \approx 10^{-5}$) to obtain "truly intrinsic" layers, with a dark conductivity activation energy E_{act} of 0.85 eV. It is known today that this is due to unintentional doping by oxygen impurities. If the oxygen content is kept below 2×10^{-18} cm^3 by using high-purity gases and other precautionary measures, then a-Si:H layers without any dopants will be "truly intrinisc" and have a dark conductivity activation energy E_{act} of 0.8 eV or more.

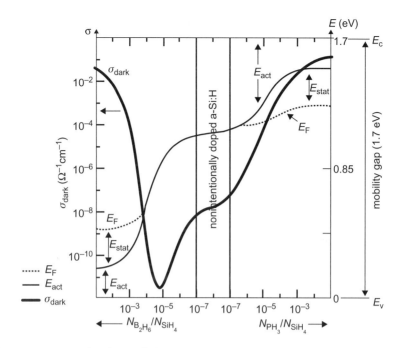

Figure 6 Measured values of dark conductivity σ_{dark} (full, thick line); measured values of the dark conductivity activation energy E_{act} (full, thin line, plotted in the figure as distance from the corresponding mobility edges E_C, E_V) and estimated position of the Fermi energy E_F for amorphous silicon layers, produced by PECVD on glass, in function of the gas-phase doping ratio N_{PH3}/N_{SiH4} (for n-type layers) and N_{B2H6}/N_{SiH4} (for p-type layers). Values of σ_{dark} and E_{act} are from [3]. To obtain the curve for the Fermi level E_F, the statistical shift has been taken into account, based on [43] (see text). The equivalent band gap of a-Si:H (or "mobility gap" as it is called here), is taken to be 1.7 eV; this corresponds to generally published values. *(Figure taken, with permission, from [42], as modified in [1].)*

3. HYDROGENATED MICROCRYSTALLINE SILICON (μc-Si:H) LAYERS

3.1 Structure of Microcrystalline Silicon

Hydrogenated microcrystalline silicon (μc-Si:H), as deposited by PECVD from a mixture of silane and hydrogen, is a mixed-phase material containing a crystalline phase (with tiny crystallites grouped into "conglomerates" or "clusters"), an amorphous phase, and voids (which are very often not real voids but just regions with a lower density [44–45]). By varying the hydrogen dilution ratio $R = [H_2]/[SiH_4]$ in the plasma deposition (where $[H_2]$ denotes the rate of hydrogen gas flow into the deposition system and $[SiH_4]$ the rate of silane gas flow), one can obtain many different types of layers: (a) at low hydrogen dilution—i.e., at low values of R, amorphous layers; (b) by slightly increasing R, layers with mainly an amorphous phase and a low concentration of very tiny crystallites (such as protocrystalline silicon layers); (c) at still higher hydrogen dilution, layers with about 50% amorphous phase and 50% crystalline phase; and (d) at very high values of R, highly crystalline layers, which tend to have a large concentration of cracks or voids and thus constitute low-density, porous material.

When deposited on a glass substrate, the μc-Si:H layers usually start off with an amorphous incubation phase and the nucleation of crystallites only begins later on. This is shown schematically in Figure 7 [46]. Within μc-Si:H solar cells, the situation is more complex, as the μc-Si:H intrinsic layer is deposited on a p-doped or an n-doped microcrystalline layer, and the latter on a rough substrate. One strives, in fact, to avoid the formation of an amorphous incubation layer (which leads to a reduction in solar cell performance) by starting off the deposition with a relatively high value of R. The value of R at which microcrystalline growth starts depends very much on the deposition parameters, such as plasma excitation frequency, substrate temperature, and deposition pressure. The values of R indicated in Figure 7 are merely given as an example and are typical of deposition at relatively high pressures (2 to 3 Torr), with a plasma excitation frequency of 13.56 MHz.

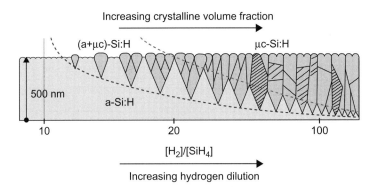

Figure 7 Range of film structures (schematic), obtained with different PECVD parameters, for films deposited on glass substrates; the dashed lines indicate the transitions between amorphous and mixed phase material, as well as between mixed phase material and highly crystalline material. *(Reproduced with permission from [46], in the form as published in [1].)*

It is generally assumed today that most of the defects (i.e., most of the recombination centres) in μc-Si:H are located at the boundaries of the conglomerates or clusters. It is also assumed that these defects are passivated by the amorphous phase. For this reason, one uses, as intrinsic layers within p–i–n- or n–i–p-type μc-Si:H solar cells, layers with about 50% crystalline volume fraction.

Let us take a closer look at the microstructure of a typical μc-Si:H layer. Figure 8 shows part of a μc-Si:H layer taken as a high-resolution transmission electron microscopic (TEM) image within a conglomerate of silicon crystallites: the latter have diameters between 10 and 20 nm and are embedded into an amorphous silicon matrix. The conglomerates themselves are separated by a varying amount of amorphous silicon, cracks and voids, and low-density material [47]. The microstructure is highly complex. In addition, the μc-Si:H layer is neither uniform nor anisotropic, because the conglomerates form cones that widen up toward the top of the layer until they touch each other as schematized in Figure 7.

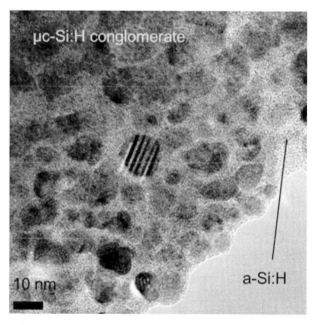

Figure 8 High-resolution TEM micrograph of a plane view taken within a microcrystalline conglomerate. Spherical nanocrystals (one of them highlighted) are embedded in amorphous tissue and constitute the microcrystalline phase itself. *(Reproduced with permission from [47].)*

The most convenient way to assess the "crystallinity" of μc-Si:H layers is to use *Raman spectroscopy*. With this technique, one investigates the local atom—atom bonding structure of a material by studying the interaction of monochromatic incoming light (photons of a given energy) with the bond vibrations in the material (phonons). The energy of the incoming photons is shifted by the energy of the phonon involved in the interaction. Due to scattering of light, one is able to collect and analyse the outgoing photons with the energy shifts. The amplitude of the scattered light is measured as a function of the shift in photon energy: this constitutes the Raman spectrum. Thereby, a unit called *wavenumber* is used, which is simply the reciprocal of wavelength and is expressed in cm^{-1}. The conversion between photon energy E, wavelength λ, and wavenumber ν is $E[eV] = 1.24/\lambda[\mu m] = 1.24 \times 10^{-4} \times \nu$ $[cm^{-1}]$. Crystalline silicon has a narrow peak in its Raman spectrum at $520\ cm^{-1}$ and, due to defective regions, a tail around 500 to $510\ cm^{-1}$ wavenumbers, whereas amorphous silicon exhibits a broad Raman signal centred at $480\ cm^{-1}$ [48]. By suitably analysing the Raman spectrum, one finds the total signal intensity I_c due to crystalline contributions and the total signal intensity I_a due to the amorphous contribution. The ratio $I_c/(I_c + I_a)$ is a semiquantitative indication for the crystalline volume fraction and is called *Raman crystallinity*. It is rather difficult and cumbersome to evaluate the actual crystalline volume fraction (this can be done, e.g., by a high-resolution TEM but requires the attribution of the various regions in the layer to the crystalline and amorphous phases, which is not an easy task). Therefore, the value of the Raman crystallinity is generally used for μc-Si:H layer optimization in connection with solar cells. The best solar cells are obtained for values of Raman crystallinity around 50% to 60%.

3.2 Optical Absorption, Gap States, and Defects in Microcrystalline Silicon

Compared with intrinsic a-Si:H layers, intrinsic μc-Si:H layers show the following striking differences.
1. Lower optical band gap (1.1 eV, similar to the band gap of crystalline silicon) is associated with the crystalline phase of the material [49].
2. Band tails are less pronounced than in a-Si:H; one may assign a value of about 30 meV to the exponential fall-off constant E_V^0 of the valence band tail [50]. The fact that the valence band tail is less pronounced in intrinsic μc-Si:H layers than in intrinsic a-Si:H layers is

probably the reason why μc-Si:H p—i—n-type solar cells can often be illuminated both from the p side as well as from the n side (see, e.g., [51]), resulting in both cases in similarly effective photocarrier collection.

3. Lower defect absorption results from midgap defects (essentially dangling bonds) [52]. The defect absorption is taken for μc-Si:H at a photon energy value of 0.8 eV [49], whereas it is taken for a-Si:H at a photon energy value of 1.2 eV. If the same calibration factor between defect absorption and defect density would apply in both materials, this would mean that the defect densities in device-quality μc-Si:H layers would be much lower than in a-Si:H layers. However, because of the mixed-phase nature of μc-Si:H layers, it is doubtful whether such a conclusion can be drawn [53]. Nevertheless, defect absorption at 0.8 eV, measured preferably by Fourier transform infrared spectroscopy (FTPS) [54] is a very convenient method for comparing the "quality" of different μc-Si:H layers, in view of their use in solar cells. The defect absorption of μc-Si:H layers has a minimum for a value of Raman crystallinity around 50% to 60%, and increases for layers with both lower and higher crystallinity. It is precisely with such layers that the solar cells with the best performances are fabricated. Figure 9 shows defect absorption (as measured by FTPS) for intrinsic μc-Si:H layers with different values of Raman crystallinity before and after degradation.

4. There is a much less pronounced light-induced degradation effect [55—58]. This is best seen in complete solar cells; here, under the standard light-soaking procedure used (50°C substrate temperature, AM1.5 light at 100 mW/cm^2 intensity; 1000 hours of light exposure), the relative degradation in the efficiency of μc-Si:H single-junction cells, with 0.5-μm-thick i layers, is only about 5%, whereas a-Si:H single-junction solar cells with 0.5-μm-thick i layers show more than 25% efficiency loss. The increase in defect absorption (at 0.8 eV) of μc-Si:H layers is also just about a factor 2, whereas the corresponding increase in defect absorption (at 1.2 eV) of a-Si:H layers is at least a factor 10. Note that the defect absorption of degraded μc-Si:H layers has a minimum for a value of Raman crystallinity around 50% to 60%. Layers with <30% crystallinity degrade strongly, whereas layers with more than 70% crystallinity hardly degrade at all but have very high values of defect absorption both in the initial and degraded states (see Figure 9).

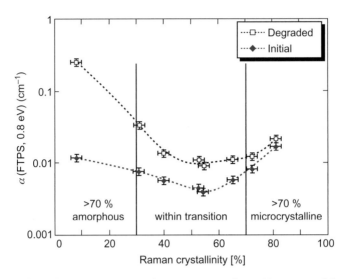

Figure 9 Defect absorption α, at a photon energy of 0.8 eV, measured by Fourier transform infrared spectroscopy (FTPS), before and after light soaking by a standard procedure (see text). *(Reproduced with permission from [56], © 2005 IEEE.)*

5. The visible range of the light spectrum has a lower optical absorption coefficient—i.e., for photon energies between 1.65 eV and 3.2 eV (i.e. for wavelengths between 750 nm and 390 nm). As a consequence, μc-Si:H solar cells have to be much thicker than a-Si:H solar cells to usefully absorb the incoming light. The lower optical absorption results from the *indirect band gap* of μc-Si:H; meaning that a phonon has to be present for a photon to be absorbed due to the rule of momentum conservation. As a consequence, fewer photons are absorbed and fewer electron−hole pairs are generated. On the other hand, in a-Si:H the rule of momentum conservation is relaxed because of the random nature of the amorphous network. The absorption of photons and the photogeneration of electron−hole pairs are correspondingly increased.

Figure 10 shows the main differences in optical properties, between typical μc-Si:H layers and typical a-Si:H layers, by displaying, in a logarithmic scale, the absorption coefficient α(hν) for the applicable spectral range—i.e., for photon energies $E_{photon} = h\nu$ from 0.7 eV to 3.5 eV—corresponding to wavelengths from 350 nm to 1750 nm.

While interpreting Figure 10, note the following:
a. The optical band gap E_g can be evaluated by extrapolation from region A in Figure 10 (the region with high absorption coefficients).

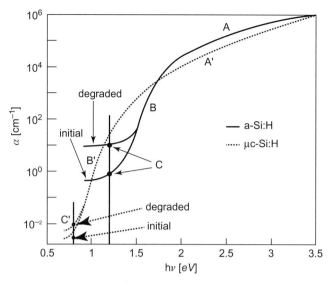

Figure 10 Plot of absorption coefficient α versus photon energy $E = h\nu$, for device-quality a-Si:H and μc-Si:H layers, indicating the three important parameters that can be evaluated from this plot: (A) optical band gap E_g; (B) Urbach energy E^0; (C) defect (dangling bond) density N_{db}. The curves for a-Si:H are drawn with thick uninterrupted lines (———); the curves for μc-Si:H with fine dashed line (- - - - -). The curves given here are purely indicative; they do not necessarily correspond to actual a-Si:H and μc-Si:H layers as used in most R and D laboratories. *(Courtesy of Michael Stückelberger, PV Lab, IMT, EPFL, Neuchâtel.)*

This is commonly done by the procedure according to Tauc et al., as shown in Figure 5.

b. The band tails (and especially the valence band tail) can be assessed from region B, where $\alpha(h\nu)$ follows an exponential curve (which is a straight line in Figure 10, because of the logarithmic scale). The exponential decrease of $\alpha(h\nu)$—i.e., the slope of the straight line in the logarithmic representation—is given by $1/E^0$, where E^0 is called the *Urbach energy*. The Urbach energy E^0 is considered in a-Si:H layers to be roughly equivalent to E_V^0, whereas in μc-Si:H layers it may, in fact, depend on both E_V^0 and E_C^0.

c. Region C in Figure 10 gives a qualitative indication for the density of midgap defects, which can be associated with dangling bonds. In μc-Si:H layers, it is not really clear where these dangling bonds are located; from transport measurements [47,59], one may presume that they are located at the boundaries of the conglomerates.

3.3 Conductivities, Doping, Impurities, and Ageing in Microcrystalline Silicon

3.3.1 Conductivities

Because of the lower mobility gap of the crystalline phase of μc-Si:H, the values of dark conductivity σ_{dark} are significantly higher than in a-Si:H; they are between 10^{-8} and 10^{-6} ($\Omega^{-1} cm^{-1}$) for Raman crystallinities between 60% and 80%, provided we have "truly intrinsic" material with a low content of impurities [60]. The photoconductivity σ_{photo} of such layers is only slightly higher than in a-Si:H, with values around 10^{-4} and 10^{-5} ($\Omega^{-1} cm^{-1}$). The photosensitivity ratio ($\sigma_{photo}/\sigma_{dark}$) can still be used as one of the criteria for layer quality. Furthermore, the dark conductivity activation energy E_{act} remains a convenient indication for the position of the Fermi level. For "truly intrinsic" material with a low content of impurities, E_{act} will be >0.5 eV. Intrinsic μc-Si:H layers with a photosensitivity ratio of 10^{-3}, a dark conductivity activation energy >0.5 eV, and a Raman crystallinity of 50% to 60% can thus be considered good candidates for the intrinsic layers of solar cells. Figure 11 shows commonly obtained values for σ_{dark} and σ_{photo} in function of Raman crystallinity.

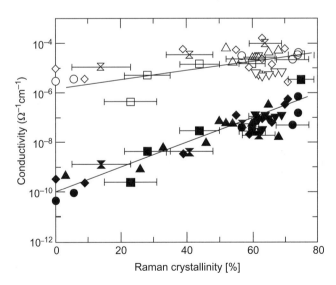

Figure 11 Dark conductivity (filled symbols) and photoconductivity (open symbols) as a function of Raman crystallinity. *(Reproduced with permission from [60], showing typical error bars.)*

3.3.2 Doping

By adding phosphine, diborane, or trimethylboron to the silane plus hydrogen gas mixture and feeding it into the deposition chamber, μc-Si:H layers can be doped in a similar way as a-Si: layers. In contrast with a-Si:H, one does not observe an additional creation of midgap defects through doping (see, e.g., [61]). As a consequence, one can push here, by doping, the Fermi level E_F almost to the mobility edges E_C, E_V, and it does not remain some 300 to 400 meV away from them as in strongly doped a-Si:H layers (see Figure 6). The conductivities obtained thereby are also considerably higher than in a-Si:H: they are about 10^2 ($\Omega^{-1}\mathrm{cm}^{-1}$), rather than just 10^{-2} ($\Omega^{-1}\mathrm{cm}^{-1}$). As a consequence thereof, the best μc-Si:H solar cells have open-circuit voltages in excess of 600 mV [17,62,63]—i.e., not very far from the theoretical limit value as given by the band gap of the crystalline phase of μc-Si:H.

3.3.3 Impurities

Undoped μc-Si:H layers with oxygen concentrations above 10^{19} cm^{-3} in general show clear n-type behaviour. Under usual deposition conditions, it is only by reducing the oxygen concentration to 2×10^{18} cm^{-3} that "truly intrinsic" layers can be deposited, with the Fermi level E_F in the middle of the gap and with dark conductivity activation energy E_{act} higher than 500 meV. By incorporating layers with low oxygen content as intrinsic layers (i layers) into p–i–n-type solar cells, one obtains solar cells with high efficiencies and with a broad spectral-response curve [16], as shown in Figure 12. Layers with low oxygen content are obtained, either by employing a gas purifier, as reported in [16], or by utilizing high-purity source gases. If the i layers are deposited at low temperatures (at temperatures below 180°C), the oxygen impurities apparently do not play the same active role as at higher temperatures (but are passivated by the hydrogen atoms), and solar cells with higher efficiencies have been obtained even with a relatively high oxygen contamination of 2×10^{19} cm^{-3} [64,65]. On the other hand, such low-temperature deposition is currently not used to fabricate the best μc-Si:H solar cells; furthermore, it is hardly compatible with the production of "micromorph" tandems (see Section 5.4). Nitrogen impurities have a similar effect as oxygen impurities [66].

3.3.4 Ageing

Because of the porous nature of μc-Si:H layers (especially of μc-Si:H layers with high crystalline volume fraction), oxygen and other impurities

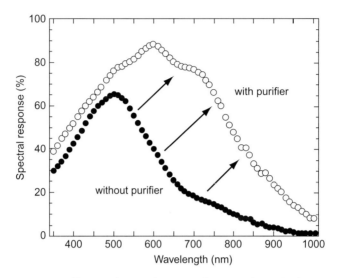

Figure 12 Quantum efficiency (spectral response) curve of a typical microcrystalline silicon solar cell, fabricated with and without gas purifier. In the latter case, a relatively high oxygen contamination of 10^{20} cm^{-3} leads to a strong deformation of the internal electric field within the i layer of the p–i–n-type solar cell by positively charged oxygen atoms acting as donors and thus to very poor collection. (Reproduced with permission from [16].)

can easily enter into these layers [19] (and even into entire solar cells if they are not encapsulated), provoking thereby a change in dark conductivity (and a reduction in cell performance). This ageing effect takes place in the dark and at room temperature over a period of days to months. By annealing in vacuum or inert gas for several hours at temperatures typically higher than 130°C, this effect can be reversed. The ageing effect is an additional difficulty when developing and characterizing μc-Si:H layers and solar cells.

4. FUNCTIONING OF THIN-FILM SILICON SOLAR CELLS WITH P–I–N AND N–I–P STRUCTURES

4.1 Role of the Internal Electric Field

All solar cells function according to the following two principles:
1. An electron and hole pair is generated by absorption of an incoming photon within a semiconductor; this is possible if the energy of the

photon $E_{photon} = h\nu = hc/e\lambda$ is larger than the band gap of the semiconductor—i.e. if $E_{photon} > E_g$, where E_g is the band gap of the semiconductor (taken to be somewhere between 1.6 and 1.85 eV for a-Si:H, depending upon deposition conditions, and 1.1 eV for μc-Si:H); ν and λ are the frequency and the wavelength of light, respectively; h is Planck's constant; c the velocity of light; and e the charge of an electron (unit charge). The majority of the photons of a given wavelength λ is only absorbed if the thickness d of the semiconductor is larger than the penetration depth d_{pen} of the photons. The penetration depth d_{pen} becomes larger as λ is increased—i.e., as E_{photon} is decreased and approaches E_g. This is especially critical for thin-film silicon solar cells, where the thickness d of the semiconductor is of the order of 1 μm. In these cells, one has to utilize special light-trapping schemes in order to absorb a sufficient part of the incoming sunlight (see Section 4.5).

2. Holes and electrons are separated by the action of an internal electric field created by a diode configuration. This can be described as follows:

   a. The majority of solar cells are built as p–n diodes; here the electric field is limited to the depletion layers—i.e., to two very narrow zones at the interface between p and n regions. Photogenerated carriers travel by diffusion up to the depletion layers and are then separated by the strong electric field prevailing there. As long as the carrier diffusion lengths are sufficiently high—i.e., as long as they are much higher than the cell thickness d, the collection losses are low, and the solar cell functions properly.

   b. In thin-film silicon solar cells, the diffusion lengths are, in general, very small; they are, in fact, often smaller than the thickness d of the solar cell. Thus, diffusion alone is not sufficient to ensure transport and collection of the photogenerated carriers. One therefore utilizes the internal electric field to assist also in the transport of the photogenerated carriers. This is only possible with a p–i–n (or n–i–p) diode configuration: here the internal electric field extends throughout the whole i layer and governs both transport and separation of the photogenerated carriers.

Figure 13 shows a comparison for the profile of the internal electric field $E(x)$ between p–n and p–i–n diodes. Note that for zero external (applied) voltage V, the integral $\int E_{int}(x)dx$ is, for both types of diodes

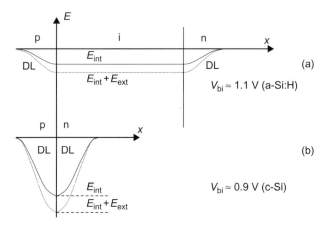

Figure 13 Internal electric field $E_{int}(x)$, for (a) p–i–n-type and (b) p–n-type diodes; E_{int} is the internal electric field for zero applied voltage, E_{ext} is the additional electric field due to an external or applied voltage V, $E = E_{int} + E_{ext}$ is the total electric field, V_{bi} is the built-in voltage of the diode, and DL is the depletion layer. The sign of the electric field is negative because it is directed from the n region to the p region (i.e., toward negative values of x). Thus, the electric field separates electrons and holes; it pushes the photogenerated holes toward the p region, where they are collected, and the photogenerated electrons toward the n region, where they are, in their turn, collected. (Reproduced from [1] with permission of the EPFL Press.)

equal to the built-in voltage V_{bi}, a parameter, which is approximately equal to 1 Volt, for all forms of silicon. If $V \neq 0$, one can write: $V = \int -E_{ext}(x)dx$; this means that if the applied voltage is negative, it gives a reverse bias to the diode and the electric field is augmented (as shown in the figure); if the applied voltage is positive, it gives a forward bias to the diode and the electric field is reduced. As V approaches the open-circuit voltage V_{oc} of the solar cell, the internal electric field becomes strongly reduced. Note that in relation to the corresponding energy gaps, V_{oc} values of p–i–n- and n–i–p-cells are fundamentally lower than the theoretical limit values found for p–n cells.

Due to the action of the internal electric field $E(x)$ in the i layer of p–i–n- and n–i–p-type solar cells, transport (and collection) of the photogenerated carriers is now governed (as long as V is not too high) by the drift length L_{drift} of both electrons and holes within the i layer. For thin-film silicon, one generally writes $L_{drift} = \mu\tau E$, where μ is the mobility of the carrier (electron or hole), τ the lifetime, and E the magnitude of the prevailing electric field. One can consider that, for a given value of E, L_{drift} will be approximately equal for both carriers (electrons and holes)

[67,68]. One finds also that, at short-circuit conditions, the drift length has a value that is about 10 times higher than the minority-carrier diffusion length [69]: this is, of course, the reason, why one uses the p−i−n (or n−i−p) configuration for all a-Si:H and for most μc-Si:H. As previously stated, the minority-carrier diffusion lengths in these materials would be too small, often quite a bit smaller than the i layer thickness d_i.

Thus, for thin-film silicon solar cells on glass substrates, the structure shown in Figure 14 is obtained. Note that for optimal performance of a-Si:H solar cells, the light has to enter the solar cell through the p layer; it is only then that the deformation of the internal electric field through trapped charge in the valence band tail can be kept negligibly small (see Section 2.2). Therefore, if the solar cell is deposited on a substrate, which is not fully transparent (or, unlike glass, does not remain for several decades fully transparent but eventually becomes yellowish like most polymers), then the deposition sequence n−i−p is used. A textured reflector

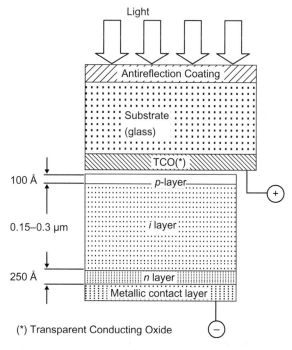

Figure 14 Typical structure of p−i−n-type amorphous silicon solar cell on glass substrate. Microcrystalline silicon solar cells have a similar structure; however, in the latter case, the i layer is much thicker—i.e., 1 to 2 μm thick. *(Reproduced from [1] with permission of the EPFL Press.)*

layer (e.g., textured silver) is deposited on the substrate followed by the n, i, and p layers; on top of the p layer, a transparent conductive oxide (TCO) and a grid for current collection are used. The n–i–p-configuration is regularly employed not only for solar cells deposited on stainless steel substrates but also on most polymer substrates.

Although it would be feasible to let the light enter through the n side in single-junction μc-Si:H solar cells (because the trapped charge in the valence band tail is much less prominent than in a-Si:H), μc-Si:H solar cells are, in practice, almost exclusively used in tandem and multijunction structures, together with a-Si:H solar cells, so that once again one prefers to let the light enter from the p side.

Because the doped layers in thin-film silicon do not have a sufficiently high conductivity as does wafer-based crystalline silicon, a TCO layer has to be used as the contact layer adjacent to the p layer—on the side where the light enters the solar cell. In practice, the TCO layer is often textured and thus contributes to light scattering (see Section 4.5). In the case of p–i–n-type solar cells (as shown in Figure 14), the thin-film silicon layers are deposited on top of this TCO layer: the TCO layer has therefore to withstand the action of a silane plus hydrogen plasma. In this case, F-doped SnO_2, Al-doped ZnO, or B-doped ZnO layers are used as TCO layers. (The SnO_2 layer tends to be reduced by the hydrogen-rich plasma used for the deposition of μc-Si:H and should in this case be covered by a thin protective layer—e.g., by a thin ZnO or a thin TiO_2 layer [47]). In the case of n–i–p-type solar cells, the TCO contact layer is deposited after the thin-film silicon layer and very often indium tin oxide (ITO) layers are used.

4.1.1 Formation of the Internal Electric Field in the i Layer

Thermal equilibrium in a p–i–n-diode is established by the formation of space charge regions in the p- and n-doped layers. The space charge is constituted by the ionized dopant atoms and is responsible for forming the internal electric field, as schematically drawn in Figure 15. In uniform doped p layers (without a p–i junction), the charge constituted by the negatively ionized acceptor atoms (density N_A^-) is neutralized by free holes (density p_f) and trapped holes (density p_t), as one has $N_A^- = p_f + p_t$ and, thus, charge neutrality. Similarly, in uniform doped n layers, (without a n/i junction) the charge constituted by the positively ionized donor atoms (density N_D^+) is neutralized by free electrons (density n_f) and trapped electrons (density n_t), as one has here $N_D^+ \, n_f + n_t$, and, thus, again charge

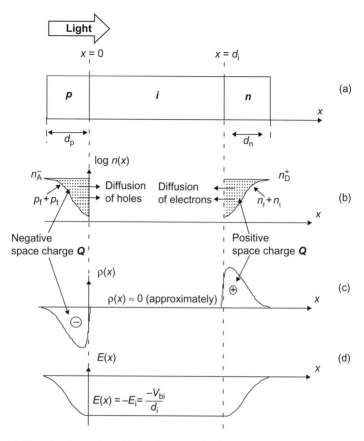

Figure 15 Sketch of a p–i–n-diode showing the formation of space charge Q in the "frontier zones" of the p and n layers (at the boundaries toward the i layer). In these frontier zones. the concentrations of free carriers (p_f, n_f) and of trapped carriers (p_t, n_t) are very low (the carriers diffuse away from these frontier zones into the i layer). In the p layer, we are therefore essentially left with the negatively ionized acceptor atoms (density N_A^-), forming a negative space charge; in the n layer, we are essentially left with the positively ionized donor atoms (density N_D^+), forming a positive space charge. *(Reproduced from [1] with permission of the EPFL Press.)*

neutrality. On the other hand, when we have a p–i junction, then that part of the p layer, which is just adjacent to the i layer, will have virtually no free holes ($p_f \approx 0$) in the valence band, because most free holes remaining in that "frontier zone of the p layer" would immediately travel by diffusion to the i layer, where their density is very much lower; now, through the processes of capture and thermal emission, there is a constant exchange between the free holes in the valence band and the trapped holes in the valence band tail and the density p_t of trapped holes can be considered roughly proportional to the density p_f of free holes (see [70]). Thus, there remain in the "frontier zone of the p layer" practically only the ionized acceptor atoms, which make up a negative space charge Q, as shown in Figure 15. Similarly there remain in the "frontier zone of the n layer" practically only the ionized donor atoms, which make up a positive space charge Q. The internal electrical field extends between these two space charge regions. If we look at an ideal p–i–n-diode, then the i layer itself will not contain any significant charge contributions and the electric field will be constant; it will have (for zero applied voltage) a value E_i equal to $-(V_{bi}/d_i)$, where d_i is the thickness of the i layer and V_{bi} is the built-in voltage of the solar cell. V_{bi} is essentially given by the sum of the two shifts in Fermi level E_F due to doping in both the p and n layers (see [1], Figure 4.33); in amorphous silicon, the mobility gap is relatively large (≈ 1.7 eV), but we can only push the Fermi level E_F by doping to a position, which is approximately 300 meV away from the mobility edges E_C, E_V so that $V_{bi} \approx 1.7$ eV $- (2 \times 0.3$ eV$) \approx 1.1$ eV. In microcrystalline silicon, the mobility gap is smaller (≈ 1.1 eV), but we can push the Fermi level E_F by doping to a position, which is just about 50 to 100 meV away from the mobility edges E_C, E_V, so that $V_{bi} \approx 1.1$ eV $- (2 \times 0.1$ eV$) \approx 0.9$ eV. The values given here are the maximum values for V_{bi} and are applicable if the p and n layers are correctly doped and sufficiently thick. (In practice, this means that the doped layers should be thicker than about 10 nm; the p layer, through which the light enters the cell, must be kept especially thin, because the light that is absorbed here is generally lost and does not contribute to photogeneration. This "parasitic" absorption can be reduced even further by using a silicon–carbon alloy as p layer with a higher band gap, especially in the case of a-Si:H solar cells. The n layer, which is at the "back" of the cell, is generally made quite a bit thicker—some 20 to 25 nm thick.)

In order to design solar cells with satisfactory performance, the internal electric field $E_i = -(V_{bi}/d_i)$, should be relatively strong, especially in a-Si:H solar cells, where in the light-soaked state (after light-induced degradation), the mobility $\times$ lifetime product $\mu\tau$ is very small and a sufficiently high drift length $L_{drift} = \mu\tau E$ is only achieved by choosing an i-layer thickness d_i in the range of 200 to 300 nm. For microcrystalline silicon solar cells, the i-layer thickness d_i can be chosen up to a few μm without any loss of efficiency [71].

4.1.2 Reduction and Deformation of the Internal Electric Field in the i Layer

The internal electric field $E(x)$ in the i layer will be deformed and reduced by additional space charge, from the following sources:

1. Ionized atoms, due to cross-contamination from dopant atoms (mainly from the doped layer deposited before the i layer—i.e., from boron atoms in the case of p—i—n-cells, or from phosphorus atoms, in the case of n—i—p-cells). For these reason one has to avoid cross-contamination and use either multichamber deposition systems or other precautionary measures [72].
2. Ionized atoms, from impurities acting as dopants, especially from oxygen and nitrogen contamination
3. Trapped carriers in the band tails, especially trapped holes in the valence band tail
4. Ionized dangling bonds [73].

The first effect is significant in both amorphous and microcrystalline silicon solar cells, the second effect is particularly important in microcrystalline silicon solar cells, the third effect will be of importance only in amorphous silicon solar cells, and the fourth effect will cause problems in degraded amorphous silicon solar cells.

For all types of p—i—n- and n—i—p-type thin-film silicon solar cells, it is of paramount importance to have a strong internal electric field and to avoid substantial reduction of this field by any of the effects listed earlier. This can be achieved by suitable design of the fabrication process and by keeping amorphous silicon solar cells sufficiently thin.

4.2 Recombination and Collection

In amorphous silicon solar cells, a large part of recombination is *bulk recombination* and takes place in the centre of the i layer due to the dangling bonds acting as recombination centres. According to [74], it is mainly the neutral dangling bonds that play an essential role in this part of recombination. One may speculate that the situation is essentially the same in microcrystalline silicon solar cells. However, it is important to realize that interface recombination in thin-film silicon solar cells can also play a significant role, as drawn schematically in Figure 16. In this case, charged (or ionized) dangling bonds will act as recombination centres. Their density can be substantially higher than the dangling bond density in the bulk of the i layer. In amorphous silicon solar cells, such interface problems arise mainly from cross-contamination—i.e.,

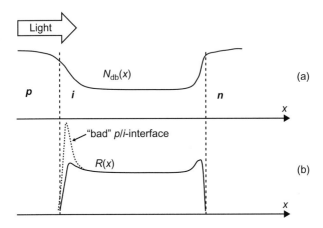

Figure 16 Schematic representation of (a) the dangling bond density $E_{db}(x)$ and (b) the recombination function $R(x)$ in the i layer of a p–i–n-type thin-film solar cell; if the cell has a problematic p–i interface (e.g., due to boron contamination from the p layer deposited before the i layer), there will be strong supplementary recombination, as indicated by the dotted line. *(Reproduced from [1] with permission of the EPFL Press.)*

from dopant atoms having diffused during the fabrication process—from one of the doped layers into the i layer. For microcrystalline silicon solar cells, there is the additional problem of crystalline growth: one often has at the beginning of the growing microcrystalline layer a layer of inferior crystallographic properties that in extreme cases can even be amorphous.

For amorphous silicon solar cells, the dangling bond density will be dramatically increased by light-induced degradation (the Staebler–Wronski effect). By keeping the solar cell very thin, i.e., by choosing an i-layer thickness d_i in the range of 200 to 300 nm, and by adopting all the other measures described earlier (see Section 2.1) one is able today to fabricate amorphous silicon solar cells with a relative efficiency loss of just 10% to 20% due to light-induced degradation.

If the recombination in a thin-film silicon solar cell becomes excessive, the resulting deficiency in photocarrier collection can be mainly identified by: (a) decrease in the fill factor FF; (b) a deficiency in the spectral-response/external quantum efficiency (EQE) curve of the cell. In case (b) it is particularly instructive to compare two EQE-curves: a first curve with no bias voltage and a second curve with a reverse bias voltage of −1 to −2 Volts. If these curves essentially do not differ, the internal electric field at no bias is sufficiently high to collect practically all photogenerated carriers. However, a difference between the two curves (see Figure 18) indicates collection problems, i.e. the internal field must be increased by an external bias to enhance the carrier collection. If the difference is seen at short wavelengths, it means that the collection problem occurs at the interface through which light enters in to the cell (at the p–i interface, for p–i–n-cells illuminated from the p side); if it occurs at longer wavelengths, it means that the collection problem occurs in the bulk of the i layer. Figure 17 represents the difference between (a) blue light and (b) red light entering into a p–i–n-type solar cell. In the case of blue light, there are photogenerated holes and electrons only near the p–i-interface, so that recombination can only take place there. We are, thus, only probing the region of the i layer near the p–i-interface. In the case of red light, holes and electrons are generated throughout the i layer and we are probing the whole i layer.

Figure 18 shows typical external quantum efficiency curves of microcrystalline p–i–n-type cells. In (a) we are looking at cells with different p layers (in the case of the "bad" p layer, it would appear that the recombination in the adjoining regions of the i layer is also increased, possible through cross-contamination or another effect). In (b) we are looking at cells with and without contamination; the contamination results in a

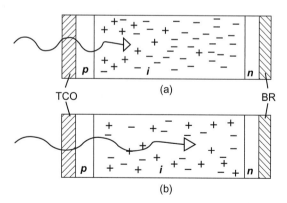

Figure 17 Light penetration and presence of photogenerated carriers (holes and electrons) within a p–i–n-type solar cell. (a) For blue, short-wavelength light, recombination only takes place near the p–i interface; (b) for red, long-wavelength light, recombination can take place throughout the i layer. TCO = transparent conductive oxide; BR = back reflector. *(Reproduced from [1] with permission of the EPFL Press.)*

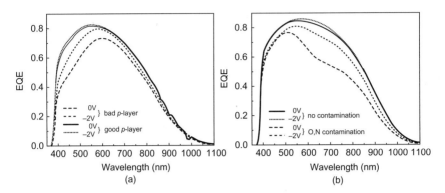

Figure 18 External quantum efficiency (EQE) curves of microcrystalline silicon solar cells with various deficiencies (see text). *(Reproduced from [1] with permission of the EPFL Press.)*

reduction or deformation of the electric field and a collection problem throughout the i layer as described earlier.

Spectral-response and EQE measurements are a powerful tool for the diagnosis of thin-film silicon solar cells. Their detailed interpretation needs, however, considerable experience, and goes well beyond the scope of the present chapter. Their main advantage is that they allow us to assign defects and shortcomings to various regions of the cell. Their "geometrical sensitivity" is excellent for the zone where the light enters into the solar cell, but it is very much reduced at the far end of the solar cell toward the back reflector (BR) in Figure 17. In order to probe the far end of the cell, it is necessary to employ a bifacial configuration, in which one can let the light enter "from the back"—i.e., through the n layer—and perform EQE measurements in this arrangement [75].

4.3 Shunts

In thin-film silicon solar cells and modules, shunts are a common problem. Shunts result in a reduction of the fill factor *FF* already at standard illumination levels of 100 mW/cm^2 light intensity. In order to distinguish them from collection problems (which also result in a reduction of the fill factor *FF* at standard light levels), it is necessary to measure the *J-V* curve of the cell or module at low light levels. If there is a substantial further drop in the *FF* as the light level is decreased to less than 1 mW/cm^2, we are facing a shunt problem, not a collection problem.

To conceptually separate the effects of collection problems and shunts, look at Figures 19 and 20. Figure 19 shows the common equivalent

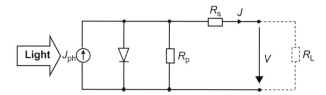

Figure 19 Common equivalent circuit universally used for all solar cells. Note that the parallel resistance R_p symbolizes both recombination–collection problems as well as the effect of shunts. R_L represents an external load. This simple equivalent circuit is used for designing electrical systems containing solar cells and modules. (Reproduced from [1] with permission of the EPFL Press.)

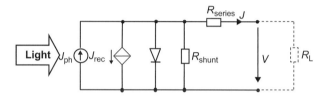

Figure 20 Modified equivalent circuit introduced by [76] in order to separate the effects of recombination–collection problems and shunts in p–i–n- and n–i–p-type thin-film silicon solar cells. The current sink J_{rec} represents current losses through recombination; the value of J_{rec} is proportional to the value of J_{ph} and depends additionally on the operating point (V) and on the "quality" of the cell; the resistor R_{shunt} represents "true" Ohmic shunts. This modified equivalent circuit is only used for the analysis of solar cells and modules with unsatisfactory performance. (Reproduced from [1] with permission of the EPFL Press.)

circuit universally used for all solar cells. This equivalent circuit is indeed very simple and therefore convenient for the design of entire electrical systems containing solar cells and modules. However, it does not provide us with physical insight into what is happening within the solar cell, especially within a thin-film silicon solar cell. In fact, the effects of shunts and of collection problems are merged together in the parallel resistance R_p. In Figure 20, a modified equivalent circuit, introduced by [76] is given: here the recombination or collection problems are symbolized by a "current sink" J_{rec} (opposite of a current source). The recombination current J_{rec} is proportional to the photogenerated current J_{ph} and will therefore result in performance reduction and in a FF reduction that is independent of light intensity. The resistance R_{shunt} represents now "true" Ohmic shunts. From the electrical diagram of Figure 20, one can see that R_{shunt} will play an increasing role as the light level, and thus, both the photogenerated current J_{ph} and the recombination current J_{rec} are decreased proportionally to the light level. On the other hand, the current through R_{shunt} (approximately V/R_{shunt}) is not reduced to the same extent, because the J-V curves at low light levels show only minimal voltage decreases compared to the current decreases.

Shunts in thin-film silicon solar cells can originate from various fabrication problems.

1. Particles may be deposited on the substrates or on the growing layers because of either a dusty environment or the formation of powder during the deposition itself (due to plasma polymerization—i.e., chemical reactions in the gas phase); after the particle falls off, a pinhole remains and a short-circuit will be formed when the final contact layer is deposited (Figure 21) and makes contact to the conducting substrate through the pinhole.
2. Mechanical operations (such as cutting of modules) may give rise to the generation of particles.
3. Laser scribing (see Section 6.3) with parameters that are not properly optimized may create partial short circuits within a module, either at locations along a scribe line or by local bridges across a scribe line.
4. Cracks may develop during the growth of the microcrystalline silicon layer (Figure 22) when a rough substrate is used and precautionary measures are not taken [45,77,78].

Shunts and collection problems can be quantified with the help of the variable intensity measurement (VIM) method. VIM analysis is based on measuring the J-V characteristics of the solar cell or module for a whole range of light intensities. An evaluation of these curves as indicated in [79,80] will enable us to determine to what extent the performance of the cell or module—and, in particular, its fill factor—are affected by shunts and to what extent they are affected by collection problems.

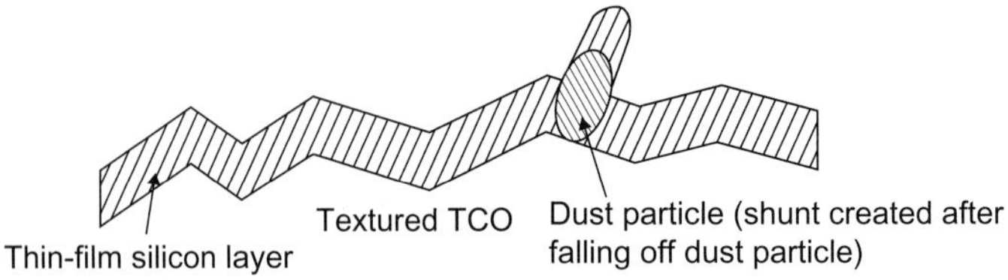

Figure 21 Schematic representation of shunt formation due to a dust particle. *(Reproduced from [1] with permission of the EPFL Press.)*

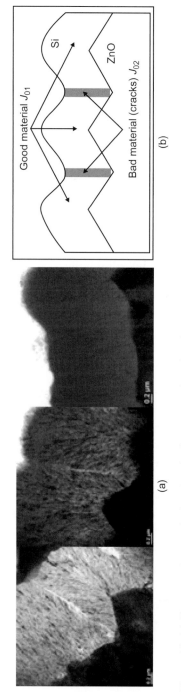

Figure 22 (a) Electron micrograph showing the formation of cracks when microcrystalline silicon layers are grown on certain rough substrates (here on a ZnO transparent contact layer) without adequate substrate surface treatment; (b) schematic drawing indicating that these cracks may not only create shunts but also "bad" diodes (diodes with high reverse saturation current $J_{02} > J_{01}$ due to porous material). *(Reproduced with permission from [1,77].)*

4.4 Series Resistance Problems

In thin-film silicon solar cells and modules, the electrical contacts are made to the front and back contact layers. Generally the contact on the "front side" (where the light enters into the cell or module) is given by a transparent conductive oxide layer. Here a compromise between optical and electrical requirements must be found: if the TCO layer is chosen to be too thin to keep the optical absorption low, it will have a high sheet resistance and will contribute in a pronounced manner to the series resistance R_{series} shown in Figure 20. If the TCO layer is too thick, it will absorb too much light, which is lost for photocarrier generation. The thickness of the back contact layer is less critical, since these layers are either highly conducting metal layers or, in case of TCO layers with additional reflector layers, can be chosen thicker for lower sheet resistance. For small individual laboratory test cells, sheet resistances usually do not cause significant electrical losses.

However, for modules that feature a monolithic series connection of cells (see Section 6.3, especially Figure 33), the earlier mentioned compromise between optical and electrical losses must be carefully taken into account.

For a cell with a photoactive width w, the relative power loss at the maximum power point is given by the expression [42]

$$(\Delta P/P)_{mp} = j_{mp} w^2 R_{sheet}/3 V_{mp}$$

where R_{sheet} denotes the sheet resistance of the TCO layer and j_{mp} and V_{mp} are current density and voltage at the maximum power point, respectively. The term $(\Delta P/P)_{mp} \times 100\%$ then gives us the reduction in fill factor due to the sheet resistance of the TCO. One can draw the following conclusions from the preceding expression:

1. It is important to use TCO layers with low sheet resistance (typically $R_{sheet} = 10\Omega$/square or lower).
2. The width w of the cell has to be kept very low (in single-junction amorphous silicon solar cells, the widths of individual cells within a solar panel are kept typically below 1 cm). However, the cell-interconnection scheme implies an area loss, since the required laser scribes to establish the interconnection represent photoinactive area losses (see Section 6.3). For a certain interconnection width Δw, the relative area loss is $\Delta w/(w + \Delta w)$, where w is the photoactive cell width—i.e., the area loss increases, opposite to the electrical loss, with decreasing cell width. Thus, there is an optimal width in which the sum of electrical and area losses reaches a minimum.

3. For cells with high current and low voltage (like single-junction microcrystalline silicon solar cells), the relative power loss will be higher; this is one reason why we do not fabricate modules with single-junction microcrystalline cells).
4. For tandem cells, and even more for triple-junction cells, the relative power loss will be lower, since the inherent series connection of the subcells results in the addition of their voltages and in a lower current corresponding to the shared absorption.

In addition to the sheet resistance of the TCO, poorly doped p and n layers will also contribute to increase the value of R_{series}. Furthermore, contact problems between the TCO (which is usually n type) and the adjacent p layer, as well as unsatisfactory tunnel or recombination junctions within tandems and multijunction cells (see Section 5.1) will also lead to an increase in R_{series}. If the cell or module is properly designed and fabricated, the loss in fill factor FF due to series resistance R_{series}, should, at standard illumination conditions, not exceed a few percentage points. If FF increases when reducing the light intensity, this is typically a sign of losses due to series resistance R_{series}. If these losses are excessive, they can be evaluated by VIM analysis [79,80].

4.5 Light Trapping

In thin-film silicon solar cells and modules, it is imperative to limit the thickness d_i of the i layer (which is usually the only layer contributing to useful absorption and photogeneration) in order to keep the internal electric field sufficiently high for effective carrier collection. A higher value of d_i, such that the solar cell could absorb a large part of the solar spectrum, would lead to a very high light-induced degradation in amorphous silicon solar cells; in microcrystalline silicon solar cells, it would mean long deposition times and high material and fabrication costs. This can be seen in Figure 23, where the absorption coefficient α and the penetration depth $d_{pen} = 1/\alpha$ of monochromatic light are plotted as a function of the wavelength λ (or of the photon energy $E_{photon} = h\nu$) of the light. Figure 23 (from [81]) compares five materials commonly used for solar cells. It can be seen that for both $CuInSe_2$ and $CdTe$ the penetration depth d_{pen} is approximately 1 μm or lower for most of the light in the visible range of the solar spectrum—i.e., for light with wavelengths $\lambda < 700$ nm. For the various forms of silicon, this is clearly not the case: d_{pen} is larger than 1 μm for $\lambda = 700$ nm, and it remains still higher than

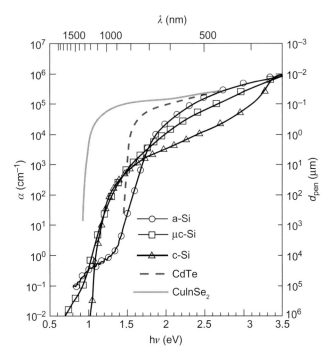

Figure 23 Absorption coefficient α and penetration depth d_{pen} in function of wavelength λ (top scale) or photon energy $E_{photon} = h\nu$ (bottom scale), for materials commonly used in solar cells: hydrogenated amorphous silicon (a-Si), hydrogenated microcrystalline silicon (with 50% to 60% Raman crystallinity; μc-Si), wafer-based crystalline silicon (c-Si), CdTe, and CuInSe$_2$. Because of the amorphous phase contained in the microcrystalline silicon layers, the curve for μc-Si lies in between the curve for a-Si and c-Si. *(Reproduced with permission from [1,81].)*

the corresponding penetration depths for CuInSe$_2$ and CdTe down to wavelengths $\lambda > 500$ nm.

In order to absorb a sufficient portion of the incoming light in a thin-film silicon solar cell, it is necessary to increase the average optical path length within the i layer by a factor m. With the help of a back reflector, one may double the optical path and reach $m = 2$. By using improved light-scattering techniques. one presently reaches substantially higher values of m—around 10 or even higher.

In p–i–n solar cells, this is currently done by using randomly textured TCO layers. Since the texture of the front TCO layer is widely replicated for the back reflector layer, scattering in reflection further increases the optical path length, particularly for the weaker absorbed wavelengths toward the red portion of the incident light.

In n–i–p solar cells, one deposits first the back reflector. By depositing a randomly textured back reflector, such as a rough silver layer [82–84], one obtains light scattering at this point, and, to some extent, also at the entry point of the light—as the layers deposited on top of the back reflector take on more or less the same form.

All present solar cells and modules use randomly textured structures for light scattering; however, from a research point of view, periodical structures for light scattering are increasingly being investigated [85–87].

The design of light-management and light-trapping schemes and their incorporation into thin-film silicon solar cells is by no means trivial. There are a whole series of questions yet to be solved:

1. Which is the best surface–interface structure (morphology) to obtain an extension of the optical path by the greatest possible factor m?
2. Given that there are multiple reflections and multiple points of scattering within a thin-film silicon solar cell, as shown in Figure 24, what is the practical limit for m?
3. How does one avoid cracks and other problems within the silicon layers when choosing a very pronounced texture? (See Figure 22.)

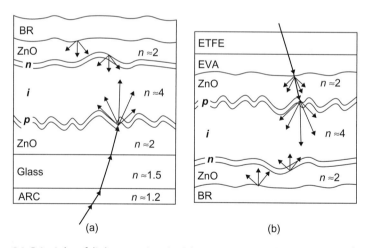

Figure 24 Principle of light trapping in (a) p–i–n-type solar cell; (b) n–i–p-type solar cell. The figure shows the case of a single-junction solar cell, with ZnO (zinc oxide) as TCO layer, a back reflector (BR). One notes the multiple interfaces at which scattering and reflection take place. For p–i–n-based modules (a), the front glass is sometimes provided with an antireflective coating (ARC), the encapsulation of the back side, usually done with EVA, is not shown. For n–i–p-based modules (b), ETFE and EVA are typical polymer foils used for encapsulation and needed for protection. *(Reproduced from [1] with permission of the EPFL Press.)*

4. Each time that light passes through a doped layer or a TCO layer, and each time light is reflected at a metal reflector (especially at a textured metal reflector), there are losses through optical absorption. How do we minimize them? (This question is particularly important for μc-Si:H solar cells, which absorb the light in the near infrared, where there are additional optical losses through free-carrier absorption—e.g., see [88].)

It can therefore be stated that improved light management and light trapping techniques hold great promise for further enhancing the performance of thin-film silicon solar cells.

5. TANDEM AND MULTIJUNCTION SOLAR CELLS

5.1 General Principles

In thin-film silicon solar cells, one so far almost exclusively uses two-terminal tandem solar cells. These devices stack two subcells, one on top of the other as indicated in Figure 25.

In the ideal case, each of the subcells absorbs the same amount of photons and therefore basically photogenerates the same electrical current. This is the case of "current matching," and it can only be achieved

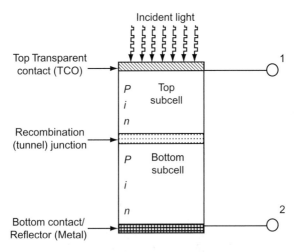

Figure 25 Principle of dual-junction stacked cell or tandem cell in the p—i—n configuration, with two electrical terminals; the light is first absorbed in the top subcell; the remaining light enters into the bottom subcell. Electrically, the two subcells are in series, so that strictly the same current circulates through each of them.

for a given spectrum of the incoming light. If the spectrum of the incident light changes, or if one of the two subcells changes its properties in a more pronounced way than the other (e.g., because of light-induced degradation or because of a difference in temperature coefficients), we will have a mismatch in the two currents; in this case, the subcell with the lower current will limit the current of the tandem. The voltage of the tandem is equal to the sum of the voltages of the two subcells.

As for the fill factor of a tandem, it is difficult to predict. In the case of current mismatch, the fill factor is artificially increased—but because of the loss in current, this does not result in a net increase of the conversion efficiency. One should therefore be very careful when trying to draw conclusions from the value of the fill factor in a tandem (or, even more so, in a multijunction cell) [89].

Why does one designate the intermediate zone between top subcell and bottom subcell as *recombination junction* or *tunnel junction*? The electrons photogenerated in the i layer of the top subcell will travel toward the n layer and then enter into the intermediate zone; similarly, the holes generated in the i layer of the bottom subcell will travel toward the p layer and then enter into the intermediate zone. To ensure current continuity, all electrons travelling downward (in Figure 25) will have to recombine with the holes travelling upward. This recombination process is in general facilitated by "tunnelling" of the electrons and holes from the conduction and valence band, respectively, of the doped layers, into imperfection states (midgap defects) localized at the interface (see Figure 26) so that one also uses the term *tunnel junction*.

The tunnel or recombination junction is composed of an n layer and a p layer; at least one of these layers (or, alternatively, the interface between these two layers) should contain a high density of midgap defects to facilitate recombination. Both layers are generally strongly doped, and, if possible, microcrystalline and not amorphous to enable tunnelling. Poorly designed tunnel or recombination junctions will contribute to increase the series resistance of the cell.

What are the advantages of using the tandem configuration instead of the single-junction configuration? There are some evident advantages.

- Currents take on roughly half the value, and voltages double the value, so that the electrical losses due to the TCO sheet resistance are reduced by roughly a factor of 4.
- The i-layer thicknesses of the subcells can be decreased, and hence the internal fields increased, while both thicknesses combined represent a

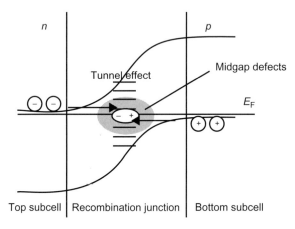

Figure 26 Principle of tunnel–recombination junction in a tandem cell. All electrons exiting from the top subcell have to recombine with the holes exiting from the bottom subcell; their recombination will be assisted by midgap defects (dangling bonds) situated within this junction and by tunnelling of carriers to the midgap defects. *(Reproduced from [1] with permission of the EPFL Press.)*

similar absorption length as in a single-junction configuration; thus, in amorphous silicon subcells, the effect of light-induced degradation will be reduced.
- The subcells can have different band gaps and, thus, may cover each a different range of the solar spectrum.

On the other hand, there is the delicate problem of current matching and a significant increase in the complexity of the fabrication process.

5.2 a-Si:H/a-Si:H Tandems

The simplest form of tandem cell is the one shown in Figure 27: an amorphous silicon top subcell is sitting on top of a bottom subcell, which consists also of amorphous silicon. If the band gaps of both subcells were exactly the same, then the i-layer thickness d_1 of the top subcell would have to be very much smaller than the i-layer thickness d_2 of the bottom subcell; this follows from the exponential absorption profile of the incident light and might lead to either an exceedingly low value of d_1 (and thus an increased tendency for the top subcell to have shunts) or a very large value of d_2 (with a corresponding undesired light-induced degradation effect in the bottom subcell). Thus, one attempts to increase the band gap of the top subcell—either by alloying with some carbon—i.e., using an a-Si,C:H layer [90]—or by changing the deposition

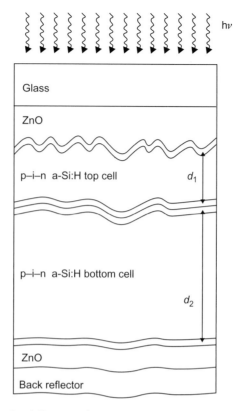

Figure 27 Sketch of a fully amorphous tandem cell (here an a-Si:H/a-Si:H tandem). *(Reproduced from [1] with permission of the EPFL Press.)*

conditions—i.e., by increasing the hydrogen dilution in the deposition plasma) and to decrease the band gap of the bottom subcell (either by alloying with some germanium—i.e., using an a-Si,Ge:H layer—or by changing the deposition conditions—i.e., by decreasing the hydrogen dilution and increasing slightly the deposition temperature). At present, the most successful a-Si:H/a-Si:H tandems do not use any alloying [20]; they can attain a total area stabilized efficiency of more than 7%.

5.3 Triple-Junction Amorphous Cells with Silicon–Germanium Alloys

Triple-junction cells with amorphous silicon–germanium alloys and modules based on this triple-junction design have been produced since the mid-1990s by the firm United Solar Ovonic. These devices are deposited on stainless steel substrates in an n–i–p sequence (see Figure 28). The light

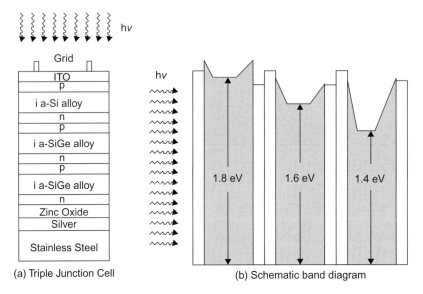

Figure 28 Principle of triple-junction cell with an amorphous silicon top subcell and middle and bottom subcells employing amorphous silicon–germanium alloys: (a) basic structure; (b) schematic band diagram showing the variation in band gap within the individual subcells as used in these cells in order to improve carrier collection. *(Reproduced from [1] with permission of the EPFL Press.)*

enters into the solar cell through the topmost p layer on which an indium tin oxide layer has been deposited as a TCO. In order to improve (reduce) the contact resistance and correspondingly reduce also R_{series}, a metal grid is deposited on top of the ITO layer. The grid has, however, the disadvantage of "shading" the solar cell, thus reducing the photoactive area by almost 10%. In the laboratory, an active-area initial efficiency of 14.6% and an active-area stabilized efficiency of 13% were achieved for small-area cells [21]. The stabilized total-area efficiency is thereby (because of the presence of the grid) only 12.1%: this is at present roughly the same value as the stabilized total-area efficiency of the best micromorph tandem cells (see Section 5.4).

The stainless steel substrates are flexible and thus lend themselves to roll-to-roll fabrication of lightweight, unbreakable modules especially suited for building-integrated photovoltaics. Module efficiencies are, however, considerably lower than the previously mentioned cell efficiencies—about the same value as for tandem a-Si:H/a-Si:H modules on glass. The relatively large discrepancy between the cell and module efficiencies are mainly caused by the method of interconnecting cells to a module (see also Sections 6.2 and 6.3). Unlike the monolithic interconnection that is applicable to nonconducting substrates like glass, relatively large-area cells on the conducting stainless steel substrate are individually connected, very much like the stringing of crystalline silicon cells. These large areas result in large total currents and hence require special precautions to reach low series resistances R_{series}. Therefore, the grid mentioned previously is absolutely necessary to balance losses by series resistance with active-area losses. In addition, the deposition rate has to be increased for economic reasons, whereby the layer quality will generally suffer, resulting in lower cell performance.

5.4 Microcrystalline—Amorphous or "Micromorph" Tandems

One of the most promising tandem cells, is the so-called micromorph tandem, containing, as shown in Figure 29, a microcrystalline silicon bottom subcell and an amorphous silicon top subcell. This structure was introduced by the Neuchâtel group in 1994 [91]. It has since then been the focus of intensive R&D efforts and stabilized small-area cell efficiencies of 12% were reported a few years back [92]. The micromorph tandem is of special interest, because the band gap combination of 1.1 eV/ 1.75 eV, as approximately given by μc-Si:H and a-Si:H, would correspond to the "ideal" band-gap combination for a tandem cell, provided (1) that collection problems can be neglected and (2) that all the photons

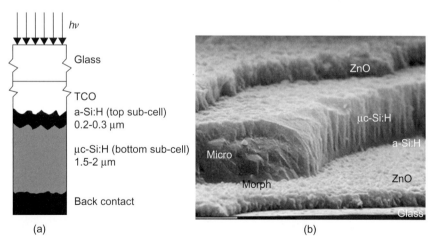

Figure 29 Microcrystalline—amorphous or micromorph tandem cell: (a) basic structure (note that the intermediate reflector between the two subcells is not drawn); (b) electron micrograph. *(Reproduced from [1] with permission of the EPFL Press.)*

with energies above the band-gap energy are usefully absorbed in the corresponding subcell [22]. These conditions are, in practice, far from being fulfilled. The amorphous top subcell suffers from the light-induced degradation effect (Staebler—Wronski effect) and therefore has to be kept as thin as possible, with an i-layer thickness d_i of 200 nm (or less). This means that the top subcell will not absorb enough light.

The remedy used here to enhance photogeneration in the *amorphous top subcell* is to provide an *intermediate reflector* (IR) between the top and the bottom subcells in such a way that the short-wavelength components of the light are reflected back into the amorphous top subcell but the long-wavelength components are (as far as possible) passed on, with little attenuation, into the microcrystalline bottom subcell. As intermediate reflector, one needs basically a material with a refractive index n lower than that of amorphous or microcrystalline silicon ($n \approx 4$). Physically, a ZnO layer, with a refractive index $n = 2$, is an excellent candidate; it was used in the first trials [93]. But ZnO is not a very convenient solution from the fabrication point of view, so later solutions, including doped silicon oxide layers, were introduced. These layers can be conveniently deposited by plasma deposition, just like the a-Si:H and μc-Si:H layers themselves (see [94—96]). The design of an efficient and spectrally selective intermediate reflector remains one of the primary research topics for micromorph tandems [97].

The *microcrystalline bottom subcell* has also to be kept as thin as possible. Here it is not the light-induced degradation, but the fabrication cost, that is the issue, The fabrication costs associated with the deposition of the microcrystalline silicon intrinsic layer are one of the main cost factors for micromorph module production. On one hand, the plasma-deposition equipment represents a very large investment. Now operation and depreciation costs for the plasma reactor are proportional to layer thickness. If the layer, which has to be deposited, is thick, the deposition process will correspondingly take a long time, thus diminishing the production throughput. Furthermore, the materials costs for gas inputs into the plasma reactor (silane and hydrogen) are also proportional to the thickness of the deposited layer. Therefore, one currently proposes to use microcrystalline silicon bottom cells with a total thickness of less than 1 μm [23]. Referring to Section 4.5, and particularly to Figure 23, one can immediately see that this will only be possible with a very effective light-trapping scheme. The multiplication factor m by which the optical path becomes longer than the intrinsic layer thickness d_i should be as high as

10, or even higher, in order to absorb a sufficiently large part of the incoming light.

Thus, one may state that most of the research problems to be solved in the coming years, in order to increase the efficiency of micromorph tandems, are related to improvements in light management (e.g., see [98]).

An interesting further development of the micromorph tandem concept is the extension to *triple-junction cells*. In these triple-junction cells, one generally retains an amorphous subcell on top and a microcrystalline subcell on the bottom. For the middle subcell, however, a number of different possibilities have been tried out.

- An amorphous silicon—germanium alloy (a-Si,Ge:H): the disadvantages are here (1) the relatively high cost of germanium and (2) the pronounced light-induced degradation present in these alloys.
- A microcrystalline silicon (μc-Si:H): the disadvantages here are (1) the low open-circuit voltage V_{oc} that can be achieved with such a subcell, (2) light-management problems (i.e., how to obtain a high enough current with a sufficiently thin μc-Si:H middle subcell?), and (3) higher fabrication costs associated with the deposition of two microcrystalline subcells.
- Unalloyed amorphous silicon (a-Si:H): the disadvantages here are linked to light-management problems and to the light-induced degradation effect.

One comes therefore to the conclusion that today there does not exist so far a clear preference for the structure of micromorph triple-junction cells. The different possibilities need further in-depth investigation, especially with respect to the light-management problem, which now truly becomes quite complex. On the other hand, if one goes for the triple-junction configuration and the increased fabrication costs associated with such a configuration, then one would hope to be able to avoid the use of intermediate reflectors.

6. MODULE PRODUCTION AND PERFORMANCE

6.1 Deposition of the Thin-Film Silicon Layers

The deposition of the thin-film silicon layers is the central step in the fabrication of solar modules based on amorphous and microcrystalline silicon. The method generally used—by all manufacturers up to now is plasma-enhanced chemical vapour deposition (PECVD) in a capacitively coupled reactor, as indicated schematically in Figure 30. For the deposition of the intrinsic layers, a mixture of silane and hydrogen is used; for the deposition of the relatively thin doped layers, diborane or trimethlyboron is added for the p layer, and phosphine is added for the n layer (see Section 2.3).

At present, the deposition of the intrinsic silicon layer is the production step that takes the largest share of total module production costs. It is also one of the most critical steps from the point of view of module performance. The high production costs for this fabrication step stem principally from two factors:

1. High investment costs necessary for the PECVD equipment. This means that depreciation and operating costs are correspondingly high. In order to reduce this part of module fabrication costs, one strives to keep the total deposition time for the intrinsic layers as low as possible by
   a. making the intrinsic layers as thin as possible (this is a question of light management—see Section 4.5) and
   b. increasing the deposition rate for the intrinsic layer without reducing layer quality; this aspect will be treated hereunder.
2. Consumption of process gases, especially of silane and hydrogen. To minimize this part of fabrication costs, one has to
   a. keep the intrinsic layers as thin as possible, as just mentioned;
   b. use a plasma-deposition process in which a large part of the silicon contained in the silane gas fed into the deposition system is actually deposited as thin-film layer; and
   c. use process parameters that allow for layer deposition with only a moderate hydrogen dilution ratio (this last criterion is specially critical for the deposition of microcrystalline silicon layers).

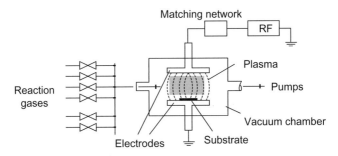

Figure 30 Capacitively-coupled plasma-deposition system. *(Reproduced from [1] with permission of the EPFL Press.)*

A capacitively coupled plasma-deposition system, as indicated in Figure 30, contains electrically neutral molecules, some of which are the growth radicals, actively contributing to layer growth. It also contains electrons and positively charged ions. Increasing the deposition rate is generally accomplished by increasing the electrical power fed into the deposition system. This not only increases the densities of growth radicals but also enhances the ion bombardment on the growing silicon layer. High-energy ion bombardment generally leads to the creation of additional defects on the growing layer and must therefore be avoided. The following are the two most common methods to avoid high-energy ion bombardment.

1. Increase the plasma excitation frequency from the standard industrial frequency of 13.56 MHz to frequencies in the *very high frequency* (VHF) range, typically to frequencies between 60 and 100 MHz [99,100]. As the plasma excitation frequency is increased, the so-called sheaths—i.e., thin regions with strong electrical fields next to each of the electrodes in Figure 30—become thinner. The voltage drop across the sheaths is reduced and, accordingly, the positive ions hitting the growing silicon surface are not accelerated to the same extent [101]. Note, however, that the design of a large-area deposition system becomes more difficult as the plasma excitation frequency is increased and the corresponding wavelength of the electrical field powering the plasma is decreased and approaches the dimensions of the electrode. As a consequence, the deposition tends to become nonuniform [102]. Uniform deposition can then only be achieved with a relatively sophisticated (and costly) design of the deposition system; one uses, e.g., lens-shaped [103] or ladder-shaped [28] electrodes.

2. An increase in the deposition pressure: at higher pressures there are more collisions between the various species in the plasma, reducing thereby the energies of the ions impinging upon the growing surface. To arrive at the highest deposition rates, relatively high pressures are combined with high excitation frequencies [104].

There are a number of other factors that have to be taken into account when designing a deposition system for thin-film silicon layers.

1. The formation of powder during deposition must be avoided. Powder is formed by premature chemical reactions in the gas phase—i.e., by so-called plasma polymerization. An increase in plasma excitation frequency, in general, will lead to reduced powder formation [105].

2. The presence of powder during deposition, on the other hand, will lead to shunts in the solar cell being fabricated. There is therefore a

need to periodically clean the deposition system; this is traditionally done by plasma etching with relatively powerful etchant gases such as SF_6 and NF_3. These gases are, however, "greenhouse gases" [106], which contribute to global warming; their use in industrial systems is therefore increasingly being avoided. Instead, one turns to alternative cleaning gases such as fluorine [107] or reverts to purely mechanical cleaning procedures—the latter are, however, cumbersome and can entail health problems for the workers involved in the process.

3. The need to minimize gas consumption and maximize gas utilisation. This factor needs at present to be investigated in more detail: there is some published work on silane utilisation [108]. On the other hand, an increase in deposition pressure seems to result generally in a strong increase of hydrogen consumption for microcrystalline layers.

4. The need to avoid the deposition of porous layers, which quite generally have inferior quality (they lead to solar cells with higher degradation and ageing effects and, in more extreme cases, to solar cells with lower initial efficiencies).

The design of economically attractive deposition systems for thin-film silicon modules is therefore a very important topic, which will have a decisive effect on the future success of this type of modules.

To conclude, one may remark that *hot wire deposition* (also called *catalytic CVD*) is increasingly being proposed and tested as an interesting alternative to plasma deposition (e.g., see [109–112]). This method is, however, at present not yet used in industrial module fabrication. Hot wire deposition completely avoids ion bombardment; instead of a plasma, one uses a filament heated to very high temperatures (around 1800°C) to obtain silane dissociation and generation of growth radicals. This method also has the advantage of a relatively high gas utilisation. Potentially it should be able to lead to the fabrication of solar modules at high deposition rates and with excellent performances. Excellent solar cells have up to now been fabricated by hot wire deposition at low deposition rates, and individual layers with excellent properties have also been fabricated at high deposition rates. But the combination of a high efficiency and a high deposition rate has so far only been demonstrated in isolated cases. Furthermore, it would seem that a certain amount of ion bombardment with low-energy ions may, in some cases, be beneficial for the fabrication of solar cells. Thus, the combination of plasma deposition and hot wire deposition [113] may be especially interesting to investigate.

6.2 Substrate Materials and Transparent Contacts

There are basically two possible configurations for thin-film silicon solar cells.

1. In a so-called superstrate configuration (indicated schematically in Figure 31), where glass is used as the support on which the solar cell is deposited and at the same time also as cover through which light

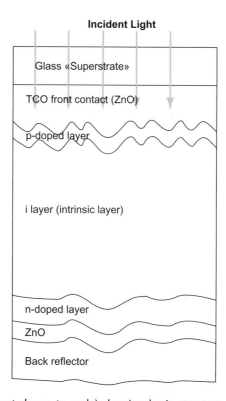

Figure 31 Sketch (not drawn to scale) showing basic structure of a single-junction thin-film silicon solar cell in the «superstrate configuration». The thickness of the glass—TCO combination is basically determined by the glass thickness, ranging from 0.5 mm to 4 mm, whereas the TCO layer thickness is typically around 1 µm. The p—i—n layer sequence and back reflector combined amount also to about 1 µm in thickness. The sketch merely illustrates the succession of different layers. Note that the ZnO between the n layer and the back reflector is used only for optical index matching and can therefore be very thin if the back reflector is an electrically conductive and highly reflecting material. If the back reflector is a dielectric layer (e.g., white paint), the adjacent ZnO has to act as a contact layer and be correspondingly thicker, subject to similar requirements as for the front TCO discussed in Section 4.4. To protect against environmental effects, various materials for encapsulation at the back of the cell structure (not shown) are being employed.

enters into the solar cell. As the glass support is on top of the finished modules (facing the incident light) this is called the *superstrate* configuration, although one generally continues to use the term *substrate* for the glass when referring to the fabrication process. In this case, one deposits first those silicon layers through which the light enters into solar cell. Here, the deposition sequence is p—i—n for amorphous silicon solar cells; it is advantageous if the light enters into the cell through the p layer and not through the n layer, as explained in Section 4.1. The main advantages of using glass are (a) subsequent deposition of the solar cell is relatively straightforward, (b) glass offers good protection against environmental influences, and (c) glass is known to remain stable for a very long time when exposed to sunlight and if adequately designed to withstand extreme weather conditions. The disadvantages of using glass are (a) weight, (b) possibility of damage through breaking, and (c) rigidity—a limitation for certain applications.

2. In the substrate configuration, which is shown schematically in Figure 32, light enters into the cell from the opposite side, and the substrate is the bottom layer. The deposition sequence, in this case, is n—i—p. The substrate does not need to be transparent; the most common substrate used here is stainless steel, a solution pioneered by the firm United Solar Ovonic. Because stainless steel is electrically conducting, it does not lend itself easily to the monolithic cell-interconnection scheme described in the Section 6.3. In fact, the solar cells produced by the firm United Solar Ovonic are connected by separate external wiring to form modules. Deposition of thin-film silicon solar cells on stainless steel has the advantage of being relatively straightforward. Increasingly, one attempts to use polymers as substrates. Here solar cell deposition is more difficult, because it is impaired by outgassing from the polymer and by temperature limitations of the latter. On the other hand, polymers are electrically isolating and can be used together with the monolithic cell-interconnection scheme, which is described later. Many polymers are initially transparent but become yellow under the influence of the UV component within sunlight. They therefore cannot be used in the same manner as glass is used in the superstrate configuration—i.e., as protective cover material. Indeed, in the substrate configuration, a protection layer or cover material must be added to rule out environmental effects. It should come in the form of flexible foil, which has (just like glass) to be

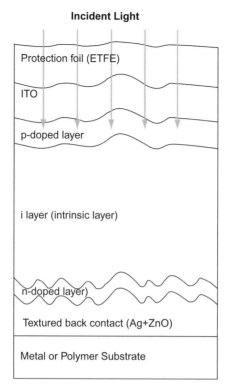

Figure 32 Sketch (not drawn to scale) showing basic structure of a single-junction thin-film silicon solar cell in the «substrate configuration». The substrate and the protection foil are each about 0.1 to 0.2 mm thick; the entire cell structure, including the ITO front contact layer and triple-junction structures, are typically about 1 μm thick. The sketch merely illustrates the succession of different layers.

transparent, mechanically stable, and UV resistant. This protection foil is one of the important cost factors for the substrate configuration. At present one mainly uses ethylene tetrafluoroethylene (ETFE) as protection foil. The substrate configuration is generally used to fabricate lightweight, flexible, and unbreakable modules that are particularly suited for transport to remote locations and for building-integrated PV (BIPV), particularly for roof integration.

Most manufacturers opt at the moment for the superstrate configuration. Here the glass is coated with a transparent conductive oxide serving as a front contact. For commercial modules, one so far employs mostly fluorine-doped SnO_2 as TCO material; SnO_2 is usually deposited by atmospheric pressure chemical vapour deposition (APCVD) "online" within the float-zone process employed for glass production [114]. This is probably the most cost-effective way of producing the glass substrate together with the transparent front contact. As mentioned in Section 4.5, the transparent contact has to be textured in order to obtain light trapping. In the case of the present SnO_2 layers, which are manufactured online, the texturing is generally not optimized for light trapping in a particular cell configuration. One is therefore currently developing highly textured SnO_2 layers with off-line processes on a pilot-line basis (e.g., see [115]). If these experimental processes lead to the desired result, one will probably attempt later to incorporate them (if possible) as an online version within the float-zone process of the glass producer. (Note that SnO_2 is chemically reduced by a high-power hydrogen-rich plasma as used for depositing μc-Si:H. Therefore, if one deposits a μc-Si:H solar cell on a SnO_2 layer, one has first to cover the latter by a thin protective layer of TiO_2 or ZnO. This consideration is of importance for the deposition of single-junction μc-Si:H cells, as studied in research laboratories, but it is irrelevant in case of micromorph tandems, where the SnO_2 is no longer exposed to the extreme deposition conditions mentioned previously but is instead protected by the amorphous top subcell, which is deposited first.)

As an alternative to SnO_2, one is at present increasingly introducing ZnO as transparent contact material [116]. ZnO acts as an effective diffusion barrier against any contaminants diffusing from the underlying glass substrate. Another advantage is that it can be deposited in a relatively easy manner by the manufacturer of the thin-film solar cell and at a lower deposition temperature than SnO_2. There are two main deposition methods for ZnO: (a) sputtering and (b) low-pressure chemical vapour deposition (LPCVD). Sputtering leads to smooth layers that have to be chemically etched in order to obtain the desired texture for light trapping. Sputtered ZnO layers are generally doped with aluminium in order to obtain the desired high conductivity. The LPCVD process leads directly to textured ZnO layers, which have a high effectiveness for light trapping. LPCVD ZnO layers are generally doped with boron.

In the substrate configuration, the TCO front contact layer is deposited at the end of the deposition process. It has correspondingly to be deposited at a relatively low temperature (typically <200°C). Therefore, SnO_2 is not an option to be considered. One can use ZnO. One can also use indium tin oxide (ITO = 0.9 In_2O_3/0.1 SnO_2). Note that ITO is never used in the superstrate configuration because the In_2O_3 would be strongly chemically reduced by the exposure to hydrogen during the initial phases of the PECVD, and thus, ITO would widely loose its transparency. Besides, indium would diffuse into the silicon layers, rendering the latter unusable.

Substrates and transparent contact layers, as well as reflecting layers, are an important part of any thin-film silicon cell. They determine to a large extent the performance (especially the current density) of the solar cell. The interaction of textured front and back layers is complex and different for the superstrate and substrate configurations. (See [117] for an analysis, which is restricted, however, to the case of single-junction microcrystalline solar cells.) These front and back layers are therefore increasingly the focus of current research efforts, as already described in Section 4.5. The focus on these parts of the solar cell is all the more warranted, as they also account for a significant part of the total fabrication cost—together with other "external features," cell-interconnection schemes (see Section 6.3), module encapsulation, and current collection schemes (e.g., metallic grids that can be helpful in reducing the series resistance).

6.3 Laser Scribing and Cell Interconnection

Individual cells have to be electrically interconnected, most often by a series connection in order to form a complete module. Whenever the thin-film silicon solar cell has been deposited on an electrically isolating material, such as glass or polymer, one uses a monolithic interconnection scheme as shown in Figure 33. (The monolithic interconnection scheme cannot be used if the thin-film silicon solar cells have been deposited on

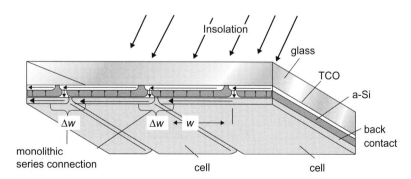

Figure 33 Monolithic series connection of thin-film silicon solar cells (photoactive cell width w, photoinactive interconnection width Δw). *(Reproduced from [1] with permission of the EPFL Press.)*

an electrically conductive substrate, like stainless steel, unless an isolating interlayer is first deposited. Such an interlayer would, however, be prone to electrical shorts, and would lead to additional complications or costs.)

In the monolithic interconnection scheme one proceeds as follows if one is using the superstrate configuration with a glass superstrate.
1. The front TCO is deposited on the glass sheet.
2. The front TCO layer is cut into stripes; this separation is best done by removing the TCO with a process called *laser scribing* [118].
3. The semiconductor layers are deposited in the sequence p–i–n (or p–i–n, p–i–n for tandem cells).
4. These semiconductor layers are now also cut into stripes of the same width as that of the TCO stripes, with scribe lines slightly offset from the previous TCO scribe lines. This separation is again preferably done by laser scribing. Thus, the grooves of the semiconductor layer lie directly above an unremoved area of the front TCO.
5. The metallic (or ZnO) back contact is deposited. This back contact layer thus makes contact with the front TCO layer through the grooves in the semiconductor layer.
6. Finally, the metallic (or ZnO) back contact is "cut up" into stripes of the same width as the previous stripes, once again with a slight offset from the previous semiconductor scribe below. Thus, one arrives at a series-connection structure, where the photoactive cell width is w (taking into account two photoinactive spacings between the three scribes) and the total width of the interconnection is Δw (the cross-section of this structure is shown in Figure 34).

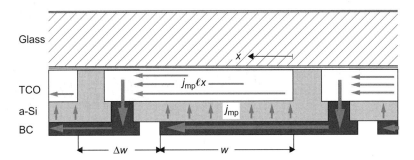

Figure 34 Monolithic series connection of thin-film silicon solar cells, shown as cross-section, with corresponding current paths (photoactive cell width w, photoinactive interconnection width Δw, cell length l, current density at the maximum power point j_{mp}, back contact BC). *(Reproduced from [1] with permission of the EPFL Press.)*

For optimal cell performance, the photoinactive interconnection width Δw should be as low as possible; it is given by the precision and reproducibility of the laser-scribing system. In the best cases, it is around 0.2 mm [119]. The interconnection area, defined by $\Delta w \times l$, is lost for photocarrier generation. Thus, there is a relative loss of incoming light, given by $(\Delta w)/(w + \Delta w)$.

This loss term (areal loss) decreases with increasing stripe width w. On the other hand, because of the sheet resistance R_{sheet} of the TCO layer, there is an increase in series resistance R_{series} and a corresponding electrical loss of the solar cell leading to a relative power loss, which is proportional to w^2, as given in Section 4.5. The sum of the relative areal and electrical losses reaches a minimum for an optimum value of w. This value is of the order of 1 cm, with slightly lower values for single-junction cells and slightly higher values for tandems.

The monolithic series connection can also be applied to the substrate configuration, as long as one is using an isolating substrate such as a polymer sheet. In this case, the laser-scribing process is, for technological reasons, a little more delicate to implement.

If the laser-scribing system is not properly adjusted, partial short circuits will be created on the module; this will lead to low values of shunt resistance R_{shunt} and to a low fill factor FF. Such a situation is definitely to be avoided.

The monolithic series connection of cells to form large-area modules is one of the key features of thin-film solar cell technologies, ensuring a higher degree of automation and an enhanced reliability, as compared to the stringing procedures used for interconnecting cells in wafer-based crystalline silicon technologies.

6.4 Module Encapsulation

Module encapsulation is indeed a key factor determining the long-term reliability of photovoltaic solar modules. It is an especially critical factor for thin-film modules. For all thin-film modules, one of the main goals of module encapsulation is to protect the semiconductor layers, as well as the transparent contact layers, against the influx of humidity. To this end, special polymer foils and dedicated sealing techniques have been developed. Many different combinations are used in commercial modules; most solutions are proprietary. Thus, at present we can only emphasize the importance of this step. For general information on this problem, the reader is referred to [120–122].

6.5 Module Performance

There are several factors influencing the performance of photovoltaic modules under actual application conditions:
1. spectrum of the incoming light,
2. intensity of the incoming light,
3. angle of incidence of the incoming light, and
4. operating temperature of the module.

The spectrum of sunlight tends to change over the course of the day (with a shift toward the red in the morning and evening). It also changes with climatic and geographical conditions (with a shift toward the blue in the presence of snow and water surfaces). In general, single-junction amorphous silicon modules perform better than crystalline silicon modules under blue shift of the incoming light and worse than the latter under red shift. Note that a blue shift is regularly observed for indoor conditions, especially for illumination with energy-saving light sources (e.g., fluorescent lamps). As for tandem and multijunction cells, they are more sensitive to changes in the incident spectrum than single-junction cells [123]. This means they must be carefully designed for the average spectrum prevailing in actual field conditions at the site of deployment.

As the intensity of the incoming light is varied from standard conditions (intensity of approximately 100 mW/cm^2) to lower values, the conversion efficiency of the cell will be reduced. The extent of this reduction depends very much on the shunt resistance R_{shunt} within the cell, and this parameter in turn depends on the fabrication process. A low value of R_{shunt} (and a poor solar cell performance at low light intensities) is an indication of a faulty manufacturing process. Amorphous silicon solar cells with well-mastered fabrication processes can exhibit a very high value of R_{shunt} and thus also excellent performance for very low light intensities, making them particularly suitable for indoor applications [124]. On the other hand, present thin-film silicon solar cells and modules are not suited for higher light intensities—i.e., for applications with sunlight concentration.

As for the angle of incidence of the incoming light, it evidently also has, for optical reasons, an influence on the efficiency of the solar module. Its influence depends on the module design, especially on the first layers, sheets, or foils through which light enters the solar module. Thus, it is almost impossible to make any general statements. Nevertheless, as the design and the fabrication of solar modules become more standardized and as their application progresses, this effect should be taken into consideration and experimentally quantified (see also [125]).

One of the most important factors that result in reduced solar module efficiency is higher module operating temperature. Figure 35 shows how various solar cell technologies show a decrease in efficiency when their operating temperature is higher than 25°C (the temperature for which the module efficiency is specified in the data sheets—it corresponds to the standard test conditions, STC). One clearly notes from Figure 35 that wafer-based crystalline silicon modules suffer a relative loss in efficiency of about 20% when their operating temperature is increased from 25°C to 75°C (curves a, b, and c in Figure 35), whereas thin-film silicon modules suffer a relative loss of only half that value (curves d and e in Figure 35). In very many applications, the solar modules are typically operating at higher temperatures—between 60°C and 80°C. This is true for applications in tropical countries, and also for BIPV in temperate countries during the summer months. Thus,

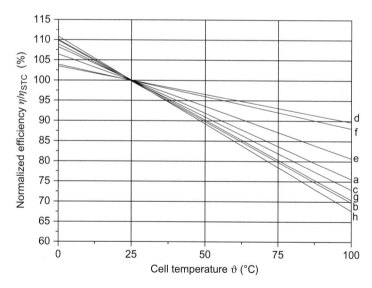

Figure 35 Effect of cell operating temperature on cell efficiency, normalized to the efficiency value at standard test conditions (STC)—i.e., to η_{STC} (at 25 °C). a, b, c = different types of crystalline wafer-based modules; d = amorphous silicon module; e = micromorph tandem module; f = CdTe module; h, g = CIGS modules. *(Courtesy of Wilhelm Durisch, PSI, Villigen, Switzerland. Reproduced from [1] with permission of the EPFL Press.)*

thin-film silicon solar modules have a "hidden" advantage over wafer-based crystalline silicon modules that does not transpire from the datasheets: their efficiency drop at higher temperatures is much less pronounced (see also [123,126]).

Because of the various effects described previously, the so-called relative performance of photovoltaic installations containing thin-film silicon solar modules is, in general, about 10% higher than that of installations with wafer-based crystalline silicon modules (see [1], Table 6.4). Note that the "relative performance" is given in kWh/kW$_p$ and denotes the total amount of electrical energy in kWh delivered by the module during the period of one year, divided by the nominal installed power in kW$_p$.

6.6 Field Experience

The field experience is a very important aspect: only with actual data from modules exposed over several years to outdoor light and weather conditions can we judge whether a given solar technology is mature enough for widespread practical deployment as an energy source.

Amorphous silicon modules have now been in operation for almost 30 years. Whereas modules produced during the 1980s did have unexpected problems and failures, the next generation of modules that came on the market in the 1990s have shown remarkably stable and reliable performance. Chapter 7 in [1] gives several examples of photovoltaic installations, in Europe, in the USA and in Brazil, with performance data for 10 years or more; in all cases, the performance is indeed quite stable after an initial phase of 1 to 2 years with distinct degradation due to the Staebler—Wronski effect (see Section 2.1). In the long run, *all* photovoltaic modules show some drop in efficiency [127]—i.e., a yearly relative decrease $\Delta\eta/\eta$ in efficiency, of about 0.5%, due partly to the imperfections of the encapsulation. For amorphous silicon modules, $\Delta\eta/\eta$ may be slightly higher, possibly about 1%. Micromorph modules (i.e., modules with microcrystalline—amorphous silicon tandem cells) have not been around for very long so that we do not have long enough field experience with these types of modules; however, from what has so far been observed, one may reasonably expect their long-term performance to be slightly better than that of amorphous silicon modules. A further reason to have full confidence in "micromorph" modules is that the fabrication processes and individual layers used here are very similar (and partly identical) to those used for amorphous silicon modules.

7. CONCLUSIONS

Thin-film silicon solar cells and modules have at present a significant disadvantage with respect to wafer-based crystalline silicon modules and even with respect to some other thin-film modules such as CIGS modules: their conversion efficiency is quite a bit lower. Commercial large-area modules do have efficiencies that are now nearing 10% for micromorph tandems. But commercial wafer-based crystalline silicon modules have efficiencies that can lie in the 15% to 18% range. And we may expect efficiencies to slightly increase over the coming years for all technological options. Thus, the relative disadvantage of thin-film silicon modules regarding efficiencies cannot be expected to disappear in the future (although it is somewhat mitigated by the better annual relative performance in kWh/kW$_p$, see Section 6.5).

However thin-film silicon modules also have significant advantages, which may become decisive in the coming years:

1. Only abundant and nontoxic raw materials are used (they share this advantage, of course, with wafer-based crystalline silicon modules).
2. If properly designed, their fabrication process is nonpolluting (just like the fabrication process for wafer-based crystalline silicon modules).
3. The temperature dependence of the PV performance is significantly lower than for wafer-based silicon modules.
4. The energy required for their fabrication is very low and can be recuperated in less than a year (they share a low energy recovery time with most other thin-film solar technologies; for wafer-based silicon technologies, the energy recovery times are about twice that of thin-film technologies).
5. The temperatures used in the fabrication process for thin-film silicon modules are around 200°C and thus much lower than those used for most other solar technologies. This advantage is of special significance for the deposition of thin solar cells on lightweight polymer substrates.
6. Large-area modules can be easily fabricated on substrates exceeding 1 m^2 area, with monolithic interconnection of the individual cells.

For all these reasons we can expect thin-film silicon to remain an interesting option for solar cells and modules in the coming decades. Research and development efforts for thin-film silicon modules are continuing worldwide and with high intensity; one may, thus, expect that significant progress will still be achieved for both efficiency and cost.

The future cost and deployment of various solar technologies depends on factors that cannot be exactly foreseen, like the cost and availability of various raw materials—and that of electricity as needed for large-scale module production, especially for crystalline silicon wafers. However, the unique combination of properties 1 to 6 listed previously is so far not matched by any other solar technology. This probably means that thin-film silicon will remain the most interesting option for at least a part of solar applications in the foreseeable future.

REFERENCES

[1] A. Shah (Ed.), Thin-Film Silicon Solar Cells, EPFL Press, 2010.
[2] R.C. Chittick, J.H. Alexander, H.F. Sterling, Preparation and properties of amorphous silicon, J. Electrochem. Soc. 116 (1969) 77−81.
[3] W.E. Spear, P.G. Le Comber, Substitutional doping of amorphous silicon, Solid State Commun. 17 (1975) 1193−1196.
[4] W.E. Spear, P.G. Le Comber, Electronic properties of substitutionally doped amorphous Si and Ge, Phil. Mag. B 33 (1976) 935−949.
[5] D.E. Carlson, C.R. Wronski, Amorphous silicon solar cells, Appl. Phys. Lett. 28 (1976) 671−673.
[6] D.E. Carlson, C.R. Wronski, A. Triano, R.E. Daniel, Solar cells using Schottky barriers on amorphous silicon, Proceedings of the 22nd IEEE Photovoltaic Specialists Conference, Baton Rouge, 1976, pp. 893−895.
[7] D.L. Staebler, C.R. Wronski, Reversible conductivity change in discharge produced amorphous silicon, Appl. Phys. Lett. 31 (1977) 292−294.
[8] S. Benagli, D. Borrello, E. Vallat-Sauvain, J. Meier, U. Kroll, J. Hötzel, et al., High-efficiency amorphous silicon devices on LPCVD−ZnO TCO prepared in industrial KAI TM-M R&D reactor, Proceedings of the 24th EC PV Solar Energy Conference, 2009, pp. 2293−2298.
[9] S. Veprek, V. Marecek, The preparation of thin layers of Ge and Si by chemical hydrogen plasma transport, Solid State Electron. 11 (1968) 683−684.
[10] S. Usui, M. Kikuchi, Properties of heavily doped glow-discharge silicon with low resistivity, J. Non-Crystalline Solids 34 (1979) 1−11.
[11] C. Wang, G. Lucovsky, Intrinsic microcrystalline silicon deposited by remote PECVD: A new thin-film photovoltaic material, in: Proceedings of the 21st IEEE Photovoltaic Specialists Conference, 1990, pp. 1614ff.
[12] M. Faraji, S. Gokhale, S.M. Choudari, M.G. Takwale, S.V. Ghaisas, High mobility hydrogenated and oxygenated silicon as a photo-sensitive material in photovoltaic applications, Appl. Phys. Lett. 60 (1992) 3289−3291.
[13] R. Flückiger, J. Meier, H. Keppner, U. Kroll, O. Greim, M. Morris, et al., Microcrystalline silicon prepared with the very high frequency glow discharge technique for p−i−n solar cell applications, Proceedings of the 11th EC PV Solar Energy Conference, 1992, pp. 617ff.
[14] J. Meier, R. Flückiger, H. Keppner, A. Shah, Complete microcrystalline p−i−n solar cell—Crystalline or amorphous cell behaviour? Appl. Phys. Lett. 65 (1994) 860−862.
[15] J. Meier, P. Torres, R. Platz, S. Dubail, U. Kroll, J.A. Anna Selvan, et al., On the way towards high effciency thin film silicon solar cells by the "micromorph" concept, Proc. Mater. Res. Soc. Symp. 420 (1996) 3−14.

[16] P. Torres, J. Meier, R. Flückiger, U. Kroll, J. Anna Selvan, H. Keppner, et al., Device grade microcrystalline silicon owing to reduced oxygen contamination, Appl. Phys. Lett. 69 (1996) 1373–1375.
[17] M.N. Van den Donker, S. Klein, B. Rech, F. Finger, W.M.M. Kessels, M.C.M. van de Sanden, Microcrystalline silicon solar cells with an open-circuit voltage above 600 mV, Appl. Phys. Lett. 90 (2007) (Paper No. 183504)
[18] B.J. Yan, G.Z. Yue, Y.F. Yan, C.S. Jiang, C.W. Teplin, J. Yang, et al., Correlation of hydrogen dilution profiling to material structure and device performance of hydrogenated nanocrystalline silicon solar cells, Proc. Mater. Res. Soc. Symp. 1066 (2008) 61–66.
[19] V. Smirnov, S. Reynolds, F. Finger, R. Carius, C. Main, Metastable effects in silicon thin films: Atmospheric adsorption and light induced degradation, J. Non-Crystalline Solids 352 (2006) 1075–1078.
[20] P. Lechner, W. Frammelsberger, W. Psyk, R. Geyer, H. Maurus, D. Lundszien, et al., Status of performance of thin film silicon solar cells and modules, Proceedings of the 23rd EC PV Solar Energy Conference, 2008, pp. 2023–2026.
[21] J. Yang, A. Banerjee, S. Guha, Triple-junction amorphous silicon alloy solar cell with 14.6% initial and 13.0% stable conversion efficiencies, Appl. Phys. Lett. 70 (1997) 2975–2977.
[22] A. Shah, M. Vanecek, J. Meier, F. Meillaud, J. Guillet, D. Fischer, et al., Basic efficiency limits, recent experimental results and novel light-trapping schemes in a-Si:H, μc-Si:H and "micromorph tandem" solar cells, J. Non-Crystalline Solids 338–340 (2004) 639–645.
[23] H. Knauss, E.L. Salabas, M. Fecioru, H.D. Goldbach, J. Hötzel, S. Krull, et al., Reduction of cell thickness for industrial micromorph tandem modules, Proceedings of the 25th EC PV Solar Energy Conference, 2010, pp. 3064–3067.
[24] T. Matsui, C.W. Chang, T. Takada, M. Isomura, H. Fujiwara, M. Kondo, Thin film solar cells based on microcrystalline silicon-germanium narrow-gap absorbers, Sol. Energy Mater. & Sol. Cells 93 (2009) 1100–1102.
[25] T. Shimizu, Staebler–Wronski effect in hydrogenated amorphous silicon and related alloy films, Jpn. J. Appl. Phys. 43 (2004) 3257–3268.
[26] E. Bhattacharya, A.H. Mahan, Microstructure and the light-induced metastability in hydrogenated amorphous silicon, Appl. Phys. Lett. 52 (1988) 1587–1589.
[27] A.H. Mahan, D.L. Williamson, B.P. Nelson, R.S. Crandall, Characterization of microvoids in device-quality hydrogenated amorphous silicon by small-angle X-ray scattering and infrared measurements, Phys. Rev. B 40 (1989) 12024–12027.
[28] H. Takatsuka, M. Noda, Y. Yonekura, Y. Takeuchi, Y. Yamauchi, Development of high efficiency large-area silicon thin film modules using VHF-PECVD, Sol. Energy 77 (2004) 951–960.
[29] W. Beyer, H. Wagner, F. Finger, Hydrogen evolution from a-Si:C:H and a-Si:Ge:H alloys, J. Non-Crystalline Solids 77–78 (1985) 857–860.
[30] W. Frammelsberger, R. Geyer, P. Lechner, D. Lundszien, W. Psyk, H. Rübel, et al., Entwicklung und Optimierung von Prozessen zur Herstellung hocheffizienter grossflächiger Si-Dünnschicht-PV-Module, Abschlussbericht, Bundesministerium für Umwelt, Naturschutz und Reaktorsicherheit, 2004, Förderkennzeichen 0329810A.
[31] J.D. Cohen, Electronic structure of a-Si:Ge:H, in: T. Searle (Ed.), Amorphous Silicon and Its Alloys, vol. 19 of EMIS Dataviews Series, INSPEC, London, 1998.
[32] F. Finger, W. Beyer, Growth of a-Si:Ge:H alloys by PECVD, in: T. Searle (Ed.), Amorphous Silicon and its Alloys (1998), vol. 19 of EMIS Dataview Series, INSPEC, London, 1998.
[33] P. Wickboldt, D. Pang, W. Paul, J.H. Chen, F. Zhong, C.C. Chen, et al., High performance glow discharge a-$Si_{1-x}Ge_x$:H of large x, J. Appl. Phys. 81 (1997) 6252–6267.

[34] J. Yang, S. Guha, Status and future perspective of a-Si:H, a-SiGe:H, and nc-Si:H thin film photovoltaic technology, in: A.E. Delahoy, L.A. Eldada, Thin film solar technology, Proceedings of the SPIE, vol. 7409, 2009, pp. 74090C1−74090C14.

[35] C.R. Wronski, Review of direct measurements of mobility gaps in a-Si:H using internal photoemission, J. Non-Crystalline Solids 141 (1992) 16−23.

[36] M. Vaneček., J. Stuchlik, J. Kocka, A. Triska, Determination of the mobility gap in amorphous silicon from a low temperature photoconductivity measurement, J. Non-Crystalline Solids 77 & 78 (1985) 299−302.

[37] J. Tauc, A. Grigorovici, A. Vancu, Optical properties and electronic structure of amorphous germanium, Physica. Status Solidi. 15 (1966) 627−637.

[38] T.M. Mok, S.K. O'Leary, The dependence of the Tauc and Cody optical gaps associated with hydrogenated amorphous silicon on the film thickness: αl experimental limitations and the impact of curvature in the Tauc and Cody plots, J. Appl. Phys. 102 (2007) (Paper No.113525).

[39] R.W. Collins, A.S. Ferlauto, G.M. Ferreira, C. Chen, J. Koh, R.J. Koval, et al., Evolution of microstructure and phase in amorphous, protocrystalline and microcrystalline silicon studied by real time spectroscopic ellipsometry, Sol. Energy Mater. & Sol. Cells 78 (2003) 143−180.

[40] A. Fontcuberta i Morral, P. Roca i Caborrocas, C. Clerc, Structure and hydrogen content of polymorphous silicon thin films studied by spectroscopic ellipsometry and nuclear measurements, Phys. Rev. B 69 (2004) (Paper No 125307)

[41] R.A. Street, Hydrogenated Amorphous Silicon, Cambridge University Press, Cambridge, 1991.

[42] A.V. Shah, H. Schade, M. Vaneček, J. Meier, E. Vallat-Sauvain, N. Wyrsch, et al., Thin-film silicon solar cell technology, Prog. Photovolt: Res. Appl. 12 (2004) 113−142.

[43] H. Overhof, P. Thomas, Electronic transport in hydrogenated amorphous semiconductors, Springer Tracts in Modern Physics, vol. 114, Springer Verlag, Berlin, 1989, pp. 26, 77.

[44] M. Python, Microcrystalline solar cells, growth, and defects, Ph. D. thesis, University of Neuchâtel, 2009, Section 3.8.

[45] M. Python, D. Dominé, T. Söderström, F. Meillaud, C. Ballif, Microcrystalline silicon solar cells: Effect of substrate temperature on cracks and their role in post-oxidation, Prog. Photovolt: Res. Appl. 18 (2010) 491−499 (see Section 3.1).

[46] R.W. Collins, A.S. Ferlauto, Advances in plasma-enhanced chemical vapor deposition of silicon films at low temperatures, Curr. Opin. Solid St. M. 6 (2002) 425−437.

[47] E. Vallat-Sauvain, A. Shah, J. Bailat, Advances in microcrystalline silicon solar cell technologies, in: J. Poortmans, V. Arkhipov (Eds.), Thin Film Solar Cells: Fabrication, Characterization, and Applications, John Wiley & Sons, Ltd, Chichester, UK, 2006 (doi: 10.1002/0470091282.ch4).

[48] E. Vallat-Sauvain, C. Droz, F. Meillaud, J. Bailat, A. Shah, C. Ballif, Determination of Raman emission cross-section ratio in microcrystalline silicon, J. Non-Crystalline Solids 352 (2006) 1200−1203.

[49] M. Vanecek, A. Poruba, Z. Remes, N. Beck, M. Nesladek, Optical properties of microcrystalline materials, J. Non-Crystalline Solids 227−230 (1998) 967−972.

[50] T. Dylla, F. Finger, E.A. Schiff, Hole drift-mobility measurements in microcrystalline silicon, Appl. Phys. Lett. 87 (2005) (Paper No. 032103).

[51] F. Finger, O. Astakhov, T. Bronger, R. Carius, T. Chen, A. Dasgupta, et al., Microcrystalline silicon carbide alloys prepared with HWCVD as highly transparent and conductive window layers for thin film solar cells, Thin Solid Films 517 (2009) 3507−3512.

[52] M. Vanecek, A. Poruba, Z. Remes, J. Rosa, S. Kamba, V. Vorlicek, et al., Electron spin resonance and optical characterization of defects in microcrystalline silicon, J. Non-Crystalline Solids 266–269 (2000) 519–523.
[53] O. Astakhov, R. Carius, F. Finger, Y. Petrusenko, V. Borysenko, D. Barankov, Relationship between defect density and charge carrier transport in amorphous and microcrystalline silicon, Phys. Rev. B 79 (2009) (Paper No 104205).
[54] A. Poruba, M. Vanecek, J. Meier, A. Shah, Fourier transform infrared spectroscopy in microcrystalline silicon, J. Non-Crystalline Solids 299–302 (2002) 536–540.
[55] F. Finger, R. Carius, T. Dylla, S. Klein, S. Okur, M. Günes, Stability of microcrystalline silicon for thin film solar cell applications, IEE Proc. Circ. Dev. Syst. 150 (2003) 300–308.
[56] F. Meillaud, E. Vallat-Sauvain, X. Niquille, M. Dubey, J. Bailat, A. Shah, et al., Light-induced degradation of thin-film amorphous and microcrystalline silicon solar cells, Proceedings of the 31st IEEE Photovoltaic Specialist Conference, Lake Buena Vista, Florida, 2005, pp. 1412–1415.
[57] Y. Wang, X. Geng, H. Stiebig, F. Finger, Stability of microcrystalline silicon solar cells with HWCVD buffer layer, Thin Solid Films 516 (2008) 733–735.
[58] F. Meillaud, E. Vallat-Sauvain, A. Shah, C. Ballif, Kinetics of creation and of thermal annealing of light-induced defects in microcrystalline silicon solar cells, J. Appl. Phys. 103 (2008) (Paper No. 054504).
[59] C. Droz, E. Vallat-Sauvain, J. Bailat, L. Feitknecht, J. Meier, X. Niquille, et al., Electrical and microstructural characterisation of microcrystalline silicon layers and solar cells, Proceedings of the 3rd World Conference on Photovoltaic Energy Conversion, Osaka, Japan, 2003, pp. 1544–1547.
[60] O. Vetterl, A. Gross, T. Jana, S. Ray, A. Lambertz, R. Carius, et al., Changes in electric and optical properties of intrinsic microcrystalline silicon upon variation of the structural composition, J. Non-Crystalline Solids 299–302 (2002) 772–777.
[61] T. Matsui, M. Kondo, A. Matsuda, Doping properties of boron-doped <<microcrystalline silicon>> from B_2H_6 and BF_3: Material properties and solar cell performance, J. Non-Crystalline Solids 338–340 (2004) 646–650.
[62] F. Finger, Y. Mai, S. Klein, R. Carius, High efficiency microcrystalline silicon solar cells with hot-wire CVD buffer layer, Thin Solid Films 516 (2008) 728–732.
[63] Y. Mai, S. Klein, X. Geng, M. Hülsbeck, R. Carius, F. Finger, Differences in the structure composition of microcrystalline silicon solar cells deposited by HWCVD and PECVD: Influence on open circuit voltage, Thin Solid Films 501 (2006) 272–275.
[64] Y. Nasuno, M. Kondo, A. Matsuda, Microcrystalline silicon thin-film solar cells prepared at low temperature using PECVD, Sol. Energy Mater. & Sol. Cells 74 (2002) 497–503.
[65] Y. Nasuno, M. Kondo, A. Matsuda, Passivation of oxygen-related donors in microcrystalline silicon by low temperature deposition, Appl. Phys. Lett. 78 (2001) 2330–2332.
[66] T. Kilper, W. Beyer, G. Bräuer, T. Bronger, R. Carius, M.N. van den Donker, et al., Oxygen and nitrogen impurities in microcrystalline silicon deposited under optimized conditions: Influence on material properties and solar cell performance, J. Appl. Phys. 105 (2009) (Paper No. 074509).
[67] N. Beck, N. Wyrsch, Ch. Hof, A. Shah, Mobility lifetime product—A tool for correlating a-Si:H film properties and solar cell performances, J. Appl. Phys. 79 (1996) 9361–9368.
[68] C. Droz, M. Goerlitzer, N. Wyrsch, A. Shah, Electronic transport in hydrogenated microcrystalline silicon: similarities with amorphous silicon, J. Non-Crystalline Solids 266 (2000) 319–324.

[69] A. Shah, R. Platz, H. Keppner, Thin film silicon solar cells: A review and selected trends, Sol. Energy Mater. & Sol. Cells 38 (1995) 501−520.

[70] A. Rose, Concepts in Photoconductivity and Allied Problems, Wiley-Interscience, John Wiley and Sons, New York, 1963 (see also <http://en.wikipedia.org/wiki/Albert_Rose/>)

[71] O. Vetterl, A. Lambertz, A. Dasgupta, F. Finger, B. Rech, O. Kluth, et al., Thickness dependence of microcrystalline silicon solar cell properties, Sol. Energy Mater. & Sol. Cells 66 (2001) 345−351.

[72] U. Kroll, C. Bucher, S. Benagli, I. Schönbächler, J. Meier, A. Shah, et al., High-efficiency p−i−n a-Si:H solar cells with low boron cross-contamination prepared in a large-area single-chamber PECVD reactor, Thin Solid Films 451−452 (2004) 525−530.

[73] M.-E. Stückelberger, A.V. Shah, J. Krc, F. Sculati-Meillaud, C. Ballif, M. Depseisse, Internal electric field and fill factor of amorphous silicon solar cells, Proceedings of the 35th IEEE Photovoltaic Specialists Conference (PVSC), Honolulu, Hawaii, 20−25 June 2010, pp. 1569−1574 (Manuscript No. 360).

[74] J. Hubin, A. Shah, E. Sauvain, Effects of dangling bonds on the recombination function in amorphous semiconductors, Phil. Mag. Lett. 66 (1992) 115−125.

[75] D. Fischer, N. Wyrsch, C.M. Fortmann, A. Shah, Amorphous silicon solar cells with low-level doped i-layers characterised by bifacial measurements, Proceedings of the 23rd IEEE Photovoltaic Specialists Conference, Louisville, 1993, pp. 878−884.

[76] J. Merten, J.M. Asensi, C. Voz, A.V. Shah, R. Platz, J. Andreu, Improved equivalent circuit and analytical model for amorphous silicon solar cells and modules, IEEE Trans. Electron. Dev. 45 (1998) 423−429.

[77] M. Python, E. Vallat-Sauvain, J. Bailat, D. Dominé, L. Fesquet, A. Shah, C. Ballif, Relation between substrate surface morphology and microcrystalline silicon solar cell performance, J. Non-Crystalline Solids 354 (2008) 2258−2262.

[78] M. Python, O. Madani, D. Domine, F. Meillaud, E. Vallat-Sauvain, C. Ballif, Influence of the substrate geometrical parameters on microcrystalline silicon growth for thin-film solar cells, Sol. Energy Mater. & Sol. Cells 93 (2009) 1714−1720.

[79] A.V. Shah, F.C. Sculati-Meillaud, Z.J. Berényi, R. Kumar, Diagnostics of thin-film silicon solar cells and solar panels/modules with VIM (variable intensity measurements), Sol. Energy Mater. & Sol. Cells 95 (2011) 398−403.

[80] F. Meillaud, A. Shah, J. Bailat, E. Vallat-Sauvain, T. Roschek, B. Rech, et al., Microcrystalline silicon solar cells: Theory and diagnostic tools, Proceedings of the 4th World Conference on Photovoltaic Energy Conversion, Kona Island, Hawaii, 2006, pp. 1572−1575.

[81] A. Shah, J. Meier, A. Buechel, U. Kroll, J. Steinhauser, F. Meillaud, et al., Towards very low-cost mass production of thin-film silicon photovoltaic (PV) solar modules on glass, Thin Solid Films 502 (2006) 292−299.

[82] A. Banerjee, S. Guha, Study of back reflectors for amorphous silicon alloy solar cell application, J. Appl. Phys. 69 (1991) 1030−1035.

[83] A. Banerjee, J. Yang, K. Hoffman, S. Guha, Characteristics of hydrogenated amorphous silicon alloy cells on a Lambertian back reflector, Appl. Phys. Lett. 65 (1994) 472−474.

[84] V. Terrazzoni Daudrix, J. Guillet, F. Freitas, A. Shah, C. Ballif, P. Winkler, et al., Characterisation of rough reflecting substrates incorporated into thin-film silicon solar cells, Prog. Photovolt: Res. Appl. 14 (2006) 485−498.

[85] V. Terrazzoni-Daudrix, J. Guillet, X. Niquille, F. Freitas, P. Winkler, A. Shah, et al., Enhanced light trapping in thin-film silicon solar cells deposited on PET and glass, Proceedings of the 3rd World Conference on Photovoltaic Energy Conversion, Osaka, Japan, 2003, pp. 1596−1600.

[86] J. Krc, M. Zeman, A. Campa, F. Smole, M. Topic, Novel approaches of light management in thin-film silicon solar cells, Proc. Mater. Res. Soc. Symp. 910 (2007) 669–680.

[87] V.E. Ferry, M.A. Verschuuren, H.B.T. Li, E. Verhagen, R.J. Walters, R.E.I. Schropp, et al., Light trapping in ultrathin plasmonic solar cells, Opt. Express 18 (2010) A237–245.

[88] S. Faÿ, J. Steinhauser, N. Oliveira, E. Vallat-Sauvain, C. Ballif, Opto-electronic properties of rough LP-CVD ZnO:B for use as TCO in thin-film silicon solar cells, Thin Solid Films 515 (2007) 8558–8561.

[89] B. Yan, G. Yue, J. Yang, S. Guha, Correlation of current mismatch and fill factor in amorphous and nanocrystalline silicon based high efficiency multi-junction solar cells, Proceedings of the 33rd IEEE Photvoltaics Specialists Conference, 2008, pp. 1038–1040.

[90] A.E. Delahoy, Recent developments in amorphous silicon photovoltaic research and manufacturing at Chronar corporation, Sol. Cells 27 (1989) 39–57.

[91] J. Meier, S. Dubail, R. Flückiger, D. Fischer, H. Keppner, A. Shah, Intrinsic microcrystalline silicon (µc-Si:H)—A promising new thin film solar cell material, Proceedings of the First World Conference on Photovoltaic Energy Conversion, 1994, pp. 409–412.

[92] K. Yamamoto, A. Nakajima, M. Yoshimi, T. Sawada, S. Fukuda, T. Suezaki, et al., A thin-film silicon solar cell and module, Prog. Photovolt: Res. Appl. 13 (2005) 489–494.

[93] D. Fischer, S. Dubail, J.A. Anna Selvan, N. Pellaton Vaucher, R. Platz, Ch. Hof, et al, The "micromorph" solar cell: Extending a-Si:H technology towards thin film crystalline silicon, Proceedings of the 25th IEEE Photovoltaic Specialists Conference, 1996, pp. 1053–1056.

[94] K. Yamamoto, A. Nakajima, M. Yoshimi, T. Sawada, S. Fukuda, T. Suezaki, et al., High efficiency thin film silicon hybrid cell and module with newly developed innovative Interlayer, Conf. Rec. 4th World Conf. Photovol. Energy Conversion 2 (2006) 1489–1492.

[95] P. Buehlmann, J. Bailat, D. Dominé, A. Billet, F. Meillaud, A. Feltrin, C. Ballif, In situ silicon oxide based intermediate reflector for thin-film silicon micromorph solar cells, Appl. Phys. Lett. 91 (2007) (Paper No.143505).

[96] D. Dominé, P. Buehlmann, J. Bailat, A. Billet, A. Feltrin, C. Ballif, Optical management in high-efficiency thin-film silicon micromorph solar cells with a silicon oxide based intermediate reflector, Phys. status solidi (Rapid Res. Lett.) 2 (2008) 163–165.

[97] T. Söderström, F.-J. Haug, X. Niquille, V. Terrazzoni, C. Ballif, Asymmetric intermediate reflector for tandem micromorph thin film silicon solar cells, Appl. Phys. Lett. 94 (2009) (Paper No. 063501).

[98] J. Bailat, et al., Recent developments of high-efficiency micromorph® tandem solar cells in KAI-M PECVD reactors, Proceedings of the 25th EC PV Solar Energy Conference, 2010, pp. 2720–2723.

[99] H. Curtins, N. Wyrsch, A.V. Shah, High-rate deposition of hydrogenated amorphous silicon—Effect of plasma excitation-frequency, Electron. Lett. 23 (1987) 228–230.

[100] H. Curtins, N. Wyrsch, M. Favre, A.V. Shah, Influence of plasma excitation frequency for a-Si:H thin film deposition, Plasma Chem. and Plasma Process. 7 (1987) 267–273.

[101] A.A. Howling, J.-L. Dorier, Ch. Hollenstein, U. Kroll, F. Finger, Frequency effects in silane plasmas for plasma enhanced chemical vapor deposition, J. Vac. Sci. Technol. A 10 (1992) 1080–1085.

[102] A.A. Howling, L. Sansonnens, C. Hollenstein, Electromagnetic sources of non-uniformity in large area capacitive reactors, Thin Solid Films 515 (2006) 5059−5064.

[103] L. Sansonnens, H. Schmidt, A.A. Howling, C. Hollenstein, C. Ellert, A. Buechel, Application of the shaped electrode technique to a large area rectangular capacitively coupled plasma reactor to suppress standing wave non-uniformity, J. Vac. Sci. Technol. A 24 (2006) 1425−1430.

[104] M. Kondo, A. Matsuda, An approach to device grade amorphous and microcrystalline silicon thin films fabricated at higher deposition rates, Curr. Opin. Solid St. M. 6 (2002) 445−453.

[105] J.-L. Dorier, Ch. Hollenstein, A.A. Howling, U. Kroll, Powder dynamics in very high frequency silane plasmas, J. Vac. Sci. Technol. A 10 (1992) 1048−1052.

[106] M.J. Prather, J. Hsu, NF3, the greenhouse gas missing from Kyoto, Geophys. Res. Lett. 35 (2008) (Paper No. L12810).

[107] M. Schottler, M. de Wild-Scholten, The carbon footprint of PECVD chamber cleaning using fluorinated gases, Proceedings of the 23rd EC PV Solar Energy Conference, 2008, pp. 2505−2509.

[108] B. Strahm, A.A. Howling, L. Sansonnens, C. Hollenstein, Optimization of the microcrystalline silicon deposition efficiency, J. Vac. Sci. Technol. A 25 (2007) 1198−1202.

[109] S. Klein, F. Finger, R. Carius, M. Stutzmann, Deposition of microcrystalline silicon prepared by hot-wire chemical-vapor deposition: The influence of the deposition parameters on the material properties and solar cell performance, J. Appl. Phys. 98 (2005) (Paper No. 024905).

[110] Q. Wang, Hot-wire CVD amorphous Si materials for solar cell application, Thin Solid Films 517 (2009) 3570−3574.

[111] K. Ishibashi, M. Karasawa, G. Xu, N. Yokokawa, M. Ikemoto, A. Masuda, et al., Development of Cat-CVD apparatus for 1-m-size large-area deposition, Thin Solid Films 430 (2003) 58−62.

[112] B. Schroeder, Status report: Solar cell related research and development using amorphous and microcrystalline silicon deposited by HW(Cat)CVD, Thin Solid Films 430 (2003) 1−6.

[113] H. Hakuma, K. Niira, H. Senta, T. Nishimura, M. Komoda, H. Okui, et al. Microcrystalline-Si solar cells by newly developed novel PECVD method at high deposition rate, Proceedings of the 3rd World Conference on Photovoltaic Energy Conversion, Osaka, Japan, 2003, pp. 1796−1799.

[114] P.F. Gerhardinger, R.J. McCurdy, Float line deposited transparent conductors-implications for the PV industry, Proc. Mater. Res. Soc. Symp. 426 (1996) 399−410.

[115] M. Kambe, A. Takahashi, N. Taneda, K. Masumo, T. Ovama, K. Sato, Fabrication of a-Si:H solar cells on high-haze SnO_2:F thin films, Proceedings of the 33rd IEEE Photovoltaic Specialists Conference (PVSC), San Diego, California, May 11−16, 2008, pp. 690−693.

[116] K. Ellmer, A. Klein, B. Rech (Eds.), Transparent conductive zinc oxide, Springer Series in Materials Science, vol. 104, Springer Verlag, Berlin, 2008.

[117] H. Sai, H.J. Jia, M. Kondo, Impact of front and rear texture of thin-film microcrystalline silicon solar cells on their light trapping properties, J. Appl. Phys. 108 (2010) (Paper No. 044505).

[118] R. Bartlome, B. Strahm, Y. Sinquin, A. Feltrin, C. Ballif, Laser applications in thin-film photovoltaics, Appl. Phys. B—Lasers Opt 100 (2010) 427−436.

[119] T. Witte, A. Gahler, H.J. Booth, Next generation of laser scribing technology for grid-parity and beyond, Proceedings of the 25th EC PV Solar Energy Conference, 2010, pp. 3213−3216.

[120] J. Pern, 2008. Module encapsulation materials, processing and testing, APP International PV Reliability Workshop, SJTU; Shanghai (China), 2008. Available as <www.nrel.gov/docs/fy09osti/44666.pdf/>.

[121] F.J. Pern, S.H. Glick, Photothermal stability of encapsulated Si solar cells and encapsulation materials upon accelerated exposures, Sol. Energy Mater. & Sol. Cells 61 (2000) 153—188.

[122] T.J. McMahon, Accelerated testing and failure of thin-film PV modules, Prog. Photovolt. 12 (2004) 235—248.

[123] W. Durisch, B. Bitnar, J.-Cl. Mayor, H. Kiess, J.-Cl. Lam, K.-H., J. Close, Efficiency model for photovoltaic modules and demonstration of the application to energy yield estimation, Sol. Energy Mater. & Sol. Cells 91 (2007) 79—84.

[124] J.F. Randall, Designing Indoor Solar Products, John Wiley and Sons, Chichester, UK, 2005.

[125] S.A. Boden, D.M. Bagnall, Sunrise to sunset optimization of thin film antireflective coatings for encapsulated, planar silicon solar cells, Prog. Photovolt: Res. Appl. 17 (2009) 241—252.

[126] W. Durisch, J.-Cl. Mayor, K.-H. Lam, J. Close, S. Stettler, Efficiency and annual output of a monocrystalline photovoltaic module under actual operating conditions, Proceedings of the 23rd European Photovoltaic Solar Energy Conference, Valencia, Spain, 2008, pp. 2992—2996.

[127] E.D. Dunlop, D. Halton, H.A. Ossenbrink, Twenty years of life and more: where is the end of life of a PV module?, Proceedings of the 31st IEEE Photovoltaic Solar Energy Specialists Conference, Lake Buena Vista, Florida, 2005, pp. 1593—1596.

CHAPTER IC-2

CdTe Thin-Film PV Modules

Dieter Bonnet
CTF Solar GmbH, Friedrichsdorf, Germany

Contents

1. Introduction	284
2. Steps for Making Thin-Film CdTe Solar Cells	285
2.1 Film Deposition	285
2.1.1 CdTe	285
2.1.2 CdS	288
2.1.3 TCO Films	288
2.1.4 Substrates	289
2.2 Improvement of Critical Regions of the CdTe Solar Cell	290
2.2.1 The p—n Heterojunction—Improvement by Activation	291
2.2.2 The Back Contact	299
2.3 Stability Issues	302
2.4 Best Performance of Cells	302
3. Making of Integrated Modules	303
3.1 Interconnection of Cells	303
3.2 Contacting	304
3.3 Lamination	305
4. Production of CdTe Thin-Film Modules	306
4.1 Generalised Production Sequence	306
4.2 Industrial Production of Modules	307
4.2.1 BP Solar Inc. (Fairfield, California, USA)	307
4.2.2 First Solar LLC (Toledo, Ohio, USA)	308
4.2.3 ANTEC Solar GmbH (Arnstadt, Germany)	309
4.3 A 10-MW Production Line	310
4.4 Environmental and Health Aspects	314
4.5 Material Resources	315
5. The Product and Its Application	316
5.1 Product Qualification	316
5.2 Examples of Installation of CdTe Modules	318
6. The Future	320
References	320

Practical Handbook of Photovoltaics.
© 2012 Elsevier Ltd. All rights reserved.

1. INTRODUCTION

CdTe is very well suited for use as active material in thin-film solar cells due to four special properties [1]:
- CdTe has an energy gap of 1.45 eV and therefore is well adapted to the spectrum of solar radiation.
- The energy gap of CdTe is 'direct,' leading to very strong light absorption.
- CdTe has a strong tendency to grow as an essentially highly stoichiometric, but p-type semiconductor film can form a p–n heterojunction with CdS. (CdS has a rather wide energy gap of 2.4 eV and grows n-type material under usual film-deposition techniques.)
- Simple deposition techniques have been developed that are suitable for low-cost production.

Current densities of as much as 27 mA cm^{-2} and open-circuit voltages of 880 mV, leading to AM 1.5 efficiencies of 18%, can be expected for CdTe cells made under a mature technology.

Figure 1 shows the typical film sequence of this cell. In the preferred arrangement, first a transparent conducting film (typically In_2O_3 or SnO_2 or a combination of both) is deposited onto glass plate used as transparent substrate. Then an n-CdS film is deposited, followed by the active p-conducting CdTe film. A special treatment improves the p–n junction between CdS and CdTe ('activation'). Finally a low-resistance contact is deposited onto the CdTe, which can be opaque.

Light enters the cell through the glass substrate. Photons transverse the TCO and CdS layers. These films are not active in the photovoltaic charge generation process, although they lead to some unwanted absorption. The

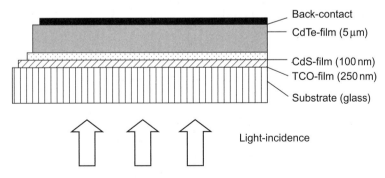

Figure 1 Film sequence of the CdTe thin-film solar cell.

CdTe film is the active absorber of the solar cell. Electron—hole pairs are generated close to the junction. The electrons are driven by the built-in field through the interface into the n-CdS film. The holes remain in the CdTe and join the pool of the holes promoting the p conduction of this material and finally have to leave the cell via the back contact. Electric power is drawn by metallic contacts attached to the TCO film and the back contact.

Due to the strong light absorption in CdTe of about 10^5 cm^{-1} for light having a wavelength below 800 nm, a film thickness of a few micrometres would be sufficient for complete light absorption. For practical reasons, a thickness of about 3–7 μm is often preferred.

Intensive research has shown that this junction can be mastered so that the following basic criteria for solar cells can be fulfilled under conditions of industrial production:
- effective generation of mobile minority-charge carriers in the CdTe film,
- efficient separation of charge carriers by means of the internal electric field of the p—n junction between n-CdS and p-CdTe,
- low loss extraction of the photocurrent by means of ohmic contacts to the TCO and back-contact films, and
- simple fabrication technologies for low-cost, high-volume production.

Solar cells of efficiencies above 16% have been made in research laboratories, and industrial efforts have led to the recent start-up of industrial production units at three private companies in the United States and Germany, each aiming at large-scale production of 100,000 m^2 per annum or more. The first large-area modules have recently surpassed the 10% efficiency mark.

2. STEPS FOR MAKING THIN-FILM CDTE SOLAR CELLS

2.1 Film Deposition

2.1.1 CdTe

Most techniques to deposit CdTe films rely on one or both of the following properties:
- If healed in vacuum up to temperatures above 600°C, CdTe sublimes congruently, liberating Cd and Te in equal amounts, the residue remaining stoichiometric CdTe.
- In CdTe films condensing on substrates kept above 400°C (or heated up to this temperature after deposition), the stoichiometric compound

is the stable solid phase. The constituting elements have a significantly higher vapour pressure than the compound.

These properties make it relatively easy to produce CdTe films suited for thin-film solar cells: no excessive care has to be taken to provide for stoichiometry, as long as the substrate temperature is sufficiently high. CdTe or Cd + Te or decomposable compounds of Cd and Te can be used as starting material. Upon arrival of Cd and Te on the substrate even in a ratio that is not 1:1, CdTe condenses (nearly) stoichiometrically as long as the substrate is heated at 400–500°C or higher during or after the actual deposition. The film quality increases with temperatures up to 600°C. At higher temperatures, the sticking coefficient decreases (resublimation). A p-doping effect is achieved due to a small natural nonstoichiometry in the form of Cd deficiencies, probably vacancies. No additional doping is used. Typical doping levels are around 10^{15} cm^{-3}. These values are somewhat low but can be tolerated in thin-film cells. Thin films are deposited at lower temperatures and therefore not necessarily at stoichiometric ratio, they can be heated to create the stoichiometric compound. This allows numerous film-deposition technologies to be applied. The only requirement is absence of disturbing impurities, which might jeopardise the native p-doping and charge-carrier lifetime. High purity (up to 99.999%) of the elements and the compound can be achieved on an industrial scale as the elements—Cd and Te—can be easily purified by standard metallurgical procedures.

Numerous film-deposition processes have been studied in the past, and all have led to good cells exhibiting efficiencies above 10%. Only a few processes have properties suited to large-scale production, though. They have been developed by industrial units, as discussed in the following.

Vacuum Deposition—Sublimation and Condensation

Solid CdTe material in the form of powder or granulate is sublimed in vacuum and condenses on the substrate maintained at elevated temperatures between 450°C and 600°C, using the basic thermodynamic properties of CdTe already mentioned. Commercial processes of a different kind have been developed that can achieve very high deposition rates (>10 µm/min) and be applied to continuous-flow in-line processes using low-cost, rugged vacuum systems. The processes do have high materials yield as the material is forced to condense only on the substrate either by close space between source and substrate (close-spaced sublimation, or

CSS) or by prevention of deposition on the walls kept at elevated temperatures above 600°C. Two factories using sublimation processes have been built in Germany and the United States. Production and sale of modules has started in both facilities.

Electrodeposition

CdTe films are formed from aqueous solutions of $CdSO_4$ and Te_2O_3 at temperatures of around 90°C. An n-type film of low electronic quality is formed. The basic reaction is as follows:

$$Cd^{24}_{(aq)} + HTeO_2^+{}_{(aq)} + 3H^+_{(aq)} + 6e^- \rightarrow CdTe_{(s)} + 2H_2O_{(liq)}$$

Grain-size enhancement, doping conversion into p-type, and improvement of electronic transport properties are achieved by thermal postannealing under the influence of Cl-based compounds. The driving electric potential is applied to the transparent conductive film on the substrate and has to be very homogeneous over the whole surface to be coated. This requires low-deposition current density, resulting in low deposition rates. This can be compensated for high throughput by coating a large number of substrates in parallel. A production plant has been built in the United States and presently is ramped up to production quantities.

Chemical Spraying

An aerosol of water droplets containing heat-decomposable compounds of Cd and Te is sprayed onto a heated substrate, forming CdTe from the liberated elements. Processes have been developed that do not require a vacuum and can be applied easily in in-line systems by using linear nozzle arrays. A pilot plant was built in the United States by an industrial venture but subsequently put up for sale and then finally abandoned.

Screen Printing

Slurries containing Cd and Te are screen printed onto the substrate and transformed into CdTe by thermal reaction under the influence of added $CdCl_2$. Due to some porosity of the films, comparatively thick layers are required for good operation of the cells. This technology is presently employed on a commercial scale with production capacity of 1 MW_p/year. Small modules are manufactured and used in consumer applications. There are some doubts about the suitability of this process for the large-scale production of high-efficiency, low-cost modules.

2.1.2 CdS

Like CdTe, CdS has the same strong tendency to form stoichiometric films; unlike CdTe, CdS films are natively n-doped by a slight nonstoichiometry. CdS films can be deposited by the same processes as CdTe because its basic properties are quite similar to those of CdTe—e.g., its tendency to sublime and condense congruently. The following processes have been studied more intensely in view of production:
- sublimation and condensation, such as close-spaced sublimation and hot-wall sublimation;
- electrodeposition; and
- screen printing.

Another process is especially suited for CdS: chemical bath deposition (CBD). In this process, a metastable solution containing Cd and S leads to spontaneous formation of thin CdS films on surfaces of substrates immersed into the solution at temperatures around 80 °C. The chemical reaction basically is as follows:

$$Cd^{2***} (complex)_{(aq)} + (NH_2)CS_{(aq)} + 2OH^-_{(aq)} \rightarrow CdS_{(s)} + H_2O_{(liq)}$$

The CdS films so formed are tightly adherent and very homogeneous even at low thickness.

A potential disadvantage on forming abrupt junctions of high photoelectronic quality between CdTe and CdS is that CdS has a significant lattice mismatch to CdTe. Fortunately postdeposition treatments, described later, allow the amelioration of the junction.

2.1.3 TCO Films

In TCO films, a compromise is achieved between high electronic conduction required for low series resistance in cells and high optical transmission for high light input and ensuing high photocurrent. Several materials are presently in use and under development for industrial application.

SnO_2

SnO_2 films can be produced by a spraying process at ambient pressure. $SnCl_4$ is dissolved in water and sprayed onto a heated substrate in air. $SnCl_4$ decomposes under reaction with oxygen to form SnO_2 films and yield HCl, which is led away. Substrates of this kind are made on a commercial basis and are presently used by several industrial solar cell manufacturers. The films typically have area resistivities of 10 Ω/square and transmission values of around 70–80%. Alternative deposition techniques are cathode sputtering of metallic Sn targets in an oxidising ambient. Although more expensive than spraying, better quality films are achieved.

ITO

Indium tin oxide films have higher performance when sputtered from an oxide target (either better conductivity or optical transmission than pure SnO_2 films). They are more expensive due to the use of In. As indium may diffuse into the CdS−CdTe film packet and lead to unwanted n-doping of CdTe during high-temperature processing, usually a thin pure SnO_2 film is deposited onto the ITO as a diffusion barrier for In.

CdSnO$_4$

This compound can be deposited by cosputtering oxides of Cd and Sn. It requires annealing processes at elevated temperatures, which are not suited for the use of cheap soda-lime glass. Future process improvements may overcome this setback. The films show better performance than ITO— i.e., higher transmission at equal resistivity or lower resistivity at equal transmission, making them an interesting option for industrial production.

ZnO:Al

This material is routinely used as transparent contact for CIS-based thin-film solar cells. It can easily be fabricated in thin-film form by sputtering a heterogeneous target containing ZnO and Al. Al acts as donor in ZnO. Unfortunately, the film loses its doping during thermal stress (>500 °C) at deposition of CdTe. There is hope that more stable films can be made eventually, as the material is more cost efficient than ITO.

2.1.4 Substrates

The most common transparent substrate to be used is glass. The cheapest glass—soda-lime glass or windowpane glass—is suitable. It exhibits, if made by the float-glass process, a very flat surface well suited to thin-film deposition. It is limited in processing temperature at 520°C, or somewhat higher if suitably suspended. It is sufficiently cheap (<$10/m^2), and can be bought cut and edge treated in virtually unlimited quantities. It is indeed used by the three production facilities that have recently become operative.

If higher temperature (which may lead to better quality films) is desired, the second option is borosilicate glass, which can be heated to temperatures above 600°C without softening. The higher cost of this material presently prevents its industrial use. Research groups have made cells of up to 16.2% efficiency on such glass.

2.2 Improvement of Critical Regions of the CdTe Solar Cell

Figure 2 shows an image obtained by scanning electron microscopy (SEM) of the broken edge of a CdTe solar cell in which the CdTe film has been made by close-spaced sublimation. Figure 3 shows the critical regions—i.e., the CdTe–CdS junction and the back-contact region—that have proved to be the most critical parts of the cell and on which efficiency depends strongly.

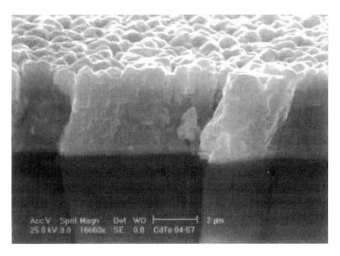

Figure 2 SEM image of the cross-section of a CdTe solar cell.

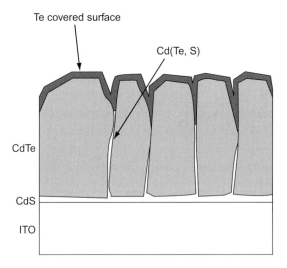

Figure 3 Schematic illustration of the key features of the CdTe solar cell.

2.2.1 The p–n Heterojunction—Improvement by Activation

Although both materials forming the junction are II–VI compounds and have a close chemical relationship, their lattice constants differ by about 5%. This leads to a significant density of interface states that can be expected to result in strong charge-carrier recombination. Junctions 'as made' indeed show low charge-carrier collection efficiency and thereby low-power efficiency of around 2%. Annealing of the system at temperatures of around 400°C leads to some improvement, but only the still not completely understood 'activation' changes the junction so far that efficiencies of up to 16% have been observed. In this activation step, the junction is annealed at temperatures of 400–500°C in the presence of Cl-containing species, generally $CdCl_2$, which is deposited onto the film stack or admitted in vapour form for a time of around a few minutes. Figure 4 shows the strong improvement of the quality of the I–V curves under this treatment [2]. This procedure in the first instance has been developed quite empirically, although it is based on historic processes for the manufacture of CdS photoconductive films [3]. Only recently has light been shed on the basic processes that take place during activation. Three essential effects go hand in hand.

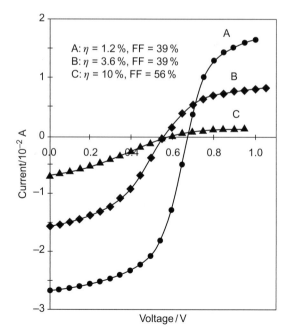

Figure 4 I–V curves of cells showing improvement by thermal treatment processes (from [2]).

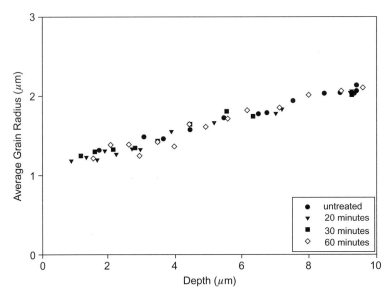

Figure 5 Grain-size distribution of a CdTe film as function of the distance from the junction after different thermal treatments [4].

Recrystallisation—Grain Growth

Generally, for polycrystalline materials in which grain growth may occur, there is a limiting grain size in which the driving force is balanced by a retarding force. The net driving force decreases as the grain radius increases. Smaller grain sizes typically are observed for films grown at lower substrate temperatures due to lower mobility of atoms during growth. Upon annealing and activation, small grains can start to grow, but only to a certain limit. If, on the other hand, larger grains are formed already upon film deposition at higher deposition temperatures, such grains will grow less, as they are already closer to the final equilibrium. This has been observed experimentally: grain size in films deposited at 500°C does not increase upon activation. Figure 5 shows that the grain diameter grows with distance from the junction—at which location nucleation of the films occurs—to an average size of 2 μm [4]. (The film has been deposited by CSS at 500°C.) It does not change even under extended treatment times with $CdCl_2$ species at 400°C. This means that for these samples the grains were at equilibrium directly after material deposition, a satisfying situation. Figure 6 shows a TEM cross-section of a CdS–CdTe film stack deposited at a substrate temperature of 525°C, indicating the typical morphology of such a system, clearly showing the

Figure 6 X-ray transmission image of a CdTe film [5]. *(Reproduced with permission from IEEE.)*

high density and three-dimensional distribution of planar defects, mainly stacking faults and twins [5]. The density, however, varies from one grain to the other.

For films deposited at significantly lower temperatures (<400°C), the films show smaller grain sizes directly after deposition, together with strong orientation with the 111-axis perpendicular to the substrate. This orientation is lost upon recrystallisation during $CdCl_2$ activation [6]. Figure 7, obtained by atomic force microscopy (AFM), shows directly three steps in the process of recrystallisation. During annealing under $CdCl_2$, new small grains form on the highly oriented films and grow into the final unoriented phase, indicating close packing. The reference film deposited at 590°C does not grow. The loss of preferential orientation upon annealing of this low-temperature film is strikingly illustrated in Figure 8 by X-ray pole diagrams by the same author [7]. The author attributes the grain growth and recrystallisation to lattice strain energy present in the films, which drives the process. In high-temperature–deposited films, no strain is present, leading to no significant recrystallisation. This strain present in low-temperature films is clearly illustrated by the author from an effective decrease of lattice constant: after (low-temperature) deposition, the lattice constant for a is 6.498 Å, which is reduced on activation to 6.481 Å, the published equilibrium value for crystalline powder.

Figure 7 Grain reconstruction in a CdTe film made at low temperature during activation (a, b, c) and unchanged film made by high-temperature CSS (d)[6]. (© American Institute of Physics.)

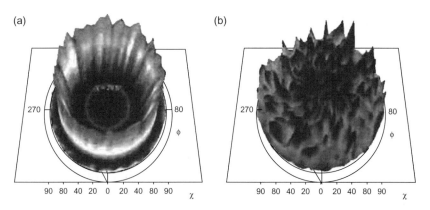

Figure 8 X-ray pole diagrams of 311 planes in a CdTe film: (a) before and (b) after activation at 400 °C V. (© American Institute of Physics.)

In another study (Figure 9), grain–size distributions were measured [8] for films deposited at relative low temperatures directly after deposition (at 340°C) and after annealing (at 580 °C) and alternatively after activation (at 430°C). The final grain-size distribution has a maximum at 1 μm.

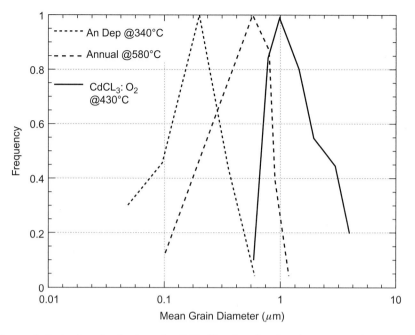

Figure 9 Grain-size distribution of a CdTe film directly after deposition, after annealing, and after activation [8]. *(Reproduced with kind permission of James & James Publ.)*

Immediately at the junction, where the film nucleates, some grain growth also has been observed for high-temperature films [9]. This region is difficult to access. More detailed knowledge may be crucial for an efficiency increase.

Interdiffusion—Intermixing

CdS and CdTe in thermal equilibrium may form mixed compounds CdS_xTe_{1-x} only for quite limited regions close to the single compounds, leaving a miscibility gap between $x = 0.16$ and $x = 0.86$ at 650 °C [10]. At lower temperatures, the gap widens. In the Te-rich region, the material shows a lower band gap than pure CdTe, an interesting feature in some II−VI compounds containing Te.

It is to be expected that CdS and CdTe intermix upon deposition at elevated temperatures to a certain degree. Intermixing at the interface can be expected to reduce the effects of lattice mismatch between CdS and CdTe. This can be analysed by secondary ion mass spectrometry (SIMS) depth profiling. SIMS analyses allow the elements to be determined quantitatively at the surface while it is removed layer by layer—e.g., by sputtering ('depth profiling'). Such experiments have shown that intermixing of CdS and CdTe is a function of substrate temperature and post-deposition $CdCl_2$ activation. The degree of intermixing indeed has been observed to increase with increase of substrate temperature. Further increase is induced by activation. Excessive interdiffusion leads to deterioration of the device performance. Spectral-response curves for devices made from CdTe films deposited at 610°C and activated for different times are shown in Figure 10 [11]. Indeed the intermixing is manifest by a longer-wavelength response due to the lower-band-gap material for small values of x, the Te-rich mixture. This goes hand in hand with increased total photocurrent. Significant amounts of sulphur (probably CdS_xTe_{1-x}) can be detected both at grain boundaries and within heavily faulted grains (as opposed to grains with low-defect density) for films deposited at temperatures between 500°C and 600°C [5]. Model calculation yielded three-dimensional distributions of S and Te that have led to isocompositional contour plots of diffused regions [12] (Figure 11). In the case shown for two adjacent grains after activation, the different grain sizes result in a different alloy profile. For the narrower grain, no pure CdTe remains at all, while the wider grain exhibits the entire range of alloy composition. The evolution of mixed regions and their progression

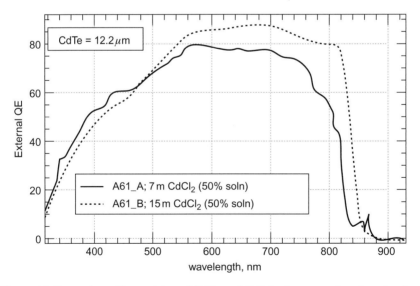

Figure 10 Illustration of intermixing of CdTe and CdS. by long-wavelength extension of sensitivity due to lower-band-gap mixed compound [11]. *(Reproduced with kind permission of the Material Research Society.)*

have been followed by X-ray diffraction studies, impressively showing the emergence of lines corresponding to mixed material. Figure 12 shows initially the line of pure CdTe and after 10, 20, and 40 minutes of activation the emergence and growth of alloy lines. Modelling of these results yielded diffusion coefficients that have been used to obtain the above three-dimensional profiles of Figure 11. Similar results have been obtained by measuring the lattice constant of films before and after activation [6]. The value for lattice constant *a* before annealing—as mentioned above—has a value of 6.498 Å (indicating stress in the CdTe film) and changes into two distinct values of 6.481 Å (relaxed CdTe) and 6.468 Å.

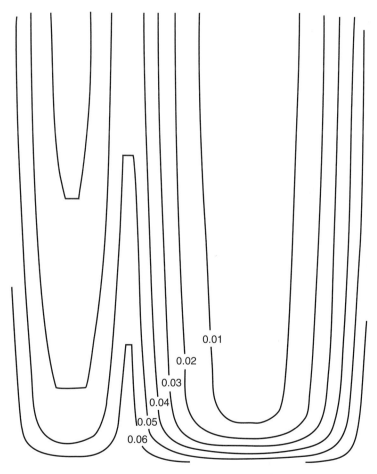

Figure 11 Simulated isocomposition lines of two grains in a CdTe/CdS film. Parameter = *x* in CdTe$_x$S$_{1-x}$ [8]. (© *American Institute of Physics.*)

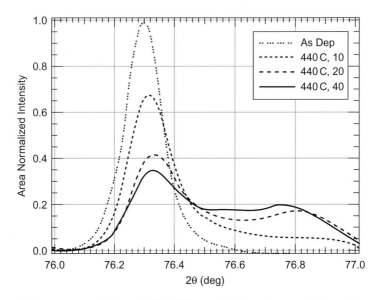

Figure 12 Emergence of mixed CdTe/CdS material after increased thermal stressing identified by X-ray diffraction [8]. (© *American Institute of Physics.*)

The latter value does not occur for films deposited onto ITO film without any CdS. These results can only be explained by the occurrence of a CdS_xTe_{1-x} species with $x = 0.2$, corresponding to 6.468 Å.

The presence of oxygen traces during CdTe film deposition usually is considered harmless, a fact that eases the requirements on vacuum equipment. Recently [13] it has been observed that oxygen leads to a reduced CdTe–CdS interdiffusion during activation, a fact that may reduce the danger of shunting of cells by CdS-enhanced grain boundaries, as illustrated in Figure 11.

Increase of Charge-Carrier Lifetime

Electrons generated as minority-carriers in the CdTe films have to reach (by diffusion) the field region at the junction and transverse the junction (by drift). Even under low-lifetime conditions, electrons generated directly at the junction can transverse the field region: the *schubweg* for an electron of 10 cm^2/Vs mobility and a lifetime of 1 ns in an internal field of 10^4 V/cm of 1 μm with at a potential difference of 1 V i:$\mu\tau E = 10^{-4}$ cm. If diffusion in the field-free region is required—for carriers created further away from the junction by red light—lifetime becomes more important. Moutinho et al. [14] have shown that CdTe films of good cells (11%) can show 2 ns lifetime.

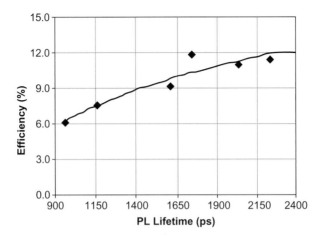

Figure 13 Increase of efficiency of a cell by increase of minority-carrier lifetime [14]. *(Reproduced with permission from IEEE.)*

They have plotted efficiency of cells made similarly as a function of lifetime (Figure 13), allowing the conclusion that in CdTe solar cells diffusion of minority-carriers also plays an important role for achieving high efficiency.

Conclusions on Activation
All these results indicate that major structural effects in CdTe solar cells occur upon the 'magic' activation process—namely, grain growth and interdiffusion, which are stronger for low-temperature deposited films than for films deposited at temperatures of 500°C or more. These ('high-temperature') films directly lead to a stable structure with less recrystallisation and grain growth required.

The main effect of activation on efficiency of the devices, although, is an electronic improvement not so much by morphological effects but by improvement of the crystalline and electronic quality of grains immediately at the junction. Evidently, the lifetime of minority carriers in CdTe (electrons) determines the charge-carrier collection efficiency of the device.

2.2.2 The Back Contact
It is well known from semiconductor physics that it is not easy to contact a low-doped p-type semiconductor of relatively high energy gap. There are two general principles for making ohmic contacts to p-type semiconductors:
1. Use of a metal of work function higher than the electron affinity + energy gap of the semiconductor in order to align the top of the valence band with the Fermi level of the metal. The electron

affinity + energy gap of CdTe is >6 eV. There is no metal of work function of >6 eV. This means that all metals lead to a blocking contact, as is illustrated in Figure 14, showing the band diagram of this situation.

2. Generation of a highly doped back-surface layer in the semiconductor. The unavoidable Schottky barrier created by the back-contact metal in the semiconductor will then be thin enough for holes to tunnel through efficiently. Figure 15 shows the band diagram for the second option, a highly p-doped surface region. Consequently, it remains to find practical solutions for this.

Efforts for high p-doping in CdTe usually fail due to a strong tendency for self-compensation of acceptors by formation of donors at elevated temperatures as used here. Furthermore, acceptors cannot be introduced by diffusion of atoms or ions from the surface, as diffusion preferentially proceeds along grain boundaries, leading to shunting of the cell by conducting grain boundaries before sufficient doping levels are achieved within the grains. In many cases, copper (an acceptor in CdTe) has been added—e.g., in graphite contacts still used for experimental contacting, which upon annealing can diffuse into the CdTe film. Its diffusion coefficient along grain boundaries is 100 times that in bulk CdTe [15]. If Cu reaches the junction, it first reduces the junction width and then compensates donors in the CdS layer. Recently, more stable contacts using Cu have been made by depositing Te—Cu double films, which can react to produce Cu_2Te compound films upon annealing, some of the Cu diffusing into the CdTe film and leading to a minute surface doping [16].

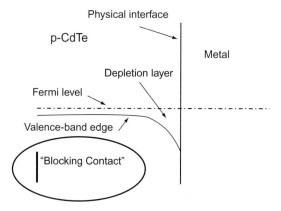

Figure 14 Energy band scheme of the metal–CdTe interface of a CdTe solar cell illustrating blocking contact.

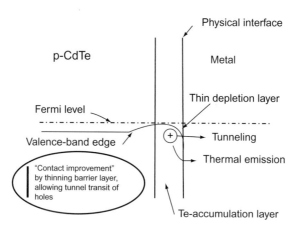

Figure 15 Energy band scheme of a metal contact to CdTe after generation of a p+ surface by Te enhancement leading to a thin barrier.

Alternative efforts in the past have been directed primarily toward three semiconductors: HgTe, ZnTe:Cu, Te and Cu_2Te [17,18]. All of them have led to unstable contacts.

It has become obvious in the course of recent work [4] that a stable back contact cannot consist simply of a metal film. A new contact system has been developed, consisting of a triple procedure [4]:

- generation of an *accumulation layer* (e.g., by suitable etching of CdTe, a Te-enhanced surface layer can be generated);
- deposition of a p-type narrow-band-gap, chemically inert semiconductor or semimetal (*buffer layer*);
- deposition of a metal film for low-resistance current collection (*metal contact*).

The role of the buffer layer essentially consists of protecting the (chemically sensitive) p-type accumulation layer from being corroded by the (reactive) metallic current-collection film. All three components had to be individually optimised and mutually adjusted. Such a triple structure can be highly stable and made by using techniques suited to large-scale production.

The etching process used proceeds into the grain boundaries, leading to 'Te caps' covering the grains as illustrated in Figure 3. This has been strikingly visualised experimentally [17] by sputter-etching away the first 30 nm of the film and analysing the new surface by scanning Auger analysis. Figure 16 clearly shows grain boundaries having excess Te in the form of lighter-coloured regions.

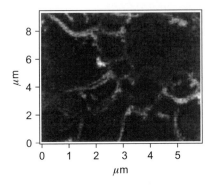

Figure 16 Te-capped CdTe grains by scanning EDX analysis of an ion milled back surface (light areas are Te-enhanced) [7]. *(Reproduced with permission from IEEE.)*

2.3 Stability Issues

Due to the material's strong ionicity (72%), the energy of the bond between Cd and Te is quite high (5.75 eV) [18]. The energy of any photon in the solar spectrum is lower than the binding energy in CdTe or CdS, so breaking of bonds must not be considered. The strong bonding leads to an extremely high chemical and thermal stability, reducing the risk of degradation of performance or any liberation of Cd to a very low level. No degradation intrinsic to the material can be expected.

The stability risks of back contacts have been virtually eliminated by the triple structure described in Section 2.2.2. Nevertheless, careful process development has to be performed to avoid stability risks from other processing steps such as influences of additives in the lamination material. Only dedicated tests of products can yield to assured stability of the product.

2.4 Best Performance of Cells

Using the most advanced techniques, record efficiencies have been achieved by a few groups, indicating the potential of the CdTe thin-film solar cell. In 1984 the magic limit of 10% efficiency was surpassed by a group at Kodak laboratories, using close-spaced sublimation (CSS) [19]. In 1993, an efficiency of 15.8% was achieved [20] by a group from the University of South Florida that used CSS again for formation of CdTe films on borosilicate glass at temperatures of around 600°C. Fine-tuning has been achieved using CdS films made by chemical bath deposition and finally applying an antireflection coating onto the glass surface positioned toward the sun. This value could be surpassed only 7 years later by a group from National Renewable Energy Laboratory (NREL), which achieved 16.5% efficiency [21]. The important advance in this work has been the use of $CdSnO_4$ deposited onto borosilicate glass.

In both recent cases, the higher deposition temperature allowed by use of (expensive) borosilicate glass has been a central issue. For industrial production, this type of glass is presently considered too expensive. Therefore, in industrial production, low-cost soda-lime glass is used, which is limited in temperature endurance.

3. MAKING OF INTEGRATED MODULES
3.1 Interconnection of Cells

Semiconductor solar cells are devices delivering open-circuit voltages of less than 1 V. As electric power for commercial applications requires higher voltages, it has proven advantageous to connect a multitude of cells in series in 'modules.' Whereas in the case of silicon solar cells, individual cells have to be series connected by conductors welded onto both sides of wafers, thin-film cells have a strong advantage to allow integrated series connection of numerous cells, which are at the same time defined in area and interconnected. If the different layers of the cells—TCO, p–n film stack, and back contact—are individually separated (scribed) into parallel stripes that overlap asymmetrically the series, then connection of one distinct cell with its neighbour can be achieved periodically for all cells so generated, as illustrated in Figure 17. After deposition of the TCO film, a first set of separation scribing lines at a periodic distance of about 1 cm is applied, typically by laser ablation. Subsequently, the p–n film sandwich is deposited and separated at the same periodicity so that this scribing line opens the TCO beneath for the back contact, which is deposited subsequently. If

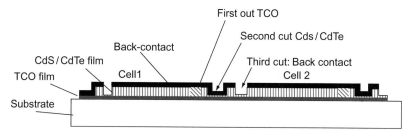

Figure 17 Interconnection principle.

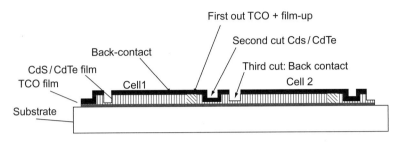

Figure 18 Modified interconnection principle.

the back contact now is separated by a similar set of lines at a small distance from its contact line to the TCO, the interconnection is achieved. Figure 17 shows the principle for a set of three cells.

In some cases, such as electrodeposition, it is not permissible to separate the TCO film before depositing the semiconductor film. Here a variation in the procedure allows the scribing of the first two lines after the deposition of the semiconductor film [22]. An insulating fill-up of the first scribing line is required in this process, as illustrated in Figure 18.

Evidently, using these techniques, the cell width can be adjusted according to technical needs or commercial requirements, On one hand, more interconnection triple scribes lead to higher loss in active area. (Typically, a scribing system requires between 0.2 and 0.3 mm.) On the other hand, wider cells will lead to increased series resistance (lower fill factor), as the current density being conducted through the TCO film will be higher, leading to higher voltage drop. For CdTe, a cell width of 9–10 mm seems to be an optimum value for TCO films of 8–10 Ω per square resistivity. (For CIS cells, due to higher current density and lower voltage of the individual cell, smaller cells—about 6 mm wide—are appropriate; conversely, for amorphous silicon, wider cells are optimal.)

3.2 Contacting

The photocurrent of a module transverses all series-connected cells and is extracted by contacts to the first and last cell; all individual cell voltages add to the total voltage. Usually, metallic conductors are attached to the free contact area of the first and last cell by conducting adhesive and are further connected by contact bands toward the point of the module, where it transverses the back sealing, typically glass. Usually the contact bands are Sn-plated Cu ribbons. Figure 19 indicates the topology of the contacting conductors. Care has to be taken to avoid shunting at any place.

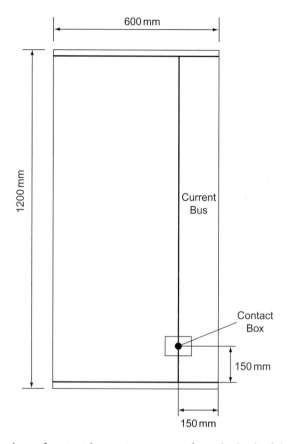

Figure 19 Topology of contact bus system, as seen from the back of the module.

3.3 Lamination

After contact-bus attachment, the module needs sealing and protection against external influences. This usually is achieved by laminating a second glass plate to the module-carrying side of the first glass plate. It is required to remove all films at a boundary region of 1–2 cm of the module to provide the required electric insulation of the module. This is usually achieved before contacting by sandblasting or laser ablation.

For historical reasons, in many cases the technology used for silicon modules is also used for CdTe modules—namely, sealing by EVA (ethyl-vinyl-acetate), which is used as a monomer, applied as a sheet, and polymerised by a thermal annealing step under vacuum. Commercial EVA is readily available. It is 'overqualified' for CdTe, as intense development work has been invested into the commercial product to stabilise it against degradation in the sunlight impinging onto silicon modules. In the case of CdTe, this special quality is not required, because sunlight directly enters the junction after having passed the glass and the lamination material is positioned below the second glass and therefore oriented away from the sun.

The cover glass so laminated to the module has a hole through which the contact bands are guided into a contact box, which is attached to the glass (cf. Figure 19). Two stable current guides with plugs are attached to the box and will be used for connecting the modules so generated to the user circuit. Evidently, all connections and contacts must be extremely well protected against water and water vapour for achieving the expected lifetime of the module.

4. PRODUCTION OF CDTE THIN-FILM MODULES

4.1 Generalised Production Sequence

The different steps to make cells and modules described in Sections 2 and 3 can be arranged into a sequence to make ready-to-use modules as a product:
1. selection of substrate glass, soda-lime glass (float glass) as substrate;
2. deposition of the transparent conductive coating (SnO_2, ITO, etc.);
3. scribing of the TCO film into parallel bands, defining and separating the cells (for some processes, this scribing can be done later after deposition of the semiconductor films);
4. deposition of CdS films of lowest possible thickness, typically around 100 nm;
5. deposition of CdTe by the process of choice for the particular product;
6. activation of the film stack by influence of $CdCl_2$ at temperatures of around 400°C;
7. application of the second scribing step, which opens the semiconductor stack for contacting the TCO film (optionally, step 3 can be applied in parallel; the scribing line afterward has to be filled up for electrical insulation);
8. application of the back-contact structure, consisting of a set of steps—e.g., etching to achieve a Te accumulation, application of a buffer layer, application of the metallic back contact;

9. separation cut to separate the back contacts of the neighbouring cells;
10. attachment of the contact-bus structure;
11. lamination with a second glass (or plastic) using a suitable plastic such as EVA or a thermoplastic film, the contacts protruding through a hole in the cover glass plate for the next step;
12. contact-box attachment in which the fragile contact bands from the module are connected to stable cables with suitable plugs for commercial application; and
13. measurement of each module's efficiency using a solar simulator.

With some modifications, this sequence is the basis of the following industrial efforts into production.

4.2 Industrial Production of Modules

A commercial product can only be manufactured with expectation of cost-covering revenues, if a factory above a certain capacity is built, using the dimensions of scale. It is generally accepted today that a capacity of around 100,000 m^2 per annum is presently appropriate. All units of significantly lower capacity should be called pilot plant as they will require more funds than any return can provide. At around 600,000 m^2 per annum a cost potential of 0.6/W_p has been estimated by a group of experts [23].

Presently three industrial units are known to actively pursue the target of large-scale production. They basically differ by the deposition technique for CdS and CdTe.

4.2.1 BP Solar Inc. (Fairfield, California, USA)

Work at BP Solar started in Great Britain in 1984 when British Petroleum took over a galvanic deposition process for CdTe thin-film solar cells from Monosolar Inc. in the United States. BP Solar continued development of the basic electrodeposition process for CdTe [24]. Upscaling work resulted in a factory built in the United States. Due to low deposition rates, parallel deposition is used: a large number (40 to 100) of plates are immersed into a tank containing a recirculation system for the continuously replenished electrolyte. A constant potential is applied to the plates via the TCO films already covered by CdS. (The CdS film is deposited by the chemical bath deposition technique, also in parallel onto a larger number of TCO-coated glasses.) The total charge applied to the plates is a measure for film thickness. A charge of about 12,000 C/m^2 leads to approximately 1.6 μm of CdTe film. Postannealing leads to strain relief, change of conduction polarity from n to p, and activation by means of a Cl compound added to the bath. The grain size grows from 0.1 to 0.2 μm to about 0.4 μm during this procedure.

Great care has to be taken to avoid lateral voltage drop over the plate surface during CdTe deposition. Te- or Cd-rich compounds can form upon deviation from this condition. To achieve good-quality films, the lowest possible surface resistivity of the TCO has to be chosen. The cell-defining and module-generating scribe application follows the principle shown in Figure 18. In order to keep the TCO film intact as long as possible, the TCO scribing lines are applied after CdTe deposition, necessitating a fill-up procedure. Module sizes are around 1 m^2, yielding a power of around 72 W under simulated terrestrial solar light, have been reported [25]. More than 150,000 m^2 of glass substrates can be processed per annum. The line can produce 0.55-m^2 and 0.94-m^2 modules. The deposition system consists of eight identical deposition tanks. Each tank is able to simultaneously coat forty substrates of 0.55-m^2 or twenty-four of 0.94-m^2. Since 2002, BP Solar has manufactured the first large area CdTe module of >10% efficiency.[1]

4.2.2 First Solar LLC (Toledo, Ohio, USA)

This company started work within the scope of a predecessor called Solar Cells Inc. around 1991 in Toledo by a senior shareholder of Glasstech Inc. In 1999, a joint venture with a finance group from Arizona was formed that led to the new name First Solar LLC [26]. The basic process for generating CdS−CdTe junctions relies on the sublimation and condensation properties of CdS and CdTe described earlier. Substrates coated by SnO$_2$ films by Libbey Owens Ford using a spray process enter a vacuum system through a loadlock and move—lying on a roller system to avoid warping—into a chamber heated to temperatures around 560°C. Vapour sources are positioned above the substrates out of which CdS and CdTe vapours emerge, are directed toward the substrates, and condense sequentially on the substrates. The CdS and CdTe material is continuously fed into the evaporators. Module interconnection of cells is performed afterward—i.e., first and second scribing lines are applied after semiconductor deposition, requiring filling of scribing line 1 (cf. Figure 18).

The standard module size of First Solar is 60 × 120 cm^2 = 0.72 m^2 [27]. Efficiencies of 6−8% can be routinely achieved. Figure 20 shows area efficiency versus sequential plate count for 3128 sequentially deposited modules.

---

[1] In 2002, the management of BP Solar decided to terminate its CdTe development and production efforts after 18 years of investment and closed down the factory.

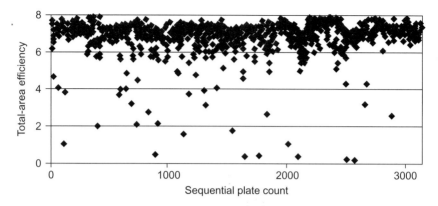

Figure 20 Total-area efficiency for 3128 sequentially deposited modules from First Solar LLC (from [27]).

4.2.3 ANTEC Solar GmbH (Arnstadt, Germany)

The technology used at ANTEC Solar is essentially based on the development started at Battelle Institute around 1970 [28]. After the closure of Battelle Institute, the know-how was transferred to ANTEC GmbH, a management buyout from Battelle, and developed into a manufacturing technology. ANTEC Solar was founded in 1996 in order to start production of CdTe thin-film PV modules. A fully automated production plant has been built in Arnstadt (Germany), and production started in 2001. The basic process used for deposition of CdTe is close-spaced sublimation. Glass substrates carrying scribed TCO films are transported in vacuum above crucibles containing CdS and CdTe at temperatures of 700°C. The semiconductor materials condense at temperatures of 500°C and form the n−p diode structure, which is activated in a $CdCl_2$ atmosphere. Module size is 60×120 cm^2; module efficiencies were about 7% and were expected to increase to 8% in 2002. In contrast to BP Solar and First Solar, the TCO is also made in the plant in the online system. Definition of interconnected cells is achieved according to Figure 17— i.e., at three different stages of the plant by laser ablation (first scribe) and mechanical ablation (second and third scribe).[2]

Figure 21 shows the total area efficiency of 2000 modules manufactured in one production run. The production plant presently employs 100 persons.

---

[2] Because of financial problems. ANTEC Solar declared insolvency in 2002. In 2003, the plant was taken over by a new owner, who successfully restarted production.

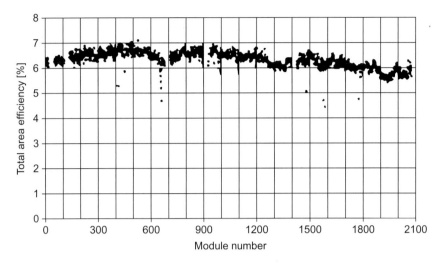

Figure 21 Total area efficiency for 2000 sequentially deposited modules from ANTEC Solar GmbH.

4.3 A 10-MW Production Line

The following uses an example to describe and illustrate the production line of ANTEC Solar GmbH.

The deposition line was conceived as a two-step production line consisting of a fully automated deposition line for integrated modules on glass substrates of 60×120 cm^2 and a semiautomatic module line for hermetical sealing, contacting, measuring, and customising of the modules into a marketable product.

The fully automatic in-line deposition procedure consists of nine steps:
1. Cleaning of the substrate (float glass).
2. Deposition of the transparent contact (ITO + SnO$_2$) at around 250°C.
3. Scribing of the TCO film for cell definition and interconnection.
4. Deposition of CdS and CdTe by CSS at around 500°C.
5. Activation (improvement of junction by annealing in Cl-containing atmosphere at around 400°C).
6. Wet-chemical etching for contact preparation (Te accumulation).
7. Scribing by mechanical tools for interconnection of cells.
8. Deposition of two-layer back contact by sputtering.
9. Scribing by mechanical tools for separation and interconnection of cells.

Most of these positions are connected by heating or cooling segments in order to present the plates to the deposition steps at the adequate temperatures They take a large part of the equipment length. The highest temperature (500°C) is reached during deposition of the semiconductor films, compatible with glass stability. The total length of the automated deposition line is 165 m. Glass plates (to become modules) are transported by automated conveyor systems and are not touched by human hands during the processes. They are collected in boxes in sets of 30 and transferred to the module line in an adjacent hall.

In the module line, the substrates with sets of interconnected cells (often called *submodules*) are contacted and sealed for convenient use in energy-generating systems. This part of the factory consists of the following procedures:

1. Cleaning of the cover glass (float glass).
2. Deposition of the contact buses onto the modules.
3. PV function testing to identify substandard modules, which can be excluded from further processing.
4. Cutting of EVA sheet to size.
5. Joining of module, EVA sheet, and cover glass.
6. Lamination (sets of six modules).
7. Fill in of contact hole in cover glass.
8. Attachment of contact box.
9. Quantitative measurement of PV performance (sets of three modules).
10. Type tag attachment.
11. Classification, selection, and packaging for dispatch.

The module line requires some manual handling and adjustments due to the heterogeneity of the processes. Figure 22 shows the geometrical arrangement of the different processes. Each line has been installed in a separate hall. Both halls are connected by an aisle for transfer of modules. Buffer stations allow the removal of partly completed modules in case of an incident in the downline stations without crashes in the upline stations. The overall primary parameter for the deposition line and also the module line has been the production speed defined at 120,000 modules per annum, which leads to an average linear transportation speed of approximately 1 m/min in the production line.

The procedures described above contain (1) state-of-the-art procedures, (2) new but simple procedures, and (3) newly developed procedures.

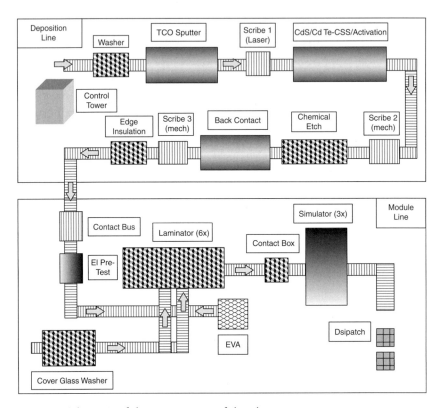

Figure 22 Schematic of the components of the plant.

Standard procedures are sputter deposition of ITO and SnO_2 metals. Dedicated equipment has been constructed by experienced equipment manufacturers. In some cases (e.g., ITO deposition), multiple sputtering targets are needed in order to achieve the required film thickness at the given transport speed of the substrates (around 1 m/min). The final steps of contacting and lamination are also using state-of-the-art techniques. Equipment can be bought custom-made on the market. An unforeseen effort in the setup and initial operation of the plant had to be dedicated to the transport of the substrate sheets through heating and cooling stages without breakage due to excessive thermal gradients. After some significant modifications, breakage under thermal stress could be virtually eliminated.

Simple but new processes are essentially the core processes for the active semiconductors, namely, CSS deposition of CdS, CdTe, and—if required—back-contact buffer semiconductors. It has been shown that

CdTe can be deposited onto stationary substrates at rates of 10 μm/min and more. For achieving the required production speed, three crucibles containing a week's supply of starting material of CdTe are used. For CdS, due to its low thickness, only one crucible is used. As mentioned, the vacuum requirements are rather low, so no high-vacuum equipment is needed. (For CdTe, oxygen incorporation to some extent indeed is considered advantageous by some researchers.) Laser scribing of the TCO film does not require basic new technologies. Equipment can be built by industry, and processing data have to be developed and adjusted. In spite of these arguments, some problems have been incurred by thermal stress conditions of crucibles and substrates. Fast heating invariably leads to breakage of glass substrates.

More involved *new processes* are activation of the junction CdS–CdTe and etching of the CdTe surface as a contact-preparation procedure. Etching requires the wetting of the substrates carrying the activated CdS–CdTe film stacks by nitric and phosphoric acid plus water solutions (NP etch), rinsing, and drying. Although being essentially simple, procedures equipment had to be developed to incorporate these processes under tightly controlled parameters into the production line to allow the target speed to be achieved. Activation requires exposition of the substrates carrying the TCO-CdS-CdTe film-stack to, e.g., $CdCl_2$ at elevated temperatures. Fortunately, traces of oxygen do not play a detrimental role. (The activation step in the laboratory can even be executed in air.) Furthermore, $CdCl_2$ at room temperature has extremely low vapour pressure. Some special chemical-engineering skill has been needed to conceive an in-line activation stage fitting into the production line. Mechanical scribing has turned out to be more involved, due to the requirement that the production line should have only one service interval per week. Scribing tools have to survive one week under tough tolerances. New solutions had to be found for recognising and identifying the first laser cuts below the CdTe film for precise positioning of the second and third scribing lines. Implementation of automatic image recognition of previously scribed lines in the TCO films for precise positioning of subsequent scribing lines has been a nontrivial task.

After the plant had been built and assembled in 1999 and the first functional tests had been passed in early 2000, debugging and process optimisation required extensive tests and equipment modifications by hardware and software manufacturers. Production started in 2001. Figure 23 shows the deposition line in its entirety.

Figure 23 View of the total deposition line.

4.4 Environmental and Health Aspects

The production of polycrystalline thin-film CdTe solar modules basically employs techniques common in chemical and microelectronic industry. The substances involved are easily manageable by standard processes. Production is possible under existing safety laws without putting staff health at risk. It is technically and economically possible to design and operate a factory with zero emissions. Workers in a production environment have been tested regarding Cd uptake and shown Cd content in blood and urine far below the threshold concern level under periodic medical scrutiny [29]. Smokers have shown a somewhat higher Cd level than do nonsmokers but still are below any threshold for concern.

A number of studies from third parties [30–33] show negligible risk under use of CdTe solar modules for the environment and humans even under irregular conditions. In case of exposure to fire, the substrate and cover glass will melt long before the CdTe decomposes, thereby including the semiconductor into the resolidifying glass. Incineration experiments conducted by BP Solar in cooperation with a fire research institution using typical household inventory plus CdTe modules have not led to detectable emissions of Cd compounds [30].

During use, a CdTe module can be compared to laminated glass similar to that used in cars. Thus, modules will not easily break and release their content. At their end of life, modules can be recycled by crushing them whole and either returning the debris to the smelters, which can inject the material into their processes without significant additional cost, or dissolving the films by liquid or gaseous etchants.

Debis Systemhaus GmbH (a subsidiary of Daimler-Chrysler AG) has established a life-cycle inventory for CTS thin-film solar modules of ANTEC Solar GmbH guided by ISO 14040 and 14041. To achieve this aim, the total energy and material flow for the module's life cycle has been accounted with help of the CUMPAN® software system. This allows, for example, the determination of the total energy required for raw materials' production and processing.

- The manufacture of 1 m^2 of CTS-module uses 126 kg of raw material and primary energy carrier and 70 kWh of electric energy.
- Under the climatic conditions of Germany, the generation of 1000 kWh of electricity per annum at a module lifetime of 30 years saves 16,244 kg of carbon dioxide and further undesired materials such as sulphuric dioxide and nitrogenic oxides.
- The production plant is built not to emit any material. Water is reprocessed and reused. (Waste heat, however, is emitted to the environment at this time.)
- A recycling process is envisaged for spent and reject modules. In view of environmental safety, economical retrieval of valuable raw materials, and securing of hazardous materials, this process can be considered satisfying.
- Compared with alternative thin-film modules (a-Si, CIS), emissions and waste during production amount to similar values.
- The emission of cadmium can be judged as low in comparison to other emission sources (e.g., coal-fired power plants, phosphate fertiliser). Even in case of accidents (e.g., fire), by reason of the small quantities of (thermally stable) material per square metre of module area, no environmentally critical emission must be contemplated.
- The total energy used for fabricating a module will be retrieved by the module within 15 months using the actual energy uptake of the ANTEC factory. From then on, the module operates in an environmentally benign way.

4.5 Material Resources

More than 99% of the weight of a CdTe modules consists of float glass, EVS, and metal connectors readily available in virtual unlimited quantities. TCO films made of SnO_2 are not considered to be limited by available Sn supplies. In used in ITO films is a more rare resource at an annual production of 120 t [34]. Furthermore, it is also used for ITO films in liquid crystal displays and the CuInSe$_2$-based thin-film solar cell. ITO can be substituted by SnO$_2$ and—possibly—ZnO, Sn and Zn being abundant metals. CdTe and CdS warrant a closer consideration. Presently, CdTe (and CdS) are offered in the required purity by five industrial enterprises. Cd is presently produced as a by-product of Zn at 20,000 t per year [34]. Due to low demand, Te is presently produced at 300 t per year as a by-product of Cu [34]. Growing future demand, according to an expert in the field [35], can be satisfied by more efficient extraction from the anodic slurries in Cu electrorefining and also by exploiting Te-rich ores in South America not yet exploited. S finally is an abundant element.

In summary, no critical material bottleneck is expected for an expanding production. The CdTe thin-film solar cell will be able to take a share in ameliorating future energy shortages and the climate change expected from burning fossil fuels.

5. THE PRODUCT AND ITS APPLICATION

The photovoltaic modules manufactured in a factory are ready to use in a suitable PV installation. CdTe modules by any one manufacturer are mass products made on large scale (100,000 per year or more). Typically, the modules are sized about 0.6×1.2 m^2. A module 60 cm wide can easily be carried under the arm by one person. Glass–glass laminates, as manufactured, for example, by First Solar and ANTEC Solar, have a weight of 16 kg. A contact box furnishes two cables carrying plugs (male and female for easy series connection) that are long-term stable. Figure 24 shows an ATF module from ANTEC Solar. A very homogeneous appearance of most thin-film modules helps organic and visually pleasing installation.

5.1 Product Qualification

A PV module is a product made for long-time deployment in a harsh environment and therefore must be furnished with a warranty for approximately 20 years of useful life. In order to provide such assurance, the international standards agencies (International Electrotechnical Commission, or IEC) have designed international norms that are valid worldwide. The norm pertaining to thin-film modules is IEC 61646,

Figure 24 View of a CdTe thin-film PV module.

'Thin film terrestrial photovoltaic modules –Design qualification and type approval.' This norm was published in 1996 and was originally designed with amorphous silicon modules in mind.

Modules to be classified as fulfilling IEC 61646 have to undergo an extensive set of tests defined in detail in the text of this norm. Briefly these are four sets, for which eight modules have to be provided, typical for the production discussed:

A. Performance tests
- Power output at standard test condition (25°C, 1 sun)
- Power output at nominal operating conditions
- Power output at low sunlight intensity (20%)

B. Endurance tests
- Long-term outdoor exposure
- Light soaking
- Temperature shock tests
- Damp-heat test

C. Mechanical tests
- Mechanical load
- Twist test
- Hail test

D. Electric tests
- Insulation test
- Water immersion test

5.2 Examples of Installation of CdTe Modules

Thin-film modules can show a highly homogeneous surface appearance and very little variation from module to module. This allows the assembly of large, highly homogeneous panels on rooftops, on facades, or on the ground. Figures 25–27 show examples of such installations. Modules in

Figure 25 Installation of CdTe modules from First Solar LLC on an office building.

Figure 26 Installation of CdTe modules from ANTEC Solar on the wall of a public administration building.

the form of glass—glass laminates can be mounted on special structures, which can be invisible and nonetheless watertight. Figure 28 shows how modules can be mounted on metal rafters that allow water drainage and also seal the interior of the roof by rubber lips.

Figure 27 Installation of CdTe modules on a private residence roof.

Figure 28 Close-up of mounting system on metal rafters for roof installation.

6. THE FUTURE

After more than 20 years of development in various industrial and academic laboratories, the CdTe thin-film solar module has entered the production stage, and experience is being gained in this process. This will definitively lead to the next generation of plant at capacities around 1,000,000 m^2 per year. It is expected that the learning curve will be transgressed quite quickly, which will lead to a mature low-cost product.

REFERENCES

[1] K. Zanio, Cadmium telluride: materials preparation, physics, defects, applications, in: Semiconductors and Semimetals, vol. 13, 1978.
[2] D. Bonnet, et al., The CdTe thin film solar cell—EUROCAD, in: Final Report to the Commission of the European Communities, Project No. JOU2-CT92−0243, 1995. Referenced data later also published in: H.M. Al Allak, et al., The effect of processing conditions on the electrical and structural properties of CdS/CdTe solar cells, in: Proceedings of the 13th European Photovoltaic Solar Energy Conference, Nice, 1995, pp. 2135−2138.
[3] R. Bube, Photoconductivity of Solids, Wiley, New York, London, 1960, p. 94
[4] D. Bonnet, et al., CADBACK: The CdTe thin film solar cell—improved back contact, in: Final Report to the European Commission, Contract No. JOR3-CT98−0218, 2002.
[5] R. Dhere, et al., Influence of CdS/CdTe interface properties on the device properties, in: Proceedings of the 26th IEEE Photovoltaic Specialists Conference, Anaheim, 1997, pp. 435−437.
[6] R.H. Moutinho, et al., Effects of CdCl$_2$-treatment on the recrystallization and electro-optical properties of CdTe films, J. Vac. Sci. Technol. A16 (1998) 1251−1257.
[7] H.R. Moutinho, et al., Alternative procedure for the fabrication of close-spaced sublimated CdTe solar cells, J. Vac. Sci. Technol. A18 (2000) 1599−1603.
[8] B. McCandless, R. Birkmire, Diffusion in CdS/CdTe thin-film couples, in: Proceedings of the 16th European Photovoltaic Solar Energy Conference, Glasgow, 2000, pp. 349−352.
[9] K. Durose, Private communication.
[10] S.-Y. Nunoue, T. Hemmi, E. Kato, Mass spectrometric study of the phase boundaries of the CdS/CdTe system, J. Electrochem. Soc. 137 (1990) 1248−1251.
[11] R.G. Dhere, et al., Intermixing at the CdS/CdTe interface and its effect on device performance, Mat. Res. Soc. Symp. Proc 426 (1966) 361−366.
[12] B. McCandless, M.G. Engelman, R.W. Birkmire, Interdiffusion of CdS/CdTe thin films: modelling X-ray diffraction line profiles, J. Appl. Phys. 89 (2001) 988−994.
[13] Y. Yan, D.S. Albin, M.M. Al-Jassim, The effect of oxygen on junction properties in CdS/CdTe solar cells, in: Proceedings of the NCPV Program Meeting, 2001, pp. 51−52.
[14] H.R. Moutinho, et al., Study of CdTe/CdS solar cells using CSS CdTe deposited at low temperature, in: Proceedings of the 28th IEEE Photovoltaic Specialists Conference, Anchorage, 2000, pp. 646−649.

[15] S.S. Hegedus, B.E. McCandless, R.W. Birkmire, Analysis of stress-induced degradation in CdS/CdTe solar cells, in: Proceedings of the 28th IEEE Photovoltaic Specialists Conference, Anchorage, 2000, pp. 535–538.
[16] S.S. Hegedus, B.E. McCandless, R.W. Birkmire, initial and stressed performance of CdTe solar cells: effect of contact processing, in: Proceedings of the NCPV Program Review Meeting, 2001, pp. 119–120.
[17] D.H. Levi, et al., Back contact effects on junction photoluminescence in CdTe/CdS solar cells, in: Proceedings of the 26th IEEE Photovoltaic Specialists Conference, Anaheim, 1997, pp. 351–354.
[18] H. Hartmann, R. Mach, B. Selle, Wide gap II–VI compounds as electronic materials, in: E. Kaldis (Ed.), Current Topics in Materials Science, Amsterdam, 1981, pp. 1–414.
[19] Y.-S. Tyan, E.A. Perez-Albuerne, Efficient thin film CdS/CdTe solar cells, in: Proceedings of the 16th IEEE Photovoltaic Specialists Conference, San Diego, 1982, pp. 794–800.
[20] J. Britt, C. Ferekides, Thin film CdS/CdTe solar cell with 15.8% efficiency, Appl. Phys. Lett. 62 (1993) 2851–2852.
[21] X. Wu, et al., 16.5% efficient CdS/CdTe polycrystalline thin film solar cell, in: Proceedings of the 17th European Photovoltaic Solar Energy Conference, Munich, 2001, pp. 995–1000.
[22] D. Rose, et al., R&D of CdTe-absorber photovoltaic cells, modules and manufacturing equipment: plan and progress to 100 MW/yr, in: Proceedings of the 28th IEEE Photovoltaic Specialists Conference, Anchorage, 2000, pp. 428–431.
[23] J.M. Woodcock, et al., A study on the upscaling of thin film solar cell manufacture towards 500 MWp per annum, in: Proceedings of the 14th European Photovoltaic Solar Energy Conference, Barcelona, 1997, pp. 857–860.
[24] D.W. Cunningham, et al., Advances in large area apollo module development, in: Proceedings of the NCPV Program Review Meeting, 2000, pp. 261–262.
[25] D.W. Cunningham, et al., Large area appollo thin film module development, in: Proceedings of the 16th European Photovoltaic Solar Energy Conference, Glasgow, 2000, pp. 281–285.
[26] A. McMaster, et al., PVMat advances in CdTe product manufacturing, in: Proceedings of the NCPV Program Review Meeting, 2000, pp. 101–102.
[27] D. Rose, R. Powell, Research and progress in high-throughput manufacture of efficient, thin-film photovoltaics, in: Proceedings of the NCPV Program Review Meeting, 2001, pp. 209–210.
[28] D. Bonnet, H. Rabenhorst, New results on the development of a thin film p-CdTe–n-CdS heterojunction solar cell, in: Proceedings of the 9th IEEE Photovoltaic Specialists Conference, Silver Springs, 1972, pp.129–131.
[29] J.R. Bohland, K. Smigielski, First solar's module manufacturing experience; environmental, health and safety results, in: Proceedings of the 28th IEEE Photovoltaic Specialists Conference, Anchorage, 2000, pp. 575–578.
[30] E.A. Alsema, B.C.W. van Engelenburg, Environmental risks of CdTe and CIS solar cell modules, in: Proceedings of the 11th European Photovoltaic Solar Energy Conference, Montreux, 1992, pp. 995–998.
[31] M.H. Patterson, A.K. Turner, M. Sadeghi, R.J. Marshall, Health, safety and environmental aspects of the production and use of cdte thin film photovoltaic modules, in: Proceedings of the 12th European Photovoltaic Solar Energy Conference, Amsterdam, 1994, pp. 951–953.
[32] P.D. Moskowitz, H. Steinberger, W. Thumm, Health and environmental hazards of CdTe photovoltaic module production, use and decommissioning, in: Proceedings of

the First World Conference on Photovoltaic Energy Conversion, Hawaii, 1994, pp. 115–118.
[33] H. Steinberger, Health and environmental risks from the operation of CdTe- and CIS thin film modules, in: Proceedings of the 2nd World Conference on Photovoltaic Solar Energy Conversion, Vienna, 1998, pp. 2276–2278.
[34] US Bureau of Mines, Mineral Commodity Summary, 1992.
[35] G. Daub, PPM Pure Metals GmbH., personal communication.

CHAPTER IC-3

Cu(In,Ga)Se$_2$ Thin-Film Solar Cells

U. Rau and H.W. Schock
Institut für Physikalische Elektronik (IPE), Universität Stuttgart, Germany

Contents

1. Introduction 324
2. Material Properties 325
   2.1 Chalcopyrite Lattice 325
   2.2 Band-Gap Energies 326
   2.3 The Phase Diagram 326
   2.4 Defect Physics of Cu(In,Ga)Se$_2$ 328
3. Cell and Module Technology 331
   3.1 Structure of the Heterojunction Solar Cell 331
   3.2 Key Elements for High-Efficiency Cu(In,Ga)Se$_2$ Solar Cells 331
   3.3 Absorber Preparation Techniques 333
      3.3.1 Basics *333*
      3.3.2 Co-Evaporation Processes *334*
      3.3.3 Selenisation Processes *335*
      3.3.4 Other Absorber Deposition Processes *337*
      3.3.5 Postdeposition Air Anneal *337*
   3.4 Heterojunction Formation 338
      3.4.1 The Free Cu(In,Ga)Se$_2$ Surface *338*
      3.4.2 Buffer Layer Deposition *339*
      3.4.3 Window Layer Deposition *340*
   3.5 Module Production and Commercialisation 341
      3.5.1 Monolithic Interconnections *341*
      3.5.2 Module Fabrication *342*
      3.5.3 Upscaling Achievements *343*
      3.5.4 Stability *344*
      3.5.5 Radiation Hardness and Space Applications *345*
4. Device Physics 346
   4.1 The Band Diagram 346
   4.2 Short-Circuit Current 348
   4.3 Open-Circuit Voltage 349
   4.4 Fill Factor 352
   4.5 Electronic Metastabilities 353
5. Wide-Gap Chalcopyrites 354
   5.1 Basics 354
   5.2 CuGaSe$_2$ 357
   5.3 Cu(In,Al)Se$_2$ 358

Practical Handbook of Photovoltaics.
© 2012 Elsevier Ltd. All rights reserved.

    5.4 CuInS$_2$ and Cu(In,Ga)S$_2$ 358
    5.5 Cu(In,Ga)(Se,S)$_2$ 359
    5.6 Graded-Gap Devices   359
6. Conclusions   361
Acknowledgements   362
References   *362*

1. INTRODUCTION

With a power conversion efficiency of 18.8% on a 0.5-cm^2 laboratory cell [1] and 16.6% for mini-modules with an area of around 20 cm^2 [2], Cu(In,Ga)Se$_2$ is established as an efficient thin-film solar cell technology. The start of production at several places provides a new challenge for research on this material. However, these recent achievements are based on a long history of research and technological development.

CuInSe$_2$ was synthesised for the first time by Hahn in 1953 [3]. In 1974, this material was proposed as a photovoltaic material [4] with a power conversion efficiency of 12% for a single-crystal solar cell. In the years 1983–84, Boeing Corp. reported efficiencies in excess of 10% from thin poly crystalline films obtained from a three-source co-evaporation process [5]. In 1987 Arco Solar achieved a long-standing record efficiency for a thin-film cell of 14.1% [6]. It took a further ten years, before Arco Solar, at that time Siemens Solar Industries (now Shell Solar), entered the stage of production. In 1998, the first commercial Cu(In,Ga)Se$_2$ solar modules were available [7]. In parallel, a process which avoids the use of H$_2$Se is being developed by Shell Solar in Germany [8] (see also Chapter IV-1). Other companies in the USA, Global Solar and ISET, plan to commercialise modules prepared on other than glass substrates. In Europe, the long-term development efforts of the EUROCIS consortium on the co-evaporation process resulted in the activity of Würth Solar with pilot production envisaged in 2003 [9,10]. In Japan, two lines for film preparation are planned by Showa Shell (selenisation by H$_2$Se) [11] and Matshushita (co-evaporation) [12].

In this chapter, we give a short overview on the present knowledge of Cu(In,Ga)Se$_2$-based heterojunction thin film solar cells. We focus on four points: (i) The description of the basic material properties such as crystal properties, phase diagram, and defect physics. (ii) Description of the cell technology starting from the growth of the polycrystalline Cu(In,Ga)Se$_2$ absorber up to device finishing by heterojunction formation and window layer deposition. This section also discusses basic technologies for module production. (iii) The electronic properties of the finished heterostructure. (iv) Finally, Section 5 discusses the photovoltaic potential of wide-gap chalcopyrites, namely CuGaSe$_2$ and CuInS$_2$, as well as that of the pentenary alloy system Cu(In,Ga)(S,Se)$_2$ and the possibility of building graded-gap structures with these alloys.

This chapter can only briefly cover those scientific issues that are relevant for photovoltaic applications. More detailed information can be found in two review articles by the present authors [13,14] as well as in references [15–19].

2. MATERIAL PROPERTIES
2.1 Chalcopyrite Lattice

CuInSe$_2$ and CuGaSe$_2$, the materials that form the alloy Cu(In,Ga)Se$_2$, belong to the semiconducting I–III–VI$_2$ materials family that crystallise in the tetragonal chalcopyrite structure. The chalcopyrite structure of, for example, CuInSe$_2$ is obtained from the cubic zinc blende structure of II–VI materials like ZnSe by occupying the Zn sites alternately with Cu and In atoms. Figure 1 compares the two unit cells of the cubic zinc

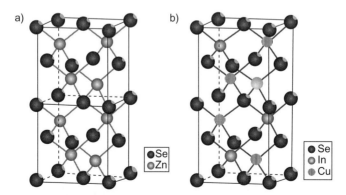

Figure 1 Unit cells of chalcogenide compounds. (a) Sphalerite or zinc blende structure of ZnSe (two unit cells); (b) chalcopyrite structure of CuInSe2. The metal sites in the two unit cells of the sphalerite structure of ZnSe are alternately occupied by Cu and In in the chalcopyrite structure.

blend structure with the chalcopyrite unit cell. Each I (Cu) or III (In) atom has four bonds to the VI atom (Se). In turn each Se atom has two bonds to Cu and two to In. Because the strengths of the I–VI and III–VI bonds are in general different, the ratio of the lattice constants c/a is not exactly two. Instead, the quantity $2-c/a$ (which is -0.01 in CuInSe$_2$, $+0.04$ in CuGaSe$_2$) is a measure of the tetragonal distortion in chalcopyrite materials.

2.2 Band-Gap Energies

The system of copper chalcopyrites Cu(In,Ga,Al)(Se,S)$_2$ includes a wide range of band-gap energies E_g from 1.04 eV in CuInSe$_2$ up to 2.4 eV in CuGaS$_2$. and even 2.7 eV in CuAlS$_2$, thus, covering most of the visible spectrum. All these compounds have a direct band gap, making them suitable for thin film photovoltaic absorber materials. Figure 2 summarises lattice constants a and band-gap energies E_g of this system. Any desired alloys between these compounds can be produced as no miscibility gap occurs in the entire system. We will discuss the status and prospects of this system in Section 5 in more detail.

2.3 The Phase Diagram

Compared with all other materials used for thin-film photovoltaics, Cu(In,Ga)Se$_2$ has by far the most complicated phase diagram. Figure 3 shows

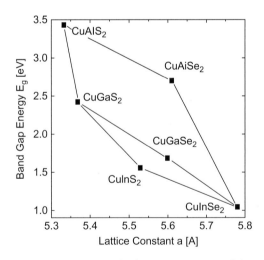

Figure 2 Band-gap energies E_g versus the lattice constant a of the Cu(In,Ga,Al)(S,Se)$_2$ alloy system.

the phase diagram of CuInSe$_2$ given by Haalboom et al. [20]. This investigation had a special focus on temperatures and compositions relevant for the preparation of thin-films. The phase diagram in Figure 3 shows the four different phases which have been found to be relevant in this range: the α-phase (CuInSe$_2$), the β-phase (CuIn$_3$Se$_5$), the δ-phase (the high-temperature sphalerite phase) and Cu$_y$Se. An interesting point is that all neighbouring phases to the α-phase have a similar structure. The β-phase is actually a defect chalcopyrite phase built by ordered arrays of defect pairs (Cu vacancies V$_{Cu}$ and In−Cu antisites In$_{Cu}$). Similarly, Cu$_y$Se can be viewed as constructed from the chalcopyrite by using Cu−In antisites Cu$_{In}$ and Cu interstitials Cu$_i$. The transition to the sphalerite phase arises from disordering the cation (Cu, In) sub-lattice, and leads back to the zinc blende structure (cf. Figure 1 (a)).

The existence range of the α-phase in pure CuInSe$_2$ on the quasi-binary tie line Cu$_2$Se-In$_2$Se$_3$ extends from a Cu content of 24 to 24.5 at.%. Thus, the existence range of single-phase CuInSe$_2$ is relatively small and does not even include the stoichiometric composition of 25 at.% Cu. The Cu content of absorbers for thin-film solar cells varies typically between 22 and 24 at.% Cu. At the growth temperature this compositional range lies within the single-phase region of the α-phase. However, at room temperature it lies in the two-phase α + β region of

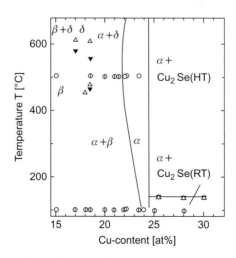

Figure 3 Quasi-binary phase diagram of CuInSe$_2$ along the tie-line that connects the binary compounds In$_2$Se$_3$ and Cu$_2$Se established by Differential Thermal Analysis (DTA) and microscopic phase analysis (After Haalboometal. [20]). Note that at 25 at.% Cu no single phase exists.

the equilibrium phase diagram [20]. Hence one would expect a tendency for phase separation in photovoltaic-grade $CuInSe_2$ after deposition. Fortunately, it turns out that partial replacement of In with Ga, as well as the use of Na-containing substrates, considerably widens the single-phase region in terms of (In + Ga)/(In + Ga + Cu) ratios [21]. Thus, the phase diagram hints at the substantial improvements actually achieved in recent years by the use of Na-containing substrates, as well as by the use of $Cu(In,Ga)Se_2$ alloys.

2.4 Defect Physics of Cu(In,Ga)Se$_2$

The defect structure of the ternary compounds $CuInSe_2$, $CuGaSe_2$, $CuInS_2$, and their alloys, is of special importance because of the large number of possible intrinsic defects and the role of deep recombination centres for the performance of the solar cells. The features that are somewhat special to the Cu-chalcopyrite compounds are the ability to dope these compounds with native defects, their tolerance to large off-stoichiometries, and the electrically neutral nature of structural defects in these materials. It is obvious that the explanation of these effects significantly contributes to the explanation of the photovoltaic performance of these compounds. Doping of $CuInSe_2$ is controlled by intrinsic defects. Samples with p-type conductivity are grown if the material is Cu-poor and annealed under high Se vapour pressure, whereas Cu-rich material with Se deficiency tends to be n-type [22,23]. Thus, the Se vacancy V_{Se} is considered to be the dominant donor in n-type material (and also the compensating donor in p-type material), and the Cu vacancy V_{Cu} the dominant acceptor in Cu-poor p-type material.

By calculating the metal-related defects in $CuInSe_2$ and $CuGaSe_2$, Zhang et al. [24] found that the defect formation energies for some intrinsic defects are so low that they can be heavily influenced by the chemical potential of the components (i.e., by the composition of the material) as well as by the electrochemical potential of the electrons. For V_{Cu} in Cu-poor and stoichiometric material, a negative formation energy is even calculated. This would imply the spontaneous formation of large numbers of these defects under equilibrium conditions. Low (but positive) formation energies are also found for the antisite Cu_{In} in Cu-rich material (this defect is a shallow acceptor which could be responsible for the p-type conductivity of Cu-rich, non-Se-deficient $CuInSe_2$). The dependence of the defect formation energies on the electron Fermi level could explain the strong tendency of $CuInSe_2$ to self-compensation and the difficulties of achieving extrinsic doping. The results of Zhang et al. [24] provide a good theoretical model of defect formation energies and defect transition energies, which exhibits good agreement with experimentally obtained data. Table 1 summarises the ionisation energies and the defect formation energies of the 12 intrinsic defects in CuInSe$_2$. The energies (bold values in Table 1) for V_{Cu}, V_{In}, Cu_I, Cu_{In}, In_{Cu}, are obtained from a first principle calculation [24] whereas the formation energies in italics (Table 1) and for the other defects are calculated from the macroscopic cavity model [25]. The ionisation energies used in Table 1 are either taken from Zhang et al. [24] or from the data compiled in reference [26]. Note that the data given in references [25,26] for the cation defects differ significantly from those computed in reference [24].

Further important results in reference [24] are the formation energies of *defect complexes* such as $(2V_{Cu}, In_{Cu})$. (Cu_{In}, In_{Cu}) and $(2Cu_I, Cu_{In})$, where Cu_I is an interstitial Cu atom. These formation energies are even lower than those of the corresponding isolated defects. Interestingly, the $(2V_{Cu}, In_{Cu})$ complex does not exhibit an electronic transition within the forbidden gap, in contrast to the isolated In_{Cu}-anti-site, which is a deep recombination centre. As the $(2V_{Cu}, In_{Cu})$ complex is most likely to

Table 1 Electronic transition energies and formation energies ΔU of the 12 intrinsic defects in CuInSe$_2$. Source: the ionisation energies in italics are derived from reference [26], and the formation energies in italics are from reference [25]. All the numbers in bold type are from reference [24].

Defect transition energies[a] and formation energies[b] (eV)

Transition	V_{Cu}	V_{In}	V_{Se}	Cu_i	In_i	Se_i	In_{Cu}	Cu_{In}	Se_{Cu}	Cu_{Se}	Se_{In}	In_{Se}
(−/0)	**0.03**	**0.17**					**0.29**					
	0.03	*0.04*	*0.04*[c]			*0.07*		*0.05*		*0.13*	*0.08*	
(−/2−)		0.41						0.58				
(2−/3−)		0.67										
(0/+)				**0.2**			**0.25**					
				0.11[d]	*0.08*	*0.07*		*0.04*	*0.06*			*0.09*
(+/2+)								0.44				
ΔU/eV	**0.60**	**3.04**		**2.88**			**3.34**	**1.54**				
	2.9	*2.8*	*2.6*	*4.4*	*9.1*	*22.4*	*1.4*	*2.5*	*7.5*	*7.5*	*5.5*	*5.0*

[a] Difference between the valence/conduction band energy for acceptor/donor states.
[b] Formation energy ΔU of the neutral defect in the stoichiometric material.
[c] Covalent.
[d] Ionic.

occur in In-rich material, it can accommodate a large amount of excess In (or likewise deficient Cu) and, at same time, maintain the electrical performance of the material. Furthermore, ordered arrays of this complex can be thought as the building blocks of a series of Cu-poor Cu−In−Se compounds such as $CuIn_5Se_5$ and $CuIn_5Se_8$ [24].

Let us now concentrate on the defects experimentally detected in photovoltaic grade (and thus In-rich) polycrystalline films. In-rich Cu(In, Ga)Se$_2$ is in general highly compensated, with a net acceptor concentration of the order of 10^{16} cm^{-3}. The shallow acceptor level V_{Cu} (which lies about 30 meV above the valence band) is assumed to be the main dopant in this material. As compensating donors, the Se-vacancy V_{Se} as well as the double donor In_{Cu} are considered. The most prominent defect is an acceptor level at about 270−300 meV above the valence band, which is reported by several groups from deep-level transient spectroscopy [27] and admittance spectroscopy [28,29]. This defect is also present in single crystals [30]. The importance of this transition results from the fact that its concentration is related to the open-circuit voltage of the device [31−33]. Upon investigating defect energies in the entire Cu(In, Ga)(Se,S)$_2$ alloy system, Turcu et al. [34] found that the energy distance between this defect and the valence band maximum remains constant when alloying CuInSe$_2$ with Ga, whereas the energy distance increases under S/Se alloying. Assuming that the defect energy is independent from the energy position of the band edges, like the defect energies of transition metal impurities in III−V and II−IV semiconductor alloys [35,36], the authors of reference [34] extrapolate the valence band offsets $\Delta E_v = -0.23$ eV for the combination CuInSe$_2$/CuInS$_2$ and $\Delta E_v = 0.04$ eV for CuInSe$_2$/CuGaSe$_2$. Recently, transient photocapacitance studies by Heath et al. [37] unveiled an additional defect state in Cu(In, Ga)Se$_2$ at about 0.8 eV from the valence band. Again, the defect energy is independent of the Ga content in the alloy. Figure 4 summarises the energy positions of bulk defects in the Cu(In$_{1-x}$Ga$_x$)Se$_2$ and the CuIn(Se$_{1-y}$S$_y$)$_2$ alloy system with the defect energy of the bulk acceptor used as a reference energy to align the valence and conduction band energy [38]. Additionally, Figure 4 shows the activation energy of an interface donor. This energy corresponds to the energy difference ΔE_{Fn} between the Fermi energy and the conduction band minimum at the buffer absorber interface [39]. Notably this energy difference remains small upon alloying CuInSe$_2$ with Ga, whereas ΔE_{Fn} increases when alloying with S.

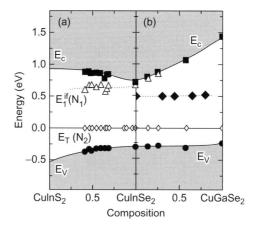

Figure 4 Band gap evolution diagram of the CuIn(Se,S)$_2$ (a) and the Cu(In,Ga)Se$_2$ (b) alloy system with the trap energy E$_T$(N$_2$, open diamonds) taken as an internal reference to align the conduction band and the valence band energies E$_c$ and E$_v$. The energy position of an additional defect state in Cu(In,Ga)Se$_2$ (full diamonds) as well as that of an interface donor (open triangles) in Cu(In,Ga)(Se,S)$_2$ is also indicated.

3. CELL AND MODULE TECHNOLOGY

3.1 Structure of the Heterojunction Solar Cell

The complete layer sequence of a ZnO/CdS/Cu(In,Ga)Se$_2$ heterojunction solar cell is shown in Figure 5. The device consists of a typically 1-μm-thick Mo layer deposited on a soda-lime glass substrate and serving as the back contact for the solar cell. The Cu(In,Ga)Se$_2$ is deposited on top of the Mo back electrode as the photovoltaic absorber material. This absorber layer has a thickness of 1–2 μm. The heterojunction is then completed by chemical bath deposition (CBD) of CdS (typically 50 nm thick) and by the sputter deposition of a nominally undoped (intrinsic) i-ZnO layer (usually of thickness 50–70 nm) and then a heavily doped ZnO layer. As ZnO has a band-gap energy of 3.2 eV it is transparent for the main part of the solar spectrum and therefore is denoted as the window layer of the solar cell.

3.2 Key Elements for High-Efficiency Cu(In,Ga)Se$_2$ Solar Cells

We first mention four important technological innovations which, during the decade 1990–2000, have led to a considerable improvement of the

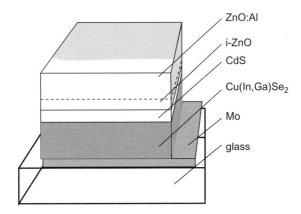

Figure 5 Schematic layer sequence of a standard ZnO/CdS/Cu(In,Ga)Se$_2$ thin-film solar cell.

efficiencies and finally to the record efficiency of 18.8% [1]. These steps are the key elements of the present Cu(In,Ga)Se$_2$ technology:

- The film quality has been substantially improved by the crystallisation mechanism induced by the presence of Cu$_y$Se ($y < 2$). This process is further supported by a substrate temperature close to the softening point of the glass substrate [40].
- The glass substrate has been changed from Na-free glass to Na-containing soda-lime glass [40,41]. The incorporation of Na, either from the glass substrate or from Na-containing precursors, has led to an enormous improvement of the efficiency and reliability of the solar cells, as well as to a larger process tolerance.
- Early absorbers consisted of pure CuInSe$_2$. The partial replacement of In with Ga [42] is a further noticeable improvement, which has increased the band gap of the absorber from 1.04 eV to 1.1–1.2 eV for the high-efficiency devices. The benefit of 20–30% Ga incorporation stems not only from the better band gap match to the solar spectrum but also from the improved electronic quality of Cu(In,Ga)Se$_2$ with respect to pure CuInSe$_2$ [21,32].
- The counter electrode for the CuInSe$_2$ absorber of the earlier cells was a 2 μm thick CdS layer deposited by Physical Vapour Deposition (PVD). This has been replaced by a combination of a 50-nm-thin CdS buffer layer laid down by chemical bath deposition [43,44] and a highly conductive ZnO window layer.

3.3 Absorber Preparation Techniques
3.3.1 Basics

The preparation of Cu(In,Ga)Se$_2$-based solar cells starts with the deposition of the absorber material on a Mo-coated glass substrate (preferably soda-lime glass). The properties of the Mo film and the choice of the glass substrate are of primary importance for the final device quality, because of the importance of Na, which diffuses from the glass through the Mo film into the growing absorber material. In the past, some processes used blocking layers such as SiN$_x$, SiO$_2$, or Cr between the glass substrate and the Mo film to prevent the out-diffusion of Na. Instead, Na-containing precursors like NaF [45], Na$_2$Se [46], or Na$_2$S [47] are then deposited prior to absorber growth to provide a controlled, more homogeneous, incorporation of Na into the film. The control of Na incorporation in the film from precursor layers allows the use of other substrates like metal or polymer foils. The most obvious effects of Na incorporation are better film morphology and higher conductivity of the films [48]. Furthermore, the incorporation of Na induces beneficial changes in the defect distribution of the absorber films [49,50].

The explanations for the beneficial impact of Na are manifold, and it is most likely that the incorporation of Na in fact results in a *variety* of consequences. During film growth, the incorporation of Na leads to the formation of NaSe$_x$ compounds. This slows down the growth of CuInSe$_2$ and could at same time facilitate the incorporation of Se into the film [51]. Also the widening of the existence range of the α-(CuInSe$_2$) phase in the phase diagram, discussed above, as well as the reported larger tolerance to the Cu/(In + Ga) ratio of Na-containing thin films, could be explained in this picture. Furthermore, the higher conductivity of Na-containing films could result from the diminished number of compensating V_{se} donors.

During absorber deposition, a MoSe$_2$ film forms at the Mo surface [52,53]. MoSe$_2$ is a layered semiconductor with p-type conduction, a band gap of 1.3 eV and weak van der Waals bonding along the c-axis. The c-axis is found to be in parallel with, and the van der Waals planes thus perpendicular to, the interface [53]. Because of the larger band gap of the MoSe$_2$ compared with that of standard Cu(In,Ga)Se$_2$ films, the MoSe$_2$ layer provides a low-recombinative back surface for the photogenerated minority carriers (electrons) in the Cu(In,Ga)Se$_2$ absorber and at the same time provides a low-resistance contact for the majority carries (holes).

Photovoltaic-grade Cu(In,Ga)Se$_2$ films have a slightly In-rich overall composition. The allowed stoichiometry deviations are astonishingly large, yielding a wide process window with respect to composition. Devices with efficiencies above 14% are obtained from absorbers with (In + Ga)/(In + Ga + Cu) ratios between 52% and 64% if the sample contains Na [48]. Cu-rich Cu(In,Ga)Se$_2$ shows the segregation of a secondary Cu$_y$Se phase preferentially at the surface of the absorber film. The metallic nature of this phase does not allow the formation of efficient heterojunctions. Even after *removal* of the secondary phase from the surface by etching the absorber in KCN, the utility of this material for photovoltaic applications is limited. However, the importance of the Cu-rich composition is given by its role during film growth. Cu-rich films have grain sizes in excess of 1 μm whereas In-rich films have much smaller grains. A model for the film growth under Cu-rich compositions comprises the role of Cu$_y$Se as a flux agent during the growth process of co-evaporated films [54]. For Cu(In,Ga)Se$_2$ prepared by selenisation, the role of Cu$_y$Se is similar [55]. Therefore, growth processes for high quality material have to go through a copper-rich stage but have to end up with an indium-rich overall composition.

3.3.2 Co-Evaporation Processes

The absorber material yielding the highest efficiencies is Cu(In,Ga)Se$_2$ with a Ga/(Ga + In) ratio of 20–30%, prepared by co-evaporation from elemental sources. Figure 6 sketches a co-evaporation set-up as used for the preparation of laboratory-scale solar cells and mini-modules. The process requires a maximum substrate temperature of about 550°C for a certain time during film growth, preferably towards the end of growth. One advantage of the evaporation route is that material deposition and film formation are performed during the same processing step. A feedback loop based on a quadrupole mass spectrometer or an atomic absorption spectrometer controls the rates of each source. The composition of the deposited material with regard to the metals corresponds to their evaporation rates, whereas Se is always evaporated in excess. This precise control over the deposition rates allows for a wide range of variations and optimisations with different sub-steps or stages for film deposition and growth. These sequences are defined by the evaporation rates of the different sources and the substrate temperature during the course of deposition.

Advanced preparation sequences always include a Cu-rich stage during the growth process and end up with an In-rich overall composition in

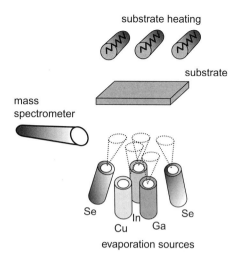

Figure 6 Arrangement for the deposition of Cu(In,Ga)Se$_2$ films on the laboratory scale by co-evaporation on a heated substrate. The rates of the sources are controlled by mass spectrometry.

order to combine the large grains of the Cu-rich stage with the otherwise more favourable electronic properties of the In-rich composition. The first example of this kind of procedure is the so called Boeing or *bilayer process* [5], which starts with the deposition of Cu-rich Cu(In,Ga)Se$_2$ and ends with an excess In rate. The most successful co-evaporation process is the so-called *three-stage process* [56]. This process starts with the deposition of In, Ga, and Se at relatively low temperatures, then uses a Cu-rich growth stage by evaporating Cu in excess at elevated temperature, and at the end again deposits only In, Ga, and Se to ensure the overall In-rich composition of the film. The three-stage process currently leads to the Cu(In,Ga)Se$_2$ solar cells with the highest efficiencies.

3.3.3 Selenisation Processes

The second class of absorber preparation routes is based on the separation of deposition and compound formation into two different processing steps. High efficiencies are obtained from absorber prepared by selenisation of metal precursors in H$_2$Se [57–59] and by rapid thermal processing of stacked elemental layers in a Se atmosphere [60]. These sequential processes have the advantage that approved large-area deposition techniques like sputtering can be used for the deposition of the materials. The Cu(In,Ga)Se$_2$ film formation then requires a second step, the selenisation.

The very first large-area modules were prepared by the selenisation of metal precursors in the presence of H_2Se more than ten years ago [6]. Today, a modification of this process yields the first commercially available $Cu(In,Ga)Se_2$ solar cells manufactured by Shell Solar Industries. This process is schematically illustrated in Figure 7. First, a stacked layer of Cu, In and Ga is sputter-deposited on the Mo-coated glass substrate. Then selenisation takes place under H_2Se. To improve the device performance, a second thermal process under H_2S is added, resulting in an absorber that is $Cu(In,Ga)(S,Se)_2$ rather than $Cu(In,Ga)Se_2$.

A variation of the method that avoids the use of the toxic H_2Se during selenisation is the rapid thermal processing of stacked elemental layers [60]. Here the precursor includes a layer of evaporated elemental Se. The stack is then selenised by a rapid thermal process (RTP) in either an inert or a Se atmosphere. The highest efficiencies are obtained if the RTP is performed in an S-containing atmosphere (either pure S or H_2S).

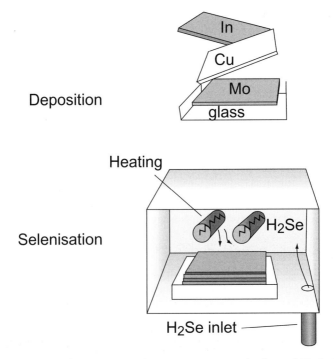

Figure 7 Illustration of the sequential process. First a stack of metal (Cu,In,Ga) layers deposited by sputtering onto a Mo-coated glass. In the second step, this slack is selenised in H_2Se atmosphere and converted into $CuIn Se_2$.

On the laboratory scale, the efficiencies of cells made by these preparation routes are smaller by about 3% (absolute) as compared with the record values. However, on the module level, co-evaporated and sequentially prepared absorbers have about the same efficiency. Sequential processes need two or even three stages for absorber completion. These additional processing steps may counterbalance the advantage of easier element deposition by sputtering.

3.3.4 Other Absorber Deposition Processes

Besides selenisation and co-evaporation, other deposition methods have been studied, either to obtain films with very high quality or to reduce the cost of film deposition on large areas. Methods that are used to grow epitaxial III−V compound films, such as molecular beam epitaxy (MBE) [61] or metal organic chemical vapour deposition (MOCVD) [62] have revealed interesting features for fundamental studies like phase segregation and defect formation, but could not be used to form the absorber material for high-efficiency solar cells.

Attempts to develop so-called low-cost processes include electrodeposition [63−66], screen printing and particle deposition [67,68]. Electrodeposition can be done either in one or two steps. The crucial step is the final film formation in a high-temperature annealing process. The recrystallisation process competes with the decomposition of the material. Therefore, process optimisation is quite difficult. Cells with high efficiencies were obtained by electrodeposition of a Cu-rich $CuInSe_2$ film and subsequent conditioning by a vacuum evaporation step of In(Se) [69]. Particle deposition by printing of suitable inks and subsequent annealing lead to absorber layers with good quality enabling the fabrication of solar cells with efficiencies over 13% [70].

3.3.5 Postdeposition Air Anneal

Air annealing has been an important process step, crucial for the efficiency especially of the early solar cells based on $CuInSe_2$. Though often not mentioned explicitly, an oxygenation step is still used for most of the present-day high-efficiency devices. The beneficial effect of oxygen was explained within the defect chemical model of Cahen and Noufi [71]. In this model, the surface defects at grain boundaries are positively charged Se vacancies V_{Se}. During air annealing, these sites are passivated by O atoms. Because of the decreased charge at the grain boundary, the band bending as well as the recombination probability for photogenerated electrons is reduced. The surface donors and their neutralisation by oxygen are important for the free Cu(In,Ga)Se$_2$ surface as well as for the formation of the CdS/Cu(In,Ga)Se$_2$ interface [72,73].

3.4 Heterojunction Formation
3.4.1 The Free Cu(In,Ga)Se$_2$ Surface

The surface properties of Cu(In,Ga)Se$_2$ thin films are especially important as this surface becomes the active interface of the completed solar cell. However, the band diagram of the ZnO/CdS/Cu(In,Ga)Se$_2$ heterojunction, especially the detailed structure close to the CdS/Cu(In,Ga)Se$_2$ interface, is still under debate (for recent reviews see references [74,75]).

The free surfaces of as-grown Cu(In,Ga)Se$_2$ films exhibit two prominent features:

1. The valence band-edge energy E_V lies below the Fermi level E_F by about 1.1 eV for CuInSe$_2$ films [76]. This energy is larger than the band-gap energy E_g^{bulk} of the bulk of the absorber material. This finding was taken as an indication for a *widening of band gap* at the surface of the film. A recent direct measurement of the surface band gap of polycrystalline CuInSe$_2$ by Morkel et al. [77] proved that the band-gap energy E_g^{surf} at the surface of the film is about 1.4 cV, i.e., more than 0.3 eV larger than $E_g^{bulk} \approx 1.04$ eV. In Cu(In$_{1-x}$,Ga$_x$)Se$_2$ alloys the distance $E_F - E_V$ was found to be 0.8 eV (almost independently of the Ga content if $x > 0$) [78].

2. The surface composition of Cu-poor CuInSe$_2$, as well as that of Cu(In,Ga)Se$_2$ films, corresponds to a surface composition of (Ga + In)/(Ga + In + Cu) of about 0.75 for a range of bulk compositions of 0.5 < (Ga + In)/(Ga + In + Cu) < 0.75 [76].

Both observations (i) and (ii) have led to the assumption that a phase segregation of Cu(In,Ga)$_3$Se$_5$, the so-called Ordered Defect Compound (ODC), occurs at the surface of the films. From the fact that bulk Cu(In,Ga)$_3$Se$_5$ exhibits n-type conductivity [79] it was argued that Cu-poor Cu(In,Ga)Se$_2$ thin films automatically generate a rectifying, buried junction. However, the existence of a separate phase on top of standard Cu(In,Ga)Se$_2$ thin films has not yet been confirmed by structural methods such as X-ray diffraction, high resolution transmission electron microscopy or electron diffraction. Furthermore, if the surface phase exhibited the weak n-type conductivity of bulk Cu(In,Ga)$_3$Se$_5$, simple charge neutrality estimations [21] show that this would not be sufficient to achieve type inversion.

Based on these arguments, another picture of the surface of Cu(In,Ga)Se$_2$ thin films and of junction formation has emerged [21,80]. Within the classical Bardeen model [81] of Fermi level pinning by electronic states at semiconductor surfaces, a density of surface states of about 10^{12} cm^{-2} eV^{-1} is sufficient to pin the Fermi level at the neutrality level of the free semiconductor surfaces. Positively charged surface donors are expected in the metal terminated (112) surface of CuInSe$_2$ owing to the dangling bond to the missing Se [71]. Thus, these surface states, rather than a distinct n-type surface phase, determine the type inversion of the surface.

Surface states play also an important role in the completed heterostructure, where they become *interface states* at the absorber/buffer interlace. *The deject layer model* [21,80] takes into account a modification of the band structure due to the Cu deficiency of the surface *as well as* the presence of positively charged surface states due to the missing surface Se. However, the defect layer model considers the surface layer not as n-type bulk material (as does the ODC model) but as a p$^+$-layer. Furthermore, the defect layer is viewed *not as the origin but rather as the consequence* of the natural surface type inversion [21,80]. Surface states are responsible for the surface band bending that leads to the liberation of Cu from its lattice sites and to Cu migration towards the neutral part of the film. The remaining copper vacancies V_{Cu}^- close to the surface result in a high density of acceptor states, i.e., the p$^+$-defect layer at the film surface. Recent photoelectron spectroscopy experiments of Klein and Jaegermann [82] suggest that the band bending occurring during junction formation leads to a loss of Cu atoms from the surface of CuInSe$_2$ and CuGaSe$_2$, whereas a similar effect was not observed in CuInS$_2$.

3.4.2 Buffer Layer Deposition

Surface passivation and junction formation is most easily achieved by the CBD deposition of a thin CdS film from a chemical solution containing Cd ions and thiourea [83]. The benefit of the CdS layer is manifold:
- CBD deposition of CdS provides complete coverage of the rough polycrystalline absorber surface at a film thickness of only 10 nm [84].
- The layer provides protection against damage and chemical reactions resulting from the subsequent ZnO deposition process.
- The chemical bath removes the natural oxide from the film surface [83] and thus reestablishes positively charged surface states and, as a

consequence, the natural type inversion at the CdS/Cu(In,Ga)Se$_2$ interface.
- The Cd ions, reacting first with the absorber surface, remove elemental Se, possibly by the formation of CdSe.
- The Cd ions also diffuse to a certain extent into the Cu-poor surface layer of the absorber material [85,86], where they possibly form Cd$_{Cu}$ donors, thus providing additional positive charges enhancing the type inversion of the buffer/absorber interface.
- Open-circuit voltage limitations imposed by interface recombination can be overcome by a low surface recombination velocity in addition to the type inversion of the absorber surface. Thus, one might conclude that interface states (except those shallow surface donors responsible for the type inversion) are also passivated by the chemical bath.

Due to the favourable properties of CdS as a heterojunction partner and the chemistry of the CBD process it is difficult to find a replacement. Avoiding CdS and the chemical bath step would be advantageous from the production point of view. On one hand, a toxic material such as CdS requires additional safety regulation; on the other hand, the chemical bath deposition does not comply with the vacuum deposition steps of an in-line module fabrication. Therefore, research and development in this area relates to two issues: (i) the search for alternative materials for a chemical deposition, and (ii) the development of ways to deposit the front electrode without an intermediate step in a chemical bath.

Promising materials to replace CdS are In(OH,S) [78], Zn(OH,S) [88,89], ZnS [90,91], and ZnSe [92–94], However, all these materials require additional precautions to be taken for the preparation of the absorber surface or front electrode deposition.

3.4.3 Window Layer Deposition

The most commonly used material for the preparation of the front electrode is ZnO doped with B or Al. The first large-area modules produced by ARCO Solar (later Siemens Solar Industries, now Shell Solar Industries) had a ZnO:B window layer deposited by chemical vapour deposition (CVD). Later production facilities at Boeing and EUROCIS use sputtering processes. Present pilot production lines also favour sputtering. As mentioned previously, an undoped i-ZnO layer with a thickness of about 50–100 nm is needed at the heterojunction in order to achieve optimum performance.

3.5 Module Production and Commercialisation
3.5.1 Monolithic Interconnections

One inherent advantage of thin-film technology for photovoltaics is the possibility of using monolithic integration for series connection of individual cells within a module. In contrast, bulk Si solar cells must be provided with a front metal grid, and each of these front contacts has to be connected to the back contact of the next cell for series connection. The interconnect scheme, shown in Figure 8, has to warrant that the front ZnO layer of one cell is connected to the back Mo contact of the next one. In order to obtain this connection, three different patterning steps are necessary. The first one separates the Mo back contact by a series of periodical scribes and thus defines the width of the cells, which is of the order of 0.5–1 cm. For Mo patterning, a laser is normally used. The

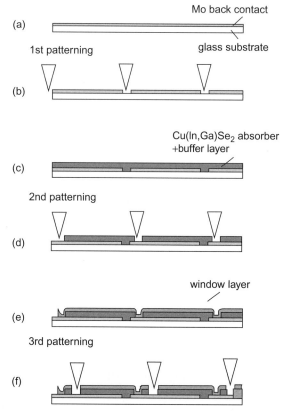

Figure 8 Deposition and patterning sequence to obtain an integrated interconnect scheme for Cu(In,Ga)Se$_2$ thin-film modules.

second patterning step is performed after absorber and buffer deposition, and the final one after window deposition (cf. Figure 8). Scribing of the semiconductor layer is done by mechanical scribing or laser scribing. The total width of the interconnect depends not only on the scribing tools, but also on the reproducibility of the scribing lines along the entire module. The length of the cells and, accordingly, that of the scribes can be more than 1 m. The typical interconnect width is of the order of 300 μm. Thus, about 3–5% of the cell area must be sacrificed to the interconnects.

3.5.2 Module Fabrication

The technologies for absorber, buffer, and window deposition used for module production are the same as those discussed above for the production of small laboratory cells. However, the challenge of producing modules is to transform the laboratory-scale technologies to much larger areas. The *selenisation* process uses as much off-the-shelf equipment and processing as possible (e.g., sputtering of the metal precursors) for fabricating Cu(In,Ga)Se$_2$ films. For *co-evaporation* on large areas, the Centre for Solar Energy and Hydrogen Research in Stuttgart (ZSW) has designed its own equipment, schematically shown in Figure 9, for an in-line co-evaporation process. Line-shaped evaporation sources allow continuous deposition of large-area, high-quality Cu(In,Ga)Se$_2$ films. The relatively high substrate temperatures that are necessary for high-quality material impose problems in handling very large area glass sheets. Future process optimisation therefore implies reduction of the substrate temperature.

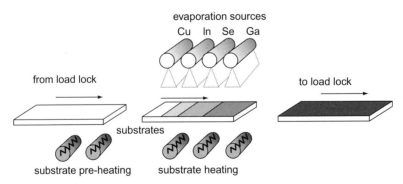

Figure 9 Sketch of an in-line deposition system for co-evaporation of Cu(In,Ga)Se$_2$ absorber films from line-sources.

A bottleneck for the production is the deposition of the buffer layer in a chemical bath. On one hand, it is not straightforward to integrate this process in a line consisting mainly of dry physical vapour deposition processes; on the other, it would be favourable for environmental reasons to replace the currently used CdS layer by a Cd-free alternative.

The ZnO transparent front electrode is put down either by CVD or sputtering. Each method has its specific advantages with respect to process tolerance, throughput, cost and film properties. The widths of the cells within the module—and therefore the relative losses from the patterning—mainly depend on the sheet resistance of the ZnO.

Module encapsulation is an important issue because module stability depends on proper protection against humidity. Low-iron cover glasses provide good protection against ambient influences. Hermetic sealing of the edges is mandatory to obtain stable modules (see the following discussion).

3.5.3 Upscaling Achievements

$Cu(In,Ga)Se_2$ has the best potential to reach more than 15% module efficiency in the near future. Mini-modules ranging in area from 20 cm^2 to 90 cm^2 that use the process sequence anticipated for larger area commercial modules have already reached efficiencies around 14–15%. Recently, Siemens Solar Industries fabricated a 1 ft × 4 ft power module (~44 watts) with an independently verified efficiency of 12.1%. Using a totally different approach to the deposition of the absorber layer, the ZSW fabricated a 30-cm × 30-cm module with a verified efficiency of 12.7%. Based on the same co-evaporation process, Würth Solar GmbH, Stuttgart, reported an efficiency of 12.5% for a module of aperture area 5932 cm^2 [95]. More results of the different processes are summarised in Table 2. Note that, because of the promising results from the laboratory scale and the first approaches of up-scaling, several companies other than those mentioned in Table 2 now plan commercial production.

A further challenge is to develop CIGS cells on flexible substrates and hence to extend their area of applications. There are ongoing efforts to produce cells on various kinds of substrates like stainless steel, polymide and at the same time to retain the performance achieved with devices on soda-lime glass [96]. For space applications it is important to reduce the weight by depositing the cells on lightweight foil substrates. Highest small-area efficiencies on polymide films formed by spin coating on a glass substrate reach 12.8% [97]. Roll to roll coating on metal foils [98] and polymer films [99,100] has already reached the stage of pilot production.

Table 2 Comparison of efficiencies η and areas A of laboratory cells, mini-modules, and commercial-size modules achieved with Cu(In,Ga)Se$_2$ thin films based on the co-evaporation and the selenisation process. NREL denotes the National Renewable Energy Laboratories (USA), ZSW is the Center for Solar Energy and Hydrogen Research (Germany), EPV is Energy Photovoltaics (USA), ASC is the Angstrom Solar Centre (Sweden)

	Laboratory cell		Mini-module		Module		
Process	η(%)	A (cm^2)	η (%)	A (cm^2)	η (%)	A (cm^2)	Laboratory/company
Co-evap.	18.8	0.45					NREL [1]
	16.1	0.5	13.9	90	12.7	800	ZSW
					12.3	5932	Würth Solar [95]
			9.6	135			EPV
	11.5	0.5	5.6	240			Global Solar[a]
			16.6	20			ASC [2]
Selenis.	>16	0.5	14.7	18	12.1	3600	Shell Solar
			14.2	50	11.6	864	Showa, Japan

[a]Flexible cells.

3.5.4 Stability

The long-term stability is a critical issue of any solar cell technology because the module lifetime contributes as much to the ratio between produced energy and invested cost as does the initial efficiency. Cu(In,Ga)Se$_2$ modules fabricated by Shell Solar Industries more than 10 years ago show until today very good stability during outdoor operation [101,102]. However, intense accelerated lifetime testing is made for the now commercially available Cu(In,Ga)Se$_2$ modules. Careful sealing and encapsulation appears mandatory, especially because of the sensitivity of Cu(In,Ga)Se$_2$ to humidity. For non-encapsulated modules, corrosion of the molybdenum contact and the degradation of zinc oxide were found to be the dominating degradation mechanisms [103] during the so-called damp heat test (1000 hours in hot (85°C) and humid (85% humidity) atmosphere). Investigations of non-encapsulated cells [104–106] unveiled further a humidity-induced degradation of the Cu(In,Ga)Se$_2$ absorber material. Despite of the sensibility of Cu(In,Ga)Se$_2$ with respect to humidity, well encapsulated modules pass the damp heat test [107].

Recent work of Guillemoles et al. [108,109] investigates the chemical and electronic stability of Cu(In,Ga)Se$_2$-based solar cells and possible fundamental instabilities of the material system, namely, interface reactions, defect metastability, and constituent element (Cu) mobility. Guillemoles et al. conclude that all reasonably anticipated detrimental interface reactions at the

Mo/Cu(In,Ga)Se$_2$, the Cu(In,Ga)Se$_2$/CdS. or the CdS/ZnO interface are either thermodynamically or kinetically limited, Furthermore, Cu mobility does not contradict long-term stability [108,109].

3.5.5 Radiation Hardness and Space Applications

One important prospective application for Cu(In,Ga)Se$_2$ cells is in space, where the main power source is photovoltaics. Satellites in low-earth orbits for communication systems require solar cells with high end-of-life efficiencies, despite the high flux of high-energy electrons and protons in that ambient. The radiation hardness of Cu(In,Ga)Se$_2$ has been recognized as early as 1984/85 [110,111], but only recently have systematic investigations on the radiation response of Cu(In,Ga)Se$_2$ solar cells been undergone using high enough electron and proton fluences to allow quantitative conclusions regarding the defect generation rates [112]. The radiation resistance of Cu(In,Ga)Se$_2$ against high-energy (0.3—3 MeV) electrons turns out to be far better than that of any other photovoltaic material [112]. The radiation hardness of Cu(In,Ga)Se$_2$ against high-energy (0.4—10 MeV) protons is also high, though the difference from other materials is not as high as in case of electron irradiation. Walters et al. [113] have analysed Cu(In,Ga)Se$_2$ in the frame of the so-called damage dose model and found that the damage coefficients for Cu(In,Ga)Se$_2$ are comparable to those of InP and considerably smaller than those of Si and GaAs.

The high mobility of Cu in the Cu(In,Ga)Se$_2$ lattice was proposed to be one important ingredient for a defect healing mechanism that could explain the high radiation resistance of Cu(In,Ga)Se$_2$ [108,114]. Recent thermal annealing experiments of electron-irradiated [115] and proton-irradiated [116] Cu(In,Ga)Se$_2$ solar cells unveiled a thermally activated healing process with an activation energy of around 1 cV leading to a complete recovery of the device performance. Illumination of the solar cell enhances this annealing process further [116,117].

The challenge for developing CIGS space cells is to reduce the weight by depositing the cells on foil substrates, and at the same time to retain the performance achieved with devices on soda-lime glass. Recently, Tuttle et al. [118] reported a Cu(In,Ga)Se$_2$ solar cell on lightweight metal foil with a power conversion efficiency of 15.2% (under AM0 illumination) and a specific power of 1235 W/kg. Other approaches to flexible and lightweight Cu(In,Ga)Se$_2$ solar cells embrace Cu(In,Ga)Se$_2$ deposited on plastic foil such as polyimide [119] or the use of a lift-off technique to remove the absorber from the glass substrate after device fabrication [120].

4. DEVICE PHYSICS

4.1 The Band Diagram

The band diagram of the ZnO/CdS/Cu(In,Ga)Se$_2$ heterostructure in Figure 10 shows the conduction and valence band energies E_c and E_v of the Cu(In,Ga)Se$_2$ absorber, the CdS buffer layer and the ZnO window. Note that the latter consists of a highly Al-doped (ZnO:Al) and an undoped (i-ZnO) layer. For the moment, we neglect the polycrystalline nature of the semiconductor materials, which in principle requires a two- or three-dimensional band diagram. Even in the one-dimensional model, important details of the band diagram are still not perfectly clear. The diagram in Figure 10 concentrates on the heterojunction and does not show the contact between the Mo and Cu(In,Ga)Se$_2$ at the back side of the absorber.

An important feature in Figure 10 is the 10–30-nm-thick surface defect layer (SDL) on top of the Cu(In,Ga)Se$_2$ absorber, already discussed in Section 3.4. The physical nature of this SDL is still under debate.

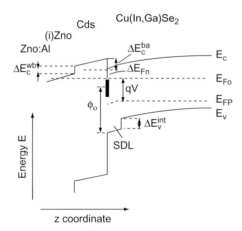

Figure 10 Band diagram of the ZnO/CdS/Cu(In,Ga)Se$_2$ heterojunction under bias voltage showing the conduction and valence band-edge energies ΔE_c and E_v. The quantities $\Delta E_c^{wb/ba}$ denote the conduction band offsets at the window/buffer and buffer/absorber interfaces, respectively. An internal valence band offset ΔE_v^{int} exists between the bulk Cu(In,Ga)Se$_2$ and a surface defect layer (SDL) on top of the Cu(In,Ga)Se$_2$ absorber film. The quantity ΔE_{Fn} denotes the energy distance between the electron Fermi level E_{Fn} and the conduction band at the CdS buffer/Cu(In,Ga)Se$_2$ absorber interface, and ϕ_n denotes the neutrality level of interface states at this heterointerface.

However, the fact that the band gap at the surface of Cu(In,Ga)Se$_2$ thin films (as long as they are prepared with a slightly Cu-poor final composition) exceeds the band-gap energy of the bulk material [76,77], has important implications for the contribution of interface recombination to the overall performance of the solar cell. A simplified approach to the mathematical description of the ZnO/CdS/Cu(In,Ga)Se$_2$ heterojunction including the consequence of the surface-band-gap widening is given in reference [121].

The most important quantities to be considered in the band diagram are the band discontinuities between the different heterojunction partners. Band discontinuities in terms of valence band offsets ΔI between semiconductor a and b are usually determined by photoelectron spectroscopy (for a discussion with respect to Cu-chalcopyrite surfaces and interfaces, see [74]). The valence band offset between a (011)-oriented Cu(In,Ga)Se$_2$ single crystal and CdS deposited by PVD at room temperature is determined as [122,123] $\Delta E_v^{ab} = -0.8$ (± 0.2) eV (and, therefore, $\Delta E_c^{ab} = E_g^{bCdS} - E_g^{CIS} + \Delta E_v^{ab} \approx 0.55$ eV, with the band gaps $E_g^{CdS} \approx 2.4$ eV and $E_g^{CdS} \approx 1.05$ eV of CdS and CuInSe$_2$, respectively). Several authors have investigated the band discontinuity between polycrystalline Cu(In,Ga)Se$_2$ films and CdS, and found values between -0.6 and -1.3 eV with a clear centre of mass around -0.9 eV, corresponding to a conduction band offset of 0.45 eV [74]. Wei and Zunger [124] calculated a theoretical value of $\Delta E_v^{ab} = -1.03$ eV, which would lead to $\Delta E_C^{ab} \approx 0.3$ eV. Recently, Morkel et al. [77] found a valence band offset between the surface of polycrystalline, Cu-poor prepared CuInSe$_2$ and chemical bath deposited CdS of $\Delta E_v^{ab} = -0.8$ eV. By combining their photoelectron spectroscopy results with measured surface band-gap energies of CdS and CuInSe$_2$ from inverse photoemission spectroscopy, the authors of reference [77] conclude that the conduction band offset ΔE_c^{ab} is actually zero. This is because the deposited CdS has $E_g = 2.2$ eV due to S/Se intermixing and the CuInSe$_2$ film exhibits a surface band gap of 1.4 eV; thus, $\Delta E_c^{ab} = E_g[\text{Cd(Se, S)}] - E_{gs}^{surf}[\text{CuInSe}_2] + \Delta E_V^{ab} \approx 0$.

The band alignment of polycrystalline CuInSe$_2$ and Cu(In,Ga)Se$_2$ alloys was examined by Schmid et al. [76,78] who found that the valence band offset is almost independent of the Ga content. In turn, the increase of the absorber band gap leads to a change of ΔE_C^{ab} from positive to negative values. The conduction band offset between the CdS buffer and the ZnO window layer was determined by Ruckh et al. to be 0.4 eV [125].

4.2 Short-Circuit Current

The short-circuit current density J_{sc} that can be obtained from the standard 100 mW cm^{-2} solar spectrum (AM1.5) is determined, on one hand, by *optical losses*, that is. by the fact that photons from a part of the spectrum are either not absorbed in the solar cell or are absorbed *without* generation of electron–hole pairs. On the other hand, not all photogenerated electron–hole pairs contribute to J_{sc} because they recombine before they are collected. We denote these latter limitations as *recombination losses*. Figure 11 illustrates how much from an incoming photon flux from the terrestrial solar spectrum contributes to the final J_{sc} of a highly efficient Cu(In,Ga)Se$_2$ solar cell [126] and where the remainder gets lost. The incoming light, i.e., that part of the solar spectrum with photon energy hv larger than the band-gap energy $E_g = 1.155$ eV of the specific absorber would

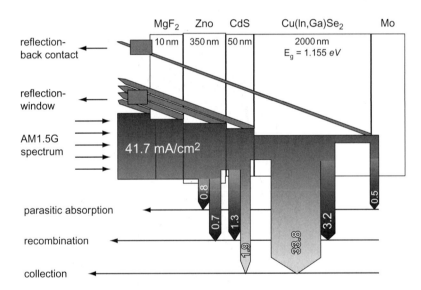

Figure 11 Optical and electronic losses of the short circuit current density J_{sc} of a high-efficiency ZnO/CdS/Cu(In,Ga)Se$_2$ heterojunction solar cell. The incident current density of 41.7 mA/cm^2 corresponds to the range of the AM 1.5 solar spectrum that has a photon energy larger than the hand gap energy $E_g = 1.155$ eV of the Cu(In,Ga)Se$_2$ absorber. Optical losses consist of reflection losses at the ambient/window, at the window/buffer, the buffer/absorber, and at the absorber/back contact interface as well as of parasitic absorption in the ZnO window layer (free carrier absorption) and at the Mo back contact. Electronic losses are recombination losses in the window, buffer, and in the absorber layer. The finally measured J_{sc} of 34.6 mA/cm^2 of the cell stems almost exclusively from the Cu(In,Ga)Se$_2$ absorber and only to a small extend from the CdS buffer layer.

correspond to a (maximum possible) J_{sc} of 41.7 mA cm^{-2}. By reflection at the surface and at the three interfaces between the MgF$_2$ anti-reflective coating, the ZnO window, the CdS buffer, and the Cu(In,Ga)Se$_2$ absorber layer we loose already 0.6 mA cm^{-2}. A further reflection loss of 0.1 mA cm^{-2} is due to those low-energy photons that are reflected at the metallic back and leave the solar cell after having traversed the absorber twice. Due to the high absorption coefficient of Cu(In,Ga)Se$_2$ for $hv > E_g$, (this portion is very small. More important are parasitic absorption losses by free-carrier absorption in the highly doped part of the ZnO window layer (0.9 mA cm^{-2}) and at the absorber/Mo interface (0.5 mA cm^{-2}). Thus, the sum of all *optical losses* amounts to 2.1 mA cm^{-2}. Note that in solar cells and modules there are additional optical losses due to the grid shadowing or interconnect areas, respectively. These losses are not discussed here.

Next, we have to consider that electron hole pairs photogenerated in the ZnO window layer are not separated. Therefore, this loss affecting photon energies $hv > E_g$ (ZnO) = 3.2 eV contributes to the *recombination losses*. As shown in Figure 11, this loss of high-energy photons costs about 0.7 mA cm^{-2}. Another portion of the solar light is absorbed in the CdS buffer layer (E_g (CdS) ≈ 2.4 eV). However, a part of the photons in the energy range 3.2 eV > hv > 2.4 eV contributes to J_{sc} because the thin CdS layer does not absorb all those photons and a part of the electron–hole pairs created in the buffer layer still contributes to the photocurrent [127]. In the present example 1.3 mA cm^{-2}, get lost by recombination and 0.8 mA cm^{-2} are collected. However, the major part of J_{sc} (33.8 mA cm^{-2}) stems from electron–hole pairs photogenerated in the absorber. Finally, the collection losses in the absorber amount to 3.2 mA cm^{-2}.

The preceding analysis shows that, accepting the restrictions that are given by the window and the buffer layer, this type of solar cell makes extremely good use of the solar spectrum. There is however still some scope for improving J_{sc} by optimising carrier collection in the absorber and/or by replacing the CdS buffer layer by an alternative material with a higher E_g.

4.3 Open-Circuit Voltage

At open circuit no current flows across the device, and all photogenerated charge carriers have to recombine within the solar cell. The possible recombination paths for the photogenerated charge carriers in the Cu(In,Ga)Se$_2$ absorber are indicated in the band diagram of Figure 12. Here we have

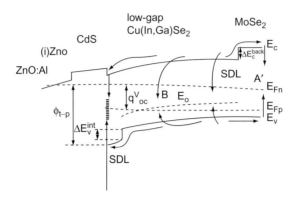

Figure 12 Recombination paths in a ZnO/CdS/(low-gap) Cu(In,Ga)Se2 junction at open circuit. The paths A represent recombination in the neutral volume, A recombination at the back contact, B recombination in the space-charge region, and C recombination at the interface between the Cu(In,Ga)Se$_2$ absorber and the CdS buffer layer. Back contact recombination is reduced by the conduction band offset ΔE_c^{back} between the Cu(In,Ga)Se$_2$ absorber and the MoSe$_2$ layer that forms during absorber preparation on top of the metallic Mo back contact. Interface recombination (C) is reduced by the internal valence band offset ΔE_v^{int} between the bulk of the Cu(In,Ga)Se2 absorber and the Cu-poor surface layer. The quantity ϕ_{bp}^* denotes the energy barrier at the CdS/absorber interface and E_T indicates the the energy of a recombination centre in the bulk of the Cu(In,Ga)Se$_2$.

considered recombination at the back surface of the absorber (A') and in the neutral bulk (A), recombination in the space-charge region (B), and recombination at the buffer/absorber interface (C). Note that due to the presence of high electrical fields in the junction region, the latter two mechanisms may be enhanced by tunnelling.

The *basic* equations for the recombination processes (A–C) can be found, for example, in [128]. Notably, all recombination current densities J_R can be written in the form of a diode law

$$J_R = J_0 \left\{ \exp\left(\frac{qV}{n_{id}k_B T}\right) - 1 \right\} \quad (1)$$

where V is the applied voltage, n_{id} the diode ideality factor, and $k_B T/q$ the thermal voltage. The saturation current density J_0 in general is a thermally activated quantity and may be written in the form

$$J_0 = J_{00} \exp\left(\frac{-E_a}{n_{id}k_B T}\right) \quad (2)$$

where E_a is the activation energy and the prefactor J_{00} is only weakly temperature dependent.

The activation energy E_a for recombination at the back surface, in the neutral zone, and in the space charge recombination is the absorber bandgap energy E_g, whereas in case of interface recombination E_a equals the barrier ϕ_{bp} that hinders the holes from the absorber to come to the buffer/absorber interface (cf. Figure 12). In the simplest cases, the diode ideality factor n_{id} is unity for back surface and neutral zone recombination as well as for recombination at the buffer/absorber interface, whereas for space charge recombination $n_{id} = 2$. Equalising the short-circuit current density J_{sc} and the recombination current density J_R in Equation (1) for the open-circuit-voltage situation (i.e., $V = V_{oc}$ in Equation (1)) we obtain with the help of Equation (2)

$$V_{oc} = \frac{E_a}{q} - \frac{n_{id}k_BT}{q}\ln\left(\frac{J_{00}}{J_{sc}}\right) \qquad (3)$$

Note that Equation (3) yields the open circuit voltage in a situation where there is single mechanism clearly dominating the recombination in the specific device. Note that Equation (3) is often used for the analysis of the dominant recombination path. After measuring V_{oc} at various temperatures T, the extrapolation of the experimental $V_{oc}(T)$ curve to $T = 0$ yields the activation energy of the dominant recombination process, e.g., $V_{oc}(0) = E_g/q$ in case of bulk recombination or $V_{oc}(0) = \phi_{bp}/q$ in case of interface recombination.

We emphasise that, in practice, measured ideality factors considerably deviate from that textbook scheme and require more refined theories (see, e.g., [14] and references therein). However, for a basic understanding of the recombination losses in thin-film solar cells, a restriction to those textbook examples is sufficient.

The band diagram in Figure 12 contains two important features that appear important for the minimisation of recombination losses in Cu(In,Ga)Se$_2$ solar cells. The first one is the presence of a considerably widened band gap at the surface of the Cu(In,Ga)Se$_2$ absorber film. This surface-band-gap widening implies a valence band offset ΔE_v^{int} at the internal interface between the absorber and the surface defect layer (discussed in Section 4.1). This internal offset directly increases the recombination barrier ϕ_{bp} to an effective value

$$\phi_{bp}^* = \phi_{bp} + \Delta E_v^{int} \qquad (4)$$

Substituting Equation (4) in Equation (3) we obtain

$$V_{oc} = \frac{\phi_{bp}}{q} + \frac{\Delta E_v^{int}}{q} - \frac{n_{id}k_B T}{q}\ln\left(\frac{J_{00}}{J_{sc}}\right) \qquad (5)$$

It is seen that the internal band offset directly adds to the open-circuit voltage that is achievable in situations where only interface recombination is present. However, the open circuit voltage of most devices that have this surface-band-gap widening (those having an absorber with a final Cu-poor composition) is then limited by bulk recombination.

The second important feature in Figure 12 is the conduction band offset ΔE_c^{back} between the Cu(In,Ga)Se$_2$ absorber and the thin MoSe$_2$ film. The back surface recombination velocity at the metallic Mo back contact is reduced from the value S_b (without the MoSe$_2$) to a value $S_b^* = S_b \exp(-\Delta E_c/k_B T)$.

4.4 Fill Factor

The fill factor FF of a solar cell can be expressed in a simple way as long as the solar cell is well described by a diode law. Green [129] gives the following phenomenological expression for the fill factor

$$FF_0 = \frac{v_{oc} - \ln(v_{oc} + 0.72)}{v_{oc} + 1} \qquad (6)$$

It is seen that, through the dimensionless quantity $v_{oc} = qV_{oc}/n_{id}k_B T$, the fill factor depends on the temperature T as well as on the ideality factor n_{id} of the diode (see also Chapter Ia-1). The fill factor FF_0 results solely from the diode law form of Equation (1). In addition, effects from series resistance R_s and shunt resistance R_p add to the fill factor losses. A good approximation is then given by

$$FF = FF_0(1 - r_s)\left[1 - \frac{(v_{oc} + 0.7)\,FF_0(1 - r_S)}{v_{oc}}\frac{}{r_P}\right] \qquad (7)$$

with the normalised series and parallel resistances given by $r_s = R_s J_{sc}/V_{oc}$ and $r_p = R_p J_{sc}/V_{oc}$, respectively. The description of Cu(In,Ga)Se$_2$ solar cells in terms of Equations (6) and (7) works reasonably well, e.g., the world record cell [1] has a fill factor of 78.6% and the value calculated from Equations (6) and (7) is 78.0% ($V_{oc} = 678$ mV, $J_{sc} = 35.2$ mA cm^{-2}, $R_s = 0.2\ \Omega$ cm^2, $R_p = 10^4\ \Omega$ cm^2, $n_{id} = 1.5$).

Factors that can further affect the fill factor are (i) the voltage bias dependence of current collection [130], leading to a dependence of $J_{sc}(V)$

on voltage V in Equation (1), (ii) recombination properties that are spatially inhomogeneous [131], or (iii) unfavourable band offset conditions at the heterointerface [132].

4.5 Electronic Metastabilities

The long time increase (measured in hours and days) of the open-circuit voltage V_{OC} of Cu(In,Ga)Se$_2$ based solar cells during illumination is commonly observed phenomenon [133,134]. In some cases it is not only V_{oc} but also the fill factor that improves during such a light soaking procedure. Consequently, light soaking treatments are systematically used to reestablish the cell efficiency after thermal treatments [107,135].

For the present day ZnO/CdS/Cu(In,Ga)Se$_2$ heterojunctions it appears established that there exist at least three types of metastablities with completely different fingerprints [136,137].

Type I: A continuous increase of the open circuit voltage during illumination and the simultaneous increase of the junction capacitance. Both phenomena are satisfactorily explained as a consequence of persistent photoconductivity in the Cu(In,Ga)Se$_2$ absorber material [138,139], i.e., the persistent capture of photogenerated electrons in traps that exhibit large lattice relaxations like the well-known DX centre in (Al,Ca)As [140]. This type of metastability affects exclusively the open circuit voltage and can vary from few mV up to 50 mV, especially if the sample has been stored in the dark at elevated temperatures.

Type II: A decrease of the fill factor after the cell has been exposed to reverse voltage bias. In extreme cases, this type of metastability leads to a hysteresis in the IV characteristics, e.g., the fill factor of an illuminated IV curve becomes dependent on whether the characteristic has been measured from negative voltages towards positive ones or vice versa. The application of reverse bias also leads to a metastable increase of the junction capacitance and to significant changes of space charge profiles as determined from capacitance vs. voltage measurements [136,137]. The type II metastability is especially important for devices with non-standard buffer/window combinations (e.g., Cd-free buffer layers) [141,142].

Type III: An increase of the fill factor upon illumination with light that is absorbed in the buffer layer or in the extreme surface region of the Cu(In,Ga)Se$_2$ absorber, i.e., the blue part of the solar spectrum. This type of instability counterbalances, to a certain extend, the consequences of reverse bias; i.e., it restores the value of the fill factor after it has been degraded by application of reverse bias.

Our overall understanding of metastabilities, especially of type II and III, is still incomplete. Fortunately, all metastabilities observed in Cu(In,Ga)Se$_2$ so far, tend to improve the photovoltaic properties as soon as the device is brought under operating conditions.

5. WIDE-GAP CHALCOPYRITES

5.1 Basics

The alloy system Cu(In,Ga,Al)(S,Se)$_2$ provides the possibility of building alloys with a wide range of band-gap energies E_g between 1.04 eV for CuInSe$_2$ up to 3.45 eV for CuAlS$_2$ (cf. Figure 2). The highest efficiency within the chalcopyrite system is achieved with the relatively low-band-gap energy E_g of 1.12 eV, and attempts to maintain this high efficiency level at $E_g > 1.3$ eV have so far failed (for a recent review, see [143]). Practical approaches to wide-gap Cuchalcopyrites comprise (i) alloying of CuInSe$_2$ with Ga up to pure CuGaSe$_2$ with $E_g = 1.68$ eV; (ii) Cu(In,Al)Se$_2$ alloys with solar cells realised up to an Al/(Al + In) ratio of 0.6 and $E_g \approx 1.8$ eV [144]; (iii) CuIn(Se,S)$_2$ [145], and Cu(In,Ga)(Se, S)$_2$ [146] alloys with a S/(Se + S) ratio up to 0.5; and (iv) CuInS$_2$ [147] and Cu(In,Ga)S$_2$ [148] alloys. Note that the approaches (i)–(iii) are realised with a final film composition that is slightly Cu-poor, whereas approach (iv) uses films that are Cu-rich. In the latter case, CuS segregates preferably at the film surface. This secondary phase has to be removed prior to heterojunction formation.

The advantage of higher voltages of the individual cells by increasing the band gap of the absorber material is important for thin-film modules. An ideal range for the band gap energy would be between 1.4 eV and 1.6 eV because the increased open circuit voltage and the reduced short circuit current density would reduce the number of necessary scribes used for monolithic integration of the cells into a module. Also, the thickness of front and back electrodes can be reduced because of the reduced current density.

Table 3 compares the solar cell output parameters of the best chalcopyrite-based solar cells. This compilation clearly shows the superiority of Cu(In, Ga)Se$_2$ with a relatively low Ga content which leads to the actual world champion device. The fact that the best CuInSe$_2$ device has an efficiency of 3% below that of the best Cu(In,Ga)Se$_2$ device is due not only to the

Table 3 Absorber band-gap energy E_g, efficiency η, open-circuit voltage V_{oc}, short-circuit current density J_{sc}, fill factor FF, and area A of the best Cu(In,Ga)Se$_2$, CuInSe$_2$, CuGaSe$_2$, Cu(I-n,Al)Se$_2$, CuInS$_2$, Cu(In,Ga)S$_2$, and Cu(In,Ga)(S,Se)$_2$ solar cells

Material	E_g (eV)	η(%)	V_{oc} (mV)	J_{sc} (mA cm^{-2})	FF (%)	A (cm^2)	Ref.
Cu(In,Gu)Se$_2$	1.12	18.8	678	35.2	78.6	0.45	[1][a]
CuInSe$_2$	1.04	15.4	515	41.2	72.6	0.38	[40][b]
CuGaSe$_2$	1.68	8.3	861	14.2	67.9	0.47	[151][a]
Cu(In,Al)Se$_2$	1.16	16.6	621	36.0	75.5	?	[157][a]
Cu(In,Ga)(S,Se)$_2$	1.36	13.9	775	24.3	74.0	0.5	[146][b]
CuInS$_2$	1.5	11.4	729	21.8	72.0	0.5	[159][a]
Cu(In,Ga)S$_2$	1.53	12.3	774	21.6	73.7	0.5	[148][a]

[a]Confirmed total area values.
[b]Effective area values (not confirmed).

less favourable band-gap energy but also to the lack of the beneficial effect of small amounts of Ga on the growth and on the electronic quality of the thin film, as discussed previously.

The difficulty of obtaining wide-gap devices with high efficiencies is also illustrated by plotting the absorber band-gap energies of a series of chalcopyrite alloys vs. the attained open-circuit voltages. Figure 13 shows that below $E_g = 1.3$ eV, the V_{OC} data follow a straight line, indicating the proportional gain of $V_{oc} = E_g/q - 0.5$ V, whereas at $E_g > 1.3$ eV the gain is much more moderate. At the high band-gap end of the scale the differences of the band-gap energies and the open-circuit voltages of CuInS$_2$ and CuGaSc$_2$ amount to 840 mV and 820 mV, respectively, whereas $E_g - qV_{oc}$ is only 434 eV in the record Cu(In,Ga)Se$_2$ device.

One reason for these large differences $E_g - qV_{oc}$ in wide-gap devices is the less favourable band offset constellation at the absorber/CdS-buffer interface. Figure 14 shows the band diagram of a wide-gap Cu chalcopyrite-based heterojunction with (Figure 14(a)) and without (Figure 14(b)) the surface defect layer. As the increase of band gap in going from CuInSe$_2$ to CuGaSe$_2$ takes place almost exclusively by an increase of the conduction band energy E_c, the positive or zero band offset ΔE_c^{ab} between the absorber and the buffer of a low-gap device (cf. Figure 12) turns into a negative one in Figure 14. This effect should be weaker for CuInS$_2$ and CuAlSe$_2$ as the increase of E_g in these cases is due to an upwards shift of E_c *and* a downwards shift of E_v. However, *any* increase of E_c implies that the barrier ϕ_{bp} that hinders the holes from the absorber from recombination at the heterointerface does not increase proportionally with the increase of the band-gap energy. Thus, in wide-gap absorbers, the importance of interface recombination

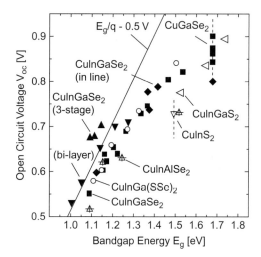

Figure 13 Open-circuit voltages of different Cu-chalcopyrite bused solar cells with various band-gap energies of the absorber layers. Full symbols correspond to Cu(In, Ga)Se$_2$ alloys prepared by a simple single layer process (squares), a bilayer process (triangles down), and the three-stage process (triangles up). Cu(In,Ga)Se$_2$ cells derived from an in-line process as sketched in Fig. 9 are denoted by diamonds. Open triangles relate to Cu(In,Ga)S$_2$, open circles to Cu(In,Ga)(S,Se)$_2$, and the crossed triangles to Cu(In,Al)Se$_2$ cells.

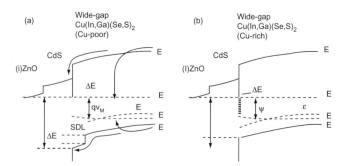

Figure 14 Energy band diagram of a ZnO/CdS/(wide-gap) Cu(In,Ga)(Se,S)$_2$ heterojunction. The band diagram (a) that includes the surface defect layer (SDL) of a Cu-poor prepared film shows that the interface recombination barrier $\phi^*_{bp} = \phi_{bp} + \Delta E_v^{int}$ is larger than the barrier ϕ_{bp} in the device that was prepared Cu-rich (b). The difference is the internal valence band offset ΔE_v^{int} between the SDL and the bulk of the absorber. The larger value of ϕ^*_{bp} reduces interface recombination.

(determined by the barrier (ϕ_{bp}) grows considerably relative to that of bulk recombination (determined by E_g of the absorber) [149]. Using the same arguments with respect to the MoSe$_2$/absorber, the back-surface field provided by this type of heterojunction back contact, as shown in Figure 12, vanishes when the conduction-band energy of the absorber is increased.

Up to now, all arguments for explaining the relatively low performance of wide-gap chalcopyrite alloys deal with the changes in the band diagram. However, recent work focuses on the differences in the electronic quality of the absorber materials. Here, it was found that the concentration of recombination active defects increases when increasing the Ga content in Cu(In,Ga)Se$_2$ above a Ga/(Ga + In) ratio of about 0.3 [32] and the S content in Cu(In,Ga)(Se,S)$_2$ alloys over a S/(Se + S) ratio of 0.3 [34]. Moreover, a recent systematic investigation of the dominant recombination mechanism in large series of Cu(In,Ga)(Se,S)$_2$ based solar cells with different compositions [150] suggests that bulk recombination prevails in all devices that were prepared with a Cu-poor final film composition. Only devices that had a Cu-rich composition (before removing Curich secondary phases) showed signatures of interface recombination. The absence of interface recombination in Cu-poor devices could be a result of the presence of the Cu-poor surface defect layer in these devices. Comparison of Figures 14(a) and (b) illustrates that interface recombination is much more likely to dominate those devices that lack this feature.

5.2 CuGaSe$_2$

CuGaSe$_2$ has a band gap of 1.68 eV and therefore would represent an ideal partner for CuInSe$_2$ in an all-chalcopyrite tandem structure. However, a reasonable efficiency for the top cell of any tandem structure is about 15%, far higher than has been reached by the present polycrystalline CuGaSe$_2$ technology. The record efficiency of CuGaSe$_2$ thin-film solar cells is only 8.3% (9.3% active area) [151] despite of the fact that the electronic properties of CuGaSe$_2$ are not so far from those of CuInSe$_2$. However, in detail, all the differences quantitatively point in a less favourable direction. In general, the net doping density N_A in CuGaSe$_2$ appears too high [152]. Together with the charge of deeper defects, the high doping density leads to an electrical field in the space-charge region, which enhances recombination by tunnelling [153]. The high defect density in CuGaSe$_2$ thin films absorbers additionally leads to a low diffusion length and, in consequence, to a dependence of the collected short circuit current density J_{sc} on the bias voltage V. Because of the decreasing width of the space charge region, $J_{sc}(V)$ decreases with increasing V affecting significantly the fill factor of the solar cell [130]. Note that, on top of the limitation by the unfavourable bulk properties, interface recombination also plays a certain role in CuGaSe$_2$ [154] because of the unfavourable band diagram shown in Figure 14. Therefore, substantial improvement of the performance requires simultaneous optimisation of bulk *and* interface properties. Notably, CuGaSe$_2$ is the only Cu-chalcopyrite material where the record efficiency of cells based on bulk crystals (with $\eta = 9.4\%$, total area) [115] exceeds that of thin-film devices.

5.3 Cu(In,Al)Se$_2$

As can be seen in Figure 2, the bandgap change within the Cu(In,Al)Se$_2$ alloy system is significant even if a small amount of Al is added to CuInSe$_2$ [144]. This fact allows to grow graded structures with only small changes in the lattice constant. Al—Se compounds tend to react with water vapour to form oxides and H$_2$Se. Furthermore, there is a tendency to phase segregations [144]. However, cells with very good performance have been achieved by a co-evaporation process of absorbers in a band gap range between 1.09 and 1.57 eV, corresponding to Al/(In + Al) ratio x between 0.09 and 0.59 [144,156]. The highest confirmed efficiency of a Cu(In,Al)Se$_2$ solar cell is 16.9% [157] (see also Table 3). This device has about the same band-gap energy as the record Cu(In,Ga)Se$_2$ device [1].

5.4 CuInS$_2$ and Cu(In,Ga)S$_2$

The major difference between CuInS$_2$ and Cu(In,Ga)Se$_2$ is that the former cannot be prepared with an overall Cu-poor composition. Cu-poor CuInS$_2$ displays an extremely low conductivity, making it almost unusable as a photovoltaic absorber material [145]. Even at small deviations from stoichiometry on the In-rich side, segregation of the spinel phase is observed [158]. Instead, the material of choice is Cu-rich CuInS$_2$. As in the case of CuInSe$_2$, a Cu-rich preparation route implies the removal of the unavoidable secondary Cu—S binary phase by etching the absorber in KCN solution [147]. Such an etch may involve some damage of the absorber surface as well as the introduction of shunt paths between the front and the back electrode. However, as shown in Table 3, the best CuInS$_2$ device [159] has an efficiency above 11%. This record efficiency for CuInS$_2$ is achieved by a sulphurisation process rather than by co-evaporation.

As we have discussed previously, interface recombination dominates the open circuit voltage V_{oc} of Cu(In,Ga)(Se,S)$_2$ devices that are prepared with a Cu-rich absorber composition (prior to the KCN etch), like the CuInS$_2$ and Cu(In,Ga)S$_2$ devices discussed here. It was found recently, that alloying CuInSe$_2$ with moderate amounts of Ga enhances the open circuit voltage V_{oc} [160]. This increase of V_{or} can counterbalance the loss of short circuit current density J_{sc} resulting from the increased band gap. Apparently, addition of Ga to CuInS$_2$ reduces interface recombination by increasing the interfacial barrier ϕ_{bp}(cf. Figure 14(b)) [150]. However, for Cu(In,Ga)S$_2$ devices prepared by the sulphurisation route, the benefit of Ga alloying is limited, because, during preparation, most of the Ga added to the precursor ends up confined to the rear part of the absorber layer and, therefore, remains ineffective at the absorber surface [160]. In contrast, when preparing Cu(In,Ga)S$_2$ with co-evaporation, Ga is homogeneously distributed through the depth of the absorber and the Ga content at the film surface is well controlled. Recent work of Kaigawa et al. [148] represents a major progress in wide-gap Cu-chalcopyrites with the preparation of Cu(In,Ga)S$_2$ solar cells with a confirmed efficiency of 12.3% at a band gap energy $E_g = 1.53$ eV. In the same publication [148] an efficiency of 10.1% is reported for a device with $E_g = 1.65$ eV. The open circuit voltage V_{oc} of this device is 831 mV, i.e., only slightly lower than V_{oc} of the best CuGaSe$_2$ cell having an efficiency of 8.3%.

5.5 Cu(In,Ga)(Se,S)$_2$

One possible way of overcoming the disadvantages of the ternary wide-gap materials CuInS$_2$ and CuGaSe$_2$ is to use the full pentenary alloy system Cu(In$_{1-x}$Ga$_x$)(Se$_{1-y}$S$_y$)$_2$ [146]. Among the materials listed in Table 3, the pentenary system is the only one with an open-circuit voltage larger than 750 mV and an efficiency above 13%, outperforming CuInS$_2$ in both these respects. The advantage of Cu(In,Ga)(Se,S)$_2$ could be due to the mutual compensation of the drawbacks of CuGaSe$_2$ (too high charge density) and that of (Cu-poor) CuInS$_2$ (too low conductivity, if prepared with a Cu-poor film composition).

5.6 Graded-Gap Devices

An interesting property of the Cu(In,Ga,Al)(S.Se)$_2$ alloy system is the possibility of designing graded-gap structures in order to optimise the electronic properties of the final device [161–164]. Such band-gap gradings are achieved during co-evaporation by the control of the elemental sources, but selenisation/sulphurisation processes also lead to beneficial compositional gradings. The art of designing optimum band-gap gradings is to push back charge carriers from critical regions, i.e., regions with high recombination probability within the device. Such critical regions are (i) the interface between the back contact and the absorber layer and (ii) the heterojunction interface between the absorber and the buffer material. Figure 15 shows a band diagram of a graded structure that fulfils the requirements for minimising recombination losses.

1. To keep the back contact region clear from electrons, one can use a Ga/In grading. The increase of the Ga/(Ga + In) ratio causes a movement of the conduction band minimum upward with respect to its position in a non-graded $CuInSe_2$ device. Thus, back surface grading leads to a gradual increase of the conduction-band energy, as illustrated in Figure 15, and therefore drives photogenerated electrons away from the metallic back contact into the direction of the buffer/absorber junction.

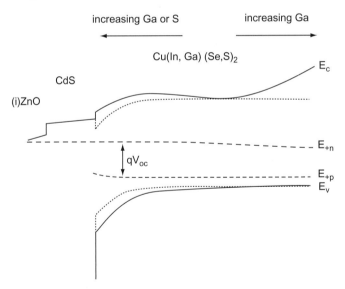

Figure 15 Band diagram of a ZnO/CdS/Cu(In,Ga)(Se,S)$_2$ heterojunction with a graded-gap absorber. The minimum band gap energy is in the quasi neutral part of the absorber. An increasing Ga/In ratio towards the back surface and an increasing Ga/In or S/Se-ratio towards the front minimise recombination in critical regions at the back contact (recombination path A'), in the space charge region (path B), and at the heterointerface (path C). The dotted lines correspond to the conduction and valence band edge energies of a non-graded device.

2. The minimisation of junction recombination, both at the point of equal capture rates of holes and electrons in the space charge region (recombination path B in Figure 15) as well as at the absorber/buffer interface (path C), requires an increased band gap towards the absorber surface. If one had the choice, one would clearly favour a decrease of the valence-band energy, as shown in Figure 15, over an increase of the conduction band energy. This favours a grading with the help of S/Se alloying, as at least a part of the increasing band-gap energy is accommodated by a decrease of the energy of the valence-band edge that should minimise interface recombination.

6. CONCLUSIONS

The objective of this chapter is the description of the recent achievements on $Cu(In,Ga)Se_2$-based solar cells as well as an account of our present understanding of the physical properties of the materials involved and the electronic behaviour of the devices. $Cu(In,Ga)Se_2$ is in a leading position among polycrystalline thin-film solar cell materials because of the benign, forgiving nature of the bulk materials and interfaces. Nevertheless, we want to guide the attention of the reader also in the direction of the work that has still to be done.

Three of the four cornerstones 1–4 for the recent achievements mentioned in Section 3.2 concern the *growth of the films*: the optimised deposition conditions and the incorporation of Na and Ga. However, no detailed model is available to describe the growth of $Cu(In,Ga)Se_2$, and especially the impact of Na, which, in our opinion, is the most important of the different ingredients available to tune the electronic properties of the absorber. A clearer understanding of $Cu(In,Ga)Se_2$ growth would allow us to find optimised conditions in the wide parameter space available, and thus to reduce the number of recombination centres and compensating donors and optimise the number of shallow acceptors. The deposition of the buffer layer, or more generally speaking, the formation of the heterojunction, is another critical issue. The surface chemistry taking place during heterojunction formation, and also during post-deposition treatments, is decisive for the final device performance. Both processes greatly affect not only the surface defects (i.e., recombination *and* charge), and therefore the charge distribution in the device, but also the defects in the bulk of the absorber. Concentrated effort and major progress in these tasks would not only allow us to push the best efficiencies further towards 20%, but would also provide a sound knowledge base for the various attempts at the commercialisation of Cu(In,Ga)Se$_2$ solar cells.

ACKNOWLEDGEMENTS

The authors thank all our colleagues at the IPE for various discussions and fruitful collaboration. We are especially grateful to J. H. Werner for his continuous support of the Cu (In,Ga)Se$_2$ research at the *ipe*. We are grateful to M. Turcu for a critical reading of the manuscript. Support of our work during many years by the German Ministry of Research, Technology, and Education (BMBF), by the Ministry of Economics, and by the European Commission is especially acknowledged.

REFERENCES

[1] M. Contreras, B. Egaas, K. Ramanathan, J. Hiltner, A. Swartzlander, F. Hasoon, et al., Progress toward 20% efficiency in Cu(In,Ga)Se$_2$ polycrystalline thin-film solar cells, Prog. Photovolt. Res. Appl. 7 (1999) 311.
[2] J. Kessler, M. Bodegard, J. Hedström, L. Stolt, New world record Cu(In,Ga)Se2 based mini-module: 16.6%. Proceedings of the 16th European Photovoltaic Solar Energy Conference, Glasgow, 2000, p. 2057.
[3] H. Hahn, G. Frank, W. Klingler, A. Meyer, G. Störger, über einige ternäre Chalkogenide mit Chalkopyritstruktur, Z. Anorg. u. Allg. Chemie 271 (1953) 153.
[4] S. Wagner, J.L. Shay, P. Migliorato, H.M. Kasper, CuInSe$_2$/CdS heterojunction photovoltaic detectors, Appl. Phys. Lett. 25 (1974) 434.
[5] R.A. Mickelsen, W.S. Chen, High photocurrent polycrystalline thin-film CdS/CuInSe$_2$ solar cell, Appl. Phys. Lett. 36 (1980) 371.
[6] K.C. Mitchell, J. Ermer, D. Pier, Single and tandem junction CuInSe2 cell and module technology. Proceedings of the 20th IEEE Photovoltaic Specialists Conference, Las Vegas, 1988, p. 1384.
[7] R.D. Wieting, CIS Manufacturing at the megawatt scale. Proceedings of the 29th IEEE Photovoltaic Specialists Conference, New Orleans, 2002, p. 480.
[8] V. Probst, W. Stetter, W. Riedl, H. Vogt, M. Wendl, H. Calwer, et al., Rapid CIS-process for high efficiency PV-modules: development towards large area processing, Thin Solid Films 387 (2001) 262.
[9] B. Dimmler, M. Powalla, H.W. Schock, CIS-based thin-film photovoltaic modules: potential and prospects, Prog. Photovolt. Res. Appl. 10 (2002) 149.
[10] M. Powalla, , E. Lotter, R. Waechter, S. Spiering, M. Oertel, B. Dimmler, Pilot line production of CIGS modules: first experience in processing and further developments. Proceedings of the 29th IEEE Photovoltaic Specialists Conference, New Orleans, 2002 p. 571.
[11] K. Kushiya, Improvement of electrical yield in the fabrication of CIGS-based thin-film modules, Thin Solid Films 387 (2001) 257.
[12] T. Negami, T. Satoh, Y. Hashimoto, S. Nishiwaki, S. Shimakawa, S. Hayashi, Large-area CIGS absorbers prepared by physical vapor deposition, Sol. Energy Mater. Sol. Cells 67 (2001) 1.

[13] U. Rau, H.W. Schock, Electronic properties of Cu(In,Ga)Se$_2$ heterojunction solar cells-recent achievements, current understanding, and future challenges, Appl. Phys. A 69 (1999) 191.
[14] U. Rau, H.W. Schock, Cu(In,Ga)Se$_2$ Solar Cells, in: M.D. Archer, R. Hill (Eds.), Clean Electricity from Photovoltaics, Imperial College Press, London, 2001, p. 270.
[15] J.L. Shay, J.H. Wernick, Ternary Chalcopyrite Semiconductors: Growth. Electronic Properties, and Applications, Pergamon Press, Oxford, 1975.
[16] L.L. Kazmerski, S. Wagner, Cu-ternary chalcopyrite solar cells, in: T.J. Coutts, J.D. Meakin (Eds.), Current Topics in Photovoltaics, Academic Press, Orlando, 1985, p. 41.
[17] T.J. Coutts, L.L. Kazmerski, S. Wagner, Ternary Chalcopyrite Semiconductors: Growth, Electronic Properties, and Applications, Elsevier, Amsterdam, 1986.
[18] A. Rockett, R.W. Birkmire, CuInSe$_2$ for photovoltaic applications, J. Appl. Phys. 70 (1991) 81.
[19] B.J. Stanbery, Copper indium selenides and related materials for photovoltaic devices, Crit. Rev. Solid State 27 (2002) 73.
[20] T. Haalboom, T. Gödecke, F. Ernst, M. Rühle, R. Herberholz, H.W. Schock, et al., Phase relations and microstructure in bulk materials and thin films of the ternary system Cu-In-Se, Inst. Phys. Conference Ser. 152E (1997) 249.
[21] R. Herberholz, U. Rau, H.W. Schock, T. Haalboom, T. Gödecke, F. Ernst, et al., Phase segregation, Cumigration and junction formation in Cu(In,Ga)Se$_2$, Eur. Phys. J. Appl. Phys. 6 (1999) 131.
[22] P. Migliorato, J.L. Shay, H.M. Kasper, S. Wagner, Analysis of the electrical and luminescent properties of CuInSe$_2$, J. Appl. Phys. 46 (1975) 1777.
[23] R. Noufi, R. Axton, C. Herrington, S.K. Deb, Electronic properties versus composition of thin films of CuInSe$_2$, Appl. Phys. Lett. 45 (1984) 668.
[24] S.B. Zhang, S.H. Wei, A. Zunger, H. Katayama-Yoshida, Defect physics of the CuInSe$_2$ chalcopyrite semiconductor, Phys. Rev B 57 (1998) 9642.
[25] H. Neumann, Vacancy formation in $A^I B^{III} C_2^{VI}$ chalcopyrite semiconductors, Cryst. Res. Technol. 18 (1983) 901.
[26] F.A. Abou-Elfotouh, H. Moutinho, A. Bakry, T.J. Coutts, L.L. Kazmerski, Characterization of the defect levels in copper indium diselenide, Solar Cells 30 (1991) 151.
[27] M. Igalson, H.W. Schock, The metastable changes of the trap spectra of CuInSe$_2$-based photovoltaic devices, J. Appl. Phys. 80 (1996) 5765.
[28] M. Schmitt, U. Rau, J. Parisi, Investigation of deep trap levels in CuInSe2 solar cells by temperature dependent admittance measurements. Proceedings of the 13th European Photovoltaic Solar Energy Conference, Nice, 1995, p. 1969.
[29] T. Walter, R. Herberholz, C. Müller, H.W. Schock, Determination of defect distributions from admittance measurements and application to Cu(In,Ga)Se$_2$ based heterojunctions, J. Appl. Phys. 80 (1996) 4411.
[30] Igalson, M., Bacewicz, R. and H.W. Schock, 'Dangling bonds' in CuInSe2 and related compounds. Proceedings of the 13th European Photovoltaic Solar Energy Conference, Nice, 1995, p. 2076.
[31] R. Herberholz, V. Nadenau, U. Rühle, C. Köble, H.W. Schock, B. Dimmler, Prospects of wide-gap chalcopyrites for thin film photovoltaic modules, Sol. Energy Mater. Sol. Cells 49 (1997) 227.
[32] G. Hanna, A. Jasenek, U. Rau, H.W. Schock, Open circuit voltage limitations in CuIn$_{1-x}$Ga$_x$Se$_2$ thin- film solar cells — dependence on alloy composition, Phys. Stat. Sol. A 179 (2000) R7.
[33] U. Rau, M. Schmidt, A. Jasenek, G. Hanna, H.W. Schock, Electrical characterization of Cu(In,Ga)Se$_2$ thin-film solar cells and the role of defects for the device performance, Sol. Energy Mater. Sol. Cells 67 (2001) 137.

[34] M. Turcu, I.M. Kötschau, U. Rau, Composition dependence of defect energies and band alignments in the $Cu(In_{1-x}Ga_x)(Se_{1-y}S_y)_2$ alloy system, J. Appl. Phys. 91 (2002) 1391.
[35] M. Caldas, A. Fazzio, A. Zunger, A universal trend in the binding energies of deep impurities in semiconductors, Appl. Phys. Lett. 45 (1984) 67.
[36] J.M. Langer, H. Heinrich, Deep-level impurities: a possible guide to prediction of band-edge discontinuities in semiconductor heterojunctions, Phys. Rev. Lett. 55 (1985) 1414.
[37] J.T. Heath, J.D. Cohen, W.N. Shafarman, D.X. Liao, A.A. Rockett, Effect of Ga content on defect states in $CuIn_{1-x}Ga_xSe_2$ photovoltaic devices, Appl. Phys. Lett. 80 (2002) 4540.
[38] M. Turcu, U. Rau, Compositional trends of defect energies, band alignments, and recombination mechanisms in the $Cu(In,Ga)(Se,S)_2$ alloy system, Thin Solid Films 431–432 (2003) 158.
[39] R. Herberholz, M. Igalson, H.W. Schock, Distinction between bulk and interface states in $CuInSe_2$/CdS/ZnO by space charge spectroscopy, J. Appl. Phys. 83 (1998) 318.
[40] L. Stolt, J. Hedström, J. Kessler, M. Ruckh, K.O. Velthaus, H.W. Schock, ZnO/CdS/$CuInSe_2$ thin-film solar cells with improved performance, Appl. Phys. Lett. 62 (1993) 597.
[41] J. Hedström, H. Ohlsen, M. Bodegard, A. Kylner, L. Stolt, D. Hariskos, et al., ZnO/CdS/Cu(In,Ga)Se/sub 2/thin film solar cells with improved performance. Proceedings of the 23rd IEEE Photovoltaic Specialists Conference, Lousville, 1993, p. 364.
[42] W.E. Devaney, W.S. Chen, J.M. Steward, R.A. Mickelson, Structure and properties of high efficiency ZnO/CdZnS/$CuInGaSe_2$ solar cells, IEEE Trans. Electron Devices ED-37 (1990) 428.
[43] R.R. Potter, C. Eberspacher, L.B. Fabick, Device analysis of $CuInSe_2$/(Cd,Zn)S solar cells. Proceedings of the18th IEEE Photovoltaic Specialists Conference, Las Vegas, 1985, p. 1659.
[44] R.W. Birkmire, B.E. McCandless, W.N. Shafarman, R.D. Varrin, Approaches for high efficiency $CuInSe_2$ solar cells. Proceedings of the 9th. European Photovoltaic Solar Energy Conference, Freiburg, 1989, p. 134.
[45] M.A. Contreras, B. Egaas, P. Dippo, J. Webb, J. Granata, K. Ramanathan, et al., On the role of Na and modifications to Cu(In,Ga)Se absorber materials using thin-MF (M = Na, K, Cs) precursor layers. Proceedings of the 26th IEEE Photovoltaic Specialists Conference, Anaheim, 1997, p. 359.
[46] J. Holz, F. Karg, H.V. Phillipsborn, The effect of substrate impurities on the electronic conductivity in CIGS thin films. Proceedings of the12th European Photovoltaic Solar Energy Conference, Amsterdam, 1994, p. 1592.
[47] T. Nakada, T. Mise, T. Kume, A. Kunioka, Superstate type $Cu(In,Ga)Se_2$ thin film solar cells with ZnO buffer layers – a novel approach to 10% efficiency. Proceedings of the 2nd. World Conference on Photovoltaic Solar Energy Conversion, Vienna, 1998, p. 413.
[48] M. Ruckh, D. Schmid, M. Kaiser, R. Schäffler, T. Walter, H.W. Schock, Influence of substrates on the electrical properties of $Cu(In,Ga)Se_2$ thin films. Proceedings of the First World Conference on Photovoltaic Solar Energy Conversion, Hawaii, 1994, p. 156.
[49] B.M. Keyes, F. Hasoon, P. Dippo, A. Balcioglu, F. Abouelfotouh, Influence of Na on the elctro-optical properties of $Cu(In,Ga)Se_2$. Proceedings of the 26th. IEEE Photovoltaic Specialists Conference, Anaheim, 1997, p. 479.

[50] U. Rau, M. Schmitt, P. Engelhardt, O. Seifert, J. Parisi, W. Riedl, et al., Impact of Na and S incorporation on the electronic transport mechanisms of Cu(In,Ga)Se$_2$ solar cells, Solid State Commun. 107 (1998) 59.
[51] D. Braunger, D. Hariskos, G. Bilger, U. Rau, H.W. Schock, Influence of sodium on the growth of polycrystalline Cu(In.Ga)Se$_2$ thin films, Thin Solid Films 361 (2000) 161.
[52] R. Takei, H. Tanino, S. Chichibu, H. Nakanishi, Depth profiles of spatially resolved Raman spectra of a CuInSe$_2$-based thin-film solar cell, J. Appl. Phys. 79 (1996) 2793.
[53] T. Wada, N. Kohara, T. Negami, M. Nishitani, Chemical and structural characterization of Cu(In,Ga)Se$_2$/Mo interface in Cu(In,Ga)Se$_2$ solar cells, Jpn. J. Appl. Phys. 35 (1996) 1253.
[54] R. Klenk, T. Walter, H.W. Schock, D. Cahen, A model for the successful growth of polycrystalline films of CuInSe$_2$ by multisource physical vacuum evaporation, Adv. Mat. 5 (1993) 114.
[55] V. Probst, J. Rimmasch, W. Stetter, H. Harms, W. Riedl, J. Holz, et al., Improved CIS thin film solar cells through novel impurity control techniques. Proceedings of the 73th European Photovoltaic Solar Energy Conf., Nice, 1995, p. 2123.
[56] A.M. Gabor, J.R. Tuttle, D.S. Albin, M.A. Contreras, R. Noufi, A.M. Hermann, High-efficiency CuIn$_x$Ga$_{1-x}$Se$_2$ solar cells from (In$_x$, Ga$_{1-x}$)$_2$Se$_3$ precursors, Appl. Phys. Lett. 65 (1994) 198.
[57] J.J.M. Binsma, H.A. Van der Linden, Preparation of thin CuInS$_2$ films via a two-stage process, Thin Solid Films 97 (1982) 237.
[58] T.L. Chu, S.C. Chu, S.C. Lin, I. Yue, Large grain copper indium diselenide films, J. Electrochem. Soc. 131 (1984) 2182.
[59] V.K. Kapur, B.M. Basol, E.S. Tseng, Lov-cost methods for the production of semiconductor films for CuInSe$_2$/CdS solar cells, Solar Cells 21 (1987) 65.
[60] V. Probst, F. Karg, J. Rimmasch, W. Riedl, W. Stetter, H. Harms, et al., Advanced stacked elemental layer progress for Cu(InGa)Se$_2$ thin film photovoltaic devices, Mat. Res. Soc. Symp. Proc. 426 (1996) 165.
[61] S. Niki, P.J. Fons, A. Yamada, R. Suzuki, T. Ohdaira, S. Ishibashi, et al., High quality CuInSe2 epitaxial films — molecular beam epitaxial growth and intrinsic properties, Inst. Phys Conference Ser. 152E (1994) 221.
[62] P.N. Gallon, G. Orsal, M.C. Artaud, S. Duchemin, Studies of CuInSe2 and CuGaSe$_2$ thin films grown by MOCVD from three organometallic sources. Proceedings of the 2nd World Conference on Photovoltaic Solar Energy Conversion, Vienna, 1998, p. 515.
[63] J.-F. Guillemoles, P. Cowache, A. Lusson, K. Fezzaa, F. Boisivon, J. Vedel, et al., High quality CuInSe$_2$ epitaxial films — molecular beam epitaxial growth and intrinsic properties, J. Appl Phys. 79 (1996) 7293.
[64] A. Abken, F. Heinemann, A. Kampmann, G. Leinkühler, J. Rechid, V. Sittinger, et al., Large area electrodeposition of Cu(In,Ga)Se$_2$ precursors for the fabrication of thin film solar cells. Proceedings of the 2nd World Conference on Photovoltaic Solar Energy Conversion, Vienna, 1998, p. 1133.
[65] D. Lincot, J.-F. Guillemoles, P. Cowache, A. Marlot, C. Lepiller, B. Canava, et al., Solution deposition technologies for thin film solar cells: status and perspectives. Proceedings of the 2nd World Conference on Photovoltaic Solar Energy Conversion, Vienna, 1998, p. 440.
[66] D. Guimard, P.P. Grand, N. Boderau, P. Cowache, J.-F. Guillemoles, D. Lincot, et al., Copper indium diselenide solar cells prepared by electrodeposition, Proceedings of the 29th IEEE Photovoltaic Specialists Conference New Orleans, 2002, p. 692.

[67] C. Eberspacher, K.L. Pauls, C.V. Fredric, Improved processes for forming $CuInSe_2$ films. Proceedings of the 2nd World Conf on Photovoltaic Solar Energy Conversion, Vienna, 1998, p. 303.
[68] C. Eberspacher, K. Pauls, J. Serra, Non-vacuum processing of CIGS solar cells. Proceedings of the 29th IEEE Photovoltaic Specialists Conference, New Orleans, 2002, p. 684.
[69] K. Ramanathan, R.N. Bhattacharya, J. Granata, J. Webb, D. Niles, M.A. Contreras, et al., Advances in the CIS research at NREL, Proceedings of the 26th IEEE Photovoltaic Specialists Conference, Anaheim, 1998, p. 319.
[70] V.K. Kapur, A. Bansal, P. Le, O.I. Asensio, Non-vacuum printing process for CIGS Solar cells on rigid and flexible substrates. Proceedings of the 29th IEEE Photovoltaic Specialists Conference, New Orleans, 2002, p. 688.
[71] D. Cahen, R. Noufi, Defect chemical explanation for the effect of air anneal on $CdS/CuInSe_2$ solar cell performance, Appl. Phys. Lett. 54 (1989) 558.
[72] U. Rau, D. Braunger, R. Herberholz, H.W. Schock, J.-F. Guillemoles, L. Kronik, et al., Oxygenation and air-annealing effects on the electronic properties of $Cu(In,Ga)Se_2$ films and devices, J. Appl. Phys. 86 (1999) 497.
[73] L. Kronik, U. Rau, J.-F. Guillemoles, D. Braunger, H.W. Schock, D. Cahen, Interface redox engineering of $Cu(In,Ga)Se_2$-based solar cells: oxygen, sodium, and chemical bath effects, Thin Solid Films 361–362 (2000) 353.
[74] R. Scheer, Surface and interface properties of Cu-chalcopyrite semiconductors and devices, Research Trends in Vacuum Sci. Technol. 2 (1997) 77.
[75] U. Rau, Role of defects and defect metastabilities for the performance and stability of $Cu(In,Ga)Se_2$ based solar cells, Jpn. J. Appl. Phys. 39(Suppl. 39–1) (2000) 389.
[76] D. Schmid, M. Ruckh, F. Grunwald, H.W. Schock, Chalcopyrite/defect chalcopyrite heterojunctions on the basis of $CuInSe_2$, J. Appl. Phys. 73 (1993) 2902.
[77] M. Morkel, L. Weinhardt, B. Lohmüller, C. Heske, E. Umbach, W. Riedl, et al., Flat conduction-band alignment at the $CdS/CuInSe_2$ thin-film solar-cell heterojunction, Appl. Phys. Lett. 79 (2001) 4482.
[78] D. Schmid, M. Ruckh, H.W. Schock, A comprehensive characterization of the interfaces in Mo/CIS/CdS/ZnO solar cell structures, Sol. Energy Mater. Sol. Cells 41 (1996) 281.
[79] M.A. Contreras, H. Wiesner, D. Niles, K. Ramanathan, R. Matson, J. Tuttle, et al., Defect chalcopyrite $Cu(In_{1-x}Ga_x)_3Se_5$ materials and high Ga-content $Cu(In,Ga)Se_2$-based solar cells. Proceedings of the 25th IEEE Photovoltaic Specialists, Washington, DC, 1996, p. 809.
[80] A. Niemeegers, M. Burgelman, R. Herberholz, U. Rau, D. Hariskos, H.W. Schock, Model for electronic transport in $Cu(In,Ga)Se_2$ solar cells, Prog. Photovolt. Res. Appl. 6 (1998) 407.
[81] J. Bardeen, Surface states and rectification at a metal semiconductor contact, Phys. Rev. 71 (1947) 717.
[82] A. Klein, W. Jaegermann, Fermi-level-dependent defect formation in Cu-chalcopyrite semiconductors, Appl. Phys. Lett. 74 (1999) 2283.
[83] J. Kessler, K.O. Velthaus, M. Ruckh, R. Laichinger, H.W. Schock, D. Lincot, et al., Chemical bath deposition of CdS on $CuInSe_2$, etching effects and growth kinetics. Proceedings of the 6th. International Photovoltaic Solar Energy Conference, New Delhi, India, 1992, p. 1005.
[84] T.M. Friedlmeier, D. Braunger, D. Hariskos, M. Kaiser, H.N. Wanka, H.W. Schock, Nucleation and growth of the CdS buffer layer on $Cu(In,Ga)Se_2$ thin films. Proceedings of the 25th IEEE Photovoltaic Specialists Conference, Washington DC, 1996, p. 845.

[85] K. Ramanathan, H. Wiesner, S. Asher, D. Niles, R.N. Bhattacharya, J. Keane, et al., High efficiency Cu(In,Ga)Se$_2$ thin film solar cells without intermediate buffer layers. Proceedings of the 2nd World Conference on Photovoltaic Solar Energy Conversion, Vienna, 1998, p. 477.

[86] T. Wada, S. Hayashi, Y. Hashimoto, S. Nishiwaki, T. Sato, M. Nishitina, High efficiency Cu(In,Ga)Se$_2$ (CIGS) solar cells with improved CIGS surface. Proceedings of the 2nd World Conference on Photovoltaic Solar Energy Conversion, Vienna, 1998, p. 403.

[87] D. Hariskos, M. Ruckh, U. Rühle, T. Walter, H.W. Schock, J. Hedström, et al., A novel cadmium free buffer layer for Cu(In,Ga)Se$_2$ based solar cells, Sol. Energy Mater. Sol. Cells 41/42 (1996) 345.

[88] K. Kushiya, T. Nii, I. Sugiyama, Y. Sato, Y. Inamori, H. Takeshita, Application of Zn-compound buffer layer for polycrystalline CuInSe$_2$-based thin-film solar cells, Jpn. J. Appl. Phys. 35 (1996) 4383.

[89] K. Kushiya, M. Tachiyuki, T. Kase, Y. Nagoya, T. Miura, D. Okumura, et al., Improved FF of CIGS thin-film mini-modules with Zn(O,S,OH)$_x$ buffer by post-depostion light soaking. Proceedings of the 26th IEEE Photovoltaic Specialists Conference, Anaheim, 1997, p. 327.

[90] T. Nakada, K. Furumi, A. Kunioka, High-efficiency cadmium-free Cu(In,Ga)Se-2 thin-film solar cells with chemically deposited ZnS buffer layers, IEEE Trans. Electron. Devices (1999) 2093ED-46 (1999) 2093.

[91] T. Nakada, M. Mizutani, 18% efficiency Cd-free Cu(In, Ga)Se$_2$ thin-film solar cells fabricated using chemical bath deposition (CBD)-ZnS buffer layers, Jpn. J. Appl. Phys. 41 (2002) L165.

[92] Y. Ohtake, K. Kushiya, M. Ichikawa, A. Yamada, M. Konagai, Polycrystalline Cu (InGa)Se$_2$ thin-film solar cells with ZnSe buffer layers, Jpn. J. Appl. Phys. 34 (1995) 5949.

[93] Y. Ohtake, M. Ichikawa, A. Yamada, M. Konagai, Cadmium free buffer layers for polycrystalline Cu(In,Ga)Se$_2$ thin film solar cells. Proceedings of the 13th European Photovoltaic Solar Energy Conference, Nice, 1995, p. 2088.

[94] M. Konagai, Y. Ohtake, T. Okamoto, Development of Cu(InGa)Se$_2$ thin film solar cells with Cd-free buffer layers, Mat. Res. Soc. Symp. Proc. 426 (1996) 153.

[95] M. Powalla, E. Lotter, R. Waechter, S. Spiering, M. Oertel, B. Dimmler, Pilot line production of CIGS modules: first experience in processing and further developments. Proceedings of the 29th IEEE Photovoltaic Specialists Conference, New Orleans, 2002, p. 571.

[96] M. Hartmann, M. Schmidt, A. Jasenek, H.W. Schock, F. Kessler, K. Ilerz, et al., Flexible and light weight substrates for Cu(In,Ga)Se$_2$ solar cells and modules. Proceedings of the 28th IEEE Photovoltaic Specialists Conference, Anchorage, 2000, p. 638.

[97] A.N. Tiwari, M. Krejci, F.-J. Haug, H. Zogg, 12.8% Efficiency Cu(In,Ga)Se$_2$ solar cell on a flexible polymer sheet, Prog. Photovolt. 7 (1999) 393.

[98] S. Wiedemann, M.E. Beck, R. Butcher, I. Repins, N. Gomez, B. Joshi, et al., Proceedings of the 29th IEEE Photovoltaic Specialists Conference, New Orleans, 2002, p. 575.

[99] L.B. Fabick, A. Jehle, S. Scott, B. Crume, G. Jensen, J. Armstrong, A new thin-film space PV module technology. Proceedings of the 29th IEEE Photovoltaic Specialists Conference, New Orleans, 2002, p. 971.

[100] G.M. Hanket, U.P. Singh, E. Eser, W.N. Shafarman, R.W. Birkmire, Pilot-scale manufacture of Cu(InGa)Se$_2$ films on a flexible polymer substrate. Proceedings of the 29th IEEE Photovoltaic Specialists Conference, New Orleans, 2002, p. 567.

[101] R.R. Gay, Status and prospects for CIS-based photovoltaics, Sol. Energy Mater. Sol. Cells 47 (1997) 19.
[102] F. Karg, D. Kohake, T. Nierhoff, B. Kühne, S. Grosser, M.C. Lux-Steiner, Performance of grid-coupled PV arrays based on CIS solar modules. Proceedings of the17th European Photovoltaic Solar Energy Conference, Munich, 2002, p. 391.
[103] M. Powalla, B. Dimmler, Process development of high performance CIGS modules for mass production, Thin Solid Films 387 (2001) 251.
[104] M. Schmidt, D. Braunger, R. Schäffler, H.W. Schock, U. Rau, Influence of damp heat on the electrical properties of $Cu(In,Ga)Se_2$ solar cells, Thin Solid Films 361–362 (2001) 283.
[105] M. Igalson, M. Wimbor, J. Wennerberg, The change of the electronic properties of CIGS devices induced by the 'damp heat' treatment, Thin Solid Films 403–404 (2002) 320.
[106] C. Deibel, V. Dyakonov, J. Parisi, J. Palm, S. Zweigart, F. Karg, Electrical characterisation of damp-heat trated $Cu(In,Ga)(S,Se)_2$ solar cells. Proceedings of the 17th European Photovoltaic Solar Energy Conference, Munich, 2002, p. 1229.
[107] F. Karg, H. Calwer, J. Rimmasch, V. Probst, W. Riedl, W. Stetter, et al., Development of stable thin film solar modules based on $CuInSe_2$, Inst. Phys. Conference Ser. 152E (1998) 909.
[108] J.-F. Guillemoles, L. Kronik, D. Cahen, U. Rau, A. Jasenek, H.W. Schock, Stability issues of $Cu(In,Ga)Se_2$-based solar cells, J. Phys. Chem. B 104 (2000) 4849.
[109] J.F. Guillemoles, The puzzle of $Cu(In,Ga)Se_2$ (CIGS) solar cells stability, Thin Solid Films 403–404 (2002) 405.
[110] C.F. Gay, R.R. Potter, D.P. Tanner, B.E. Anspaugh, Radiation effects on thin film solar cells, Proceedings of the 17th IEEE Photovoltaic Specialists Conference, Kissimmee, 1984, p. 151.
[111] R.A. Mickelsen, W.S. Chen, B.J. Stanbery, H. Dursch, J.M. Stewart, Y.R. Hsiao, et al., Development of $CuInSe_2$ cells for space applications, Proceedings of the 18th IEEE Photovoltaic Specialists Conference, Las Vegas, 1985, p. 1069.
[112] A. Jasenek, U. Rau, Defect generation in $Cu(In,Ga)Se_2$ heterojunction solar cells by high-energy electron and proton irradiation, J. Appl. Phys. 90 (2001) 650.
[113] R.J. Walters, G.P. Summers, S.R. Messenger, A. Jasenek, H.W. Schock, U. Rau, et al., Displacement damage dose analysis of proton irradadiated CIGS solar cells on flexible substrates. Proceedings of the 17th European Photovoltaic Solar Energy Conference, Munich, 2002, p.2191.
[114] J.-F. Guillemoles, U. Rau, L. Kronik, H.W. Schock, D. Cahen, $Cu(In,Ga)Se_2$ solar cells: device stability based on chemical flexibility, Adv. Mat. 8 (1999) 111.
[115] A. Jasenek, H.W. Schock, J.H. Werner, U. Rau, Defect annealing in $Cu(In,Ga)Se_2$ heterojunction solar cells after high-energy electron irradiation, Appl. Phys. Lett. 79 (2001) 2922.
[116] S. Kawakita, M. Imaizumi, M. Yamaguchi, K. Kushia, T. Ohshima, H. Itoh, et al., Annealing enhancement effect by light illumination on proton irradiated $Cu(In,Ga)Se_2$ thin-film solar cells, Jpn. J. Appl Phys. 41(2) (2002) L797.
[117] A. Jasenek, U. Rau, K. Weinert, H.W. Schock, J.H. Werner, Illumination-enhanced annealing of electron irradiated $Cu(In,Ga)Se_2$ solar cells. Proceedings of the 29th IEEE Photovoltaic Specialists Conference, New Orleans, 2002, p. 872.
[118] J.R. Tuttle, A. Szalaj, J. Keane, A 15.2% AM0/1433 W/kg thin-film $Cu(In,Ga)Se_2$ solar cell for space applications. Proceedings of the 28th IEEE Photovoltaic Specialists Conference, Anchorage, 2000, p. 1042.
[119] M. Hartmann, M. Schmidt, A. Jasenek, H.W. Schock, F. Kessler, K. Herz, et al., Flexible and light weight substrates for $Cu(In,Ga)Se_2$ solar cells and modules.

Proceedings of the 28th IEEE Photovoltaic Specialists Conference, Anchorage, 2000, p. 838.
[120] A.N. Tiwari, M. Krejci, F.J. Haug, H. Zogg, 12.8% Efficiency Cu(In,Ga)Se$_2$ solar cell on a flexible polymer sheet, Prog. Photov. 7 (1999) 393.
[121] M. Turcu, U. Rau, Fermi level pinning at CdS/Cu(In,Ga)(Se,S)$_2$ interfaces: effect of chalcopyrite alloy composition, J. Phys. Chem. Solids 64 (2003), p. xxx.
[122] A.J. Nelson, C.R. Schwerdtfeger, S.-H. Wei, A. Zunger, D. Rioux, R. Patel, et al., Theoretical and experimental studies of the ZnSe/CuInSe$_2$ heterojunction band offset, Appl. Phys. Lett. 62 (1993) 2557.
[123] T. Löher, W. Jaegermann, C. Pettenkofer, Formation and electronic properties of the CdS/CuInSe$_2$ (011) heterointerface studied by synchrotron-induced photoemission, J. Appl. Phys. 77 (1995) 731.
[124] S.-H. Wei, A. Zunger, Band offsets at the CdS/CuInSe$_2$ heterojunction, Appl. Phys. Lett. 63 (1993) 2549.
[125] M. Ruckh, D. Schmid, H.W. Schock, Photoemission studies of the ZnO/CdS interface, J. Appl. Phys. 76 (1994) 5945.
[126] K. Orgassa, Q. Nguyen, I.M. Kötschau, U. Rau, H.W. Schock, J.H. Werner, Optimized reflection of CdS/ZnO window layers in Cu(In,Ga)Se2 thin film solar cells. Proceedings of the 17th European Photovoltaic Solar Energy Conference, Munich, 2002, p. 1039.
[127] F. Engelhardt, L. Bornemann, M. Köntges, Th. Meyer, J. Parisi, E. Pschorr-Schoberer, et al., Cu(In,Ga)Se$_2$ solar cells with a ZnSe buffer layer: interface characterization by quantum efficiency measurements, Prog. Photovolt. Res. Appl. 7 (1999) 423.
[128] R.H. Bube, Photolectronic Properties of Semiconductors, Cambridge University Press, Cambridge, UK, 1992.
[129] M.A. Green, Solar Cells, University of New South Wales, Sydney, Australia, 1986. p. 96
[130] W.N. Shafarman, R. Klenk, B.E. McCandless, Device and material characterization of Cu(InGa)Se$_2$ solar cells with increasing band gap, J. Appl. Phys. 79 (1996) 7324.
[131] U. Rau, M. Schmidt, Electronic properties of ZnO/CdS/Cu(In,Ga)Se$_2$ solar cells – aspects of heterojunction formation, Thin Solid Films 387 (2001) 141.
[132] A. Niemegeers, M. Burgelman, A. De Vos, On the CdS/CuInSe$_2$ conduction band discontinuity, Appl. Phys. Lett. 67 (1995) 843.
[133] M.N. Ruberlo, A. Rothwarf, Time-dependent open-circuit voltage in CuInSe$_2$/CdS solar cells: Theory and experiment, J. Appl. Phys. 61 (1987) 4662.
[134] R.A. Sasala, J.R. Sites, Time dependent voltage in CuInSe$_2$ and CdTe solar cells. Proceedings of the 23rd IEEE Photovoltaic Specialists Conference, Louisville, 1993, p. 543.
[135] K. Kushia, M. Tachiyuki, T. Kase, I. Sugiyama, Y. Nagoya, D. Okumura, et al., Fabrication of graded band-gap Cu(InGa)Se$_2$ thin-film mini-modules with a Zn(O, S,OH)$_x$ buffer layer, Sol. Energy Mater. Sol. Cells 49 (1997) 277.
[136] U. Rau, A. Jasenek, R. Herberholz, H.W. Schock, J.-F. Guillemoles, D. Lincot et al., The inherent stability of Cu(In,Ga)Se$_2$-based solar cells. Proceedings of the 2nd World Conference on Photovoltaic Energy Conversion, Vienna, 1998, p. 428.
[137] P. Zabierowski, U. Rau, M. Igalson, Classification of metastabilities in the electrical characteristics of ZnO/CdS/Cu(In,Ga)Se$_2$ solar cells, Thin Solid Films 387 (2001) 147.
[138] U. Rau, M. Schmitt, J. Parisi, W. Riedl, F. Karg, Persistent photoconductivity in Cu(In,Ga)Se$_2$ heterojunctions and thin films prepared by sequential deposition, Appl. Phys. Lett. 73 (1998) 223.

[139] Th. Meyer, M. Schmidt, F. Engelhardt, J. Parisi, U. Rau, A model for the open circuit voltage relaxation in Cu(In,Ga)Se$_2$ heterojunction solar cells, Eur. Phys. J. App. Phys. 8 (1999) 43.
[140] D.V. Lang, R.A. Logan, Large-lattice-relaxation model for persistent photoconductivity in compound semiconductors, Phys. Rev. Lett. 39 (1977) 635.
[141] A.K. Delahoy, A. Ruppert, M. Contreras, Charging and discharging of defect states in CIGS/ZnO junctions, Thin Solid Films 161–162 (2000) 140.
[142] U. Rau, K. Weinert, Q. Nguyen, M. Mamor, G. Hanna, A. Jasenek, et al., Mat. Res. Soc. Symp. Proc. 668 (2001). p. H9.1.1
[143] S. Siebentritt, Wide gap chalcopyrites: material properties and solar cells, Thin Solid Films 403–404 (2002) 1.
[144] P.D. Paulson, M.W. Haimbodi, S. Marsillac, R.W. Birkmire, W.N. Shafarman, CuIn$_{1-x}$Al$_x$Se$_2$ thin films and solar cells, J. Appl. Phys. 91 (2002) 10153.
[145] T. Walter, A. Content, K.O. Velthaus, H.W. Schock, Solar-cells based on CuIn(Se,S)$_2$, Sol. Energy Mater. Sol. Cells 26 (1992) 357.
[146] T.M. Friedlmeier, H.W. Schock, Improved voltages and efficiencies in Cu(In,Ga)(S,Se)2 solar cells. Proceedings of the 2nd. World Conference on Photovoltaic Solar Energy Conversion, Vienna, 1998, p. 1117.
[147] R. Scheer, T. Walter, H.W. Schock, M.L. Fearhailey, H.J. Lewerenz, CuInS$_2$ based thin film solar cell with 10.2% efficiency, Appl. Phys. Lett. 63 (1993) 3294.
[148] R. Kaigawa, A. Neisser, R. Klenk, M.-Ch Lux-Steiner, Improved performance of thin film solar cells based on Cu(In,Ga)S$_2$, Thin Solid Films 415 (2002) 266.
[149] R. Klenk, Characterisation and modelling of chalcopyrite solar cells, Thin Solid Films 387 (2001) 135.
[150] M. Turcu, O. Pakma, U. Rau, Interdependence of absorber composition and recombination mechanism in Cu(In,Ga)(Se,S)$_2$ heterojunction solar cells, Appl. Phys. Lett. 80 (2002) 2598.
[151] V. Nadenau, D. Hariskos, W. Schock H., CuGaSe$_2$ based thin-film solar cells with improved performance, Proceedings of the 14th European Photovoltaic Solar Energy Conference Barcelona, 1997, p. 1250.
[152] A. Jasenek, U. Rau, V. Nadenau, H.W. Schock, Electronic properties of CuGaSe$_2$-based heterojunction solar cells. Part II. Defect spectroscopy, J. Appl. Phys. Vol. 87 (2000) 594.
[153] V. Nadenau, U. Rau, A. jasenek, H.W. Schock, Electronic properties of CuGaSe$_2$-based heterojunction solar cells. Part I. Transport analysis, J. Appl. Phys. 87 (2000) 584.
[154] J. Reiß, J. Malmström, A. Werner, I. Hengel, R. Klenk, M.Ch. Lux-Steiner, Current Transport in CuInS$_2$ solar cells depending on absorber preparation, Mat. Res. Soc. Symp. Proc. 668 (2001), p. H9.4.1
[155] J.H. Schön, M. Klenk, O. Schenker, E. Bucher, Photovoltaic properties of CuGaSe$_2$ homodiodes, Appl. Phys. Lett. 77 (2000) 3657.
[156] W.N. Shafarman, S. Marsillac, P.D. Paulson, M.W. Haimbodi, R.W. Birkmire, Material and device characterization of thin film Cu(InAl)Se$_2$ solar cells. Proceedings of the 29th IEEE Photovoltaic Specialists Conference, New Orleans, 2002, p. 519.
[157] S. Marsillac, P.S. Paulson, M.W. Haimbodi, R.W. Birkmire, W.N. Shafarman, High-efficiency solar cells based on Cu(InAl)Se$_2$ thin films, Appl. Phys. Lett. 81 (2002) 1350.
[158] T. Walter, H.W. Schock, Structural and electrical investigations of the anion-exchange in polycrystalline CuIn(S,Se)2 thin-films, Jpn. J. Appl. Phys. 32(3) (1993) 116.
[159] K. Siemer, J. Klaer, I. Luck, J. Bruns, R. Klenk, D. Bräunig, Efficient CuInS$_2$ solar cells from a rapid thermal process (RTP), Sol. Energy Mater. Sol. Cells 67 (2001) 159.

[160] I. Hengel, A. Neisser, R. Klenk, M.C. Lux-Steiner, Ga/Cds/ZnO − solar cells, Thin Solid Films, 361−362, Current transport in $CuInS_2$, 2000, p. 458.
[161] J.L. Gray, Y.J. Lee, Numerical modeling of graded band gap CIGS solar cells. Proceedings of the First World Conference on Photovoltaic Solar Energy Conversion, Hawaii, 1994, p. 123.
[162] A. Dhingra, A. Rothwarf, Computer simulation and modeling of graded bandgap CuInSe2/CdS based solar cells, IEEE Trans. Electron Devices ED-43 (1996) 613.
[163] A.M. Gabor, J.R. Tuttle, M.H. Bode, A. Franz, A.L. Tennant, M.A. Contreras, et al., Band-gap engineering in $Cu(In,Ga)Se_2$ thin films grown from $(In,Ga)_2Se_5$ precursors, Sol. Energy Mater. Sol. Cells 41 (1996) 247.
[164] T. Dullweber, U. Rau, M. Contreras, R. Noufi, H.W. Schock, Photogeneration and carrier recombination in graded gap $Cu(In. Ga)Se_2$ solar cells, IEEE Trans. Electron. Dev. ED-47 (2000) 2249.

CHAPTER IC-4

Progress in Chalcopyrite Compound Semiconductor Research for Photovoltaic Applications and Transfer of Results into Actual Solar Cell Production

Arnulf Jäger-Waldau

European Commission, DG JRC, Institute for Energy Renewable Energies Unit, 21027 Ispra, Italy

Contents

1. Introduction		374
2. Research Directions		375
3. Industrialisation		379
3.1 Technology Transfer		381
3.2 Chalcopyrite Cell Production Companies		382
3.2.1	Ascent Solar Technologies Incorporated (USA)	383
3.2.2	AVANCIS GmbH and Co. KG (Germany)	383
3.2.3	DayStar Technologies (USA)	383
3.2.4	Global Solar Energy Inc. (USA, Germany)	384
3.2.5	Honda Soltec Co. Ltd. (Japan)	384
3.2.6	International Solar Electric Technology (USA)	384
3.2.7	Jenn Feng New Energy (Taiwan)	385
3.2.8	Johanna Solar Technology GmbH (Germany)	385
3.2.9	Miasolé (USA)	385
3.2.10	Nanosolar (USA)	385
3.2.11	Odersun AG (Germany)	386
3.2.12	Shandong Sunvim Solar Technology Co., Ltd. (PRC)	386
3.2.13	Solar Frontier (Japan)	386
3.2.14	Solarion AG (Germany)	387
3.2.15	Solibro GmbH (Germany)	387
3.2.16	Solo Power Inc. (USA)	388
3.2.17	Solyndra (USA)	388
3.2.18	Sulfurcell Solartechnik GmbH (Germany)	388
3.2.19	Sunshine PV Corporation (Taiwan)	389

Practical Handbook of Photovoltaics.
© 2012 Elsevier Ltd. All rights reserved.

    3.2.20 Tianjin Tai Yang Photo-Electronic Technology Co. Ltd. (PRC)   389
    3.2.21 Würth Solar GmbH (Germany)   389
4. Conclusions and Outlook   390
References   390

1. INTRODUCTION

Since a number of years photovoltaics continues to be one of the fastest-growing industries with growth rates well beyond 40% per annum. This growth is driven not only by the progress in materials and processing technology, but by market support programmes in a growing number of countries around the world [1].

Since 1988, world wide solar cell production has increased from about 35 MW to 11.5 GW in 2009 or more than 340 times (Figs. 1 and 2). Between 1988 and the mid 1990s, PV production exhibited a moderate growth rate of about 15%. With the introduction of the Japanese 70,000 roof programme in 1997, the growth rate more than doubled for the end of the decade. Since the introduction of the German feed-in tariff scheme for PV generated electricity in 2000, total PV production increased more than 30-fold, with annual growth rates between 40% and 80%.

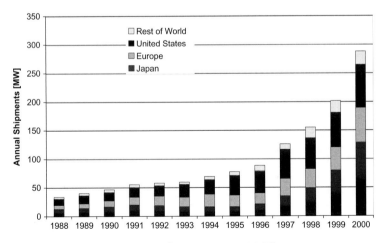

Figure 1 World solar cell production from 1988 to 2000 [2].

Wafer-based silicon solar cells is still the main technology and had around 80% market shares in 2009, but thin-film solar cells are continuously increasing their share since 2005. Back then, production of thin-film solar modules reached for the first time more than 100 MW per annum. Since then, the *Compound Annual Growth Rate* (CAGR) of thin-film solar module production was even beyond that of the overall industry, increasing the market share of thin-film products from 6% in 2005 to 10% in 2007 and 16—20% in 2009.

Amongst the different thin film technologies, the family of chalcopyrites has so far demonstrated the highest efficiency for solar cells in the laboratory as well as in production. This review gives an overview of the long-term R&D efforts to realise this as well as the industry activities to commercialise chalcopyrite solar cells. The specific details of research for chalcopyrite solar cells will be left to the specialised Chapter Ic-3.

2. RESEARCH DIRECTIONS

The chalcopyrite compound family has two subclasses, which are based on the use of elements from group I, III, and VI or II, IV, and V of

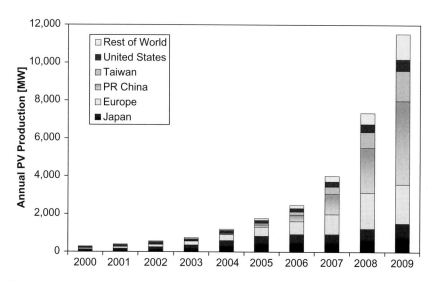

Figure 2 World solar cell production from 2000 to 2009 [1,3—5].

the periodic table and are derived from the III—V and II—VI compounds, respectively. The first ternary semiconductor device was already made in the 1920s. It was a point-contact diode made out of a mineral chalcopyrite ($CuFeS_2$) [6]. In the early 1950s the search for new semiconductor materials led to investigations of the first III—V materials focused on finding materials with a high carrier mobility and an amphoteric doping [7,8]. Soon the search was broadened to ternary compounds and the artificial synthesis of the two subgroup families led to intensive studies of their physical properties [9—12]. In the 1960s and 70's electrical, optical and structural properties of the $(Cu,Ag)(Al,Ga,In)(S,Se,Te)_2$ as well as the $(Zn,Cd)(Si,Ge,Sn)(P,As)_2$ family were investigated [13—15].

The large number of the ternary compounds and their corresponding variety of properties like the variation of band-gaps ranging from 0.26 eV ($CdSnAs_2$) to 3.5 eV ($CuAlS_2$), the possibility to substitute certain elements like In by Ga or Se by S, as well as the possibility to synthesise the material n- or p-type, resulted in the development of a number of novel devices at the end of the 1960s and early 1970s [14,16,17].

Initially they were explored for their non-linear optical properties [14]. Chalcopyrites were investigated with respect to their use a frequency mixing devices like the parametric oscillator, harmonic generator as well as up and down converters. Frequency mixing materials are $ZnGeP_2$, $CdGeAs_2$, $AgGaS_2$, $AgGaSe_2$, $ZnSiAs_2$, $AgGaTe_2$, and Proustite (Ag_3AsS_3) [18]. Another non-linear optical device application investigated was the use of chalcopyrites as electro-optic modulators.

The fact that a large number of chalcopyrites exhibit direct bandgaps and relatively narrow luminescent bands (sometimes at low temperatures only) made them an interesting candidate to obtain stimulated emission/laser action. It could be demonstrated that stimulated emission could be achieved in six of the ternary compounds by brute force pumping with either a high-voltage electron beam [19,20] or laser pumping [21—23]. The chalcopyrites in question were $CdSiAs_2$, $CdSnP_2$, $CdSnP_2$:Ag, $CuGaS_2$, $AgGaS_2$, and $AgGaSe_2$.

The ability of some of the ternary compounds to be synthesised with both n- and p-type conduction and a low resistivity was used to form the first homojunctions in $CdSnP_2$ [24], $CuInS_2$ [23], and $CuInSe_2$ [25] by dopant diffusion or stoichiometry variation and vapour annealing. The stoichiometry variation and vapour anneal technique was used to produce the first ternary homodiode, which could be characterised in detail [26]. A n-type $CuInSe_2$ crystal was treated with a short annealing in Se vapour to make a junction a short distance below the crystal surface. Then the crystal was etched to a mesa structure. For ±1 V a rectification ratio of 300:1 was obtained. Electroluminescence could be obtained with a forward bias centred on 1.34 µm and with 10% internal quantum efficiency at 77 K.

The first ternary heterodiode was reported in 1970 based on a n-type $CdSnP_2$ single crystal combined with p-type Cu_2S [27]. Despite a lattice mismatch of around 5% the device exhibited a large photovoltaic response under white light illumination. The focus of investigations changed after the first $CuInSe_2$/CdS solar cell device with 12.5% efficiency was produced at the Bell Laboratories in 1974 [28]. This device was based on a p-type $CuInSe_2$ single crystal onto which a thin n-type CdS layer was deposited. The result was outstanding at the time and by far better than the results achieved with thin film devices, which at that time reached around 6% [29,30]. Until the early 1980s there was a limited interest in this type of solar cell [31,32] until the first polycrystalline $CuInSe_2$ thin-film solar cell with an efficiency of 9.4% was made in 1980 [33].

The fact that the representatives of the $Cu(In,Ga)(S,Se)_2$ family have different bandgaps ranging from 1.04 eV ($CuInSe_2$) over 1.54 eV ($CuInS_2$) to 1.68 eV for $CuGaSe_2$ and that the band-gaps can be engineered by substituting In with Ga and Se with S led to the development of $Cu(In,Ga)Se_2$ or $Cu(In,Ga)(S,Se)_2$ with slightly higher band-gaps of around 1.2 eV. This increase in the band-gap enhanced the voltage performance for this heterojunction solar cell.

In the early 1980s significant progress was made when the Boeing group invented the bilayer process, where a Cu-rich and In-rich layer of $CuInSe_2$ is deposited sequentially [34]. The process was to co-evaporate Cu, In, and Se onto a molybdenum-coated glass substrate. The In-rich layer close to the heterojunction interface prevented the formation of copper nodules. In addition, the Boeing group increased the band-gap of the window layer by using $Cd_xZn_{1-x}S$ instead of CdS and $Cu(In,Ga)Se_2$ instead of $CuInSe_2$ which led to the then record efficiency of 14.6%. During this time the EUROCIS activities, a series of EU-funded projects, started in Europe. In August 2010 the Centre for Solar and Hydrogen Research in Stuttgart, Germany (ZSW), reported the latest world record of a 20.3% efficient $Cu(In,Ga)Se_2$ small, area solar cell, a further increase from the 20.1% reported earlier that year [35].

In the following years research activities focused on the increase of small size solar cell efficiencies as well as on the up-scaling of the active solar cell areas in order to move the system to a production stage.

Despite a wide range of production methods, there are three methods which have dominated both research and large-scale production:
1. co-evaporation of elements,
2. selenisation/sulphidisation of elemental precursor layers, and
3. stacked elemental layer (SEL) processing.

The first method was used by Mickelson and Chen [33], in 1981 to produce the first 10% all-thin-film cell based on the use of $CuInSe_2$. The second method was initiated by Basol and Kapur in 1989 [36]. The third method was developed at Newcastle Polytechnic, now Northumbria University, by Carter et al. [37].

Over the years, many groups across the world have developed CIGS solar cells with efficiencies in the range 15−19%, depending on different growth procedures. The most commonly used substrate is glass, but significant efforts are being made to develop flexible solar cells on polyimide [38−46] and metal foils [47−59]. Highest efficiencies of 17.6%, 17.7%, and 17.9% have been reported for CIGS cells on polyimide [60], ceramics, [61] and metal foils [55], respectively.

Besides the research activities focusing on the understanding of the fundamental material and device properties and the industrialisation of $Cu(In,Ga)Se_2$ or $Cu(In,Ga)(S,Se)_2$, the compound family was investigated for using different members either for multijunction solar cell concepts (e.g., $CuGaSe_2$) and concentrator concepts (single and tandem). The highest concentrator efficiency was reported in 2002 with 21.5%, but the result was recalibrated to 21.8±1.5% in the efficiency tables [62]. The first concepts end experimental result of thin-film tandem solar cells based on $CuInSe_2$ in combination with $CuGaSe_2$ or $(CdHg)Te$ were presented as early as 1985 [63,64]. A combination of thin-film silicon and $CuInSe_2$ 4-terminal tandem cell (4 cm^2) with 15.6% efficiency and a 30- × 30-cm^2 module with 12.3% aperture efficiency were reported as early as 1988 [65].

Another rather recent approach for the development of a multijunction device is the application of a dye sensitised cell (DSC) on top of the CIGS. In 2006 an efficiency of 15.09% was reported for a DSC/CIGS two terminal device [66].

$CuInS_2$ has a direct band-gap of 1.54 eV, near the optimum for solar energy conversion. Research activities on this material for photovoltaic applications already started in the late 1970s, when a photoelectrochemical cell with about 3% efficiency was realised [67]. Early works on $CuInS_2$ solar cells included the formation of two-source evaporated thin-film homojunctions with 3.3−3.6% efficiency [68] as well as heterojunctions with sprayed tin oxide/CuInS$_2$ [69,70] and sprayed (ZnCd)S/CuInS$_2$ [71] with 2.0% to 2.9% efficiency.

In 1986 a 9.7% electrochemical solar cell using a n-type CuInS$_2$ crystal was realised at the Hahn-Meitner-Institute in Berlin using a I$^-$/I$_3^-$ T-HCl redox electrolyte [72]. This was a first proof that the material could be used as a good solar cell absorber material.

The next step was to realise this efficiency with CuInS$_2$ thin films. In 1988 a 7.3% efficient ZnO/thin CdS/CuInS$_2$ cell was reported [73], and the first CuInS$_2$ thin-film solar cell exceeding 10% followed in 1993 [74]. The record efficiency of CunS$_2$ was reported with 11.4% in 2001 [75], and a further increase in efficiency was due to the widening of the band gap by adding Ga resulting in Cu(In,Ga)S$_2$ with 12.9% efficiency [76].

One of the latest developments is the investigation of chalcopyrites as material for thin-film intermediate solar cells [77].

3. INDUSTRIALISATION

The early days of PV manufacturing were dominated by investments or buy-outs of the big oil companies in small start-up or research and development companies, and with this move they took control of the research and patents. Therefore, it is no surprise that already from very early on scientific findings in the field of chalcopyrite solar cells were protected by patents as early as 1982. Table 1 gives an overview on the early US patents covering chalcopyrite solar cells.

As already mentioned earlier, the first CuInSe$_2$ solar cell with 9.4% efficiency was made by the Boeing research lab [33]. The patents held by Boeing made it clear that they were active in the field of space applications. They actively participated in the Polycrystalline Thin-Film Programme at the National Renewable Energy Laboratory (NREL), formerly the Solar Energy Research Institute, but did not move beyond mini-modules into a terrestrial module manufacturing stage.

Despite the fact that Matsushita Electric and Yazaki Corporation were quite active to develop CIS solar cells in the 1990s and earlier this decade, no manufacturing activities emerged. However, it is interesting to note that despite the fact that the number of patents filed in the second half of the last decade has increased sharply, these two companies still hold the largest number of patents in this field followed by Shell and Fujielectric.

Table 1 Early US patents on CIS

Company	Period of granting	Topics	Reference
The Boeing Company, Seattle, WA	1982–1993	Window layer, multi-junction, module design, deposition, interconnection, light weight	[78]
Atlantic Richfield Company (Arco Solar), Los Angeles, CA	1984–1991	Co-deposition by magnetron sputtering, window layer, multi-junction cells	[79]
Matsushita Electric Industrial Co., Ltd., Kadoma, Japan	1990–1998	Precursor deposition, deposition equipment, window layer, co-evaporation, precursor pastes, module sealing	[80]
Siemens Solar and other Siemens companies	1992–1997	Sequential deposition; sodium control, module design, window layers, TCO, multi junction	[81]
Yazaki Corporation, Tokyo, Japan	1996–2001	Deposition system, electro-deposition	[82]

Since 1985 about 10% of all patens filed in the field of thin film solar cells deal with the chalcopyrite compounds CIS/CIGS.

The Atlantic Richfield Company (ARCO) was one of the early movers to work on the commercialisation of CIS solar modules for terrestrial applications. The research work at the company to develop the CIS module technology to take advantage of CIS cell performance resulted in the achievement of a 30- × 30-cm^2 module with an aperture efficiency of 11.7% by 1988 [83]. In 1987 Arco Solar started to co-operate with Siemens Solar in the field of thin film solar cells (a-Si and CIS) and in February 1989 Siemens Solar completely takes over Arco Solar. Already in 1991, the company could present stability data of their mini-modules obtained by outdoor exposure testing for over 1000 days and accelerated lifetime testing [84].

Despite the promising results reported by Siemens Solar in 2001, it took another 6 years until the first commercial 5 and 10 W modules were available on the market in 1997. An expansion of the product portfolio to 20- to 40-W modules took place in 1998. At that stage a 3 MW pilot plant was operational.

In 1994 about 80 companies with a total production capacity of 130 MW existed worldwide and their activities ranged from research to production of solar cells. About half of them were actually manufacturing. Another 29 companies were involved in module production only. Out of the solar cell companies, 41 companies used crystalline silicon, 2 ribbon silicon, 19 amorphous silicon, 3 CdTe, 5 CIS (of which one was in Europe), and 10 companies worked on other concepts like III–V concentrator cells or spherical cells.

Since then, the total number of PV companies has increased manifold, and more than 300 companies with more than 450 manufacturing plants with a total 2010 production capacity exceeding 36, 11.5 GW thin films, were announced worldwide [1].

More than 200 companies are involved in thin-film solar cell activities, ranging from basic R&D activities to major manufacturing activities, and over 150 of them have announced the start or increase of production. More than 100 companies are silicon based and use either amorphous silicon or an amorphous/microcrystalline silicon structure. Over 30 companies announced using $Cu(In,Ga)(Se,S)_2$ as absorber material for their thin-film solar modules, whereas 9 companies use CdTe and 8 companies go for dye and other materials.

In 2009, a total of 166 MW of CIGSSe solar modules was manufactured worldwide, representing about 1.5% of the annual production. Compared to the announced capacity expansions in other technologies, the expansion to 3.4 GW for CIGSSe looks small, but seems realistic (Fig. 3).

3.1 Technology Transfer

A crucial part in the industrialisation of the $Cu(In,Ga)(Se,S)_2$ material family for thin-film solar cells was and still is played by a number of research institutions, which do not only conduct basic material research but focus on process development and up-scaling issues as well.

Besides a larger number of university institutes, which are very active in fundamental material and/or device research there are a few research institutes, which are working on pre-pilot or pilot stage development of process steps for the commercialisation of the $Cu(In,Ga)(Se,S)_2$ material family for thin-film solar cells. The importance of this research activities should not be underestimated as it was and is the foundation of a number of thin-film companies which are in the process of commercialisation the $Cu(In,Ga)(Se,S)_2$ technology.

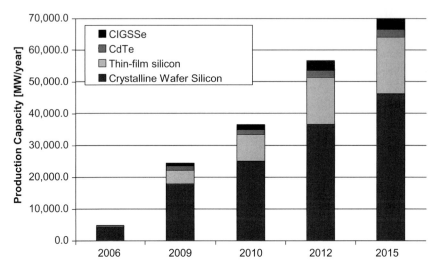

Figure 3 Actual and planned PV production capacities of different solar cell technologies [1].

The most well-known institutes are the Ångström Solar Center, Uppsala University, Sweden; the Helmholtz Zentrum Berlin (former Hahn-Meitner-Institute), Germany; the Institute of Energy Conversion (IEC), USA; the National Institute of Advanced Industrial Science and Technology (AIST), Japan; the National Renewable Energy Laboratory (NREL), USA; and the Zentrum für Sonnenenergie- und Wasserstof-Forschung (ZSW), Germany.

3.2 Chalcopyrite Cell Production Companies

Compared with the silicon PV or the leading CdTe company, those companies manufacturing $Cu(In,Ga)(S,Se)_2$ solar cells are still small or at an early stage, but the anticipated growth rates are significant, and it will be interesting to see how this market segment develops. The following chapter gives a short description of these companies and the basic manufacturing process they are using. This listing does not claim to be complete, especially due to the fact that the availability of information or data for some companies were very fragmentary. The given manufacturing process info are only those publicly available. The capacity, production, or shipment data are from the annual reports or financial statements of the respective companies or the cited references.

3.2.1 Ascent Solar Technologies Incorporated (USA)
The company was established in 2005 to manufacture CIGS thin-film solar modules with a roll-to-roll process. According to the company, it is on track to commence full-scale production on their 1.5 MW pilot line by the end of 2008. A 30-MW production line was completed in 2009, and it is planned to ramp up to full capacity during 2010. For 2012 the company plans to increase production capacity to 110 MW.

3.2.2 AVANCIS GmbH and Co. KG (Germany)
In 2002 Shell Solar acquired Siemens Solar including the CIS technology and continued with the development of the manufacturing technology. In 2006 Shell Solar sold the silicon production facilities to Solarworld and formed a joint venture company for the development, production, and marketing of the next generation of CIS together with Saint-Gobain called AVANCIS.

The manufacturing sequence starts with the soda-lime glass covered by a silicon nitride (SiN) alkali-barrier layer and the molybdenum (Mo) back electrode as a substrate. The first step in the CIGSSe absorber formation is the elemental precursor film deposition consisting of DC-magnetron sputtering of Cu−In−Ga:Na and thermal evaporation of selenium (Se). The second step is the reaction of the elemental precursor stack to form the CIGSSe semiconductor. Rapid thermal processing (RTP) is conducted in an infrared heated furnace capable of high heating rates in a sulphur containing ambient. The CdS buffer layer is deposited in a chemical bath deposition process, the ZnO:Al window layer by magnetron sputtering [85].

In 2008 commercial production started in the new factory with an initial annual capacity of 20 MW in Torgau, Germany. In 2009 Saint-Gobain took over the shares of Shell and started the construction of a second CIS factory with a total capacity of 100 MW in Torgau. Production in 2009 is estimated at 15 MW.

3.2.3 DayStar Technologies (USA)
The company was founded in 1997 and conducted an Initial Public Offering in February of 2004. Products are *LightFoil*^{TM} and *TerraFoil*^{TM} thin-film solar cells based on CIGS. In addition, DayStar has its patented ConcentraTIR^{TM} (Total Internal Reflection) PV module, which has been

designed to incorporate a variety of cell material components, including wafer-Si, Spheral Si, thin-film CIGS, and a-Si.

3.2.4 Global Solar Energy Inc. (USA, Germany)

GSE is located in Tucson and was established in 1996. In 2006, German module manufacturer, SOLON AG, acquired a 19% stake in Global Solar Energy Inc. The remaining 81% are owned by an European venture capital investor. The company is producing thin-film photovoltaic CIGS solar cells for use in solar products, as well as installing and managing large solar photovoltaic systems.

A molybdenum back contact layer is deposited onto the stainless steel film by means of sputtering. All the steps follow a roll-to-roll procedure. Copper, indium, gallium, and selenium raw materials are deposited by means of vaporisation sources whose design is primarily a result of proprietary know-how. A very thin buffer layer is then applied onto the CIGS layer. The top layer of the solar cell is a transparent but conductive layer, which is sputtered.

According to the company, the new 40-MW plant was opened in March 2008 and 35-MW plant in Germany opened in the autumn of 2008. GSE aims for 175-MW production capacity in 2010 [86]. In 2009, about 20-MW production was estimated.

3.2.5 Honda Soltec Co. Ltd. (Japan)

Honda R&D Co. Ltd. developed a CIGS thin-film module with a power output of 112 W. To commercialise the product, Honda Soltec Co. Ltd was established on 1 December 2006. Since June 2007, the company has been selling 125-W modules produced by Honda Engineering Co. Ltd. and announced that the mass production at the Kumamoto Plant, with an annual capacity of 27.5 MW, started its production in November 2007 [87].

3.2.6 International Solar Electric Technology (USA)

In 1985 the company was established in California. Since then ISET has conducted advanced technology R&D on Copper Indium Gallium Selenide (CIGS) solar cells. In 2006 ISET began with the development of a production facility for manufacturing printed CIGS photovoltaic modules. Current annual pilot plant manufacturing capacity is given with 3 MW and according to the company an expansion to 30 MW is in the planning stage.

3.2.7 Jenn Feng New Energy (Taiwan)

The company belongs to the Jenn Feng Group, an industry group which manufactures power tools, HID lights, LED lights, and solar modules based on silicon and CIGS. In addition, the company manufactures production lines for CIGS. The company developed its own nano-particle "Chemical Wet Process" for CIGS absorb layer deposition. According to the company the first production line with 30-MW capacity started mass production in December 2009, and an expansion with a second line is planned for 2010.

3.2.8 Johanna Solar Technology GmbH (Germany)

In June 2006 the company started to build a factory for copper indium gallium sulphur selenide (CIGSSE) thin-film technology in Brandenburg/Havel, Germany. The technology was developed by Prof. Vivian Alberts at the University of Johannesburg. The molybdenum back contact is deposited by DC-magnetron sputtering. After the first of three scribing steps for integrated series connection of the cells several alternating layers of a copper–gallium alloy and elemental indium are DC-magnetron sputtered. This precursor layer stack is exposed to a reactive atmosphere in which the absorber is formed [88].

The company built up a production line with a nominal capacity of 30 MW. In March 2008 the company granted a licence to the Chinese company Shandong Sunvim Solar Technology Co. Ltd. for the construction of a thin-film solar module production plant. In November 2008 the solar cell production started and in August 2009, the Robert Bosch GmbH purchased the company.

3.2.9 Miasolé (USA)

The company was formed in 2001 and produces flexible CIGS solar cells on a continuous, roll-to-roll production line. The company has installed two 20 MW production lines in its Santa Clara facility. In January 2010, the company announced that it installed a second 20-MW line and another 60-MW are planned, bringing the total capacity to 100 MW at the end of 2010 [89].

3.2.10 Nanosolar (USA)

The company was founded in 2001 and is based in Palo Alto. It is a privately held company with financial-backing of private-technology-investors. According to the company, Nanosolar developed nanotechnology and high-yield high-throughput process technology for a proven thin-film solar device technology based on GIGS. The company made headlines when it announced on 21 June 2006 that it has secured $100 million in funding and intends to build a 430 MW thin-film factory [90]. In September 2009 the company announce the completion of its European 640 MW panel-assembly factory [91]. Production for 2009 was estimated at 12 MW.

3.2.11 Odersun AG (Germany)

The company was founded in 2002 and developed a unique thin-film technology for the production of copper-indium-sulphide—based solar cells. The technology uses consecutive reel-to-reel processes. In the first process, a Cu-tape chemically cleaned followed by a variety of rinsing processes. Then indium is electrochemically deposited on the front side of the tape only. A solid Cu-In-S layer is formed by partial conversion of the In—Cu precursor into the CISCuT-absorber when the tape is exposed to reactive gaseous sulphur inside a sulphurisation reactor. Then the absorber layer surface must be treated with a KCN-solution to remove $Cu_{2-x}S$ from the surface before the tape is annealed. A wide-band-gap p-type CuI buffer layer is chemically deposited before a TCO stack is deposited by DC sputtering as a transparent front contact [92].

The main investor is Doughty Hanson Technology Ventures, London, and the company has signed an agreement with Advanced Technology and Materials Co. Ltd., which is listed on the Shenzhen Stock Exchange to co-operate in August 2004. The first production line was inaugurated on 19 April 2007. On 26 March 2008 the company laid the cornerstone for its 30-MW expansion project. The first 20-MW phase of this expansion was inaugurated in June 2010.

3.2.12 Shandong Sunvim Solar Technology Co., Ltd. (PRC)

The company is a subsidiary of Sunvim Group Co., Ltd. and is located in Gaomi, China. The company took a licence for manufacturing CIGSSe solar cells from Johanna Solar and its first 30-MW production line became operational in 2010. According to the company, an increase of the production capacity to 240 MW is planned without a date specified.

3.2.13 Solar Frontier (Japan)

Solar Frontier is a 100% subsidiary of Showa Shell Sekiyu K.K. In 1986 Showa Shell Sekiyuki started to import small modules for traffic signals, and started module production in Japan, co-operatively with Siemens (now Solar World). The company developed CIS solar cells and completed the construction of the first factory with 20-MW capacity in October 2006. The baseline device structure is a ZnO:B (BZO) window by a metal-organic chemical vapor deposition/Zn(O, S, OH)$_x$ buffer by a chemical bath deposition/CIS-based absorber by a sulphurisation after selenization (SAS)/Mo base electrode by a sputtering/glass substrate with three patterns for interconnections [93].

Commercial production started in FY 2007. In August 2007 the company announced the construction of a second factory with a production capacity of 60 MW to be fully operational in 2009 [94]. In July 2008 the company announced to open a research centre *"to strengthen research on CIS solar powered cell technology, and to start collaborative research on mass production technology of the solar modules with Ulvac, Inc."* [95]. The aim of this project is to start a new plant in 2011 with a capacity of 900 MW. In 2010 the company changed its name to Solar Frontier, and production was given with 43 MW [96].

3.2.14 Solarion AG (Germany)

In 2000 the company was founded as a spin-off the Institute for Surface Modification (IOM) in Leipzig, Germany, with the aim to produce flexible CIGS solar cells. On a flexible polyimide web a molybdenum layer is sputtered. Then the elements Cu, In, and Ga are coevaporated, and Selenium is supplied by an ion beam source. A CdS buffer layer is deposited using chemical bath deposition and i-ZnO and ZnO:Al layers are sputter on the web as front contact [97].

In 2003 the company starts with the construction of the roll-to-roll pilot plant and the transfer of the laboratory process into an industrial roll-to-roll technology, which results in the manufacturing of first modules in 2009. According to the company they are in the planning stage for an industrial-scale manufacturing plant for flexible solar cells with a production capacity greater than 20 MW per year.

3.2.15 Solibro GmbH (Germany)

The company was established in early 2007 as a joint venture between Q-Cells AG (67.5%) and the Swedish Solibro AB (32.5%). In 2009 the company became a 100% subsidiary of Q-Cells. The company develops thin-film modules based on a Copper Indium Gallium Diselenide (CIGS) technology. The CIGS layer is deposited onto a molybdenum sputter-coated glass substrate using a co-evaporation method (PVD). Then a CdS buffer layer is deposited by chemical bath deposition before a ZnO layer as front contact is deposited.

A first production line in Thalheim, Germany, with a capacity of 30 MWp, started test production in April 2008. The line was expanded to 45 MW. A second line, with 90 MW was built in 2009 and the ramp-up started in the second half of 2009 and will continue throughout 2010. Solibro produced about 14 MW in 2009.

3.2.16 Solo Power Inc. (USA)

The company was founded in 2006 and is a California-based manufacturer of thin-film solar photovoltaic cells and modules based on CIGS. The company developed its own electrochemical roll-to-roll process. In June 2009, the company received certification under ANSI/UL 1703 standard. In April 2010, NREL confirmed that flexible CIGS solar panels manufactured on Solo Power's pilot production line have achieved aperture conversion efficiencies of 11% [98]. According to the company, current capacity is 10 MW, with an expansion of 75–100 MW in planning.

3.2.17 Solyndra (USA)

The company was founded in 2005 and produces PV modules using their proprietary cylindrical CIGS modules and thin-film technology. The company operates a state-of-the-art 300,000-square-foot factory, which would allow production of up to 100 MW. For 2009 a production of 30 MW was reported [99]. In 2010 the company announced to increase their total production capacity to 300 MW by 2011 [100].

3.2.18 Sulfurcell Solartechnik GmbH (Germany)

The company was incorporated in June 2001 to commercialise the $CuInS_2$ technology developed at the Hahn-Meitner-Institute (now "Helmholtz-Zentrum Berlin"), Berlin and is jointly owned by its founders and investing partners. In 2004, the company set up a pilot plant to scale up the copper indium sulphide (CIS) technology developed at the Hahn-Meitner-Institut, Berlin. Molybdenum-coated soda-lime glass is used as substrate material. The absorber deposition process starts with sputtering of copper and indium precursor layers and a subsequent rapid thermal anneal (RTP) in a sulphur containing atmosphere. The absorber is completed by a window layer consisting of a CdS buffer layer and a sputtered ZnO front contact [101].

First prototypes were presented at the 20th PVSEC in Barcelona in 2005. Production of CIS modules started in December 2005 and in 2006 the company had sales of 0.2 MW. For 2007, a production increase to 1 MW and in 2008, an increase to 5 MW was planned. The first 35-MW expansion phase was completed in October 2009. The new production site can be expanded to 75 MW.

In September 2010, the company announced the production of the first CIGSe solar modules and plans to convert part of their production facilities to this new product [102].

3.2.19 Sunshine PV Corporation (Taiwan)
The company was founded in May 2007 and is a subsidiary of Solartech Energy located in Hsinchu Industrial Park. According to the company their production capacity for their 120-W modules will reach 7 MW at the end of 2010. An expansion to 90 MW is scheduled to be completed in 2013.

3.2.20 Tianjin Tai Yang Photo-Electronic Technology Co. Ltd. (PRC)
The company is a joint-stock company with integrated research and manufacturing and was founded in June 2007. The aim is to industrialise the results of the CIGS thin film solar cell research of Nankai University performed under the national "863 key project" and the "Tianjin key scientific project."

According to the company it is foreseen to expand the 1.5-MW pilot line to a large-scale production facility in 2010 with a single line capacity of 12 MW by 2012. At the same time, a production line for flexible substrate-based CIGS thin-film solar cells will be developed, which should lead to a 100-MW production capacity.

3.2.21 Würth Solar GmbH (Germany)
Würth Solar GmbH and Co. KG was founded in 1999 with the aim of building up Europe's first commercial production of CIS solar modules. The company is a joint venture between Würth Electronic GmbH and Co. KG and the Centre for Solar and Hydrogen Research (ZSW). Pilot production started in the second half of the year 2000, a second pilot factory followed in 2003, increasing the production capacity to 1.3 MW. The Copper Indium Selenide (CIS) thin-layer technology was perfected in a former power station to facilitate industrial-scale manufacture. The company uses a co-evaporation method to deposit the CIGSe absorber on molybdenum-coated glass.

In August 2008 the company announced the successful ramp-up of their production facilities to 30 MW [103]. For 2009 a production volume of 30 MW is estimated.

In July 2010 the company announced the exclusive licensing of their technology to Manz Automation, an equipment manufacturer, who will be offering integrated production solutions to third parties.

4. CONCLUSIONS AND OUTLOOK

Since the invention of the first CIS solar cell in 1974, the technology has come a long way with respect to efficiency increase and commercialisation.

Thin film CIGS solar cells have exceeded 20% efficiency in 2010 and are almost at the same level as polycrystalline solar cells now. The ability to combine the $Cu(In,Ga)(S,Se)_2$ materials family internally as well as with other materials to form multijunctions leaves the pathway open to further efficiency increases, especially for concentrator concept. However, to continue this road, further fundamental research to understand the basic material properties of the respective components as well as the interface behaviours is of great importance.

Despite the current relative low market share, the future of CIGSSe manufacturing looks very promising and the anticipated production capacity of 3.4 GW by 2015 would represent about 5% of the announced total production capacity. On the other hand, such a capacity would be sufficient to supply about 10% of the then anticipated annual world market, provided that the cost structure is taking full advantage of the advantages of this thin-film technology.

According to a number of market studies, CIGSSe should benefit from the commercialisation of high throughput manufacturing technologies in the next two to three years and it is expected, that manufacturing costs could be in the range of $0.65/W to $0.55/W by 2015, with the lower end very close to the expected cost structure ($0.55/W) of the current CdTe cost market leader.

REFERENCES

[1] A. Jäger-Waldau, PV Status Report 2010, Office for Official Publications of the European Union, EUR 24344 EN, ISBN:978-92-79-15657-1.

[2] P.D. Maycock, P.V. News, 2003, ISSN 0739-4829.
[3] Paula Mints, Manufacturer shipments, capacity and competitive analysis 2009/2010, in: Navigant Consulting Photovoltaic Service Program, Palo Alto, CA.
[4] Paula Mints, The PV industry's black swan, in: Photovoltaics World, March 2010.
[5] P.V. News, Greentech Media, May 2010, ISSN 0739-4829.
[6] E.T. Wherry, Radio detector minerals, Am. Mineral. 10 (1925) 28−31.
[7] N.A. Goryunova, A.P. Obukhov, Studies on AIIIBV type (InSb, CdTe) compounds, in: Presentation at the Seventh All-Union Conference on Properties of Semiconductors, Kiev, 1950.
[8] H. Welker, Über neue halbleitende Verbindungen, Z. Naturforsch. 7a (1952) 744−749.
[9] H. Hahn, G. Frank, W. Klingler, A.-D. Meyer, G. Störger, Über einige ternäre Chalkogenide mit Chalcopyritestruktur, Z. Anorg. Allg. Chem. 271 (1953) 153−170.
[10] H.L. Goodman, R.W. Douglas, New semiconducting compounds of diamond type structure, Physica 20 (1954) 1107−1109.
[11] H.L. Goodman, A new group of compounds with diamond type (chalcopyrite) structure, Nature 179 (1957) 828−829.
[12] H. Pfister, Kristallstruktur von ternären Verbindungen der Art $A^{II}B^{IV}C^{V}$, Acta Crystallogr 11 (1958) 221−224.
[13] E. Parthé, Crystal Chemistry of Tetrahedral Structures, first ed., Gordon and Breach, New York, 1964.
[14] J.L. Shay, J.H. Wernick, The science of the solid state, in: B.R. Pamplin (Ed.), Ternary Chalcopyrite Semiconductors: Growth, Electronic Properties and Applications, vol. 7, Pergamon Press, New York, 1975.
[15] B.R. Pamplin, T. Kiyosawa, K. Masumoto, Ternary chalcopyrite compounds, Prog. Cryst. Growth Charact. 1 (1979) 331−387.
[16] N.A. Goryunova, Composite Diamond-Like Semiconductors, (R) Sowetskoje Radio, Moscow, 1968.
[17] L.I. Berger, V.D. Prochukhan, Ternary Diamond-Like Semiconductors, Consultants Bureau, New York, 1969.
[18] R.C. Smith, Device applications of the ternary semiconducting compounds, J. Phys 36(C3) (1975) 89−99.
[19] M. Berkovskii, A. Goryunova, V.M. Orlov, S.M. Ryvkin, I. Sokolova, V. Tsvetkova, G.P. Shpen'kov, $CdSnP_2$ laser excited with an electron beam, Sov. Phys.-Semicond. 2 (1969) 1027−1028.
[20] G.K. Averkirvak, N.A. Goryunova, V.D. Prochukhan, S.M. Ryvkin, M. Serginov, Yu.G. Shreter, Stimulated recombination radiation emitted by CdSiAs2, Sov. Phys.-Semicond. 5 (1971) 151−152.
[21] J.L. Shay, W.D. Johnston, E. Buehler, J.H. Wernick, Plasmaron coupling and laser emission in Ag-doped $CdSnP_2$, Phys. Rev. Lett. 27 (1971) 711−714.
[22] J.L. Shay, L.M. Schiavone, E. Buehler, J.H. Wernick, Spontaneous and stimulated emission spectra of $CdSnP_2$, J. Appl. Phys. 43 (1972) 2805−2810.
[23] J.L. Shay, B. Tell, H.M. Kasper, Visible stimulated emission in ternary chalcopyrite sulfides and selenides, Appl. Phys. Lett. 19 (1971) 366−368.
[24] E. Buehler, J.H. Wernick, J.L. Shay, The CdP_2-Sn system and some properties of $CdSnP_z$ crystals, Mater. Res. Bull. 6 (1971) 303−310.
[25] J. Parkes, R.D. Tomlinson, M.J. Hampshire, The fabrication of p and n type single crystals of $CuInSe_2$, J. Cryst. Growth 20 (1973) 315−318.
[26] P. Migliorato, B. Tell, J.L. Shay, H.M. Kasper, Junction electroluminescence in $CuInSe_2$, Appl. Phys. Lett. 24 (1971) 227−228.

[27] N.A. Goryunova, A.V. Anshon, I.A. Kaprovich, E.I. Leonov, V.M. Orlov, On some properties of $CdSnP_2$ in strong electrical field, Phys. Status Solidi (a) 2 (1970) K 117–K120.
[28] S. Wagner, J.L. Shay, P. Migliorato, H.M. Kasper, $CuInSe_2$/CdS heterojunction photovoltaic detectors, Appl. Phys. Lett. 25 (1974) 434–435.
[29] L.L. Kazmerski, M.S. Ayyagari, G.A. Sanborn, F.R. White, Growth and properties of vacuum deposited $CuInSe_2$ thin films, J. Vac. Sci. Technol. 13 (1976) 139–144.
[30] L.L. Kazmerski, Ternary compound thin film solar cells, in: G. Holah (Ed.), Ternary Compounds 1977, Institute of Physics, London, 1977, pp. 217–228.
[31] L.L. Kazmerski, F.R. White, M.S. Ayyagari, Y.J. Juang, R.P. Patterson, Growth and characterization of thin-film compound semiconductor photovoltaic heterojunctions, J. Vac. Sci. Technol. 14 (1977) 65–68.
[32] J.J. Loferski, J. Shewchun, B. Roessler, R. Beaulieu, J. Piekoszewski, M. Gorska, et al., Investigation of thin film cadmium sulfide/mixed copper ternary heterojunction photovoltaic cells, In: Proceedings of the 13th IEEE PVSC, Washington, D.C., 1978, pp. 190.
[33] R.A. Mickelsen, W.S. Chen, Development of a 9.4% efficient thin film $CuInSe_2$/CdS solar cell. In: Proceedings of the 15th IEEE PVSC, Orlando, 1981, pp. 800–803.
[34] R.A. Mickelsen, W.S. Chen, Polycrystalline thin film $CuInSe_2$ solar cells. In: Proceedings of the 16th IEEE PVSC, San Diego, 1982, pp. 781–784.
[35] M.A. Green, K. Emery, Y. Hishikawa, W. Warta, Solar cell efficiency tables (version 36), Prog. Photovolt.: Res. Appl. 18 (2010) 346–352; (Z.S.W., Press Release, 23 August 2010.).
[36] B.M. Basol, V.J. Kapur, Deposition of $CuInSe_2$ films by a two-stage process utilizing E-beam evaporation, IEEE Trans. Electr, Dev 37 (1990) 418–421.
[37] M.J. Carter, I.I.'Amm, A. Knowles, H. Ooumous, R. Hill, Laser processing of compound semiconductor thin films for solar cell applications. In: Proceedings of the 19th IEEE Photovoltaics Specialist Conference, New Orleans, 1987, pp. 1275–1278.
[38] B.M. Basol, V.K. Kapur, C.R. Leidholm, A. Halani, K. Gledhill, Flexible and light weight copper indium diselenide solar cells on polyimide substrates, Sol. Energy Mater. Sol. Cells 43(1) (1996) 93–98.
[39] A.N. Tiwari, M. Krejci, F.-J. Haug, H. Zogg, 12.8% efficiency $Cu(In,Ga)Se_2$ solar cell on a flexible polymer sheet, Prog. Photovolt.: Res. Appl. 7(5) (1999) 393–397.
[40] G.M. Hanket, U.P. Singh, E. Eser, W.N. Shafarman, R.W. Birkmire, Pilot-scale manufacture of $Cu(InGa)Se_2$ films on a flexible polymer substrate, in: Proceedings of the 29th IEEE Photovoltaic Specialists Conference, May 2002, pp. 567–569.
[41] R. Birkmire, E. Eser, S. Fields, W. Shafarman, $Cu(InGa)Se_2$ solar cells on a flexible polymer web, Prog. Photovolt.: Res. Appl. 13(2) (2005) 141–148.
[42] S. Ishizuka, H. Hommoto, N. Kido, K. Hashimoto, A. Yamada, S. Niki, Efficiency enhancement of $Cu(In,Ga)Se_2$ solar cells fabricated on flexible polyimide substrates using alkali-silicate glass thin layers, Appl. Phys. Express 1(9) (2008) 1–3.
[43] P. Gečys, G. Račiukaitis, M. Gedvilas, A. Selskis, Laser structuring of thin-film solar cells on polymers, EPJ Appl. Phys. 46(1) (2009) 12508.
[44] L. Zhang, Q. He, W.-L. Jiang, C.-J. Li, Y. Sun, Flexible $Cu(In, Ga)Se_2$ thin-film solar cells on polyimide substrate by low-temperature deposition process, Chin. Phys. Lett. 25(2) (2008) 734–736.
[45] R. Caballero, C.A. Kaufmann, T. Eisenbarth, M. Cancela, R. Hesse, T. Unold, A. Eicke, R. Klenk, H.W. Schock., The influence of Na on low temperature growth of CIGS thin film solar cells on polyimide substrates, Thin Solid Films 517(7) (2009) 2187–2190.

[46] H. Zachmann, S. Heinker, A. Braun, A.V. Mudryi, V.F. Gremenok, A.V. Ivaniukovich, M.V. Yakushev, Characterisation of Cu(In,Ga)Se$_2$-based thin film solar cells on polyimide, Thin Solid Films 517(7) (2009) 2209−2212.

[47] F. Kessler, D. Rudmann, Technological aspects of flexible CIGS solar cells and modules, Sol. Energy 77(6) (2004) 685−695.

[48] T. Satoh, Y. Hashimoto, S. Shimakawa, S. Hayashi, T. Negami, CIGS solar cells on flexible stainless steel substrates, in: Proceedings of the Conference Record of the 28th IEEE Photovoltaic Specialists Conference, 2000, p. 567.

[49] K. Herz, F. Kessler, R. Wächter, M. Powalla, J. Schneider, A. Schulz, et al., Dielectric barriers for flexible CIGS solar modules, Thin Solid Films 403−404 (2002) 384−389.

[50] D. Herrmann, F. Kessler, K. Herz, M. Powalla, A. Schulz, J. Schneider, U. Schumacher, High-performance barrier layers for flexible CIGS thin-film solar cells on metal foils, in: Proceedings of the Materials Research Society Symposium L: Compound Semiconductor Photovoltaics, 763, San Francisco, California, USA, 2003, pp. 287−292.

[51] D.R. Hollars, R. Dorn, P.D. Paulson, J. Titus, R. Zubeck, Large area Cu(In,Ga)Se$_2$ films and devices on flexible substrates made by sputtering. In: Proceedings of the Materials Research Society Spring Meeting, San Francisco, vol. 865, April 2005, pp. 477−482, F14.34.1.

[52] K. Otte, L. Makhova, A. Braun, I. Konovalov, Flexible Cu(In,Ga)Se$_2$ thin-film solar cells for space application, Thin Solid Films 511−512 (2006) 613−622.

[53] S. Ishizuka, A. Yamada, P. Fons, S. Niki, Flexible Cu(In,Ga)Se$_2$ solar cells fabricated using alkali-silicate glass thin layers as an alkali source material, J. Renew. Sustain. Energy 1 (2008) 013102.

[54] R. Wuerz, A. Eicke, M. Frankenfeld, F. Kessler, M. Powalla, P. Rogin, O. Yazdani-Assl, CIGS thin-film solar cells on steel substrates, Thin Solid Films 517(7) (2009) 2415−2418.

[55] T. Yagioka, T. Nakada, Cd-free flexible Cu(In,Ga)Se$_2$ thin film solar cells with ZnS (O,OH) buffer layers on Ti foils, Appl. Phys. Express 2(7) (2009) 072201.

[56] D. Brémaud, D. Rudmann, M. Kaelin, K. Ernits, G. Bilger, M. Döbeli, H. Zogg, A.N. Tiwari, Flexible Cu(In,Ga)Se$_2$ on Al foils and the effects of Al during chemical bath deposition, Thin Solid Films 515(15) (2007) 5857−5861.

[57] S. Ishizuka, A. Yamada, K. Matsubara, P. Fons, K. Sakurai, S. Niki, Development of high-efficiency flexible Cu(In,Ga)Se$_2$ solar cells: a study of alkali doping effects on CIS, CIGS, and CGS using alkali-silicate glass thin layers, Curr. Appl. Phys. 20(2 Suppl. 1) (2009) S154−S156.

[58] C.Y. Shi, Y. Sun, Q. He, F.Y. Li, J.C. Zhao, Cu(In,Ga)Se$_2$ solar cells on stainless-steel substrates covered with ZnO diffusion barriers, Sol. Energy Mater. Sol. Cells 93(5) (2009) 654−656.

[59] M.S. Kim, J.H. Yun, K.H. Yoon, B.T. Ahn, Fabrication of flexible CIGS solar cell on stainless steel substrate by co-evaporation process, Diffusion Defect Data B 124−126 (2007) 73−76.

[60] A. Chirilă, P. Bloesch, A. Uhl, S. Seyrling, F. Pianezzi, S. Buecheler, et al., Progress towards the development of 18% efficiency flexible CIGS solar cells on polymer film, In: Proceedings of the 25th European Photovoltaic Solar Energy Conference/ 5th World Conference on Photovoltaic Energy Conversion, Valencia, Spain, 6−10 September 2010, pp. 3403−3405.

[61] S. Ishizuka, A. Yamada, K. Matsubara, P. Fons, K. Sakurai, S. Niki, Alkali incorporation control in Cu(In,Ga)Se2 thin films using silicate thin layers and applications in enhancing flexible solar cell efficiency, Appl. Phys. Lett. 93 (2008)124105-1−124105-3.

[62] J. Ward, K. Ramanathan, F. Hasoon, T. Coutts, J. Keane, T. Moriarty, R. Noufi, A 21.5% efficient Cu(In,Ga)Se2 thin-film concentrator solar cell, Prog. Photovolt.: Res. Appl. 10 (2002) 41–46; M.A. Green, K. Emery, Y. Hishikawa, W. Warta, Solar cell efficiency tables (version 36), Prog. Photovolt.: Res. Appl. 18 (2010) 346–352.
[63] W. Arndt, H. Dittrich, F. Pfister, H.W. Schock, CuGaSe2-ZnCdS and CuInSe2-ZnCdS thin film solar cells for tandem systems, in: Proceedings of the Sixth E.C. Photovoltaic Solar Energy Conference, London, UK, 15–19 April 1985, pp. 260–264.
[64] R.W. Birkmire, J.E. Phillips, L.C. DiNetta, J.A. Meakin, Thin film tandem solar cells based on CuInSe2, in: Proceedings of the Sixth European Photovoltaic Solar Energy Conference, London, UK, 15–19 April 1985, pp. 270–274.
[65] K. Mitchell, C. Eberspacher, J. Ermer, D. Pier, P. Milla, Copper indium diselenide photovoltaic technology. In: Proceedings of the 8th E.C. Photovoltaic Solar Energy Conference, 9–13 May 1988, Florence, Itlay, pp. 1578–1582.
[66] P. Liska, K.R. Thampi, M. Grätzel, D. Brémaud, D. Rudmann, H.M. Upadhyaya, A.N. Tiwari, Nanocrystalline dye-sensitzied solar cell/copper indium gallium selenide thin-film tandem showing greater than 15% conversion efficiency, Appl. Phys. Lett. 88 (2006) 203103.
[67] M. Robbins, K.J. Bachmann, V.G. Lambrecht, F.A. Thiel, J. Thomson Jr., R.G. Vadimsky, S. Menezes, A. Heller, B. Miller, CuInS$_2$ liquid junction solar cells, J. Electrochem. Soc. 125 (1978) 831.
[68] L.L. Kazmerski, G.A. Sanborn, CuInS$_2$ thin film homojunction solar cells, J Appl. Phys. 48 (1977) 3178–3180.
[69] P.R. Ram, R. Thangaraj, A.K. Sharma, O.P. Agnihotri, Totally sprayed CuInSe$_2$/Cd(Zn)S and CuInS2/Cd(Zn)S solar cells, Solar Cells 14 (1985) 123–131.
[70] A.N. Tiwari, D.K. Pandya, K.L. Chopra, Sol. Energy Mater. 15 (1987) 121–133.
[71] A.N. Tiwari, D.K. Pandya, K.L. Chopra, Fabrication and analysis of all-sprayed CuInS$_2$/ZnO solar cells, Sol. Cells 22 (1987) 263–273.
[72] H.J. Lewerenz, H. Goslowsky, K.-D. Husemann, S. Fiechter, Efficient solar energy conversion with CuInS$_2$, Nature 321 (1986) 687–688.
[73] K.W. Mitchell, G.A. Pollock, A.V. Mason, 7.3% efficient CuInS$_2$ solar cell, in: Proceedings of the 20th IEEE Photovoltaic Specialists Conference, Las Vegas, USA, 26–30 September 1988, pp. 1542–1544.
[74] R. Scheer, T. Walter, H.W. Schock, M.L. Fearheiley, H.J. Lewerenz, CuInS$_2$ based thin film solar cell with 10.2% efficiency, Appl. Phys. Lett. 63 (1993) 3294–3296.
[75] K. Siemer, J. Klaer, I. Luck, J. Bruns, R. Klenk, D. Bräunig, Efficient CuInS$_2$ solar cells from a rapid thermal process (RTP), Sol. Energy. Mater. Sol. Cells 67 (2001) 159–166.
[76] R. Kaigawa, A. Neisser, R. Klenk, M Ch., Lux-Steiner, Improved performance of thin film solar cells based on Cu(In,Ga)S$_2$, Thin Solid Films 415 (2002) 266–271.
[77] D. Fuertes Marrón, A. Martí, A. Luque, Thin-film intermediate band chalcopyrite solar cells, Thin Solid Films 517 (2009) 2452–2454.
[78] 15 US patents owned by The Boeing Company, Seattle, WA: US patent numbers: 4335266 (1982), RE31968 (1985), 4684761 (1987), 4703131 (1987), 4867801 (1989), 5078804 (1992), 5141564 (1991), 5261969 (1994).
[79] 12 US patents owned by Atlantic Richfield Company, Los Angeles, CA: US patent numbers: 4465575 (1984); 461109 (1986), 4612411 (1986) 4638111 (1987), 4798660 (1989), 4915745 (1990), 5045409 (1991).
[80] 13 US patents owned by Matsushita Electric Co., Ltd., Kadoma, Japan: US patent numbers: 4940604 (1990), 5445847 (1995), 5474622 (1995), 5500056 (1996), 5567469 (1996), 5633033 (1997), 5714391 (1998), 5725671 (1998), 5728231 (1998).
[81] 14 US patents owned by Siemens Solar or other Siemens companies: US patent numbers: 5078803 (1992), 5103268 (1992), 5474939 (1995), 5512107 (1996), 5578503 (1996), 5580509 (1996), 5626688 (1997).

[82] 20 US patents owned by Yazaki Corporation, Tokyo, Japan: US patent numbers: 5501786 (1996), 5695627 (1998), 5772431 (1998), 5935324 (1999), 6036822 (2000), 6207219 (2001).
[83] K. Mitchell, C. Eberspacher, J. Ermer, D. Pier, Single and tandem junction CuInSe$_2$ cell and module technology, in: Proceedings of the 20th IEEE Photovoltaic Specialists Conference, Las Vegas, USA, 26—30 September 1988, pp. 1384—1389.
[84] D.E. Tarrent, Al. R. Ramos, D.R. Willett, R.R. Gay, CuInSe$_2$ module environmental durability, in: Proceedings of the 22nd IEEE Photovoltaic Specialists Conference, Las Vegas, USA, 7—11 October 1991, pp. 553—556.
[85] T. Dalibor, S. Jost, H. Vogt, R. Brenning, A. Heiß, S. Visbeck, et al., Advanced CIGSSe device for module efficiencies above 15%, in: Proceedings of the 25th European Photovoltaic Solar Energy Conference and Exhibition/5th World Conference on Photovoltaic Energy Conversion, Valencia, Spain, 6—10 September 2010, pp. 2854—2857.
[86] Global Solar Energy, Press Release, 6 March 2008.
[87] Honda, Press Release, 12 November 2007.
[88] V. Probst, F. Hergert, B. Walther, R. Thyen, G. Batereau-Neumann, B. Neumann, et al., High performance CIS modules: status of production and development at Johanna solar technology, in: Proceedings of the 24th European Photovoltaic Solar Energy Conference and Exhibition, Hamburg, Germany, 21—25 September 2009, pp. 2455—2459.
[89] Miasolé, Press Release, 14 January 2010.
[90] Nanosolar, Press Release, 21 June 2006.
[91] Nanosolar, Press Release, 9 September 2009.
[92] M. Winkler, J. Griesche, I. Konovalov, J. Penndorf, J. Wienke, O. Tober, CISCuT—solar cells and modules on the basis of CuInS$_2$ on Cu-tape, Sol. Energy 77 (2004) 705—716.
[93] K. Kushiya, Y. Tanaka, H. Hakuma, Y. Goushi, S. Kijima, T. Aramoto, Y. Fujiwara, Interface control to enhance the fill factor over 0.70 in a large-area CIS-based thin-film PV technology, Thin Solid Films 517 (2009) 2108—2110.
[94] Showa Shell, Press Release, 15 August 2007.
[95] Showa Shell Sekiyu, Press Release, 17 July 2008.
[96] Osamu Ikki, PV Activities in Japan, volume 16, no. 6, June 2010.
[97] C. Scheit, H. Herrnberger, A. Braun, M. Ehrhardt, K. Zimmer, Interconnection of flexible CIGS thin film solar cells, in: Proceedings of the 25th European Photovoltaic Solar Energy Conference and Exhibition/Fifth World Conference on Photovoltaic Energy Conversion, Valencia, Spain, 6—10 September 2010, pp. 3414—3417.
[98] SoloPower, Press Release 27 April 2010.
[99] PV News, May 2010, Greentech Media, 2010, ISSN 0739-4829.
[100] Solyndra, Press Release, 17 June 2010.
[101] A. Neisser, A. Meeder, F. Zetzsche, U. Rühle, C. Von Klopmann, R. Stroh, et al., Manufacturing of large-area CuInS$_2$ solar modules—from pilot to mass production, in: Proceedings of the 24th European Photovoltaic Solar Energy Conference and Exhibition, 21—25 September 2009, Hamburg, Germany, pp. 2460—2464.
[102] Sulfurcell, Press Release, 6 September 2010.
[103] Würth Solar GmbH, Press Release, 04 August 2008.

PART ID

Space and Concentrator Cells

CHAPTER ID-1

GaAs and High-Efficiency Space Cells

V.M. Andreev
Ioffe Physico-Technical Institute, St. Petersburg, Russia

Contents

1. Historical Review of III—V Solar Cells — 399
2. Single-Junction III—V Space Solar Cells — 403
   2.1 Solar Cells Based on AlGaAs—GaAs Structures — 403
   2.2 Solar Cells with Internal Bragg Reflector — 404
   2.3 GaAs-Based Cells on Ge Substrates — 406
3. Multijunction Space Solar Cells — 407
   3.1 Mechanically Stacked Cells — 407
   3.2 Monolithic Multijunction Solar Cells — 409
Acknowledgements — 412
References — 412

1. HISTORICAL REVIEW OF III—V SOLAR CELLS

Since the first solar-powered satellites Vanguard 1 and Sputnik 3 were launched in the spring of 1958, solar cells had become the main source of energy on the spacecraft. The first space arrays were based on single crystal silicon solar cells with an efficiency of about 10%. During the 1960s and 1970s, improvements in Si cell design and technology, such as fabrication of 'violet' cells with an increased short-wavelength photosensitivity, back-surface field formation, application of photolithography to make optimal front grid lingers, reduction of optical losses owing to front-surface texturing, and improvement of the antireflection coating, allowed an increase of efficiency of up to 14% (1 sun, AM0). In the last two decades, the Si space cell efficiencies were increased by up to 18%. These advanced Si cells are used for space missions that do not strictly require III—V cells with both higher efficiency and better radiation stability [1,2].

Practical Handbook of Photovoltaics.
© 2012 Elsevier Ltd. All rights reserved.

At the beginning of the 1960s, it was found that GaAs-based solar cells with the Zn-diffused p–n junction ensured the better temperature stability and higher radiation resistance. One of the first scaled applications of the temperature-stable GaAs solar cells took place on the Russian spacecraft Venera 2 and Venera 3 launched in November 1965 to the 'hot' planet Venus. The area of each GaAs solar array constructed by the Russian Enterprise KVANT for these spacecraft was 2 m^2. Then the Russian moon cars Lunokhod 1 and Lunokhod 2 were launched in 1970 and 1972 with GaAs 4 m^2 solar arrays in each. The operating temperature of these arrays on the illuminated surface of the Moon was about 130°C. Therefore, silicon-based solar cells could not operate effectively in these conditions. GaAs solar arrays have shown efficiency of 11% and have provided the energy supply during the lifetime of these moon cars.

The first AlGaAs–GaAs solar cells with passivating wide-band-gap windows were created in 1970 [3,4]. In the following decades, by means of the liquid-phase-epitaxy (LPE) of AlGaAs–GaAs heterostructures [4–12], their AM0 efficiency was increased up to 18–19% [10–14] owing to the intensive investigations in the fields of physics and technology of space solar cells [15–18]. Those investigations were stimulated and supported by ambitious space programmes in the former Soviet Union and in the United States [1,2,18]. Owing to the high efficiency and improved radiation hardness of the AlGaAs–GaAs solar cells, the LPE technology was utilised in the high-scale production of AlGaAs–GaAs space arrays for the spacecraft launched in the 1970s and 1980s. For example, an AlGaAs–GaAs solar array with a total area of 70 m^2 was installed in the Russian space station MIR launched in 1986. During 15 years in orbit, the array degradation appeared to be lower than 30%, despite being under severe conditions such as appreciable shadowing, effects of numerous dockings, and the ambient environment of the station. At that time, it was the largest demonstration of the AlGaAs–GaAs solar cell advantages for space applications. The further improvement of the LPE technology [19,20] led to increased efficiencies of 24.6% (AM0, 100 suns) on the base of the heterostructures with an ultra-thin window AlGaAs layer and a back-surface-field layer.

Since the late 1970s, AlGaAs–GaAs heterostructures have been produced by the metal organic chemical vapour deposition (MOCVD) technique using metal organic compounds of Group III elements and hydrides of Group V elements [21,22]. The advantage of MOCVD is a possibility to fabricate multilayer structures in high-yield reactors with layers of a specified composition and a precise thickness varying from 1 to 10 nm to several microns. AlGaAs—GaAs heterostructures with an ultrathin (0.03 μm) top window layer and with a back-surface wide-band-gap barrier were fabricated by MOCVD for space cells. AlGaAs—GaAs 4-cm^2 1-sun space solar cells with efficiencies of 21% [23] and 21.7% [24] were fabricated on the base of these structures.

Enhanced light absorption was provided in the cells with an internal Bragg reflector [25—27]. This dielectric mirror increases the effective absorption length of sunlight within the long-wavelength part of photoresponse spectrum and allows the base layer to be made thinner. In this case, the cell efficiency is more tolerant to a reduction of the carrier diffusion length; as a result, these cells are more radiation resistant [27].

Owing to the fact that MOCVD is capable of producing single crystal layers on silicon and germanium substrates, it has a potential for fabrication of low-cost, high-efficiency III—V solar cells on these substrates. Growing GaAs on Si of a sufficient quality is not possible as a result of the mismatch of 4% between Si and GaAs. However, there is progress in improving the GaAs—Si structure quality by using special structures and growth techniques: strained superlattice, thermo cyclic growth, and cyclic structure annealing.

Ge is a quite good lattice match to the GaAs material. Therefore, epitaxial growth of GaAs with a high quality was realised by MOCVD, and this is now the basic technique for growth of multilayer AlGaAs—GaAs—Ge single-junction and GaInP—GaAs—Ge multijunction solar cells. This method provides a good crystal quality of epitaxial structures on Ge substrates, high productivity, and reproducibility.

Among other single junction cells, the InP-based cells are rather promising for space applications owing to that fact that InP has a higher radiation resistance [2,28] than GaAs. However, there are some obstacles for the scaled application of InP-based cells in space arrays. First, there is no lattice-matched wide-band-gap window for InP to make stable passivation of the front surface. Second, it is difficult to grow this material to a high quality on the Ge and Si substrates due to lattice mismatches that are as high as 8% between InP and Si and 4% between InP and Ge. Third, InP is a quite expensive material. In spite of these obstacles, there are some possible applications of InP cells in the arrays for satellites expected to be launched toward the intermediate orbits, high Earth orbits, or orbit transfer that are characterised by very high radiation fluencies.

Multijunction (cascade) cells ensured the further increase of III−V solar cell efficiencies. Despite a large number of theoretical studies of cascade solar cells [1,29−32], their efficiencies remained low enough for a long time, since the ohmic and optical losses in available designs were unacceptably large. Monolithic and mechanically stacked tandem cells with increased efficiencies were developed in the beginning of the 1990s. In mechanically stacked tandems with GaAs top cells and GaSb (or InGaAs) bottom cells [33−35,38,41−43], efficiencies of about 30% were achieved under the concentrated AM0 sunlight. Monolithic cascade cells have been developed and fabricated by MOCVD on the structures GaInAs−InP [40], Si−AlGaAs [44,64], AlGaAs−GaAs [45,42], GaAs−Ge [47−51], GaInP−GaAs [36,52,53] GaInP−GaAs−Ge [54−60], GaInP−GaInAs [61], and GaInP−GaInAs−Ge [57,62,63] heterostructures. Table 1 presents the best reported efficiencies of the cells based on the different structures under one sun and concentrated AM0 sunlight illumination.

Table 1 Reported efficiencies for III−V space solar cells under AM0 conditions at $T = 25-28°C$

Cell material	Cell type	Sunlight concentration	Area, cm^2	Eff., %	Ref.
AlGaAs/GaAs	single junction	Unconcentrated	4	21.7	[24]
GaAs/Si	single junction		4	18.3	[64]
InP	single junction		4	19.9	[2]
AlGaAs/GaAs	monolithic 2-junction		0.5	23.0	[2]
GaInP/GaAs	monolithic 2-junction		4	27.2	[56]
GaInP/GaAs/Ge	monolithic 3-junction		4	29.3	[56]
GaInP/GaInAs/Ge	monolithic 3-junction		4	29.7	[57]
GaInP/GaAs/Ge	monolithic 3-junction		26.6	29.0	[56]
AlGaAs/GaAs	single junction	×100	0.07	24.6	[41]
GaAs/GaSb	mechanical stack 2-Junction	×100	0.05	30.5	[2]
GaAs/Ge	monolithic 2-junction	×9	0.136	23.4	[50]
GaInP/GaInAs/Ge	monolithic, 3-junction	×7.6	4.1	31.1	[58]
GaInP/GaAs/GaSb	monolithic/mechanical stack 3-junction	×15	circuit	34.0	[65]

2. SINGLE-JUNCTION III–V SPACE SOLAR CELLS
2.1 Solar Cells Based on AlGaAs–GaAs Structures

Among different investigated heterostructures based on III–V heterojunctions appropriate for fabrication of single-junction solar cells, AlGaAs–GaAs heterostructures have found the first application due to the well-matched lattice parameters of GaAs and AlAs, and because GaAs has an optimal band gap for effective sunlight conversion. In the first solar cells based on AlGaAs–GaAs heterojunctions [3,4], the basic narrow-band-gap material was GaAs. A wide-band-gap window was made of AlGaAs close in the composition to AlAs, which is almost completely transparent to sunlight, making solar cells very sensitive in the short-wavelength range of the sun spectrum. Such a cell is illuminated through the window, and the light with photon energy exceeding the band-gap value of GaAs is absorbed in it, while the generated minority carriers are separated by the p–n junction field located in GaAs. Because of the close lattice parameters of the contacting materials, the interface in AlGaAs–GaAs heterojunctions is characterised by a low density of surface states, providing a highly effective accumulation of carriers.

The composition n-GaAs–p-GaAs–p-AlGaAs (Figure 1(a)) was the first widely used heterostructure. The structures were grown by LPE

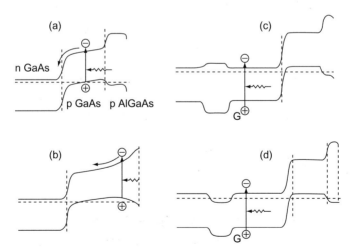

Figure 1 Band diagrams of AlGaAs–GaAs heterostructures developed for space solar cells: (a) p-AlGuAs–p-GaAs–n-GaAs; (b) structure with a graded p-AlGaAs front layer; (c, d) structures with a back-surface field made of n-AlGaAs (c) and n^+ GaAs (d).

[3−14] or by MOCVD [21,22]. For example, a 0.5-μm-thick p-GaAs layer is either grown epitaxially (MOCVD) or formed by zinc or beryllium diffusion during the growth (LPE) of the Al_xGa_{1-x} As solid solution doped with one of these impurities. The diffusion produces a quasi-electric field that arises as a result of the acceptor concentration gradient (Figure 1(a)), which enhances the effective diffusion length of electrons generated by light in the p-GaAs layer. Using the LPE technique, the highest conversion efficiencies of about 19% (1 sun; AM0) have been obtained in the structures with the wide-band-gap layer of smaller thickness [10−14]. A way of enhancing the short-wavelength photosensitivity is to use in the front layer a solid solution of graded composition with the band gap (Figure 1(b)) increasing toward the illuminated surface [10]. The strong built-in electric field significantly enhances the value of the effective electron diffusion length and suppresses the surface recombination of the electron−hole pairs generated near the surface by short-wavelength light.

Introduction of a potential barrier at the back surface of the cell photoactive region assists the collection of minority carriers generated in the base. This barrier is made either by growing a buffer layer (Figure 1(d)) of n^+-GaAs doped to a level exceeding that in the active layer or by growing an n-AlGaAs layer (Figure 1(c)). The 1-sun AM0 efficiency of 21.7% and the concentrator AM0 efficiency of 24.5% at 170 suns have been measured [24] in the cells based on these structures grown by MOCVD with the n-$Al_xGa_{1-x}As$ layer as the back-surface barrier. Similar results have been obtained during development of a low-temperature LPE modification for the growth of AlGaAs−GaAs structures [19] that resulted in fabrication of high-efficiency solar cells with a structure shown in Figure 2(a). Silicone prismatic covers optically eliminate the gridline obscuration losses in concentrator cells. Owing to the high crystal quality of the LPE material and optimised optical parameters, a high quantum yield is obtainable in a wide spectral range. The AM0 efficiency of these cells with a prismatic cover was 24.6% (Figure 2(b)) under 103 suns at 25°C.

2.2 Solar Cells with Internal Bragg Reflector

The Bragg reflector (BR) made of semiconductor layers is widely used in lasers and other optical devices. By using a multiple layer composed of two materials with different refractive indices, nearly 100% reflectance can be achieved over a restricted wavelength range. The thickness of each of the two materials is chosen for quarter-wavelength reflection for the given wavelength. These multilayer dielectric stacks selectively reflect a part of the unabsorbed photons, providing a second pass through the photoactive region, thereby increasing the photocurrent.

Epitaxial (MOCVD) Bragg reflectors in solar cells [25–27] were based (Figure 3(a)) on the pairs of Al_xGa_{1-x} As and GaAs layers. By increasing the number (N) of pairs, the BR reflectance increases, asymptotically tending to unity and reaching 96% at $N = 12$ [26,27]. A reflector of this

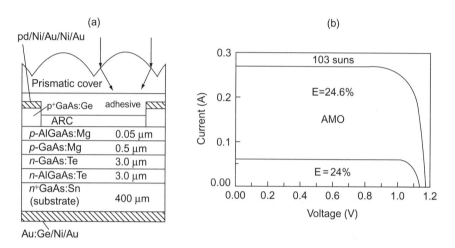

Figure 2 (a) Schematic diagram and (b) illuminated I–V curves of the LPE grown single-junction AlGaAs–GaAs concentrator solar cell with a prismatic cover [19]. The cell area is 0.07 cm².

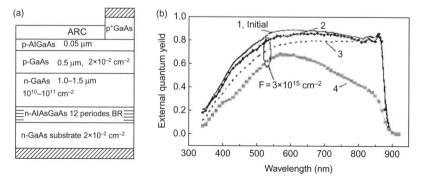

Figure 3 Schematic cross-section of a AlGaAs–GaAs solar cell with (a) internal Bragg reflector (BR) and (b) spectral responses, which are shown for the cells with BR (1–3) and without BR (4) before (1) and after (2, 3, 4) irradiation by 3.75 MeV electrons with fluence of 3×10^{15} cm^{-2}.

type allows us to increase the effective absorption within the long-wavelength part of the photosensitivity spectrum and to make the base layer thinner. The cell efficiency in this case becomes more tolerant to decreasing the diffusion lengths caused by the high-energy particle irradiation. The solar cell structure with a Bragg reflector (Figure 3(a)) was grown [26,27] by MOCVD using equipment with a low-pressure horizontal reactor. BR was optimised for reflectance in the 750–900 nm spectral region and consisted of 12 pairs of AlAs and GaAs layers with a thickness of 72 nm for AlAs layers and 59 nm for GaAs. The photocurrent density of 32.7 mA/cm^2 (AM0, 1 sun, 25°C) and the efficiency of 23.4% (AM0, 18 suns, 25°C) were registered for this cell. These values are fairly good, taking into account the smaller thickness of the n-GaAs base layer. The long-wavelength response of the cell with a 1.5 μm n-GaAs layer is nearly the same as for the cell with a 3 μm n-GaAs layer without BR. Reduction of the base thickness improves the cell radiation resistance. Figure 3(b) shows spectral responses of the cells with BR (curves 1–3) and without BR (4). Base thickness in the cells with BR was reduced to 1.1 μm (curves 1, 2) and to 1.3 μm (3), and the base doping level was reduced to 10^{15}cm^{-3} (curves 1, 2) and to 7×10^{15}cm^{-3} (curve 3). The base thickness of 3.5 μm and the base doping level of 10^{17} cm^{-3} were in the cell without BR (curve 4). It is seen from Figure 3 (b) that reduction of the base thickness and the base doping level in the cells with a Bragg reflector allowed an increase in the radiation resistance. The remaining power factor was 0.84–0.86 in these cells after 1-MeV electron irradiation with a fluence of 10^{15} cm^{-2}.

2.3 GaAs-Based Cells on Ge Substrates

Intensive investigations of single- and dual-junction GaAs-based heterostructures MOCVD grown on Ge substrates were carried out [47–50]. Ge is less expensive and more mechanically strong than GaAs. Therefore, Ge substrates can be thinned down to 100–150 μm, and cells can be made larger, reducing the weight and cost of space arrays. Owing to a good lattice matching between Ge and GaAs, the structures based on GaAs can be grown with a good crystal quality. An increase of the output voltage was observed in GaAs–Ge cells with a photoactive Ge–GaAs interface. A disadvantage of this photoactive Ge results from mismatching of the photocurrents generated in Ge and in GaAs active regions. Usually, the photocurrent from a Ge subcell was lower. Therefore, the I–V curve of these cells was 'kinked' with reduced FF and efficiency [48,49,51].

Reproducible growth conditions were developed to form GaAs cells on Ge substrates with an inactive interface [49,51]. MOCVD equipment produced by EMCORE (USA) and AIXTRON (Germany) provides the GaAs-based solar cell structures on Ge substrates with high productivity (up to 0.2 m^2 per run in AIX-3000 reactor) and with excellent uniformity and reproducibility of the cell performance. AlGaAs—GaAs structure production on the 4-inch Ge wafers allows fabricating the large-area (up to 36 cm^2) cells with an average efficiency of 19% (1 sun, AM0, 25°C).

3. MULTIJUNCTION SPACE SOLAR CELLS
3.1 Mechanically Stacked Cells

Optimum band gaps for multijunction solar cells were calculated in a number of works [1,2,29—40]. Figure 4 illustrates the band gaps and lattice constants for III—V compounds and their solid solutions. The hatched areas represent the theoretical optimum E_g values for the current matched bottom and top cells: $E_{g1} = 1.65-1.8$ eV for the top cells and $E_{g2} = 1.0-1.15$ eV for the bottom cells. The theoretical one-sun AM0 efficiency for tandem solar cells with these band-gap values is about 32.5% (Figure 5, curve 1).

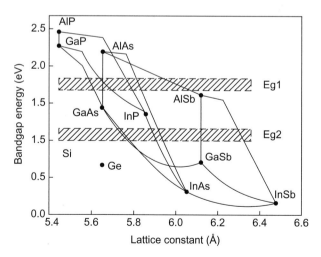

Figure 4 Energy band gap versus lattice constant for Ge, Si, III—V compounds and ternary solid solutions. Hatched boxes correspond to the highest efficiency current-matched two-junction solar cells.

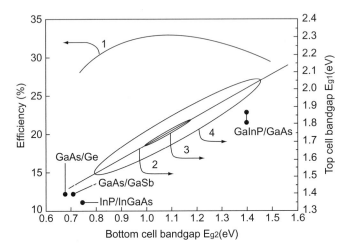

Figure 5 Curve 1: one sun AM0 efficiency (curve 1) for two-junction, two-terminal solar cells as a function of E_{g2} under conditions of matched photocurrents in the top and bottom cells. Line 2: relationship between E_{g1} and E_{g2} values, for which the condition of matched photocurrents is fulfilled. Curve 3: iso-efficiency contour ensuring the highest AM0 efficiencies of about 32.5%. Curve 4: iso-efficiency contour, for which efficiency decreases down to 30%. The bold points show the handgaps E_g, mid E_{g2} of GaAs−Ge, GaAs−GaSb, InP−InGaAs, and GaInP−GaAs tandems.

Mechanical stacks ensure more possibilities for the material choice than monolithic tandems owing to the use of lattice- and current-mismatched combinations of semiconductors. An obstacle for high-scale applications of stacks in space is a more complicated design. It is one reason why stacks are only used in concentrator array applications. In spite of the higher equilibrium temperature in the cells operating under concentrated sunlight, theoretical efficiencies of multijunction, mechanically stacked cells under 10−100 suns at temperatures of 70−80°C are higher by 2−4% than 1-sun solar cells at 25°C. The use of sunlight concentration for space arrays offers additional advantages: better cell shielding from space radiation and a potentially low cost of concentrator arrays owing to the low semiconductor material consumption.

The first high-efficiency, mechanically stacked tandems consisted of AlGaAs−GaAs infrared-transparent top cells and GaSb infrared-sensitive bottom cells were fabricated [35]. GaAs-based cells were made transparent to the infrared part ($\lambda > 0.9~\mu$m) of sunlight in these stacked cells as n-GaAs substrate doping level was reduced to 10^{17} cm^{-3}. GaSb cells for stacks were fabricated mainly using Zn diffusion. Maximum photocurrent densities in GaSb cells for the AM0-spectrum were about 30 mA/cm^2 behind a GaAs filter, and efficiencies of about 6% (AM0, 100×) were achieved behind a transparent GaAs cell. The best efficiencies of 29–30% (AM0, 100×) were obtained in two-junction stacks based on AlGaAs–GaAs top and GaSb (or InGaAs) bottom cells [35,41], which have promise for use in space concentrator arrays [66,67].

The further efficiency increase up to 34% (AM0, 25°C, 15 suns) was obtained in the triple-junction, mechanically stacked voltage-matched circuits based on the monolithic GaInP–GaAs two-junction top cell and the GaSb bottom cell [65]. The efficiencies of 27.5% in the GaInP–GaAs top cell and 6.5% in the bottom cell were obtained at 15 suns AM0 concentration. To obtain two-terminal circuits, seven GaSb cells were connected in series ensuring output voltage (V_{mp}) of 7 × 0.375 V = 2.63 V, slightly exceeding V_{mp} of 2.4 V for the InGaP–GaAs two-junction cells that were connected in parallel.

Efficiency of 30% is expected in the minimodules based on these stacks in the ultralight stretched lens array [66] characterised by overall efficiency of 26% and array power density of 350 W/m^2. The further efficiency increase is expected in four-junction stacks based on the monolithic two-junction GaInP–GaAs top cells and the monolithic two-junction AlGaAsSb–GaSb bottom cells (or with another type of the cascade bottom cells).

3.2 Monolithic Multijunction Solar Cells

In monolithic AlGaAs–GaAs tandems consisting of an $Al_{0.37}Ga_{0.63}As$ (E_g = 1.93 eV) upper cell and a GaAs lower cell were grown by MOCVD [45]. The component cells were electrically connected by a metal contact fabricated during the postgrowth processing. The efficiency of 25.2% measured under AM0 1-sun illumination was achieved in AlGaAs–GaAs–InGaAsP three-junction cells consisting of a monolithic AlGaAs–GaAs tandem mechanically stacked with an InGaAsP (E_g = 0.95 eV) single-junction cell.

Monolithic two-terminal GaAs–Ge tandem space concentrator cells with efficiency of 23.4% (9 suns, AM0, 25°C) were developed [50]. MOCVD growth of n-GaAs formed a bottom cell in Ge owing to simultaneous diffusion of As and Ga into Ge. Series resistance of the n$^+$GaAs-p$^+$Ge tunnel junction formed on the interface limited the effective operation of these tandem cells to 10 suns only.

As is seen in Figures 4 and 5, silicon is a material with an optimum band gap for the fabrication of bottom cells for two-junction cells with a theoretical efficiency exceeding 30%. Wide-band-gap cells in these tandems, however, can be made only from such materials as AlGaAs, GaInP, and GaPAs, which are not lattice-matched to silicon. A considerable advance was realised in the fabrication of GaAs-based epitaxial layers on Si substrates [44,64]. The results obtained hold a promise for high-efficiency monolithic cascade cells on Si substrates, costing less than those on Ge substrates.

Cascade cells based on GaInP–GaAs heterostructures were at first proposed and fabricated at National Renewable Energy Laboratory (NREL) [52,53]. Then this technology was successfully applied for high-scale production of space arrays based on dual- and triple-junction GaInP–GaAs–Ge in Spectrolab [54–58], Tecstar [59], and Emcore [60]. Figure 6 shows two of the developed triple-junction cell structures consisting of a (Al)GaInP top cell connected in series by tunnel junction to a GaAs (Figure 6(a)) or InGaAs (Figure 6(b)) middle cell, connected in turn by tunnel junction with a bottom Ge cell. A 1-sun AM0 efficiency as high as 29.3% was achieved in Spectrolab [56] in a three-junction

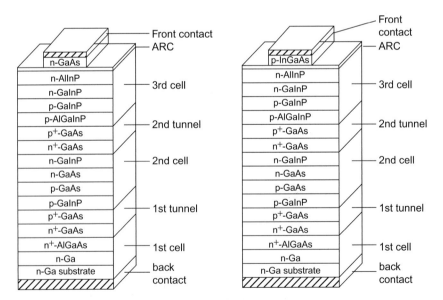

Figure 6 Cross-section of the developed triple-junction solar cells: (a) (Al) GaInP–GaAs–Ge cascade cell (n-on-p) with GaAs-based first and second-tunnel junctions and second cell; (b) (Al)GaInP–(In)GaAs–Ge cascade cell (p-on-n) with InGaAs-based first-tunnel junction and second cell, and (Al)GaInP-based second-tunnel junction.

GaInP−GaAs−Ge cell. Large-area (26.6 cm^2) three-junction cells have reached 29% AM0 efficiency. The high efficiencies of these cells are compatible with the high radiation hardness. The power-remaining factor P/P$_0$ = 0.83 at 10^{15} e$^-$/cm^2 was measured in typical GaInP−GaAs−Ge cells, ensured mainly by the high radiation resistance of the top GaInP cell.

As mentioned above, the subcells with E$_g$ = 1−1.15 eV ensure a higher theoretical efficiency in cascade cells. Suitable Ge substrates can only be employed for the growth of lattice-mismatched GaInAs epilayers of an optimal composition. GaInAs layers on GaAs or Ge substrates of a satisfactory quality were grown and used for a subsequent epitaxial growth of GaInP−GaInAs cascade structures [61−63]. Efficiency of 27.3% (1 sun, AM0, 28°C) was measured [63] in the triple junction Ga$_{0.43}$In$_{0.57}$P−Ga$_{0.92}$In$_{0.08}$As−Ge cells with the 0.5% lattice-mismatch to Ge substrate, similar to high-efficiency conventional lattice-matched GaInP−GaAs−Ge cells. The record efficiency of 29.7% [57] at 1 sun AM0 was achieved in a triple-junction cell based on the GaInP−GaInAs−Ge structure with improved band-gap control: the band gap of the GaInP top cell and tunnel junction layers, the band gap of the GaInAs structure, and simultaneous reduction of dislocation density in the structure. Owing largely to this very high efficiency at the start of life, the prototype cells have demonstrated end-of-life AM0 efficiency of more than 24.4% after irradiation with 1-MeV electrons at fluence of 1×10^{15}e$^-$/cm^2.

The next step for the efficiency increase was proposed in [36]: the development of monolithic four-junction (Al)GaInP−GaAs−GaInNAs−Ge cells that contain a 1-eV GaInNAs subcell lattice matched to GaAs between the Ge and GaAs subcells. However, GaInNAs layer has not been obtained with parameters acceptable for incorporation in such a four-junction cells until now in spite of the intensive investigations of this material. The further efficiency increase was predicted for monolithic five-junction cells based on the (Al)GaInP−GaInP−GaAs−GaInAs−Ge structures with the lattice-mismatched GaInAs layers in a second (from Ge substrate) cell [62].

More recently, new III−V low-dimensional structures based on supperlattice and multiquantum wells [68,69], as well as metallic and quantum dot intermediate bands [70,71], were proposed for solar cells. The main idea in these works is to use low-dimensional heterostructures in order to extend the sunlight absorption to longer wavelengths and to conserve the high output voltage corresponding to the wide-band-gap bulk semiconductor. It was predicted that these structures could obtain higher theoretical efficiencies than the multijunction solar cells reviewed earlier. However, new materials, new technologies, and maybe new approaches should be developed to realise these predictions.

ACKNOWLEDGEMENTS

The author expresses his thanks to colleagues from the Photovoltaics Laboratory of the Ioffe Physico-Technical Institute for the help and valuable discussions, Zh. I. Alferov for the continuous interest and support, and to all researchers for the permissions to use the copyright material.

REFERENCES

[1] D. Flood, H. Brandhorst, Space solar cells, in: T.J. Coutts, J.D. Meakin (Eds.), Current Topics in Photovoltaics, Vol. 2, Academic Press, New York, London, 1987, pp. 143–202.
[2] S.G. Bailey, D.J. Flood, Space photovoltaics, Prog. Photovolt: Res. Appl. 6(1) (1998) 1–14.
[3] Antonio L. Luque López and Viacheslav M. Andreev, *Concentrator Photovoltaics* (Springer, Heidelberg; 2010).
[4] Zh.L Alferov, V.M. Andreev, M.B. Kagan, I.I. Protasov, V.G. Trofim, Solar cells based on heterojunction p-AlGaAs-n-GaAs, Sov. Phys. Semicond. 4(12)) (1970).
[5] H.J. Hovel, J.M. Woodall, High-efficiency AlGaAs-GaAs solar cells, Appl Phys. Lett. 21 (1972) 379–381.
[6] V.M. Andreev, T.M. Golovner, M.B. Kagan, N.S. Koroleva, T.A. Lubochevskaya, T.A. Nuller, et al., Investigation of high efficiency AlGaAs–GaAs solar cells, Sov. Phys. Semicond. 7(12) (1973).
[7] V.M. Andreev, M.B. Kagan, T.L. Luboshcvskaia, T.A. Nuller, D.N. Tret'yakov, Comparison of different heterophotoconverters for achievement of highest efficiency, Sov. Phys. Semicond. 8(7) (1974).
[8] Zh.L. Alferov, V.M. Andreev, G.S. Daletskii, M.B. Kagan, N.S. Lidorenko, V.M. Tuchkevich, Investigation of high efficiency AlAs-GaAs heteroconverters. Proceedings of the World Electrotechnology Congress, Moscow, 1977, Section 5A, report 04.
[9] H.J. Hovel, Solar cells, in: R.K. Willardson, A.C. Beer (Eds.), Semiconductors and Semimetals, vol. 11, Academic Press, New York, London, 1975.
[10] J.M. Woodall, H.J. Hovel, An isothermal etchback–regrowth method for high efficiency $Ga_{1-x}Al_xAs$–GaAs solar cells, Appl. Phys. Lett. 30 (1977) 492–493.
[11] V.M. Andreev, V.R. Larionov, V.D. Rumyantsev, O.M. Fedorova, Sh.Sh. Shamukhamedov, P-AlGaAs–pGaAs–nGaAs solar cells with efficiencies of 19% at AM0 and 24% at AM1 .5, Sov. Tech. Phys. Lett. 9(10) (1983) 537–538.
[12] H.J. Hovel, Novel materials and devices for sunlight concentrating systems, IBM J. Res. Develop. 22 (1978) 112–121.
[13] E. Fanetti, C. Flores, G. Guarini, F. Paletta, D. Passoni, High efficiency 1.43 and 1.69 eV band gap $Ga_{1-x}Al_xAs$–GaAs solar cells for multicolor applications, Solar Cells 3 (1981) 187–194.

[14] R.C. Knechtly, R.Y. Loo, G.S. Kamath, High-efficiency GaAs solar cells, IEEE Trans. Electron Dev. ED-31(5) (1984) 577–588.
[15] H.S. Rauschenbach, Solar Cell Array Design Handbook. The Principles and Technology of Photovoltaic Energy Conversion, Litton Educational Publishing, Inc., New York, 1980.
[16] A. Luque, Solar Cells and Optics for Photovoltaic Concentration, Adam Hilger, Bristol, Philadelphia, 1989.
[17] L.D. Partain (Ed.), Solar Cells and Their Application, John Wiley & Sons, 1995.
[18] P.A. Iles, Future of Photovoltaics for space applications, Prog. Photovolt: Res. Appl. 8 (2000) 39–51.
[19] V.M. Andreev, A.B. Kazantsev, V.P. Khvostikov, E.V. Paleeva, V.D. Rumyantsev, M. Z. Shvarts, High-efficiency (24.6%, AM0) LPE Grown AlGaAs/GaAs Concentrator Solar Cells and Modules. Proceedings of the First World Conference on Photovoltaic Energy Conversion, Hawaii, 1994, pp. 2096–2099.
[20] V.M. Andreev, V.D. Rumyantsev, A^3B^5 based solar cells and concentrating optical elements for space PV modules, Sol. Energy Mater. Sol. Cells 44 (1996) 319–332.
[21] R.D. Dupuis, P.D. Dapkus, R.D. Vingling, L.A. Moundy, High-efficiency GaAlAs/GaAs heterostructure solar cells grown by metalorganic chemical vapor deposition, Appl. Phys. Lett. 31 (1977) 201–203.
[22] N.J. Nelson, K.K. Jonson, R.L. Moon, H.A. Vander Plas, L.W. James, Organometallic- sourced VPE AlGaAs/GaAs concentrator solar cells having conversion efficiencies of 19%, Appl. Phys. Lett. 33 (1978) 26–27.
[23] J.G. Werthen, G.F. Virshup, C.W. Ford, C.R. Lewis, H.C. Hamaker, 21% (one sun, air mass zero) 4 cm^2 GaAs space solar cells, Appl. Phys. Lett. 48 (1986) 74–75.
[24] S.P. Tobin, S.M. Vernon, S.J. Woitczuk, C. Baigar, M.M. Sanfacon, T.M. Dixon, Advanced in high-efficiency GaAs solar cells. Proceedings of the 21st IEEE Photovoltaic Specialists Conference, 1990, pp. 158–162.
[25] S.P. Tobin, S.M. Vernon, M.M. Sanfacon, A. Mastrovito, Enhanced light absorption in GaAs solar cells with internal Bragg reflector. Proceedings of the 22nd IEEE Photovoltaic Specialists Conference, 1991, pp. 147–152.
[26] V.M. Andreev, V.V. Komin, I.V. Kochnev, V.M. Lantratov, M. Z. Shvarts, High-efficiency AlGaAs–GaAs Solar Cells with Internal Bragg Reflector. Proceedings of the First World Conference on Photovoltaic Energy Conversion. Hawaii, 1994, pp. 1894–1897.
[27] M.Z. Shvarts, O.I. Chosta, I.V. Kochnev, V.M. Lantratov, V.M. Andreev, Radiation resistant AlGaAs/GaAs concentrator solar cells with internal Bragg reflector, Sol. Energy Mater. Sol. Cells 68 (2001) 105–122.
[28] M. Yamaguchi, Space solar cell R&D activities in Japan. Proceedings of the 15th Space Photovoltaic Research and Technology Conference, 1997, pp. 1–10.
[29] C.C. Fan, B.-Y. Tsaur, B.J. Palm, Optimal design of high-efficiency tandem cells. Proceedings of the 16th IEEE Photovoltaic Specialists Conference, San Diego, 1982, pp. 692–698.
[30] M.A. Green, Solar Cells, Prentice-Hall Inc., New Jersey, 1982.
[31] M.F. Lamorte, D.H. Abbott, Computer modeling of a two-junction, monolithic cascade solar cell, IEEE Trans. Electron. Dev. ED-27 (1980) 231–249.
[32] M.B. Spitzer, C.C. Fan, Multijunction cells for space applications, Solar Cells 29 (1990) 183–203.
[33] L.M. Fraas, High-efficiency III–V multijunction solar cells, in: L.D. Partain (Ed.), Solar Cells and Their Applications, John Wiley & Sons, 1995, pp. 143–162.
[34] R.K. Jain, D.J. Flood, Monolithic and mechanical multijunction space solar cells, J. Solar Energy Eng. 115 (1993) 106–111.

[35] L.M. Fraas, J.E. Avery, J. Martin, V.S. Sundaram, G. Giard, V.T. Dinh, et al., Over 35-percent efficient GaAs/GaSb tandem solar cells, IEEE Trans. Electron. Dev. ED-37 (1990) 443–449.
[36] S.R. Kurtz, D. Myers, J.M. Olson, Projected performance of three- and four-junction devices using GaAs and GaInP. Proceedings of the 26th IEEE Photovoltaic Specialists Conference, Anaheim, 1997, pp. 875–878.
[37] M. Yamaguchi, Multi-junction solar cells: present and future. Technical Digest 12th International Photovoltaic Solar Energy Conference, 2001, pp. 291–294.
[38] A.W. Bett, F. Dimroth, G. Stollwerk, O.V. Sulima, III–V compounds for solar cell applications, Appl. Phys. A69 (1999) 119–129.
[39] M. Yamaguchi, A. Luque, High efficiency and high concentration in photovoltaics, IEEE Trans. Electron Devices ED-46(10) (1999) 41–46.
[40] M.W. Wanlass, J.S. Ward, K.A. Emery, T.A. Gessert, C.R. Osterwald, T.J. Coutts, High performance concentrator tandem solar cells based on IR-sensitive bottom cells, Solar Cells 30 (1991) 363–371.
[41] V.M. Andreev, L.B. Karlina, A.B. Kazantsev, V.P. Khvostikov, V.D. Rumyantsev, S. V. Sorokina, et al., Concentrator Tandem Solar Cells Based on AlGaAs/GaAs–InP/InGaAs (or GaSb) Structures. Proceedings of the First World Conference on Photovoltaic Energy Conversion, Hawaii, 1994, pp. 1721–1724.
[42] V.M. Andreev, V.P. Khvostikov, E.V. Paleeva, V.D. Rumyantsev, S. V. Sorokina, M.Z. Shvarts, et al., Tandem solar cells based on AlGaAs/GaAs and GaSb structures. Proceedings of the 23d International Symposium on Compound Semiconductors, 1996, pp. 425–428.
[43] V.M. Andreev, R&D of III–V compound solar cells in Russia. Technical Digest 11th International Photovoltaic Solar Energy Conference, 1999, pp. 589–592.
[44] M. Umeno, T. Kato, M. Yang, Y. Azuma, T. Soga, T. Jimbo, High efficiency AlGaAs/Si tandem solar cell over 20%. Proceedings of the First World Conference on Photovoltaic Energy Conversion, Hawaii, 1994, pp. 1679–1684.
[45] B.-C. Chung, G.F. Virshup, M. L. Ristow, M.W. Wanlass, 25.2%-efficiency (1-sun, air mass 0) AlGaAs/GaAs/InGaAsP three-junction, two-terminal solar cells, Proceedings of the 22nd IEEE Photovoltaic Specialists Conference, Las Vegas, 1991, pp. 54–57.
[46] V.M. Andreev, V.P. Khvostikov, V.D. Rumyantsev, E.V. Paleeva, M.Z. Shvarts, Monolithic two-junction AlGaAs/GaAs solar cells, Proceedings of the 26th IEEE Photovoltaic Specialists Conference, Anaheim, 1997, pp. 927–930.
[47] M.L. Timmons, J.A. Hutchley, D.K. Wagner, J.M. Tracy, Monolithic AlGaAs/Ge cascade cell. Proceedings of the 21st IEEE Photovoltaic Specialists Conference, Kissimmee, 1988, pp. 602–606.
[48] S.P. Tobin, S.M. Vernon, C. Bajgar, V.E. Haven, L.M. Geoffroy, M.M. Sanfacon, et al., High efficiency GaAs/Ge monolithic tandem solar cells. Proceedings of the 20th IEEE Photovoltaic Specialists Conference, Las Vegas, 1988, pp. 405–410.
[49] P.A. Iles, Y.-C.M. Yeh, F.N. Ho, C.L. Chu, C. Cheng, High-efficiency (>20% AM0) GaAs solar cells grown on inactive Ge substrates, IEEE Electron Device Lett. 11(4) (1990) 140–142.
[50] S. Wojtczuk, S. Tobin, M. Sanfacon, V. Haven, L. Geoffroy, S. Vernon, Monolithic two-terminal GaAs/Ge tandem space concentrator cells. Proceedings of the 22nd IEEE Photovoltaic Specialists Conference, Las Vegas, 1991, pp. 73–79.
[51] P.A. ILes, Y.-C.M. Yeh, Silicon, gallium arsenide and indium phosphide cells: single junction, one sun space, in: L.D. Partain (Ed.), Solar Cells and Their Applications, John Wiley & Sons, 1995, pp. 99–121.

[52] J.M. Olson, S.R. Kurtz, A.E. Kibbler, P. Faine, Recent advances in high efficiency GaInP$_2$/GaAs tandem solar cells. Proceedings of the 21st IEEE Photovoltaic Specialists Conference, Kissimmee, 1990, pp. 24−29.
[53] K.A. Bertness, S.R. Kurtz, D.J. Friedman, A.E. Kibbler, C. Kramer, J.M. Olson, High-efficiency GaInP/GaAs tandem solar cells for space and terrestrial applications. Proceedings of the First World Conference on Photovoltaic Energy Conversion, Hawaii, 1994, pp.1671−1678.
[54] P.K. Chiang, D.D. Krut, B.T. Cavicchi, K.A. Bertness, S.R. Kurtz, J.M. Olson, Large area GaInP/GaAs/Ge multijunction solar cells for space application. Proceedings of the First World Conference on Photovoltaic Energy Conversion, Hawaii, 1994, pp. 2120−2123.
[55] P.K. Chiang, J.H. Ermer, W.T. Niskikawa, D.D. Krut, D.E. Joslin, J.W. Eldredge, et al., Experimental results of GaInP$_2$/GaAs/Ge triple junction cell development for space power systems. Proceedings of the 25th IEEE Photovoltaic Specialists Conference, Washington, D.C., 1996, pp. 183−186.
[56] R.R. King, N.H. Karam, J.H. Ermer, M. Haddad, P. Colter, T. Isshiki, et al., Next-generation, high-efficiency III−V multijunction solar cells. Proceedings of the 28th IEEE Photovoltaic Specialists Conference, Anchorage, 2000, pp. 998−1005.
[57] R.R. King, C.M.Fetzer, P.C. Colter, K.M. Edmondson, J.H. Ermer, H.L. Cotal, et al., High-efficiency space and terrestrial multijunction solar cells trough bandgap control in cell structures. Proceedings of the 29th IEEE Photovoltaic Specialists Conference, New Orleans, 2002, pp. 776−781.
[58] A. Stavrides, R.R. King, P. Colter, G. Kinsey, A.J. McDanal, M.J. O'Neill, et al., Fabrication of high efficiency, III−V multi-junction solar cells for space concentrators. Proceedings of the 29th IEEE Photovoltaic Specialists Conference, New Orleans, 2002, pp. 920−922.
[59] P.K. Chiang, C.L. Chu, Y.C.M. Yeh, P. Iles, G. Chen, J. Wei, et al., Achieving 26% triple junction cascade solar cell production. 2000. Proceedings of the 28th IEEE Photovoltaic Specialists Conference, Anchorage, 2000, pp. 1002−1005.
[60] H.Q. Hou, P.R. Sharps, N.S. Fatemi, N. Li, M.A. Stan, P.A. Martin, et al., Very high efficiency InGaP/GaAs dual-junction solar cell manufacturing at Emcore Photovoltaics. Proceedings of the 28th IEEE Photovoltaic Specialists Conference, Anchorage, 2000, pp. 1173−1176.
[61] A.W. Bett, F. Dimroth, G. Lange, M. Meusel, R. Beckert, M. Hein, et al., 30% monolithic tandem concentrator solar cells for concentrations exceeding 1000 suns. Proceedings of the 28th IEEE Photovoltaic Specialists Conference, 2000, Anchorage, pp. 961−964.
[62] F. Dimroth, U. Schubert, A.W. Bett, J. Hilgarth, M. Nell, G. Strobl, et al., Next generation GaInP/GaInAs/Ge multijunction space solar cells. Proceedings of the 17th European Photovoltaic Solar Energy Conference, Munich, 2001, pp. 2150−2154.
[63] R.R. King, M. Haddad, T. Isshiki, P. Colter, J. Ermer, H. Yoon, et al., Metamorphic GaInP/GaInAs/Ge solar cells. Proceedings of the 28th IEEE Photovoltaic Specialists Conference, Anchorage, 2000, pp. 982−985.
[64] M. Yamaguchi, Y. Ohmachi, T. O'Hara, Y. Kadota, M. Imaizumi, S. Matsuda, GaAs-on-Si solar cells for space use. Proceedings of the 28th IEEE Photovoltaic Specialists Conference, Anchorage, 2000, pp. 1012−1015.
[65] L.M. Fraas, W.E. Daniels, H.X. Huang, L.E. Minkin, J.E. Avery, M.J. O'Neill, et al., 34% efficient InGaP/GaAs/GaSb cell-interconnected-circuit for line-focus concentrator arrays. Proceedings of the 17th European Photovoltaic Solar Energy Conference, Munich, 2001, pp. 2300−2303.

[66] M.J. O'Neill, A.J. McDanal, P.J. George, M.F. Piszczor, D.L. Edwards, D.T. Hoppe, et al., Development of the ultra-light stretched lens array. Proceedings of the 29th IEEE Photovoltaic Specialists Conference, New Orleans, 2002, pp. 916–919.

[67] V.M. Andreev, V.R. Larionov, V.M. Lantratov, V.A. Griükhes, V.P. Khvostikov, V.D. Rumyantsev, et al., Space concentrator module based on short focus linear Fresnel lenses and GaAs/GaSb tandem stacks. Proceedings of the 28th IEEE Photovoltaic Specialists Conference, Anchorage, 2000, pp. 1157–1160.

[68] M.A. Green, Prospects for photovoltaic efficiency enhancement using low-dimensional structures, Nanotechnology 11 (2000) 401–405.

[69] N.J. Ekins-Daukes, J.M. Barnes, K.W.J. Barnham, J.P. Connolly, M. Mazzer, J.C. Clark, et al., Strained and strain-balanced quantum well devices for high-efficiency tandem solar cells, Sol. Energy Mater. Sol. Cells 68 (2001) 71–87.

[70] A. Luque, A. Marti, A metallic intermediate band high efficiency solar cell, Prog. Photovolt: Res. Appl. 9 (2001) 73–86.

[71] A. Marti, L. Cuadra, A. Luque, Partial filling of a quantum dot intermediate band for solar cells, IEEE Trans. Electron Devices ED-48(10) (2001) 2394–2399.

CHAPTER 1D-2

High Efficiency III−V Multijunction Solar Cells

Simon P. Philipps, Frank Dimroth, and Andreas W. Bett
Fraunhofer Institute for Solar Energy Systems ISE, Freiburg, Germany

Contents

1. Introduction 417
2. Special Aspects of III−V Multijunction Solar Cells 419
   2.1 Fields of Application and Reference Conditions 419
   2.2 Band-Gap Choice 421
   2.3 Band Gap versus Lattice Constant 423
   2.4 Tunnel Diodes 424
   2.5 Characterisation 425
   2.6 Design of Concentrator Solar Cells 425
3. III−V Solar Cell Concepts 427
   3.1 Lattice-Matched Triple-Junction Solar Cells on Ge 429
   3.2 Quantum Well Solar Cells 430
   3.3 Upright Metamorphic Growth on Ge 430
   3.4 Inverted Metamorphic Growth 433
   3.5 Bifacial Growth 434
   3.6 III−V on Si 434
   3.7 More Than Three Junctions 436
   3.8 Other Approaches 439
4. Conclusions 440
Acknowledgements 440
References 440

1. INTRODUCTION

Solar cells made of III−V semiconductors reach the highest efficiencies of any photovoltaic technology so far. The materials used in such solar cells are composed of compounds of elements in groups III and V of the periodic table. Figure 1 shows the development of record efficiencies of III−V multijunction solar cells under concentrated sunlight over the last two decades. An impressive increase from about 32% in the early 1990s to

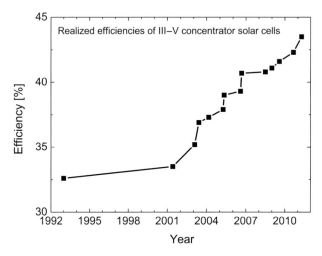

Figure 1 Development of best-realized efficiencies of III–V multijunction concentrator solar cells. Data are based on the Solar Cell Efficiency Tables, in which record efficiencies have regularly been published since 1993 [1]. The latest edition considered here is reference [2].

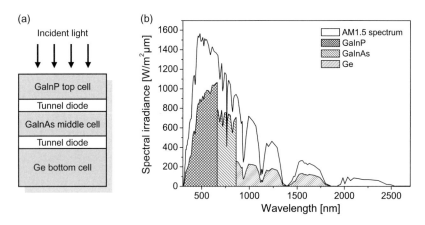

Figure 2 (a) Schematic structure of a monolithic GaInP–GaInAs–Ge triple-junction solar cell, which represents the state-of-the-art approach for III–V multijunction solar cells. (b) Spectral irradiance of the AM1.5 spectrum together with the parts of the spectrum that can be used by a triple-junction solar cell.

more than 43% in 2011 has been achieved. The prerequisite for such high efficiencies is the ability to stack solar cells made of different III–V semiconductors. This enables an efficient use of the solar spectrum. Figure 2(a) shows a scheme of a typical triple-junction solar cell. Three subcells consisting of GaInP, GaInAs, and Ge are stacked on top of each other and are series interconnected by tunnel diodes. The key to efficient use of the solar spectrum is that each subcell has a higher band gap than the one below it. In this way, each subcell absorbs light from a spectral range closest to its band gap, hence reducing thermalisation losses (Figure 2(b)). Moreover, transmission losses can also be reduced if the lowest band gap of the stack has a lower band gap than do the conventional single-junction solar cells.

This approach results in a device with only one positive and one negative contact. As the subcells are connected in series within the multijunction solar cell, the total current is limited by the lowest current generated by one of the subcells. Therefore, current matching of the subcells is a central design aspect for III−V multijunction solar cells, which will be discussed in more detail in this chapter. In recent years, III−V multijunction solar cells have usually been grown by metal-organic vapour phase epitaxy (MOVPE) reactors, resulting in favourable economics of growth as well as high crystal quality. Large-area commercial MOVPE reactors are available from different companies.

III−V multijunction solar cells are used in different applications, the most prominent being satellites and space vehicles as well as terrestrial concentrator systems. Record efficiencies of 34.2% (AM0, 1367 W/m^2) [3] and 43.5% (AM1.5d, 418 suns) [2] have already been realized in these fields. Intensive research is ongoing to further optimize the cell structures in order to achieve even higher efficiencies. Various approaches for III−V multijunction solar cells are currently investigated. This chapter summarizes the state of the art as well as recent trends of these devices. The first part describes special features of III−V multijunction solar cells in comparison to conventional single-junction solar cells. The second part discusses some of the different approaches and designs for III−V solar cells.

2. SPECIAL ASPECTS OF III−V MULTIJUNCTION SOLAR CELLS

III−V multijunction solar cells differ from conventional single-junction solar cells in several aspects. This section introduces these features.

2.1 Fields of Application and Reference Conditions

III−V multijunction solar cells are used in different applications. They have become the state-of-the-art photovoltaic power generator for satellites and space vehicles [4]. This development was driven by the fact that

III—V multijunction solar cells are particularly suitable for specific needs in space. They offer high reliability, a high power-to-mass ratio, excellent radiation hardness, small temperature coefficients, and the possibility to operate at high voltage and low current [5]. Despite their higher production costs compared to silicon solar cells, III—V multijunction solar cells are integrated into flat-plate modules for space applications. This becomes feasible as the determining measure for cost in space applications are €/kg rather than €/W_p as in terrestrial applications. The different measure originates from the consideration of launch costs as well as of spacecraft attitude control [6]. Due to their higher power-to-mass ratio (W/kg), flat-plate modules of III—V multijunction solar cells are beneficial under these cost considerations.

The use of III—V multijunction solar cells in flat-plate modules on Earth would currently be too expensive. However, the expensive cell area can be reduced by using a cost-efficient concentrating optic. In recent years, many companies have implemented this idea by placing III—V multijunction solar cells into terrestrial concentrator systems. Most of these systems use high concentration factors above 400, which enables a significant cost reduction and also leads to higher efficiencies. An extensive overview about CPV can be found in reference [7]. For a recent review on the status and forecast of CPV efficiencies, see reference [8].

In addition to their use in space and terrestrial concentrator systems, III—V solar cells are also used in several niche applications. One of these is thermophotovoltaics (TPV). In such systems, light from an emitter other than the sun is converted to electricity by photovoltaic cells. The emitter can either be a flame or a material that is heated to a temperature between 1000°C and 1500°C [9] by the sun or by burning fuel. As the emitted spectrum is shifted toward longer wavelengths compared to the spectrum of the sun, TPV photovoltaic cells need to have a rather small band gap. Materials like Ge, GaSb, and InGaAs(Sb) are suitable. TPV could, for example, enable the use of industrial waste heat for electricity generation and offer advantages like high-power density outputs, in-phase supply and demand, and potential low cost [10]. A detailed overview about TPV can be found in references [11,12].

III—V photovoltaic cells are also used as laser power converters, which convert light emitted by a laser into electricity [13—18]. This is a promising alternative to using copper wires as source and load can be electrically isolated giving improved safety, e.g., in explosive areas as well as reducing the influence of electromagnetic pulses and interferences. Possible applications of these power-by-light systems are sensor applications in industrial monitoring as well as medical diagnostic tools [18]. Recently, an optically powered camera video link was realized, which proves that complex information can be transmitted in such a system [19].

Depending on the intended field of application III−V multijunction solar cells are rated with different reference spectra. For space applications the reference spectrum AM0 (with a total irradiance of 1367 W/m^2) is used [20]. For terrestrial concentrator applications the AM1.5d spectrum (1000 W/m^2) is applicable, which only takes direct irradiance into account. Due to the concentration of the incident light a concentration ratio needs to be indicated for measurements under AM1.5d. The unit of this factor is 'suns' with '1 sun' corresponding to unconcentrated incident light. For solar cells in conventional flat-plate modules on Earth the global AM1.5 g spectrum is applicable. Both terrestrial reference spectra are currently defined in the norm ASTM G173 03 [21].

2.2 Band-Gap Choice

One of the benefits of using III−V semiconductors for multijunction solar cells is the wide flexibility in band-gap combinations that can be realized. Thus, the first decision to be made when designing a III−V multijunction solar cell is the number of junctions and band-gap combinations to be chosen. Ignoring possible restrictions due to other material properties such as lattice constant, achievable material quality and availability, the question of choosing the optimal band-gap combinations comes down to optimizing in which parts the solar spectrum should be divided by the multijunction solar cell (see Figure 2). This determines on one hand the current densities of the subcells and on the other hand their voltages.

Different models are used for this optimisation (for an overview see Kurtz et al. [22]). One widely used method is the 'detailed balance approach' suggested by Shockley and Queisser [23], which allows us to calculate the theoretical conversion efficiency of a solar cell with a given band-gap energy under a defined spectrum. Only radiative recombination is considered because this is the only unavoidable recombination mechanism. In addition, ideal solar cells are assumed that have an external quantum efficiency of unity and behave according to the one-diode model. As a rule of thumb, between 70% and 80% of the theoretical efficiencies can be achieved in reality. This approach was, for example, implemented for

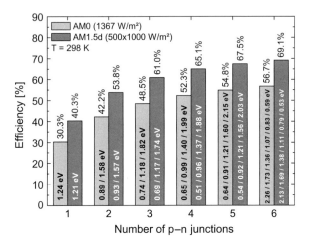

Figure 3 Theoretical efficiency limit versus number of p–n junctions under the reference spectrum AM0 (1367 W/m²) for space applications [20] as well as under the reference spectrum AM1.5d (500 × 1000 W/m²) for concentrator solar cells [21]. The calculation was carried out with the program etaOpt [24], which implements the detailed balance approach of Shockley and Queisser [23].

single and multijunction solar cells in the program etaOpt [24], which is also available for download on the Website of Fraunhofer ISE. Figure 3 shows the maximum efficiencies (AM1.5d, 500) suns calculated with etaOpt for different numbers of p–n junctions. The numbers in the bars indicate the optimal band-gap combination. The efficiency increases with the number of subcells. However, the gain of efficiency for any additional junction gets smaller with increasing number of junctions.

In recent years a trend toward evaluating the potential of different band-gap combinations and solar cell concepts for terrestrial concentrator applications in terms of energy yield is observable, e.g., [25–29]. This is motivated by the fact that the main value of interest in real applications is not the efficiency under a reference spectrum but the annual energy production under realistic operating conditions. Since multijunction solar cells are known to be sensitive to changes in the solar spectrum, e.g., [25,30–34], spectral variation throughout day and year should be taken into account when calculating the annual energy production of these solar cells. Owing to this aspect the potential of several solar cell concepts discussed in this chapter is evaluated with the energy-harvesting efficiency which is the total energy produced by a cell in a year divided by the incident solar energy at the investigated location [26].

2.3 Band Gap Versus Lattice Constant

Realizing III−V multijunction solar cells with an optimal band-gap combination can be a challenging task. To understand the problems it is important to note that each (III−V) semiconductor is characterized by a band gap as well as a characteristic lattice constant. Figure 4 shows the relation between band gap and lattice constant for various semiconductors which are important for III−V solar cells. The band-gap combination of the most common $Ga_{0.50}In_{0.50}P$−$Ga_{0.99}In_{0.01}As$−Ge triple-junction solar cell (Figure 9) is indicated as an example. As all materials in this structure nearly have the same lattice constant, the approach is called *lattice matched*. In contrast to this devices with material combinations that are not lattice matched to each other are called *lattice mismatched* or *metamorphic*.

Growing layers with different lattice constant on top of each other causes the formation of dislocations which need to be confined in order to ensure high material quality. As a high number of dislocations can already be expected for a relatively low lattice mismatch of 1−2%, it is obvious that particular care has to be taken when realizing metamorphic structures.

Another approach for realizing multijunction solar cells is to fabricate individual solar cells with different band-gap energies, which are then

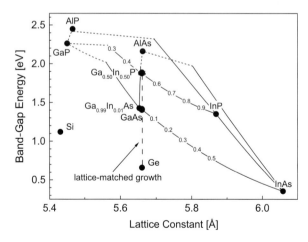

Figure 4 Band gap as a function of lattice constant for exemplary semiconductors. Ternary compounds are indicated by lines between binary crystals. Solid lines refer to direct band-gap semiconductors, and broken lines mark indirect band-gap semiconductors. The band-gap combination of the most common lattice-matched $Ga_{0.50}In_{0.50}P$−$Ga_{0.99}In_{0.01}As$−Ge triple-junction solar cell is marked.

stacked mechanically (e.g., [35]). The main advantages of this approach are that the individual cells do not need to be lattice-matched and that each solar cell can be contacted individually. Mechanically stacked III–V multijunction solar cells show good efficiencies (e.g., [36–39]). However, the complexity of fabrication and assembly as well as the higher material costs due to the multiple substrates usually lead to higher overall costs than for monolithic multijunction solar cells. Therefore, this approach has been used less frequently in recent years.

2.4 Tunnel Diodes

The heart of the subcells in most multijunction solar cells is realized as a thin n-doped emitter on a thick p-doped base layer. Stacking such n-on-p junctions would lead to p-on-n diodes in between the subcells, which would block current flow. Thus, solutions for the interconnection of the subcells need to be implemented.

A suitable interconnector must have a low electrical resistivity and a high optical transmissivity, and it has to be integrated into the structure. Esaki interband tunnel diodes [40] have become standard for this purpose. They are realized through thin highly doped p-on-n diodes between the subcells. Figure 5(a) shows the band diagram of an exemplary GaAs–GaAs tunnel diode. Due to the high doping levels, the quasi-Fermi level (E_F) on

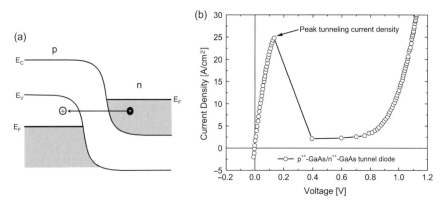

Figure 5 (a) Schematic band diagram of a tunnel diode. (b) Measured IV curve of a GaAs–GaAs tunnel diode with a peak tunnelling current density of above 25 A/cm². (Both graphs after [41]. Reproduced with permission, © 2008 John Wiley & Sons, Ltd.)

each side of the p−n junction moves into the valence (E_V) and conduction (E_C) band, respectively. If a small positive voltage is applied to the junction full states on the n side become aligned with empty states on the p side (as shown in Figure 5(b)). Based on quantum mechanical principles, charge carriers can now tunnel through the barrier. The right graph in Figure 5 shows an exemplary current−voltage curve of a tunnel diode. High current densities flow at low voltages. However, the current flow falls off strongly after the characteristic peak tunnelling current density as full states on the n side are no longer aligned with empty states on the p side in the corresponding voltage range. At significantly higher voltages, the tunnel diode behaves like a conventional p−n junction, leading to another increase in current. As the targeted operating range of the tunnel diode is between 0 V and the voltage at which the peak tunnelling current density is reached, tunnel diodes for multijunction solar cell need to have a sufficiently high peak tunnelling current density. Realizing such tunnel diodes is a key issue in the development process of monolithic multijunction solar cells especially if high concentration levels are targeted.

2.5 Characterisation

The series interconnection of the subcells, the wide absorption range, and possibly high concentration factors pose additional challenges to the experimental characterisation of III−V multijunction solar cells compared with conventional single-junction solar cells. One example is the measurement of the external quantum efficiency (EQE). Due to the close proximity and the series interconnection of the subcells, interactions between the subcells can lead to measurement artefacts [42−45]. With adequate measurement routines, the artefacts can be eliminated. A detailed description of the characterisation of III−V multijunction solar cells can be found in reference [46].

2.6 Design of Concentrator Solar Cells

A particular challenge arises for III−V concentrator solar cells as various CPV systems exist today. Therefore, concentrator solar cells with different sizes and geometries are requested. Figure 6 shows an example of a test wafer with different solar cell designs. Apart from the different sizes and geometries, the solar cells differ in the structure of the front contact grid.

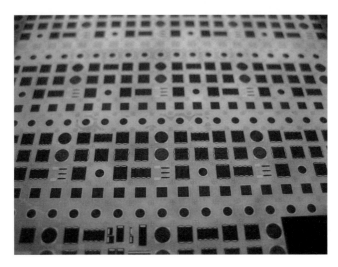

Figure 6 Picture of a solar cell wafer with concentrator solar cells of different geometry and grid design.

The grid structure has a significant influence on the solar cell performance and should be optimized for the concentration ratio of the particular CPV system. The optimisation is usually supported through numerical modelling (e.g., [47–51]).

The potential of grid optimisation on the efficiency of a GaAs single-junction solar cell is visualized in Figure 7 [51]. While the measured solar cell parameters that are linked to the epitaxial layer structure and the anti-reflective coating were kept constant, the grid design and the solar cell size were optimized for different concentration levels. Efficiency-versus-concentration curves are exemplarily shown for optimized grids for 100, 450, and 1000 suns. The solid line indicates the efficiency that can be reached with the optimal grid and size for each concentration level. Note that a practical minimum size of 1 mm^2 is defined here. For comparison, a curve without this size restriction is also shown. The global maximum of 29.09% is reached at 450 suns.

Other challenges for III–V concentrator solar cells include inhomogeneous light profiles caused by the concentrating optics, which can lead to significantly different current densities throughout the solar cell (e.g., [52]). In addition, the tunnel diodes within multijunction solar cells need to be capable of supporting the high current densities (see preceding discussion).

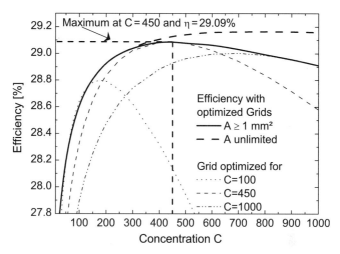

Figure 7 Simulated efficiency versus concentration ratio for GaAs solar cells with grid designs optimized for 100, 450, and 1000 suns. In addition, the highest overall efficiency for grid-optimized solar cells with limited (A ≥ 1 mm²) and unlimited cell area is shown. The global efficiency maximum of 29.09% is reached at 450 suns. Note that only the grid structure and the cell size were optimized, whereas the semiconductor layer structure and the antireflection coating of the solar cell were not changed. *(After [51]. Reproduced with permission,* © *2010 John Wiley & Sons, Ltd.)*

3. III–V SOLAR CELL CONCEPTS

To further increase the efficiency of III–V multijunction solar cells, various concepts are currently being investigated by research groups around the world. Figure 8 shows the approaches that will be discussed in the following section. Other recent overviews can be found in references [5,53–55]. Note that the schematic drawings are strongly simplified. In reality, each subcell consists of many layers, tunnel junctions are placed between the subcells, metamorphic buffer layers are composed of several layers, and antireflective coatings are placed on top of the device (see Figure 9).

The main focus of research nowadays is on III–V multijunction solar cells with three or more junctions, so this chapter emphasizes these concepts. Note, however, that single- and dual-junction solar cells are still also being investigated. What we learn from these simpler devices might help us optimize more complex approaches. For III-V single-junction concentrator solar cells a record efficiency of 29.1% (AM1.5d,

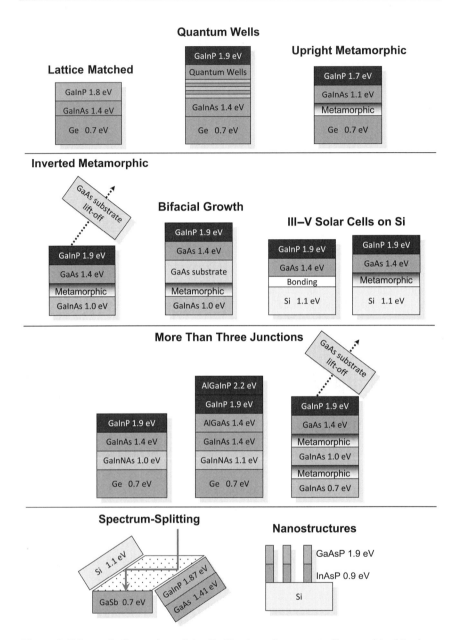

Figure 8 Schematic illustration of the III—V solar cell concepts discussed in this chapter. Subcells and metamorphic buffers are indicated. Note that some concepts have also been realized with different materials and band-gap combinations.

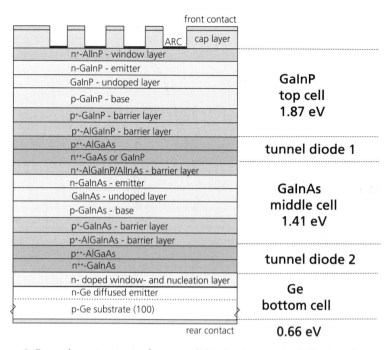

Figure 9 Exemplary structure of a monolithic lattice-matched III−V triple-junction solar cell. The band gap of the subcells decreases from top to bottom. The subcells are interconnected in series by tunnel diodes within the device.

117 suns) was achieved by Fraunhofer ISE with a crystalline GaAs solar cell. With the same material, Alta Devices recently realized a thin-film single-junction solar cell with an efficiency of 28.1% under AM1.5g [2,56]. A record value of 32.6% under 1000 suns (AM1.5d) was achieved by the UPM Madrid with a monolithic $Ga_{0.51}In_{0.49}P-GaAs$ dual-junction solar cell [57]. Note that all recent efficiency records can be found in Green et al. [2]. A historical overview of III−V solar cells can be found in Chapter Id-1 in this book and in references [58,59].

3.1 Lattice-Matched Triple-Junction Solar Cells on Ge

The state-of-the-art III−V solar cell in space and concentrator applications is the lattice-matched $Ga_{0.50}In_{0.50}P-Ga_{0.99}In_{0.01}As-Ge$ triple-junction solar cell, which is shown exemplarily in Figure 9. In this device, all materials have nearly the same lattice constant (see Figure 4), which facilitates achieving good material quality. A champion efficiency of 41.6% (AM1.5d,

364 suns) has already been achieved [60], hence showing the mature status of the lattice-matched approach.

However, the band-gap combination of this device leads to a not optimal split of the solar spectrum, causing a strong excess current in the Ge bottom cell. Therefore, various approaches try to overcome the strong current mismatch by increasing the absorption in the upper two cells (see sections 3.2 and 3.3) or by integrating an additional junction between the GaInAs and the Ge subcell (see section 3.7).

3.2 Quantum Well Solar Cells

The deficiency of high excess current in the bottom cell of the lattice-matched triple-junction solar cell can be partly reduced by implementing quantum wells (QWs) into the middle cell. These can be realized by thin alternating layers of $GaAs_yP_{1-y}$ and $Ga_xIn_{1-x}As$ (Figure 10(a)) [61]. Quantum wells with a band-gap energy lower than 1.4 eV extend the absorption of the middle cell toward longer wavelengths [62,63]. Thus, the current of the middle cell is increased at the expense of a reduction of the excess current in the Ge bottom cell. The lowest transition energy in the quantum wells and the number of wells determine the current density, which is achievable for the middle cell. However, the open-circuit voltage of QW solar cells is lower than for a cell structure without wells [62]. This drawback can be compensated by the increased current density due to the QWs. Theoretical calculations showed that an overall gain in energy-harvesting efficiency between 3% and 9% relative to the lattice-matched triple-junction solar cell is possible [28]. In addition, QWs are seen as a possibility to tune solar cells for specific spectral conditions [29]. Moreover, it was reported that QW solar cells may lead to higher radiation hardness in space [64].

Triple-junction solar cells with quantum wells have already been experimentally realized. Figure 10(b) shows a sample $Ga_{0.50}In_{0.50}P-Ga_{0.99}In_{0.01}As-Ge$ solar cell with 40 QWs in the middle cell [65]. The quantum wells lead to an increase of the EQE of the middle cell linked with a decrease of the bottom cell's EQE. Therefore, a better current-matching of the subcells can be achieved. For a recent overview about quantum well solar cells, see references [66,67].

3.3 Upright Metamorphic Growth on Ge

As discussed previously, the band-gap combination of the lattice-matched triple-junction is not optimal because it leads to large excess current in the Ge

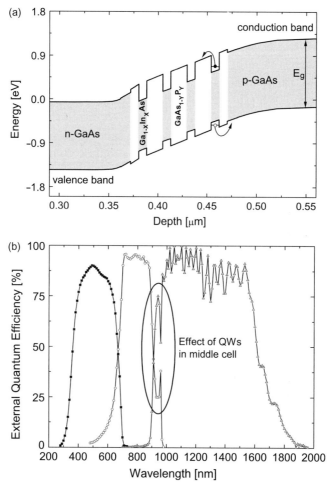

Figure 10 (a) Schematic band diagram of four QWs in a GaAs solar cell (Reprinted with permission from [68]. Copyright 2010, American Institute of Physics.). (b) Measured external quantum efficiency of a $Ga_{0.50}In_{0.50}P-Ga_{0.99}In_{0.01}As-Ge$ solar cell with 40 QWs in the middle cell *(after [65] Reproduced with permission, © 2010 IEEE.)*. The EQE in the band-gap region of the middle cell increases, leading to a corresponding reduction of the EQE of the bottom cell.

bottom cell. This deficiency originates from the large band-gap difference between the $Ga_{0.99}In_{0.01}As$ middle cell (1.41 eV) and the Ge bottom cell (0.66 eV). Calculations show that lower band gaps for the top and middle cells lead to a higher theoretical efficiency under AM1.5d [69,70] and to higher energy yields [26,27]. As the band gaps of $Ga_yIn_{1-y}P$ and $Ga_xIn_{1-x}As$ decrease with increasing In content, these materials can also be used to realize a more optimal band-gap combination. However, an additional technical challenge arises as the lattice constant increases with higher In content (lower band gap, see Figure 4). The monolithic growth of materials with different lattice constants leads to misfit dislocations that deteriorate the material quality. Therefore, lattice-mismatched or metamorphic approaches require adequate strategies to mitigate the effect of dislocations. This is achieved through the implementation of adequate buffer structures between the Ge bottom cell and the GaInAs middle cell [71,72]. These buffer structures increase the lattice constant gradually and hence reduce or confine dislocations as shown in Figure 11(a). A comparison of the external quantum efficiency of a lattice-matched solar cell and a metamorphic triple-junction solar cell is shown in Figure 11(b). The absorption range of the upper two subcells in the metamorphic structure are extended toward longer wavelengths. This leads to higher current generation compared to the lattice-matched structure.

Efficiencies above 40% have already been realized with two variants of this approach [70,73], hence proving that the misfit dislocations due to the differences in lattice constant can be handled successfully within the structure. It is also noteworthy that the triple-junction solar cell in reference [70] is well current-matched under AM1.5d. Further improvements can be expected based on the significantly higher theoretical potential of

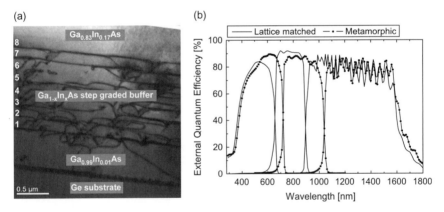

Figure 11 (a) Cross-sectional transmission electron micrograph (TEM) of a step-graded $Ga_{1-x}In_xAs$ buffer layer grown on Ge (after [70], reprinted with permission. Copyright 2009, American Institute of Physics.). The In content is increased in seven steps from 1% to 17% (1–7) followed by another layer with 20% (8), which helps to fully relax the buffer. (TEM measured at the Christian-Albrechts-University in Kiel, Germany). (b) Comparison of the external quantum efficiency of a lattice-matched $Ga_{0.50}In_{0.50}P–Ga_{0.99}In_{0.01}As–Ge$ and a metamorphic $Ga_{0.35}In_{0.65}P–Ga_{0.83}In_{0.17}As–Ge$ solar cell *(after [74]. Reprinted with the permission of Cambridge University Press.).* The band-gap energies of the two upper subcells of the metamorphic approach are shifted toward longer wavelengths (lower energies).

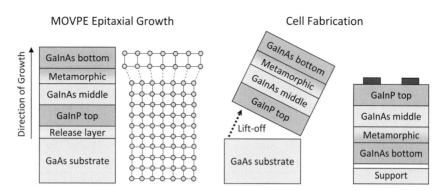

Figure 12 Scheme of the production process of IMM solar cells. The epitaxial structure is grown inverted to the conventional growth direction. The first two subcells can hence be grown lattice matched to the GaAs or Ge substrate. A metamorphic buffer gradually increases the lattice constant for the subcell that is grown last. The substrate is then removed using liftoff techniques. Finally, the epitaxial structure is processed to solar cells. *(Illustration inspired by [79].)*

the metamorphic approach in comparison to the lattice-matched approach [26,27].

3.4 Inverted Metamorphic Growth

A challenge of the upright metamorphic concept on Ge is that the lattice mismatch is introduced early in the growth process. Thus, imperfections of the grading buffer can affect the material quality of both upper subcells. To reduce this possible deficiency, research on inverted metamorphic (IMM) growth has been intensified in recent years. In the IMM approach, the growth direction is inverted, hence the top cell is grown first followed by the other subcells (see Figure 12). The substrate is later removed from the top cell using liftoff techniques. This approach allows growing lattice matched on the substrate first, while buffer layers are postponed to later growth phases. Another possible advantage is that the bottom cell is grown epitaxially rather than being created by diffusion into the Ge substrate, hence allowing higher flexibility for the bottom cell band gap. Geisz et al. [75] presented an IMM structure with an efficiency of 40.8% (AM1.5d, 326 suns). The $Ga_{0.51}In_{0.49}P$ top cell (1.83 eV) was grown lattice matched on a GaAs substrate. A buffer is then used to increase the lattice constant by 0.3% for the $Ga_{0.96}In_{0.04}As$ middle cell (1.34 eV) followed by another buffer for the $Ga_{0.63}In_{0.37}As$ bottom cell (0.89 eV, 2.6% misfit). Several other IMM designs are also investigated, e.g., in [76–79]. With subcells of GaInP (1.9 eV), GaAs (1.4 eV) and GaInAs (1.0 eV) Cornfeld et al. [76] reached an efficiency of 32% under AM0 (1367 W/m^2), while Takamoto et al. [79] achieved an efficiency of 35.8% under AM1.5g (1000 W/m^2).

A possible drawback for the IMM approach compared to Ge-based structures is the higher complexity of cell processing connected with substrate removal, which may lead to higher production costs and low yield. Yet these disadvantages might be counterbalanced by lower material costs if the substrate can be reused. Another benefit of the IMM structure in particular for space application is their possible lower weight and the chance to realize flexible devices.

3.5 Bifacial Growth

Another approach that deviates from the conventional growth direction is bifacial growth [80], which recently led to an efficiency of 42.3% (AM1.5d, 406 suns) [81]. The specific feature is that subcells are grown on both sides of a GaAs substrate. First, a graded buffer is grown on the backside of a GaAs wafer followed by a GaInAs bottom cell (0.95 eV). The wafer is then flipped within the MOVPE reactor, and a tunnel junction, a GaAs middle cell (1.42 eV), another tunnel junction, and a GaInP top cell (1.89 eV) are grown. The advantage of this approach is that the two upper cells can be grown lattice matched to the GaAs substrate. In addition, the thick substrate protects the upper two cells from dislocations originating from the lattice-mismatched GaInAs bottom cell. The complexity of cell processing is similar to upright metamorphic approaches.

3.6 III−V on Si

Since the early 90s research efforts have been ongoing to grow III−V solar cells on silicon substrates (1.1 eV), e.g., [82−84]. The interest in this field has increased in recent years in order to replace the expensive Ge substrate of today's lattice-matched triple-junction solar cells with lower-cost Si. Moreover, the higher band gap of Si compared with Ge leads to an increase of the theoretical energy-harvesting efficiency by 3% [28].

The main challenges here are to overcome the 4.1% difference in lattice constant between Si and GaAs and to handle the difference in thermal expansion coefficient. In general, two different approaches need to be distinguished: direct growth on the Si substrate and wafer bonding. The first approach requires strategies to ensure sufficient material quality in the III−V subcells. As it is extremely challenging to achieve good material quality for direct GaAs growth on Si (e.g., [83,85]) different

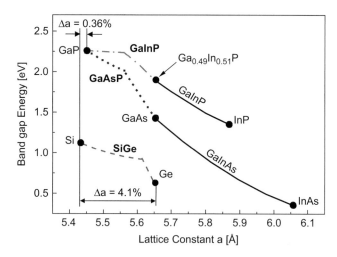

Figure 13 Different strategies to overcome the 4.1% difference in lattice constant between GaAs and Si. One option is to first create a Ge layer either directly or through the use of SiGe compounds to transfer from Si to Ge (e.g., [87,90,91]). Other options are to use $Ga_{1-x}In_xP$ or $GaAs_xP_{1-x}$ buffers on a GaP nucleation (e.g., [92–95]). *(Graphic from [94]. Reproduced with permission.)*

buffer layers are investigated to increase the lattice constant gently. Several of the investigated strategies are shown in Figure 13. GaAs or GaInP solar cells on Si substrates have already been demonstrated (e.g., [86–89]). Research efforts are continuing to realize III–V triple-junction solar cells on Si with efficiencies close to the potential of this approach.

The necessity of a suitable buffer can be avoided by using dilute nitrides (GaIn)(NAsP). This material system offers a wide band-gap range from 1 eV to 2 eV and allows choosing materials that are lattice matched to Si. Yet this promising approach has been limited so far by the short minority-carrier diffusion length of the dilute nitrides (e.g., [96–98]).

The use of wafer-bonding [53,99,100] relaxes the challenge of realizing a transition buffer as the silicon bottom cell and the upper subcells (e.g., a GaInP–GaAs dual-junction solar cell) are grown independently and then combined through wafer bonding. After the bonding process, the substrate of the upper two solar cells is removed by using liftoff techniques such as ion implantation [101] and laser [102], stress-induced [103], or wet chemical liftoff [104]. Both the liftoff process and the bonding lead to technological challenges. However, promising result have already been achieved (e.g., [105–107]). Recently, a GaInP–GaAs–Si triple-junction solar cell with an efficiency of 23.3%

under AM1.5d (24 suns) has been demonstrated at Fraunhofer ISE [108]. It was realized through wafer bonding of a GaInP–GaAs dual-junction and a silicon solar cell.

3.7 More Than Three Junctions

As shown in Figure 3, the theoretical efficiency limit of a multijunction solar cell under reference conditions increases with the number of junctions. For applications with rather stable spectral conditions such as satellites in Earth orbit, this trend directly translates into higher efficiencies. Yet under real operating conditions on Earth or other planets like Mars [34], spectral changes throughout the day and year need to be taken into account. To evaluate the effect of an additional junction on the energy yield, the energy-harvesting efficiency can be studied, for example, using a model as presented in reference [26]. Figure 14 shows the energy-harvesting efficiency and the optimal band-gaps for multijunction solar cells with up to six p–n junctions for the varying spectral conditions at three exemplary locations [28]. For all three locations, the energy-harvesting efficiency increases steadily from three to six junctions. Thus, even under varying spectral conditions, a higher power output for solar

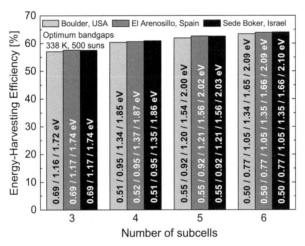

Figure 14 Calculated energy-harvesting efficiency of ideal multijunction solar cells with three to six junctions for three different locations on Earth. Calculations are performed for a concentration factor of 500 suns and a cell temperature of 338 K. Spectral influences of the concentrator optics are not taken into account. The optimum combination of band-gap energies is given for each number of junctions. *(After [28]. Reproduced with permission, © 2010 IEEE.)*

cells with increasing number of subcells can be expected. However, the benefits of adding more junctions become smaller: adding a fourth junction leads to 5% higher efficiencies on average, whereas a fifth junction only adds another 3% to the energy-harvesting efficiency. It needs to be taken into account that the model assumes ideal solar cells. Therefore, it is an open question if the small theoretically expected gain in energy yield for solar cells with more than four junctions can be achieved in reality. Another important point to note is that the ideal band-gap energies at the three locations only vary by 10–20 meV. This indicates that it is not necessary to develop a different solar cell for each operating site.

One straightforward approach for III–V multijunction solar cells with more than three junctions is to grow lattice matched on Ge substrates. Starting from the standard lattice-matched $Ga_{0.50}In_{0.50}P-Ga_{0.99}In_{0.01}As-Ge$ solar cell, theoretical calculations show that a 1-eV fourth junction placed above the Ge bottom cell could boost the efficiency significantly. Research efforts are ongoing to realize such a junction with the quaternary alloy GaInNAs [60,96,109–111]. Yet up to now, the subcell suffers from a low minority-carrier diffusion length, leading to current limitation in a four-junction solar cell. One way to work around this deficiency is to use a five- or six-junction configuration with a smaller current of each subcell. Such a device can be realized by adding subcells of AlGaInP and AlGaInAs. Figure 15 (a) shows the internal quantum efficiency of such a six-junction solar cell, which was recently realized at Fraunhofer ISE [111]. Note that the complex device is composed of more than 40 semiconductor layers, leaving large room for ongoing optimisation. The IV characteristic of the six-junction solar cell under AM0 is compared with those of triple- and four-junction solar cells in Figure 15(b). The voltage increases with the number of junctions, while the current decreases. The IV characterisation of the four- and six-junction solar cells was performed under a novel flash simulator with six independently variable light channels, which allows the spectrum to be adjusted for each junction.

An option for a lattice-matched five-junction solar cell is to include junctions of AlGaInP and AlGaInAs into a lattice-matched triple-junction solar cell [112,113]. The main benefit of this structure is that the thickness of the subcells is significantly reduced. Thus, a low minority-carrier lifetime has a comparably smaller impact, which improves the radiation hardness of the device. In addition, only well-known materials are used in this structure. Another approach, which is lattice matched to Ge, is the AlGaInP–AlGaInAs–GaInAs–Ge (1.95 eV, 1.66 eV, 1.39 eV, and 0.72 eV, respectively)

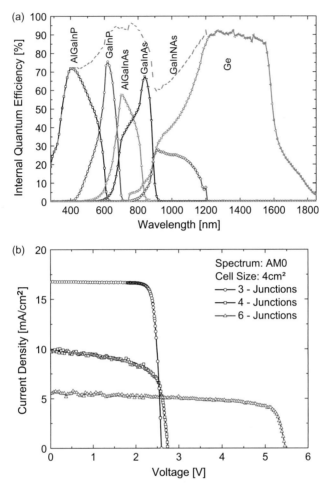

Figure 15 (a) Measured internal quantum efficiency of an AlGaInP–GaInP–AlGaInAs–GaInAs–GaInNAs–Ge six-junction solar cell. (b) Comparison of measured IV characteristic under AM0 of the six-junction solar cell with a GaInP–GaAs–Ge triple-junction and an AlGaInP–GaInAs–GaInNAs–Ge four-junction solar cell. *(Both graphs after [111]. Reproduced with permission.)*

four-junction solar cell realized by King et al. [60]. A prototype device already achieved an efficiency of about 37% (AM1.5d, 500 suns) in preliminary measurements.

A different route to realizing III–V solar cells with more than three junctions is to use inverted metamorphic [77,114] or semiconductor-bonding techniques [115]. Cornfeld et al. [116] presented an inverted metamorphic four-junction solar cell that recently reached an in-house measured efficiency above 34% under AM0 [3]. A bottom cell of GaInAs (0.7 eV) was added to a formerly developed inverted metamorphic triple-junction solar cell [76]. Another metamorphic buffer was implemented to accommodate the additional 2% lattice mismatch. Moreover, designs for inverted metamorphic solar cells with five and six junctions have also been proposed [3,60].

Law et al. [115] realized a 31.7% (AM0) five-junction solar cell using a direct semiconductor-bonding technique. The three upper subcells—AlGaInP (2.0 eV), AlGa(In)As (1.7 eV), and Ga(In)As (1.4 eV)—were grown inverted on GaAs or Ge substrates, while a InP substrate was used for the upright growth of the two lower subcells consisting of GaInPAs (1.1 eV) and GaIn(P)As (0.8 eV). The epitaxial wafers are then bonded together using semiconductor-bonding techniques. An advantage of this approach is that all subcells are grown lattice matched to the corresponding substrate. Recently, an AM0 efficiency of around 31% was presented for a semiconductor-bonded four-junction solar cell with a band-gap combination of 1.9—1.4—1.0—0.73 eV [78].

3.8 Other Approaches

The challenge of current matching in monolithic multijunction solar cells can be circumvented by using optical beam splitting (for a detailed review, see Imnes et al. [117]). In a spectrum-splitting architecture, spectrally selective filters are used to split the incident solar radiation and to direct the light toward different single- or dual-junction solar cells, which are individually designed for the corresponding wavelength range. This allows realizing optimal band-gap combinations without the usual constraints of lattice or current matching. A light-splitting approach was already experimentally realized in 1978 by Moon et al. [118] and has found increasing interest again in recent years due to progress in dichroic filter technology [119—122]. Theoretical calculations of the energy-harvesting efficiency showed that an optical beam splitting approach with a GaInP—GaAs (1.87—1.41 eV), a Si (1.10 eV), and a GaSb (0.7 eV) solar cell has a theoretical potential that is 18% higher than theoretical values for a lattice-matched GaInP—GaInAs—Ge solar cell [28]. Another recent theoretical study also presented the result that operating efficiencies of solar cell arrays with two to four cells are expected to be higher in a spectral-splitting approach than for series interconnection [123]. Several spectral-beam splitting receivers have recently been realized showing promising efficiencies [120—122,124,125]. The devices are still in a prototype stage.

Further optimisation of these rather complex structures as well as steps toward mass production must be undertaken.

Another approach for optimal band-gap combinations are nanowire solar cells in which materials with different lattice constants can be combined in high material quality. The small diameter of the wires allows the strain from the lattice mismatch to relax by an expansion or contraction of the nanowires to the side [126]. Another advantage of this approach is that expensive III−V substrate material can be saved by growing the III−V nanowires on Si substrates or solar cells [127,128]. Nanowire solar cells are still in the development phase. However, theoretical [126,129] and experimental [130,131] investigations of different solar cell designs show promising results.

Several other approaches use nanostructures for photovoltaics. Some of these could also be used for III−V solar cells. An overview can be found in reference [132].

4. CONCLUSIONS

III−V multijunction solar cells have already reached efficiencies above 40% and a further increase toward 50% can be expected. These devices are widely used in space applications and terrestrial concentrators as well as in several niche applications. The great choice of III−V semiconductor materials allows various design options, several of which are currently being investigated, while others remain to be explored in the future. Material availability, quality, and manufacturing complexity will finally determine the most successful approaches for achieving optimum performance.

ACKNOWLEDGEMENTS

The authors thank all members of the department III−V—Epitaxy and Solar Cells at Fraunhofer ISE for their contributions to the work presented here. The authors especially appreciate the very helpful contributions of Stephanie Essig, Peter Kailuweit, René Kellenbenz, Vera Klinger, Tobias Roesener, Marc Steiner, and Wolfram Wettling.

REFERENCES

[1] M.A. Green, K. Emery, D.L. King, S. Igari, W. Warta, Solar cell efficiency tables (Version 01), Prog. Photovolt: Res. Appl. 12 (1993) 55−62.

[2] M.A. Green, K. Emery, Y. Hishikawa, W. Warta, E.D. Dunlop, Solar cell efficiency tables (Version 38), Prog. Photovolt: Res. Appl. 19 (2011) 565–572.
[3] P. Patel, D. Aiken, A. Boca, B. Cho, D. Chumney, M.B. Clevenger, et al., Experimental results from performance improvement and radiation hardening of inverted metamorphic multi-junction solar cells. Proceedings of the 37th IEEE Photovoltaics Specialists Conference, Seattle, Washington, 2011, in press.
[4] S. Bailey, R. Raffaelle, Space solar cells and arrays, in: A. Luque, S. Hegedus (Eds.), Handbook of Photovoltaic Science and Engineering, John Wiley & Sons, Ltd., Chichester, West Sussex, UK, 2011, pp. 365–401.
[5] D.J. Friedman, J.M. Olson, S. Kurtz, High-efficiency III–V multijunction solar cells, in: A. Luque, S. Hegedus (Eds.), Handbook of Photovoltaic Science and Engineering, John Wiley & Sons, Ltd., Chichester, West Sussex, UK, 2011, pp. 314–364.
[6] S.G. Bailey, R. Raffaelle, K. Emery, Space and terrestrial photovoltaics: synergy and diversity, Prog. Photovolt: Res. Appl. 10 (2002)399–306
[7] A. Luque, V.M. Andreev, Concentrator Photovoltaics, Springer Verlag, Heidelberg, Germany, 2007. p. 345
[8] P. Pérez-Higuerasa, E. Muñoz, G. Almonacida, P.G. Vidala, High Concentrator PhotoVoltaics efficiencies: present status and forecast, Renew. Sustain. Energy Rev. 15 (2011).
[9] V.M. Andreev, Solar cells for TPV converters. in: Next Generation Photovoltaics: High Efficiency Through Full Spectrum Utilization, Institute of Physics Publishing, St. Petersburg, Russia, 2004, pp. 246–273.
[10] T.J. Coutts, A review of progress in thermophotovoltaic generation of electricity, Renew. Sustain. Energy Rev. (1999) 77–184.
[11] V. Andreev, V. Khvostikov, A. Vlasov, Solar thermophotovoltaics, in: A. Luque, V. Andreev (Eds.), Concentrator Photovoltaics, Springer Verlag, Heidelberg, Germany, 2007, pp. 175–198.
[12] T. Bauer, Thermophotovoltaics: Basic Principles and Critical Aspects of System Design, Springer, Berlin, Germany, 2011. p. 222
[13] S.J. Wojtczuk, Long-wavelength laser power converters for optical fibers. Proceedings of the 26th IEEE Photovoltaic Specialists Conference, Anaheim, CA, 1997, pp. 971–974.
[14] D. Krut, R. Sudharsanan, W. Nishikawa, T. Issiki, J. Ermer, N.H. Karam, Monolithic multi-cell GaAs laser power converter with very high current density. Proceedings of the 29th IEEE Photovoltaics Specialists Conference, New Orleans, 2002, pp. 908–911.
[15] V. Andreev, V. Khvostikov, V. Kalinovsky, V. Lantratov, V. Grilikhes, V. Rumyantsev, et al., High current density GaAs and GaSb photovoltaic cells for laser power beaming. Proceedings of the 3rd World Conference on Photovoltaic Energy Conversion, Osaka, Japan, 2003, pp. 761–764.
[16] J.G. Werthen, Powering next generation networks by laser light over fiber. Proceedings of the conference on optical fiber communication, San Diego, CA, 2008, pp. 2881–2883.
[17] E. Oliva, F. Dimroth, A.W. Bett, GaAs converters for high power densities of laser illumination, Prog. Photovolt: Res. Appl. 4 (2008) 289–295.
[18] J. Schubert, E. Oliva, F. Dimroth, W. Guter, R. Loeckenhoff, A.W. Bett, High-voltage GaAs photovoltaic laser power converters, IEEE Trans. Electron. Dev. 56 (2009) 170–175.
[19] G. Böbner, M. Dreschmann, C. Klamouris, M. Hübner, M. Röger, A.W. Bett, et al., An optically powered video camera link, IEEE Photon. Tech. Lett. 20 (2008) 39–41.

[20] ASTM E490-00a, Standard Solar Constant and Zero Air Mass Solar Spectral Irradiance Tables, American Society for Testing and Materials, 2000.
[21] ASTM-G173, Standard Tables for Reference Solar Spectral Irradiances: Direct Normal and Hemispherical on 37° Tilted Surface, ASTM, 2008.
[22] S.R. Kurtz, D. Myers, W.E. McMahon, J. Geisz, M. Steiner, A comparison of theoretical efficiencies of multi-junction concentrator solar cells, Prog. Photovolt: Res. Appl. 16 (2008) 537–546.
[23] W. Shockley, H.J. Queisser, Detailed balance limit of efficiency of p–n junction solar cells, J. Appl. Phys. Vol. 32 (1961) 510–519.
[24] G. Létay, A.W. Bett, EtaOpt—a program for calculating limiting efficiency and optimum bandgap structure for multi-bandgap solar cells and TPV cells. Proceedings of the 17th European Photovoltaic Solar Energy Conference, Munich, Germany, 2001, pp. 178–181.
[25] G.S. Kinsey, K.M. Edmondson, Spectral response and energy output of concentrator multijunction solar cells, Prog. Photovolt: Res. Appl. 17 (2009) 279–288.
[26] S.P. Philipps, G. Peharz, R. Hoheisel, T. Hornung, N.M. Al-Abbadi, F. Dimroth, et al., Energy harvesting efficiency of III–V triple-junction concentrator solar cells under realistic spectral conditions, Sol. Energy Mater. Sol. Cells 94 (2010) 869–877.
[27] S.P. Philipps, G. Peharz, R. Hoheisel, T. Hornung, N.M. Al-Abbadi, F. Dimroth, et al., Energy harvesting efficiency of lll–V multi-junction concentrator solar cells under realistic spectral conditions. Proceedings of the 6th International Conference on Concentrating Photovoltaic Systems, Freiburg, Germany, 2010, pp. 294–298.
[28] F. Dimroth, S.P. Philipps, E. Welser, R. Kellenbenz, T. Roesener, V. Klinger, et al., Promises of advanced multi-junction solar cells for the use in CPV systems. Proceedings of the 35th IEEE Photovoltaics Specialists Conference, Honolulu, Hawai, 2010, pp. 1231–1236.
[29] M. Norton, A. Dobbin, A. Phinikarides, T. Tibbits, G.E. Georghiou, S. Chonavel, Field performance evaluation and modelling of spectrally tuned quantum-well solar cells. Proceedings of the 37th IEEE Photovoltaics Specialists Conference, Seattle, Washington, 2011, in press.
[30] S.R. Kurtz, J.M. Olson, P. Faine, The difference between standard and average efficiencies of multijunction compared with single-junction concentrator cells, Sol. Energy Mater. Sol. Cells 30 (1991) 501–513.
[31] P. Faine, S.R. Kurtz, C. Riordan, J.M. Olson, The influence of spectral solar irradiance variations on the performance of selected single-junction and multijunction solar cells, Sol. Cells 31 (1991) 259–278.
[32] K. Araki, M. Yamaguchi, Influences of spectrum change to 3-junction concentrator cells, Sol. Energy Mater. Sol. Cells 75 (2003) 707–714.
[33] G. Peharz, G. Siefer, A.W. Bett, A simple method for quantifying spectral impacts on multi-junction solar cells, Sol. Energy 83 (2009) 1588–1598.
[34] R. Hoheisel, S.P. Philipps, A.W. Bett, Long-term energy production of III–V triple-junction solar cells on the Martian surface, Prog. Photovolt: Res. Appl. 18 (2010) 90–99.
[35] A.W. Bett, F. Dimroth, G. Stollwerck, O.V. Sulima, III–V compounds for solar cell applications, Appl. Phys. 69 (1999) 119–129.
[36] L.M. Fraas, J.E. Avery, J. Martin, V.S. Sundaram, G. Girard, V.T. Dinh, et al., Over 35-percent efficient GaAs/GaSb tandem solar cells, IEEE Trans. Electron. Dev. 37 (1990) 443–449.
[37] V.M. Andreev, L.B. Karlina, A.B. Kazantsev, V.P. Khvostikov, V.D. Rumyantsev, S.V. Sorokina, et al., Concentrator tandem solar cells based on AlGaAs/GaAs-InP/

InGaAs(or GaSb) structures. Proceedings of the 1st World Conference on Photovoltaic Energy Conversion, Hawaii, USA, 1994, pp. 1721−1724.

[38] T. Takamoto, E. Ikeda, T. Agui, H. Kurita, T. Tanabe, S. Tanaka, et al., InGaP/GaAs and InGaAs mechanically stacked triple junction solar cells. Proceedings of the 26th IEEE Photovoltaic Specialists Conference, Anaheim, California, USA, 1997, pp. 1031−1034.

[39] A.W. Bett, C. Baur, R. Beckert, F. Dimroth, G. Létay, M. Hein, et al., Development of high-efficiency mechanically stacked GaInP/GaInAs-GaSb triple-junction concentrator solar cells. Proceedings of the 17th European Photovoltaic Solar Energy Conference, Munich, Germany, 2001, pp. 84−87.

[40] L. Esaki, New phenomenon in narrow Germanium p-n junction, Phys. Rev. 109 (1958) 603−604.

[41] M. Hermle, G. Létay, S.P. Philipps, A.W. Bett, Numerical simulation of tunnel diodes for multi-junction solar cells, Prog. Photovolt: Res. Appl. 16 (2008) 409−418.

[42] M. Meusel, C. Baur, G. Létay, A.W. Bett, W. Warta, E. Fernandez, Spectral response measurements of monolithic GaInP/Ga(In)As/Ge triple-junction solar cells: measurement artifacts and their explanation, Prog. Photovolt: Res. Appl. 11 (2003) 499−514.

[43] C. Baur, M. Hermle, F. Dimroth, A.W. Bett, Effects of optical coupling in III−V multilayer systems, Appl. Phys. Lett. 90 (2007) 192109/1−192109/3.

[44] G. Siefer, C. Baur, A.W. Bett, External quantum efficiency measurements of germanium bottom subcells: measurement artifacts and correction procedures. Proceedings of the 35th IEEE Photovoltaic Specialists Conference Honolulu, HI, 2010, pp. 704−707.

[45] J.-J. Li, S.H. Lim, C.R. Allen, D. Ding, Y.-H. Zhang, Combined effects of shunt and luminescence coupling on external quantum efficiency measurements of multijunction solar cells. Proceedings of the 37th IEEE Photovoltaics Specialists Conference, Seattle, Washington, 2011, in press.

[46] A.W. Bett, F. Dimroth, G. Siefer, Multijunction concentrator solar cells, in: A. Luque, V. Andreev (Eds.), Concentrator Photovoltaics, Springer Verlag, Heidelberg, Germany, 2007, pp. 67−87.

[47] A.R. Moore, An optimized grid design for a sun-concentrator solar cell, RCA Review 40 (1979) 140−151.

[48] P. Nubile, N. Veissid, A contribution to the optimization of front-contact grid patterns for solar cells, Solid-State Electron. 37 (1994) 220−222.

[49] I. Rey-Stolle, C. Algora, Modeling of the resistive losses due to the bus-bar and external connections in III−V high-concentrator solar cells, IEEE Trans. Electron. Dev. 49 (2002) 1709−1714.

[50] B. Galiana, C. Algora, I. Rey-Stolle, I.G. Vara, A 3-D model for concentrator solar cells based on distributed circuit units, IEEE Trans. Electron. Dev. 52 (2005) 2552−2558.

[51] M. Steiner, S.P. Philipps, M. Hermle, A.W. Bett, F. Dimroth, Validated front contact grid simulation for GaAs solar cells under concentrated sunlight, Prog. Photovolt: Res. Appl. 19 (2010) 73−83.

[52] C. Algora, Very-high-concentration challenges of III−V multijunction solar cells, in: A. Luque, V. Andreev (Eds.), Concentrator Photovoltaics, Springer Verlag, Heidelberg, Germany, 2007, pp. 89−111.

[53] D.C. Law, R.R. King, H. Yoon, M.J. Archer, A. Boca, C.M. Fetzer, et al., Future technology pathways of terrestrial III−V multijunction solar cells for concentrator photovoltaic systems, Sol. Energy Mater. Sol. Cells 94 (2008) 1314−1318.

[54] D.J. Friedman, Progress and challenges for next-generation high-efficiency multijunction solar cells, Curr. Opin. Solid State Mater. Sci. 14 (2010) 131−138.

[55] S.P. Philipps, W. Guter, E. Welser, J. Schöne, M. Steiner, F. Dimroth, et al., Present status in the development of III−V multi-junction solar cells, in: L. Luque, A. Martí, A.B. Cristóbal (Eds.), New Concepts for a Next Generation of Photovoltaics, Springer, to be published, Berlin, Germany, 2011.

[56] B.M. Kayes, H. Nie, R. Twist, S.G. Spruytte, 27.6% conversion efficiency, a new record for single-junction solar cells under 1 sun illumination. Proceedings of the 37th IEEE Photovoltaics Specialists Conference, Seattle, Washington, 2011, in press.

[57] I. Garcia, I. Rey-Stolle, B. Galiana, C. Algora, A 32.6 percent efficient lattice-matched dual-junction solar cell working at 1000 suns, Appl. Phys. Lett. 94 (2009) 053509.

[58] P.R. Sharps, M.A. Stan, D.J. Aiken, F.D. Newman, J.S. Hills, N.S. Fatemi, High efficiency multi-junction solar cells − past, present, and future. Proceedings of the 19th European Photovoltaic Solar Energy Conference, Paris, France, 2004, pp. 3569−3574.

[59] Z.I. Alferov, V.M. Andreev, V.D. Rumyantsev, III−V heterostructures in photovoltaics, in: A. Luque, V. Andreev (Eds.), Concentrator Photovoltaics, Springer Verlag, Heidelberg, Germany, 2007, pp. 25−50.

[60] R. King, A. Boca, W. Hong, D. Larrabee, K.M. Edmondson, D.C. Law, et al., Band-gap-engineered architectures for high-efficiency multijunction concentrator solar cells. Proceedings of the 24th European Photovoltaic Solar Energy Conference and Exibition, Hamburg, Germany, 2009, pp. 55−61.

[61] K.W.J. Barnham, I. Ballard, J.P. Connolly, N.J. Ekins-Daukes, B.G. Kluftinger, J. Nelson, et al., Quantum well solar cells, Physica 14 (2002) 27−36.

[62] K. Barnham, I. Ballard, J. Barnes, J. Connolly, P. Griffin, B. Kluftinger, et al., Quantum well solar cells, Appl. Surf. Sci. 113−114 (1997) 722−733.

[63] N.J. Ekins-Daukes, K.W.J. Barnham, J.P. Connolly, J.S. Roberts, J.C. Clark, G. Hill, et al., Strain-balanced GaAsP/InGaAs quantum well solar cells, Appl. Phys. Lett. 75 (1999) 4195−4197.

[64] R.J. Walters, G.P. Summers, S.R. Messenger, A. Freundlich, C. Monier, F. Newman, Radiation hard multi-quantum well InP/InAsP solar cells for space applications, Prog. Photovolt: Res. Appl. 8 (2000) 349−354.

[65] R. Kellenbenz, R. Hoheisel, P. Kailuweit, W. Guter, F. Dimroth, A.W. Bett, Development of radiation hard $Ga_{0.50}In_{0.50}P/Ga_{0.99}In_{0.01}As/Ge$ spase solar cells with multi quantum wells. Proceedings of the 35th IEEE Photovoltaic Specialists Conference, Honolulu, Hawai, 2010, pp. 117−122.

[66] K.W.J. Barnham, I.M. Ballard, B.C. Browne, D.B. Bushnell, J.P. Connolly, N.J. Ekins-Daukes, et al., Recent progress in quantum well solar cells, in: L. Tsakalakos (Ed.), Nanotechnology for Photovoltaics, CRC Press, Boca Raton, Florida, USA, 2010, pp. 187−210.

[67] J.G.J. Adams, B.C. Browne, I.M. Ballard, J.P. Connolly, N.L.A. Chan, A. Ioannides, et al., Recent results for single-junction and tandem quantum well solar cells, Prog. Photovolt: Res. Appl. (2011), in press.

[68] P. Kailuweit, R. Kellenbenz, S.P. Philipps, W. Guter, A.W. Bett, F. Dimroth, Numerical simulation and modeling of GaAs quantum-well solar cells, J. Appl. Phys. 107 (2010) 064317-1−064317-6.

[69] F. Dimroth, R. Beckert, M. Meusel, U. Schubert, A.W. Bett, Metamorphic $Ga_yIn_{1-y}P/Ga_{1-x}In_xAs$ tandem solar cells for space and for terrestrial concentrator applications at C>1000 suns, Prog. Photovolt: Res. Appl. 9 (2001) 165−178.

[70] W. Guter, J. Schöne, S.P. Philipps, M. Steiner, G. Siefer, A. Wekkeli, et al., Current-matched triple-junction solar cell reaching 41.1% conversion efficiency under concentrated sunlight, Appl. Phys. Lett. 94 (2009) 223504-1−223504-3.

[71] A.W. Bett, C. Baur, F. Dimroth, J. Schöne, Metamorphic GaInP-GaInAs layers for photovoltaic applications, Mater. Res. Soc. Symp. Proc. 836 (2005) 223−234.

[72] J. Schöne, E. Spiecker, F. Dimroth, A.W. Bett, W. Jäger, Misfit dislocation blocking by Dilute Nitride intermediate layers, Appl. Phys. Lett. 92 (2008) 081905.
[73] R.R. King, D.C. Law, K.M. Edmondson, C.M. Fetzer, G.S. Kinsey, H. Yoon, et al., 40% efficient metamorphic GaInP/GaInAs/Ge multijunction solar cells, Appl. Phys. Lett. 90 (2007) 183516-1−183516-3.
[74] F. Dimroth, S. Kurtz, High-efficiency multijunction solar cells, MRS Bull. 32 (2007) 230−234.
[75] J.F. Geisz, D.J. Friedman, J.S. Ward, A. Duda, W.J. Olavarria, T.E. Moriarty, et al., 40.8% efficient inverted triple-junction solar cell with two independently metamorphic junctions, Appl. Phys. Lett. 93 (2008) 123505-1−123505-3.
[76] A.B. Cornfeld, M. Stan, T. Varghese, J. Diaz, A.V. Ley, B. Cho, et al., Development of a large area inverted metamorphic multi-junction (IMM) highly efficient AM0 solar cell. Proceedings of the 33rd IEEE Photovoltaic Specialists Conference, San Diego, 2008, pp. 26/1−5.
[77] H. Yoon, M. Haddad, S. Mesropian, J. Yen, K. Edmondson, D. Law, et al., Progress of inverted metamorphic III−V solar cell development at Spectrolab. Proceedings of the 33rd IEEE Photovoltaic Specialists Conference, San Diego, USA, 2008, pp. 25/1−6.
[78] J. Boisvert, D. Law, R. King, D. Bhusari, X. Liu, A. Zakaria, et al., Development of advanced space solar cells at Spectrolab. Proceedings of the 35th IEEE Photovoltaic Specialists Conference, Honolulu, Hawaii, USA, 2010, pp. 123−127.
[79] T. Takamoto, T. Agui, A. Yoshida, K. Nakaido, H. Juso, K. Sasaki, et al., World's highest efficiency triple-junction solar cells fabricated by inverted layers transfer process. Proceedings of the 35th IEEE Photovoltaics Specialists Conference, Honolulu, Hawai, 2010, pp. 412−417.
[80] S. Wojtczuk, P. Chiu, X. Zhang, D. Derkacs, C. Harris, D. Pulver, et al., InGaP/GaAs/InGaAs 41% concentrator cells using bi-facial epigrowth. Proceedings of the 35th IEEE Photovoltaics Specialists Conference, Honolulu, Hawai, 2010, pp. 1259−1264.
[81] P. Chiu, S. Wojtczuk, C. Harris, D. Pulver, M. Timmons, 42.3% efficient InGaP/GaAs/InGaAs concentrators using bifacial epigrowth. Proceedings of the 37th IEEE Photovoltaics Specialists Conference, Seattle, Washington, 2011, in press.
[82] M. Yamaguchi, C. Amano, Efficiency calculations of thin-film GaAs solar cells on Si substrates, J. Appl. Phys. 58 (1985) 3601−3606.
[83] S.F. Fang, K. Adomi, S. Iyer, H. Morkoç, H. Zabel, C. Choi, et al., Gallium arsenide and other compound semiconductors on Silicon, J. Appl. Phys. 68 (1990) R31−58.
[84] A.W. Bett, K. Borgwarth, C. Schetter, O.V. Sulima, W. Wettling, GaAs-on-Si solar cell structures grown by MBE and LPE. Proceedings of the 6th International Photovoltaic Scinece and Engineering Concerence, New Delhi, India, 1992, pp. 843−847.
[85] R.K. Ahrenkiel, M.M. Al Jassim, B. Keyes, D. Dunlavy, K.M. Jones, S.M. Venon, et al., Minority carrier lifetime of GaAs on silicon, J. Electrochem. Soc. 137 (1990) 996−1000.
[86] K. Hayashi, T. Soga, H. Nishikawa, T. Jimbo, M. Umeno, MOCVD growth of GaAsP on Si for tandem solar cell application. Proceedings of the 35th IEEE Photovoltaics Specialists Conference, Waikoloa, Hawai, 1994.
[87] S.A. Ringel, J.A. Carlin, C.L. Andre, M.K. Hudait, M. Gonzalez, D.M. Wilt, et al., Single-junction InGaP/GaAs solar cells grown on Si substrates with SiGe buffer layers, Prog. Photovolt: Res. Appl. 10 (2002) 417−426.
[88] M.R. Lueck, C.L. Andre, A.J. Pitera, M.L. Lee, E.A. Fitzgerald, S.A. Ringel, Dual junction GaInP/GaAs solar cells grown on metamorphic SiGe/Si substrates with high open circuit voltage, IEEE Electron Dev. Lett. 27 (2006) 142−144.

[89] J.F. Geisz, J.M. Olson, M.J. Romero, C.S. Jiang, A.G. Norman, Lattice-mismatched GaAsP solar cells grown on Silicon by OMVPE. Proceedings of the 4th World Conference on Photovoltaic Energy Conversion, Waikoloa, Hawaii, USA, 2006, pp. 772–775.
[90] L. Colace, G. Masini, F. Galluzzi, G. Assanto, G. Capellini, L.D. Gaspare, et al., Metal–semiconductor–metal near-infrared light detector based on epitaxial Ge/Si, Appl. Phys. Lett. 72 (1998) 3175–3177.
[91] J.-S. Park, M. Curtin, J. Bai, M. Carrol, A. Lochtefeld, Growth of Ge thick layers on Si(001) substrates using reduced pressure chemcial vapor deposition, Jpn. J. Appl. Phys. 45 (2006) 8581–8585.
[92] T.J. Grassman, M.R. Brenner, M. Gonzalez, A.M. Carlin, R.R. Unocic, R.R. Dehoff, et al., Characterization of metamorphic GaAsP/Si materials and devices for photovoltaic applications, IEEE Trans. Electron. Dev. 57 (2010) 3361–3369.
[93] H. Döscher, T. Hannappel, In situ reflection anisotropy spectroscopy analysis of heteroepitaxial GaP films grown on Si(100), J. Appl. Phys. 107 (2010) 123523.
[94] T. Roesener, H. Döscher, A. Beyer, S. Brückner, V. Klinger, A. Wekkeli, et al., MOVPE growth of III-V solar cells on Silicon in 300 mm closed coupled showerhead reactor. Proceedings of the 25th European Photovoltaic Solar Energy Conference and Exhibition, Valencia, Spain, 2010, pp. 964–968.
[95] K. Volz, A. Beyer, W. Witte, J. Ohlmann, I. Németha, B. Kunert, et al., GaP-nucleation on exact Si (001) substrates for III/V device integration, J. Cryst. Growth 315 (2011) 37–47.
[96] M. Kondow, K. Uomi, A. Niwa, T. Kitatani, S. Watahiki, Y. Yazawa, GaInNAs: a novel material for long-wavelength-range laser diodes with excellent high-temperature performance, Jpn. J. Appl. Phys. 35 (1996) 1273–1275.
[97] J.F. Geisz, D.J. Friedman, III-N-V semiconductors for solar photovoltaic applications, Semicond. Sci. Technol. 17 (2002) 769–777.
[98] K. Volz, W. Stolz, J. Teubert, P.J. Klar, W. Heimbrodt, F. Dimroth, et al., Doping, electrical properties and solar cell application of GaInNAs, in: E. Ayse (Ed.), Dilute III–V Nitride Semiconductors and Material Systems, Springer Berlin Heidelberg, Heidelberg, 2008, pp. 369–404.
[99] Q.-Y. Tong, U. Gösele, Semiconductor Wafer Bonding: Science and Technology, John Wiley & Sons, Inc., New York, New York, USA, 1999. p. 297.
[100] M. Reiche, Semiconductor wafer bonding, Phys. Status Solidi 203 (2006) 747–759.
[101] J.W. Mayer, Ion implantation in semiconductors. International Electron Devices Meeting, 1973, pp. 3–5.
[102] M.K. Kelly, O. Ambacher, R. Dimitrov, R. Handschuh, Stutzmann, Optical process for liftoff of group III-nitride films, Phys. Status Solidi (a) 159 (1997) R3–R4.
[103] F. Dross, J. Robbelein, B. Vandevelde, E. Van Kerschaver, I. Gordon, G. Beaucarne, et al., Stress-induced large-area lift-off of crystalline Si films, Appl. Phys. A: Mater. Sci. Process. 89 (2007) 149–152.
[104] P. Demeester, I. Pollentier, P. De Dobbelaere, C. Brys, P. Van Daele, Epitaxial lift-off and its applications, Semicond. Sci. Technol. 8 (1993) 1124–1135.
[105] A. Fontcuberta i Morral, J.M. Zahler, H.A. Atwater, S.P. Ahrenkiel, M.W. Wanlass, InGaAs/InP double heterostructures on InP/Si templates fabricated by wafer bonding and hydrogen-induced exfoliation, Appl. Phys. Lett. 83 (2003) 5413.
[106] J.M. Zahler, K. Tanabe, C. Ladous, T. Pinnington, F.D. Newman, H.A. Atwater, High efficiency InGaAs solar cells on Si by InP layer transfer, Appl. Phys. Lett. 91 (2007) 012108.
[107] M.J. Archer, D.C. Law, S. Mesropian, M. Haddad, C.M. Fetzer, A.C. Ackerman, et al., GaInP/GaAs dual junction solar cells on Ge/Si epitaxial templates, Appl. Phys. Lett. 92 (2008) 103503.

[108] K. Dreyer, E. Fehrenbacher, E. Oliva, A. Leimenstoll, F. Schätzle, M. Hermle, et al., GaInP/GaAs/Si triple-junction solar cell formed by wafer bonding. DPG Spring Meeting, Dresden, Germany, 2011.
[109] K. Volz, D. Lackner, I. Nemeth, B. Kunert, W. Stolz, C. Baur, et al., Optimization of annealing conditions of (GaIn)(NAs) for solar cell applications, J. Cryst. Growth 310 (2008) 2222−2228.
[110] S. Kurtz, S.W. Johnston, J.F. Geisz, D.J. Friedman, A.J. Ptak, Effect of nitrogen concentration on the performance of $Ga_{1-x}In_xN_yAs_{1-y}$ solar cells. Proceedings of the 31st IEEE Photovoltaic Specialists Conference, Orlando, Florida, USA, 2005, pp. 595−598.
[111] S. Essig, E. Stämmler, S. Rönsch, E. Oliva, M. Schachtner, G. Siefer, et al., Dilute nitrides for 4- and 6-junction space solar cells. Proceedings of the 9th European Space Power Conference, Saint Raphaël, France, 2011, in press.
[112] F. Dimroth, U. Schubert, A.W. Bett, J. Hilgarth, M. Nell, G. Strobl, et al., Next generation GaInP/GaInAs/Ge multi-junction space solar cells. Proceedings of the 17th European Photovoltaic Solar Energy Conference, Munich, Germany, 2001, pp. 2150−2154.
[113] F. Dimroth, M. Meusel, C. Baur, A.W. Bett, G. Strobl, 3−6 junction photovoltaic cells for space and terrestrial concentrator applications. Proceedings of the 31st IEEE Photovoltaic Specialists Conference, Orlando, Florida, USA, 2005, pp. 525−529.
[114] D.J. Friedman, J.F. Geisz, A.G. Norman, M.W. Wanlass, S.R. Kurtz, 0.7-eV GaInAs junction for a GaInP/GaAs/GaInAs(1 eV)/GaInAs(0.7 eV) four-junction solar cell. Proceedings of the 4th World Conference on Photovoltaic Energy Conversion, Waikoloa, Hawaii, USA, 2006, pp. 598−602.
[115] D.C. Law, D.M. Bhusari, S. Mesropian, J.C. Boisvert, W.D. Hong, A. Boca, et al., Semiconductor-bonded III−V multi-junction space solar cells. Proceedings of the 34th IEEE Photovoltaic Solar Energy Conference, Philadelphia, USA, 2009, pp. 2237−2239.
[116] A.B. Cornfeld, D. Aiken, B. Cho, A.V. Ley, P. Sharps, M. Stan, et al., Development of a four sub-cell inverted metamorphic multi-junction (IMM) highly efficient AM0 solar cell. Proceedings of the 35th IEEE Photovoltaics Specialists Conference, Honolulu, Hawaii, USA, 2010, pp. 105−109.
[117] A.G. Imenes, D.R. Mills, Spectral beam splitting technology for increased conversion efficiency in solar concentrating systems: a review, Sol. Energy Mater. Sol. Cells 84 (2004) 19−69.
[118] R.L. Moon, L.W. James, H.A. Vander Plas, T.O. Yep, G.A. Antypas, Y. Chai, Multigap solar cell requirements and the performance of AlGaAs and Si cells in concentrated sunlight. Proceedings of the 13th IEEE Photovoltaic Specialists Conference, Washington DC, USA, 1978, pp. 859−867.
[119] A.G. Imenes, D. Buie, D. McKenzie, The design of broadband, wide-angle interference filters for solar concentrating systems, Sol. Energy Mater. Sol. Cells 90 (2006) 1579−1606.
[120] L.M. Fraas, J.E. Avery, H.X. Huang, L. Minkin, E. Shifman, Demonstration of a 33% efficient Cassegrainian solar module. Proceedings of the 4th World Conference on Photovoltaic Energy Conversion, Waikoloa, Hawaii, USA, 2006, pp. 679−682.
[121] A. Barnett, D. Kirkpatrick, C. Honsberg, D. Moore, M. Wanlass, K. Emery, et al., Short communication: accelerated publication very high efficiency solar cell modules, Prog. Photovolt: Res. Appl. 17 (2009) 75−83.
[122] D. Vincenzi, A. Busato, M. Stefancich, G. Martinelli, Concentrating PV system based on spectral separation of solar radiation, Phys. Status Solidi (a) 206 (2009) 375−378.

[123] E.R. Torrey, P.P. Ruden, P.I. Cohen, Performance of a split-spectrum photovoltaic device operating under time-varying spectral conditions, J. Appl. Phys. 109 (2011) 074909.
[124] B. Mitchell, G. Peharz, G. Siefer, M. Peters, T. Gandy, J.C. Goldschmidt, et al., Four-junction spectral beam-splitting photovoltaic receiver with high optical efficiency, Prog. Photovolt: Res. Appl. 19 (2010) 61–72.
[125] J.D. McCambridge, M.A. Steiner, B.L. Unger, K.A. Emery, E.L. Christensen, M.W. Wanlass, et al., Compact spectrum splitting photovoltaic module with high efficiency, Prog. Photovolt: Res. Appl. 19 (2011) 352–360.
[126] P. Kailuweit, M. Peters, J. Leene, K. Mergenthaler, F. Dimroth, A.W. Bett, Numerical simulations of absorption properties of InP nanowires for solar cell applications, Prog. Photovolt: Res. Appl. (2011).
[127] T. Martensson, C.P.T. Svensson, B.A. Wacaser, M.W. Larsson, W. Seifert, K. Deppert, et al., Epitaxial III-V nanowires on silicon, Nano Lett. 4 (2004) 1987–1990.
[128] W. Wei, X.-Y. Bao, C. Soci, Y. Ding, Z.-L. Wang, D. Wang, Direct heteroepitaxy of vertical InAs Nanowires on Si substrates for broad band photovoltaics and photodetection, Nano Lett. 9 (2009) 2926–2934.
[129] A. Kandala, T. Betti, A.F.i. Morral, General theoretical considerations on nanowire solar cell designs, Phys. Status Solidi (a) 206 (2009) 173–178.
[130] H. Goto, K. Nosaki, K. Tomioka, S. Hara, K. Hiruma, J. Motohisa, et al., Growth of Core–Shell InP nanowires for photovoltaic application by selective-area metal organic vapor phase epitaxy, Appl. Phys. Express 2 (2009) 035004.
[131] M. Heurlin, P. Wickert, S. Fält, M.T. Borgström, K. Deppert, L. Samuelson, et al., Axial InP nanowire tandem junction grown on a Silicon substrate, Nano Lett. 11 (2011) 2028–2031.
[132] L. Tsakalakos (Ed.), Nanotechnology for Photovoltaics, CRC Press, Boca Raton, Florida, USA, 2010.

CHAPTER ID-3

High-Efficiency Back-Contact Silicon Solar Cells for One-Sun and Concentrator Applications

Pierre J. Verlinden
AMROCK Pty Ltd, McLaren Vale, Australia

Contents

1. Introduction	449
2. Concentrator Applications of IBC Solar Cells	450
3. Back-Contact Silicon Solar Cells	453
3.1 IBC Solar Cells	453
3.2 Front-Surface-Field, Tandem-Junction, and Point-Contact Solar Cells	456
4. Modelling of Back-Contact Solar Cells	458
5. Perimeter and Edge Recombination	462
6. Manufacturing Process for Back-Contact Solar Cells	464
7. Stability of Back-Contact Cells	464
8. Toward 30% Efficiency Silicon Cells	467
9. How to Improve the Efficiency of Back-Contact Solar Cells	468
9.1 Reduce Emitter Saturation Current Density	470
9.2 Demonstrate Low-Contact Resistance	470
9.3 Reduce the Cell Thickness	471
9.4 Improve Light Trapping	471
9.5 Shrink Geometries	471
9.6 Reduce Series Resistance	472
9.7 Target Performance	472
10. Conclusions	472
Acknowledgements	473
References	474

1. INTRODUCTION

Interdigitated back-contact (IBC) silicon solar cells were originally designed for concentrating photovoltaic (CPV) applications. When the first IBC design concept was introduced by R.J. Schwartz [7], the principal

Practical Handbook of Photovoltaics.
© 2012 Elsevier Ltd. Asll rights reserved.

objective was to circumvent the limited usage of conventional silicon solar cells with top and bottom contacts for CPV application. In a conventional solar cell design, very low series resistances are difficult to obtain due to the practical sheet resistance of the front-side emitter that must be maintained to keep good quantum efficiency. Therefore, conventional silicon solar cells are not suitable for high concentration ratio—for example, greater than about ×50. Following the introduction of IBC cells and their modelling [7,8], the technology to fabricate such a cell, which required a particularly clean process to maintain high bulk lifetime and low surface recombination velocity, was subsequently developed by the universities of Stanford, Louvain, and Marseilles [4,10,11,19] and later commercialised by companies such as SunPower Corporation [1,2] and Amonix Inc [37]. Over time, the IBC solar cell design became the best silicon solar cell design for CPV applications and, to this date, is still the most efficient one, with efficiencies up to 28.3% in laboratory [2,3].

Although the first applications for IBC solar cells were dense-array and Fresnel lens CPV [1,15], SunPower Corporation also commercialised a version of the IBC solar cell for high-value one-sun applications [13,29–32]. For example, the silicon solar cells powering the Honda Dream solar race car and the NASA Helios unmanned airplane were basically modified and simplified versions of the concentrator IBC solar cells. These solar cells were, however, still too expensive, by more than two orders of magnitude, to be used in flat-plate PV modules for terrestrial applications. A cost-effective flat-plate PV module using IBC silicon solar cells required the development of a low-cost fabrication process, including a simplified process flow [36], lower-cost silicon Czochralski (CZ) substrates, cost-effective lithography steps, and electroplated metallisation [33]. Over the years, the efficiency of one-sun IBC silicon solar cells increased from about 21% to 24.2% [34,35], making its design the most efficiency solar cell design to date for both CPV and one-sun applications in large-volume manufacturing.

2. CONCENTRATOR APPLICATIONS OF IBC SOLAR CELLS

Concentrating sunlight for photovoltaic conversion has always been a very attractive solution. Since one can easily acknowledge that the cost of photovoltaic energy conversion is driven by the fabrication cost of the solar cells, particularly the cost of the semiconductor material, it becomes obvious that much less expensive concentrating lenses or mirrors can replace the expensive solar cells area. The concentration ratio can be increased several hundred fold to the point where the cost of fabricating the solar cells becomes a small part (typically 10% to 15%) in the overall PV system cost. However, this benefit does not come without complexity and cost. Concentrating PV systems need solar trackers with a tracking precision that increases with concentration ratio and a more expensive module design with a well-engineered and low-cost passive or active cooling system for the solar cells.

Concentration not only increases the energy productivity of the solar cell material and device but also increases its efficiency, since both current and voltage increase with the light intensity. However, in order to compare efficiencies of concentrator PV systems with flat-plate PV systems, we have to remember that concentrator systems only use direct sunlight, about 85% of the incident power density on a clear sunny day. Therefore, one could think that a 20% efficient concentrator system would produce about the same amount of energy as a 17% efficient flat-plate system if they are both mounted on the same tracking system. However, CPV systems usually benefit from additional advantages—for example, a lower temperature coefficient due to higher voltage—than one-sun solar cells and, in most cases, a lower cell temperature than in flat-plate modules [6].

The series resistance of the cell limits the concentration ratio to which the solar cells can be used and the efficiency advantage of concentration systems over flat plates. To collect a current that is, for example, at a 500× concentration ratio, almost 20 A/cm^2 from a solar cell that has an open-circuit voltage of 0.800 V requires a series resistance that is less than 0.001 Ω cm^2. To achieve such low series resistance, the concentrator solar cell requires a well-engineered double-level, solderable metallisation scheme [4,5] (see Figure 1).

The carrier recombination in commercial one-sun solar cells for flat-plate application is usually dominated by bulk Shockley–Read–Hall (SRH) or surface recombination. In high-efficiency (>18%) solar cells or when medium-concentration (<100×) is applied, the carrier recombination is usually dominated by junction recombination. Auger recombination usually dominates in very high-concentration silicon solar cells [11]. Auger recombination occurs when an electron from the conduction band recombines with a hole from the valence band, giving its energy to another electron. This is the opposite mechanism of impact ionisation.

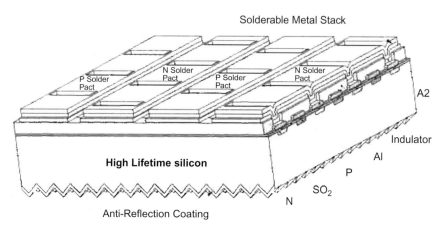

Figure 1 Structure of IBC silicon solar cell.

The Auger recombination rate increases as the cube of the carrier density and is generally the dominant recombination mechanism when the carrier concentration exceeds 10^{17} cm^{-3}. In order to reduce the Auger recombination rate, concentrator solar cells must be as thin as possible (typically 120 μm or less) and therefore require a good light trapping.

Because the solar cell cost represents only around 10% of the total concentrator PV system cost, a high-efficiency solar cell provides a great leverage for reducing the levelised cost of solar electricity (LCOE); the higher the concentration ratio, the greater the leverage. The most efficient silicon solar cell, both for laboratory cells and for production scale, is currently the IBC and, in particular, the point-contact (PC) solar cell [5–9,37]. The structure of such a cell is shown in Figure 1. It has attained a conversion efficiency of 28.3% [2,3] in the laboratory and 27.6% at 92× (AM1.5D, 10 W/cm^2, 25°C) at the production scale [37]. At present, it is the most efficient silicon solar cell for CPV applications. Figures 2 and 3 show examples of concentrator silicon solar cells manufactured by SunPower Corporation of Richmond, California. Two examples of a reflective concentrator dish built by Solar Systems Pty Ltd of Abbotsford, Australia, and using high-efficiency concentrator silicon solar cells, are shown in Figures 4 and 5. The 20-m^2 concentrator system represented in Figure 5 was the first silicon-based concentrator system to reach an overall system efficiency of 20% under normal operating conditions [6]. Figure 6 shows a picture of the cells at the receiving point of such a concentrator system in operation.

Figure 2 A point-contact silicon solar cell for Fresnel concentrator application. *(Courtesy of SunPower Corporation.)*

Figure 3 A point-contact silicon solar cell for dense-array concentrator application. *(Courtesy of SunPower Corporation.)*

3. BACK-CONTACT SILICON SOLAR CELLS
3.1 IBC Solar Cells

Schwartz et al. [7,8] introduced the IBC design in 1975. The main reason that the IBC solar cell design is particularly suitable for high concentration is that both metal contacts are made on the back surface of the cell (Figure 1), so there is no shadowing effect on the front side. The trade-off between the shadowing effect and the series resistance is eliminated, which is particularly interesting for concentration applications. Furthermore, the design of the front side (optical side) and the back (electrical side) can easily be optimised separately. In a conventional solar

Figure 4 Picture of a 25-kW dish concentrating PV system with point-contact silicon solar cells. *(Courtesy of Solar Systems Ply Ltd.)*

Figure 5 A dish concentrating PV system with point-contact silicon solar cells. This PV system, installed in White Cliff, Australia, demonstrated 20% overall efficiency under normal operating conditions [6]. *(Courtesy of Solar Systems Ply Ltd.)*

cell, the doping of the front-side emitter is a trade-off between series resistance and efficiency: a lighter-doped front-side emitter would improve the quantum efficiency and reduce the emitter saturation current density, but it would increase the series resistance of the cell. On the other hand, in IBC solar cells, the front side could be optimised for

Figure 6 A dense-array receiver with point-contact silicon solar cells. *(Courtesy of Solar Systems Ltd Ply.)*

maximum quantum efficiency, reduced recombination, and reduced sublinearity without compromising the series resistance of the cell.

The main requirements for obtaining a high efficiency with IBC cells are the following:

- Long recombination lifetime in the bulk. After solar cell fabrication, the diffusion length must be at least 5 times longer than the solar cell thickness. For this reason, float-zone (FZ), high-resistivity, n-type substrates with carrier lifetimes greater than 5 ms are generally preferred. However, recent improvements in bulk lifetime control in n-type CZ wafer production have made this type of substrate the wafer of choice for large-volume manufacturing. The typical solar cell thickness is 100–150 μm for CPV, while 150 to 200 μm is usual for one-sun applications.
- Low front-surface recombination velocity. Ideally, the front-surface recombination velocity must be much smaller than the ratio of the minority-carrier diffusion constant, D_n, D_p, or D_a, and the thickness of the cell, W, is typically less than 10 cm/sec. The front surface must be passivated with a thin, thermally grown, silicon-dioxide layer, grown in a very clean environment before depositing an antireflective coating (ARC) of silicon nitride.

3.2 Front-Surface-Field, Tandem-Junction, and Point-Contact Solar Cells

Over the years, several variants in the IBC design have been introduced. The front-surface-field (FSF) solar cell has a high–low (n+/n or p+/p) junction on the front side of the IBC solar cell (Figure 7(a)). The function of the front-surface field is to reduce the effective front-surface recombination velocity for the carriers generated in the bulk of the device. It behaves

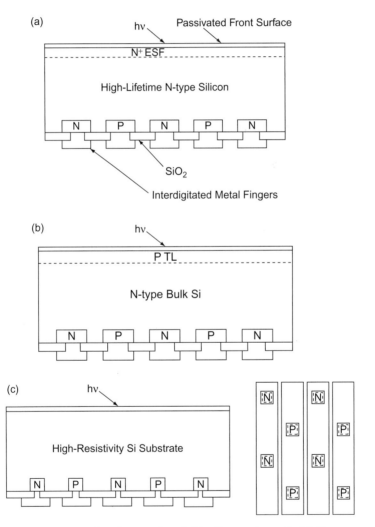

Figure 7 (a) Structure of a front-surface-field silicon solar cell. (b) Structure of a tandem-junction silicon solar cell. (c) Structure of a point-contact silicon solar cell.

the same way as a back-surface field (BSF) in a conventional n+/p/p+ solar cell. The front-surface recombination current density can be expressed either as a function of the FSF saturation current density, J_{0+}, or as a function of the effective surface recombination velocity, S_{eff},

$$J_{fs,\ rec} = \frac{N_D p}{n_i^2} J_{0+} = q p S_{eff} \qquad (1)$$

where p is the minority-carrier concentration at the front surface.

It is, however, important to note that if the solar cell operates in high-level injection (HLI), the effective recombination velocity S_{eff} increases as the injection level increases and the effect of the FSF decreases. Indeed, in high-level injection, the front-surface recombination current density, $J_{fs,rec}$, can be expressed by

$$J_{fs,rec} = \frac{n^2}{n_i^2} J_{0+} = q n S_{eff} \qquad (2)$$

where n is the carrier concentration (note that in HLI the electron and hole concentrations are equal, $n = p$) at the front surface of the cell and n_i is the intrinsic carrier concentration. As we can see from Equation (2), the front-surface recombination current increases as n^2 and the effective surface recombination velocity increases as the injection level increases:

$$S_{eff} = \frac{n}{q n_i^2} J_{0+} \qquad (3)$$

These characteristics result in a more or less important sublinearity of the responsivity of the cell, depending on the value of J_{0+}. The responsivity of a solar cell is, for a given spectrum, the total photogenerated current per unit of incident power. The responsivity has the dimension of amp/watt. The typical value of responsivity for high-efficiency concentrator solar cells is 0.395 A/W. If it varies with the concentration ratio, the solar cell is called *sublinear* or *superlinear*. Therefore, an IBC cell with a doped front-side surface (tandem-junction or FSF) is usually sublinear at HLI and is therefore reserved to one-sun or low-concentration applications (<10×).

Tandem-junction (TJ) solar cells are identical to FSF solar cell, except that a floating p–n junction replaces the high-low junction at the front side of the cell (Figure 7b). The front-surface recombination current density, $J_{fs,rec}$, is determined by Equations (1) or (2) in the same way as in the

FSF solar cell, and there is practically no difference between a TJ and a FSF solar cell. The only differences reside in
- the risk of shunting of the tandem junction at the edge of the cell, which would decrease the voltage across the floating junction and decrease the performance of the device, and
- the additional space charge recombination current at a very low intensity.

The designer choice between a TJ and a FSF solar cell is fully determined by the ability to make a front-side emitter with the lowest emitter saturation current density, J_{0+}, as possible. For these reasons, a FSF design with an n-type (phosphorous-doped) high–low emitter on n-type substrates is usually preferred.

PC solar cells are IBC cells designed for high concentration ratios with a reduced emitter area and a reduced metal contact area on the back of the cell (Figure 7(c)) to decrease the emitter recombination as much as possible. For one-sun or low-concentration applications (<5×), since the dominant recombination mechanisms are SRH and surface recombinations, a simple IBC or FSF design is preferred, the back-surface emitter coverage fraction being kept at around 80% to 90%, and the back base diffusion being kept at around 10% or 20%, with a pitch (width of unit cell) of 200 µm to 2000 µm. However, the metal-to-semiconductor contact area always has a very low coverage fraction, typically less than 1%, depending on the metal–Si contact resistance. For medium concentration ratios between ×5 and ×200, the back-surface emitter coverage fraction could be reduced to typically around 10%, whereas the metal-to-semiconductor contact area is typically 5% of the back surface of the cell. For medium to high concentration ratios, the unit cell width is reduced to about the value of the thickness of the cell.

4. MODELLING OF BACK-CONTACT SOLAR CELLS

The simplest way to model IBC solar cells is to consider it as a back-surface–illuminated p–n junction solar cell. As an example, we can use the integral approach as proposed by R.M. Swanson [9,10,12]. For a precise simulation of an IBC or PC solar cell, we refer the reader to the model developed in these papers that includes quasi-3D effects close to the back contacts. In the following, we give a simple formulation.

The current, I, at the terminals of a solar cell can be written in the form

$$I = I_{ph} - I_{b,rec} - I_{s,rec} - I_{em,rec} \qquad (4)$$

where I_{ph}, is the photogenerated current (the maximum photogenerated current within the silicon material with a defined thickness, including reflection losses and light trapping), $I_{b,rec}$ is the sum of all the bulk recombination currents (including SRH and Auger recombination), $I_{s,rec}$ is the sum of all the surface recombination currents (including front side, back, and edge surface recombination), and $I_{em,rec}$ is the sum of all the emitter recombination currents (including back emitters, the front TJ or FSF emitter if it exists, and also recombination at the contacts). The recombination currents increase as the carrier concentration increases and therefore increase as the terminal voltage increases. Eventually, at the open-circuit condition, the sum of all the recombination currents will be equal to the photocurrent, I_{ph}, and the terminal current will be equal to zero. Therefore, at V_{oc}:

$$I_{ph} = I_{b,rec} - I_{s,rec} - I_{em,rec} \qquad (4)$$

Calculating the different components of the recombination current is the most interesting and most powerful approach to simulate, optimise, or analyse the performance of solar cells. In particular, it provides a good method for a power-loss analysis. Table 1 gives a summary of the different relevant recombination mechanisms in silicon, their controlling parameters, typical values in high-efficiency one-sun or concentrator silicon solar cells, and the corresponding recombination currents.

If an IBC solar cell is to be modelled in a simple manner as a back-surface−illuminated conventional solar cell, one has to realise the following:
- In most of the device, the electron and holes are flowing in the same direction, from the front side of the cell toward the back, where the collecting n-type and p-type junctions are.
- When the cell is under HLI and the electrical neutrality in the bulk is to be maintained, the electron and hole concentrations are equal throughout the device and are flowing with the same ambipolar diffusion constant, Da = 2DnDp/(Dn + Dp), where Dn and Dp are the electron and hole diffusion constants.
- In most of the device, from the front side to very close to the emitter area where 3D current flow starts to appear, the current flow is almost unidirectional, perpendicular to the surface:
  - in low-level injection (LLI) and for an n-type substrate,

Table 1 Recombination mechanisms and the corresponding recombination currents in n-type silicon solar cells under high-level injection (HLI) and low-level injection (LLI)

Carrier recombination mechanism	Controlling parameter	Unit	Typical values	Recombination current density (A/cm²) HLI	LLI
Bulk (trap assisted)	τ_B, τ_n or τ_p	s	1–10 ms	$\frac{qnW}{\tau_B}$	$\frac{q(p-p_0)W}{\tau_p}$
Surface (trap assisted)	S	cm/s	1–4 cm/s	qnS	$q(p \times p_0)S$
Emitter	J_0	A/cm²	50–200 fA/cm²	$\frac{n^2}{n_i^2}J_0$	$\frac{(p-p_0)N_D}{n_i^2}J_0$
Auger	C_n, C_p or C_A	cm⁶/s	1.66×10^{-30}	$qn^3 C_A W$	$\frac{q(p-p_0)}{N_D^2 C_n W}$

The ambipolar Auger coefficient value is from Sinton [11,12]. In this chapter, τ_B represents the bulk lifetime, either τ_n or τ_p in LLI. When we are considering HLI, we have assumed that $\tau_B = \tau_p + \tau_n$, where τ_n and τ_p are the electron and the hole lifetimes.

$$J_p = -qD_p \frac{dp}{dx} \qquad (6)$$

- and in HLI,

$$J_n = -J_p = qD_a \frac{dn}{dx} = qD_a \frac{dp}{dx} \qquad (7)$$

where J_n and J_p are the electron and hole current densities, and therefore the total current along the wafer J_T is negligible far away from the collecting junctions:

$$J_T \approx 0 \qquad (8)$$

Since carrier concentration at the cell back is determined by the terminal voltage of the solar cell:
- in LLI,

$$p_{back} = \frac{n_i^2}{N_D} \exp\left(\frac{qV}{k_B T}\right) \qquad (9)$$

- and in HLI,

$$n_{back} = n_i \exp\left(\frac{qV}{2k_B T}\right) \qquad (10)$$

we can use Equation (4) and the expression of recombination currents in Table 1 to calculate the cell current at any bias voltage. As a simple example, we can take a high-efficiency IBC or PC solar cell under HLI with long carrier lifetime ($\tau_B = \tau_n + \tau_p \gg W^2/D_a$, where W is the thickness of the cell) and low surface recombination velocity ($S_0 \ll D_a/W$, where S_o is the real or effective surface recombination velocity at the front surface).

In short-circuit condition, we can consider that, in the first approximation,

- $n_{back} = 0$;
- the emitter and surface recombination currents at the back of the cell can be neglected, as well as Auger recombination; and
- the electron and hole concentrations linearly decrease from front to back.

Equation (4) then becomes

$$J_{sc} = J_{ph} - \frac{qn_0 W}{2\tau_B} - qn_0 S_0 = qD_a \frac{n_0}{W} \qquad (11)$$

where J_{sc} is the short-circuit current density, n_0 is the front-surface electron concentration, and currents were replaced by current densities in view of the effectively planar geometry of the cell as a result of the perpendicular direction of the current. A few iterations are necessary to determine the short-circuit current and the front-surface carrier concentration.

It becomes immediately apparent that if the thickness of the cell, W, increases, the front-surface carrier concentration, n_0, increases, which, in turn, results in increased bulk and front-surface recombination currents.

In open-circuit condition, we can consider that the electron and hole concentrations are constant throughout the device. The open-circuit voltage is given by

$$V_{oc} = 2 \frac{k_B T}{q} \ln\left(\frac{n}{n_i}\right) \qquad (12)$$

after solving the following equation:

$$J_{ph} = \frac{qnW}{\tau_b} + qn^3 C_a W + qnS_0 + qnS_{back}(1 - A_n - A_p) + \frac{n^2}{n_i^2}(A_n J_{0n} + A_p J_{0p}) \qquad (13)$$

where J_{0n} and J_{0p} are the saturation current densities at the n and p junctions; A_n and A_p are the n-type and p-type emitter coverage fractions, respectively; and S_{back} is the surface recombination velocity at the back of the cell.

The maximum power point and efficiency of the solar cell can be determined the same way, considering that the back-carrier concentration is given by Equation (10) and that the carrier concentration is still linearly distributed from front to back. Equation (4) becomes

$$J_{mp} = J_{ph} - \frac{qn_{avg}W}{\tau_B} - q\int_0^W n(x)^3 C_a dx - qn_0 S_0$$
$$- qn_{back}S_{back}(1 - A_n - A_p) - \frac{n_{back}^2}{n_i^2}(A_n J_{0p} + A_p J_{0p}) = qD_a \frac{n_0 - n_{back}}{W} \quad (14)$$

where $n_{avg} = (n_0 + n_{back})/2$ is the average carrier concentration and

$$n(x) = n_{back} + (n_0 - n_{back})(1 - \frac{x}{W}) \quad (15)$$

Several iterations are necessary to reach the solution.

This is a very simple way to analyse the IBC or PC solar cells. It allows for a quick determination of the carrier concentration in the device and an analysis of the different recombination mechanisms. Note that Equation (1) or (2) must be used for the front-side recombination current expression in FSF or TJ solar cells. For a more precise modelling, including quasi-3D effects at the back surface of the solar cell, the reader is referred to the Swanson model [9,10].

IBC solar cells for one-sun applications can be easily modelled using the LLI Equations (1), (6), (9) and the LLI recombination currents in Table 1. However, one needs to remember that, in most high-efficiency solar cells, the device is in LLI at short-circuit condition but actually in HLI at open-circuit condition. At the maximum power point, the device is often at the limit of HLI.

5. PERIMETER AND EDGE RECOMBINATION

So far, we have not considered the edge recombination. It has been recently demonstrated that, in most high-efficiency silicon solar cells, one of the most important recombination mechanisms is a recombination current at the unpassivated surface at the edge of the silicon die [13]. Two cases need to be considered here:
- aperture-illuminated solar cells (e.g., cells for Fresnel lens modules, Figure 2), and
- totally illuminated solar cells (e.g., cells for one-sun application or CPV cells for dense-array receivers, dish, and thermophotovoltaic [TPV] applications, Figure 3).

In order to reduce the edge recombination, the aperture-illuminated IBC solar cells must have its active area as far as possible from the edge of the silicon die—in theory, at a distance of least three times the ambipolar diffusion length at the considered carrier concentration. In practice, economical considerations prevent manufacturers from increasing this distance too much, so a typical distance between the edge of the silicon die and the nearest emitter at the back of the cell is about 500 μm. For totally illuminated solar cells, there is an optimal distance that can be calculated as explained in [13]. The width of the border region is optimal when the illuminated border region generates just enough current to supply the recombination current at the edge of the silicon die. A wider or narrower border region than the optimal width would result in a lower efficiency. If d is the width of the border region, P the cell perimeter, and p the average carrier density at the middle of the cell, at the maximum power point (V_m, I_m) and at the considered concentration ratio, the current generated by the border region is

$$I_{border} = dPJ_{sc} \tag{16}$$

and the edge recombination current is

$$I_{edge} = qPWD_a p/d \tag{17}$$

The border width is optimal when

$$I_{border} = I_{edge} \tag{18}$$

and

$$d = \sqrt{\frac{qWD_a p}{J_{SC}}} \tag{19}$$

In LLI, use D_p instead of D_a in Equations (17) and (19).

For a fixed geometry and cell thickness, the carrier concentration and short-circuit current are more or less proportional to the concentration ratio. Therefore, the optimal border width is generally independent of the concentration ratio. However, since the average carrier concentration increases with the thickness of the cell, the optimal border width is roughly proportional to the thickness of the cell. For a typical IBC solar cell for dense-array application at ×400 and with a thickness of 100 μm ($n \approx 10^{17}$ cm^{-3}), the optimal border width is about 135 μm.

Equation (19) is valid if the edge is unpassivated (assuming infinite recombination velocity) such as in silicon solar cells diced with a dicing saw. New techniques to passivate the edges of solar cells are needed, but so far there has been no satisfactory development in low-temperature passivation of silicon surfaces after dicing.

6. MANUFACTURING PROCESS FOR BACK-CONTACT SOLAR CELLS

The typical six-mask process flow to manufacture IBC solar cells for CPV application and with a double-level metallisation scheme is presented in Figure 8, although actual processes may significantly differ in practice. A typical low-cost process for one-sun IBC solar cells requires only three lithography steps: junction, contact, and metal [33,36]. Developing a low-cost lithography technology has been the key driver for making commercial flat-plate PV modules with IBC cells a reality. On the other hand, when very high efficiencies are desired, additional photolithography steps may be required for
- inverted pyramid texturisation instead of random texture and
- local thinning instead of uniform wafer thinning.

7. STABILITY OF BACK-CONTACT CELLS

In the same way as many other high-efficiency silicon solar cells, the IBC cells are subject to efficiency degradation. If FZ wafers are used, the degradation is limited to the silicon—silicon dioxide interface [1,14—16]. The degradation of the interface could be due to the loss of hydrogen atoms passivating the dangling bonds at the silicon—silicon dioxide interface or the creation of new interface states. The degradation of the interface is manifested by:
- an increase of the surface recombination velocity, which, in turn, results in a reduction of the short-circuit current over the entire range of concentration ratio, or
- an increase of the front emitter current J_0 that results in an increase of the sublinearity (the cell responsivity decreases as the concentration

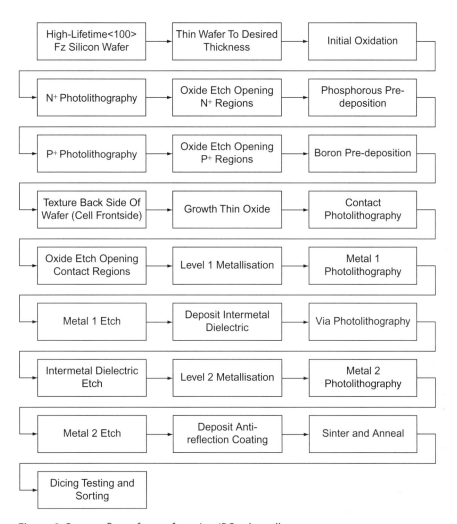

Figure 8 Process flow of manufacturing IBC solar cells.

ratio increases) if the front surface is lightly doped or inverted due to charges in the dielectric layer.

Both mechanisms will result in a significant decrease in efficiency. The degradation has been observed so far due to the following individual or combined conditions:
- UV light, which creates hot electrons that can overcome the potential barrier at the silicon—silicon dioxide interface and create new interface states;

- elevated temperature (>100°C), which allows hydrogen atoms to escape from the interface; and
- mechanical stress.

It is also assumed that the following conditions enhance the degradation:
- highly concentrated sunlight, which creates a large concentration of carriers at the surface of the solar cell;
- the presence of moisture; and
- the presence of atomic hydrogen, which could break an existing Si−H bond to form a hydrogen molecule H_2.

There is still much research to be done to understand the degradation mechanisms and to find a solution to this issue.

The easiest way to prevent the strong influence of the degradation of the surface recombination velocity to solar cell efficiency is to isolate the surface from the rest of the cell with FSF or a floating tandem junction. Unfortunately, although it makes the cell more stable, the front-side emitter has a significant impact on the sublinearity of the solar cell [1,15], which makes this solution undesired for concentration applications. In any case, the front emitter must be lightly doped in such way that the front-emitter saturation current, J_{0+}, is kept as low as possible. Indeed, we can see from Equation (2) that the greater the front emitter J_{0+}, the more significant the sublinearity of the solar cell will be.

There are other ways to ensure stability of the IBC solar cells:
- During the solar cell fabrication, it is very important that the front passivating oxide layer is grown with the smallest interface state density. 1,1,1-trichloroethane (TCA), dichloroethane (DCA), trichloroethylene (TCE), or HCl oxide growth, as well as an aluminium anneal, should be avoided for the final oxidation as they have been recognised as giving very unstable Si−SiO_2 interfaces [16,17].
- UV light with wavelengths below 400 nm should be filtered from the solar spectrum, either at the solar cell level or at the concentrator module level. For example, the ARC can include a UV filter such as TiO_2 or a very thin polysilicon layer [14,16].
- The front surface of the solar cell should be coated with a thin layer of silicon nitride (as part of the ARC) or another material, hopefully compatible with the ARC, to form a great hydrogen and moisture diffusion barrier (for example, a very thin amorphous silicon or polysilicon layer) [16].

Another mechanism of degradation has been observed in IBC solar cells [38]. Since the front surface of the cells is undoped (in the case of PC cells, for example) or very lightly doped (in the case of FSF or TJ cells), it can easily be brought into depletion by electric charges. When the front surface is depleted, the front-surface recombination current significantly increases and the efficiency decreases. The electric charges can be easily reversed in the factory and the original module performance can be reestablished. As an example of solution to this problem, the positive DC leg of an array of flat-plate PV modules containing n-type IBC or FSF solar cells should be earthed in order to prevent the front surface from being negatively charged and to prevent the front surface of the cells from being depleted. With the positive DC leg being earthed, the front surface of the cells stay positively charged and stay in accumulation. For the same reason, floating PV array schemes and transformerless inverters are not to be used with PV systems using IBC cells.

The two mechanisms of degradation described above are not exclusive to IBC cells. They are observed in any solar cell presenting a lightly doped front surface; the methods of prevention are identical for all cells.

8. TOWARD 30% EFFICIENCY SILICON CELLS

In 1985, Gray and Schwartz presented a paper titled 'Why Don't We Have a 30% Efficient Silicon Solar Cell?' [18]. At that time, the highest reported efficiency for silicon solar cell was 22% under concentrated sunlight. In that paper, the authors concluded that 30% efficiency would be attainable with a silicon solar cell if

- good light trapping was developed that did not degrade the open-circuit voltage of the solar cell,
- the contact and grid series resistance could be reduced enough to allow for a high concentration ratio,
- the current crowding effects are reduced by a judicious design, and
- a novel heterojunction or hetero-interface contact with small contact resistance and low emitter saturation current was developed.

A few years later, the first three steps to attain the 30% efficiency target had been addressed and resolved: the PC solar cell with a thin substrate and long carrier lifetime, with a textured and passivated front

Table 2 Strategy for 30% efficient PC silicon solar cells for CPV applications

Step	Parameter	Target value	Acceptable value	Unit
1	Emitter J_o	3×10^{-14}	5×10^{-14}	A/cm^2
2	Emitter contact resistance	10^{-6}	10^{-4}	Ω cm^2
3	Cell thickness with manufacturable yield	30	80	μm
4	Light trapping: number of passes for long wavelengths	50	35	passes
5	Unit cell geometry	25	40	μm
6	Total series resistance	5×10^{-4}	10^{-3}	Ω cm^2

surface, proved to be the best design for high-efficiency concentrator solar cell [19,20]. In 1989, the highest reported efficiency for silicon PC solar cell was 28.3% [21]. The same year, Swanson responded to Gray's paper in a publication titled "Why We Will Have a 30% Efficient Silicon Solar Cell" [22] in which he demonstrated that all the pieces were in place to fabricate a 30% efficient silicon solar cell. The last milestone to reach the 30% breakthrough—namely, the development of polysilicon emitters with a low contact resistance and low emitter saturation currents—had just been demonstrated at Stanford [23,24]. Swanson announced, "Within one year, cells will be reported with efficiency in excess of 30%" [22].

It is believed that 30% efficient silicon CPV solar cells are possible in a manufacturing environment. In order to achieve this goal, polysilicon or heterojunction emitter technology needs to be implemented, along with several other improvements to the existing technology (thinner cells, improved light trapping, and reduced dimension geometries).

9. HOW TO IMPROVE THE EFFICIENCY OF BACK-CONTACT SOLAR CELLS

Campbell and Green [25] discussed the efficiency limit of silicon solar cells under concentrated sunlight. They showed that the limit of efficiency for a silicon cell is between 30% and 35%. These very high efficiencies have never been demonstrated so far with a single-junction silicon solar cell. Commercially available PC concentrator cells have efficiencies around 26.5%. In order to reach such high efficiencies, the recombination mechanisms such as trap-assisted SRH recombination, in the bulk or at the surface, and emitter recombination must be negligible compared to Auger recombination. The Auger recombination is intrinsic to the material and, for example, cannot be improved by using a purer starting material or a cleaner fabrication process. Therefore, the only way to reduce the Auger recombination rate inside the cell is to reduce its thickness. However, using a very thin silicon solar cell requires the use of a very effective scheme for light trapping. For example, Campbell and Green [25] suggest that the optimum cell would be less than 1 μm thick and could reach 36–37% under concentration.

By comparison, the recombination in the PC solar cell at ×250 (25 W/cm^2) is almost equally dominated by emitter (40%) and Auger (40%) recombination. The surface and bulk SRH recombination represents less than 20% of the overall recombination rate. In the PC design, the emitter coverage fraction has already been reduced to the minimum technologically acceptable, and the only way to further reduce the emitter recombination rate is to reduce the emitter saturation current density J_0. Therefore, in order to reach 30% or greater efficiency, the following strategy needs to be used:

1. Reduce the emitter recombination by reducing the emitter saturation current density from 2×10^{-13} to 5×10^{-14} A/cm^2. Polysilicon heterojunction emitters with less than 5×10^{-14} A/cm^2 as saturation current density have already been demonstrated [23,24].
2. Demonstrate low contact resistance between the bulk and the emitters. A desired value for the contact resistance between bulk and emitter is less than 10^{-5} Ω cm^2.
3. Reduce the cell thickness in order to reduce the Auger recombination rate. For a practical concentrator for which the angle of incidence of the light to the cell is between 0 and 30 degrees, the optimal thickness is about 30 μm [25]. This is very difficult to achieve in a manufacturing environment. Manufacturing techniques, such as local thinning, need to be implemented.
4. Improve light trapping by designing new textured surfaces on both sides of the cell.
5. Reduce the unit cell geometry down to a dimension equal to or smaller than the thickness of the cell.
6. Reduce the external series resistance (metal and interconnect).

9.1 Reduce Emitter Saturation Current Density

The saturation current density of an emitter J_0 represents the sum of all the recombination mechanisms inside the emitter. It includes the SRH, surface, contact, and Auger recombination mechanisms, as well as heavy doping effects such as band-gap narrowing. The best diffused emitter J_0 that can be achieved is around 2×10^{-14} A/cm^2. However, such an emitter is very transparent, which means that if a metal contacts it, its saturation current density dramatically increases about 100-fold. For contacting PC solar cell, the emitter must be opaque and the best saturation current density of such emitter is around 2×10^{-13} A/cm^2.

An ideal emitter for silicon should be transparent to majority carriers and a mirror for minority carriers. Therefore, it should have a wide band gap such that an additional potential barrier will appear in the minority band when it is doped (see Figure 9). Although there are many semiconductors with a larger band gap than silicon, so far none of the large band-gap materials have proved to be ideal for contacting silicon. Indeed, the potential barrier at the emitter heterojunction blocks the minority carriers; they still can recombine at the heterojunction interface due to the presence of interface traps.

There are only two materials that were reported to emitters with low J_0: polysilicon emitters and semi-insulating polysilicon (SIPOS) emitters, both used with an interfacial thin oxide layer [23,24,26,27]. However, some very promising results have recently been reported on nonconcentrator solar cells with amorphous silicon emitters [28] that do not require an interfacial thin oxide to block the minority carriers.

9.2 Demonstrate Low-Contact Resistance

The growth of a thin (10 Å to 20 Å) interfacial layer of silicon dioxide is critical to obtain low emitter saturation current density with polysilicon. However, in order to achieve a low contact resistance, it is necessary to

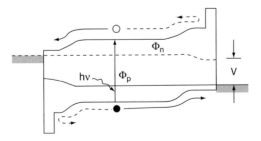

Figure 9 Band diagram of a p–i–n solar cell with heterojunction emitters.

anneal the polysilicon emitter to break up the oxide layer on about 1% of the contact area [23]. The trade-off between J_0 and a high contact resistance is highly dependent on the oxide thickness and the anneal (or breakup) step.

9.3 Reduce the Cell Thickness

Once the emitter recombination mechanism has been reduced, it is possible to reduce the Auger recombination rate by using a thinner substrate for the solar cell. Solar cells made on 100-μm-thick wafers can be manufactured in large volumes, but thinner cells present a real challenge for manufacturing. Since both sides of the cells must be processed for texturing, passivation, and antireflection coating, the thinning of the wafers cannot just be done by lapping the wafers at the end of the process. Solar cells of 30–80-μm thickness must be fabricated on locally thinned substrates.

9.4 Improve Light Trapping

When very thin silicon solar cells are designed, the light-trapping property of the cells becomes a significant factor for the efficiency. Currently available cells have only one-side texturisation. Unfortunately, this is not enough for cells thinner than 100 μm. In this case, a double side texturisation or perpendicular grooving must be designed in order to ensure good light trapping. In current PC solar cells, the analysis of the internal quantum efficiency near the edge of the band gap shows that the effective number of passes for long-wavelength light is greater than 30 [1]. The light trapping can be improved with double side texture, perpendicular slats, or parquet grooves in order to attain as many as 50 passes of the light.

9.5 Shrink Geometries

Currently available PC solar cells have a significant current crowding loss and large internal series resistance accounting for about 0.002 Ω cm^2, which represents about 3% power loss at 25 W/cm^2 or 250$\times$. The unit cell dimension should be reduced from the current 140 μm to 40 μm and to 25 μm if possible. The metallisation and intermetal dielectric technology usually prevents the shrinking of the unit cell. In order to be able to shrink the unit-cell dimension to 40 μm or even 25 μm, a state-of-the-art plasma etching process for the metal, as well as a plasma-enhanced chemical vapour deposition (PECVD), SiO$_2$-based dielectric as intermetal dielectric layer must be adopted.

Table 3 Comparison of performance of a commercial PC solar cell (HECO335 from SunPower Corporation) cell with the projected new high-performance cell at ×100 and ×275 (AM1.5D, 10 and 27.5 W/cm^2, $T_c = 25°C$). The values at ×275 concentration are in parentheses

Parameter	Commercially available PC solar cell	New high-performance cell	Unit
Responsivity	0.403 (0.379)	0.416 (0.416)	A/W
Open-circuit voltage	807 (825)	826 (845)	mV
Fill factor	82.6 (79)	87 (85.9)	%
Efficiency	26.8 (24.7)	29.9 (30.2)	%

9.6 Reduce Series Resistance

Shrinking the geometries of the unit cell in order to reduce the internal series resistance may have a negative effect on the external series resistance. This problem is due to the fact that, if the gap between metal lines is kept constant, the metal coverage fraction decreases with decreasing the dimensions of the unit cell. A new metallisation design, similar to what has been proposed in previous papers [4,5], must be implemented in order to reduce the series resistance. As in step 9.5, the new metal design requires a plasma etching process for metal and a SiO$_2$-based dielectric layer. The target is to reduce the total series resistance to less than 0.001 Ω cm^2, or about a 2.5% power loss at ×500 concentration.

9.7 Target Performance

Table 3 shows the target performance of a 30% efficient silicon solar cell compared to the commercially available point-contact solar cells. Figure 10 shows the improvement in efficiency based on the above strategy. Of course, the full benefit of the polysilicon emitter with low J_0 is observed when all the other improvements (thinner cells, improved light trapping, and shrunk geometries) are in place.

10. CONCLUSIONS

Interdigitated-back-contact and point-contact silicon solar cells have been demonstrated to be the most efficient and most suitable silicon solar cells for one-sun and high-concentration applications. Commercially

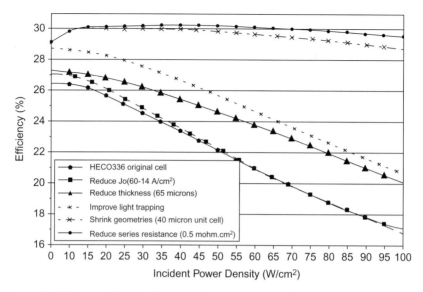

Figure 10 Efficiency (AM1. 51, 25°C) of the point-contact silicon solar cell versus incident power density after each step of this strategy.

available PC solar cells have demonstrated efficiencies up to 27.6% (at 9.2 W/cm^2, AM1.5D, 25°C) in large-volume production. The best large-area IBC silicon solar cell for one-sun applications has an efficiency of 24.2%. Commercially available large flat-plate PV modules with IBC silicon solar cells have demonstrated efficiencies greater than 21%. This chapter has described the structure of IBC, FSF, TJ, and PC cells, as well as the process to fabricate them. A simple model for the simulation, optimisation, and analysis of the different recombination mechanisms in one-sun and concentrator IBC cells has been presented. Finally, we have presented a plan for the development of 30% efficiency concentrator PC solar cells.

ACKNOWLEDGEMENTS

The author would like to thank A. Terao and S. Daroczi of SunPower Corporation and J. Lasich of Solar Systems Pty Ltd for supplying pictures of the point-contact solar cells and the concentrator PV systems.

REFERENCES

[1] P.J. Verlinden, R.M. Swanson, R.A. Crane, K. Wickham, J. Perkins, A 26.8% Efficient concentrator, point-contact solar cell, Proceedings of the 13th European Photovoltaic Solar Energy Conference, Nice, 1995, pp. 1582–1585.

[2] R.A. Sinton, R.M. Swanson, Development efforts on silicon solar cells, Final Report, Electric Power Research Institute, Palo Alto, CA, February 1992, pp. 2–44.
[3] R.M. Swanson, R.A. Sinton, N. Midkiff, D.E. Kane, Simplified designs for high-efficiency concentrator solar cells. Sandia Report, SAND88–0522, Sandia National Laboratories, Albuquerque, NM, July 1988.
[4] P.J. Verlinden, R.M. Swanson, R.A. Sinton, D.E. Kane, Multilevel metallization for large area point-contact solar cells, Proceedings of the 20th IEEE Photovoltaic Specialists Conference, Las Vegas, 1988, pp. 532–537.
[5] P.J. Verlinden, R.A. Sinton, R.M. Swanson, High-efficiency large-area back contact concentrator solar cells with a multilevel interconnection, Int. Journal of Solar Energy 6 (1988) 347–365.
[6] P.J. Verlinden, A. Terao, D.D. Smith, K. McIntosh, R.M. Swanson, G. Ganakas, et al., will we have a 20%-efficient (ptc) photovoltaic system?, Proceedings of the 17th European Photovoltaic Solar Energy Conference, Munich, 2001, pp. 385–390.
[7] R.J. Schwartz, M.D. Lammert, IEEE International Electron Devices Meeting. Washington DC, 1975, pp. 350–351.
[8] M.D. Lammert, R.J. Schwartz, IEEE Trans. Electron Devices ED-24(4) (1977) 337–342.
[9] R.M. Swanson, Point contact silicon solar cells, theory and modeling, Proceedings of the 18th IEEE Photovoltaic Specialist Conference, Las Vegas, 1985, pp. 604–610.
[10] R.M. Swanson, Point contact solar cells, modeling and experiment, Solar Cells Vol. 7(1) (1988) 85–118.
[11] R.A. Sinton, Device Physics and Characterization of Silicon Point-Contact Solar Cells, Ph.D. Thesis, Stanford University, Stanford, CA, 1987.
[12] R.A. Sinton, R.M. Swanson, Recombination in highly injected silicon, IEEE Trans. Electron Devices ED-34(6) (1987) 1380.
[13] R.A. Sinton, P.J. Verlinden, R.M. Swanson, R.A. Crane, K. Wickham, J. Perkins, Improvements in silicon backside-contact solar cells for high-value one-sun applications, Proceedings of the 13th European Photovoltaic Solar Energy Conference, Nice, 1995, pp.1586–1589.
[14] P.E. Gruenbaum, J.Y. Gan, R.R. King, R.M. Swanson, Stable passivations for high-efficiency silicon solar cells, Proceedings of the 21st IEEE Photovoltaic Specialists Conference, Kissimmee, 1990, pp. 317–322.
[15] P.J. Verlinden, R.M. Swanson, R.A. Sinton, R.A. Crane, C. Tilford, J. Perkins, et al., High-efficiency point-contact silicon solar cells for fresnel lens concentrator modules, Proceedings of the 23rd IEEE Photovoltaic Specialists Conference, Louisville, 1993, pp. 58–64.
[16] P.E. Gruenbaum, Photoinjected Hot-Electron Damage at the Silicon/Silicon Dioxide Interface in Point-Contact Solar Cells, Stanford University, Stanford, 1990.
[17] M. Cudzinovic, T. Pass, A. Terao, P.J. Verlinden, R.M. Swanson, Degradation of surface quality due to anti-reflection coating deposition on silicon solar cells, Proceedings of the 28th IEEE Photovoltaic Specialists Conference, Anchorage, 2000, pp. 295–298.
[18] J.L. Gray, R.J. Schwartz, Why don't we have a 30% efficient silicon solar cell?, Proceedings of the 18th IEEE Photovoltaic Specialists Conference, Las Vegas, 1985, pp. 568–572.
[19] P. Verlinden, F. Van de Wiele, G. Stehelin J.P. David, An interdigitated back contact solar cell with high efficiency under concentrated sunlight, Proceedings of the 7th European Photovoltaic Solar Energy Conference, Seville,1986, pp. 885–889.
[20] R.A. Sinton, et al., 27.5 Percent silicon concentrator solar cells, IEEE Electron Device Letters EDL-7(10) (1986) 567–569.

[21] R.A. Sinton, R.M. Swanson, An optimization study of si point-contact concentrator solar cells, Proceedings of the 19th IEEE Photovoltaic Specialists Conference, New Orleans, 1987, pp. 1201–1208.
[22] R.M. Swanson, Why we will have a 30% efficient silicon solar cell. Proceedings of the 4th International Photovoltaic Science and Engineering Conference, Sydney, 1989, pp. 573–580.
[23] J.Y. Gan, Polysilicon Emitters for Silicon Concentrator Solar Cells, Ph.D. Thesis, Stanford University, Stanford, 1990.
[24] J.Y. Gan, R.M. Swanson, Polysilicon emitters for silicon concentrator solar cells, Proceedings of the 21st IEEE Photovoltaic Specialists Conference, Kissimmee, 1990, pp. 245–250.
[25] P. Campbell, M. Green, The limiting efficiency of silicon solar cells under concentrated sunlight, IEEE Trans. Electron Devices ED-33(2) (1986) 234–239.
[26] L.A. Christel, Polysilicon-contacted P+ emitter for silicon solar cell applications, Sandia National Laboratories Report SAND87–7021, 1987.
[27] Y.H. Kwark, R.M. Swanson, SIPOS heterojunction contacts to silicon, Sandia National Laboratories Report SAND85–7022, 1985.
[28] H. Sakata, K. Kawamoto, M. Taguchi, T. Baba, S. Tsuge, K. Uchihashi, et al., 20.1% Highest efficiency large area (101 cm^2) HIT cell, Proceedings of the 28th IEEE Photovoltaic Specialists Conference, Anchorage, 2000, pp. 7–12.
[29] P.J. Verlinden, R.M. Swanson, R.A. Crane, 7000 High-efficiency cells for a dream, Prog. Photovolt − Res. Appl. 2 (1994) 143–152.
[30] P.J. Verlinden, R.M. Swanson, R.A. Crane, High efficiency silicon point-contact solar cells for concentrator and high-value one-sun applications, Proceedings of the 12th EC Photovoltaic Solar Energy Conference, Amsterdam, 1994, pp. 1477–1480
[31] P.J. Verlinden, R.A. Crane, R.M. Swanson, T. Iwata, K. Handa, H. Ogasa, et al., A 21.6% Efficient photovoltaic module with backside contact silicon solar cells, Proceedings of the 12th EC Photovoltaic Solar Energy Conference, Amsterdam, 1994, pp. 1304–1307.
[32] C.Z. Zhou, P.J. Verlinden, R.A. Crane, R.M. Swanson, 21.9% Efficient silicon bifacial solar cells, Proceedings of the 26th IEEE Photovoltaic Specialists Conference, Anaheim 1997, pp. 287–290.
[33] W.P. Mulligan, et al., Manufacture of solar cells with 21% efficiency, Proceedings of the 19th European Photovoltaic Solar Energy Conference, 2004, pp. 387.
[34] D. De Ceuster, P.J. Cousins, D. Rose, D. Vicente, P. Tipones, W. Mulligan, Low cost high volume production of >22% efficiency solar cells, Proceedings of the 22nd European Photovoltaic Solar Energy Conference, 2007, pp. 816–819.
[35] P.J. Cousins, D.D. Smith, H.-C. Luan, J. Manning, T.D. Dennis, A. Waldhauer, et al., Generation 3: Improved performance at lower cost, Proceedings of the 35th IEEE Photovoltaic Specialists Conference, Honolulu, 2010, pp. 275–278.
[36] R.A. Sinton, R.M. Swanson, Simplified backside-contact solar cells, IEEE Trans. Electron Dev. 37(2) (1990) 348–352.
[37] A. Slade, V. Garboushian, 27.6% efficient silicon concentrator cell for mass production, Technical Digest of 15th International Photovoltaic Science and Engineering Conference, Shanghai, October 2005, pp. 701.
[38] SunPower Corporation Press Release, <http://investors.sunpowercorp.com/release-detail.cfm?ReleaseID = 179402/>, August 22, 2005.

PART IE

Dye-Sensitized and Organic Solar Cells

CHAPTER IE-1

Dye-Sensitized Photoelectrochemical Cells

Anders Hagfeldt[a,*], Ute B. Cappel[a], Gerrit Boschloo[a], Licheng Sun[b], Lars Kloo[c], Henrik Pettersson[d] and Elizabath A. Gibson[e]

[a]Department of Physical and Analytical Chemistry, Uppsala University, Box 259, SE-75105 Uppsala, Sweden
[b]Department of Chemistry, KTH—Royal Institute of Technology, Teknikringen 30, SE-10044 Stockholm, Sweden; State Key Laboratory of Fine Chemicals, Dalian University of Technology (DUT), Dalian 116012, China
[c]Department of Chemistry, KTH—Royal Institute of Technology, Teknikringen 30, SE-10044 Stockholm, Sweden
[d]Swerea IVF AB, Box 104, SE-431 22 Mölndal, Sweden
[e]School of Chemistry, University of Nottingham, University Park, Nottingham NG7 2RD, United Kingdom
*This is a joint publication from the Centre for Molecular Devices, KTH-Royal Institute of Technology, Teknikringen 30, SE−10044 Stockholm, Sweden.
To whom correspondence should be addressed: E-mail: Anders.Hagfeldt@fki.uu.se

Contents

1. Introduction 480
2. Photoelectrochemical Cells 483
   2.1 Levels of Energy and Potential 484
   2.2 The Semiconductor 485
   2.3 The Electrolyte 487
   2.4 The Semiconductor—Electrolyte Junction 488
   2.5 Light-Induced Charge Separation 491
   2.6 Dye Sensitization 493
   2.7 Basic Operational Principles of Photoelectrochemical Cells 495
3. Dye-Sensitized Solar Cells 496
   3.1 Overview of Current Status and Operational Principles 497
   3.2 Overview of the Different Electron-Transfer Processes 499
      3.2.1 Reactions 1 and 2: Electron Injection and Excited State Decay *501*
      3.2.2 Reaction 3: Regeneration of the Oxidized Dyes *502*
      3.2.3 Reaction 4: Electron Transport Through the Mesoporous Oxide Film *503*
      3.2.4 Reaction 5 and 6: Recombination of Electrons in the Semiconductor with Oxidized Dyes or Electrolyte Species *504*
      3.2.5 Reaction 7: Transport of the Redox Mediator and Reactions at the Counterelectrode *505*
   3.3 Characterization of DSC Devices 506
      3.3.1 Efficiency Measurements *507*
      3.3.2 External and Internal Quantum Efficiencies *509*
      3.3.3 The DSC Toolbox *512*

Practical Handbook of Photovoltaics.
© 2012 Elsevier Ltd. All rights reserved.

3.3.4　The Stark Effect in DSC　　　　　　　　　　　　　　525
　3.4　Development of Material Components and Devices　　　527
　　　3.4.1　Mesoporous Oxide Working Electrodes　　　　　　528
　　　3.4.2　Dyes　　　　　　　　　　　　　　　　　　　　　529
　　　3.4.3　Electrolytes　　　　　　　　　　　　　　　　　　531
　　　3.4.4　Counterelectrodes　　　　　　　　　　　　　　　532
　　　3.4.5　Development of Modules　　　　　　　　　　　　533
4.　Future Outlook　　　　　　　　　　　　　　　　　　　　535
References　　　　　　　　　　　　　　　　　　　　　　　　536

1. INTRODUCTION

The magnitude of the change required in the global energy system will be huge. The challenge is to find a way forward that simultaneously addresses issues of energy supply and saving, climate change, security, equity, and economics. The mean global energy consumption rate was 13 terawatt (TW) in the year 2000. Assuming a kind of "business-as-usual" scenario with rather optimistic but reasonable assumptions of population growth and energy consumption, the projection is that the global energy demand in 2050 will at least be double. Solar energy will play a pivotal role in the paradigm shift that is needed for energy supply. After fusion, solar energy has the largest potential to satisfy future global needs for renewable energy sources. From the 1.7×10^5 TW of solar energy that strikes Earth's surface, a practical terrestrial global solar potential value is estimated to be about 600 TW. Thus, using 10% efficient solar farms, about 60 TW of power could be supplied. The umbrella of solar-energy conversion encompasses solar thermal, solar fuels, solar-to-electricity (photovoltaic, PV) technology, and the great many subcategories below those. Photovoltaics, or solar cells, are fast growing both with regards to industrialization and research. Globally, the total PV installation is around 40 gigawatt (GW), and an annual growth rate of 45% has been experienced over recent years. Solar cell technologies can be divided into three generations. The first is established technology such as crystalline silicon. The second includes the emerging thin-film technologies that have just entered the market, while the third generation covers future technologies that are not yet commercialized. A link for PV updates is www.solarbuzz.com, and our own contribution for a review of PV technologies with special emphasis on the materials science aspects is reference [1].

When comparing different photovoltaic technologies, a figure of merit is the production cost per peak watt of solar electricity produced. For so-called second-generation thin-film solar cells, production costs down to and even below $1/W_{peak}$ are reported. To be competitive for large-scale electricity production, new PV technologies thus need to aim at production costs below $0.5/W_{peak}$. To give an example, this means a cost of 70 $/m^2 at a module efficiency of 14%. The dye-sensitized solar cell (DSC) is a molecular solar cell technology that has the potential to achieve production costs below $0.5/W_{peak}$.

DSC is based on molecular- and nanometre-scale components. Record cell efficiencies of 12%, promising stability data and energy-efficient production methods, have been accomplished. In the present table of record solar cell efficiencies [2], in which the solar cell area must be at least 1 cm^2, the record is held by the Sharp company in Japan at 10.4 ± 0.3% [3]. The record for a DSC module is 9.2% achieved by Sony, Japan [4]. As selling points for the DSC technology, the prospect of low-cost investments and fabrication and short energy-payback time (<1 year) are key features. DSCs offer the possibilities to design solar cells with a large flexibility in shape, colour, and transparency. Integration into different products opens up new commercial opportunities for niche applications. Ultimately, the comparison of different energy sources is based on the production cost per kWh—i.e., the cost in relation to energy production. For DSC technology, it is advantageous to compare energy cost rather than cost per peak watt. DSCs perform relatively better compared with other solar cell technologies under diffuse light conditions and at higher temperatures.

DSC research groups have been established around the world with the most active groups in Europe, Japan, Korea, China, and Australia. The field is growing fast, which can be illustrated by the fact that about two or three research articles are being published every day. From a fundamental research point of view, we can conclude that the physical chemistry of several of the basic operations in the DSC device remain far from fully understood. For specific model and reference systems and controlled conditions, there is a rather detailed description in terms of energetics and kinetics. It is, however, still not possible to predict accurately how a small change to the system—e.g., replacing one component or changing the electrolyte composition—will affect DSC performance. With time, the chemical complexity of DSCs has become clear, and the main challenge for future research is to understand and master this complexity, in particular at the oxide–dye–electrolyte interface. It is noteworthy that such a fundamental effect as the Stark shift of the dye molecules at the interface took more than 15 years to identify and describe. This was independently made by ourselves [5] and Meyer et al. [6], USA. A challenging but realizable goal for the present DSC technology is to achieve efficiencies above 15%. We have for many years known where the main losses in the state-of-the-art DSC device are—that is, the potential drop in the regeneration process and the recombination loss between electrons in the TiO_2 and acceptor species in the electrolyte. With our breakthrough of using one-electron transfer redox systems such as co-complexes, in combination with a dye, which efficiently prevents the recombination loss, we may now have found the path to increase the efficiency significantly [7]. With the recent world record of 12.3 by Grätzel and co-workers [8] using co-complexes, the main direction of the research field is now to explore this path.

The industrial interest in DSCs is strong with large multinational companies such as BASF and Tata Steel in Europe and Toyota, Sharp, Panasonic, Sony, Fujikura, and Samsung in Asia. These companies present encouraging results, in particular with regard to upscaling with world record minimodule efficiencies above 9% (Sony and Fujikura). In this context, we note that world record efficiencies are not the same as stable efficiencies obtained after durability tests. Reported stable module efficiencies vary significantly in the literature and are difficult to judge. Best values in the literature are about 5%, although presentations at conferences report better results. Several companies are dedicated to setting up manufacturing pilot lines. G24i is a company based in Cardiff, Wales, that focuses on consumer electronics. On its Web site (www.g24i.com), such niche products are now for sale. Companies that sell material components, equipment, and consultancy services have increased and are growing.

The structure of this chapter basically follows the paper by McEvoy in the first edition of *Practical Handbook of Photovoltaics: Fundamentals and Applications* [9]. Our aim is to provide the reader with a general description of photoelectrochemical cells with a specific focus on DSCs. There are several recent reviews on DSCs, and the reader is referred to these papers for further information [10–22]. Instead of trying to add another review article to the field, we want to make this chapter more of a handbook. We present the general concepts and principles of PEC and put more emphasis on different types of characterization methods for DSC devices rather than provide detailed overviews of, for example, materials development. This also means that the reference list is more of a general summary of the field, including mainly books, review articles, and some original papers.

2. PHOTOELECTROCHEMICAL CELLS

The photovoltaic effect was first observed by A.-E. Becquerel almost 200 years ago [23] when he illuminated a thin layer of silver chloride coated on a sheet of Pt immersed in an electrolytic solution and connected to a counterelectrode. The Becquerel device would at present be classified as a photoelectrochemical (PEC) cell, see, for example, reference [9]. The contact of the semiconductor with the electrolyte, in which the conduction mechanism is the mobility of ions rather than of electrons or holes, forms a photoactive junction functionally equivalent to those later discovered for solid-state photovoltaic devices. Unlike their solid-state counterparts, however, semiconductor—liquid junctions are versatile in that solar energy can be used to drive chemical reactions—for example, for the production of chemical fuels and self-cleaning and anti-fogging surfaces. For an interesting and recent review on the origins of photoelectrochemistry, see reference [24].

The first boom of photoelectrochemistry came with the famous 1972 paper of Fujishima and Honda, who reported the use of a TiO_2 photoanode in an electrochemical cell to split water into H_2 and O_2 [25]. As pointed out by Rajeshwar [24], perhaps the first instance of the use of a light-responsive electrode for the photo-assisted decomposition of water is the 1960 paper 'Decomposition of Water by Light' [26]. The promising results of PEC cells as solar-energy converters together with the 1973 oil crisis triggered a dramatic growth of research in this area. From an industrial point of view, the biggest success so far of the Fujishima—Honda paper was probably the development of photocatalytic TiO_2 systems for environmental remediation, self-cleaning materials, and antifogging surfaces [24,27,28]. For solar-energy conversion, the intense activities on PEC systems from the mid-1970s to the mid-1980s was followed by a considerable slowing down of the progress [24]. The second boom of photoelectrochemistry for solar-energy conversion occurred in the early 1990s starting with the seminal paper by O'Regan and Grätzel in 1991 [29]. The very unexpected finding that a high-surface-area electrode could be used to dramatically increase the performance of dye-sensitized photoelectrochemical cells caused a paradigm shift in the understanding of how an efficient solar cell can be prepared and operate. The principle of dye-sensitized solar cells has also become a part of the core chemistry and energy science teaching and research. Textbooks have sections or chapters dealing with DSC [30,31]. Laboratory kits have been developed for educational purposes making use of dyes from all possible natural sources such as blackberries, spinach, blueberries, tea, and wine. Not only photoelectrochemistry but also energy science, photochemistry, materials science, and transition metal coordination chemistry have significantly benefitted from DSC research.

The basic structure of a PEC cell is a working electrode (semiconductor), an electrolyte, and a counterelectrode (normally a metallic material with a low overpotential for reduction–oxidation of the electrolyte). The heart of the device is the semiconductor–electrolyte interface (SEI) of the working electrode. In the following, we go through the basic features of this interface, starting with the individual components followed by the description of the SEI and finishing with the phenomenon of dye sensitization.

2.1 Levels of Energy and Potential

The semiconductor–electrolyte junction is an 'interface' between physics and chemistry. In solid-state physics and electrochemistry, one normally uses the energy scale with vacuum as reference for the former and the potential scale with the standard hydrogen electrode (SHE) or normal hydrogen electrode (NHE) as reference for the latter. The electrochemical potential of electrons in a semiconductor is normally referred to as the *Fermi level*, E_F, and in an electrolyte solution it is often referred to as the *redox potential*, $E^0_{F,redox}$ in energy scale and V^0_{redox} in potential scale. At equilibrium, the electrochemical potentials (or the Fermi levels) of the semiconductor and electrolyte will be equal. For most purposes in electrochemistry, it is sufficient to reference the redox potentials to the NHE (or any other more practical reference system), but it is sometimes of interest to have an estimate of the absolute potential (i.e., versus the potential of a free electron in vacuum). For example, it may be of interest to estimate relative potentials of semiconductors, redox electrolytes, and solid hole conductors in solid-state DSC based on their work functions. Taking a redox system dissolved in an electrolyte as an example, the absolute energy of a system is shifted against the conventional scale by the free energy E_{ref}, according to

$$E_{abs} = E_{ref} - E^0_{F,redox} \tag{1}$$

For the NHE, E_{ref} is estimated to be -4.6 ± 0.1 eV[63]. With this value, the standard potentials of other redox couples can be expressed on the absolute scale.

2.2 The Semiconductor

In nondegenerate semiconductors, the equilibrium Fermi level is given by

$$E_F = E_c + k_B T \ln\left(\frac{n_c}{N_c}\right) \qquad (2)$$

where E_c is the energy at the conduction band edge, $k_B T$ is the thermal energy, n_c is the density of conduction band electrons, and N_c is the effective density of conduction band states. With respect to vacuum, E_c is given by the electron affinity E_A, as shown in Figure 1. The ionization energy, I, in the same figure determines the position of the valence band, E_V, whereas the distance between the vacuum level and the equilibrium Fermi level is the work function, ϕ. Thus, the positions of the energy bands can be predicted from electron-affinity values.

However, these values are very sensitive to the environment, and measurements of absolute and relative energies in vacuum must be carefully interpreted and analyzed in terms of their relevance to DSCs. In the field of semiconductor electrochemistry, the standard approach of determining the so-called flatband potential of a semiconductor, V_{fb}, which estimates the work function of the semiconductor in contact with the specific electrolyte, is Mott-Schottky analysis of capacitance data [32]. This approach is based on the potential-dependent capacitance of a depletion layer at the semiconductor surface. For a DSC, such behaviour is not expected to be

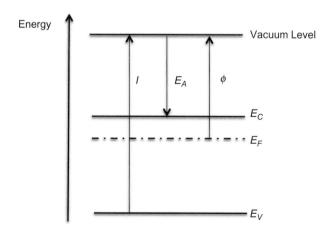

Figure 1 Positions of energy bands for a semiconductor with respect to vacuum level.

observed for the ~20-nm anatase nanocrystals that are expected to be fully depleted [10,14]. Instead, cyclic voltammetry and spectroelectrochemical procedures have been used to estimate E_c. These methods also give information of the density of states (DOS) of the semiconductor.

For a recent review on the measurements of E_c, and the density of states in mesoporous TiO_2 films, see reference [14]. Fitzmaurice has reviewed the first spectroelectrochemical measurements of E_c for transparent mesoporous TiO_2 electrodes [33]. Using an accumulation-layer model to describe the potential distribution within the TiO_2 nanoparticle at negative potentials [34], and assuming that E_c remains fixed as the Fermi level is raised into accumulation conditions, E_c values in organic and aqueous electrolytes were estimated. At present there is an extensive compilation of data that shows that E_c is not that well defined. Many electrochemical and spectroelectrochemical studies indicate that mesoporous TiO_2 films possess a tailing of the DOS (trap states) rather than an abrupt onset from an ideal E_c. Nevertheless, the E_c values estimated in the early work by Fitzmaurice and co-workers are still used today, at least qualitatively, to discuss, for example, the energy-level matching between the conduction band edge of the oxide and the excited state of the dye.

The position of the conduction band edge depends on the surface charge (dipole potential). The pH dependence of E_c for mesoporous TiO_2 films in aqueous solutions follows a Nernstian behaviour with a shift of 59 mV/pH unit due to protonation–deprotonation of surface titanol groups on TiO_2 (see references [14,34]). In nonaqueous solutions, E_c can be widely tuned by the presence of positive ions, cations. This effect is greatest with cations possessing a large charge-to-radius ratio. For example, E_c has been reported to be -1.0 V vs. SCE in 0.1 M $LiClO_4$ acetonitrile electrolyte and ~ -2.0 V when Li^+ was replaced by a large cation, tert-butyl ammonium. The large variation of E_c with different 'potential-determining' ions can be explained by cation-coupled reduction potentials for TiO_2 acceptor states, due to surface adsorption or insertion into the anatase lattice [14]. This cation-dependent shift in E_c is used to promote photo-induced electron injection from the surface-bound sensitizer. For this to occur, E_c must be at a lower energy than the excited state of the sensitizer, S^+/S^*. In contrast, one would like E_c to be at as high energy as possible to achieve a high photovoltage. Thus, there is a compromise for the position of E_c in order to attain an efficient electron injection while maintaining a high photovoltage. Additives in the electrolyte are normally used to fine-tune the energy-level matching of E_c and S^+/S^*. The effect of the additive—most often 4-tert-butylpyridine (tBP) is used—can be studied by measuring the relative shifts of E_c, depending on surface charge, and by measuring the electron lifetime. The shift of E_c is measured by charge-extraction methods and electron lifetime measurements as described in the toolbox section.

2.3 The Electrolyte

The electrochemical potential of electrons, or the Fermi level, for a one-electron redox couple is given by the Nernst equation and can be written as

$$F_{F,redox} = E^0_{F,redox} + k_B T \ln\left(\frac{c_{ox}}{c_{red}}\right) \quad (3)$$

where c_{ox} and c_{red} are the concentrations of the oxidized and reduced species of the redox system. Besides the Fermi energy, we also need a description of the energy states being empty or occupied by electrons. The electronic energies of a redox system are shown in Figure 2 and are based on the model developed by Gerischer [32,35–37].

In this energy scale, E^0_{red} corresponds to the energy position of occupied electron states and E^0_{ox} to the empty states. They differ from the Fermi level $E^0_{F,redox}$ by the so-called reorganization energy, λ. The reorganization energy is the energy involved in the relaxation process of the

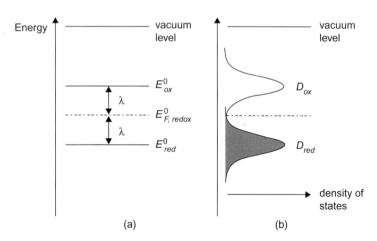

Figure 2 (a) Electron energies of a redox system using vacuum as a reference level. E^0_{red} = occupied states, E^0_{ox} = empty states, $E^0_{F,redox}$ = Fermi level of the redox couple. (b) Corresponding distribution functions. *(Taken from Figure 4 in reference [35].)*

solvation shell around the reduced species following transfer of an electron to the vacuum level. For the reverse process—i.e., electron transfer from vacuum to the oxidized species—there is an analogous relaxation process. It is normally assumed that λ is equal for both processes. The electron states of a redox system are not discrete energy levels but are distributed over a certain energy range due to fluctuations in the solvation shell surrounding the molecule. This is indicated by the distribution of energy states around E^0_{red} and E^0_{ox} (Figure 2b). D_{red} is the density of occupied states (in relative units) represented by the reduced component of the redox system and D_{ox} the density of empty states represented by the oxidized component. Assuming a harmonic oscillation of the solvation shell, the distribution curves, D_{red} and D_{ox}, are described by Gaussian functions:

$$D_{red} = D^0_{red} \exp\left[-\frac{(E - E^0_{F,\,redox} - \lambda)^2}{4kT\lambda}\right] \quad (4)$$

$$D_{ox} = D^0_{ox} \exp\left[-\frac{(E - E^0_{F,\,redox} + \lambda)^2}{4kT\lambda}\right] \quad (5)$$

D^0_{red} and D^0_{ox} are normalizing factors such that $\int_{-\infty}^{\infty} D(E) de = 1$. The half-width of the distribution curves is given by

$$\Delta E_{1/2} = 0.53 \lambda^{1/2} \text{ eV} \quad (6)$$

Accordingly, the widths of the distribution function depend on the reorganization energy, which is of importance for the kinetics of electron-transfer processes at the oxide–dye–electrolyte interface. Typical values of λ are in the range from a few 10ths of an eV up to 2 eV. In Figure 2b, the concentrations of reduced and oxidized species are equal ($D_{red} = D_{ox}$). Changing the concentration ratio varies the redox potential ($E^0_{F,\,redox}$), according to the Nernst equation.

2.4 The Semiconductor–Electrolyte Junction

If a semiconductor is immersed in the electrolyte, chemical equilibrium will be established—i.e., the Fermi levels of the semiconductor and the electrolyte will adjust to each other and

$$E_F = E_{F,\,redox} = qV_{redox} \quad (7)$$

where q is the elementary charge and V_{redox} the electrochemical (redox) potential of the electrolyte. From now on, we limit the discussion to n-type doped semiconductors keeping in mind that the inverse but analogous situation occurs with p-type semiconductors. For an n-type semiconductor with the initial Fermi level higher than the redox potential, the equilibration of E_F and $E_{F,\,redox}$ occurs by transfer of electrons from the semiconductor to the electrolyte. This will produce an electrical field in the solid—i.e., the so-called space charge layer (also referred to as the *depletion layer* since the region is depleted of majority carries) (see Figure 3).

For an n-type semiconductor the space charge layer is positive. The electric field is represented by curvature of the conduction and the valence bands (band bending) and indicate the direction that the free carriers will migrate in the field. The width of the space charge layer was derived by Gärtner [38] to be

$$w = \sqrt{\frac{2\varepsilon\varepsilon_0(V - V_{fb})}{qN_D}} \tag{8}$$

where ε is the relative dielectric constant, ε_0 the permittivity of free space, N_D the donor concentration (or dopant density), and V_{fb} and V the electrostatic potentials of the semiconductor surface and the bulk, respectively. Thus, $V - V_{fb}$ gives the band bending. While the space charge layer normally extends from a few nanometres to micrometres, the main part of the potential drops takes place in the solution within a few Ångströms of the surface in the so-called Helmholtz layer. This is particularly the case in solutions with high ionic strengths.

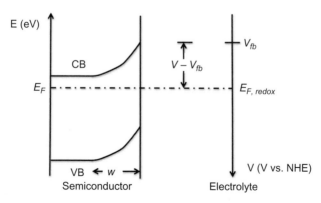

Figure 3 Energy-versus-distance diagram for an n-type semiconductor–liquid junction in equilibrium. For an n-type semiconductor, the space charge region, *w*, is positively charged due to depletion of conduction band electrons.

For a semiconductor, the space charge layer formed gives rise to a capacitance C_{SC} in this region, which is usually much smaller than that of the adjacent Helmholtz layer in the electrolyte, C_H. Since the total capacity is given by

$$\frac{1}{C} = \frac{1}{C_H} + \frac{1}{C_{SC}} \qquad (9)$$

it follows that $1/C \approx 1/C_{SC}$. Thus, any variation in an externally applied voltage, V, often changes only the potential drop within the semiconductor—i.e., the Fermi level in the semiconductor—and so the band bending is changed. The energy (or potential) difference between the conduction band edge position at the semiconductor surface and Fermi level of the electrolyte is therefore not affected by the external voltage. The flatband potential, V_{fb}, is the potential at which the semiconductor bands are flat (zero space charge), and is measured with respect to a reference electrode with a well-defined electrode potential.

The above-described situation is different in a colloidal semiconductor or in a nanocrystalline network. The potential distribution in a spherical semiconductor particle has been derived by semiconductor nanoparticles, the total potential drop within the semiconductor becomes

$$V - V_{fb} = \frac{kT}{6q}\left(\frac{r_0}{L_D}\right)^2 \qquad (10)$$

where r_0 is the radius of the particle and $L_D = (\varepsilon\varepsilon_0 kT/2q^2 N_D)^{0.5}$ is the Debye length, which depends on the number of ionized dopant molecules per cubic centimetre, N_D. From this equation, it is apparent that the electrical field in semiconductor nanoparticles is usually small and that high dopant levels are required to produce a significant potential difference between the surface and the center of the particle. For example, in order to obtain a 50 mV potential drop in a TiO_2 nanoparticle with $r_0 = 6$ nm, a concentration of 5×10^{19} cm^{-3} of ionized donor impurities is necessary. Undoped TiO_2 nanoparticles have a much smaller carrier concentration and the band bending within the particles is therefore negligibly small.

This chapter focuses on dye-sensitized solar cells, which are based on a mesoporous semiconductor oxide film. When the dye-sensitized mesoporous solar cell was first presented, perhaps the most puzzling phenomenon was the highly efficient charge transport through the nanocrystalline TiO_2 layer. Early on it was realized that the mesoporous electrodes were very much different compared to their compact analogues described above because (1) the inherent conductivity of the film is very low, (2) the small size of the individual colloidal particles does not support a built-in electrical field, and (3) the oxide particles and the electrolyte-containing pores form interpenetrating networks whose phase boundaries produce a junction of huge contact area. These films are thus best viewed as an ensemble of individual particles through which electrons can percolate by hopping from one crystallite to the next. Charge transport in mesoporous systems is still under keen debate today [21,39–41] and will be described in Section 3.2.3. Common characteristics in the operations of mesoporous oxide electrodes are the filling of trap states and that the charge-separation process after photoexcitation is mainly governed by the kinetics at the SEI.

The effect of an external field in a mesoporous semiconductor film is then different compared to the discussion above for a doped compact semiconductor in which the applied potential changed the band bending of the semiconductor. Since there is no macroscopic space charge layer formed in a mesoporous electrode, the applied potential will change the Fermi level at the back contact (conducting substrate–mesoporous oxide interface) throughout the mesoporous film with the result of a different electron concentration of the electrode compared to zero bias. The term *flatband potential* is therefore not appropriate for a mesoporous electrode. The term *position of the conduction band edge*, E_c, is then used for conventional DSC systems.

2.5 Light-Induced Charge Separation

The depletion layer at the interface between a compact semiconductor and a liquid medium plays an important role in light-induced charge separation. The local electrostatic field present in the space charge layer serves to separate the electron–hole pairs generated by illumination of the semiconductor. For n-type materials, the direction of the field is such that holes migrate to the surface where they undergo a chemical reaction, while the electrons drift through the bulk to the back contact of the semiconductor and subsequently through the external circuit to the counterelectrode. Charge carriers which are photogenerated in the field-free space of the semiconductor can also contribute to the photocurrent. In solids with low defect concentration, the lifetime of the electron–hole pairs is long enough to allow for some of the minority carriers to diffuse to the depletion layer before they undergo recombination.

In the case of semiconductor nanoparticles, the band bending is small and charge separation occurs via diffusion. The absorption of light leads to the generation of electron–hole pairs in the particle, which are oriented in a spatially random fashion along the optical path. These charge carriers subsequently recombine or diffuse to the surface where they undergo chemical reactions with suitable solutes or catalysts deposited on the surface of the particles. Since in nanoparticles the diffusion of charge carriers from the interior to the particle surface can occur more rapidly than their recombination, it is feasible to obtain quantum yields for photoredox processes approaching unity. Whether such high efficiencies can really be achieved depends on the rapid removal of at least one type of charge carrier—i.e., either electrons or holes—upon their arrival at the interface. This underlines the important role played by the interfacial charge-transfer kinetics.

The charge-separation process in mesoporous electrodes was measured independently for mesoporous TiO_2 films [42] and electrodeposited CdS and chemically deposited CdSe [43]. High quantum yields were measured in all cases. By illuminating the mesoporous film both through the electrolyte and through the transparent conducting oxide (TCO) substrate, the dependence of the quantum yield as a function of depth in the semiconductor film could be monitored. From these measurements, a qualitative model to describe the photocurrent generation in nanocrystalline films was presented. The electrolyte penetrates the whole mesoporous film up to the surface of the back contact, and a semiconductor–electrolyte junction occurs thus at each nanocrystal, much like a normal nanoparticle system. During illumination, light absorption in any individual nanoparticle will generate an electron–hole pair. Assuming that the kinetics of charge transfer to the electrolyte is much faster for one of the charges (holes for TiO_2) than for the recombination processes, the other charge (electrons) can create a gradient in the electrochemical potential between the particle and the back contact. In this gradient, the electrons (for TiO_2) can be transported through the interconnected colloidal particles to the back contact, where they are withdrawn as a current. The charge separation in a mesoporous semiconductor electrode therefore does not need to depend on a built-in electric field but is mainly determined by kinetics at the SEI. The creation of the light-induced electrochemical potential for the electrons in TiO_2 also explains the building up of a photovoltage. Early on in DSC research it was concluded that charge transport in mesoporous films is dominated by diffusion of the charge carriers [44].

2.6 Dye Sensitization

The history of dye-sensitized photoelectrochemical cells goes back to the early days of colour photography. For further insights into this interesting parallel between sensitization in photography and photoelectrochemistry, readers are referred to references [9,12,45]. In 1873, Vogel, professor of "photochemistry, spectroscopy and photography" in Berlin, established empirically that silver halide emulsions could be sensitized to red and even infrared light by suitably chosen dyes [46]. The concept of dye enhancement was carried over already by 1887 from photography to the photoelectric effect by Moser [47] using the dye erythrosine, again on silver halide electrodes. That the same dyes were particularly effective for both processes was recognized among others by Namba and Hishiki [48] at the 1964 International Conference on Photosensitization in Solids in Chicago, a seminal event in the history of dyes in the photosciences. It was also recognized there that the dye should be adsorbed on the semiconductor surface in a closely packed monolayer for maximum sensitization efficiency [49]. On that occasion, the theoretical understanding of the processes was clarified, since until then it was still disputed whether the mechanism was a charge transfer or an Auger-like energy-coupling process. With the subsequent work of Gerischer and Tributsch [50,51] on ZnO, there could be no further doubt about the mechanism, and it was evident that the process involved the excitation of the dye from its charge-neutral ground state to an excited state by the absorption of the energy of a photon followed by a charge transfer from the excited dye to the semiconductor.

Excitation of a sensitizer dye molecule, S, for example by the absorption of the energy of a photon, can promote an electron from the highest occupied molecular orbital (HOMO) to the lowest unoccupied molecular level (LUMO). Therefore, as far as absorption of light is concerned, the HOMO−LUMO gap of a molecule is fully analogous to the band gap of a semiconductor. It defines the response to incident light and consequently the optical absorption spectrum. At the same time, the absolute energy level of the excited state of S, approximately the LUMO, can determine the energetics of the permitted relaxation processes of the excited molecule. When it lies above the conduction band edge of a semiconductor substrate, relaxation may take the form of emission of an electron from the dye into the semiconductor, leaving that molecule in a positively charged oxidized state. The Fermi level of the oxidation potential of S is $E^0_{F,redox}(S/S^+)$. An excited sensitizer, S^*, is more easily reduced or oxidized, because of the excitation energy ΔE^* stored in the molecule. The Fermi levels of excited molecules, $E^*_{F,redox}(S^*/S^+)$, can be estimated by adding ΔE^* from the redox energy level of the molecule in the ground state. The stored excitation energy ΔE^* corresponds to the energy of the $0-0$ transition between the lowest vibrational levels in the ground and excited states—that is, $\Delta E^* = \Delta E_{0-0}$. ΔE_{0-0} can be estimated by the photoluminescence (PL) onset or from the intersection of the absorption and PL spectra. If difficulties arise in measuring the PL spectrum, another way to estimate ΔE_{0-0} is from the absorption onset of the dyes adsorbed on the oxide at a certain percentage (e.g., 10%) of the full amplitude at the absorption maximum.

Introducing the corresponding distribution functions of the occupied and empty states for the most relevant reactions in a photoelectrochemical cell, we arrive at a so-called Gerischer diagram for an excited-state electron injection from surface-bound sensitizers into the density of states of a TiO$_2$ film (Figure 4). In Figure 4, the distribution functions of the empty and occupied states for the ground state and excited state are drawn with

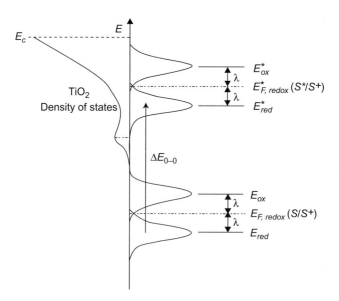

Figure 4 Gerischer diagram of DSC with iodide–triiodide electrolyte. The level of the unstable reaction intermediate didiodide (I_2^-) is indicated. *(Adapted from reference [14] and reproduced by permission of the Royal Society of Chemistry; and reference [35], copyright 1985, with permission from Elsevier.)*

equal areas, indicating that the concentrations of the different species are the same. Differences in concentration will lead to different Fermi levels, $E^0_{F,redox}(S/S^+)$ and $E^*_{F,redox}(S^*/S^+)$, and thus, different driving forces for electron injection and for regeneration of the oxidized dyes by the electrolyte. The actual concentrations of the different species indicated in Figure 4 will depend on several factors such as the Fermi levels of the semiconductor and electrolyte, dye loading, extinction coefficients, and light intensity.

2.7 Basic Operational Principles of Photoelectrochemical Cells

DSC is an example of a PEC cell that converts solar energy to electricity. The basic operational principles of such a device are described in Section 3.1. Since a PEC can be integrated into an electrolytic cell, it is also an option that electrochemistry can give rise to a photoelectrolytic or a photosynthetic effect converting solar energy to chemistry. Figure 5 is an example of an energy-level diagram for photoelectrolysis [32,35] in which water is split into its elements as reported by Fujishima and Honda [25]. In Figure 5, light is absorbed by an n-type semiconductor electrode in contact with a water-based electrolyte.

The counterelectrode on the right-hand side of the figure is a metal, and the whole configuration is similar to a photoelectrochemical photovoltaic cell. The two electrodes are short-circuited by an external wire. In the case of an n-type semiconductor electrode, the holes created by light excitation must react with H_2O, resulting in O_2 formation,

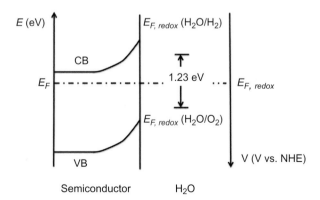

Figure 5 Energy-level diagram for photoelectrolysis in which water is split into its elements.

whereas at the counterelectrode H_2 is produced. The electrolyte can be described by two redox potentials (Fermi levels)—$E^0_{F,redox}(H_2O/H_2)$ and $E^0_{F,redox}(H_2O/O_2)$—that differ by 1.23 eV. The two reactions can obviously only occur if the band gap is >1.23 eV, the conduction band in energy being above $E^0_{F,redox}(H_2O/H_2)$ and the valence band below (positive) $E^0_{F,redox}(H_2O/O_2)$. Since multielectronic steps are involved in the reduction of and oxidation of H_2O, certain overvoltages occur for the individual processes that lead to losses. In general, it can be stated that it is usually easy to produce hydrogen with n-type semiconductor electrodes; the real problem is the oxidation of H_2O. Much research is at present taking place and being launched worldwide to address this "holy grail" of photoelectrochemistry, also with the use of molecular systems such as ruthenium complexes for light absorption and oxygen evolution [52,53].

3. DYE-SENSITIZED SOLAR CELLS

Attempts to develop dye-sensitized photoelectrochemical cells had been made before [12,45,54,55] the breakthrough of O'Regan and Grätzel in 1991 [29]. The basic problem was the belief that only smooth semiconductor surfaces could be used. The light-harvesting efficiency (LHE) for a monomolecular layer of dye sensitizer is far less than 1% of the AM 1.5 spectrum. Attempts to harvest more light by using multilayers of dyes were in general unsuccessful. Indications of the possibilities to increase the roughness of the semiconductor surface so that a larger number of dyes could be adsorbed directly to the surface and simultaneously be in direct contact with a redox electrolyte had also been reported before 1991. For example, Matsumura et al. [56] and Alonso et al. [57] used sintered ZnO electrodes to increase the efficiency of sensitization by rose bengal and related dyes. But the conversion yields from solar light to electricity remained well below 1% for these systems. Grätzel, Augustynski, and co-workers presented results on dye-sensitized fractal-type TiO_2 electrodes with high surface area in 1985 [58]. For DSC, there was thus an order-of-magnitude increase when O'Regan and Grätzel in 1991 reported efficiencies of 7–8% [29]. With regards to stability, a turnover number of 5×10^6 was measured for the ruthenium-complex sensitizer. This was followed up by the introduction of the famous N3 dye, giving efficiencies around 10% [59]. For more than a decade, the ruthenium complex N3, $Ru(L_{bip})_2(NCS)_2$ with L_{bip} being a

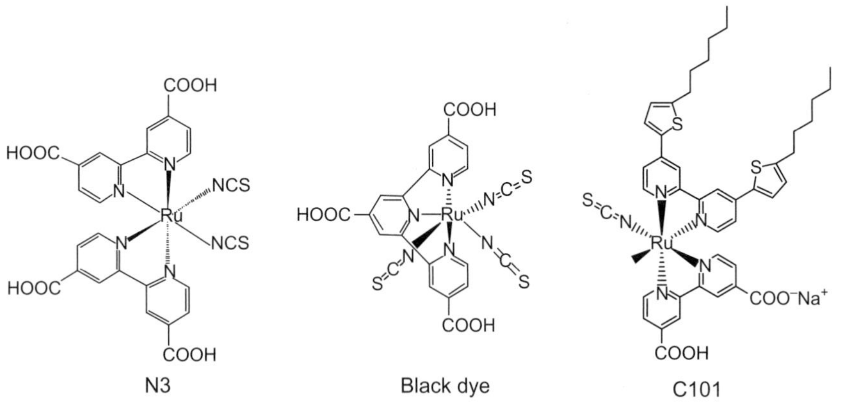

Figure 6 Molecular structures of N3, the black dye, and C101.

dicarboxylated bipyridyl ligand, its salt analogue N719, and the so-called black dye $RuL_{ter}(NCS)_3$ with L being a ter-pyridyl ligand, were state-of-the art sensitizers. Recently developed Ru complexes such as C101 show now higher performances both in terms of efficiency and stability [20,60]. The molecular structures of N3, the black dye, and C101 are shown in Figure 6.

3.1 Overview of Current Status and Operational Principles

Since the initial work in the beginning of the 1990s, a wealth of DSC components and configurations have been developed. At present, several thousands of dyes have been investigated. Fewer, but certainly hundreds of electrolyte systems and mesoporous films with different morphologies and compositions have been studied and optimized. For DSCs at present, in the official table of world record efficiencies for solar cells, the record is held by the Sharp company in Japan at 10.4±0.3% [61]. A criterion to qualify for these tables is that the solar cell area is at least 1 cm^2. For smaller cells, conversion efficiencies above 12% have been reached using the so-called C101 sensitizer as the sensitizer and with Co-complex based electrolytes [8], see Section 3.4.3.

A schematic of the interior of a DSC showing the principle of how the device operates is shown in Figure 7.

The typical configuration is as follows. At the heart of the device is the mesoporous oxide layer composed of a network of TiO$_2$ nanoparticles, which have been sintered together to establish electronic conduction. The layer is in the sintering step also deposited on a transparent

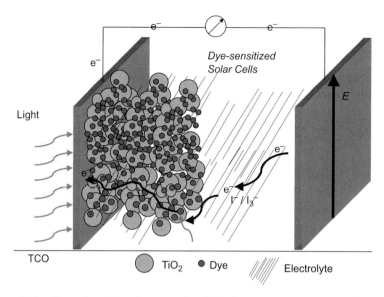

Figure 7 A schematic of the interior of a DSC showing the principle of how the device operates.

conducting oxide (TCO) substrate forming an ohmic contact. Typically, the mesoporous film thickness is ca 10 μm and the nanoparticle size 10–30 nm in diameter; the porosity is 50–60%. The mesoporous layer is deposited on a transparent conducting oxide on a glass or plastic substrate. A typical scanning electron microscopy (SEM) image of a mesoporous TiO_2 film is shown in Figure 8.

Attached to the surface of the nanocrystalline film is a monolayer of the charge-transfer dye. Photoexcitation of the latter results in the injection of an electron into the conduction band of the oxide leaving the dye in its oxidized state. The dye is restored to its ground state by electron transfer from the electrolyte, usually an organic solvent containing the iodide–tri-iodide redox system. The regeneration of the sensitizer by iodide inhibits the recapture of the conduction band electron by the oxidized dye. The I_3^- ions formed by oxidation of I^- diffuse a short distance (<50 μm) through the electrolyte to the cathode, which is coated with a thin layer of platinum catalyst where the regenerative cycle is completed by electron transfer to reduce I_3^- to I^-. For a DSC to be durable for more than 15 years outdoors, the required turnover number is 10^8, which is attained by the ruthenium complexes mentioned above.

Figure 8 SEM picture of a typical mesoporous TiO_2 film.

The voltage generated under illumination corresponds to the difference between the electrochemical potential of the electron at the two contacts, which for DSCs is generally the difference between the Fermi level of the mesoporous TiO_2 layer and the redox potential of the electrolyte. Overall, electric power is generated without permanent chemical transformation.

As mentioned earlier, a huge number of material components, dyes, mesoporous and nanostructured electrodes, electrolytes, and counterelectrodes have been synthesized and developed for DSC applications. The material component variations of a DSC device are therefore endless. It is important to keep in mind that the description above, and the one below in the next section, of the conventional DSC device with a mesoporous TiO_2, a Ru-complex sensitizer, I^-/I_3^- redox couple, and a platinized TCO counterelectrode is only valid for this particular combination of material components. As soon as one of these components is modified or completely replaced by another component, the picture has changed; energetics and kinetics are different and need to be determined for the particular system at hand. To generalize a result in DSC research is therefore difficult and can many times be misleading.

3.2 Overview of the Different Electron-Transfer Processes

The basic electron-transfer processes in a DSC, as well as the potentials for a state-of-the-art device based on the N3 dye adsorbed on TiO_2

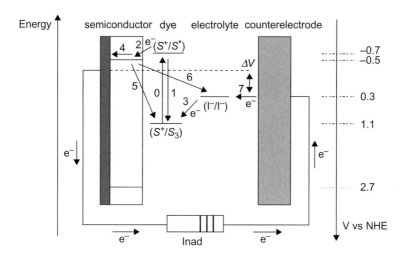

Figure 9 Simple energy-level diagram for a DSC. The basic electron-transfer processes are indicated by numbers (1–7). The potentials for a DSC based on the N3 dye, TiO$_2$, and the I$^-$/I$_3^-$ redox couple are shown. *(Taken from Figure 5 in reference [21].)*

and I$^-$/I$_3^-$ as redox couple in the electrolyte, are shown in Figure 9. The corresponding kinetic data for the different electron-transfer processes taking place at the oxide–dye–electrolyte interface are summarized in Figure 10.

Besides the desired pathway of the electron-transfer processes (processes 2, 3, 4, and 7) described in Figure 9, the loss reactions 1, 5, and 6 are indicated. Reaction 2 is electron injection from the excited state of the dye to the semiconductor, 3 is regeneration of the oxidized dye by the electrolyte, 4 is electron transport through the mesoporous oxide layer, and 7 is the reduction of the electrolyte at the counterelectrode. Reaction 1 is direct recombination of the excited dye reflected by the excited state lifetime. Recombination reactions of injected electrons in the TiO$_2$ with either oxidized dyes or with acceptors in the electrolyte are numbered 5 and 6, respectively. In principle, electron transfer to I$_3^-$ can occur either at the interface between the nanocrystalline oxide and the electrolyte or at areas of the TCO contact that are exposed to the electrolyte. In practice, the second route can be suppressed by using a compact blocking layer of oxide deposited on the anode by spray pyrolysis [62,63]. Blocking layers are mandatory for DSCs that utilize one-electron redox systems or for cells using solid organic hole-conducting media [21].

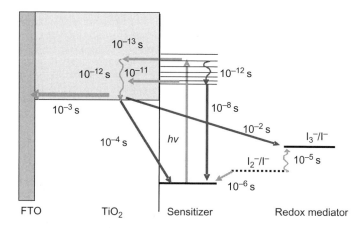

Figure 10 Summary of the kinetic data for the different electron-transfer processes depicted in Figure 9. *(Taken from Figure 6 in reference [21].)*

3.2.1 Reactions 1 and 2: Electron Injection and Excited State Decay

One of the most astounding findings in DSC research is the ultrafast injection from the excited ruthenium complex in the TiO_2 conduction band. Although the detailed mechanism of this injection process is still under debate, it is generally accepted that a fast, femtosecond (fs) component is observed for these types of sensitizers, directly attached to an oxide surface [64–67]. This would then be one of the fastest chemical processes known to date. There is a debate for a second, slower, injection component on the picosecond (ps) timescale. The reason for this could, on one hand, be due to an intersystem crossing of the excited dye from a singlet to a triplet state. The singlet state injects on the fs time scale whereas the slower component arises from the relaxation time of the singlet to triplet transition and from a lower driving force in energy between the triplet state and the conduction band of the TiO_2 [67]. Another view of the slower component is that it is very sensitive to sample condition and originates from dye aggregates on the TiO_2 surface [66]. For DSC device performance, the time scales of the injection process should be compared with direct recombination from the excited state of the dye to the ground state. This is given by the excited state lifetime of the dye, which for typical ruthenium complexes used in DSC, is 30–60 ns [11]. Thus, the injection process itself has not generally been considered to be a key factor limiting device performance. Koops et al. observed a much slower electron injection in a complete DSC device with a time constant of around 150 ps. This would then be slow enough for kinetic competition between electron injection and excited state decay of the dye, with potential implications for the overall DSC performance [68]. These results are debated since they are obtained with a single-photon counting technique that is a nondirect measurement of the injection process. A slower injection in the sub-ps time regime could be even more severe for organic dyes, for which the excited state decay time could be less than nanoseconds. Research on the electron injection process will thus continue to be an important topic in DSC research and needs to be extended to other classes of sensitizers beside the ruthenium complexes.

3.2.2 Reaction 3: Regeneration of the Oxidized Dyes

The interception of the oxidized by the electron donor, normally I^-, is crucial for obtaining good collection yields and high cycle life of the sensitizer. For a turnover number (cycle life of the sensitizer in the DSSC device) above 10^8 (which is required for a DSC lifetime of >15 years in outdoor conditions), the lifetime of the oxidised dye must be longer than 100 s if the regeneration time is 1 μs [69]. This is achieved by the best performing ruthenium complexes such as C101.

A large number of sensitizers are efficiently regenerated by iodide, as follows from their good solar cell performance. Most of these sensitizers have oxidation potentials that are similar to or more positive than that of the standard N3 sensitizer $Ru(L_{bip})_2(NCS)_2$ ($V^0_{redox} = +1.10$ V vs. NHE). Because the redox potential of the iodide−triiodide electrolyte with organic solvent is about +0.35 V versus NHE, the driving force ΔG^0 for regeneration of N3 is 0.75 eV. It is of interest to estimate how much driving force is needed. Kuciauskas et al. [70] investigated regeneration kinetics of a series of Ru and Os complexes and found that $Os(L_{bip})_2(NCS)_2$ with a $\Delta G^0 = 0.52$ eV is not (or is very slowly) regenerated by iodide, while $Os(L_{bip})_2(CN)_2$ ($\Delta G^0 = 0.82$ eV) is regenerated. The black dye, $RuL_{ter}(NCS)_3$, with $\Delta G^0 = 0.60$ eV, shows rapid regeneration [71]. Clifford et al. [72] studied regeneration in a series of Ru sensitizers and found that $Ru(L_{bip})_2Cl_2$ with $\Delta G^0 = 0.46$ eV gave slow regeneration (>100 μs), leading to a low regeneration efficiency. The results suggest that 0.5 to 0.6 eV driving force is needed for regeneration of Ru complex sensitizers in iodide−triiodide electrolyte. The need for such a large driving force comes probably from the fact that the initial regeneration reaction involves the I^-/I_2^- redox couple, having a more positive potential than I^-/I_3^- [73].

Fast regeneration kinetics are also found for the one-electron redox mediators. Cobalt(II)-bis[2,6-bis(1′-butylbenzimidazol-2′-yl)pyridine] (Co(dbbip)$_2^{2+}$) gave regeneration times of some microseconds and regeneration efficiencies of more than 0.9 [74,75]. Ferrocene and phenothiazine gave rapid regeneration, while cobalt(II) bis(4,4′-di-tert-butyl-2,2′-bipyridine) was slow [76]. Interestingly, mixtures of this Co complex with ferrocene and phenothiazine were efficient in dye-sensitized solar cells, suggesting that a mix of redox mediators can be a viable approach in DSCs [76]. Very rapid dye regeneration was observed in the case of the solid-state DSCs where the redox electrolyte is replaced by the solid hole conductor spiro-MeOTAD [77]. Bach et al. found that hole injection from the oxidized Ru(L$_{bip}$)$_2$(SCN)$_2$ dye to the spiro-MeOTAD proceeds over a broad timescale, ranging from less than 3 ps to a few nanoseconds [78].

Very recently, several papers have been published dealing with the regeneration of oxidized dyes with the iodide—tri-iodide electrolyte [73,79,80].

3.2.3 Reaction 4: Electron Transport Through the Mesoporous Oxide Film

The mesoporous semiconductor electrode consists of numerous interconnected nanocrystals. Because these particles are typically not electronically doped and surrounded by ions in the electrolyte, they will not have an internal electrical field and will not display any significant band bending. Electrons photoinjected into the nanoparticles from the dye molecules are charge compensated by ions in the electrolyte. Photocurrent will be detected in the external circuit once the electrons are transferred into the conducting substrate. The gradient in electron concentration appears to be the main driving force for transport in the mesoporous oxide films—that is, electron transport occurs by diffusion [cf. 14,21,41]. Because the electrons in the mesoporous electrode are charge compensated by ions in the electrolyte, the diffusion processes of electrons and ions will be coupled through a weak electric field. This will affect transport of charge carriers. The measured electron diffusion can thus be described by an ambipolar diffusion model [81,82].

In contrast to the notion that electron transport occurs by diffusion, it is observed that the electron transport depends on the incident light intensity, becoming more rapid at higher light intensities [83,84]. This can be explained by a diffusion coefficient that is light-intensity dependent or, more correctly, dependent on the electron concentration and Fermi level in the TiO$_2$. The measured value of the diffusion coefficient is orders of magnitude lower than that determined for single-crystalline TiO_2 anatase (<0.4 cm^2 s^{-1}) [85]. These observations are usually explained using a multiple trapping model [84,86–89]. In this model, electrons are considered to be mostly trapped in localized states below the conduction band, from which they can escape by thermal activation. Experiments suggest that the density and energetic location of such traps is described by an exponentially decreasing tail of states below the conduction band [86,88]. The origin of the electron traps remains obscure at present: they could correspond to trapping of electrons at defects in the bulk, grain boundaries, or surface regions of the mesoporous oxide or to Coulombic trapping due to local field effects through interaction of electrons with the polar TiO_2 crystal or with cations of the electrolyte [90–92].

3.2.4 Reaction 5 and 6: Recombination of Electrons in the Semiconductor with Oxidized Dyes or Electrolyte Species

During their relatively slow transport through the mesoporous TiO_2 film, electrons are always within only a few nanometres distance of the oxide–electrolyte interface. Recombination of electrons with either oxidised dye molecules or acceptors in the electrolyte is therefore a possibility. The recombination of electrons with the oxidised dye molecules competes with the regeneration process, which usually occurs on a timescale of submicroseconds to microseconds. The kinetics of the back electron-transfer reaction from the conduction band to the oxidised sensitizer follow a multiexponential time law, occurring on a microsecond to millisecond timescale, depending on electron concentration in the semiconductor and, thus, the light intensity. The reasons suggested for the relatively slow rate of this recombination reaction are (1) weak electronic coupling between the electron in the solid and the Ru(III) centre of the oxidised dye, (2) trapping of the injected electron in the TiO_2, and (3) the kinetic impediment due to the inverted Marcus region [93]. Application of a potential to the mesoporous TiO_2 electrode has a strong effect [94–97]. When the electron concentration in the TiO_2 particles is increased, a strong increase in the recombination kinetics is found. Under actual working conditions, the electron concentration in the TiO_2 particles is rather high and recombination kinetics may compete with dye regeneration.

Recombination of electrons in TiO_2 with acceptors in the electrolyte is, for the I^-/I_3^- redox system, generally considered to be an important loss reaction, in particular under working conditions of the DSC device when the electron concentration in the TiO_2 is high. The kinetics of this reaction are determined from voltage decay measurements and normally referred to as the electron lifetime. Lifetimes observed with the I^-/I_3^- system are very long (1—20 ms under one-sun light intensity) compared with other redox systems used in DSCs, explaining the success of this redox couple. The mechanism for this recombination reaction remains unsettled but appears to be dominated by the electron trapping—detrapping mechanism in the TiO_2 [98]. Recently, a lot of attention has been drawn to the effects of the adsorbed dye on the recombination of TiO_2 electrons with electrolyte species. There are several reasons: first, adsorption of the dye can lead to changes in the conduction band edge of TiO_2 because of changes in surface charge. This will lead to a larger driving force for recombination. Second, dyes can either block or promote reduction of acceptor species in the electrolyte [99]. The size of the oxide particle, and thus the surface-to-volume ratio, is also expected to have a significant effect on electron lifetime [100,101].

3.2.5 Reaction 7: Transport of the Redox Mediator and Reactions at the Counterelectrode

Transport of the redox mediator between the electrodes is mainly driven by diffusion. Typical redox electrolytes have a high conductivity and ionic strength so that the influence of the electric field and transport by migration is negligible. In viscous electrolytes such as ionic liquids, diffusion coefficients can be too low to maintain a sufficiently large flux of redox components, which can limit the photocurrent of the DSC [102].

In the case of the iodide—tri-iodide electrolyte, an alternative and more efficient type of charge transport can occur when high mediator concentrations are used: the Grotthus mechanism. In this case, charge transport corresponds to formation and cleavage of chemical bonds. In viscous electrolytes, such as ionic liquid-based electrolytes, this mechanism can contribute significantly to charge transport in the electrolyte [102—105]. In amorphous hole conductors that replace the electrolyte in solid-state DSCs, charge transport takes place through hole hopping. In the most investigated molecular hole conductor for DSCs, spiro-MeOTAD, mobility is increased 10-fold by the addition of a Li salt [106]. Resistance, however, in the hole-transporting layer can be a problem in sDSCs.

At the counterelectrode in standard DSCs, triiodide is reduced to iodide. The counterelectrode must be catalytically active to ensure rapid reaction and low overpotential. Pt is a suitable catalyst as iodine (tri-iodide) dissociates to iodine atoms and iodide upon adsorption, enabling a rapid one-electron reduction. The charge-transfer reaction at the counterelectrode leads to a series resistance in the DSC, the charge-transfer resistance R_{CT}. Ideally, R_{CT} should be $\leq 1\ \Omega\ cm^2$ to avoid significant losses. A poor counterelectrode will affect the current–voltage characteristics of the DSC by lowering the fill factor.

3.3 Characterization of DSC Devices

In this section, we describe the basic solar cell measurements—i.e., the determination of solar-to-electrical energy conversion efficiency, η, and the quantum efficiency. Then there are the so-called DSC toolbox techniques that are useful in order to obtain information on the energetics of the different components and on the kinetics of the different charge-transfer processes in complete DSC devices measured under normal solar cell operating conditions. As mentioned previously, there are a huge number of material components and combinations, which can be used to prepare a DSC device. As examples of results from the characterizations methods described in this chapter we will therefore limit the possibilities to a few systems, including a comparison between liquid and solid-state DSC [107]. This comparison demonstrates one of the main use of the DSC toolbox—namely, comparing different material components to investigate how properties of individual components affect the complete DSC device. The results refer to a DSC device built using an organic sensitizer, D35, with a liquid I^-/I_3^- electrolyte or a solid-state hole conductor (spiro-MeOTAD). The molecular structures of D35 and spiro-MeOTAD are shown in Figure 11.

For the liquid cell, a platinized fluorine-doped tin oxide TCO substrate is used as counterelectrode and for the solid-state DSC (sDSC) an evaporated silver layer on top of the hole conductor is used. Mesoporous TiO_2 films, 1.8-μm thick, screen printed on dense TiO_2 blocking layers, were used as working electrodes. The mesoporous TiO_2 films were treated with a $TiCl_4$ solution [108]. The electrolyte concentrations were 0.05 M I_2, 0.5 M LiI, and 0.5 M 4-tert butyl pyridine (tBP) in 3-methoxy propionitrile (MPN), while the spiro-MeOTAD solution used for spin coating consisted of 150 mg spiro-MeOTAD per ml of chlorobenzene with 15 mM LiTFSI and 60 mM tBP added.

With these examples of results, it should be kept in mind that they do not represent optimized efficiencies of the DSC devices but are used for general descriptions of the methods and for comparing different material components.

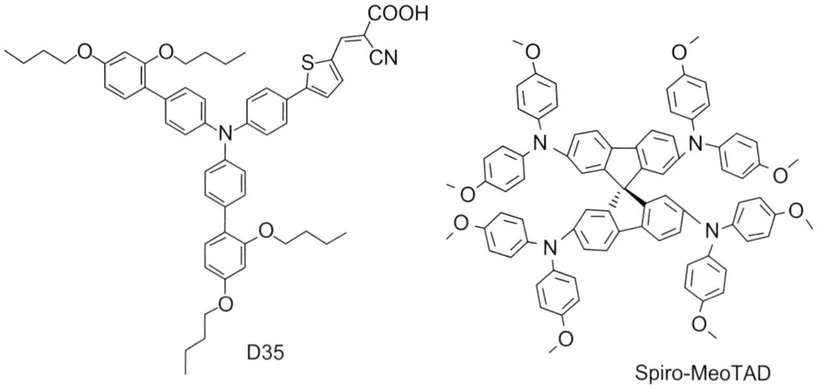

Figure 11 Molecular structures of the organic dye D35 and hole conductor spiro-MeOTAD.

3.3.1 Efficiency Measurements

Current–voltage measurements (IV-measurements) under illumination are used to determine the power conversion efficiencies of solar cells. A lamp with a spectrum simulating the solar AM1.5 spectrum is used for illumination and calibrated to an intensity of 1000 W m^{-2} for measurements at one-sun intensity. A Newport solar simulator of class B was used for the results presented below. A voltage is then applied between the working and counter-electrode of the solar cell and the current output is measured. The voltage range should include the voltage at which the current is zero (the open-circuit voltage, V_{OC}) and 0 V, at which the short-circuit current density (J_{SC}) is measured. The resulting current–voltage curve is usually referred to as an *IV curve*. The conditions for measuring the current–voltage characteristics of a DSC device should be carefully checked so that the determined efficiencies represent "steady-state" efficiencies. The IV characteristics of DSC can be quite sensitive to scan rate, preconditioning of the cell (which potential is applied and for how long), as well as changes occurring after repeated scans—see, for example, discussions in reference [107].

Measurements can also be carried out in the dark, and the measured data is accordingly called a *dark current curve*. Figure 12 shows an example of IV curves under illumination and in the dark for the solid and the liquid-electrolyte cell with the D35 dye.

The efficiency of a solar cell, η, is given by

$$\eta = \frac{P_{max}}{P_{in}} = \frac{(J \cdot V)_{max}}{P_{in}} \tag{11}$$

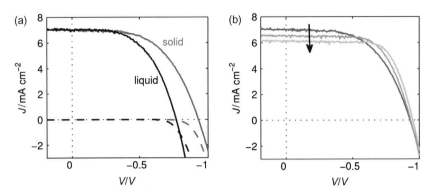

Figure 12 *IV* curves of a solid-state DSC (grey) and a liquid-electrolyte DSC (black) with D35 as sensitizer under one-sun illumination (solid line) and in the dark (dashed line). *(Taken from Figure 5.12 in reference [107].)*

where P_{in} is the illumination intensity and P_{max} is the maximum power output of the solar cell at this light intensity. To describe the efficiency of a solar cell in terms of V_{OC} and J_{SC}, a quantity called the *fill factor* (FF) is introduced, relating P_{max} to V_{OC} and J_{SC}:

$$ff = \frac{(J \cdot V)_{max}}{J_{SC} \cdot V_{OC}} \quad (12)$$

The efficiency can then be written as

$$\eta = \frac{J_{SC} \cdot V_{OC} \cdot ff}{P_{in}} \quad (13)$$

For the example in Figure 12, the efficiencies of the liquid DSC and sDSC were 2.9% and 3.6%, with V_{OC} of 0.77 V and 0.93 V, J_{SC} of 7.0 and 7.0 mA/cm² and FF of 0.54 and 0.55, respectively. Thus, the solid-state DSC has a higher V_{OC} than the liquid-electrolyte DSC, which is the reason for the higher efficiency of the sDSC. It should be noted that the film thickness is only 1.8 μm, so the optical density of the film is relatively low, reducing the overall efficiency. What was also observed in reference [107] and shown in Figure 12b is that in consecutive scans, the short-circuit current of the solid-state cell decreases, while the fill factor and the overall efficiency increase, demonstrating how care must be taken in measuring *IV* curves for DSCs, in particular for sDSCs.

With regards to illumination of the DSC cell, the cells should be masked. A mask size that is 1mm on each side bigger than the active area is recommended in reference [109]. Using thin TCO glass (ca 1 mm) and a device size of at least 5 × 5 mm is also recommended. This will reduce optical artefacts that can enhance or diminish the apparent power-conversion efficiency. To be qualified in the official table of world record efficiencies for photovoltaics, the solar cell area must be at least 1 cm^2. For efficiency measurements of solar cells in general, we refer to [110]. For DSC specifically, we summarize the discussions above according to [111].

1. Scan rate must be slow, and the *IV* curve should be scanned in both directions to determine if it is slow enough.
2. The efficiency depends on premeasurement state. The temperature should be 25°C, and there should be a one-sun light bias at P_{max}. Procedures that approximate this can be used. It should be noted that V_{OC} or J_{SC} might not give the same results as preconditioning at P_{max}.
3. Light piping from outside the defined area is possible. The magnitude of this should be determined by looking at J_{SC} with or without an aperture.
4. Quantum efficiency measurements (see below) may be nonlinear, and the results should be checked as function of light intensity. It is recommended that these measurements should be done with bias light and preferably chopped monochromatic beam.

3.3.2 External and Internal Quantum Efficiencies

The *incident photon to current conversion efficiency* (IPCE), sometimes also called the *external quantum efficiency* of the solar cell, describes how many of the incoming photons at one wavelength are converted to electrons:

$$\text{IPCE}(\lambda) = \frac{\text{electrons out}(\lambda)}{\text{incident photons}(\lambda)} = \frac{J_{SC}(\lambda)}{q\Phi(\lambda)} = \frac{hc}{q} \cdot \frac{J_{SC}(\lambda)}{\lambda \cdot P_{in}(\lambda)}$$
$$= 1240 \cdot \frac{J_{SC}(\lambda)[\text{A cm}^{-2}]}{\lambda[\text{nm}] \cdot P_{in}(\lambda)[\text{W cm}^{-2}]} \quad (14)$$

where J_{SC} is the short-circuit current density, Φ the photon flux, P_{in} the light intensity at a certain wavelength λ, q the elementary charge, and h and c Planck's constant and speed of light, respectively.

IPCEs are made typically using a xenon or halogen lamp coupled to a monochromator. The photon flux of light incident on the samples is measured with a calibrated photodiode, and measurements are typically made at 10- or 20-nm wavelength intervals between 400 nm and the absorption threshold of the dye. Since the DSC is a device with relatively slow relaxation times, it is important to make sure that the measurement duration for a given wavelength is sufficient for the current to be stabilized (normally 5–10 seconds). If it is observed that IPCE depends on light intensity, then the measurements should be made with additional bias light to ascertain that IPCE is determined at relevant light-intensity conditions. The reasons for light-intensity–dependent IPCE may be that the charge-collection efficiency (process 4 in Figure 9) increases with light intensity due to faster electron transport, or that there are mass transport limitations in the electrolyte, decreasing IPCE with light intensity.

The magnitude of the IPCE spectrum depends on how much light is absorbed by the solar cell and how much of the absorbed light is converted to electrons, which are collected:

$$IPCE(\lambda) = LHE(\lambda) \cdot \varphi_{inj}(\lambda) \cdot \varphi_{reg} \cdot \eta_{CC}(\lambda) \qquad (15)$$

where LHE is the light-harvesting efficiency and equal to $1-10^{-A}$ with A being the absorbance of the film; φ_{inj} and φ_{reg} the quantum yields for electron injection and dye regeneration, respectively; and η_{CC} the charge-collection efficiency.

IPCE spectra of the liquid and solid-state DSCs sensitized with D35 are shown in Figure 13 [107].

The spectra are slightly different in shape, although the same TiO_2 thickness and the same dye were used. The spectrum of the solid-state DSC is lower at around 380 nm and higher at the red edge of the spectrum than the spectrum of the liquid-electrolyte DSC. These differences

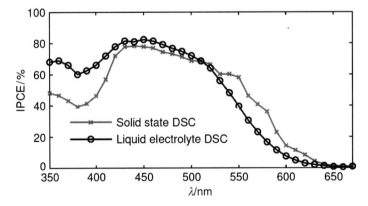

Figure 13 IPCE spectra of the solid-state DSC and the liquid-electrolyte DSC with D35 as sensitizer. *(Taken from Figure 5.14 in reference [107].)*

can be explained with Equation (15): at around 380 nm, spiro-MeOTAD absorbs strongly causing a filtering effect at this wavelength in the solid-state device compared to the liquid-electrolyte DSC and therefore decreasing the IPCE. *LHE* at the absorption maximum of D35 was close to 1 for the devices, resulting in IPCE maxima of 80%. However, at longer wavelengths, light harvesting was incomplete. In the solid-state DSC, the reflecting back contact increased the light harvesting at these wavelengths and therefore also the IPCE.

The short-circuit current of a solar cell can be calculated by integrating over the product of the IPCE and the AM 1.5 solar spectrum:

$$J_{SC} = \int IPCE(\lambda) q \phi_{ph,AM1.5}(\lambda) d\lambda \qquad (16)$$

where $\phi_{ph,AM1.5}$ is the photon flux in AM1.5 at wavelength λ. For the DSC presented in Figure 13, the integrated J_{SC} were determined to be 7.75 mA cm^{-2} for the solid-state DSC and 7.4 mA cm^{-2} for the liquid-electrolyte cell. These currents are slightly higher than the currents determined in the *IV* measurements. For the solid-state DSC, this might be due to the fact that the IPCE measurement was carried out prior to the *IV* measurements, so the analysis of the data must be checked according to the discussions above.

From a fundamental viewpoint, the so-called *absorbed photon to current conversion efficiency* (APCE) values provide further insight into the properties of the device. APCE (or internal quantum efficiency) shows how efficiently the numbers of absorbed photons are converted into current. APCE is obtained by dividing the IPCE number by the *LHE* (0–100%). The IUPAC name for *LHE* is *absorptance*. Thus,

$$APCE = \frac{IPCE}{LHE} = \varphi_{inj}(\lambda) \cdot \varphi_{reg} \cdot \eta_{CC}(\lambda) \qquad (17)$$

Quantitative *in situ* measurement of the light-harvesting efficiency of complete dye solar cells is complicated because of light scattering by the mesoporous oxide film and light absorption by the other cell components. For fundamental studies, it is therefore advisable to use transparent mesoporous TiO$_2$ films. There are several descriptions of the procedures to obtain *LHE* in the literature—see, for example, [112–115]. As an example, we refer to the work by Barnes et al. [114] on relatively transparent TiO$_2$ films.

To take into account reflective and absorption losses in the DSC that are not attributable to the dye—oxide system, measurement of the conducting glass, platinized counterelectrode layer, and electrolyte are made. In the example of Barnes et al., a platinized counterelectrode and a liquid I^-/I_3^- electrolyte were used. We follow their notation and denote reflectance from the glass by R, $(1 - T_{Pt})$ for the absorption due to the Pt layer and $(1 - T_I)$ for the absorption in the electrolyte between the TiO_2 and the Pt layer. The fraction of light transmitted by complete cells with dye ($T_{dye/TiO2}$) and without dye (T_{TiO2}) and of film thickness d gives the absorption coefficient of the dye coated mesoporous film according to

$$\alpha(\lambda) = -\frac{1}{d}\ln\frac{T_{dye/TiO_2}(\lambda)}{T_{TiO_2}(\lambda)} \tag{18}$$

The assumptions underlying Equation (18) are that there is an exponential variation of light intensity with position in the mesoporous film and that the dye molecules do not scatter a significant fraction of light—i.e., $T_{dye/TiO2}/T_{TiO2} \approx [T_{dye/TiO2}/(1 - R_{dye/TiO2})]$. Since iodine in the electrolyte, filling the oxide pores, absorbs light, one needs to take this into account by determining its absorption coefficient, $\alpha_I(\lambda)$ by measuring transmission with and without iodine in the electrolyte and estimating the porosity of the films. In the work by Barnes et al. [114], relatively nonscattering TiO_2 films were used and scattering effects were negligible for $\lambda > 480$ nm. To describe scattering effects, the optical measurements should be made with an integrating sphere and more sophisticated and appropriate models that include the scattering properties such as Kubelka–Munk [116].

Integration of the generation rate of excited dye states as a function of position x in the film across the thickness of the oxide film gives the *LHE* for illumination from the working electrode side, LHE_{WE}, and from the counterelectrode–electrolyte side, LHE_{CE} [114]:

$$\eta_{lh,PE}(\lambda) = \frac{(1-R)\alpha(1-e^{-(\alpha+\alpha_I)d})}{\alpha + \alpha_I} \tag{19}$$

$$\eta_{lh,CE}(\lambda) = \frac{(1-R)T_{Pt}T_I\alpha(1-e^{-(\alpha+\alpha_I)d})}{\alpha + \alpha_I} \tag{20}$$

3.3.3 The DSC Toolbox

The dye-sensitized solar cell is a complex, highly interacting system. To understand the precise working mechanism of the DSC and to optimize its performance, it is important to map the energetics of the different components and interfaces and the kinetics of the different electron-transfer

reactions for *complete DSC devices working under actual operating conditions*. The so-called toolbox of characterization techniques is used to in situ investigate the kinetics of different reactions in DSC devices. These studies are particularly fruitful, as the interactions between different components can be studied. Toolbox methods are continuously being developed by several research groups, and for two recent reviews we refer to references [21,40]. In this section, we specifically discuss the following techniques:
- photovoltage and photocurrent as a function of light intensity;
- small-modulation photocurrent and photovoltage transients to investigate electron transport and recombination;
- steady-state, quantum efficiency, measurements to determine injection efficiency, collection efficiency, and electron diffusion length;
- electron-concentration measurements;
- determination of the internal potential (quasi-Fermi level) in the mesoporous electrode; and
- photo-induced absorption spectroscopy to obtain information on recombination reactions and regeneration of the oxidized dye by the electrolyte.

A set of very powerful toolbox techniques is based on *electrochemical impedance spectroscopy* (EIS). The reader is referred to the works of Bisquert and co-workers on this topic, and as examples of references we propose [39,40,117]. In EIS, the potential applied to a system is perturbed by a small sine-wave modulation, and the resulting sinusoidal current response (amplitude and phase shift) is measured as a function of modulation frequency. The impedance is defined as the frequency domain ratio of the voltage to the current and is a complex number. For a resistor (R), the impedance is a real value, independent of modulation frequency, while capacitors (C) and inductors (L) yield an imaginary impedance with values that vary with frequency. The impedance spectrum of an actual system—that is, the impedance measured in a wide range of frequencies—can be described in terms of an equivalent circuit consisting of series- and parallel-connected elements R, C, L, and W, which is the Warburg element that describes diffusion processes. Using EIS, the following parameters can be obtained: series resistance, charge-transfer resistance of the counterelectrode, diffusion resistance of the electrolyte, the resistance of electron transport and recombination in the semiconductor, and the chemical capacitance of the mesoporous electrode. One of the advantages of impedance spectroscopy is that it allows simultaneous characterization of electron transport in the mesoporous oxide and of recombination of the electrons from the oxide to the hole-conducting medium. The transport and interfacial transfer of electrons in the mesoporous oxide layer can be modelled using a distributed network of resistive and capacitive elements in the form of a finite transmission line.

In the following descriptions of the toolbox techniques, we again use examples of liquid and solid-state DSCs with D35 as sensitizer.

Photovoltage and Photocurrent as a Function of Light Intensity

The short-circuit current and open-circuit voltage of a DSC can be determined as a function of light intensity. Ideally, the short-circuit current should increase linearly with light intensity. Considering Equation (15), this will be the case if the effective injection efficiency, regeneration efficiency, and collection efficiency are independent of light intensity. Plots of J_{SC} versus light intensity for the example cells are shown in Figure 14 [107]. The gradient of the log-log plot was 1.07 for the solid-state cell and 0.99 for the liquid-electrolyte cell, showing that the currents increased almost linearly with light intensity.

If the contacts in a DSC are ohmic, the open circuit is given by the difference between the Fermi level of the semiconductor, E_F, and the Fermi level of the redox electrolyte or hole conductor, $E_{F,\,redox}$:

$$qV_{OC} = E_F - E_{F,redox} \qquad (21)$$

For the liquid-electrolyte cell, the concentrations in the electrolyte do not vary significantly between the dark condition and under illumination, and therefore $E_{F,\,redox}$ can be treated as a constant (a typical value for

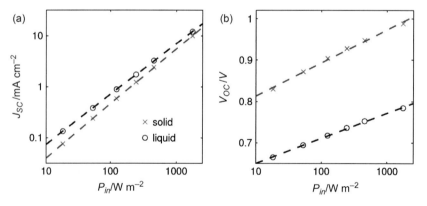

Figure 14 Light-intensity dependence of (a) J_{SC} and (b) V_{OC} for the two DSCs. Dashed lines indicate fits to the data. *(Taken from Figure 5.15 in reference [107].)*

iodide—tri-iodide is 0.31 V vs. NHE) (see Equation (3)). The voltage change at different illumination intensities therefore only depends on $\ln n_{CB}$. If n_{CB} increases linearly with light intensity, then the voltage should increase by 59 mV per decade of increase in light intensity at 298 K [cf. 40]. In the example shown here (Figure 14b), the slope of a plot of V_{OC} vs. log P_{in} is indeed 60 mV. Normally, however, DSCs are nonideal to some extent, and generally the photovoltage varies by more than 59 mV per decade of intensity (values as high as 110 mV/decade are not uncommon). The origin of this nonideality is not understood, but it has important consequences for the interpretation of many of the techniques discussed in this chapter. An empirical nonideality factor, m (>1), is often used to account for nonideality so that the intensity dependence of the photovoltage is given by

$$\frac{dU_{photo}}{d\log_{10}I_0} = m\frac{2.303 k_B T}{q} \quad (22)$$

In the case of solid-state DSCs, the relative concentrations of oxidised spiro-MeOTAD and ground-state spiro-MeOTAD are not as well known as the ratio of I_3^- and I^- in the liquid electrolyte. Therefore, $E_{F,\ redox}$ in the dark is not as well known as for the liquid-electrolyte cell. If C_{ox} is very small, then the concentration of holes injected under illumination might significantly influence the $E_{F,\ redox}$ of spiro-MeOTAD and therefore V_{OC}. The slope of a plot of V_{OC} vs. log P_{in} for the solid-state cell (Figure 14b) was 80 mV. This higher value could be due to an effect of the change of concentration of oxidised spiro-MeOTAD under illumination, but other nonideality effects such as recombination of electrons to the hole conductor via surface states also may be important.

Small-Modulation Photocurrent and Photovoltage Transients to Investigate Electron Transport and Recombination

Information about the transport and lifetimes of charge carriers in the DSC can be obtained by monitoring the current and voltage transients following a small modulation of the light intensity. Figure 15a shows normalised photocurrent transients of the solid-state DSC and the liquid-electrolyte DSC measured under short-circuit conditions at the same bias light intensity [107]. Due to the small amplitude of the modulation, the transients can be reasonably well fitted by a single exponential decay:

$$J(t) = J_{SC} + \Delta J \cdot e^{-t/\tau_{resp}} \quad (23)$$

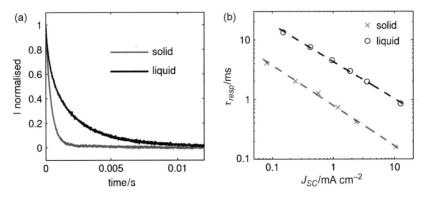

Figure 15 (a) Normalised current decays of the two DSCs at the same bias light intensity (560 W m^{-2}) following a small downward modulation of the light intensity. (b) Photocurrent response times obtained from fits at different light intensities as a function of J_{SC}. Dashed lines indicate a log-log fit to the data. *(Taken from Figure 5.17 in reference [107].)*

where τ_{resp} is the characteristic time constant of the decay. A plot of τ_{resp} measured at different bias light intensities can be seen in Figure 15b. The photocurrent response times of the solid-state DSC are significantly faster than the corresponding time constants for the liquid-electrolyte DSC at the same short-circuit current.

The time constant of the current decay depends on how fast the system adjusts to the decreased injection of charges into the TiO$_2$. It therefore depends on the recombination rate and on the rate at which charge carriers can be transported out of the cell. Considering that both holes and electrons have to be transported out of the cell independently, it will depend on the limiting one of these two types of charge transport. For a liquid-electrolyte DSC, this is the electron transport in the TiO$_2$. The transport of I_3^- to the counterelectrode can usually be ignored because an excess of I_3^- is available in the electrolyte. τ_{resp} can then be related to the transport time of electrons in the TiO$_2$ (τ_{tr}) and to the electron lifetime (τ_e) by

$$\frac{1}{\tau_{resp}} = \frac{1}{\tau_{tr}} + \frac{1}{\tau_e} \qquad (24)$$

For optimised liquid-electrolyte cells under short-circuit conditions, τ_e is much larger than τ_{tr}, and the photocurrent response time becomes a direct measure of the electron-transport time in the TiO$_2$. The chemical diffusion coefficient, D_e, can be calculated from the electron-transport time using Equation (25):

$$D_e = \frac{d^2}{C \cdot \tau_{tr}} \qquad (25)$$

where d is the thickness of the mesoporous oxide film and C is a constant with a value of about 2.5, which depends slightly on absorption coefficient of the film and direction of illumination [118–119].

For the sDSC, it seems possible that the charge transport could be limited by the transport of holes to the back contact if the concentrations of oxidised spiro-MeOTAD is very low [see reference 107 and references therein]. However, in such a case an increase in τ_{resp} would be expected compared to a comparable liquid-electrolyte DSC. The reason for the shorter transport time for sDSC is unclear.

One method to measure τ_e is to measure the photovoltage decay, the V_{OC} decay method [21,120]. Here the open-circuit potential of a DSC is monitored as function of time when the light is switched off. The electron lifetime is calculated from the slope of the V_{OC} transient:

$$\tau_e = -\frac{kT}{q}\left(\frac{dV_{OC}}{dt}\right)^{-1} \qquad (26)$$

An advantage of this method is that the lifetime can be determined in a wide potential range with one single measurement. It can also be performed without using light by application of a negative potential before open-circuit decay [121].

The charge-collection efficiency, η_{CC}, can be estimated from electron transport and lifetime measurements as follows [122]:

$$\eta_{CC} = 1 - \frac{\tau_{resp}}{\tau_e} = \frac{1}{1 + \tau_{tr}/\tau_e} \qquad (27)$$

For a correct calculation of the charge-collection efficiency, the transport time and lifetime must be measured at the same quasi-Fermi level (electron concentration) in the mesoporous electrode. If τ_{resp} (measured at short circuit) and τ_e (measured at open circuit) are used to calculate η_{CC}, then the resulting value will underestimate the true η_{CC} value under short-circuit conditions. To determine the latter, τ_e must be determined under conditions where the quasi-Fermi level in the mesoporous oxide is equal to that under short-circuit conditions. This can be achieved by (1) measuring the quasi-Fermi level under short-circuit conditions (see the following discussion) and determining τ_e from the relation between τ_e and V_{OC} or (2) measuring the extracted charge under short-circuit conditions and determining τ_e from the relation between τ_e and the extracted charge (see below). Electrochemical impedance spectroscopy [123] and photovoltage rise or decay measurements (see the following discussion) can be used to determine τ_{tr} and τ_e simultaneously at the same quasi-Fermi level. The charge-collection efficiency can also be determined from IPCE using Equation 15.

The electron diffusion length, L, in the DSC is closely related to the charge-collection efficiency. L is a wavelength independent parameter, whereas η_{CC} depends on wavelength. Södergren et al. derived expressions for the IPCE of mesoporous DSCs as a function of the diffusion length, absorption coefficient, and film thickness, assuming quantitative electron injection [44], see the following discussion. In recent works by Halme et al. [113] and Barnes et al. [114] these relations were used to determine the electron diffusion length in DSCs under various conditions. Dynamic, small amplitude methods (impedance spectroscopy, electron-transport, and lifetime measurements) can also be used to determine the electron diffusion length [124,125].

$$L = \sqrt{D_e \tau_e} \qquad (28)$$

where D_e is the effective electron diffusion coefficient. Interestingly, the values of L determined in this way were at least a factor 2 larger than those obtained from IPCE measurements [126]. Bisquert and Mora-Sero [127] demonstrated in a simulation study that this can be attributed to the fact that recombination kinetics in DSCs are nonlinearly dependent on the electron concentration in the conduction band, whereas linearity is assumed in the IPCE method. Recently, a careful comparison has been made between values of the diffusion length obtained by impedance spectroscopy and by front- or rear-side IPCE measurements using bias light, showing that the methods agree if measurements are made close to open circuit to ensure that the electron concentration across the film is almost constant [128].

An alternative technique devised to measure electron-transport time in DSCs is a photovoltage rise method [129]. Here the cell is kept under open-circuit conditions, and the characteristic time constant for photovoltage rise is measured after application of a short light pulse superimposed on a constant bias illumination. The electron-transport time is calculated from the rise time using the capacity values of the mesoporous oxide and the substrate—electrolyte interface, which have to be measured independently. The electron lifetime is determined in the same experiment from the voltage decay. The advantage of this method is that it can be performed under conditions where the resistance-capacitance (RC) time constant of the DSC, arising from the series resistance of the mesoporous oxide and conducting substrate and the large capacity of the mesoporous film, does not limit the transport measurement.

Steady-State (Quantum Efficiency) Measurements to Determine Injection Efficiency, Collection Efficency, and Electron Diffusion Length

Under steady-state conditions, the rate of trapping and detrapping must be equal so the presence of trap states will not have any influence on, for example, the electron diffusion length or the charge-collection efficiency. Thus, steady-state measurements avoid some of the complications involved in finding the electron diffusion coefficient, L, by perturbation frequency domain or transient methods. A standard steady-state method is measurement of the IPCE spectrum. Comparison of spectral IPCE measurements taken for opposite illumination directions is not only a good diagnostic test for detecting low η_{CC} but also provides basis for quantitative estimation of L by the diffusion model. The derivation can be found in references [40,44,113,114]. The model assumes that the electron transport occurs via diffusion, the recombination reactions are first order in the electron concentration, D_e and τ are independent of x and $n(x)$, and that under short-circuit conditions, extraction of electrons at the substrate contact is fast enough to keep the excess electron concentration at the contact close to the dark equilibrium value.

The ratio between the IPCE, or APCE, spectra, which depends only on $\alpha(\lambda)$, d and L since $\eta_{inj}(\lambda)$ can be assumed independent of the direction of illumination, is expressed as

$$\frac{IPCE_{CE}(\lambda)}{IPCE_{WE}(\lambda) \cdot T_{Pt} \cdot T_I} = \frac{APCE_{CE}(\lambda)}{APCE_{WE}(\lambda)} = \frac{\eta_{CC,CE}}{\eta_{CC,WE}}$$
$$= \frac{\sinh(d/L) + L\alpha(\lambda)\cosh(d/L) - L\alpha(\lambda)e^{\alpha(\lambda)d}}{(\sinh(d/L) - L\alpha(\lambda)\cosh(d/L) + L\alpha(\lambda)e^{-\alpha(\lambda)d})} \cdot e^{-\alpha(\lambda)d}$$

(29)

Two special cases can be considered with respect to uniform or highly nonuniform electron generation. For uniform generation, obtained in the limit of weak light absorption, η_{CC} becomes

$$\eta_{CC} = \frac{\tanh(d/L)}{d/L} \quad (1/\alpha \gg d), \qquad (30)$$

and independent of the illumination direction. In the opposite case $1/\alpha \ll d$, corresponding to high LHE, η_{CC} approaches 100% for the WE illumination, whereas for the CE illumination it becomes

$$\eta_{CC} = \frac{1}{\cosh(d/L)} \quad (1/\alpha \ll d, \text{ CE illumination}) \tag{31}$$

The main predictions from the preceding equations can be summarized as [113]:

1. For constant L and α, η_{CC} decreases with d for the both directions of illumination.
2. For L much larger than d, η_{CC} approaches 100%, irrespective of α and the direction of illumination.
3. For uniform electron generation obtained in the limit of weak light absorption, η_{CC} becomes equal at the both illumination directions, irrespective of d and L.
4. For constant d and L, η_{CC} increases with α at the WE illumination; at the CE illumination, the trend is opposite.

Using the experimental data for $APCE_{CE}/APCE_{WE}$, d, and α, L can be estimated directly from Equation (29) for each λ and d. With the estimated L, η_{CC} can be calculated from Equations (30) and (31), and η_{inj} subsequently as

$$\eta_{inj} = \frac{APCE_{WE}}{\eta_{CC,WE}} = \frac{APCE_{CE}}{\eta_{CC,CE}} = \tag{32}$$

It should be noted that a fit to the IPCE ratio, Equation (29), is independent of calibration errors in the IPCE measurement (the ratio can be normalized to unity at long wavelengths if the losses due to Pt and iodine are uncertain) [114].

Electron Concentration Measurements

The total concentration of electrons in the mesoporous oxide film under solar cell operating conditions can be determined using different methods [21,40]. In charge-extraction methods, the light is switched off and all remaining charge in the film is extracted as a current during a certain period and integrated to obtain the charge. Under short-circuit conditions, this simply corresponds to the integration of the photocurrent transient recorded after the light is switched off [130,131]. Starting from open-circuit conditions, the cell connection has to switch from open to short-circuit simultaneously with switching off of the light. When the V_{OC} is allowed to decay in the dark for different periods before charge extraction, a complete charge–potential curve can be obtained by repeated experiments [132]. When charge-extraction measurements are used, it should be realized that not all electrons will be extracted due to recombination losses and limitations in extraction time (electron transport becomes very slow at low electron concentrations).

In an alternative method, the capacity of the mesoporous film is measured rather than the charge [133]. At a certain open-circuit potential obtained by bias illumination, a light pulse is added and the resulting voltage rise is measured. The photocurrent transient resulting from the light pulse is measured separately under short-circuit conditions and used to calculate the injected charge induced by the light pulse. The capacity is calculated from the ratio of injected charge and voltage change. Integration of capacity with respect to the range of open-circuit potentials gives the charge–potential relation.

Finally, near-infrared transmission measurements can be done to determine the total electron concentration in mesoporous dye-sensitized solar cells [134]. An optical cross section of 5.4×10^{18} cm^2 was determined at 940 nm for electrons in mesoporous TiO_2, corresponding to a decadic extinction coefficient of 1400 M^{-1} cm^{-1}, in good agreement with spectroelectrochemical studies on mesoporous TiO_2 (1200 M^{-1} cm^{-1} at 700 nm) [135].

Approximate electron concentrations in the mesoporous TiO_2 film of a DSC under operating conditions (one sun) are on the order of 10^{18}–10^{19} cm^{-3} [134,136] corresponding to about 4–40 electrons per TiO_2 particle, assuming spherical 20-nm sized particles.

Determination of the Internal Potential (Quasi-Fermi Level) in the Mesoporous Electrode

Two methods have been developed to determine the quasi-Fermi level in mesoporous oxide electrodes. The first method is a switching method and can be performed on standard DSCs [136,137]. The DSC is illuminated under short-circuit conditions (or kept at a certain applied potential), when simultaneously the light is switched off and the cell is switched to open circuit while the potential is measured. The potential will rise to reach a value that is similar to the quasi-Fermi level of the electrons in the mesoporous film, as it was present in the illuminated film. Considering that there was a gradient in the Fermi level, the resulting value will be an average value. Model calculations suggest that even under short-circuit conditions, the

Fermi level under illumination conditions is relatively flat in the whole mesoporous film, except for about 1 μm directly adjacent to the conducting substrate, so that the measured value gives a good indication of the Fermi level in most of the film. Additionally, the rise of the potential gives information on the electron transport in the mesoporous film, and the following decay of the potential information on the electron lifetime.

In the second method, the deposition of an additional titanium electrode on top of a mesoporous TiO_2 film was made [138,139]. The surface of the metal is passivated by thermal oxidation and does not lead to an additional recombination pathway for electrons in the TiO_2. The potential of the additional electrode can be measured under operating conditions and directly gives the quasi-Fermi level at the outside of the mesoporous TiO_2 film. Both methods give similar results and show that in standard DSCs, illuminated at one sun, the quasi-Fermi level under short circuit in the mesoporous TiO_2 is located about 0.5−0.6 V negative of the redox potential of the electrolyte.

Photo-Induced Absorption Spectroscopy to Obtain Information on Recombination Reactions and Regeneration of the Oxidized Dye by the Electrolyte

Transient absorption spectroscopy (TAS) is a common tool for the study of the kinetics of DSCs [96,140−144]. Usually, the change in absorption of a visible probe light by a dyed TiO_2 film is measured as a function of time following excitation of the dye by a short laser pulse (the pump). This technique is also referred to as *laser flash photolysis*, and kinetics can be measured on different timescales, depending on the equipment and the laser pulse length. When using a femtosecond laser to generate both the pump and the probe pulse, kinetics in the femto- and picosecond time scales can be resolved (electron injection and regeneration in solid-state DSCs). When lasers with nanosecond long pulses are used, the probe light is often supplied by a separate lamp, and kinetics on the nano- and microsecond timescales can be resolved (regeneration in liquid-electrolyte DSCs and recombination). To create a detectable concentration of transient species, TAS often requires the use of high light intensities in the pump.

As a toolbox technique, we have developed photo-induced absorption (PIA) spectroscopy [145,146]. In PIA, the difference in absorption between a pump light being on and off is measured as a function of modulation frequency and probe wavelength. As the on time of the pump is relatively long in these experiments (55 ms at a 9 Hz modulation), light intensities similar to the light intensities under operating conditions of the solar cell can be used. The kinetics of slower processes in the DSC ($t > 10^{-5}$ s) can be followed using PIA—e.g., recombination of electrons in the oxide with oxidized dyes or electrolyte species. It is very useful in qualitative studies, for instance, to check whether a dye is injecting electrons into the oxide after photoexcitation and whether a dye is regenerated when in contact with a redox electrolyte. PIA can also be measured over a large wavelength range and the measurements are relatively fast (acquisition of a spectrum often takes only 10 minutes) and very easy to set up and use.

A schematic diagram of the PIA setup in our laboratory is shown in Figure 16. A white probe light, provided by a 20-W tungsten-halogen lamp, is focused onto the sample by a series of optics. Superimposed at the sample is the pump light used for excitation. Typical excitation intensities at the sample are between 8 and 25 mWcm^{-2}, and typical probe intensities are between 10% of 1 sun and 40% of 1 sun. The pump light can be chosen between a blue LED (Luxeon Star, 1W, Royal Blue, 460 nm), a green LED-pumped laser at 530 nm, or a red-diode laser at 640 nm. The pump light is square-wave modulated (on−off) with a function generator.

After transmission through the sample, the probe light is focused onto a monochromator and detected by a UV-enhanced silicon photodiode, allowing for measurements between 400 and 1100 nm. The detector is connected to a current amplifier and a lock-in amplifier. The lock-in

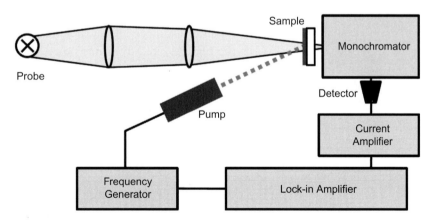

Figure 16 Schematic diagram of the PIA setup. *(Taken from Figure 5.9 in reference [107].)*

amplifier is locked to the modulation frequency, such that $\Delta T/T$ in phase with the modulation frequency (the "real" part of $\Delta T/T$, $Re(\Delta T/T)$) and $\Delta T/T$ out of phase with the modulation frequency (the "imaginary" part of $\Delta T/T$, $Im(\Delta T/T)$) are measured. In this way, only signals occurring at the modulation frequency are detected, which greatly enhances the signal to noise ratio. The setup allows the measurement of ΔA at different wavelengths, resulting in the measurement of a photo-induced absorption spectrum at a given frequency and the measurement of ΔA at different frequencies (frequency-resolved PIA at a given λ). Additionally, ΔA can also be measured as a function of time—i.e., individual time traces following the rise and decay of ΔA can be obtained.

For the theory of PIA measurements we refer to references [107,145,146]. As the concentrations of transient species (and therefore their absorbances) depend on recombination rates, only processes with a relatively long lifetime (milliseconds to seconds) can be resolved with PIA. A sample of, for example, dyed TiO_2 therefore only shows a signal of oxidised dye molecules if recombination times are sufficiently long. This is often the case, and PIA can show if effective injection is happening. When a redox electrolyte or hole conductor are added to such a sample and the regeneration of oxidised dye is fast and efficient, the "lifetime" of the oxidised dye becomes very short. Consequently, the concentration of oxidised dye is decreased by many orders of magnitude, and the PIA signal of oxidised dye will be negligible. Instead, a signal due to holes and electrons will be observed. PIA can therefore show if the regeneration of oxidised dye molecules is efficient. An example of such a study is presented in Figure 17. In this figure we used a perylene organic dye as sensitizer, the ID28 dye [107] and spiro-MeoTAD as hole conductor. Without the hole conductor the photoexcited dye injects electron in the TiO_2 and the spectrum of the oxidized dye is observed (solid line). With the hole conductor (dashed line) there are no remaining features of the oxidized dye, indicating an efficient regeneration of the oxidized dye.

To use PIA for quantification of the injection and regeneration efficiencies, however, is difficult. First, one needs to choose a wavelength at which the Stark effect (this effect is described below) of the dye is not observed. Then one needs to determine the rate constant of recombination and the order of recombination by measuring and fitting frequency-resolved PIA at different light intensities. Finally, any values of injection and regeneration efficiencies will include any recombination that occurs on faster time scales than the PIA measurement.

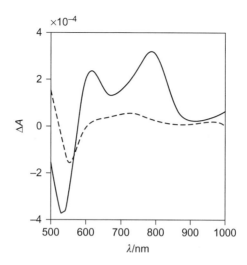

Figure 17 Example of a PIA study using (a) the ID28 perylene dye on TiO$_2$ and (b) with the addition of the spiro-MeOTad hole conductor [107].

3.3.4 The Stark Effect in DSC

Electric fields can induce a change in the transition energies of molecules (ΔE) [147–149]. This is called the *Stark effect, electroabsorption,* or *electrochromism*. Recently, it has been shown that a potential drop across dye molecules in DSC systems upon electron injection into TiO$_2$ leads to a Stark shift of the absorption spectra of dyes [5,6,77,150]. In very general terms, the change in transition energy due to an external electric field ($\rightarrow F$) is given by

$$\Delta E = - \Delta \vec{\mu} \cdot \vec{F} - \frac{1}{2} \vec{F} \cdot \Delta \alpha \cdot \vec{F} \qquad (33)$$

where $\Delta \mu$ is the change in dipole moment and $\Delta \alpha$ is the change in polarisability due to the transition. This equation is valid for electronic transitions as well as for other transitions. One distinguishes the first-order Stark effect, which is linear in the electric field, and the second-order Stark effect, which is quadratic in the electric field. Considering dye-sensitized solar cells, we are interested in how the Stark effect affects electronic transitions of dye molecules that are adsorbed to titanium dioxide surfaces. The effect will be induced by electrons injected into the titanium dioxide. We can define a main direction of the electric field, which is normal to the titanium dioxide surface, assuming for simplification that

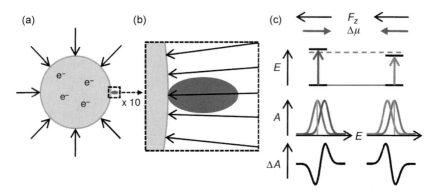

Figure 18 (a) Electric field direction outside a titanium dioxide particle with a 20-nm diameter into which electrons have been injected. A dye molecule with a length of 2 nm and a width of 1 nm is included. (b) Enlargement of the dye molecule on the surface. The electric field lines around the dye molecule are almost parallel. (c) Representation of the first-order Stark shift when $\Delta\mu_z$ is antiparallel to F_z and $\Delta\mu_z$ is parallel to F_z. *(Taken from Figure 4.2 in reference [107].)*

the positive countercharges are uniformly distributed at a certain distance away from the TiO$_2$ particle (Figure 18a). On average, dye molecules adsorbed to the surface should therefore experience an almost uniform electric field in one direction (Figure 18b) and we can rewrite Equation (33) in a one-dimensional form (chosen to be along the z-axis here):

$$\Delta E = -\Delta\mu_z \cdot F_z - \frac{1}{2}\Delta\alpha \cdot F_z^2 \tag{34}$$

However, it is the change in absorbance (ΔA) due to an electric field rather than the change of transition energy, which is of practical interest. An expression for ΔA can be derived from a Taylor expansion of A around E at $F_z = 0$:

$$\Delta A = \frac{dA}{dE}\Delta E + \frac{1}{2}\frac{d^2A}{dE^2}\Delta E^2 + \ldots \tag{35}$$

Substituting ΔE from Equation 34 into this equation, one can obtain an expression for the terms of ΔA linear and quadratic with the electric field:

$$\Delta A = -\frac{dA}{dE}\Delta\mu_z F_z + \frac{1}{2}\left(\frac{d^2A}{dE^2}\Delta\mu_z^2 - \frac{dA}{dE}\Delta\alpha\right)F_z^2 \tag{35}$$

Organic dyes used in DSCs are usually designed to have a large change in dipole moment upon excitation and should therefore show a first-order Stark effect, given by the linear term of Equation (35). A representation of this first-order Stark effect is shown in Figure 18c. For organic donor-acceptor (push-pull) dyes designed for n-type DSCs, $\Delta\mu_z$ usually points away from the TiO_2 surface (electron density is moved towards the TiO_2 surface upon excitation). The transition energies are therefore expected to increase upon electron injection into the TiO_2, and the absorption spectrum is expected to shift to shorter wavelengths. The Stark effect has been observed in different types of measurements of DSC systems, such as in the following [107]:

- photo-induced absorption and transient absorption measurements where the Stark effect of ground-state dye molecules was observed both in absence and presence of a redox mediator after photo-induced electron injection [5,6,77];
- electroabsorption measurements where an external electric field was applied to a monolayer of dye molecules adsorbed to a flat TiO_2 substrate [5,77]; and
- spectroelectrochemistry, where charges were injected into a dye-coated mesoporous TiO_2 film surrounded by a supporting electrolyte by applying a negative potential to the film [107 and refs. therein].

3.4 Development of Material Components and Devices

Since the initial work in the beginning of the 1990s, a wealth of DSC components and configurations have been developed. Perhaps a key concept for the future success of DSC is *diversity*. At present, several thousands of dyes have been investigated as well as numerous types of mesoporous films with different morphologies and compositions. The last year has seen some very interesting breakthroughs in the use of alternative redox systems, and this field is now opened up after almost 20 years of I^-/I_3^- dominance. With such a diversity to explore, the DSC technology can be expected to progress rapidly, be it through design of new materials and combinations based on fundamental insights or by *evolution*—i.e., trial and error—or better, with the use of combinatorial approaches.

In the following sections, we briefly overview the development of the material components and devices in DSC and in general refer to the many recent reviews on these topics.

3.4.1 Mesoporous Oxide Working Electrodes

The key to the breakthrough for DSCs in 1991 was the use of a mesoporous TiO_2 electrode with a high internal surface area to support the monolayer of a sensitizer. Typically, the increase of surface area by using mesoporous electrodes is about a factor 1000 in DSCs. TiO_2 still gives the highest efficiencies, but many other metal oxide systems have been tested, such as ZnO, SnO_2, and Nb_2O_5. Besides these simple oxides, ternary oxides, such as $SrTiO_3$ and Zn_2SnO_4, have been investigated, as well as core-shell structures, such as ZnO-coated SnO_2. For recent reviews on the development of nanostructured metal oxide electrodes for DSC, the reader is referred to references [1,21,151–154]. During the last few years, large efforts have been paid to optimize the morphology of the nanostructured electrode, and a large range of nanostructures has been tested from random assemblies of nanoparticles to organized arrays of nanotubes and single-crystalline nanorods. These studies are motivated by the expectation of an improved and directed charge transport along the rods and tubes and by an improved pore filling of hole conductor materials for solid-state DSC. General reviews for preparation techniques and structures are, for example, Chen et al. for TiO_2 [155] and Ozgur et al. for ZnO [156].

TiO_2 is a stable, nontoxic oxide that has a high refractive index ($n = 2.4-2.5$) and is widely used as a white pigment in paint, toothpaste, sunscreen, self-cleaning materials, and food (*E171*). Several crystal forms of TiO_2 occur naturally: rutile, anatase, and brookite. Rutile is the thermodynamically most stable form. Anatase is, however, the preferred structure in DSCs, because it has a larger band gap (3.2 vs. 3.0 eV for rutile) and a higher conduction band edge energy, E_c. This leads to a higher Fermi level and V_{OC} in DSCs for the same conduction band electron concentration.

For state-of-the-art DSCs, the employed architecture of the mesoporous TiO_2 electrode is as follows [21 and refs. therein]:

1. A TiO_2 blocking layer (thickness ~50 nm), coating the FTO plate to prevent contact between the redox mediator in the electrolyte and the FTO, prepared by chemical bath deposition, spray pyrolysis, or sputtering.
2. A light absorption layer consisting of a ~10-µm thick film of mesoporous TiO_2 with ~20-nm particle size that provides a large surface area for sensitizer adsorption and good electron transport to the substrate.

3. A light scattering layer on the top of the mesoporous film, consisting of a ~3-μm porous layer containing ~400-nm-sized TiO_2 particles. Voids of similar size in a mesoporous film can also give effective light scattering.
4. An ultrathin overcoating of TiO_2 on the whole structure, deposited by means of chemical bath deposition (using aqueous $TiCl_4$), followed by heat treatment.

3.4.2 Dyes

As one of the crucial parts in DSCs, the photosensitizer should fulfill some essential characteristics:

1. The absorption spectrum of the photosensitizer should cover the whole visible region and even the part of the near-infrared (NIR).
2. The photosensitizer should have anchoring groups ($-COOH$, $-H_2PO_3$, $-SO_3H$, etc.) to strongly bind the dye onto the semiconductor surface.
3. The excited state level of the photosensitizer should be higher in energy than the conduction band edge of n-type DSCs, so that an efficient electron-transfer process between the excited dye and conduction band of the semiconductor can take place. In contrast, for p-type DSCs, the HOMO level of the photosensitizer should be at more positive potential than the valence band level of the p-type semiconductor.
4. For dye regeneration, the oxidized state level of the photosensitizer must be more positive than the redox potential of electrolyte.
5. Unfavourable dye aggregation on the semiconductor surface should be avoided through optimization of the molecular structure of the dye or by addition of coadsorbers that prevent aggregation. Dye aggregates can, however, be controlled (H- and J-aggregates), leading to an improved performance compared with a monomer dye layer as, for example, described in reference [157].
6. The photosensitizer should be photostable, and electrochemical and thermal stability are also required.

Based on these requirements, many different photosensitizers including metal complexes, porphyrins, phthalocyanines, and metal-free organic dyes have been designed and applied to DSCs in the past decades. Ru-complexes have since the early days given the best results. Osmium and iron complexes and other classes of organometallic compounds such as phthalocyanines and porphyrins have also been developed. These dyes have recently been reviewed [21,158]. Metal-free organic dyes are catching up, showing efficiencies close to 10% with the use of indoline dyes. Moreover, chemically robust organic dyes showing promising stability results have recently been developed by several groups. Organic dyes for DSCs have been reviewed in references [21,159].

During the last 2 to 3 years, the advent of heteroleptic ruthenium complexes furnished with an antenna function has taken the performance of the DSC to a new level [21 and refs. therein]. An example of these dyes is C101 (see Figure 6). Compared with the classical N3, N719, and black dye, its extinction coefficients is higher and the spectral response is shifted to the red. Efficiencies above 12% have been reported [20]. The C101 sensitizer furthermore maintains outstanding stability at efficiency levels more than 9% under light soaking at 60°C for more than 1000 h [60]. This is achieved by the molecular engineering of the sensitizer but also very importantly by the use of robust and nonvolatile electrolytes, such as ionic liquids and adequate sealing materials.

A common line of development for organic dyes in DSCs is to synthesise so-called D-π-A dyes that consist of an electron donor (D), a conjugated linker (π), and an electron acceptor (A). The idea behind this concept is that in such molecules, most of the electron density of the HOMO will be located on the donor, while most of the density of the lowest unoccupied molecular orbital (LUMO) will be located on the acceptor. The dyes therefore show intramolecular charge transfer from the donor to the acceptor upon excitation. For typical n-type dyes used in standard DSCs, the anchoring group is located close to or is even part of the acceptor. This ensures that the excited state of the dye is located closer to the TiO_2 surface than the hole remaining on the dye after electron injection. This should help to supress recombination between electrons and oxidised dye molecules and make the hole easily accessible for the redox mediator, which is beneficial for regeneration. A variation of this theme is employed in p-type DSCs, which use a p-type semiconductor (NiO) as a photocathode [160–162]. In these DSCs, a hole is injected into the semiconductor and therefore the anchoring group will now be on the donor of the sensitizer, facilitating hole injection. A p-type DSC can be combined with the conventional n-type DSC into a tandem DSC device. See references [1,21] for recent reviews on these systems.

The donor, acceptor, and linker of the molecule can then be independently varied to tune the properties of the molecules [21]. Typically used donors are triphenylamine, indoline, or coumarin units. Examples of acceptors, which include anchoring groups, are cyanoacrylic acid and rhodanine-3-acetic acid. Another modification often made to organic dyes is to introduce alkyl chains to the linker or donor part of the molecule. These can prevent the formation of dye aggregates or inhibit electron and hole recombination by protecting the TiO$_2$ surface. An example of such a dye based on a triphenylamine donor is D35 [163,164] (Figure 11), which has been successfully employed in both liquid-electrolyte and solid-state DSCs.

3.4.3 Electrolytes

There are many demands on the electrolyte or hole conducting material. They should be chemically stable; provide good charge-transport properties; not dissolve the dye from the oxide surface or the oxide, counterelectrode, and substrates; and be compatible with a suitable sealing material to avoid losses by evaporation or leakage. Several types of electrolyte and hole conductor materials have been developed over the years.

The standard redox electrolyte for DSC ever since the very start in 1991 now comprises the iodide−tri-iodide (I^-/I_3^-) redox couple in an organic solvent. It has good solubility and a suitable redox potential, and it provides rapid dye regeneration. A serious disadvantage of this redox mediator is that a significant part of the potential is lost in the regeneration of the oxidized dye due to intermediate reactions [73]. To develop dye-sensitized solar cells with efficiencies significantly larger than 12%, our strategy was to use one-electron redox couples or hole conductors instead of I^-/I_3^-. Unfortunately, the use of one-electron mediators in DSC nearly always leads to strongly increased recombination between electrons in TiO$_2$ and the oxidized part of the redox couple, which seriously limits the solar cell efficiency. Recently, however, we obtained a breakthrough with cobalt polypyridine-based mediators in combination with the organic dye, D35. Careful matching of the steric bulk of the mediator and the dye molecules minimizes the recombination between electrons in TiO$_2$ and Co(III) species in the electrolyte and avoids mass transport limitations of the redox mediator. The organic sensitizer D35, equipped with bulky alkoxy groups, efficiently suppresses recombination, allowing the use of cobalt redox mediators with relatively small steric bulk. The best efficiency obtained for a DSC sensitized with D35 and employing a [Co(bpy)$_3$]$^{3+/2+}$-based electrolyte was 6.7% at full sunlight (1000 W m^{-2} AM1.5G illumination) [7], which is more than a doubling of previously published record efficiencies using similar cobalt-based mediators. We could thus show that the dye layer itself could efficiently suppress the recombination of electrons in the TiO_2 with oxidized species in the electrolyte opening up the possibilities to use alternative redox couples. With the recent world record of 12.3% using Co-complexes by Grätzel and co-workers [8,165] the main direction of the research field is now to explore this path. Interesting results are also currently presented using other redox systems such as organic redox couples [166,167], ferrocenes [168], and Cu-complexes [169].

Alternative redox systems have been developed over the years such as Br^-/Br_3^- [170], $SCN^-/(SCN)_3^-$ [171], $SeCN^-/(SeCN)_3^-$ [172], and organic redox systems such as TEMPO [173]. For practical applications, it is preferable to use a low-volatility electrolyte system simplifying production processes, encapsulation methods, and materials. Ionic liquids and gel electrolytes are also normally based on I^-/I_3^- [174]. Solid-state DSC [175] have developed quickly recently using the spiro-MeOTAD hole conductor with efficiencies above 5% and recently above 7% [8], but interesting results are also obtained with conducting polymers such as P3HT [176] and inorganic p-type semiconductors such as CuSCN [177] and CuI [178]. The electrolyte—hole conductor systems have recently been reviewed in reference [21].

3.4.4 Counterelectrodes

Counterelectrodes for DSCs with iodide—triiodide electrolytes can be rather easily prepared by deposition of a thin catalytic layer of platinum onto a conducting glass substrate. Without platinum, conducting tin oxide (SnO_2:F) glass is a very poor counterelectrode and has a very high charge-transfer resistance, more than $10^6\ \Omega\ cm^2$, in a standard iodide—tri-iodide electrolyte [179]. Pt can be deposited using a range of methods such as electrodeposition, spray pyrolysis, sputtering, and vapour deposition. Best performance and long-term stability has been achieved with nanoscale Pt clusters prepared by thermal decomposition of platinum chloride compounds [180]. In this case, very low Pt loadings (5 $\mu g\ cm^{-2}$) are needed so that the counterelectrode remains transparent. Charge-transfer resistances of less than 1 $\Omega\ cm^2$ can be achieved.

Alternative materials are, for example, carbon, which is suited as a catalyst for the reduction of tri-iodide [181] and provides good conductivity [182]. Films prepared on TCO substrates from carbon powders and carbon nanotubes showed good catalytic activity for I_3^- reduction as well as good conductivity [183–186]. Conducting polymers have been successfully developed, PEDOT doped with toluenesulfonate anions in particular [187–189]. Recently, metal (Co, W, Mo) sulfides have been identified as suitable catalysts for the I^-/I_3^- redox couple [190–192].

3.4.5 Development of Modules

The manufacturing, reliability, performance, and stability of a DSC module are more complex compared with the test cell situation due to the larger size. Moreover, interconnected cells in a DSC module may interact through, for example, mismatched performance of the cells or unwanted mass transport of electrolyte between adjacent cells. The dominating DSC module designs can be divided into five categories [21] and as schematically shown in Figure 19. In the figure caption, the term *sandwich* is used to define a device structure carrying the working and the counterelectrodes on two substrates. The term *monolithic* is used to define a device structure carrying the working and counterelectrode on one and the same substrate. Figure 20 is an example of a monolithic module prepared at Swerea IVF AB, Sweden. In the present table of record solar cell efficiencies a DSC module efficiency of 9.2% is listed, achieved by Sony, Japan [4].

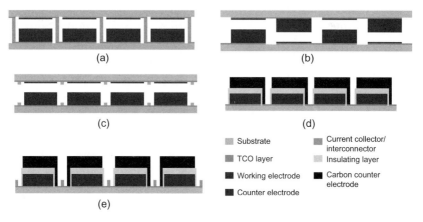

Figure 19 Schematic cross sections of examples of constructions of the five categories of DSC modules: (a) sandwich Z-interconnection; (b) sandwich W-interconnection; (c) sandwich current collection; (d) monolithic serial connection; (e) monolithic current collection. Each module consists of four working and counterelectrodes. The proportions of the module layers and substrates are not drawn to scale but have been adjusted to illustrate the device constructions. *(Taken from Figure 21 in reference [21].)*

Figure 20 A 900-cm² demonstrator consisting of 12 parallel-connected modules that are connected in series. The cell width is 8 mm, and the amount of active area is 75%.

As described above, there are many different DSC module designs, substrates, and manufacturing combinations. According to us, it is still too early to identify a winning combination. Clearly, the monolithic concept offers a cost advantage because only one substrate is used. Likewise, the use of flexible substrates, polymers or metal foils, allow for rapid manufacturing using, for example, roll-to-toll processes. However, the highest module efficiencies have been obtained on glass-based sandwich modules. The performance of the described DSC module designs may be increased by improved device components—for example, TCO glass and dye, or closer packaging of cells, or a less inactive area. The latter can, for the designs utilizing current collectors or interconnects, be obtained by using an inert combination of electrical conductor and electrolyte. Various companies have in the last years reported such DSC systems. Unfortunately, the DSC technology is not so trivial that the highest module efficiency, the fastest production methods, or the lowest materials cost necessarily provides the best module solution. The big challenge is to produce reliable devices that are long-term stable to fulfill the requirements of the designed application. It may thus be that a winning combination springs from the most functional encapsulation method.

4. FUTURE OUTLOOK

With time, the chemical complexity of DSCs has become clear, and the main challenge for future research is to understand and master this complexity, in particular at the oxide−dye−electrolyte interface. Thus, for future research, it will be important to carefully select several reference systems that emphasize different key aspects of the device and characterize these systems in depth with all the different techniques we have at hand. From comparisons and modelling, we may then find a better generality of our fundamental understanding.

A challenging but realizable goal for the present DSC technology is to achieve efficiencies above 15%. The new one-electron−based redox couples may significantly lower the voltage drop between the redox system and the oxidized dye, thus pointing out the direction to reach above 15%. This will require electron-transfer studies to clarify how much driving force is needed in the regeneration step and how this can be optimized by tuning of the redox potential by modification of the redox complexes and the oxidation potential of the dye. Moreover, the blocking of the recombination reaction of electrons in the oxide with electrolyte species is an important task for further design of dye molecules and surface passivation methods.

Regarding stability of DSC cells and modules, the conditions for accelerated testing have not yet been standardized. Until now, 1000-h light-soaking and 1000-h high-temperature storage tests have mainly been performed to compare materials and to show feasibility of the technology. As the technology advances, the outcome of accelerated testing will also be used to start calculating acceleration factors and estimating product life. Since this is not straightforward, intensified research and development is urgently required to define the procedures for relevant accelerated testing of dye-sensitized solar devices. Especially important is collecting outdoor test results from different locations and application-relevant conditions. The outcome needs to be related to results from accelerated testing to define the key tests for accelerated testing of DSC modules.

For the DSC technology, it is advantageous to compare energy cost rather than cost per peak watt. DSCs perform relatively better compared with other solar cell technologies under diffuse light conditions and at higher temperatures. An overall goal for future research will thus be to collect data and develop models to make fair judgments of the DSC technology with regards to energy costs.

With the ever-increasing industrial development of the DSC technology, we anticipate some exciting years to come with possible introduction of niche applications such as consumer electronics and successful development of manufacturing processes.

REFERENCES

[1] E.L. Gibson, A. Hagfeldt, Solar energy materials, in: D.W. Bruce, D. O'Hare, R.I. Walton (Eds.), Energy Materials, John Wiley & Sons, Ltd., 2011978-0-470-99752-9, Chapter 3
[2] M.A. Green, K. Emery, Y. Hishikawa, W. Warta, Prog. Photovoltaics 16 (2008) 435.
[3] Y. Chib, A. Islam, K. Kakutani, R. Komiya, N. Koide, L. Han, High efficiency dye-sensitized solar cells. *Technical Digest*, 15th International Photovoltaic Science and Engineering Conference, Shanghai, October 2005; 665.
[4] (i) M.A. Green, K. Emery, Y. Hishikawa, W. Warta, Prog. Photovolt: Res. Appl. 18 (2010) 346. (ii) M. Morooka, K. Noda, 88th Spring Meeting of the Chemical Society of Japan, Tokyo, 26 March 2008.
[5] U.B. Cappel, S.M. Feldt, J. Schoeneboom, A. Hagfeldt, G. Boschloo, J. Am. Chem. Soc. 132 (2010) 9096.
[6] S. Ardo, Y. Sun, A. Staniszewski, F.N. Castellano, G.J. Meyer, J. Am. Chem. Soc. 132 (2010) 6696.
[7] S.M. Feldt, E.A. Gibson, E. Gabrielsson, L. Sun, G. Boschloo, A. Hagfeldt, J. Am. Chem. Soc. 132 (2010) 16714.
[8] M. Grätzel, Keynote lecture, 3rd Hybrid and Organic Photovoltaics Conference, May 15–18, 2011, Valencia, Spain.
[9] A. McEvoy, Photoelectrochemical cells, in: T. Markvart, L. Castaner (Eds.), Practical Handbook of Photovoltaics: Fundamentals and Applications, Elsevier, 20031856173909, Chapter IIe. 1.
[10] A. Hagfeldt, M. Grätzel, Chem. Rev. 95 (1995) 49.
[11] A. Hagfeldt, M. Grätzel, Acc. Chem. Res. 33 (2000) 269.
[12] M. Grätzel, Nature 414 (2001) 338.
[13] M. Grätzel, Inorg. Chem. 44 (2005) 6841.
[14] S. Ardo, G.J. Meyer, Chem. Soc. Rev. 38 (2009) 115.
[15] L.M. Peter, J. Phys. Chem. C 111 (2007) 6601.
[16] L.M. Peter, Phys. Chem. Chem. Phys. 9 (2007) 2630.
[17] J. Bisquert, D. Cahen, G. Hodes, S. Ruhle, A. Zaban, J. Phys. Chem. B 108 (2004) 8106.
[18] B.C. O'Regan, J. Durrant, Acc. Chem. Res. 42 (2009) 1799.
[19] K. Kalyanasundaram, Dye-Sensitized Solar Cells, EFPL Press, 2010. ISBN 10: 143980866X.
[20] M. Grätzel, Acc. Chem. Res. 42 (2009) 1788.
[21] A. Hagfeldt, G. Boschloo, L. Sun, L. Kloo, H. Pettersson, Chem. Rev. 110 (2010) 6595.
[22] G.J. Meyer, ACS Nano 4 (2010) 4337.
[23] A.E. Becquerel, Memoire sur les effets électriques produits sous l'influence des rayons solaires, C. R. Acad. Sci., Paris 9 (1839) 561.
[24] K. Rajeshwar, J. Phys. Chem. Lett. 2 (2011) 1301.

[25] A. Fujishima, K. Honda, Nature 238 (1972) 37.
[26] M. Kallmann, M.D Pope, Nature 188 (1960) 935.
[27] K. Rajeshwar, M.E. Osugi, W. Chanmanee, C.R. Chenthamarakshan, M.V.B. Zanoni, P. Kajitvichyanukul, et al., J. Photochem. Photobiol. C: Photochem. Rev 9 (2008) 15.
[28] K. Hashimoto, H. Irie, A. Fujishima, Jpn. J. Appl. Phys. 44 (2005) 8269.
[29] B. O'Regan, M. Grätzel, Nature 353 (1991) 737.
[30] R. Memming, Semiconductor Electrochemistry, Wiley VCH, Weinheim, Germany, 2001.
[31] P. Würfel, Physics of Solar Cells: From Principles to New Concepts, Wiley-VCH, Weinheim, Germany, 2005.
[32] R. Memming, Semiconductor Electrochemistry, Wiley VCH, Weinheim, 2001.
[33] D. Fitzmaurice, Sol. Energy Mat. Solar Cells 32 (1994) 289.
[34] G. Rothenberger, D. Fitzmaurice, M. Grätzel, J. Phys. Chem. 96 (1992) 5983.
[35] R. Memming, Prog. Surf. Sci. 17 (1984) 7.
[36] H. Gerischer, Z. Phys. Chem. 26 (1960) 223.
[37] H. Gerischer, in: Eyring, Henderson, Jost (Eds.), Physical Chemistry, vol. VIA, Academic Press, New York, 1970, p. 463.
[38] W.W. Gärtner, Phys. Rev. 116 (1959) 84.
[39] J. Bisquert, F. Fabregat-Santiago, Impedance spectroscopy: A general introduction and application to dye-sensitized solar cells, in: K. Kalyanasundaram (Ed.), Dye-Sensitized Solar Cells, EPFL Press, 2010ISBN-10:143980866X
[40] A. Hagfeldt, L. Peter, Characterization and modelling of dye-sensitized solar cells: A toolbox approach, in: K. Kalyanasundaram (Ed.), Dye-Sensitized Solar Cells, EPFL Press, 2010ISBN-10:143980866X
[41] P.R.F. Barnes, A.Y. Anderson, J.R. Durrant, B.C. O'Regan, Phys. Chem. Chem. Phys. 13 (2011) 5798.
[42] A. Hagfeldt, U. Bjorksten, S. -E. Lindquist, Sol. Energy Mat. Sol. Cells 27 (1992) 293.
[43] G. Hodes, I.D.J. Howell, L.M. Peter, J. Electrochem. Soc. 139 (1992) 3136.
[44] S. Södergren, A. Hagfeldt, J. Olsson, S. -E. Lindquist, J. Phys. Chem. 98 (1994) 5552.
[45] A.J. McEvoy, M. Grätzel, Sol. Energy Mater. Sol. Cells 32 (1994) 221.
[46] W. West, Photogr. Sci. Eng. 18 (1974) 35.
[47] J. Moser, Monatsh. Chem. 8 (1887) 373.
[48] S. Namba, Y. Hishiki, J. Phys. Chem. 69 (1965) 774.
[49] R.C. Nelson, J. Phys. Chem. 69 (1965) 714.
[50] H. Gerischer, H. Tributsch, Ber. Bunsenges. Phys. Chem. 72 (1968) 437.
[51] H. Tributsch, Ph.D. thesis, Techn. Hochschule München, Germany, 1968.
[52] G.F. Cf. Moore, G.W. Brudvig, Annu. Rev. Condens. Matter Phys. 2 (2011) 303.
[53] A. Magnuson, M. Anderlund, O. Johansson, P. Lindblad, R. Lomoth, T. Polivka, et al., Acc. Chem. Res. 42 (2009) 1899.
[54] Same as reference 50.
[55] H. Gerischer, M.E. Michel-Beyerle, F. Rebentrost, H. Tributsch, Electrochim. Acta 13 (1968) 1509.
[56] M. Matsumura, Y. Nomura, H. Tsubomura, Bull. Chem. Soc. Jpn. 50 (1977) 2533.
[57] N. Alonso, M. Beley, P. Chartier, V. Ern, Rev. Phys. Appl. 16 (1981) 5.
[58] J. Desilvestro, M. Grätzel, L. Kavan, J. Moser, J. Augustynski, J. Am. Chem. Soc. 107 (1985) 2988.
[59] M.K. Nazeeruddin, A. Kay, I. Rodicio, R. Humphry-Baker, E. Mueller, P. Liska, et al., J. Am. Chem. Soc. 115 (1993) 6382.
[60] D.B. Kuang, C. Klein, S. Ito, J.E. Moser, R. Humphry-Baker, N. Evans, et al., Adv. Mater. 19 (2007) 1133.

[61] Same as reference 2.
[62] B. O'Regan, D.T. Schwartz, J. Appl. Phys. 80 (1996) 4749.
[63] T.W. Hamann, R.A. Jensen, A.B.F. Martinson, H. van Ryswyk, J.T. Hupp, Energy Environ. Sci. 1 (2008) 66.
[64] J.B. Asbury, R.J. Ellingson, H.N. Ghosh, S. Ferrere, A.J. Nozik, T.Q. Lian, J. Phys. Chem. B. 103 (1999) 3110.
[65] G. Ramakrishna, D.A. Jose, D.K. Kumar, A. Das, D.K. Palit, H.N. Ghosh, J. Phys. Chem. B 109 (2005) 15445.
[66] D. Kuang, S. Ito, B. Wenger, C. Klein, J. -E. Moser, R. Humphry-Baker, et al., J. Am. Chem. Soc. 128 (2006) 4146.
[67] G. Benkö, J. Kallioinen, J.E.I. Korppi-Tommola, A.P. Yartsev, V. Sundström, J. Am. Chem. Soc. 124 (2002) 489.
[68] S.E. Koops, B. O'Regan, P.R.F. Barnes, J.R. Durrant, J. Am. Chem. Soc. 131 (2009) 4808.
[69] P. Wang, B. Wenger, R. Humphry-Baker, J.E. Moser, J. Teuscher, W. Kantlehner, et al., J. Am. Chem. Soc. 127 (2005) 6850.
[70] D. Kuciauskas, M.S. Freund, H.B. Gray, J.R. Winkler, N.S. Lewis, J. Phys. Chem. B 105 (2001) 392.
[71] C. Bauer, G. Boschloo, E. Mukhtar, A. Hagfeldt, J. Phys. Chem. B 106 (2002) 12693.
[72] J.N. Clifford, E. Palomares, M.K. Nazeeruddin, M. Grätzel, J.R. Durrant, J. Phys. Chem. C 111 (2007) 6561.
[73] G. Boschloo, A. Hagfeldt, Acc. Chem. Res. 42 (2009) 1819.
[74] H. Nusbaumer, J. -E. Moser, S.M. Zakeeruddin, M.K. Nazeeruddin, M. Grätzel, J. Phys. Chem. B 105 (2001) 10461.
[75] H. Nusbaumer, S.M. Zakeeruddin, J. -E. Moser, M. Grätzel, Chem. —Eur. J. 9 (2003) 3756.
[76] S. Cazzanti, S. Caramori, R. Argazzi, C.M. Elliott, C.A. Bignozzi, J. Am. Chem. Soc. 128 (2006) 9996.
[77] U.B. Cappel, A.L. Smeigh, S. Plogmaker, E.M.J. Johansson, H. Rensmo, L. Hammarström, et al., J. Phys. Chem. C 115 (2011) 4345.
[78] U. Bach, Y. Tachibana, J.E. Moser, S.A. Haque, J.R. Durrant, M. Grätzel, et al., J. Am. Chem. Soc 121 (1999) 7445.
[79] A.Y. Anderson, P.R.F. Barnes, J.R. Durrant, B.C. O'Regan, J. Phys. Chem. C 115 (2011) 2439.
[80] J.G. Rowley, G.J. Meyer, J. Phys. Chem. C 115 (2011) 6156.
[81] N. Kopidakis, E.A. Schiff, N.G. Park, J. van de Lagemaat, A.J. Frank, J. Phys. Chem. B 104 (2000) 3930.
[82] D. Nistér, K. Keis, S. -E. Lindquist, A. Hagfeldt, Sol. Energy Mater. Sol. Cells 73 (2002) 411.
[83] F. Cao, G. Oskam, P.C. Searson, J. Phys. Chem. 100 (1996) 17021.
[84] L. Dloczik, O. Ileperuma, I. Lauermann, L.M. Peter, E.A. Ponomarev, G. Redmond, et al., J. Phys. Chem. B 101 (1997) 10281.
[85] L. Forro, O. Chauvet, D. Emin, L. Zuppiroli, H. Berger, F. Lévy, J. Appl. Phys. 75 (1994) 633.
[86] J. Van de Lagemaat, A.J. Frank, J. Phys. Chem. B 104 (2000) 4292.
[87] J. Bisquert, V.S. Vikhrenko, J. Phys. Chem. B 108 (2004) 2313.
[88] A.C. Fisher, L.M. Peter, E.A. Ponomarev, A.B. Walker, K.G.U. Wijayantha, J. Phys. Chem. B 104 (2000) 949.
[89] J. Bisquert, J. Phys. Chem. B 108 (2004) 2323.
[90] K. Westermark, A. Henningsson, H. Rensmo, S. Södergren, H. Siegbahn, A. Hagfeldt, Chem. Phys. Lett. 285 (2002) 157.

[91] L.M. Peter, Acc. Chem. Res. 42 (2009) 1839.
[92] E. Hendry, M. Koeberg, B.C. O'Regan, M. Bonn, Nano Lett. 6 (2006) 755.
[93] J.E. Moser, M. Grätzel, Chem. Phys. 176 (1993) 493.
[94] S.A. Haque, Y. Tachibana, D.R. Klug, J.R. Durrant, J. Phys. Chem. B 102 (1998) 1745.
[95] D. Kuciauskas, M.S. Freund, H.B. Gray, J.R. Winkler, N.S. Lewis, J. Phys. Chem. B 105 (2001) 392.
[96] B. O'Regan, J. Moser, M. Anderson, M. Grätzel, J. Phys. Chem. 94 (1990) 8720.
[97] S.G. Yan, J.T. Hupp, J. Phys. Chem. 100 (1996) 6867.
[98] L.M. Peter, J. Phys. Chem. C 111 (2007) 6601.
[99] B.C. O'Regan, J.R. Durrant, Acc. Chem. Res. 42 (2009) 1799.
[100] K. Zhu, N. Kopidakis, N.R. Neale, J. van de Lagemaat, A.J. Frank, J. Phys. Chem. B 110 (2006) 25174.
[101] S. Nakade, Y. Saito, W. Kubo, T. Kitamura, Y. Wada, S. Yanagida, J. Phys. Chem. B 107 (2003) 8607.
[102] N. Papageorgiou, Y. Athanassov, M. Armand, P. Bonhote, H. Pettersson, A. Azam, et al., J. Electrochem. Soc. 143 (1996) 3099.
[103] R. Kawano, M. Watanabe, Chem. Commun. (2005) 2107.
[104] M. Zistler, P. Wachter, P. Wasserscheid, D. Gerhard, A. Hinsch, R. Sastrawan, et al., Electrochim. Acta 52 (2006) 161.
[105] N. Yamanaka, R. Kawano, W. Kubo, N. Masaki, T. Kitamura, Y. Wada, et al., J. Phys. Chem. B 111 (2007) 4763.
[106] H.J. Snaith, M. Grätzel, Appl. Phys. Lett. 89 (2006) 262114.
[107] U. Cappel, Characterization of organic dyes for solid-state dye-sensitized solar cells. PhD Thesis., Uppsala University, Sweden. ISBN 978-91-554-8042-4 (http://uu.diva-portal.org/smash/record.jsf?pid = diva2:406937).
[108] The $TiCl_4$ treatment leads to the deposition of an ultrapure TiO_2 shell (approximately 1 nm) on the mesoporous TiO_2 (which may contain impurities or have carbon residues at the surface). The procedure leads to increased dye adsorption, lowers the acceptor levels in TiO_2 in energy, which can improve the injection efficiency, and it improves electron lifetime significantly.
[109] S. Ito, T.N. Murakami, P. Comte, P. Liska, C. Grätzel, M.K. Nazeeruddin, et al., Thin Solid Films 516 (2008) 4613.
[110] K.A. Emery, Measurement and characterization of solar cells and modules, in: A. Luque, S. Hegedus (Eds.), Handbook of Photovoltaic Science and Engineering, 2nd ed., John Wiley & Sons, W. Sussex, U. K., 2011, pp. 797−840. Chap. 18
[111] Correspondence with Keith Emery, National Renewable Energy Laboratory, Golden, Colorado, USA.
[112] W. Kubo, A. Sakamoto, T. Kitamura, Y. Wada, S. Yanagida, J. Photochem. Photobiol. A 164 (2004) 33.
[113] J. Halme, G. Boschloo, A. Hagfeldt, P. Lund, J. Phys. Chem C. 112 (2008) 5623.
[114] P.R.F. Barnes, A.Y. Anderson, S.E. Koops, J.R. Durrant, B.C. O'Regan, J. Phys. Chem. C 113 (2009) 1126.
[115] Y. Tachibana, K. Hara, K. Sayama, H. Arakawa, Chem. Mat. 14 (2002) 2527.
[116] A.B. Murphy, Appl. Optics 46 (2007) 3133.
[117] F. Fabregat-Santiago, G. Garcia-Belmonte, I. Mora-Seró, J. Bisquert, Phys. Chem. Chem. Phys. 13 (2011) 9083.
[118] J. Van de Lagemaat, A.J. Frank, J. Phys. Chem. B 105 (2001) 11194.
[119] S. Nakade, T. Kanzaki, Y. Wada, S. Yanagida, Langmuir 21 (2005) 10803.
[120] A. Zaban, M. Greenshtein, J. Bisquert, ChemPhysChem 4 (2003) 859.
[121] J. Bisquert, A. Zaban, M. Greenshtein, I. Mora-Sero, J. Am. Chem. Soc. 126 (2004) 13550.

[122] J. Van de Lagemaat, N.G. Park, A.J. Frank, J. Phys. Chem. B 104 (2000) 2044.
[123] F. Fabregat-Santiago, J. Bisquert, G. Garcia-Belmonte, G. Boschloo, A. Hagfeldt, Sol. Energy Mater. Sol. Cells 87 (2005) 117.
[124] J. Bisquert, V.S. Vikhrenko, J. Phys. Chem. B 108 (2004) 2313.
[125] A.C. Fisher, L.M. Peter, E.A. Ponomarev, A.B. Walker, K.G.U. Wijayantha, J. Phys. Chem. B 104 (2000) 949.
[126] P.R.F. Barnes, A.Y. Anderson, S.E. Koops, J.R. Durrant, B.C. O'Regan, J. Phys. Chem. C 113 (2008) 1126.
[127] J. Bisquert, I. Mora-Sero, J. Phys. Chem. Lett. 1 (2009) 450.
[128] J.R. Jennings, F. Li, Q. Wang, J. Phys. Chem. C 114 (2011) 14665.
[129] B.C. O'Regan, K. Bakker, J. Kroeze, H. Smit, P. Sommeling, J.R. Durrant, J. Phys. Chem. B 110 (2006) 17155.
[130] G. Boschloo, A. Hagfeldt, J. Phys. Chem. B 109 (2005) 12093.
[131] J. Van de Lagemaat, N. Kopidakis, N.R. Neale, A.J. Frank, Phys. Rev. B (2005) 71.
[132] N.W. Duffy, L.M. Peter, R.M.G. Rajapakse, K.G.U. Wijayantha, Electrochem. Commun. 2 (2000) 658.
[133] B.C. O'Regan, S. Scully, A.C. Mayer, E. Palomares, J. Durrant, J. Phys. Chem. B 109 (2005) 4616.
[134] T.T.O. Nguyen, L.M. Peter, H. Wang, J. Phys. Chem. C 113 (2009) 8532.
[135] G. Boschloo, D. Fitzmaurice, J. Phys. Chem. B 103 (1999) 2228.
[136] G. Boschloo, L. Häggman, A. Hagfeldt, J. Phys. Chem. B 110 (2006) 13144.
[137] G. Boschloo, A. Hagfeldt, J. Phys. Chem. B 109 (2005) 12093.
[138] K. Lobato, L.M. Peter, U. Wurfel, J. Phys. Chem. B 110 (2006) 16201.
[139] K. Lobato, L.M. Peter, J. Phys. Chem. B 110 (2006) 21920.
[140] J. Kallioinen, G. Benkö, V. Sundström, J.E.I. Korppi-Tommola, A.P. Yartsev, J. Phys. Chem. B 106 (4396) (2002).
[141] R. Eichberger, F. Willig, Chem. Phys. 141 (1990) 159.
[142] Y. Tachibana, J.E. Moser, M. Grätzel, D.R. Klug, J.R. Durrant, J. Phys. Chem. 100 (1996) 20056.
[143] B. Wenger, M. Grätzel, J. -E. Moser, J. Am. Chem. Soc. 127 (2005) 12150.
[144] S.E. Koops, P.R.F. Barnes, B.C. O'Regan, J.R. Durrant, J. Phys. Chem. C 114 (2010) 8054.
[145] G. Boschloo, A. Hagfeldt, Chem. Phys. Lett. 370 (2003) 381.
[146] G. Boschloo, A. Hagfeldt, Inorg. Chim. Acta 361 (2008) 729.
[147] J. Stark, Nature 92 (1914) 401.
[148] M. Leone, A. Paoletti, N. Robotti, Phys. Perspect. 6 (2004) 271.
[149] G.U. Bublitz, S.G. Boxer, Annu. Rev. Phys. Chem. 48 (1997) 213.
[150] S. Ardo, Y. Sun, F.N. Castellano, G.J. Meyer, J. Phys. Chem. B 114 (2010) 14596.
[151] T.W. Hamann, R.A. Jensen, A.B.F. Martinson, H. Van Ryswyk, J.T. Hupp, Energy Environ. Sci. 1 (2008) 66.
[152] M. Pagliaro, G. Palmisano, R. Ciriminna, V. Loddo, Energy Environ. Sci. 2 (2009) 838.
[153] R. Jose, V. Thavasi, S. Ramakrishna, J. Am. Ceram. Soc. 92 (2009) 289.
[154] Q.F. Zhang, C.S. Dandeneau, X.Y. Zhou, G.Z. Cao, Adv. Mater. 21 (2009) 4087.
[155] X. Chen, S.S. Mao, Chem. Rev. 107 (2007) 2891.
[156] U. Ozgur, Y.I. Alivov, C. Liu, A. Teke, M.A. Reshchikov, S. Dogan, et al., J. Appl. Phys. (2005) 98.
[157] J.R. Mann, M.K. Gannon, T.C. Fitzgibbons, M.R. Detty, D.F. Watson, J. Phys. Chem. C 112 (2008) 13057.
[158] N. Robertson, Angew. Chem., Int. Ed. 45 (2006) 2338.
[159] A. Mishra, M.K.R. Fischer, P. Bauerle, Angew. Chem., Int. Ed. 48 (2009) 2474.

[160] J.J. He, H. Lindstrom, A. Hagfeldt, S.E. Lindquist, J. Phys. Chem. B 103 (1999) 8940–8943.
[161] P. Qin, H.J. Zhu, T. Edvinsson, G. Boschloo, A. Hagfeldt, L.C. Sun, J. Am. Chem. Soc. 130 (2008) 8570–8571.
[162] (i) L.L. Pleux, A.L. Smeigh, E. Gibson, Y. Pellegrin, E. Blart, G. Boschloo, et al., Energy Environ. Sci. 4 (2011) 2075. (ii) F. Odobel, L.L. Pleux, Y. Pellegrin, E. Blart, Acc. Chem. Res. 43 (2010) 1063.
[163] D.P. Hagberg, X. Jiang, E. Gabrielsson, M. Linder, T. Marinado, T. Brinck, et al., J. Mater. Chem. 19 (2009) 7232.
[164] X. Jiang, T. Marinado, E. Gabrielsson, D.P. Hagberg, L.C. Sun, A. Hagfeldt, J. Phys. Chem. C 114 (2010) 2799.
[165] H.N. Tsao, C. Yi, T. Moehl, J-H. Yum, S.M. Zakeeruddin, M.K. Nazeeruddin, et al., ChemSusChem. 4 (2011) 591.
[166] M. Wang, N. Chamberland, L. Breau, J. -E. Moser, R. Humphry-Baker, B. Marsan, et al., Nature Chem. 2 (2010) 385.
[167] H. Tian, Z. Yu, A. Hagfeldt, L. Kloo, L. Sun, J. Am. Chem. Soc. 133 (2011) 9413.
[168] T. Daeneke, T. -H. Kwon, A.B. Holmes, N.W. Duffy, U. Bach, L. Spiccia, Nature Chem. 3 (2011) 211.
[169] Y. Bai, Q. Yu, N. Cai, Y. Wang, M. Zhang, P. Wang, Chem. Commun. 47 (2011) 4376.
[170] (i) S. Ferrere, A. Zaban, B.A. Gregg, J. Phys. Chem. B 101 (1997) 4490. (ii) Z.S. Wang, K. Sayama, H. Sugihara, J. Phys. Chem. B 109 (2005) 22449.
[171] (i) G. Oskam, B.V. Bergeron, G.J. Meyer, P.C. Searson, J. Phys. Chem. B 105 (2001) 6867. (ii) P. Wang, S.M. Zakeeruddin, J. -E. Moser, R. Humphry-Baker, M. Grätzel, J. Am. Chem. Soc. 126 (2004) 7164. (iii) B.V. Bergeron, A. Marton, G. Oskam, G.J. Meyer, J. Phys. Chem. B 109 (2005) 937.
[172] Same as reference 171.
[173] Z. Zhang, P. Chen, T.N. Murakami, S.M. Zakeeruddin, M. Grätzel, Adv. Funct. Mater. 18 (2008) 341.
[174] (i) M. Gorlov, L. Kloo, Dalton Trans. (2008) 2655.(ii)B. Li, L.D. Wang, B.N. Kang, P. Wang, Y. Qiu, Sol. Energy Mater. Sol. Cells 90 (2006) 549.
[175] (i) J. Kruger, R. Plass, L. Cevey, M. Piccirelli, M. Grätzel, U. Bach, Appl. Phys. Lett. 79 (2001) 2085. (ii) H.J. Snaith, A.J. Moule, C. Klein, K. Meerholz, R.H. Friend, M. Grätzel, Nano Lett. 7 (2007) 3372.
[176] K.J. Jiang, K. Manseki, Y.H. Yu, N. Masaki, K. Suzuki, Y.L. Song, et al., Adv. Funct. Mater. 19 (2009) 2481.
[177] (i) B. O'Regan, D.T. Schwartz, J. Appl. Phys. 80 (1996) 4749. (ii) G. Kumara, A. Konno, G.K.R. Senadeera, P.V.V. Jayaweera, D. De Silva, K. Tennakone, Sol. Energy Mater. Sol. Cells (69) (2001) 195.
[178] K. Tennakone, G. Kumara, A.R. Kumarasinghe, K.G.U. Wijayantha, P.M. Sirimanne, Semicond. Sci. Technol. 10 (1995) 1689.Q.B. Meng, K. Takahashi, X.T. Zhang, I. Sutanto, T.N. Rao, O. Sato, et al., Langmuir 19 (2003) 3572.
[179] A. Hauch, A. Georg, Electrochim. Acta 46 (2001) 3457.
[180] N. Papageorgiou, W.F. Maier, M. Grätzel, J. Electrochem. Soc. 144 (1997) 876.
[181] A. Kay, M. Grätzel, Sol. Energy Mater. Sol. Cells 44 (1996) 99.
[182] H. Pettersson, T. Gruszecki, R. Bernhard, L. Häggman, M. Gorlov, G. Boschloo, et al., Prog. Photovoltaics 15 (2007) 113.
[183] T.N. Murakami, S. Ito, Q. Wang, M.K. Nazeeruddin, T. Bessho, I. Cesar, et al., J. Electrochem. Soc. 153 (2006) A2255.
[184] E. Ramasamy, W.J. Lee, D.Y. Lee, J.S. Song, Appl. Phys. Lett. 90 (2007) 173103.
[185] K. Imoto, K. Takahashi, T. Yamaguchi, T. Komura, J.-i. Nakamura, K. Murata, Sol. Energy Mater. Sol. Cells 79 (2003) 459.

[186] K. Suzuki, M. Yamaguchi, M. Kumagai, S. Yanagida, Chem. Lett. 32 (2003) 28.
[187] L. Bay, K. West, B. Winther-Jensen, T. Jacobsen, Sol. Energy Mater. Sol. Cells 90 (2006) 341.
[188] Y. Saito, T. Kitamura, Y. Wada, S. Yanagida, Chem. Lett. (2002) 1060.
[189] Y. Saito, W. Kubo, T. Kitamura, Y. Wada, S. Yanagida, J. Photochem. Photobiol., A 164 (2004) 153.
[190] M. Wang, A.M. Anghel, B. Marsan, Ha, N. -L. Cevey, N. Pootrakulchote, S.M. Zakeeruddin, et al., J. Am. Chem. Soc. 131 (2009) 15976.
[191] M. Wu, X. Lin, A. Hagfeldt, T. Ma, Chem. Commun. 47 (2011) 4535.
[192] M. Wu, X. Lin, A. Hagfeldt, T. Ma, Angew. Chem. Int. Ed. 50 (2011) 3520.

CHAPTER IE-2

Organic Solar Cells

Clare Dyer-Smith and Jenny Nelson
Imperial College London, Blackett Laboratory, Prince Consort Road, London SW7 2AZ, UK

Contents

1. Introduction	544
1.1 Moving Toward the Market	544
2. Organic Electronic Materials	544
2.1 Excitons in Organic Semiconductors	545
3. Principles of Device Operation	548
3.1 Device Architectures	549
3.2 Bilayer Architecture	550
3.3 Bulk Heterojunction Architecture	551
4. Optimising Solar Cell Performance	552
4.1 Optimising Light Harvesting	553
4.2 Optical Trapping	554
4.3 Optimising Voltage Generation	554
4.4 Optimising Electrodes for Voltage Generation	555
4.5 Organic Tandem Solar Cells	556
4.6 Optimising Microstructure of Bulk Heterojunctions	557
4.7 Blend Composition	557
4.8 Annealing	558
4.9 Blend Additives	558
4.10 Alternative Approaches	559
5. Production Issues	559
5.1 Substrates	560
5.2 Stability	560
5.3 Degradation of Materials	560
5.4 Degradation of Morphology	561
5.5 Degradation of Electrodes	561
5.6 Encapsulation	562
5.7 Deposition Processes	562
6. Conclusions	563
References	564

Practical Handbook of Photovoltaics.
© 2012 Elsevier Ltd. All rights reserved.

1. INTRODUCTION

The discovery of a photovoltaic effect from organic materials [1–2] has led to an intense stimulation of research interest. Organic semiconductor materials offer the potential for low-cost and low-temperature preparation methods, mechanical flexibility, colour tuning, and the ability to be deposited on a wide variety of substrates, resulting in a widening of the range of applications of photovoltaic technology. The expected low capital cost for manufacture of organic photovoltaic (OPV) modules makes OPV technologies scalable and therefore suitable for distributed power applications.

Organic solar cell efficiencies have undergone a drastic acceleration in performance since the last edition of this book. At the time of this writing, the record power-conversion efficiency for solar cells with an all-organic active layer is 8.3%; this record has been achieved both for a polymer–fullerene blend deposited from solution[3] and for a blend consisting of two nonpolymeric semiconductors, deposited under vacuum[4]. These results represent almost a doubling of efficiency in the past 5 years.

1.1 Moving Toward the Market

Despite the impressive recent advances in device performance, further improvements, particularly in the performance of cost ratio through advancements in efficiency (with a goal of $1 per watt generated) and in device stability, are required to convert these laboratory-scale achievements into commercial reality. However, the prospect of cheap organic solar power is coming closer, and a small number of OPV products have already been brought to market, mostly for charging applications (e.g., Konarka's Power Plastic range[5]). The acceleration in this field in the commercial sector is evident in the increasing number of newly launched companies and existing materials or device manufacturers who have added organic photovoltaics to their portfolios.

The remainder of this chapter outlines the current understanding of organic photovoltaics and the key challenges that must be addressed in order to deliver on the promise of plastic solar power.

2. ORGANIC ELECTRONIC MATERIALS

The materials used in organic solar cells are classed as organic semiconductors as a consequence of their ability to absorb light and conduct

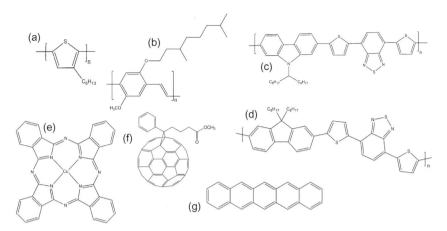

Figure 1 Selected organic photovoltaic materials: (a) poly-3-hexyl thiophene (P3HT); (b) poly[2-methoxy-5-3(3′,7′-dimethyloctyloxy)-1-4-phenylene vinylene] (MDMO-PPV); (c) poly[N-9′-hepta-decanyl- 2,7-carbazole-alt-5,5-(4′,7′-di-2-thienyl-2′,1′,3′-benzothiadiazole)] (PCDTBT); (d) poly((9,9-dioctylfluorene)-2,7-diyl-alt-[4,7-bis(3-hexylthien-5-yl)-2,1,3-benzothiadiazole]-2′,2″-diyl) (F8DTBT); (e) copper phthalocyanine (CuPc); (f) phenyl-C61-buytric acid methyl ester (PCBM); (g) pentacene.

charge, either within the molecules (such as in conjugated polymers) or through a molecular network. Conjugated polymers that have been studied in photovoltaic applications include polythiophenes, poly-phenylene-vinylenes (PPVs), polyfluorenes, and polycarbazoles, whereas nonpolymeric ('small molecule') organic semiconductors used in organic photovoltaic devices include functionalised fullerenes, phthalocyanines, perylene derivatives, and pentacene. Figure 1 shows examples of these conjugated materials. The desirable properties of visible absorption and charge transport in conjugated organic molecules arise from the network of π bonding between unsaturated atoms (largely carbon). The carbon atoms in the conjugated sections of the molecule form strong σ bonds using three sp^2 orbitals, whereas the p orbitals lying perpendicular to the plane of the molecule are able to form π bonds, which are less tightly localised than the strong σ bonds. Increasing conjugation across several atoms allows delocalisation of the π orbitals (in conjugated polymers, this typically extends over 2–10 repeat units), which may support the conduction of charge.

2.1 Excitons in Organic Semiconductors

The primary photogenerated state arising from light absorption in organic electronic materials is an *exciton*, a quasiparticle consisting of an electron in the lowest unoccupied molecular orbital (LUMO) and a hole in the highest occupied molecular orbital (HOMO). Organic semiconductors primarily differ from inorganic semiconductors in the greater localisation of the charge carriers and in the stronger Coulombic attraction between opposite charges. The localisation of the molecular orbitals in organic solids is a consequence of the van der Waals interactions between molecules being much weaker than the bonds within the molecules (in contrast to crystalline solids, where the intermolecular bonds are relatively strong). As a result, the so-called Frenkel excitons in organic materials are more localized and much less extensive than the excitons in crystalline inorganic semiconductors (known as Mott–Wannier excitons), which exist across several lattice sites (Figure 2) The Coulombic attraction between charges, known as the *exciton binding energy* (E_B), is given by Equation 1, in which e is the electronic charge, ε is the dielectric permittivity of the medium, ε_0 is the vacuum permittivity and r is the electron–hole separation.

$$E_B = \frac{e^2}{4\pi\varepsilon\varepsilon_0 r} \qquad (1)$$

In inorganic semiconductors with typical dielectric constants, ε, of greater than 10, the binding energy is around 0.01 eV. The small binding energy in these excitons means that the electron and hole are readily separated at typical operating temperatures (the thermal energy $k_B T$ at room temperature is around 0.025 eV). Organic materials usually have a smaller dielectric constant of around 2–4 and a smaller separation between electron and hole within the localized Frenkel exciton. As a result, the exciton binding energies are typically much larger (around 0.3–0.5 eV), requiring

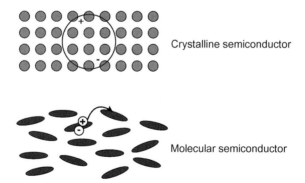

Figure 2 Mott–Wannier excitons in crystalline inorganic semiconductors are delocalised over several lattice sites and are easy to separate. Frenkel excitons in organic semiconductors are strongly localised and move by hopping.

the presence of a driving force in addition to the thermal energy to generate free-charge carriers. Organic solar cells therefore require a heterojunction—that is, an interface between two different types of material (the electron donor and the electron acceptor) that provides a difference in free energy that can drive charge separation in order to function. The details of charge separation at the heterojunction, and the considerations for optimum device performance, are given in Sections 3 and 4.

The migration of excitons in organic materials occurs by a relatively slow process of hopping between localised sites. The energies of excitons on these sites may vary as a result of disorder in the material, and the hopping may be downhill in energy or thermally activated [6]. The mean free path taken by an exciton within its radiative lifetime is known as the *exciton diffusion length*, L_D, and in organic semiconductors is on the order of 5–20 nm.

Another consequence of the localisation of π orbitals in organic semiconductors is that charges in these materials tend to experience strong coupling to the lattice and are usually described as positive and negative polarons rather than free holes and electrons. As a result, charge transport in organic semiconductors is qualitatively different from that in inorganic semiconductors, and it proceeds by a mechanism of hopping between sites, with the hopping rate determined by the electronic coupling between sites and by the difference in the energy of the initial and final states. The inherent disorder in the electronic energy levels of organic semiconductors usually leads to a variety of hopping rates and some charge trapping. Carrier mobilities in organic semiconductors therefore tend to be rather low compared with their inorganic counterparts (typically 10^{-4} cm^2 V^{-1} s^{-1} to 10^{-1} cm^2 V^{-1} s^{-1}), which limits the useful thickness of devices since recombination can more effectively compete with charge collection in thicker layers. Moreover, the hopping mechanism of charge transport gives rise to temperature-dependent carrier mobility, leading to temperature-dependent effects in device performance.

To summarise, the following are the key differences between organic and inorganic semiconductors are that in organic semiconductors:
1. Light absorption occurs in narrower spectral bands.
2. The molecular orbitals are more localized on account of weak intermolecular interactions. Excitons, resulting from the absorption of light, move by hopping.
3. The dielectric constant of the materials is low and the exciton binding energy is high as a result. A heterojunction is required to separate the

exciton into free charges. This is distinct from the situation in inorganic semiconductors, for which it can be assumed that every absorbed photon immediately generates a pair of separate charges.
4. Charge transport occurs by hopping and is usually dispersive. As a result, carrier mobilities are generally quite low.

3. PRINCIPLES OF DEVICE OPERATION

In organic solar cells, photogeneration can occur only at a heterojunction between two materials, where the additional free energy difference resulting from the energy level offset of the materials provides the energy to overcome the Coulombic binding between electron and hole. This driving force, sometimes referred to as ΔG, is usually defined as the difference between the energy of the singlet exciton and the charge separated state, with the latter defined as the difference between the donor ionisation potential and the acceptor electron affinity. The latter two quantities represent the energy of the donor HOMO and acceptor LUMO, respectively. The presence of donor and acceptor components promotes charge separation in two ways: (1) by ensuring that the process is energetically 'downhill' and (2) by allowing the localisation of the electron and hole on different molecules, increasing their spatial separation and decreasing the likelihood that they will recombine.

Figure 3 shows a simple band-diagram schematic of charge separation in an organic solar cell. Light absorption in the active layer is followed by the diffusion of the exciton through the material until it either decays or reaches an interface where it can undergo charge separation to form polarons that can produce a current if they are able to exit the cell at the anode and cathode before recombining with one another. The open-circuit voltage is believed to be controlled by the frontier orbital energy offset between the donor and acceptor; this is in contrast with an inorganic p−n junction cell where it is determined by the difference in the doping levels of n and p regions in the device. In an organic solar cell, asymmetry in the electrode work functions assists in sweeping the charges out of the device (particularly important in a blend device; see the following discussion) but is not required to generate a photovoltage.

Charge separation at the donor−acceptor interface was initially believed to be a single-stage process driven by the excess energy in the

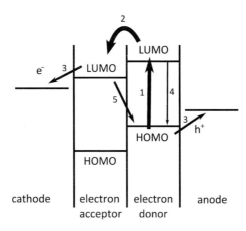

Figure 3 Schematic of solar cell operation. Light absorption in the active layer (1) is followed by the diffusion of the exciton through the material until it either decays (4) or reaches an interface where it can undergo charge separation (2) to form polarons that can produce a current if they are able to exit the cell (3) before recombining with one another (5).

exciton (ΔG), with a minimum ΔG of 0.3–0.6 eV required to efficiently separate charges [7–8]. However, it is now believed by many researchers that the separation of the exciton to give spatially separate polarons in the donor and acceptor proceeds via a charge transfer state in which the electron and hole are located on different molecules at the interface but still experience a significant binding energy [9–13]. In some cases, interfacial charge transfer states are observable in absorption [14–16] and in emission [17–20]. The mechanism by which this intermediate state is separated into charges is still under some debate in the literature, with some researchers using a model of hot charge-transfer state dissociation [13,21].

Since charge separation can only occur at the heterojunction, the exciton diffusion length L_D introduced in the previous section is an important parameter in organic solar cells since it places a limit on the volume fraction of the donor and the acceptor domains that can usefully generate charges. Excitons generated outside this volume (that is, at a distance greater than L_D from the heterojunction) will tend to decay before they are able to reach the donor–acceptor interface and be separated.

3.1 Device Architectures

An 'ideal' architecture for a photovoltaic blend comprises domains sufficiently small to maximize exciton dissociation but sufficiently large to

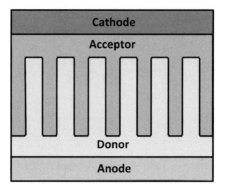

Figure 4 An example of an 'ideal' heterojunction structure, with domains of a similar size to the exciton diffusion length and continuous pathways to the electrodes to prevent losses through recombination at the electrodes and shunt pathways.

promote charge separation and allow the charges to reach the electrodes. Moreover, the preferential concentration of donor material close to the anode and acceptor close to the cathode can help to minimise losses through shunt pathways and the surface recombination losses that occur when a charge reaches the 'wrong' electrode. Figure 4 shows a schematic representation of one 'ideal' active layer structure that satisfies these criteria.

In practice, the most common types of organic photovoltaic devices are based on either planar (bilayer) or mixed (bulk) heterojunctions.

3.2 Bilayer Architecture

In a bilayer solar cell (Figure 5), the donor and acceptor layers are deposited sequentially, by vacuum deposition [22], spin coating from orthogonal solvents [23–24], or deposition of one layer onto the other by a process such as lamination [25–27] or stamp transfer [28]. The bilayer offers the ability to tune the properties of the donor and the acceptor both to maximise charge separation and allow absorption in both layers by suitable matching to the solar spectrum. It also allows good charge transport from the interface to the contacts. Aside from these properties, bilayer devices have proved useful model systems to test and study fundamental device behaviour.

Bilayer efficiencies have reached around 5%, with the best devices using C_{60} as an electron acceptor and a phthalocyanine (typically, copper phthalocyanine CuPc) as the electron donor. Recent encouraging results have been achieved using pentacene [29] or subphthalocyanine [30] as the

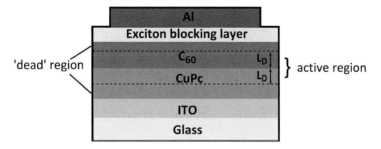

Figure 5 Architecture of a bilayer solar cell using copper phthalocyanine (CuPc) as the electron donor and buckminsterfullerene (C_{60}) as the electron acceptor. The dashed lines demarcate the active region of the cell. Excitons generated outside this region will decay before reaching the donor−acceptor interface.

electron donor or by using long-wavelength absorbers such as tin or lead phthalocyanine, either alone or in ternary systems containing two light-absorbing components [31−32].

Bilayer solar cells present a major drawback: the charge separation efficiency is limited by the small interfacial area, a consequence of the limited diffusion length of the exciton in organic materials [6,33−34], which requires that a photogenerated exciton must be formed close to a donor−acceptor interface to have a chance of separating before it decays to its ground state. This limits the thickness of the exciton-generating layer in a bilayer device and hence prevents the use of thick films with greater light absorption. In general, the better-performing bilayer structures are based on small molecules rather than polymers, possibly because the greater crystalline order in these materials offers improved exciton diffusion. The bilayer concept has been further improved in vacuum-deposited cells by including a mixed donor−acceptor layer with the aim of increasing the available interfacial area for charge dissociation and approaching the morphology shown in [35]. The intermixing of donor and acceptor is the basis of the bulk heterojunction, which is described next.

3.3 Bulk Heterojunction Architecture

In order to overcome the limitation on device thickness due to the small exciton diffusion lengths in organic materials, the bulk heterojunction concept has been widely adopted. The basis of the bulk heterojunction is the intimate mixing of the donor and acceptor in a single layer, resulting

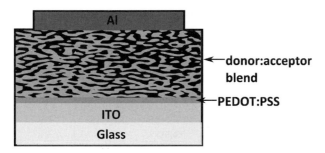

Figure 6 Bulk heterojunction blend. To maximize exciton dissociation, the domain sizes should be comparable to the exciton diffusion length.

in a film with a high interfacial surface area (Figure 6). Various types of bulk heterojunction system have been demonstrated, including polymer—fullerene blends (the main focus of this section), polymer—polymer blends, block copolymers, and inorganic—organic hybrid layers.

The high interfacial surface area and the smaller donor and acceptor domains in a bulk heterojunction compared to a bilayer ensure that excitons are produced close enough to an interface that they may reach it before decaying. This greatly increases the probability of exciton dissociation and allows the use of thicker films with increased harvesting of light. However, the intermixing of the two phases into small domains tends to decrease the order in the molecular packing and means that continuous pathways are harder to produce, and isolated domains, which cause recombination losses, may exist. In addition, the larger interfacial area encourages recombination, and poor control of the phase morphology means that shunt pathways may exist. Balancing the increased photogeneration yield against the increased recombination losses requires the optimisation of the microstructure of the blend film (Section 4).

4. OPTIMISING SOLAR CELL PERFORMANCE

High power-conversion efficiency requires the efficient harvesting of visible light, the efficient conversion of photogenerated potential energy into photovoltage at the electrodes, and a high fill factor (which represents good photocurrent generation in forward bias). The total power-conversion efficiency η of a solar cell can be expressed as a product of three key parameters: (1) the short-circuit current J_{sc}, (2) the open-circuit voltage V_{oc}, and

(3) the fill factor FF (Equation (2), where P_s is the radiant power incident on the cell):

$$\eta = \frac{J_{sc} V_{oc} FF}{P_s} \qquad (2)$$

The short-circuit current density J_{sc} measures the number of charges exiting the cell per unit time and area, and it has been shown to correlate strongly with the yield of photogenerated charges, provided the charge mobility is above around 10^{-5} cm^2 V^{-1} s^{-1} [36]. As such, for most organic semiconductors of interest for photovoltaics, the main parameters affecting J_{sc} are the light harvesting by the cell, a function of the absorption profile of the materials, absorption profile of the materials, the film thickness and optical confinement effects, the blend microstructure, and the efficiency of charge generation at the donor–acceptor heterojunction. The photovoltage is mainly determined by the offset of the energy levels at the donor–acceptor heterojunction and by bimolecular recombination. Bimolecular recombination also influences the fill factor, which is further determined by the parasitic resistances in the device (series resistance and shunt pathways). Bimolecular recombination in organic solar cells is strongly dependent upon charge-carrier density and energetic disorder (necessitating a model of device operation distinct from that of inorganic solar cells [37]) and the blend morphology.

4.1 Optimising Light Harvesting

Although organic semiconductors frequently have high absorption coefficients ($\sim 10^5$ cm^{-1}), their absorption tends to occur in fairly narrow bands (Figure 7). The available photon flux provided by the sun peaks at around 600 nm and contains large numbers of photons in the red or infrared. Polymers or acceptors with absorption in this long-wavelength region are desirable for increased light harvesting.

Materials with increased absorption at long wavelengths are desirable for OPV applications and are mostly based upon low-band-gap polymers [38] (many containing some charge-transfer character), although phthalocyanine or subphthalocyanine dyes [39], or inorganic nanoparticles [40], have also been employed to enhance light harvesting at longer wavelengths in dye-sensitised and hybrid organic-inorganic solar cells. Decreasing the material band gap to increase the number of photons absorbed has one disadvantage: this may also decrease the driving force for charge separation and lead to a decrease in the yield of photogenerated charges.

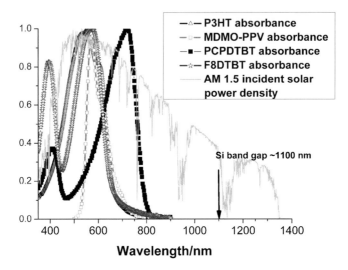

Figure 7 The absorption spectra of conjugated polymers are typically narrow and may not overlap well with the available solar emission spectrum. By comparison, Si has a band gap corresponding to an absorption edge at 1100 nm.

4.2 Optical Trapping

Because the layer thicknesses employed in organic solar cells are comparable to the wavelength of light, interference effects are important, and these can be manipulated to enhance optical absorption in the active layer. Light trapping can be achieved by the use of grating structures [41], microlenses [42–43], and pattern metal electrodes to induce plasmonic effects [44].

4.3 Optimising Voltage Generation

The photovoltage generated in a donor–acceptor OPV device is dependent upon the alignment of the energy levels of the donor and acceptor in the cell and also upon the rate of bimolecular recombination, which is discussed in the next section. One strategy to increase V_{oc} is to increase the energy level offset at the interface by either lowering the donor HOMO or raising the acceptor LUMO. Theoretical studies have suggested that modifying the device energy levels in this way could result in device efficiencies of 10% or more [45]. Maintaining good device performance does, however, depend upon the effect of these changes upon the charge-generation efficiency (and hence the short-circuit current). Increasing the donor ionisation potential (Figure 8) increases the voltage

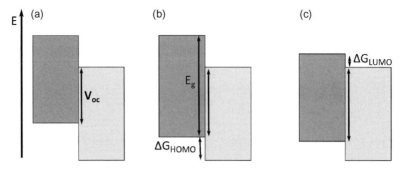

Figure 8 Increasing the open-circuit voltage (a) by increasing the polymer ionisation potential (b) results in an increased polymer band gap and decreased driving force for hole transfer, both of which may reduce short-circuit current. Maintaining the same band gap (c) results in a decreased driving force for electron transfer, with the same effect.

but may reduce J_{sc}, either by increasing photon energy required to excite the molecule or by reducing the driving force for charge separation. In polymer–fullerene cells, an alternative is to reduce the electron affinity of the fullerene. The use of multiply substituted fullerenes leads to an increase in open-circuit voltage with no loss of charge generation in poly-3-hexylthiophene (P3HT) blends [46–47] but may adversely affect the blend morphology and reduce carrier mobility [48].

In some polyfluorene–fullerene blends, increasing the frontier orbital offset between the donor HOMO and the acceptor LUMO may also lead to the formation of triplet states that are detrimental to device performance [49–51] and are of concern for the long-term stability of devices (Section 5).

4.4 Optimising Electrodes for Voltage Generation

In order to extract the electrochemical potential generated by the absorbed light, the interface between the electrodes and the active layer must be optimised to provide an ohmic contact. Since the Fermi levels of organic materials are seldom easily altered by doping, achieving good contact to the active layer requires careful selection of the electrode materials. In a P3HT–PCBM device, the transport level for holes (the polymer HOMO) is 4.8 eV and for electrons (the PCBM LUMO) is around 4 eV (estimates range from 4.3 eV to 3.7 eV) below the vacuum level. In a 'normal' device configuration (Figure 6), this requires the use of a low-work-function metal as the cathode. This electrode is typically based on aluminium, although its work function of

4.3 eV typically requires the insertion of a thin layer of a lower-work-function material such as LiF or barium or calcium metal in order to extract electrons more efficiently [52]. The anode is normally based on indium-doped tin oxide (ITO), which has a work function of around 4.7 eV and is normally coated with a conducting layer of polyethylenedioxythiophene–poly(4-styrenesulfonate) (PEDOT–PSS), which increases the work function to 5.0 eV. While the electrode materials listed above are considered to be optimal for P3HT–PCBM blends, we are likely to need higher-work-function anodes or lower-work-function cathodes to contact to the new low-band-gap active materials discussed previously.

4.5 Organic Tandem Solar Cells

As discussed earlier, decreasing the band gap of the light-absorbing component increases the number of photons and leads to improved photocurrent generation. However, the excess energy in photons with energy greater than the band gap will be lost to thermalisation, reducing the attainable photovoltage. In order to avoid this, it is possible to use low-band-gap semiconductors in combination with higher-band-gap semiconductors in a tandem structure to achieve a higher voltage. The tandem architecture is shown in Figure 9 and requires the semiconductor materials to be deposited sequentially in layers. At first, this was a challenge for OPV because of the difficulty in processing several layers from solution without damaging the layers below, although much recent progress has been made, and all-solution processed tandem solar cells are now possible [23,53–54]. The intermediate layer in an organic tandem cell may be made of a metal such as gold or metal oxide [55] or based on the conducting polymer PEDOT–PSS [23,56]; its role is to provide a

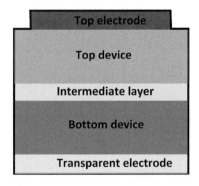

Figure 9 Tandem organic solar cell device architecture.

recombination region for charges and to ensure the alignment of the quasi-Fermi levels of electrons and holes in the two (or more) devices so that the photovoltages generated by the separate cells add. As usual, the photocurrent generated by the top and bottom cells should be equal.

A detailed review of tandem organic solar cells is given in [57].

4.6 Optimising Microstructure of Bulk Heterojunctions

The structure of organic blends, both on the microscopic length scale and the exact structure of the interface, is extremely important in determining the charge-generation efficiency and fill factor in organic solar cells, but it has proved challenging to control. The degree of intermixing between materials depends upon many factors, including the blend ratio, the surface energies of the blend components (influencing the relative affinity of each material for the substrate, the air, and each other), and the device processing conditions. This section presents a summary of ways in which the morphology can be influenced and some strategies to improve performance.

4.7 Blend Composition

The blend composition is known to have a large effect: in blends of MDMO-PPV with PCBM, the optimum device performance is achieved using a blend containing 80% PCBM by weight [58–59]. This result, while initially unexpected on the basis that the majority of the light absorption occurs in the polymer, is explained by the surprising observation that increased PCBM content in the film actually increases the hole mobility [59–60]. Moreover, the formation of large PCBM domains is also thought to provide an improvement to charge photogeneration. In a study of polyfluorene–fullerene blend films[19], the increased charge separation efficiency in blends with larger PCBM domains was ascribed to an increase in the dielectric constant and local electron mobility, as the crystalline domains increased in size. Influence over domain sizes can be achieved by changes to the solvent- and spin-coating conditions [58,61–62] or by the use of blend additives [63–65] to increase control over the blend microstructure.

A great deal of research has been carried out into blends of P3HT with PCBM, and this system has until quite recently been held as a benchmark blend due to the possibility of achieving power-conversion efficiencies of more than 5%. P3HT–PCBM also provides an important demonstration of the ways in which the microstructure not only strongly influences the blend's optoelectronic properties and ultimately device performance but also may be controlled. In P3HT–PCBM blends, the optimum blend composition requires 40—50% PCBM by weight, a much lower amount than is required for MDMO-PPV. In this system, the polymer crystallinity is high, and the formation of P3HT crystalline domains serves to expel the fullerene and decrease the level of intermixing between the materials. This results in less trapping of the fullerene in the polymer phase, and high fullerene content is not then required to produce a percolating network for electrons. The crystalline P3HT also has a reasonably high hole mobility and allows balanced charge transport. A study of the phase behaviour in P3HT—PCBM blends concluded that the composition required for optimum device performance is slightly below the eutectic point of the mixture, resulting in a blend with high interfacial area *and* optimal charge transport properties [66]. This was correlated both with photoluminescence quenching and the yield of photogenerated charges in a subsequent study [67].

4.8 Annealing

Post-treatment processing by annealing, either by heating above the glass transition temperature of the polymer or by placing in a solvent atmosphere, is known to be a necessary step to achieve the optimum performance in P3HT—PCBM blends [68—71]. Annealing results in an increased ordering in the blend as a result of crystallisation of the polymer, which is followed by clustering of the fullerene molecules. This results in improved charge-transport properties and charge-generation yield in the film [68], as well as an increase in the red-light absorption by the polymer. Vertical phase segregation has been observed upon annealing, which improves charge collection at the electrodes [72—73]. Annealing also has the effect of causing a decrease in the polymer ionisation potential (as a result of increased polymer crystallinity and the resulting increase in delocalisation of the HOMO) [68]. Increased crystallinity may also promote charge separation by an increase in the spatial extent of the exciton or in the local charge-carrier mobility, thus improving charge photogeneration yield [19].

4.9 Blend Additives

In recent years, a selection of molecular additives have been used in an attempt to optimise the morphology of polymer blends, particularly in systems where, unlike P3HT, the morphology cannot be so strongly influenced by annealing as a result of the lower crystallinity of the polymer. High performances have been achieved in polymer—PCBM [74—75] and small-molecule—PCBM [76] blends by the addition of

1,8-octanedithiol to the blend solution. The dithiol additive is believed to increase the phase segregation in the blend and provide better percolation pathways for electrons [77]. In blends of the polymer PCDTBT with the fullerene $PC_{70}BM$, the addition of diiodooctane to the blend has been shown to reduce the yield of triplet formation, thereby improving the charge-generation yield in the film [63].

4.10 Alternative Approaches

Inorganic—organic hybrid devices offer another route to control of the active-layer morphology. In these devices, the electron acceptor is based upon a nanostructured metal oxide film, similar to the microporous metal oxides commonly employed in dye-sensitised solar cells. Devices utilising nanostructured metal oxide layers have achieved efficiencies of a few percent, with a range of structures including nanowires, nanorods, and ordered nanocrystals being formed by these materials [78–80].

Block copolymers are another way to manipulate the blend morphology and may offer an advantage over binary blends of better long-term stability in morphology. Block copolymers consist of two or more different polymer 'blocks' joined end to end, allowing two materials that would ordinarily be immiscible to be forced into close contact and giving rise to unique phase behaviour and well-defined morphologies [81]. They can be used either as a sacrificial template for photoactive materials (for example, to produce nanostructured oxide films, which may be filled by a conjugated polymer [82] or used in dye-sensitised solar cells [83]) or in the case of a conjugated block copolymer, as a single active-layer component [84–85]. Block copolymers have been shown to improve the long-term stability of morphology in P3HT–PCBM blends when used as a compatibilising agent, decreasing the tendency of the blend to phase segregate over time [86]. In the long term, block copolymers are promising candidates in that the length scale of phase segregation is 'written in' to the material during synthesis, and so the solar cell performance should be less sensitive to variations in processing.

5. PRODUCTION ISSUES

The previous sections have dealt with the research challenges involved in increasing power-conversion efficiencies in organic solar cells, mainly at the laboratory scale. Now we address some of the key issues facing the production of these devices for commercial applications.

5.1 Substrates

In order to realise the potential of organic solar cells in unique applications and their mass production by roll-to-roll processes, flexible substrates are required. The most widely used material is ITO-coated poly(ethylene terephthalate) (PET).

At present, the transparent conductor used as the basis for OPV is indium-doped tin oxide. This material is limited by its transmission of visible light (80–90), high resistivity (typically 10–100 Ohm-cm^2) and susceptibility to mechanical stress, a problem for the production of flexible devices. Indium is also a limited and expensive resource, limiting the cost reductions available by using organic semiconductors in solar cells. As a result, alternatives to ITO are being sought. These include the alternative oxide material fluorine-doped tin oxide (FTO), which is cheaper but cannot be deposited on plastic substrates. Organic conducting coatings—for example, high-conductivity PEDOT–PSS or vapour-phase–polymerised PEDOT (VPP-PEDOT)—single-walled carbon nanotubes [87] and metal electrodes in an inverted geometry [88] have also been demonstrated as ITO alternatives in OPV devices.

5.2 Stability

One of the major challenges facing the field of organic photovoltaics is the limited stability of devices. Although impressive recent stability results have been achieved in accelerated degradation tests, organic solar cells do not yet achieve lifetimes of more than a few thousand hours (equivalent to a few years insolation in northern latitudes [89]), limiting their scope to small-scale consumer products rather than large-scale building-integrated applications. There are many sources of degradation for organic solar cells, mostly caused by the ingress of water and oxygen to the cell or by reactions at the electrodes. A detailed review of solar cell degradation processes and the measurement techniques used to characterise them is given in [90].

5.3 Degradation of Materials

Organic materials are susceptible to photodegradation, particularly when in contact with water or oxygen. Oxygen doping of conjugated polymers is well known, and the formation of triplet states upon photoexcitation offers another degradation pathway through the resulting reactivity to the triplet ground state of atmospheric oxygen. The chemical degradation of the materials depends on their molecular structure; the presence of reactive groups either within the polymer repeat units (such as the vinyl linkages in PPV-type polymers) or at the ends of the polymer chain. The development of materials with decreased reactivity is central to solving this problem [91].

5.4 Degradation of Morphology

In many types of bulk heterojunction solar cell, the morphology for optimal device performance is a nonequilibrium structure (it is 'frozen in' by the fast drying of the solvent during deposition). Over time, the structure may evolve toward a more stable molecular arrangement with an accompanying decrease in device performance. Most strategies to prevent this are based on materials designed to produce a thermodynamically stable morphology in the first place, although other approaches including the use of cross-linkable polymers have been shown to reduce the degradation in device performance over time [92].

5.5 Degradation of Electrodes

The electrodes and the electrode-active layer interfaces are another source of instability. Metal electrodes, particularly the low-work-function metals used for devices in a standard configuration, may react with atmospheric oxygen. Using an inverted device architecture (Figure 10) allows higher-work-function metals to be used or even conducting polymer anodes that can be used with printed silver ink cathodes, avoiding the need for metal evaporation and resulting in fully solution-processable devices [93]. Inverted devices also require an additional layer, frequently TiO_2, at the cathode. Although inverted-device performances are so far limited to less than 5% [94], they are gaining interest for their relative ease of

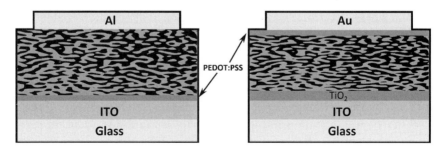

Figure 10 Normal (left) and inverted (right) devices.

manufacture. ITO may undergo chemical degradation, causing indium atoms to leach into the active layer, and it has limited mechanical strength cracking when flexed. The PEDOT−PSS layer is also susceptible to the rapid uptake of water and migration of PSS ions over time. Alternatives to PEDOT−PSS include metal oxides such as MoO_3, V_2O_5[95], and NiO[96] and solution-processible graphene oxide [97].

5.6 Encapsulation

At the laboratory scale, the fabrication of devices is frequently carried out in an inert environment to avoid the atmospheric degradation of the organic materials. However, at a manufacturing plant level, maintaining oxygen-free environments would result in a significant cost increase for the product. Encapsulation will be essential in prolonging the shelf life of commercially produced OPV devices and is already routine in the production of organic light-emitting diodes (OLEDs) [98].

The main requirement for encapsulating materials is that they should have low permeability to oxygen and water. Glass is a reliable material for encapsulation, leading to lifetimes of several thousand hours, but it negates the key advantage of organic solar cells, their flexibility, and the resulting low production cost. Flexible devices require encapsulation to be performed by applying a flexible plastic layer with barrier layers applied to reduce the permeability to ambient degrading agents. This approach has been demonstrated using, for example, poly(ethylene naphthalate) films coated with alternating layers of silicon oxide and organosilicon compounds that improve the lifetime of MDMO-PPV−PCBM and P3HT−PCBM cells to more than 3000 hours and 6000 hours, respectively [99−100].

5.7 Deposition Processes

Most photovoltaic devices produced at the laboratory scale are prepared by spin coating, but this technique is not amenable to scaling up to production levels. Large-scale device manufacture will require scalable and high-throughput production techniques, which for flexible substrates can be performed in a roll-to-roll manner.

A range of printing and deposition techniques have been shown to be effective in preparing organic solar cells, including printing techniques such as screen printing [101], ink-jet printing [102], slot-die coating [103−104], and gravure printing [105−106], deposition in vacuo or spray coating [107−109]. An overview of some of the different processes is given in Figure 11.

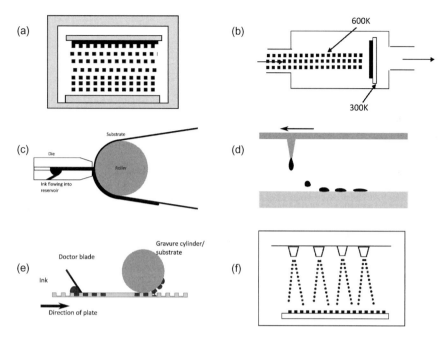

Figure 11 Deposition processes amenable to solar cell production: (a) thermal evaporation, (b) organic molecular beam deposition, (c) slot-die coating, (d) ink-jet printing, e) gravure printing, (f) spray coating.

6. CONCLUSIONS

Organic solar cells have moved from low-efficiency laboratory-scale devices to the first commercial products in less than a decade. The highest efficiencies are still mostly achieved using bulk heterojunction devices made from conjugated polymers and fullerene acceptors, although other classes of organic solar cell are beginning to compete. The main challenges for future development of the technology are to continue the improvements to device efficiency by materials design to achieve high light harvesting, high photovoltage, control of blend microstructure, the development of new electrodes to compete with and ultimately supplant the ubiquitous yet expensive ITO, and improvement to the long-term stability of devices. Roll-to-roll printing and coating processes employing flexible substrates are highly compatible with organic solar cells with the main challenges for production being in morphology control and in increasing device stability through a combination of materials design and device encapsulation.

REFERENCES

[1] H. Spanggaard, F.C. Krebs, A brief history of the development of organic and polymeric photovoltaics, Sol. Energy Mater. Sol. Cells 83(2–3) (2004) 125–146.
[2] M. Volmer, Die verschiedenen lichtelektrischen Erscheinungen am Anthracen, ihre Beziehungen zueinander, zur Fluoreszenz und Dianthracenbildung, Annalen der Physik 345(4) (1913) 775–796.
[3] Konarka. Konarka's power plastic achieves world record 8.3% efficiency certification from national energy renewable laboratory (NREL). 2010 Nov. 29, 2010 [cited 2011 22 February]; Press release]. Available from: <http://www.konarka.com/index.php/site/pressreleasedetail/konarkas_power_plastic_achieves_world_record_83_efficiency_certification_fr/>.
[4] Heliatek, Heliatek and IAPP achieve production-relevant efficiency record for organic photovoltaic cells. 11 November 2010.
[5] Konarka. http://www.konarka.com/index.php/power-plastic/about-power-plastic/. 2011 [cited 2011 10 March].
[6] O.V. Mikhnenko, et al., Temperature dependence of exciton diffusion in conjugated polymers, J. Phys. Chem. B 112(37) (2008) 11601–11604.
[7] H. Ohkita, et al., Charge carrier formation in polythiophene/fullerene blend films studied by transient absorption spectroscopy, J. Am. Chem. Soc. 130(10) (2008) 3030–3042.
[8] J.J.M. Halls, et al., Charge and energy transfer processes at polymer/polymer interfaces: A joint experimental and theoretical study, Phys. Rev. B 60(8) (1999) 5721–5727.
[9] C. Deibel, T. Strobel, V. Dyakonov, Role of the charge transfer state in organic donor-acceptor solar cells, Adv. Mater. (2010). p. n/a.
[10] J. Lee, et al., Charge transfer state versus hot exciton dissociation in polymer-fullerene blended solar cells, J. Am. Chem. Soc. 132(34) (2010) 11878–11880.
[11] M. Hallermann, S. Haneder, E. Da Como, Charge-transfer states in conjugated polymer/fullerene blends: below-gap weakly bound excitons for polymer photovoltaics, Appl. Phys. Lett. (2008) 05330793 (2008) 053307.
[12] D. Veldman, S.C.J. Meskers, R.A.J. Janssen, The energy of charge-transfer states in electron donor-acceptor blends: insight into the energy losses in organic solar cells, Adv. Funct. Mater. 19(12) (2009) 1939–1948.
[13] S.K. Pal, et al., Geminate charge recombination in polymer/fullerene bulk heterojunction films and implications for solar cell function, J. Am. Chem. Soc. 132(35) (2010) 12440–12451.
[14] J. Benson-Smith, et al., Formation of a ground-state charge-transfer complex in polyfluorene/[6,6]-phenyl-C61 butyric acid methyl ester (PCBM) blend films and its role in the function of polymer/PCBM solar cells, Adv. Funct. Mater. 17(3) (2007) 451–457.
[15] L. Goris, et al., Observation of the subgap optical absorption in polymer-fullerene blend solar cells, Appl. Phys. Lett. 88(5) (2006) 052113.
[16] K. Vandewal, et al., The relation between open-circuit voltage and the onset of photocurrent generation by charge-transfer absorption in polymer: Fullerene bulk heterojunction solar cells, Adv. Funct. Mater. 18(14) (2008) 2064–2070.
[17] C. Yin, et al., Relation between exciplex formation and photvoltaic properties of PPV polymer-based blends, Sol. Energy Mater. Sol. Cells 91 (2007) 411–415.
[18] P.E. Keivanidis, et al., Delayed luminescence spectroscopy of organic photovoltaic binary blend films: probing the emissive non-geminate charge recombination, Adv. Mater. 22(45) (2010) 5183–5187.
[19] D. Veldman, et al., Compositional and electric field dependence of the dissociation of charge transfer excitons in alternating polyfluorene copolymer/fullerene blends, J. Am. Chem. Soc. 130 (2008) 7721–7735.

[20] S. Westenhoff, et al., Charge recombination in organic photovoltaic devices with high open-circuit voltages, J. Am. Chem. Soc. 130(41) (2008) 13653–13658.
[21] R.D. Pensack, J.B. Asbury, Barrierless free carrier formation in an organic photovoltaic material measured with ultrafast vibrational spectroscopy, J. Am. Chem. Soc. 131(44) (2009) 15986–15987.
[22] S.R. Forrest, Ultrathin organic films grown by organic molecular beam deposition and related techniques, Chem. Rev. 97(6) (1997) 1793–1896.
[23] J. Gilot, M.M. Wienk, R.A.J. Janssen, Double and triple junction polymer solar cells processed from solution, Appl. Phys. Lett. 90(14) (2007) 143512–143513.
[24] D. O'Brien, et al., Use of poly(phenyl quinoxaline) as an electron transport material in polymer light-emitting diodes, Appl. Phys. Lett. 69(7) (1996) 881–883.
[25] W. Wiedemann, et al., Nanostructured interfaces in polymer solar cells, Appl. Phys. Lett. 96(26) (2010) 263109–263113.
[26] J.B. Kim, et al., Reversible soft-contact lamination and delamination for non-invasive fabrication and characterization of bulk-heterojunction and bilayer organic solar cells, Chem. Mater. 22(17) (2010) 4931–4938.
[27] M. Granstrom, et al., Laminated fabrication of polymeric photovoltaic diodes, Nature 395(6699) (1998) 257–260.
[28] T.A.M. Ferenczi, et al., Planar heterojunction organic photovoltaic diodes via a novel stamp transfer process, J. Phys.: Condens. Matter 20(47) (2008) 475203.
[29] S. Yoo, B. Domercq, B. Kippelen, Efficient thin-film organic solar cells based on pentacene/C[sub 60] heterojunctions, Appl. Phys. Lett. 85(22) (2004) 5427–5429.
[30] H.H.P. Gommans, et al., Electro-optical study of subphthalocyanine in a bilayer organic solar cell, Adv. Funct. Mater. 17(15) (2007) 2653–2658.
[31] J. Dai, et al., Organic photovoltaic cells with near infrared absorption spectrum, Appl. Phys. Lett. 91(25) (2007) 253503–253513.
[32] B.P. Rand, et al., Organic solar cells with sensitivity extending into the near infrared, Appl. Phys. Lett. 87(23) (2005) 233508–233513.
[33] S.R. Scully, M.D. McGehee, Effects of optical interference and energy transfer on exciton diffusion length measurements in organic semiconductors, J. Appl. Phys. 100 (2006) 3.
[34] S. Athanasopoulos, et al., Trap limited exciton transport in conjugated polymers, J. Phys. Chem. C 112(30) (2008) 11532–11538.
[35] J. Xue, et al., A hybrid planar-mixed molecular heterojunction solar cell, Adv. Mater. 17(1) (2005) 66–71.
[36] T.M. Clarke, et al., Analysis of charge photogeneration as a key determinant of photocurrent density in polymer: fullerene solar cells, Adv. Mater. 22(46) (2010) 5287–5291.
[37] C.G. Shuttle, et al., Charge-density-based analysis of the current–voltage response of polythiophene/fullerene photovoltaic devices, Proc. Natl. Acad. Sci. U.S.A. 107 (38) (2010) 16448–16452.
[38] C. Winder, N.S. Sariciftci, Low bandgap polymers for photon harvesting in bulk heterojunction solar cells, J. Mater. Chem. (2004) 1077–108614 (2004) 1077–1086.
[39] J.-J. Cid, et al., Molecular cosensitization for efficient panchromatic dye-sensitized solar cells, Angew. Chem. Int. Ed. 46(44) (2007) 8358–8362.
[40] J.A. Chang, et al., High-performance nanostructured inorganic–organic heterojunction solar cells, Nano Lett. 10(7) (2010) 2609–2612.
[41] M. Niggemann, et al., Diffraction gratings and buried nano-electrodes–architectures for organic solar cells, Thin Solid Films 451–452 (2004) 619–623.
[42] S.D. Zilio, et al., Fabrication of a light trapping system for organic solar cells, Microelectronic Eng. 86(4–6) 1150–1154.

[43] K. Tvingstedt, et al., Trapping light with micro lenses in thin film organic photovoltaic cells, Opt. Express 16(26) (2008) 21608–21615.
[44] T.H. Reilly, et al., Surface-plasmon enhanced transparent electrodes in organic photovoltaics, Appl. Phys. Lett. 92(24) (2008) 243304–243313.
[45] M.C. Scharber, et al., Design rules for donors in bulk-heterojunction solar cells—towards 10% energy-conversion efficiency, Adv. Mater. 18(6) (2006) 789–794.
[46] M.A. Faist, et al., Effect of multiple adduct fullerenes on charge generation and transport in photovoltaic blends with poly(3-hexylthiophene-2,5-diyl), J. Polym. Sci. Part B: Pol. Phys. 49(1) (2010) 45–51.
[47] M. Lenes, et al., Fullerene bisadducts for enhanced open-circuit voltages and efficiencies in polymer solar cells, Adv. Mater. 20(11) (2008) 2116–2119.
[48] M. Lenes, et al., Electron trapping in higher adduct fullerene-based solar cells, Adv. Funct. Mater. 19(18) (2009) 3002–3007.
[49] J.J. Benson-Smith, et al., Charge separation and fullerene triplet formation in blend films of polyfluorene polymers with [6,6]-phenyl C61 butyric acid methyl ester, Dalton Trans. 2009 (2009) 10000–10005.
[50] S. Cook, et al., Singlet exciton transfer and fullerene triplet formation in polymer-fullerene blend films, Appl. Phys. Lett. (2006) 10112889 (2006) 101128.
[51] C. Dyer-Smith, et al., Triplet formation in fullerene multi-adduct blends for organic solar cells and its influence on device performance, Adv. Funct. Mater. 20(16) (2010) 2701–2708.
[52] C.J. Brabec, et al., Effect of LiF/metal electrodes on the performance of plastic solar cells, Appl. Phys. Lett. 80(7) (2002) 1288–1290.
[53] J.Y. Kim, et al., Efficient tandem polymer solar cells fabricated by all-solution processing, Science 317(5835) (2007) 222–225.
[54] O. Hagemann, et al., All solution processed tandem polymer solar cells based on thermocleavable materials, Sol. Energy Mater. Sol. Cells 92(11) (2008) 1327–1335.
[55] A.G.F. Janssen, et al., Highly efficient organic tandem solar cells using an improved connecting architecture, Appl. Phys. Lett. 91(7) (2007) 073519–073623.
[56] A. Hadipour, et al., Solution-processed organic tandem solar cells, Adv. Funct. Mater. 16(14) (2006) 1897–1903.
[57] T. Ameri, et al., Organic tandem solar cells: a review, Energy Environ. Sci. 2(4) (2009) 347–363.
[58] J.K.J. van Duren, et al., Relating the morphology of poly(p-phenylene vinylene)/methanofullerene blends to solar-cell performance, Adv. Funct. Mater. 14(5) (2004) 425–434.
[59] V. Mihailetchi, et al., Compositional dependence of the performance of poly(p-phenylene vinylene):methanofullerene bulk-heterojunction solar cells, Adv. Funct. Mater. 15(5) (2005) 795–801.
[60] S. Tuladhar, et al., Ambipolar charge transport in films of methanofullerene and poly(Phenylenevinylene)/methanofullerene blends, Adv. Funct. Mater. (2005) 1171–118215 (2005) 1171–1182.
[61] E. Moons, Conjugated polymer blends: Linking film morphology to performance of light emitting diodes and photdiodes, J. Phys.: Condens. Matter 14 (2002) 12235–12260.
[62] D.M. DeLongchamp, et al., Variations in semiconducting polymer microstructure and hole mobility with spin-coating speed, Chem. Mater. (2005) 5610–561217 (2005) 5610–5612.
[63] D.D. Nuzzo, et al., Improved film morphology reduces charge carrier recombination into the triplet excited state in a small bandgap polymer-fullerene photovoltaic cell, Adv. Mater. 22(38) (2010) 4321–4324.

[64] J.S. Moon, et al., Effect of processing additive on the nanomorphology of a bulk heterojunction material, Nano Lett. 10(10) (2010) 4005–4008.
[65] J.K. Lee, et al., Processing additives for improved efficiency from bulk heterojunction solar cells, J. Am. Chem. Soc. 130(11) (2008) 3619–3623.
[66] C. Müller, et al., Binary organic photovoltaic blends: a simple rationale for optimum compositions, Adv. Mater. 20 (2008) 3510–3515.
[67] P.E. Keivanidis, et al., Dependence of charge separation efficiency on film microstructure in poly(3-hexylthiophene-2,5-diyl): 6,6-phenyl-C-61 butyric acid methyl ester blend films. J. Phys. Chem. Lett. 1(4) 734–738.
[68] T.M. Clarke, et al., Free energy control of charge photogeneration in polythiophene/fullerene solar cells: the influence of thermal annealing on p3ht/pcbm blends, Adv. Funct. Mater. 18(24) (2008) 4029–4035.
[69] L. Nguyen, et al., Effects of annealing on the nanomorphology and performance of poly(alkylthiophene):fullerene bulk-heterojunction solar cells, Adv. Funct. Mater. 17(7) (2007) 1071–1078.
[70] M. Reyes-Reyes, K. Kim, D.L. Carroll, High-efficiency photovoltaic devices based on annealed poly(3-hexylthiophene) and 1-(3-methoxycarbonyl)-propyl-1-phenyl-(6,6)C_{61} blends, Appl. Phys. Lett. 87(8) (2005) 083506.
[71] P.E. Keivanidis, et al., Dependence of charge separation efficiency on film microstructure in poly(3-hexylthiophene-2,5-diyl):[6,6]-phenyl-C61 butyric acid methyl ester blend films, The J. Phys. Chem. Lett. 1(4) (2010) 734–738.
[72] Y. Kim, et al., Device annealing effect in organic solar cells with blends of regioregular poly(3-hexylthiophene) and soluble fullerene, Appl. Phys. Lett. 86 (2005) 063502.
[73] M. Campoy-Quiles, et al., Morphology evolution via self-organization and lateral and vertical diffusion in polymer:fullerene solar cell blends, Nat. Mater. (2008) 158–1647 (2008) 158–164.
[74] N.E. Coates, et al., 1,8-octanedithiol as a processing additive for bulk heterojunction materials: Enhanced photoconductive response, Appl. Phys. Lett. 93(7) (2008) 072105–072113.
[75] I.-W. Hwang, et al., Carrier generation and transport in bulk heterojunction films processed with 1,8-octanedithiol as a processing additive, J. Appl. Phys. 104(3) (2008) 033706–033709.
[76] H. Fan, et al., Efficiency enhancement in small molecule bulk heterojunction organic solar cells via additive, Appl. Phys. Lett. 97(13) (2010) 133302–133303.
[77] M. Morana, et al., Bipolar charge transport in PCPDTBT-PCBM Bulk-heterojunctions for photovoltaic applications, Adv. Funct. Mater. 18(12) (2008) 1757–1766.
[78] P. Ravijaran, et al., Hybrid polymer/zinc oxide photovoltaic devices with vertically oriented ZnO nanorods and an amphiphilic molecular interface layer, J. Phys. Chem. B 110 (2006) 7635–7639.
[79] T. Rattanavoravipa, T. Sagawa, S. Yoshikawa, Photovoltaic performance of hybrid solar cell with TiO_2 nanotubes arrays fabricated through liquid deposition using ZnO template, Sol. Energy Mater. Sol. Cells 92(11) (2008) 1445–1449.
[80] J. Bouclé, et al., Hybrid solar cells from a blend of poly(3-hexylthiophene) and ligand-capped TiO_2 nanorods, Adv. Funct. Mater. 18(4) (2008) 622–633.
[81] A.K. Khandpur, et al., Polyisoprene-polystyrene diblock copolymer phase diagram near the order-disorder transition, Macromolecules 28(26) (1995) 8796–8806.
[82] K.M. Coakley, M.D. McGehee, Photovoltaic cells made from conjugated polymers infiltrated into mesoporous titania, Appl. Phys. Lett. 83(16) (2003) 3380–3382.
[83] E.J.W. Crossland, et al., Block copolymer morphologies in dye-sensitized solar cells: probing the photovoltaic structure–function relation, Nano Lett. 9(8) (2008) 2813–2819.

[84] L. Bu, et al., Monodisperse co-oligomer approach toward nanostructured films with alternating donor–acceptor lamellae, J. Am. Chem. Soc. 131(37) (2009) 13242–13243.
[85] Q. Zhang, et al., Donor–acceptor poly(thiophene-block-perylene diimide) copolymers: synthesis and solar cell fabrication, Macromolecules 42(4) (2009) 1079–1082.
[86] K. Sivula, et al., Amphiphilic diblock copolymer compatibilizers and their effect on the morphology and performance of polythiophene:fullerene solar cells, Adv. Mater. 18(2) (2006) 206–210.
[87] J. van de Lagemaat, et al., Organic solar cells with carbon nanotubes replacing In[sub 2]O[sub 3]:Sn as the transparent electrode, Appl. Phys. Lett. 88(23) (2006) 233503–233513.
[88] B. Zimmermann, et al., ITO-free wrap through organic solar cells–A module concept for cost-efficient reel-to-reel production, Sol. Energy Mater. Sol. Cells 91(5) (2007) 374–378.
[89] J.A. Hauch, et al., Flexible organic P3HT:PCBM bulk-heterojunction modules with more than 1 year outdoor lifetime, Sol. Energy Mater. Sol. Cells 92(7) (2008) 727–731.
[90] M. Jørgensen, K. Norrman, F.C. Krebs, Stability/degradation of polymer solar cells, Sol. Energy Mater. Sol. Cells 92(7) (2008) 686–714.
[91] S. Hee Kim, et al., Long-lived bulk heterojunction solar cells fabricated with photo-oxidation resistant polymer, Sol. Energy Mater. Sol. Cells 95(1) (2011) 361–364.
[92] S. Miyanishi, K. Tajima, K. Hashimoto, Morphological stabilization of polymer photovoltaic cells by using cross-linkable poly(3-(5-hexenyl)thiophene), Macromolecules 42(5) (2009) 1610–1618.
[93] F.C. Krebs, All solution roll-to-roll processed polymer solar cells free from indium-tin-oxide and vacuum coating steps, Org. Electron. 10(5) (2009) 761–768.
[94] S.K. Hau, H.-L. Yip, A.K.-Y. Jen, A review on the development of the inverted polymer solar cell architecture, Polym. Rev. 50(4) (2010) 474–510.
[95] V. Shrotriya, et al., Transition metal oxides as the buffer layer for polymer photovoltaic cells, Appl. Phys. Lett. 88(7) (2006) 073508–073513.
[96] M.D. Irwin, et al., p-Type semiconducting nickel oxide as an efficiency-enhancing anode interfacial layer in polymer bulk-heterojunction solar cells, Proc. Natl. Acad. Sci. U.S.A. 105(8) (2008) 2783–2787.
[97] S.-S. Li, et al., Solution-processable graphene oxide as an efficient hole transport layer in polymer solar cells, ACS Nano 4(6) (2010) 3169–3174.
[98] A.B. Chwang, et al., Thin film encapsulated flexible organic electroluminescent displays, Appl. Phys. Lett. 83(3) (2003) 413–415.
[99] G. Dennler, et al., A new encapsulation solution for flexible organic solar cells, Thin Solid Films 511–512 (2006) 349–353.
[100] C. Lungenschmied, et al., Flexible, long-lived, large-area, organic solar cells, Sol. Energy Mater. Sol. Cells 91(5) (2007) 379–384.
[101] S.E. Shaheen, et al., Fabrication of bulk heterojunction plastic solar cells by screen printing, Appl. Phys. Lett. 79(18) (2001) 2996–2998.
[102] C.N. Hoth, et al., Printing highly efficient organic solar cells, Nano Lett. 8(9) (2008) 2806–2813.
[103] J. Alstrup, et al., Ultra fast and parsimonious materials screening for polymer solar cells using differentially pumped slot-die coating, ACS Appl. Mater. Interfaces 2 (10) (2010) 2819–2827.
[104] Y. Galagan, et al., Technology development for roll-to-roll production of organic photovoltaics. Chemical Engineering and Processing: Process Intensification. In Press, Corrected Proof.

[105] M.M. Voigt, et al., Gravure printing for three subsequent solar cell layers of inverted structures on flexible substrates, Sol. Energy Mater. Sol. Cells 95(2) (2011) 731–734.

[106] P. Kopola, et al., High efficient plastic solar cells fabricated with a high-throughput gravure printing method, Sol. Energy Mater. Sol. Cells 94(10) (2010) 1673–1680.

[107] L.-M. Chen, et al., Multi-source/component spray coating for polymer solar cells, ACS Nano 4(8) (2010) 4744–4752.

[108] C. Girotto, et al., Exploring spray coating as a deposition technique for the fabrication of solution-processed solar cells, Sol. Energy Mater. Sol. Cells 93(4) (2009) 454–458.

[109] D. Vak, et al., Fabrication of organic bulk heterojunction solar cells by a spray deposition method for low-cost power generation, Appl. Phys. Lett. 91(8) (2007) 081102–081103.

PART IIA

Photovoltaic Systems

CHAPTER IIA-1

The Role of Solar-Radiation Climatology in the Design of Photovoltaic Systems

John Page
Emeritus Professor of Building Science, University of Sheffield, UK, Fellow of the Tyndall Centre for Climate Change at Tyndall North, UMIST, Manchester, UK

Contents

1.	Introduction	575
2.	Key Features of the Radiation Climatology in Various Parts of the World	577
	2.1 Some Definitions and Associated Units	577
	2.2 The Variability of Solar Radiation and the Implications for Design	578
	2.3 Cloudless Sky Global Solar Radiation at Different Latitudes	579
	2.3.1 Partially Clouded Conditions	*584*
	2.4 Overcast Sky Conditions	586
	2.5 Sequences of Global-Radiation Data	586
	2.6 Simplified Radiation Climatologies	587
	2.6.1 Typical Components of Simplified Radiation Climatologies	*587*
	2.6.2 Examples of Simplified Radiation Climatologies for Europe	*588*
	2.6.3 Basic Solar-Radiation Climatology of the Humid Tropics	*590*
	2.6.4 Solar Radiation in Desert-Type Climates	*597*
	2.6.5 Special Issues in Mountainous Areas	*598*
	2.7 Radiation Climatology of Inclined Planes for Photovoltaics Applications	601
	2.8 Interannual Variability	602
	2.8.1 Estimating Daily Radiation from Sunshine Data	*603*
	2.9 Conclusions Concerning Basic Radiation Climatology	604
3.	Quantitative Processing of Solar Radiation for Photovoltaic Design	605
	3.1 Introduction	605
	3.2 Assessing Solar Radiation on Inclined Planes: Terminology	606
	3.3 An Example: Analysis of the Components of Slope Irradiation for London	606
	3.4 Design Conclusions	609
4.	The Stochastic Generation of Solar-Radiation Data	610
	4.1 The Basic Approach and the Implicit Risks	610
	4.1.1 General Principles	*610*
	4.2 Estimating Time Series of Daily Global Radiation from Monthly Means using KT-Based Methods	612
	4.3 Improvements in the Stochastic Estimation of Daily Solar Radiation in the SoDa Project	613

Practical Handbook of Photovoltaics.
© 2012 Elsevier Ltd. All rights reserved.

		4.3.1 Validation of Daily Irradiation Generation Models	613
	4.4	Generating Hourly Values of Global Radiation from Daily Values	614
		4.4.1 Generating Mean Daily Profiles of Irradiance from Specific Values of Daily Radiation	614
		4.4.2 Profiling of the Global Irradiance	615
		4.4.3 Stochastic Generation of Hourly Values of Global Irradiation	615
	4.5	Splitting the Global Radiation to Diffuse and Beam	616
	4.6	Assessment of Progress	616
5.	Computing the Solar Geometry		618
	5.1	Angular Movements of the Sun Across the Seasons	618
	5.2	Time Systems Used in Conjunction with Solar Geometry and Solar-Radiation Predictions	618
	5.3	Conversion of Local Mean Time (LMT) to Local Apparent Time (Solar Time)	620
		5.3.1 Example	621
	5.4	Trigonometric Determination of the Solar Geometry	621
		5.4.1 Key Angles Describing the Solar Geometry	621
		5.4.2 Climatological Algorithms for Estimating Declination	623
		5.4.3 The Calculation of Solar Altitude, Azimuth Angles, and Astronomical Day Length	624
	5.5	The Calculation of the Angle of Incidence and Vertical and Horizontal Shadow Angles	625
	5.6	Establishing the Accurate Noon Declination and the Accurate Solar Geometry	627
6.	The Estimation of Hourly Global and Diffuse Horizontal Irradiation		627
	6.1	The Estimation of the Extraterrestrial Irradiance on Horizontal Planes	627
	6.2	The Daily Clearness Index (KT_d Value)	628
	6.3	The Estimation of Mean Daily Profiles of Global Solar Irradiation from Monthly Means	629
		6.3.1 Method 1: Estimating the Monthly Mean Daily Irradiance Profile of $G_h(t)$ from the Monthly Mean Extraterrestrial Irradiance Profile	629
		6.3.2 Method 2: Estimating the Monthly Mean Daily Profile of $G_h(t)$ from the Clear-Sky Irradiance Profile	630
	6.4	The Estimation of Hourly Diffuse Radiation on Horizontal Surfaces	630
		6.4.1 Introduction	630
		6.4.2 Estimating the Monthly Average Daily Diffuse Horizontal Irradiation from the Monthly Average $(KT_d)_m$ Value	630
		6.4.3 Estimating the Daily Diffuse Horizontal Irradiation from the Daily Global Irradiation	631
		6.4.4 Estimating the Monthly Mean Daily Profile of $D_h(t)$	632
7.	The Estimation of the All Sky Irradiation on Inclined Planes from Hourly Time Series of Horizontal Irradiation		632
	7.1	Estimating the Components of Slope Irradiation from First Principles	632
	7.2	Direct Beam Irradiation on Inclined Planes	633

7.3	The Estimation of the Hourly Diffuse Irradiation on Inclined Surfaces from the Hourly Horizontal Diffuse Irradiation	634
7.4	The Estimation Process for Sunlit Sun-Facing Surfaces for Solar Altitudes Below 5.7°	636
7.5	Ground-Reflected Irradiation	636
8.	Conclusion	637
	Acknowledgements	638
	References	638
	Appendix: Solar Energy Data for Selected Sites	642

1. INTRODUCTION

Climate and solar radiation impact on issues for both the system supply side and the system demand side. Designers need both solar data and temperature data. Temperature affects the performance of photovoltaic devices per se. It also has a strong bearing on the demands of the energy required for heating and cooling. Relating supply and demand within any renewable energy structure requires study of the interrelationships between supply and demand. One has to establish the resources that need to be devoted to energy storage to achieve an acceptably reliable energy supply from an intermittent supply resource. Ideally, one needs long-term time series of solar-radiation data and temperature data for each specific site at the hourly level. Such data are relatively rare, so in recent years statistical approaches have been developed to help fill the gap. Long series of daily data are needed for sizing and modelling of stand-alone systems (Chapters 11a-3, 11a-4). Effective statistical approaches have to recognise the links between solar-radiation data and temperature data.

Section 2 aims to provide the reader with a brief general overview of the main features of solar-radiation climatology that are important for photovoltaics design in different parts of the world. Section 3 discusses in more detail some of the currently available quantitative techniques available for compiling solar-radiation design data for specific sites. The goal is to help optimise future photovoltaic design through the application of improved climate knowledge. Sections 4 through 7 explain in detail the quantitative procedures that are suitable for developing quantitative solar-radiation data in the forms needed for the study of the performance of photovoltaic systems.

Attention is drawn throughout to important sources of systematic climate data that are already readily accessible to designers. These modern approaches, aiming to provide the user with specialised meteorological data for specific sites, usually try to deliver information to users through user-friendly PC-based methodologies. The most successful now use advanced CD-ROM—based computational toolboxes as a matter of routine. For example, the digitally based fourth *European Solar Radiation Atlas* (ESRA) [1] provides an advanced stand-alone CD-ROM toolbox. The ESRA toolbox includes coupled applications modules that enable users to address in a user-friendly way the design of standalone and grid-connected PV systems for any site in the mapped area. The efficient climate data supply from the ESRA database is enhanced to meet user needs with a wide range of supporting algorithms. Global coverage based on CD-ROM approaches is available in the Meteonorm system marketed by Meteotest of Switzerland [2—5]. RetScreen® International [6] developed between Canada and NASA and enhanced with support from UNEP provides another example of a CD-ROM—based design tool. This contains photovoltaic applications tools. A UK CD-ROM—based achievement is the Chartered Institution of Building Services Engineers (CIBSE) Guide J, Weather, solar and illuminance data [7]. It is published on a CD-ROM in PDF format. Additional supporting quality-controlled hour-by-hour observed climate databases may be purchased for the UK. The PDF format provides opportunities for the intelligent user to extract published tabulated data into spreadsheets to enhance their practical utility in design, for example, for facilitating intersite interpolation and for creating site-specific design graphics.

More recently developed systems provide user-friendly live interactive IT methodologies that operate through the World Wide Web. After development, which is costly, such systems allow users to specify their own climate data requirements directly through the Internet and then receive back nearly immediate application-oriented answers to their questions for any part of the globe. Designated server sites interacting within organised computational networks offer users intelligently programmed access to the use of various distributed and interlinked data-bank systems to provide advanced solar-radiation data and other climate design data. The user-requested results are delivered back through the internet. The EU-funded SoDa project provides a good example of such Web-based approaches [8]. The EU-based Satel-Light program provides another Web-based example, covering the area 34°N to 68°N, 20°W to 50°E with data for 5 years (1996 to 2000) and specialising in illuminance data information and daylighting design [9,10].

Extensive use is now made of digital mapping techniques in such programmes. Such maps have the advantage that they can be accessed in very user-friendly ways using the computer mouse to extract climate data associated with any specific pixel from the database resource. The data associated with any pixel are arranged in arrays spatially structured to correspond with the mapped pixels (see, for example, [1,8]). Both systems enable users to work directly from the maps themselves. The mouse is used to gain computational access to detailed design databanks associated with any specific pixels held in the database. These input data are then used in structured algorithmic systems to supply outputs that match declared user needs like time series of hourly radiation on inclined surfaces of any selected orientation.

This chapter does not attempt to reproduce the advanced technological studies that underlie these new approaches. Readers should consult the previously listed publications to find out more about them. These new approaches include the use of satellite technology to estimate radiation data for regions of the world with no ground observational resources for measuring solar radiation [1,8]. These processed satellite data then become part of the organised database resource.

The world is a big place. No one can be familiar with more than a small part of it. There are great dangers in photovoltaics design if designers attempt to guess the properties of the radiation climate at unfamiliar places when assessing design risks. As globally based climatological tools are becoming more widely available, designers should make themselves aware of the powers of the new tools for design assessment in renewable energy design. Such tools should not be looked on as luxuries. They deliver quality-controlled programmed results very efficiently to users. This chapter draws heavily on the resources available in the previously mentioned publications.

2. KEY FEATURES OF THE RADIATION CLIMATOLOGY IN VARIOUS PARTS OF THE WORLD

2.1 Some Definitions and Associated Units

The solar short-wave radiation falling on a horizontal surface from Sun and sky combined is known as the *global short-wave radiation*. The global short-wave radiation flux—i.e., radiant energy flow per unit time—is known as the *irradiance*, symbol G. The most commonly used unit is watts per square metre (Wm^{-2}). The integral of irradiance flux over any period is called the *irradiation*. Typical integration periods are the hour, which yields the hourly global irradiation, G_h (units MJ m^{-2} h^{-1} or Wh m^{-2} h^{-1}); the day, G_d (units MJ m^{-2} d^{-1} or Wh m^{-2} d^{-1}); and the month, G_m (units MJ m^{-2} per month or Wh m^{-2} per month). The monthly daily mean irradiation is written as $(G_d)_m$ in this chapter. The subscripts indicate the integration period. The global radiation may be split into its two components: beam and diffuse. The method of indicating the integration period for these irradiation data corresponds to that used for global irradiation, yielding B_h, B_d, B_m, and $(B_d)_m$ as the symbols for the horizontal beam irradiation, and D_h, D_d, D_m, and $(D_d)_m$ as the symbols for the horizontal diffuse irradiation over the indicated periods—h hour, d day, and m month—and monthly mean. The symbols for irradiance, G, B, and D carry no time subscripts, so making a clear distinction between irradiance and irradiation.

Solar-radiation data are often presented in a dimensionless form called the *clearness index*. The clearness index, often called the *KT* value, is obtained by dividing the global irradiation at the surface on the horizontal plane by the corresponding extraterrestrial irradiation on a horizontal plane for the same time period. Three integration time periods are in common use. The hourly clearness index KT_h is the ratio G_h/G_{oh}. The daily clearness index KT_d is the ratio G_d/G_{od}. The monthly mean clearness index, KT_m, is $(G_d)_m/(G_{od})_m$; G_{oh}, G_{od}, and $(G_{od})_m$ are the corresponding extraterrestrial global irradiation quantities for the time integration periods as indicated by the subscripts.

The solar constant I_o is the extraterrestrial irradiance of the solar beam at mean solar distance. The accepted value is 1367 Wm^{-2}. The distance of Earth from the Sun varies slightly according time of year. The symbol used for the correction to mean solar distance is ε. It is dimensionless. The extraterrestrial irradiance normal to beam is therefore εI_o Wm^{-2}.

2.2 The Variability of Solar Radiation and the Implications for Design

Solar photovoltaic designers are essentially facing the design problem of achieving optimal performance from an input of a natural energy resource of great intrinsic variability. Three basic sky conditions can be identified: cloudless skies, partially clouded skies, and overcast skies. Cloud cover is the primary cause of the variability in solar-radiation energy supply from one minute to another and from one day to another. The typical patterns of cloud cover vary according to geographic locations. In desert climates, there may be no cloud cover day after day. In high-latitude maritime climates, overcast skies may persist day after day in winter with only rare breaks of sunshine. Statistics of cloud cover and daily sunshine, which are widely available, yield qualitative information that can help inform design. Such statistics are available in a systematic mapped form for Europe [9]. Reliable design, however, needs quantitative data about the actual solar-radiation energy fluxes. The starting point is to understand the basic radiation climatology for the three basic conditions: clear sky, partially cloudy skies, and overcast skies.

2.3 Cloudless Sky Global Solar Radiation at Different Latitudes

The global irradiation is made up of two components: the horizontal beam irradiation and the horizontal diffuse irradiation coming from the hemispherical sky dome. The cloudless sky global irradiation may be calculated using the improved ESRA clear-sky model [1,11]. The model requires as inputs the latitude of the site, the date in the year that determines the solar geometry, and the clarity of the atmosphere. The clarity of the sky is described by an index known as the *Linke turbidity factor*. Dust, human-made pollution, and water vapour in the cloudless atmosphere deplete the clear-sky beam irradiation and increase the clear-sky diffuse irradiation. Figure 1 plots the calculated clear-sky irradiance normal to the solar beam as a function of the sea level solar altitude for different values of the Linke turbidity factor. Figure 2 plots the corresponding values of the diffuse irradiance on horizontal surfaces. Representative spectra of the clear-sky global and diffuse irradiance are shown in Figure 3.

By combining the information in Figures 1 and 2 with the geometric information about the solar altitude, a clear-day irradiance plot can be produced. Figure 4 shows the calculated distribution of the hourly global, beam, and diffuse irradiation under clear skies at latitude 45°N on 30 April. The Linke turbidity factor for this example has been set at 3. This represents quite a clear sky. The direct beam dominates. Under cloudless sky conditions, the diffuse irradiation forms typically about 10–20% of the global radiation. If one performs the same calculation day by day throughout the year and then integrates the hourly values, one can estimate the annual pattern of the clear-day daily irradiation at this latitude at any reference level of atmospheric clarity. Figure 5 shows the

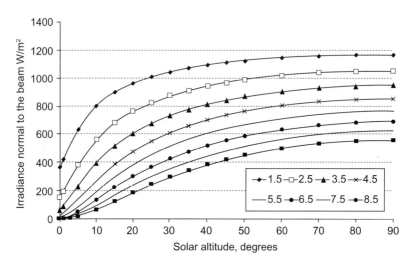

Figure 1 *European Solar Radiation Atlas* clear-sky beam-irradiance model [1,11] estimates of the irradiance normal to the beam in Wh m^{-2} at sea level at mean solar distance for a range of Linke turbidity factors of 1.5–8.5. In practice, the sea-level Linke turbidity factor is seldom below about 3.5. Values above 6 are common in desert areas due to dust in the atmosphere.

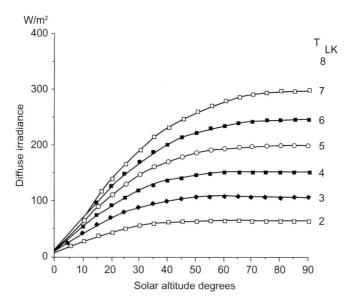

Figure 2 *European Solar Radiation Atlas* clear-sky irradiance model [1,11] estimates of the diffuse irradiance on a horizontal plane in Wh m^{-2} at sea level at mean solar distance for Linke turbidity factors between 2 and 8. Note the diffuse irradiance increases as the Linke turbidity factor increases and the beam irradiance decreases.

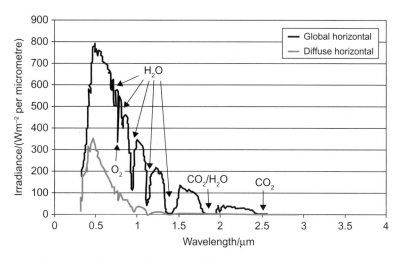

Figure 3 Calculated clear-sky global and diffuse spectral irradiances on a horizontal surface for a solar altitude of 30°. The integrated global irradiance was 476 Wm^{-2}, and the integrated diffuse irradiance was 114 Wm^{-2}. Note the dominance of diffuse radiation in the ultraviolet and the contrasting low clear-sky diffuse spectral irradiance in the infrared region. Water vapour in the atmosphere is an important absorbing agent in the infrared region. *Source: Reference [7].*

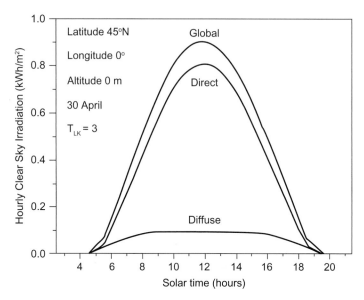

Figure 4 Estimated hourly horizontal irradiation under cloudless skies at sea level at latitude 45°N, showing the split into hourly direct and diffuse irradiation with a fixed Linke turbidity factor of 3. *Source: Reference [14].*

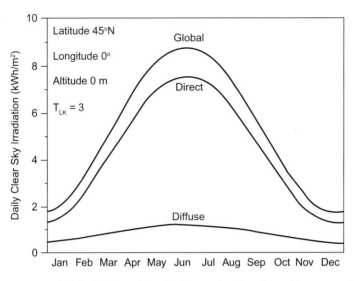

Figure 5 Estimated daily horizontal irradiation under cloudless skies at sea level in different seasons, latitude 45°N, showing the split into daily direct and diffuse irradiation with a fixed Linke turbidity factor of 3. *Source: Reference [14].*

result. The interseasonal variations due to the changing day length and the associated changes in solar altitude with time of year are very evident. Figure 6 shows the substantial impact of the Linke turbidity factor on the clear-sky daily global radiation at latitude 45°N. It is very rare to encounter a turbidity below about 2.5 at sea level. Large cities have typical turbidity factors around 4.5 in summer. Extremes of dust in the atmosphere may raise the Linke turbidity factor to 6.0 or more. The diffuse proportion of the daily irradiation, which is not plotted, increases as the sky clarity decreases. The hour-by-hour diffuse proportion also becomes greater as the solar-altitude angle gets lower. The available clear-sky global radiation becomes extremely low in December at high latitudes the closer one gets to the polar circle and the noon Sun is low in the sky. Finally Figure 7 plots the influence of latitude on the clear-day irradiation as a function of the day number. As the knowledge of the Linke turbidity factor is so important for making solar-energy estimates, considerable efforts have had to be made within the SoDa program [8,12,13] to develop a reliable global database of Linke turbidity factor values.

Note in Figure 7 the bunching of the curves in June around the summer solstice in the northern hemisphere. The equator values show

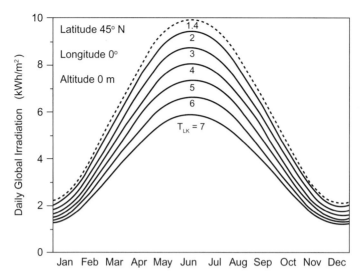

Figure 6 Estimated daily horizontal global irradiation under cloudless skies at sea level, latitude 45°N. The combined Imparts of season and of Linke turbidity factor. *Source: Reference [14].*

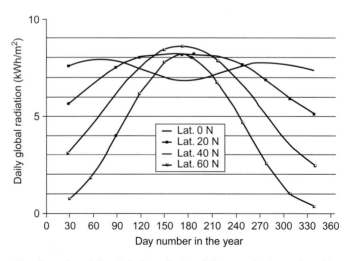

Figure 7 Cloudless day daily global irradiation falling on horizontal surfaces at four different latitudes calculated using a Linke turbidity factor of 3.5. *Derived from Reference [7].*

the maximum values there occur at the equinoxes, 21 March and 23 September. Figure 7 effectively defines the near-maximum global radiation likely to be observed on horizontal planes on any date. Considerable improvements in daily beam irradiation received per unit collector area can be achieved by tilting its surface to face the Sun. This helps offset some of the adverse impacts of latitude on solar-radiation availability in winter. The quantitative estimation of the cloudless sky irradiation falling on inclined planes is discussed later.

2.3.1 Partially Clouded Conditions

The cloud interrupts the direct beam, so the direct beam irradiance is very variable in partially clouded climates. Dramatic changes in beam irradiance can occur within a few seconds. Skies with a lot of scattered cloud tend to be relatively bright, so the associated diffuse radiation received during such periods can be quite large. However, the short-term variations in the diffuse irradiance are normally not as great as in the case of the beam. The beam irradiance may fall from $750\ \mathrm{Wm}^{-2}$ to zero within seconds and then, within less than a minute, increase to more than $750\ \mathrm{Wm}^{-2}$. Figure 8 shows some detailed observed radiation records for four European sites for both global and diffuse radiation on a continuous basis [from 15]. The four days were selected as being representative of different types of intermittent cloudiness. The top curve gives the global flux. The bottom curve gives the diffuse flux. When the curves coincide, the weather is overcast $(G = D)$. At Vaulx-en-Velin, near Lyon, on 29 September 1994, the morning was cloudy followed with an afternoon of broken cloud. At Nantes on 9 May 1994, very rapid changes were experienced in the afternoon. This is characteristic with broken cloud under windy conditions. At Nantes on 22 July 1994, the morning was overcast and the afternoon sunny. In Athens on 25 October 1994, there was some thin cloud throughout most of the day followed by a brief overcast period. The day and night periods are indicated using night as black. This presentation makes the changes in day length evident. The numbers at the top right-hand side give the temperatures. The daily global radiation is obtained by integrating the area under the curve. This yields a single variable G_d, thus suppressing all knowledge of the detailed variations. The first question is, 'Is the detail important for design?' The second question, if it is important, is, 'Can the detail be statistically regenerated for design studies?' The answer to the first question is yes, because one is dealing with nonlinear systems. The answer to the second question

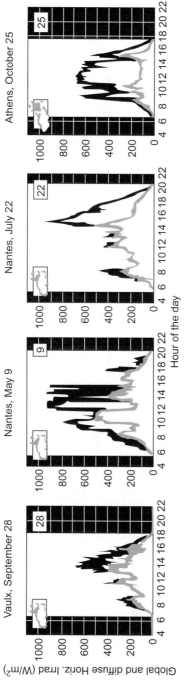

Figure 8 Variation of the global and diffuse horizontal irradiance with time of day for four European sites under conditions of intermediate cloudiness. The bottom line is the diffuse component; the top line is the global component. When the two lines merge into one line, it is overcast. Vaulx and Nantes reveal a cloudy morning followed by a sunnier afternoon. The irradiance at Nantes on 9 May 1994 is very variable. *Source: Reference [15]*.

is that stochastic data-generation methods are available, but they are complex. Stochastic methods are discussed in more detail in Section 4.

2.4 Overcast Sky Conditions

There is no direct beam during overcast periods. The radiation on overcast days is not steady. The height, type, and depth of clouds influence the atmospheric transmission. Changes in the cloud cover make the irradiation on overcast days variable, but the minute-to-minute variations in global irradiance tend to be far less than with partially cloudy skies.

Kasten and Czeplak [16] studied the impact of cloud cover type on the global irradiation at Hamburg. They found that the overcast-hour diffuse radiation from a given cloud type was roughly a linear function of the sine of the solar altitude. There were strong differences between the different cloud types. They expressed the overcast-hour horizontal radiation data with different cloud types as a fraction of the corresponding clear-sky global horizontal radiation data found at Hamburg. Their findings are given in Table 1. This table provides a way to link typical overcast-day values to the clear-day values discussed in Section 2.3. The impact of rain clouds is especially evident. Such days result in serious storage battery drawdown. The practical risk is that such days often occur in prolonged runs of adverse weather rather than as single days in isolation.

2.5 Sequences of Global-Radiation Data

The experience gained in the production of ESRA 2000 [1] has indicated the great importance of having daily time series of global radiation for making sensibly reliable design decisions. Otherwise, the day-to-day variability patterns in solar radiation due to weather change will be lost. The daily global radiation can be profiled to match the clear-day profile to estimate hourly irradiation values with an acceptable loss in accuracy. The monthly mean values cannot be used with accuracy to simulate daily profiles. Figure 9 shows

Table 1 Overcast cloud transmittance relative to clear-sky values

Cloud type	Relative transmittance
Cirrus, cirrostratus, and cirrocumulus	0.61
Altocumulus and altostratus	0.27
Cumulus	0.25
Stratus	0.18
Nimbostratus	0.16

Source: Reference [16].

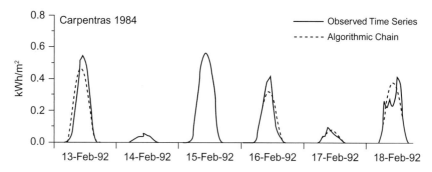

Figure 9 Using an average profile model to generate hourly data, starting with the respective lime series of daily global-irradiation values G_d. *Source: Reference [14]*.

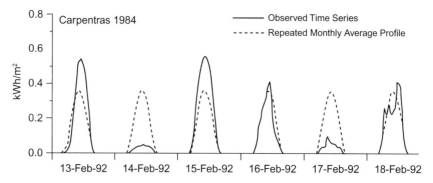

Figure 10 Using a monthly average hourly global-radiation profile model to generate hourly data starting with monthly mean daily global radiation $(G_d)_m$. *Source: Reference [14]*.

the advantages of using a profiled daily time-series model to simulate performance compared with Figure 10 showing the consequences of using a profiled monthly average model. The differences between the methodologies become even greater when one moves on to estimating slope irradiation.

Another issue is that there is an autocorrelation between the global radiation on successive days because of the common persistence of cyclones and anticyclones over several days. Strings of dull days may therefore occur, which may put a serious strain on battery storage resources. These statistical issues are discussed in Section 4.

2.6 Simplified Radiation Climatologies

2.6.1 Typical Components of Simplified Radiation Climatologies

It is useful to have some way of profiling in acceptably simple ways the solar-radiation climatology found in different parts of the world. As climate data are compiled on a month-by-month basis, it is sensible to adopt the calendar month as the basic timescale division in climatological summaries. The most important solar-radiation climatological value is the long-term monthly mean daily global solar radiation $(G_d)_m$. This should be an average value compiled over several years. For example, ERSA 2000 used the 10-year period 1981–90. It is very useful to have this daily value split into the monthly beam component $(B_d)_m$ and the monthly diffuse component $(D_d)_m$. As diffuse radiation is only measured at a few sites, it usually has to be estimated using the methods described later in Section 6. As knowledge of extremes is useful, it is also valuable to have representative monthly maximum values of daily global irradiation, $G_d max$. The third *European Solar Radiation Atlas* [23] used the mean of the 10 monthly maximum measured values in each 10-year data series for each site to obtain $G_d max$. SoDa 2003 generates the clear $G_d max$ values from terrain height and its global database of monthly mean Linke turbidity factors. However, there is a risk of overestimation in $G_d max$ in using the SoDa approach when applied to climates where cloudless days have a near-zero probability of occurrence—for example, in the rainy season months in equatorial Africa and in the monsoon months in countries like India. It is also useful to tabulate representative minimum global values $G_d min$. The *European Solar Radiation Atlas* [23] used the mean of the 10 monthly minimum measured daily values in each 10 year-data series. Sunshine observation data can be usefully added. As the day length varies so much with season, especially at high latitudes, it is useful to express the monthly mean daily bright sunshine in hours as a fraction of the astronomical day length. This ratio is called the *relative sunshine duration*. Experience with the various European Atlases has also indicated the value of including the monthly mean daily clearness index $(KT_d)_m$ in simplified radiation climatological summaries. Many of the algorithms used in data preparation, described in Section 3 onward of this chapter, use this dimensionless quantity as an input.

2.6.2 Examples of Simplified Radiation Climatologies for Europe

Figure 11 presents monthly mean climatological summaries for a selected range of European sites prepared by the author using reference [1]. Figure 11(a) immediately makes evident the large latitudinal gradient in monthly mean daily global radiation in winter. In midsummer, this strong latitudinal gradient is absent. This winter gradient reflects the extraterrestrial values shown in Figure 11(a) by the dashed line. A better understanding of exactly what is happening is often achieved by examining the

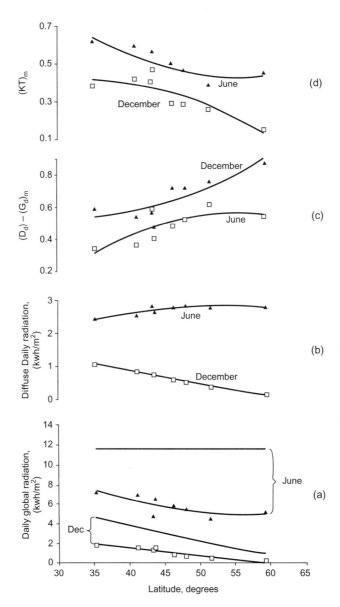

Figure 11 Monthly mean climatology of selected European sites 1981–90 in June and December. Prepared from reference [1]. (a) Monthly mean daily global-radiation (points and full lines) and the radiation outside the atmosphere (dashed lines); (b) monthly mean diffuse radiation on horizontal surfaces (all in kWh m^{-2}); (c) ratio of mean daily diffuse radiation on horizontal surfaces to the corresponding global values; (d) monthly mean KT values. The graph corresponds to solar-radiation data shown in the Appendix.

dimensionless $(KT_d)_m$ values. These values are shown in Figure 11(d). For example, the range of $(KT_d)_m$ values in December ranges from 0.150 at Stockholm (latitude 59.35°N) to 0.468 at Nice (latitude 43.65°N). The Sun is scraping the horizon in Sweden in December. The weather is often overcast. In contrast, Nice is favoured by both latitude and its relatively sunny winter climate due to the protection offered by the mountains immediately to the north. In summer, London (latitude 51.52°N) has a cloudier summer climate than Stockholm because of its maritime location and so receives less radiation than more northerly Stockholm, which has a more continental type of summer.

Figure 11(b) presents the corresponding information about monthly mean daily diffuse irradiation falling on horizontal surfaces. The diffuse radiation is relatively constant across Europe in the period May to August and can be less at southerly latitudes than at northerly latitudes. There are large variations however in the relative proportion of beam irradiation. Figure 11(c) sets down the month-by-month values of the ratio $(D_d)_m/(G_d)_m$. Of course, $(B_d)_m/(G_d)_m = 1 - (D_d)_m/(G_d)_m$. While Figures 11(b) and (c) reveal big variations in the diffuse radiation, they show that the diffuse radiation is an important resource even in relatively sunny climates. In London, more than 60% of the incoming resource is diffuse radiant energy, so proper attention must be attached to diffuse irradiation analysis. Only the beam radiation can be focused successfully.

Figure 12(a) to 12(f) plots month by month the monthly mean global and diffuse radiation for the period 1981–90 for six of the eight European sites included in Figure 11. Figure 12 contains one desert site, Sde Boker 12(h), and one humid tropical site, Ilorin in Nigeria. A systematic description of the climatology of solar radiation in Europe may be found in the ESRA User Guidebook of reference [1], which identifies three roughly defined climatic categories—maritime climates, continental climates, and Mediterranean climates—and adds a separate category of high mountainous climates. Photovoltaic designers are strongly encouraged to strengthen their understanding of the basic dynamics of climate in the regions in which they plan to work before embarking on detailed quantitative studies. Such knowledge strengthens understanding of the basic strategic issues that should underlie the detailed engineering design decisions.

2.6.3 Basic Solar-Radiation Climatology of the Humid Tropics

The humid tropics are located close to the equator. The Tropic of Cancer at latitude 22.5°N and the Tropic of Capricorn at latitude 22.5°S

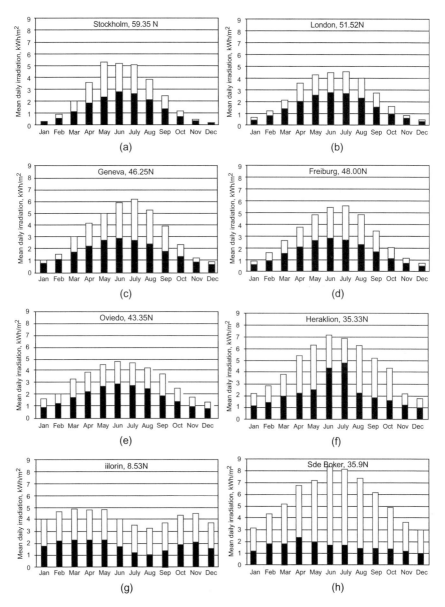

Figure 12 (a)–(f) Monthly mean global and diffuse radiation on horizontal surfaces for six European sites, for a humid tropical location (g), and for a desert region (h). The numerical data on which these graphs are based can be found in the Appendix. The shaded bands represent the diffuse irradiation magnitude, and the clear bands represent the beam-irradiation magnitude on horizontal surfaces.

give a broad definition to the zone. The equatorial radiation climate is quite different in nature to the climate produced by the varying patterns of cyclones and anticyclones that dominate the typical high-latitude solar-radiation climates. For a start, the month-to-month variations in day length are very small close to the equator. There are no long summer evenings. There are no long winter nights. So the hours during which indoor lighting is needed do not vary much across the year. This has a bearing on patterns of photovoltaic energy demand.

The noon solar elevation of the Sun at low latitudes remains fairly high throughout the year with the consequence that, in the absence of cloud cover, the global solar energy varies very little across the year. Figure 7 in Section 2.3 shows clearly the contrast between the annual pattern of clear-sky daily irradiation at latitude 0°N and the corresponding patterns found at higher latitudes. However, cloudless conditions are rare in equatorial areas.

The wet and dry seasons are usually well-defined periods in tropical regions. There is always a considerable amount of cloud during the wet seasons. There is also often quite a lot of cloud in the drier seasons, too. The daily pattern of irradiation is often dominated by convective clouds. The clouds build up as the surface temperatures increase. These clouds frequently obstruct the Sun, and some produce large amounts of rainfall. This rainfall is often associated with violent thunderstorms. The normal pattern of insolation is therefore an oscillating mix of periods of bright sunshine intermingled with periods of heavy obstruction of the solar beam. The diffuse radiation frequently dominates throughout the year.

The variations in the amount and type of cloud cover associated with the different seasons is the main factor influencing the monthly mean global solar radiation available at different times of the year. The patterns of radiation experienced reflect the impacts of the general circulation of Earth's atmosphere in tropical regions. Surface warming by the Sun induces a global atmospheric circulation pattern of ascending air above the equatorial zone. Air from both the northern and southern hemispheres flows horizontally in at the bottom at low levels to drive this vast vertical circulation system in the equatorial zone. This air ascends in massive convection cells that move upward, cooling and shedding moisture as they ascend. Vertical clouds that may grow to considerable heights in the atmosphere often form, producing heavy rain. The air that has ascended, losing much of its initial moisture on the way up through rainfall, then flows away from the equatorial region toward higher latitudes, moving horizontally at relatively high levels close to the stratosphere. This circulating air eventually descends heated adiabatically by recompression to reach the surface of the land warm and dry. Such descending dry air overlies the surface of the great deserts of the globe and desiccates them. These regions of descending dry air are essentially regions of high sunshine and small or virtually nonexistent rainfall. Earth's rotation causes a distortion of the convection cells, so, for example, the returning air from the Sahara moving into the west of Africa arrives predominately from a northeasterly direction.

The zone where the air masses from the south and north merge to form the equatorial convection cells is known as the *intertropical convergence zone* (ITCZ). The position of the intertropical convergence zone oscillates north and south as the solar declination changes with season. This movement produces the typical patterns of wet and dry seasons found in tropical climates. There are northern and southern limits to the movement of the intertropical convergence zone. If the latitude of any site is close to the latitude of that limit, a single short rainy season will be experienced. Closer to the equator, a pattern of two wet seasons is usually experienced as the rains come up from the equator and produce the first rainy season. The ITCZ then passes on to a region further from the equator and the rains yield, only to return later with a second season of rain as the intertropical convergence zone falls back later in the year toward the other hemisphere. The northern and southern hemisphere tropical rainy seasons are typically about 6 months out of phase from each other, reflecting the annual movements of the ITCZ back and forward across the equator.

Global and Diffuse Irradiance Estimates Under Partially Clouded Conditions

Maximum diffuse radiation values tend to occur during hours of broken thin cloud. The diffuse radiation from partially clouded skies is normally substantially higher than the diffuse radiation from clear skies and from typical overcast skies. *It is thus an invalid scientific process to attempt to calculate monthly mean diffuse radiation by linear interpolation between estimated overcast sky and estimated clear-sky diffuse radiation values.* The typical hourly sunshine amount producing the highest diffuse irradiation conditions is about 0.3 to 0.5 hours within the hour. Surprisingly high global-radiation levels can occur under thin high cloud conditions, but the radiation is very diffuse dominated.

An interesting study by Gu et al. [17] relates to the cloud modulation of surface solar irradiance at a pasture site in southern Brazil. They report broken cloud fields typical of humid tropical climates create mosaic radiative landscapes with interchanging cloud shaded and sunlit areas. While clouds attenuate solar-radiation incident on cloud shaded areas, sunlit ground surfaces may actually receive more irradiance than under a clear sky due to light scattering and reflection from neighbouring clouds. The authors analysed a high-resolution time series of radiation measurements made in the wet season of 1999 in southern Brazil. Surface solar irradiance frequently (more than 20% of the time) exceeded clear-sky levels and occasionally surpassed the extraterrestrial radiation. They found the clouds created a bimodal frequency distribution of surface irradiance, producing an average of approximately 50% attenuation for about 75% of the time and 14% enhancement for about 25% of the time, respectively, as compared to the corresponding clear-sky levels of irradiance. The study is helpful in drawing attention to the actual patterns of irradiance likely to be found on solar cell systems across the humid tropics.

Examples of the Radiative Climatology of a Typical Humid Region: Kumasi, Ghana

Figure 13 presents observed climatological data for Kumasi in Ghana in West Africa from [18]. Kumasi, located at latitude 6.72°N, longitude 1.6°W at a height of 287 m, is nearly at the centre of the rain-forest belt in Ghana. The values of the monthly mean daily clearness index $(KT_d)_m$ reveal the fairly small annual range characteristic of humid tropical climates. The lowest values occur in the rainy season. The double peak in the rainfall in the rainy season in June and September is evident (see Figure 12(g)). For most of the year, wet, moist air moves in from the South Atlantic. However, for a short period around December, the wind tends to blow off the desert regions to the north. These winds, known locally as *Harmattan winds*, are dry and very dusty, especially in December and January. In this season, there is very little rainfall for a rain-forest area. The Linke turbidity factor is very high in the Harmattan season. It may reach 6 or 7. There is a lot of cloud cover throughout the year. The monthly mean values of $(KT_d)_m$ range from 0.32 in August to 0.47 in February to May and in November. The typical values of $KT_d max$ are around 0.60, only rising in November to 0.70. In August, no cloudless

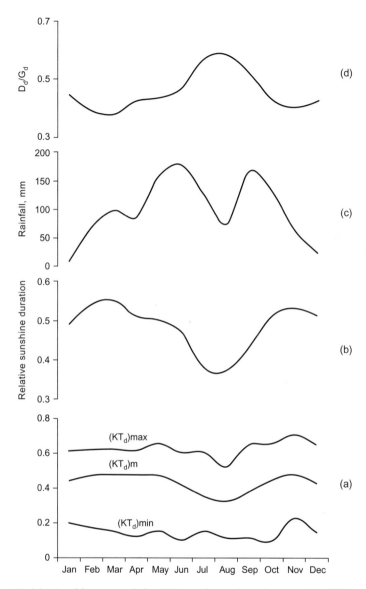

Figure 13 (a) Monthly mean daily $(KT_d)_m$ values plotted alongside (KT_d)max and (KT_d)min. The monthly mean daily relative sunshine duration values S_d/S_{od} (b), the rainfall (c), and the ratios $(D_d)_m/(G_d)_m$ estimated using Equation (1). The KT_d values are for Kumasi in Ghana, West Africa; the rainfall data, sunshine duration, and D_d/G_d are for Ilorin, Nigeria, with a similar climate. *Source: References [18] and [19].*

days occur and the KT_dmax value is as low as 0.52. The values of KT_dmin are lowest in the wet season, standing at around 0.10.

Ilorin, Nigeria

The next example is for Ilorin in Nigeria, located at 8.53°N, 4.57°E at a height of 375 m, and is based on data from [19], with a very similar radiation climate to that found at Kumasi, though being slightly further north. The Ilorin study included relative sunshine duration data. Figure 12(g) shows the rainy season causes a strong drop in the relative sunshine duration. The impact of the Harmattan is again evident. Figure 14 gives the monthly percentage cumulative frequency of the KT_d values for the months of January, June, and November. There are significant differences in the cumulative frequency curves for different months. In June, 50% of the days have a KT_d value of less than 0.36. In November, 50% of the days have a KT_d value below 0.50. The seasonal implications for battery storage are obvious. Udo [19] showed that the cumulative frequency curves for Ilorin and Kumasi have a much smaller spread of values than typical cumulative frequency curves for higher latitudes. This small spread reflects the low observed values of KT_dmax. Also, the observed values of KT_dmin in the humid tropics are high compared with those found in high-latitude climates.

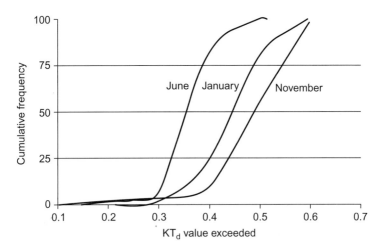

Figure 14 Monthly percentage cumulative frequency of the daily clearness index KT_d for Ilorin, Nigeria. *Source: Reference [19].*

Estimation of Diffuse Radiation in the Humid Tropics

The diffuse proportion of the monthly mean global irradiation can be estimated approximately using the following simple formula developed by Page [20] and widely verified since in many papers:

$$(D_d)_m/(G_d)_m = 1.00 - 1.13(KT_d)_m \tag{1}$$

Monthly mean $(D_d)_m/(G_d)_m$ estimates for Kumasi using Equation (1) are shown in Figure 13(d).

2.6.4 Solar Radiation in Desert-Type Climates

The main hot desert climates of the world lie in those latitudinal regions where the air from the equatorial zone transported aloft at the equator descends to the surface in a very dry condition. Such climates are characterised by long hours of sunshine, few clouds, and very little rain. As there is no water available in the centre of large deserts like the Sahara, human activity is usually confined to the fringes of such deserts. In some areas, water from nearby mountains can flow underground considerable distances and re-emerge at the surface to provide the water needed for life in areas without rain. The desert atmosphere, however, can contain a lot of sand raised by the wind. Consequently, the actual irradiance levels are not as high as those found in many cooler climates with clearer local atmospheres. The Linke turbidity factors are often above 6. High temperatures must not be assumed to provide high irradiation levels from the cloudless sky. Figure 6 in Section 2.3 shows the reductions expected at latitude 45°N as the Linke turbidity factor increases. The clear-day daily global radiation may be nearly halved if the Linke turbidity factor rises to 7. The colour of the desert sky is an important indicator of the amount of dust. The blue colour may become weak as dust increases. In heavy pollution conditions, the Sun's disc may become quite red when it is below 30° elevation and cut out altogether when the Sun is below 15°. The sky may take on a brown hue. Dust deposition on photovoltaic collectors is likely to impact adversely on performance. Arrangements for achieving effective periodic cleaning of solar panels are essential in desert regions.

Example: The Solar-Radiation Climatology of the Negev Region of Israel

This example has been selected to bring out the main features of the solar-radiation climatology of a relatively dry and sunny area. The data presented by Ianetz et al. for the Negev have been selected [21]. This

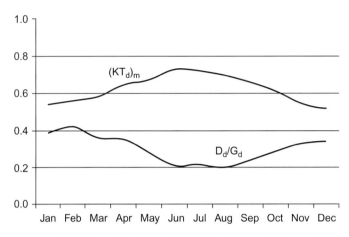

Figure 15 Ratios of $(KT_d)_m$ and D_d/G_d for Sde Boker. *Source: Reference [21].*

data set includes measured beam-irradiation data as well as measured global irradiation data. Figure 12(h) shows the monthly mean irradiation data for Sde Boker at 30.90°N, 34.8°E at 470 m. As the latitude in this case is around 30°N, one finds the expected substantial increase from winter to summer. Figure 15 presents the same data in dimensionless forms, as the monthly mean values of the daily clearness index $(KT_d)_m$ and as the monthly mean daily diffuse divided by the monthly mean global irradiation ratio $(D_d)_m/(G_d)_m$. The very high monthly mean values of $(KT_d)_m$ found in the June to August period may be noted. Every day is virtually cloudless. The $(KT_d)_m$ ratios are less in the winter period but are still relatively high compared with the humid tropics. The diffuse data are of special interest as they show there is still a lot of diffuse radiation even in very sunny climates.

2.6.5 Special Issues in Mountainous Areas

Stand-alone photovoltaics applications are becoming quite common at high-level mountain farms and isolated mountain recreational retreats in countries like Norway, Austria, and Switzerland. This study, which shows the influence of site height and cloud cover on horizontal surface irradiation in mountainous regions, is drawn from published Austrian data. Figure 16 demonstrates the influence of site elevation and cloud cover on the typical horizontal surface irradiation at different elevations in the Austrian Alps during different months of the year. Detailed examination

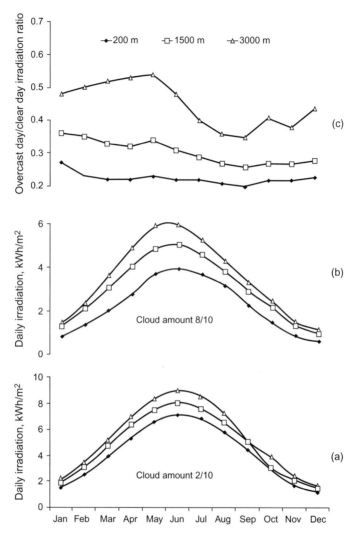

Figure 16 Influence of cloud and site elevation on global solar radiation. (a) The mean daily global irradiation for cloud amounts 2/10. (b) The mean daily global irradiation for cloud amounts 8/10. (c) Overcast-day to clear-day irradiation ratio at three elevations. A case study from the Austrian Alpine area. Units are kWh m^{-2} d^{-1}. *Source: Reference [22].*

of Figure 16 shows there are good climatic reasons favouring the use of photovoltaics in mountainous regions. However, designers habituated to sea level may only partially appreciate these issues, unless they make more focused studies of the radiation climates of such mountainous regions.

Figure 16 will now be discussed in detail. Figure 16(a) plots the daily irradiation for cloudless and near cloudless conditions. The increase with elevation is evident. The radiation obviously decreases with increase in cloud cover as Figure 16(b) shows. However, there is a substantial elevational effect. The ratio of the overcast-day values to the clear-day values is shown in Figure 16(c). The typical radiation on overcast days at the 200 III level is about 22% of the radiation on cloudless days. At low levels, this reduction is slightly less in the snowy month of January. (Snow in this region is more persistent at higher elevations.) However, in general, the mean overcast sky irradiation reduction at 200 m is relatively constant from month to month. There are, of course, substantial day-to-day variations in overcast sky radiation due to cloud type.

Let us now examine the effect of height on the irradiation on cloud free days. The ratios of the monthly values at 1500 m and 3000 m to the 200 m monthly values are presented to make the comparison simpler. The clear-sky daily radiation increases with site elevation. The increases are greatest in midwinter when the Sun is lowest and least in midsummer when the Sun is highest. The increases are quite substantial, especially at 3000 m—for example, a 33% increase in January and a 21% increase in June at 3000 m.

Examining next the impact of site elevation on the corresponding overcast sky data, it is clear that the global radiation on overcast days is substantially greater at higher elevations when compared with the low-level values. The differences are greatest in the mountain snowy period February to May. In March, the ratio on overcast days is 1.79 at 1500 m and 3.00 at 3000 m compared with 1.17 and 1.25 on clear days. Two effects are at work. First, as one gets higher, the water content of the clouds above the site gets less. Second, as one gets higher, snow is more likely to be lying, especially in winter. Snow increases the diffuse radiation through back-scattering. Some of the reflected energy returns from the atmosphere. The increases become less as the summer melt takes place, with the overcast-day ratios falling to 1.48 at 1500 m and 2.13 at 3000 m in August. The consequence is the overcast-day to clear-day irradiation ratios at 1500 m and 3000 m are substantially greater than those found at lower elevations.

Figure 16 cannot, of course, bring out the mix of days with different cloud amounts at different mountain sites. However, as is well known, many mountain climates have a high proportion of sunshine in winter months and then become cloudier in summer as the moisture content of the air rises.

The preceding factors favour stand-alone photovoltaics in mountainous regions, because the conditions of high sunshine coincide with the seasons of low sun, thus easing the issue of battery storage capacity requirements in winter. The atmosphere is not so opaque on cloudy days, reducing the energy-storage risks associated with runs of cloudy days.

2.7 Radiation Climatology of Inclined Planes for Photovoltaics Applications

While national meteorological services normally concentrate on the provision of horizontal surface data, the practical utilisation of solar energy normally demands detailed design knowledge of the irradiation on inclined planes. Choice of a favourable collector orientation helps reduce costs by increasing the radiant flux per unit collector area, so enabling a greater energy-collection efficacy to be achieved per unit of investment. When the exposure of any site used for harnessing solar energy photovoltaically allows, the most favourable orientations tend to be south facing in the northern hemisphere and north facing in the southern hemisphere. However, there are some climates where the afternoons are systematically cloudier than the forenoons. In these cases, the true north—south orientation rule may be inappropriate. The appropriate choice of collector tilt and collector orientation is always an important decision. The most appropriate choice is strongly influenced by latitude. A common recommendation is that the ideal exposure is an equator-facing surface with a tilt equal to the latitude. Accepting this recommendation, this section presents variations in the monthly mean irradiation on equator-facing surfaces on planes with a tilt equal to the latitudes using data from the third *European Solar Radiation Atlas* [23].

Figure 17 shows the monthly mean daily inclined surface irradiation data for a south-facing surface with a tilt equal to the latitude predicted for 18 selected European sites during 3 selected months plotted against latitude. It is clear from Figure 17 that latitude makes a very big impact on expected inclined surface radiation in December. There is very little solar energy available around midwinter for latitudes above 45°. Tilting the collector plane cannot compensate for the longer solar path length through the atmosphere and the greater cloudiness. In December, there is about 4 times as much solar radiation on a south-facing slope around

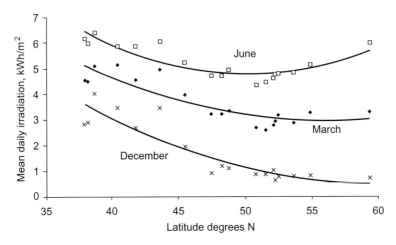

Figure 17 Monthly mean daily global irradiation on a south-facing plane with a slope equal to the latitude plotted against site latitude for 18 European cities of different latitudes in 3 selected months. Period 1966–75. Data source: Reference [23]. The ESRA 2000 toolbox [1] contains the tools to enable such calculations for slopes of any orientation and slope in the geographical range covered. This range is from 30°W to 70°E and 29°N to 75°N. The SoDa–IS [8] system provides similar computational systems for the whole world.

latitude 40°N in Europe than is available for sites above latitude 50°N. The latitudinal slope is less in March than it is in December. However, there is about three times as much solar-radiation energy available at high latitudes in March compared with December. At midsummer, the curve becomes U-shaped, with a minimum around latitude 50°N. The longer days close to the arctic circle and the more continental summer climate found in Sweden increase the plotted midsummer daily gains at these high latitudes. While there are significant variations between sites—say, Nice and Venice at nearly the same latitudes—the latitudinal effects broadly dominate solar-radiation availability in Europe.

2.8 Interannual Variability

Most of the short-wave radiation variability on cloudless days is the direct result of the apparent movements of the Sun as seen from Earth. However, some of the clear-sky variability is associated with the varying amounts of water vapour and aerosols in the atmosphere in different seasons in different parts of the world. Imposed on this basic variability is the variability that results from the day-to-day variations of weather as cyclones and anticyclones pass over the site. These variations are expressed in terms of variations in cloud type and cloud amount in different months. In high latitudes, such disturbances often last three or five days. The dynamics of climatology drives variations in both the total amount of solar energy and in the balance between beam and diffuse radiation received from the sky. There is also usually considerable interannual variability. This interannual variability in some cases is linked to fundamental oceanic oscillations in sea temperature—for example, the impacts of El Niño felt around the world. In Northern Europe, the North Atlantic pressure oscillation is important, determining the typical tracks of weather fronts in different years. There are also the occasional impacts of major volcanic eruptions. The consequence of such influences is that there is often a persistence of either radiatively favourable or radiatively unfavourable weather across succeeding months in particular years. These experiences leave behind human memories of excessively hot summers and miserably cold summers, as well as bright sunny winters and dark overcast winters. Risk analysis of stand-alone photovoltaic systems' performance is affected by the impacts of such interannual variations in climate. There is a danger in placing too much reliance on short-term records as the basis of risk assessment. When long-term radiation observations are missing, sunshine data can provide an alternative approach. The method below is based on the use of the Angström regression equation (see Equation (2) in the next section) in conjunction with monthly mean sunshine data selected on a year-by-year basis. It is based on the fourth *European Solar Radiation Atlas* [1].

2.8.1 Estimating Daily Radiation from Sunshine Data

The daily global irradiation can be estimated from sunshine data using the Angström formula. Monthly mean sunshine data are first converted into the monthly mean relative sunshine duration, σ_m, by dividing the observed monthly sunshine data, S_m, by the corresponding monthly mean maximum possible sunshine, S_{om}. S_{om} can be determined from the sunset hour angle, ω_s, using Equation (11) in Section 5.4.4, as $2\omega_s/15$ hours.

The linear relationship between the monthly mean daily clearness index defined as $(G_d)_m/(G_{od})_m$ and the monthly mean relative sunshine duration is known as the *Angström regression formula*. The following equation is used to estimate the monthly mean value of $(G_d)_m$ from σ_m:

$$(G_d)_m/(G_{od})_m = a_m + b_m \sigma_m \tag{2}$$

where $(G_{od})_m$ is the monthly mean daily extraterrestrial irradiation on a horizontal plane, and a_m and b_m are site-dependent monthly regression coefficients. G_{od} is calculated using Equation (19) in Section 6.1. The site-dependent Angström regression coefficients a_m and b_m used in [1] were extracted on a month-by-month basis. They were derived for the period 1981–90 from the quality-controlled daily series of global irradiation and sunshine duration data. Such data were available for more than 500 sites in the ESRA mapped region. The geographical range coverage is from 30°W to 70°E, 29°N to 75°N. The values are stored in two text files on the ESRA CD-ROM, one containing values of a_m and the other the values of b_m. By assuming that these values still hold for nearby sites with the same type of climate, one can make a good approximation of the global irradiation provided there are no major impacting geographical features in the region of interest. These include mountain ranges, large water bodies, etc., from which significant climatic gradients can originate.

While site-dependant linear regression coefficients are widely available from the literature for many parts of the world (publications of the national meteorological services, atlases, and solar-energy journals), the coefficients are not usually derived using the same methodologies, as adopted in the ESRA 2000 project. Caution has to be used in drawing international comparisons based on published Angström coefficients. If a site lies outside the ESRA area and a short-term series of daily global radiation is available, then the short-term series can be used can be used month by month to estimate site-applicable values of a_m and b_m, following the detailed procedures reported in [1]. These values can then be applied to the long-term records of observed monthly mean sunshine to explore year-to-year variability.

2.9 Conclusions Concerning Basic Radiation Climatology

It is always wise to aim to achieve a sound understanding of the basic climatic influences driving the local solar radiation at any point of the globe before proceeding to any detailed quantitative technical analysis of system designs. Assessing risk reliably is a difficult task in the design of renewable energy systems. Always try to obtain long-term records where possible.

3. QUANTITATIVE PROCESSING OF SOLAR RADIATION FOR PHOTOVOLTAIC DESIGN

3.1 Introduction

Photovoltaic system designers usually have to aim to extract the maximum economic return from any investment over the lifetime of the installation. One way of doing this is to try to increase the incoming energy density by tilting the solar cell panels toward the Sun. However, the optimum tilt for increasing the daily energy density varies with time of year. The value of the collected energy to the user also varies with time of year, particularly for stand-alone photovoltaic systems. For example, if the electrical energy is to be used for lighting, the nighttime period during which the electric lighting is needed will vary across the year. So choosing the optimal tilt and orientation of collectors requires careful thought. Design also has to deal with the assessment of the effects of site obstruction on collection performance. Partial shading of photovoltaic panels is also undesirable. So a proper understanding of the geometry of solar movements is needed. The irradiation on inclined surfaces can only be calculated if the global irradiation can first be split into its beam and diffuse radiation components at the hourly level. The designer is handling a three-dimensional energy-input system. The capacity to be able to interrelate the solar geometry and the energy fluxes is usually important.

National meteorological services aim to provide generalised data for the regions they cover. Their work has two main components:
- short-range and long-range weather forecasting, and
- providing climatological advice and data.

Climate data needs for solar-radiation data were historically met using ground-based observation records. More recently, ever-increasing use is being made of satellite observed data to generate ground-level information. Satellite data, for example, form a crucial component of the radiation database material available to users of the SoDa—IS 2003. The ESRA 2000 [1] solar-radiation maps are based on a combination of satellite-observed and ground-observed data.

Photovoltaics design depends on successfully harnessing the available climatological information to the detailed task of design. The gap between what data that national meteorological services can provide and what systems designers actually need is currently often quite wide. This part of the chapter considers some of the currently available methodologies for providing photovoltaic designers with quantitative information that they need to address various design tasks.

Performance has to be assessed in the context of risk. Supply and demand have to be matched through appropriate energy-storage strategies. This places especial value on the availability of time series of solar-radiation data preferably linked to temperature data. The diurnal temperature range is strongly associated with the daily horizontal radiation received. The daily mean temperature is usually dominated by the source of the air passing over the site. Polar air can bring low daily mean temperatures in association with large solar-radiation gains and consequently associated big diurnal swings of temperature.

3.2 Assessing Solar Radiation on Inclined Planes: Terminology

In the case of inclined planes, there are three components of the incident radiation: beam radiation, diffuse radiation from the sky, and radiation reflected from the ground. The ground-reflected component depends on the albedo of the ground surface. The albedo is the surface reflectivity averaged over the whole solar spectrum. Typical values of the albedo are 0.2. Snow cover can raise this to 0.6−0.8, according to the age of the snow. The terminology used here for inclined plane irradiance is $B(\beta,\alpha)$ for the direct-beam irradiance falling on a surface with tilt β and azimuth angle α. $D(\beta,\alpha)$ is used for the diffuse radiation received from the sky and $R(\beta,\alpha)$ for the diffuse radiation reflected from the ground alone. Irradiation quantities for different integration periods are indicated by adding the following subscripts: h for hour, d for day, and m for month. This practice conforms with the terminology of Section 2. The supplementary subscript cs refers to cloudless sky data.

3.3 An Example: Analysis of the Components of Slope Irradiation for London

A specific graphical study is next presented in order to clarify the nature of the decision-making issues concerned in deciding collector orientation and tilt. This example is based on the use of numerical monthly mean irradiation data and clear-day irradiation values data for London (Bracknell). These data have been extracted from reference [7].

Figure 18(a) shows the estimated monthly means of the daily irradiation on south-facing inclined planes of different inclinations at Bracknell. These data were processed from observed hour-by-hour global and diffuse horizontal surface irradiation data for the period 1981−92. (UK readers should

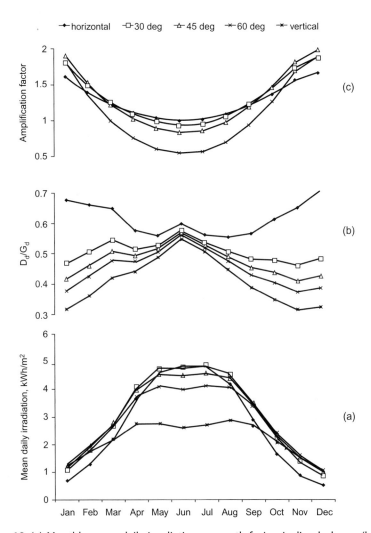

Figure 18 (a) Monthly mean daily irradiation on south-facing inclined planes. (b) The ratio of diffuse to daily global radiation on the inclined plane. (c) The total radiation amplification factor over the horizontal surface. London Bracknell (1981–92). Units are Wh m^{-2} d^{-1}. latitude 51.38°N, longitude 0.78°W. Site elevation 73 m. Ground albedo 0.2. *Source: CIBSE Guide] [7].*

note that there are corresponding tables for west, southwest, southeast, and east surfaces as well in reference [7] for this site, as well as for Edinburgh and Manchester Aughton, all valuable for photovoltaic design purposes in the UK.) The large variation found in the global radiation from winter to summer at high latitudes has already been discussed in Section 2. Figure 18(b) presents the corresponding ratios of the daily slope diffuse radiation to the daily global radiation on that inclined plane. Figure 18(c) compares the estimated total irradiation values on the various south-facing inclined planes with the corresponding observed global-radiation values on the horizontal plane. The ratio *monthly mean irradiation on the inclined plane divided by monthly mean irradiation on horizontal plane* has been termed the *monthly mean daily amplification factor* for that specific slope and orientation. It is usually a design aim to optimise the amplification factor in the context of supply and demand. The monthly mean amplification factors are different for beam radiation, diffuse radiation and total radiation. An important question is, 'When is the energy most valuable?' Choice of collector slope and orientation becomes especially important for the periods of the year when the solar-radiation energy supply is low and the energy demand high.

Figure 18(c) shows the maximum tabulated global amplification factor in January is 1.90 on a south-facing plane with a tilt of 60°. This tilt may be compared with latitude of Bracknell of 51.38°N. The December value is slightly higher at 1.98. It also occurs on a south-facing plane with a tilt of 60°. The maximum tabulated total amplification factor value in June is 1.01 on a south-facing plane with a tilt of 30°. A vertical south-facing surface in June has a monthly mean amplification factor of only 0.56 compared with 1.79 in January.

Figure 19 summarizes the predictions for July and December for south-facing slopes of different inclination. The annual mean curve is also included.

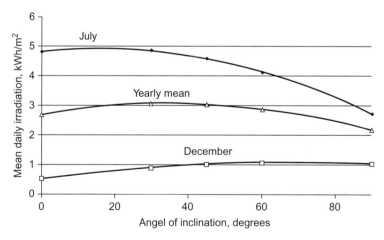

Figure 19 Mean daily irradiation on a south-facing inclined surface in London in July and December and the mean yearly value as a function of the angle of inclination.

3.4 Design Conclusions

This analysis has shown that the interactions between global irradiation and inclined plane irradiation are very complex. This complexity makes it difficult to make sound decisions about photovoltaic design without having simulation methodologies available to explore the issues of optimisation of collector exposure efficiently under different energy-demand scenarios.

The monthly mean energy density gains achieved through tilting collection systems vary considerably from month to month. Designers have to decide how to best match supply and demand across the whole annual cycle, taking account of the economics of energy storage in bridging between periods of favourable and unfavourably energy supply. Decisions about tilt angle are critical.

Compared with horizontal exposures, inclining photovoltaic collectors appropriately toward the Sun usually increases the radiation density available on sunny days and reduces it on overcast days. Statistically, the variance range in the daily irradiation is increased as a surface is tilted toward the equator.

Collecting diffuse radiation efficiently is often as important as collecting beam energy efficiently, especially in high-latitude cloudy climates.

Satisfactory time series of hourly slope irradiation can be developed from time series of observed daily horizontal irradiation. It follows that observed daily global-radiation data are needed to aid effective simulation. For example, the fourth *European Solar Radiation Atlas* [1] contains a 10-year time series of quality-controlled daily solar-radiation observations for the period 1981–90 for nearly 100 European sites. These data can be harnessed to help improve design understanding using suitable simulation models.

User-friendly tools for preparing hourly time series of inclined surface beam and diffuse irradiation data already exist within [1], which covers Europe, and [8], which covers the whole world. The development of these tools has had to address algorithmically a number of issues, including the generation of the solar geometry, the estimation of clear-sky radiation, the estimation of monthly mean global radiation, the estimation of diffuse radiation, and conversion radiation from horizontal surfaces to inclined planes, and reflections from the ground. Included are tools for examining the geometric impact of obstructions. Considerable attention has had to be given to the development of stochastic tools for the generation of irradiation time series. These statistical approaches are described in Section 4.

4. THE STOCHASTIC GENERATION OF SOLAR-RADIATION DATA

4.1 The Basic Approach and the Implicit Risks

4.1.1 General Principles

Stochastic models for generating time series of solar-radiation data use various statistical approaches to generate solar-radiation data. One type of stochastic model commonly used creates time series of daily global-radiation data statistically from values of the monthly mean daily global irradiation. A different type of stochastic model is used for generating time series of hour-by-hour irradiation statistically from day-by-day time series of daily global-radiation data. Using the principle of the chaining of stochastic models, the day-by-day models can be used first to process the relatively widely available monthly mean daily irradiation data to provide statistically generated representative time series of daily radiation data. This day-by-day output then provides the starting statistical daily time series needed as input to drive a second statistical stage to generate, hour by hour, the correspondingly more detailed time series of hourly global-radiation data. However, before moving to the hour-by-hour stochastic stage, mean daily profiles of hourly solar-irradiation data have to be generated for each day from the generated daily global irradiation time series using some suitable algorithm. Once the daily profile of the mean hourly global irradiation associated with each specific daily irradiation value has been established, the hour-by-hour stochastic models, which generate statistically the departures from the expected mean profile for the day, can then be applied. This process converts the stochastically generated daily time series of global-radiation data into a stochastically generated hourly time series of global-radiation data for the simulated period. The most common simulated periods are a month or a whole year. The stochastic generation of a time series of hourly global-irradiation data falling on a horizontal surface thus involves three basic stages. The next step for practical applications is to estimate the split of the generated hour-by-hour global-irradiation time-series data into their hour-by-hour beam and diffuse components. This information is essential to make estimates of the irradiation falling on inclined planes. This adds a fourth stage involving the application of another appropriate algorithm to achieve the split.

The definitions of most of the dimensionless quantities used here have been given already in Section 2.1. Following the SoDa project [8], a new objective dimensionless quantity has been added for this section. This is the clear-sky KT value, $KT_{d,cs}$. It is simply defined as $G_d/G_{d,cs}$, where $G_{d,cs}$

is the estimated representative daily clear-sky irradiation on day d. The SoDa project has a facility for estimating $G_{d,cs}$ for any place in the world.

The Problem of Data Generalisation

The various models in use attempt to generalise the underlying statistical relationships in order to build successful stochastic models. The aim often is to generate acceptable time-series data covering the radiative characteristics of a wide range of climates. The methodology used needs to reproduce the distribution of the probabilities as well as preserve the degree of autocorrelation between succeeding days [24]. A fine day has a greater probability of being followed by another reasonably fine day than being followed by an overcast day. This is due to the typical persistence of anti-cyclonic weather, which statistically typically lasts several days. There is also the corresponding day-to-day persistence in overcast-day irradiation, because cyclonic disturbances also typically endure more than one day.

Limitations Imposed by Variations in Global Climatology

The various stochastic models proposed do not usually achieve their prediction goals successfully for all geographic locations. Unfortunately, many authors working at specific places tend to believe the aim is to find universal solutions. In following this ambitious aim, they often reveal a lack of detailed knowledge of the complexities of radiation climatology in various parts of the globe. So they risk becoming scientifically exposed through their lack of global-climate knowledge.

Most of the available stochastic models have tended to be derived predominately from midlatitude observed data. The underlying statistical derivation processes are influenced by the predominant use of observed data from the midrange of latitudes in their generation. This geographically limited approach unfortunately has tended to make the many of the commonly available generalised stochastic models less reliable at high latitudes and at low latitudes than they are at midlatitudes. It is therefore important to assess the type of observed data being used in any statistical model development process against the actual latitude of use. Great benefits can still be obtained, however, from the use of stochastic techniques for generating realistic time series of solar-radiation data for assessing the performance of photovoltaic systems in various parts of the world, provided these underlying difficulties are recognised.

It is a simple question to ask, 'Has this specific stochastic model been validated against radiation observations in my part of the world?' If not, proceed with care.

4.2 Estimating Time Series of Daily Global Radiation from Monthly Means using *KT*-Based Methods

There is normally considerable variation in the global irradiation from day to day, especially at high latitudes. The alternation of anticyclonic and cyclonic conditions causes large variations in both cloud and sunshine amounts. This results in a highly variable supply in the available daily solar energy in such regions. This is especially true for areas exposed to a strong maritime influence. The desert climates, in contrast, show much less day-to-day variation in irradiance. Some humid tropical climates have some cloud nearly every day. In such regions, cloudiness often develops progressively across the day due to vertical convection as the ground warms. Other climates are linked to strongly defined wet and dry seasons—for example, the monsoon climates of Southeast Asia. There still remains a lot of work to be done in developing sound stochastic models covering the wide range of climates found across the world.

The model of Aguiar et al. [24] was implemented within METEONORM Version 3.0 [3]. This commonly used algorithm converts the monthly mean global radiation into a statistically generated time series using a library of Markov transition matrices to generate the daily global-radiation series. The aim is to make the statistically generated series indistinguishable statistically from the observed time series. This approach provides a way of obtaining radiation sequences for locations where such sequences have not been measured.

The method is based on two observations [24]:
1. There is a significant correlation between radiation values for consecutive days.
2. The probability of occurrence of dimensionless daily radiation KT_d values is the same for months with the same monthly mean KT_m values at different locations in the world.

The basic algorithm generates a statistical time series of daily values of KT_d. The global radiation G_d is then estimated from KT_d as $KT_d - G_{od}$, where G_{od} is the daily extraterrestrial value for day d. The method depends on the availability of a suitable library of monthly Markov transition matrices (MTMs), each covering a defined range of KT_m values. The library [24] of MTMs was built up using daily data from five Portuguese stations, plus Polana in Mozambique, Trappes and Carpentras in France, and Macau. There were 300 months of data.

There are a number of shortcomings of the KT_m approach for the stochastic generation of daily global-radiation time series. The difficulties at higher latitudes have already been mentioned. Workers at midlatitudes do not always realise how significant change of latitude can be on the KT_d max values in winter. This factor was not taken into account adequately in the Aguiar et al. [24] midlatitude studies. The earlier Aguiar approach also produces difficulties at sites elevated well above sea level. For these reasons, the SoDa project has developed the alternative MTM methodology outlined in the next section.

4.3 Improvements in the Stochastic Estimation of Daily Solar Radiation in the SoDa Project

The following important methodological improvement has been recently introduced in the EU-supported SoDa program [8,13]. The SoDa project has produced world maps of the monthly representative Linke turbidity factors across the globe automatically adjusted to site level. The SoDa computational processes also provide the user with accurate information on terrain elevation. These maps and their associated data bases are conjoined to SoDa clear-sky modelling programs so an objective process is now available for estimating the representative value of the clear-sky daily global irradiation $G_{d,cs}$ on any day. The whole system of MTMs was therefore changed from a KT_m clearness index basis to a clear-sky clearness index basis. Formulated like this, the maximum value of $KT_{d,cs}$ ($= 1$) must correspond automatically to the clear-sky model predictions used. $KT_{d,cs}$ is calculated as the ratio $G_d/G_{d,cs}$.

This change required the daily MTM tables to be completely revised to match the new formulation. Data from 121 stations in the world were used [13].

4.3.1 Validation of Daily Irradiation Generation Models

Two sets of validation studies are needed. The first set has to verify that there is a reasonable representation of the overall probability profile. This is normally achieved by accumulating the observed daily time series into bins of defined KT_d or $KT_{d,cs}$ width and comparing these values with the corresponding predicted values, binned in exactly the same manner. A reasonable match should be obtained. The match will never be perfect. A second basic check is to compare the one-day time-lapse autocorrelation values between prediction and observation.

The new model was validated at 12 sites and compared the KT estimates against the KT_{cs} observations. Generally, the new model was found to be slightly better. The probability profile distributions generated from the two methods are very similar. Generally the clear-day KT_{cs} model shows better results for North American sites than for European sites. This result can be explained by the different geographical choice of sites used for making the original Aguiar model and the new METEONORM/SoDa model developed by Meteotest of Switzerland in conjunction with other EU SoDa project partners.

While the results are very similar in numeric detail, it has to be underlined that good convergence with the clear-day observations is only achieved with the new version. The close interaction with the clear-sky model is helpful especially when looking at clear-day extremes. Additionally, the new Meteotest/SoDa model is easier to use (requires no minimum and maximum classes) and runs generally faster (fewer loops). The autocorrelation also seems to be predicted better with the new $KT_{m,cs}$ model [13].

4.4 Generating Hourly Values of Global Radiation from Daily Values

4.4.1 Generating Mean Daily Profiles of Irradiance from Specific Values of Daily Radiation

The knowledge of the mean daily profiles is essential for many simulation applications. It is also the first step in the generation of hourly values of global radiation from daily values of clearness index or clear-sky clearness index. The model of Aguiar and Collares-Pereira [25] was used in METEONORM Version 3.0 [3] to generate the daily profile. Unfortunately, this earlier model is very badly suited for high latitudes. In METEONORM Version 4.0 [4] the hourly integration model of Gueymard [26] was introduced. Gueymard validated different models for 135 stations. One of them was the model of Collares-Pereira and Rabl [27] that was used for ESRA [1]. Gueymard as a result presented a corrected version of Collares–Rabl model that was better adapted to high latitudes. Gueymard's new model and the corrected model of Collares–Rabl showed about the same quality.

In a report by Page [28], a firm recommendation was made to use another daily profile of beam irradiance. This report pointed out that the profile model of Collares-Pereira and Rabl [27] was unsafe, principally for low solar altitudes, a point especially important for high latitudes, where low solar altitudes are much more dominant in solar-radiation simulations. Page proposed the hypothesis that the average day's beam-irradiance profile should mirror the clear-day beam profile exactly in the form

$$B_h = B_d \cdot \frac{B_{h,sc}}{B_{d,cs}} \qquad (3)$$

where B_d is the daily beam irradiance, $B_{h,cs}$ the clear-sky hourly beam irradiance, and $B_{d,cs}$ the daily clear-sky beam irradiance.

The clear-day hourly beam irradiance and the daily clear-day beam irradiation can be calculated with any standard clear-day formulae. The SoDa project has adopted the improved ESRA clear-day model for this calculation [8,11]. The disadvantage of the model described by Equation (3) is that the daily beam value has to be known first. This value can be calculated, for example, with Erbs's model as discussed in reference [1]. Page suggested [28] the use of a new model for daily beam values using a kind of Angström formulae with sunshine data. In order to escape this problem in the estimation of daily beam values, Meteotest postulated the alternative method for calculating the global mean daily profile directly. This avoids the difficulties mentioned. As the diffuse profile can be calculated from the improved ESRA model, the global profile can be easily calculated as sum of the two components.

4.4.2 Profiling of the Global Irradiance

The following new method for generating mean daily irradiation profiles was proposed and has been incorporated as a standard part of the SoDa stochastic software. The global-radiation clear-sky profile is used to calculate the global-radiation profile from the daily global irradiation using

$$G_h = G_d \cdot \frac{G_{h,cs}}{G_{d,cs}} \qquad (4)$$

where G_d is the daily global horizontal irradiance, $G_{h,cs}$ the clear-sky hourly global irradiance, and $G_{d,cs}$ the daily clear-sky global irradiation [29].

The advantage of this model is that daily values of beam or diffuse do not have to be known in advance. The model, called here *Remund–Page* (RP), relates perfectly to the upper edge of the distribution, the clear-sky profile. This calculation is always needed as the first step in the SoDa chain of algorithms. A short validation of the model has been made and will be published in due course by Meteotest.

4.4.3 Stochastic Generation of Hourly Values of Global Irradiation

The stochastic generation of hourly values of global irradiance in the SoDa program is based on the TAG (time-dependent, autoregressive Gaussian) model of reference [25]. This model, also used in METEONORM, consists of two parts. The first part calculates an average daily profile. The second part simulates the intermittent hourly variations by superimposing an autoregressive procedure of the first-order (AR(1)-procedure) [30]. The improved chain of stochastic models is implemented in a user-accessible way in SoDa [8] and also in the new version of METEONORM [5].

4.5 Splitting the Global Radiation to Diffuse and Beam

Two well-researched models for estimating the fraction of the hourly diffuse radiation in the hourly global radiation, D_h/G_h, are the models of Perez [31], as used in METEONORM Version 4.0 [4], and of Skartveit et al. as used in the Satel-Light project [32]. The disadvantage of the use of a model like that of Perez or Skartveit is that the daily values of beam or diffuse the hourly values cannot be determined without stochastically generating hourly values. So the beam and diffuse values depend to a small extent on random numbers. The use of mean daily profiles to calculate the beam and diffuse profile is not really possible, because both the Skartveit and Perez models depend on the hourly variations from one hour to the next. Mean profiles and hourly values with variations do not give the same result.

One advantage of the Skartveit model is that a correction for the impact of high-ground albedos is available for this model. There is also a formula to calculate the hourly variation of global irradiation if this value is not known. The original model was adjusted to observed data from Bergen, Norway. The Perez model is more widely used and is still the standard model. It can be adapted to both US and EU climates.

4.6 Assessment of Progress

In the past, it has been usual to use the KT_m values as the key inputs, taking no detailed account of the impact of local atmospheric transmission variations when assembling data from a wide range of sites. It is now possible to take better account of local variations in atmospheric transmission and, also, of the impacts of site elevation on the division between clear-day beam and diffuse radiation, because of the advances achieved in the SoDa project.

The earlier stochastic work of Aguiar et al. [24] on the estimation of daily global-radiation time series has proved very valuable, especially in middle latitudes. It has been available to users of the Meteonorm system for some time as a fully working user-friendly programme in METEONORM Version 3.0 [3]. This model, however, has weaknesses in dealing with high-latitude sites, because the impact of latitude on KT max values is not adequately treated. The new Meteotest model for the generation of daily global-radiation time series addresses these issues of clear-day data generation with greater precision using the SoDa global database of Linke turbidity factor data. The impacts of site elevation also are now treated more scientifically.

The models for generating the daily profiles of irradiance proposed by Collares-Pereira and Rabl [27] and subsequently used in ESRA 2000 have been shown to be unreliable at higher latitudes. A new improved method for estimating the daily radiation profile has been created based on the use of improved ESRA clear-sky model implemented in SoDa—IS and in METEONORM Version 5.0 [5].

An improved group of stochastic modelling procedures has been developed as part of the work on the SoDa project. The new approach is based of the use of the clear-sky KT factor. This approach has required the development of new Markov transition matrices based on the clear-sky KT factor. This new stochastic approach has been implemented in a user-friendly way as a key element in the SoDa computing chains (visit www.soda-is.com for practical implementation).

The hour-by-hour stochastic model of Aguiar and Collares-Pereira [25] was used in METEONORM Version 3.0 [3]. However, this model has significant weaknesses at high latitudes, especially in winter. Version 4 of METEONORM [4] introduced the hourly integration model of Gueymard [26]. This helped to rectify weaknesses in high-latitude estimates.

These improvements have been implemented in a systematic way with the SoDa—IS and embedded within a global computational structure for generating hourly and daily time series of radiation data.

A new version of METEONORM 5.0 has been published [5]. This version will contain the algorithmic improvements reported here and in other papers associated with the SoDa project.

More studies are needed on the stochastic generation of solar-radiation data for tropical areas, especially for humid and cloudy climates. The physical patterns of tropical weather generation are essentially different from those encountered at high latitudes. Convectively generated cloud often plays an important role. Additionally, there are often considerable asymmetries of global radiation between morning and afternoon.

Statistically based solar-radiation data-generation modules are likely to become increasingly important in photovoltaic system design. The models available to users are likely to continue to improve in scientific quality.

5. COMPUTING THE SOLAR GEOMETRY

5.1 Angular Movements of the Sun Across the Seasons

The calculation of the solar geometry is often needed. This knowledge is essential to assist both in the choice of the most effective tilt and orientation of the solar cells and also for consideration of the detailed impacts of overshadowing obstructions. This section describes the calculation of the movements of Sun as seen from any point on Earth's surface.

The daily solar path depends on the latitude of the site and the date in the year. The basic input variable in the trigonometric estimation of the solar geometry is the solar declination. The *solar declination* is the angle between the direction of the centre of the solar disc measured from Earth's centre and the equatorial plane. The declination is a continuously varying function of time. The summer solstice—i.e., the longest day—occurs when the solar declination reaches its maximum value for that hemisphere. The winter solstice—i.e., the shortest day—occurs when the declination reaches its minimum value for that hemisphere. The declination on any given day in the southern hemisphere has the opposite sign to that in the northern hemisphere on that same day. While any day in the year may be studied, most design manuals select one or two days in each month for detailed reference calculations—for example, in the *European Solar Radiation Atlas* [1] and the CIBSE Guide [7]. The mid-month values adopted here are given in Table 2.

The passage of days is described mathematically by numbering the days continuously through the year to produce a Julian day number, J: 1 January, $J = 1$; 1 February, $J = 32$; 1 March, $J = 57$ in a nonleap year and 58 in a leap year; and so on. Each day in the year can be then be expressed in an angular form as a day angle, J', in degrees by multiplying J by 360/365.25. The day angle is used in the many of the trigonometric expressions that follow.

5.2 Time Systems Used in Conjunction with Solar Geometry and Solar-Radiation Predictions

Different time systems are in use. Legal clock time differs from solar time. Solar time is determined from the movements of the Sun. The moment when the Sun has its highest elevation in the sky is defined as solar noon. Solar noon at any place defines the instant the Sun crosses the north–south meridian. It is then precisely due south or precisely due north. It is strongly

Table 2 Equation of time (EOT) (expressed in both hours and minutes) and solar declination values at monthly design dates based on monthly mean declination. Values for 21 June and 22 December are also given. The declination angles are stated for the northern hemisphere; change the sign for the southern hemisphere

Date	Jan. 17	Feb. 15	Mar. 16	Apr. 15	May 5	Jun. 11	Jun. 21	Jul. 17	Aug. 16	Sep. 16	Oct. 16	Nov. 15	Dec. 11	Dec. 22
EOT, hour	−0.163	−0.241	−0.157	−0.006	+0.061	+0.009	−0.029	0.099	−0.070	+0.094	+0.250	+0.252	+0.106	−0.195
EOT, min	−9.8	−14.5	−9.4	−0.3	+3.7	+0.5	−1.8	5.9	−4.2	+5.6	+15.0	+15.1	+6.4	−11.7
δ degrees	−20.71	12.81	−1.80	+9.77	+18.83	+23.07	+23.4.3	+21.16	+13.65	+2.89	−8.72	−18.37	−22.99	23.46

recommended that all calculations of the solar geometry are carried out in solar time. This time system is usually called *local apparent time* (LAT). Solar time is converted to its angular form, the *solar hour angle*, for trigonometric calculations of the solar path. The solar hour angle is referenced to solar noon. Earth rotates through 15° in each hour. The standard convention used is that the solar hour angle is negative before solar noon. So 14:00 hours LAT represents a solar hour angle of 30°, and 10:00 hours LAT represents a solar hour angle of −30°.

5.3 Conversion of Local Mean Time (LMT) to Local Apparent Time (Solar Time)

Local mean time (LMT), often called *clock time* or *civil time*, differs from LAT, often called *solar time*. The difference depends on the longitude of the site, the reference longitude of the time-zone system in use at that site and the precise date in the year. Most climate observations are made in synoptic time. This is Greenwich Mean Time (GMT) in the UK. Hour-by-hour solar irradiation and bright sunshine observations have been important exceptions. They are usually observed in and summarised in LAT. There is also the complication of summer time.

As climatic data are typically provided in two different systems of time, it is sometimes important to be able to interrelate the two time systems, especially in the case of simulation studies. The solar irradiance is a discontinuous function at sunrise and sunset. The other synoptic weather variables are continuous. So, it is more accurate to interpolate synoptic values into solar time than it is to interpolate irradiation values into synoptic time when compiling consistent data sets for simulation.

The conversion of time systems requires knowledge both of the longitude of the site and the reference longitude of the time system being used. The conversion also requires the application of the equation of time, which accounts for certain perturbations in Earth's rotation around its polar axis. In the UK, LMT is GMT in winter and British Summer Time (BST) in summer. Many countries in the western portion of the European Union use West European Time (WET) and its summer-time variants. Based on longitude 15°E, these nearby countries have time systems one hour ahead of GMT and BST (it is EU policy that all member states change to and from summer time on the same dates). Countries sometimes change their time-reference longitudes: for example, Portugal switched to WET to align with the European Union and was then forced by formal public protest to switch back to GMT. Telephone directories and up-to-date airline timetables are useful sources of current time-zone systems in use in different zones of different countries. UK readers should note that the latest UK BT telephone directories contain an excellent colour map showing in detail world local time systems. On the Web, www.worldtimezone.com/index24.html also provides this information.

The equation of time is the difference in time between solar noon at longitude 0° and 12:00 GMT on that day. The calculation of the equation of time requires as input the day number in the year. This is then converted into a day angle in the year as explained in Section 5.1. The equation of time is calculated as

$$\text{EOT} = -0.128 \sin(J' - 2.80°) - 0.165 \sin(2J' + 19.7°) \text{ hours} \quad (5)$$

Then

$$\text{LAT} = \text{LMT} + (\lambda - \lambda_R)/15 + \text{EOT} - c \text{ hours} \quad (6)$$

where J' is the day angle in the year in degrees; λ is the longitude of the site, in degrees, cast positive; λ_R is the longitude of the time zone in which the site is situated, in degrees, east positive; and c is the correction for summer time, in hours. Values of the equation of time for selected dates may be found in Table 2.

5.3.1 Example
Estimate the time of occurrence of solar noon at Belfast on 4 August in BST. The longitude of Belfast is 6.22° W. The time system is BST. The time-zone reference longitude is Greenwich 0.00°. A westward displacement of 6.22° yields a value of $(\lambda - \lambda_R)/15 = (-6 - 13/60)/15$ hours $= -24.9$ minutes. From Equation (5), the equation of time on 4 August is -5.9 minutes. LAT is 12:00. In LMT, solar noon will occur at $12:00 - (-24.9 - 5.9)$ minutes— i.e., 12:32 GMT or at 13:32 BST.

5.4 Trigonometric Determination of the Solar Geometry
5.4.1 Key Angles Describing the Solar Geometry
Two angles are used to define the angular position of the Sun as seen from a given point on Earth's surface (Figure 20): solar altitude and solar azimuth.
- Solar altitude (γ_S) is the angular elevation of the centre of the solar disc above the horizontal plane.
- Solar azimuth (α_S) is the horizontal angle between the vertical plane containing the centre of the solar disc and the vertical plane running

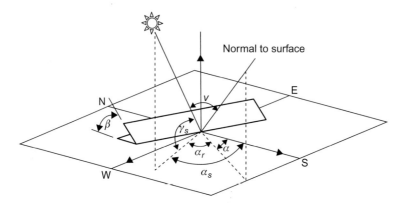

Figure 20 Definition of angles used to describe the solar position (γ_s and α_s), the orientation and tilt of the irradiated plane (α and β), the angle of incidence (ν), and the horizontal shadow angle (α_1). *Source: Reference [7].*

in a true north–south direction. It is measured from due south in the northern hemisphere, clockwise from the true north. It is measured from due north in the southern hemisphere, anticlockwise from true south. Values are negative before solar noon and positive after solar noon.

Four other important solar angles are the following.

- The *solar incidence angle on a plane of tilt α and slope β* ($\nu(\beta,\alpha)$) is the angle between the normal to the plane on which the Sun is shining and the line from the surface passing through the centre of the solar disc. The cosine of $\nu(\beta,\alpha)$ is used to estimate the incident beam irradiance on a surface from the irradiance normal to the beam.
- The *vertical shadow angle*, sometimes called the *vertical profile angle* (γ_p), is the angular direction of the centre of the solar disc as it appears on a drawn vertical section of specified orientation (see Figure 21).
- The *wall solar azimuth angle*, sometimes called the *horizontal shadow angle* (α_F) is the angle between the vertical plane containing the normal to the surface and the vertical plane passing through the centre of the solar disc. In other words, it is the resolved angle on the horizontal plane between the direction of the Sun and the direction of the normal to the surface (see Figure 21).
- The *sunset hour angle* (ω_s) is the azimuth angle at astronomical sunset. It is a quantity used in several algorithmic procedures.

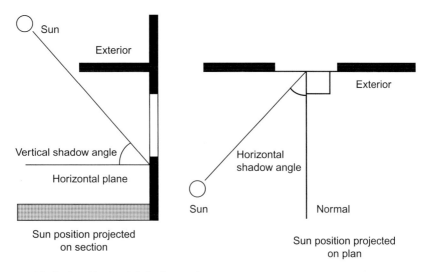

Figure 21 Definition of vertical shadow angle γ_p and the horizontal shadow angle α_F. Source: Reference [7].

5.4.2 Climatological Algorithms for Estimating Declination

The solar year is approximately 365.24 days long. The calendar is kept reasonably synchronous with the solar-driven seasons through the introduction of leap years. This leap-year cycle means that the precise declination for any selected day varies according to the position of the day within the four-year leap cycle. For calculations involving climatological radiation data averaged over several years, it is appropriate to use long-term mean value formulae to estimate the declination. In contrast, if observed data for specifically identifiable days in specific years are available, it is logical to use more accurate declination formulae to calculate the declination for that specific day and longitude—for example, in developing the solar geometry to use with simulation tapes based on observed time series. Usually, the noon declination values are used. This gives sufficient accuracy for practical calculations. The following equation is used in the *European Solar Radiation Atlases* to compute representative mean values of the declination based on a 365-day year:

$$\delta = \sin^{-1} 0.3978 \ \sin[J' - 1.400 + 0.0355 \ \sin(J' - 0.0489)] \quad (7)$$

where J' is the day angle in radians (Julian day number $\times 2\pi/365.25$).

Monthly mean declination values can be derived for each month by integrating the daily declination values over each month and taking the mean. Each mean can be associated with a monthly mean representative design date. Table 2 tabulates the recommended climatological values of the mean declination for use at the monthly mean level. Table 2 includes the day in the month when the representative declination is closest to the monthly mean value.

More accurate formulae for calculating the solar declination for a specific time at a specific place in a specific year are available; for example, the Bourges algorithm [33] is recommended for use by engineers to obtain an accurate assessment of the declination on specific days in specific years [1,7] (see Section 5.6 for details). Solar cell designers will not normally require this greater accuracy unless they are using refined beam-focusing systems.

5.4.3 The Calculation of Solar Altitude, Azimuth Angles, and Astronomical Day Length

These two angles, which are defined in Section 5.4.1, are dependent on the time of day, t, as measured in hours LAT on the 24-hour clock. For solar trigonometric calculations that follow, time is expressed as an hour angle, ω, where

$$\omega = 15(t - 12) \text{ degrees} \tag{8}$$

where t is the solar time in hours (i.e., LAT). The solar altitude angle γ_s is obtained from

$$\sin\gamma_s = \sin\phi \sin\delta + \cos\phi \cos\delta \cos\omega \tag{9}$$

$$\gamma_s = \sin^{-1}(\sin\gamma_s)$$

where ϕ is the latitude of the location. The solar azimuth angle is obtained from

$$\cos\alpha_s = (\sin\phi \sin\gamma_s - \sin\delta)/(\cos\phi \cos\gamma_s)$$

$$\sin\alpha_s = \cos\delta \sin\omega/\cos\gamma_s \tag{10}$$

If $\sin \alpha_s < 0$, then $\alpha_s = -\cos^{-1}(\cos \alpha_s)$; if $\sin \alpha_s > 0$, then $\alpha_s = \cos^{-1}(\cos \alpha_s)$.

Both formulae in Equation (10) are needed to resolve the azimuth angle into the correct quadrant in computer programs. The sunset hour

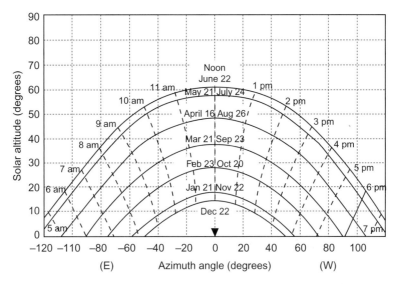

Figure 22 Sun-path diagram for latitude 52°N. The times of day, defined by near vertical lines, are expressed in LAT. Note that, in the earlier and later parts of the day, the Sun lies to the north of a south-facing facade during the period between the spring and autumn equinoxes. *Source: Reference [7].*

angle, ω_s, is calculated by setting the solar altitude to zero in Equation (9). It follows that ω_s may be calculated as

$$\omega_s = \cos^{-1}(-\tan\phi \tan\delta) \qquad (11)$$

The astronomical day length is therefore $2\omega_s/15$ hours. The day length is an important design variable. At high latitudes, it changes considerably with time of year. The sun-path diagram for latitude 52°N is shown in Figure 22. The annual variation in day length (shown in Figure 23) is much less at lower latitudes than at higher latitudes. The astronomical day length is 12 hours every day on the equator. These facts have a strong bearing of the length of time during which night-time electric lighting is needed at different locations, impinging strongly on the demand side in the analysis of photovoltaic system performance.

5.5 The Calculation of the Angle of Incidence and Vertical and Horizontal Shadow Angles

The determination of the exact geometry of the solar movements across the sky is important for many aspects of practical photovoltaic design. The

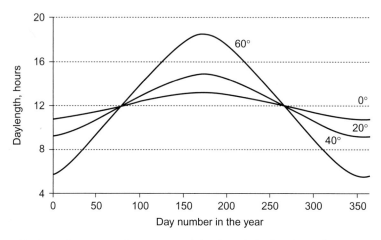

Figure 23 Annual variation of astronomical day length for four latitudes.

size of the array required to meet a given load can be optimised through appropriate geometric decisions. Knowledge of the solar geometry is also important in the assessment of the periods of partial and total overshadowing. Collection performance may be considerably reduced when collectors are placed within the detailed potentially obstructive structure of their surrounding townscape and landscape. It is often useful to present the geometry on the same projections as those used in standard engineering design.

The angle of incidence of the solar beam, $\nu(\beta,\alpha)$, on a surface of tilt β and surface azimuth angle α is calculated from the solar altitude and solar azimuth angles. The wall solar azimuth angle α_F has to be calculated first using Equation (12). The sign convention adopted for wall solar azimuth angle is the same as that used for the solar azimuth angle. The angle is referenced to due south in the northern hemisphere and to due north in the southern hemisphere. Values to the east of the north–south meridian are negative. Values to the west are positive.

$$\alpha_F = \alpha_s - \alpha \qquad (12)$$

where α_s is the solar azimuth angle and α is the azimuth angle of the surface. If $\alpha_F > 180°$, then $\alpha_F = \alpha_F - 360°$; if $\alpha_F < 180°$, then $\alpha_F = \alpha_F + 360°$.

Once the wall solar azimuth angle has been calculated, the angle of incidence $\nu(\beta,\alpha)$ can be calculated using Equation (13). A negative value implies the Sun is behind the surface and then the value is usually set to zero:

$$\nu(\beta, \alpha) = \cos^{-1}(\cos\gamma_s\ \cos\alpha_F\ \sin\beta + \sin\gamma_s\ \cos\beta) \qquad (13)$$

A particularly useful geometric parameter for engineering design purposes is the vertical shadow angle (see Figure 21) because it aligns with standard engineering projection practice. It may be calculated as

$$\gamma_p = \tan^{-1}(\tan\gamma_s/\cos\alpha_F) \tag{14}$$

If $\gamma_p < 0$, $\gamma_p = 180 + \gamma_p$. For a vertical surface, if $\gamma_p > 90°$, then the Sun is falling on the opposite parallel vertical surface. The horizontal shadow angle is identical with α_F.

5.6 Establishing the Accurate Noon Declination and the Accurate Solar Geometry

Sometimes when working with data for specific days or making live observations, it is useful to use a more accurate formula for estimating the declination. The Bourges formula [33] is simple to use. It is adequate to accept the noon declination as representative for the whole day:

$$\delta_{noon} = 0.3723 + 23.2567 \sin\omega_t + 0.1149 \sin 2\omega_t - 0.1712 \sin 3\omega_t$$
$$- 0.7580 \cos\omega_t + 0.3656 \cos 2\omega_t - 0.2010 \cos 3\omega_t \text{ degrees} \tag{15}$$

where

$$\omega_t = \omega_o(J - t_1) \text{ radians}, (\omega_o - 2\pi/\gamma/365.2422 \text{ radians},$$
$$t_1 = -0.5 - \lambda/360 - n_o \text{ days}, n_o = 78.8946 + 0.2422(Y - 1957)$$
$$- \text{INT}[(\gamma - 1957)/4]$$

in which λ is the longitude (east positive), Y is the year in full, INT is the integer part number of the expression, and J is the Julian day running from 1 to 365 in nonleap years and from 1 to 366 in leap years.

6. THE ESTIMATION OF HOURLY GLOBAL AND DIFFUSE HORIZONTAL IRRADIATION

6.1 The Estimation of the Extraterrestrial Irradiance on Horizontal Planes

Knowledge of the diurnal pattern of extraterrestrial irradiation is needed for establishing the profile of the daily radiation at the surface. The extraterrestrial irradiance normal to the solar beam at the mean solar distance (solar constant, $I_o = 1367\ \text{Wm}^{-2}$, has already been introduced in Section 2.1). The correction needed to allow for the variation of Sun–Earth distance from its mean value, ε, can be conveniently expressed as a function of the day angle J' in radians:

$$\varepsilon = 1 + 0.03344\cos(J' - 0.048869) \tag{16}$$

The extraterrestrial irradiation values incident on a horizontal surface can be obtained by numerical integration over time of the extraterrestrial irradiance, given by

$$I_0 = \varepsilon\ \sin\gamma_s$$

But it is also possible [1] and simpler to make a direct integration, resulting in simplified formulae where, instead of the solar altitude angle γ_s, the three basic angles (latitude ϕ, solar hour angle ω in radians, and declination δ, all defined in Section 5) are used as inputs:

- In the general case,

$$G_{0(1\ 2)} = I_0\varepsilon(T/2\pi)[\sin\phi\ \sin\delta(\omega_2 - \omega_1) + \cos\phi\ \cos\delta(\sin\omega_2 - \sin\omega_1)] \tag{17}$$

- In the case of hourly values ($1\omega_1 - \omega_2 1 = \pi/12$), provided γ_s is positive or 0 throughout the hour,

$$G_{oh} = I_0\varepsilon(T/2\pi)[\sin\phi\ \sin\delta(\pi/12) + \cos\phi\ \cos\delta(\sin\omega_2 - \sin\omega_1)] \tag{18}$$

- In the case of daily values, the sunrise hour angle values and sunrise hour angle values are used ($\omega_1 = -\omega_s$, $\omega_2 = \omega_s$ inputs in radians [see Section 5, Equation (11) for the calculation of ω_s]),

$$G_{od} = I_0\varepsilon(T/\pi)\cos\phi\ \cos\delta(\sin\omega_s - \omega_s\cos\omega_s) \tag{19}$$

The parameter T is the duration of a rotation of Earth around its axis; an average value of 86,400 s (i.e., 24 hours) can be used in practice.

6.2 The Daily Clearness Index (KT_d Value)

The daily clearness index is the ratio of the global irradiance at the surface, falling on a horizontal plane, to the corresponding extraterrestrial global irradiance on the horizontal plane. The clearness index value may be extracted at the daily level KT_d or the hourly level KT_h. The daily extraterrestrial irradiation G_{od}, Equation (19), is required to estimate the widely used daily clearness index KT_d. Typical values of KT_d are around 0.68 to 0.72 under cloudless conditions, with lower values at high latitudes in winter. The monthly mean daily clearness index is indicated by $(KT_d)_m$. It is a mapped variable in ESRA [1]. Section 2.8.1 outlines how to calculate the monthly mean daily radiation from sunshine data using the monthly mean daily relative sunshine duration.

6.3 The Estimation of Mean Daily Profiles of Global Solar Irradiation from Monthly Means

Hour-by-hour irradiation data are essential if the irradiation on inclined planes is to be estimated. Two well-described methodologies are available for estimating hourly global-irradiation data from daily data. One method is based on the use of the diurnal profile of extraterrestrial irradiance. The other is based on the diurnal profile of clear-day global irradiance at the surface.

6.3.1 Method 1: Estimating the Monthly Mean Daily Irradiance Profile of G$_h$(t) from the Monthly Mean Extraterrestrial Irradiance Profile

The mean daily profile of the hourly global horizontal irradiation $G_{hm}(t)$ is estimated from the respective monthly mean daily sum $(G_d)_m$ using the model of Collares-Pereira and Rabl [27]. This model is set down as Equation (20) below:

$$G_{hm}(t) = (G_d)_m r_{om}[a + b\cos(\omega(t))] \qquad (20)$$

with seasonally varying coefficients a and b:

$$a = 0.4090 + 0.5016 \; \sin[(\omega_s)_m - \pi/3]$$
$$b = 0.6609 - 0.4767 \; \sin[(\omega_s)_m - \pi/3]$$

where r_{om} is the monthly mean ratio of hourly to daily extraterrestrial radiation, which is a function of the hour angle $\omega(t)$ (see Section 5.4.3, Equation (8)) and the monthly average sunset hour angle $(\omega_s)_m$ for the month in question (see Section 5.4.3, Equation (11)):

$$r_{om}(t) = \frac{\pi}{24}\left[\frac{\cos\omega(t) - \cos(\omega_s)_m}{\sin(\omega_s)_m - (\omega_s)_m \cos(\omega_s)_m}\right] \qquad (21)$$

This model does not perform well at high latitudes [13]. See also Section 4 for further discussion of this point.

6.3.2 Method 2: Estimating the Monthly Mean Daily Profile of $G_h(t)$ from the Clear-Sky Irradiance Profile

This method provides more accurate estimates of the value of the hourly surface global irradiation, G_h. However, it needs accurate information if available about the clarity of the atmosphere at the site under consideration in order to estimate $G_{c,cs}$ [8,13]. More discussion is included in Section 4 and Equation (4).

The clear-sky model requires the user to input the site elevation and the Linke turbidity factor. (The Linke turbidity factor and the terrain elevation are mapped variables on a global scale within the SoDa–IS 2003 project that make it easy to implement Equation (4) of Section 4 at any selected site.) A comparison by Meteotest found the rmse of the hourly values of KT_h ($=G_h/G_{oh}$) generated with this second model tested against 10 worldwide sites was 0.072 compared with 0.095 using Method 1, i.e., the Collares-Pereira and Rabl model. Method 2 has the added value that it avoids all risk of generating negative values close to sunrise and sunset, a fault encountered sometimes in the first method.

6.4 The Estimation of Hourly Diffuse Radiation on Horizontal Surfaces

6.4.1 Introduction

The estimation of the irradiation on inclined surfaces can only be achieved through knowledge of hourly values of beam and diffuse radiation. Unfortunately, few observed data sets contain observations of diffuse irradiation. It is frequently necessary to estimate the monthly mean and daily diffuse irradiation algorithmically from the observed monthly mean and daily global radiation. Once this split has been achieved, the monthly mean data and daily data can then be converted to representative daily profiles. The hourly diffuse profile approximately follows the extraterrestrial hourly profile. By using these facts, representative hourly values of diffuse horizontal irradiance can be developed, which are then used to estimate inclined surface values.

6.4.2 Estimating the Monthly Average Daily Diffuse Horizontal Irradiation from the Monthly Average $(KT_d)_m$ Value

ESRA [1] uses the following third-order polynomial models to estimate the monthly mean daily diffuse horizontal irradiation $(D_d)_m$ as a fraction of the monthly mean global irradiation $(G_d)_m$ from the monthly mean clearness index $(KT_d)_m$:

$$\frac{(D_d)_m}{(G_d)_m} = c_0 + c_1(KT_d)_m + c_2(KT_d)_m^2 + c_3(KT_d)_m^3 \qquad (22)$$

Table 3 Coefficients of the third-order polynomial for estimation of monthly average diffuse horizontal irradiation from $(KT_d)_m$ values for use in Equation (22)

Latitude band	Season	c_0	c_1	c_2	c_3	$(KT_d)_m$ validation range
61°N > φ 56°N	Winter	1.061	−0.397	−2.975	2.583	[0.11,0.50]
	Spring	0.974	−0.553	−1.304	0.877	[0.24,0.50]
	Summer	1.131	−0.895	−1.616	1.555	[0.26,0.60]
	Autumn	0.999	−0.788	−0.940	0.788	[0.21,0.51]
56°N > φ > 52°N	Winter	1.002	−0.546	−1.867	1.490	[0.14,0.49]
	Spring	1.011	−0.607	−1.441	1.075	[0.22,0.59]
	Summer	1.056	−0.626	−1.676	1.317	[0.29,0.64]
	Autumn	0.969	−0.624	−1.146	0.811	[0.23,0.53]
φ < 52°N	Winter	1.032	−0.694	−1.771	1.562	[0.15,0.51]
	Spring	1.049	−0.822	−1.250	1.124	[0.23,0.61]
	Summer	0.998	−0.583	−1.392	0.995	[0.27,0.63]
	Autumn	1.019	−0.874	−0.964	0.909	[0.22,0.55]

Source: Reference [1]. Winter = November to February; spring = March, April; summer = May to August; autumn = September and October.

The coefficients c_0, c_1 and c_2 depend on latitude and season. The coefficients of the polynomial are listed in Table 3.

It was verified during mapping of the diffuse irradiation over the ESRA geographical range that the discontinuities of the estimated value of $(D_d)_m$ at the two interfaces of the latitude bands selected 56°N and 52°N are very small [1].

6.4.3 Estimating the Daily Diffuse Horizontal Irradiation from the Daily Global Irradiation

A different formula is recommended in ESRA [1] for estimating daily D_d values from daily G_d values. The *overall* test findings were that the models of [34] and [35] were the most appropriate for this purpose. This Erbs et al. model allows for seasonal variations (to some extent). So it was selected as the most appropriate current algorithm in ESRA [1]. For sunset hour angle $\omega_s < 81.4°$:

$$\frac{D_d}{G_d} = \begin{cases} 1.0 - 0.2727 KT_d + 2.4495 KT_d^2 - 11.9514 KT_d^3 + 9.3879 KT_d^4, & \text{for } KT_d < 0.715 \\ 0.143, & \text{for } KT_d \geq 0.715 \end{cases}$$

(23)

For sunset hour angle $\omega_s \geq 81.4°$:

$$\frac{D_d}{G_d} = \begin{cases} 1.0 - 0.2832 KT_d + 2.5557 KT_d^2 - 0.8448 KT_d^3, & \text{for } KT_d < 0.722 \\ 0.175, & \text{for } KT_d \geq 0.722 \end{cases}$$

(24)

6.4.4 Estimating the Monthly Mean Daily Profile of $D_h(t)$

The final step is to obtain the respective average daily profile of the hourly diffuse horizontal irradiation $D_{hm}(t)$. This is estimated from the respective daily sum (D_d) found using Section 6.1, Equation (18). This method adopts the simple model of Liu and Jordan [35], developed based on North American data:

$$D_{hm}(t) = (D_d)_m r_{om}(t) \qquad (25)$$

and $r_{om}(t)$ is found from Equation (18). Note that because the global- and diffuse-radiation profiles are generated independently, it is necessary on rare occasions to reset some diffuse radiation values close to sunrise and sunset hours to make sure that $D_{hm}(t) \leq G_{hm}(t)$. When this happens, the $D_{hm}(t)$ profile must be suitably renormalised to yield the input daily sum $(D_d)_m$.

7. THE ESTIMATION OF THE ALL SKY IRRADIATION ON INCLINED PLANES FROM HOURLY TIME SERIES OF HORIZONTAL IRRADIATION

7.1 Estimating the Components of Slope Irradiation from First Principles

This section provides guidance on the conversion of observed hourly time series of global and diffuse irradiation on horizontal surfaces into the corresponding hourly components of irradiation on inclined planes. The method assumes the availability of an hourly time series of global and diffuse horizontal irradiation derived from a reliable meteorological source. If observed data on diffuse irradiation are not available, they have to be filled first using the algorithmic approaches given in Section 6. If an observed time series is to be used for simulation, any gaps in the horizontal data should first be filled using appropriate techniques [7]. The data may be generated stochastically (see Section 4). Then there should be no gaps.

The tilt of the irradiated surface from the horizontal plane is defined as β. This needs to be stated in radians for some of the algorithms used. The azimuth angle of the surface is α (see Figure 20 in Section 5). The azimuth angle is measured from due south in the northern hemisphere and from due north in the southern hemisphere. Directions to the west of north–south are positive, east is negative.

The inclined surface total short-wave irradiation for any hour is obtained by summation of the three slope irradiation components; the hourly slope beam component $B_h(\beta,\alpha)$, the hourly slope sky diffuse component $D_h(\beta,\alpha)$, and the hourly slope ground reflected component $R_{gh}(\beta,\alpha)$. Each component has to be separately estimated:

$$G_h(\beta, \alpha) = B_h(\beta, \alpha) + D_h(\beta, \alpha) + R_{gh}(\beta, \alpha) \tag{26}$$

The associated hour-by-hour solar geometry has to be developed first, using the algorithms in Section 5 provide a matching time series of solar geometric data, in order to start the process. When observed data for specific hours are involved, an accurate formula for the declination should be used. The Bourges formula given in Section 5 is accurate and simple in engineering use.

7.2 Direct Beam Irradiation on Inclined Planes

Knowing the midhour observational times of the input data, the detailed solar geometry can be established using Section 5, making appropriate adjustments in the sunrise and sunset hour. The astronomical sunrise and sunset times must be calculated each day first to do this. The angle of incidence of the beam on the plane is also a required geometrical input (see Equation (13)).

The estimation of the slope beam irradiance from the beam normal irradiance is straightforward once the cosine of the angle of incidence of the solar beam for the inclined plane under consideration has been established:

$$\begin{aligned}(\beta, \alpha) &= B_n \cos\nu(\beta, \alpha) \quad \text{for } \cos\nu(\beta, \alpha) > 0 \\ B(\beta, \alpha) &= 0 \quad \text{otherwise}\end{aligned} \tag{27}$$

However, this process is dealing with observed irradiation data summarised on an hour-by-hour basis. The Sun is not necessarily on the selected surface for the whole of the summation period. Outside the sunrise and sunset hour, the hourly mean beam normal irradiation B_{nh} may be estimated as

$$B_{nh} = (G_h - D_h)/\sin\gamma_s \tag{28}$$

where γ_s is the midhour solar altitude. In the sunrise and sunset hour, account must be taken of the proportion of the hour when the Sun is above the horizon. Then

$$B_{nh} = (G_h - D_h)\Delta T/\sin\gamma_s \tag{29}$$

where γ_s is now the solar altitude at a time halfway between sunrise and the end of the sunrise hour or halfway between the beginning of the sunset hour and sunset, and ΔT is the length of time during the sunrise or sunset hour during which the Sun is above the horizon.

The accurate estimation of beam irradiation on inclined planes using standard observational time periods of 1 hour also raises complications, because undetected in the hour-by-hour calculation process the Sun may move off the surface during the hour in question. A simple irradiation approximation can be achieved for hours when the Sun is on the surface for the whole hour by using the midhour angle of incidence and setting the hourly irradiation equal to the midhour irradiance. Thus,

$$B_h(\beta, \alpha) = B_{nh} \cos \nu(\beta, \alpha) \tag{30}$$

The ESRA methodology [1] reduces the risk of errors by calculating at 6-minute intervals throughout each hour.

7.3 The Estimation of the Hourly Diffuse Irradiation on Inclined Surfaces from the Hourly Horizontal Diffuse Irradiation

This section is based on research by Muneer [36]. It includes some recent small improvements both by Muneer himself and by the CEC ESRA team [1]. The method requires, as inputs, hourly values of global and diffuse horizontal irradiation, G_h and D_h. The method has been tested and selected as a recommended algorithm in the CEC *European Solar Radiation Atlas* project. It gave better results in tests using a set of European-wide observed inclined surface data than the currently widely used Perez model [31]. European tests showed, for Sun-facing surfaces, that the model yields similar values to the Perez model at the day-by-day hourly level. It yielded better results at the monthly mean level. It gave more accurate values for surfaces facing away from the Sun than the algorithm developed by Perez et al. [31], for both monthly mean hourly values and daily hour-by-hour values.

The algorithms distinguish between potentially sunlit surfaces and surfaces that are not potentially sunlit. For Sun-facing surfaces, there are different algorithms for low sun (below 5.7°) and high sun (above 5.7°). For potentially insolated surfaces, a distinction is also made between overcast hours and nonovercast hours. An overcast sky hour is defined as an hour having $G_h - D_h < 5$ Wh m^{-2}. A modulating function K_b is first calculated as

$$K_b = B_h/(\varepsilon I_0 \sin \gamma_s) = (G_h - D_h)/(\varepsilon \times 1367 \sin \gamma_s) \tag{31}$$

where I_o is the solar constant (1367 Wm^{-2}), ε is the correction to mean solar distance on day J, and γ_s is the solar altitude angle. These angles have been defined previously. K_b expresses the horizontal beam irradiance as a ratio to the extraterrestrial horizontal irradiance, corrected to mean solar distance. Then a diffuse function $f(\beta)$ for slope β is calculated. β *must be expressed in radians.* This function is defined as

$$f(\beta) = \cos^2(\beta/2) + [2b/\pi(3+2b)][\sin\beta - \beta\cos\beta - \pi\sin^2(\beta/2)] \quad (32)$$

where b takes the following values: shaded surface, 5.73; sunlit surface under overcast sky, 1.68; sunlit surface under nonovercast sky, −0.62.

As an improvement, for certain specific areas where appropriate observed data exist, Muneer has suggested an alternative way of evaluating $2b/\pi(3+2b)$ to be applied to the sunlit surface cases only.

Northern Europe considered represented by Bracknell:

Replace $2b/\pi(3+2b)$ with $0.00333 - 0.4150K_b - 0.6987K_b^2$ \quad (33)

Southern Europe considered represented by Geneva:

Replace $2b/\pi(3+2b)$ with $0.00263 - 0.7120K_b - 0.6883K_b^2$ \quad (34)

Equation (32) for a vertical surface reduces to

$$f(\beta) = 0.5 + [2b/\pi(3+2b)](1 - 0.5\pi) \quad (35)$$

For a sunlit vertical surface, under the nonovercast sky conditions, using $b = -0.62$:

$$f(\beta) = 0.5 + [2(-0.62)/\pi(3 - 2\times 0.62)](1 - 0.5\pi) = 0.628$$

If the vertical surface is not potentially sunlit—i.e., $\cos v(\beta,\alpha)$ is zero or less, then Equation (35) must be evaluated using $b = 5.73$, and one obtains for a vertical surface

$$f(\beta) = 0.5 + [2\times 5.73/\pi(3 + 2\times 5.73)](1 - 0.5\pi) - 0.357$$

For *a potentially sunlit surface under an overcast sky,* using $b = 1.68$, one obtains for a vertical surface

$$f(\beta) = 0.5 + [2\times 1.68/\pi(3 + 2\times 1.68)](1 - 0.5\pi) = 0.404$$

For *sunlit surfaces,* with $\gamma_s > 5.7$ degrees, $D_h(\beta,\alpha)$ is found using the following formula:

$$\begin{aligned}D_h(\beta,\alpha)/D_h &= f(\beta)(1 - K_b) + K_b\cos v(\beta,\alpha)/\sin\gamma_s \\ &= 0.628(1 - K_b) + K_b\cos v(\beta,\alpha)/\sin\gamma_s \text{ for a vertical surface}\end{aligned} \quad (36)$$

where D_h is the hourly diffuse irradiation on the horizontal plane, and $\cos v(\beta,\alpha)$ is the angle of incidence on the surface.

For a potentially sunlit surface with an overcast sky $K_b = 0$ and for all solar altitudes:

$$D_h(\beta,\alpha)/D_h = f(\beta) = 0.404 \quad \text{for a vertical surface} \tag{37}$$

For *surfaces in shade*—i.e., $\cos v(\beta,\alpha) = 0$—and for all solar altitudes:

$$D_h(\beta,\alpha)/D_h = f(\beta) = 0.357 \quad \text{for a vertical surface} \tag{38}$$

As there is not a perfect conjunction in the functions at the instant the Sun changes from being just on the surface to being just off the surface, a small adjustment was made in the program for the CECESRA project to ensure continuity in estimation. Linear interpolation was used to achieve a smooth transition in this zone, using a zone of 20° width either side of the on–off Sun switch position.

7.4 The Estimation Process for Sunlit Sun-Facing Surfaces for Solar Altitudes Below 5.7°

The above Muneer algorithm is not suitable for sunlit surfaces below a solar elevation of 5.7°. In this region, Muneer has proposed applying an adjustment based on the Temps and Coulson algorithms. The original Temps and Coulson algorithm [37] for cloudless skies is

$$D_c(\beta,\alpha)/D_c = \cos^2(\beta/2)[1 + \sin^3(\beta/2)][1 + \cos^2 v(\beta,\alpha)\sin^3(90 - \gamma_s)] \tag{39}$$

This formula was modified by Klucher [38] to become an all-sky model by introducing a modulating function, F2. F2 was set by Klucher as $1 - (D_h/G_h)^2$. Muneer substituted K_b in Equation (39), giving

$$D_c(\beta,\alpha)/D_c = \cos^2(\beta/2)[1 + K_b\sin^3(\beta/2)][1 + K_b\cos^2 v(\beta,\alpha)\sin^3(90 - \gamma_s)] \tag{40}$$

7.5 Ground-Reflected Irradiation

Three assumptions are implicit in the method adopted for making estimates of the irradiation reflected from the ground. The first is that the ground surface reflects isotropically. The second assumption is that the ground surface is fully irradiated. The third is that there is no reflected energy other than that directly reflected from the level ground. A ground albedo of 0.2 has been adopted in the ESRA [1] and CIBSE Guide J [7] tables. This value can be changed if better knowledge exists about the ground cover in different months. Table 4 provides typical values of the

Table 4 Typical albedo values for various ground types

Surface type	Albedo
Grass (July, August, UK)	0.25
Lawns	0.18–0.23
Dry grass	0.28–0.32
Uncultivated fields	0.26
Bare soil	0.17
Macadam	0.18
Asphalt	0.15
Concrete, new before weathering	0.55
Concrete, weathered industrial city	0.20
Fresh snow	0.80–0.90
Old snow	0.45–0.70
Water surfaces for different values of solar altitude	
$\gamma_s > 45°$	0.05
$\gamma_s = 30°$	0.08
$\gamma_s = 20°$	0.12
$\gamma_s = 10°$	0.22

ground albedo for different types of surface. The most important albedo change is that due to snow cover. Table 4 also demonstrates the directional reflection characteristics of water surfaces at different solar altitudes. Late-afternoon water-reflected gains can be important in some locations.

Additional information about ground-reflected irradiation in the UK may be found in [39]. The reflected irradiance $R_{gh}(\beta, \alpha)$, which is assumed independent of orientation α, is simply estimated as

$$R_{gh}(\beta, \alpha) = \gamma_g \rho_g G_h \qquad (41)$$

where ρ_g is the ground albedo; r_g is ground slope factor, with $r_g = (1 - \cos \beta)/2$; and G_h is the hourly irradiation falling on the ground.

8. CONCLUSION

All the detailed climatological algorithmic procedures discussed here are accessible as pre-programmed computational chains in user-friendly forms in ESRA [1] and SoDa–IS [8]. ESRA is a European-based resource whereas SoDa–IS is an international resource. In addition, considering specifically PV built-in applications modules, the SoDa–IS contains (a) a daylighting service for SoDa; (b) an advanced user application—a grid-connected PV system; (c) an advanced user application—solar home system (PV nongrid connected); and (d) a daily temperature information database covering the domain latitude 34°N to 68°N longitude 20°W to 50°W on a grid of 5 minutes of arc between 1996 and 2001. The ESRA computational toolbox contains a facility for the assessment of PV grid-connected systems and a PV stand-alone system with batteries. The SoDa–IS system also contains a facility for estimating ultraviolet irradiation on inclined planes coupled to the internationally based solar-radiation data-generation facility. This facility could be useful in the assessment of the long-term environment exposure risks to PV systems. The content of the METEONORM CD-ROM [5] is closely related to SoDa–IS and incorporates many facilities that have evolved from the participation of METEOTEST in the European Commission supported program SoDa–IS, including the estimation of UV irradiation. It also incorporates the latest advanced stochastic models to aid the effective simulation of climate variability thus enabling effective performance risk assessment of photovoltaic systems at any place across the world.

ACKNOWLEDGEMENTS

The material in this chapter has drawn very heavily on research sponsored by the Commission of the European Communities. It reflects more than 30 years of EC research support to the systematic study of solar-radiation climatology in which the author has been deeply involved. Two especially important EU actions in which the author has participated have been:

- the fourth *European Solar Radiation Atlas* (EU Contract JOULE II Project Number JOU2-CT-94-00305), and
- Integration and Exploitation of Networked Solar Radiation Databases for Environmental Monitoring (SoDa), funded in part under the Information Societies Technology Programme (EU Contract IST-1999-122245 SoDa). The SoDa project was led by Professor Lucien Wald, Ecole des Mines de Paris, BP 207, 06904 Sophia Antipolis cedex, France.

On the personal level, I would especially like to express my thanks to my key Swiss collaborator in the SoDa project, *Jan Remund* of Meteotest, Bern, Switzerland, and to my two academic colleagues who have worked with me on the EU SoDa programme in the Department of Physics at UMIST Manchester, Dr Ann Webb and Richard Kift. I thank them for their considerable help and advice.

REFERENCES

[1] ESRA, Les Presses de l'Ecole des Mines de Paris, vol. 2. Database and exploitation software contains photovoltaics applications module, K. Scharmer, J. Grief, co-ordinators, European Solar Radiation Atlas, fourth ed., 2000.

[2] J. Remund, S. Kunz, METEONORM—a comprehensive meteorological database and planning tool for system design. Proceedings of the 13th European Photovoltaic Solar Energy Conference, Nice, 1995, pp. 733–735.
[3] J. Remund, S. Kunz, METEONORM Version 3.0. Nova Energie GmbH, Schachenallee 29, CH-5000 Aarrau, Switzerland, 1997.
[4] J. Remund, S. Kunz, METEONORM Version 4.0. Nova Energie GmbH, Schachenallee 29, CH-5000 Aarrau, Switzerland, 1999.
[5] J. Remund, S. Kunz, METEONORM Version 5.0. METEOTEST, Fabrikstrasse 14, 3012 Bern, Switzerland, 2003.
[6] RETSCREEN®INTERNATIONAL, Prepared by CANMET Energy Technology Centre, Varennes Canada on behalf of National Resources Canada, 2003, Visit <http:/www.retscreen.net/>. These programs include a PV module.
[7] CIBSE Guide J Weather, Solar and Illuminance Data, Chartered Institution of Building Services Engineers, 222 Balham High Road, London SW12 9BS, UK, 2002.
[8] SoDa–IS, Integration and exploitation of networked solar radiation databases for environmental monitoring. Either contact the welcome Web site at www.helioclim.net or consult www.soda-is.com. Contains photovoltaic applications modules, 2003.
[9] SATELLIGHT, Fontoynont, M., Dumortier, D., Heinemann, D., Hammer, A., Olseth, J., Skartveit, A., Ineichen, P., Reise, C., Page, J., Roche, L., Beyer, H.G. and Wald, L. Processing of Meteosat data for the production of high-quality daylight and solar radiation data available on a web server. Application to Western and Central Europe. JOR3-CT95-0041, 1995.
[10] M. Fontoynont, D. Dumortier, D. Heinemann, A. Hammer, J. Olseth, A. Skartveit, et al., Satellight: a WWW server which provides-high quality daylight and solar radiation data for Western and Central Europe. Proceedings of the 9th Conference On Satellite Meteorology and Oceanography, vol. 1, Darmstadt, Germany, EUM P22, 1998, pp. 434–435.
[11] C. Rigollier, O. Bauer, L. Wald, On the clear sky model of the ESRA with respect to the heliostat method, Sol. Energy 68 (2000) 33–48.
[12] J. Remund, M. Levevre, T. Ranchin, L. Wald, Constructing maps of the Linke turbidity factor, SoDa project deliverable D5-2-1, Internal Document, 2002, Meteotest, Bern. SoDa project working paper for European Commission. Refer also Remund, J., Wald L. Lefevre M., Ranchin T. and Page J. 2003 Worldwide Linke Turbidity Information. Proceedings of World Solar Congress, Götebourg, Sweden, June 14–19, 2003. In press.
[13] J. Remund, J.K. Page, Integration and exploitation of networked Solar radiation Databases for environmental monitoring (SoDa Project). Advanced parameters: WP 5.2b: Deliverable D5-2-2 and D5-2-3. Chain of algorithms: short- and long-wave radiation with associated temperature prediction resources. SoDa project working paper for European Commission, 2002.
[14] Aguiar and Page, 2000. Chapter 3 in reference [1].
[15] D. Dumortier, Prediction of air temperatures from solar radiation. SoDa project Working Partnership WP-5-2.c, deliverable 5-2-4. CNRS-ENTPE. Department Genie Civil et Batiment, Rue Maurice Audin, 69518 Vaulx-en-Velin, near Lyons, France, 2002. E-mail: dominique.dumortier@entpe.fr.
[16] F. Kasten, G. Czeplak, Solar radiation and terrestrial radiation dependent on the amount and type of cloud, Sol. Energy 24 (1980) 117–189.
[17] L. Gu, J.D. Fuentes, M. Garstang, J. Tota da Silva, R. Heitz, J. Sigler, H.H. Shugart, Cloud modulation of surface solar irradiance at a pasture site in southern Brazil, Agr. Forest Meteorol. 106 (2001) 117–129.
[18] F.O. Akuffo, A. Brew-Hammond, The frequency distribution of daily global irradiation at Kumasi, Sol. Energy 50 (1993) 145–154.

[19] S.O. Udo, Sky conditions at Ilorin as characterized by clearness index and relative sunshine, Sol. Energy 69 (2000) 45–53.
[20] J.K. Page, The estimation of monthly mean values of daily total short wave radiation on vertical and inclined surfaces from sunshine records for latitudes 40°N to 40°S. Proceedings of the UN Conf. on New Sources of Energy, Rome, vol. 4, 1964, pp. 378–390.
[21] A. Ianetz, I. Lyubansky, I. Setter, E.G. Evseev, A.I. Kudish, A method for characterization and intercomparison of sites with regard to solar energy utilization by statistical analysis of their solar radiation data as performed for three sites in the Israel Negev region, Sol. Energy 69 (2000) 283–294.
[22] I. Dirmhirn, Quoted by P. Valko, Swiss Meteorological Institute, Zurich, 1983, in IEA programme to develop and test solar heating and cooling systems, Task V Use of existing meteorological information for solar energy applications. Final draft. A literature search has showed that US DoE funded I. Dirmhirn to produce the report Solar energy potential, ultraviolet radiation, temperature and wind conditions in mountainous areas, DoE Grant Number EG-77-S-07-1656, completed 1982.
[23] W. Palz, J. Grief (Eds.), European Solar Radiation Atlas, 3rd Edition. Solar Radiation on Horizontal and Inclined Surfaces, Commission of the European Communities, Springer-Verlag, Berlin, Heidelberg, New York, 1996.
[24] R. Aguiar, M. Collares-Pereira, J.P. Conde, A simple procedure for generating sequences of daily radiation values using a library of Markov transition matrices, Sol. Energy 40(3) (1988) 269–279.
[25] R. Aguiar, M. Collares-Pereira, TAG: A time-dependent auto-regressive, Gaussian model, Sol. Energy 49(3) (1992) 167–174.
[26] C. Gueymard, Prediction and performance assessment of mean hourly global radiation, Sol. Energy 68(3) (2000) 285–303.
[27] M. Collares-Pereira, A. Rabl, The average distribution of solar radiation—correlations between diffuse and hemispherical and between daily and hourly insolation values, Sol. Energy 22 (1979) 155–164.
[28] J. Page, ESRA Task II Technical Report Number 14. Internal project report, 2000.
[29] J. Remund, L. Wald, J. Page, Chain of algorithms to calculate advanced radiation parameters. Proceedings of World Solar Congress, Göteborg, Sweden, June 14-19, 2003, In the press. This paper contains the new Markov transition matrices based on the used of the clear sky clearness index.
[30] E.P. Box, G.M. Jenkins, Time series analysis, Forecasting and control, Holden Day, San Francisco, 1970.
[31] R. Perez, R. Seals, P. Ineichen, R. Stewart, D. Menicucci, A new simplified version of the Perez Diffuse Irradiance Model for tilted surfaces, Sol. Energy 39(3) (1987) 221–231.
[32] A. Skartveit, J.A. Olsen, M.E. Tuft, An hourly diffuse fraction model with correction for variability and surface albedo, Sol. Energy 63 (1998) 173–183.
[33] B. Bourges, Improvement in solar declination calculations, Sol. Energy 35 (1985) 367–369.
[34] K. Erbs, Duffie, Estimation of the diffuse radiation fraction for hourly, daily and monthly-average global radiation, Sol. Energy 28 (1982) 293–302.
[35] B. Liu, R. Jordan, The interrelationship and characteristic distributions of direct, diffuse and total solar radiation, Sol. Energy 4 (1960) 1–19.
[36] T. Muneer, Solar radiation model for Europe, Building Serv. Eng. Res. Technol. 11 (4) (1990) 153–163.
[37] R.C. Temps, K.L. Coulsen, Solar radiation incident on slopes of different orientations, Sol. Energy 19 (1977) 179–184.

[38] T.M. Klucher, Evaluation of models to predict insolation on inclined surfaces, Sol. Energy 23 (1979) 111–114.

[39] G.S. Saluja, T. Muneer, Estimation of ground-reflected radiation for the United Kingdom, Building Serv. Eng. Res. Technol. 9(4) (1988) 189–196.

US DATA MAY BE FOUND IN THE FOLLOWING REFERENCES:

[40] NSRDB, User's Manual, National Solar Radiation Data Base (1961–1990), vol. 1, National Renewable Energy Laboratory, Golden, CO, USA, 1992.

[41] NSRDB, Final Technical Report, National Solar Radiation Data Base (1961–1990), vol. 2, National Renewable Energy Laboratory, Golden. CO, USA, 1995 (NERL/TP-463-5784).

[42] J.D. Garrison, A programme for calculation of solar energy collection by fixed and tracking collectors, Sol. Energy 73 (2003) 241–255.

APPENDIX: SOLAR ENERGY DATA FOR SELECTED SITES

Monthly mean daily global radiation on horizontal surfaces (Wh m^2)

	Jan.	Feb.	Mar.	Apr.	May	Jun.	July	Aug.	Sep.	Oct.	Nov.	Dec.	Year
Stockholm, 59.35°N	292	892	1947	3583	5289	5175	5072	3847	2431	1156	453	144	2523
London, 51.52°N	683	1222	2119	3561	4294	4481	4572	4022	2719	1617	853	486	2552
Freiburg, 48.00°N	869	1539	2603	3731	4736	5406	5528	4789	3411	2031	1122	708	3039
Geneva, 46.25°N	975	1553	2978	4153	4933	5819	6083	5178	3847	2228	1133	803	3307
Nice, 43.65°N	1758	2325	3658	4708	5906	6556	6681	5903	4514	2917	1878	1506	4026
Oviedo, 43.35°N	1589	2039	3272	3864	4497	4753	4678	4217	3681	2492	1717	1311	3176
Porto, 41.13°N	1928	2511	4069	5036	6189	6931	6683	6200	4781	3258	2011	1533	4261
Heraklion, 35.33°N	2189	2819	3797	5392	6319	7136	6853	6228	5153	3369	2203	1781	4437
Sde Boker, 30.90°N	3175	4361	5169	6767	7231	8333	8106	7381	6211	4908	3689	3056	5699
Ilorin, 8.58°N	4031	4631	4867	4803	4806	4156	3511	3253	3733	4317	4478	3725	4192

Monthly mean daily diffuse radiation on horizontal surfaces (Wh m^{-2})

	Jan.	Feb.	Mar.	Apr.	May	Jun.	July	Aug.	Sep.	Oct.	Nov.	Dec.	Year
Stockholm, 59.35°N	226	573	1147	1849	2348	2774	2613	2116	1358	720	317	126	1347
London, 51.52°N	483	823	1381	2023	2563	2773	2693	2286	1574	1001	577	369	1545
Freiburg, 48.00°N	610	976	1543	2109	2625	2826	2699	2324	1691	1156	721	511	1649
Geneva, 46.25°N	679	1027	1612	2142	2641	2791	2614	2299	1709	1225	769	577	1673
Niee, 43.65°N	800	1147	1647	2147	2542	2643	2462	2167	1680	1291	887	716	1677
Oviedo, 43.35°N	835	1172	1702	2208	2652	2820	2741	2436	1822	1332	917	742	1781
Porto, 41.13°N	903	1239	1674	2153	2498	2526	2475	2126	1715	1354	993	828	1707
Heraklion, 35.33°N	1146	1454	1939	2219	2506	2431	2423	2237	1835	1590	1225	1052	1838
Sde Boker, 30.90°N	1236	1836	1875	2400	2053	1747	1750	1472	1486	1414	1225	1058	1629
Ilorin, 8.58°N	1799	1805	1842	2035	2091	1949	2004	1893	1919	1829	1796	1578	1878

CHAPTER IIA-2

Energy Production by a PV Array

Luis Castañer[a], Sandra Bermejo[a], Tom Markvart[b], and Katerina Fragaki[b]
[a]Universidad Politecnica de Catalunya, Barcelona, Spain
[b]School of Engineering Sciences, University of Southampton, UK

Contents

1. Annual Energy Production	645
2. Peak Solar Hours: Concept, Definition, and Illustration	646
3. Nominal Array Power	648
4. Temperature Dependence of Array Power Output	650
5. Module Orientation	651
5.1 Fixed Tilt Arrays	651
5.2 Arrays with Tracking	652
6. Statistical Analysis of the Energy Production	653
7. Mismatch Losses and Blocking/Bypass Diodes	655
Acknowledgement	658
References	658

1. ANNUAL ENERGY PRODUCTION

The power produced by a PV system depends on a range of factors which need to be examined when the system is designed; it is also useful to assess the accuracy of simplified treatments where these factors are ignored or neglected. Such analysis is conveniently carried out by looking at the total annual energy produced by the system. A recent comprehensive study has identified seven factors influencing the annual performance of PV modules [1] and a brief summary of the main conclusions follows.

- *Cumulative solar irradiance.* Long-term irradiance profiles depend on surface orientation and possibly tracking. This factor depends on the location and varies between a reduction by about 25% for a vertical surface to over 30% increase for two axis tracking, in comparison with

Practical Handbook of Photovoltaics.
© 2012 Elsevier Ltd. All rights reserved.

a latitude-tilt fixed system. The effect of module orientation is considered further in Section 5 and is analysed in detail in Part 1.
- *Module power rating at standard test conditions.* Analysis of several PV technologies has shown that for the same power rating, all technologies were equivalent in terms of the expected annual energy production within 5% calculation error.
- *Operating temperature.* Analysis of various technologies and sites shows that the annual production can be reduced due to the operating temperature by a factor between 2% and 10%, depending on the module design, wind speed, mounting technique and ambient temperature. The effect of operating temperature is discussed quantitatively in Section 4.
- *Maximum power point voltage dependence on irradiance level.* a-Si and CdTe modules tend to have a value of the maximum power point voltage larger at low irradiance levels than at the standard 1-sun conditions. This fact can result in an additional 10% increase in annual energy production.
- *Soiling.* Soiling may account for up to a 10% of reduction of the annual energy production.
- *Variation in solar spectrum.* It is found that the effects of the hourly variation of the solar spectrum almost cancel out in a yearly basis. Amorphous silicon technology has the highest sensitivity to this effect, but the observed changes usually remain under 3%.
- *Optical losses when the sun is at a high angle of incidence (AOI).* The optical losses are due to the increased reflectance of the cover glass of the PV modules for AOI greater than approximately 60°. However, the effect on a long-term basis is relatively small (typically under 5%) although it may have larger effect on a seasonal basis (close to 10% for a vertical inclination).

2. PEAK SOLAR HOURS: CONCEPT, DEFINITION, AND ILLUSTRATION

The initial approximate analysis and design of a PV system is usually based on Peak Solar Hours (PSH): a convenient definition of the equivalent of one day. This concept is particularly useful for the first-order sizing of flat-plate (non-concentrating) arrays which operate under global radiation (see IIIa-3). The magnitude of Peak Solar Hours is equal to the length of an equivalent day with a constant irradiance equal to the 1-sun intensity (1 kW/m²), resulting in the same value of the daily radiation. This parameter has units of time, and when given in hours, it has the same numerical value as the total daily radiation in kWh/m²-day. Accordingly, the total generation of a PV array exposed to solar radiation a whole year can be estimated as

$$E_A = \sum_{i=1}^{365}(PSH)_i \, P_0 \qquad (1)$$

where $(PSH)_i$ is the value of the parameter PSH for day i and P_0 is the nominal array power under standard or reference conditions.

For arrays operating at the maximum power point, the normalised instantaneous power output P_A/P_{max} depends on temperature and irradiance, and Equation (1) is therefore only an approximation. A better estimate of the annual generation can be obtained by a model which uses actual values of the ambient temperature, and estimates the cell operating temperature using the $NOCT$ concept (see Section 4) [3] and the cell efficiency temperature coefficient [2]. Figure 1 shows the error involved in the use of Equation (1) rather than by taking into account the full temperature and irradiance dependence of the array power output. As can be seen, the average annual

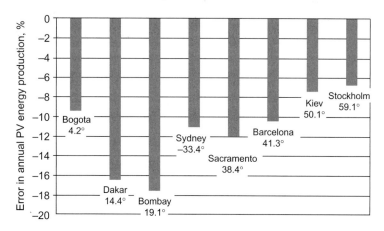

Figure 1 Error involved in the use of Equation (1) to calculate the annual energy output by a PV array at several locations worldwide. Numbers by locations indicate the latitude. *Source of the radiation and temperature data: Meteonorm program [11].*

errors lie between −2% and −15%, similar to the experimentally observed values discussed in Section 1. Clearly, the error is larger in locations where the yearly radiation and average temperature are higher.

3. NOMINAL ARRAY POWER

Results in Figure 1 are specific for arrays operating at the maximum power point This usually includes all systems with an inverter and larger DC systems, with power rating in excess of 1 kW or so. In these systems, the voltage is controlled by the maximum power point tracker (MPPT) to follow the optimum maximum power point voltage V_m.

The situation is somewhat different in smaller stand-alone systems where the array voltage is controlled by the load typically consisting of a charge regulator and a battery (Figure 2). Here, the array delivers power at a voltage close to the battery voltage V_{bat} and the reduction in array output shown in Figure 1 is compensated by a lower nominal array power instead of P_{max}. The design of a standard PV module consisting of 36 crystalline silicon cells has evolved from the need to charge a 12 V battery. In practical usage, the module then operates in the linear part of its I–V characteristic and supplies approximately the same current I_{sc} as at short circuit. The power P_A delivered by the array to the battery and load in parallel connection is then

$$P_A \cong V_{bat} I_{sc} \quad (2)$$

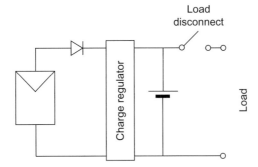

Figure 2 A typical configuration of a stand-alone photovoltaic system. The load disconnect is usually incorporated in the charge regulator.

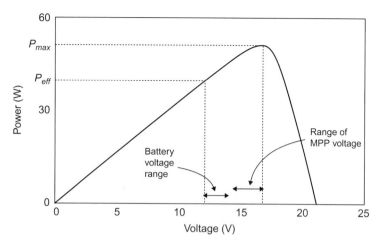

Figure 3 The power output from a typical crystalline silicon PV module, showing two characteristic power levels: P_{max}, the power at the maximum power point, and P_{eff}, the effective power output at the nominal battery voltage.

which takes into account the actual bias point of the module and the instantaneous value V_{bat} of the battery voltage. The voltage drop at the blocking diode which is not included in Equation (2) is discussed in more detail in Section 7. Losses in the charge regulator can often be neglected as most modern regulators have switches in place of diodes, and hence, the voltage drop is small.

The average value of P_A—the effective module power rating—is often approximated by nominally setting V_{bat} equal to 12 V:

$$P_{eff} = \langle P_A \rangle \cong 12 I_{sc} \tag{3}$$

Equation (3)—*valid for a system without maximum power point tracking*—is a useful approximation for the analysis of the long-term energy balance in stand alone systems, and for the development of sizing procedures, as discussed in Chapters IIa-3 and IIa-4. The accuracy of Equation (2) in a specific application can be verified for the actual daily load profile with the help of detailed system models.

As a practical example, BP Solar module with manufacturers rating of 85 W_p, I_{sc} = 5 A and maximum power point values V_m = 18 V, I_m = 4.72 A produces P_{eff} = 12 × 5 = 60 W, which is a reduction of about 29% on P_{max} at STC.

4. TEMPERATURE DEPENDENCE OF ARRAY POWER OUTPUT

The principal effect of temperature on the PV array output comes from the temperature dependence of the open-circuit voltage (see Figure 4) which can be described by

$$V_{oc}(T_c) = V_{oc}(STC) + \frac{dV_{oc}}{dT}(T_c - 25) \tag{4}$$

where T_c is the cell temperature and dV_{oc}/dT is the temperature coefficient. If an accurate measured value is not known, the following theoretical expression

$$\frac{dV_{oc}}{dT} = \frac{V_{oc} - E_{g0} - \gamma k_B T_c}{T_c} \times \text{number of cells in the module} \tag{5}$$

can be used, where the energy gap E_{g0} (extrapolated linearly to 0 K) and the thermal energy $k_B T_c$ are given in electronvolts, and $\gamma = 3$ for silicon [2]. For a typical module of 36 crystalline silicon cells, Equation (4) gives a temperature coefficient of approximately -80 mV/°C. The cell temperature T_c can be estimated from the ambient temperature T_a and the

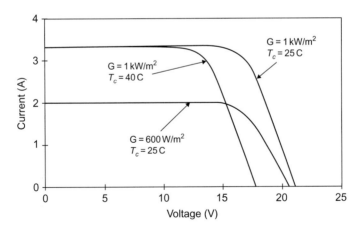

Figure 4 The IV characteristic of a typical crystalline silicon PV module as a function of irradiance G and temperature. T_a denotes the ambient temperature, and T_c the is temperature of the solar cells.

irradiance G with the use of a parameter called the Nominal Operating Cell Temperature (NOCT)

$$T_c = T_a + \frac{NOCT - 20}{800} G \qquad (6)$$

where NOCT is expressed in °C and the irradiance G in W/m² [3]. If NOCT is not known, 48°C is recommended as a reasonable value which describes well most of the commonly used PV modules.

Under most circumstances, the temperature dependence of the short-circuit current can be neglected (for example, the temperature coefficient of the BP585 module is equal to 0.065%/°C).

The measurement of the temperature coefficient of PV modules is discussed in Chapter III-2. A theoretical discussion of the temperature dependence of solar cell parameters including other materials can be found in [4]. A scheme that can be used to calculate the temperature dependence of the power output at maximum power is discussed by Lorenzo [5]. A simplified version which uses the assumption on temperature-independent fill factor is given in [6]. At maximum power, for example, BP Solar give the value of −0.5%/°C for the module BP585.

5. MODULE ORIENTATION
5.1 Fixed Tilt Arrays

The output from the PV generator depends on solar radiation incident on the inclined panels of the PV array. Methods to analyse solar radiation data are discussed extensively in Part I; software packages which were developed for PV applications are reviewed in Chapter IIa-4. In some applications, the array orientation will be constrained by the nature of the support system: for example, the orientation of a building-integrated array will normally be dictated by the orientation of the roof or facade where the array is to be installed. For a freestanding PV array, the most important consideration in deciding the array orientation is to maximise the energy collection by the inclined PV panels. This will frequently depend on the seasonal nature of the load. The typical examples include

- Some applications require energy only during the summer months. This may be the case, for example, for many irrigation systems.

Table 1

Season	Optimum angle of inclination
Summer	Latitude $-15°$
Winter	Latitude $+15°$
Year average	Latitude

- If load is to be powered over the entire year, energy supply during the winter months is likely to be key to satisfactory operation, and the panel orientation should ensure optimum energy capture during the month with the lowest daily radiation. In the Northern hemisphere, this is usually be the month of December.
- The annual average is sometimes used in locations with little variation of daily solar radiation during the year. The PV system battery is then used partly as a seasonal energy storage, and care should be exercised to choose an appropriate type of battery for this purpose (see Chapter IIb-2).

Practical 'rules-of-thumb' for panel inclination are summarised in Table 1 [7]. The use of the recommended values in Table 1 is normally justified by the simplicity and by the fact that the energy produced by the array is not very sensitive to the precise angle of inclination. For best results, however, one of the software packages discussed in Chapter IIa-4 and Part I should be consulted.

5.2 Arrays with Tracking

PV arrays which track the sun can collect a higher amount of energy than do those installed at a fixed tilt. The use of tracking is common for concentrator arrays which—at least for appreciable concentration ratios—collect only the direct (beam) radiation. The relationship between the annual solar radiation captured by a tracking system and a fixed-tilt panel inclined at the angle of latitude for a number of locations across the world is illustrated in Figure 5 [8]. It is seen that, on a yearly basis, the energy capture by a tracking flat-plate system is increased by more than 30% over a fixed array at latitude inclination. At the same time we note, however, that a tracking concentrator system will collect more energy than will a flat-plate system only in locations with predominantly clear skies. Practical methods of estimating the solar energy available to concentrator systems are discussed further in Chapter IId-1.

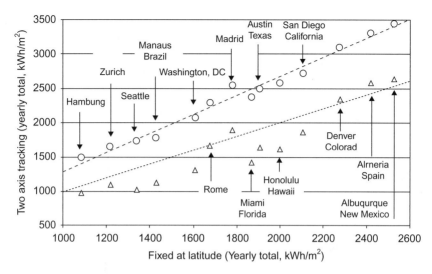

Figure 5 Comparison of energy collected by a panel at a fixed tilt equal to the angle of latitude, with two axis tracking. Circles are measured values of global radiation, triangles correspond to direct beam. The dashed line is a fit to the global radiation points, corresponding to an increase of 33.5%. The dash-dot line shows solar radiation on the fixed tilt plane at latitude inclination *(adapted from reference 18])*.

6. STATISTICAL ANALYSIS OF THE ENERGY PRODUCTION

In a number of situations—for example, when considering the rating of an inverter in grid-connected systems (see Chapter IIa-4)—it is useful to know the statistical distribution of power produced by the array. To this end, let us consider the (average) number of hours that solar irradiance on the array falls in a small interval δG between G and $G + \delta G$. This number of hours is proportional to δG and will be denoted $h_1(G)\delta G$. Clearly

$$\int_0^\infty h_1(G)dG = H \qquad (7)$$

where H is the total number of hours in a year which can be taken equal to 8760. The integral of the incident energy $Gh_1(G)\delta G$, expressed in kWh. is then equal to the sum of Peak Solar Hours over the whole year:

$$\int_0^\infty Gh_1(G)dG = \sum_{i=1}^{365} PSH_i \qquad (8)$$

The statistics of power generation by the array can be analysed in a similar fashion. If $h_P(P)\delta P$ denotes the number of hours the array generates power between P and $P + \delta P$, the energy produced in this interval is equal to $Ph_P(P)\delta P$. The integral of this function is the total energy produced in a year E_A:

$$\int_0^\infty Ph_1(P)dP = E_A \qquad (9)$$

The energy E_A was already discussed in Sections 1 and 2 which also analysed the validity of approximating E_A by the sum (1).

The functions $Gh_I(G)$ and $Ph_P(P)$ are important for the sizing of inverters and will be discussed further in Chapter IIa-4, Section 4. In practice, these functions can be obtained from a representative sample of solar radiation and array power output data by sorting the results in classes and counting the energy generated in the whole year when the irradiance or power output belongs to each class. This values can also be plotted in absolute values [1] or as a percentage [3]. A detailed statistical analysis of the irradiance for different parts of the world can be found in Part I.

For comparison purposes, it is usual to consider the array output function $Ph_P(P)$ as a function of the ratio $P_{max}/P_{max}(STC)$, where $P_{max}(STC)$ is the peak power rating of the array at STC, as illustrated in Figure 6

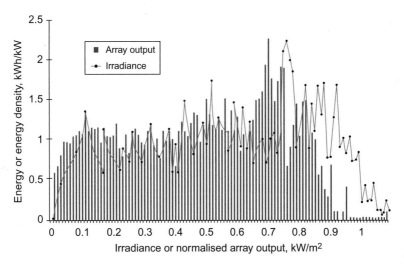

Figure 6 The array output energy $Ph_P(P)$ and solar radiation $Gh_I(G)$ as functions of the ratio $P_{max}/P_{max}(STC)$ and G, respectively. *Source of data: STaR Facility, University of Southampton.*

using observed data in the Northern Europe. The function $Gh_I(G)$ is also shown. At lower latitudes, the peak of both graphs is shifted towards higher values at the expense of energy output at lower power levels.

7. MISMATCH LOSSES AND BLOCKING/BYPASS DIODES

A number of issues arises in an array consisting of several series or parallel connected modules. Mismatch losses may occur, for example, due to non-uniform illumination of the array or because different modules in the array have different parameters. As a result, the output power from the array will be less than the sum of the power outputs corresponding to the constituent modules. Worse still, some cells may get damaged by the resulting excess power dissipation by what is called a *hot spot formation*.

The PV array in the dark behaves as a diode under forward bias and, when directly connected to a battery, will provide a discharge path for the battery. These reverse currents are traditionally avoided by the use of blocking (or string) diodes (Figure 7). Blocking diodes also play a role in preventing excess currents in parallel connected strings. The mismatch losses which result from shading a part of a series string are illustrated in Figure 8, which shows the I–V characteristics of five series connected solar cells. When one cell is shaded, the current output from the string is determined by the current from the shaded cell. At or near the short circuit, the shaded cell dissipates the power generated by the illuminated

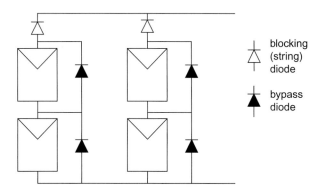

Figure 7 An array consisting of two strings, each with a blocking diode. Each module is furnished with a bypass diode. In practice, it recommended that bypass diode are used for every series connection of 10–15 cells [2].

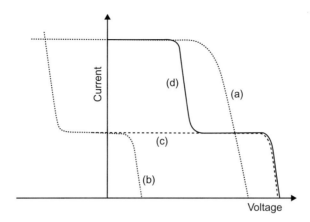

Figure 8 The I–V characteristic of a series string, with four illuminated and one shaded cell. (a) Four illuminated cells. (h) One shaded cell with a bypass diode. (c) Four illuminated cells and one shaded cell, no diode. (d) Four illuminated cells and one shaded cell, with a bypass diode across the shaded cell.

cells in the string; if the number of cells is substantial, the resulting heating may damage the glass, the encapsulant or the cell. This problem can be alleviated by the use of bypass diodes. It should be noted, however, that the resulting I–V characteristic now has two local maxima, an effect that may affect adversely the maximum power point tracking.

The use of blocking diodes has been subject of some discussion and their use should be evaluated in each specific situation, focusing on the trade off between the power losses due to voltage drops across the diode, and the losses through reverse currents in the dark if the diodes are omitted. With the use of modern charge regulators and inverters which disconnect the array in the dark, the blocking diodes may become redundant in any case.

As an illustration, Figure 9 compares the losses incurred with and without the use of blocking diodes, in an array shown in Figure 9(a) of a stand-alone system with a battery, without a maximum power point tracker. If no diodes are connected and one of the strings is in dark and the other is illuminated with the irradiance in Figure 9(d), the total power dissipated in the dark string is shown in Figure 9(b). It can be seen that the power dissipated by the dark string never reaches more than some 200 mW, which is less that 0.1% of the nominal peak power of the array. When a blocking diode is included to avoid dissipation in the dark

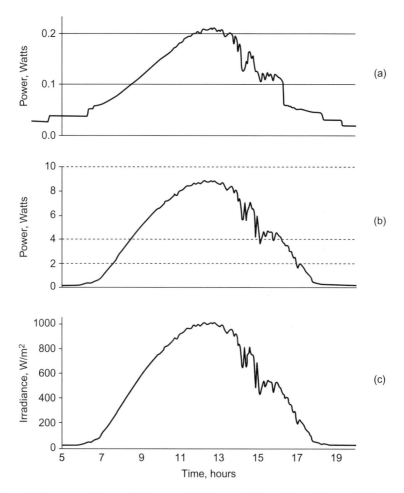

Figure 9 The power losses during one day in one of the strings of array in (a) resulting from reverse currents through the string in the dark (shaded) if the blocking diode is omitted (b). The power dissipated in the blocking diode of one string (c). (d) shows the irradiance used in the modelling. Each module of the array (a) consists of 32 cells in series with nominal power of 45.55 Wp under standard AM1.5, 1 kW/m² irradiance.

string, the power dissipated is reduced to the level of tenths of milliwatts. The power dissipated by the diode itself, however, is much higher and reaches several watts, as shown in Figure 9(c).

This has different implications for grid-connected and stand-alone systems. Grid-connected systems normally have MPPT features and the power lost in the diode reduces the available power generation, thus reducing the overall system efficiency. In a stand-alone system without MPPT, the operating point at the load is set by the battery voltage and—unless the diode connection brings the operating point beyond the maximum power point—the energy supplied to the load remains the same. The energy dissipated in the diode comes from the additional energy produced by the PV array.

In low-voltage applications there are, however, concerns about potential safety hazards if no fuses or blocking diodes are used [9], especially under fault or other unusual operating conditions. These issues have been addressed by simulation and experimental work which conclude that fuses may not be the best solution to the problem and that blocking diodes may be more reliable.

Recommendations as to the installation of blocking diodes in grid connected systems for a number of countries in the Task 5 of the International Energy Agency can be found in reference [10].

ACKNOWLEDGEMENT

We are grateful to Santiago Silvestre for help with modelling and useful comments.

REFERENCES

[1] D.L. King, W.E. Boyson, J.A. Kratochvil, Analysis of factors influencing the annual energy production of photovoltaic systems, Proceedings of the 29th IEEE Photovoltaic Specialist Conference, New Orleans, 2002, pp. 1356–1361.
[2] S.R. Wenham, M.A. Green, M.E. Watt, Applied Photovoltaics, Centre for Photovoltaic Devices and Systems, UNSW.
[3] R.G. Ross, M.I.Smockler, Flat-plate solar array project, Jet Propulsion Laboratories Publ. No. 86–31, 1986.
[4] J.C.C. Fan, Theoretical temperature dependence of solar cell parameters, Sol. Cells 17 (1986) 309–315.
[5] E. Lorenzo, Solar Electricity, Progensa, Seville, 1994.
[6] L Castaner, Photovoltaic systems engineering, in: T. Markvart (Ed.), Solar Electricity, second ed., John Wiley & Sons, Chichester, 2000, Chapter 4.
[7] T. Markvart (Ed.), Solar Electricity, second ed., John Wiley & Sons, Chichester, 2000, Chapter 2.
[8] E.C. Boes, A. Luque, Photovoltaic concentrator technology, in: T.B. Johansson, et al. (Eds.), Renewable Energy: Sources of Fuel and Electricity, Earthscan, London, 1993, p. 361.
[9] J.C. Wiles, D.L. King, D.L. 1997. Blocking diodes and fuses in low-voltage PV systems, Proceedings of the 26th IEEE Photovoltaic Specialist Conference, Anaheim, pp. 1105–1108.
[10] PV System Installation and Grid-Interconnection Guidelines in Selected IEA Countries. International Energy Agency Report IEA PVPST5–04–2001.
[11] METEONORM version 4.0, Meteotest, Switzerland, <http://www.meteonorm.com/>

CHAPTER IIA-3

Energy Balance in Stand-Alone Systems

Luis Castañer[a], Sandra Bermejo[a], Tom Markvart[b], and Katerina Fragaki[b]

[a]Universidad Politecnica de Catalunya, Barcelona, Spain
[b]School of Engineering Sciences, University of Southampton, UK

Contents

1. Introduction 659
2. Load Description 663
    2.1 Electricity Consumption by Lighting and Electrical Appliances 664
        2.1.1 Daily Energy-Balance Dynamics 666
3. Seasonal Energy Balance 668

1. INTRODUCTION

An important part of stand-alone PV system design is concerned with the balance between energy produced by the PV array and energy consumed by the load. Any short-term mismatch between these two energy flows is compensated by energy storage, usually in the form of a rechargeable battery. Other types of storage—for example, water pumped into a tank—may also be encountered in certain applications. Considerations of energy balance cover a number of characteristic time scales and can be discussed in terms of the energy stored in the battery (in other words, the battery state of charge), which is shown schematically in Figure 1.

The energy balance displays various cycles that occur with different degree of regularity [1]. In the daily cycle, the battery is charged during the day and discharged by the nighttime load or at any time that energy consumption exceeds supply (see Section 2.2). The depth of discharge in

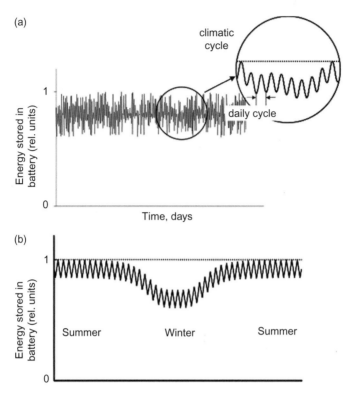

Figure 1 A schematic diagram of the characteristic times scales of energy balance in stand-alone PV systems, as indicated by the energy stored in the battery: (a) the daily and climatic cycles; (b) the seasonal cycle. (adapted from [1]).

the daily cycle varies from application to application, but for systems without a backup generator, it is always fairly shallow.

Superimposed on the daily cycle is a climatic cycle that is due to variable climatic conditions. The climatic cycle occurs when the daily load exceeds the average design value of the daily energy supply from the PV generator. In some systems (usually in applications where reliability is not paramount), the battery may act as seasonal energy storage, and the climatic cycle then extends over a substantial part of the season.

Figure 2 illustrates these phenomena by plotting the battery voltage (a measure of the state of charge) and the energy flow in and out of the battery as measured by the charge (integrated current) removed from the battery. The data correspond to an 11-day climatic cycle observed during the operation of an experimental PV system in the south of England in December 2002. [2] The system parameters are plotted alongside the solar

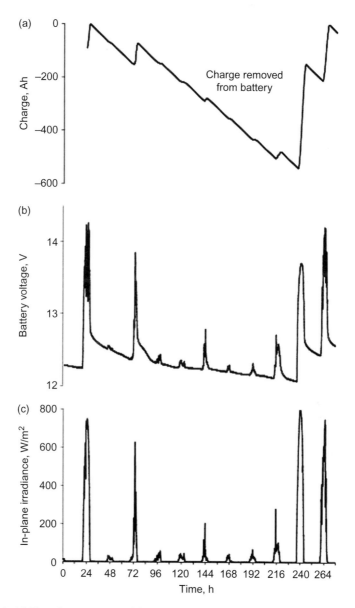

Figure 2 (a) The charge removed from the battery and (b) the battery voltage in a stand-alone PV system during a 'climatic cycle.' The in-plane irradiance is shown in (c). Each point corresponds to a 5-minute average of measured data [2].

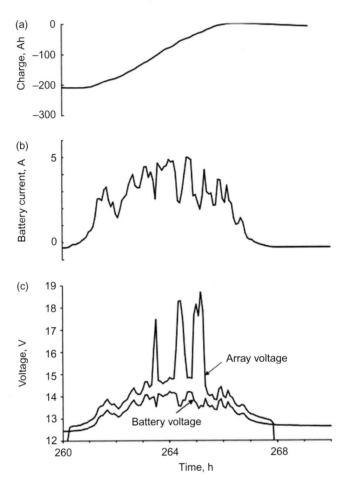

Figure 3 (a) The charge removed from the battery, (b) the battery current, and (c) the battery and array voltage during the last day of the climatic cycle in Figure 2. Intervals where the array is disconnected by the series regulator as the battery approaches full charge are clearly visible [2].

irradiance in the plane of the array. The system voltages, and the battery current and charge during one day of this cycle, are shown in more detail in Figure 3.

For a regular daily load, the daily and long-term energy balance become effectively decoupled and can be discussed separately. The daily energy balance and the calculation of the mismatch between solar radiation and the load that needs to be covered by the battery are discussed in Section 2. The long-term aspects of energy balance can be analysed using the time series of daily solar radiation (measured or synthetic; see Chapter IIa-1) to produce a statistical picture of the energy supply by the system as a function of its configuration. This aspect of PV system design is related to sizing and is discussed in detail in Chapter IIa-4. An example of the seasonal cycle in battery operation and its relationship to system design is discussed briefly in Section 5 of this chapter.

2. LOAD DESCRIPTION

Stand-alone systems can only be sized effectively for predictable loads, and random load patterns are likely to result in uncertain reliability of supply by the PV system. We should note, however, that the load coverage will be improved if there is a possibility of adjusting or disconnecting nonessential loads.

Load description is recognised in the international standards (see, for example, [3,4]), which describe accurate procedures for load determination. For example, the standard [4] contains the following recommendations:

1. Describe all loads by a voltage and current with a starting and finishing time for a period of 24 hours.
2. Describe the AC loads separately and combine with the inverter efficiency.
3. The following types of load data are considered:
    - momentary current, or a current lasting less than 1 minute, which is associated with the starting or surge of certain loads;
    - running current, or a current drawn by the load once the initial transient has come to an end;
    - parasitic currents;
    - load duration;
    - load coincidence, or the simultaneous occurrence of loads; and
    - load voltage, or the range of maximum and minimum voltage.

The meaning of some of these parameters is illustrated in Figure 4.

It is seen that there may be loads having a momentary current of duration less than 1 minute and a running current of known duration. There may be several occurrences per day, as shown in Figure 4(a), or conventional loads having only a running rated current and a run time as shown in Figure 4(b). The loads may be coincident or noncoincident in time

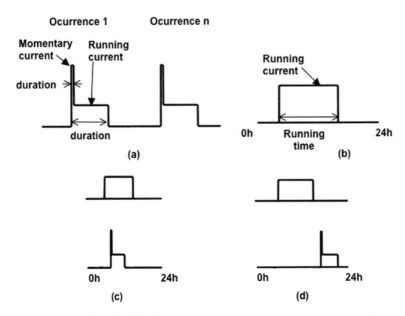

Figure 4 (a) Individual load having a momentary and running currents and a number *n* of occurrences, (b) load used a number of hours per day, (c) coincident loads, and (d) noncoincident loads. Elaborated from IEEE standard 1144-1996.

(see Figure 4(c) and (d), respectively). There may also be load profiles requiring a longer period of time than 24 hours that need to be described; in this case, an average and a maximum daily load should be given.

This information allows the derivation of the load profile, or at least the typical load profile for a 24-hour-period load. This is performed by multiplying each load current by the duration and then adding all the products to find the daily ampere-hour load. If the duration of the momentary load is not known, then 1 minute should be taken as recommended in reference [3]. The load profile also provides the necessary information about the maximum discharge rates that need to be sustained by the battery.

2.1 Electricity Consumption by Lighting and Electrical Appliances

Table 1 gives a summary of the load values for the main domestic appliances. The principal results on energy consumption for lighting in the

Table 1 Values of typical energy end use. Source of data: lighting—reference [8]; other appliances—reference [9]. Asterisk indicates estimated value.

Appliance	Average rated power (W)	Average usage (hrs/day)	Annual energy consumption (kWh/yr)
Lighting			
Bedroom	94	1.0	36
Closet	66	1.3	31
Dining room	165	2.3	136
Hall	78	1.7	49
Family room	106	2.0	77
Garage	103	1.9	71
Kitchen	95	3.2	109
Living room	124	2.4	109
Master bedroom	93	1.2	41
Outdoor	110	2.9	116
Utility room	84	2.4	74
Bathroom	138	1.9	96
Other	103	1.5	55
All	105	2.0	78
Other Appliances			
Refrigerator			649
Freezer			465
Washing machine	0.375(kWh/load)	4 loads per week★	78
Dishwasher	0.78 (kWh/load)	One load per day★	283
Electric oven	2300	0.25★	209
Coffee machine (drip)			301
Microwave			120
Toaster			50
Vacuum cleaner			14
Audio equipment			36
TV	100	5★	182
PC			25
Printer (ink-jet)			28
Satellite dish			96
VCR			158
Video games			49
Clothes iron			53
Hair dryer			40

residential sector can be found in references [5,6,7,8,9]. Most of these studies conclude that 50-W to 150-W incandescent lamps that are on for at least 3 hours a day can be cost-effectively replaced by compact fluorescent lamps. This affects the results shown in Table 1 as an incandescent lamp can be substituted by a fluorescent lamps of approximately a factor 2.8 to 3.9 lower rating (i.e., a 60-W incandescent lamp can be substituted by a 20-W compact fluorescent lamp; more details can be found in Table 4−1 of reference [8]). The average total consumption per household in the USA used for lighting is about 1800 kWh/year [8]. About one-third less energy is used during months with high solar radiation, and this lighting consumption does not appear to depend on the population demographics. More information about typical values of the power of common appliances can be found in reference [10].

There are, however, indications that the owners of off-grid houses have a tendency to adapt their pattern of energy use, leading to a clear reduction in comparison with baseline load figures [11], which are in the range of 25% for the larger systems that have been analysed. Most of the studies of the performance of PV systems agree that an adequate estimation of the load is essential to reach reasonable values of the performance ratio.

2.1.1 Daily Energy-Balance Dynamics

Power has to be supplied to the load according to the consumption patterns of the application, and energy may be required at different times than when it is generated by the PV array. This creates a mismatch in the energy flow between the PV system and the load.

The daily energy balance between PV energy supply and a typical domestic load is illustrated in Figure 5 on the example of a PV system operating in Sacramento, California. Figure 5(a) and (b) show the ambient temperature and solar-radiation data for a typical day of the year, with PV array at latitude tilt. These data have been generated statistically by the METEONORM software [12]. Figure 5(c) shows the profile of energy consumption in a typical household obtained as described in the IEEE standard [3]. If the array size is chosen so that the total energy produced by the array is equal to the energy consumed by the load, the resulting hourly energy balance (energy produced minus energy consumed) is shown in Figure 5(d). The cumulative energy balance (integral of (d) from 0 to time t) is shown in Figure 5(e).

Energy Balance in Stand-Alone Systems 667

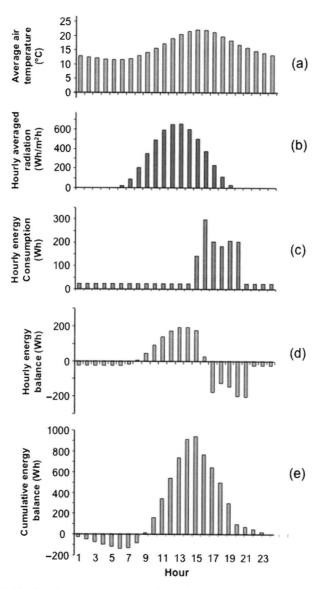

Figure 5 (a) The hourly temperature profile and (b) solar radiation for a typical day in a year in Sacramento, California; (c) typical hourly load profile; (d) hourly energy balance; (e) cumulative energy balance.

This result shows that, although there is complete energy balance at the end of the day, there is considerable imbalance between the energy supply and the load for a number of hours during the day when the load consumes more energy than supplied by the PV generator. If this situation is encountered in a stand-alone system, this mismatch has to be supplied by energy storage or a backup generator. This is the daily energy storage that was introduced in Section 1.

3. SEASONAL ENERGY BALANCE

Considerations regarding the seasonal energy balance are key to the design of stand-alone systems and depend critically on the latitude where the system is installed. The gently varying profile of daily solar radiation during the year in a tropical location near the equator should be contrasted with the wide differences between summer and winter at high latitudes (see Section 2 in Chapter IIa-1). The potential mismatch between the PV energy supply and demand is potentially much larger than commonly encountered in the daily energy balance, and a careful analysis is needed to determine the principal parameters of the system: the size of the array and the battery capacity. These parameters, in turn, affect the likely reliability of energy supply by the PV system. This analysis is the principal aim of sizing, which is discussed in some detail in Chapter IIa-4.

In concluding this chapter, we take up an issue that was introduced in Section 1 and consider the profile of energy stored in the battery during the seasonal energy cycle. This analysis illustrates the intimate relationship between the configuration of a stand-alone PV system and the behaviour of battery state of charge over prolonged periods of time. This behaviour, in turn, has a considerable impact on the battery lifetime. By way of example, Figure 6 shows the battery state of charge during the year for two PV systems operating in the south of England. The results, obtained by modelling, refer to high-reliability systems with a similar reliability of supply; detailed system configurations can be found in Chapter IIa-4. For clarity, the daily charge and discharge cycle has been omitted from the graphs, which show the maximum state of charge reached by the battery in each particular day.

The system in Figure 6(a) has 16 days of autonomy; this is a large battery but not uncommon in remote industrial systems at these latitudes.

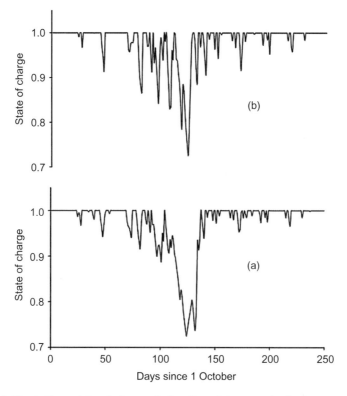

Figure 6 The battery state of charge during the winter months for two stand-alone PV systems operating in the south of England with different system configurations but a similar reliability of energy supply.

Figure 6(b) corresponds to a system with an array 60% larger than the system in Figure 6(a) but with only 7.2 days of autonomy. Although the minimum depth of discharge is similar, there are important differences in the detailed battery behaviour. We note, in particular, that the duration of the longest climatic cycle for the system in (b) is approximately half the duration observed for the system in Figure 6(a). The implications of these profiles for the optimal type of battery and for the battery life will be considered again in Chapter IIb-2.

From the fact that the irradiance and temperature are random functions, when the analysis of the operation of a stand-alone PV system is performed by means of stochastic time series of irradiance and ambient temperature, it soon becomes clear the importance that the size of the system has on several operating parameters such as (a) loss of load (LOL)

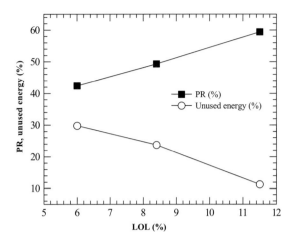

Figure 7 Plots of PR and the percentage of unused energy as a function of the LOL probability for a stand-alone system 2 kWh/day consumption. Radiation data from Barcelona (Spain) at optimal inclination.

probability, (b) performance ratio (PR), and (c) percentage of unused energy due to battery full state of charge. We have used PVSYST to explore such relationships for a small-sized system working to provide energy supply to a 2 kWh/day load in a geographical location in Barcelona (northeastern Spain). Figure 7 shows a plot of the PR and of the percentage of unused energy as a function of LOL. It can be seen that both follow opposite trends, indicating that if a smaller value of LOL is desired, and hence a more reliable system, then the unused energy fraction increases, thereby producing a reduction of the PR.

Figure 8 shows the values of the PV array size and battery size for the same example as in Figure 7 showing that the reduction of the LOL is accomplishable by increasing the size of the system (array and battery).

An alternative approach to defining system reliability is to use practices familiar from other braches of engineering (used, for example, to assess the effect of extreme winds on building structures or in the design of flood-protection measures) where extreme values are considered as function of certain recurrence intervals. Notions such as 50-year wind or 100-year flood may be familiar concepts, and one can similarly design a stand-alone PV system to operate without loss of load over a certain period of time. Clearly, a system designed to operate over 10 years will have a higher reliability (by virtue of a larger PV array or a larger battery

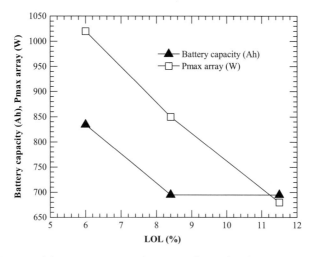

Figure 8 Array and battery size as a function of LOL for the same example as in Figure 2.7

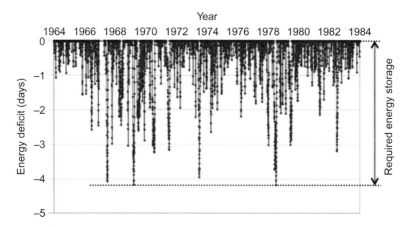

Figure 9 The daily deficit in the energy store predicted by modelling PV system operation for Bulawayo, Zimbabwe, showing also the size of energy storage required for 20-year continuous operation. Energy storage sized for a shorter period will be smaller, but the reliability of supply will be lower.

bank) than a system designed for 5-year operation (Figure 9) [13]. Such a philosophy has recently been adopted, for example, by the World Health Organization in defining the photovoltaic power system standards for solar vaccine-refrigeration systems [14].

REFERENCES

[1] T. Markvart (Ed.), Solar Electricity, second ed., John Wiley and Sons, 2000 (Sec. 4.4.2.)
[2] K. Fragaki, Advanced Photovoltaic System Design. MPhil Thesis, University of Southampton, 2003, unpublished.
[3] IEEE standard 1013-2000: IEEE Recommended practice for sizing lead-acid batteries for Photovoltaic Systems, IEEE, 2001.
[4] IEEE standard 1144-1996: IEEE Recommended practice for sizing nickel-cadmium batteries for stand-alone PV systems, IEEE, 1997.
[5] From Table 1 in <http://homeenergysaver.lbl.gov/hes/aboutltg.html/>
[6] <http://hes.lbl.gov/hes/aboutapps.html/>
[7] M.C. Sanchez, Miscellaneous electricity use in US residences, University of California, Berkeley, June 1997, MS thesis.
[8] L.S. Tribwell, D.I. Lerman, Baseline residential lighting energy use study, Final report, Tacoma Public Utilities, Tacoma, Washington, May 29th, 1996 (<www.Nwcouncil.org/comments/documents/CFUselighting.doc/>)
[9] T.P. Wenzel, J.G. Koomey, G.J. Rosenquist, M. Sanchez, J.W. Handford, Energy data sourcebook for the US residential sector. Report LBNL-40297, September 1997 (available at <www.osti.gov/gpo/servlets/> as pdf document nr. 585030).
[10] Energy use of some typical home appliances. Energy efficiency and renewable energy network of the US Department of Energy (in <www.eren.doe.gov/consumerinfo/refbriefs/ec7.html/>).
[11] Energy use patterns in off-grid houses, Canadian Housing Information Center, Technical Series 01-103 (<www.cmhc-schl.gc.ca/publications/en/rh-pr/tech/01-103-e.html/>).
[12] METEONORM version 4.0, Meteotest, Switzerland, <www.meteonorm.com/>.
[13] T. Markvart, K. Fragaki, J.N. Ross, PV system sizing based on observed time series of solar radiation, Sol. Energy 80 (2006) 46−50.
[14] <www.who.int/immunization_standards/vaccine_quality/pqs_e03_pv1.2.pdf/>

CHAPTER IIA-4

Review of System Design and Sizing Tools

Santiago Silvestre
Universidad Politecnica de Catalunya, Barcelona, Spain

Contents

1. Introduction 673
2. Stand-Alone PV Systems Sizing 674
   2.1 Sizing Based on Energy Balance 674
   2.2 Sizing Based on the Reliability of Supply 676
   2.3 Sizing of Solar Pumping Systems 680
3. Grid-Connected PV Systems 681
4. PV System Design and Sizing Tools 684
   4.1 Sizing Tools 684
   4.2 Simulation Tools 687
References 692

1. INTRODUCTION

System modelling forms a key part of the photovoltaic system design. It can provide answers to a number of important issues such as the overall array size, orientation, and electrical configuration; it can also determine the size of various subsystems such as the battery or the inverter (see the standard [1] for a detailed discussion of the relevant terminology used to describe PV systems). The design criteria will vary, depending on the nature of the application. In stand-alone systems, the consideration of energy production to meet the load is paramount; the reliability of supply and economic considerations may also be important. A critical aspect of stand-alone system design is sizing, which is discussed in Section 2.

Practical Handbook of Photovoltaics.
© 2012 Elsevier Ltd. All rights reserved.

The applications of grid-connected systems vary from small-building integrated systems to PV power stations. Modelling tools are available to provide solar-radiation data, assess possible shading effects (which may be particularly important at an urban site), and produce the resulting electrical layout of the array. Further considerations will include the restrictions imposed by the connection to the local utility, which is discussed in Chapter IIc-1. Economic aspects may include an investigation of financial support mechanisms and of the economic impact of local electricity generation, including possible revenue for electricity exported to the utility through feed-in tariffs or net metering (see, for example, [2,3]).

2. STAND-ALONE PV SYSTEMS SIZING

Sizing is one of the most important tasks during the design of a stand-alone PV system. The sizing procedure will determine the power rating of the PV array and the battery storage capacity needed to power the required load; the electrical configuration of the array may also be considered at this stage. More sophisticated sizing procedures will ensure that the reliability of power delivered to the load is appropriate for a given application and optimise the cost of the system. Sections 2.1 and 2.2 below outline the two most frequently used sizing methods: using energy balance and using a more complex procedure that invokes the reliability of supply [4–7]. An example of sizing a water-pumping system is discussed in Section 2.3.

2.1 Sizing Based on Energy Balance

The essential features of photovoltaic system sizing can be understood in terms of the daily energy balance between the daily load and the energy delivered by the array, which is considered in some detail in Chapter IIa-3. The input side of this balance—the expected energy production by the PV array—is determined principally by the solar radiation at the site, and it can be conveniently discussed using the concept of *peak solar hours* (PSHl see Chapter IIa-2). Depending on the application, the appropriate value of PSH to use may be an average over the entire year or a part of the year.

Typical examples were considered in Section 5.1 of Chapter IIa-2. The appropriate value of PSH should correspond to the critical period of the system operation, be it the month with the lowest solar radiation or the month with the highest load. A similar consideration governs also the array inclination. The annual average value of PSH is sometimes used when daily solar radiation does not vary a great deal throughout the year and the battery can be used partly as a seasonal buffer.

The second fundamental parameter is the typical daily load demand. Detailed aspects of the load, including the analysis of the daily energy balance, are considered in Chapter IIa-3. Here we shall only need the total amount of energy consumed in one day, to be denoted by L.

The values of PSH and L determine the (average) daily balance between the energy supply and the required nominal power rating of the PV array,

$$P_0 = \frac{PSH}{L} \tag{1}$$

which, in turn, gives the total number of PV modules N:

$$N = \frac{P_0}{P_{\text{mod}}} \tag{2}$$

where P_{mod} is the power produced by one module under standard conditions, which is equal to $P_{\max}$ or P_{eff} for systems with and without a maximum power point tracker, respectively (see Chapter IIa-2).

Once the operating DC voltage has been specified, this sizing procedure gives the array configuration in terms of the number of modules to be connected in series and in parallel (Figure 1). This argument applies for systems without maximum power point tracking, where the nominal battery voltage V_{bat} is usually set equal to 12 V or 24 V, as discussed in Chapter IIa-2. The security (sizing) factor SF is inserted to allow for additional losses, such as the accumulation of dirt on the modules, or to increase the system performance. It plays a similar role as the dimensionless array size C_A, which will be introduced in Section 2.2.

A separate argument is used to size the battery. The critical parameter is the number of 'days of autonomy' (denoted by C_S) that the system is required to operate without any energy generation. C_S is related to the battery capacity C_n (in energy units—i.e., the battery capacity in Ah multiplied by the voltage) by

$$C_n = C_S \frac{L}{DOD_{\max}} \tag{3}$$

where $DOD_{\max}$ is the maximum allowed depth of discharge of the battery. If appropriate, seasonal storage can be added to the battery capacity

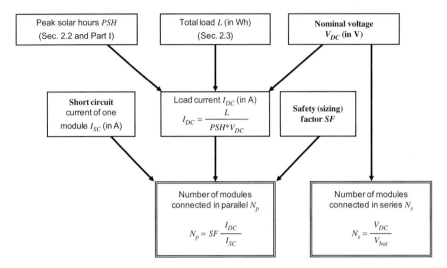

Figure 1 Sizing based on energy balance.

Table 1 Autonomy

Latitude of site (°)	Days of autonomy
0–30	5–6
30–50	10–12
50–60	15

C_n but is not included in Equation (3). In the energy-balance method, the number of autonomy days is determined from experience in the field rather than by a rigorous theoretical argument. BP Solar [8], for example, recommends the values shown in Table 1. A more rigorous method for the battery size is offered by sizing procedures based on the reliability of supply, which are discussed in Section 2.2.

The determination of battery size completes the sizing of the system. One of the software packages discussed in Section 4 can be now used to analyse the system performance in more detail. This is recommended particularly for 'professional' (or remote industrial) systems where strict requirements on system performance may be imposed by the nature of the application.

2.2 Sizing Based on the Reliability of Supply

The reliability of electricity supply is an important factor in PV system design, and this is also reflected in some sizing procedures. One way to

Table 2 LLP

Application	Recommended LLP
Domestic	
Illumination	10^{-2}
Appliances	10^{-1}
Telecommunications	10^{-4}

quantify the reliability of supply is by a parameter known as the *loss-of-load probability* (LLP), defined as the ratio between the estimated energy deficit and the energy demand over the total operation time of the installation. Other names for the same concept have also been used: *load coverage rate* (LCR) [9], *loss of power probability* (LOPP) [10], or *loss of power supply probability* (LPSP) [11]. The recommended values of LLP for various applications are shown in Table 2 [12]. Reviews of the methods based on reliability of supply can be found in [6,7,12,13]; an elegant analytical method that uses the random walk methodology was developed by Bucciarelli [14,15].

Sizing methods of this type are frequently used in applications where high reliability is required. To be statistically significant, long time series of solar-radiation data are required that can usually be obtained only in the form of synthetic solar-radiation data [7]. A more pragmatic approach in such circumstances may be to determine system configurations that would be expected to deliver energy to load without interruption over a certain specified period of time; such configurations can be determined with the help of system modelling [15]. Despite the apparent differences, this methodology and the methods based on LLP result in a similar formalism.

The results of sizing based on system reliability are often displayed in a graphical form. To this end, the expected daily energy produced by the PV generator is expressed in a dimensionless form [6],

$$C_A = \frac{P_0 \; PSH}{L} \qquad (4)$$

where P_0 is the nominal power of the array, and the value of PSH, corresponding to in-plane daily radiation, is usually taken for the 'worst month' of the year (see Section 2.1). The array size C_A (4) and the number of autonomy days C_S (3), are then considered as coordinates in a Cartesian coordinate system and the required system configurations are displayed as a locus of points in the C_S–C_A plane.

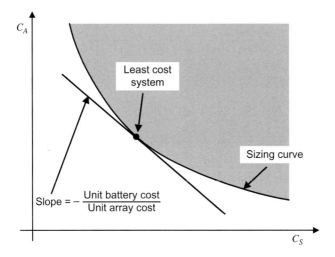

Figure 2 The sizing curve, showing also the system with minimum initial outlay cost.

It turns out (see, for example [12,13,16]) that all system configurations (C_S, C_A) that deliver energy with a given reliability of supply lie on a line called the *sizing curve* (Figure 2; the term *isoreliability line* is also used [12]); all configurations with a higher reliability lie in the shaded region in the graph bounded by the sizing curve. A different reliability of supply (a different value of LLP or a different period of time required for expected operation without shedding load) will yield a different sizing curve.

A further bonus of this method is that it allows an insight into the system economics. Indeed, it can be shown that the least-cost system is determined by a point on the sizing curve where the tangent slope equals the ratio of unit battery cost to the unit array cost. It should be emphasised, however, that this calculation is based solely on capital costs, as it only takes into account components that are installed at the start of the system operation; later replacements are not included. This concerns, in particular, the batteries that have to be replaced several times during the lifetime of PV modules.

Egido and Lorenzo [6] developed a comprehensive library of data that allows the determination of LLP for a wide range of locations over the Iberian peninsula. A sizing curve based on measured solar-radiation data is illustrated in Figure 3 for a PV system operating in the south of England for 10 years without shedding load. Once the load is specified, this universal curve provides an immediate description of all high-reliability PV

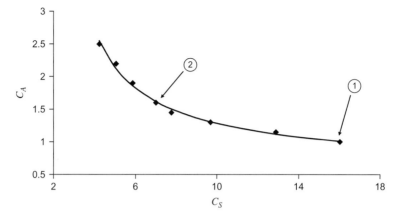

Figure 3 Example of a sizing curve using solar-radiation data for Bracknell, near London. Panel inclination = 65 degrees; period of data = 1981–1990; source of data = reference [70]. Points are results obtained by modelling; line is a fit to the points with the expression

$$C_A = \frac{1}{0.4514 \ln C_s - 0.2601}$$

system configurations by converting the dimensionless coordinates C_S and C_A into battery capacity and array size with the use of Equations (3) and (4). It is seen that taking the array size as predicted by the energy-balance method and $C_A(=S_F) = 1$; the sizing curve in Figure 3 gives 16 days of autonomy, quite close to the BP-recommended value of 15 days. If the array size is increased by 15% (in other words, we take $C_A(=S_F) = 1.15$), then the battery can be reduced to 13 days of autonomy without affecting the supply reliability; a larger array increase would give a more substantial battery reduction.

The consideration of different system configurations has implications for the system cost and affects also another important feature of PV system operation: the pattern of seasonal battery charge and discharge cycles that may affect battery life. This point was already touched upon and is illustrated in Figure 6 of Chapter IIa-3. This figure illustrates that although all points on the sizing curve correspond to systems with identical reliability of supply, the pattern of battery charge and discharge cycles will be different along the curve. The two system configurations on the sizing curve marked by 1 and 2 in Figure 3 correspond to graphs (a) and

(b) in Figure 6 of Chapter IIa-3, respectively. Thus, it is seen that increasing the array size along the sizing curve thus shortens the duration of the climatic cycles and helps reduce possible degradation of the battery. A full discussion of battery life can be found in Part IIb-2 and in the standards [17,18].

2.3 Sizing of Solar Pumping Systems

Water pumping for irrigation and water supply for rural communities represents an important area of stand-alone PV systems. These systems usually consist of a photovoltaic generator, a source of water, a water-storage tank, and a DC pump. The role of batteries is here played by the water-storage tank, and the electric power load demand L is now replaced by daily water demand. If expressed in Wh/day, this represents the energy needed to pump the required volume of water demanded by the user into the storage tank. These considerations show that PV pumping systems can be sized in a similar way to PV systems with other applications [19–21]. Some computer simulation programs developed to determine the performance of PV pumping systems are also available in the literature [22,23].

In particular, the PV system-sizing method based on loss-of-load probability can be also used to design PV pumping systems [19,21]; LLP now corresponds to the ratio of energy deficit and energy demand, as water pumped volume, over the system's working period. To introduce the LLP sizing method into water-pumping applications, Equation (2) can be used to calculate the PV generator capacity, C_A. In this case, the energy load demand L represents the daily load or water demand. Equation (3) can then be rewritten as Equation (5) [19] to estimate the storage capacity C_s that represents the necessary volume of the water-storage tank:

$$C_s = \frac{C_U}{L} \qquad (5)$$

where C_U (in units of energy per day) is the energy required to pump the entire tank full of water.

Using the desired values of LLP for the PV pumping system, it is possible to obtain sizing curves that give the required pairs of C_A and C_s values to determine the sizes of the water tank and PV generator that represent the best compromise between cost and reliability, in the same way as for the system-sizing methods considered in Section 2.2.

3. GRID-CONNECTED PV SYSTEMS

The array architecture in grid-connected systems should be considered alongside the DC characteristics of the inverter, including the maximum input current, the nominal and minimum input voltage, and the maximum power tracking range. Sizing of grid-connected systems is considered, for example, in [24–26]. This analysis can be carried out by analogy with the energy-balance arguments that were used in Section 2.1 to develop a sizing procedure for stand-alone PV systems. The AC energy produced, E_{AC}, by the PV system in one day, say, can be estimated by using an equation similar to (1):

$$E_{AC} = \eta \; P_0 PSH \tag{6}$$

where P_0 is the nominal power of the array at STC, PSH is the average value of peak solar hours at the specific location, and η is the inverter efficiency, which in general depends on the output power. This dependence can be expressed in terms of the inverter self-consumption and load-dependent losses as [27,28]

$$\eta = \frac{p_{out}}{p_{out} + k_0 + k_1 p_{out} + k_2 p_{out}^2} \tag{7}$$

where

$$p_{out} = \frac{P_{AC}}{P_I}$$

is the instantaneous AC output power, P_{AC}, normalized to the nominal AC output power of the inverter P_I. The parameter k_0 represents the self-consumption factor, which is independent of the output power. The linear and quadratic terms in the denominator take into account losses that are linear in the load power such as voltage drops, whereas the ohmic losses are taken into account by the quadratic term. Experimentally measured inverter efficiency curves can be easily fitted to this equation and parameter values can be extracted (see, for example, [28]). This procedure allows a comparison between different inverters and offers insight into the origin of the dominant losses by modelling [29]. A typical form of the functional dependence of the inverter efficiency of the output power is shown in Figure 4.

The reduction of inverter efficiency at low power should be allowed for in the choice of the power rating of the inverter. It is generally

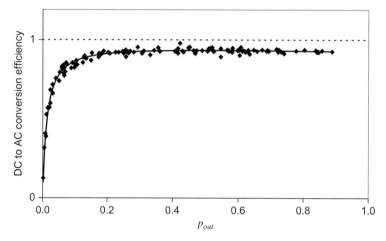

Figure 4 Typical inverter efficiency during operation in a grid-connected PV system as a function of the output power. Data: STaR Facility, University of Southampton. Line is a fit using Equation (7) with $k_0 = 0.013$; $k_1 = 0.02$; $k_2 = 0.05$.

recognised that advantage can be gained by choosing the power rating of the inverter P_I smaller than the nominal power P_0 of the array. Based on inverter efficiency, this represents, in fact, a trade-off between the self-consumption and inverter losses that scale down as the inverter is rated smaller, and the energy lost at the upper limit of PV generation close to the STC conditions. Thus, the choice of optimum inverter size depends on the statistics of the array output, which is related to the latitude of the site as discussed in Section 6 of Chapter IIa-2. For example, the recommended value of the ratio P_I/P_0 is between 0.65 and 0.8 for countries in Northern Europe and 0.75 to 0.9 is more suitable in mid-European latitudes. In Southern Europe, the suggested figure is 0.85 to 1 [28,30].

Whilst system efficiency is important, care should also be taken to avoid long periods of inverter operation under overload conditions. Nofuentes and Almonacid [30] suggest a size that exceeds the DC power supply to the inverter for 99% of the time (Figure 5). The inverter power rating determined using this '99% criterion' is usually somewhat larger than the value based on the efficiency argument.

Once the P_I/P_0 ratio and the inverter have been selected for the grid-connected PV system, the sizing of the DC part of the system, the PV generator, can be addressed.

The number of parallel strings Npg, of the PV generator, can be calculated by taking into account the maximum input allowed by the

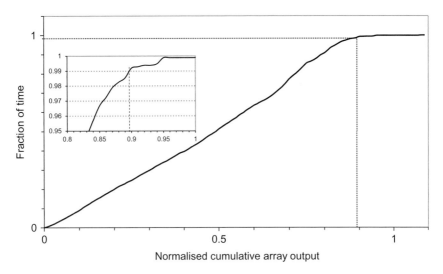

Figure 5 The cumulative array output obtained by integrating the graph of $G\,h_I(G)$ in Figure 6, of Chapter IIa-2, illustrating inverter sizing by the '99% criterion.'

inverter as shown in Equation (8). The value of this maximum current, Imax, is an usual parameter given by inverter manufacturers:

$$Npg < \frac{Im\,ax}{Im\,pp} \qquad (8)$$

where Impp represents the output current of the PV module at the maximum power point in STC conditions.

The number of PV modules in series, Nsg, by string, can be directly deduced from the selected P_I/P_O ratio, where the power rating of the inverter P_I is also known, as shown in Equation (9):

$$Nsg = \frac{P_0}{Npg} \qquad (9)$$

The product of the number of PV modules in series, Nsg, by the PV module output voltage at the maximum power point must also accomplish Equation (10) to ensure that the inverter will be working inside the maximum power point tracking voltage window:

$$V\min_{MPPT} \leq N_{sg}Vmpp \leq V\max_{MPPT} \qquad (10)$$

where $V\min_{MPPT}$ and $V\max_{MPPT}$ are, respectively, the minimum and maximum inverter input voltages for a correct tracking of the PV generator maximum power point tracking, and Vmpp is the maximum power voltage of the PV modules forming the PV generator.

Equation (10) must be verified for different temperatures of work, taking into account the temperature dependence of the PV module given by the manufacturer.

4. PV SYSTEM DESIGN AND SIZING TOOLS

A wide variety of software tools now exist for the analysis, simulation, and sizing of photovoltaic systems. These tools present different degrees of complexity and accuracy, depending on the specific tasks for which each tool has been developed. This chapter presents an overview of the available software tools. The following list of software tools described is not exhaustive due to the large numbers that exist but can help PV system designers and installers to examine the available PV system design and analyse the options from different points of view. It is usual to distinguish between sizing tools (which determine the component size and possibly also configuration) and simulation or modelling tools, which analyse the system output and performance once its specifications are known.

Examples of these sizing and simulation tools are given in Sections 4.1 and 4.2, accompanied by a brief description of each program and the source from which it can be procured. The source indicated in the references should be contacted for more detailed information on specific capabilities, operating environment requirements, report functions, technical support, and licensing or user restrictions.

4.1 Sizing Tools

Sizing tools allow photovoltaic systems to be dimensioned, taking into account energy requirements, site location, and system costs (Table 3). Most of these software tools are relatively simple and help users automatically solve energy-balance calculations considering different combination of PV system components, including batteries, modules, and loads. These tools are usually implemented using spreadsheets at different level of complexity and offer a first approach for the evaluation of specific PV system applications. In their basic form, these tools are easy to use, and some solar cell manufacturers and vendors of PV system components offer this

Table 3 System-sizing programs

Program	Source	Description
MaxDesign (Version 3.1)	SolarMax [31]	Max Design is an interactive photovoltaic system design tool for all SolarMax inverters. It offers design variants, finding out which configurations are the best for a specific PV system. Results include dimensioning, yield calculation, and analysis of DC line losses.
NSol (Version 4.6)	Fear The Skunk [32]	It includes modules for stand-alone PV, PV–generator hybrids, and grid-tied PV. The stand-alone version includes the loss-of-load (LOL) probability statistical analysis.
PVGIS	European Commission [33]	Web application to estimate the performance of PV systems located in Europe.
PVWATTS (Version 2)	NREL [34]	An Internet-accessible tool that calculates the electrical energy produced by a grid-connected PV system for locations within the United States and its territories.
RETScreen (Version 4)	CANMET Energy Diversification Research Laboratory. [35]	RETScreen International is a renewable energy decision-support and capacity-building tool. Each RETScreen renewable-energy-technology model, including hybrid PV systems, is developed within an individual Microsoft Excel spreadsheet workbook file. The workbook file is in turn composed of a series of worksheets. These worksheets have a common format and follow a standard approach for all RETScreen models. In addition to the software, the tool includes product, weather, and cost

(continued)

Table 3 (continued)

Program	Source	Description
		databases; an online manual; a Web site; project case studies; and a training course.
SAM (Version 2010.11.9)	NREL [36]	SAM makes performance predictions for grid-connected solar, small wind, and geothermal power systems and economic estimates for distributed energy and central generation projects.
Simulate your PV system	Helyos Technology [37]	Online sizing estimation of stand-alone systems using the energy-balance method. Allows sizing a PV system in response to the annual energy demand. Offers an interesting economic analysis. Available for locations in the Italian republic.
SINVERT	Siemens [38]	A dimensioning program for determining the best possible configuration for a PV system with Siemens SINVERT inverters.
Solar.configurator (Version 2.5.2)	Fronius [39]	Allows the analysis of the different configuration possibilities: PV module types, strings, tilt angles, and wirings.
Solar Pro (Version 3.0)	Laplace System Co., Kyoto, Japan [40]	Program enables users to gauge the effects of shade from adjacent buildings or objects so as to determine the optimum positioning and scale of the photovoltaic modules and system. It also allows for estimating the I-V curve of solar cell modules based on the electrical characteristics of various module designs, and it calculates the electrical yield of systems based on geographic location and atmospheric conditions over a wide array of possibilities.

(continued)

Table 3 (continued)

Program	Source	Description
Sunny Design (Version 1.55)	SMA [41]	Sunny Design allows definition of the PV generator, definition of system location, module type, orientation, and power generated or number of photovoltaic module and the determination of the typical investor (SMA). Results help determine the characteristics of performance of a year of service.
Solar Sizer	Center for Renewable Energy and Sustainable Technology (CREST) and Solar Energy International [42]	Design of residential photovoltaic systems. Provides information on energy usage, storage, and production as well as initial cost, annual costs, and life-cycle cost. Program includes solar-radiation data for 240 sites in North America and 50 around the world.

NREL = National Renewable Energy Laboratory, USA.
SMA = SMA Solar Technology AG.

kind of tool to potential customers (often via the Internet) to enable customers to adapt PV system components to their products. Similar sizing tools are also available that focus on specific studies for complete sets of modules, batteries, inverters, electronic power conditioning components, and loads from the principal manufacturers in the photovoltaic market.

More sophisticated sizing tools are also available in the market, some offering the possibility of optimising the size of each PV system component. More detailed analysis can then be carried out of the energy flows in the PV system and the determination of critical periods along the year. Of particular interest is often the deficit of energy associated with these periods of time and minimising the final cost of the system for a specific application.

4.2 Simulation Tools

More specialised simulation tools are also available on the market, which offer, in addition to sizing, the chance to model PV system components or to carry out accurate simulations of the PV system behaviour and a more accurate design. A selection of such simulation and modelling tools is given in Table 4.

Apart from these specialised simulation tools, the possibility of modelling and simulation of PV systems and components using programs such as Matlab [53] or Pspice [54] should be given a serious consideration. Most electrical and electronic engineers today will be familiar with these open-architecture research tools that include advanced mathematical manipulation toolboxes, which can be helpful in the accurate simulation of PV systems.

Pspice is one of the most popular standard programs for analogue and mixed-signal simulation. The possibilities of modelling the different components of a PV system (including solar cells, PV modules, and batteries) has been shown in the literature. Furthermore, experimental measurements have been successfully compared with Spice simulations [55–59]. Long-term simulation of PV systems using Pspice has also been demonstrated [24]. System designers are generally interested in these long-term simulations provided that they are fast enough to allow easy interaction, such as changes of parameter values, to refine the adjustment of the system size at the design stage.

Matlab is a powerful technical computing environment that can be complemented by a wide set of associated toolboxes offered by Mathworks, the developer of Matlab. Matlab can also be combined with the Simulink interface, a friendly modular graphical environment of simulation, resulting in a very powerful modelling and simulation platform. As in the case of Pspice, models developed in the Matlab or Simulink environment are available for main components of PV systems, and very accurate simulations of the PV system behaviour can be carried out [60–63]. The output data can be easily manipulated to extract further realistic values of the model parameters of PV system components. Matlab capabilities can help PV system designers gain deep understanding of the PV system behaviour under realistic circumstances of system operation. Furthermore, Matlab could be an interesting tool in the field of automatic supervision and fault detection of PV systems [63].

Finally, general purpose programming languages—for example, C, Fortran, and Pascal—have been also used for PV system simulation and modelling. Some computer models and calculation algorithms specially developed for the simulation of PV systems have been presented by many

Table 4 PV system simulation software

DDS-CAD PV	Data Design System [43]	Allows planning and simulation of complete photovoltaic systems (roof-mounted, roof and facade, and open-space installation). Software includes modelling of shadows cast by adjacent objects using photorealistic 3D representation.
Hybrid2	NREL and University of Massachusetts [44]	Hybrid2 allows a detailed long-term performance and economic analysis of a wide variety of hybrid power systems: wind, PV, and diesel hybrid systems.
INSEL (Version 8)	INSEL [45]	INSEL provides plenty of well-validated simulation models that support the solution of problems in renewable energy systems. It includes a module for PV system simulation (module 10) that covers grid-connected and stand-alone applications.
PV F-CHART	University of Wisconsin Solar Energy Laboratory [46]	A PV system analysis and design program. The program provides monthly averaged performance estimates for each hour of the day. System types: utility interface systems, battery-storage, and stand-alone systems; tracking options are also included. Main features: weather data for more than 300 locations (can also be added), hourly load-power—demand profiles for each month, statistical load variation, buy—sell cost differences, time-of-day rates for buy—sell, and life-cycle economics.
PV★SOL Expert	Valentin Energy Software [47]	A program for the design, planning, and simulation of PV systems. Program features include grid-connected and stand-alone systems; any number of panels set up at varying angles; shading

(*continued*)

Table 4 (continued)

		from the horizon and other objects; use of different makes of PV modules and inverters within the system; module, string, and system inverters; 3D visualization manufacturer and weather databases; determination of electricity consumption through profiles; full flexibility in inputting charge rates for electricity use and supply to utility-detailed information on power production and consumption; and costs calculation.
PVSYST 5.31	CUEPE [48]	PVSYST is a PC software package for the study, sizing, simulation, and data analysis of complete PV systems. It is suitable for grid-connected, stand-alone, and DC-grid (public-transport) systems, and it offers an extensive meteorological and PV-components database. Program features include multilanguage interface (English, French, German, Spanish, and Italian); the possibility to define multi-PV fields; and improved definition of inverters, allowing consideration of multi-MPPT devices and the simulation of PV systems with heterogeneous orientations.
PV-DesignPro	Maoui Solar Soft. Corporation [49]	PV-DesignPro is designed to simulate photovoltaic energy system operation on an hourly basis for one year based on a user-selected climate and system design. Three versions of the PV-DesignPro program are included on the CD-ROM: PV-DesignPro-S for standalone

(*continued*)

Table 4 (continued)

		systems with battery storage, PV-DesignPro-G for grid-connected systems with no battery storage, and PV-DesignPro-P for water-pumping systems.
Solmetric PV designer	Solmetric [50]	This software enables the user to lay out strings of solar PV modules on a roof surface and calculate the energy production of the system. Results are corrected for the impacts of partial shading. This software is designed for residential and smaller commercial systems of 15 kW or less.
HOMER (Version 2.81)	NREL [51]	HOMER (Hybrid Optimisation Model for Electric Renewable) is a computer model that simulates and optimises stand-alone electric hybrid power systems. It can consider any combination of wind turbines, PV panels, small hydro power, generators, and batteries. The design optimisation model determines the configuration, dispatch, and load-management strategy that minimises life-cycle costs for a particular site and application.
TRNSYS	Solar Energy Lab, University of Wisconsin-Madison [52]	This tool was first created to study passive solar heating systems. Nowadays has been revised and includes main components of PV systems.

CUEPE = Centre universitaire d'étude des problèmes de l'énergie de l'Université de Genève.
NREL = National Renewable Energy Laboratory, USA.

authors [64–69]. These tools offer total freedom of programming to create an accurate model of a specific PV system and its components in return, however, for substantial efforts in the development and implementation of these tools.

REFERENCES

[1] ANSI/IEEE Standard 928-1986: IEEE Recommended Criteria for Terrestrial Photovoltaic Power Systems, Institute of Electrical and Electronics Engineers, New York, NY.
[2] E.T. Morton, M.J. Peabody, Feed-in tariffs: Misfits in the federal and state regulatory regime? The Electricity J. 23(8) (2010) 17−26.
[3] K. Sedghisigarchi, Residential solar systems: Technology, net-metering, and financial payback. Proceedins of the IEEE Electrical Power & Energy Conference, Montreal, 2009, pp. 978−1-4244-4509-7/09.
[4] T. Markvart, L. Castañer, M.A. Egido, Sizing and reliability of stand-alone PV systems. Proceedings of the 12th European Photovoltaic Solar Energy Conference, Amsterdam, 1994, pp. 1722−1724.
[5] L. Castañer, Photovoltaic engineering, in: T. Markvart (Ed.), Solar Electricity, John Wiley and Sons, Chichester, 1994, pp. 74−114.
[6] M.A. Egido, E. Lorenzo, The sizing of stand-alone PV-systems. A review and a proposed new method, Sol. Energy Mat. Sol. Cells 26 (1992) 51−69.
[7] E. Lorenzo, L. Narvate, On the usefulness of stand-alone PV sizing methods, Prog. Photovolt: Res. Appl. 8 (2000) 391−409.
[8] <www.bpsolar.com/>
[9] E. Negro, On PV simulation tools and sizing techniques: A comparative analysis towards a reference procedure. Proceedings of the 13th European PV Solar Energy Conference, Nice, 1995, pp. 687−690.
[10] W. Cowan, A performance test and prediction method for stand-alone/battery systems looking for quality assurance. Proceedings of the 12th European Solar Energy Conference, Amsterdam, 1994, pp. 403−407.
[11] I. Abouzahr, R. Ramakumar, Loss of power supply probability of stand-alone photovoltaic systems: A closed form solution approach, IEEE Trans. Energ. Convers. EC-6 (1991) 1−11.
[12] Lorenzo, E. 1994. Solar Electricity. Engineering of Photovoltaic Systems, Progensa, Sevilla.
[13] J.P. Gordon, Optimal sizing of stand-alone photovoltaic power systems, Sol. Energy 20 (1987) 295.
[14] L.L. Bucciarelli, Estimating loss-of-power probabilities of stand-alone photovoltaic conversion systems, Sol. Energy 32 (1984) 205.
[15] L.L. Bucciarelli, The effect of day-to-day correlation in solar radiation on the probability of loss of power in a stand-alone photovoltaic energy systems, Sol. Energy 36 (1986) 11.
[16] T. Markvart, W. He, J.N. Ross, A. Ruddell, J. Haliday, B. Rodwell, et al., Battery charge management for minimum-cost photovoltaic systems. Proceedings of the 16th European Photovoltaic Solar Energy Conference, Munich, 2001, pp. 2549−2552.
[17] IEEE Standard 1013-2000: IEEE Recommended practice for sizing lead−acid batteries for Photovoltaic Systems, Institute of Electrical and Electronics Engineers, New York.
[18] IEEE Standard 1144-1996: IEEE Recommended practice for sizing nickel−cadmium batteries for stand-alone PV systems, Institute of Electrical and Electronics Engineers, New York.
[19] P. Díaz, M.A. Egido, Sizing PV Pumping systems method based on loss of load probability. Proceedings of the 2nd World Conference on Photovoltaic Solar Energy Conversion, Vienna, 1998, pp. 3246−3249.

[20] O.C. Vilela, N. Fraidenraich, A methodology for the design of photovoltaic water supply systems, Prog. Photovolt: Res. Appl. 9 (2001) 349–361.
[21] A. Hadj Arab, F. Chenlo, M. Benghanem, Loss-of-load probability of photovoltaic water pumping systems, Sol. Energy 76(6) (2004) 713–723.
[22] A.A. Ghoneim, Design optimization of photovoltaic powered water pumping systems, Energ. Convers. Manag. 47(11–12) (2006) 1449–1463.
[23] A.I. Odeh, Y.G. Yohanis, B. Norton, Influence of pumping head, insolation and PV array size on PV water pumping system performance, Sol. Energy 80 (2006) 51–64.
[24] L. Castañer, S. Silvestre, Modelling Photovoltaic Systems Using Pspice, John Wiley and Sons, Chichester, 2002.
[25] R. Messenger, J. Ventre, Photovoltaic Systems Engineering, CRC Press LLC, 2000.
[26] G. Nottona, V. Lazarovb, L. Stoyanova, Optimal sizing of a grid-connected PV system for various PV module technologies and inclinations, inverter efficiency characteristics and locations, Renew. Energy 35(2) (2010) 541–554.
[27] H. Laukamp, Wechselrichter in Photovoltaik-Anlagen, Proceedings of the 4th Symposium on Photovoltaic Energy Conversion, Staffelstein, 1989.
[28] M. Jantsch, H. Schmidt, J. Schmid, Results of the concerted action on power conditioning and control. Proceedings of the 11th Photovoltaic Solar Energy Conference, Montreux, 1992, pp. 1589–1593.
[29] T. Schilla, Development of a Network Model of a PV Array and Electrical System for Grid Connected Applications, University of Northumbria, 2003, PhD Thesis.
[30] G. Nofuentes, G. Almonacid, Design tools for the electrical configuration of architecturally integrated PV in buildings, Prog. Photovolt: Res. Appl. 7(1999) (1999) 475–488.
[31] <www.solarmax.com/pub/downloads.php?lng = es&mc = espana#cat31/>
[32] <www.nsolpv.com/products.htm/>
[33] <http://re.jrc.ec.europa.eu/pvgis/apps4/pvest.php/>
[34] <www.nrel.gov/rredc/pvwatts/version2.html/>
[35] <www.retscreen.net/ang/home.php/>
[36] <https://www.nrel.gov/analysis/sam/>
[37] <www.thermosolar.it/index.php?lang = eng/>
[38] <www.automation.siemens.com/mcms/solar-inverter/en/solar-inverters-sinvert/software/Pages/home.aspx/>
[39] <www.fronius.com/cps/rde/xchg/fronius_international/hs.xsl/83_8594_ENG_HTML.htm/>
[40] <www.lapsys.co.jp/english/index.html/>
[41] <www.sma.de/de/service/downloads.html/>
[42] <oikos.com/esb/54/solarsizer.html/>
[43] <www.dds-cad.net/143x2 × 0.xhtml/>
[44] <www.ceere.org/rerl/rerl_hybridpower.html/>
[45] <www.insel.eu/index.php?id = 73&L = 1/>
[46] <www.fchart.com/pvfchart/>
[47] <www.valentin.de/en/>
[48] <www.pvsyst.com/5.2/index.php/>
[49] <www.mauisolarsoftware.com/>
[50] <www.solmetric.com/pvdesigner.html/>
[51] <www.homerenergy.com/>
[52] <www.trnsys.com/about.htm/>
[53] <www.mathworks.com/>
[54] <www.cadence.com/products/orcad/pages/default.aspx/>

[55] L. Castañer, D. Carles, R. Aloy, S. Silvestre, SPICE simulation of PV systems. Proceedings of the 13th European Photovoltaic Solar Energy Conference, Nice, 1995, pp. 950–952.
[56] A. Moreno, S. Silvestre, J. Julve, L. Castañer, Detailed simulation methodology for PV Systems. Proceedings of 2nd World Conference on Photovoltaic Energy Conversion, Vienna, 1998, pp. 3215–3218.
[57] A. Moreno, J. Julve, S. Silvestre, L. Castañer, SPICE macromodeling of photovoltaic systems, Prog. Photovolt: Res. Appl. 8 (2000) 293–306.
[58] S. Silvestre, A. Boronat, A. Chouder, Study of bypass diodes configuration on PV modules, Appl. Energy 86 (2009) 1632–1640.
[59] S. Silvestre, D. Guasch, A. Moreno, J. Julve, L. Castañer, A comparison on modelling and simulation of PV systems using Matlab and Spice. Technical Digest of 11th International Photovoltaic Science and Engineering Conference, Hokkaido, Japan, 1999, pp. 901–902.
[60] S. Silvestre, D. Guasch, U. Goethe, L. Castañer, Improved PV battery modelling using Matlab. Proceedings of the 17th European Photovoltaic Solar Energy Conference, Munich, 2001, pp. 507–509.
[61] S. Silvestre, D. Guasch, A. Moreno, J. Julve, L. Castañer, Characteristics of solar cells simulated using Matlab. Proceedings of the CDE-99, Madrid, Spain, 1999, pp. 275–278.
[62] D. Guasch, S. Silvestre, Dynamic battery model for photovoltaic applications, Prog. Photovolt: Res. Appl. 11 (2003) 193–206.
[63] A. Chouder, S. Silvestre, Automatic supervision and fault detection of PV systems based on power losses analysis, Energ. Convers. Manag. Vol. 51 (2010) 1929–1937.
[64] S. Wakao, T. Onuki, K. Ono, R. Hirakawa, T. Kadokura, J. Wada, The analysis of PV power systems by computational simulation. Proceedings of the 2nd World Conference on Photovoltaic Solar Energy Conversion, Vienna, 1998, pp. 3262–3265.
[65] A. Hamzeh, Computer aided sizing of stand-alone photovoltaic systems considering markedly variations of solar insolation over the year. Proceedings of the 2nd World Conference on Photovoltaic Solar Energy Conversion, Vienna, 1998, pp. 3250–3253.
[66] G.T. Klise, J.S. Stein, Models Used to Assess the Performance of Photovoltaic Systems, Sandia National Laboratories, 2009, Report SAND2009-8258.
[67] C. Tammineedi, J.R.S. Brownson, K. Leonard, Modeling improved behavior in stand-alone PV systems with battery–ultracapacitor hybrid systems. SOLAR 2010 Conference Proceedings of the American Solar Energy Society, 2010.
[68] E. Karatepe, M. Boztepe, M. Colak, Development of a suitable model for characterizing photovoltaic arrays with shaded solar cells, Sol. Energy 81(8) (2007) 977–992.
[69] E. Karatepe, M. Boztepe, M. Colak, Neural network based solar cell model, Energ. Convers. Manag. 47(9–10) (2006) 1159–1178.
[70] K. Sharmer, J. Giref, coordinators, European Solar Radiation Atlas, fourth ed. Les Press de l'Ecole des mines de Paris, Paris, 2000.

PART IIB

Balance of System Components

CHAPTER IIB-1

System Electronics

J. Neil Ross
Department of Electronics and Computer Science, University of Southampton, UK

Contents

1. Introduction — 697
2. DC to DC Power Conversion — 698
   2.1 The Buck Converter — 698
   2.2 The Boost Converter — 701
   2.3 The Buck-Boost or Flyback Converter — 702
   2.4 More Advanced Topologies for DC to DC Converters — 703
3. DC to AC Power Conversion (Inversion) — 703
   3.1 Single-Phase Inverters — 703
   3.2 Three-Phase Inverters — 707
   3.3 Isolation — 708
4. Stand-Alone PV Systems — 708
   4.1 Battery Charge Control — 708
   4.2 Maximum Power Point Tracking — 711
   4.3 Power Inversion — 713
5. PV Systems Connected to the Local Electricity Utility — 714
6. Available Products and Practical Considerations — 715
   6.1 Charge Controllers — 715
   6.2 Inverters — 717
7. Electromagnetic Compatibility — 718
References — 719

1. INTRODUCTION

The very simplest PV system requires no electronic control or power conditioning. A PV array with a suitably chosen number of cells in series charges a battery. The battery sustains supply when there is insufficient solar energy, but it also helps to maintain the supply voltage within limits. Such an approach has the merit of simplicity but has severe limitations. There is no control to limit the charge supplied to the battery, other than that imposed by the battery voltage and the open circuit voltage of the PV array, and there is no means of limiting the depth of discharge of the battery. The power supplied to the load will be direct current (DC), and the voltage may fluctuate substantially according to the state of charge. Also, there is no way of controlling the voltage across the PV array to ensure that it is providing its maximum power.

In order to overcome these limitations power electronic circuits are used to control the battery charging current, transform the voltage (DC to DC conversion) and to convert the direct current to alternating current, AC (inversion). In this chapter some of the techniques of DC to DC conversion and power inversion will be described briefly and then their use within stand-alone and grid-connected PV systems will be considered. In Section 5 the types of charge controllers and power conditioners available commercially are reviewed briefly.

2. DC TO DC POWER CONVERSION

For electronic power conversion it is essential that high efficiency be maintained, both to avoid wasting power and to avoid excessive heat dissipation in the electronic components. For this reason all practical power conversion circuits are built around energy storage components (inductors and capacitors) and power switches. The power switches used depend on the level of power to be converted or controlled. MOSFETs (metal oxide field effect transistors) are usually used at relatively low power (up to a few kW) or IGBTs (insulated gate bipolar transistors) at higher powers. At one time the use of thyristors was common, but these have been generally superseded, except at the very highest power levels.

There are three basic circuit topologies for DC to DC converters: the buck, or forward, converter; the boost converter; and the buck boost, or flyback, converter. More complex circuits exist, but these are sufficient to illustrate the principles involved. Many good texts on power electronics have been written. Good wide-ranging books, which cover the basic principles of power conversion and some applications, are Mohan et al. [1] and Rashid [2].

2.1 The Buck Converter

The buck converter, in its basic form, is shown in Figure 1. The key components are the inductor, L; the switch, S; and the diode, D. The

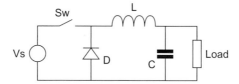

Figure 1 Basic circuit of the buck converter.

capacitor, C, stores charge and maintains a smooth output voltage as the switch is cycled on and off. The switch may be a MOSFET or an IGBT which can be turned on or off rapidly. The source voltage, V_s, must be greater than the load voltage V_l.

While the switch is on the current in the inductor the current in the inductor will increase at a rate given by

$$\frac{dI}{dt} = \frac{V_s - V_I}{L}$$

When the switch is turned off the current in the inductor will still flow but it will be diverted from the switch to the diode, a process referred to as commutation. The polarity of the voltage across the inductor is now reversed and the current in the inductor will decay. Neglecting the voltage drop across the diode, the rate of change of current is given by

$$\frac{dI}{dt} = \frac{V_I}{L}$$

The switch is turned on and off cyclically with a period T, and the current builds up and decays. If the current in the inductor does not decay to zero before the switch turns on again the converter is said to be operating in *continuous current* mode, as illustrated in Figure 2. In this case the load voltage depends only on the source voltage and the duty ratio, D:

$$V_I = DV_s$$

When operating in continuous current mode the current flowing from the inductor to the load is also continuous. This reduces the ripple current in the capacitor, enabling the use of a smaller value capacitor. However the current drawn from the supply is discontinuous.

In *discontinuous current* mode the inductor current falls to zero between switching cycles. In this case the load voltage depends in a more complex way on duty ratio and load current. Figure 3 shows how the load voltage varies with load current, at fixed duty ratio. At high load current the

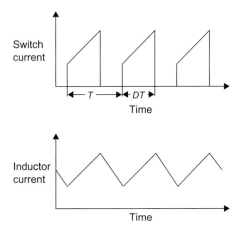

Figure 2 The currents in the switch and the inductor for a buck converter in continuous current.

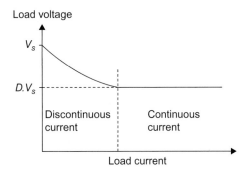

Figure 3 The variation of load voltage with load current for a buck converter at constant duty ratio.

inductor current is continuous and the load voltage constant (for an ideal loss-less converter). When the converter moves to discontinuous mode at low load current the voltage rises. Continuous conduction mode is generally preferred, as it is more efficient.

In either continuous or discontinuous operation the load voltage may be controlled at any value less than the source voltage by varying the duty ratio D. Usually the operating frequency is kept constant and the pulse width (switch on time) varied. This is referred to as pulse width modulation (PWM). An alternative, used occasionally, is to hold the switch on time constant and vary the frequency (pulse frequency modulation, or

PFM). At zero load current the voltage will rise to the source voltage even at very low duty ratio; hence, most switching converters specify a minimum operating current at which proper voltage control can be maintained.

The switching frequency depends on the type and speed of the switch used. For MOSFET switches a switching frequency of greater than 100 kHz would be common. For IGBT switches, which are much slower to turn off, the frequency might be around 25 kHz.

The efficiency of this type of circuit may be high, typically the load power will be 90%, or more, of the power drawn from the source. Power is dissipated in the switch when it is on, but provided it has a low on resistance this power will be small. Power is also dissipated when the switch changes state (at this time both the voltage across the switch and the current through it may be substantial). These switching losses become more important as the system voltage rises, but are minimised by ensuring that the switching transition is as fast as possible. Similar considerations also apply to the diode.

It is important to note that a practical converter will require a suitable electrical filter at the input, between the array and the converter. This is necessary to make the current drawn from the PV array reasonably constant. This not only avoids problems of radiated interference from the converter but also is necessary to store energy and ensure that the array is supplying power continuously.

2.2 The Boost Converter

The buck converter reduces the voltage, a circuit topology that increases the voltage, the boost converter, is shown in Figure 4. Again, the transistor is turned on and off in a cyclic manner. While it is on the current in the inductor builds up, then when the switch opens the voltage across the inductor changes sign and the current flows through the diode to the load. Again a capacitor is used to keep the load voltage constant.

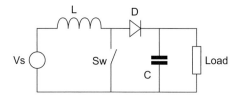

Figure 4 The boost converter.

As with the buck converter PWM is used to control the transformation ratio. The load voltage, in continuous current operation, is given by

$$V_I = \frac{1}{1-D} V_s$$

The boost converter in continuous conduction draws a continuous source current, but the load current will be discontinuous. This continuous source current may be advantageous for PV applications as it reduces the filtering required between the PV array and the converter.

Again the converter may be operated in discontinuous mode, when the load voltage will depend on both duty ratio and switching current. With the boost (or buck-boost) converter the voltage may rise to a very high value if there is no load current, unless active measures are taken to prevent this happening.

2.3 The Buck-Boost or Flyback Converter

The third basic topology is shown in Figure 5. This converter has the feature that the voltage may be increased or decreased, depending on the switching duty ratio. The output voltage is however of opposite sign to the input voltage. While the switch is on the current in the inductor increases and energy is stored. When the switch turns off the voltage across the inductor is reversed and the energy transferred to the load via the diode. In this case, for continuous inductor current, the load voltage is given by

$$V_I = \frac{-D}{1-D}$$

In this case, even in continuous current mode, both source and load currents are discontinuous.

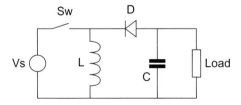

Figure 5 The buck-boost converter.

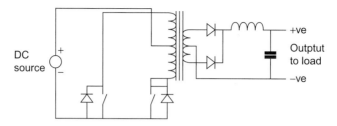

Figure 6 An example of an isolated DC to DC power converter.

2.4 More Advanced Topologies for DC to DC Converters

The basic converters described above have a number of limitations. The load is not electrically isolated from the source, which may be a serious limitation in higher voltage applications. Also efficient operation requires very rapid switching. This is because during the transition from ON to OFF or OFF to ON, in all the above circuits the switch may experience both large voltages and large currents. This implies that significant losses may occur during switching. A variety of *resonant converters* have been developed to reduce these losses (see, for example, references [1] or [2]).

Isolation of source from load may be achieved using a high-frequency transformer. Both the buck converter and the flyback converter may be isolated in this way. A common circuit for a moderate-power, isolated converter is shown in Figure 6. This is in fact a derivative of the buck converter with a transformer and diodes between the power switch and the inductor. The transformer may be made small and light by using a high switching frequency, typically of order 100 kHz at power levels around 1 kW. The transformer may also be used to vary the voltage ratio (in addition to the PWM) and hence this buck-derived converter, usually called a forward converter, may step the voltage up or down.

3. DC TO AC POWER CONVERSION (INVERSION)

3.1 Single-Phase Inverters

In essence a single-phase inverter is just a switching circuit which reverses the polarity of the supply on a cyclic basis. There are two basic configurations, the half-bridge and the full-bridge, which are shown in Figure 7.

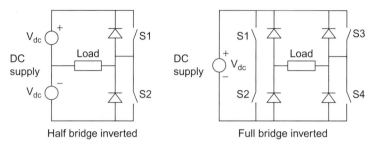

Figure 7 Single-phase inverter circuits.

The switching devices shown here as simple switches will, at low- or medium-power levels, be MOSFETs, IGBTs, or bipolar transistors, depending on the voltage and power. The diodes across the switches are necessary to allow for reactive loads where the current will continue to flow even if all the switches are open, or try to flow through the switch in the reverse direction, which, the electronic switch may not permit.

The switches of the inverter may simply be switched alternately at the required frequency for the AC to obtain a *squarewave* (S1 and S4 ON, then S2 and S3 ON) as shown in Figure 8(a). This simple switching scheme has the merit of simplicity, but no control of load voltage is possible and the resulting waveform will have a high harmonic content. *Modified square-wave* inverters adjust the switching scheme so that the two pairs of switches do not operate simultaneously, but there is a phase shift between them. The consequence is that there are intervals when both sides of the bridge are high or low and the net waveform is as shown in Figure 8(b). The equivalent sine wave has the same r.m.s. value as the modified square wave. This is sometimes referred to as single pulse width modulation. It may be used to reduce and control the r.m.s. load voltage, or it may be used to reduce the harmonic distortion, by making the waveform somewhat closer to a sine wave [2]. These modified square-wave inverters are usually, rather misleadingly, called *modified sinewave* by the manufacturers.

To reduce harmonic distortion and control the load voltage many modern inverters for PV and similar applications use high frequency pulse width modulation to synthesis a sinusoidal output. A variety of PWM schemes may be used; only one will be described to illustrate the principle.

The switches are operated in pairs to alternately provide a positive or negative voltage as in Figure 8(a), but at a much higher frequency than

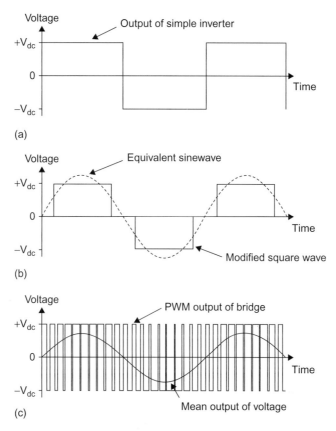

Figure 8 Output of single phase inverter with (a) simple square wave switching and (b) a PWM, quasi-sinusoidal output.

the desired AC output. By varying the duty ratio, the mean output voltage may be positive or negative. By modulating the duty ratio in an appropriate way, as illustrated in Figure 8(c), the output voltage averaged over the switching period may be made to vary sinusoidal manner. The modulation frequency has been reduced in Figure 8(c), relative to the fundamental output frequency, for clarity. With a high modulation frequency, and using a suitable filter, the modulation frequency components may be removed, leaving a power frequency output with low harmonic distortion. Varying the maximum and minimum duty ratio, the amplitude of the power frequency output may be controlled.

The inverters considered so far are voltage source inverters; that is, the DC source is at a constant voltage. Such an inverter will act as an alternating voltage source on the AC side. It is also possible to have a current source inverter in which the DC voltage source is replaced by a constant direct current source. The inverter bridge must be modified slightly as shown in Figure 9, and the switches must be operated so that current flowing from the source has a suitable path. Zero output current is obtained by having both switches in a leg in the ON state to divert current from the load. The current source inverter approximates an ideal alternating current source. In practice the constant DC current source is supplied by a large inductance, which keeps the current constant on a short time scale. On a longer time scale a control system matches the source voltage to the load voltage to keep the current in the inductor constant.

If the inverter is to supply power independently of any other source, then the inverter must control the load voltage and a voltage source inverter is the obvious choice. However, if the power is to be fed into a local power network where the voltage is controlled by other, more substantial sources of power then a current source inverter may be more appropriate.

An alternative to the current source converter is to use a voltage source inverter with PWM and with a fast control loop determining the current supplied to the load. In this case only a small value inductive filter is necessary at the output side of the inverter to keep the current approximately constant over the switching period and give the control loop time to respond. This approach is much more appropriate to small inverters as it avoids the large and heavy inductance.

A PV array acts as a source, which is neither an ideal current source nor an ideal voltage source. It approximates a current source if the voltage is well below the open-circuit voltage and a voltage source if the voltage is close to the open-circuit voltage. In order to provide a stable output voltage, or current, feedback control is necessary. Also required is energy storage to keep the source voltage (or current) stable during the power

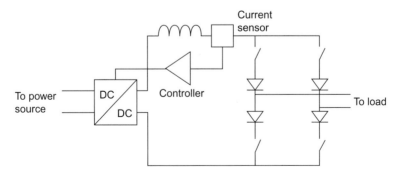

Figure 9 A current source converter.

frequency cycle, as the load current and source current vary. This may be achieved with a battery or capacitor for a voltage-fed inverter or a series inductor for a current-fed inverter.

If the load is not a pure resistance, but has a reactive component so that the load voltage and current are not in phase, or if the load is non-linear and generates harmonics, then the instantaneous power will flow through the inverter in both directions. (The reversal of power flow implies a reversal of the current on the DC side of the inverter.) This is not a problem for a stand-alone system with battery storage; however, if no battery storage is used, a large value capacitor across the input is required to absorb the cyclic return of current through the inverter if the reactive power is non-zero.

3.2 Three-Phase Inverters

At higher power levels it is usual to generate and distribute power using three phases. A three-phase inverter is usually based on the circuit of Figure 10. The three pairs of switches are switched in a cyclic manner with a phase shift of 120° between each pair. Using a simple square-wave switching scheme the corresponding waveforms are as in Figure 11, which shows the voltages at each phase with respect to the negative end of the DC source, and one of the interphase voltages. The other interphase voltages look the same but with a phase shifts of 120° and 240°. As with a single-phase inverter PWM may be used to produce a quasi-sinusoidal output and/or control the output voltage or current.

For high-power applications the preferred switches may be thyristors or gate turn-off thyristors (GTOs) since these devices are able to operate at high voltage and very high current. Circuits based on these devices are considered in many textbooks on power electronics [2,3].

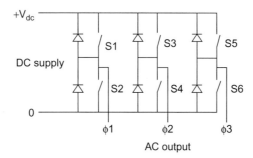

Figure 10 The basic circuit of a three-phase inverter.

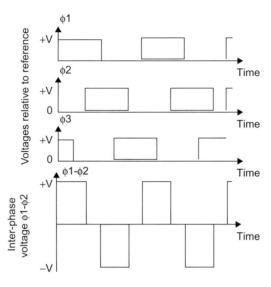

Figure 11 The voltage waveforms at $\phi 1$, $\phi 2$, and $\phi 3$ for simple square wave switching and one of the interphase voltages, $\phi 1 - \phi 2$.

3.3 Isolation

Generally it will be necessary to have electrical isolation between the PV array and the load, especially where the load side of the inverter is connected to the local electricity supply. This can be done simply with a power frequency transformer between the inverter and the load. The disadvantage of this approach is that power frequency transformers are relatively large heavy and expensive. An alternative approach is to provide isolation between the array and the inverter with a DC to DC converter isolated using a high frequency transformer (as in Figure 6). The high-frequency transformer approach has the advantages of small size and weight, at the expense of increased complexity.

4. STAND-ALONE PV SYSTEMS

4.1 Battery Charge Control

Small stand-alone PV systems generally use battery backup, and the roles of the system electronics are to control the charging of the battery and if necessary to change the voltage or to convert the DC to AC for the load.

The most widely used batteries are lead—acid batteries (Chapter IIb-2) in one of their various forms. This is because they are relatively cheap, have relatively good energy storage density, and can be robust and reliable. Their disadvantage is that they do need some care in controlling the charging and discharging if good life and a large number of charge/discharge cycles are to be obtained.

A typical, simple, ideal, charging cycle for a lead—acid battery is as illustrated in Figure 12. Initially the battery is charged at constant current (the bulk charge phase) until the voltage reaches some predefined value then the voltage is held constant while the charging current decays (tapered charge phase). After a suitable time, the charging voltage is reduced, or removed completely to avoid excessive gassing and loss of electrolyte. This ideal charging sequence can never be achieved with a PV system where the available power is constantly changing. The best the controller can do is to limit the peak charging current if necessary during the bulk charge phase, limit the voltage during the tapered charge phase and cut off the charge if the battery is deemed fully charged. Some charging schemes, rather than cutting off the charge altogether reduce the charging voltage by 5—10% to provide a trickle charge with a voltage low enough to avoid significant gassing. With some types of lead—acid cell it is also desirable to overcharge the battery occasionally to promote gassing and stir up the electrolyte.

With a PV array there are two basic methods by which the voltage or current may be controlled [4]. A series regulator introduces resistance in series with the array (Figure 13(a)), reducing the load current but allowing the array voltage to rise towards its open-circuit value, and a shunt

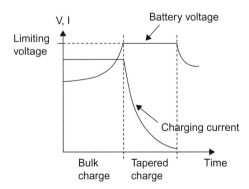

Figure 12 The typical ideal charging cycle for a lead acid battery.

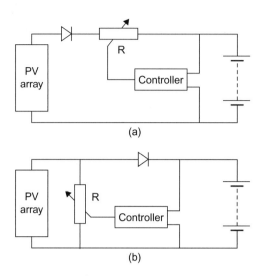

Figure 13 (a) Series regulation of battery charge current; (b) shunt regulation.

regulator which dumps current from the array pulling down both load and array voltage (Figure 13(b)). The diodes are necessary to avoid the battery being discharged either through the array, or through the shunt regulator. Such regulators may be implemented using transistors as variable resistance elements. This approach requires the transistors to dissipate a large amount of power and is not generally used. The much more popular approach is to use the transistors as switches which turn on and off in such a way as to reduce the mean current flowing to the battery (Figure 14). Again either a series (a) or a shunt connection is possible (b). In the series-switching regulator, the switch makes and breaks the current to the battery. The mean switch current depends on the fraction of the time for which the switch is on. If the switching is rapid the voltage across the battery will be almost constant as the battery voltage responds quite slowly to changes in current. The excess power from the PV array is dissipated in the cells of the array. The shunt switching regulator operates in a similar manner. The diode prevents the battery from discharging through the switch and effectively acts as a series switch which is off when the controlled switch is on. Again excess power is dissipated within the PV array.

Provision should also be made for cutting off the supply to the load if the state of charge of the battery falls below an acceptable level otherwise permanent damage to the battery may result. This requires some method of measuring or estimating the state of charge of the battery. This information is

System Electronics

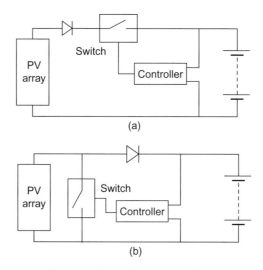

Figure 14 (a) Series switching regulator; (b) shunt switching regulator.

important to know when to shed load, it is also needed to establish when to discontinue charge. The state of charge is frequently deduced from the terminal voltage, either off load or with a known load. However this method is not very reliable as the voltage depends on many factors such as temperature and the recent charge/discharge history of the battery. For flooded lead−acid cells the electrolyte specific gravity may be used, but this is not easy to measure using electronic techniques and may be modified by stratification of the electrolyte or loss of water due to excessive gassing. To try to overcome this ampere hour counting may be used to establish how much charge has been removed or added to the battery and hence deduce the state of charge. This method will drift, but may be recalibrated if there are times when the battery is known to be fully charged.

4.2 Maximum Power Point Tracking

The power supplied by a PV array varies with load voltage as shown in Figure 15. The voltage for optimum power transfer varies quite strongly with array temperature and more slowly with intensity of illumination. To ensure the optimum power transfer a DC to DC converter may be located between the array and the battery and used to optimise the power transfer. Any of the DC to DC converters described above may be used, depending on the array and battery voltages. The boost converter is probably the preferred configuration as it can draw current continuously at

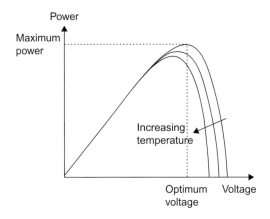

Figure 15 Power supplied by PV array as function of voltage, at constant illumination.

the input side, minimising the need for filtering and energy storage on the array side of the converter.

In order to ensure that the power converter is operating at the optimum voltage transformation ratio, some form of feedback control is required. This requires a method of identifying the voltage at which maximum power is obtained from the array and a controller on the power converter to maintain the array voltage at, or close, to this value. There are several approaches to estimating the optimum operating condition. The most general is to sweep the array current, measure the array voltage and current, and deduce the maximum power point. Such an approach is however of little value if the array illumination and temperature are varying.

A more practical approach is to vary the array current and deduce if increasing the current increases or decreases the power delivered, and then deduce how the converter control must be changed so as to move the operating point in the direction of the maximum power point. A wide range of these hill-climbing algorithms have been proposed. Possibly the simplest is the perturb and observe (P&O) algorithm [5]. Periodically a small variation is made to the operating point of the power controller. If this results in increased power the next perturbation will be in the same direction, if not the direction will be reversed. Another hill-climbing scheme is to periodically vary the load current while measuring the instantaneous power and hence deduce the local gradient of the P-I curve. It is worth noting that the source current will be modulated by the input ripple current of the converter, and this modulation may be used

[6] as the stimulus. The control algorithm moves the operating point to set this gradient to zero.

An approach to maximum power point tracking which requires minimal control electronics has been described by Boehringer [7]. This technique avoids the need to measure the power; only the current and voltage are needed. The scheme was originally proposed for use in a space on a small satellite. The scheme is simple and elegant but may be more appropriate to space applications where the system is not required to operate with large changes of illumination (see, however, [8] for a discussion of an application of this technique to grid-connected systems).

A simple approach to finding the maximum power point is to use the property of silicon solar cells that the optimum power point is always about 70–80% of their open-circuit voltage [9]. By simply disconnecting the array periodically for a few milliseconds and measuring the open circuit voltage, the optimum array voltage may be found by multiplying the open circuit voltage by a factor k of about 0.75. A small error in the choice of k is relatively unimportant since power varies slowly with voltage near the turning point. (A similar approach is also possible using the short-circuit current.) A similar concept uses a monitor cell to deduce the open-circuit voltage.

There has also been interest in the use of neural networks [10]. In this case a monitoring cell is used to deduce the open-circuit voltage, as above, but the converter is controlled using a neural network which has been trained so that it can predict the optimum voltage of the installed array.

A recent comparison of various schemes [5] showed that the 'perturb and observe' scheme was simple to implement and could achieve high efficiency even under conditions of fluctuating illumination.

4.3 Power Inversion

Where the stand-alone PV system is to deliver AC for electrical appliances, an inverter must be used. The appropriate type of converter for this type of application is voltage fed, with the battery controlling the input voltage to the converter. The inverter control system must maintain the supply voltage and frequency within acceptable limits. It must also cut the supply if battery charge falls to an unacceptably low value. Simple square wave inverters are now rarely used. Modified sine wave (really modified square wave, as described in Section 3.2) or sine-wave converters are preferred because their better voltage waveform is required by many types of load and the cost difference is small. Modified sine-wave converters still generate significant harmonics and are unsuitable for certain types of load, for example some types of electronic equipment. For this reason sine-wave converters may be preferred. The main disadvantages of sine wave inverters are their somewhat higher cost and they may provide lower efficiency when compared with modified sine-wave inverters (because of increased switching frequency, and, hence, switching losses). However, these differences are being significantly reduced.

5. PV SYSTEMS CONNECTED TO THE LOCAL ELECTRICITY UTILITY

When a PV system is connected to the electrical utility the inverter must be operated synchronously with the utility, the voltage frequency and phase of the generation must match that of the utility. Inverters for this application are likely to be controlled so as to appear at the utility interconnection as a current source with a sinusoidal waveform synchronous with the utility voltage. This may be achieved using a current fed inverter, but, as noted earlier, would in practice be achieved using a suitable PWM feedback control system to ensure that the load current is sinusoidal and to control its value. Control is quite complex. The utility voltage fixes the load voltage of the inverter, but the inverter must implement maximum power point tracking to ensure that the input voltage from the array is at the optimum value. This criterion determines the optimum r.m.s. value for the load current.

For medium-sized PV systems there seems to be a movement towards the use of multiple inverters, each connected to its own segment of the array, rather than a single, larger inverter. Such an approach could lead to better reliability because of the redundancy. It may also reduce wiring costs if the AC voltage is much higher than the DC source voltage, as is usually the case, by reducing the need to carry large DC currents over significant distances.

There are two key issues that arise as the result of connecting small power generators to the public utility supply. These are power quality and safety. As far as power quality is concerned there are a number of points to consider. The power supplied must be sinusoidal with minimal harmonic distortion, there must be minimal injection of high-frequency switching components and the net power factor of the local generation and load should be as close as possible to unity. Good inverter design should ensure that the frequency spectrum of the injected power is acceptable and that the output from the inverter has no DC content, since a small direct current may cause saturation of the distribution transformer. The issue of power factor is more complex. If the inverter is controlled so as to generate the sinusoidal current in phase with the voltage it will not generate reactive power. Thus, while the inverter will supply some or all the power needed by the load, the reactive current for the load will be drawn from the utility. Hence, the local generation will degrade the power factor seen by the utility.

A particular safety issue raised by the connection of local generation to the utility is what happens when there is a loss of supply from the utility. If the loss of connection is not detected, and the inverter shut down, the local network may remain energised. If the inverter is able to supply sufficient power to match the load this situation may continue for some time. This is referred to as islanding. Islanding presents a potential safety hazard to utility and other personnel and also potential problems if the connection is remade while the inverter is supplying power. Basic protection measures such as shutting down operation if the voltage or frequency lie outside normal values will usually be sufficient to ensure that islanding is detected and the inverter shut down. However, there are circumstances where such measures are inadequate, notably with multiple inverters, a matched resonant load or rotating machine. Other measures that may be taken to improve the prevention of islanding include monitoring the frequency for rapid changes and measuring the impedance of the network by injecting a suitable current impulse.

A detailed overview of islanding and other issues which arise when a PV generator is connected to the utility supply can be found in Chapter IIc-1.

6. AVAILABLE PRODUCTS AND PRACTICAL CONSIDERATIONS

6.1 Charge Controllers

The role of the charge controller is to ensure that the battery is not overcharged. There is a bewildering array of commercial devices from a large number of manufacturers, with current ratings from a few amps up to hundreds of amps, and operating voltages generally in multiples of 12 V up to 48 V (nominal battery voltage).

Most commercially available charge controllers for PV applications use a switched series regulator to control the charging current Figure 14(a). The most common control scheme is pulse width modulation. The power to the battery is switched on and off at a constant frequency, with the duty ratio varied, to control either the mean current to the battery or the charging voltage of the battery. This scheme is similar to a buck converter, but with a PV power source the current is limited, so a series inductor to limit the peak current is unnecessary, and the load voltage is smoothed by the battery. The control algorithm depends on the type of battery and most charge controllers provide a number of settings to accommodate different voltages and types of lead—acid battery. A typical control scheme would allow continuous current until the battery reaches a predetermined voltage and then the duty ratio of the PWM is reduced to limit the battery voltage. Some controllers use a three stage charging algorithm with a *bulk charge phase*, where the charging current is the maximum available, a *taper phase* where the voltage is held constant, and a *float phase* where the battery voltage is held constant at a reduced value. Many controllers will have control algorithms that will allow different control regimes and/or settings to accommodate different battery types.

A few controllers use switched shunt regulation. In this case a transistor switch bypasses the current from the PV array (Figure 14(b)). Again PWM may be used to control the mean load current. A variant on the shunt controller uses switches to divert power from charging the battery to a diversion load, again using PWM. The diversion load might be a water heater or similar. This enables the energy not used for charging the battery to be used usefully, rather than dissipated as heat in the PV panel. It is not clear that either series or shunt regulation is to be preferred. Both can achieve high overall efficiency.

In both series and shunt controllers the switches used are field effect transistors (FETs) with an ON state resistance of only a few mΩ. Thus, the voltage drop across the switch when it is on is usually small, avoiding unnecessary power dissipation, and in the case of a series regulator ensuring that overall efficiency is high. Control schemes other than PWM are possible and are used by some suppliers. One proprietary scheme, Flexcharge™, uses the way in which the battery voltage rises and falls as the current is switched ON and OFF to control the switching. The switch supplies current to charge the battery until the battery voltage reaches a desired upper limit. The current is then switched off until the voltage has fallen to a lower limit. This procedure charges the battery with pulses of current which become shorter and less frequent as the battery approaches full charge.

Charge controllers incorporating maximum power point tracking are available, but much less common than simple charge controllers. To be useful a maximum power point tracker must have an efficiency of greater than 90%. This calls for careful design and the cost of such systems is significantly greater than for a basic charge controller. The gain in system efficiency however may be significant; gains of up to 30% are claimed. Such large gains will only be achieved when the battery is in a low state of charge (low voltage) and the PV array is cold (high voltage). Average efficiency gains are however likely to be significantly less (of the order of 10%). The cost of the maximum power point tracker must be weighed against the cost of extra array area.

6.2 Inverters

Inverters can broadly be split into two groups: those for stand-alone systems where the inverter controls the voltage and frequency of the *AC*, sometimes referred to as *off grid*, and those intended for use to feed power into an AC supply which is connected to the power utility, often referred to as *grid tied* or *grid connected*. The difference between the two groups is largely in the control algorithm which is generally implemented using a microprocessor. A significant number of dual application systems are available where the controller can implement either voltage control for stand-alone operation or current control for operation connected to the public utility or other generators.

Inverters available for stand-alone applications are readily available from as low as about 60 W up to several kW. They fall into two categories, modified sine wave or sine wave. Modified sine wave inverters generate a modified square wave as discussed in Section 3.1 and illustrated in Figure 8(b). This waveform may have relatively low total harmonic distortion and is adequate for many applications. The advantage of modified sine-wave inverters is lower cost than true sine-wave inverters; however some types of load will not function correctly with this type of waveform (e.g., thyristor or triac controllers may experience problems due to the ill-defined zero crossing point).

Sine-wave inverters produce a waveform much closer to the ideal sinusoidal waveform. Typically the total harmonic distortion will be less than 3%. The waveform is much smoother than for modified sine wave and is suitable for all types of load. The only disadvantages are potentially lower efficiency because of the high switching frequency and risk of the emission electromagnetic interference (again due to the high switching frequency). This latter problem may be avoided with good design and adequate filtering. In practice the efficiencies claimed by manufacturers for sine-wave inverters are not significantly less than for the simpler type.

Grid-connected inverters are rather more constrained than those used in standalone systems. Because they are connected to the public utility network they must meet all the safety and other criteria imposed by the local electricity utility. Thus, they will be expected to generate low levels of harmonics and low levels of high frequency noise due to the switching. In addition, as discussed earlier, they must be able to detect the loss of power from the utility and shut down. This islanding protection is required to ensure safety under the condition of loss of supply from the utility. Commercially available inverters use a variety of schemes to detect loss of utility connection (a detailed review of this topic can be found in Chapter IIc-1). It is not usual to supply battery storage for grid-connected systems.

Inverters are available that will operate with either current control or voltage control. With suitable storage such a system can provide both utility connected and stand-alone operation. However, if the system is used to provide backup power in the event of loss of utility supply, islanding protection is still required. Following loss of supply the inverter must be shut down and the local network isolated from the utility before the inverter may be restarted in voltage control mode.

7. ELECTROMAGNETIC COMPATIBILITY

Converters and inverters using PWM are sources of electromagnetic interference because of the rapidly changing voltages and currents. Measures must be taken to limit the emissions from such systems in order to meet regulatory requirements and to avoid interference with other electronic equipment [11]. To minimise radiation care is necessary with the internal layout to minimise the area of loops with large, or rapidly changing, circulating currents. In addition a carefully designed screened enclosure will reduce radiated emissions further. Interference may also be

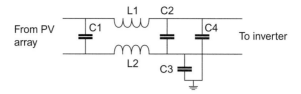

Figure 16 Example of filter to reduce the transmission of electromagnetic interference along power lines.

conducted up the power leads into and out of the converter or inverter. This is minimised by the use of filters, a typical example of a filter for the out input of an inverter is shown in Figure 16. Careful design of the filters is necessary to obtain low levels of interference. It is also essential to ensure that the earthing arrangements do not provide loops around which the unwanted high-frequency currents may circulate.

REFERENCES

[1] N. Mohan, T.M. Undeland, W.P. Robbins, Power Electronics, Converters, Applications and Design, 2nd ed., John Wiley & Sons, New York, 1995.
[2] M.H. Rashid, Power Electronics, Circuits, Devices and Applications, 2nd ed., Prentice Hall, 1993.
[3] C.W. Lander, Power Electronics, 3rd ed., McGraw-Hill, London, 1993.
[4] F. Lasnier, T.G. Ang, Photovoltaic Engineering Handbook, Adam Hilger, Bristol, 1990.
[5] D.P. Hohm, M.E. Ropp, Comparative study of maximum power point tracking algorithms, Prog. Photovolt: Res. Appl. 11 (2003) 47–62.
[6] P. Midya, P.T. Krein, R.J. Turnbull, R. Reppa, J. Kimball, Dynamic maximum power point tracker for photovoltaic applications 1996, Proceedings IEEE Annual Power Electronic Specialists Conference (PESC), 1996, pp. 1710–1716.
[7] A.F. Boehringer, Self-adapting dc converter for solar spacecraft power supply, IEEE Trans. Aerospace Electronic Systems 4 (1968) 102–111.
[8] T. Schilla, Development of a Network Model of a PV Array and Electrical System for Grid Connected Applications, PhD Thesis, University of Northumbria, 2003.
[9] J.H.R. Enslin, M.S. Wolf, D.B. Snyman, and W. Swiegers, Integrated photovoltaic maximum power point converter, IEEE Trans. Ind. Electron. 44 (1997) 769–773.
[10] T. Hiyama, S. Kouzuma, T. Imakubo, T.H. Ortmeyer, Evaluation of neural network based real time maximum power tracking controller for PV system, IEEE Trans. Energy Conversion 10 (1995) 543–548.
[11] B.E. Keiser, Principles of Electromagnetic Compatibility, 5th ed., Artech House Inc, MA, USA, 1985.

CHAPTER IIB-2

Batteries in PV Systems

David Spiers
Naps Systems, Naps-UK, Abingdon, UK

Contents

1. Introduction 722
2. What Is a Battery? 723
3. Why Use a Battery in PV Systems? 724
4. Battery Duty Cycle in PV Systems 726
   - 4.1 How Are Rechargeable Batteries Normally Used? 726
   - 4.2 How Does This Differ from PV Systems? 726
5. The Battery as a 'Black Box' 727
   - 5.1 Battery Performance Definitions and Summary 729
     - 5.1.1 Definitions of Capacity, Efficiency and Overcharge 729
     - 5.1.2 Discharge Rate and Charge Rate 730
     - 5.1.3 Battery Capacity Is Not Fixed 730
     - 5.1.4 Depth of Discharge and State of Charge 731
     - 5.1.5 Self-Discharge Rate 732
     - 5.1.6 Cycle Life 732
     - 5.1.7 Maximum Lifetime 733
   - 5.2 Example of Simple 'Black Box' Battery Calculations 733
     - 5.2.1 Basic Data 734
     - 5.2.2 Calculations 734
6. The Battery as a Complex Electrochemical System 737
7. Types of Battery Used in PV Systems 740
8. Lead–Acid Batteries 741
   - 8.1 Basics of Lead–Acid Batteries 741
   - 8.2 Types of Lead–Acid Battery 742
   - 8.3 Construction of Lead–Acid Batteries 743
     - 8.3.1 Plate Type 743
     - 8.3.2 Grid Alloy 743
     - 8.3.3 Grid Thickness 744
   - 8.4 Sealed Lead–Acid Batteries 744
     - 8.4.1 AGM Type 745
     - 8.4.2 Gel Type 745
   - 8.5 Mass-Produced and Industrial Batteries 745
   - 8.6 How the Capacity Varies 746
     - 8.6.1 Capacity Depends on the Discharge Rate 746
     - 8.6.2 Capacity Reduces at Low Temperatures 747
     - 8.6.3 Capacity Depends on the End Voltage 748

Practical Handbook of Photovoltaics.
© 2012 Elsevier Ltd. All rights reserved.

8.7	Acid Density	749
	8.7.1 Acid Density Falls During Discharge	*749*
8.8	Acid Stratification	750
8.9	Freezing	752
	8.9.1 Acid Freezing Point	*752*
	8.9.2 Battery Freezing Points	*753*
8.10	Sulphation and Deep Discharge	753
9.	Nickel—Cadmium Batteries	754
10.	How Long Will the Battery Last in a PV System?	756
	10.1 Factors Affecting Battery Life and Performance in PV Systems	757
	10.2 Cycle Life Can Be Misleading	758
	10.3 Battery Float Lifetime	758
	10.4 Sealed Battery Lifetimes	759
	10.5 Examples of Predicted Battery Lifetimes	760
	10.6 Comparison of Predicted Battery Lifetime and Field Data	762
	10.7 Battery Lifetimes Summary	765
11.	Selecting the Best Battery for a PV Application	765
12.	Calculating Battery Size for a PV System	767
	12.1 Select the Appropriate Voltage	767
	12.2 Define Maximum Depths of Discharge (DOD)	767
	12.3 Define Maximum Depths of Discharge (DOD)	768
	12.3.1 We Now Calculate Four Capacities	*768*
13.	Looking After the Battery Properly	770
	13.1 System Design Considerations	770
	13.1.1 Charge Control	*770*
	13.1.2 Internal Healing	*771*
	13.1.3 Battery Environment	*771*
	13.2 Commissioning	773
	13.3 Maintenance, Replacement, and Disposal	773
14.	Summary and Conclusions	774
Acknowledgements		775
References		775
Further Reading		776

1. INTRODUCTION

Batteries are used in most stand-alone PV systems, and are in many cases the least understood and the most vulnerable component of the system. Most design faults (e.g., undersizing the PV array or specifying the wrong type of controller) and operating faults (e.g., use of more daily electrical energy than was designed for or simply some breakdown in the PV array or charge controller) will lead to ill health, if not permanent failure, of the battery. In these cases, the battery is then often blamed for the failure of the system to deliver the promised regular amount of electrical energy, and the type of battery used in such a failed system can acquire, quite unfairly, a bad reputation as a 'PV battery.'

The aim of this chapter is to present the reader with enough information to understand how important it is to specify an appropriate type of battery, and with sufficient capacity, for satisfactory use in a PV system. We start by considering the battery as a 'black box' with certain properties, and for many purposes this is sufficient to design a satisfactory PV system as long as we understand that these 'black box' properties can change according to certain external conditions. To understand why these properties change the way they do, it is necessary to appreciate some general principles of how batteries are made and how they work, and this material forms the second part of this chapter. Finally, some guidelines are given on how to specify a battery correctly for a PV system and how to look after it properly.

2. WHAT IS A BATTERY?

To some people, a battery is a small object that you insert into a radio or personal music system to provide some electrical energy. When it has run out of energy, it is thrown away (often irresponsibly). This is perhaps the best-known example of a nonrechargeable, or primary, battery. Many of these are in the form of cylindrical cells, and you need to put 2, 3, 4, or more of them into your device to make it work. This illustrates another essential feature of batteries: like PV cells, they are essentially low-voltage devices, and you have to connect several in series to get a useful voltage (which is often 4.5, 6, or 9 V for small portable electrical consumer products). Even if a primary battery is supplied as a single package (for example, the familiar small 9-V battery with snap-on connectors), it nearly always consists of more than one cell inside.

To other people, a battery is the device in their mobile phone, laptop computer, cordless drill, electric toothbrush, etc. that needs recharging from time to time, often at the most inconvenient moment. These are examples of rechargeable, or secondary, batteries that in earlier times were often referred to as accumulators. In these types of devices, the batteries are often used until they have run out of energy to deliver, but instead of being thrown away are 're-filled' with electricity by connecting them to a main charger. After several of these 'cycles,' the battery will start to show signs of wearing out, and eventually it will not provide a satisfactory interval between recharges and will have to be replaced.

Most people prefer to forget that there is a battery inside their car, until of course it goes wrong—which often means that the car refuses to start, typically on the first cold morning of the winter. Actually, there is a 12-V rechargeable battery in every car that provides the electrical power for starting the engine and is recharged quite quickly by a generator that is driven by the car engine. This battery also powers the lights when needed, providing a more constant voltage than if the lights were powered from an engine-driven generator alone (so the lights do not change intensity as the engine speed varies). It also provides electrical power when the engine is not running. In many ways, this is quite similar to how a rechargeable battery is used in a PV system—it provides electricity when there is no power from the PV array (i.e., at night), it is recharged when power from the PV array is available (i.e., during the day), and it stabilises the voltage when there is power from the PV array. In several other ways, these two types of duty are quite different.

Few people realise just how widespread batteries are depended upon in other aspects of daily life. In offices, building plant rooms, power stations, etc., batteries are constantly waiting in 'stand-by' mode to provide emergency electrical power if there is a power cut. They are kept at full charge by a small charging current (called *float charge*), and they only undergo a discharge (and recharge) in the event of emergency power being required. In contrast, electric forklift trucks and other electric vehicles have a large rechargeable battery that is discharged almost fully every working day and recharged every night.

3. WHY USE A BATTERY IN PV SYSTEMS?

There are three main functions that a battery performs in a PV system:
- *It acts as a buffer store to eliminate the mismatch between power available from the PV array and power demand from the load.* The power that a PV

module or array produces at any time varies according to the amount of light falling on it (and is zero at nighttime). Most electrical loads need a constant amount of power to be delivered. The battery provides power when the PV array produces nothing at night or less than the electrical load requires during the daytime. It also absorbs excess power from the PV array when it is producing more power than the load requires.

- *The battery provides a reserve of energy (system autonomy)* that can be used during a few days of very cloudy weather or, in an emergency, if some part of the PV system fails.
- *The battery prevents large, possibly damaging, voltage fluctuations.* A PV array can deliver power at any point between a short circuit and an open circuit, depending on the characteristics of the load it is connected to. In a nominal 12-V system, for example, this means anything between 0 V and around 20 V is possible from the PV array. Many loads cannot operate over such a wide range of voltages. Placing a battery between the PV array and the load ensures that the load will not see anything outside the range of voltages that the battery can experience—in the case of a 12-V system from around 9.5 V under deep discharge to around 16 V under extreme charging conditions.

These functions are needed in most PV applications. It is, for example, difficult to think of a viable PV lighting system that does not contain a battery. Perhaps it is easier to list those PV systems where a battery is not commonly used:

- *In grid-connected systems*, where the PV array produces AC electricity that is either used inside the building or is exported to the main electricity supply network when there is an excess. Note that this only works at present without disturbing the grid because the amount of PV electricity produced is so small compared to the grid's capacity. If grid-connected PV becomes extremely widespread, some form of electrical storage will be needed to accommodate it.
- *In water-pumping systems*, where a PV array is connected directly to an electric pump of suitable characteristics. The pump speed varies with amount of power available from the PV array at any time, and the mismatch between water supply and demand is smoothed out by storing water in a tank. This is a cheaper way to store energy than in batteries.
- *In other cases of directly driven motor loads*, such as fans and sometimes compressors in refrigeration units. As with the PV driven pumps, the speed varies with the PV array output (and is zero at night).

4. BATTERY DUTY CYCLE IN PV SYSTEMS

PV systems demand that batteries are used in a different way to any other type of battery application.

4.1 How Are Rechargeable Batteries Normally Used?

As we have seen above, batteries are mostly used in one of three modes:
- Regular deep cycling (as in electric vehicles and consumer devices)
- Standby use (kept at full charge in case of an emergency)
- Starting, lighting, ignition (SLI) for road vehicles. Starting requires a very high current for a short time, but the depth of discharge is small.

In all of the above normal duty cycles, the battery is given a full charge after discharging. Full charging means that some adequate 'overcharge' is given, often over quite a long period of time.

4.2 How Does This Differ from PV Systems?

The first difference to note in PV systems is that there is a limited (and variable) amount of charging energy available from the PV array, and it is by no means guaranteed that the battery will be fully charged at the end of each day.

If the load is used every day, then the battery will undergo a daily cycle of discharge and recharge, but this is often a shallow cycle compared to normal cycling duty. Most stand-alone PV systems are specified to have a certain number of days of autonomy reserve in the battery—depending on the reliability required and the weather pattern, this can be anything from 3 to 20 days, and it is normal to specify that there is some charge left in the battery after this autonomy reserve has been used (often 20%). So, simplistically, we can estimate the range of daily cycling as follows:
- For 3 days autonomy, the battery needs a storage capacity equal to $(3/0.8) = 3.75$ times the daily load. The maximum daily cycle is then $(1/3.75) = 26.7\%$ (if the load is all consumed at night).
- For 20 days autonomy, the same reasoning leads to a battery capacity of $(20/0.8) = 25$ times the daily load and a maximum daily cycle of $(1/25) = 4\%$.

If the load is continuous throughout the 24-hour period, these average daily depths of discharge will be less, as for some of the time the battery will be charging when the load is being powered. We can therefore say that the typical range of daily cycling is of the order of 2–30% per day.

One exception to this is when a PV array is coupled with a diesel generator and the diesel generator provides the backup autonomy reserve instead of the battery. In these cases the batteries can indeed perform a deep cycle every day, and they are unlikely to give more than 2–3 years' service (1 year in such a duty cycle requires 365 deep cycles; 2 years, 730 deep cycles; and 3 years, 1095 deep cycles).

In addition to shallow daily cycles, the battery may be required to provide seasonal storage during winter or some rainy season like a monsoon. This is basically a deeper and much longer cycle superimposed on the daily cycling pattern. The battery may be in a lower state of charge than normal for several weeks.

Similarly, after a deep discharge to supply system autonomy in an emergency, the battery may take several weeks to recover its full charge—there may only be a small excess of daily energy available from the PV array to restore this charge that has been withdrawn.

Figure 1 illustrates these various aspects of cycle duty in PV systems.

5. THE BATTERY AS A 'BLACK BOX'

It is very convenient (and tempting) to treat a rechargeable battery as a 'black box,' or perhaps, more graphically, a 'black bucket,' into which electricity can be filled and withdrawn as needed. Like a bucket, a battery has a certain storage capacity, and when it is full, there is no point in trying to fill it further. Also like a bucket, when it is empty, there is no point in trying to empty it further. If you want to store more water in a bucket, or electricity in a battery, you have to use either a larger capacity bucket or battery, or more than one of the same size.

If you use ampere-hours (Ah) as the unit of electricity being stored, then as a first approximation, you can use the 'black box' or 'black bucket' approach. So if you want to store enough reserve electricity to power a 1 A load continuously for 5 days (120 hours), you are going to need a storage capacity of 120 ampere-hours (120 Ah). To determine whether that means using a battery with 120 Ah written on the label requires either understanding batteries in more detail or simply learning a set of rather strange empirical rules about this black box that stores electricity.

If you want to just follow 'black box rules' for specifying batteries for PV systems, then the following are the main ones to be aware of. It soon

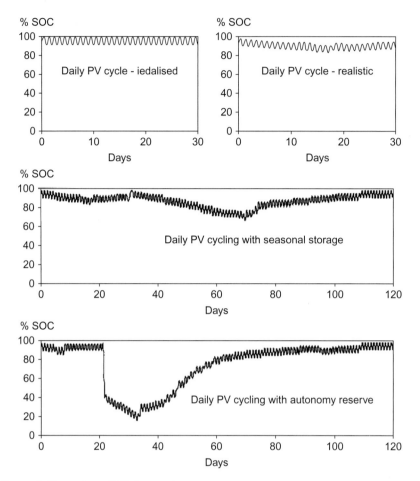

Figure 1 Illustration of PV duty cycles.

becomes clear that the idea of a bucket into which electricity can be filled and withdrawn is a little simplistic.
- You must choose a battery of the appropriate voltage. Simply specifying that this is a 120-Ah battery does not tell you whether it should be 6, 12, 24, 48 V, etc. In most PV systems, to get the voltage you require means connecting 2, 6, or 12 V units of the required capacity in series.
- The storage capacity of a battery is not fixed. It depends mainly on how fast you try to extract it (i.e., the discharge current) and the temperature. How this varies depends on the exact battery type, and you need to consult a data sheet.

- It is not a good idea to try to extract the full capacity of the battery, even under worst-case conditions. Under most circumstances, you should aim not to extract more than 80% of the available capacity at any time.
- The more capacity you try to extract from a battery every day, the more quickly it will wear out. Just how quickly depends on the type of battery and its cycle life. For a given daily load, the smaller the battery capacity, the shorter its life will be, unless something else causes it to fail earlier than the wearing out due to this charge and discharge cycling.
- The battery voltage on charging depends to some extent on how fast you are charging it. If the current is particularly high, the voltage may rise more than the charge controller is expecting to find for a fully charged battery. It may disconnect the PV charge current (or reduce it) before the battery is fully charged.
- If you operate a battery at high temperatures (in this case, 'high' means more than about 20°C on average), it can wear out much faster than anything stated on a data sheet, or indeed much faster than the daily cycling and cycle life might predict.
- In general, the longer you wish a battery to last in a PV system, the more expensive it will be. But it is not always the case that the most expensive battery you can think of will give the longest life.

5.1 Battery Performance Definitions and Summary

If we consider a battery to be simply a 'black box' that stores electricity, there are various properties that we need to define in order to describe its performance.

5.1.1 Definitions of Capacity, Efficiency and Overcharge

- The amount of electrical energy stored is measured in watt-hours (Wh) or kilowatt-hours (kWh). The energy efficiency of a rechargeable battery is

$$\frac{\text{energy in Wh discharged}}{\text{energy in Wh required for complete recharge}}$$

and is usually around 70–80%.

- Battery capacity is measured in ampere-hours (Ah). The charge efficiency or Ah efficiency is

$$\frac{\text{Ah discharged}}{\text{Ah required for complete recharge}}$$

and is around 95% for a lead—acid battery, somewhat lower for a nickel—cadmium battery.

The energy efficiency of a battery is lower than the Ah efficiency because batteries discharge at a lower voltage than they charge at. Since the Ah efficiency is close to 1, it is considerably more convenient to work in Ah when balancing how much charging is required to replace a certain amount of discharge in PV (and indeed other) calculations. However, since the Ah efficiency for a full recharge is always at least slightly less than 1, somewhat more Ah have to be delivered to the battery than are consumed in the actual charging process. This additional charge, or overcharge, is consumed by other, unwanted, chemical reactions within the battery. In lead—acid and nickel—cadmium batteries, these are normally the production of oxygen gas from water at the positive electrode, and, in open batteries, the production of hydrogen gas from water at the negative electrode too.

5.1.2 Discharge Rate and Charge Rate

Discharge and charge rates are convenient scales to compare currents at which batteries are charged, independent of battery capacity. They are expressed as a number of hours, e.g., the 10-hour rate, the 240-hour rate, etc. The current to which they correspond is the appropriate total discharge capacity divided by the number of hours:

$$\text{Rate} = \frac{\text{Capacity (Ah)}}{\text{Time (h)}}$$

For example, C/10 (10-hour rate) is a current equal to the rated capacity in Ah divided by 10.

5.1.3 Battery Capacity Is Not Fixed

Unfortunately, the capacity of a battery is not constant, and we have to be very careful to understand the way it varies.

The Nominal or Rated capacity of a battery (in Ah) is defined as the maximum Ah a fully charged battery can deliver under certain specified conditions. These conditions include

- the voltage to which the battery is discharged (the end voltage),
- the current (or rate) at which the discharge is carried out and
- the battery temperature.

In particular, the discharge rate has to be carefully stated along with any capacity, since, for example, a battery rated at 100 Ah at the 10-hour rate will give 10 hours' discharge at 10 A, but normally less than 1 hour's discharge at 100 A and normally more than 100 hours' discharge at 1 A. Capacity increases at lower discharge currents (longer discharge rates) and decreases at higher discharge currents (shorter discharge rates).

At low temperatures, capacities of all batteries are reduced. If the PV system requires a certain amount of autonomy backup in a month when the battery will experience a low temperature, then allowance for this has to be made in specifying the nominal battery capacity.

The end voltage obviously affects the amount of capacity that can be delivered. If a battery is discharged down to a lower voltage, it will of course give more capacity.

These capacity variations are illustrated in more detail later on in the chapter when the different types of batteries used in PV systems have been described. The exact variation of capacity depends on the type of battery, although the above trends are always true (Figure 5).

5.1.4 Depth of Discharge and State of Charge

Depth of Discharge (DOD) is the fraction or percentage of the capacity which has been removed from the fully charged battery. Conversely, the State of Charge (SOC) is the fraction or percentage of the capacity is still available in the battery. It is similar to considering whether a bucket (or drinking glass) is half empty or half full.

The following table shows the simple relationship between these two scales:

SOC (%)	DOD (%)
100	0
75	25
50	50
25	75
0	100

However, these states of charge/depth of discharge scales are normally referred to the nominal capacity (e.g., the capacity at the 10-hour rate). For lower discharge currents you may see references to a DOD of more than 100%. This simply means that the battery can produce more than 100% of its nominal capacity at discharge rates lower than the nominal rate.

5.1.5 Self-Discharge Rate

Self-discharge is the loss of charge of a battery if left at open circuit for an appreciable time. For example, a primary battery that has been sitting on the shelf of a shop for some years will not have its full capacity remaining (if any). For rechargeable batteries, the self-discharge rate is normally quoted as a percentage of capacity lost per month when starting with a fully charged battery, but it must be stated along with a battery temperature. In many cases it will double for each 10°C rise in battery temperature. In most calculations for PV batteries, the self-discharge rate of the preferred battery types is low (between 1% and 4% per month at 20–25°C) and self-discharge requires so little additional charging compared to the load (or even the control electronics) that it can be neglected.

5.1.6 Cycle Life

Cycling describes the repeated discharging and recharging process that a battery undergoes in service. One cycle equals one discharge followed by one recharge. Cycle life is a measure of how many cycles a battery can deliver over its useful life. It is normally quoted as the number of discharge cycles to a specified DOD that a battery can deliver before its available capacity is reduced to a certain fraction (normally 80%) of the initial capacity.

The cycle life depends very much on the depth of each cycle, and this is described in more detail in the section on battery lifetimes. However, it can be mentioned here that if cycle life is experimentally measured at a high DOD, at lower DODs the product of (number of cycles $\times$ the DOD) is approximately constant; i.e., the 'capacity turnover' is about the same for lower DODs.

Care should be taken when analysing cycle life that is published by battery manufacturers. It is normally measured at relatively high currents (short discharge times) and the DOD quoted often refers to the capacity available at that short discharge time. As a concrete example, one manufacturer quotes a cycle life of 400 cycles of 50% DOD for their product. Closer inspection of the details shows that this was done at the 5-hour rate of discharge, and that the DOD quoted refers to the capacity at this rate. The rated capacity of this battery is quoted at the 20-hour rate, and the 5-hour-rate capacity is 85% of the rated capacity. Although the capacity turnover is $400 \times 50\% = 200$ for the 5-hour-rate figures, in terms of actual Ah turnover, it is only $400 \times 50\% \times 85\% = 170$ relative to the rated

(20-hour) capacity. That means that one might reasonably expect only 340 cycles at 50% DOD relative to the rated 20-hour capacity, not 400.

In cycle life tests, batteries are given a full recharge after each discharge. In PV systems, the recharging is not so thorough. As a safety factor, it is therefore prudent to derate the cycle life somewhat when using it to estimate lifetimes in PV systems. A figure of 80% of the tested cycle life is often used. Thus, our above example battery, which started at 400 cycles at '50% DOD' and was reduced to 340 cycles at 'true 50% DOD,' would now be counted on to give only 272 such cycles under PV conditions.

In the early days of PV system design, the 'quick fix' for increasing battery life when these were found to be disappointingly low in some cases was to look for a battery with an increased cycle life. Unfortunately, cycle life is not the only factor that determines the lifetime of batteries in PV systems, and in some cases this change actually led to an even shorter lifetime.

5.1.7 Maximum Lifetime

Batteries that are used in standby ('float') applications are not cycled regularly. Using cycle life to estimate their lifetime is not at all useful. Instead the float service lifetime is stated as a certain number of years (at given temperature and float voltage) before the available capacity of a standby battery drops to 80% of its initial value. The float service life of a lead−acid battery approximately halves for every 10°C rise in temperature above the stated value, because it is most often determined by internal corrosion processes that follow such a temperature dependence. The lifetime of a shallow cycling PV battery will often be similar to that in float service under the same conditions, and this gives an upper limit to the expected lifetime if the cycle life data predicts a longer (and perhaps unbelievable) value. Because the battery is not kept in such a precisely fully charged condition as on float or standby duty, it is again prudent to derate its maximum lifetime for PV life estimation purposes (again 80% is often used).

5.2 Example of Simple 'Black Box' Battery Calculations

In practice, you need to know the detailed characteristics of the battery type you are considering in order to work out accurately the size of battery required for a given PV system, and also its possible lifetime. You do not actually need to know why these characteristics vary the way they do for different battery types, but it does help if you appreciate something about the general principles of how battery properties depend on how they are made. These are summarised in the rest of this chapter. However, here is a hypothetical example to show how the battery properties influence the size and selection, and also the basic way to determine the battery size from the PV system characteristics.

5.2.1 Basic Data

A PV system has to run a daily load of 40 Ah per day at 24 V nominal. The load is lighting, and so is all consumed during nighttime.

- The PV array produces a maximum current of 9 A charging current under full sun conditions (1 kW/m^2 solar irradiance).
- The battery should provide at least 4 days of autonomy (i.e., run the load for 4 days if the PV system fails).
- The annual average temperature of the battery can be assumed to be around 20°C in this location.

  Two types of lead—acid batteries can be considered:
- Type A: Available as 12 V, 100 Ah nominal only. Cycle life: 200 deep (80%) cycles. Expected maximum life: 5 years at 20°C.
- Type B: Available as 12 V, 50, 100, 150, 200 Ah nominal. Cycle life: 1200 deep (60%) cycles. Expected maximum life: 12 years at 20°C.

Type A is typical of a 'good' flat-plate battery, and type B is typical of an industrial tubular-plate battery. Per kWh, type B is about 3 times more expensive than is type A.

5.2.2 Calculations

For explanations of many of these steps, you will need to read the later parts of this Chapter. Remember that the details will change for other battery types, and even between manufacturers for the same battery type.
Autonomy Requirement

- 4 days × 40 Ah/day = 160 Ah storage needed.
- If we specify this to extract no more than 80% of the battery capacity, we actually require (160/0.8) = 200 Ah actual capacity.
- This is extracted at a low rate of discharge (4 days to 80% capacity is equivalent to 5 days or 120 hours to extract 100% of the capacity). From their data sheets, battery type A will give 1.1 × more than will its rated capacity at this low rate, and battery type B 1.4 ×. Therefore the minimum rated capacity needed will be type A: (200/1.1) = 182 Ah; type B: (200/1.4) = 143 Ah. We do not need to correct these

capacities for other operating temperatures, since the annual average temperature of the battery is around 20°C. Particularly if this operating temperature was lower, we would need to correct for the capacity variation with temperature.
- Batteries do not come in such precise capacities as the above. The nearest actual sizes are 200 Ah for type A (actually 2×100 Ah batteries in parallel) and 150 Ah for type B.

Lifetime Limit

Since the annual average temperature of the battery is around 20°C, if the batteries were operated as standby batteries, the highest lifetime expected would be 5 years for type A and 12 years for type B. However, in PV use we derate this to about 80% of these values, giving maximum expected daily lifetimes of 4 years and about 10 years (actually 9.6), respectively. If the expected battery temperature would be higher than 20°C, then we would need to reduce these lifetime limits.

Now we check the daily cycling. Since it is all at night, the batteries will be discharged by 40 Ah each night. This is equivalent to $(40/200) = 20\%$ of the nominal capacity for type A and $(40/150) = 26.7\%$ for type B.

If the load was continuous, or at least did not all occur at nighttime, the daily cycling would be less than the above, since, for some of the time, the battery would not be discharging whilst the load was being delivered.
- Battery type A gives a maximum 'capacity turnover' of $200 \times 80\% = 160$ times its nominal capacity. Scaling this to the 20% of nominal capacity it will cycle every day gives $(160/0.2) = 800$ cycles. At one cycle per day, this is $(800/365) = 2.2$ years. If we are not happy with this short battery lifetime, and want to increase it to the maximum value of 4 years, then we have to increase the capacity, such that the daily cycling over 4 years is no more than 160 times the nominal capacity, i.e., so that it is no more than $(160/(4 \times 365))$ or about 11%. For a 40-Ah daily load, this means a minimum of $(40/(160/(4 \times 365)))$, which actually comes to 365 Ah. As the battery only comes in 100 Ah units, this means we have to specify 400 Ah of nominal capacity to achieve a 4 year life (i.e., 4×100 Ah in parallel).
- Battery type B gives a maximum 'capacity turnover' of $1200 \times 60\% = 720$ times its nominal capacity. Scaling this to the 26.7% of nominal capacity it will cycle every day gives $(720/0.267) = 2700$ cycles. At one cycle per day, this is $(2700/365) = 7.4$ years. If we wanted to get

the maximum lifetime of this battery (9.6 years), then we would need to increase the capacity somewhat (by 9.6/7.4), but the lifetime due to the cycle limit is probably sufficient in this case.

Check the Maximum Charge Rate

For proper controller function, it is a good idea that the maximum charge current is no more than the 10 hour rate for open batteries such as these. The maximum charging current from the PV array is 9 A.

- For the type A 200-Ah battery, the maximum charge rate is (200/9) = 22.2 hours.
- For the type A 400-Ah battery, the maximum charge rate is (400/9) = 44.4 hours.
- For the type B 150-Ah battery, the maximum charge rate is (150/9) = 16.7 hours.

All of these are acceptable, and there is no need to increase the battery capacity to satisfy this condition.

Remember to Specify the Voltage!

The system (PV array and load) is 24 V nominal. If we are using 12-V batteries, we need two in series. If we are using more than one battery in parallel, each parallel group must be two 12-V units in series.

Examine the Options and Bear the Cost in Mind

Our three possible options for this example are then

- Option 1: 24-V, 200-Ah type-A battery (two parallel strings of two 12-V, 100-Ah batteries in series). Expected lifetime: a little over 2 years. The lowest cost option.
- Option 2: 24-V, 400-Ah type-A battery (four parallel strings of two 12-V, 100-Ah batteries in series). Expected lifetime: 4 years. Cost relative to option 1: double.
- Option 3: 24-V, 150-Ah type-B battery (two 12-V, 150-Ah batteries in series). Expected lifetime: 7–8 years. Cost relative to option 1: 2.25 times (3 times the cost per kWh and 75% of the actual kWh of option 1).

Unfortunately, it is tempting to specify the first option, either to produce just the lowest cost quotation or perhaps out of ignorance of any better option. However, the user is likely to be disappointed with the actual battery lifetime and may not be able to afford a replacement battery after 2 years (especially if, like many real cases, this is an overseas aid project that has assumed there are no recurring costs).

If the other two alternatives are compared, then option 3 (paying 2.25 the lowest battery price and getting around 3.5 times the battery life) is better than option 2 (paying twice the lowest battery price and getting a little less than twice the battery life).

6. THE BATTERY AS A COMPLEX ELECTROCHEMICAL SYSTEM

Electrochemistry can be described very generally as the study of what happens when electrical conductors are placed in an electrolyte (which is a material that conducts electricity via ions, not electrons). Electrolytes are commonly thought of as liquids, and common examples that come to mind are sulphuric acid, brine (or salt solution), and more complex solutions such as seawater or blood. However, they do not have to be water-based solutions; they may be a solution of something dissolved in a nonaqueous solvent, or a pure liquid (or melted solid) or even a special type of solid. If two conductors (now we will call them electrodes) are placed in the same electrolyte, they may develop a voltage difference that can be measured, or they may have a current forced through them, or they may simply be short-circuited together. All these cases are covered by the realms of electrochemistry, and the most familiar practical examples include corrosion processes, electrolysis, electroplating, and battery and fuel cell technology. At their most basic levels,

- corrosion occurs when two dissimilar materials in the same electrolyte are connected electronically and reactions occur at each material.
- electrolysis is what happens when two electrodes in the same electrolyte have a current passed through them. One of the most familiar examples is the production of hydrogen and oxygen gas when current is passed through a water-based electrolyte (but not through pure water, which is a very poor conductor of ions). Electroplating is a special form of electrolysis where a metal is deposited smoothly at one of the electrodes.

In a battery, two dissimilar materials in the same electrolyte can produce a voltage and current that can drive an electrical circuit or load. It can be thought of as the reverse of electrolysis. A fuel cell works on the same principle, except that the dissimilar materials are gases, normally hydrogen and oxygen, which are fed to the electrodes and produce electrical energy. The principle here can be thought of as the reverse of water electrolysis.

One common feature of all electrochemical processes is that chemical reactions take place at both electrodes. At one electrode, electrons are removed from the active material, which is a reaction that chemists call oxidation, and injected into the electrical circuit. At the other electrode, electrons from the electrical circuit are added to the active material, a reaction that chemists call reduction. If the process is a 'driven' one, such as electrolysis, the oxidation occurs at the electrode that is made electrically positive and reduction occurs at the negative electrode. However, if the process is a 'driving' one, such as a battery discharging, oxidation occurs at the electrode, which becomes electrically negative and reduction occurs at the positive electrode.

Whilst the basic bulk properties of different material determine what may or may not occur in theory in an electrochemical process, the actual events that take place in electrochemistry happen at surfaces and interfaces and follow special rules of their own. Small amounts of other substances at these surfaces and interfaces can affect the electrochemical behaviour in a totally disproportionate way, either beneficially or detrimentally. Even today, some of these effects are imperfectly understood theoretically, and although practical electrochemists know what works and what does not, they may not be sure exactly why. There is still an element of art in all the main electrochemical disciplines, including battery technology.

A practical battery basically consists of the following items:
- Positive and negative electrodes. Often called *plates*, these hold the different active material on some form of conducting support.
- An electrolyte (normally a liquid)
- Separators to stop the electrodes touching
- A container
- Positive and negative connections to the external circuit

On discharge, the active material in the electrodes is chemically changed and electrical energy released, the positive active material being reduced and the negative active material being oxidised. The voltage of the battery reflects the different 'chemical potentials' of the two active materials when they react with the electrolyte—simplistically speaking, the more reactive the two materials, the higher will be the voltage of a battery they compose. Of course, when we say 'reactive' here, what we really mean is 'reactive when the battery circuit is completed, but otherwise totally unreactive toward the electrolyte, toward the conducting supports, or toward anything like air or other gases that may be present inside the battery.' If these unwanted reactions go on when the battery is not being used, the battery will self-discharge and will not be a very good electrical storage device.

We may be lucky enough to find a pair of active materials and an electrolyte that give us the required combination of high reactivity when we want it (meaning a good working voltage) and a lack of unwanted reactivity toward the surroundings (meaning a low rate of self-discharge). This alone, however, does not guarantee that we have viable battery system. During the oxidation and reduction reactions, the active materials change into different chemical forms, and to be sure that our desired reactions continue to consume the active materials as we desire, certain properties of the products that are formed have to be right. They must not, for instance, be too low in density, since this could cause the active material to swell too much and block any further contact between the remaining active material and the electrolyte. Nor must they form an electrically insulating 'crust' on the surface of the remaining active material or at the point where the remaining active material is in contact with the electrically conducting support. If any of these occurs at either electrode, the battery discharge will quickly come to a halt with little of the active material consumed (i.e., there will be little capacity delivered).

It may be apparent from the above that even getting the right mix of material properties to make a primary, or nonrechargeable, battery is not that easy. An even larger combination of special materials properties are required to make a rechargeable system, since, on charging, the active material in the electrodes has to be chemically changed back to the original state using electrical energy. That means not just back to the same chemical form, but that the same basic shape (or morphology) must be maintained so that electrolyte access is possible to make the reactions happen. And it also means that active material must not be lost from the electrodes, nor must it become insulated from the main conductors that carry the current in and out of the battery electrodes. It is, in fact, these last two processes that largely control what the cycle life of the battery will be.

7. TYPES OF BATTERY USED IN PV SYSTEMS

The above should explain why there are very few rechargeable battery systems compared to primary (nonrechargeable) types. Examples of rechargeable battery systems are
- Lead—acid
- Nickel—cadmium (Ni—Cd)
- Nickel—iron
- Nickel—metal hydride (Ni—MH)
- Rechargeable lithium of various types, especially lithium-ion

This chapter has as its main focus the use of batteries in relatively large stand-alone PV systems. Only lead—acid and to a small extent nickel—cadmium batteries are used in this type of PV system.

Nickel—iron batteries are rarely used in any application and suffer from a particularly high self-discharge rate that makes them unsuitable for most PV applications.

Small portable electronic devices such as laptop computers, mobile phone and personal music players have in recent years all benefited from the rapid development and commercialisation of lithium-ion and nickel—metal hydride (Ni—MH) batteries. Of course, solar-powered versions of these will also use such modern battery types. However, lithium-ion batteries need some rather sophisticated protection circuitry that is not easy to adapt to the changing nature of PV charge currents without some loss of overall charging efficiency. This protection is required at the cell level rather than the battery level, and so this type of batteries has to be supplied as complete battery packs with integrated protection circuitry and not just individual cells or blocks. This restricts the flexibility of their use in PV systems that are built as custom-made power sources for higher daily energy supply, as opposed to solar-powered portable appliances and gadgets.

There is now an EU ban on the use of nickel—cadmium (Ni—Cd) batteries, which applies only to portable (i.e., small and sealed) batteries, and even then there are some exceptions. The EU Directive (2006/66/EC) [5] specifically defines Ni—Cd batteries for PV use as industrial batteries that are not part of this ban, but they have to be recycled properly, not through domestic waste channels. Most of the (small) demand for large Ni—Cd batteries in remote-area PV systems is in any case in countries that have no such ban on Ni—Cd batteries.

Ni−Cd batteries are only used in larger PV systems when there are extremely high (>40°C) or extremely low (< −10°C) operating temperatures. The operating range of Ni−MH batteries is −20°C to +40°C. Ni−Cd batteries can operate down to −50°C and up to at least +50°C. Ni−MH batteries are therefore not a substitute for Ni−Cd batteries in the extreme temperature conditions where the latter are required. The price per kWh of Ni−MH batteries is similar to that of Ni−Cd batteries and much higher than that of lead−acid batteries. Therefore, Ni−MH batteries are unlikely to be preferred over lead−acid batteries for the majority of PV applications.

Lithium-ion batteries are being developed for electric vehicles, so they are becoming available in battery packs considerably larger than those used in portable electronics equipment, and thus of possible interest for some of the larger PV systems. The main advantage they have for electric vehicles is a superior weight energy density (kWh per kg). This, however, comes at a price premium. In 2010, the factory price per kWh of large lithium-ion batteries for electric cars was at least US$400 per kWh [6]. This is more than double the factory gate price of a high-quality industrial lead−acid battery as used in large PV systems at present. The higher weight energy density is of no real advantage in stationary PV systems. At least for the foreseeable future, there is no sign of lithium-ion batteries becoming lower in cost per kWh than lead−acid batteries and so there is no real prospect yet of lead−acid batteries losing their dominant position in the stand-alone PV market to them.

There are other rechargeable battery types under development for such future battery applications as electric vehicles or load levelling. These are not commercially available yet, except in some limited cases. There is nothing yet to suggest that any of these would have the required properties or price to be competitive in PV systems, but we can always hope.

8. LEAD−ACID BATTERIES

8.1 Basics of Lead−Acid Batteries

A lead−acid battery or cell in the charged state has positive plates with lead dioxide (PbO_2) as active material, negative plates with high surface area (spongy) lead as active material, and an electrolyte of sulphuric acid solution in water (about 400–480 g/1, density 1.24–1.28 kg/l). On discharge, the lead dioxide of the positive plate and the spongy lead of the negative plate are both converted to lead sulphate.

The basic reactions are

$$\text{Overall}: PbO_2 + Pb + 2H_2SO_4 \underset{\text{charged}}{\overset{}{\Leftrightarrow}} 2PbSO_4 + 2H_2O \;\text{(discharged)}$$

$$\text{At the positive plate}: PbO_2 + 3H^+ + HSO_4^- + 2\bar{e} \underset{\text{charged}}{\overset{}{\Leftrightarrow}} PBSO_4 + 2H_2O \;\text{(discharged)}$$

$$\text{At the negative plate}: Pb + HSO_4^- \underset{\text{charged}}{\overset{}{\Leftrightarrow}} PbSO_4 + H^+ + 2e^- \;\text{(discharged)}$$

Note that the electrolyte (sulphuric acid) takes part in these basic charge and discharge reactions, being consumed during discharge and regenerated during charge. This means that the acid concentration (or density) will change between charge and discharge. It also means that an adequate supply of acid is needed at both plates when the battery is discharging in order to obtain the full capacity.

The lead–acid battery system has a nominal voltage of 2.0 V/cell. The typical end voltage for discharge in PV systems is 1.8 V/cell, and the typical end voltage for charging in PV systems varies between 2.3 and 2.5 V/cell, depending on battery, controller, and system type. The relation of open circuit voltage to SOC is variable but somewhat proportional. However, if charging or discharging is interrupted to measure the open-circuit voltage, it can take a long time (many hours) for the battery voltage to stabilise enough to give a meaningful value.

8.2 Types of Lead–Acid Battery

Many different types of lead–acid battery are manufactured for different uses. The PV market today is not large enough to warrant the manufacture of a radically different lead–acid battery design from the standard products that are made in higher volumes for other uses, although some slightly modified 'solar' battery types are available. We can basically classify lead–acid batteries in two ways:
- Open or 'sealed' construction
- Mass-produced or 'industrial' types

There is a wide range of lead–acid battery types to choose from when designing a PV system, and there is always a trade-off between battery cost and expected lifetime. In order to understand the differences between the various types of lead–acid battery available, it is necessary to understand a little about how they differ in construction.

8.3 Construction of Lead–Acid Batteries

8.3.1 Plate Type

Lead–acid batteries for PV systems have one of the following types of plate:

- Pasted flat plates. The most common form of lead–acid battery plate is the flat plate or grid. It can be mass-produced by casting or it can be wrought. This is what is in car batteries. The active material is applied to the grids by pasting and drying.
- Tubular plates. These are used in the positive plates of some larger industrial lead–acid batteries. Cycle life is longer because the active material is more firmly retained in woven tubes. The spines that carry the current are more protected against corrosion. So-called positive plate batteries actually have a tubular positive plate and flat negative plate.

Figure 2 illustrates these two plate types.

8.3.2 Grid Alloy

This is vitally important to achieve a satisfactory battery life. Pure lead is too weak to use as a conventional grid material except in special battery constructions. Alloy additives (antimony or calcium) are mainly used to strengthen the grid, but these primary additives can have bad effects on

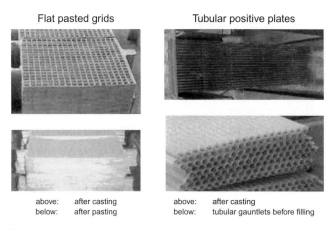

Figure 2 Flat plates and tubular plates.

cycling, corrosion, or water consumption, so secondary additives are used also.
- High-antimony alloys (5—11%) give high water consumption as the battery ages but give good cycle life. Their use these days is mostly restricted to traction batteries.
- Low antimony alloys (1—3%) plus other elements such as selenium or arsenic are widely used in open lead-batteries, but cannot be used in 'sealed' batteries because a small amount of hydrogen gas is always produced.
- Calcium alloys (0.06—0.9%) are used for sealed batteries but pure lead-calcium alloys give poor cycle life and poor recovery from a deep discharge. The addition of tin to the alloy corrects this.
- Other alloys are used by some manufacturers but have similar properties to low antimony or lead-calcium-tin alloys.

It is important to check that the grid alloys are suitable for a battery selected for PV use.

8.3.3 Grid Thickness

Especially for the positive grid where corrosion mostly happens, the thicker the grid, the longer the battery life in most cases. However, the thicker the grid, the more expensive the battery is likely to be. For tubular plates, the same applies to the spine thickness.

8.4 Sealed Lead—Acid Batteries

There is a fundamental difference between open (vented) or 'sealed' (valve-regulated) lead—acid batteries. In open batteries, overcharge results in the conversion of water into hydrogen and oxygen gases, which are lost to the atmosphere. Water has to be added to these batteries from time to time to make up this loss. In valve-regulated batteries, overcharge results in oxygen gas production at the positive plate, but because the space between the plates is not completely filled with acid, the oxygen gas can reach the negative plate, where it is re-converted back to water. This recombination of oxygen gas can only proceed at a certain rate. If the charging current is too high, then oxygen gas pressure will build up inside the cell, and eventually the safety valve will release oxygen (and some acid spray) into the atmosphere. This will result in permanent loss of water.

There are two types of 'sealed' (more correctly, valve-regulated) lead—acid battery, the so-called AGM and gel types.

8.4.1 AGM Type

This type of battery uses an absorbent glass mat (AGM) between tightly packed flat plates. All the acid is absorbed in the glass mat separator, but the pores of the glass mat are not completely filled. The empty (or part-empty) pores provide a pathway for oxygen gas, formed at the positive plate during charging, to move to the negative plate for recombination. AGM batteries were mainly developed for a good high-current (short discharge time) performance. They contain very little acid, which means that they are very susceptible to water losses that especially occur at high temperatures. In contrast, they have a good resistance to being frozen solid, since there is space for expansion within the AGM.

8.4.2 Gel Type

In lead—acid gel batteries, the sulphuric acid is mixed with finely divided silica, which forms a thick paste or gel. The freshly mixed gel is poured into the cell container before it sets. As the gel dries, microscopic cracks form that allow the passage of gas between the positive and negative plates required for the recombination process. This formation of cracks may occur during the early part of a gel battery's service life, so both hydrogen and oxygen can be given off from a new battery through the safety valve. Attention should be paid to the manufacturer's instructions concerning this, especially regarding ventilation requirements.

Unlike AGM batteries, gel batteries can be made with either flat or tubular positive plates. The gel provides a better means of heat conduction from the plates to the cell walls than in AGM batteries, so heat produced on overcharge is lost more efficiently. The sustained high-current capability (both charge and discharge) is not as good for gel batteries as for AGM batteries, but this is not normally a problem for PV use. At high operating temperatures, they will suffer to some extent from water loss, but since there is more acid than in an equivalent AGM battery, the lifetime reduction will not be so severe.

8.5 Mass-Produced and Industrial Batteries

The mass-produced lead—acid batteries are basically of the type used in cars for starting, lighting and ignition (SLI) use. They have relatively thin flat plates that are optimised for producing the high currents needed to start a vehicle engine. Conversely, the thin plates do not lead to a long lifetime in any other application that involves either cycling or operation at elevated temperatures. Most SLI batteries are of the open type. For trucks and boats, batteries with thicker plates are produced, and with some modification these may make a reasonable PV battery for light duty (infrequent or shallow cycling, no high-temperature operation). Sealed types with moderately thick flat plates are made for such uses as golf carts, invalid carriages and general leisure uses. Although not strictly speaking mass-produced on the scale that SLI batteries are produced, they are capable of being made in reasonably large volumes and at relatively low cost.

Industrial batteries are made for two general applications: float (or standby) duty and deep cycling (especially traction batteries for forklift trucks, etc). In Europe especially, the tubular plate construction is often used for both types. The tubular-plate standby battery type, whether open or gelled, is often the battery of choice for larger PV systems where the highest possible lead−acid battery lifetime is required. The tubular plate battery is also available in deep cycling or 'solar' versions from some manufacturers. Unfortunately the techniques used to increase the cycle life nearly always reduce the standby life at the same time, and this generally means that they give a lower PV service lifetime unless the cycling requirement is unusually deep.

Industrial lead−acid batteries with very thick flat plates are also made, especially for standby use. They are available as open, gelled, or AGM types. The thickness of their flat plates determines their maximum lifetime on standby duty, and it is common to see such industrial batteries described by the manufacturer as 5-year design life, 10-year design life, or even 20-year design life. These of course refer to the design life under optimum operating conditions, not those in PV systems.

Higher capacity industrial lead−acid batteries are mostly available only as 2-V single cells. These are assembled into batteries of the required voltage. Lower capacity industrial lead−acid batteries are normally available as 12- or 6-V units, known as *blocks* or *monoblocks*.

8.6 How the Capacity Varies

In this section graphs are used to show typical variations of capacity for different battery types. Manufacturers' data have been used to draw these graphs. It is not recommended that the figures in these graphs are used for actual PV system design, rather that similar information is requested from the intended battery supplier and that specific data are used.

8.6.1 Capacity Depends on the Discharge Rate
Lead−acid batteries with a lot of free acid and thicker plates gain more capacity at low discharge rates. The reason is that, at high rates, fresh acid

cannot get inside the plates fast enough and the capacity is limited by the amount of acid available. At lower rates, acid can diffuse to the plates during the discharge and maintain a more adequate supply of acid to the discharge reactions.

The AGM type of sealed batteries have considerably less free acid than do open or gel batteries, and therefore do not show much capacity increase at low discharge rates. The same is true for other types of flat-plate battery in which there is not much free acid. In nickel–cadmium batteries, the electrolyte does not participate in the reaction, and there is not the same low-rate capacity increase.

Figure 3 illustrates the typical increase at low rates for different battery types used in PV systems. Please note that these are average figures for selected ranges of commercially available batteries and that the actual variation for a particular type of battery of a specified capacity should be used in accurate calculations.

8.6.2 Capacity Reduces at Low Temperatures

The capacity (to a given end voltage) of all lead–acid batteries decreases at low temperatures. This is due to a combination of factors, including increased resistance and decreased diffusion rates in the electrolyte. The

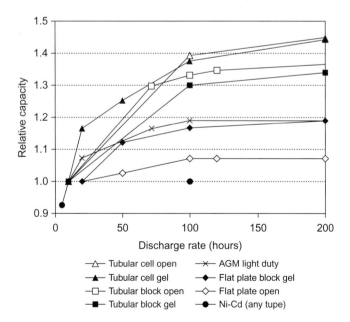

Figure 3 Capacity variation with discharge rate.

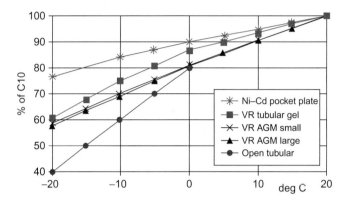

Figure 4 Capacity (10-h) reduction at low temperatures.

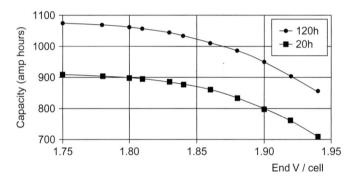

Figure 5 Capacity variation with end voltage. Typical values for large AGM type.

latter effect means that lead–acid batteries with a large reserve of acid tend lose more capacity at low temperatures than those with a smaller acid volume.

Rated capacities are normally stated for 20°C operating temperature. If a battery is required to provide an autonomy reserve capacity at a lower operating temperature, it is normal practice to increase the specified rated capacity to take into account the reduced capacity at the worst-case temperature.

Capacities increase slightly at operating temperatures above 20°C, but it is not normal to reduce the specified battery capacity on account of this.

8.6.3 Capacity Depends on the End Voltage

Capacities stated in manufacturers' data sheets should specify the end voltage to which that capacity applies. For batteries used in PV systems, this

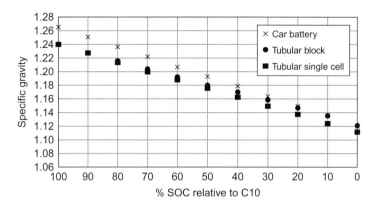

Figure 6 Typical specific gravity (20°C) change on discharge.

is often between 1.75 V and 1.85 V per cell. When comparing two different batteries, ensure that capacities to the same end voltage are compared. Obviously, the lower the end voltage, the greater will be the available capacity.

8.7 Acid Density

The concentration (or 'strength') of acid is normally conveniently measured by its density (kg/1), or, more precisely, by its density at 20°C. Specific gravity (SG) is a measure of density relative to that of water. It is the same as the value in kg/1, but without the units (i.e., 1.24—1.28). In some cases, the specific gravity is quoted as a number 1000 times the true value (e.g. 1240 instead of 1.24) (Figure 6).

Most batteries used in temperate climates are supplied with acid of around 1.24 SG. For cold climates, and particularly for starter batteries, a higher acid SG of 1.26—1.28 is often used. Among other things, this provides a greater resistance to freezing. Batteries for hot climates may be supplied with somewhat lower SG acid than the usual 1.24. Lower SG starting acid means lower internal corrosion rates, but it also means slightly lower capacity.

Batteries with gelled electrolyte normally have the same acid starting SG as do their open counterparts. Sealed AGM type batteries generally have a much higher acid starting SG of around 1.30.

8.7.1 Acid Density Falls During Discharge

Note the use of the term 'starting SG' in the above paragraphs. This means the SG of the acid that is supplied with a new, fully charged battery. Because sulphuric acid is consumed in the discharge reactions, and regenerated in the charging reactions, the acid density falls as the battery is discharged. A fully discharged lead—acid battery will have an average SG of around 1.05—1.15, depending on the battery type and the rate of discharge. Note that we have to use the term 'average SG' now, since there is no guarantee that the acid will be completely mixed. As shown in the next section, the acid will have areas of higher and lower than average density in different parts of the battery and at different points in the discharge and recharge cycle.

8.8 Acid Stratification

Stratification describes the tendency of the acid in lead—acid batteries to form layers of different density on cycling. Denser acid tends to form at the bottom of the battery, especially in the space below the plates. This can cause increased corrosion. A layer of less dense acid tends to form at the top of the battery, especially in the area above the plates. Batteries that are regularly deep discharged and then fully recharged, e.g., as in fork lift trucks, tend to suffer most from the build-up of lower density acid at the bottom. In contrast, batteries in PV systems that only experience regular shallow cycling and which are not 100% recharged every time lend to suffer more from the buildup of a less dense acid layer at the top. Figure 7 gives a simplified picture of how stratification develops in a lead—acid battery.

Stratification of an open battery in a PV system can be largely eliminated by stirring up the acid quite thoroughly by giving from time to time more overcharge than normal ('boost charging'). This produces gas bubbles that tend to carry the denser acid at the bottom of the battery upward.

In AGM batteries, the acid is present as a liquid absorbed in a glass mat. Stratification will occur in these, although not to such a large extent as in an open battery since the denser acid experiences more resistance to downwards flow. However, because gassing to stir up the acid is not possible with such batteries, once stratification has occurred, it is difficult to remove. For this reason, larger (and especially taller) AGM batteries are often operated in the horizontal position to minimise the buildup of stratification.

In gel batteries, the denser acid has a much higher resistance to downwards flow, and stratification is not so pronounced as in open batteries. It

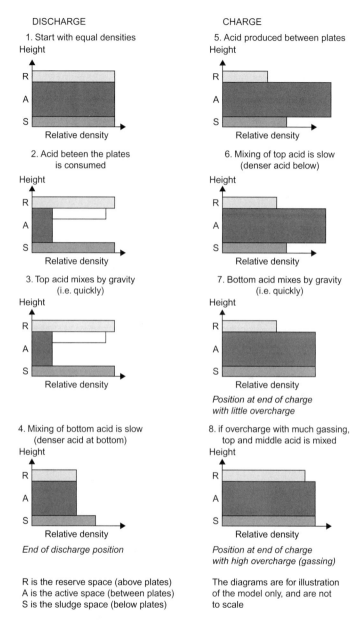

Figure 7 Principle of stratification. R denotes the reserve space (above plates), A is the active space (between plates), and S is the sludge space (below plates). The diagrams are for illustration of the model only and are not to scale.

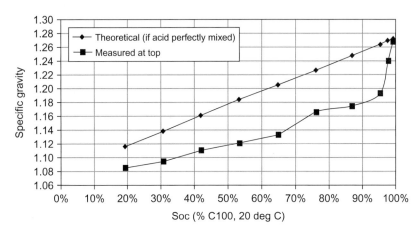

Figure 8 Illustration of stratification: acid density measured at the top of a flat-plate battery during recharging.

does, however, occur to a small extent, and, as with AGM batteries, cannot be reversed by periodic boost charging.

Stratification in open batteries can give misleading hydrometer readings of acid density when the acid is withdrawn from the top of a cell. Figure 8 shows some actual measurements on a battery discharged to 80% of its capacity and then recharged. Until the battery is more or less 100% recharged, the acid density reading taken at the top of a cell gives a quite false reading, suggesting that the battery is not as charged as it actually is. For this reason, a low hydrometer reading should not be taken to indicate a fault in a PV system, as it does not necessarily mean that the battery is in a low state of charge. In contrast, if one can be sure that the battery has received a full charge, then a suitably high hydrometer reading can confirm this.

8.9 Freezing

Since the acid in a lead—acid battery becomes more 'watery' as the battery is discharged, we can expect the freezing point to be raised. This can become dangerous if the battery is operated in subzero temperatures.

8.9.1 Acid Freezing Point

Figure 9 shows the freezing point of sulphuric acid of different SGs. We can see from this that the freezing point of 'normal starting SG acid' (1.24) is around −45°C, and if higher SG acid is used for colder climates (1.26−1.28), the freezing point is lowered to between −60°C and

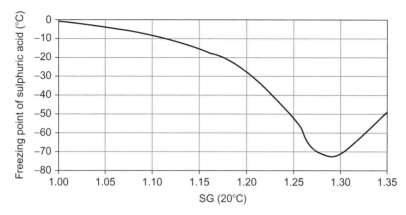

Figure 9 Freezing point of sulphuric acid.

−70°C. The acid in a fully discharged battery (SG 1.10−1.15, say) has a much higher freezing point of −10°C to −15°C.

8.9.2 Battery Freezing Points

Obviously, a discharged battery with low acid density will be more at risk of freezing than a fully charged one. However, since available capacity falls at low temperatures, the average acid density of a battery discharged at low temperatures will probably not fall to such low levels as 1.1−1.15. The most dangerous freezing risk is for a battery that has been discharged at normal temperatures (giving its full capacity, and so having a low acid density), and then left uncharged as the temperature falls to below zero.

Stratification can raise the freezing point of the acid above the plates. The 'theoretical' and 'local' freezing points for the stratified battery in Figure 8 are shown in Figure 10. Calculation of the freezing point of the battery from theoretical acid densities that assume the acid is completely mixed can be misleading for a partly recharged battery, since the weaker acid at the top has a much higher freezing point than the rest.

8.10 Sulphation and Deep Discharge

Sulphation is a process that can reduce the capacity of lead−acid batteries permanently if they are kept at a low SOC for prolonged periods without recharging. Small lead sulphate crystals initially formed during discharge may recrystallise on standing into larger pieces that are not easy to recharge. The reason for this recrystallisation is that as the acid concentration falls to a low level. the lead sulphate becomes slightly soluble in the acid. Another consequence of this slight solubility of lead sulphate in the weaker acid is

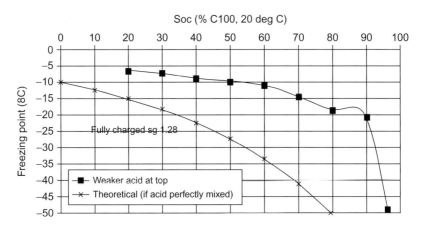

Figure 10 Freezing point of acid at the top of a deeply discharged stratified flat-plate battery during discharge.

that some fine and 'whiskery' lead crystals (dendrites) can be electroplated back on the negative plate on recharging, and in extreme cases these can cause a disastrous short circuit to the positive plate. On deep discharge a thin layer of an electrically insulating type of lead sulphate can also form on the grid material (passivation). Just how serious a problem this is depends on the exact nature of the grid alloy.

The ability of lead–acid batteries to recover from a very deep discharge is something that depends on the exact nature of the battery, as grid alloy type, additives, etc. will affect all the above problems of sulphation, dendrites and passivation. Careful selection of the battery type and the recharging conditions in a PV system, can give more or less full recovery of a lead–acid battery from a deep discharge, even if the battery has been in a deeply discharged condition for some weeks [2]. However, use of an inappropriate battery type, or an inappropriate means of PV recharging from such a condition, can result in total battery failure.

In view of the above, it is normal to restrict the maximum depth of discharge of a lead–acid battery in a PV system to around 80%, in order that the worst problems arising from deep discharge are not encountered.

9. NICKEL–CADMIUM BATTERIES

Nickel–cadmium (Ni–Cd) batteries in the charged state have positive plates with nickel oxy-hydroxide (NiOOH) as active material, negative plates with finely divided cadmium metal as active material, and an electrolyte of potassium hydroxide (KOH) in water (20–35% by weight). On discharge, the NiOOH of the positive plate is converted to Ni(OH)$_2$ and the cadmium metal of the negative plate is converted to Cd(OH)$_2$.

The basic reactions are

$$\text{Overall}: 2\text{NiOOH} + \text{Cd} + 2\text{H}_2\text{O} \underset{\text{discharged}}{\overset{\text{charged}}{\Leftrightarrow}} 2\text{NI(OH)}_2 + \text{Cd(OH)}_2$$

$$\text{At the positive plate}: \text{NiOOH} + \text{H}_2\text{O} + e^- \underset{\text{discharged}}{\overset{\text{charged}}{\Leftrightarrow}} \text{Ni(OH)}_2 + \text{OH}^-$$

$$\text{At the negative plate}: \text{Cd} + 2\text{OH}^- \underset{\text{discharged}}{\overset{\text{charged}}{\Leftrightarrow}} \text{Cd(OH)}_2 + 2e^-$$

Note that in the nickel–cadmium battery, there is no involvement of the KOH electrolyte in the charge or discharge reactions. This means that the electrolyte concentration does not change on charging and discharging, nor does the discharge reaction need to have an adequate supply of ions from the electrolyte to ensure that full capacity is reached. Both of these are in contrast to the behaviour of the lead–acid battery.

The nickel–cadmium battery system has a nominal voltage of 1.2 V/cell. The typical end voltage for discharge in PV systems is 0.9–1.0 V/cell, and the typical end voltage for charging in PV systems varies between 1.45 and 1.6 V/cell, depending on battery, controller, and system type. There is no relationship between open-circuit voltage and SOC.

In PV systems, nickel–cadmium batteries are usually only selected in preference to lead–acid batteries when operation is at very low (subzero) or very high (over 40°C) temperatures, where lead–acid acid batteries may suffer from freezing or a much reduced lifetime respectively. Industrial open-type nickel–cadmium batteries are typically 3–4 times more expensive per kWh of energy stored than are industrial open types of lead–acid batteries.

Although a single nickel–cadmium cell can be discharged fully (to 0 V) without harm, it is not advisable to allow a complete battery to discharge to very low voltages. This is because some cells will inevitably have less capacity than do others, and if the battery discharge exceeds their capacity limit, the low-capacity cells can be driven into reverse polarity (i.e., will have a voltage less than 0 V), which can shorten their life. It is therefore usual to specify that a nickel–cadmium battery in a PV system has a maximum DOD of 90%.

Industrial nickel–cadmium batteries used in PV systems are normally of the open type designed for standby use at low discharge rates. They may be of the pocket-plate or fibre-plate type. There is worldwide pressure to ban nickel–cadmium batteries because of the toxic waste problem, and this has already happened in the EU [5] for small consumer-type sealed batteries, for which alternative battery types are available. However, for larger batteries, there is no alternative system at present with similar properties, and it is difficult to see how these can be banned before such an alternative system is available. It should be borne in mind that any nickel–cadmium battery specified for a PV system has to be disposed of correctly at the end of life (by returning to the manufacturer for recycling or through an approved battery recycling organisation).

The memory effect is a phenomenon that is observed in some types of Ni–Cd batteries in shallow cycle service, but not in the open pocket-plate type used in the larger stationary PV systems which this chapter is about. The memory effect describes a battery's loss of its ability to deliver its full capacity at its normal voltage under regular shallow cycling, without a full discharge. The remaining capacity that has not been used regularly will be available, but at a lower voltage. The cause of this memory effect is thought to be due to the formation of large crystals in the cadmium electrode in the presence of a large surface area of nickel metal. It therefore occurs mostly in sintered plate Ni–Cd batteries (both open and vented) but not in the pocket-plate or fibre-plate type used in larger stand-alone PV systems in extreme temperature conditions.

Most industrial nickel–cadmium standby batteries are supplied with 20% KOH electrolyte as standard. The freezing point of this is −25°C. If the reason for choosing a nickel–cadmium battery rather than lead–acid is to prevent freezing problems, this freezing point may not be sufficiently low, and it may be necessary to use 30% KOH electrolyte, which has a freezing point of −58°C.

10. HOW LONG WILL THE BATTERY LAST IN A PV SYSTEM?

This rather simple question has been the subject of much confusion within the PV industry, especially in the early years. The author has been fortunate to have been able to work with an able team of battery researchers for several years to investigate the really important factors that affect battery life, and for more details of this work the reader is directed to the original published papers [2−4].

10.1 Factors Affecting Battery Life and Performance in PV Systems

In order to obtain the fullest possible life of a battery in a PV system, we first have to avoid the following:

Disasters, such as
- Manufacturing faults
- User abuse
- Accidents

Among other things, this needs attention to such things as
- Choosing a reliable and trusted battery manufacturer.
- Providing proper documentation and supervision or training for commissioning, operation, and maintenance.
- Being careful (try to avoid dropping a metal spanner across the battery terminals).
- Being lucky!

Next we need to be sure that we design the PV system so that the following possible problems are avoided:
- Sulphation
- Stratification
- Freezing

These we can do by ensuring a full charge (at least periodically), restricting the maximum depth of discharge to an appropriate level and providing as rapid a recharge as possible following a deep discharge. Much of this is done by choosing an appropriate battery, sizing it properly, and providing an appropriate method of charge control. Buying individual PV system components and putting them together without proper system design is probably the best way to ensure that one of the above will be a problem.

If we do all the above, we should be in a position to get the maximum service life of the battery in our PV system, and that should now be controlled by one of these two main characteristics:
- Cycle life
- Grid corrosion (temperature dependent)

Just which one this will be depends on the exact circumstances. It is not always the cycle life.

10.2 Cycle Life Can Be Misleading

Let us take, for example, a battery that is stated to give a thousand 80% cycles. This is equivalent to 'turning over' 800 times the capacity in cycling over the life of the battery. For lower depths of discharge, we can, to a good and conservative approximation, scale this figure [3,4]. So, scaling this for one cycle per day predicts a battery life of the following:

Daily DOD (%)	Cycles	Years
80	1000	2.7
40	2000	5.5
20	4000	11
10	8000	22
5	16,000	44
2	40,000	110
1	80,000	219

Whilst we may believe the first few lines of this table, we start to predict some pretty unbelievable lifetimes for very shallow cycling. Something else must cause the end of the battery life before it wears out due to cycling here. The most logical case to look at here is the limiting case of a battery that does no cycles at all, i.e., is used on standby (or float charge) duty.

10.3 Battery Float Lifetime

Batteries that are used for noncycling, non-PV duty (standby batteries) are continuously 'float-charged,' only being discharged when a mains power failure occurs. For lead–acid batteries in this case, the life-limiting process is the corrosion of the positive plate that occurs slowly during charging. This corrosion rate approximately doubles for each 10°C rise in temperature for a lead–acid battery. Correspondingly, the lifetime on float charging approximately halves for each 10°C rise in temperature.

Expected lifetimes in such float or standby applications can be 5, 10, or even 20 years, so accelerated tests to establish such figures are the only practical option. These involve measuring the float charge lifetime at one or more elevated temperatures and extrapolating to a lower standard temperature.

A shallow cycling PV battery does not experience a constant charging voltage, although the voltage averaged over a day does not differ very much from a typical float charge voltage. Therefore, to a first approximation, we can take the limit to the battery lifetime in a PV system to be no more than what one would expect on float charging at the same temperature. If the cycle life predicts a longer lifetime than this, we can expect the temperature-dependent corrosion process to be the life-determining process.

10.4 Sealed Battery Lifetimes

There can be additional life-limiting factors for sealed lead–acid batteries, especially at high operating temperatures, of which the most common are

- Water or acid loss, caused by one of three main processes:
  (a) Less than 100% recombination (some oxygen from the positive plate is not reduced back to water at the negative plate on overcharge, and the gas may be vented)
  (b) Positive grid corrosion (which converts the lead grid to an oxide)
  (c) Water vapour loss through case
- Negative capacity loss

Self-discharge of the negative plate is a chemical, rather than electrochemical process, and can only be balanced in the battery by some positive grid corrosion current or recombination that is less than 100% efficient. Both of these processes consume water, so if you avoid the gradual discharging of the negative plate, it may be at the expense of losing water. Fortunately, the self-discharge of the negative plate is very slow unless there are some undesirable impurities present in the battery.

For more details of these additional factors for sealed lead–acid batteries, the reader is directed to reference [4] and the references cited therein.

Water loss is less of a problem in gel batteries than in the AGM type, simply because there is more acid to start with. For the AGM type of battery, the capacity is often limited by the amount of acid, and any loss of water or acid in these batteries will reduce the available capacity. In the gel batteries, especially the industrial types with tubular plates, there is the same volume of acid as in the corresponding open type of battery, and it will at least take much longer for any loss of water or acid to affect the capacity in a gel battery than an AGM type. For this reason, gel batteries are preferred to the AGM type in PV systems where high operating temperatures are expected.

There is no simple method (yet) of predicting the effect of water and acid loss on the expected lifetime of an AGM type of sealed battery. For the gel batteries, the indication from actual field lifetimes in PV systems is that the lifetime can be predicted quite well from the cycle life or float life, as for open batteries.

10.5 Examples of Predicted Battery Lifetimes

The depth of daily cycling varies considerably between different types of PV systems. In systems with a large autonomy reserve, such as in telecommunications, the cycling will be quite shallow, whereas in systems with less autonomy, such as lighting systems, the daily cycling will be higher. As shown, earlier, the typical range of daily cycling is of the order of 2—30%, except in some hybrid PV-diesel systems, where it can be much deeper. Similarly, there can be quite a wide range of operating temperatures experienced by batteries in PV systems, depending on the climate at the site and also the nature of the battery enclosure. For predicting the lifetime limit due to corrosion, the annual average battery temperature should be considered, and in a well-designed system this should be close to (probably a few degrees above) the average annual ambient temperature.

Figure 11 illustrates the different limits for typical PV systems of different types. We plot the corrosion-limited lifetime against battery temperature, and also display the cycle limit for different numbers of days of autonomy on the same graph.

The upper graph shows the situation in a 'professional' PV system with a continuous load and tubular-plate batteries, typical of a remote telecommunications or cathodic protection PV system. The number of days of autonomy required for such systems is normally at least 5, and is often much higher. The combination of the shallow cycling and the good cycle life of the tubular-plate battery means that the cycle life hardly ever limits the battery life at operating temperatures above 20°C. Instead, the corrosion limit of 12 years at 20°C, 6 years at 30°C, or 3 years at 40°C is what normally applies for the battery life in such systems.

The middle graph illustrates a much different case, typical of many rural lighting PV systems. For reasons of cost, a flat-plate battery is often specified. A good but low-cost flat-plate battery may give around 200 deep (80%) cycles, and this is what is illustrated in this graph. Daily cycling is considerably deeper in such a system than in the first example, not just because less days of autonomy are usually specified for such systems (2—5 is typical), but also because all the discharge occurs at night. Now we see that the predicted battery life is around 1 year for 2 days

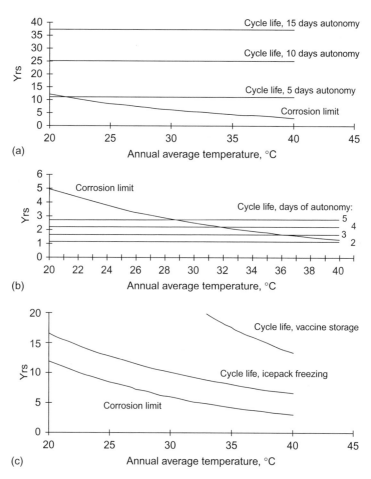

Figure 11 Predicted corrosion and cycle life limits in different types of PV system. (a) Professional PV system with continuous load, tubular batteries. (b) PV rural lighting system with flat-plate batteries. (c) Vaccine Refrigerator CFS, tubular batteries

autonomy, whatever the temperature, around 2 years for 3–4 days autonomy at temperatures below about 32°C and around 3 years for 5 days autonomy at temperatures below about 30°C. Above these temperature limits, the corrosion limit applies, giving an even shorter life. If a normal car battery is used in such a PV system, it will give even less life than shown in the graph, due a shorter cycle life than the example battery. The message here is quite clear—relatively cheap batteries with a short cycle life can give a very disappointing life in PV lighting systems, especially if a low capacity (3 days autonomy or less) is also specified.

The lower graph shows the interesting case of a PV vaccine refrigerator system with tubular-plate batteries. Here the daily load, and thus the daily depth of discharge, will vary with the ambient temperature (the refrigerator consumes more electrical energy if it is working in a hot location), and will also vary according to how it is used (e.g., if ice packs are frozen regularly, it will consume more than if it is just storing vaccine. Early examples of PV vaccine refrigerator systems were often installed in some very hot locations, and the resulting battery life was disappointing (less than 5 years), even though good-quality tubular-plate batteries were used. The initial reaction was to look for batteries with a better cycle life—e.g., traction batteries—but the graph clearly shows that it is not the cycle life but the temperature-dependent corrosion that is causing the short battery life at high operating temperatures. Siting the battery in a cooler location is the answer to obtaining a longer battery life in this case.

10.6 Comparison of Predicted Battery Lifetime and Field Data

So far, all the above reasoning is theoretical. Checking such calculations against actual field data is not easy, especially when the PV systems are often in extremely remote places. However, we have published what data are available from PV systems that we have supplied over a period that covers from the early 1980s to 1999 [3,4], and in general there is quite good agreement between those battery lifetimes we have managed to establish and what we have predicted.

Tables 1 and 2 are taken from [4] and show the situation in 1999 when the PV systems were last checked. Unfortunately various changes in different companies have meant that it was not possible to trace the history of most of these batteries beyond 1999 to determine their ultimate lifetime.

Table 1 shows results from relatively large single installations. Only three of the batteries in this table had been reported as having been replaced because they had failed, and their lifetimes were more or less in agreement with the predicted value. Of the remaining 10 entries, 4 had equalled or exceeded the predicted lifetime, and 1 was definitely still operating within its predicted lifetime. For the remaining 5, which had not failed inside their predicted lifetime in 1995, there were unfortunately no firm updated data, although if these unattended and very remote systems were not still working in 1999, then we would probably have been notified.

Table 1 Battery lifetimes: single PV systems with open tubular-plate batteries

Location	Type	Autonomy	Installed	Battery replacement	Last reported working	Service life (years)	Predicted years
Argentina	telecom	5 days	1984	—	1995	>11	10–12
Saudi Arabia	telecom	10 days	1984	—	1995	>11	8
Djibouti	ac system	? 4 days	1985	1994		9	6
Jordan	telecom	20 days	1987	—	1995	>8	8–10
Oman	telecom	10 days	1987	1994		7	7–8
Oman	telecom	5 days	1987	—	1995	>8	7–8
Venezuela	telecom	10 days	1988	—	1995	>7	10–11
Uganda	lighting	6 days	1989	—	1999	>10	10
Kenya	demo	8 days	199U	—	1999	>9	11
Uganda	freezer	4 days	1990	1998		8	9
Guinea	telecom	5 days	1992	—	?1999	>7	7–8
Chad	telecom	3 days	1993	—	?1999	>6	4–6
Vietnam	telecom	5 days	1993	—	?1999	>6	6–8

Table 2 Battery lifetimes: PV-powered vaccine refrigerators

Location	Fridge type	Installed	Capacity (Ah)	Battery replacement	Last reported working	Service life (years)	Predicted years
Ghana	1	1986	360	1993		7	5–6
Maldives	1	1987	360	–	1995	>8	5
Bolivia	2	1989	200	–	1998	>10	12
Tanzania	1	1991	300	1996–98		5–7	6–8
Ethiopia	2	1992	200		1999	>7	9–11
Ethiopia	2	1992	200		1999	>7	5–7

In Table 2, all the entries except for Bolivia refer to several vaccine refrigerator systems (ranging from 15 to 150 in a particular country), so these are better statistical cases than the single systems in Table 1. Of the two locations where batteries had needed replacement at the time of reporting, this was within 1 year of the predicted lifetime.

All the entries in Tables 1 and 2 are for open tubular-plate lead—acid batteries. It is less easy to track the lifetime of flat-plate batteries that are more commonly used in domestic PV systems. However, evidence from the Nordic country cottage market for PV systems indicates that the lifetime model works for those too.

10.7 Battery Lifetimes Summary

We can summarise the above as follows:
- In most industrial PV systems, positive grid corrosion rather than cycle life limits the battery lifetime when tubular-plate batteries are used.
- If flat-plate batteries are used in Solar Home Systems (mostly PV lighting), the cycle life often does limit the battery lifetime, especially when a low battery capacity (low autonomy) is specified.
- It is sometimes less easy to predict the lifetime of sealed batteries due to 'dry out' and negative capacity loss, especially for the AGM type.

It is necessary to know both the cycle life at some deep discharge depth and the life on standby duty at some reference temperature for a particular battery type in order to make an accurate estimate of the lifetime in a particular PV system. Such information should be obtained from the battery supplier and suitable derating factors applied to them for PV system use. Specific PV system details required for the battery lifetime calculation are the average daily depth of discharge and the average battery temperature. These can only be estimated accurately after the PV system has been specified in some detail.

11. SELECTING THE BEST BATTERY FOR A PV APPLICATION

There is a wide range of lead—acid battery types to choose from when designing a PV system and there is always a trade-off between cost and expected lifetime.

The preceding examples of lifetime calculations show that if the daily cycling is relatively deep (i.e., less than 5 days autonomy), an open flat-plate battery with a cycle life equivalent to 200 deep (75—80%) cycles can give a disappointingly short life of about 1—2 years. A car battery, with even shorter cycle life, will give even an even shorter and more disappointing service life in such a PV system. For a lifetime of 5 years or more in such relatively deep cycling conditions, there are only two real alternatives: specify a much larger battery capacity (i.e., increase the autonomy) to reduce the daily depth of discharge or specify a battery with a longer cycle life, such as one with tubular plates.

In contrast, if the daily cycling is very shallow (i.e., autonomy more than 5 days for a 'day and night' load), the battery life is likely to be limited by the internal corrosion processes, and this is highly temperature-dependent. At high annual average operating temperatures, this is almost certain to limit the ultimate lifetime of a battery in such a PV system. Here the selection of a battery with at least 10 years' design lifetime on standby duty at 20°C is recommended, which basically means an 'industrial' battery. A tubular-plate lead—acid battery is suitable for most of the annual average operating temperatures that might be experienced in hot climates in a well-designed system. If the operating temperature is extremely high (say averaging close to 40°C) and a battery life of more than 5 years is required, then the two alternatives are really only either to reduce the annual average operating temperature somehow (e.g., better shelter or battery box design) or to use a nickel—cadmium battery.

Provision of purified water (distilled or deionised) for regular topping up of batteries may be problematical in very remote areas. In these cases, 'sealed' lead—acid batteries are often specified. For high operating temperatures, the gel type is preferable as it is less likely to suffer from 'dry-out.'

In small PV systems, where the battery capacity required may be only 40 Ah or less, the full range of battery types is not available. Often, small AGM-type sealed lead—acid batteries are used for these, especially those that have an enhanced cycle life or float life (depending on the daily depth of cycling of the system). It should be noted that normal methods of PV charge control are not possible for sealed nickel—cadmium, nickel—metal hydride or lithium batteries, and these are rarely specified for smaller PV systems.

If the battery is expected to experience very low temperatures, such that there is a danger of freezing of a lead—acid battery, this must be taken into account at the battery specification stage. Unless there is a chance of a battery being very deeply discharged at normal temperatures and then subjected to subzero temperatures, the position is not quite as bad as may first be thought. If the battery undergoes discharge at subzero temperatures, its capacity is reduced and its freezing point will not be so high as if it had been discharged more fully. Often, the specification of a higher acid density than normal, plus some additional oversizing of the capacity, is enough to avoid freezing problems in cold climates. Most AGM types of sealed lead−acid battery have a higher starting acid density (around 1.30 SG) than do others, and these therefore will have a generally lower freezing point. However, if really low temperatures (say, below −30°C) are expected when the battery will be discharged to any great extent, then a nickel−cadmium battery is the safest (but most expensive) choice. If this is specified, it must be supplied with the correct electrolyte (30% KOH) to avoid freezing problems.

12. CALCULATING BATTERY SIZE FOR A PV SYSTEM

Earlier, some rough calculations of required battery capacity were presented. We now list the full process of correctly calculating the capacity required for a particular battery type in a specific PV system.

12.1 Select the Appropriate Voltage

This is defined by the load (and PV array) nominal voltage unless some dc/dc converter is present in the system. This sets the number of cells or blocks that must be connected in series.

12.2 Define Maximum Depths of Discharge (DOD)

These must be defined for each battery type according to the mode of operation.
- The maximum DOD for autonomy reserve is normally set at 80% for a lead−acid battery, although somewhat less may be specified for lead−acid batteries, which are known to be particularly susceptible to sulphation problems.
- The maximum daily depth of discharge may either be set arbitrarily (e.g., a figure of 20−30% is common), or it may be worked out from

the known daily cycle, the cycle life of the battery in question and the required lifetime (if cycling is the limiting factor).
- For seasonal storage (if used) a maximum depth of discharge needs to be set. If a lead–acid battery is not to be fully charged for some weeks, it is inadvisable to discharge it to more than about 30% DOD, for example.

12.3 Define Maximum Depths of Discharge (DOD)

Define the maximum charge rate.
- For open batteries in most PV systems, a charge rate faster than the 10-hour rate is not recommended, as the voltage will rise very quickly toward the end of charge and most types of PV charge controller will interpret this as a sign of full charge being reached when in fact it has not.
- For sealed batteries, another consideration is the highest overcharge current that can be sustained with efficient gas recombination, and this is temperature-dependent. A guideline maximum charge rate of 20 hours is often used for sealed lead–acid batteries at normal operating temperatures, but a lower limit may be specified for very low temperatures.

12.3.1 We Now Calculate Four Capacities

For seasonal storage (if used), the amount of storage required to make up for the shortfall in PV array production in certain months is the 'seasonal Ah' requirement:

$$C1 = \frac{\text{seasonal Ah}}{\text{maximum seasonal DoD}}$$

(If there is no seasonal storage requirement, $C1 = 0$.)

For autonomy storage, we need to specify a certain number of days for which we wish the battery to supply the load under emergency conditions. Strictly speaking, the worst time this could occur is when the seasonal storage (if any) has just been used. If the site could encounter subzero conditions, we may wish to modify the normal DOD limit here to avoid battery freezing problems:

$$C2 = \frac{\text{Average daily load Ah} \times \text{days of autonomy} + \text{seasonal Ah}}{\text{maximum DoD (adjusted to prevent freezing if necessary)}}$$

For the capacity required to fulfil the daily cycling requirement, we need to know if the load is required only at night, only during the day, continuously for 24 hours, or some other combination. Night-time only loads mean that the daily Ah discharged will be equal to the total daily load. For other cases, the daily Ah discharged will be less than the daily load and some correction factor is needed. For continuous loads, we can expect the battery to be undergoing charging for 6–8 hours in a typical day, so the daily Ah of discharge would be daily load in Ah multiplied by a factor of between (16/24) to (18/24):

$$C3 = \frac{\text{daily Ah discharged}}{\text{maximum daily DoD}}$$

Finally we need to ensure that the maximum charging rate is not exceeded. For this, we need to know the maximum current that the PV array will produce under maximum sunlight conditions:

$$C4 = \frac{\text{battery maximum C rate (in hours)}}{\text{maximum array current in A}}$$

The required battery capacity, before any corrections are made, is whichever is the highest of C1 to C4.

We now correct the chosen capacity for temperature and discharge rate.

- If the battery will experience average daily temperatures below 20°C in any month, then a correction should be made for the reduced capacity available at such temperatures. This means increasing the actual capacity specified.
- If, as is normal, the load will be delivered at a lower discharge rate than the normal 10- or 20-hour rate at which the standard capacity is specified, the capacity available at this rate is higher than the rated capacity. The 'data sheet' capacity we require can then be somewhat less than the capacity calculated above.
- The battery manufacturer may be able to supply data to correct for low temperature and low discharge rates at the same time, but do not use such data unless they specifically refer to the particular battery model. If manufacturer's specific data are not available, then an approximation is to make both corrections using the deviations from the standard capacity given at a standard discharge rate and a temperature close to 20°C.

Final Specification of the Battery

Having arrived at a calculated battery capacity, we now have to see which is the closest battery model with the desired capacity. Often two or more parallel batteries are used to improve system reliability, especially in larger systems, and if parallel batteries are used, the capacity per parallel string is the total capacity calculated divided by the number of strings. Finally, we need to remember the number of cells or blocks that are needed in series. If parallel strings are used, each string must contain this number of series cells or blocks.

13. LOOKING AFTER THE BATTERY PROPERLY

Maintenance is vitally important for obtaining the maximum lifetime from a battery, but even the highest quality maintenance will not produce the maximum benefit if the battery is operating under the best possible conditions. Therefore, the first thing to ensure is that the PV system has been designed with the battery's good health in mind.

13.1 System Design Considerations

13.1.1 Charge Control

Solar electricity is expensive and variable, and conventional charging methods that ensure 100% recharge are not possible. In most cases, the full available current from the PV array is transferred to the battery until the voltage level rises to a certain level, indicating that full charge is almost achieved. Then the PV array current is either cut off or reduced. This action is called *charge regulation* or, more correctly, *voltage regulation*. We need to ensure that the battery is as fully charged as practically possible, but on the other hand we do not want to cause problems from excessive overcharging.

Overcharge

Overcharge is the excess Ah delivered to recharge the battery. Some overcharge is necessary to achieve full charge and to prevent sulphation.
- In PV systems, 1–4% overcharge is common.
- In conventional non-PV cycling systems, at least 10% overcharge is common.

The energy delivered as overcharge causes gassing. In open batteries this results in water loss. In sealed batteries, overcharge results in heat being generated inside the battery. Gassing starts before full charge is reached and increases as charging progresses. Some gassing is needed in open batteries to stir up the acid and reduce stratification, but not such an excessive amount that would consume too much water.

Water Loss

In an open battery, each Ah of overcharge causes a loss of approximately 0.3 ml of water from each battery cell. We can estimate the time it takes to consume the 'acid reserve' above the plates from this factor, as in this example calculation: A 1500 Ah (C10)/2265 Ah (C100) tubular-plate single cell (2 V) in a typical telecom system, when new, is filled with 28.8 litres of 1.24 SG acid and the acid volume over the plates is 4.68 litres (16% of the acid volume). The number of days to lose this amount of water (i.e., to when the acid level falls to the top of plates) is

- At 2% overcharge per day: 464 days
- At 3% overcharge per day: 309 days

The average acid SG increases to 1.28 if the level falls to the top of the plates.

A maximum 1-year interval between water additions is appropriate at overcharge levels below 3% in this case.

13.1.2 Internal Healing

All batteries produce significant amounts of heat on overcharge. In contrast, heat generation during charge and discharge is relatively low in lead—acid batteries at the low rates of charge and discharge encountered in most PV systems. In a sealed battery on overcharge, no net chemical changes occur, and all the input overcharge energy is turned into heat. In an open battery on overcharge, we can think of the 'excess' energy due to overvoltage of the gas-producing reactions being turned into heat. In practice, this means any voltage in excess of 1.4 V/cell.

Example calculations of the internal heating produced by 24-V (12-cell) batteries at 2.4 V/cell at the end of charge, being charged at an average current of 10A:

- Open battery: $12 \times (2.4 - 1.4) \times 10 = 120$ W internal heating produced
- Sealed battery: $12 \times 2.4 \times 10 = 288$ W internal heating produced

13.1.3 Battery Environment

Ventilation

Ventilation is needed, not just to get rid of gases, but to lose heat, especially the internal heat produced on overcharge. Sealed batteries in tightly enclosed surroundings will, in particular, overheat, often disastrously.

On charge, open batteries will produce hydrogen and oxygen gases. Hydrogen will concentrate at the top of the battery enclosure if adequate ventilation is not given. If the concentration of hydrogen in air exceeds 4%, there is an explosion hazard, and the ventilation of open batteries should be designed to prevent this happening.

Sealed lead—acid batteries produce small amounts of hydrogen due to internal corrosion and some sealed batteries (especially gel batteries) produce normal amounts of gas at the beginning of their life. Sealed batteries also require some ventilation, although the amount required is more dependent on cooling needs than on the removal of hydrogen.

Temperature Control

The battery in a PV system may need protection from high temperatures or low temperatures, or both. This has to be done without consuming significant amounts of extra energy (which would lead to a larger PV system being needed). Some key guidelines for battery environments in different climates are as follows:

For a very hot climate,
- Avoid direct sunlight on the battery enclosure, by shading it if necessary.
- Use light-coloured enclosures if possible.
- Allow plenty of air circulation by providing sufficient air space and ventilation.
- Do not use a heavily insulated enclosure—this retains internal heating.
- Do not be afraid to use an 'oversize' shelter. This will allow more air volume inside and have a higher surface area for heat losses.

For a very cold climate,
- Reverse the above guidelines, but be careful of summer temperatures.
- Use a highly insulated enclosure.
- Use as little air space as possible around the battery (but do not restrict necessary ventilation).
- Use any available heat source (e.g., waste heat from a backup generator), and even consider using passive solar techniques.

The most difficult type of climate to design a battery enclosure for is one that has hot summers and subzero winter temperatures. In some cases, the only real option here is to arrange for some changes to the battery's thermal environment between winter and summer.

13.2 Commissioning

Some batteries supplied for PV systems will only give their full capacity and lifetime if they are commissioned correctly. This particularly applies to dry-charged batteries, where acid of the correct purity needs to be added carefully to each cell and a proper commissioning charge given. This is a longer charge than normal, and the load should not be switched on until the battery has reached full charge. Dry-charged batteries should not be stored for too long, nor in excessive heat or humidity, before they are commissioned.

When batteries of any type are supplied for PV use in very remote areas, it is often inevitable that they will have been stored for some time. Even if they are not supplied dry charged, they should be given a thorough charge (from the PV array) before use.

13.3 Maintenance, Replacement, and Disposal

There is no such thing as a truly 'maintenance-free' battery. Even sealed batteries, which do not need water additions, should be regularly inspected and have their terminals cleaned and, if necessary, tightened.

For open batteries, the main maintenance requirement is to add distilled or demineralised water periodically. For an estimate of how to judge the safe period between water additions, see the earlier section on water loss. As mentioned in an earlier section, a low specific gravity reading does not mean that the battery or cell is necessarily undercharged—it may simply be due to stratification. However, if one cell does give a quite different reading to the others, especially if the battery is more or less fully charged, then that may indeed be taken as signifying a faulty cell. Similarly, a considerably different voltage to that of other cells or blocks in the same series connected circuit can also indicate a fault condition, as can one cell being considerably hotter than the rest.

Therefore, periodic maintenance procedures for all batteries should attempt to identify faulty cells or blocks. This may be done by specific gravity measurement (not possible for sealed batteries), voltage checks, or case temperature checks.

If really defective cells or blocks are detected (e.g., those with internal short circuits or zero capacity), they should be replaced whenever practical. However. it is not recommended to mix aged and new cells or blocks in one series string. If replacement of some cells or blocks is contemplated, it is best to renew all the elements in one series group. If the battery consists of more than one parallel group, there is scope to reconfigure the cells or blocks so that those of similar aged characteristics are in the same series string. If any cells or blocks are replaced, the new elements should be clearly marked. If a series-parallel battery arrangement is changed, all the parallel strings should be fully charged separately before making the parallel connection.

Dead batteries should be disposed of responsibly and not left lying around where they can be a source of pollution. Lead—acid and nickel—cadmium batteries can be recycled and have some scrap value, and if possible their return to the local battery company or the original supplier should be arranged. Indeed, recycling of industrial nickel—cadmium batteries is now mandatory in many countries.

14. SUMMARY AND CONCLUSIONS

The rechargeable batteries used in PV systems are required to perform under conditions that are different to the more conventional battery applications for which they are designed. In particular, the charging and discharging they undergo is not entirely regular or predictable, being subject to variations in the weather. Different types of PV system require different amounts of daily discharging, but in most cases this cycling is relatively shallow. The cycling capability of a battery is an important factor in determining its PV system lifetime, but it is not the only one. The operating temperature, and the battery's resistance to internal corrosion, is equally important.

The main property of a battery—its capacity—is not fixed, but varies with temperature, discharge current and other factors. To specify the battery capacity correctly, the above system-specific factors must be known as well as the maximum fraction of the capacity that can be extracted safely under different conditions.

Rechargeable batteries are complex electrochemical devices that depend on a large number of material properties being 'just right' to function correctly. There are therefore not that many basic types in general use, and even fewer of these used regularly in PV systems. The majority of batteries used in PV systems are lead—acid, but there are several distinct types of these, each with their own set of properties. Perhaps the most obvious classification is into 'sealed' and 'open' types, but within each of these categories there are also many variations. The chemistry of the lead—acid battery means that there are certain specific problems to be avoided, such as stratification, freezing and sulphation. However, the relatively low cost and general availability of lead—acid batteries means that they are used in all but the most demanding PV system environments.

It is often the case that choosing a more expensive lead—acid battery will result in longer PV service life than if a 'cheap' one is used, but it is not always the most expensive battery that gives the longest PV system life. Use of low-capacity, low-cost, flat-plate lead—acid batteries similar to truck starter batteries can result in a disappointingly short lifetime in many solar lighting systems in hot climates. In general, the best PV service lifetime is given by a tubular-plate industrial lead—acid battery, although they are not available in the very low capacities required for small PV systems.

Whatever type of battery is chosen for a particular PV system, it must be sized correctly, placed in the best possible environment, and maintained correctly if it is to have a good chance of reaching its predicted service lifetime.

It is impossible to present anything but an outline of the required information in a chapter of this size. For more comprehensive information, the reader is directed to [1] and to the recommended literature for further reading.

ACKNOWLEDGEMENTS

The author would like to thank
- Naps Systems, for permission to publish various battery test results and conclusions.
- Colleagues at Naps Systems and the former Neste Corporate R&D for many hours of painstaking battery tests and interpretation, plus stimulating discussions about what the results might mean.
- Many major battery companies for answering detailed technical questions with patience, and for allowing the author to photograph parts of the their production process, including the photographs shown in Figure 2.

REFERENCES

[1] D.J. Spiers, J. Royer, Guidelines for the Use of Batteries in Photovoltaic Systems. Joint publication, NAPS, Vantaa, Finland and CANMET, Varennes, Canada, 1998.
[2] D.J. Spiers, A.A. Rasinkoski, Predicting the service lifetime of lead/acid batteries in photovoltaic systems, J. Power Sources 53 (1995) 245—253.

[3] D.J. Spiers, A.A. Rasinkoski, Limits to battery lifetime in photovoltaic applications, Solar Energy 58 (1996) 147–154.
[4] D.J. Spiers, Understanding the factors that limit battery life in PV systems. Proceedings of the World Renewable Energy Congress VI (WREC2000), Elsevier Science Ltd., Vol. II, 2000, pp. 718–723.
[5] Directive 2006/66/EC of the European Parliament and of the Council of 6 September 2006 on batteries and accumulators and waste batteries and accumulators and repealing Directive 91/157/EEC <http://eurlex.europa.eu/LexUriServ/LexUriServ.do?uri = CELEX:32006L0066:EN:NOT/>
[6] See, for example, J.L. Petersen, The Cruel Realities of EV Range, Batteries International, 77 (2010) 8–10

FURTHER READING

M. Barak, Lead–acid batteries, in: M. Barak (Ed.), Electrochemical Power Sources, Institute of Electrical Engineers, UK, 1980, Chapter 4.

Bechtel National Inc., Handbook for Battery Energy Storage in Photovoltaic Power Systems, Department of Energy, USA, 1979.

H. Bode, Lead–Acid Batteries, John Wiley & Sons, New York, USA, 1977.

T. Crompton, Battery Reference Book, Butterworths, UK, 1990.

V.S. Donepudi, W. Pell, J.W. Royer, Storage Module Survey of the IEA-SHCP Task 16, PV in Buildings, Canadian Solar Industry Association, Ottawa, Ont. Canada, 1993.

U. Falk, Alkaline storage batteries, in: M. Barak (Ed.), Electrochemical Power Sources, Institute of Electrical Engineers, UK, 1980, Chapter 5.

R. Foster, S. Harrington, S. Durand, Battery and Charge Controller Workshop for Photovoltaic Systems, PV Design Assistance Center, Sandia National Laboratories, Albuquerque, NM, USA, 1994.

M. Hill, S. McCarthy, PV Battery Handbook, Hyperion Energy Systems Ltd., Ireland, 1992.

PV Design Assistance Center, Stand-Alone Photovoltaic Systems. A Handbook of Recommended Design Practices, Sandia National Laboratories, Albuquerque, NM. USA. 1990.

S. Roberts, Solar Electricity; A Practical Guide to Designing and Installing Small Photovoltaic Systems, Prentice Hall, New Jersey, 1991.

S. Strong, The Solar Electric House: A Design Manual for Home-Scale Photovoltaic Power Systems, Rodale Press, Emmaus, PA, USA, 1987.

PART IIC

Grid-Connected Systems

CHAPTER IIC-1

Grid Connection of PV Generators: Technical and Regulatory Issues

Jim Thornycroft[a] and Tom Markvart[b]

[a]Halcrow Group Ltd, Burderop Park, Swindon, UK
[b]School of Engineering Sciences, University of Southampton, UK[*]

Contents

1. Introduction	780
2. Principal Integration Issues	783
2.1 Safety	783
2.2 Power Quality	783
2.3 DC Injection	784
2.4 Radio Frequency Suppression	784
3. Inverter Structure and Operating Principles	784
4. Islanding	785
4.1 Passive Methods	787
4.1.1 Voltage and Frequency Detection	787
4.1.2 ROCOF Technique or Phase Jump Detection	788
4.2 Active Methods	788
4.2.1 Variation of Frequency	788
4.2.2 The Measurement of Impedance	788
4.2.3 Variation of Power and Reactive Power	789
5. Regulatory Issues	789
5.1 Technical Connection Guidelines	789
5.2 Type Testing	790
5.2.1 Draft IEC Standard 62776	790
5.3 Parallel Connection Agreement	798
5.4 Tariff Agreement	798
5.5 Post- or Pre-notification	803
Acknowledgements	803
References	803

[*] With acknowledgement to IEA PV-PS Task 5.

Practical Handbook of Photovoltaics.
© 2012 Elsevier Ltd. All rights reserved.

1. INTRODUCTION

The "grid connection" of PV systems is a fast growing area, with a vast potential for domestic and industrial locations. A grid-connected PV system provides an individual or a business with the means to be its own power producer, as well as contributing to an environmentally friendly agenda. The key to grid-connected systems is that they work in parallel with the already established Electricity Supply Network, and a number of their features are determined by this connection to the utility supply.

Traditionally, the connection of generators into the supply system has been done at high voltage (HV), in the kV range. Due to the restricted numbers, extensive experience and interconnected nature of the HV system, such generators can be accommodated into the distribution network using one-off assessments and are large enough to justify the cost of the assessment and reinforcement measures. Photovoltaic generators on domestic roofs generally connect to the utility supply at low voltage (LV). Together with other generators which are connected to the low-voltage network, they are classed as distributed generation (often called 'embedded generation' in the UK). Such connections present new issues for the distribution network operators which have traditionally distributed power 'downwards' from a relatively few generators connected at high voltage (Figure 1).

One possible way of integrating the PV generator within a domestic installation is shown in Figure 2. In this scheme, the PV electricity is used to supply the consumer demand and any excess is metered and exported to the utility network. Other schemes are possible where, for example, the PV generator (after DC/AC conversion and the export meter) is connected directly to the incoming service cable. All the generated electricity is thus exported to the network, and the consumer unit is connected by a separate line through the import meter in the usual manner. This scheme is beneficial to the consumer if the utility price for PV electricity is higher than the usual domestic tariff.

For PV systems, a significant issue is the grid connect inverter used as part of the PV system. The inverter converts the DC power produced by the PV generators to alternating current (AC) in order that the generator may be connected and synchronised to the utility network. Previous regulations and recommendations for the connection of generators were not written with small inverter interfaces in mind, and so a new framework is

Grid Connection of PV Generators: Technical and Regulatory Issues 781

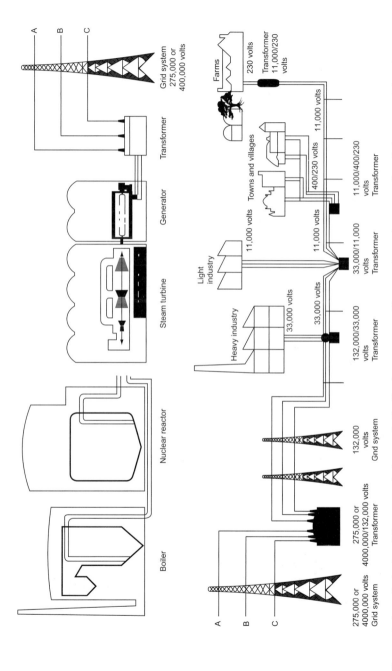

Figure 1 A typical electricity transmission and distributions system, illustrated on the example of the UK public electricity supply (adapted from R. Cochrane, *Power to the People*, CEGB/Newness Books, 1985). All but the largest PV generators are usually connected at 400/230 V.

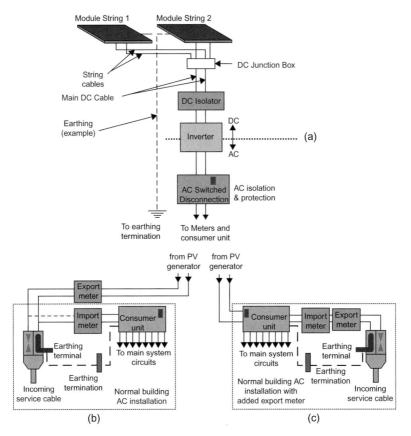

Figure 2 Typical layout of a domestic PV installation. (a) The PV generator. (b) Connection scheme of the meters and consumer units suitable for countries (notably Spain) where all PV power is exported to the network. (c) Connection scheme for countries (for example, UK) where only the excess power not used on-site is exported to the network. In the case (c), a PV generation meter is often installed on the AC side of the inverter. Combined export/import meters are now also available as a 'dual register' unit.

emerging which will allow the PV generators to connect to the utility network safely but without undue complexity and cost.

Although there are differences in the grid-connection procedures which are at present applied in different countries and by different utility companies, these procedures share a number of common attributes. The network operator will generally require that the connection of a PV generator conforms to the relevant codes of practice and engineering recommendations, particularly with respect to safety. There must be adequate protection for both the supply network and the inverter. The power quality will also have to be sufficient not to affect adversely the utility equipment and other users connected to the network.

This chapter gives an overview of the principal grid connection issues and the existing codes of practice and engineering recommendations, drawing on the work of the Task 5 of the Photovoltaic Power Systems (PVPS) Programme of the International Energy Agency. Further details can be found on their website (http://www.iea-pvps.org) with copies of reports downloadable from www.oja-services.nl/iea-pvps/products/home.htm.

2. PRINCIPAL INTEGRATION ISSUES

The integration of PV systems to electricity networks is covered at the top level in the standard [1] which groups the issues into two main categories: safety and power quality. DC injection and radio frequency suppression are also important considerations.

2.1 Safety

Safety of personnel and protection of equipment are the most important issues concerning a grid-connected PV generator. When the source of power is disconnected from the network section to which the inverter is connected, the inverter is required to shut down automatically within a given time. An inverter that remains generating into such a network is termed to be in *island operation*. Such operation could critically affect the safety of electricity supply staff and the public, as well as the operation and integrity of other equipment connected to the network [2]. The risks involved depend on the type of network and how it is operated.

With the advent of distributed PV generation, islanding has become one of the central issues of concern for the utility supply industry. For this reason, it forms the principal subject of this chapter, and is discussed in detail in Section 4.

2.2 Power Quality

Any generator or load connected to the 'mains' network may affect the quality of the waveform. Alterations to the voltage or frequency, or variations which affect the 'shape' of the sine wave, such as harmonics and flicker, are all important. Power factor and fault current contribution are also relevant considerations. Fortunately, with inverter connected PV systems, neither of these is a particular problem. Most modern inverters can adjust their power factor to suit the network. Fault level contributions are low as there is little stored energy in an inverter in contrast with the rotation energy of a spinning generator.

2.3 DC Injection

Another aspect of inverter generators is the possibility of DC injection into the network. The network operators will not allow the presence of DC current in any significant amount and impose a limit close to zero on the DC current produced by the inverter. An inverter with an isolation transformer at the output to the utility is designed not to produce any DC current but this issue may arise for inverters with high frequency transformers which are now beginning to appear on the market.

2.4 Radio Frequency Suppression

Inverters now generally operate at switching frequencies of 20 kHz or higher, and may cause some interference in the RF region. This is usually avoided by the use of appropriate filtering and shielding.

3. INVERTER STRUCTURE AND OPERATING PRINCIPLES

Inverters are discussed in detail in Chapter IIb-1, and only a brief review of the relevant aspects will be given here. Inverters for utility connection can be broadly classified into two types: single-phase inverters and three-phase inverters. Detection of islanding is much easier in a three-phase than a single-phase inverter, although inverters that are rated at a power below 5 kW are mostly connected to single-phase networks.

Most modern grid-connect PV inverters use self-controlled power switches (e.g., MOSFET, IGBT) and generally use pulse width modulation (PWM) control signals for producing an AC output. Previous thyristor based systems were turned off using the 'zero crossing' of alternating current from the mains. In either case, inverters need to synchronise with the utility network and switch off on islanding.

Figure 3 shows a schematic structure of a typical PV inverter. The input DC power enters the maximum power point tracker aiming to keep the DC power input extracted from the array as high as possible.

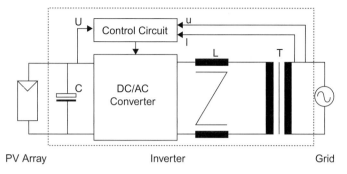

Figure 3 Schematic diagram of a typical PV inverter with low-frequency isolation.

This DC power is converted to output AC power by the DC/AC converter. The basic function of the control circuit is to control the DC/AC converter to produce a sine-wave output (voltage or current) which is synchronised with the grid and has a low distortion. In order to reduce the weight and the price, some more recent PV inverters use a high frequency transformer. Here, the DC/DC converter converts the input DC voltage to a higher DC level to meet the voltage requirement of the DC/AC converter which is directly connected to the grid. Resonant technology can be used in this structure for higher efficiency.

4. ISLANDING

The main technical issue for utilities has been the issue of 'islanding' where the power to a local area of the electricity network could potentially be maintained live by a distributed generator, even where the main power station supply is switched off or lost in a fault. Alternatively, islanding can be considered as the condition in which a portion of the utility system, which contains both load and generation, is isolated from the remainder of the utility system and continues to operate.

Island operation is potentially undesirable both from the safety point of view—where a circuit may be assumed to be 'dead' when it is not, and from the point of view of power quality—where the standards normally guaranteed to customers may not be met. The risks of islanding have been researched in some detail under IEA Task 5 and the results are available in several IEA reports [2–4].

Traditionally, the problem would have been solved by providing an accessible lockable switch for each generator, but clearly this is impracticable once numbers rise above a very few. As a result, the functionality of the accessible switch has now normally been replaced by automatic anti-islanding protection circuits within each inverter [2].

Various methods of anti-islanding detection and prevention have been developed and adopted across different countries, as described in Table 1. In general, there is a move not to be prescriptive on how islanding is detected leaving the choice of the islanding detection method to the inverter manufacturer. Instead, a standardised method of the testing of these devices is being produced through international standards [5]. Some

Table 1

Methods	Characteristics
Passive methods	
Over Voltage	Switches off when voltage rises above a preprogrammed 'window' limit-very common
Under Voltage	As above but when value falls below preprogrammed limit—very common
Over Frequency	As above but for frequency—very common
Under Frequency	As above but for frequency—very common
Frequency Variation Rate	ROCOF (rate of change of frequency)—a more sophisticated method that is triggered by unusual changes in frequency—common
Voltage Phase Jump	Also known as power factor detection—monitors phase difference between inverter output voltage and current for sudden change
Third Harmonic Voltage Detection	Harmonics will typically increase if the low impedance of the grid is lost
Active methods	
Frequency Shift	A method of ensuring that if the mains is lost the inverter trips on its frequency limit. Various methods have been designed to achieve this
Impedance Measurement	One method injects a pulse into the network to detect if it is present—some difficulties in multi-inverter situations and with power quality degradation
Harmonic Impedance Measurement	Where a harmonic frequency is deliberately injected to detect response
Utility methods	
Utility Communications	Requires continuous signal from utility to keep generator connected (either power line or other communications)

differences may, however, remain to take into account the differences in electricity distribution networks in different countries.

In the countries participating in the IEA PVPS, for instance, anti-islanding requirements have been evolving for many years and today they still vary considerably from country to country. Some countries such as the Netherlands require only the out of frequency and out of voltage windows. Other countries such as Germany and Austria require a specific method based on sudden impedance changes and described as ENS or MSD. Standards are also sometimes adopted that require inverters to detect and shut down within a variable amount of time that is determined by the out of tolerance condition that exists on the island or even on the utility grid.

However, a growing trend is to require that utility-interactive inverters be type tested for the purpose using a standard test circuit and test method that has been determined to be a worst-case condition. This allows a single inverter to be tested rather than requiring multiple inverter tests. There is a move to standardise the test, as far as possible, under IEC.

A general overview of the possible range of islanding detection techniques is given in Table 1. Some of these techniques (for example, those that are implemented by the utility) are outside the scope of the present book. We shall therefore focus on detection methods integrated within the PV system (e.g. inside or close to the inverter). These techniques can be further divided into active and passive measures depending on the type of monitoring, as discussed in the following sections. Further detail can be found in the comprehensive IEA report [6].

4.1 Passive Methods

Passive islanding detection methods detect the characteristic features of the islanding mode, and the operation of the inverter is stopped. Typical measured values are voltage, frequency, and phase. This method is suitable for integration into the inverter. Passive techniques do not by themselves alter the operation of the power system in any way; they detect loss of grid by deducing it from measurements of system parameters. Passive techniques are suitable for all types of generators.

4.1.1 Voltage and Frequency Detection

The simplest way to achieve loss of mains protection is to use standard voltage and frequency protection to identify the islanding condition. This method works in most cases except where the load exactly matches the output from the generator. Voltage and frequency detection is implemented as basic protection on most inverters. Other methods can be added to this such as ROCOF below.

4.1.2 ROCOF Technique or Phase Jump Detection

A technique known as ROCOF [6] measures the rate of change frequency to determine whether the connection to the mains has been lost. Sometimes it is also referred as phase jump detection. The relay monitors the system frequency and is arranged to ignore the slow changes in frequency, which normally occur on the grid system, but to respond to the relatively rapid changes of frequency when the grid is disconnected.

4.2 Active Methods

Active methods for detecting the island introduce deliberate changes or disturbances to the connected circuit and then monitor the response to determine if the utility grid with its stable frequency, voltage, and impedance is still connected. If the small perturbation is able to affect the parameters of the load connected within prescribed limits, the active circuit causes the inverter to shut down.

4.2.1 Variation of Frequency

This is sometimes referred to as frequency shift. The output frequency of the inverter when the utility is lost is made to diverge, by giving bias to the inverter free running frequency (making it increase or decrease). Islanding is then detected when the frequency veers outside the allowed frequency band. This is a very efficient way to detect islanding in a single generator. Its effectiveness is being evaluated in situations when several generators work together. The effect of the technique on the mains when large numbers of these generators are connected is also subject to further research.

4.2.2 The Measurement of Impedance

In one implementation of this method, current is injected into the utility to determine the impedance of the utility line. The circuit is designed to detect significant changes in impedance over a short period of time such as would occur if the utility were disconnected. This has been adopted in the ENS or MSD systems, but the effect on power quality and the effect when there are multiple units is still being evaluated. An example of waveform generated by an inverter using the ENS impedance measuring technique is shown in Figure 4.

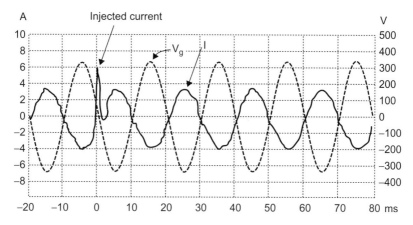

Figure 4 A typical waveform produced by an inverter using the impedance measuring technique of islanding detection.

4.2.3 Variation of Power and Reactive Power

Islanding can also be detected from the frequency change on islanding as a result of continuous variation of inverter reactive power output. Alternatively, islanding can be detected by monitoring the voltage change in response to a continuous variation of inverter active power output.

5. REGULATORY ISSUES

To facilitate the installation of photovoltaic generators (particularly small domestic installations), regulations are being developed and are in place in a number of countries that relate to connection of PV generators. The technical, legal, and commercial aspects of these regulations in various countries are discussed below. We shall also touch upon the associated procedures for the type approval of inverters and/or the allied interfacing equipment.

5.1 Technical Connection Guidelines

The grid connection of PV systems is covered at the top level in the standard [1]. This document is in the process of being revised to include developments since its introduction several years ago. Although the main features are likely to remain in place, the precise values of the parameters are specific to each country, and a copy of the latest national regulations should therefore be obtained through the local electricity company. Current detailed information and contacts are given in the IEA report [7] and are summarized here in Tables 2 and 3.

Until relatively recently, only a minority of countries had PV-specific standards, but today most that are looking to implement PV systems have developed guidelines for the grid interconnection of PV inverter systems. PV systems using static inverters are technically different from rotating generators and this fact has been generally recognised in these new guidelines.

Whilst it is evident that most countries now have specific PV standards, these standards differ from country to country and a harmonisation is intended as the next step. Activities at the IEC level have started work in this direction, with the main focal point IEC TC82 Working Groups WG3 and WG6. The output of the work from Task 5 to IEC TC82 has been essential for some of the present activities in these groups, and several members of Task 5 are also working as IEC experts. At the same time, it is clear that due to differing technical boundary conditions in different countries (earthing philosophy, layout of the grid, etc.) it will be difficult to achieve total harmonisation.

The status of this regulatory process in a number of countries that took part on the Task 5 of the Photovoltaic Power System Programme of the International Energy Agency is shown in Table 2.

5.2 Type Testing

The large number and relatively small size of many grid connected PV systems will impact on current network operator procedures. Through the pressure of numbers, each system cannot be considered on a 'one-off' basis. Projects also cannot be required to bear a connection cost (charge levied by a network operator) that is out of proportion to the cost of the systems themselves. This has lead to the concept of 'type verification' where a representative test is carried out on a sample of a product, and then this does not have to be repeated for every new installation. The situation on Type Testing is shown for various countries in Table 3.

5.2.1 Draft IEC Standard 62776

It has already been mentioned that international tests for inverter anti-islanding are being developed in the International Electrotechnical Commission Technical Committee 82—Solar Photovoltaic Energy Systems, to be titled 'Testing Procedure of Islanding Prevention Measures

Table 2 Guidelines for selected countries. Data from [7] and [9].

Country	Authorisation procedure
Australia	Installations are authorised by the local electricity utility. Usually required: — a copy of inverter conformance certificates — a licensed electrician to connect the PV system to the grid Commissioning tests include visual inspection and tests of the anti-islanding measures
Austria	Installations are authorised by the local electric utility. Usually required: — a copy of the declaration of CE conformity of the inverter — a copy of the type-testing of the built-in anti-islanding measure ENS Only a licensed electrician may connect the PV system to the grid Commissioning tests include visual inspection and tests of the anti-islanding measures.
Germany	Approval by the grid operator is required. Installation work may be performed only by an authorised (skilled and licensed) installer. The grid operator usually demands single phase circuit diagram of installation — a copy of the declaration of CE conformity of the inverter — a copy of the type test approval of the anti-islanding ENS — protocol of commissioning by installer Tests are not regulated. Generally simple functional tests of inverter and ENS are conducted. Documentation of commissioning procedure is requested.
Italy	Installations are authorised by the local electric utility. Usually required: — a copy of the declaration of CE conformity of the inverter — a copy of the type-testing of the built-in anti-islanding measure ENS Only a licensed electrician may connect the PV system to the grid Commissioning tests include visual inspection and tests of the anti-islanding measures.
Japan	Installations are authorised by the local electric utility. Usually required: — a copy of electrical layout of the system — a copy of the declaration of confirmation of the requirements for grid interconnection guideline — a copy of the type-testing certification by JET Only a licensed electrician may connect the PV system to the grid Commissioning tests include visual inspection and insulation tests.
Netherlands	The connection of a generator to the power network has to be approved by the network operator. However, this requirement

(continued)

Table 2 (continued)

Country	Authorisation procedure
	does not apply when dealing with small and medium sized PV systems on dwellings. Permission is asked for all large PV systems. By law, the network operator cannot refuse the connection of a PV system (or any other renewable energy source).
	All PV systems have to comply with regular and special standards described in more detail in the IEA report. Official procedures for commissioning are not available. Specialised consultancy firms often assist owners of large and/or special PV system in authorisation and acceptance testing.
Portugal	The licensing of an Independent Power Producer (IPP) plant requires the following documents to be submitted to the Government's Directorate General of Energy (DGE):
	1. Normalised formal request to the Ministry of Economy (Energy)
	2. Liability term, by which the installation conforms to the regulations in force.
	3. Technical information provided by the utility (grid company) regarding:
	— Interconnection point
	— Maximum and minimum short circuit power at the interconnection point
	— Type of neutral connection
	— Automatic reclosing devices available or to be installed
	4. Detailed design of the whole system (generators, transformers, protective devices, connection line, local grid, etc.).
	The final decision (approval) is from the Minister of Economy or the Director
	General of Energy (DGE), depending on the system installed power ($P > 1$ MVA or $P < 1$ MVA. respectively).
	Before starting its regular operation, the power plant must be inspected by the DGE ($P > 10$ MVA) or the Ministry's Regional Delegation ($P < 10$ MVA). which will deliver an 'Operating Licence'.
Switzerland	All installations have to be authorised by the local electric utility. Usually required:
	— description of the inverter and the solar modules
	— form for grid connection of the electric power utility
	Only a licensed electrician may connect the PV system to the grid. Commissioning tests include visual inspection and tests of the anti-islanding measures.

(continued)

Table 2 (continued)

Country	Authorisation procedure
	If an installation exceeds 3.3 kVA on one phase or 10 kVA on three phase an approval of the Eidgenossische Starkstrominspektorat (ESTI) is required. Usually required: — description of the inverter and the solar modules — form for grid connection of the ESTI Commissioning tests include visual inspection of the lightning protection and the grid connection. Tests of the anti-islanding measures.
United Kingdom	Installations need to be agreed with the local Distribution Network Operator (DNO). For larger units the DNO may require on-site commissioning tests and may also want the opportunity to witness the commissioning tests. However, the use of a type approved inverter for smaller units greatly simplifies the process. A connection (operating) agreement also needs to be in place with the local DNO and an appropriate supply (tariff) agreement with an electricity supplier if any export settlement is required.
Spain	For P < 100 kVA, the local electric utility requires: — Data concerning ownership and location of PV system — Single phase circuit layout of the installation — Proposal for the interconnection point to the grid — Technical characteristics of the main components of the installation: Nominal PV power, inverter characteristics, description of protection devices and connection elements A contract must be formalised between the local electric utility and the PV system owner, fixing economical and technical relationships as can be the interconnection point to the grid and the interconnection conditions. An authorized installer must inspect and test the system before it is connected to the grid.
USA	Permits for installation are required. Drawings, specifications, equipment lists and layout are normal. The local authority having jurisdiction (AHJ) inspects installations. Utility-interactive systems are required to use listed hardware and components. Generally, only a licensed electrician may wire and connect the PV system to the grid. Some states now certify PV installers. Inspections include visual checks before interconnection is approved.

Table 3 The accepted standardised islanding test and institutions authorised to perform the type approval tests in selected countries. Data from [7] and [9]

Country	Guideline/standard	Scope	Protection Settings
Australia	Australian Guidelines for grid connection of energy systems via inverters[a]	All energy sources that are connected to the electricity system via inverters 0–10 kVA phase to neutral 0–30 kVA 3 phase.	• Over Voltage (2 s) • Under Voltage (2 s) • Over frequency (2 s) • Under Frequency (2 s) • At least one active method of detecting AC disruptions
Austria	ÖVE/Önorm E2750[b]	PV systems—both stand alone and grid connected. Without specific approval the maximum power is limited to 5 kWp DC (4.6 kVA) per phase for small installations.	ENS (impedance measurement method): if grid impedance is >1.75 Ω OR grid impedance jump >0.5 Ω is detected then inverter must disconnect within 5 seconds.[c]
Germany	DIN VDE0126 (draft)	No general limit for power generation capacity: up to 4.6 kVA (inverter) or 5 kWp (PV generator) single phase connection is permitted (VdEW Guideline).	Islanding prevention device ENS (also called MSD) has become the de-facto standard for new systems. It monitors grid voltage, grid frequency, grid impedance, and ground leakage current and disconnects the inverter, if one parameter is out of bounds.[c]
	VdEW guideline of the Association of Electric Power Companies	Above 5 kWp three-phase connection is requested. All systems need approval by local grid operator.	
Italy	CEI 11–20[d]	All types including photovoltaic systems. The standard gives an indication of 50 kVA for LV systems and 8 MVA for MV systems (HV systems are not covered by this standard).	Voltage and frequency window.

Japan	See note[e]	Both active and passive methods should be installed. Any kind of active method and passive method is acceptable. e.g., *Active methods*: Frequency shift, power variation, load variation. *Passive methods*: voltage phase jump, frequency variation rate. third harmonic voltage detection. Required protection: • Overvoltage • Undervoltage • Overfrequency • Underfrequency • Overload protection or maximum current
Netherlands	1: Supplementary conditions for decentralized generators low-voltage level 2: Guidelines for the electrical installation of grid connected photo-voltaic (PV) systems[f]	1: Covers all types of generators connected to the LV network. Simple protection is required for generators below 5 kVA. This class is intended for small generators in residential applications, e.g. PV systems and micro CHP. 2: This standard is intended for grid connected PV systems only. Clauses referring to the AC side may also be used for other small types of generators like micro CHP.
Portugal	No specific standard for grid connected PV systems. Law 168/99[g]	When the connection to the grid is made at the low voltage level (up to 1 kV). the power cannot exceed 4% of the minimum short-circuit power at the interconnection point, with an upper limit of 100 kW.[h] No reference is made (in the IPP law) to this phenomenon. The co-ordination between the IPP and the electrical grid must consider situations of grid disconnection, for maintenance and repair, in order to ensure the necessary safety conditions,

(*continued*)

Table 3 (continued)

Country	Guideline/standard	Scope	Protection Settings
Switzerland	ESTINr. 233.0690; VSE Sonderdruck Abschnitt 12[i]	PV systems—both stand alone and grid connected.	Recommended: measurement of network voltage (one phase) and shutdown in case of frequency shift (e.g. in the US, Japan and Holland)[j]
		PV installations up to 3.3 kVA on one phase or up to 10 kVA on 3 phase do not require an approval of the ESTI.	
United Kingdom	G77/1 (2002)[k]	PV inverters for grid connected systems. Up to 5 kVA, for a single installation.	• Over Voltage • Under Voltage • Over frequency • Under Frequency • A recognised loss of mains technique, such as vector shift or frequency shift. Active techniques that distort the waveform beyond harmonic limits or that inject current pulses are not allowed.
Spain	No specific standard for grid connected PV systems. Law 17599. Real Decreto (RD) 1663/2000	Nominal power, <100 kVA and low voltage level, Connection Voltage to the grid <1 kV 3 phase inverter is required for interconnection to the grid of systems with power rating; >5 kW	Disconnection time 5 sec – Manual line breaker – Automatic disconnection and reconnection to the grid – Overfrequency – Underfrequency – Overvoltage – Undervoltage

USA	NFPA 70; IEEE 929-2000; UL1741[l]	IEEE929 covers interconnection requirements when connected to the utility, and defines small systems as up to 10 kW. Some utilities still use their internally generated distributed generation interconnection requirements.	Requirements for allowable islanding are given in IEEE929 and UL1741. Methods are not specified.

Abbreviations: LV = low voltage; MV = Medium Voltage; HV = High Voltage; CHP = combined heat and power.
[a] Available from http://ee.unsw.edu.au/~std_mon/html.pages/inverter.passed.html.
[b] ÖVE/Önorm E 2750 Photovoltaische Energieerzeugunsanlagen—Sicherheits-anforderungen (Photovoltaic power generating systems—safety requirements).
[c] Older requirement for single phase inverters (still possible as alternative to ENS): 3 phase undervoltage relay and single phase overvoltage relay required.
[d] CEI 11–20—Impianti di produzione di energia elettrica e gruppi di continuità collegati a reti di I e II categoria (Power production plants and uninterruptable power systems connected to 1st and 2nd grids). Available form Comitato Elettrotecnico Italiano CEI; Viale Monza, 261; 20126 Milano; e-mail: mcei@ceiuni.it.
[e] Technical Guidelines for the Grid Interconnection of Dispersed Power Generating Systems (only available in Japanese). Available from for grid-interconnection guideline: http://www.energy-forum.co.jp/index.htm. For photovoltaic generation system standard: JIS (Japan Industry Standard). Japanese Standards Association, 4-1-24 Akasaka, Minato-ku, Tokyo 107–8440; see also the English page of JSA web site http://www.jsa.or.jp/eng/index.htm
[f] Available from EnergieNed; Utrechtseweg 310: P.O. Box 9042; 6800 Arnhem; tel +31 263 56 94 44; fax +31264 46 0146.
[g] Law 168/99 (last revision of all the legal and technical framework to IPP), first introduced in 1988.
[h] For higher voltage levels, see the IEA report.
[i] ESTI Nr. 233.0690 'Photovoltaische Energieerzeugungsanlagen—Provisorische Sicherheitsvorschriften' (Photovoltaic power generating systems—safety requirements draft); VSE Sonderdruck Abschnitt 12: 'Werkvorschriften über die Erstellung von elektr. Installation' Elektrische Energieerzeugungsanlagen; Completes VSE 2.8d-95; available from: Eidgn. Starkstrominspektorat ESTI; Luppmenstr. 1:8320 Fehraltorf. Verband Schweizerischer Elektrizitätswerke VSE: Gerbergasse 5: Postfach 6140; 8023 Zürich.
[j] Methods used but not recommended include ENS (impedance measurement method) and the measurement of phase-to-phase and phase-to-neutral voltage for single phase inverters feeding into three phase networks.
[k] G77/1 (2002)—Recommendations for the Connection of Inverter-Connected Single-Phase Photovoltaic (PV) Generators up to 5 kVA to Public Distribution Networks. Available from Electricity Association, Millbank, London. Work is underway on G83/1 (2003) which it is planned will supersede G 77/1 when published.
[l] NFPA 70—National Electrical Code; IEEE 929–2000 Recommended Practice for Utility-interface of Photovoltaic Systems. UL1741 UL Standard for Static Inverters and Charge Controllers for Use in Photovoltaic Power Systems.

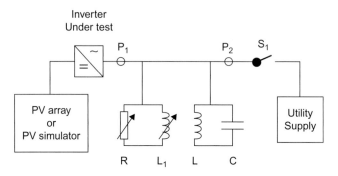

Figure 5 The test circuit for islanding detection.

for Grid Connected Photovoltaic Power Systems' [5]. An example test circuit which is being discussed is shown in Figure 5.

The circuit has a resonant load to model the situation of multiple inverters. In preparation for the test, the resistive part of the load and the impedance of the resonant circuit are changed to match the active and reactive power produced by the inverter. The switch to the utility supply is then opened and the duration of inverter operation before shutting down is recorded. Depending on the regulations, the tests may be repeated at several values of the resistive load. The main item for debate between countries is the 'Q' factor of the resonant load which determines how sensitive the circuit is to the disruption, and also whether motors or similar loads should be included as part of the local load.

On issuing of a Type Test certificate (or 'listing' in the USA), simplified procedures and checks can be adopted by the utility companies when requests for connection of these type tested units are made.

5.3 Parallel Connection Agreement

A Parallel Connection Agreement is generally necessary for connection and operation even if export of power to the main network is not required. This covers the legal aspects and liability of connection. Requirements in selected countries are shown in Table 4.

5.4 Tariff Agreement

The Tariff Agreement embodies the commercial terms for payment for exported energy and varies from country to country and electricity company to electricity company. Examples of arrangements for metering and 'profiling' for PV electricity in various countries are summarised in Table 5. In a

Table 4 The connection procedures in selected countries. Data from [7] and [9]

Country	Accepted standardised Islanding test	Authority/institute authorised to perform such a test and issue a certificate
Australia	Australian Guidelines for grid connection of energy systems via inverters. Appendix B	Not specified.
Austria	The ENS (impedance measurement) test method is nationally accepted.	Berufsgenossenschaft für Feinmechanik und Elektrotechnik. Cologne. Germany Bundesforschungs- und Prüfzentrum Arsenal. Vienna, Austria
Germany	The ENS (impedance measurement) test method as defined in VDE 0126 is empowered and nationally accepted.	Authorisation is not regulated, theoretically each laboratory could conduct the type approval test. In fact type tests are performed by: Berufsgenossenschaft für Feinmechanik und Elektrotechnik, Koln, Germany, hv@bgfe.de and by: TÜV Rheinland. Koln, Germany, mail@de.tuv.com
Japan	Japanese Standard	Japan Electrical Safety & Environment Technology Laboratories 5-14-12 Yoyogi, Shibuya-ku, Tokyo, Japan

(*continued*)

Table 4 (continued)

Country	Accepted standardised islanding test	Authority/institute authorised to perform such a test and issue a certificate
Netherlands	Correct function of the protection and proper settings have to be guaranteed by the manufacturer. For AC Modules a special KEMA-KEUR (K150) safety certification is required. The functioning and settings of the protections are included.	KEMA Registered Quality P.O. Box 9035 6800 ET Arnhem The Netherlands
Switzerland	The ENS (impedance measurement) test method is nationally not accepted.	Berufsgenossenschaft für Feinmechanik und Elektrotechnik, Cologne, Germany TÜV Rheinland. Cologne. Germany Bundesforschungs- und Prüfzentrum Arsenal. Vienna, Austria Manufacturers themselves, or STaR Facility.
United Kingdom	Part of the G77/1 Type Approval tests in G77/1 Appendix.	SES, University of Southampton. Southampton SO 17 1BJ.
USA	IEEE929 and UL1741 contain test setup and requirements for listing and utility interconnection recommended practices. The procedure was written so tests conducted on single inverters would apply to multiple inverters connected together.	Qualified electrical testing laboratories that are recognised as having the facilities to test as required by the NEC.

Table 5 Legal and tariff situation for electricity exported to the network by small grid-connected PV generators in member countries of IEA PVPS Task V. Data from [7] and [9].

Country	Legal and tariff situation
Australia	Utilities are not obliged to buy electricity produced by PV systems.
	Tariffs paid for electricity fed into the grid vary with Utility.
Austria	Utilities are obliged to buy electricity produced by PV systems.
	Tariffs paid for electricity fed into the grid vary locally; a recent change in law has generally raised the rates in most parts of Austria.
	Kates are between 0.04 and 0.55 euro/kWh.
Denmark	Experimentally, for a 4-year period, normal households may use meters that can run in both directions. This means that they buy and sell electricity at the same price. Over one year there must not be a negative consumption.
	Companies and similar have to use separate meter for energy fed into the grid.
Germany	Utilities are obliged to buy electricity produced by PV systems (Law of Privilege for Renewable Energies).
	Tariffs are defined in this law. They are fixed for 20 years and depend on the year of installation. Price for systems built in 2001 is 0.99DM/kYVh.
	Metering is not regulated, it depends on local grid operator. Usually energy fed to the grid is measured using a separate meter. The meter belongs to the grid operator and is rented for some 20 to 50 Euro annually to the independent Power Producer.
Japan	Utilities are voluntarily buying electricity produced by PV systems. It is not legally regulated.
	Tariffs paid for electricity fed into the grid vary locally, however the rate is almost the same as electricity from the utility. Rates are around 23 Yen/kWh for a low voltage customer.
	A separate meter for energy fed into the grid is required.
Netherlands	Utilities normally accept decentralised generators in their networks. The new Dutch legislation on electrical energy requires a zero-obstruction policy from network operators towards renewable energy.
	Pay back rates and the necessity of a net-export kWh-meter are subject to the contract between the network operator and the owner of the PV system.

(continued)

Table 5 (continued)

Country	Legal and tariff situation
Portugal	Utilities are obliged to buy electricity produced by PV systems and all the other IPP systems according to the Law 168/99. providing the technical conditions are met. Tariffs paid for electricity fed into the grid are specified in the Law 1 68/99: Rates are between 0.055 and 0.065 euro/kWh. Different meters for the energy supplied by the IPP and for the energy consumed from the utility grid are mandatory.
Switzerland	Utilities are obliged to buy electricity produced by PV systems. Tariffs paid for electricity fed into the grid vary locally; the minimum rate for the electricity produced is 0.15 SFr./kWh (around 0.10 Euro/kWh). Separate meter is used in most installations. Over 100 power utilities in Switzerland offer their clients the possibility to buy solar power. If a PV plant is installed within a green pricing model, tariffs are paid from 0.80 to 1.20 SFr/kWh (around 0.50–0.75 Euro/kWh). A separate net meter is mandatory.
United Kingdom	Electricity suppliers are not obliged to buy electricity produced by PV systems. Tariffs paid for electricity led into the grid vary with supplier, however the rate is normally less than that for imported units. Some suppliers now offer the same price. A separate meter for energy fed into the grid is required.
Spain	Utilities are obliged to buy electricity produced by PV systems at the rate of approximately 2.5 times the selling price.
USA	Utilities are obliged to buy electricity produced by PV systems but net metering laws are only recently providing guidelines for protection equipment required by the utilities, insurance requirements imposed on the PV owner, etc. Tariffs paid for electricity fed into the grid vary widely and depend on state laws and utility. Various net metering laws apply that vary from state to state.

number of countries (for example, Germany, Spain, and Switzerland) the utilities are obliged by law to pay a premium price for PV electricity. Further information on year by year changes is available from an annual survey report published by IEA Task 1 [8]. In some countries this benefit is enhanced further by the fact that all the PV generated power is exported direct to the utility (as mentioned in Section 1).

5.5 Post- or Pre-notification

The administration of processing a large number of network connections is also of issue to network operators. Once a Type Verification process is in place, there is no need for on-site tests, and hence, the installer is responsible for correct connection. A system requiring preinstallation or postinstallation notification can then be established.

ACKNOWLEDGEMENTS

The authors are grateful to the International Energy Agency for permission to publish sections of IEA Reports and, in particular, the work of Bas Verhoeven, Christoph Panhuber and Ward Bower. We also thank Rod Hacker, Martin Cotterell, Ian Butterss, Ray Arnold, Weidong He and Santiago Silvestre for useful input and discussions.

REFERENCES

[1] IEC Standard 61727, Photovoltaic Systems—Characteristics of the Utility Interface, 1995.
[2] Utility Aspects of Grid Connected Photovoltaic Power Systems, IEA Report PVPS T5−01:1998, International Energy Agency, 1998.
[3] Probability of Islanding in Utility Networks Due to Grid Connected Photovoltaic Power Systems, IEA Report PVPS T5−07:2002, International Energy Agency, 2002.
[4] Risk Analysis of Islanding of PV Power Systems Within Low Voltage Distribution Networks, IEA Report PVPS T5−08:2002. International Energy Agency, 2002.
[5] Draft IEC Standard 62116. Testing Procedure of Islanding Prevention Measures for Grid Connected Photovoltaic Power Generating Systems.
[6] Evaluation of Islanding Detection Methods for Photovoltaic Utility-Interactive Power Systems, IEA Report PVPS T5−09:2002, International Energy Agency, 2002.
[7] PV System Installation and Grid Connection Guidelines in Selected IEA Countries, Report PVPS T5−04:2001, International Energy Agency, 2001.
[8] Trends in Photovoltaic Applications in Selected IEA Countries between 1992 and 2000, IEA Report PVPS Tl-10:2001, International Energy Agency, 2001.
[9] Santiago Silvestre, personal communication. An overview of the Spanish legal framework for the grid connection of photovoltaic generators including contract forms can be found at <http://www.solarweb.net/fotovoltaica.php>

CHAPTER IIC-2

Installation Guidelines: Construction

Bruce Cross
PV Systems Ltd, Cardiff, UK

Contents

1. Roofs — 806
   1.1 Roof Types — 806
      1.1.1 Integral — *806*
      1.1.2 Over Roof — *807*
      1.1.3 Tiles — *807*
   1.2 Substructure — 808
   1.3 New Build vs. Retrofit — 808
   1.4 Mechanical Strength — 809
   1.5 Loading — 810
   1.6 Fixing Systems — 810
   1.7 Weatherproofing — 810
      1.7.1 Sublayer Membranes — *810*
   1.8 Interfaces with Traditional Roof Types — 810
   1.9 Standards — 811
2. Facades — 811
   2.1 Facade Types — 811
      2.1.1 Rain Screen — *811*
      2.1.2 Curtain Wall — *812*
   2.2 Substructure — 813
   2.3 New Build vs. Retrofit — 814
   2.4 Mechanical Strength — 814
   2.5 Weatherproofing — 814
   2.6 Cooling — 814
   2.7 Maintenance — 815
   2.8 Site Testing — 816
3. Ground-Mounted Systems — 817
References — 817

1. ROOFS

1.1 Roof Types

1.1.1 Integral

An integral roof has the potential advantage of saving of material of the roof envelope that has been replaced. However, there are arduous requirements for water-tightness of roofs, which must be met. This can be achieved with regard to traditional roofing practice (which is country specific) and realistic engineering design (see Figure 1). A general overview of this field as well as examples of projects can be found in references [1–5].

The solutions that have been proved fall into the following categories:
- Interlocking panel systems, which either use panels that mimic roofing tiles with the PV element embedded in the surface or have a frame bonded to the PV panel which provides the sealing interlock.
- Adaptations of standard face-sealed sloping glazing systems, where the PV may be built into a double-glazed sealed unit. These are particularly suitable to buildings where the PV is visible from the inside, and must also provide a degree of transparency for internal lighting needs.
- Internally drained, secondarily sealed systems, where the high cost of a face-sealed system cannot be justified.

Figure 1 Corncroft development, Nottingham, UK.

1.1.2 Over Roof

The over-roof mounting of PV panels has been the normal practice in early installations. It is simple in concept, and has been proven provided that the attachment through the traditional roof is performed well. A range of standard frame systems have been tried, and provided that durable materials (hot dip galvanised steel, aluminium, or stainless steel) are used, have been successful. The time taken to install these varies greatly, and has the largest impact on the final cost of the mounting. The mounting brackets are generally most successful when they are standard roofing products, rather than 'special PV' made items, and should be rigid engineered mounts rather than the flexible strap type of fixing sometimes used for solar thermal collector mountings (Figure 2).

1.1.3 Tiles

PV roof tiles have been manufactured in several countries. The advantage of using a traditional roof product is that normal building trade practice can be used, and there is little resistance to the concept from the naturally conservative building trade. However, tiles are small components and a large number are required for an installation—this implies are large number of interconnections, and also a mix of building trades is needed (e.g., electricians must work on a roof, or roofers must perform an electrical function). The use of tiles or slates frequently requires special tiles for edges, angled valleys, chimney joins, ridges, etc., and slates are generally cut to shape on site, and nail holes made to suit. This implies that the PV tile must link seamlessly into the standard roofing material (Figure 3).

Figure 2 Domestic PV systems for Mr. Treble. Farnborough, UK.

Figure 3 Electraslate PV slate system.

1.2 Substructure

Most types of roof have been used for a PV system at some time. The overall construction must be capable of taking the additional load of the PV (or indeed survive the additional uplift when the PV replaces a much heavier roof surface such as concrete tiles). A more arduous requirement may be the local loads of a frame mounted system that has relatively few attachment points. The structure must also be able to accommodate any fixed, or temporary, access structures for maintenance or repair. A transparent or semi-transparent roof has a different set of requirements as it has to provide all the elements of the roof within a single system. This will also have to meet any aesthetic requirements from below as well as from outside. Thus an integral wiring pathway within the mounting system is an advantage (Figure 4).

1.3 New Build vs. Retrofit

In a niche market such as PV at present, there is a need to be able to fit PV to both new and existing houses. In many countries the building stock is old and not being replaced, so the mechanisms for retrofitting PV are necessarily being developed. The costs of fitting PV to new buildings is significantly lower than for a retrofit due to access and adaptation at design stage being zero cost and the one-off costs being absorbed in the larger building project costs (Figure 5).

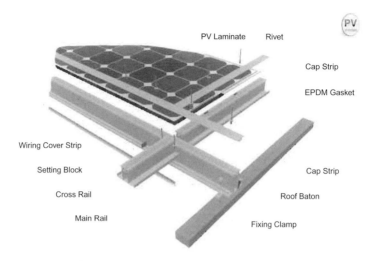

Figure 4 RIS PV roof integration system.

Figure 5 PV retrofits for New Progress Housing, UK.

1.4 Mechanical Strength

The PV elements of the roof have to fulfil the requirements of wind loading, snow loading, fire resistance, and possible traffic for maintenance. This means that a PV panel made for ground mounting may not always be suitable for a BIPV application. The grab zone of a standard PV laminate is small, and the glass thickness may also be inadequate. A purpose designed laminate or alternatively, a mounting system designed to transform the standard laminate into a building component may often be required to meet local codes.

1.5 Loading

Many standard PV laminates are fairly lightweight in roofing terms. The panels themselves may only weigh 5 kg/m^2, and, say, another 5 kg/m^2 for an aluminium mounting structure. However, a double-glazed panel with a double glass front PV in a structural roofing system may add up to 40 kg/m^2.

1.6 Fixing Systems

Traditional roofs are fitted together with nails and screws. High-technology fixings are rare in roofing, and PV systems that require precision fitting with specialist components will be expensive compared to those that are adapted for the trades already in use on site.

1.7 Weatherproofing

The requirement for the roof is to resist the ingress of water, and also to resist the loss of heat from the building, and to provide the degree of protection against fire for the type of building in which it is used. The external surface will have to resist degradation from UV, wind, and rain for 30–60 years. This can be achieved for roofs with traditional materials, but is hard to demonstrate for new materials. Hence, most PV on roofs has a glass external surface.

1.7.1 Sublayer Membranes

Traditional interleaving roof coverings, such as tiles, slates, etc., have a need for a secondary membrane to stop water penetration in times of extreme weather. This requirement remains where PV equivalents are used. The high value of such roofs makes it a good investment to use high-specification membranes for additional security.

1.8 Interfaces with Traditional Roof Types

The join between the specialist PV part of a roof, and the traditional area of a roof, is an area for greatest care. Here a complete engineered system meets with a system developed over centuries and implemented by a craftsperson. If the understanding between these two is not clear, there is potential for the project to fail (Figure 6).

Figure 6 Flashing detail.

1.9 Standards

There are requirements for all building components to meet certain standards. For new products these take time to develop, and these are not yet in place for PV products. There are IEC standards that cover the operation of a PV product in itself, but not yet one that covers its operation in a building. Early moves have been made to establish a European test standard against which products can be tested in order to receive a CE mark, but this goal is still several years away [6]. The USA has several UL tests against which products can be tested (Figure 7).

2. FACADES

2.1 Facade Types

2.1.1 Rain Screen

The simplest type of facade integration is the rain screen. Here the object is to keep the direct impact of the rain from the waterproofing layer, which is some distance behind. The intermediate space is ventilated with the ambient air, and an allowance is made for moisture to drain out from the space. Rain screen panels are generally a simple sheet product within a frame that hooks onto the lattice substructure. PV has been successfully embedded into such frames using the PV laminate to replace the sheet material (Figure 8).

Figure 7 Shell PV roof undergoing 'prescript' test in Solar Simulator at Cardiff, UK.

Figure 8 Bowater House PV facade, Birmingham, UK.

2.1.2 Curtain Wall

In a curtain wall the external surface is the waterproof layer, and hence, all parts of the structure behind are considered dry. This is not to say that there is no chance of moisture, as condensation must still be considered. Normal practice would be to allow a small amount of air movement behind the outer skin, but in the case of PV it should be increased, if possible, to provide some cooling from the rear surface of the PV. The advantage of the curtain wall is that it allows a continuous skin incorporating all the facade elements—windows, PV, and blank panels within a proven design. These systems are complex and expensive without the PV and so the additional cost may be more readily absorbed into such a facade (Figure 9). It should be noted that the use of terms 'rain screen' and 'curtain wall' varies internationally.

2.2 Substructure

The substructure for a facade may use any type of normal construction material. The extra requirements caused by having PV as the external skin are that temperatures may be higher (larger expansion may occur) and that a path must be found for the cabling and access provided to any connections or marshalling boxes within the building envelope. These cable

Figure 9 William Jefferson Clinton Peace Centre PV facade, Enniskillen, Northern Ireland, UK.

paths must be accessible for maintenance and must not provide a path for moisture ingress into the building fabric.

2.3 New Build vs. Retrofit

A retrofit facade has more potential areas of risk than a new-build. The uncertainties of dimensions may require significant site work, and the locations for cabling and electrical plant may have to be defined late in the project, when the old materials are removed. In a new-build facade there is likely to be a long time delay between fitting the PV elements as part of the external envelope, and installation and wiring of the electrical plant. The design of the system must ensure that these operations can be separated (e.g., external stringing may not be accessible for checking at the time the system is finally commissioned).

2.4 Mechanical Strength

Building designs often require the use of large glazing elements. This has the effect of increasing the glass thickness, which, in a PV panel, reduces output. In many areas the facade must withstand impact from foot traffic, and may have to provide a degree of security also. Standard laminates rarely have sufficient thickness, so more expensive custom made panels are needed. In general BIPV facade panels are 50–80% more expensive than standard modules (Figure 10).

2.5 Weatherproofing

The PV is just one element of a facade system and this total system must provide the weatherproofing. Care is required that the cabling and the increased expansion allowance do not compromise the performance of the mounting system.

2.6 Cooling

The output of a PV panel decreases with rising temperature. It is therefore advantageous to provide for the flow of air over the rear surface of the PV if at all possible. For rain screen systems there are usually ventilation slots around the periphery of each panel, so these need only to be checked for adequate passage of air. However, for curtain walls, it will be necessary to allow an entry for air at the lowest part of the facade and to allow the exit of air from the top, if the maximum output is to be achieved. This design requires care if the waterproofing of the system is

Figure 10 Office of the Future PV facade, Building Research Establishment. UK.

not to be compromised. An EU project to assist this design process is underway (pvcool-build.com). When a fully face-sealed double-glazed transparent facade utilises PV, there is frequently no opportunity to ventilate the rear, and so the loss of performance (5–10%) must be accepted.

2.7 Maintenance

An important part of the design of the building is its future maintainability. Physical cleaning is not any more arduous for PV than for glass panels in general. Access should be provided for inspection and testing to any cable marshalling box, and a system should be in place to allow the testing, and possible replacement of any PV module in the system. In a large facade this may require simultaneous access externally, and internally, to test and to identify a fault.

2.8 Site Testing

There are standards for site testing the water tightness of facade systems. These have been developed by the centre for window and cladding technology [7]. A high-pressure water jet is used at susceptible joints and the system is inspected for water penetration. An addition to this test is to perform a high-voltage isolation test immediately after, in order to evaluate whether any water has compromised the DC cabling or the PV modules (Figure 11).

Figure 11 Pluswall PV facade, UK.

3. GROUND-MOUNTED SYSTEMS

When PV is mounted directly on a frame, pole, or concrete block, the requirements are to provide a rigid attachment that will resist gravitational, wind or impact forces, while protecting the PV from undue twisting or deflection. A typical frame uses standard sections which are assembled on-site.

Since such systems are not usually part of a building and perform no weather protection function, there are few constraints on their design. However, car parking shelters, and entrance canopies are favoured locations for PV and often fall into the scope of building codes by association.

REFERENCES

[1] S. Roaf, Ecohouse—A Design Guide, Architectural Press, Elsevier Science, Oxford, 2001.
[2] F. Sick, T. Erge, PVs in Buildings—A Design Handbook for Architects and Engineers, International Energy Agency, Solar Heating and Cooling Programme, Task 16, 1996.
[3] M.S. Imamura, P. Helm, W. Palz (Eds.), Photovoltaic System Technology: A European Handbook, U.S. Stephens & Associates, Bedford, 1992.
[4] Max Fordham & Partners in association with Feilden Clegg Architects, Photovoltaics in Builidings. A Design Guide, Department of Trade and Industry, UK, 1999, Report ETSU S/P2/00282/REP.
[5] Studio E Architects, Photovoltaics in Buildings. BIPV Projects, Department of Trade and Industry, UK, 2000, Report ETSU S/P2/00328/REP.
[6] Prescript, Final Report on EU Project, European Commission, Brussels, 2000, JOR3-CT97−0132.
[7] Test Methods for Curtain Walling, second ed., Centre for Window and Cladding Technology, Bath University, 1996.

CHAPTER IIC-3

Installation Guidelines: Electrical

Martin Cotterell
Sundog Energy

Contents

1. Introduction — 820
2. Codes and Regulations — 820
3. DC Ratings (Array Voltage and Current Maxima) — 821
   3.1 Maximum Voltage — 821
   3.2 Maximum Current — 822
      3.2.1 Designing for Current and Voltage Variations — 822
      3.2.2 Voltage Multiplication Factor: Mono- and Multicrystalline Silicon Modules — 822
      3.2.3 Current Multiplication Factor: Mono- and Multicrystalline Silicon Modules — 823
      3.2.4 Multiplication Factors: Other Modules — 823
      3.2.5 Module Effects—Initial Period of Operation — 823
4. Device Ratings and Component Selection — 823
5. Array Fault Protection — 824
   5.1 Fault Scenarios: Earth Faults — 824
   5.2 Line—Line Faults — 825
   5.3 Fault Analysis — 825
      5.3.1 String Cable—Fault Current — 825
      5.3.2 Array Cable—Fault Current — 826
      5.3.3 PV Module—Reverse Current — 826
   5.4 String Cable Protection — 827
   5.5 Array Cable Protection — 827
   5.6 Module Reverse-Current Protection — 827
   5.7 String Overcurrent Protection — 828
   5.8 Blocking Diodes — 829
   5.9 Fault Protection with an Additional Source of Fault Current — 830
6. Earthing Arrangements — 830
   6.1 Earthing of Live Conductors — 830
   6.2 Earthing of Exposed Conductive Parts — 831
7. Protection by Design — 832
8. Labelling — 834
References — 834

Practical Handbook of Photovoltaics.
© 2012 Elsevier Ltd. All rights reserved.

1. INTRODUCTION

The design of a PV system will draw much from standard practice as detailed in national and international wiring standards. However, certain aspects of the design and installation of PV systems can call for a slightly different approach. This chapter covers some of the key issues specific to PV installations. It covers those areas that need particular attention when designing a PV system or specifying PV system components.

The specific requirements of PV systems are the result of three key characteristics.
- A PV array does not have a fixed voltage and current output—rather one that varies considerably under different operating conditions. It is important to determine the expected current and voltage maxima that may be generated by the PV array in order to select suitably rated system components and configurations.
- A PV module is a current-limiting device—PV arrays require a slightly different approach when designing suitable fault protection.
- PV modules cannot be switched off and, due to the current limiting nature, faults may not be detected or cleared. Precautions need to be taken to mitigate shock and fire hazards during installation, operation, and maintenance. This includes careful consideration of grounding of the PV system.

2. CODES AND REGULATIONS

While local codes retain a very significant role in many countries, since the first edition of this book, PV systems are now better addressed within international electrical codes:
- IEC 60364 *Low voltage Electrical Installations* is designed to harmonize national wiring standards. Many European wiring regulations, such as BS7671 in the UK, closely follow the format of IEC 60364. Part 7-712 of IEC 60364 *Solar Photovoltaic (PV) Power Supply Systems* [1] contains requirements that supplement modify or replace the general requirements of IEC 60364—specifically for PV systems.
- Work is also underway within IEC to produce a dedicated PV array standard—IEC 62548 *Requirements for Photovoltaic (PV) Arrays*. When

issued, it is expected that this document will provide a significant resource and reference for PV system design.
- While not a design document, IEC62446 *Grid-Connected PV Systems—Minimum Requirements for System Documentation, Commissioning Tests, and Inspection* [2] is also an important document for the PV system designer, as it sets out requirements a PV system will need to meet on completion.

3. DC RATINGS (ARRAY VOLTAGE AND CURRENT MAXIMA)

When considering the voltage and current carrying requirements of the PV DC system, the maximum values that will originate from the PV array need to be assessed. The maximum values are derived from the two key PV module ratings: the open-circuit voltage (Voc) and the short-circuit current (Isc).

The values of Voc and Isc provided by the manufacturer on the module label or data sheet are those at standard test conditions (STC): irradiance of 1000 W/m^2, air mass 1.5, and cell temperature of 25°C. As described in Chapter Ia-2 the operation of a module outside of STC affects the values of Voc and Isc, and these variations must be taken into consideration when specifying the voltage or current rating of all components of the DC system.

Cell temperatures well above and below the STC value of 25°C can be expected in real systems. Cell temperatures are affected by ambient air temperatures, absorbed heat from the Sun, cooling from the wind, and the degree of ventilation and air movement behind the modules.

Irradiance can also vary considerably from the STC value of 1000 W/m^2. While irradiance below STC values will be common, a clear atmosphere and cloud-enhancement effects can raise the irradiance above the STC value.

3.1 Maximum Voltage

The maximum voltage produced by a PV array occurs when the modules are open circuit. The Voc of a silicon PV cell will increase above the nominal STC value if the cell temperature drops below the STC temperature of 25°C. The amount of variation with respect to temperature depends upon the module type and cell material.

For example, a module with a temperature coefficient of 130 mV/°C would see an increase in Voc of 3.25 V at 0°C. For a series string of 10 such modules, this would result in an increased string voltage of 32.5 V.

3.2 Maximum Current

The maximum current generated by a PV module occurs under short-circuit conditions. The Isc is affected most significantly by irradiance fluctuations and to a lesser extent by temperature changes.

Irradiance may increase above the STC value of 1000 W/m^2, taking the short-circuit current above the nominal rated value. This can occur around midday when the Sun is at its highest, in very clear atmospheric conditions, and from enhancement from cloud or snow reflections. A potential increase in Isc of 125% is typically assumed in many codes [1,3].

3.2.1 Designing for Current and Voltage Variations

In order to simplify component selection, the standard approach to accommodating the potential variation in voltage and current due to deviations from STC conditions, is to apply a multiplication factor to the STC values of Voc and Isc quoted on the module label or data sheet. The resultant voltage and current values are then taken as a minimum requirement for all parts of the DC system.

3.2.2 Voltage Multiplication Factor: Mono- and Multicrystalline Silicon Modules

A multiplication factor of 1.15 is commonly applied to Voc, but multiplication factors greater than this are used in some national codes, particularly in regions where low ambient temperatures may occur.

The following table, taken from the US National Electric Code [4], shows Voc multiplication factors for crystalline silicon PV modules, according to the lowest expected ambient temperature.

Minimum ambient temperature	Voc multiplication factor
4°C to 0°C	1.10
−1°C to −5°C	1.12
−6°C to −10°C	1.14
−11°C to −15°C	1.16
−16°C to −20°C	1.18
−21°C to −25°C	1.20
−26°C to −30°C	1.21
−31°C to −35°C	1.23
−36°C to −40°C	1.25

3.2.3 Current Multiplication Factor: Mono- and Multicrystalline Silicon Modules

A multiplication factor of 1.25 is applied to Isc in most cases, is stated in IEC 60364-7-712 [1], and is commonly used in many national codes. However, multiplication factors above this value may be applied in some cases.

3.2.4 Multiplication Factors: Other Modules

As discussed in Part Ic some types of PV modules have temperature coefficients considerably different to those constructed of crystalline silicon cells. The effects of increased irradiance may also be more pronounced. In such cases, the multiplication factors used for crystalline silicon modules will not be appropriate.

In such cases, specific calculations of worst case Voc and Isc must be performed, based on manufacturer's data and guidance. These calculations should allow for the lowest possible ambient temperature and an irradiance level of at least 1250 W/m^2.

3.2.5 Module Effects—Initial Period of Operation

Some module types such as amorphous silicon systems have an electrical output that is significantly elevated during the first weeks of operation. This increase is additional to that produced by temperature or irradiance variation.

Where modules are subject to this effect, the initial phase increase (see editors' note above) in voltage and current quoted by the manufacturer must be considered first, before any other multipliers are applied.

4. DEVICE RATINGS AND COMPONENT SELECTION

The calculations in Chapter IIa-4 enable a system designer to assess the voltage and current maxima that may occur under normal operation. Following these calculations the designer will then ensure that
- module maximum operating voltage is not exceeded (the number of modules in string is selected to ensure that the string voltage will not exceed the maximum voltage rating of modules within the string),
- electronic components such as inverters will remain at all times within their operating range (e.g., maximum string open-circuit voltage will not exceed inverter maximum DC rating),

- all cables have suitable voltage and current ratings (after any derating factors applied),
- all switch gear is rated for DC operation at the voltage and current maxima calculated, and
- any overcurrent devices, any means of disconnection, and all other system components are suitably rated for DC operation at the voltage and current maxima calculated.

It is important to note that the DC rating of array components, particularly overcurrent protection devices and switch gear, must be checked. Some devices may not be suitable or rated for DC operation, and some devices may have DC ratings lower than their nominal AC rating.

Splitting an array into a number of smaller subarrays to reduce the maximum shock voltage is sometimes considered at this stage. Such a reduction in voltage may enable the use of lower-rated system components.

5. ARRAY FAULT PROTECTION

The following analysis is directed toward faults arising where the PV array provides the only source of fault current, such as in a typical grid-connected system with no battery. For a system with a battery or other source of fault current, see also Chapter IIb-2.

The conventional means of automatic disconnection of supply under fault conditions is to rely upon a circuit protective device such as a fuse to clear the fault current. However, the current limiting nature of a PV circuit means that conventional protection measures are often difficult to apply, and electric arcs can develop that will not be detected or cleared by a protective device.

The short-circuit current of a module is little more than the operating current, so in a single-string system, a circuit fuse would simply not detect or operate to clear a short-circuit fault. However, in systems with multiple strings, the prospective fault current may be such that overcurrent protective devices can be utilised. Hence, the selection of overcurrent protective measures depends upon the system design, the number of strings, and the fault scenario.

5.1 Fault Scenarios: Earth Faults

Earth faults may occur for a number of reasons. For example, a fault to earth in PV cabling systems may arise due to insulation damaged during installation, subsequent impact or abrasion damage to the cable sheath, or vermin damage. Earth faults may also develop within PV modules, for example, due to damage to the module backsheet during installation.

In an ungrounded system (where none of the PV array conductors have been intentionally connected to earth), two earth faults are required to create a fault current. Such an occurrence may be rare and can be minimised by good system design and installation practice. However, such faults can be difficult to detect and may develop over time. Additionally, it is often stated that where one earth fault has developed, a second is likely to follow.

For grounded systems where one of the supply conductors is connected to earth, then only one earth fault may be necessary in order for a fault current to flow.

5.2 Line–Line Faults

While generally less likely, other fault circuits can develop. Such faults are often classed as line–line faults where string or array output cables become shorted together. Such faults may occur as a result of cable insulation failure (from impact, abrasion, or installation damage) or from faults developing in the DC junction box, DC disconnect, or inverter (through mechanical damage, water ingress, or corrosion).

5.3 Fault Analysis

For the purposes of examining fault conditions, an example of an earth fault is considered. Such a fault provides a case with the potential for both cable overloading and module reverse currents. Figure 1 depicts a PV system consisting of three parallel connected strings, with the PV string cables commoned in separate DC junction boxes. Under normal conditions, the maximum current in each string cable is equal to the module Isc.

In this floating system with no reference to earth, two earth faults are required to create a fault current. For example, if a long-standing earth fault in the PV array main negative cable is combined with a new fault at a point along a PV string (marked X on the diagram), then fault currents will flow.

5.3.1 String Cable–Fault Current

The maximum fault current in any part of the string cabling in the system of Figure 1 is I_R — flowing to point X and fed from the two nonfaulty strings. With no current flowing to any load circuit, I_R is the sum of the

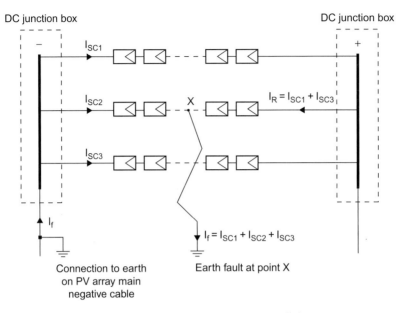

Figure 1 Schematic diagram of a PV system with three parallel strings.

short-circuit current flowing from the two other strings: $I_R = 2 \times Isc$. For a system of n strings,

$$\text{Maximum string cable fault current} = (n-1) \times Isc$$

5.3.2 Array Cable–Fault Current

The maximum fault current in the array cables in the system of Figure 1 is I_F – flowing in the PV array main negative cable. This fault current cannot exceed $3 \times Isc$. For an array of n strings,

$$\text{Maximum array cable fault current} = n \times Isc$$

5.3.3 PV Module–Reverse Current

In the illustration given, a number of the PV modules in the faulty string have a reverse current flowing through them. The magnitude of this current is the same as that flowing through the string cable:

$$\text{Maximum PV module reverse current} = (n-1) \times Isc$$

5.4 String Cable Protection

From the fault analysis above, it can be seen that for a system of n parallel connected strings, with each formed of m series connected modules and with no cable overcurrent protection devices installed, string cables need to be rated at a minimum of

$$\text{Voltage} = \text{Voc (max)} \times m$$

$$\text{Current} = \text{Isc (max)} \times (n - 1)$$

It is important that the voltage and current values used in this calculation take into account potential variations due to factors such as temperature and irradiation. This can be achieved by using the multiplication factors described in Section 3 of this chapter.

To take into account factors such as cable installation method and grouping, cable derating factors also need to be considered. These may reduce the effective cable current-carrying capacity.

This approach provides a means of cable overcurrent protection by ensuring that the maximum potential fault current can be safely accommodated by the string cable in question. This method does not clear the fault: it simply prevents a fire risk from overloaded string cables.

5.5 Array Cable Protection

For a system with only one subarray, no special measures need to be taken to protect the main array cables from fault currents. This is because the maximum fault current ($n \times$ Isc) is the same as the peak design current (also $n \times$ Isc). The PV array main cables should be sized for the peak design current, and hence, these cables will not be overloaded under fault conditions.

For a system with multiple subarrays, depending on the system architecture, there is the potential for overcurrent in subarray main cables (multiple subarrays feeding fault currents into a faulted array). In such cases, overcurrent protection can be achieved by the use of suitably rated overcurrent protective devices installed on each subarray cable.

5.6 Module Reverse-Current Protection

While string cable sizes can be increased as the number of parallel connected strings (and the potential fault current) increases, the ability of a module to withstand the reverse current must also be considered. Where currents exceed the modules maximum reverse current rating, there is the potential for damage to the affected modules and also a fire risk.

IEC61730-2 *Photovoltaic (PV) Module Safety Qualification—Part 2: Requirements for Testing* [5] includes a reverse-current overload test. This reverse-current test is part of the process that enables the manufacturer to provide the maximum overcurrent protection rating or *maximum series fuse*. Fault currents above the maximum series fuse rating present a safety risk and must be addressed within the system design.

The maximum module reverse current to be experienced under fault conditions depends on the number of parallel connected strings (see Figure 1): $I_R = (n-1) \times Isc$.

Overcurrent protective measures need to be considered where I_R is greater than the maximum series fuse rating of the module. The test being overcurrent protection required where

$$(n-1) \times Isc > \text{module } \textit{maximum series fuse rating}$$

5.7 String Overcurrent Protection

The installation of string overcurrent protection provides a means to prevent excessive fault currents in systems with multiple parallel connected strings. This can provide protection against excessive module reverse currents and may also enable the use of smaller string cables. See Figure 2.

For floating (ungrounded) systems, to provide full protection, overcurrent protection is required in both the positive and negative conductors. In Figure 1, a protective device in the positive leg of the string cabling would be required to clear the fault. However, if the earth fault at X were combined instead with an earth fault in the main positive cable (rather than in the negative cable as shown), this could only be cleared by a protective device in the negative leg.

The trip value I_{TRIP} of the overcurrent protective device needs to be carefully selected. It needs to be large enough to prevent nuisance tripping and small enough to ensure it clears the potential fault current, as well as ensuring it is less than or equal to the maximum series fuse value specified by the manufacturer. A common rule for the selection of the device is to ensure

$$I_{TRIP} > 1.5 \times Isc\ stc$$

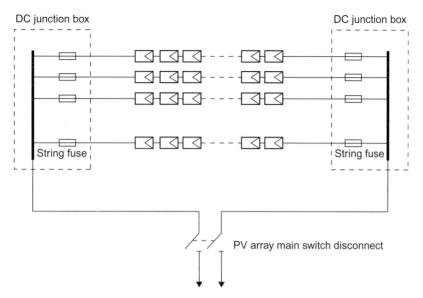

Figure 2 Typcial array schematic with multiple strings.

$$I_{TRIP} \leq 2.4 \times I_{sc}\ stc$$

$$I_{TRIP} \leq \text{Maximum series fuse value}$$

Fuses are the most common overcurrent protective devices selected by system designers. Type gPV fuses according to IEC60269-6 [6] are specifically designed for use in PV systems. String fuses should be selected to ensure they are rated at the maximum system voltage V_{oc} (max).

The use of circuit breakers may also be acceptable provided they are DC rated, meet the above criteria, and are rated for operation with currents in either direction.

5.8 Blocking Diodes

Blocking diodes are sometimes specified in a PV array to prevent reverse currents, though they should not be seen as a substitute for an overcurrent protective device. One application is to prevent losses in a battery system at night, but most modern charge controllers usually prevent such reverse currents in any case. Chapter IIa-2 should be consulted for a further discussion of this topic.

5.9 Fault Protection with an Additional Source of Fault Current

Systems with another source of fault current may need additional fault protection to that outlined above. With PV systems, a typical fault current source will be a battery, although other fault current sources may also be present.

Standard circuit design is to be applied for fault protection of other fault current sources. With a battery system, it is important to ensure that any fuse or circuit breaker is rated to interrupt the very high fault currents that a battery can generate under short-circuit conditions. Standard AC fuses are sometimes seen on battery systems; these fuses not only are often not DC rated but also may be incapable of interrupting the high battery fault currents.

6. EARTHING ARRANGEMENTS

Earthing practice varies considerably around the world and the local code must be applied when designing a PV system. However, any connection to earth within a PV system needs special consideration because, as discussed above, the behaviour of a PV system under fault conditions may differ to that of a conventional AC mains circuit.

Earthing decisions will be driven by factors including
- local code requirements;
- functional reasons—some PV technologies require a connection to earth (see Chapter IIb-1);
- the nature of the application circuit or device; and
- lightning or surge-protection requirements.

6.1 Earthing of Live Conductors

A connection to earth of one of the live conductors within the PV DC circuit is most likely to be the result of a local code requirement or be driven for functional reasons by the PV technology being used.

There are a variety of possible system architectures and earthing scenarios. These can be broadly summarized as
- no earth connection,
- hard-wired connection of positive or negative conductor to earth,
- centre-tapped array—with or without earth connection, and
- high impedance connection of positive or negative conductor to earth (for functional reasons).

The system designer must ensure that the requirements of local codes and manufacturers of both PV modules and the equipment to which PV arrays are connected are taken into account in determining the most appropriate earthing arrangements.

In grid-connected systems, a connection to earth of either of the conductors of the PV array is typically not permitted in a system without at least simple separation between the PV and AC main circuits. Functional grounding of one of the conductors of the PV array may be allowed in a system without at least simple separation, where residual current detection is provided that limits the residual current in the array to 30 mA.

In the case of PV systems connected to an inverter, it is to be noted that IEC62109-2 (*Safety of Power Convertors for Use in Photovoltaic Power Systems—Part 2: Particular Requirements for Inverters*) [7] includes requirements according to the type of earthing arrangement (and inverter topology). These include minimum inverter isolation requirements, array ground-insulation resistance measurement requirements and array residual-current detection and earth-fault alarm requirements.

Where there is a hard-wired connection to earth, there is the potential for significant fault currents to flow if an earth fault occurs somewhere in the system. A ground fault interrupter and alarm system can interrupt the fault current and signal that there has been a problem. The interrupter (such as a fuse) is installed in series with the ground connection and selected according to array size. IEC 62548 provides guidance on selection of the interrupter and also the requirements for the alarm. It is important that the alarm is sufficient to initiate action, as any such earth fault needs to be immediately investigated and action taken to correct the cause.

A high-impedance connection to earth of one of the current carrying conductors may be specified where the earth connection is required for functional reasons. The high-impedance connection fulfills the functional requirements while limiting fault currents.

6.2 Earthing of Exposed Conductive Parts

Exposed conductive parts are those parts of the PV system that are not energised during normal operation, such as the array frame. There are three main reasons why a connection to earth of the exposed conductive parts of a PV array may be considered:

1. Protective earthing (to provide a path for fault currents)
2. Lightning, surge, EMI protection

3. Equipotential bonding (to avoid uneven potentials across different parts of an installation)

An earth connection may provide one or more of the above functions.

The selection and installation of the earth conductor will need to be performed with regard to local codes and will depend upon the function, or combination of functions, it is performing.

7. PROTECTION BY DESIGN

Faults are best prevented by good design and the appropriate selection of system components coupled with careful installation.

The DC wiring of an array should be specified to minimise the risk of faults occurring. This can be achieved by the use of single-core insulated and sheathed cables (sometimes referred to as *double-insulated cable*, Figure 3), single-core basic insulated cable with an earthed armour or screen (Figure 4), or insulated cable where the cables of opposite poles are laid in separate ducts or conduits (Figure 5).

Array cables should also be selected to ensure that they are properly rated for the environment in which they will be installed. For example,

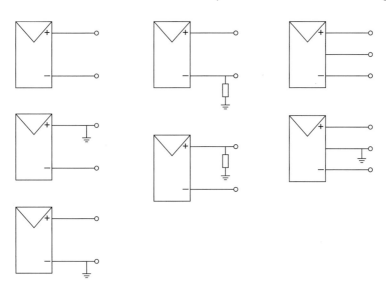

Figure 3 Common PV system architectures.

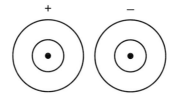

Figure 4 Single core insulated and sheathed cables.

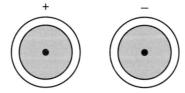

Figure 5 Single core basic insulated cable with an earthed armour or screen.

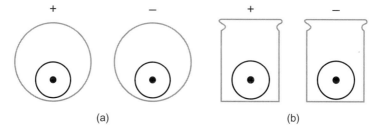

Figure 6 Insulated cable where the cables of opposite poles are laid in separate ducts (a) or conduits (b).

module interconnect cables must have sufficient UV and temperature ratings. Cables specifically designed for PV applications are now readily available and are the obvious choice for the PV system specifier.

The short-circuit protection afforded by double-insulated cables also needs to be maintained in the construction and makeup of DC junction boxes and switch assemblies (see Figure 6). Short-circuit protection may be provided by

- fabrication of the enclosure from nonconductive material;
- positive and negative buses adequately separated within the enclosure, by a suitably sized insulating plate, or by separate positive and negative junction boxes; and
- cable and terminal layout such that short circuits during installation and subsequent maintenance are extremely difficult.

8. LABELLING

A PV array circuit cannot be turned off: terminals will remain live at all times during daylight hours. It is important to ensure that anyone opening any enclosure within the DC system is fully aware of this fact.

It is good practice to ensure that all parts of the system are suitably labelled to warn of this hazard—e.g., 'Danger, Contains Live Parts During Daylight.' All labels should be clear, easily visible, and constructed to last and remain legible as long as the enclosure.

REFERENCES

[1] IEC standard 60364—Part 7–712. Electrical Installations of Buildings. Special Installations or Locations—Solar Photovoltaic (PV) Power Supply Systems.
[2] IEC62446, Grid-Connected PV Systems—Minimum Requirements for System Documentation, Commissioning tests, and Inspection.
[3] Photovoltaics in Buildings. Guide to the Installation of PV Systems, DTI/Pub URN 02/788, UK, 2002.
[4] NFPA 70. National Electrical Code (NEC). National Fire Protection Association, USA.
[5] IEC61730-2, Photovoltaic (PV) Module Safety Qualification—Part 2: Requirements for Testing.
[6] IEC60269-6, Low-Voltage fuses—Part 9: Supplementary Requirements for Fuse-Links for the Protection of Solar Photovoltaic Energy Systems.
[7] IEC62109-2, Safety of Power convertors for Use in Photovoltaic Power Systems—Part 2: Particular Requirements for Inverters.

EDITOR'S NOTE

There should be an awareness of electrical hazards to personnel, particularly installation personnel who may be required to work at height, where even minor electrical shock can lead to further accidents. Obviously where there is light, there is a potential developed by PV modules and arrays, particularly in "islanding" where disconnection from the grid does not eliminate shock hazards.

PART IID

Space and Concentrator Systems

CHAPTER IID-1

Concentrator Systems

Gabriel Sala
Instituto de Energia Solar, Universidad Politecnica de Madrid, Spain

Contents

1. Objectives of PV Concentration 837
2. Physical Principles of PV Concentration 839
3. Description of a Typical Concentrator: Components and Operation 843
4. Classification of Concentrator Systems 845
5. Tracking-Control Strategies 851
6. Applications of C Systems 854
7. Rating and Specification of PV Systems 854
   7.1 Rating of C Modules 856
   7.2 Specifications That Must Be Required From C Systems 858
8. Energy Produced by a C System 859
9. The Future of Concentrators 860
References 861

1. OBJECTIVES OF PV CONCENTRATION

The aim of combining solar cells with concentration systems is an attempt to reduce the cost of the electrical energy produced. Once this principal goal has been stated several aspects of this statement can be discussed in more detail.

On one hand, PV concentrators reduce the fraction of the cell's cost in the total system cost, by substituting the area of expensive cells by less expensive collectors. This has been historically the first step in the development of concentrators, using mainly silicon concentrator solar cells (Figure 1).

On the other hand, concentration has more recently been seen as the only way to accept, into the commercial sphere, new generations of highly sophisticated solar cells. At the same time, these are very efficient,

Practical Handbook of Photovoltaics.
© 2012 Elsevier Ltd. All rights reserved.

Figure 1 The Ramón Areces array (Madrid, 1980). One of the first two axis tracking European PV concentrators made following the ideas of the early Sandia I and II (USA, 1978). It uses patented hybrid glass-silicone Fresnel lenses and 50 mm diameter Si cells at 38×.

but also very expensive. Without the adoption of concentration as the operating mode these extremely expensive cells cannot be utilised. But by increasing the intensity of light (irradiance or W/m^2) even exceeding 1000 times over the standard one-sun solar radiation, the cell cost per unit of output power becomes acceptable. The higher efficiency then compensates for the cell cost achieving an overall cost reduction of the system.

Since cost reduction is the prime objective of PV concentration, let us examine a simple expression for the cost of energy produced by a concentrator system [1,2]:

$$C_{kWh}\left(\frac{\$}{kWh}\right) = \frac{Array + \frac{Cell}{X}}{Collectede\ Energy.\eta_{op}.\eta_{cell}.PR} Return\ rate \quad (1)$$

where *array* is the cost of the optics, the tracking structure, the heat sink, and of the driving control system, per m^2; *cell* is the cost per unit area of the cells, per/m^2; X is the geometrical concentration (i.e., the ratio of the area of the collector to the area of the solar cells); *collected energy* is the direct radiation incident on the collector optics (depends on tracking strategy) in kWh/m^2 per year; η_{op} is the optical efficiency of the collector (i.e., the ratio of power incident on the receiver divided by the power incident on the collector); η_{cell} is the efficiency of the cell at standard test conditions (STC); and *PR* is the performance ratio. This accounts for the losses at operating conditions of the cells over the losses at STC. Typically, PR is 0.70, including all balance-of-system

Table 1 Typical values of PV concentrator cost components [1,2,11]

System concentration level	Solar cell type	Collector (euros/m²)	Cell cost (euros/m²)	X	η_{opt} (%)	η_{cell} (%) at STC
Low concentration	Silicon	150	200–500	10–40	80	20
High concentration	Silicon	200	15,000	300	80	26
Very high concentration	III–V single-junction	300–400	40,000	1000	80	28
Very high concentration	III–V multijunction	300–400	80,000	>2000	80	35

(BOS) loses. A discussion of the balance of system components is deferred to Section 3. *Return rate* is the annuity over 20 years of the total capital invested (usually taken as 6% per year in Europe).

From this expression we understand that the cell cost contribution becomes less significant as the concentration level increases. It is also clear that the efficiencies are the key figures for cost reduction because they affect all components of the generator including the land needed to deploy the collectors. Typical figures for each term in Equation (1) are shown in Table 1 for different systems.

2. PHYSICAL PRINCIPLES OF PV CONCENTRATION

Solar cells generate a current (in A) proportional to the total power of light absorbed (in W). Assuming that the spectrum of light is kept the same, the electrical current supplied by the cell exactly at short circuit will follow a linear law up to a very high incident light power, i.e., up to several thousand times the normal sunlight [3,4].

The intensity of light incident on a surface is commonly expressed by the value of the *irradiance* (W/m²)—that is, the power received per unit of area normal to the light rays. The effective concentration level, C, is usually defined as the ratio of the averaged irradiance on the PV receiver divided by the direct irradiance on the collector optics, and is measured as

$$C = + \frac{I_{sc}(under\ concentration)\ (A)}{I_{sc}(@1000 W/m^2)(A)} = \frac{I_{sc}(C)}{I_{sc}(1)} \qquad (2)$$

where I_{sc} is the short circuit current of the receiver cell under concentrated light or under the standard irradiance of 1000 W/m², as indicated in brackets. The irradiance of 1000 W/m² is usually referred to as '1 sun.'

We can now define the optical efficiency, η_{op} as

$$C = \eta_{op} X \tag{3}$$

For single-junction cells (for example, cells made from crystalline silicon or III–V semiconductors) the variation of the spectral composition of natural sunlight can cause variations in the cell current of about 3%— a value which is insignificant for practical purposes.

In contrast, the output voltage increases as a result of the larger current relative to 1-sun conditions as

$$\Delta V = \frac{kT}{q} \ln C = \frac{kT}{q} \frac{I_{sc}(C)}{I_{sc}(1)} \tag{4}$$

where ΔV is the increase of voltage, kT/q is the thermal voltage ~0.026 V at 25°C, InC is the natural (Napier) logarithm of the effective concentration, $Isc(C)$ and $Ics(1)$ are the short-circuit currents at C and 1 sun, respectively.

Combining Equations (2) and (4) we deduce that the efficiency of a cell should be higher under concentration than at 1 sun. However, the unavoidable series resistance of solar cells dissipates power according to the expression:

$$P_{loss}(C) = I_{mp}^2(C) R_s = C_{mp}^2(1).Rs = C^2.P_{loss}(1) \tag{5}$$

In other words, for 10× concentration, the losses due to R_s are 100 times larger than at 1 sun. For this reason, special contact grid patterns must be designed to reduce the series resistance of C cells at least C times to maintain the same ratio of power losses to the power generated. The series resistance is the main physical parameter that prevents the use of cells designed for flat modules to operate efficiently under concentration.

The curve of Figure 2 provides a simple method for the determination of the series resistance of a cell, receiver, or array, by measuring the negative slope of the curve at high concentration levels. The series resistance is given approximately by

$$R_s = \frac{\eta(C_2) - \eta(C_1)}{C_1 - C_2} \cdot \frac{I_{mp}^2(1)}{A, E(1)} \tag{6}$$

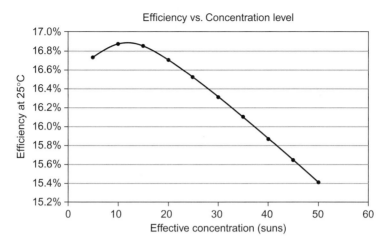

Figure 2 Variation of cell efficiency as a function of the concentration level. The maximum is obtained for 15 suns. The continuous drop after the maximum is caused by series resistance losses [4]. (Obtained by the author on a Euclides-Tenerife cell type.)

where $\eta\,(C_1)$ and $\eta\,(C_2)$ are two values of efficiency of the curve to the right of the maximum, C_1 and C_2 are the abscissas of these points, A_r is the area of the receiving cell, $E(1)$ is the average irradiance at 1 sun ($= 0.1$ W/cm^2), and I_{mp} is the cell current at the maximum power point at 1 sun ($I_{mp} = 0.95\ I_{sc}$).

Figure 3 shows a set of I–V curves for a C cell under concentration. Figure 4 compares the I–V curves for a conventional cell under 1 sun and under concentration.

An implicit drawback of concentration is that the density of power which has to be dissipated as heat increases proportionally with C while the surface area of the receiver which is available to exchange the heat with the surrounding air is fixed. Special heat sinks are therefore required to hold the cell temperatures sufficiently low to maintain an acceptable value of the efficiency. Figure 5 shows the heat sink of the Madrid EUCLIDES prototype.

The series resistance constitutes the most serious limit to concentration for classical silicon n$^+$pp$^+$ structures because the base cannot be made sufficiently thin, and this resistance term cannot be reduced with any grid design. However, this limitation does not apply to the silicon Back Point Contact solar cells which maintain effective operation up to 400× (see Chapter Id-3). In contrast, recent III–V cells have been

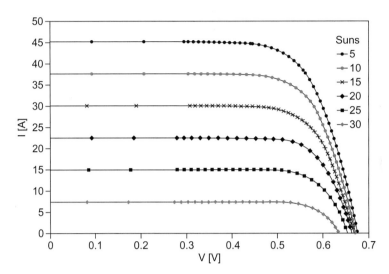

Figure 3 I–V curves of the EUCLIDES cells at several concentrations obtained with a multiflash test [11]. Note the increase of V_{oc} with concentration and the linearity of I_{sc} as a function of the concentration level (expressed in suns).

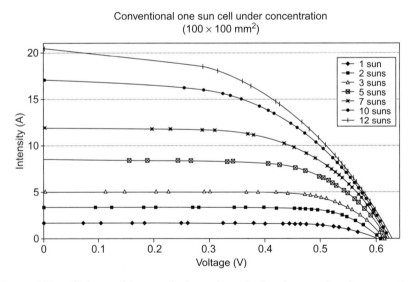

Figure 4 The efficiency of 1 sun cells drops dramatically when used under concentration due to series resistance: the efficiency of 14.6% at 1 sun becomes 10.3% at 12 suns.

Figure 5 Close view of the EUCLIDES receiver (Madrid). You can see the focused light on a module glued to an optimal thin fin heat sink made of aluminium.

operated at concentrations up to 5900× and many experiments in the range 500× to 2000× are currently under way.

The secret to withstanding such intensities with neither series resistance problems nor thermal limitations is to use very small solar cells, of area in the range 1–2 mm^2 (Figure 6).

3. DESCRIPTION OF A TYPICAL CONCENTRATOR: COMPONENTS AND OPERATION

A PV concentrator consists of two principal elements: the optical collector and the cell receiver (Figure 7). The collector can be a mirror, a lens, or a combination of both. The cells must be designed for the irradiance level which will be received by the focus from the collector, and they must be thermally bonded to a heat sink in order to remove the energy not converted into electricity. The wires must also be of sufficient thickness to carry the large currents generated. The thermal bond to heat sink must be compatible with a good electrical insulation which is sometimes a difficult compromise.

The collector, receiver and heat sink as a whole are commonly known as the 'concentrator module.' This is the smallest part of the system that

Figure 6 (a) Silicon concentrator cell sized 5 cm in diameter able to operate at 50×. This cell was actively cooled for testing purposes. Project MINER (Madrid, 1982). (b) GaAs solar cell with AlGaAs window made at IES (Madrid, 1995). This device demonstrates that operation over 1000× can be achieved without thermal problems if the cell is sufficiently small (0.1 cm^2, operating at 1290 suns, 23% efficiency).

Figure 7 Concentrator test setup showing the principal subsystems: lens, cell, and heat sink (here actively cooled with water). Note that the cell requires suitable thermal and electrical connectors (IES, 1978).

includes all direct current generating components. The rest of the C-system components are called the *balance of system* (BOS).

The module must, in general, be directed towards the sun in order to cast the focussed direct light beam on the receiver. As a result, the C systems require a mobile structure to point them continuously towards the sun. Concentrator systems, except those with a high acceptance angle, cannot collect diffuse radiation, but this drawback is compensated for by the orientation of the collector towards the sun at all times.

As an example of a concentrator PV System, Figures 8–10 illustrate the principles of the single-axis tracking parabolic trough concentrator, and show the main components.

4. CLASSIFICATION OF CONCENTRATOR SYSTEMS

The classification of concentrator systems is a complex task because of the number of different criteria according to which this can be done: the concentration level, the optical component (lenses or mirror), the

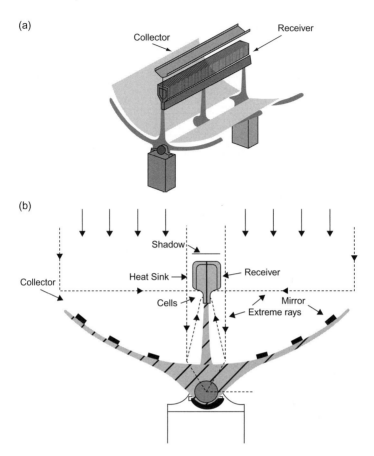

Figure 8 (a) Example of one C array composed by two C modules as used in EUCLIDES system. The collector is a line-focus parabolic mirror. The receiver consists of a heat sink and a linear cell assembly. (b) Frontal view of the EUCLIDES array showing the sun rays reflecting in the collector and striking the receiver.

shape of the optics (point focus, linear, etc.), the cell material or the cell structure (silicon, III−V semiconductors, single-junction, multijunction), or the tracking strategy (two-axis, single-axis, stationary, quasi-stationary, etc.).

Among the large variety of concentrator systems that have been developed for research and demonstration, five specific combinations are sufficient to represent the tendencies and prototypes which are of interest today. These are listed in Table 2, and a picture of each is shown in Figures 10−14. The typical combinations of materials, concentration levels, cells tracking, etc. used for each type are presented in Table 2.

Figure 9 The picture shows a section of a EUCLIDES receiving module before final encapsulation and sealing. (Courtesy of BP Solar Ltd.)

Figure 10 A complete EUCLIDES Module. Output wires emerge from the back allowing continuity at the receiver plane. (Courtesy of BP Solar Ltd.)

The experience during the last 20 years has not been sufficient in determining the level of concentration or which collector or tracking driver might be best. This lack of conclusion is probably due to the scarcity and diversity of experiments that have been carried out. However, the road to achieving competitiveness with the conventional energy sources should be based on a high overall system efficiency in which the cell cost share is below 15% of the total system cost.

The low- and medium-level concentration systems (from $2\times$ to $40\times$) can reach probably this cost objective with inexpensive 20% efficient cells (crystalline silicon cells). The new multijunction and other third-generation cells can reach these goals at over $1000\times$ concentration, if the cell cost

Table 2

Reference concentrator type	Optics	Cell assembly	Cell type	Concentration ratio	Cooling	Tracking
Point focus on a single solar cell	Fresnel lens	One single cell or several cells with spectral beam splitting	Unijunction silicon or unijunction III–V or multijunction	$50 < Xg < 500$ for silicon cells. >500 for all other cells	Passive[a]	Two axis
Large area point focus systems	Big or medium size parabolic dish or central tower power plant	Parquet of cells	Unijunction silicon or unijunction III–V or multijunction	$150 < Xg < 500$	Active	Two axis
Linear systems	Linear lens or parabolic trough	Linear array of cells	Silicon	$15 < Xg < 60$	Passive	One axis for parabolic troughs. Two axis for lenses
Static systems	Non-imaging device	Usually linear array of cells	Silicon	$1.5 < Xg < 10$	Passive	No tracking
Compact minipoint focus systems	Small lens or small parabolic dish or RXI[b] device	One single cell	Unijunction silicon or unijunction III–V or multijunction	$Xg > 800$	Passive	Two axis

[a] Active cooling cannot be discarded in systems with high concentration ratio.
[b] RXI: refractive and reflective with total internal reflection.

Figure 11 Example of a single-axis tracking linear concentrator with a reflective collector. The trough is oriented north to south. EUCLIDES Plant (Tenerife. 1998).

Figure 12 Prototypes of large-area point-focus parabolic dishes. A plant based on this concept will be built in Australia by Solar Systems PTY Ltd. 2002. (Courtesy of Solar Systems Ply Ltd.)

remains, as before, below 15% of total system cost. Currently, there are 34% efficient cells, operating at 400×, but suitable for concentrations up to 2000×. However, their cost is still too high, and there is limited experience with outdoor systems. Notwithstanding, the miniconcentrators which are

Figure 13 An early prototype of a static concentrator using bifacial cells. The acceptance angle is ±23°, thus allowing collection of the sun light along its path for the entire year. Bifacial cells provide twice the concentration level for the same acceptance angle than the monofacial cell (IES, 1978).

Figure 14 XRI concentrator allowing 1000 + x geometric concentration with ±2.5° acceptance angle. It is envisaged that such compact miniconcentrators will provide adequate optics for high-efficiency third-generation microcells (IES, 1994).

used in these very high concentration cells allow the genuine modularity of PV systems. This can be the key to the market penetration of concentrators as they need not compete with the conventional energy generators in the same way as the larger C systems do.

5. TRACKING-CONTROL STRATEGIES

PV concentrators converts only the light which comes directly from the solar disc because only the incident set of rays within a limited solid angle can reach the receiver. The laws of thermodynamics state that the maximum achievable concentration level is linked to the acceptance angle of a concentrator by [5]:

$$C \leq \frac{n^2}{\sin^2\theta} \text{ for point-focus C system} \quad (4)$$

$$C \leq \frac{n}{\sin\theta} \text{ for linear C system} \quad (5)$$

where θ is the half-acceptance angle and n the index of refraction of the transparent material surrounding the solar cell. The actual limits, however, are several times lower for many practical concentrators.

The limited acceptance angle of concentrators requires that the collector is oriented towards the solar disc. According to Equations (4) and (5), the accuracy of such orientation is related to the concentration level and the optical design. As a result, the concentrators must be built on mechanically driven structures which are able to turn the collector to face the sun (Figure 15).

If the collector surface must be normal to the sun, 'two-axis' tracking is required. As an alternative, single-axis strategies achieve good performance with linear reflective concentrators or with low gain linear lenses (Figure 16).

Table 3 shows the energy collected by tracking systems and by a static flat panel tilted at the angle of latitude towards the equator.

Another subject of interest concerns the strategies which can be employed so that the mobile mechanical structure tracks the sun every moment of the day with a given accuracy (for example, under $\pm 0.2°$). Two principal methods are commonly adopted:
- Control system based on direct detection of the sun position.
- Control system based on the theoretical position of the sun according to astronomical equations and time.

The direct method is based on sensors that can generate a null signal when pointing directly towards the sun, and some proportional error signal in case of misorientation [6,7]. A classical servomechanism circuit drives the motors to correct the array position continuously or by steps.

Figure 15 A close-up view of a large tracking driver for a single-axis tracking system. The mechanism can carry up to 2000 N and provide a position accuracy of ±0.05°. The accuracy of the driver is derated by the mechanical deformation of the structure, allowing only ±0.2° (EUCLIDES-THERMIE Plant, Tenerife, 1998).

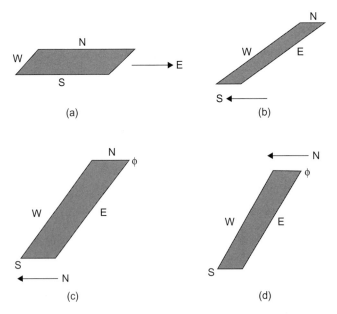

Figure 16 Single-axis tracking strategies: single-axis tracking is only compatible with mirrors, but lenses can be used if geometric concentration is well below 10×.

Table 3 Daily energy collected by an array in Madrid (yearly averages)

Radiation (kWh/m² year)	Static tilted 30° to noon[a]	1 axis E/W tracking	1 axis N/S tracking	Polar axis tracking	Two axis tracking
Global	5.20	5.61	6.24	6.87	7.08
Direct (beam)	—	3.48	4.46	4.90	5.15

[a] Due south in the northern hemisphere and north in the southern hemisphere.

Problems with these systems are encountered on cloudy days, and additional circuits are required to turn back the array during the night and to correct the possible miscalibration of the sensors.

The other principal method is based on the astronomical equations [8, 9] and requires just a good clock signal which can currently be supplied by a GPS connection. Another requirement is an accurate zero reference for each axis and a precise orientation of the axis.

The most onerous requirements of both methods can be eliminated if the electric power output of the system is used as a feedback to adjust and optimise the sun tracking. With such correcting capability the control unit becomes an adaptive system. In this case the concentration array is able to self-learn the sun path, change the values of the circuit variables, and correct errors.

A proposed third tracking method used for very low concentration (2–10X) consists of a thermo-driving mechanism that is at the same time a sensor and a driver. It consists of two metallic cylinders painted black which are located at the extremes of the array. When the array is not normal to sun the cylinders are at different temperature because one of them collects more light, generating a hydraulic force that corrects the orientation until both cylinders are at same temperature. This family of trackers are called *passive*, and those using driving motors are called *active* trackers.

At present, the solar tracking-control circuits and tracking structures are being developed and built principally by prototype makers. However, some companies (INSPIRA, SOLARTRACK, etc.) produce general purpose tracking-control systems and mechanisms. [7,10].

Tracking is considered as a drawback by contractors because
(a) It adds cost.
(b) It reduces system reliability.
(c) It limits the applications (in houses, on roofs, for example).

However, points (a) and (b) will probably become negligible when experience is gained and quality control is applied to these subsystems.

6. APPLICATIONS OF C SYSTEMS

The low (or even non-existent) penetration of C systems in the market 25 years after the demonstration of the first modern prototype shows that concentrator systems are not suitable for the current applications of photovoltaics, isolated professional applications, rural electrification, grid-connected home systems, etc. (see Part V for further examples).

The cost reduction achieved with early concentrators was due to their use in medium and large plants (between 100 kW$_p$ and 2000 kW$_p$). These figures are today considered as very large in terms of PV solar electricity production, and can only be realised in grid-connected plants that must compete with the conventional sources of electricity.

However, the demonstrated cost in the range of 3 euros/W$_p$ which is in line with the government incentives can change this situation in the short term. In many developing countries, the energy cost is high enough to make PV concentrators (with price in the range of 3 euros/W$_p$) competitive in comparison with diesel generators because of their lower maintenance and the price stability. Concentrators are often welcome in developing countries because more than 50% of the total cost is a direct local cost, but simplicity and reliability must be assured.

A new set of applications are beginning to appear based on the new high-efficiency concentrator systems which use multijunction solar cells operating at efficiencies of 30% or more. These arrays are as thin as a conventional flat module, with the same physical aspect and weight. The only drawback is the tracking, but since the 'panels' are so light the systems can easily enter the current modular market of few hundreds watts if reliable and very cheap trackers become available.

Figures 17 and 18 shows the largest and most representative PV concentrators PV plants deployed in the world.

7. RATING AND SPECIFICATION OF PV SYSTEMS

There are no standards and testing methods to rate concentrator cells, receivers, modules and arrays which are generally accepted at present. However, several methods have been adopted and widely used during the last 20 years.

Figure 17 Example of a concentrator array field. Seven of the fourteen parabolic troughs constituting the EUCLIDES demonstration plant installed in Tenerife are shown in this picture (480 kW$_p$). The plant includes a monitoring station and seven inverters, one for each pair of arrays. The project is supported by the JOULE-THERMIE Programme.

Figure 18 Example of an array field: 100 kW system at CSW Solar Park in Texas made by ENTECH Inc. The plant is composed of four arrays. Each array comprises 72 refractive Concentrator Modules with 220 m^2 total aperture. (Courtesy of Entech Inc.)

The rating of concentrator solar cells requires that the cell is connected by thick wires and plates which are able to carry the high currents that flow under concentrated light. At the same time, a heat sink is required to avoid any damage to the cell and to maintain as closely as possible the reference temperature during the test.

The objective of the rating is to obtain the I–V curve of the cell under reference conditions at the nominal or other concentration levels.

To plot the I–V curve under illumination requires a powerful light source, but the sources of uniform, high intensity illumination are scarce and very expensive. The most frequently recommended light source for concentrator cell measurement is a flash bulb. With a high power flash lamp (3000–6000 W) it is possible to illuminate 1 m^2 at 30× or 1 cm^2 at 2500× with an uniformity better than 2%. The lamp spectrum and its linear evolution along its life can be determined. It allows an interpolation of the actual spectrum between the spectra at the beginning of life and at the end of life.

To plot the complete I–V curve, one point per each flash discharge must be recorded. In addition, several I–V pairs can be obtained during the decay of each flash (lasting several milliseconds), and a complete set of curves at different concentration levels can thus be acquired within a single run. The use of a flash instead of continuous light makes the control of cell temperature much easier, and does not required additional precautions.

Such measurements were used to obtain the plots in Figures 2 and 3. These plots should be compulsory to specify any solar cell for C applications.

The test of single-junction cells for practical use is not highly sensitive to the precise details of the light spectrum. The error in the figures obtained with natural sunlight or with a flash instead of a correct reference spectrum (whatever this may be) is always under 3%. Furthermore, if one assumes the linearity of the short-circuit current as a function of the total power incident on the cell, the concentration level of any measured I–V curve is given simply and accurately by the ratio $C = J_{sc}(C)/I_{sc}$ (1).

On the other hand, the rating of multijunction cells requires a careful knowledge of the light spectrum impinging on the cell (note that optical elements modify the incoming spectrum), and also the spectral response of each cell in the stack. The calibration of multijunction cells thus requires specialised laboratories.

7.1 Rating of C Modules

The C modules include the optical collector, the receiving cells and the heat sink. Except in the case of very small modules (for example, mini-concentrators or microcells), the amount of uniform light necessary to illuminate the whole collector requires outdoor testing, and consequently sun tracking. The test should be carried out with the collector normal to sun on a clear day with a direct irradiance higher that 750 W/m². (If this condition is fulfilled but there are thin clouds in the sky, do not test during such days: you will waste your time!)

Once the module is installed onto a two-axis tracker, the following parameters must be measured and calculated for the correct and complete characterisation of the module (see Tables 4(a, b, c)).

The calibration of an array is identical to the calibration of a module except that the array cannot point to the sun if the system does not have two-axis tracking, as in the case of a horizontal parabolic trough or a static concentrator. In these cases, Table 4 remains valid if we substitute for B (W/cm²) the quantity:

$$B_{eff} - B \cdot \cos \gamma \qquad (6)$$

where γ is the angle between the actual direct beam and the normal to the collector surface.

Table 4(a) Parameters to measure

Symbol	Name and units	Testing equipment
B	Direct irradiance (W/m²)	Pyrheliometer
T_{amb}	Ambient temperature (°C)	Thermometer
T_{cell}	Cell temperature in operation (°C)	Thermocouple[a]
$I_{sc}(C)$	Short circuit current under concentration C at 1 sun (A)	Ammeter
$V_{oc}(C)$	Open-circuit voltage under concentration C at T_{cell} (V)	Voltmeter
$I-V(C)$ curve	Complete I–V curve under concentration C at T_{cell} (I. V) pairs	I–V curve tracer (transient or continuous)

[a]Cell temperature can be deduced from V_{oc} if you have information of V_{oc} and dV/dT at given cell irradiance and temperature.

Table 4(b) Parameters calculated from measurement. Parameters required prior to concentrator measurements

Calculated parameter	Measured data and calculating formulas	Measurement conditions
$P_{max}(C)$	Maximum power at T_{cell}	At B and T_{cell}
η = module overall efficiency	$P_{max}(C)/(B \times A_c)$	At B and T_{cell}
η_{ap} = optical efficiency	$I_{sc}(C)/(X \times I_{sc}(1))$	At B
η_{rec} = receiver efficiency	Overall efficiency/optical efficiency = η/η_{op}	At B and T_{cell}

Table 4(c) Parameters required previously to concentrator measurements

Symbol	Name and units	Test conditions
Ac	Collector area (cm^2)	—
$I_{sc}(1)$	Receiver short circuit current at 1 sun (A)	At B
$V_{oc}(1, T_{ref})$	Open circuit voltage at known cell temperature and illumination (V)	At B and T_{cell}
R_{th}	Thermal resistance dependence vs. wind (provided by the manufacturer or measured indoors) (°C W^{-1} cm^2)	At B and T_{cell}
R_s	Cell series resistance (Ω)	At B and T_{cell}

The efficiency of linear or static concentrator arrays can be different due to the different performance of the optical elements under different angles of incidence of sunrays. As a result, any comparative measurements must be carried out at an identical angle of incidence of the light.

Furthermore, due to possible deformation of the structure supporting the modules, the efficiency can vary with array orientation. This fact must also be taken into account if a complete picture of the system performance under all real operating conditions is required, rather than only with the system 'centred at noon.'

The effect of structure deformation on the optical elements decreases the current at the maximum power point of the I–V curve. This effect is known as an 'optical mismatch' and is caused by the unequal illumination of the cells in a series connection or by the dispersion of the I–V characteristics of the receiver.

7.2 Specifications That Must Be Required From C Systems

A PV concentrator is a complex system made of very different components: optical materials, a mechanical structure, control circuits, cells, receivers, heat sinks, insulators, wires, connectors, etc. Each component requires a set of specifications, but an approved and accepted list of minimum required specifications does not yet exist. The same can be said about the qualification of C modules, systems, and subsystems.

Currently, the IEEE has approved a qualification procedure for receivers and concentration modules (IEEE P1513) that will be probably adopted by IEC (IEC 62108). In Europe, a project (entitled C-RATING) with support from the European Commission aims to prepare a draft on specifications, testing methods and modelling for cells, modules, and systems (see www.ies-def.upm.es/c-rating/). In addition, an initiative to develop qualification procedures for tracking-control units and tracking drivers is about to start. However, it will take about four years before any qualification method can be accepted worldwide.

In the meantime, standards, specifications, and qualification tests approved for flat modules, with suitable modifications, are being used as a compromise.

8. ENERGY PRODUCED BY A C SYSTEM

The prediction of the energy produced by a concentrator should, in principle, be easier than for a flat panel because the irradiance distribution during the day for a concentrator is more uniform because of sun tracking. However, concentrators use only the direct sunlight, and the availability of historical data for direct beam irradiance (W/m^2) and radiation (kWh/m^2 day) is limited. There are computer codes and models that generate direct irradiance data based on global radiation, hourly average global irradiance, etc. Other data come from an analysis of meteorological satellite images. All these models are complicated to use and have not been verified.

One approximation that gives good results is based on the knowledge of two numbers: (W_{day}), the monthly averaged direct daily radiation on a tracking array (two-axis, single-axis north or south, etc.) in kWh/m^2 day, and (S), averaged number of sunshine hours (as recorded with a Campbell-Stokes sunshine recorder).

A simple and effective model can then be obtained by assuming that the concentrator receives a uniform and constant irradiance equal to

$$B_{\it{eff}} = \frac{\textit{Total direct daily radiation}}{\textit{Sun hours} - 2} = \frac{(W_{day})}{(S) - 2} \qquad (7)$$

during the (S) − 2 hours around the solar noon.

Another simple method consists of assuming that the beam irradiance normal to the sun is always 800 W/m^2. The operating time is then given by:

$$S(hours) = \frac{\textit{Total direct daily radiation}}{0.8 \ kW/m^2} = \frac{(W_{day})}{0.8 \ kW/m^2} \qquad (8)$$

which is also centred around noon.

Regarding ambient temperature, it can be assumed that it follows a daily oscillatory variation related to the horizontal solar radiation but delayed by 2 hours relative to noon. As a result, the average daylight ambient temperature is given by

$$\langle T_{amb} \rangle = \langle T_{av} \rangle + 2 \frac{\langle T_{max} \rangle - \langle T_{av} \rangle}{\pi} \quad (9)$$

where $\langle T_{max} \rangle$ and $\langle T_{av} \rangle$ are the maximum and average daily temperatures.

Using such site data and the system specifications outlined in Table 4, one can calculate the expected daily averaged I–V curve of the array for any day from the I–V curve at reference conditions by using the following formulas. The average cell temperature for each day is

$$\langle T_{cell} \rangle = \langle T_{amb} \rangle + R_{th} \cdot B_{\mathit{eff}} \cdot \eta_{op}(1 - \eta_{rec}) \quad (10)$$

$$I_{day}(model) = I(ref) \cdot \frac{C(model)}{C(ref)} \quad (11)$$

$$V(model) = V(ref) + (E_G - V(ref))\left(1 - \frac{T_{model}}{T_{ref}}\right)$$
$$+ \frac{kT_{ref}}{q} \ln \frac{C(model)}{C(ref)} - R_s(I(model) - I(ref)) \quad (12)$$

where E_g is the band gap of the cell material, q the electronic charge, and η_{rec} the receiver efficiency.

Once the corrected I–V curve is known, the power output of the array can be calculated for any moment of time. The energy produced in a day is then:

$$P_{day}\left(\frac{kWh}{m^2 \, day}\right) = P_{max}(model) \cdot (\langle S \rangle - 2) \quad (13)$$

Figure 19 shows a table of measured reference data and model results for a set of given site values. The measured and corrected I–V curves are shown (courtesy of Project C-RATING, EC Thermie Programme).

9. THE FUTURE OF CONCENTRATORS

High conversion efficiency of PV systems is the key to low cost. It is currently accepted that the progress of technology based on the learning curve of conventional cell technology cannot reach the goal of 0.1 euros per kWh. Thus, innovative effort is necessary to increase the cell efficiency.

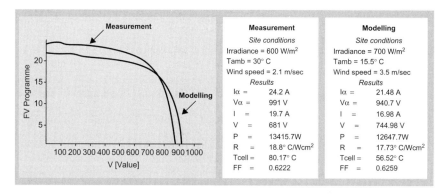

Figure 19 The I–V values, measured for given site conditions, can be used to deduce the I–V curve for any other set of conditions, by using Equations (9) and (11). (Results obtained by the code for modelling concentration systems within the project C-RATING, EC 5th FV Programme.)

In recent years, third-generation cells are being mentioned with the promise of conversion efficiencies in the range of 40%; efficiencies of 34% have already been demonstrated experimentally. The required efficiency can therefore be reached, but the use of such highly expensive cells requires operation under high concentration levels, in excess of 1000×. This can be compared with demonstrated efficiencies of 24 and 20% which have been accomplished using the second generation cells (usually based on silicon) and systems, respectively. The total cost, including BOS, is then in the range of 3 euros/W_p, and compatible with the energy cost objectives.

In summary, two scenarios of PV progress have been outlined, based on the use of concentrators. It is not certain, however, what arrangement will be better for large-scale applications, or if one single type of concentrator is optimal for all applications and climatic regions.

REFERENCES

[1] E.G. Boes, A. Luque, Photovoltaic concentrator technology, in: T.B. Johansson, H. Kelly, A.K.N. Reddy, R.H. Williams (Eds.), Renewable Energy: Sources for Fuel and Electricity, Earthscan, London, 1993.

[2] G. Sala, PV concentrator alternatives towards low energy costs, Proceedings of the PV in Europe: From PV technology to energy solutions, Rome, 2002.

[3] E.T. Zirkle, R.C. Dondero, C.E. Backus, D.V. Schorder, The superlinear behaviour of short-circuit current in silicon concentrator cells at concentrations up to 1400 suns, Proceedings of the Seventh European Photovoltaic Solar Energy Conference, Seville, 1986.

[4] G.L. Araújo, J.M. Ruiz, Variable injection analysis of solar cells, in: A. Luque (Ed.), Solar Cells and Optics for Photovoltaic Concentration, Adam Hilger, Bristol and Philadelphia, 1989, pp. 32—74.
[5] W.T. Welford, R. Winston, The Optics of non-imaging Concentrators, Academic Press, New York, 1978.
[6] J.A. Castle, F. Ronney, 10 kW$_p$ photovoltaic concentrator system design, Proceedings of the Thirteenth IEEE Photovoltaic Specialists Conference, Washington DC, 1978, pp. 1131—1138.
[7] A. Luque, G. Sala, A. Alonso, J.M. Ruiz, J. Fraile, G.L. Araújo, J. Sangrador, M.G. Agost, J. Eguren, J. Sanz, E. Lorenzo, Project of the "Ramón Areces" concentrated photovoltaic power station, Proceedings of the Thirteenth IEEE Photovoltaic Specialists Conference, Washington DC, 1978, pp. 1139—1146.
[8] A.B. Maish, M.L. O'Neill, R. West, D.S. Shugar, Solartrack controller developments for today's application, Proceedings of the Twenty-fifth IEEE Photovoltaic Specialist Conference, Washington DC, 1996, pp. 1211—1214.
[9] J.C. Arboiro, G. Sala, A constant self-learning scheme for tracking system, Proceedings of the Fourteenth European Photovoltaic Solar Energy Conf., Barcelona, 1997, pp. 332—335.
[10] G. Sala, I. Antón, J.C. Arboiro, A. Luque, E. Camblor, E. Mera, M.P. Gasson, M. Cendagorta, P. Valera, M.P. Friend, J. Monedero, S. Gonzalez, F. Dobon, The 480 kW$_p$ EUCLIDES™-THERMIE power plant: Installation, set-up and first results. Proceedings of the Sixteenth European Photovoltaic Solar Energy Conference, Glasgow, 2000, pp. 2072—2077.
[11] I. Anton, R. Solar, G. Sala, D. Pachón, IV testing of concentration modules and cells with nonuniform light patterns, Proceedings of the Seventeenth European Photovoltaic Solar Energy Conference, Munich, 2001, pp. 611—614.
[12] M. Yamaguchi, A. Luque, High efficiency and high concentration in photovoltaics, IEEE Trans. on Electron Devices ED-46(10) (1999).

CHAPTER IID-2

Operation of Solar Cells in a Space Environment

Sheila Bailey[a] and Ryne Raffaelle[b]

[a]Photovoltaic and Space Environments Branch, NASA Glenn Research Center, USA
[b]National Center for Photovoltaics, National Renewable Energy Laboratory (NREL), 1617 Cole Blvd., Golden, CO 80401-3305

Contents

1. Introduction — 863
2. Space Missions and their Environments — 865
   2.1 The Air Mass Zero Spectrum — 865
   2.2 The Trapped Radiation Environment — 867
   2.3 Solar Flares — 868
   2.4 The Neutral Environment — 869
   2.5 The Paniculate Environment — 870
   2.6 Thermal Environment — 870
3. Space Solar Cells — 872
   3.1 Radiation Damage in Space Solar Cells — 873
4. Small Power Systems — 875
5. Large Power Systems — 877
References — 879

1. INTRODUCTION

The beginning of the Space Age brought about a perfect application for the silicon solar cell developed at Bell laboratory in 1953. Sputnik was battery powered and remained active only a little over a week. The US launched the first successful solar powered satellite, Vanguard 1, seen in Figure 1, on March 17, 1958 [1]. The solar powered transmitter lasted six years before it is believed that the transmitter circuitry failed.

Vanguard I had eight small panels with six p on n silicon solar cells, each 2 cm × 0.4 cm, connected in series. Each panel output was approximately

Practical Handbook of Photovoltaics.
© 2012 Elsevier Ltd. All rights reserved.

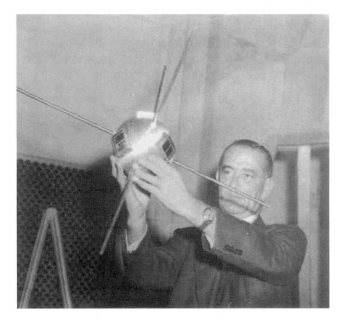

Figure 1 3/12/1957, Senator Lyndon B. Johnson, chairman of the US Senate Preparedness subcommittee, holds the tiny 6.5-inch American test satellite.

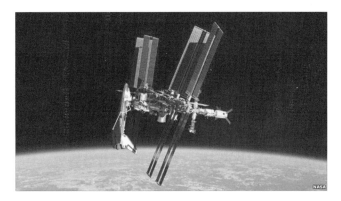

Figure 2 The International Space Station (compare shuttle for size).

50 mW with a cell efficiency of ~8%. This can be contrasted with the International Space Station (ISS), see Figure 2, which has the largest photovoltaic power system ever present in space, with 262,400 n on p silicon solar cells, each 8 cm × 8 cm, with an average efficiency of 14.2% on 8

US solar arrays (each ~34 m × 12 m) [2]. This can generate about 110 kW of average power, which after battery charging, life support, and distribution, can supply 46 kW of continuous power for research experiments on ISS.

2. SPACE MISSIONS AND THEIR ENVIRONMENTS

Space missions are defined by their trajectories. For Earth orbiting missions these are roughly classified as low Earth orbit (LEO), 300–900 km, mid-Earth orbit (MEO), and geosynchronous (GEO), 35,780 km. The orbit's size, defined by the semimajor axis; shape, defined by the eccentricity; orientation, defined by the orbital plane in space (inclination and right ascension of the ascending node); and the orbit within the plane, defined by the argument of perigee, determine the space environment the spacecraft will encounter. NASA missions also involve interplanetary flight both toward and away from the Sun, planetary fly-bys, and orbiting other planets, each with their own unique set of environments. In addition both the moon and Mars may be sites for future human visits and have their own individual conditions for surface power.

At the heart of our solar system is the Sun, which is both the source of the solar irradiance which a solar cell converts to electricity and the solar wind which is primarily a stream of protons and electrons moving with a mean velocity by Earth of ~400–500 km/s with a mean density of approximately $5/cm^3$. In addition, the Sun is a dynamic body exhibiting facula, plages, spicules, prominences, sunspots, and flares over time. The only other source of radiation is galactic cosmic rays which emanate from beyond our solar system. These consist of about 85% protons, about 14% alpha particles, and about 1% heavier nuclei [3]. The differential energy spectra of the cosmic rays near the Earth tend to peak around 1 GeV/nucleon and the total flux of particles seen outside the magnetosphere at the distance of the Earth from the Sun (i.e., 1 AU) is approximately 4 per square centimeter per second.

2.1 The Air Mass Zero Spectrum

The spectral illumination that is available in space is not filtered by our atmosphere and thus is quite different from what is incident on Earth's surface (see Figure 3). Space solar cells are designed and tested under an Air

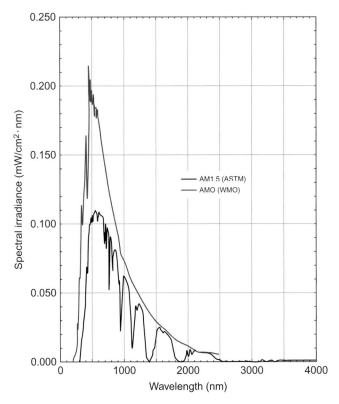

Figure 3 The Air Mass Zero (AMO) spectrum (WMO) and the Air Mass 1.5 (ASTM) spectrum.

Mass Zero (AMO) spectrum. This is in contrast to an Air Mass 1.5 as reduced by 1.5 times the spectral absorbance of the Earth's atmosphere, which is the standard condition for testing terrestrial solar cells. Thus, cells intended for use in space will be optimized for a somewhat different spectrum. The change in spectral distribution will typically result in a decrease in overall cell efficiency, even though the intensity of light is somewhat higher (i.e., 1367 W/m^2 in space as compared to 1000 W/m^2 on Earth). A 12% efficient silicon solar cell as measured under AM1.5 condition on Earth would translate into an approximately 10% cell as measured under AM0.

As seen below the Sun resembles a blackbody with a surface temperature of 5800 K with a peak of spectrally emitted energy at 480 nm. Approximately 77% of the emitted energy lies in the band from 300 to 1200 nm. The total energy received from the Sun per unit area perpendicular to the Sun's rays at the mean Earth–Sun distance (1 AU) is called the solar constant. The current accepted value of the solar constant is 1367 W/m². The solar intensity of course varies in time. However since space solar cells are calibrated in near-space, the variation in the value of the solar constant primarily affects the predicted solar cell operational temperature in orbit.

2.2 The Trapped Radiation Environment

The solar wind, solar flares and galactic cosmic rays all consist of charged particles (electrons, protons, and ions). These interact with a planetary magnetic field. Some planets have a very weak magnetic field or no magnetic field. Jupiter has a very large magnetic field. Jupiter, Saturn, and Uranus have trapped radiation belts. The Earth's magnetic field is 0.3 gauss at the surface on the equator and does change over time even reversing polarity every 10,000 years. The Earth's magnetic poles do not coincide with the poles determined by the axis of rotation, with approximately an 11 degree difference. The total magnetic field of the magnetosphere is determined by the internal magnetic field of the planet and the external field generated by the solar wind. These interact with each other and provide the complex asymmetric pattern of the geomagnetic cavity. Charged particles gyrate around and bounce along magnetic field lines, and are reflected back and forth between pairs of conjugate mirror points in opposite hemispheres. At the same time electrons drift eastward around the Earth while protons and heavy ions drift westward. These regions of trapped charged particles are called the Van Allen belts. An illustration of the regions of trapped particles can be seen in Figure 4, where L is a dimensionless ratio of the Earth's radius equal to the radial distance divided by the cos 2Λ, where Λ is the invariant latitude.

An example of the number of trapped particles as a function of energy for both Earth and Jupiter can be seen in Figure 5 [4]. The Jupiter data were provided by Pioneers 10 and 11 and Voyagers 1 and 2 during their encounters with the planet.

The models that are used for the trapped electron and proton environment at Earth were developed by the US National Space Science Data Center at NASA's Goddard Space Flight Center from available radiation measurements from space. The most recent models in use, AP8 for protons [5] and AE8 for electrons [6], permit long-term average predictions of trapped particle fluxes encountered in any orbit and currently constitute the best estimates for the trapped radiation belts, although they have been noted to overestimate the radiation in certain low Earth orbits.

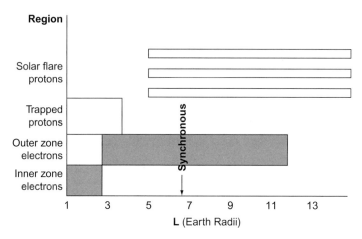

Figure 4 Regions of trapped particles as a function of distance in Earth radii from the Earth's centre.

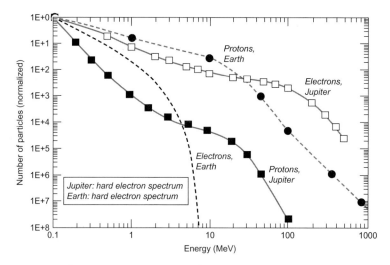

Figure 5 Normalized energy spectrum of trapped electron and proton radiation environment at Jupiter, compared to Earth.

2.3 Solar Flares

Solar activity, as measured by the number of sunspots, follows an 11-year cycle between maxima. The cycle has an active 7-year period during which solar flare events are probable and a quiescent 4-year period during which solar flare events are rare. The last peak in activity occurred in

2000. The most recent model, JPL 91, allows the spacecraft designer to predict the proton integral fluence as a function of confidence level and exposure time [7]. The exposure time must be correlated to the solar maxima. The calculated integral fluence is at 1 AU from the Sun. For missions at other radial distances from the Sun the fluence should be modified by $1/R^2$ to produce the most probable estimate. These solar flare proton events are associated with coronal mass ejections on the Sun. They occur at highly localized places on the Sun and rotate with the Sun. Because they follow the field lines of the interplanetary magnetic field, only when the ejection occurs along the line from the Sun that intersects the Earth will the protons propagate immediately to the Earth, arriving approximately an hour after the flare. These protons are highly anisotropic and therefore variations in proton flux can be as large as 100 from the same flare at different points of the orbit. Protons can also reach Earth when ejection occurs away from a field line but then the protons must diffuse through the solar corona before they propagate to Earth. This takes longer, up to 10 hours, and the flux tends to become isotropic. The trapped proton population is also significantly effected by solar activity, particularly in lower Earth orbits. The increased energy output of the Sun expands the atmosphere and increases proton densities. There is a region of trapped particles close to the surface of the Earth called the South Atlantic Anomaly. For low altitude, low inclination orbits the South Atlantic Anomaly may be the most significant source of radiation.

The electron environment is also influenced by the solar cycle. Magnetic solar storms can raise the electron flux in the outer orbits by an order of magnitude over a short time period. This short-term variation in electron flux is usually more significant for spacecraft charging effects in GEO. AP8 and AE8 mentioned above have data for both solar maximum and solar minimum. A more recent model based on the CRRES spacecraft launched in July 1990 found differences between the AP8 and AE8 predictions to be as high as a factor of 3 orders of magnitude particularly at low altitudes due to the highly variable proton belts [8]. At high altitudes AE8 predictions are typically higher than CRRES predictions.

2.4 The Neutral Environment

The density, composition, pressure, and temperature change dramatically as a function of altitude. Orbits lower than 200 km are generally not stable due to atmospheric drag. At 300 km only proportionately 23% of the sea-level molecular nitrogen remains and only 10% of molecular oxygen. 80% of the atmosphere at 300 km is highly reactive atomic oxygen. Atomic oxygen erodes polymers and composites that might be used in array substrates and also silver interconnects between solar cells [9]. The atmospheric density is very small above 800 km.

2.5 The Particulate Environment

The particulate environment is composed of both naturally occurring meteoroids and man-made space debris. The particles of most concern to space arrays are between 10^{-3} and 10^{-6} g, since those below 10^{-6} g do not have sufficient energy to cause significant damage and those above 10^{-3} are less frequently encountered [10]. Meteoroids, whose origin is either asteroids or comets, have an average velocity of 20 km/s and their density varies with the Earth's position around the Sun. Debris has of course become more problematical as the number of launched spacecraft and their relative time in orbit has increased. The relative damage of orbital debris, except for very large objects, is less due to the reduced difference in orbital velocities for the debris that was created in that orbit. The flux of meteoroids and orbital debris has been observed for a variety of orbits [11]. Damage to the solar arrays of Mir and the Hubble Space Telescope were noted in primarily the erosion and cracking of the cover glass on the array and the erosion of the substrate rear surface thermal control coating.

2.6 Thermal Environment

The temperature of a solar cell in space is largely determined by the intensity and duration of its illumination [12]. In the case of the US array on the ISS, the operating temperature of the silicon solar cells is as high as 55°C while under illumination, and as low as −80°C when in eclipse. Similarly, as spacecraft venture further away from the Sun their average temperatures will decrease. Likewise, as they move closer to the Sun their average temperatures will rise. The average illuminated temperature at the orbit of Jupiter is −125°C, whereas at the average orbit of Mercury the temperature is 140°C.

The orbital characteristics of a space mission are also a major source of thermal variation for the associated photovoltaic arrays. The relative amount of illumination versus eclipse time and the rate of change in the temperature vary dramatically will the orbital path. The orbital path will also affect the fraction of incident solar radiation returned from a planet or albedo. The average albedo from the Earth is 0.34, but can range anywhere from 0.03 (over forests) to 0.8 (over clouds) [7].

The available power generated by a solar cell is directly related to its operating temperature. An increase in temperature will result in a reduction in output power. Although there will be a slight increase in the short-circuit current with increasing temperature, it will be overshadowed by the decrease in the open-circuit voltage. A GaAs solar cell will experience a decrease of about 0.05 mW/cm^2 per °C.

The degradation of solar cell performance as a function of temperature is expressed in terms of temperature coefficients. There are several different temperature coefficients used to describe the thermal behavior of solar cells. They are based in terms of the change in a characteristic cell measurement parameter (i.e., I_{sc}, V_{oc}, I_{mp}, V_{mp}, or η) as a function of the change in temperature. The difference in the measured value at the desired temperature and a reference temperature is used to determine the coefficient. The International Space Organization (ISO) standard reference measurement is taken at 25°C. For most space solar cells, the change in output is fairly linear from -100°C to 100°C.

Temperature coefficients are often expressed as a normalized number. For example, if the case of the efficiency temperature coefficient the normalized value would be expressed as

$$\beta = \frac{1}{\eta}\frac{d\eta}{dT} \qquad (1)$$

or the fractional change in efficiency with temperature. Representative temperature coefficients for the various types of cells used in space are given in Table 1. The temperature coefficient is inversely related to band gap and negative for the majority of space solar types.

Table 1 Measured temperature coefficients for various types of solar cells used in space [26]

Cell Type	Temp (°C)	η (28°C)	$1/\eta d\eta/dT$ ($\times 10^{-3}$°C^{-1})
Si	28–60	0.148	-4.60
Ge	20–80	0.090	-10.1
GaAs/Ge	20–120	0.174	-1.60
2-j GaAs/Ge	35–100	0.194	-2.85
InP	0–150	0.195	-1.59
a-Si	0–40	0.066	-1.11 (non-linear)
CuInSe$_2$	-40–80	0.087	-6.52

3. SPACE SOLAR CELLS

The first 30 years of space solar cell development focused on the of silicon solar cells, although it was known even in the early days that better materials existed [13]. The concept of a tandem cell was also proposed in the early days to enhance the overall efficiency. An optimized three-cell stack was soon to follow with a theoretical optimum efficiency of 37% [14]. However, it was 40 years later before a multijunction solar cell flew in space. Today silicon cells still fly in space but the cell of choice is a multijunction solar cell. Table 2 shows some of the best efficiencies for small area cells and the comparison to an AM0 efficiency.

A variety of cell types are listed in Table 2 because, while commercial satellites use silicon or dual or triple junction GaInP/GaAs/Ge. there is a marked interest in military applications of thin film cells. NASA also has planned missions in which a large specific power (kW/kg) and lower cost would be beneficial. The advantages of thin film solar cells are their large specific power when deposited on a flexible, lightweight substrate with a

Table 2 Measured Global AM1.5 and measured or *estimated AM0 efficiencies for small area cells

Cells	Efficiency (%) Global AM1.5	Efficiency (%) AM0	Radio AM0/ AM1.5	Area (cm²)	Manufacturer
c-Si	22.3	21.1	0.95	21.45	Sunpower [15]
Poly-Si	18.6	17.1*	0.92	1.0	Georgia Tech/ HEM[16]
c-Si film	16.6	14.8*	0.89	0.98	Astropower [17]
GaAs	25.1	22.1*	0.88	3.91	Kopin [17]
InP	21.9	19.3*	0.88	4.02	Spire [17]
GaInP(1.88ev)	14.7	13.5	0.92	1.0	ISE [18]
GaInP/GaAs/Ge	31.0	29.3	0.95	0.25	Spectrolab [18]
$Cu(Ga,In)Se_2$	18.8	16.4*	0.87	1.04	NREL [15]
CdTe	16.4	14.7*	0.90	1.131	NREL [15]
a-Si/a-Si/a-SiGe	13.5	12.0	0.89	0.27	USSC [15]
Dye-sensitized	10.6	9.8*	0.92	0.25	EPFL [15]

*These are based on cells measured under standard conditions, courtesy or Keith Emery, NREL. The calculated efficiency uses the ASTM E490−2000 reference spectrum and assumes that the fill factor does not change for the increased photocurrent. Quantum efficiencies corresponding to the table entries were used in the calculations.

suitably lightweight support structure. Thin film solar cells are currently lower in efficiency and require a larger area for the same power levels; however, trade studies have identified several potential applications [19].

3.1 Radiation Damage in Space Solar Cells

Radiation degradation in space is a complex issue. The degradation is dependent on the type of particle, energy and fluence, shielding, and cell design (layer thickness, number of junctions, etc.). In addition ground based radiation measurements use monoenergetic, unidirectional beams of particles (electrons or protons) and the simulation of the space solar environment, especially for multijunction cells, is difficult. Historically the Jet Propulsion Laboratory (JPL) has provided the format for determining radiation damage in silicon and gallium arsenide space solar cells [20,21]. The elements needed to perform degradation calculations are degradation data under normal 1 MeV electron irradiation, the effective relative damage coefficients for omnidirectional space electrons and protons of various energies with various cover-glass thickness, and the space radiation data for the orbit of interest. As discussed in the section on the trapped radiation environment, AP8 and AE8 are current NASA models for trapped radiation. The models were based on observations from 43 satellites from 1958 to 1970 for AP8 and from 1958 to 1978 for AE8. They can return an integral or differential omnidirectional flux for a set of energies. The integral flux is the number of particles with energy greater than or equal to the input energy. The models calculate a numerical derivative of the integral flux to obtain the differential flux. For AP8 the energy range is 0.1 to 400 MeV with McIlwain L number ranging from 1.1 to 6.6. For AE8 the energy range is 0.04 to 7 MeV with an L number from 1.1 to 11. The models calculate a numerical derivative of the integral flux to obtain the differential flux. As mentioned in the section on solar flares the models permit a choice of solar maximum or solar minimum. AP8 and AE8 provide the largest coverage for Earth orbiting spacecraft and are internationally available. Other models exist with narrower coverage: the US Air Force model from the CRRES data, an ESA model based on the SAMPEX spacecraft, and a Boeing model based on TIROS/NOAA satellites.

In recent years the Naval Research Lab (NRL) has developed a model of displacement damage dose based on the nonionizing energy loss (NIEL) [22]. The NIEL gives the energy dependence of the relative damage coefficients, and because the NIEL is a calculated quantity, the NRL method enables the analysis of a solar cell response to irradiation by a spectrum of particle energies, as encountered in space, based on only one or two ground measurements. The Solar Array Verification and Analysis Tool (SAVANT) [23,24] computer program being developed at NASA Glenn Research Center combines the NRL method with the NASA space environment models to produce a user-friendly space solar array analysis tool. Equator-S and COMETS satellite data have been analyzed using SAVANT. In the Equator-S mission, the model was successful in predicting the onboard degradation of both GaAs/Ge and $CuInSe_2$ solar cells. This is the first time that the model has been applied to a thin film technology. SAVANT and the onboard measurements agreed to within a few percent over the entire mission. The COMETS mission used GaAs/Ge solar cells as its main power source, and SAVANT accurately modeled the power output of the arrays for the bulk of the mission lifetime [25]. SAVANT is not currently available to the public.

Normalized power as a function of altitude for solar cells in a 60° orbit for 10 years with 75 μm cover glass can be seen in Figure 6. Note that by using normalized power the higher radiation resistance of $CuInSe_2$ can be seen. Also, the larger particle flux can be noted for MEO orbits.

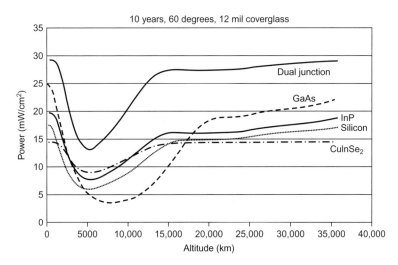

Figure 6 Normalised power versus altitude for a 60° orbit for 10 years with 75-μm cover glass. *(Graph courtesy of Tom Morton, Ohio Aerospace Institute.)*

4. SMALL POWER SYSTEMS

There has been considerable development over the last several years in the development of small spacecraft. Many so-called microsats or satellites whose total mass is less than 100 kg have been deployed. In fact, satellites whose total mass is less than 10 kg (i.e., nanosats) and even satellites weighing less that 1 kg (i.e., picosat) have already been tested in the space environment (see Figure 7). Consequently, the demands on the space power community to develop appropriately sized power systems for these new classes of satellites have arisen.

It is true that a premium has always been placed on the efficiency of space power systems and specifically the photovoltaic arrays. Specific power (W/kg) or power per mass is one of the most important figures of merit in judging a power system. The higher the specific power, the less the spacecraft mass that has to be dedicated to the power system and the more that can be used for the scientific mission. This is especially true in the case of a small power system. There is an economy of scale savings that can sometimes be recouped on a larger satellite. As power components are reduced in size, so too is there capacity. In the case of a solar cell this translates into their ability to gather light.

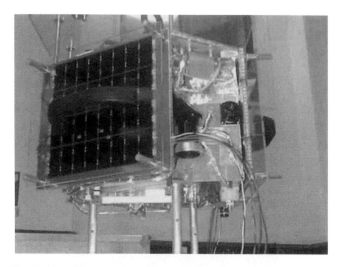

Figure 7 The SNAP-1 Surrey Nanosatellite Applications Platform was a 6-kg satellite with imager and propulsion. *(Picture courtesy of NASA and Surrey Satellite Technology Ltd.)*

A number of approaches have been used to meet the demands of small satellites. One such approach is the development of integrated power supplies. These supplies combine both power generation and storage into single devices. In fact, NASA, the Naval Research Labs, and others are working to develop monolithically grown devices that combine monolithically interconnect module (MIM) solar cells or micro-sized solar arrays with lithium-ion energy storage. A first demonstration of this

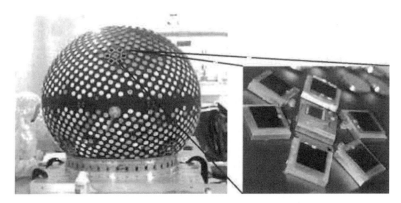

Figure 8 Photograph of Starshine 3 satellite with a magnified region of the integrated power supply.

Figure 9 Aerospace Corporation PowerSphere nanosatellite. *(Picture courtesy of the Aerospace Corporation.)*

concept, although not truly monolithic, was flown on the Starshine 3 satellite in 2002 (see Figure 8) [26].

Another approach is to integrate the photovoltaics to the satellite is such a way as to ensure light absorption. One method of this approach is to have photovoltaics incorporated to the skin of the spacecraft. This return to body mounted panels is very similar to the way the first small satellites where powered back in the space programs infancy. Another method is to tether a spherical array to the small spacecraft. The primary example of this approach is the so-called power sphere concept developed by Aerospace Corporation in collaborations with ILC Dover and others (see Figure 9).

5. LARGE POWER SYSTEMS

On the other end of the spectrum from nanosats in terms of the size of photovoltaic arrays used is the proposed development of Space Solar Power (SSP) systems. The intent is to develop systems that are capable of generating up to gigawatts of power. The proposed uses of these systems have been such things as beaming power to the Earth, Moon, or Mars or even to serve as an interplanetary refueling station. These type of large power systems may play a key role in future manned missions to Mars. Several different concepts have been proposed, but they all have the common element of an extremely large area of solar cells. The proposed systems employ solar arrays which have a total area in the neighborhood of several football fields. One such SSP concept is the NASA Sun Tower shown in Figure 10.

The largest space solar array that has been deployed to date is the United State Solar Array which is being used to power the International Space Station (ISS) (see Figure 2). As completed the ISS is powered by 262,400 (8 cm $\times$ 8 cm) silicon solar cells with an average efficiency of 14.2% on 8 US solar arrays (each ~ 34 m $\times$ 12 m) [2]. This will generate about 110 kW of average power. An additional 20 kW of solar power is also provided by arrays developed by Russia.

Another example of large solar power systems which although are not truly in space but share many of the same requirement are high-altitude airships and aerostats. Lockheed Martin with ITN Energy Systems, Linstrand Balloons Ltd, and others are developing high altitude airships

Figure 10 Sun tower. *(Picture courtesy of NASA.)*

Figure 11 An artist's conception of the Linstrand Bulloons Ltd proposed high altitude airship. *(Picture courtesy of Linstrand Balloons Ltd.)*

that incorporate large array solar arrays to produce power (see Figure 11), An airship with a surface area on the order of $10,000\ m^2$ would only need a small portion of its surface to be covered by solar cells to achieve a daytime power production over 100 kW.

REFERENCES

[1] R.L. Easton, M.J. Votaw, Vanguard IIGY Satellite (1958 Beta), Rev. Sci. Instrum. 30 (2) (1959) 70−75.
[2] L. Hague, J. Metcalf, G. Shannon, R. Hill, C. Lu, Performance of International Space Station electric power system during station assembly, Proceedings of the Thirty-first Intersociety Energy Conversion Engineering Conference, 1996, pp. 154−159.
[3] E. Stassinopoulos, J. Raymond, The space radiation environment for electronics, Proc. IEEE 11 (1988) 1423−1442.
[4] S. Kayali, Space radiation effects on microelectronics, NASA Jet Propulsion Laboratories Course. See material on JPL web site at: <http://nppp.jpl.nasa.gov/docs/Raders_Final2.pdf/>.
[5] D.M. Sawyer, J.I. Vette, AP8 trapped proton environment for solar maximum and solar minimum, Report NSSDA 76−06, National Space Science Data Center, Greenbelt, MD, 1976.
[6] J.I. Vette, AE8 trapped electron model, NSSDC/WDC-A-R&S 91−24, National Space Science Data Center, Greenbelt, MD, 1991.
[7] J. Feynman, G. Spitale, J. Wang, S. Gabriel, Interplanetary Proton Fluence Model: JPL 1991, J. Geophys. Res. 98(A8) (1993) 13281.
[8] H.B. Garrett, D. Hastings, The Space Radiation Models, Paper No. 94−0590, 32nd AIAA Aerospace Sciences Meeting, Reno, Nevada, 1994.
[9] R.C. Tennyson, Atomic oxygen and its effect on materials, In: The Behaviour of Systems in the Space Environment, Kluwer Academic, 1993, pp. 233−257.
[10] Solar Array Design Handbook, vol. 1, JPL, 1976.
[11] TRW Space & Technology Group, TRW Space Data, N. Barter (Ed.), Fourth ed., 1996.
[12] A.L. Fahrenbruch, R.H. Bube, Fundamentals of Solar Cells, Academic Press, Boston, 1983, Chapter 2.
[13] J.J. Loferski, Theoretical considerations governing the choice of the optimum semiconductor for the photovoltaic solar energy conversion, J. Appl. Phys. 27 (1956) 777.
[14] E.D. Jackson, Areas for improvement of the semiconductor solar energy converter. Trans. Conference On the Use of Solar Energy Tucson, Arizona, 1955, Vol. 5, p. 122.
[15] K. Bücher, S. Kunzelmann, The Fraunhofer ISE PV Charts: Assessment of PV Device Performance, Report EUR 18656 EN, Joint Research Center, 1998, pp. 2329−2333.
[16] M. Green, K. Emery, K. Bücher, D. King, S. Igari, Solar cell efficiency tables (version 11), Prog. Photovolt: Res. Appl. 6 (1998) 35−42.
[17] M. Green, K. Emery, D. King, S. Igari, W. Warta, Solar cell efficiency tables (version 18), Prog. Photovolt: Res. Appl. 9 (2001) 87−293.
[18] R. King, M. Haddad, T. Isshiki, P. Colter, J. Ermer, H. Yoon, et al., Metamorphic GaInP/GaInAs/Ge solar cells, Proceedings of the Twenth-eighth IEEE Photovoltaic Specialist Conference, Anchorage, 2000, pp. 982−985.
[19] D. Murphy, M. Eskenazi, S. White, B. Spence, Thin-film and crystalline solar cell array system performance comparisons, Proceedings of the Twenty-ninth IEEE Photovoltaic Specialists Conference, 2000, Anchorage, pp. 782−787.
[20] H. Tada, J. Carter, B. Anspaugh, R. Downing, Solar Cell Radiation Handbook, third ed., JPL Publication, 1982, pp. 82−69.
[21] B. Anspaugh, GaAs Solar Cell Radiation Handbook, JPL Publication 96−9, 1996.
[22] G.P. Summers, E.A. Burke, M.A. Xapsos, Displacement damage analogs to ionizing radiation effects, Radiation Measurements 24(1) (1995) 1−8.

[23] S. Bailey, K. Long, H. Curtis, B. Gardner, V. Davis, S.Messenger, et al., Proceedings of the Second World Conference on Photovoltaic Solar Energy Conversion, Vienna, 1998, pp. 3650—3653.
[24] T. Morton, R. Chock, K. Long, S. Bailey, S. Messenger, R. Walters, et al., Techical Digest 11th International Photovoltaic Science and Engineering Conference, Hokkaido, 1999, pp. 815—816.
[25] S. Messenger, R. Walters, G. Summers, T. Morton, G. La Roche, C. Signorini, et al., A displacement damage dose analysis of the COMETS and Equator-S space solar cell flight experiments, Proceedings of the Sixteenth European Photovoltaic Solar Energy Conference, Glasgow, 2000, pp. 974—977.
[26] P. Jenkins, T. Kerslake, D. Scheiman, D. Wilt, R. Button, T. Miller, et al., First results from the starshine 3 power technology experiment, Proceedings of the Twenty-ninth IEEE Photovoltaic Specialists Conference, New Orleans, 2002, pp. 788—791.

CHAPTER IID-3

Calibration, Testing, and Monitoring of Space Solar Cells

Emilio Fernandez Lisbona
ESA-Estec, Noordwijk, The Netherlands

Contents

1. Introduction	882
2. Calibration of Solar Cells	883
2.1 Extraterrestrial Methods	883
2.1.1 High-Altitude Balloon	*883*
2.1.2 High-Altitude Aircraft	*884*
2.1.3 Space Methods	*884*
2.2 Synthetic Methods	884
2.2.1 Global Sunlight	*884*
2.2.2 Direct Sunlight	*885*
2.2.3 Solar Simulator	*885*
2.2.4 Differential Spectral Response	*886*
2.3 Secondary Working Standards	886
3. Testing of Space Solar Cells and Arrays	886
3.1 Electrical Tests	887
3.1.1 Electrical Performance	*887*
3.1.2 Relative Spectral Response	*889*
3.1.3 Reverse Characterisation	*890*
3.1.4 Capacitance Characterisation	*890*
3.2 Environmental Tests	891
3.2.1 Radiation Testing	*892*
3.2.2 Ultraviolet Radiation	*893*
3.2.3 Atomic Oxygen (ATOX)	*893*
3.2.4 Thermal Cycling	*894*
3.2.5 Vacuum	*894*
3.2.6 Micrometeoroids	*894*
3.2.7 Electrostatic Discharge	*895*
3.2.8 Humidity	*895*
3.3 Physical Characteristics and Mechanical Tests	895
4. Monitoring of Space Solar Cells and Arrays	895
4.1 Flight Experiments	895
4.2 Monitoring of Solar Array Performance in Space	901
4.3 Spacecraft Solar Array Anomalies in Orbit	901

Practical Handbook of Photovoltaics.
© 2012 Elsevier Ltd. All rights reserved.

	4.3.1	European Communication Satellite (ECS) and Maritime European	
		Communication Satellite (MARECS)	901
	4.3.2	X-Ray Timing Explorer (XTE)	905
	4.3.3	CPS Navstars 1–6	905
	4.3.4	Pioneer Venus Orbiter SA	905
4.4	Postflight Investigations on Returned Solar Arrays		906
	4.4.1	Hubble Space Telescope Solar Array 1	906
	4.4.2	EURECA	907
	4.4.3	MIR Solar Array	908
Acknowledgements			908
References			908

1. INTRODUCTION

Solar energy is the main power source technology for most spacecraft since the 1960s. A total failure of the solar array (SA) performance will lead to complete mission loss. SA behaviour in the space environment has to be predicted in order to assure endurance during mission life.

The SA electrical performance is a basic parameter that needs to be predicted for mission life, tested on the ground, and monitored continuously in space. Electrical performance (EP) at beginning of life conditions is measured on the ground to check power output prediction, based on performance measurements of single solar cells and before their integration on the SA. These measurements are performed with solar simulators, having adjusted their light intensity to standard AMO illumination conditions with suitable reference solar cells. Reference cells are space calibrated using different methods that will be described in Section 2.

Endurance of the SA to the space environment has to be simulated by ground environmental testing. Different mechanical and environmental tests, together with electrical tests for degradation assessment, are performed at the different steps of development, manufacture, and integration of SA components and intermediate assemblies. An overview of these tests is given in Section 3, mainly focused on tests at solar cell levels.

Monitoring the performance of the SA in orbit is essential to validate the predicted behaviour during the mission, and this provides valuable data for verification of ground testing and further SA design improvements. Section 4 deals with the monitoring of spacecraft SA in orbit. Flight experiments are conducted to assess the performance and behaviour in space of novel solar cell or SA integration technologies. Two other important sources of data are unpredicted anomalies in orbit and investigations carried out on SAs returned from space.

2. CALIBRATION OF SOLAR CELLS

Standard solar cells are used to set the intensity of solar simulators to standard illumination conditions, in order to electrically characterise solar cells with similar spectral response. Space calibration methods of solar cells can be extraterrestrial when performed outside the atmosphere or synthetic if they are carried out on the ground, using natural sunlight or indoor simulated illumination [1]. To prevent continuous handling operations of the expensive extraterrestrial/synthetic cells, so-called secondary working standard solar cells are calibrated for routine electrical performance testing in industry and testing laboratories.

2.1 Extraterrestrial Methods

Two calibration methods are the main suppliers of extraterrestrial standards: the high-altitude balloon and the high-altitude aircraft. Both methods require minimum data correction due to the small residual air mass at the altitude where the calibration is performed.

2.1.1 High-Altitude Balloon

Calibrations are performed on board stratospheric balloons flying at altitudes of around 36 km, where the illumination sun conditions are very close to AM0. Cells to be calibrated are directly exposed to the sun, mounted on supports with sun trackers. Currently, two institutions, JPL-NASA in the USA [2] and CNES in France [3,4], are conducting, on a yearly basis, these calibration campaigns. The main differences between the two calibration institutes are the position of the cells, which in the case of JPL-NASA is mounted on the balloon apex and in the case of CNES, is a gondola hanging from the balloon. Both institutes correct calibrated data, taking into account the effect of temperature and the variation of illumination due to the Earth—Sun distance variation over the year. CNES also corrects its calibrated data, taking into account the effect of the residual atmosphere.

2.1.2 High-Altitude Aircraft

Calibrations are performed on board of an aircraft capable of flying at altitudes of 15—16 km. Cells are mounted at the end cap of a collimating tube on a temperature controlled plate. NASA Glenn Research Centre is currently conducting more than 25 flights per calibration campaign using a Gates Learjet 25 equipped even with a spectroradiometer to measure the solar spectrum at that altitude. Data are corrected for the ozone absorption, the geocentric distance and extrapolated to the air mass value of zero [5].

2.1.3 Space Methods

The most realistic environment on which calibration of solar cells can be performed is indeed outside the atmosphere. The first constraint of these methods is their relatively high cost compared with the other two extraterrestrial methods and their lower level of maturity.

- Space shuttle: On board the space shuttle, the Solar Cell Calibration Experiment (SCCE) was conducted in two flights in 1983/84, where solar cells from different agencies, institutions and space solar cell industries around the world were calibrated and returned back to Earth [6].
- Photovoltaic Engineering Testbed: This is a NASA-developed facility flown in the International Space Station, where after exposure and calibration of cells in the space environment, they are returned back to Earth for laboratory use [7].
- Lost Twin: This is an ESA-proposed method, based on the flight of several solar cells on a nonrecoverable spacecraft. Cells nearly identical to the flight ones are kept on Earth. The orbiting cells are calibrated and these calibrated values are given to their respective twin cells.

2.2 Synthetic Methods

There are two methods carried out under natural sunlight conditions.

2.2.1 Global Sunlight

The cells to be calibrated and a pyranometer are placed on a horizontal surface, where simultaneous readings of spectral irradiance over the sensitivity range of the pyranometer and short-circuit current of the cells are recorded in global sunlight. The calibration site environmental conditions need to fulfil several requirements relating to global and diffuse irradiance levels, solar elevation, unobstructed view over a full hemisphere, etc. The calibrated short circuit current of the cell is calculated by means of the following formula:

$$I_{sc} = I_{sg} \frac{\int (k_2 E_{g\lambda}) d\lambda}{E_{glob}} \frac{\int (k_1 S_\lambda) E_{s\lambda} d\lambda}{\int (k_1 S_\lambda)(k_2 E_{g\lambda}) d\lambda}$$

where $k_1 S_\lambda$ is the absolute spectral response of the cell, $k_2 E_{g\lambda}$ the absolute spectral irradiance of the sun at the calibration site, $E_{S\lambda}$ the AM0 spectral irradiance, E_{glob} the pyranometer irradiance reading, and I_{sg} the measured short circuit current of the cell.

The final calibration value is the average of three calibrations of three different days. The former RAE (UK) performed for several years global sunlight calibrations at Cyprus [8] and presently INTA-SPASOLAB (Spain) is performed on a yearly basis in Tenerife [9,10].

2.2.2 Direct Sunlight
The cells to be calibrated are placed on the bottom plate of a collimation tube, a normal incidence pyrheliometer and a spectroradiometer are kept pointing to direct sunlight while measurements of short-circuit current, total irradiance and spectral irradiance are recorded. Several conditions need to be fulfilled by the calibration site and its environment, i.e., certain irradiance level, stable cell short-circuit readings, ratio of diffuse to direct irradiance, etc. The calibrated short circuit current of the cell is calculated by means of the following formula:

$$I_{sc} = \frac{I_{sd} \int E_{d\lambda} d\lambda \int E_{s\lambda} S_\lambda d\lambda}{E_{dir} \int E_{d\lambda} S_\lambda d\lambda}$$

where I_{sd} is the measured short circuit current, E_{dir} is the total solar irradiance, $E_{d\lambda}$ is the spectral solar irradiance, $E_{s\lambda}$ is the AM0 spectral irradiance, and s_λ is the relative spectral response of the cell to be calibrated.

The calibrated short circuit current value is the average of three calibrations performed in three different days. CAST (China) presently performs calibrations following this method [11].

The following two methods are carried out under simulated sunlight.

2.2.3 Solar Simulator
The cell to be calibrated is illuminated by means of a steady-state solar simulator adjusted to 1 AM0 solar constant with a previously calibrated cell or a suitable detector. The spectral irradiance of the solar simulator is measured with a spectroradiometer, and the relative spectral response of

the cell is measured separately. The calibrated short circuit current of the cell is calculated as follows:

$$I_{sc} = I_{sm} \frac{\int Es\lambda s_\lambda d\lambda}{\int Em\lambda s\lambda d\lambda}$$

where I_{sm} the short circuit current and $E_{m\lambda}$ the spectral irradiance, both measured under the solar simulator. NASDA (Japan) regularly performs calibrations following this method [12].

2.2.4 Differential Spectral Response

The calibrated short circuit current of the cell is calculated with its absolute spectral response together with the reference AM0 solar spectral irradiance. The absolute spectral response is obtained as follows: first, the relative spectral response of the cell to be calibrated and then for certain wavelengths the absolute differential spectral response, is determined by the ratio of the cell short-circuit current to irradiance measured by a standard detector. This method was developed and is frequently presently used by PTB (Germany) for solar cell calibration [13].

2.3 Secondary Working Standards

Secondary working standard (SWS) solar cells are used to set intensity of solar simulators to standard conditions for routine measurements of identical (same spectral response) solar cells during acceptance or qualification testing. For the EP characterisation of SA, panels or coupons, SWSs are preferred for reference. SWSs are calibrated using standards obtained by the methods defined above and a continuous or pulsed light source. The measured data are corrected by means of the spectral response of both cells and the spectral irradiance of the light source and the standard AM0 spectrum, following the spectral mismatch correction method [14]. This secondary calibration method also gives relations between calibrated solar cells by different methods [15].

3. TESTING OF SPACE SOLAR CELLS AND ARRAYS

In order to assess the behaviour of solar cells and solar arrays for a specific space mission or environment, several tests need to be conducted at different hardware levels and phases of a project.

- Solar cells:
  - Development: To know their performance, their endurance to the space environment and therefore decide on the most appropriate solar cell candidate for the the specific application.
  - Design: Measured solar cell data at different environmental conditions is necessary for an accurate power prediction during the mission and therefore a suitable sizing of the solar array.
  - Qualification: Verify that the solar cells manufactured in the production line meet a set of requirements defined by the specific space mission [17].
  - Acceptance: To provide cell performance and physical data: essential for their further integration in the solar array electrical network.
- Higher levels of solar array components integration: The so-called photovoltaic assemblies (test specimens with all the components existing and integrated as in the solar array electrical network) are also tested in the development and qualification phases.
- Solar array level: Tests are performed in development phases and in the qualification phase of the flight hardware. These tests are required to see whether or not the solar array is integrated with the spacecraft body.

Tests on solar cells and solar arrays can be split in three types: Electrical, Environmental, and Mechanical/Physical characteristics. The following sections deal with these types of tests, focusing chiefly for their application to solar cells assemblies (SCAs); however, when relevant, their application to higher levels of solar array integration or other solar array components is described.

3.1 Electrical Tests

3.1.1 Electrical Performance

The objective of this test is to assess the corresponding electrical parameters of the solar cells and to provide data for solar generator design. The electrical current of solar cells under 1 Solar Constant AM0 equivalent illumination shall be measured and recorded at a certain voltage. A solar cell test set up consists basically of a continuous or pulsed light source, a load connected across the cell's terminals and electrical current and voltage measurement equipment. During the measurement, the temperature of the cell junction is kept at a constant temperature (25°C) and a four-point probe measurement of the cell is used in order to minimise the effects of lead and contact resistances [16].

Solar simulators need to meet certain requirements on their light beam spectrum, uniformity, and stability for optimum EP measurements of photovoltaic devices [1,17]:
- Spectrum: Maximum allowable deviations of spectral energy in certain wavelength regions of the standard AM0 spectrum define the solar simulator spectral quality classification. The spectral irradiance is measured with spectroradiometers [18] or special filtered solar cells [19].
- Uniformity: Uniformity of the irradiance on the test area is a critical parameter for accurate measuring of panels or SA.
- Stability: The light beam stability has to be maintained under certain values, especially when no simultaneous correction is done when measuring the EP.

Continuous or pulsed light sources are used to simulate solar illumination in laboratories or test facilities:
- Continuous solar simulators are mostly based on xenon short arc lamps where the beam is filtered and collimated to achieve the previously mentioned requirements. They are mainly used for the electrical characterisation of solar cells and small coupons. Large-area continuous solar simulators based on argon discharge lamps are used to electrically characterised solar cells or panels [20]. Multisource solar simulators are required for measuring multijunction (Mj) solar cells, in order to set equivalent AM0 illumination conditions on each subcell [21].
- Pulsed solar simulators are based on xenon large arc lamps where the beam usually is not filtered to meet the above requirements on the test plane. Either solar cells or large panels can be electrically characterised, being not heated during the test, but special techniques are needed for measuring slow response cells [22,23]. When measuring Mj solar cells, a better matching of the AM0 is needed, precisely in the near infrared spectral range, where xenon large arc lamps have less radiant energy [24].

Reference cells, either primary or secondary standards, are used to set the intensity of solar simulators to standard illumination conditions. For Mj solar cells, either so-called component cells (Mj cell structures with only one active junction) [25] or methods based on mismatch factor are followed to set standard illumination conditions on each cell junction [26].

Under standard illumination conditions and constant temperature the current voltage curve of the photovoltaic device is traced by polarising at different voltages. The shape and magnitude of the I−V curve depends on the junction characteristics, shunt and series resistance, and on total radiant energy converted, regardless of wavelength composition [27].

Table 1 Typical EP parameters and temperature coefficients of some space solar cells. Abbreviations: Sj = single junction; Dj = double junction; Tj = triple junction

Solar cell technology	I_{sc} (mA/cm^2)	V_{oc} (mV)	P_{max} (mW/cm^2)	η (%)	$dIsc/dT$ (mA/cm^2/°C)	$dVoc/dT$ (mV/°C)	dP_{max}/dT (mW/cm^2/°C)
Si BSR	37.0	595	17.5	13.0	0.02	−2.20	−0.080
Si BSFR	39.0	610	19.0	14.0	0.03	−2.00	−0.075
Sj GaAs/Ce	32.0	1030	26.5	19.5	0.02	−1.85	−0.050
Dj GaInP/GaAs/Ge	16.3	2350	31.5	23.0	0.01	−5.50	−0.065
Tj GaInP/GaAs/Ge	16.5	2560	41.5	26.0	0.01	−6.50	−0.085

However, for Mj solar cells, wavelength composition of the radiant energy affects the shape of the I−V curve [28].

Temperature coefficients of solar cell electrical parameters can be calculated from experimental data, by measuring the device EP at different temperatures [29,30] (Table 1).

The solar cell EP behaviour under different angles of incidence is of most importance for SA designs with curved substrates and operation of planar SA at high tilt levels. The potential angle of incidence-dependent effects are the cosine function, Fresnel reflectivity, cover-glass coatings and filters, solar cell multilayer antireflecting coating, extreme angle effects, and end-of-life (EOL) behaviour. Assessments of these effects for each SCA component combinations are needed for SA performance prediction [31].

3.1.2 Relative Spectral Response

Relative Spectral Response is the short-circuit current density generated by unit of irradiance at a particular wavelength as a function of wavelength. Relative spectral response provides valuable data for improving solar cells under development, for the calculation of performance measurement errors, and for solar simulator verifications. It is measured by illuminating with a narrow bandwidth (monochromator or narrow band filters) light source (pulsed or continuous) the solar cell [32], at different wavelengths in its sensitivity range, while measuring the cell short circuit current and the irradiance with a sensor. A cell with known spectral response can be used as reference, replacing the irradiance sensor [1].

To measure spectral response of Mj solar cells, each junction needs to be characterised separately by light biasing (filtered light or variable

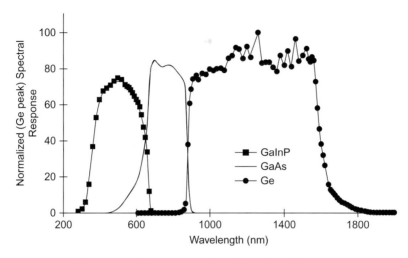

Figure 1 Spectral response of a proton-irradiated Tj solar cell.

intensity lasers) of the nonmeasured junctions and by voltage biasing, to measure in short circuit conditions the subcell junction under test [33] (Figure 1).

3.1.3 Reverse Characterisation

The reverse voltage behaviour of solar cells is needed for the prediction of shadowing and hot-spot phenomena on solar cell strings. Reverse-biased cells may experience excessive heating, minor permanent loss of power output, or permanent short-circuit failure [34]. Generally, single and Mj gallium arsenide solar cells are more sensitive to reverse bias than silicon cells [35] as seen in Figure 2, requiring the insertion of by-pass diodes on each cell for effective protection. Testing apparatus and procedures are similar to the EP ones, but current and power limitations are needed to avoid cell breakdown.

3.1.4 Capacitance Characterisation

The dynamic behaviour of solar cells may introduce specific requirements on the subsequent solar array regulator. Therefore, the capacitance of solar cells needs to be characterised following two different methods:
- Small signal or frequency domain method: This is the measured high-frequency impedance around a certain bias point. Tests are performed with voltage biasing and in darkness [36].

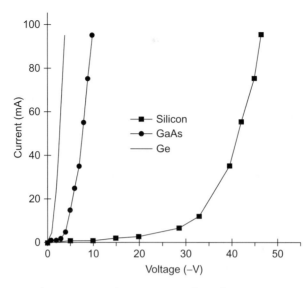

Figure 2 Reverse characteristics of some space solar cells.

- Large signal or time domain method: The rise of solar cell voltage between two operational points gives the solar cell capacitance by applying the formula $C = I_{sc}(t_2 - t_1)/(V_2 - V_1)$ [37], where t_2, t_1 and V_2, V_1 are the time and voltages associated with these operational points.

3.2 Environmental Tests

Environmental tests are performed to check solar array endurance to the different surroundings to which it is exposed during its complete lifetime. The most damaging environments are depicted here:

- Ground operations: Solar arrays are exposed to possible physical damage during manufacturing, integration, handling, and transportation activities. During long storage periods, solar array components maybe corroded by humidity. Tests are performed at component, solar cell, and SCA levels.
- Launch: Vibration, shocks, acceleration, and acoustic fields affect the solar array in this phase, producing high mechanical stress levels that could produce physical damage either just after testing or in orbit. Vibration, shock, and acoustic tests are performed usually at higher levels of solar array integration; panel, wing, and spacecraft level.

- Space: Particles, temperature, vacuum, and micrometeoroids are the main factors degrading solar arrays in space. Each factor affects different solar array components and interfaces. Tests are mainly conducted at solar cell, SCA level, and coupon level.

EP and visual inspection tests are performed before and after exposure of photovoltaic devices to any environmental tests. The main environments affecting solar array performance are described in the following sections in more detail.

3.2.1 Radiation Testing

The radiation environment in space basically comprises electrons and protons of different spectral energies. Solar cells are permanently damaged by these particles; displacement damage is produced in the cells' crystalline structure, reducing the minority carrier diffusion length and lifetimes in the cells' base region, driving a degradation of the cells' electrical parameters. For medium- and high-radiation environment missions, solar cell particle degradation is the key parameter for solar array sizing. Cover glasses and adhesives can be darkened by radiation reducing the array performance, by transmission losses and operational temperature increases.

Two methods are followed to predict the performance of solar cells under the space radiation environment: JPL method based on reducing all proton/electron energies from a certain space environment to an equivalent normal incidence and mono-energetic irradiation, usually 1 MeV electrons [38−40] and NRL model based in the displacement damage dose methodology [41] (see Figure 3).

Solar cell radiation testing is performed on solar cell or SCAs at electron and proton irradiation facilities:

- Electrons are produced by Van der Graaff generators. Typical electron energies range from 0.6 up to 2.5 MeV and flux between 10^9 up to 1.5×10^{12} $e^-/cm^2/s$. Cells are irradiated under vacuum or inert gas conditions.
- Low-energy protons (<2MeV) are produced by hydrogen ionising chambers and mass separators. Tandem Van der Graaff generators produced protons with energy from 2 MeV to 10 MeV and cyclotrons and synchrocyclotrons from 10 MeV to 50 MeV and 50 MeV to 155 MeV, respectively. Cells are always irradiated under vacuum conditions.

In general, the crystalline damage and performance degradation of irradiated solar cells is not stable for certain types. Recovery or further degradation phenomena are observed after annealing at temperatures higher

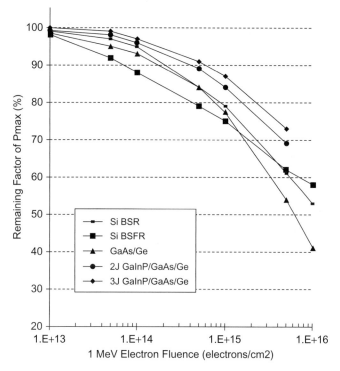

Figure 3 Power degradation of space solar cells under 1 MeV electron particles.

than 20°C and exposure to sunlight [42,43], suggesting performance of photon irradiation and annealing testing after particle irradiation.

3.2.2 Ultraviolet Radiation

Ultraviolet radiation can darken certain types of coated solar cell cover glass and adhesives, reducing the sunlight transmission to the solar cells, increasing solar array temperature and therefore lowering its EP [44]. Cracks on solar cell cover glasses may increase cell current degradation by 2% more for EOL [45]. Tests are conducted in vacuum chambers on cover glasses, SCAs, or solar array coupons, at solar cell operational temperature in orbit, using UV light sources based on xenon arc, high-pressure mercury, or halogen arc discharge lamps [46].

3.2.3 Atomic Oxygen (ATOX)

For low-Earth orbits (between 180 km and 650 km), the presence of ATOX is a main cause of erosion of silver solar cell interconnectors [47,48] and the kapton foil glued to the support structure outer layer [49]. ATOX durability testing on components or solar array coupons is performed in plasma asher chambers, being air raw material that becomes a plasma of atomic oxygen and other particles [50]. ATOX dose is determined using uncoated kapton samples whose erosion is known from flying data.

3.2.4 Thermal Cycling
The temperature cycling experienced by solar arrays in orbit (eclipses) is the cause with time of fatigue cracking of harnesses, bus-bars, interconnector material, and interconnector solder/weld joints [51], and also the cause of increased series resistance in the solar cell and interconnector interface. Temperature cycling tests are performed following two methods on either solar array coupons or SCAs:
- Vacuum: So-called *thermal vacuum* or *vacuum thermal cycling* provides a good simulation of the space environment, but not of the temperature rate decay in orbit [52].
- Ambient pressure: Fast temperature change rates are achieved with circulating inert gas chambers. Cost and test duration are considerably reduced compared to vacuum chambers, but failures tend to happen earlier than with vacuum chambers [53].

3.2.5 Vacuum
A space vacuum might vaporise metals (Mg, Cd, and Zn) and also volatile materials like adhesives. Thermal vacuum is a standard test performed at component and up to solar array level, for endurance testing of components and interfaces. Chambers as described in [52] are commonly available in the space photovoltaic industry and test houses. Failures coming from wrong manufacturing process or contamination of materials are also quickly revealed with these tests [54].

3.2.6 Micrometeoroids
More frequent impacts from micrometeoroids and space debris (between 10^{-6} and 10^{-3} g) mainly erode cover glass and solar array exposed coatings, with small solar array performance degradation due to optical losses. Predictions are in agreement with in-orbit degradation [55] and permanent loss of solar array sections by impacts on harnesses, though these are rare [56]. Hypervelocity impacts of particles are simulated with plasma drag accelerators [57,58] and light gas guns [59].

3.2.7 Electrostatic Discharge

Dielectric solar array surfaces, mainly solar cell cover glass and kapton layers, are subject to electrostatic charging due to geomagnetic substorm activity or by the spacecraft surrounding plasma. Subsequent sudden electrostatic discharge (ESD) effects may permanently damage solar array components [60]. Cover glasses are coated with conductive coatings (i.e., ITO) and grounded [61] to lighten charging and to give an equipotential surface for scientific field measurements. Tests are conducted at coupon level to check adequacy of components and interfaces [62] and at component level for the survival of the conductive coating to the mission environment [17].

3.2.8 Humidity

Accelerated humidity/temperature testing of solar cells is conducted to check the stability of solar cell contacts and anti-reflection coatings for long storage periods [63]. GaAs solar cells with AlGaAs window layers are submitted to this test in order to assure the effective protection of the antireflection coating to the corrosion of this window layer [64,65].

3.3 Physical Characteristics and Mechanical Tests

Several tests are presented in this section, not only tests to check mechanical characteristics as adhesion of coatings, contacts or interconnectors, but also measurements of some physical characteristics needed for solar array sizing or essential inputs to other solar array analysis (Mass Budget, Thermal Analysis, etc). A summary is depicted in Table 2.

4. MONITORING OF SPACE SOLAR CELLS AND ARRAYS

4.1 Flight Experiments

Several flight experiments have been conducted with solar cells/coupons in order to verify their endurance to the space environment and ground radiation testing assessment. Most of the flight experiments measure main electrical parameters of the cells and coupons (I_{sc}, V_{oc}, and power at certain voltage, full I–V curve), sun aspect angle, and the operational temperature. In Table 3 some of the most relevant recent flight experiments are listed together with the publication reference, dates of data

Table 2 Mechanical/physical characterisation tests on space solar cells

Test name or phys./charact.	Purpose	Test method	Requirements
Visual inspection	Find solar cell or component obvious defects	Unaided eye or low magnification ($5 \times - 10 \times$)	Several defects are not allowed at component level, relaxation criteria for higher levels of integration exist.
Interconnector adherence	Interconnector weld and cell contact adhesion	Pull test	Maximum pull force value and breakage mode.
Coating adherence	Contact and coatings adhesion	Tape peel test	Percentage of delaminated area below certain value.
Solar cell dimensions	External dimensions and contacts disposition	Microscope	Maximum dimensions provided by solar array electrical network or cell manufacturers.
Solar cell weight	Data for solar array weight budget	Balance	Maximum weight provided by solar array or cell manufacturers.
Solar cell flatness	Solar cell flatness	Profile microscope	Maximum bow provided by solar array electrical network manufacturer.
Contact thickness	Data for interconnector integration	X-ray spectroscopy	Maximum thickness provided by solar array electrical network or cell manufacturers.

Contact surface roughness	Data for interconnector integration	Roughness tester
Hemispherical emittance	Emitted energy by the cell	Infrared spectrophotometer	Maximum thickness provided by solar array electrical network or cell manufacturers. Maximum emittance provided by solar array or cell manufacturers.
Solar absorptance	Absorbed/incident energy to the cell	Solar spectrometer	Maximum absorptance provided by solar array or cell manufacturers.

Table 3 Summary of flight experiments

Experiment	Ref.	Dates	Orbit	Cell/coupon types	Main conclusions for each cell/coupon type
Equator-S	[66].	Dec. 97, May 98	Equatorial	1. GaAs/Ge	1. Degradation according to modelling,
			500 km/	2. MBE Dj GaInP$_2$/GaAs/GaAs	2. High radiation tolerance.
			67,000 km	3. UT[a] GaAs	3. High radiation tolerance—thick cover.
			High	4. Si NRS[b]/BSF	4. Confirms ground radiation-testing data.
			radiation	5. CIGS	5. Improvement performance by light soaking effect in orbit. Degradation according to modelling.
				6. CIS	6. Low energy protons heavily damaged uncovered cells. Degradation according to modelling.
PASP-Plus	[68–70]	Aug. 94, Aug. 95	70° Elliptic	1. Si Planar & ISS[c]	1. Less degradation than predicted with modelling (*).
			362 km/	2. GaAs	2. Degradation according to modelling (*).
			2552 km	3. Dj AlGaAs/GaAs	3. GaAs degradation according to modelling.
			High	4. Dj GaAs/CIS	4. GaAs degradation less than predicted with modelling.
			radiation	5. InP	5. Less degradation than predicted with modelling.
				6. a-Si	

ETS-VI (SCM)	[71]	Nov. 94, Jun. 96	Elliptic 8550 km/ 38,700 km High radiation	1. Sj GaAs/Si 2. GaAs/GaAs 3. Si BSFR 4. Si BSR	6. Positive P_{max} temp/coeff. with increasing temp. 1. and 2. GaAs/Si cells more resistant to radiation damage than GaAs/GaAs. 3. and 4. BSR are more resistant to radiation than BSFR, as in ground tests.
ETS-V (SCM)	[72]	Sep. 87, Sep. 97	GEO 150°E	1. GaAs LPE 2. GaAs MOCVD 3. Si BSFR 4. Si BSR 5. Si NRS/BSFR	
EURECA (ASGA)	[73]	Aug. 92, Jun. 93	LEO 510 km Circular	1. GaAs/Ge MOCVD 2. GaAs LPE 3. GaAs MOCVD	1. GaAs/Ge cells showed higher operation temperature than GaAs/GaAs. 2. No degradation of solar cells during the flight confirmed with postflight ground measurements. 3. Postflight analysis studied the effects on coupon components (interconnectors) of LEO environment (Large number of thermal cycles and ATOX).

(continued)

Table 3 (continued)

Experiment	Ref.	Dates	Orbit	Cell/coupon types	Main conclusions for each cell/coupon type
LIPS-III	[74–77]	May 87, Aug. 93	LEO 1100 km 60° circular	1. Si BSR and BSFR 2. a-Si:H 3. CuInSe$_2$ 4. GaAs MOCVD and LPE 5. Sj AlGaAs/GaAs	1. BSR more resistant than BSFR. Degradation according to modelling. 2. Photodegradation main cell degradation mechanism (40% in power). 3. Extremely high radiation resistant. 4. and 5. Degradation according to modelling.
UoSAT-5	[78]	Jul. 91, Jul. 95	770 km polar Sun-sync. Low radiation	1. InP 2. ITO/InP 3. GaAs/Ge 4. Si High Eta	1. Small degradation according to modelling. 2. Anomalous degradation in voltage. 3. Small degradation according to modelling. 4. Higher degradation than 1. and 3. but fits models.

For other abbreviations and standard solar cell terms see text or Chapter Ia-1.
*Parasitic current collected by these coupons correlates with ground testing and prediction models.
[a] UT = ultra thin.
[b] NRS = nonreflective.
[c] Silicon cells of the International Space Station Array

acquisition, orbit (apogee/perigee), cell/coupon types and main conclusions achieved.

Some flight experiments are, at the time of this writing, in preparation:
- Mars array technology experiment (MATE): Several solar cell technologies shall be sent to Mars surface for checking their performance and endurance, together with instrumentation for the Mars surface (sun spectrum, dust, temperature, etc.) characterisation [79].
- Concentrator solar cell array technology flight experiment: Assessment of the performance of reflective concentrators with Mj solar cells [80].

4.2 Monitoring of Solar Array Performance in Space

Monitoring of solar arrays in space is mainly needed to verify that their performance meets the spacecraft power requirements for planned operations and that the design performance predictions for the complete mission are met. Reliable preflight data based on ground performance measurements, solar cell qualification tests, and power budget calculations, based on qualification tests, are needed initially for an accurate performance evaluation in orbit. For flight data acquisition, temperature sensors, operational and short circuit sensors, and operational and open circuit voltage sensors are required, together with precise attitude and orbit data. Their quantity and precision drives the flight data quality [81]. Flight data are converted to standard conditions (1 Solar Constant and 25°C) for comparison with predicted data. In-orbit performance of some recent spacecraft is shown in Table 4, which includes relevant literature references, dates of evaluated data, orbit, SA type (array layout, power and solar cell type), and main conclusions achieved.

4.3 Spacecraft Solar Array Anomalies in Orbit

Another source of data for improving solar array design comes regrettably from anomalies experienced by spacecraft SA in orbit. Investigations of the failure mechanism in-orbit are much more complicated due to the small quantities of data often available. However, some anomalies in orbit could be acceptably explained; a few of them are depicted here in the following sections.

4.3.1 European Communication Satellite (ECS) and Maritime European Communication Satellite (MARECS)

After 1.5 years in GEO both SA (virtually identical, two wings of three rigid panels each with silicon solar cells) started to suffer partial loss of power [60]. The failures seemed to be short-circuits between the cell network and panel

Table 4 In-orbit performance of recent spacecraft

Spacecraft (design life)	Ref.	Dates	Orbit	SA type	Main conclusions
SOHO (2.5 years)	[82]	Dec. 95, Dec. 01	LG1[a] 1.5×10^6 km from Earth	2 wings × 2 rigid panels EOL^b power 1.4 kW Si 2 Ω cm BSR	1. Solar array design and good margin between working and P_{max} point allowed mission extension. 2. Less SA radiation degradation than predicted. 3. SA recovery after sun flares degradation.
SPOT 1 (3 years)	[83]	Feb. 86, Feb. 98	LEO^c Sun-synchronous	2 wings × 1 flexible panel EOL^b power 1 kW Si 1 Ω cm	1. Several loss factors have been over evaluated on the design, allowing a longer SA life. 2. SPOT 1 database shall improve EOL^b performance predictions of coming LEO^c spacecraft.
HS 601 HP C1 (15 years)	[84]	Aug. 97, Mar. 98	GEO^d	2 wings × 3 rigid panels BOL^c power 9.5 kW Dj $GaInP_2/GaAs/Ge$	1. SA power in orbit is 1.1% less than predicted. 2. Systematic errors may be the source of this discrepancy: calibration of flight balloon standard and the calibration of ground performance testing.
	[85]	Dec. 81, Dec. 94	GEO^d		

INTELSAT-V (7 years)		Feb. 88, Nov. 88 GEO[d]	2 wings × 3 rigid panels EOL[b] power 1.5 kW Si BSR	1. SA power is 8–10% higher than predicted (solar flares) and 4–6% higher than predicted (no solar flares) for 13 spacecraft.
CS-3A (7 years)	[86]	Feb. 88, Nov. 88 GEO[d]	2 body mounted panels BOL[c] power 0.85 kW Sj GaAs	1. SA power is 1–5% higher than predicted.
HIPPARCOS (3 years)	[87]	Feb. 90, Jun. 93 GTO[f]	3 deployed panels EOL[b] power 325 W Si BSR 10 Ω cm	1. SA power is according to radiation degradation modelling.
IRS-1A (3 years)	[88]	Mar. 88, Mar. 95 LEO[e]	6 deployed panels EOL[b] power 0.7 kW Si BSR 10 Ω cm	1. The silver mesh interconnector survived more than 35000 cy. 2. Effects of ATOX negligible on interconnectors. 3. Power degradation due to radiation matches with modelling.
Space Telescope SA (5 years)	[89]	Apr. 90, Dec. 93 LEO[e] 600 km	2 flexible wings × double roll-out 2 year life power 4.4 kW Si BSFR 10 Ω cm	1. Degradation performance is in agreement with most design loss factors. 2. Radiation fluence below initial prediction. 3. Random failures main degradation SA mechanism.

(continued)

Table 4 (continued)

Spacecraft (design life)	Ref.	Dates	Orbit	SA type	Main conclusions
JCSAT (10 years)	[90]	Mar. 89, May 91	GEO[d]	2 telescopic cylindrical EOL[b] power 1.7 kW K7 and K3 Si	1. 1–2% less BOL[e] in-orbit performance than predicted probably due to reference standards for performance ground testing, 2. Increase performance over time probably due to radiation model more severe than in orbit. 3. Some solar flares did not produce any damage on the SA. No explanation.

For other abbreviations and standard solar cell terms see text or Chapter Ia-1.
[a] La Grangian Point 1.
[b] End of life.
[c] Low–Earth orbit.
[d] Geosynchronous orbit.
[e] Beginning of life.
[f] Geosynchronous transfer orbit.

structure. These failures continued intermittently until the end both missions, however, for ECS the power losses were recovered. Several potential failure modes were identified: imperfections of the Kapton insulation layer or embedded particles between layers, insulation breakdown by electrostatic discharge, thermal cycling, corona effects, micrometeoroids, or a combination of all of them. None of the potential failure modes could be identified as being responsible for the ECS and MARECS anomalies; however, several weak points in the SA design were identified, investigations continued in the direction of the most probable failure mode (ESD) [91], and some improvements were proposed aiming to lower the risk of these failures: designs should be adapted to incorporate sufficient margins in areas where uncertainties exist, parallel cell strings sections instead of single-string sections, and more stringent tests in manufacturing and acceptance for early failure detection.

4.3.2 X-Ray Timing Explorer (XTE)

The XTE spacecraft was launched in December 1995. SA is composed of two wings of three rigid panels each, with silicon solar cells. Shortly after launch, the array showed discontinuous current drops, consistent with the loss of a part of a cell, when coming out from eclipse. The failure mechanism seems to be cell cracks not detected in ground inspections that became open in orbit due to the temperature gradients. These cracks were probably produced during the extensive tap tests, performed to detect SCA to substrate delaminations. During testing on the ground, following the same activities as for the flight SA, the qualification panel showed these effects, giving high confidence to this theory [92].

4.3.3 CPS Navstars 1–6

Six GPS Navstars satellites were placed in 20,000 km circular orbits from 1980. Mission lifetime for each spacecraft was five years and silicon solar cells K4 or K6 were in the SAs. After two years in orbit all spacecraft suffered an unexpected additional degradation of 2.5%. Investigations carried out in optical reflectors surfaces of one of the spacecraft revealed traces of contamination covering all spacecraft external surfaces. These contaminants mainly come from the outgassing of materials from the spacecraft, leading to reflectivity degradation of the cover glasses [93].

4.3.4 Pioneer Venus Orbiter SA

Pioneer Venus orbiter was a spin-stabilised (5 rpm) cylindrical spacecraft that operated in a high-eccentric near-polar orbit around Venus for more than eight years. After two years orbiting, power drops correlated with string losses were observed depending on the vehicle rotating angle. This suggested failures on strings due to reverse bias of cells (no shunt diodes protected the strings) produced by cyclic shadows made by the magnetometer boom cast, not predicted and unavoidable for the mission success. Ground tests were not conclusive that the cyclic reverse bias operation ended in cell breakdowns. Therefore, other interactions, as the ATOX environment in the Venus upper atmosphere, could favour the SA degradation [94].

4.4 Postflight Investigations on Returned Solar Arrays

Returned SAs from space are valuable opportunities to assess their predicted behaviour in the space environment. Few SAs have been returned to Earth and a brief summary of their investigation programmes and the major conclusions are outlined in subsequent paragraphs.

4.4.1 Hubble Space Telescope Solar Array 1

One wing of the Hubble Space Telescope SA was retrieved from space in December 1993, after more than 3.5 years operating in a low-Earth orbit, The SA of the Hubble Space Telescope consisted of two wings of a double roll-out concept using two flexible solar cell blankets on each wing. The 48760 Silicon BSFR solar cells should provide the required 4.4 kW after two years in operation.

The postflight investigation programme carried out between 1994 and 1995 [95] had the following main objectives:
- Assess the effect of different LEO interaction and environments as: thermal fatigue, ATOX, meteoroid and space debris damage, contaminations, UV, etc.
- Explain the anomalies experienced in orbit

During the investigation programme the SA was submitted to several tests like detailed visual inspections, EP and health checks, wipe testing, etc. The SA mechanisms were also mechanically tested to study their deploy/retract performance, and finally the SA was totally disassembled for detailed investigation of all its components.

The main conclusions of the postflight investigation programme related to the SA blankets are the following:
- SA performance: 5% more power than predicted after 3.6 years in orbit, despite several anomalies (string shorts) that reduced the power by 6.7%. Random failures are the main contributors to SA

degradation. SA overall degradation excluding failures was less than predicted, mainly as radiation model used was pessimistic.
- Solar cell interconnectors: No fatigue effects on interconnection loops were detected as expected from preflight qualification data.
- Harness: Fatigue effects were evident on flexible data harnesses, but no full detachments were found.
- Adhesives for ATOX protection: Darkening due to UV could increase SA operational temperature.
- Micrometeoroids: More than 4000 impacts were detected on the SA, but none of them produced permanent short circuits. The loss factor applied in the design is in full agreement with the results of the observed degradation (1.8%).

4.4.2 EURECA

The European Retrievable Carrier (EURECA) was launched in July 1992 (500-km orbit) and completely retrieved in July 1993 by the Space Shuttle. The SA consists of two interchangeable wings of five rigid panels (~ 100 m^2) each providing initially 5 kW. Silicon BSFR 10-Ω cm solar cells of two sizes were used to manufacture the charge and load array networks. The solar array postflight investigation programme had the objectives of studying LEO environment effects and mainly the anomalies faced during the mission [96]. Main conclusions are depicted here:
- Failures by fatigue (inadequate bend radii in the stress relief loop) in the Wiring Collecting Panels (WCPs) were responsible for open circuits on solar cell strings. WPCs were never tested in a flight representative configuration, as it was not possible to detect in advance the weakness of this design.
- A short-circuit on the load array produced current from the battery during eclipse to the solar array (no blocking diodes were placed between the SA and battery circuits). A large burn mark was found at the suspected location of the short circuit after retrieval.
- Kapton FEP (Fluorinated Ethylene-Propylene) coatings of the cable insulations were completely eroded in X-ray/UV direction.
- The exposed side of MoAg interconnectors was oxidised and eroded by ATOX.
- Adhesives for ATOX protection were also darkened (top surfaces converted to SiO$_2$) and all surfaces investigated showed contamination of carbon or silicone.

4.4.3 MIR Solar Array

In January 1998, a segment of the MIR solar array was retrieved by the space shuttle. The segment, composed of eight panels, spent 10.5 years in a 380 km orbit. The panel design is exclusive; a laminated sandwich of cover glass, glass cloth, silicon solar cells (11% efficiency), glass cloth and optical solar reflectors (OSRs) [97]. Two postflight investigation programmes have been conducted in the USA and Russia. The main conclusions are the following:

- Hot spots are the main reason for the 50% power degradation of the solar array. By-pass diodes were not installed on the panel, relying especially on solar cell screening for handling full reverse currents. High temperatures during the hot spots destroyed separate commutation bundles in the circuits of serial connected solar cells [98,99].
- Large SiO_x contaminations were found on all exposed surfaces of the panel, due to outgassing of silicone adhesives, resulting in a total power loss of only 0.72%.
- The meteoroid and space debris impact produced less than 1% power loss.
- The temperature increase over life was 7°C, due to an increase of the emittance and decrease of the absorptance.
- Solar cells not influenced by the hot spots had only 10—15% power degradation.

ACKNOWLEDGEMENTS

To my wife, Maria Jesus, for her constant support and patience. My colleagues C. Signorini and R. Crabb (ESA-Estec) and T. J. Gomez (Spasolab) for their good advice and helpful comments about the contents of this chapter. All my colleagues at Estec, especially of the solar generator section, Spasolab and the space photovoltaic community for fruitful discussions about these subjects. I am grateful to ESA for its support and permission to publish this work.

REFERENCES

[1] ISO/DIS 15387: Space Systems—Single-Junction Space Solar Cells—Measurement and Calibration Procedures.
[2] B.E. Anspaugh, et al., Results of the 2001 JPL balloon flight solar cell calibration program. JPL Publication 02—004, 2002.
[3] V. Pichetto, et al., Casolba calibration of solar cells using balloon flight, in: Proceeding of the 29th IEEE Photovoltaic Specialists Conference, New Orleans, 2002.
[4] M. Roussel, et al., Calibration de cellules solaires hors atmosphère, in: Proceeding of the 4th European Space Power Conference, ESA SP-210, 1984, pp. 257—264.

[5] P. Jenkins, et al., Uncertainty analysis of high altitude aircraft air mass zero solar cell calibration, in: Proceeding of the 26th IEEE Photovoltaic Specialists Conference, Anaheim, 1997, pp. 857–860.
[6] E.G. Suppa, Space Calibration of solar cells. Results of 2 shuttle flight missions, in: Proceeding of the 17th IEEE Photovoltaic Specialists Conference, Orlando, 1984, pp. 301–305.
[7] G.A. Landis, et al., Calibration and measurement of solar cells on the international space station: A new test facility, in: Proceeding of the 36th Intersociety Energy Conversion Conference, 2001, pp. 229–231.
[8] M.A.H. Davies, C. Goodbody, The calibration of solar cells in terrestrial sunlight, in: Proceeding of the 2nd European Space Power Conference, ESA SP-320, 1991, pp. 583–587.
[9] L. Garcia-Cervantes, et al., Ground level sunlight calibration of space solar cells, in: Proceeding of the 5th European Space Power Conference, ESA SP-416, 1998, pp. 615–620.
[10] L. Garcia, et al., Uncertainty analysis for ground level sunlight calibration of space solar cells at Tenerife, in: Proceeding of the 17th European Photovoltaics Solar Energy Conference, Munich, 2001, pp. 2259–2262.
[11] Y. Yiqiang, et al., Calibration of AM0 reference solar cells using direct normal terrestrial sunlight, in: Proceeding of the 9th Asia/Pacific Photovoltaic Science and Engineering Conference, 1996.
[12] O. Kawasaki, et al., Study of solar simulator method and round robin calibration plan of primary standard solar cell for space use, in: Proceeding of the 1st World Conference on Photovoltaic Energy Conversion, Hawaii, 1994.
[13] J. Metzdorf, et al., Absolute indoor calibration of large area solar cells, in: Proceeding of the 5th European Sympousium on Photovoltaic Generators in Space, ESA SP-267, 1986, pp. 397–402.
[14] ASTM E973M-96, Test method for determination of the spectral mismatch parameter between a Photovoltaic device and a Photovoltaic reference cell.
[15] A. Gras, et al., Terrestrial secondary calibration analysis, in: Proceeding of the 16th European Space Power Conference, 2000, pp. 1011–1014.
[16] A. Gras, et al., Generic test procedure for solar cell testing, in: Proceeding of the 3rd European Space Power Conference, ESA WPP-054, 1993, pp. 743–748.
[17] ESAPSS-01–604, Generic specification for silicon solar cells, 1988.
[18] C.H. Seaman, et al., The spectral irradiance of some solar simulators and its effect on cell measurements, in: Proceeding of the 14th IEEE Photovoltaic Specialists Conference, San Diego, 1980, pp. 494–499.
[19] G.S. Goodelle, et al., Simulator spectral characterization using balloon calibrated solar cells with narrow band pass filters, in: Proceeding of the 15th IEEE Photovoltaic Specialists Conference, Orlando, 1981, pp. 211–217.
[20] T. Thrum, et al., Characterizing state of the art solar panels—A new approach for large area testing, in: Proceeding of the 28th IEEE Photovoltaic Specialists Conference, Anchorage, 2000, pp. 1320–1323.
[21] L.C. Kilmer, A more accurate, higher fidelity dual source AM0 solar simulator design, in: Proceeding of the 4th European Space Power Conference, ESA SP-369, 1995, pp. 671–675.
[22] J.J. Sturcbecher, et al., The mini-flasher: a solar array test system, Sol. Energy Mater. Sol. Cells 36 (1994) 91–98.
[23] W. Lukschal, et al., A pulsed solar simulator for electrical performance tests of space solar cells/arrays, in: Proceeding of the 1st European Space Power Conference, ESA SP-294, 1989, pp. 689–693.

[24] J.E. Granata, et al., Triple-junction GaInP$_2$/GaAs/Ge solar cells, production status, qualification results and operational benefits, in: Proceeding of the 28th IEEE Photovoltaics Specialists Conference, Anchorage, 2000, pp. 1181−1184.
[25] A. Gras, et al., Analysis for multi-junction solar cell measurements at Spasolab, in: Proceeding of the 6th European Space Power Conference, ESA SP-502, 2002, pp. 577−580.
[26] K. Emery, et al., Procedures for evaluating multi-junction concentrators, in: Proceeding of the 28th IEEE Photovoltaic Specialists Conference, Anchorage, 2000, pp. 1126−1130.
[27] H.S. Rauschenbach, Solar Cell Array Design Handbook, Litton Educational Publishing, 1980.
[28] R. Adelhelm, et al., Matching of multi-junction solar cells for solar array production, in: Proceeding of the 28th IEEE Photovoltaic Specialists Conference, Anchorage, 2000, pp. 1336−1339.
[29] D.L. King, et al., Temperature coefficients for PV modules and arrays: Measurement methods, difficulties and results, in: Proceeding of the 26th IEEE Photovoltaic Specialists Conference, Anaheim, 1997, pp. 1183−1186.
[30] R. Adelhelm, et al., Temperature coefficients of tandem solar cells under appropriate spectra, in: Proceeding of the 14th European Photovoltaics Solar Energy Conference, Barcelona, 1997.
[31] D.R. Burger, et al., Angle of incidence corrections for GaAs/Ge solar cells with low absorptance coverglass, in: Proceeding of the 25th IEEE Photovoltaic Specialists Conference, Washington DC, 1996, pp. 243−246.
[32] J.C. Larue, Pulsed measurement of solar cell spectral response, in: Proceeding of the 2nd European Photovoltaic Solar Energy Conference, West Berlin, 1979, pp. 477−486.
[33] D.L. King, et al., New methods for measuring performance of monolithic Mj solar cells, in: Proceeding of the 28th IEEE Photovoltaic Specialists Conference, Anchorage, 2000, pp. 1197−1201.
[34] H.S. Rauschenbach, et al., Breakdown phenomena on reverse biased silicon solar cells, in: Proceeding of the 9th IEEE Photovoltaic Specialists Conference, Silver Springs, 1972, pp. 217−225.
[35] W.R. Baron, et al., GaAs solar cell reverse characteristics, in: Proceeding of the 19th IEEE Photovoltaic Specialists Conference, New Orleans, 1987, pp. 457−462.
[36] D. Schwander, Dynamic solar cell measurement techniques: new small signal measurement techniques, in: Proceeding of the 6th European Space Power Conference, ESA SP-502, 2002, pp. 603−608.
[37] P. Rueda, et al., Mj GaAs solar cell capacitance and its impact upon solar array regulators, in: Proceeding of the 6th European Space Power Conference, ESA SP-502, 2002, pp. 29−34.
[38] H.Y. Tada, et al., The Solar Cell Radiation Handbook, JPL publication, 1982, pp. 82−69.
[39] B.E. Anspaugh, GaAs Solar Cell Radiation Handbook, JPL publication, 1996, pp. 96−9.
[40] D.C. Marvin, Assessment of Mj solar cell performance in radiation environments. Aerospace Report TOR-2000 (1210)-1. The Aerospace Corporation, 2000.
[41] R.J. Walters, et al., Analysis and modelling of the radiation response of Mj space solar cells, in: Proceeding of the 28th IEEE Photovoltaic Specialists Conference, Anchorage, 2000, pp. 1092−1097.
[42] R. Crabb, Photon induced degradation of electron and proton irradiated silicon solar cells, in: Proceeding of the 10th IEEE Photovoltaic Specialists Conference, Palo Alto, 1973, pp. 396−403.

[43] H. Fischer, et al., Investigation of photon and thermal induced changes in silicon solar cells, in: Proceeding of the 10th IEEE Photovoltaic Specialists Conference, Palo Alto, 1973, pp. 404–411.
[44] G.S. Goodelle, et al., High vacuum UV test of improved efficiency solar cells, in: Proceeding of the 11th IEEE Photovoltaic Specialists Conference, Scottsdale, 1975, pp. 184–189.
[45] A. Meulenberg, et al., Evidence for enhanced UV degradation to cracked coverslides. XV Space Photovoltaic Research and Technology, 1997, pp. 213–218.
[46] J. Matcham, et al., Effects of simulated solar-UV radiation on solar cell efficiency and transparent cell components, in: Proceeding of the 5th European Space Power Conference, ESA SP-416, 1998, pp. 643–650.
[47] L. Gerlach, et al., Advanced solar generator technology for the Eureca low earth orbit, in: Proceeding of the 18th IEEE Photovoltaic Specialists Conference, Las Vegas, 1985, pp. 78–83.
[48] A. Dunnet, et al., Assessment of ATOX erosion of silver interconnects on Intelsat VI. F3, in: Proceeding of the 2nd European Space Power Conference, ESASP-320, 1991, pp. 701–706.
[49] B.A. Banks, et al., Protection of solar array blankets from attack by low earth orbital atomic oxygen, in: Proceeding of the 18th IEEE Photovoltaic Specialists Conference, Las Vegas, 1985, pp. 381–386.
[50] S.K. Ruthledge, et al., Atomic oxygen effects on SiO_x coated kapton for photovoltaic arrays in low earth orbit, in: Proceeding of the 22nd IEEE Photovoltaic Specialists Conference, Las Vegas, 1991, pp. 1544–1547.
[51] D. Richard, A rational approach to design and test a space photovoltaic generator, in: Proceeding of the 15th IEEE Photovoltaic Specialists Conference, Orlando, 1981, pp. 554–559.
[52] W. Ley, DFVLR facility for thermal cycling tests on solar cells panel samples under vacuum conditions, in: Proceeding of the 12th IEEE Photovoltaic Specialists Conference, Baton Rouge, 1976, pp. 406–412.
[53] J.C. Larue, et al., Accelerated thermal cycling of solar array samples, in: Proceeding of the 1st European Symposium on Photovoltaic Generators in Space, ESASP-140, 1978, pp. 57–64.
[54] L. Norris Blake III, Lessons learned about fabrication of space solar arrays from thermal cycle failures, in: Proceeding of the 25th IEEE Photovoltaic Specialists Conference, Washington DC, 1996, pp. 329–332.
[55] L. Gerlach, et al., HST-SA1: electrical performance evaluation. Hubble Space Telescope Solar Array Workshop, ESA WPP-77, 1995, pp. 257–264.
[56] J.F. Murray, et al., Space environment effects on a rigid panel solar array, in: Proceeding of the 22nd IEEE Photovoltaic Specialists Conference, Las Vegas, 1991, pp. 1540–1543.
[57] K.G. Paul, et al., Post-Flight particle impacts on HST solar cells. Hubble Space Telescope Solar Array Workshop, ESA WPP-77, 1995, pp. 493–500.
[58] H.W. Brandhorst Jr., et al., Hypervelocity impact testing of stretched lens array modules, in: Proceeding of the 6th European Space Power Conference, ESA SP-502, 2002, pp. 585–590.
[59] E. Schneider, Micrometeorite impact on solar panels, in: Proceeding of the 5th European Symposium on Photovoltaic Generators in Space, ESA SP-267, 1986, pp. 171–174.
[60] K. Bogus, et al., Investigations and conclusions on the ECS Solar Array in orbit power anomalies, in: Proceeding of the 18th IEEE Photovoltaic Specialists Conference, Las Vegas, 1985, pp. 368–375.

[61] T.G. Stern, et al., Development of an electrostatically clean solar array panel, in: Proceeding of the 28th IEEE Photovoltaic Specialists Conference, Anchorage, 2000, pp. 1348–1351.
[62] A. Bogorad, et al., Electrostatic discharge induced degradation of solar arrays, in: Proceeding of the 22nd IEEE Photovoltaic Specialists Conference, Las Vegas, 1991, pp. 1531–1534.
[63] C.J. Bishop, The fundamental mechanism of humidity degradation in silver-titanium contacts, in: Proceeding of the 8th IEEE Photovoltaic Specialists Conference, Seattle, 1970, pp. 51–61.
[64] P.A. Iles, et al., The role of the AlGaAs window layer in GaAs heteroface solar cells, in: Proceeding of the 18th IEEE Photovoltaic Specialists Conference, Las Vegas, 1985, pp. 304–309.
[65] K. Mitsui, et al., A high quality AR coating for AlGaAs/GaAs solar cells, in: Proceeding of the 17th IEEE Photovoltaic Specialists Conference, Orlando, 1984, pp. 106–110.
[66] G. La Roche, et al., Evaluation of the flight data of the Equator-S mini-modules, in: Proceeding of the 16th European Photovoltaic Solar Energy Conference, Glasgow, 2000, pp. 945–950.
[67] S.R. Messenger, et al., A displacement damage dose analysis of the Comets and Equator-S space solar cell experiments, in: Proceeding of the 16th European Photovoltaic Solar Energy Conference, Glasgow, 2000, pp. 974–977.
[68] H. Curtis, et al., Final results from the PASP-Plus flight experiment, in: Proceeding of the 25th IEEE Photovoltaic Specialists Conference, Washington DC, 1996, pp. 195–198.
[69] V.A. Davis, et al., Parasitic current collection by PASP PLUS solar arrays. XIV Space Photovoltaic Research and Technology, NASA CP-10180, 1995, pp. 274–285.
[70] D.A. Guidice, High voltage space-plasma interactions measured on the PASP Plus test arrays. XIV Space Photovoltaic Research and Technology, NASA CP-10180, 1995, pp. 286–295.
[71] M. Imaizumi, et al., Flight degradation data of GaAs-on-Si solar cells mounted on highly irradiated ETS-VI, in: Proceeding of the 28th IEEE Photovoltaic Specialists Conference, Anchorage, 2000, pp. 1075–1078.
[72] T. Aburaya, et al., Analysis of 10 years' flight data of solar cell monitor on ETS-V, Sol. Energy Mater. Sol. Cells 68 (2001) 15–22.
[73] C. Flores, et al., Post-flight investigation of the ASGA solar cell experiment on Eureca, in: Proceeding of the 1st World Conference on Photovoltaic Energy Conversion, Hawaii, 1994, pp. 2076–2081.
[74] R.M. Burgess, et al., Performance analysis of CuInSe$_2$ and GaAs solar cells aboard the LIPS-III flight Boeing lightweight panel, in: Proceeding of the 23rd IEEE Photovoltaic Specialists Conference, Louisville, 1993, pp. 1465–1468.
[75] J.R. Woodyard, et al., Analysis of LIPS-III satellite a-Si:H alloy solar cell data, in: Proceeding of the 25th IEEE Photovoltaic Specialists Conference, Washington DC, 1996, pp. 263–266.
[76] J.G. Severns, et al., LIPS-III. A solar cell test bed in space, in: Proceeding of the 20th IEEE Photovoltaic Specialists Conference, Las Vegas, 1988, pp. 801–807.
[77] H. Kulms, et al., Results of the MBB LIPS-III experiment, in: Proceeding of the 21st IEEE Photovoltaic Specialists Conference, Orlando, 1990, pp. 1159–1163.
[78] C. Goodbody, et al., The UoSAT-5 solar cell experiment—Over 4 years in orbit, in: Proceeding of the 25th IEEE Photovoltaic Specialists Conference, Washington DC, 1996, pp. 235–238.
[79] D.A. Scheiman, et al., Mars array technology experiment (MATE), in: Proceeding of the 28th IEEE Photovoltaic Specialists Conference, Anchorage, 2000, pp. 1362–1365.

[80] J.K. Jain, et al., Concentrator solar array technology flight experiment, in: Proceeding of the 29th IEEE Photovoltaic Specialists Conference, New Orleans, 2002, pp. 1362–1365.
[81] K. Bogus, et al., Comparative evaluation of the in-orbit performance of ESA's satellite solar generators, in: Proceeding of the 3rd European Space Power Conference, ESA WPP-054, 1993, pp. 529–535.
[82] P. Rumler, et al., SOHO power system performance during 6 years in orbit, in: Proceeding of the 6th European Space Power Conference, ESA SP-502, 2002, pp. 141–146.
[83] A. Jalinat, et al., In orbit behaviour of SPOT 1,2 and 3 solar arrays, in: Proceeding of the 5th European Space Power Conference, ESA SP-416, 1998, pp. 627–631.
[84] J.S. Fodor, et al., In-orbit performance of Hughes HS 601 solar arrays, in: Proceeding of the 2nd World Conference on Photovoltaic Energy Conversion, Vienna, 1998, pp. 3530–3533.
[85] A. Ozkul, et al., In-orbit performance characteristics of Intelsat-V solar arrays, in: Proceeding of the 1st World Conference on Photovoltaic Energy Conversion, Hawaii, 1994, pp.1994–1997.
[86] N. Takata, et al., In-orbit performance of CS-3A spacecraft GaAs solar array, in: Proceeding of the 1st European Space Power Conference, ESA SP-294, 1989, pp. 823–828.
[87] R.L. Crabb, A.P. Robben, In-flight Hipparcos solar array performance degradation after three and a half years in GTO, in: Proceeding of the 3rd European Space Power Conference, ESA WPP-054, 1993, pp. 541–549.
[88] S.E. Puthanveettil, et al., IRS-1A Solar array—in-orbit performance, in: Proceeding of the 4th European Space Power Conference, ESA SP-369, 1995, pp. 583–585.
[89] L. Gerlach, et al., Hubble Space Telescope and EURECA solar generators a summary of the post flight investigations, in: Proceeding of the 4th European Space Power Conference, ESA SP-369, 1995, pp. 5–20.
[90] S.W. Gelb, et al., In-orbit performance of Hughes HS 393 solar arrays, in: Proceeding of the 22nd IEEE Photovoltaic Specialists Conference, Las Vegas, 1991, pp. 1429–1433.
[91] L. Levy, et al., MARECS & ECS anomalies: attempt for insulation defect production in kapton, in: Proceeding of the 5th European Symposium on Photovoltaic Generators in Space, ESASP-267, 1986, pp. 161–169.
[92] E.M. Gaddy, et al., The Rossi X-Ray timing explorer XTE solar array anomaly, XV Space Photovoltaic Research and Technology, 1997, pp. 144–153.
[93] D.C. Marvin, et al., Anomalous solar array performance on GPS, in: Proceeding of the 20th IEEE Photovoltaic Specialists Conference, 1988, pp. 913–917.
[94] L.J. Goldhammer, et al., Flight performance of the Pioneer Venus orbiter solar array, in: Proceeding of the 19th IEEE Photovoltaic Specialists Conference, New Orleans, 1987, pp. 494–499.
[95] Proceedings of the Hubble Space Telescope SA workshop. ESA WPP-77, 1995.
[96] EURECA The European retrievable carrier, Technical Report ESA WPP-069, 1994.
[97] R.J. Pinkerton, MIR returned solar array, in: Proceeding of the 36th Intersociety Energy Conversion Conference, 2001, pp. 217–222.
[98] V.A. Letin, Optical, radiation and thermal cycling losses of power solar array returned from orbital station MIR after 10.5 years of operation, in: Proceeding of the 6th European Space Power Conference, ESA SP-502. 2002, pp. 7.13–718.
[99] A.B. Grabov, et al., A terrestrial investigation of material's degradation mechanisms in silicon solar cells, which returned from MIR space station after ten years exploitation, in: Proceeding of the 6th European Space Power Conference, ESA SP-502, 2002, pp. 733–740.

PART IIE

Case Studies

CHAPTER IIE-1

Architectural Integration of Solar Cells

Rafael Serra i Florensa[a] and Rogelio Leal Cueva[b]

[a]ETSAB, Universitat Politècnica de Catalunya, Barcelona, Spain
[b]GESP, Parc Cientific de Barcelona, Spain

Contents

1. Introduction	918
2. Architectural Possibilities for PV Technology	919
2.1 Architectural PV Energy System	919
2.2 Surface Availability in Buildings	921
2.3 Solar Energy Availability in Buildings	923
3. Building-Integrated Photovoltaics (BIPVs)	923
3.1 Multifunctionality of PV Modules	923
3.2 PV Mounting Techniques	924
3.2.1 Flat Roofs	925
3.2.2 Sloped Roofs	925
3.2.3 Facade	926
3.2.4 Independent Structures	926
4. Aesthetics in PV Technology	926
4.1 Shape	929
4.2 Size	930
4.3 Colour	931
4.4 Texture and Patterns	931
4.5 Translucency	932
4.6 Point of View	933
5. Built Examples	934
5.1 Case Study 1: The British Pavilion, Expo '92	934
5.1.1 Description	934
5.1.2 PV System	934
5.1.3 Architectural Energy System	936
5.2 Case Study 2: ECN, Buildings 31 and 42	936
5.2.1 Description	936
5.2.2 Building 31	936
5.2.3 Building 42	937
5.3 Case Study 3: Pompeu Fabra Public Library	937
5.3.1 Description	938
5.3.2 PV System	938

Practical Handbook of Photovoltaics.
© 2012 Elsevier Ltd. All rights reserved.

5.4 Case Study 4: Mont Cenis Conference Centre   938
    5.4.1 Description   940
    5.4.2 PV System   940
References   940
Further Reading   941

1. INTRODUCTION

Architecture may be defined as a complex whole of interrelated systems that produce satisfactory spaces for human activity. The architectural conception of buildings is commonly made from matter and geometry, where energy usually plays a secondary role for architects who tend to interpret buildings by what can easily be represented in drawings or photographs. Energy in architecture is difficult to represent and, therefore, briefly studied in architectural projects. Only when the existence of energy is revealed in visual elements and components will the issue become important for architects.

The integration of energy-related systems in architecture tends to be roughly simplified, and unfortunately, often considered as components that should be hidden from view. Furthermore, as so often happens in the history of architecture, the incorporation of new technologies is realised in a shameful way, hidden as if they were an offence to good taste.

This is particularly evident in the integration of natural energy collector technologies in buildings, especially solar photovoltaic technology. When integration is considered, it is often thought of as covering up the presence of the collector surfaces, hiding them behind the railings of flat surfaces or putting them on top of inclined roofs with the same inclination as the roofs.

The final result is aesthetically inadequate, the camouflage tends not to be absolute and in many cases the added elements are visible with a dreadful appearance. The preexisting constrictions frequently force collector surfaces to be badly oriented or to be partially shaded at a certain period of the day or year. This is one more example of the existing difficulty in architecture to integrate new technologies.

It would make sense to consider photovoltaic systems as one more element in the building that is being designed. Choosing a proper location, not only from the technical point of view, but also from the aesthetic point of view to ensure that the module and its supports have the maximum quality. They should be interpreted as one more element in architecture that contributes to the formal quality of the whole.

2. ARCHITECTURAL POSSIBILITIES FOR PV TECHNOLOGY

Architecture allows for solar energy to be used when it is incident on building exteriors. The impact of this energy may be positive or negative in buildings, depending on the climatic characteristics of the site and the time of the year. Therefore, the use of solar energy to generate electricity should be done in a way that the building may absorb heat during cold weather and be protected from it during hot weather.

On the other hand, buildings are vast energy consumers and use it for a varied range of applications. This energy is normally generated in complex, expensive and polluting systems. The collection of solar energy *in situ* results in a tempting complement to the powerful supply grids that are compulsory, allowing us to reduce the external supply or even eliminating it in many cases.

Making a comparison with natural ecosystems, buildings in our urban ecosystems must carry out a similar function to trees in the woods. This way, a building would collect energy in its outer surface and be able to conveniently transform and accumulate it to be used for its own needs. This way of approaching the issue would reduce dependence on external energies; the urban grids would be minor, functioning more like a balance of energy than a supplier.

2.1 Architectural PV Energy System

The city of today is a complex ecosystem, crossed by intense flows of matter, energy, and information. A great part of the load that these ecosystems impose upon the planet is the result of deficient management of these flows, and in particular, the energy flows.

Solar energy, collected by photovoltaic systems, is actually one of today's most promising energy sources. Its integration in buildings located in urban areas could play an important role when analysing its use on a large scale.

The quantity of solar radiation that is incident on buildings in cities represents a huge volume of energy that is only used in a minor proportion, partially due to it being ignored. Also, high-density building in cities has turned solar-exposed urban structures into compact building forms, where buildings shade mutually (Figures 1 and 2).

There is an emerging awareness about the possibilities of using natural energies in architecture, but there is very little awareness about how urban building forms may contribute to this happening. A determined action in this sense would not only reduce fossil fuel consumption, but would also considerably improve the quality of the urban environment.

It is especially important to consider the most favourable strategies to obtain an efficient use of solar energy that is often being wasted. It is particularly important to develop and use collector surfaces that may be integrated in buildings, using them to improve their functionality and aesthetics.

Buildings could be considered to be open ecosystems since various types of energy penetrate them. Some are natural energy, as solar radiation. Others are artificial energy, as electricity, gas, etc. The thermal equilibrium of the building depends on the input and output (loss) flows of energy. Among artificial energy, electrical energy is more and more important every day. It is a high-quality type of energy, offers great flexibility, and is used in multiple types of applications even though it causes high environmental damage in its origin. In order to improve the architectural energy

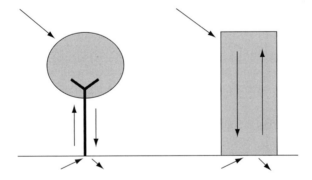

Figure 1 Urban and natural ecosystems.

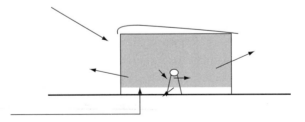

Figure 2 Architectural energy system.

performance and the environment in general, we should reduce electricity consumption from artificial sources. Photovoltaic technology allows us to integrate solar collectors in buildings that will reduce consumption drastically while producing clean electrical energy.

2.2 Surface Availability in Buildings

There are four main possibilities to locate solar energy collectors in architectural projects (see Figures 3 and 4).

A. Flat roof. A common solution for PVs on buildings is on rooftops, usually locating the collector surface on existing rooftops on additional structures.

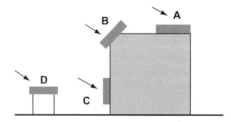

Figure 3 Surface availability.

Figure 4 Flat PV roof at Gouda. *(© BEAR Architecten. Reprinted with permission).*

B. *Sloped surface.* Collector surfaces may be linked to external inclined surfaces in buildings. This may result in a correct, aesthetic, and functional solution if the inclined surface is well oriented and has the appropriate inclination.

There is, however, the danger of generating these rooftops 'artificially,' without corresponding to spaces appropriate to the building, having a negative aesthetic effect and imposing unjustifiable supplementary economic costs. In this case, it could be a better solution to design a support structure for the collectors, integrating it to the forms of the building, but not trying to pretend to that functional spaces are being covered.

C. *Facade.* Photovoltaic modules may be located on the facade of buildings, replacing or complementing a part of the outer coating.

In this case, the main problem is the verticality of the facades. The annual efficiency of a 90° inclination surface would be around 35% below optimal in southern European countries and around 20% for northern European countries (see Figure 8). However, losses during the winter would be lower than during the summer. The advantages that may result from facade integration, including the protection to the building from excessive solar radiation that modules may offer, make this solution advisable in certain cases (Figure 5).

In other occasions, collector surfaces form auxiliary facade elements, as blinds, railings, and others. In these cases, the inclination of the modules may be optimised, and even be adaptable to the solar path during different times of the year.

Figure 5 Sloped PV roofs at Amersfoort.

D. Independent structures. It is possible to incorporate collector surfaces to independent structures, such as gazebos, shadings, and others, as part of the architectural whole of a project. In this case, greater freedom of shape and situation may improve the orientation of the surfaces. Also, the independence from the building may improve the formal impact of the whole, as long as the design of the element is adequate.

2.3 Solar Energy Availability in Buildings

The amount of solar energy that is incident to the external coating in buildings is related to several factors. These are basically shadings, orientation and meteorological conditions. Figure 8 shows the incident solar energy on a collector surface in southern Europe, compared to the optimum according to its orientation. The collector surfaces in buildings must be free of shading from elements such as trees, chimneys, light posts, neighbouring buildings, and others.

3. BUILDING-INTEGRATED PHOTOVOLTAICS (BIPVS)

Photovoltaic modules may become part of external coatings in buildings not only as energy generators, but also as external building elements capable of reducing energy consumption. To upgrade photovoltaic modules from energy generators to aesthetic and functional building elements, it requires the collaboration of a multidisciplinary team and the introduction of additional design concepts. Full awareness of the functionality of PV systems and of architectural quality is very important in order to create a multifunctional PV element that complies simultaneously with practical and aesthetic needs in buildings. Creativity will be a determinant factor when combining disciplines to achieve an appealing result. Both multifunctionality and aesthetics are the most important factors in BIPVs (Figures 6, 7 and 8).

3.1 Multifunctionality of PV Modules

A key property of BIPVs is that they perform various tasks simultaneously. They are active components that maximise energy production, and they are passive external building elements that contribute to minimising energy consumption (Table 1). Thus, sharing a number of functions in one same element forms a crucial part of the energy strategy in buildings.

Figure 6 PV facade at Sonnepark Dornbirn, Austria. *(Source: IEA Task VIIPV Database.* © *stromaufwärts. Reprinted with permission.)*

It is possible to use conventional PV modules to achieve multifunctionality, but then, the architectural possibilities are limited and the aesthetics are compromised. More convenient are multifunctional PV modules, which are designed to comply with mechanical, aesthetic, and functional requirements in buildings. There is a wide range of multifunctional PV modules in the market. Also, modules with special requirements may be custom-made by certain manufacturers. All BIPV should comply with local building regulations (see also Chapter IIc-2) [1].

3.2 PV Mounting Techniques

There are various ways of mounting PV modules to buildings. They vary according to the location of the PV array in the building, the size of the array, and the additional functions the PV modules must carry out for the energy strategy of the building.

Figure 7 An independent PV structure: the giant solar cube in southern California. *(Photo courtesy of Steven Strong, Solar Design Associates, Harvard, MA, USA.)*

3.2.1 Flat Roofs

A typical solution for installing PV modules in flat roofs is to fix the modules to a mounting system that is heavy enough to be able to stay in place without having to be fixed by bolts and nuts to the building, avoiding perforation. This mounting system is usually made of concrete, or there are solutions where a lightweight, easy-to-transport case may be filled in at the site with any heavy material such as stones (Figure 9).

3.2.2 Sloped Roofs

Aluminium sections are often mounted to the roof to provide a primary structure to which the PV modules may be attached. Another common solution for mounting PVs in a sloped roof is as PV tiles. In this type of

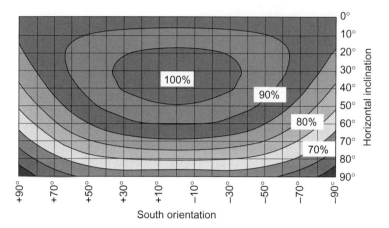

Figure 8 Solar energy availability in southern Europe. *(Source: Teulades i Façanes Multifuncionals.)*

solution, the PV cells are encapsulated in modules that act as traditional tiles (Figure 10).

3.2.3 Facade

In facades, as on rooftops, when the system is large, it is convenient to preassemble the modules so that they may be mounted in groups. This way, time and money are saved. On the other hand, PV modules installed as louvres have a special mounting system that usually offers a variety of sizes, fixing possibilities, solar cell types, and densities (Figure 11).

3.2.4 Independent Structures

Photovoltaics in independent structures are normally mounted in a conventional way. If they are to be integrated to the architectural whole of the building, as part of a landscape, then special attention should be paid to the aesthetic qualities of the mounting structure (Figure 12).

4. AESTHETICS IN PV TECHNOLOGY

The aesthetic properties of PV technology is crucial for its acceptance and implementation in the built environment. It is therefore important to harmonise the visual impact of PVs with the architectural language in buildings. There are various factors that determine the visual

Table 1 Multifunctionality of BIPVs

PVs as:	Multifunctionality	Figure	Explanation
Active components	Hybrid PV–thermal system		The produced heat in the PV modules may be transported in a heat-transfer medium (air or water) to be used for minimising energy consumption for heating during winter. During summer, the heat may be discharged to the atmosphere, cooling the PV cells and increasing their efficiency.
Passive components	Natural daylight		Semitransparent PV modules produce electricity whilst allowing natural daylight into the building. Energy consumption is reduced in artificial lighting. The heat introduced to the building by daylight should be controlled during hot weather.
	Shading		PV modules may produce electricity whilst protecting the building from excessive solar radiation. The cooling energy loads are reduced in hot weather by reducing solar incidence on the building.
	Natural ventilation		If well planned, may be used to cool the building, reducing cooling energy loads during hot weather. Natural ventilation cools the PV modules, increasing their performance.

Figure 9 PV rooftop mounting system. (© *Econergy International. Reprinted with permission.)*

Figure 10 Installing a sloped PV roof. *(Source: IEA Task VII PV Database. © MSK Corporation. Reprinted with permission.)*

Figure 11 Installing PV facade. *(© Teulades i Façanes Multifuncionals. Reprinted with permission.)*

impact of PV technology. Mainly, these factors are the elements that compose the PV modules and their configuration. It is essential to explore the aesthetic potential these factors have in order to enhance the visual value of PV systems.

Figure 12 Installing an independent PV structure. (© *Teulades i Façanes Multifuncionals. Reprinted with permission.*)

It is in the manufacturing process that opportunities are found to transform the visual impact of the PV modules. A few of the properties that may be modified are shape, size, colour, texture, and translucency (Table 2) [2].

4.1 Shape

The possibility to achieve various shapes of PV cells and modules increases the chances of adapting this technology onto architectural projects where there are various types of external surfaces not necessarily flat nor square. There are certain buildings in which the specific shapes of PV modules is the most convenient solution for fitting the system properly in place.

There are basically two factors that determine the shape of PV modules:
1. The photovoltaic wafers
2. The front and back cover of the module

Conventional crystalline-silicon wafers are square in shape if they are multicrystalline and circular if they are single-crystalline. Single-crystalline circular wafers are often trimmed to semisquares so that space is optimised in the module. During the manufacturing process, it is possible to obtain

Table 2 Visual determinants in PV technology

Property	Determined by
Shape	Semiconductor material
	Front and back covers
Size	Number of cells
	Gap between the cells
Colour	Antireflective coating (front side)
	Back cover (back side)
Texture	Semiconductor material
	Gap between cells
	Contact grid
Translucency	Semiconductor material
	Gap between cells
	Front and back covers

Table 3 Typical sizes for multicrystalline silicon PV modules

Nominal Power (W)	Area (m^2): n	H*W*T (mm)	No. of cells
50	0.5	940*500*50	36
120	1.10	1100*990*50	72

other forms of wafers as triangles, hexagons, and others. Various shapes of thin-film PV modules may also be achieved by depositing the semiconductor material to a glass substrate with any type of shape.

The shape of the module where the wafers are encapsulated may be any shape the glass industry may offer. Aesthetically, most architects would prefer self-similitude where the shape of the PV module is the same shape as the cells (i.e., triangular modules with triangular cells, hexagonal cells in hexagonal modules, etc.) [3] (Figure 13).

4.2 Size

Photovoltaic technology is characterised for being modular. A wide range of nominal power may be achieved in modules by interconnecting solar cells in series and/or in parallel. The nominal power for building-integrated PV modules is normally between 50 W and 200 W, but this may vary according to the specific project. The size of a crystalline silicon PV module is mostly determined by the rated power of the solar cells that are encapsulated in it (Table 3). Amorphous silicon PV modules require more area per Wp than do crystalline silicon PV modules (Table 4).

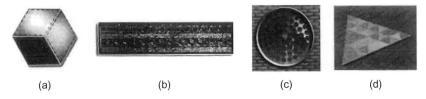

(a)   (b)   (c)   (d)

Figure 13 PV shapes, (a–c) Modules demonstrated in the EU BIMODE Project [4]. (© BP Solar. Reprinted with permission.) (d) The triangular module developed by one of the authors (R.L.C.) at the University of Southampton, UK.

Table 4 Typical sizes for amorphous silicon PV modules

Nominal Power (W)	Area (m^2)	H*W*T (mm)	No. of cells
50	0.8	1200*650*50	71
128	2.6	5800*450*30	22

Source: Photon Magazine

4.3 Colour

The colour of crystalline silicon solar cells is determined by the width of the antireflective coating, a thin layer of silicon nitride that prevents reflection of solar energy from the cells. The colours we normally see in solar cells (i.e., dark grey for single crystalline, dark blue for multicrystalline) are produced by the antireflective coating thickness that allows the highest efficiencies. By varying the thickness of the antireflective coating, we achieve new colours that add to the aesthetic possibilities of PV technology but compromise the efficiency of the cells (Figure 14).

4.4 Texture and Patterns

The texture of the solar cells is determined by the type of technology from which the solar cell is made of. Single crystalline solar cells have a homogenous texture; the texture of multicrystalline solar cells is characterised by multifaceted reflections. Amorphous silicon has a different texture from crystalline silicon cells as a result of the thin-film manufacturing process.

Patterns in solar cells are determined by the contact grid, which extends across the cell in order to collect the produced electricity. It is necessary to consider the minimum obstruction possible on the front side

Figure 14 Coloured monocrystalline (a) and multicrystalline (b) silicon solar cells manufactured by Solartec. The efficiency of monocrystalline cells ranges from 11.8% (silver) to 15.8% (dark blue, standard). *(© Solartec. Reprinted with permission.)*

in order to allow the maximum solar energy to reach the semiconductor material. Although most creative designs of contact grid are on a experimental stage, some may be found in the market. The designs shown in Figure 16 are an alternative to the standard H-grid pattern shown in Figure 15.

4.5 Translucency

The translucency of PV modules may be achieved in various ways. The front and back covers play an important role. The front cover is always transparent in order to allow the maximum solar energy through to the cells. The back cover may be opaque, translucent, or transparent, depending on the natural lighting needs of the building. The gap between the cells may determine the semitransparency of the modules by letting light through. Further, crystalline silicon solar cells may be microperforated to achieve translucency. Thin-film PV modules made of amorphous silicon or dye-sensitised cells can also be used to produce semitransparent or translucent cells [7] (Figure 17).

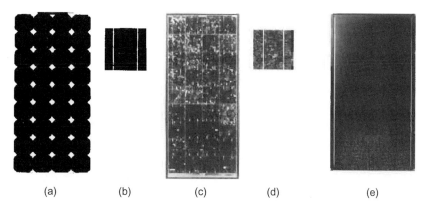

Figure 15 Texture in photovollaics. Single-crystal modules (a) and solar cells (b). The pseudosquare shape in (a) is more usual for the single crystal than the true square in (c). Multicrystalline modules (c) and solar cells (d). Amorphous modules (e). *(© BP Solar. Reprinted with permission.)*

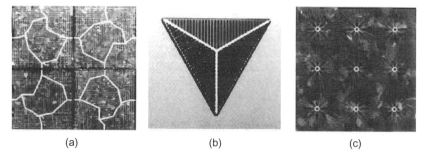

Figure 16 Patterns in photovoltaics. (a) One of the patterns developed by the Atomic Institute Vienna [5, 6]. (© Photon magazine. Reprinted with permission.) (b) The triangular cell developed by one of the authors (R.L.C.) at University of Southampton, UK. (c) ECN Pinup. *(Source: ECN homepage.)*

4.6 Point of View

The distance and the angle from where the PVs are seen from is a determinant factor for the visual impact they produce. PV modules differ in appearance from the front and from the back. The front cover, seen from the outside of the building, shows the semiconductor material with an antireflexive coating and the contact grid. The back side, on the other hand, is seen from the interior of the building. There is also a difference whether it is seen from near or from far, as the following images show (Figure 18).

Figure 17 Translucency in PVs. (© *Teuludes i Façanes Multifuncionals.* Reprinted with permission.)

5. BUILT EXAMPLES

The following projects show how building-integrated photovoltaics enrich the architectural value of buildings, applying new concepts of sustainability to the urban landscape.

5.1 Case Study 1: The British Pavilion, Expo '92 [8]

Location: Seville, Spain.
Date of construction: 1992 (Figure 19).

5.1.1 Description
An example of bioclimatic architecture in a hot and dry climate that combines traditional cooling strategies with high-tech PVs.

5.1.2 PV System
- Nominal power: 46 kWp
- Electricity generated: 70 MWh/year (estimated)
- Multifunctionality: PV elements shade the building whilst supplying energy for the cooling system.

Figure 18 PV back and front, far and close. (a) *Source: IEA Task VII PV Database.* © *Energie-Forum Innovation.* (b) (© *Teulades i Façanes Multifuncionals. Reprinted with permission.*)

Figure 19 The British Pavilion, Expo '92. *(© BP Solar. Reprinted with permission.)*

5.1.3 Architectural Energy System

A large water wall applies traditional cooling principles. Apart from giving the glass a translucent quality, the water limits the glass temperature to around 24 degrees, providing a cooler internal environment. The energy for the water pumps is supplied by the photovoltaic modules integrated to the sun sails on the flat roof.

5.2 Case Study 2: ECN, Buildings 31 and 42 [9]

Location: Netherlands Energy Research Foundation, Petten, the Netherlands.
Date of construction: 2000–2001 (Figure 20).

5.2.1 Description

PV technology is architecturally integrated in buildings 31 and 42 of the ECN, showing the proper integration of this technology into both new and existing buildings.

5.2.2 Building 31
PV System
- Nominal power: 72 kWp

Figure 20 ECN, Buildings 31 and 42. *(© BEAR Architecten.)*

- Electricity generated: 57.60 MWh/year
- Multifunctionality: PV blinds that allow natural daylight and solar gains whilst shading.

Architectural Energy System
An existing building is covered in PV sunshades that prevent the building from overheating during summertime whilst diffusing natural daylight to the interior. The renovation of the building is expected to reduce the primary energy demand by 75%.

5.2.3 Building 42
PV System
- Nominal power: 43 kWp
- Electricity generated: 34.40 MWh/year
- Multifunctionality: Glass—glass PV modules that allow natural daylight and solar gains whilst shading.

Architectural Energy System
A PV-covered conservatory, which is the connecting space between the two buildings, shades the building preventing overheating in summer at the same time that allows natural ventilation and a maximum use of daylight.

5.3 Case Study 3: Pompeu Fabra Public Library [10]
Location: Mataró, Spain.
   Date of Construction: 1993 (Figure 21).

Figure 21 Pompeu Fabra Public Library. *(© Teulades i Façanes Multifuncionals.)*

5.3.1 Description

A pioneer demonstration project where a hybrid PV–thermal multifunctional system was installed in the facade and rooftop.

5.3.2 PV System

- Nominal power: 53 kWp
- Electricity generated: 45 MWh/year
- Multifunctionality: An air chamber in the PV modules acts as a thermal collector system whilst allowing natural daylight, solar gains, and shading.

Architectural Energy System

The south-facing vertical facade of the building and the skylights are equipped with a multifunctional hybrid PV–thermal system. The produced electricity is exported to the electric grid. The produced thermal energy is dissipated to the surroundings in summer, and used by the heating system during winter, thus reducing cooling and heating loads, respectively. Natural daylight is introduced to the building through the PV skylights and south facade, reducing artificial lighting in the building.

5.4 Case Study 4: Mont Cenis Conference Centre [11]

Location: Herne Sondingen, Germany.
 Date of Construction: 1999 (Figure 22).

Architectural Integration of Solar Cells

Figure 22 The Mont Cenis Conference Centre in Herne Sondingen. (a) The front elevation. (b) The nonuniformly spaced solar cells in the roof modules resemble clouded sky.

5.4.1 Description

The old coal mining site of Mont Cenis, in Herne-Sodingen has now been developed into an energy and environmentally conscious conference centre. The buildings are protected by a carefully designed membrane containing photovoltaic modules that resemble clouds in the sky and create a Mediterranean climate in the site.

5.4.2 PV System

- Nominal power: 1000 kWp
- Electricity generated: 750 MWh/year
- Multifunctionality: PV modules in the external membrane shade the buildings whilst allowing solar gains, natural daylight, and natural ventilation to the conference centre area.

Architectural Energy System

A series of architectural sustainability concepts have been applied to this project. There is plenty of vegetation, openings in the membrane allow for natural ventilation, and the translucent PV modules create a comfortable natural lighting effect. Rain water is collected and the building has an efficient water use system. The structure is made from wood grown near the site reducing the energy content in energy and materials.

During winter, the heat received by solar irradiation in the building is used, reducing heating loads. During the summer, natural ventilation is backed up by a mechanical system creating a breeze that maintains the site in a comfortable atmosphere.

REFERENCES

[1] T. Markvart (Ed.), Solar Electricity, John Wiley & Sons, 2000.
[2] F. Sick, T. Erge, Photovoltaics in Buildings—A Design Handbook for Architects and Engineers, International Energy Agency, Solar Heating and Cooling Programme, Task 16, 1996, pp. 87 and 88.
[3] N. Pearsall, Results of European wide consultation of architects, In: Architecture and Photovoltaics—The Art of Merging, January 28, 1997, Leuven, Belgium EUREC Agency.
[4] A. Schneider, J. Claus, J.P. Janka, H. Costard, T.M. Bruton, R. Noble, et al., Development of Coloured Solar Modules for Artistic Expression with Solar Facade Designs Within the 'BIMODE' Project. Proceedings of the Sixteernth European Photovoltaic Solar Energy Conference, Glasgow, 2000.
[5] M. Radike, J. Summhammer, A. Breymesser, V. Sclosser, Optimization of artistic contact patterns on mSi solar cells. Proceedings of the Second World Conference on Photovoltaic Solar Energy Conversion, Vienna, 1998, p. 1603.
[6] M. Radike, J. Summhammer, Electrical and shading power losses of decorative PV Front Contact Patterns, Prog. Photovolt. Res. Appl. 7 (1999) 399–407.

[7] A. Hänel, State of the art in building integrated photovoltaics, European Directory of Sustainable and Energy Efficient Building, 1999, pp. 80–87.
[8] O. Humm, P. Toggweiler, Photovoltaics in Architecture, Birkhaüser Verlag, Basel, 1993.
[9] T. Reijenga, PV—Integration in solar shading (renovation) and PV—Integration in atrium glazing (New Building), ECN 31 and 42, Petten (NL).
[10] A. Lloret, The Little Story of the Pompeii Fabra Library of Mataró, Editoreal Mediterranea; Lloret, A. 1996. Móduls Fotovoltaics Multifuncionals a la Biblioteca de Mataró. Eficiencia Energetica no. 137, April/June, 1996.
[11] Entwicklungsgesellschaft Mont-Cenis (Brochure with description of the project) Stadt Herne/Montan Grustuckgesselschaft, Energieland NRW, Model Project: Mont-Cenis Energy Park, Future Energies, 1998.

FURTHER READING

A Thermie Programme Action. 1998. Integration of Solar Components in Buildings. Greenpeace. 1995. Unlocking the Power of our Cities—Solar Power and Commercial Buildings.
Greenpeace, Building Homes with Solar Power, London, 1996.
A. Hänel, State of the Art in Building Integrated Photovoltaics, European Directory of Sustainable and Energy Efficient Building, 1999.
D. Lloyd Jones, The Solar Office: A Solar Powered Building With a Comprehensive Energy Strategy, European Directory of Sustainable and Energy Efficient Building, 1999.
Max Fordham & Partners in association with Feilden Clegg Architects, Photovoltaics in Buildings—A Design Guide, Department of Trade and Industry, UK, 1999, Report No ETSU S/P2/00282/REP.
National Renewable Energy Laboratory, Solar Electric Buildings—An Overview of Today's Applications, US, Department of Energy, Washington, USA, 1996, DOE/GO-10096–253.
D. Niephaus, N. Mosko, The polyfacial PV power plant concept based on triangle solar cells and bypass support modules, Proceedings of the Fourteenth European Photovoltaic Solar Energy Conference, Barcelona, 1997, pp. 1589–1592.
Photovoltaic Solar Energy—Best Practice Stories, European Commission, Directorate General XVII Energy, 1997.
E. Shaar-Gabriel, Berlin Turns to the Sun. Solar Projects in the German Capital, European Directory of Sustainable and Energy Efficient Building, 1999.
S.J. Strong, Power Windows—Building Integrated Photovoltaics, IEEE Spectrum, October, 1996.
Studio E. Architects, *Photovoltaics in Buildings—BIPV Projects*, Report ETSU S/P2/00328/REP, Department of Trade and Industry, London, 2000.
Ten Hagen & Stam, *Building with Photovoltaics*, International Energy Agency, the Netherlands, ISBN 9071694372, 1995.
P. Toggweiler, *Integration of Photovoltaic Systems in Roofs, Facades and the Built Environment*, European Directory of Sustainable and Energy Efficient Building, 1998, pp. 42–45.
E. Van Zee, Building Power Stations—Will New Residential Areas Be the Power Stations of the 21st Century? *Renewable Energy World*, July, 1998.

CHAPTER IIE-2

Solar Parks and Solar Farms

Philip R. Wolfe
WolfeWare Limited, UK

Contents

1. What Is a Solar Park?   944
   1.1 A Brief History of Solar Parks   944
2. Design Issues for Solar Parks   946
   2.1 Site Selection   946
       2.1.1 Optimising the Load Factor   946
       2.1.2 Optimising Power Density   947
       2.1.3 Local Electricity Network and Usage   947
       2.1.4 Location and Land Usage   947
   2.2 The Solar Arrays   948
       2.2.1 PV Technology Selection   948
       2.2.2 Fixed Arrays   949
       2.2.3 Tracking Arrays   950
       2.2.4 Array Mountings   951
   2.3 Power Conditioning   953
3. Solar Park Project Development Issues   953
   3.1 Planning Consent   954
       3.1.1 Rationale for Development and Site Location   954
       3.1.2 Landscape and Visual Change   955
       3.1.3 Light Reflection and Aviation   955
       3.1.4 Surface Water Drainage, Flooding, and Ground Conditions   956
       3.1.5 Noise, Vibration, Traffic, and Air Quality   956
       3.1.6 Ecology and Nature Conservation   956
       3.1.7 Archaeology and Cultural Heritage   957
       3.1.8 Community Engagement and Legacy Issues   957
   3.2 Connection   958
   3.3 Financing   958
4. Regulatory Issues for Solar Parks   959
   4.1 Energy and Environment Policy   959
   4.2 Incentive Schemes   959
   4.3 Consenting   960
5. The End Game   960
   5.1 Grid Parity   961
   5.2 The Future of Solar Parks   962
Acknowledgements   962

Practical Handbook of Photovoltaics.
© 2012 Elsevier Ltd. All rights reserved.

Since the first edition of this book, the deployment of photovoltaic systems has mushroomed dramatically. Much of this has happened in Europe, stimulated by the feed-in tariffs that many countries have adopted.

The feed-in tariff mechanism is well suited to rooftop and other building-integrated applications as described in Chapter IIe-1. Additionally, it has created a new market for merchant power generation using stand-alone systems.

1. WHAT IS A SOLAR PARK?

No surprises here! A solar park is to photovoltaics what a wind farm is to wind power—a large-scale generating station generally designed to supply bulk power to the electricity grid. Many people call them *solar farms*, though some only use this expression for installations on agricultural land. This chapter uses *solar park* throughout and will clarify when alluding specifically to greenfield sites.

Before PV reached the price level where it could realistically compare with grid electricity, it was used mainly for stand-alone off-grid use and specialised applications where its high reliability or architectural qualities enhanced its value.

When feed-in tariffs made it economically competitive with grid power, they opened up a new market for merchant generating stations. While many of these may be rooftop installations, the easiest place to install large systems is on the ground.

1.1 A Brief History of Solar Parks

The first megawatt-scale merchant PV installation, and arguably the first solar park was installed by the Atlantic Richfield subsidiary Arco Solar in 1982. It was a two-axis tracking system at Hesperia, California (Figure 1), followed in 1983 by a 5-MW installation at Carrizo Plain (Figure 2).

At the time, PV was being used primarily for stand-alone off-grid projects. Arco was convinced that large-scale grid deployment was the future and undertook these groundbreaking projects to illustrate that potential, despite the relatively high prevailing cost at the time. The 5-MW system was eventually decommissioned in 1994, by which time

Figure 1 The 1-MW system built in 1982 by Arco Solar at Hesperia, California, was arguably the world's first solar park.

Figure 2 The 5-MW system at Carrizo Plain built by Arco Solar in 1983 and decommissioned in 1994 used two-axis tracking and mirrors to achieve low concentration on the rows of solar modules in the middle.

it could not possibly have recovered its cost. Nonetheless, it proved the ability of PV to integrate with the grid and produce power to utility standard.

Meanwhile, primarily in Japan and Europe, grid-connected schemes of several hundred kilowatts capacity were also installed under subsidised programmes.

Figure 3 Solar farming. The German feed-in tariffs made-larger scale photovoltaic systems like this rooftop installation in Dardesheim economic.

More widespread deployment was then stimulated by the introduction of feed-in tariffs initially in Germany (Figure 3) and then progressively in other markets in Europe and around the world.

2. DESIGN ISSUES FOR SOLAR PARKS

Before getting into detail about the technological and design issues associated with this way of deploying photovoltaics, a review of one of the project development issues should be helpful. Other project drivers are then examined further in Section 3.

2.1 Site Selection

Several factors influence the selection of sites for solar parks: some technical and PV specific; others related to regulatory, funding, and operational factors. All of these may impact on the cost and therefore the financial returns of the project in various ways.

2.1.1 Optimising the Load Factor

Clearly, the best financial returns will be obtained by maximising the delivery of the system—the annual kilowatt hours per kilowatt peak of capacity. This will inform not only the design of the system as discussed in Section 2 but also the site selection. The radiation level is obviously one key parameter. In most regions of the world, this tends to vary only marginally with horizontal distance (unlike wind, for example, which can be very site specific). There are, however, parts of the world where local microclimates can make sites particularly advantageous or undesirable.

Similarly, sites that are shaded at certain times of day or year by mountains, buildings, or vegetation are best avoided.

2.1.2 Optimising Power Density

The orientation is also important in defining the size of site for a particular energy output. In the northern hemisphere, sites that slope down to the south allow the solar arrays to be closer spaced (as described in Section 2.2) and therefore require less hectares per megawatt of capacity. For similar reasons, unevenly contoured sites will lead to less-efficient packing density.

While the power density does not affect the array capacity required for a given output (unlike the load factor discussed in Section 2.1.1), it does affect the size of the site. Therefore, it has an impact on balance-of-system costs, such as cabling and site rental, if not the primary solar array cost.

2.1.3 Local Electricity Network and Usage

Depending on the applicable energy payment regime, it is generally preferable to locate solar generation stations where they can supply at least part of their energy to local users. This is because that part of the energy can be supplied at the retail rather than grid price of electricity.

Solar parks on or near industrial estates, farms, and other electricity consumers can often therefore show better returns than those that purely supply wholesale power to the grid.

In almost every case, they will need connection to the wider electricity network, so the closeness and the capacity of this network becomes another key determinant. If the distribution or the transmission network to which the installation would be connected is relatively far away or weak and in need of reinforcement, this can load substantial additional costs on the project and make it unviable.

2.1.4 Location and Land Usage

Another key issue of course is the geographical location itself and the current use of the land. This can impact, in particular, the speed and success of planning consent (as further discussed in Section 4.1).

Siting solar parks on so-called brownfield sites, such as former industrial or military estates, quarries, and landfills, can often enhance the prospects of planning approval. Many authorities would consider this a good way of remediating such land.

Using agricultural land is more contentious. Solar parks do not fully cover the land, so there is often considerable space between the solar arrays, which can continue to be used for other purposes, provided these do not shade or dirty the arrays. The grazing of sheep in solar parks is a good complementary application, as that will prevent the growth of vegetation that could create shadows over time. However, many agricultural activities—intensive farming in particular—become impossible in solar farms. To prevent unnecessary conflict for land use between energy and food, therefore, solar parks should generally not be sited on prime agricultural land.

Having found a suitable site, it's then time to design the system. Many factors are generic to PV systems in general and are therefore covered elsewhere in this book. I shall deal here with the specifics of large-scale solar parks.

2.2 The Solar Arrays

The most suitable PV technology and configuration will depend on the application and location.

2.2.1 PV Technology Selection

No particular technology is fundamentally better suited for solar park applications. Both mono- and polycrystalline silicon are widely used, and thin films have also been applied to solar parks (Figure 4).

As in other applications, the relationship between output and cost is the primary determinant. If it becomes clear that certain technologies achieve predictably higher yields (in terms of kilowatt hours per kilowatt peak) than others in certain radiation regimes, then that will also impact the choice.

At present, it seems that the technology distribution broadly mirrors the wider market, with crystalline silicon the most widely used.

Concentrator cells as described in Chapter IId-1 may also be appropriate in certain instances, especially where there is a high proportion of direct radiation. For the reasons identified there, these will almost invariably use tracking systems as described below.

There are also several choices of how the arrays can be mounted.

Figure 4 Solar park at Salmdorf, Germany, using thin-film solar cells.

2.2.2 Fixed Arrays

The simplest mounting option is a fixed array tilted at a prescribed pitch angle roughly toward the equator.

The pitch angle is typically selected to maximise the annual output of the system. In subtropical areas, the angle is typically close to the angle of latitude, but in more temperate climates it is less than the latitude—to maximise summer output and the collection of diffuse radiation.

Direct radiation, of course, arrives at different angles, depending on the time of day and year, and as described in Chapter IIa-2 the intensity is proportional to the cosine of the angle between its direction and the normal to the solar array. Therefore, fixed arrays only rarely achieve the maximum possible output, so tracking arrays (described in the next section) can sometimes improve project returns.

In climates with a lower proportion of direct radiation, the benefits are less marked, because the higher proportion of diffuse radiation—that filtered through clouds, for example—is multidirectional. It is therefore normal for installations in temperate climates, such as northern Europe, to use fixed-pitch arrays.

Additionally, the selected pitch angle needs to give some consideration to the spacing between subarrays. Large sites use many rows of subarrays that, particularly in winter, may shade each other if too closely spaced. Lower tilt angles reduce the height difference between the top and bottom edges of the subarrays and therefore allow closer spacing and thus lower area-based costs.

2.2.3 Tracking Arrays

The solar arrays can use mounting structures that change the inclination of the solar modules to face toward the Sun.

The most comprehensive form is two-axis tracking where the arrays can be controlled to follow the Sun precisely and continuously (Figure 5). This requires a drive system to orientate the arrays and realign them at the end of the day. This usually uses some form of preprogrammed logic control to identify the correct direction, though some trackers sense the Sun's direction by constantly optimising the output. Depending on location, the system output can be increased by as much as 50% by dual-axis trackers, compared to a fixed system.

A mechanically less-complex option is the single-axis tracker. There are two options:
1. Daytime trackers (Figure 6) are fixed at a specific pitch in the plane of latitude but follow the azimuth angle of the Sun.
2. Seasonal trackers ((Figure 7) are adjusted only a few times per year to optimise output against the height of the Sun in the sky.

The daytime trackers again need continuous control and require power. Depending on location, they can achieve performance improvements of around 30%. Seasonal trackers achieve lower gains, seldom more than 10%, but are simpler to implement, and the adjustments can even be made manually typically 2 to 4 times per annum. However, the steeper angles used in winter lead to the need for wider spacing between subarray rows, as noted in Section 2.2.2.

Figure 5 Dual-axis trackers follow the Sun by rotating in both the horizontal and vertical planes.

Figure 6 Daytime or polar single-axis trackers like these at the solar park at Nellis Air Force Base in the USA tilt to follow the Sun as it moves from east to west during the day.

Figure 7 Seasonal or horizontal single axis trackers are adjusted only a few times per year to match the height of the Sun in the sky.

Both the structure and the controls required for tracker systems are more complex and expensive than for fixed-pitch arrays. They also typically require a small amount of power. They can therefore be justified only when the benefit of the increased output exceeds this additional outlay.

2.2.4 Array Mountings

Although the array mounting for a solar park is fundamentally the same as any other ground-mounted arrays, there are a few considerations worth

Figure 8 Pile-driven mountings or ground screws like those shown here minimise the site disturbance when locating array mountings.

bearing in mind, and a number of new mounting options have been evolved specifically for this application. Most of these comments relate to fixed arrays, because trackers (apart from single-axis seasonal trackers) need more substantial mounting to carry the more complex structure and mechanical loading, so they present fewer options.

Solar parks are stand-alone generating stations usually for merchant power generation, rather than systems servicing identified loads such as building-integrated or professional applications. The latter typically use concrete plinths to mount the arrays firmly for a very long design life. Solar parks, on the other hand, are configured to minimise their impact on the surrounding environment, minimise the incremental water-runoff and flood-risk issues, and facilitate decommissioning at the end of their service life.

A range of screwed and pile-driven mountings have therefore been developed (Figure 8), especially for greenfield sites to minimise the disturbance to the site and so that they can be removed entirely at the end of the project life. Another way of minimising disruption to the subsoil is to use ballasted systems (Figure 9) that sit on the existing site without penetrating the surface. Clearly, these need to be designed with sufficient weight and rigidity to withstand the expected wind loading. It may also be a suitable approach for corrosive or contaminated soil or to minimise disruption to the habitat of underground wildlife.

Figure 9 This ballasted support system at Ferrassières sits on the ground without any foundations or fixings.

2.3 Power Conditioning

Fundamentally, the requirements for power conditioning in solar parks are similar to those for any other PV system delivering an AC output detailed in Chapter IIb-1. However, the large size of the systems gives rise to additional options.

For example, the system designer can use a single large inverter or alternatively a number of parallel-connected *string inverters*. No single preferred approach to this option has yet gained the ascendency, but either way a level of redundancy is preferable to limit the impact of any electronic malfunctions on system performance. Clearly, string inverters inherently offer a degree of redundancy while centralised inverters can do the same by configuring the power conversion in a number of discrete modules.

3. SOLAR PARK PROJECT DEVELOPMENT ISSUES

We have seen that the technology used in solar parks is common to that applied in PV systems for other applications. Many of the distinctive factors are connected with the nontechnical issues facing project developers.

In most parts of the world, solar parks require approval before they can be constructed. They may need specific consents connected with national policies on solar energy or renewables in general, as further discussed in Section 4. They will often also need approval for their siting and construction and to connect to the electricity network as described in the next two subsections.

3.1 Planning Consent

Many issues will affect the view that planning authorities will take to proposed solar park developments. The key ones from experience in Europe and the UK in particular are the following.

3.1.1 Rationale for Development and Site Location

Planners will expect to understand why the solar park is being proposed in the local context, the contribution it will make to national and local policies, energy supply, the social economy, and the environment. This will typically include an overview of the capacity and output of the scheme measured in terms of local electricity consumption. The potential for local jobs during construction and the life of the project should also be considered.

Planners will also wish to ensure that excessive land is not given over to solar parks. They will therefore prioritise sites taking into account criteria such as the existing land use and the local energy off-take as discussed in Section 2.1. There may also be local issues that influence this, including electricity security of supply.

Agricultural land can be contentious for solar farms, because of the visual impact and possible loss of productive land. Developments should therefore be directed first toward less-productive land. Because of the spacing of solar arrays within a solar farm, it is possible to retain agricultural usage of the land on which they are sited. Although arable application would generally prove difficult, grazing remains possible within array fields and indeed is particularly synergistic as it prevents plant growth that may, in due course, shadow the solar arrays. For similar reasons, array fields can also play a part in nature conservation plans, as further described below.

There may be further issues about locating solar parks in sensitive and restricted areas such as heritage sites, conservation areas, sites of special scientific interest, and areas of outstanding natural beauty. In such circumstances, the project developer should take particular care to select the site to avoid or minimise impacts.

3.1.2 Landscape and Visual Change

Because there are no adverse environmental impacts, planning considerations tend to focus on the physical and visual impacts of solar systems—principally, the solar array.

Designs can minimise such visual impacts by limiting the height of the solar arrays, even at their steepest angle, so that they are often no higher than surrounding hedgerows. Rows of arrays can be arranged to follow local topography (Figure 10). The arrangement of discrete array blocks at staggered intervals and with spacing in between breaks up any potential visual monotony and avoids undue wind concentration or turbulence.

Generally, there is a need to ensure solar facilities are adequately secured to reduce theft and vandalism risk and protect passersby. The requirements will vary from site to site, and designs should be produced to avoid unacceptable landscape or visual impact. This includes utilising existing hedges, minimising the height of security fencing, using natural features, and taking appropriate measures that ensure continued access by larger mammals such as badgers and foxes.

3.1.3 Light Reflection and Aviation

Solar panels are designed to absorb irradiation and therefore to minimise the amount of light reflected. PV solar panels use an antireflective coating to maximise the light capture of the solar cells. They are thus responsible

Figure 10 Arrays can be designed to follow the local topography, like this solar park in Ehrfurt, Germany.

for only limited levels of either glint or glare—substantially less reflective than glass houses, for example.

Because systems avoid adverse effects from reflected light, they conform to the air navigation requirements of most countries that I am aware of. Many systems are indeed located on airfields.

3.1.4 Surface Water Drainage, Flooding, and Ground Conditions

Solar panels drain to the existing ground, and (except in unusual circumstances where substantial concrete plinths are used) runoff should be no greater for the developed site than it was for the predeveloped site.

If suitably designed, solar farms can be located in flood plains; in fact, they offer a good utilisation of such areas.

Solar arrays are typically installed using screw-piled, piled, or ballast-supported systems as described in Section 2.2.4. They therefore cause minimal adverse impact on the ground and no contamination.

3.1.5 Noise, Vibration, Traffic, and Air Quality

Solar cells are inert solid state devices that convert light into electricity. The systems therefore produce virtually no noise and no emissions.

The inverters require some cooling, so there is a slight fan noise perceptible only if standing immediately adjacent to the housing. Otherwise, there are no moving parts, except in tracking systems as described in Section 2.2.3.

The panels do not give rise to any emissions that will have an impact on air quality.

During the construction stage, each MW of capacity may generate a few dozen movements for the delivery of the equipment. Thereafter, there will be visits by service personnel typically twice per annum. The systems need no refuelling or other routine provisions, so they will generate no additional traffic movements in the area.

3.1.6 Ecology and Nature Conservation

Solar farms have the potential to increase the biodiversity value of a site if the land was previously intensively managed. Sheep grazing or an autumn cut with removal of grass cuttings could increase the botanical diversity of the site. The solar park at Kobern-Gondorf in Germany, for example, is used as a nature reserve for endangered species of flora and fauna.

Solar parks typically present no negative environmental impact to the surrounding area and wildlife. They make a substantial positive contribution, of course, toward efforts to achieve a reduction in CO_2 emissions, and this has a positive impact on ecology generally.

3.1.7 Archaeology and Cultural Heritage

The support structures for the solar arrays typically protrude up to 1.5 metres into the ground. In a minority of cases of known archaeological interest, therefore, it might be necessary to commission a geophysical evaluation of the site or an appropriate programme to ensure that any archaeological interests are safeguarded and mitigated if necessary. It is possible as mentioned in Section 2.2.4 to employ ballasted support systems, which do not penetrate the ground.

The impact of solar park developments on local heritage assets will be very site dependent. One of the first solar parks consented in the UK, for example (Figure 11), was designed by Ownergy on a previous airfield site and adjudged 'a very effective re-use of the heritage assets, in character with the innovative technological history of the site.'

3.1.8 Community Engagement and Legacy Issues

In any new development, it is advisable to inform local communities about the characteristics and benefits of the technology as part of the engagement process.

Because solar systems are silent, clean, and unobtrusive, they have proved to be one of the most acceptable of the renewable-energy technologies and substantially less contentious than, for example, wind power.

Figure 11 Westcott Solar Park, one of the first in the UK, locates the solar array on the runway of this former airfield.

The lack of traffic movements also avoids the objections sometimes raised against thermal technologies like anaerobic digestion and biomass.

Attention normally centres on the potential visual impact. Because, as detailed in Section 3.1.4, there are no substantial reflection issues and well-designed schemes maintain the height of the systems below that of surrounding hedges and landscaping, proposed developments have met with widespread support from local communities.

Solar facilities developed on agricultural ground are 'reversible,' allowing the site to be easily restored to a more-intensive agricultural use. Intrusive development, such as trenching and foundations, are minimised. Some planning consents apply the condition that the land be permanently reinstated to its original condition at the end of the project life. Much of the equipment can also be recycled at the end of its useful life.

3.2 Connection

The requirements for connection approvals vary substantially from country to country and most are common to other forms of distributed generation too.

While there is little that I can say generically about this issue, solar park developers should consider this issue very early in the programme, as in many cases the availability, consenting requirements, or cost of the grid connection can be a make-or-break issue for the project.

3.3 Financing

As this section deals with the fundamental issues that define whether a solar park can viably be constructed, it is useful to consider the third prerequisite for a successful project—the money.

Clearly, beyond the specific incentive mechanisms discussed in Section 4.1, there is nothing PV specific about constructing an engineering project that achieves the necessary level of return for its investors. However, the successful development of solar farms in Europe and elsewhere has been enhanced by or generated specific funding options worthy of brief mention.

A number of funds now running to hundreds of millions of dollars have been created specifically to invest in larger-scale solar installations. These make a return to their investors from the dependable income stream these projects can generate through the incentive payments and electricity sales. This has in turn created opportunities for companies who develop projects on their balance sheets to then sell them, thus liberating funds for more new projects.

Some countries have also established infrastructure banks—such as the KfW in Germany and, hopefully, the proposed Green Infrastructure Bank in the UK—making them able to offer funding for such projects when it is not available on suitable terms from traditional finance providers.

4. REGULATORY ISSUES FOR SOLAR PARKS

Although I have started this chapter at the microlevel looking at system components and then moved on to project-level considerations, it is usually the macro issues at the national policy level that provide the main driver for solar parks and farms. We will look at these now.

4.1 Energy and Environment Policy

Environmental and energy policy is, of course, a primary driver for the increased deployment of solar parks and renewables in general. Many countries have specific targets for renewable generation. In the European Union, this is underpinned by the Renewable Energy Directive under which all states have presented national renewable-energy action plans, many of which include specific objectives for PV in general and, in some cases, solar parks in particular.

These policies are supported by a range of incentive mechanisms as described in Section 4.2 and supportive planning policies as discussed briefly in Section 4.3.

Additionally, many countries have policies that aid the connection of systems to national and local electricity grids. These include a legislative right to connect and, in many cases, *priority despatch* whereby renewable generators are guaranteed the ability to export power to the grid. In cases of high availability and low demand, therefore, it is the nonrenewable sources that are constrained first.

4.2 Incentive Schemes

Solar generation is heading toward *grid parity* where it can be economic without subsidy as discussed in Section 5. At present costs and with fossil-based energy still artificially cheap in most parts of the world, PV can look uncompetitive, so many countries and states have introduced incentive schemes to encourage deployment.

Most schemes apply to all PV applications (not just solar parks) and many to renewables more widely. I do not propose to deal with them all here, merely to highlight why tariff-based measures in particular have been so successful in bringing forward this type of application.

Until feed-in tariffs were introduced in Germany in the early 1990s, most support for PV installations had been in the form of R&D funding and capital grants toward the cost of installations. European programmes, for example, had supported a percentage of the capital cost of the installation, capped at a specified maximum in euros per peak watt. This provided little incentive to optimise the output of the system, and many installations performed relatively poorly. Also, the incentives were often linked to specific applications where alternative power costs were high, such as isolated communities or desalination plants, rather than merchant power into the grid. This approach did not act as a very strong driver to cost reduction.

Tariffs, on the other hand, reward the power delivered by the system. The most-efficient, lowest-cost systems benefit most under this regime. Therefore, efficient systems whose cost is partly covered by a predictable stream of tariff income are able to sell power competitively on the merchant market.

4.3 Consenting

Solar PV systems are very benign, as described in Section 3.1. Many countries have therefore introduced proactive policies to simplify or streamline the consenting process. The UK and other countries have designated some solar systems as *permitted development* that do not require planning approval, though consent is still normally required for large solar parks.

Similarly, many regional and district strategies encourage solar and other renewables, imposing a *presumption in favour* as the default response to planning applications.

5. THE END GAME

You will have read throughout this book how great the potential is for PV to reduce its cost. As a semiconductor technology, there is almost limitless potential for cost improvements with both increasing volume and developing technology. There is a strong and realistic expectation that photovoltaics will achieve grid parity in coming years—perhaps even by the time you read this!

5.1 Grid Parity

Grid parity is the level at which the cost of PV power matches that of the main network. This is really 'comparing apples with pears.' Traditional energy costs are made up largely of the cost of fuel, with a lesser element of plant cost—in some cases, the capital equipment has already been fully depreciated. Solar power has no refuelling cost, so it is derived by amortising the relatively higher capital cost over the useful output of the plant. As capital costs decline with volume, so PV power will become cheaper. Meanwhile, fossil fuels, emissions costs, and therefore traditional power prices are expected to rise.

Additionally, solar power generates at relatively small scale (even a 50-MW plant—large by PV standards—is one to two orders of magnitude smaller than a typical coal-fired plant). It is also typically located close to where the power is used. This means that in many cases the electricity can be supplied directly to users. In such cases, therefore, the energy is competing with the retail, rather than the wholesale, electricity price and therefore reaches parity earlier (Figure 12).

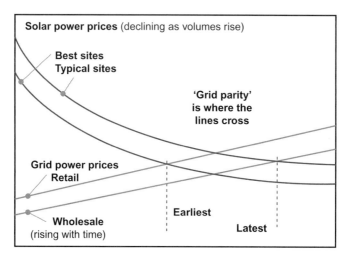

Figure 12 This diagram shows when grid parity is achieved as solar system prices fall while traditional electricity prices rise.

Any new energy source needs to offer the prospect of achieving grid parity if it is to become a serious contributor to the energy mix. Although often marginalised because of relatively higher present costs, photovoltaics is likely to be one of the first technologies to achieve grid parity in certain locations. When it does, it will no longer need incentives such as feed-in tariffs and will be able to compete in the open market. Its high reliability, long life, and benign environmental credentials should make it a very attractive—probably even the dominant—source of electrical power.

5.2 The Future of Solar Parks

What will be the future of solar parks in this scenario?

I personally don't see centralised fossil generation being replaced solely by centralised PV generation. A key benefit of the technology is its modularity and scalability, so decentralised applications such as rooftop systems delivering retail electricity will be especially attractive.

However, grid parity against wholesale electricity prices will also provide a market for grid-feed solar parks. Expect these to become very widespread, especially on otherwise unproductive land and where the interspacing between solar arrays can be put to good use.

ACKNOWLEDGEMENTS

The author thanks those companies mentioned herein and those whose installations are featured in this chapter.

CHAPTER IIE-3

Performance, Reliability, and User Experience

Ulrike Jahn
Institut für Solarenergieforschung GmbH, Hameln/Emmerthal (ISFH), Germany

Contents

1. Operational Performance Results 964
   1.1 Overview of Performance Indicators 964
   1.2 Summary Performance Results 967
   1.3 Results from 'Low Yield Analysis' 969
      1.3.1 Deviation from PV Module Specifications 970
      1.3.2 Shading 970
      1.3.3 Defects of DC Installations 971
      1.3.4 Inverter Problems 972
2. Trends in Long-Term Performance and Reliability 972
   2.1 Trends from German 1000-Roofs-PV-Programme 973
   2.2 Trends from Demonstration Projects in Switzerland 974
   2.3 Trends from PV Programmes in Italy 976
   2.4 Rise in PV System Performance 977
3. User Experience 978
   3.1 Small Residential PV System in Germany 978
   3.2 PV Sound Barrier in Switzerland 979
   3.3 PV Power Plant in Italy 980
   3.4 Lessons Learnt 981
Acknowledgements 982
Appendix. Specifications of Performance Database of IEA PVPS 983
References 984

Practical Handbook of Photovoltaics.
© 2012 Elsevier Ltd. All rights reserved.

1. OPERATIONAL PERFORMANCE RESULTS

1.1 Overview of Performance Indicators

This section focuses on the evaluation of the technical performance of photovoltaic (PV) systems. To a great extent existing evaluation procedures are based on the European Guidelines and the IEC Standard 61724. In addition, there are recommendations on guidelines at national level, aiming to harmonise procedures on data collection, data processing, and presentation of data. The increasing awareness of the importance of the PV technology and its potential has resulted in a worldwide acceptance of impressive research and investment programmes.

The *European Guidelines for the Assessment of Photovoltaic Plants* had been prepared by the European Solar Test Installation of the Joint Research Centre (JRC) in Ispra resulting in Document A [1], Document B [2], and Document C [3]. These guidelines have played an important role in the preparation and realisation of the IEC Standard 61724 [4]. The Standard titled 'Photovoltaic System Performance Monitoring—Guidelines for Measurement, Data Exchange and Analysis,' first published in April 1998, expresses an international consensus on the subject of PV system performance monitoring and analysis. The document has the form of guidelines for international use published in the form of standards and is accepted by national committees.

Various derived parameters related to the system's energy balance and performance can be calculated from the recorded monitoring data using sums, averages and ratios over reporting periods τ such as days, months, or years. The irradiation quantities H_I are calculated from the recorded irradiance G_I. The electrical energy quantities are calculated from their corresponding measured power parameters over the reporting period τ, whereas E_A is the energy from the PV array, E_H the DC energy input to the inverter, and E_{IO} the AC energy output from the inverter. Key parameters for PV system performance evaluation are given in Table 1.

For the comparison of PV systems, normalised performance indicators are used: e.g., energy yields (normalised to nominal power of the array P_O), efficiencies (normalised to PV array area A_a), and performance ratio (normalised to in-plane irradiation H_I). The most appropriate performance indicators of a PV system are:

- The final PV system yield Y_f is the total system output energy $E_{use,\tau}$ delivered to the load per day and kilowatt peak of installed PV array.

Table 1 Derived parameters for performance evaluation according to Standard ICC 61724

Parameter	Symbol	Equation	Unit
Global irradiation, in plane of array	H_I	$\int G_I dt$	kWh/m^2
Array yield	Y_A	E_A/P_O	$kWh/(kWp \star day)$
Final yield	Y_f	$E_{use,\tau}/P_O$	$kWh/(kWp \star day)$
Reference yield	Y_r	$\int G_I dt / G_{STC}$	$kWh/(kWp \star day)$
Array capture losses	L_c	$Y_r - Y_A$	$kWh/(kWp \star day)$
System losses	L_s	$Y_A - Y_f$	$kWh/(kWp \star day)$
Mean array efficiency	η_{Amean}	$E_A / \int \tau G_I \cdot A_a dt$	%
Efficiency of the inverter	η_t	E_{IO}/E_H	%
Overall PV plant efficiency	η_{tot}	$E_{use,\tau}/\int \tau G_I \cdot A_a dt$	%
Performance Ratio	PR	Y_f/Y_r	—

The reference yield Y_r is based on the in-plane irradiation H_I and represents the theoretically available energy per kilowatt peak of installed PV per day.

- The performance ratio PR is the ratio of PV energy actually used to the energy theoretically available (i.e., Y_f/Y_r). It is independent of location and system size and indicates the overall effect of losses on the array's nominal power due to module temperature, incomplete utilisation of irradiance and system component inefficiencies or failures.

The normalised losses are calculated by subtracting yields (see Table 1) and also have units of [$kWh/kWp \star day$]. The array capture losses L_c are caused by operating cell temperatures higher than 25°C (thermal losses) [5] and by miscellaneous causes such as

- low irradiance;
- wiring losses, string diode losses;
- partial shading, contamination, snow covering, non-homogenous irradiance;
- maximum power point tracking errors;
- reduction of array power caused by inverter failures or by fully charged accumulators; and
- spectral losses, losses caused by glass reflections.

System losses L_s are gained from inverter conversion losses in grid-connected PV systems and from accumulator storage losses in stand-alone PV systems.

Figure 1 shows exemplarily the annual performance indices of a large grid-connected PV power system operating in southern Italy since 1995.

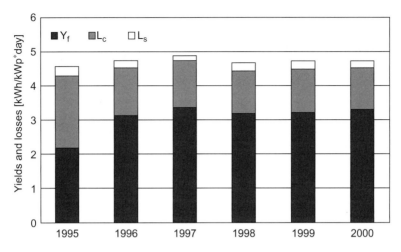

Figure 1 Annual performance results of 3.3 MWp PV system in Serre, Italy, in terms of final energy yield (Y_f), capture losses (L_c) and system losses (L_s).

This standard presentation of energy yields and losses allows us to deduce malfunctions of PV systems. High capture losses and low final yields occurred in the first year of operation in 1995, corresponding to a low performance ratio of smaller 0.5 and low system availability as shown in Figure 2 for the same system and monitoring period. Both figures are related by the equation: $PR = Y_f/Y_r = Y_f/(Y_f + L_c + L_s)$.

Because of the given definition of L_c and L_s (see Table 1), an inverter malfunction or failure in grid-connected PV plants will result in a remarkable rise of capture losses L_c. This quantity is a very good indicator for system problems occurring in grid-connected PV plants. If the grid-connected system fails completely, the values of Y_A, Y_f, and PR will drop to zero, while capture losses will rise towards Y_r and system losses become negligible. In the case of PV plant Serre during the first half of 1995, the inverters had to be adjusted to make them compatible with the frequent disturbances of the local medium voltage grid [6]. Thus, unreliable inverters and complete system breakdowns were responsible for high capture losses and relatively low system losses in 1995 as shown in Figure 1.

The higher the PR, the better the system uses its potential. A low PR value means production losses due to technical or design problems. For the performance assessment of stand-alone PV systems, a high PR value does not always mean that the system is operating in the best conditions. If the system is undersized for the considered application, the PV system will show very high value of PR, but the user will not be supplied with

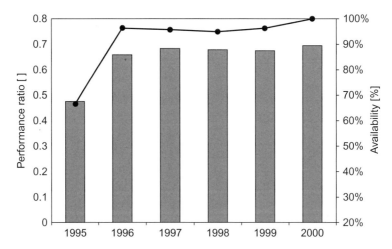

Figure 2 Annual performance results of 3.3 MWp PV system in Serre, Italy, in terms of performance ratio (PR) and system availability.

electricity. For stand-alone systems (SAS), the value of PR is user consumption dependent. If the consumption level is not correlated to the potential of the PV array, the PR will reach low values due to high capture losses. It has been shown that the PR widely used for grid-connected PV systems, cannot be used alone to describe the quality of operation of SAS [7]. Different attempts have been made to introduce new parameters for the performance assessment of SAS [8,9].

In this chapter, all results, experiences, and lessons learnt are gained from the analysis of grid-connected PV systems. The investigated PV systems have an installed capacity between 1 kWp and 3 MWp adapted to various applications and located worldwide. The article aims at illustrating the operational behaviour of different grid-connected PV systems and at presenting the summary results in standard quantities allowing cross-comparison between the systems.

1.2 Summary Performance Results

From the performance analysis of 260 grid-connected PV systems supplied by the Performance Database of the International Energy Agency [10] (see Appendix), it was learnt that the average annual yield (Y_f) fluctuates only slightly from one year to another and has typical average values for one country (700 kWh/kWp for the Netherlands, 730 kWh/kWp for Germany, 790 kWh/kWp for Switzerland, and up to $Y_f = 1470$ kWh/kWp for Israel). However, there is a considerable scattering around these average values for

Table 2 Annual final energy yields in different countries for comparison

Country	Systems analysed	Range of final yield in kWh/(kWp*year)	Average final yield in kWh/(kWp*year)	Reference
Netherlands	10	400–900	700	[12]
Germany	88	400–1030	730	[13]
Switzerland	51	450–1400	790	[14]
Italy	7	450–1250	864	[15]
Japan	85	490–1230	990	[16]
Israel	7	740–2010	1470	[17]

individual systems ranging from 400 kWh/kWp to 1030 kWh/kWp in Germany and from 450 kWh/kWp to 1400 kWh/kWp for plants in Switzerland [11].

Table 2 shows the annual final yields in terms of average, minimum and maximum values for PV systems in six different countries. While the different value of average final yield in each country can generally be explained by the difference in mean irradiation, the broad range of annual yields of factor two to three is quite significant and has system specific reasons.

High energy yields are due to
- well-operating systems,
- high component efficiencies,
- optimum orientations of PV arrays, and
- well-maintained PV systems.

Reduced energy yields are caused by:
- failures of system or components,
- shading of PV arrays,
- frequent inverter problems,
- long repair times,
- bad orientations of PV arrays (e.g., facades), and
- high module temperatures.

The performance and yield of a PV system depend on all 'gain' and 'loss' factors and to which extend the loss factors can be avoided during project planning, plant installation and operation.

Figure 3 shows the distribution of annual performance ratios calculated from 993 annual datasets of 309 grid-connected PV systems operating in 14 different countries during 1989 and 2002 (for data used see Appendix). The annual performance ratio (PR) differs significantly from plant to plant and ranges between 0.4 and 0.85 with an average value of 0.67 for all 309 PV systems.

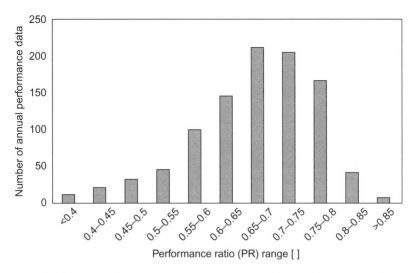

Figure 3 Distribution of annual performance ratios of 309 grid-connected PV systems (993 annual data sets) operating in 14 countries from 1989 to 2002.

The broad range of performance ratios was also found for other research programmes, such as the Thermie programme of the European Commission [18,19], the German 1000-Roofs-PV-Programme [20–22], the Japanese residential PV monitoring programme [23], and the PV programmes of the International Energy Agency [24,25].

Former investigations on annual PR have shown that for well operating grid-connected PV systems during 1990 to 1999, a PR value between 0.6 and 0.8 can be expected [26]. It was found that well-maintained PV systems show PR values of typically 0.75 at an availability of higher 98%. Taking into account realistic efficiency values of the improved inverter types of today, optimum values of annual PR between 0.81 and 0.84 may be achieved [27]. A tendency of increasing annual PR values during the past five years has been observed.

1.3 Results from 'Low Yield Analysis'

Within the German 1000-Roof's-Photovoltaic-Programme more than 2000 grid-connected PV residential systems of 1 to 5 kWp were installed during 1991 and 1995. This programme was accompanied by a monitoring and evaluation programme, which delivered remarkable results and experiences [28].

The low yield analysis was carried out at 17 selected PV systems to find the quantified reasons of reduced yields, which cannot be explained

by system losses and tolerances due to system components. The selection criteria for the systems under investigation were the performance ratio (annual value lower than 0.60) and the final energy yield (75% less than the average value in that region) in addition to practical considerations, which allow technical inspections and four to six weeks monitoring campaigns at the selected sites. The detailed investigations of this project resulted in four additional reasons for very low yields and performances of grid-connected PV systems:
- Deviation from manufacturer's module specifications
- Shading due to trees, buildings or walls
- Defects of DC installation
- Inverter problems

1.3.1 Deviation from PV Module Specifications

During the 1000-Roofs-PV-Programme it became clear that various types of PV modules show significant differences in their performance [29,30]. On one hand, there is a typical variance due to large-scale series production. On the other hand, a specific deviation of the PV module nominal power from the quoted rating was found depending on the manufacturer. This deviation of real measured power from the rated power was varying from manufacturer to manufacturer and generally appeared to be negative (typical value: -10%). During the investigations of the 17 selected PV systems, the differences of the monitored and to Standard Test Conditions (STC) converted module power and the power rating according to the manufacturer's module specifications ranged from -5% to -26% [31].

As a consequence, module manufacturers learned the lessons by declaring their PV modules for a quality check according to performance measurements. Additionally, standards organisations aimed at reduced measurement tolerances as well as at harmonisation of test equipment used by different manufacturers and authorised institutions.

1.3.2 Shading

Partial shading of the PV array leads to a significant reduction of the energy yield of a PV system [32]. The quantity of the reduction of the system yield is depending on the geometry of the shading object and that of the PV array. If shading of one or more PV strings occurs during a clear and sunny day, the string currents of the shaded strings will be heavily reduced. During the annual course, the winter months are much more influenced by shading effects.

Out of the 17 PV plants under investigation, 10 PV systems were shaded to a higher or lower degree. Simulation calculations on the effect of PV array shading were carried out in order to gain quantitative figures [31]:
- Case 1: Partial shading of five of the six strings during morning hours by a tall (18 meters) leaf tree resulted in a reduced yield of 4%.
- Case 2: Partial shading of PV array by roof of neighbouring building as well as by surrounding trees lead to a reduced yield up to 10%.
- Case 3: Significant shading of the array by multiple trees located close to the house and exceeding the PV rooftop resulted in an annual reduction loss of higher 20%.

The analysis of the measured data and the calculations of the simulation programmes lead to the following results:
- Shading has reduced the annual energy yield up to 25% for an individual PV plant.
- Each plant has to be investigated individually as there are no typical shading conditions.
- From the very beginning of the project planning and concept making, the selection of shading free sites and optimum design of PV strings should be taken care of in order to avoid or minimise losses due to shading.

1.3.3 Defects of DC Installations

It is useful to perform technical inspections right after commissioning the PV system to detect and dissolve problems. During system operation failures due to bad installation techniques may occur, which often remain unnoticed. This may affect all components of the DC system. For the 17 PV systems under investigation, the following problems were found [31]:
- Defect clamp of a switch caused a loose contact and thus a partial failure of the string current.
- Defective screwed contacts inside the PV junction box caused the relevant strings to malfunction.
- Defect string diodes were caused by lightning.
- Cracked PV modules due to bubbles inside the laminate were observed and results in a reduced string current of about 60%.
- Defect connectors for PV roof tiles caused failures of strings.
- Defect monitoring equipment was responsible for failing string currents.
- Defect string fuses were registered in many plants.

The problems and defects encountered mainly caused failures of one string, in some cases that of several strings. For a PV system of five strings, the reduced yield of one failing string is −20%, for example. It is advisable to check the strings currents and open circuit voltage on a regular basis (e.g., yearly maintenance). For the concept of new and improved PV systems, the use of switching elements and string fuses should be avoided and the installation technique should be as simple as possible.

1.3.4 Inverter Problems

The following inverter problems were identified:
- Partial or complete failure of the inverter due to hardware and software problems.
- Non-optimum maximum power point (MPP) tracking due to the permanent MPP tracking process of the inverter caused by fluctuations of irradiation and by partial shading of the PV array.
- Bad power regulation of inverters that operate with power/current limitation in case of limitation level being far below the rated power of the inverter.
- Fixed voltage operation may evolve adaptation losses, if an unsuitable level of inverter voltage is chosen.

It was found that the user of the PV system often discovers the problems related to inverters, but he is unable to estimate the significance of a particular failure. Except for complete inverter failures, the energy losses due to non-optimum operating inverters were found to be below 10%.

2. TRENDS IN LONG-TERM PERFORMANCE AND RELIABILITY

For the wide dissemination of PV technology, the results of long-term performance and reliability of PV systems are important. Financing schemes for PV such as fed-in tariffs require sound figures about real energy yields and performance data. This section deals with the questions:
- What is the technical performance of grid-connected PV systems in different countries?
- What are the significant differences?
- What is the long-term operational behaviour of PV systems?

- Does the average performance of PV systems increase for new installations?
- What is the reliability of PV systems and their components?
- Does the number of failures decrease?
- What are the trends in the technical performance of PV systems?

Performance data from 309 grid-connected PV systems in 14 countries are exported from the Performance Database of IEA (see Appendix) and 993 annual performance data are presented for operational years between 1989 and 2002. Monthly input data come from different research and demonstration programmes in each of the countries, whereas the quality of the collected data depends on the level of monitoring (analytical or global), the kind of data acquisition system used, the availability of the data, the conditions of the programme, and the reliability of the data source.

For the interpretation of the results and the comparison between different countries, the amount of available data has to be considered; e.g., Germany is presented by 326 annual data sets, Switzerland by 244 data sets, and Italy by 81 annual data sets.

2.1 Trends from German 1000-Roofs-PV-Programme

The 48 PV systems under investigation are part of the 1000-roofs-PV-project in northern Germany and were installed between 1991 and 1993. These residential rooftop systems have an installed capacity between 1 and 5 kWp (average 2.3 kWp) and have operated for more than ten years. Global monitoring data are available from the standard monitoring and evaluation programme for the years 1993 to 2000. All PV rooftops are equipped with a calibrated silicon sensor (NES Si161) to measure the in-plane irradiation at the site of the PV array.

Figure 4 shows the annual performance ratios of 48 PV systems operating since the early 1990s. The annual PR differs significantly from plant to plant and shows a broad range for all operating years (e.g., in 1993: PR = 0.44 to 0.82). The average annual PR of 0.67 (48 plants) in 1993 drops to an average value of 0.62 (12 plants) in 2000. The average performance values show negative tendency and energy yields have been reduced by 8% between 1993 and 2000, although some defect inverters were replaced against new types with higher efficiencies.

The registered failures at 21 PV systems during 10 years of operation show that defects and problems were first decreasing (1992 to 1995), but started to increase during last years (1999 and 2000). Considering a total number of 47 failures in 210 operational years, a statistical failure will occur

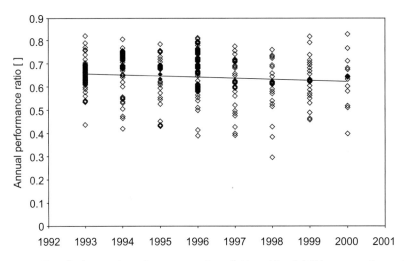

Figure 4 Trend of annual performance ratios of 48 residential PV systems in northern Germany installed between 1991 and 1993.

every 4.5 years per plant. Inverters are contributing with 63%, PV modules with 15% and other system components with 23% to the total failures.

It can be concluded that a negative tendency in terms of performance and yields was observed for these early installations (1991–1993) from the rooftop programme during eight years of operation. Although the installers provided maintenance, some PV plants were facing severe failures of inverters and other components, which led to significant reductions of the annual energy yields. New PV installations (after 2000) in Germany revealed that they reach higher component efficiencies (e.g., inverter) and high performance ratios (>0.80), but further investigations are required to confirm the increased quality of the newer systems in Germany.

2.2 Trends from Demonstration Projects in Switzerland

The 51 grid-connected demonstration plants in Switzerland are characterised by a broad range of applications (power plant, building-integrated PV, rooftop, sound barrier, freestanding) and by analytical monitoring campaigns. PV systems of 1 to 560 kWp installed power were continuously installed between 1989 and 1999 and monitored during 1991 and 2001. Due to online monitoring data availabilities of higher 95% were achieved. The average annual yield of 51 plants and 244 operational years result in 790 kWh/kWp and the average PR is 0.68.

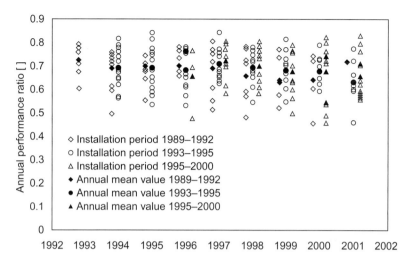

Figure 5 Trend of annual performance ratios of 51 grid-connected PV systems in Switzerland grouped in three parts according to their installation years.

Figure 5 shows the distribution of performance ratios between 1993 and 2001, which indicates a negative tendency of performance ratios. The range of PR values per year is rather high, e.g., in 2001 the plants take up PR values between 0.45 and 0.82. Well-performing plants have reach annual 0.82 in conjunction with annual inverter efficiency of higher than 97%.

In Figure 5 the 244 annual PR values are grouped in three parts according to the period of plant installation:

1. The performance of early installations (1989 to 1992, presented by a square) has decreased between 1993 (PR = 0.72) and 2000 (PR = 0.64) due to failures of components and system. During their first operational years, they have performed well. It is assumed that the design criteria were high yields and favourable orientation conditions, which led to in optimum power plants and often freestanding PV systems.
2. PV plants installed between 1993 and 1995 (presented by a circle) show less decrease of average PR during eight operational years. They have lower PR values in the beginning (J 994: PR = 0.69) and drop down to 0.63 in 2001. In this group of installations, many building integrated plants (BIPV) and facades are in operation, which reach lower PR values although they apply improved components (inverter).
3. The PV systems installed after 1995 (presented by a triangle) show low average PR values (1996: PR = 0.66), but this value of annual

PR keeps nearly constant. The range of annual PR in 2001 is still very broad. During this period, PV design criteria were focussed on building requirements and architectural design criteria, which has led to lower yields and performances compared to the plants of the first group (primarily technical design criteria).

From this example, reduced performance factors are identified:
- System and component failures are the main factors for reduced PR values.
- High module temperatures of BIPV and facades have a negative impact on yields and PR (energy losses of up to 10% [33]).
- Partial shading of the PV array is consciously accepted.
- Unfavourable array orientations given by the building structure play a negative role for the energy yield.

The overall performance of PR = 0.68 for the 51 Swiss PV systems remains below the expectations and efforts to build well-integrated *and* well-performing PV systems. BIPV substituting the building skin has a double value and must fulfil the compromise between architectural building aspects and technical performance of the system.

2.3 Trends from PV Programmes in Italy

Large PV power systems were installed since 1983 with capacities between 100 kWp and 3.3 MWp and deliver long-term performance data to 2002. New PV systems from the Italian PV-Roof-Top-programme were built since 2000 and monitoring data are available for two years (2001−2002).

Figure 6 shows the annual PR of 29 systems and 81 data sets from both programmes and a positive tendency of PR development between 1992 (average 0.60) and 2002 (average 0.74). The very low PR values of the early plants, which were designed as experimental and demonstration plants, can be explained by experimental activities and measurement campaigns as well as by frequent failures and long repair times for replacement of inverters and DC components [15].

The performance results of very large, but later installed plants (e.g., Serre, Figure 2) show good PR values and system availability (in 2000: PR = 0.7 and 100% availability) confirming their design values. The new PV systems from the Italian rooftop programme reach high performance values due to improved components and know-how. The Italian programme is accompanied by intensive PV training courses for designer and installers. The outcome of these efforts can be seen in the positive learning curve of long-term PV plant operation in Italy.

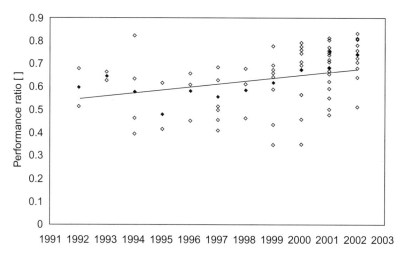

Figure 6 Trend of annual performance ratios of 29 grid-connected PV systems in Italy installed between 1983 and 2002.

2.4 Rise in PV System Performance

Will the overall performance of new installations automatically increase as a consequence of higher efficiencies of PV modules and inverters available at the PV market? The 51 examples with a high proportion of BIPV in Switzerland showed that this is not a must if other design criteria will be applied and technical performance is not the only goal for a PV system.

Considering the 298 grid-connected PV plants in 14 countries from the Performance Database, a clear answer is given in Figure 7. It shows the distribution of 962 annual PR values, which are grouped into two installation periods: all PV systems installed before 1995 have their maximum in the PR range of 0.65 to 0.7 and an average PR of 0.65 for 587 annual performance data.

The newer installations since 1995 have their maximum in the range 0.75–0.8 with an average value of PR = 0.70 for 375 annual data sets. This is a significant rise in PV system performance and reliability gained in these 14 countries during the past seven years of installation.

Using the same system data, it was observed that
- Measured inverter efficiencies also tend to have higher values.
- Measured array efficiency has slightly improved.
- System availability seems to improve with the years.

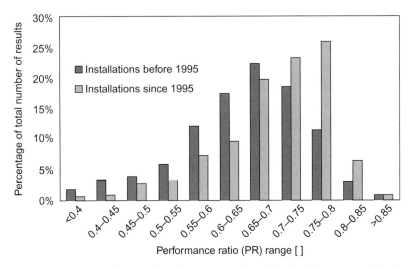

Figure 7 Distribution of annual performance ratios of 298 grid-connected PV systems (962 annual data sets) in 14 countries for two installation periods: (a) The average annual PR of PV plants installed before 1995 is 0.65. (b) The average annual PR of PV plants installed since 1995 is 0.70.

3. USER EXPERIENCE

3.1 Small Residential PV System in Germany

Long-term performance results for a 3.6 kWp PV rooftop system in northern Germany are presented in Figure 8. The monthly PR values during nine years of operation are shown in Figure 8. In its first years, the PV system obtains good annual PR values of 0.67 in 1992 and 0.71 in 1993, while the values continuously drop down in the following years to end up with to 0.54 in 2000.

The sudden decrease of monthly PR is well correlated with the frequent exchange of the inverter: seven exchanged inverters in nine years! And the installer checked the complete system after each replacement. The overall decline of performance of this system can be explained by

- frequent inverter failures,
- very long repair times (1–3 months for each replacement), and
- bad adaptation of the inverter to the given PV array.

The user did not notice how low the annual energy yields were during the past years due to a lack of information and data for comparison.

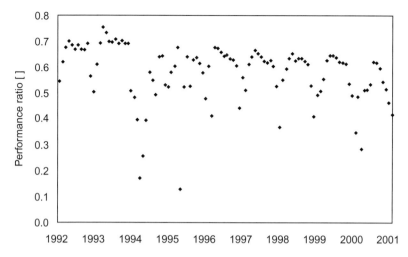

Figure 8 Monthly performance results of 3.6 kWp residential PV system in northern Germany installed in 1991.

He actively reduced the energy consumption of his household (four to six persons) and succeeded in obtaining a solar fraction of 60–70% per year independent of the continuously reduced yields.

3.2 PV Sound Barrier in Switzerland

The pilot and demonstration project of 104 kWp was built as the first PV sound barrier in 1989 and is analytically monitored for 13 years. The large PV plant along a Swiss motorway is regularly inspected and maintained. As shown in Figure 9, the PV system performs well over 10 years (annual PR = 0.73 to 0.80) except for the first year, when PV modules had to be replaced [34]. Severe operational losses in 2000 and 2001 are due to component failure of the inverter and the delays in the replacements of these components.

Climbing plants had to be removed from the PV array surfaces to avoid losses due to shading. In total, 33 modules had to be replaced (21 damaged, 12 stolen). The glass surfaces of the PV modules have never been cleaned and no significant decrease in array efficiency ($\eta_A = 8.8\%$) was observed [34].

This successful project achieves an average annual PR of 0.68 and an average energy yield of 950 kWh/kWp for 13 years of operation.

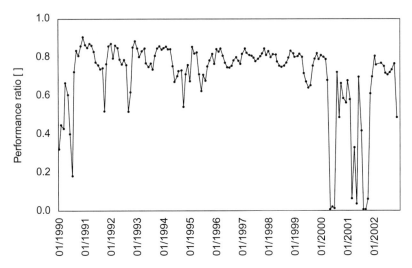

Figure 9 Monthly performance results of 104 kWp PV sound barrier in Switzerland installed in 1989.

3.3 PV Power Plant in Italy

The 100-kWp PV system is retrofitted on the flat roof of a parking shelter belonging to a research centre in southern Italy and is monitored and inspected since 1991 [35]. The first prototype system shows low performance over 11 years (average PR of 0.57) and has performance constraints due to frequent failures. Between 1992 and 2002 about 40 outage events have been recorded corresponding to an average plant availability of 87%. Figure 10 shows the annual PR and availability for 1992 to 2002.

In 1997 and 1998, performance losses were caused by array partial shutdown due to a serious failure of the junction box. The replacement of the AC switchgear was responsible for the total shutdown of the system during summer months [36].

The plant does not perform according to specifications and expectations except for the years 1999 and 2001 with high annual PR and availabilities: significant energy losses during the 11 years of operation were caused by [36]
- inverter failures in summer months,
- long repair times for replacement,
- prototype nature of the inverter, and
- DC circuits' poor sizing.

The PV plant produces about 0.5% of the electricity demand of the research centre.

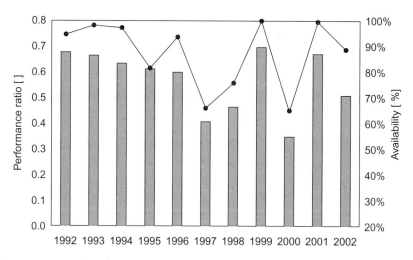

Figure 10 Annual performance ratios and availability of 100 kWp PV system in Italy installed in 1991.

3.4 Lessons Learnt

While many PV systems perform according to plan, a common performance experience was that in terms of final energy yield, the system did not meet the expectations. Dominating performance constraints were the poor reliability of inverters, long repair times, and shading problems.

From PV systems that provided well-documented failure reports, it was learnt that the power conditioner units often caused problems over the first year of installation. After problems of the utility interactive inverters had been solved, the system operated well for many years. A second high-failure rate period of inverters was observed after 8–10 years of system operation. The replacement of 'old' power conditioner units caused long repair times and thus high operational losses. Good contacts between system user, installer, and manufacturer of the system components helped to reduce delays in the replacement of the faulty parts. Step by step, the reliability will improve as the inverter technology matures. It has been suggested as an interim solution to design PV systems that allow for quick replacement of the inverter [37].

The energy losses due to partial shading by nature (e.g., trees) and by parts of the building structure are generally underestimated and cannot be eliminated after installation has been completed. In the planning stage partial shading can be avoided or reduced by the PV system designer in cooperation with the architect or project planner. The negative impact of unavoidable shading can be minimised by suitable electrical wiring configuration. Keeping shaded modules in one electrical string in the array will reduce shading losses.

It was further observed that some inverters deliver low energy output because the inverter is not well adapted to the PV array, the arrays are partly shaded or PV arrays with different orientations are connected to one central inverter. Under these conditions, the MPP unit of the inverter will not function properly and leads to energy losses, which are a matter of poor system design rather than one of poor DC/AC power conversion.

The following is recommended in the planning stage to improve PV system performance:
- Avoid shading and shading losses. Use suitable electrical wiring configuration of the PV array to minimise unavoidable losses.
- Optimise the orientations of PV arrays and avoid that PV arrays of different orientations are connected to one central inverter.
- Make DC installation technique as simple as possible: avoid switching elements and string fuses.
- Consider the reliability of the power conditioner units: ensure that repair parts are available from the manufacturer in 10 years time and allow for easy replacement of faulty units.
- Optimise the efficiency of the inverter performance by suitable sizing of the PV array and adapting to operating conditions.
- Give realistic figures of expected energy yields to the PV system user by understanding and explaining the factors that influence the performance (e.g., reflection losses, module temperature, PV array coverage of snow or dirt, module mismatch, DC installation losses, inverter efficiency).

The summary of lessons learnt might give rise to the wrong impression of frequent negative results of PV system performance. The reason of focussing and presenting examples of low performance is that one learns how to improve from the negative examples. For the overall performance assessment, long-term performance results have shown a positive learning curve in terms of improved system energy yields and reliability of components. Many PV systems have demonstrated excellent performance and are meeting the expectations of their users and designers.

ACKNOWLEDGEMENTS

Special thanks go to my colleague Mr Wolfgang Nasse for his support and for preparing the figures of this chapter. The data and information on the Swiss and Italian case studies provided by Mr Luzi Clavadetscher, TNC Consulting, and by Mr Salvatore Castello, ENEA, are greatly appreciated. Thanks are due to all colleagues of IEA PVPS Task 2 who contributed their PV system data to the Performance Database.

APPENDIX. SPECIFICATIONS OF PERFORMANCE DATABASE OF IEA PVPS

Performance Database

IEA International Energy Agency
Photovoltaic Power Systems Programme

The Performance Database of Task 2 of the Photovoltaic Power Systems Programme (PVPS) of the International Energy Agency (IEA) is designed to provide experts, industry, utilities, manufacturers, system designers, installers and schools with suitable information on the operational performance, reliability and design of photovoltaic (PV) systems and components. The benefit of the Performance Database lies in sharing technical information focusing on long-term performance and reliability of PV systems and providing tools for practical and educational purposes.

Database Programme

- CD-ROM at price of 20 EUR
- Internet download for free at www.task2.org (47 MB)
- English language
- First release in July 2001
- Last Update in May 2003
- Financed by the German Federal Ministry of Economy and Technology (BMWi)

Database Contents

- Information on 372 photovoltaic systems in IEA countries worldwide
- Grid-connected, off-grid, and hybrid photovoltaic systems of 1 kWp up to 3 MWp
- General system information (size, system type, mounting, location, cost, photo)
- System configuration and component data
- Monitoring data (values of monthly energies, irradiation and temperature)
- Calculated data (monthly and annual values of performance indicators)

Programme Specifications

- PC of 64 MB RAM, 100 MB hard disk space
- Windows 95/98, NT4.0, 2000, XP
- Excel 97/2000 for reports and data exports
- Filter, selection, and easy navigation through the database
- Import and export tools

Information and CD-ROM are available from:

Reinhard Dahl (Operating Agent)
Projekttraeger Juelich – PTJ ERG
Forschungszentrum Juelich GmbH
D-52425 Juelich
Germany
Fax: +49 (0) 24 61 - 61 28 40
Email: r.dahl@fz-juelich.de
Website: www.task2.org

Operational performance, maintenance and sizing of photovoltaic power systems and subsystems

REFERENCES

[1] Commission of the European Communities, Guidelines for the Assessment of Photovoltaic Plants, Document A, Photovoltaic system monitoring, issue 4.2, Joint Research Centre, 1993.
[2] Commission of the European Communities, Guidelines for the Assessment of Photovoltaic Plants, Document B, Analysis and presentation of monitoring data, issue 4.3, Joint Research Centre, 1997.
[3] Commission of the European Communities, Guidelines for the Assessment of Photovoltaic Plants, Document C, Initial and periodic tests on photovoltaic plants, Joint Research Centre, 1997.
[4] International Standard IEC 6 1724: Photovoltaic system performance monitoring — Guidelines for measurement, data exchange and analysis. first ed., International Electrotechnical Commission (IEC), Geneva, April 1998.
[5] H. Haeberlin, C. Beutler, Normalized representation of energy and power for analysis of performance and online error detection in PV systems. Proceedings of the Thirteen European PV Solar Energy Conference, Nice, 1995, pp. 934—937.
[6] S. Castello, IEA PVPS Task 2 case study on PV plant Serre, written communication, Italy, 2002.
[7] D. Mayer, PV stand-alone systems in France. In: Report IEA-PVPS 12—01:2000, Analysis of photovoltaic systems, 2000, pp. 52—58.
[8] A. Sobirey, H. Riess, P. Sprau, Matching factor — a new tool for the assessment of stand-alone systems. Proceedings of the Second World Conference on Photovoltaic Solar Energy Conversion, Vienna.
[9] D. Mayer, Performance assessment for PV stand-alone systems. In: Report IEA-PVPS 12—03:2002, Operational performance, reliability and promotion of photovoltaic systems, 2002, pp. 39—42.
[10] W. Nasse, Performance Database of PVPS Task 2. In: Report IEA-PVPS 12—03:2002, Operational performance, reliability and promotion of photovoltaic systems, 2002, pp. 7—10.
[11] U. Jahn, D. Mayer, et al, International Energy Agency PVPS Task 2 — Analysis of the operational performance of the IEA database PV systems. Proceedings of the Sixteenth European Photovoltaic Solar Energy Conference, Glasgow, 2000, pp. 2673—2677.
[12] K.V. Otterdijk, PV systems in the Netherlands. In: Report IEA-PVPS 12—01:2000, Analysis of photovoltaic systems, 2000, pp. 92—100.
[13] U. Jahn, PV systems in Germany. In: Report IEA-PVPS 12—01:2000, Analysis of photovoltaic systems, 2000, pp. 59—72.
[14] L. Clavadetscher, Experiences from long-term PV monitoring in Switzerland. In: Report IEA-PVPS 12—03:2002, Operational performance, reliability and promotion of photovoltaic systems, 2002, pp. 23—28.
[15] S. Castello, S. Guastella, M. Guerra, Operational performance know-how and results of PV system analysis in Italy. In: Report IEA-PVPS T2—03:2002, Operational performance, reliability and promotion of photovoltaic systems, 2002, pp. 11—15.
[16] K. Sakuta, T. Sugiura, Results and future plans for monitoring residential PV systems in Japan. In: Report IEA-PVPS T2—03:2002, Operational performance, reliability and promotion of photovoltaic systems, 2002, pp. 17—21.
[17] D. Faiman, PV systems in Israel. In: Report IEA-PVPS T2—01:2000. Analysis of photovoltaic systems, 2000, pp. 73—76.
[18] W.B. Gillett, W. Kaut, D.K. Munro, G. Blaesser, G. Riesch, Review of recent results and experience from CEC PV demonstration and THERMIE programmes. Proceedings of the Eleventh European Photovoltaic Solar Energy Conference, Montreux, 1992, pp. 1579—1582.

[19] G. Blaesser, G. Riesch, D.K. Munro, W.B. Gillett, Operating experience with PV system components from the THERMIE programme. Proceedings of the Twelfth European Photovoltaic Solar Energy Conference Amsterdam, 1994, pp. 1159–1162.
[20] U. Jahn, J. Grochowski, D. Tegtmeyer, U. Rindelhardt, G. Teichmann, Detailed monitoring results and operating experiences from 250 grid-connected PV systems in Germany. Proceedings of the Twelfth European Photovoltaic Solar Energy Conference, Amsterdam, 1994, pp. 919–922.
[21] K. Kiefer, T. Koerkel, A. Reinders, E. Roessler, E. Wiemken, 2050 Roofs in Germany – Operating results from intensified monitoring and analysis through numerical modelling. Proceedings of the Thirteenth European Photovoltaic Solar Energy Conference, Nice, 1995, pp. 575–579.
[22] B. Decker, U. Jahn, Performance of 170 grid-connected PV plants in Northern Germany – Analysis of Yields and optimisation potentials, Sol. Energy 59(4–6) (1997) 127–133.
[23] K. Otani, Year 2000 report book for PV systems, J. Jpn. Inst. Energy 80(3) (2001) 116–122 (in Japanese).
[24] M.V. Schalkwijk, T. Schoen, H. Schmidt, P. Toggweiler, Overview and results of IEA-SHCP Task 16 demonstration buildings. Proceedings of the Thirteenth European Photovoltaic Solar Energy Conference, Nice, 1995, pp. 2141–2144.
[25] U. Jahn, Results of performance analysis. In: Report IEA-PVPS T201:2000, Analysis of photovoltaic systems, 2000, pp. 5–7.
[26] U. Jahn, W. Nasse, et al. Analysis of the Operational Performance of the IEA Database PV systems. Proceedings of the Sixteenth European Photovoltaic Solar Energy Conference, Glasgow, 2000, pp. 2673–2677.
[27] U. Rindelhardt, Photovoltaische Stromversorgung, Teubner Verlag, 2001.
[28] T. Erge, et al., The German 1000-Roofs-PV Programme – a resume of 5 years pioneer project for small grid-connected PV systems. Proceedings of the Second World Conference on Photovoltaic Solar Energy Conversion, Vienna, 1998.
[29] H.A. Ossenbrink, Comparison of PV modules – data sheet specifications and reality. Proceedings of the Ninth Symposium PV Solarenergie, 1994pp. 419–425 (in German).
[30] U. Jahn, D. Tegtmeyer, J. Grochowski, Results from the 1000-Roofs-PV-Programme – Measurements of PV arrays. Proceedings of the Tenth Symposium PV Solarenergie, 1995, pp. 69–77 (in German).
[31] J. Grochowski, B. Decker, Reasons for low energy yields. In: Report 1000-Daecher-Mess- und Auswerteprogramm, Fraunhofer Institute for Solar Energy Systems, Germany, 1998, pp. 30–42 (in German).
[32] J. Grochowski, U. Jahn, B. Decker, J. Offensand, First results from the Low Yield Analysis and optimization potentials – a project within the 1000-Roofs-PV-programme. Proceedings of the Thirteenth European Photovoltaic Solar Energy Conference, Nice, 1995, pp. 356–359.
[33] T. Nordmann, L. Clavadetscher, Understanding temperature effects on PV system performance. Proceedings of the Third World Conference on Photovoltaic Energy Conversion, Osaka, 2003.
[34] L. Clavadetscher, IEA PVPS Task 2 case study on PV plant Domat, written communication, Switzerland, 2002.
[35] L. Barra, S. Castello, C. Messana, Design and development of a standard 100 kW PV plant. Proceedings of the Tenth European Photovoltaic Solar Energy Conference, Lisbon, 1991.
[36] S. Castello, IEA PVPS Task 2 case study on PV plant Cassacia, written communication, Italy, 2002.
[37] P. Drewes, 2002. Electrical design issues. In: Report IEA-PVPS Task 7:2002, Building with Solar Power, October 2002.

CHAPTER IIE-4

Solar-Powered Products

Philip R. Wolfe
WolfeWare Limited, UK

Contents

1. The Genesis of Solar-Powered Products   988
   1.1 First Steps   988
   1.2 Integration of PV in Products   988
   1.3 Thin-Film Cells   989
2. Stand-Alone Consumer Products   989
   2.1 Indoor Products   990
       2.1.1 Calculators   990
       2.1.2 Watches and Clocks   991
       2.1.3 Chargers   991
       2.1.4 Accent Lighting   991
       2.1.5 Air Fresheners   992
       2.1.6 Other Indoor Products   992
   2.2 Garden and Stand-Alone Consumer Products   992
       2.2.1 Lighting and Markers   992
       2.2.2 Aquatic Products   994
       2.2.3 Deterrents and Other Products   994
   2.3 Household Products   996
       2.3.1 Ventilation and Air Conditioning   996
       2.3.2 Lighting and Other Products   997
   2.4 Portable and Transportable Products   997
       2.4.1 Products for Vehicles   997
       2.4.2 Portable Electronics   999
3. Solar Products for Grid Connection   1001
   3.1 Roofing Products   1001
   3.2 Cladding and Facade Products   1002
4. Nonconsumer Products   1003
   4.1 Commercial Applications   1003
   4.2 Transportation   1003
   4.3 Other Industrial Applications   1003
5. Designing PV for Products   1004
   5.1 Electrical Characteristics   1005
   5.2 Product Dimensions   1005
   5.3 Mechanical and Structural Considerations   1006
6. Solar Products of the Future   1006
Acknowledgements   1007

Practical Handbook of Photovoltaics.
© 2012 Elsevier Ltd. All rights reserved.

1. THE GENESIS OF SOLAR-POWERED PRODUCTS

Solar cells have two great advantages that other energy sources cannot offer:
- They can be used anywhere on the face of the planet—the only requirement is incoming light.
- They can generate power at any scale from milliwatts to gigawatts.

This is why the development of photovoltaics 'broke the mould' of traditional energy technology. For the first time, power generation can be built into products at a practical level.

1.1 First Steps

Solar cells have been used for applications on Earth since the 1970s. Most early applications used solar systems of discrete components as described in Part IV. However, it soon became apparent that PV offers unique benefits in integrating the power generation into the very product that it is powering.

It is hard to say which came first, but by the mid-1980s, quite a range of products incorporated integrated solar generation. Marine buoys had been developed with solar cells around the edges to power a flashing light on the top. Solar battery chargers were available, and soon solar-powered calculators entered the mass market.

The advent of thin-film solar cells made it easier to incorporate photovoltaics into a wider range of products, and the number of applications mushroomed. Selections of the most common are described further in this chapter.

1.2 Integration of PV in Products

A huge proportion of the products we use today involve some form of electricity consumption. Traditionally, this was provided by internal batteries, which needed to be replaced or recharged, or by connection to an external power supply such as the grid.

The advent of the solar cell made it possible for the first time to build power generation into products, making them permanently independent of external energy sources. This is particularly beneficial in remote locations (including outer space, of course), but even in more accessible places it can save time, money, and inconvenience.

Solar cells generate power only when exposed to light, so power storage must be included when the product needs power at other times. In many cases, however, this is not necessary. Many solar calculators are designed so that, as long as there is enough light to see the display, there is enough to power the device. Some solar ventilators operate on the basis that they are needed only when it is hot, and this is normally when plenty of sunlight is available. Many solar pumps also are only needed when the sun is shining, or they can pump into a reservoir during the day to use the water at any time.

The solar cells generate the power necessary to operate the product, in terms of both current and voltage. In general, the current is related to the size of the solar cell array, while the voltage is related to the number of cells used. This is further described in Section 5.1, which shows how the size of the solar cell can be matched to the power required by the product. For this reason, solar cells are best suited to products with low power consumption. Recent advances in energy efficiency of a wide range of appliances have therefore brought ever more products into the realms that are suitable for 'solarisation.'

1.3 Thin-Film Cells

The development of thin-film solar cells gave great new impetus to the solar products business. First, thin-film cells are less expensive, because they use less material of substantially lower cost. More important, however, they allow connections between cells to be made within the manufacturing process in an arrangement known as *monolithic interconnection*. As detailed in Part II, this means that the series connections needed to obtain higher voltages are compact and cheap, compared to crystalline solar cells, where series connections need further manufacturing operations to solder or weld neighbouring cells together (see Figure 1).

Solar products typically operate at voltages between 2 and 48 V, requiring therefore at least four and as many as 100 series connections. Thin-film cells have made this easy. They are therefore increasing the range of products that can practically be solarised.

2. STAND-ALONE CONSUMER PRODUCTS

The major benefit of solar products is that they do not need a separate power source, so many of today's solar products are designed to

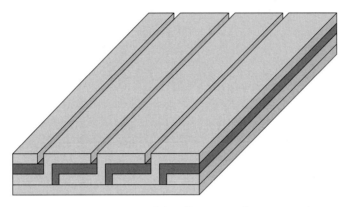

Figure 1 Monolithic interconnection of thin-film solar cells on an insulating substrate. The front contact of one cell connects to the rear contact of the neighbouring cell.

operate independently of other electricity supplies. This is commonly known as *stand-alone operation*.

2.1 Indoor Products

Several products have been designed for use in interior light levels as described below. Many other products also operate indoors, although the solar cell is exposed to external daylight such as those described in Section 2.3 and 3.

Thin-film solar cells perform better than their crystalline counterparts at low light levels, and can be tailored to optimise the output under the typical indoor light spectrum, which is different from sunlight. For this reason, almost all indoor products use thin-film solar cells, usually of amorphous silicon.

2.1.1 Calculators

Millions of solar calculators have been produced, probably accommodating more individual photovoltaic generators than any other single application. Many adults and most schoolchildren are so familiar with solar calculators that this 'eternal' energy source is taken for granted.

The solar cell array looks like a small black panel, is typically less than 1×4 cm, and is enough to power the calculator even under dim internal lighting. It typically comprises four monolithically interconnected amorphous silicon cells, many of which are produced in factories in Japan and Taiwan.

Figure 2 Solar watch incorporating a round array of four amorphous silicon solar cells.

2.1.2 Watches and Clocks

Solar watches have also been mass-produced and are widely available. The solar array may be a small rectangle, but many designs also use a circular cell, which can be incorporated within the watch face (Figure 2).

Again, amorphous silicon is usually used in clocks and watches, because of its performance in interior lighting levels.

2.1.3 Chargers

Any parent knows how expensive it is to replace batteries in the plethora of portable music centres and game consoles to which children seem to be permanently attached.

The usage and disposal of primary batteries can, of course, be avoided by the use of rechargeable equivalents. Many suppliers have now developed solar chargers for maintaining this type of battery, with the added advantage that it can be used anywhere without connection to a mains power socket. Ideally, these chargers are placed on windowsills to maximise the light level and reduce the recharge time.

2.1.4 Accent Lighting

Primary interior lighting is still too power hungry to be viably powered with PV—though the advent of high-efficiency LED lighting may in due course change this.

Specialist low-power lighting, however, especially when intermittent, can be suitable. Look in the shops in December, for example, and you'll find solar-powered Christmas tree lights.

2.1.5 Air Fresheners

Powered air fresheners are becoming very widely used not only in public buildings but also in the home, and many now incorporate electronic sensing. The modest overall power consumption of these devices makes them another ideal applications for solar cells tailored to internal lighting.

2.1.6 Other Indoor Products

Any electrical or battery powered product can be solar powered by incorporating solar cells into it.

In addition to those mentioned above, many have been designed with internal use in mind. A few examples are
- digital weighing scales for domestic and industrial use,
- other solar-powered meters,
- a huge number of kits and toys,
- smoke alarms,
- desktop gadgets and executive toys,
- units for rotating plants and shop window displays, and
- even a solar-powered food whisk and solar toothbrush.

The number of such products is doubtless increasing even as I write.

2.2 Garden and Stand-Alone Consumer Products

Where solar products are designed for external use, the major market is for garden products, especially lighting and aquatic items.

2.2.1 Lighting and Markers

Lighting technology has developed rapidly in recent years, particularly in terms of the power efficiency of the light source. New high-intensity LEDs in particular give high light output, yet consume very little power.

Solar garden lights are now increasingly available in most major retailers as an alternative to mains-powered garden lighting. The major benefit of having a built-in power source is that there is no need to dig up the garden to provide electric cabling to the units. Solar lights, of course, incorporate a rechargeable battery so that the power can be stored during the day and used at night. Many also have a circuit that automatically turns the light on at nightfall either using a separate photocell or simply

Figure 3 Solar garden light.

measuring the solar cell output and, when this falls to zero, switching the light on.

Solar lighting products first came to prominence in the mid- 1980s with offerings from Siemens and Brinkmann, amongst others. Probably the first thin-film powered light at that time was the Chronar WalkLite range, which achieved significant sales despite some initial technical problems. A bewildering range of lights is now available, mainly in four distinct groups.

- Area illumination: these lights are solar equivalents of the mains-powered lighting, which has been available for years. The lights are designed to illuminate a patio, feature, or area of the garden.
- Decorative accent lighting: typically less bright than the area lights, accent lighting is used to provide a point of interest and to cast a glow in the immediate vicinity. Often several lights will be used to create an overall effect (Figure 3).
- Marker lights: solar markers, usually with LEDs as the light source, are used to mark steps, paths, and obstacles.
- Security lighting: again, these are a direct equivalent to mains-powered devices. A sensor circuit is typically used to turn the lights

on when someone passes, for example. As the operational period is quite small, since the light switches off again after several seconds, the overall power consumption is low despite the brightness of the light.

2.2.2 Aquatic Products

Solar products for pools and garden water features have enjoyed tremendous growth in popularity over recent years. Many of these products particularly suit solarisation because they are required to operate only when the sun is out. This means that solar fountains, for example, do not need batteries but can be directly driven from the solar panel. I now find that I hear the water droplets as the fountain outside my window starts up, and think, 'Oh the sun must have come out!'

Most products in this category are pumps and derivatives thereof. There are many solar-driven pumps for waterfalls, fountains, and a wide range of other aquatic applications. 'Jug and bowl' features are particularly popular as these give all the attraction and sound of moving water while being sufficiently compact to use on patios and balconies.

A floating fountain system was an innovation in the late 1990s first promoted by Solar Trend. Because this had the solar panel built in to the product, it needed no external wires and was therefore free to float around in ponds and pools (Figure 4).

Again, the range of products in this subsector has expanded with the increasing acceptance of photovoltaic power. The principle that anything electrical can be solar powered has extended to include pond aeration systems and several swimming pool accessories, including a floating solar ioniser.

The ability of PV to make power available, where it would not otherwise be possible has led to several new concepts, too. In particular there is now a wide range of floating solar lights on the market, taking advantage of the independence of a built-in power source (Figure 5).

2.2.3 Deterrents and Other Products

Making power available where it would otherwise be too difficult or too expensive has given rise to several new products outside the aquatics sector.

Solar power is now one of the primary accepted ways of powering electric fences, which are mainly used for agricultural purposes but also in larger gardens. Other forms of animal deterrent, such as ultrasonic vibration generators, have also be developed and solarised. And to keep

Figure 4 The integral solar cells power this fountain, which is therefore free to float around the pool.

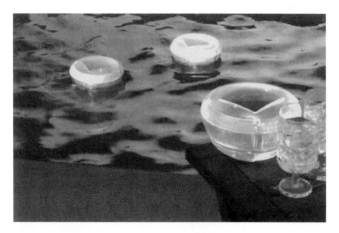

Figure 5 Floating solar light from Intersolar Group.

the lawn tidy without effort, an automatic lawnmower has been developed that also features a built-in solar cell array.

Solar products are now widely used in greenhouses and garden sheds. As most are derivatives of products used in household applications, these are described in the following section.

2.3 Household Products

There is an increasing range of discrete household products now powered by solar cells, and some are described below. An important new trend is the development of architectural products incorporating solar cells that generate power to feed into the building's traditional electricity distribution system, and these are described further in Section 3.

2.3.1 Ventilation and Air Conditioning

Ventilation is another application particularly well suited to solar power because the need is highest when the sunshine is strongest (Figure 6).

When Intersolar Group first acquired the SolarVent, originally developed for use on boats (see Section 2.4.1), it soon adapted the product for household use in conservatories and greenhouses. As an extension to the concept, the company has now patented a range of building ventilation designs incorporating solar power.

Solar-powered air conditioning is something of a 'holy grail' for the PV industry as the match between supply and demand is so good. However, this is an application that has not yet been widely exploited, mainly because of the high power consumption of traditional air-conditioning systems. Systems based on evaporative coolers have been offered in the market, but

Figure 6 The SolarVent Turbo designed to fit into conservatory roofs as seen from inside.

these require an abundant water supply, and I am not aware that the takeup has been strong.

The development of a more power-efficient air-conditioner system, coupled with a solar power supply, will one day make someone very rich!

2.3.2 Lighting and Other Products

PV's main advantage is its ability to act as an independent power source without the need to replace batteries or wire into an established supply. An electricity supply is available throughout most houses in the developed world, and so the benefits of solar power are less marked. It is most widely used, therefore, in more remote rooms and outhouses, without a ring main.

Similarly, in the developing world where an estimated 2 billion people do not have access to a grid network, solar power is often the primary source of electricity. These areas have championed a wide range of products and applications for PV, especially the following:

- Lighting, particularly using low-power fluorescent technology and increasingly high-intensity LEDs. Various solar lanterns and stand-alone products have been developed, while the more advanced distributors in many countries offer package systems.
- Solar-powered refrigeration both for domestic and medical uses.
- Televisions powered by solar panels for educational and recreational usage.

Solar products to feed into existing household electricity distribution are described further in Section 3.

2.4 Portable and Transportable Products

The independence of solar power makes it ideal for transportable products. Because there is already such a bewildering selection, I will highlight here only the most widely adopted or interesting products.

2.4.1 Products for Vehicles

Probably the first outdoor solar consumer product was the SolarVent, first developed by Solar Ventilation Ltd in the UK. This comprised a round monocrystalline solar cell cut into sections that were then soldered together in series to achieve the required voltage to directly drive a DC motor. A fan is attached to the motor and passes air through the unit whenever sufficient daylight falls on the cell (Figure 7).

Figure 7 A marine SolarVent with schematic of the airflow.

The SolarVent was originally designed for use in boats and has subsequently been developed for other transport uses such as caravans in addition to the household applications already mentioned.

A massively successful follow-on product is the AutoVent first developed by Intersolar in 1987 and subsequently patented. This applies the solar ventilation principle to parked cars, without any modification to the vehicle itself. The AutoVent is designed to fit securely into the window of the vehicle and is calculated to change the air inside about every 15 minutes. More than 2 million AutoVents have been sold, with North America, Japan, and Australasia being the major markets (Figure 8).

Vehicle applications naturally lend themselves to solar power because they need to be independent of a fixed electricity source. Boats, caravans, and cars all include electrical devices, and many are conveniently solar powered, especially for use when the vehicle is unattended and the engine (its main power provider) is off.

Amongst the many applications are alarms and deterrents (including gas alarms for boats), lights, coolers, and pumps. Solar panels also are

Figure 8 An AutoVent Turbo fitted in the front window of a parked car. The thin-film solar cell array inside the window powers a fan that exhausts stale air out over the top of the window glass.

often used for topping up the vehicle's primary battery, and a wide range of chargers is available for cars, caravans, recreational vehicles, and boats.

2.4.2 Portable Electronics

Hands up anyone who doesn't carry some form of portable electrical device? Not many, I'm sure.

Portable phones, music, and electronic books have now become ubiquitous, partly because of ever more energy-efficient processing power. Probably the first to go beyond simple battery power was the Vreeplay radio, which started with a manual windup version and then progressed to solar power (Figure 9).

I had expected more of these devices to incorporate solar cells by now, but actually integrated solar cells are still relatively rare, perhaps because these devices are generally carried in pockets or cases out of the light or perhaps because they are already covered by charger applications as mentioned below. Nokia, and probably others, have looked at integrating solar cells into mobile phone housings, but I haven't seen any on the market. There are some exceptions, though, including the first mass PV-powered product, the calculator mentioned in Section 2.1.1. A more recent example is the wireless keyboard with integral solar cells (Figure 10).

Figure 9 The Freeplay range of radios was originally developed to utilise the windup technology developed by Trevor Baylis, but many now incorporate solar cells too. Photo courtesy of the Freeplay Group.

Figure 10 The eminent computer peripherals company Logitech has incorporated solar cells into its Wireless Solar Keyboard K750.

Of course, most portable electronic and electrical products are battery powered. A wide range of solar chargers also exists to replenish rechargeable batteries, which can be used in any such product (see Section 2.1.3). Some of these chargers are specifically tailored to particular types of battery (notably mobile phones). Others include integral batteries so that they can be used as a primary power source for mobile phones and handheld game stations in particular.

Their need for independent power makes handheld personal products particularly suited to solar energy. Again the range is almost endless, perhaps the largest being lanterns and torches, especially keychain versions. Hats incorporating solar fans seemed like a gimmick at first but have now spawned many variants.

3. SOLAR PRODUCTS FOR GRID CONNECTION

Most of the solar products described so far incorporate the solar cell array directly in the product being powered. An exciting recent development is the incorporation of solar cells in other products in such a way that they can then be connected to contribute to the overall electricity supply of buildings, for example.

Most such systems operate in conjunction with a traditional grid supply. The system draws a proportion of power from the grid when the solar generation is less than the local consumption, and it can feed power back into the grid when there is a surplus, as described in Chapter IIc-1.

Originally, such architectural solar systems were individually designed for each building, or standard solar systems were retrofitted to existing buildings, as described in Chapters IIc-2 and IIe-1. Recently, however, products have been specially designed for building use to include a solar cell array.

3.1 Roofing Products

The most obvious approach is to incorporate solar cells into roofing products, and it is now possible to obtain solar tiles, slates, shingles, and even pantiles. Initially, solar tiles of standard sizes had low operating voltages of 6 V and less, so they required specialist installation due to the series- and parallel connection arrangements to achieve the required DC voltage, typically in the range of 24–120 V.

A further breakthrough came with the use of monolithically interconnected thin-film cells in solar roofing products. These make a high voltage per unit area viable and therefore enable each product to operate at the designated system voltage. Only parallel connections are needed, and it is possible to achieve this through simple interconnections that require no specialist electrical expertise and can be undertaken by established roofing contractors (Figure 11).

Figure 11 The uniform appearance of the thin-film solar cells means that the four rows of solar slates being installed in Cardiff are almost indistinguishable from the rows of standard synthetic slate above and below them. The glass section at the top incorporates the interconnections and is covered by the row above when installation is complete.

Figure 12 Each facade panel used at the first-floor level of this building in the Department of the Built Environment at Nottingham University incorporates about 1 m^2 of thin-film solar cells but look almost identical to standard glazing panels.

3.2 Cladding and Facade Products

A similar approach has been adopted for other products used in building envelopes, such as cladding panels and facade elements for use on vertical walls (Figure 12).

4. NONCONSUMER PRODUCTS

I have attempted in this chapter to focus primarily on solar products for which individuals may have a use. There are many applications of PV in other areas, such as those described in Part IV. Many of these, too, have led to the design of integrated solar products. Some of the main ones are described briefly in the following.

4.1 Commercial Applications

Many of the products described above or their industrial equivalents are used in commercial situations such as hotels, parks, shops, and public buildings.

Additionally, several solar products have been designed specifically for commercial applications, particularly for displays in shops and for roadside advertising.

4.2 Transportation

Solar-powered navigational aids are now commonplace in our ports, rivers, and in-shore waters and will be familiar to any sailor. Solar aircraft-warning lights are often used on towers, masts, and high buildings on land, too.

Road and rail transport have given rise to many solar products, especially signals, track circuits, streetlights (Figure 13), parking meters, traffic and hazard signs, bus shelters, and the ubiquitous roadside emergency telephones.

Solar products in commercial vehicles are increasingly common, including solar roofs to power refrigerated trucks and the ventilation of delivery vans. This led coincidentally to the invention of the AutoVent. During a trip to Japan, where we were selling SolarVents to Nippon Fruehauf for use in produce-delivery vans, Simon Pidgeon and I were sweltering in the traffic when I idly suggested, 'If only we could fit a SolarVent to this taxi.' And bingo!

4.3 Other Industrial Applications

Wireless telecommunications is a natural application for a wireless source of electricity such as PV. In addition to the emergency telephones mentioned above, there are now too many solar-powered telecommunications products to mention here.

Figure 13 Solar-powered roadside equipment such as this streetlight is now very common.

The range also extends to telemetry products such as SCADA devices, remote meteorological and hydrology systems, which are used particularly in the oil and gas and utility industries.

5. DESIGNING PV FOR PRODUCTS

The PV industry is relatively young, and there is still plenty of evidence that it is still learning about solar product design. I see too many products where the solar cell looks as if it has been stuck on as an afterthought or where a PV manufacturer seems to have said, 'What else can I do with my solar panels?' In many cases, the solar array is too intrusive or incongruous.

This is primarily a design issue, though marketing clearly has a part to play as some products cannot possibly have been conceived with any sane consumer in mind. In my view, a good photovoltaic product would not look substantially different from the nonsolar version unless there is a good reason to highlight the PV aspect.

Thin films have made such holistic design easier, as the cells are largely uniform in appearance and without the shiny contact grids and tabs used

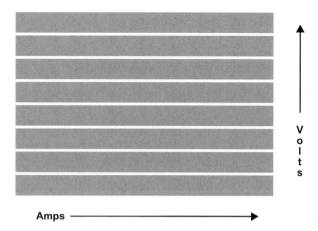

Figure 14 Monolithically interconnected thin-film plate. The voltage is proportional to the number of series cells and therefore the width. The current is proportional to the cell area and therefore the length.

in most crystalline cells. However, there are some instances where, for example, the mottled appearance of multicrystalline cells can be used to good visual effect.

5.1 Electrical Characteristics

Operationally, the solar cell array is there to fulfill a defined electrical function. This can usually be reduced to a specified operating voltage and an expected peak daily or annual current output. Where the solar cell is used as a trigger to switch the product on in the dark (as described in Section 2.2.1), the electrical characteristic at low light level is also important.

The voltage is proportional to the number of series-connected cells, while the current is related to the cell area. In monolithically interconnected thin-film arrays, these factors can bear a direct relationship to the dimensions of the solar cell array (Figure 14).

5.2 Product Dimensions

Ideally, the solar array size should be defined by the product, not the other way round, though in many cases some compromise is necessary. Where space is limited, more efficient cells will need to be used.

The electrical characteristics may be a first determinant of the size of a thin-film cell array as indicated previously. Existing multijunction amorphous silicon cells, for example, give about 50 V per metre of width and about 1 A per metre of length.

5.3 Mechanical and Structural Considerations

The most important feature of solar product design is the environmental protection of the solar cell. PV devices are highly reliable solid-state semiconductors. Their operating life is extremely long. The eventual failure of the device is more likely to relate to the connections or moisture-induced degradation than to the photovoltaic element itself.

All cells are susceptible to moisture in particular and should be packaged for weather resistance. In most designs, protection is applied to the solar cell first by spraying, adhering, or printing a protective film or by lamination. The product design should then provide secondary protection, especially if the product is to be used outdoors.

These aspects of design are a science in themselves, and there is a gulf separating good solar products from some of the cheap replicas, which inevitably follow the success of a rapidly growing market.

6. SOLAR PRODUCTS OF THE FUTURE

This has given a flavour of the massive range of solar product applications that have already been introduced. As I said at the start, anything that is or could be electrically operated can be solar powered. The eventual range is limitless.

The genesis of new solar products is expected to accelerate in the future, driven primarily by ever-lower solar cell costs but also by new advances in solar cell technology and by more power-efficient products generally. Improvements in high-efficiency LED technology have stimulated the solar lighting market tremendously. I expect continuing advances in consumer electronics to make this a major growth area for PV products in the medium-term future.

What of the long term? The key issue will become that of storage. Without reliable low-cost power storage, it will still be necessary to operate in conjunction with other generation sources, and it is therefore likely that grid-connected products will lead the way. I hope an efficient, cheap, and preferably thin-film battery will be developed. Imagine a credit-card—sized unit with solar cells on one face, a battery in the middle, and display on the other face as a fully self-contained personal computer, television, and communications system.

The applications, as many have said before, are limited only by our imagination.

ACKNOWLEDGEMENTS

The author thanks those companies that are mentioned herein and whose products and developments are featured in this chapter.

PART III

Testing, Monitoring, and Calibration

CHAPTER III-1

Characterization and Diagnosis of Silicon Wafers, Ingots, and Solar Cells

Andrés Cuevas[a], Daniel Macdonald[a], and Ronald A. Sinton[b]
[a]School of Engineering, The Australian National University, Canberra, ACT 0200
[b]Sinton Instruments, Boulder, CO, USA

Contents

1. Introduction	1012
2. Factors Affecting Carrier Recombination	1012
3. Measurement of the Minority-Carrier Lifetime	1015
3.1 Lifetime-Testing Methods	1015
3.2 Interpretation of Effective Lifetime Measurements	1021
3.2.1 Surface Recombination Velocity	*1021*
3.2.2 Emitter Saturation Current Density	*1024*
3.3 Lifetime Measurements in Silicon Ingots	1027
3.4 Lifetime Instabilities and Measurement Artifacts	1028
3.4.1 Lifetime Changes in Boron-Doped CZ Silicon and Iron-Contaminated Silicon	*1028*
3.4.2 High Apparent Lifetime at Low Excess Carrier Densities	*1029*
4. Relationship Between Device Voltage and Carrier Lifetime	1030
5. Applications to Process Monitoring and Control of Silicon Solar Cells	1031
5.1 Resistivity Measurements	1031
5.2 Process Monitoring via Lifetime Measurements	1032
5.2.1 Measurements of Carrier Lifetime in an Ingot or Brick	*1033*
5.2.2 Lifetime Measurements on Bare Wafers	*1033*
5.2.3 Effective Lifetime After Emitter Formation	*1034*
5.3 Device-Level Characterization	1035
5.3.1 Standard Diode Analysis of Suns—V_{oc} Curves	*1036*
5.3.2 Monitoring Contact Formation	*1037*
5.3.3 Fill Factor and Series Resistance Measurements	*1038*
6. Conclusions	1040
Acknowledgements	1040
References	1041

Practical Handbook of Photovoltaics.
© 2012 Elsevier Ltd. All rights reserved.

1. INTRODUCTION

Monitoring the fabrication process of a solar cell entails measurements of mechanical, optical, and electronic properties of the silicon wafers and the different layers and coatings formed on them. This chapter focuses on electronic characterization, with emphasis on measurements of the effective minority-carrier lifetime and their interpretation to extract relevant information about the silicon material, the emitter regions, and the surface passivation. The chapter also includes a brief description of techniques based on photo and electroluminescence that can provide high-resolution images of ingots, wafers, and devices. A direct measurement of the open-circuit voltage under variable illumination gives the Suns–V_{oc} characteristics, which provide most of the parameters needed to model solar cells at the device level, including diode saturation current, ideality factor, shunt resistance, and the upper bound on fill factor. The application of these techniques to process monitoring and control is described for the case of silicon solar cells, illustrating the correlation between the effective lifetime measured at various stages of fabrication and device performance. Final I-V curve testing at one-sun illumination and 25°C temperature is routinely integrated in solar cell production lines following standard methods that are described elsewhere in this book.

2. FACTORS AFFECTING CARRIER RECOMBINATION

Once created within a semiconductor, photogenerated electrons and holes last, on average, a finite time called the *lifetime*. When carriers are continuously generated, as in a solar cell, the value of the lifetime determines the stable population of electrons and holes. This population should desirably be as high as possible because it determines the voltage produced by the device. A second, equally important aspect of the lifetime is that it is directly related to the *diffusion length*, which is the average distance that carriers can travel from the point of generation to the point of collection (the p–n junction). The relationship between L and τ is $L = \sqrt{(D_n \tau_n)}$, with a value for the diffusion coefficient of $D_n = 27$ cm^2/s, for electrons (minority carriers) in 1-Ωcm p-type silicon. The diffusion length should be greater than the wafer thickness or the longest generation depth to ensure a high short-circuit current. Since the lifetime determines both the voltage and the current of the device, its characterization is of the utmost importance.

The lifetime should not be assumed to be a constant single value. It can change considerably, depending on the process history of the sample (including possible contamination and thermal degradation) and measurement conditions (injection level and temperature), and it can degrade or recover when the sample is exposed to light or annealed at certain temperatures. The measured, apparent lifetime can also be affected by nonrecombination mechanisms such as carrier trapping. Figure 1 shows, as an example, the lifetime measured over a broad range of carrier-density injection levels for two float-zone (FZ) silicon wafers, one of which was purposely contaminated with iron. The behavior of the lifetime can be understood in most cases with the assistance of the theoretical models that describe the most important physical mechanisms. It should be kept in mind that the measured lifetime, $\tau_{\it{eff}}$, may include several of them simultaneously.

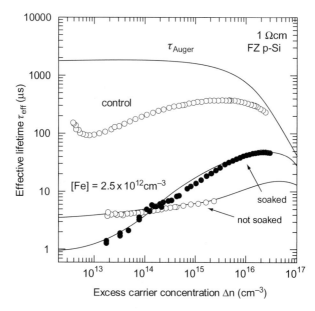

Figure 1 Measured effective-carrier lifetime τ_{eff} as a function of the excess carrier concentration Δn for a 1-Ωcm boron-doped FZ silicon wafer contaminated with iron before and after light soaking, which dissociates iron−boron pairs. An uncontaminated control wafer is also shown. Both had the surfaces passivated with PECVD SiN.

In practical silicon wafers, the most important recombination losses occur through crystallographic defects and impurities that create energy levels within the band gap. The effect of such recombination centers can be described by the Shockley–Read–Hall model, which predicts that the injection-level dependence of the lifetime, τ_{SRH}, is a function of the dopant density, N_A; recombination center density, N_{SRH}; defect energy level, E_T; and capture cross-sections:

$$\frac{1}{\tau_{SRH}} = \frac{N_A + \Delta n}{\tau_{p0}(n_1 + \Delta n) + \tau_{n0}(N_A + p_1 + \Delta n)} \quad (1)$$

In this expression, which applies to p-Si, $\Delta n = \Delta p$ is the excess carrier density (assuming negligible trapping), and τ_{n0} and τ_{p0} are the fundamental electron and hole lifetimes, which are related to the recombination center density, the thermal velocity $v_{th} = 1.1 \times 10^7$ cms^{-1}, and the capture cross-sections via $\tau_{n0} = 1/(v_{th}\sigma_n N_{SRH})$ and $\tau_{p0} = 1/(v_{th}\sigma_p N_{SRH})$. The magnitudes n_1 and p_1 are given by

$$n_1 = N_C \exp\left(\frac{E_T - E_C}{kT}\right); \quad p_1 = N_V \exp\left(\frac{E_C - E_G - E_T}{kT}\right) \quad (2)$$

The values for the effective densities of states at the conduction and valence band edges are $N_C = 2.86 \times 10^{19}$ and $N_V = 3.10 \times 10^{19}$ cm^{-3}. For recombination centers located near the middle of the energy gap, Equation (1) predicts an increased lifetime between very low and very high injection level from τ_{n0} to $\tau_{n0} + \tau_{p0}$. Nevertheless, every recombination centre is characterized by a distinctive set of parameters E_T, τ_{n0}, and τ_{p0}. This leads to different injection and temperature dependences of the lifetime and can be used to identify the characteristic signature of each centre. The diversity of possible lifetime curves and the determination of E_T, τ_{n0}, and τ_{p0} based on the injection level or temperature dependence of the lifetime are reviewed in [1]. Figure 1, which is part of an injection-level study for the case of iron–boron pairs in silicon [2], shows a good fit of Equation (1) to the experimental data. The light-soaked results in which the iron–boron pairs are dissociated show a widely varying lifetime in the low-injection range of interest for solar cells.

The upper limit to lifetime that may be measured for an otherwise perfect silicon sample is determined by two intrinsic mechanisms: Coulomb-enhanced Auger and radiative band-to-band recombination, with Auger being the most restrictive. The following empirical expressions for the intrinsic lifetime, incorporating both of these mechanisms, for p-type and n-type silicon are widely used [3]:

$$\frac{1}{\tau_{intrinsic}} = (\Delta n + N_A)(6 \times 10^{-25} N_A^{0.65} + 3 \times 10^{-27} \Delta n^{0.8} + 9.5 \times 10^{-15})$$
(3)

$$\frac{1}{\tau_{intrinsic}} = (\Delta n + N_D)(1.8 \times 10^{-24} N_D^{0.65} + 3 \times 10^{-27} \Delta n^{0.8} + 9.5 \times 10^{-15})$$
(4)

It should be emphasized that these two expressions are based on fitting experimental measurements of well surface-passivated long-lifetime samples, and they should not be taken as absolute physical limits. With improved surface passivation, slightly higher lifetimes have already been measured, although an alternative empirical expression has not yet been proposed. For perspective, Equation (3) is plotted in Figure 1 for a 1-Ωcm boron-doped wafer. Even though the intrinsic lifetime "limit" is quite high, it has a strong dependence on carrier density at high injection and should therefore be taken into account when using injection-level dependence techniques to discriminate between different recombination mechanisms, as described in the following sections.

3. MEASUREMENT OF THE MINORITY-CARRIER LIFETIME

As mentioned above, the minority-carrier lifetime that can be measured experimentally in a given sample is an effective parameter that encompasses several different recombination and transport mechanisms, including surface recombination. Accordingly, it is labeled here τ_{eff}. For a detailed characterization of silicon wafers and devices, it is essential to discriminate between the various physical mechanisms and factors that affect the effective lifetime. Several well-established methods to perform such analysis are described in the following sections.

3.1 Lifetime-Testing Methods

There are three basic approaches to measure the carrier lifetime, depending on the way that an excess of carriers is generated and varied in the semiconductor. In the *transient decay method,* the photogeneration is terminated abruptly and the rate at which carriers disappear, dn/dt, is measured, together with the excess electron concentration itself, Δn (note that every photon generates one electron−hole pair, so that $\Delta n = \Delta p$). If no current is flowing from the device, then the rate of carrier density change is equal to the recombination rate:

$$\frac{d(\Delta n)}{dt} = -\frac{\Delta n}{\tau_{\mathit{eff}}} \tag{5}$$

Equation (5) implies an exponential decay of the carrier density with time, which means that 37% of electrons are still present after one lifetime, decreasing to 5% after three lifetimes. This method, classically known as *photoconductance decay* (PCD) [4], is quite robust, because it is based on the measurement of the relative change of Δn with time. It is, nevertheless, advisable to measure the absolute value of Δn at which the lifetime has been determined, since the lifetime is, in general, strongly dependent on the carrier injection level.

In the *steady-state method* [5], a constant generation rate of known value is maintained, and the lifetime is determined from the balance between generation and recombination:

$$G_L = \frac{\Delta n}{\tau_{\mathit{eff}}} \tag{6}$$

This simple expression assumes a uniform generation rate across the thickness of the sample and also a uniform excess carrier density, Δn. Nevertheless, in cases where they may be nonuniform, Equation (6) is still applicable, with Δn representing an average value. If the illumination varies slowly, quasi−steady-state conditions prevail within the semiconductor, leading, for example, to the *quasi−steady-state photoconductance method* (QSSPC) [6].

$$\tau_{\mathit{eff}} = \frac{\Delta n(t)}{G_L(t) - \frac{\partial \Delta n}{\partial t}} \tag{7}$$

Equation (7) is applicable to the three experimental situations and is frequently referred to as the *generalized* method [7]. An absolute measurement of the excess carrier density, Δn, is required to determine the lifetime in the steady state and QSSPC methods. In addition, the photogeneration rate needs to be determined accurately as well. The latter is measured using a photodetector (for example, a calibrated solar cell).

The detector gives the total photon flux incident on the surface of the wafer. For the standard AM1.5G solar spectrum commonly referred to as *one-sun* intensity (100 mW cm^{-2}) the number of photons per second and cm^2 with energy above the energy band gap of silicon (photon energy above 1.12 eV) is $N_{ph} = 2.7 \times 10^{17}$ cm^{-2}. This photon flux gives, multiplied by the electronic charge, an upper limit of the current density for a silicon solar cell, 43.5 mA cm^{-2}. Silicon wafers absorb only a fraction of these photons, depending on the reflectivity of the front and back surfaces, possible faceting of those surfaces, and the thickness of the wafer. The value of the absorption fraction for a polished, bare silicon wafer of thickness 250 μm is $f_{abs} \approx 0.6$. If the wafer has an optimized antireflection coating, such as a 70-nm-thick silicon nitride or titanium oxide layer, $f_{abs} \approx 0.9$, while a textured wafer with antireflection coating can approach $f_{abs} \approx 1$. The uncertainties associated with the determination of f_{abs} can be kept very small by using tables or graphs that can be calculated using optical models. The generation rate per unit volume G_L can then be evaluated from the incident photon flux and the wafer thickness:

$$G_L = \frac{N_{ph} f_{abs}}{W} \tag{8}$$

For all three testing methods, the excess electron concentration, Δn, needs to be measured. There are several techniques to do this based on different properties of semiconductor materials. The simplest and most common of them is to measure the conductance of the wafer and the way it changes with illumination and time. The excess photoconductance for a wafer of thickness W is given by

$$\Delta \sigma_L = qW(\mu_n + \mu_p)\Delta n \tag{9}$$

Typical mobility values for 1-Ωcm silicon in low injection are $\mu_n = 1100$ cm^2 V^{-1} s^{-1} for electrons and $\mu_p = 400$ cm^2 V^{-1} s^{-1} for holes. The electron and hole mobilities are functions of the dopant density and injection level, as has been amply documented in the literature [8].

A straightforward way to measure the photoconductance is to apply electrical contacts to the silicon (typically an ingot or a wafer), forcing an electric current through it and measuring the resulting voltage drop [9]. This can also be done in a contactless fashion by using an inductively coupled radio-frequency circuit that produces a voltage proportional to the conductivity of the wafer. This method, implemented in commonly used QSSPC testing instruments, has the added advantage that the relationship between conductance and voltage is practically linear over a broad range [10]. Typical microwave-detected PCD systems, μ-PCD, are based on directing a microwave beam to the silicon wafer and measuring the reflected microwave power, which is proportional to the conductance of the wafer [11,12], although only under carefully optimized configurations [13]. Schroder [14] gives a comprehensive description of these methods, their theoretical background and practical implementation.

The excess carrier density can also be directly determined by probing the sample with infrared light and measuring the amount of free-carrier absorption [15,16] or by using an infrared camera to visualize the free-carrier IR absorption or emission for the entire wafer, which is the basis for the infrared lifetime mapping [17] or carrier density imaging [18] techniques.

Excess carriers may also be detected via photons emitted during radiative band-to-band recombination. Although the fraction of recombination events occurring via this mechanism is small, sensitive detectors have allowed accurate techniques based on Electroluminescence (EL) and Photoluminescence (PL) to be developed. EL is based on the application of a forward bias to generate carriers within a finished device. It can be used, for example, to image spatially resolved diffusion lengths of minority carriers [19,20]. PL uses illumination to generate the excess carriers, and so does not require the presence of a metallized p−n junction. It can therefore be applied at the wafer level, even without surface passivation. The PL emission rate is proportional to $\Delta n(N_{dop} + \Delta n)$ but also depends sensitively on the optical properties of the test sample [21]. This makes absolute calibration of the technique somewhat complicated for the steady-state case, which is the mode commonly used for imaging applications. Nevertheless, by direct comparison to an excess carrier density measured on the same sample via another technique, such as the QSSPC, it is possible to obtain a calibrated, rapid, high-resolution image of the effective lifetime across a sample [22]. The high spatial resolution, on the order of 100 μm, makes the technique ideal for process monitoring, allowing cracks, scratches and contaminated regions to be identified.

Additional methods, also described in [14], such as the surface photovoltage (SPV) and the short-circuit current response, measure the minority-carrier diffusion length. Both rely on the formation of a surface space charge region to collect minority carriers. The SPV approach achieves this by chemically treating the surface so that a surface charge is created that forms the space charge region. A transparent contact is used to measure the voltage associated to the collection of carriers by the space charge region. The latter is usually in the mV range to maintain linearity and simplify the analysis. The formation of a liquid electrolyte—semiconductor junction that allows the extraction of the short-circuit current is also possible and constitutes the basis for the Elymat technique [14]. When a junction exists in the device, as in a finished solar cell, it may be preferable to measure the spectral response of the short-circuit current, an analysis of which can give information on the bulk diffusion length, surface recombination velocity and optical light trapping [23].

Depending on the specific details of their practical implementation, in particular the magnitude of the excitation used, lifetime-testing methods can be broken down into two categories, large-signal and small-signal methods. In R&D laboratories, μ-PCD often uses short pulses to create a relatively small number of carriers in the semiconductor in order to operate in the low-injection range of the detector where the signal can be proportional to the excess carrier density [13]. The resulting small signal is then separated from the background conductance provided by the doping or a bias light. It should be kept in mind that the direct result of measurements using a bias light are, in general, different from the true recombination lifetime. Converting the small-signal lifetime into the actual lifetime requires measuring the former as a function of injection level, followed by integration [24,25]. Industrial μ-PCD systems often use a large-signal approach, which include excitation carrier densities comparable to or higher than the doping levels [26]. Using such testing conditions, the reported values can differ substantially from the actual lifetime since they are a convolution of the lifetime with the sensor nonlinearity [13] as well as the nonlinearity of the photoconductance Equation (9). The microwave-detected photoconductance (MDP) technique is similar to the μ-PCD methodologies, except that it allows for a long excitation pulse [27].

A comparison of contactless lifetime measurement techniques is shown in Table 1. All of them can be used to measure the excess carrier density, and therefore Equations (5) and (9) form a common basis for their analysis, which enables comparisons and complementarities between them.

Table 1 Contactless sensing methods commonly used for measuring the excess carrier concentration and lifetime in silicon

Technique	Excess carrier-density sensing method	Pros/Cons
RF-QSSPC	Eddy-current sensing of photoconductance. Conversion to Δn using known mobility function	Simple calibration, valid over very wide range of carrier density. Requires mobility and photogeneration model or measurement. Low spatial resolution. Trapping and depletion region modulation effects at low carrier density.
RF-transient PCD	Same as QSSPC.	Same as QSSPC, except that no calibration of photogeneration is required.
ILM/CDI	IR free-carrier absorption or emission. Calibration using samples of known majority-carrier density.	High-resolution imaging capability. Surface texture complicates interpretation. Subject to trapping and DRM effects.
μ-PCD/MDP	Microwave sensing of photoconductance. Carrier density set by bias light, long-excitation pulse into SS conditions, or by injecting known number of photons in a very short pulse.	High-resolution mapping capability. Nonlinear detection of photoconductance in some injection-level or dopant ranges; skin depth comparable to sample thickness in some cases. DRM and trapping effects at low carrier density.
Photoluminescence	Radiative band to band emission. Models for coefficient of radiative emission and reabsorption, or calibration via another method such as QSSPC.	Used in both nonimaging and high-resolution imaging applications. Signal depends on doping, photon reabsorption, surface texture, detector EQE, and wafer thickness. Unaffected by trapping and DRM effects, it can be applied at very low carrier densities.

3.2 Interpretation of Effective Lifetime Measurements
3.2.1 Surface Recombination Velocity

The *effective lifetime*, τ_{eff}, is the net result of summing up all the recombination losses that occur within the different regions that constitute a given silicon sample. The ability to separate out the different recombination mechanisms and identify where the major losses occur within the wafer or the solar cell device is an essential part of the characterization and diagnosis process. For example, if the sample is a silicon wafer, the surfaces usually have a significant impact on the measured effective lifetime. In the special case that the excess carrier density is uniform, the *effective lifetime* can be expressed mathematically in a relatively simple form:

$$\frac{1}{\tau_{\mathit{eff}}} = \frac{1}{\tau_{\mathrm{bulk}}} + \frac{S_{\mathit{front}} + S_{\mathit{back}}}{W} \qquad (10)$$

where W is the thickness of the wafer and the *surface recombination velocities*, S_{front} and S_{back} represent in simplified form the recombination occurring at the front and back sides of the wafer. Equation (10) indicates that measurements of the effective lifetime can be used to study surface recombination. The latter is due to Shockley–Read–Hall processes and can vary with the carrier injection level, in much the same way as the bulk lifetime [28], making a separation between surface and bulk recombination difficult. Nevertheless, it is always possible to determine an upper bound for S by assuming $\tau_{\mathit{bulk}} = \infty$ (or $\tau_{\mathit{bulk}} = \tau_{\mathit{intrinsic}}$) in Equation (10), which works quite well when wafers with a very high bulk lifetime are used. Conversely, a lower bound for the bulk carrier lifetime can be obtained by making $S = 0$ in Equation (10), which applies especially to wafers with a high-quality passivation on both sides. A complete separation between surface and bulk recombination, is possible by using several (at least two) wafers with different thicknesses and identical bulk and surface properties [14].

The assumption of a uniform carrier density that underlies Equation (10) can be self-consistently checked for transient methods if the measured effective lifetime is much greater than the transit time of carriers across the wafer.

$$\tau_{\mathit{eff}}(\Delta n) \gg \frac{W^2}{2D} \qquad (11)$$

When this condition is met, the transient method becomes independent of the details of the light pulse excitation, such as wavelength and duration. By waiting a transit time before analyzing the data, excess carriers can spread and distribute themselves evenly across the wafer, even if they were nonuniformly generated. The diffusion coefficient for low-injection conditions in p-type material, is D_n, but in the general case it becomes injection dependent [29]:

$$D = \frac{(n+p)D_n D_p}{nD_n + pD_p} \quad (12)$$

The transit time depends quadratically on the wafer thickness. In addition, it also depends on the type and doping density of the wafer through the diffusion coefficient, which can vary in the range of 9–30 cm^2/s for typical n- and p-type silicon wafers. For symmetric wafers with the same surface recombination velocity on both sides, the requirement of uniform carrier density can be relaxed for the QSSPC method [30].

Equation (10) is a simplified expression, and would predict a zero effective lifetime when the surface recombination velocity is very high. In reality there is a limit on how low the effective lifetime can be because electrons and holes have to diffuse toward the surfaces in order to recombine, which is a relatively slow mechanism. For homogeneous steady-state photogeneration, such as that produced by infrared light, or for a transient decay measurement, the minimum effective lifetime is, respectively, given by the following expressions:

$$\tau_{eff(S=\infty)QSSPC} = \tau_{bulk}\left(1 - \frac{2L_n}{W}\tanh\left(\frac{W}{2L_n}\right)\right) \quad (13)$$

$$\tau_{eff(S=\infty)PCD} = \frac{W^2}{\pi^2 D_n} \quad (14)$$

For a typical p-type 300-μm-thick wafer and an electron diffusion coefficient $D_n = 27$ cm^2 s^{-1}, the lifetime that can be expected for a transient decay measurement is $\tau_{eff} = 3.4$ μs. The steady-state surface-limited lifetime is a function of the bulk lifetime τ_{bulk} and diffusion length L_n, and also of the wavelength spectrum of the light source. In the limit $L_n \gg W$, Equation (13) simplifies to $\tau_{eff(QSSPC)} = W^2/(12D_n)$.

The dependence of the measured effective lifetime on the surface recombination velocity is shown in Figure 2 for the example of a 300-μm-thick, 1-Ωcm p-type wafer and four different values of the bulk lifetime. The graph indicates the lifetime that would be measured for this wafer with either 400-nm or 1000-nm light using a steady-state

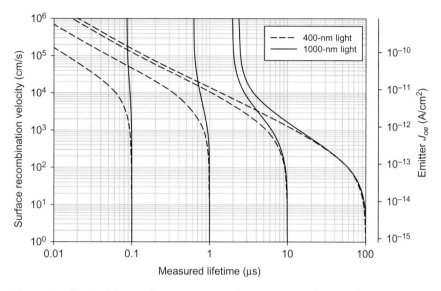

Figure 2 Effective lifetime for a 300-μm, 1-Ωcm p-type wafer as a function of the surface recombination velocity. The curves, modelled with PC1D for steady-state conditions, correspond to wafers with 0.1-, 1-, 10-, and 100-μs bulk lifetimes. Two sets of curves are shown, for 400-nm and 1000-nm monochromatic illumination, respectively.

photoconductance method. The curves were calculated for minority-carrier densities in low injection conditions using the PC1D computer software package [31]. Transient methods, using a short pulse with evaluated with Equation (1), will determine the lifetime approximately as shown by the 1000 nm curves, except that in the limit of high surface recombination the difference between the measured effective lifetime and the actual bulk lifetime is reduced by 22%, as follows from Equation (14). Several main features can be identified on this plot.

1. The measured effective lifetime is equivalent to the actual bulk lifetime if the surfaces are well passivated, independent of the wavelength of illumination (lower part of Figure 2). The demands on surface passivation can, nevertheless, be relaxed when determining relatively low bulk lifetimes. For example, for 0.1 μs bulk lifetime surface recombination velocities up to 1000 cm/s can be tolerated. On the other hand, to accurately determine high bulk lifetimes requires very good passivation. For example, the surface recombination velocity should be less than 10 cm/s to determine $\tau_{bulk} > 100$ μs.

2. If the surface recombination velocity is greater than 10^5 cm/s, typical of unpassivated silicon wafers, infrared light (for example, 1000 nm) measurements are more indicative of the bulk lifetime. The sensitivity to τ_{bulk} is excellent if the latter is lower than 2 μs and a one-to-one correspondence can be made between the IR measured effective lifetime and the actual bulk lifetime. The sensitivity is still reasonable to resolve bulk lifetimes up to 10 μs, but is poor for higher lifetimes. Importantly, however, a pass-fail test for τ_{bulk} up to 10 μs can be established for bare, unpassivated wafers by using infrared light. For example, a measured lifetime of 2 μs indicates an actual bulk lifetime greater than 10 μs. A more detailed study is given in [32].
3. When the surface recombination velocity is greater than 1000 cm/s (upper portion of Figure 2), the effective lifetime is significantly different for blue and IR light, and this allows the bulk lifetime and the surface recombination velocity to be uniquely determined by measuring with both wavelengths [33]. For example, from Figure 2, a wafer that measures 0.65 μs with 1000-nm light and 0.05 μs with 400-nm light has a bulk lifetime of 1 μs and a surface recombination velocity of 1×10^5 cm/s. This method exploits the fact that the photogeneration from blue light is very sensitive to the front surface recombination velocity, while the IR light penetrates deep into the bulk of the wafer. Note that the higher lifetime range (above 10 μs) cannot be discriminated even using this method, because the relatively high SRV above 1000 cm/s completely masks bulk recombination.

3.2.2 Emitter Saturation Current Density

Highly doped or *emitter* regions in silicon solar cells can be formed by thermal diffusion, laser doping or ion implantation. They are commonly characterized by a *saturation current density* parameter, J_{oe}, that encompasses recombination within the bulk of the thin heavily doped region as well as the recombination at the emitter doped surface. If both sides of a wafer are diffused and the excess carrier density is uniform, then the effective lifetime can be expressed as

$$\frac{1}{\tau_{eff}} - \frac{1}{\tau_{intrinsic}} = \frac{1}{\tau_{SRH}} + \left[J_{oe(front)} + J_{oe(back)} \right] \frac{(N_A + \Delta n)}{qn_i^2 W} \quad (15)$$

In this expression, N_A is the dopant density of the wafer, assumed here to be p-type, while $J_{oe(front)}$ and $J_{oe(back)}$ are the *saturation current densities* that characterize the front and back emitter regions, respectively. At 25°C, $qn_i^2 = 12$ C cm^{-6}.

It is straightforward to consider the case when one of the surfaces is diffused (hence characterized by an emitter saturation current) and the other is not (hence characterized by a surface recombination velocity) by combining Equations (10) and (15). In fact, both concepts are equivalent in low injection, $S_{\mathit{eff}} \approx J_o N_A/qn_i^2$. The scale on the right of Figure 2 shows the emitter-saturation current densities that correspond to the surface recombination velocities on the left axis, for a 1-Ωcm wafer. Equations (10) and (15) assume an approximately uniform carrier density across the wafer. The region of applicability of this assumption is visually displayed in Figure 2 as the region where the 400-nm curves and the 1000-nm curves overlap.

Usually, symmetrically diffused wafers are used to determine the saturation current density. It is possible to determine an upper bound for J_{oe} by performing a measurement in low injection and assuming $\tau_{bulk} = \infty$ (or, even better, $\tau_{bulk} = \tau_{intrinsic}$) in Equation (15), if the wafer doping and thickness are known. It is worth noting that a given J_{oe} places a bound on the carrier lifetime that may be observed experimentally, especially for relatively high dopant densities [34,35].

The fact that the emitter recombination term has a different dependence on the carrier injection level than the bulk (SRH and intrinsic) recombination terms allows determination of the J_{oe} by examining the injection level dependence of the lifetime, particularly in the high-injection regime. This method, widely used in conjunction with high resistivity, low N_A wafers, was first proposed by Kane and Swanson [36].

Figure 3 shows a family of curves for different values of emitter saturation current density and a bulk doping of 1.5×10^{16} cm^{-3} (1 Ωcm). The data are plotted as inverse measured lifetime, following the form of Equation (15). This equation indicates that for measurements performed at minority-carrier densities above the dopant density, the emitter saturation current density will be given by the slope of the line for each curve in Figure 3. The variations in lifetime due to the SRH recombination in the $1/\tau_{bulk}$ term occur primarily at carrier densities less than the doping level, so that they do not significantly affect the slope of the data above the doping level.

There is a limited range for which the use of Equation (15) for extracting J_{oe} is valid. For very high-injection levels, the near-surface recombination due to the doped emitter becomes very high, and the

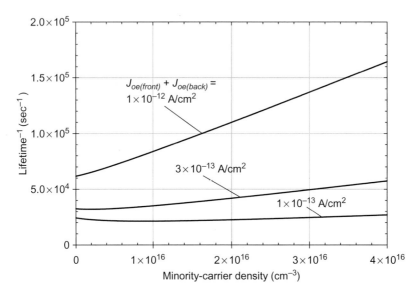

Figure 3 The inverse of the measured lifetime for a 1-Ωcm p-type wafer with three different emitters. Calculation from Equation (15), assuming a low-injection lifetime of 50 μs and subtracting out the Auger contribution to the recombination.

effective lifetime decreases strongly. The limit imposed by the finite diffusivity of carriers toward the surface discussed above in Equation (11) applies also to the case of very high J_{oe} (in high-injection conditions the ambipolar diffusivity, $D_A \approx 18 \text{ cm}^2 \text{ s}^{-1}$, substitutes the minority-carrier diffusion coefficient, D_n). Because of this, the highest J_{oe} that can be discriminated is, assuming equal front and back diffusions:

$$J_{oe} \ll \frac{6\, q n_i^2 D_A}{W[N_A + \Delta n]} \approx \frac{1300}{W[N_A + \Delta n]} \qquad (16)$$

This indicates that lower substrate doping, thinner wafers or both can be used to extend the range of J_{oe} values that can be determined and optimize the range of data that can be used. The thinner and more lightly doped the wafer, the wider the range of minority-carrier densities that can be fit to Equation (15).

For optimal measurements of silicon wafers after emitter diffusion, it is advisable to exclude short wavelengths in order to minimize the fraction of light absorbed within the emitter. For a detailed discussion of the qualifications for this data analysis for various cell geometries, lifetimes, and J_{oe} values, see references [34,36]. For wafers with high recombination on one or both surfaces, a measurement strategy can be devised using front illumination, back illumination, IR, and blue light to determine the values for recombination at the surfaces and in the bulk [33,37].

3.3 Lifetime Measurements in Silicon Ingots

The minority-carrier lifetime in a silicon ingot or brick can be measured using most of the techniques in Table I, following the same methodology described above. Two significant differences with respect to the case of relatively thin wafers are that we now have a semi-infinitely thick bulk and a very high surface recombination. At the same time, the excitation and sensing of carriers takes place near the surface of the sample only. One challenge is to define the average density that appears in Equations (5) through (9). A weighted average carrier concentration, Δn_{avg}, has been proposed [38] in order to take into the analysis only those sections of the crystal that have a significant light-induced excess carrier concentration, while areas that have no excess carriers (such as the back section of a thick bulk sample) are automatically discarded from the analysis:

$$\Delta n_{avg} = \frac{\int_0^\infty \Delta n^2 dx}{\int_0^\infty \Delta n dx} \qquad (17)$$

An effective width, W_{eff}, for the sample can also be defined as the total excess carrier concentration divided by the average carrier concentration as determined in Equation (17):

$$W_{eff} = \frac{\left(\int_0^\infty \Delta n dx\right)^2}{\int_0^\infty \Delta n^2 dx} \qquad (18)$$

This effective width for the carrier density distribution allows one to transform a measurement of bulk thick silicon into a standard wafer measurement, as far as the lifetime analysis is concerned. The total carrier concentration is then the product of n_{avg} and W_{eff}:

$$\int_0^\infty \Delta n dx = \Delta n_{avg} W_{eff} \qquad (19)$$

The above functions can be determined by computer simulation of the carrier density profile, which is dependent on the diffusion length in the material as well as the distribution of light wavelengths incident on the sample. The problem can also be solved analytically for the important case of steady-state illumination with monochromatic light and infinite surface recombination [39]:

$$\tau_{eff} = \frac{\tau_{bulk}}{\alpha L + 1} \qquad (20)$$

$$W_{eff} = 2\left[L + \frac{1}{\alpha}\right] \qquad (21)$$

where L is the diffusion length of carriers in the sample and α is the absorption coefficient for the light. Equation (20) permits the calculation of the bulk lifetime of thick samples based on the measurement of effective lifetime at a specific wavelength. It indicates that the measured effective lifetime is a lower bound on the bulk lifetime, which can approach the bulk lifetime only for absorption depths in the silicon that are greater than the carrier diffusion length in the material. Equation (21) offers a figure of merit for the practical sensors, which should ideally have a sensing depth exceeding W_{eff}.

Another special case of interest is the transient decay that follows the termination of an initial steady-state illumination, essentially a PCD measurement using a long excitation pulse. In this case, the effects of surface recombination diminish with time after the excitation pulse is extinguished, as the carriers diffuse deeper into the silicon bulk. The carrier lifetime will asymptotically approach the actual bulk lifetime after the surface becomes depleted of excess carriers. Since the effective width in this case is changing with time, the average carrier density, given by Equation (17), needs to be evaluated numerically [39,40]. The ability of a given instrument to measure the true bulk lifetime depends on its sensing depth, as the carrier density profile moves into the bulk away from the sensor. The advantage of the transient method for measuring bulk lifetimes is that it is far less sensitive to the surface recombination velocity than the steady-state approach.

3.4 Lifetime Instabilities and Measurement Artifacts
3.4.1 Lifetime Changes in Boron-Doped CZ Silicon and Iron-Contaminated Silicon

For p-type silicon wafers, the lifetime can change pronouncedly upon illumination or heating. It is important to be aware of the possibility of lifetime instabilities while performing measurements; on the other hand, such changes constitute useful tools for process diagnosis, particularly to check for contamination. Two especially significant cases of carrier lifetime instability have been documented extensively and are described here as examples: one is due to the presence of iron, the other to the presence of oxygen, both forming complexes with boron. While iron is avoidable, oxygen is innate to CZ grown silicon (except advanced CZ methods using magnetic confinement of the melt), which typically has an oxygen concentration in the vicinity of 7×10^{17} cm^{-3}. The effect only occurs when both boron and oxygen are present, and is more severe the higher the concentration of both [41]. The conversion efficiency of 1-Ωcm CZ silicon cells has been reported to degrade by 4% relative after about 50 hours of one-sun illumination, while for 0.45 Ωcm the degradation is greater (about 7.5%), and for 10 Ωcm it is lower (1% or less). The lifetime of typical 1-Ωcm boron-doped silicon degrades by more than a factor of ten after 5 hours of exposure to one-sun illumination, approaching a final lifetime of 10–20 µs [42]. Recent work has focused on methods that have been found to put this defect into a seemingly stable, low-recombination state by use of certain temperature-illumination histories [43].

Similar levels of efficiency degradation of 3-4% relative (about 0.5% absolute) after 30 minutes, one-sun illumination can occur due to the presence of iron [44]. The change in lifetime upon exposure is a well established method to determine the concentration of interstitial iron in silicon [45]. There are, however, significant differences between FeB pairs and BO complexes: (1) in the case of Fe, an anneal at 210°C degrades the lifetime in low injection, in much the same way as illumination [46], whereas in the case of BO complexes, the lifetime recovers after annealing at moderate temperatures. (2) The shape of the injection level dependent lifetime curve changes differently: the curves before and after degradation cross over in the case of Fe and FeB (see Figure 1), while they remain approximately parallel in the case of BO. (3) An additional differentiating feature is that the lifetime of Fe contaminated silicon recovers after long-term storage in the dark, whereas that of BO-limited silicon does not.

3.4.2 High Apparent Lifetime at Low Excess Carrier Densities
In some cases an abnormally high apparent lifetime is observed at very low carrier densities. The phenomenon affects transient-decay and steady-state photoconductance measurements, as well as infrared absorption or emission, but not photo or electro-luminescence measurements [47]. The

abnormality is evidenced by a strong *increase* of the apparent lifetime as the carrier injection level *decreases*. For some materials, particularly multicrystalline silicon, the effect can obscure the true carrier recombination even in the carrier-density range corresponding to the operating point for solar cells, generally between 1×10^{13} cm^{-3} and 1×10^{15} cm^{-3}. This behavior can be due to three different physical mechanisms, the trapping of minority carriers [48], grain boundary potentials that impede lateral conductance in the dark [49,50], or, in samples with junctions, Depletion Region Modulation [51,52]. For many practical situations, simple corrections can be used to retrieve the actual recombination lifetime [53].

4. RELATIONSHIP BETWEEN DEVICE VOLTAGE AND CARRIER LIFETIME

The output voltage of a solar cell is principally determined by the excess minority-carrier density at the junction edge of the quasi-neutral base region, defined to be at the plane $x = 0$ in the following approximate expression:

$$\Delta n(0)[N_A + \Delta n(0)] = n_i^2 \exp\left(\frac{V}{kT/q}\right) \quad (22)$$

In open-circuit conditions, this minority-carrier density is the result of a balance between photogeneration and recombination. The conditions are identical to those of the steady-state method to measure the lifetime, described in Section 3. It is therefore possible to obtain Δn from a photoconductance measurement and determine an *implicit*, or expected voltage. Complete characteristic curves can be obtained by plotting the implicit voltage as a function of the illumination, expressed in unit of *suns* (1 sun = 1 kW/m^2).

Direct measurements of the voltage are possible as soon as a junction exists in the wafer. By taking data at a range of illumination intensities, Suns–V_{oc} curves can be constructed and analyzed to obtain diode saturation currents, ideality factors, shunts and ideal efficiencies [54]. Applications of this are shown in Section 5. A generalized expression to obtain the Suns–V_{oc} curves from voltage measurements under quasi–steady-state illumination or during a transient open-circuit decay (OCVD) has also been developed [55].

Conversely, solving for Δn as a function of voltage and substituting this value in Equation (7) converts any voltage measurement into an effective lifetime value. Note that the equation for the lifetime uses an average carrier density, which in some cases can be substantially lower than the value of $\Delta n(0)$ obtained from the voltage through Equation (22) [37]. Temperature control, knowledge of the intrinsic carrier density n_i, and the dopant density are necessary to obtain the lifetime from a voltage measurement, and vice versa. In contrast, photoconductance lifetime measurements require knowledge of carrier mobilities and are relatively insensitive to temperature.

Photoluminescence can provide a contactless probe of the electrochemical potential, or implicit voltage, within a sample, since the PL signal is proportional to the left-hand side of Equation (22). This method, sometimes called $Suns-PL$ [56], is analogous to the $Suns-V_{oc}$ methodology described earlier, without requiring junction formation or probing.

5. APPLICATIONS TO PROCESS MONITORING AND CONTROL OF SILICON SOLAR CELLS

In-process monitoring of the electronic properties of the silicon wafer using minority-carrier measurements has been extensively used for high-efficiency solar cell optimization at research laboratories, and this can be useful for industrial production environments as well. Several possible industrial process-control applications are detailed below for single- and multicrystalline solar cells. The main electronic properties to be measured include resistivity, carrier lifetime, and device voltage. It is possible to measure the resistance of the substrate, doped layers, and metallization at each stage in the process. The effective lifetime of minority carriers in the silicon can also be monitored at each stage using, for example, photoconductance measurements, or via calibrated PL images. Finally, after junction formation, measurements can be made of the voltage, which is a clear indicator of final device performance.

5.1 Resistivity Measurements

The resistivity of the silicon bulk material can be measured at the ingot or wafer levels using contacting or noncontacting arrangements [14,57]. Knowledge of the resistivity and, subsequently, the net doping density, is very important in the optimization of the fabrication process and the interpretation of solar cell performance, since both the voltage and the current depend on it. The lifetime and diffusion length are also correlated to the resistivity of the substrate, with lower resistivities usually being accompanied by lower lifetimes. A higher resistivity, on the other hand, makes the device more dependent on the quality of the surfaces.

Given that residual iron is almost omnipresent in solar grade silicon, and interstitial iron forms pairs with ionized acceptors at a rate that is proportional to the acceptor concentration, techniques based on lifetime testing have also been developed to accurately determine the acceptor concentration in p-type silicon [58].

The sheet resistance of the emitter (usually phosphorus) diffusion is another essential control parameter. It is correlated to the dopant density profile, which determines the emitter saturation current density, the emitter quantum efficiency, the contact resistance properties, and the tolerance to shunt formation during firing. The sheet resistance can be measured by the four-point-probe method [14] or by contactless inductive coupling [10] if the wafer bulk resistivity is subtracted. Resistance measurements can also be used to monitor the thickness of metal deposited at the rear of the wafers (an alternative is to measure the weight differential) and the resulting alloying of the aluminum to form a highly doped p-type BSF layer.

5.2 Process Monitoring via Lifetime Measurements

In R&D laboratories, most of the fabrication steps can be characterized in detail using lifetime measurements. Frequently, such detailed studies are done under ideal conditions to achieve unambiguous results. For example, investigations of mc-Si wafers versus position in the original cast brick have often been made by etching off the surface and subsequently applying a high-quality surface passivation, in order to determine the effect of each process step on bulk lifetime without confounding effects of the surface [59,60]. Alternatively, the surface can be etched back and then passivated using a liquid, such as HF [61] or iodine in ethanol [60,62]. Satisfactory low-temperature passivation has also been obtained with corona-charged photoresist [63] and polymer films [64]. Many studies have also been performed on wafers with the surfaces as they exist after each step in the process [65,66]. Such measurements do not require special wafer preparation and are ideal for industrial process control in production lines. Especially on multicrystalline material, high-resolution mapping of the wafer has been used to identify the response of different grains and grain boundaries to gettering and hydrogenation [67,68]. These spatially resolved methods include microwave-PCD (µ-PCD), modulated free-carrier absorption, EL and PL, surface photovoltage, IR carrier density imaging, microwave phase-shift techniques and light beam induced current (LBIC) [69].

5.2.1 Measurements of Carrier Lifetime in an Ingot or Brick

Cast mc-Si ingots usually present a low minority-carrier lifetime near the top, bottom and side surfaces, due to contact with the crucible and impurity segregation. Once the ingot is sawn into bricks, the minority-carrier lifetime is usually mapped in order to determine the region of potentially good wafers. Two-dimensional high-resolution µ-PCD measurements have often been used to characterize mc-Si bricks and CZ ingots using commercially available instruments. More recently, QSSPC as well as Photoluminescence techniques have been applied to measure carrier lifetimes in bricks [39,70]. The initial lifetimes for as-grown material are, however, not perfectly correlated with final solar cell efficiency. It has been shown, for example, that some regions of cast bricks that might be discarded based on initial lifetime measurements recover lifetime during the high-temperature steps in cell processing due to gettering effects [66,71]. Many of these low lifetime regions can also result in good solar cells after hydrogen bulk passivation procedures. Despite these complexities, a better understanding of the initial silicon lifetime and the effects of the process on the various regions of ingots and bricks is expected to result in better strategies for optimizing the silicon growth and choosing the regions of silicon to submit to the expensive sawing, wafer cleaning, and cell fabrication processes. Rejecting material at this early stage has great value. Tight tolerances for incoming wafers at the beginning of cell fabrication will also permit better optimization and control of the fabrication process.

5.2.2 Lifetime Measurements on Bare Wafers

In principle, measurements on bare wafers should not be necessary if the ingot or the brick has been measured prior to sawing. In practice, a measurement at the wafer level can indicate if a sufficient thickness of silicon was removed at the saw-damage etch step. In addition, many silicon solar cell manufacturers do not fabricate the wafers, but purchase them, and it is of great interest for both the vendor and the manufacturer to have the capability to measure the minority-carrier lifetime of the individual bare wafers. This is not simple, since the measured effective lifetime of a wafer without surface passivation can be very low.

For typical commercial solar cells, the efficiency is relatively constant for bulk lifetimes greater than 5 μs, corresponding to diffusion lengths greater than 116 μm. Therefore, the main point of a lifetime measurement at the stage of bare wafers is to determine if the wafer has a minimum required lifetime of 2–5 μs. Wafers with greater lifetime than this clearly "pass" and wafers with lower lifetime "fail." This type of pass/fail test is possible even in the absence of surface passivation by using infrared light, as discussed in section 3.2. This methodology has been discussed in detail in several recent publications [32] in which a one-to-one correlation was shown between the measured bulk lifetimes and the final cell efficiencies. Photoluminescence imaging has also been applied to wafer sorting, using the spatial resolution of lifetime to give information on the defect distribution across the wafer [72].

As seen in Figure 2, measurements on bare wafers will generally result in effective lifetimes in the range of 0.1 to 2 μs. When using methods subject to trapping-like effects, the low levels of photoconductance corresponding to these lifetimes makes it critically important to determine the lifetime with a data analysis that removes the artifacts that come from trapping, as discussed in Section 3.4.2. Due to the dependence of the lifetime on injection level described in Section 2, all measurements of lifetime should be performed at the same minority-carrier injection level. This injection level should preferably be chosen to be relevant to the maximum power operating point of the finished solar cell.

5.2.3 Effective Lifetime After Emitter Formation

An ideal point to measure the effective lifetime is after forming the emitter, since at this stage it can represent a close prediction of the final solar cell efficiency. The emitter region of a solar cell, usually formed by a phosphorus diffusion, acts as a surface passivation. Amorphous silicon heterojunction technology (a-Si:H) offers an even clearer, and frequently more effective, example of surface passivation. In normal operating conditions of low injection, an emitter region is seen from the base of the cell as a surface recombination velocity, whose value is determined by the emitter-saturation current density J_{oe} that characterizes recombination losses within the emitter region (see Section 3.2.2). This correspondence between SRV and J_{oe} is shown in Figure 2 for a 1-Ωcm p-type wafer. It can be noted that J_{oe} values below 10^{-12} Acm^{-2} provide a reasonable passivation (SRV < 1000 cm/s). Yet such $J_{oe} = 10^{-12}$ Acm^{-2} would limit the measurable effective lifetime to approximately 12 μs in 1-Ωcm material, even if the bulk lifetime may be much higher. It should be noted that such limit is less restrictive for higher resistivity wafers. This surface-like role of the emitter regions applies to both aluminum-alloyed BSF regions and to phosphorus- or boron-diffused regions, as well as to a-Si:H heterojunctions. Representative values for industrial phosphorus (n$^+$ region) and aluminium (p$^+$ region) diffusions are 8×10^{-13} Acm^{-2} and 5×10^{-13} Acm^{-2}, respectively.

The techniques discussed in Section 3.2.2 can be applied to determine both the bulk lifetime and the emitter saturation current density. It is advisable to use special test wafers with a lower substrate doping to maximize the range of carrier injection data that can be used for the analysis and extend the range of J_{oe} values that can be determined. Once J_{oe} is measured, its contribution to the low injection range of the effective lifetime can then be subtracted. Therefore data taken in two minority-carrier density injection ranges allow both the bulk lifetime in the relevant range of cell operation and the emitter saturation current density to be uniquely determined. The exact analysis will depend on the technology used for the emitter formation, since this step may result on a diffused or heterojunction region at the front only, or at both surfaces. If emitter-surface passivation is done in a subsequent, the same methodology discussed above can be used to determine the improvements to the bulk lifetime and the emitter saturation current density that may result from that process step. This provides an ideal way to characterize the effectiveness of emitter passivation by, for example, oxidation, silicon nitride deposition, or Al_2O_3 deposition.

5.3 Device-Level Characterization

The most common characterization used in industry consists of the final I-V curve test of the solar cell under simulated one-sun illumination, sometimes supplemented by a measurement of the I-V curve in the dark. A classical interpretation of these measurements with the traditional double-exponential model leads to the determination of saturation current densities, ideality factors, and shunt and series resistances. These methods are, nevertheless, of limited usefulness as a diagnostic and control tool. In the past, research laboratories have used $I_{sc} - V_{oc}$ measurements to gain further insight into the device. Very early in the history of solar cell development, it was realized that the I_{sc}–V_{oc} curve contained information about the fundamental diode characteristic free from series resistance effects [73]. The Suns–V_{oc} method [74] is a convenient implementation of the same idea, where the device is kept in open-circuit at all times, but its short-circuit current is not actually measured. Instead, the incident light intensity, which varies with time in a quasi–steady-state fashion, is measured with a calibrated reference solar cell. New methods for determining the local series and shunt resistances as well as the local saturation current density, to produce an image of these parameters across the surface of the solar cell, have been developed recently. A selection of them is described below.

5.3.1 Standard Diode Analysis of Suns–V_{oc} Curves

Suns–V_{oc} measurements are possible early in the fabrication process, by probing the wafer after junction formation. This allows the qualification of basic materials and device properties in terms of potential device performance, before the "back-end" processing may complicate the interpretation. The illumination intensity can be expressed in units of standard solar irradiance, that is, *Suns*, or converted to units of current by using as a scaling factor either the short-circuit current of the cell (if it has been measured separately) or an estimate of it based on the modeled photogeneration in the sample.

An example of Suns–V_{oc} data is shown in Figure 4 in the form of the standard semi-logarithmic diode characteristic curve. Process control using the analysis of Figure 4 should optimize the voltage at 0.05 to 0.1 suns, corresponding to the maximum power operating voltage of the solar cell. Similarly to dark I–V curves, Suns–V_{oc} curves can be interpreted by means of a recombination current density and a shunt resistance. The data in Figure 4 can be separated into recombination and shunt current densities, as shown in the plots. The recombination current is usually expressed, following Shockley's terminology, as a pre-exponential factor called the *saturation current density* and an exponential term of the junction voltage affected by an ideality factor. It is important to realize that the saturation current density obtained from the analysis of I–V or Suns–V_{oc} curves is a global parameter that reflects all the recombination processes within the device, encompassing recombination within the emitter and base regions, as well as at the surfaces and within any space charge regions. Another important observation is that, whereas a satisfactory fit to the data is usually possible using a double-exponential model,

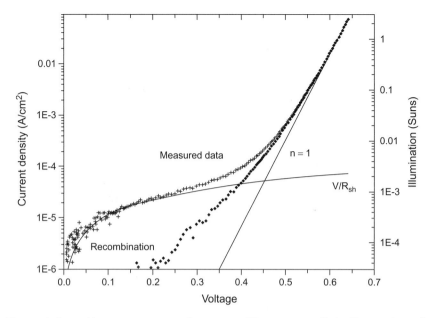

Figure 4 Suns–V_{oc} measurement of a p-type CZ monocrystalline silicon solar cell plotted in a semi-logarithmic scale, with a separation into components due to the shunt and the diode recombination current. A line for the ideal diode equation with unity ideality factor is shown for reference. The measured data is illumination (right scale) vs. voltage. The left scale is constructed from (Suns * J_{sc}).

the transition from ideality 1 to 2 (or greater) factors is often due to the variability of the bulk lifetime or the surface recombination velocity [75], and not to recombination in the space charge region. In the case shown in Figure 4, the deviation from unity ideality factor can be attributed to the B:O defect, which produces lower lifetimes at lower carrier density.

5.3.2 Monitoring Contact Formation

Measuring the contact resistance between the metal and the semiconductor usually requires separate experiments and special test structures. The Corescan instrument [76] provides detailed information about the contact resistance and the emitter sheet resistance on the finished solar cell. The technique, which is destructive, is based on scanning a probe directly on the silicon and the metal fingers to map the voltage drop versus position across the illuminated solar cell. Such voltage maps can be used to optimize the front-grid metallization and diagnose metal contact problems [77].

The metallization and sintering process steps can also be monitored for voltage loss, shunts, and contact problems using the Suns–V_{oc} technique. Although not immediately obvious, data taken under open-circuit conditions can be valuable for monitoring the properties of the solar cell contacts. Metal contacts can be modeled as a Schottky potential barrier, which is formed by most metals on silicon, in parallel with some form of leakage current. The leakage might be from the metal locally doping or spiking the silicon or from thermally assisted tunneling through the potential barrier. Under one-sun conditions, for a well-formed contact, the Schottky barrier is effectively shorted by the leakage. However, on a poorly formed contact at a sufficiently high light intensity the Schottky diode will build up a voltage opposing the junction voltage by generating a current that the leakage is unable to fully shunt. Another example is when aluminum is used to penetrate through a phosphorus diffusion on the back of the solar cell. Insufficient firing of the Al results in an opposing voltage at high light intensities from the parasitic back p–n junction. Schottky type contacts usually result in low fill factors and efficiencies. This poor contact effect is indicated as an ideality factor less than unity for the global solar cell, or even a voltage that decreases at high-illumination intensity. By monitoring the open-circuit voltage at light intensities significantly higher than the nominal operating conditions, this effect can be used to anticipate and solve problems in the contact formation. This method is both a good diagnostic and a process optimization technique [78].

5.3.3 Fill Factor and Series Resistance Measurements

The global series resistance of the device can be measured by a number of methods. It is important to emphasize that not all them provide the relevant value of series resistance. Methods based on measurements of the device under illumination are more realistic than measurements in the dark [79]. Wolf and Rauschenbach [73] described how a comparison of I_{sc}–V_{oc} data with the final I-V curve of the solar cell can then be used to determine the series resistance very precisely. They also demonstrated a method based on measuring the I-V curve at two or more different illumination intensities. An experimental comparison between these methods, plus the Suns–Voc method described below, can be found in reference [80].

An alternative presentation of Suns–V_{oc} data is shown in Figure 5. By using the superposition principle, a photovoltaic *pseudo* I–V curve can be

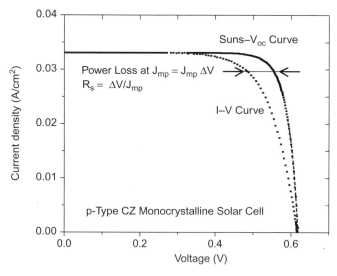

Figure 5 The same Suns–V_{oc} data as in Figure 4, plotted as a photovoltaic pseudo I–V curve and compared to the actual I–V curve taken on the finished solar cell.

constructed from the open-circuit voltage measurements of Figure 4. At each value of the voltage, the *pseudo*, or implied current is given by

$$J_{pseudo} = J_{sc}(1 - suns) \qquad (23)$$

This yields the familiar photovoltaic I-V curve format, permitting the customary interpretation of fill factor, efficiency and shunt. The parameters that matter most to solar cell performance are now visually obvious. For example, it can immediately be seen if the shunt is having a major effect on the maximum power point or not, and the upper bound on fill factor and efficiency (without series resistance effects) is clearly displayed.

Eventually, by comparing the Suns–V_{oc} curve with the actual output I–V curve of the finished solar cell, the series resistance can be determined with precision. Since the *pseudo* I–V curve from Suns–V_{oc} has the shunt and ideality factors fully included, the differences between the two curves can be clearly attributed to series resistance. The latter is simply given by the voltage difference between the two curves at the knee, in the vicinity of the maximum power point, divided by the current. As an example, the same data in Figure 4 are shown in Figure 5. The curves follow each other except near the maximum power point, where series resistance effects are greater. The solar cell in Figures 4 and 5 has a large series resistance loss that would generally imply that different parts of the solar cell are operating at different voltages at the maximum power point. This can complicate the separation of the diode recombination current from series resistance and shunting. The use of the Suns—V_{oc} curve to obtain the diode characteristics and the shunting independently from series resistance effects is especially useful under these circumstances. Electroluminescence imaging [19] can provide similar information as the Suns—V_{oc} data, with the added feature of being spatially resolved, since the luminescence signal measured is proportional to the junction voltage. Photoluminesence can provide similar information to the EL signal, but more weighted toward the average bulk lifetime rather than the junction voltage. Due to the fact that the current flow in the device is different during an EL and PL measurement, a comparison between both can be used to give information on the series resistance in the device. In addition, PL measurements at two different operating conditions can be used to produce a high-resolution map of series resistance [72].

6. CONCLUSIONS

Characterizing material quality and monitoring the impact of each fabrication step is crucial for the PV industry to develop high efficiency solar cell technologies. A broad range of characterization techniques has been developed in recent years, many of them fast enough for in-line process control. The advent of advanced techniques that give a detailed image of the electronic quality of silicon ingots, wafers, and devices provides a wealth of information. More than ever, it is imperative to interpret such information with sound physics to understand the limitations and applicability of each technique. In this chapter the common principles underlying the most important characterization methods has been outlined and possible applications of these methods to monitoring solar cell fabrication have been described. This review is not exhaustive, but should be sufficient to illustrate the range of options available for silicon material, surface, and device characterization, a range that can be expected to broaden in the future.

ACKNOWLEDGEMENTS

The authors would like to thank Keith Forsyth for the data presented in Figures 4 and 5.

REFERENCES

[1] S. Rein, T. Rehrl, W. Warta, S.W. Glunz, Lifetime spectroscopy for defect characterization: systematic analysis of the possibilities and restrictions, J. Appl. Phys. 91 (2002) 2059.
[2] D. Macdonald, A. Cuevas, J. Wong-Leung, Capture cross sections of the acceptor level of iron-boron pairs in p-type silicon by injection-level dependent lifetime measurements, J. Appl. Phys. 89 (2001) 7932.
[3] M.J. Kerr, A. Cuevas, General parameterization of Auger recombination in crystalline silicon, J. Appl. Phys. 91 (2002) 2473.
[4] D.T. Stevenson, R.J. Keyes, Measurement of carrier lifetimes in germanium and silicon, J. Appl. Phys. 26 (1955) 190.
[5] R.H. Bube, Photoconductivity of Solids, Wiley, New York, 1960.
[6] R.A. Sinton, A. Cuevas, Contactless determination of current-voltage characteristics and minority-carrier lifetimes in semiconductors from quasi—steady-state photoconductance data, Appl. Phys. Lett. 69 (1996) 2510.
[7] H. Nagel, C. Berge, A. Aberle, Generalized analysis of quasi—steady-state and quasi-transient measurements of carrier lifetimes in semiconductors, J. Appl. Phys. 86 (1999) 6218.
[8] D.B.M. Klaassen, A unified mobility model for device simulation—I. Model equations and concentration dependence, Solid-State Electron. 35 (1992) 953.
[9] ASTM, Designation F-28-75. Measuring the Minority Carrier Lifetime in Bulk Germanium and Silicon, American Society for Testing of Materials, 1981.
[10] G.L. Miller, D.A.H. Robinson, J.D. Wiley, Contactless measurement of semiconductor conductivity by radio frequency free carrier power absorption, Rev. Sci. Instrum. 47 (1976) 799.
[11] S. Deb, B.R. Nag, Measurement of carriers in semiconductors through microwave reflection, J. Appl. Phys. 33 (1962) 1604.
[12] M. Kunst, G. Beck, The study of charge carrier kinetics in semiconductors by microwave conductivity measurements, J. Appl. Phys. 60 (1986) 3558.
[13] M. Schofthaler, R. Brendel, Sensitivity and transient response of microwave reflection measurements, J. Appl. Phys. 77 (1995) 3162.
[14] D.K. Schroder, Semiconductor Material and Device Characterization, third ed., John Wiley and Sons, 2006.
[15] N.J. Harrick, LIfetime measurements of excess carrier kinetics in semiconductors, J. Appl. Phys. 27 (1956) 1439.
[16] S.W. Glunz, W. Warta, High resolution lifetime mapping using modulated free-carrier absorption, J. Appl. Phys. 77 (1995) 3243.
[17] R. Brendel, M. Bail, B. Bodman, Analysis of photoexcited charge carrier density profiles in Si wafers by using an infrared camera, Appl. Phys. Lett. 80 (2002) 437.
[18] J. Isenberg, S. Riepe, S.W. Glunz, W. Warta, Imaging method for laterally resolved measurement of minority carrier densities and lifetimes: measurement principle and first applications, J. Appl. Phys. 93 (2003) 4268.
[19] T. Fuyuki, H. Kondo, T. Yamazaki, Y. Takahashi, Y. Uraoka, Photographic surveying of minority carrier diffusion length in polycrystalline silicon solar cells by electroluminescence, Appl. Phys. Lett. 86 (2005) 262108.
[20] P. Wurfel, T. Trupke, T. Puzzer, E. Schaffer, W. Warta, S.W. Glunz, Diffusion lengths of silicon solar cells from luminescence images, J. Appl. Phys. 101 (2007) 123110.
[21] T. Trupke, R.A. Bardos, M.D. Abbott, Self-consistent calibration of photoluminescence and photoconductance lifetime measurements, Appl. Phys. Lett. 87 (2005) 184102.
[22] T. Trupke, R.A. Bardos, M.C. Schubert, W. Warta, Photoluminescence imaging of silicon wafers, Appl. Phys. Lett. 89 (2006) 044107.

[23] P.A. Basore, Extended spectral analysis of internal quantum efficiency, 23rd IEEE Photovoltaic Specialists Conference, Louisville, 1993, p. 147.
[24] R. Brendel, M. Wolf, Differential and actual surface recombination velocities, 13th European Photovoltaic Solar Energy Conference, Nice, 1995, p. 428.
[25] J. Schmidt, Measurement of differential and actual recombination parameters on crystalline silicon wafers, IEEE Trans. Electron Dev. 46 (1999) 2018.
[26] SEMI PV9-1110—Test method for excess charge carrier decay in PV silicon materials by non-contact measurements of microwave reflectance after a short illumination pulse, 2010.
[27] K. Lauer, A. Laades, H. Ubensee, H. Metzner, A. Lawerenz, Detailed analysis of the microwave-detected photoconductance decay in crystalline silicon, J. Appl. Phys. 104 (2008) 104503.
[28] A.G. Aberle, S.J. Robinson, A. Wang, J. Zhao, S.R. Wenham, M.A. Green, High-efficiency silicon solar cells: fill factor limitations and non-ideal diode behaviour due to voltage-dependent rear surface recombination velocity, Prog. Photovoltaics 1 (1993) 133.
[29] S.K. Gandhi, Semiconductor Power Devices, Wiley Intersicence, 1977.
[30] J. Brody, A. Rohatgi, A. Ristow, Review and comparison of equations relating bulk lifetime and surface recombination velocity to effective lifetime measured under flash lamp illumination, Sol. Energy Mater. Sol. Cells 77 (2003) 293.
[31] P.A. Basore, D.A. Clugston, PC1D V5.3, University of New South Wales, Sydney, Australia, 1998.
[32] K. Bothe, R. Krain, R. Falster, R. Sinton, Determination of the bulk lifetime of bare multicrystalline silicon wafers, Prog. Photovoltaics 18 (2010) 204.
[33] M. Bail, R. Brendel, Separation of bulk and surface recombination by steady state photoconductance measurements, 16th European PVSEC, Glasgow, 2000, p. 98.
[34] A. Cuevas, The effect of emitter recombination on the effective lifetime of silicon wafers, Sol. Energy Mater. Sol. Cells 57 (1999) 277.
[35] C. Reichel, F. Granek, J. Benick, O. Schultz-Wittmann, S.W. Glunz, Comparison of emitter saturation current densities determined by injection-dependent lifetime spectroscopy in high and low injection regimes, Prog. Photovoltaics Res. Appl. (2010) n/a.
[36] D.E. Kane, R.M. Swanson, Measurement of the emitter saturation current by a contactless photoconductivity decay method, 18th IEEE PVSC, Las Vegas, 1985, p. 578.
[37] A. Cuevas, R.A. Sinton, Prediction of the open-circuit voltage of solar cells from the steady-state photoconductance, Prog. Photovoltaics 5 (1997) 79.
[38] S. Bowden, R.A. Sinton, Determining lifetime in silicon blocks and wafers with accurate expressions for carrier density, J. Appl. Phys. 102 (2007) 124501.
[39] J.S. Swirhun, R.A. Sinton, M.K. Forsyth, T. Mankad, Contactless measurement of minority carrier lifetime in silicon ingots and bricks, Prog. Photovoltaics Res. Appl. (2010) n/a.
[40] N. Schüler, T. Hahn, K. Dornich, J.R. Niklas, B. Gründig-Wendrock, Theoretical and experimental comparison of contactless lifetime measurement methods for thick silicon samples, Sol. Energy Mater. Sol. Cells 94 (2010) 1076.
[41] J. Schmidt, A.G. Aberle, R. Hezel, Investigation of carrier lifetime instabilities in Cz-grown silicon, 26th IEEE Photovoltaic Specialists Conference, Washington, 1997, p. 13.
[42] S.W. Glunz, S. Rein, W. Warta, J. Knobloch, W. Wettling, On the degradation of CZ-silicon solar cells, 2nd World Conference on Photovoltaic Energy Conversion, Vienna, 1998, p. 1343.
[43] A. Herguth, G. Hahn, Kinetics of the boron-oxygen related defect in theory and experiment, J. Appl. Phys. 108 (2010) 114509.

[44] J.H. Reis, R.R. King, K.W. Mitchell, Characterization of diffusion length degradation in Czochralski silicon solar cells, Appl. Phys. Lett. 68 (1996) 3302.
[45] D. Macdonald, L.J. Geerligs, A. Azzizi, Iron detection in crystalline silicon by carrier lifetime measurements for arbitrary injection and doping, J. Appl. Phys. 95 (2004) 1021.
[46] G. Zoth, W. Bergholz, A fast, preparation-free method to detect iron in silicon, J. Appl. Phys. 67 (1990) 6764.
[47] R.A. Bardos, T. Trupke, M.C. Schubert, T. Roth, Trapping artifacts in quasi—steady-state photoluminescence and photoconductance lifetime measurements on silicon wafers, Appl. Phys. Lett. 88 (2006) 053504.
[48] D. Macdonald, A. Cuevas, Trapping of minority carriers in multicrystalline silicon, Appl. Phys. Lett. 74 (1999) 1710.
[49] H.P. Maruska, A.K. Ghosh, A. Rose, T. Feng, Hall mobility of polycrystalline silicon, Appl. Phys. Lett. 36 (1980) 381.
[50] R.A. Sinton, J. Swirhun, M. Forsyth, T. Mankad, J. Nyhus, L. Camel, The effects of subbandgap light on QSSPC measurements of lifetime and trap density: what is the cause of trapping? Proceedings of the 25th European Photovoltaic Solar Energy Conference, Valencia, 2010.
[51] M. Bail, M. Schulz, R. Brendel, Space-charge region-dominated steady-state photoconductance in low-lifetime Si wafers, Appl. Phys. Lett. 82 (2003) 757.
[52] P.J. Cousins, D.H. Neuhaus, J.E. Cotter, Experimental verification of the effect of depletion-region modulation on photoconductance lifetime measurements, J. Appl. Phys. 95 (2004) 1854.
[53] D. Macdonald, R.A. Sinton, A. Cuevas, On the use of a bias-light correction for trapping effects in photoconductance-based lifetime measurements in silicon, J. Appl. Phys. 89 (2001) 2772.
[54] R.A. Sinton, A. Cuevas, A quasi—steady-state open-circuit voltage method for solar cell characterization, 16th European PVSEC, Glasgow, Scotland, 2000, p. 1152.
[55] M.J. Kerr, A. Cuevas, R.A. Sinton, Generalized analysis of quasi—steady-state and transient decay open circuit voltage measurements, J. Appl. Phys. 91 (2001) 399.
[56] T. Trupke, R.A. Bardos, M.D. Abbott, J.E. Cotter, Suns-photoluminescence: contactless determination of current-voltage characteristics of silicon wafers, Appl. Phys. Lett. 87 (2005) 093503.
[57] R.A. Sinton, A. Cuevas, M. Stuckings, Quasi—steady-state photoconductance, a new method for solar cell material and device characterization, Proceedings of the 25th Photovoltaic Specialists Conference, Washington, D.C., U.S.A., 1996, p. 457.
[58] D. Macdonald, A. Cuevas, L.J. Geerligs, Measuring dopant concentrations in compensated p-type crystalline silicon via iron-acceptor pairing, Appl. Phys. Lett. 92 (2008) 202119.
[59] D. Macdonald, A. Cuevas, F. Ferrazza, Response to phosphorus gettering of different regions of cast multicrystalline silicon ingots, Solid-State Electron. 43 (1999) 575.
[60] A. Rohatgi, V. Yelundur, J. Jeong, A. Ebong, D. Meier, A.M. Gabor, et al., Aluminium-enhanced PECVD SiNx hydrogenation in silicon ribbons, Proceedings of the 16th European Photovoltaic Solar Energy Conference, Glasgow, U.K., 2000, p. 1120.
[61] E. Yablonovitch, D.L. Allara, C.C. Chang, T. Gmitter, T.B. Bright, Unusually low surface-recombination velocity on silicon and germanium substrates, Phys. Rev. Lett. 57 (1986) 249.
[62] T.S. Horanyi, T. Pavelka, P. Tutto, In situ bulk lifetime measurement on silicon with a chemically passivated surface, App. Surf. Sci. 63 (1993) 306.
[63] J. Schmidt, A.G. Aberle, Easy-to-use surface passivation technique for bulk carrier lifetime measurements on silicon wafers, Prog. Photovoltaics 6 (1998) 259.

[64] D. Biro, W. Warta, Low temperature passivation of silicon surfaces by polymer films, Sol. Energy Mater. Sol. Cells 71 (2002) 369.
[65] M. Stocks, A. Cuevas, A. Blakers, Process monitoring of multicrystalline silicon solar cells with quasi—steady state photoconductance measurements, Proceedings of the 26th IEEE Photovoltaic Specialists Conference, Anaheim, 1997, p. 123.
[66] G. Coletti, S.D. Iuliis, F. Ferrazza, A new approach to measure multicrystalline silicon solar cells in a production process, 17th European Photovoltaic Solar Energy Conference, 2001, p. 1640.
[67] P. Geiger, G. Kragler, G. Hahn, P. Fatch, E. Bucher, Spatially resolved lifetimes in EFG and string ribbon silicon after gettering an hydrogenation steps, 29th IEEE Photovoltaic Specialists Conference, New Orleans, 2002.
[68] J. Tan, A. Cuevas, D. Macdonald, T. Trupke, R. Bardos, K. Roth, On the electronic improvement of multi-crystalline silicon via gettering and hydrogenation, Prog. Photovolt. Res. Appl. 16 (2008) 129.
[69] W. Warta, Defect and impurity diagnostics and process monitoring, Sol. Energy Mater. Sol. Cells 72 (2002) 389.
[70] T. Trupke, J. Nyhus, R. Sinton, J.W. Weber, Photoluminescence imaging on silicon bricks, 24th European Photovoltaic Solar Energy Conference, Hamburg, 2009, p. 1029.
[71] A. Cuevas, D. Macdonald, M.J. Kerr, C. Samundsett, A. Sloan, A. Leo, et al., Evidence of impurity gettering by industrial phosphorus diffusion, 28th IEEE Photovoltaic Specialists Conference, Anchorage, Alaska, 2000, p. 108.
[72] H. Kampwerth, T. Trupke, J.W. Weber, Y. Augarten, Advanced luminescence based effective series resistance imaging of silicon solar cells, Appl. Phys. Lett. 93 (2008) 202102.
[73] M. Wolf, H. Rauschenbach, Series resistance effects on solar cell measurements, Adv. Energy Convers. 3 (1963) 455.
[74] R.A. Sinton, A. Cuevas, A quasi-steady open-circuit voltage method for solar cell characterisation, Proceedings of the 16th European Photovoltaic Solar Energy Conference, Glasgow, U.K., 2000, p. 1152.
[75] D. MacDonald, A. Cuevas, Reduced fill factors in multicrystalline silicon solar cells due to injection-level dependent bulk recombination lifetimes, Prog. Photovoltaics 8 (2000) 363.
[76] A.S.H. van der Heide, A. Schonecker, G.P. Wyers, W.C. Sinke, 16th European Photovoltaic Solar Energy Conference, Glasgow, 2000, p. 1438.
[77] A.S.H. van der Heide, J.H. Bultman, J. Hoornstra, A. Schonecker, G.P. Wyers, W.C. Sinke, Optimizing the front side metallization process using the Corescan, 29th IEEE Photovoltaic Specialists Conference, New Orleans, 2002.
[78] S.W. Glunz, J. Nekarda, H. Mäckel, A. Cuevas, Analyzing back contacts of silicon solar cells by Suns—Voc measurement at high illumination densities, 22nd European Photovoltaics Solar Energy Conference, Milan, Italy, 2007, p. 849.
[79] G.L. Araujo, A. Cuevas, J.M. Ruiz, Effect of distributed series resistance on the dark and illuminated characteristics of solar cells, IEEE Trans. Electron Dev. ED-33 (1986) 391.
[80] D. Pysch, A. Mette, S.W. Glunz, A review and comparison of different methods to determine the series resistance of solar cells, Sol. Energy Mater. Sol. Cells 91 (2007) 1698.

CHAPTER III-2

Standards, Calibration, and Testing of PV Modules and Solar Cells

Carl R. Osterwald
National Renewable Energy Laboratory

Contents

1. PV Performance Measurements — 1046
   - 1.1 Introduction — 1046
   - 1.2 Radiometry — 1046
   - 1.3 Instrumentation and Solar Simulation — 1049
   - 1.4 Temperature — 1051
   - 1.5 Multijunction Devices — 1052
   - 1.6 Other Performance Ratings — 1053
   - 1.7 Potential Problems and Measurement Uncertainty — 1054
2. Diagnostic Measurements — 1054
3. Commercial Equipment — 1056
4. Module Reliability and Qualification Testing — 1057
   - 4.1 Purpose and History — 1057
   - 4.2 Module Qualification Tests — 1059
     - 4.2.1 Thermal Cycling Sequence — 1059
     - 4.2.2 Damp-Heat Sequence — 1060
     - 4.2.3 UV Exposure, Thermal Cycling, and Humidity-Freeze Sequence — 1060
   - 4.3 Reliability Testing — 1060
   - 4.4 Module Certification and Commercial Services — 1061
5. Module Degradation Case Study — 1061
- Acknowledgments — 1063
  - Permissions — 1063
- References — 1064

1. PV PERFORMANCE MEASUREMENTS

1.1 Introduction

When we refer to the performance of a PV cell or module, the most important parameter is, of course, the maximum power point P_{max} (see fundamentals in Chapter Ia-1), which is usually determined by varying the forward bias voltage across the device under test while illuminated. The light I−V curve is then recorded by measuring the current and voltage. Unfortunately, determination of P_{max} is complicated because it is a function of the total and spectral irradiance incident upon the device, the spatial and temporal uniformity of the irradiance, and the temperature of the device. It can even be a function of the voltage sweep rate and direction, as well as the voltage biasing history. Another important parameter, the power-conversion efficiency η, defined as the power out divided by the power in, is proportional to P_{max} and also to two quantities: radiometric (the total incident irradiance, G) and physical (the device area, A)—i.e., $\eta = P_{max}/A \cdot G$. Accurate determination of PV performance requires knowledge of the potential measurement problems and how these problems are influenced by the specific device to be tested. This section covers common PV measurement techniques and shows how potential problems and sources of error are minimized.

The first solar cell applications were for satellite power systems, so it was important for designers to know how much power could be expected from an individual solar cell in Earth orbit (i.e., when illuminated by extraterrestrial solar irradiance). This could not be determined exactly for two reasons: (1) the precise nature of the extraterrestrial irradiance could only be estimated, and (2) sunlight at Earth's surface is filtered by the atmosphere. Therefore, space solar cell performance could not be measured in a laboratory. By the early 1970s, a similar situation existed for terrestrial applications, but for a different reason—the total and spectral irradiance vary continuously due to the effects of Earth's atmosphere. Thus, in both cases, it was essentially impossible to independently verify the efficiency measurements made by any laboratory [1].

1.2 Radiometry

The problems with measurement comparisons led to the concept of standard reporting conditions (SRC, also referred to as standard test conditions, STC), which consist of the device temperature, the total irradiance, and the spectral irradiance under which PV performance measurements are made or corrected to [1,2]. Using SRC allows performance comparisons to be made within the error uncertainty limits of the measurements involved [2]. Table 1 lists the SRCs for space and terrestrial applications, and Figure 1 is a plot of the extraterrestrial and global spectral irradiances, which are the result of measurements made by solar resource satellites, high-altitude balloons, and aircraft, combined with ground measurements and atmospheric models [4—11].

Table 1 Standard reporting conditions

Application	Reference spectral irradiance	Total irradiance W/m^2)	Temperature (°C)
Low–Earth orbit	ASTM E 490 [1]	1366.1	28
Terrestrial global (nonconcentrator)	IEC 60904-3 [3] or ASTM G 159 [2]	1000	25

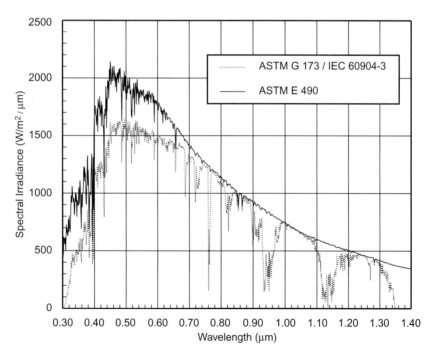

Figure 1 Reference spectral irradiances for air mass 0 (AM0) and air mass 1.5 hemispherical (global) [1–3].

For space applications, it is easy to identify a reference spectral irradiance, which is the extraterrestrial solar irradiance at a distance of one astronomical unit from the Sun. Identification of a reference spectral irradiance for terrestrial applications is a much more difficult subject because of the effects of the atmosphere on sunlight, which are a function of the path length through the atmosphere [7]. A way of describing an atmospheric path is called the *relative optical path length*, which is commonly simplified to *air mass* (AM). To a first approximation, the air mass is the secant of the solar zenith angle. Thus, AM1 indicates the Sun is directly overhead, and AM1.41 is a 45° zenith angle. Because air mass is actually a relative path length referenced to sea level, it can be less than 1 as the altitude increases above sea level. The extraterrestrial irradiance is commonly called *air mass zero* (AM0) because at the top of the atmosphere, the path length is zero.

Because solar cells convert light to electricity, radiometry is a very important facet of PV metrology. Radiometric measurements have the potential to introduce large errors in any given PV performance measurement because radiometric instrumentation and detectors can have total errors of up to 5% even with careful calibration [12,13]. Other errors can be introduced through means as subtle as misunderstanding the objective of a particular measurement or its potential limitations. Broadband radiometers, such as pyranometers and pyrheliometers can be difficult to use in PV measurements because of the large spectral response differences. The calibration and use of radiometric instruments such as pyrheliometers, spectroradiometers, and pyranometers is a subject that is beyond the scope of this chapter, but these instruments are important for PV measurements. A number of standards for radiometric instrumentation are available [14−19].

It should be emphasized that the standardized spectral irradiances in Figure 1 cannot be reproduced exactly in the laboratory. In addition, although light sources such as xenon arc lamps can approximate these curves (i.e., solar simulators), a method of setting the total irradiance is needed. These limitations have been overcome by the so-called reference cell method. Using this method, an unknown device is tested in a solar simulator for which the total irradiance has been set with a calibrated reference cell that has the same or similar spectral response as the test device. The output level of the simulator is adjusted until the short-circuit current, I_{sc}, of the reference cell is equal to its calibration. With the total irradiance established, the reference cell is replaced with the device to be tested and the performance measurement can then be made [20−22].

When considering the reference cell method, another problem becomes immediately obvious: how can it be used to test an unknown cell if a matched reference cell is not available? This problem was solved by calculating the error in a measurement of I_{sc} caused by spectral response differences between the reference and unknown devices, and by spectral irradiance differences of the solar simulator from the desired standard spectrum [23–25]; this error is now called the *spectral mismatch error*, M. It is important to note that once M is known, it can be used to correct PV current measurements for spectral error by dividing by M. The PV performance standards mentioned above all rely on the reference cell method with spectral mismatch corrections [20–22].

The measurement problem is then reduced to obtaining calibrated reference cells. Historically, reference cells calibrated in sunlight have been called *primary cells*, and cells calibrated in solar simulators are called *secondary cells*. Historically, for space applications, primary reference cell calibrations have relied on I_{sc} measurements under spectral conditions as close to AM0 as possible with locations such as high-altitude balloons and aircraft and manned spacecraft [26–28]. Primary terrestrial calibrations require stable, clear-sky conditions with total irradiances measured with an absolute cavity radiometer [29,30]. Other primary terrestrial calibration methods are used by national laboratories worldwide [31]. Reference [32] is a standard for secondary reference cell calibrations, and reference [33] describes calibrations and use of reference modules.

1.3 Instrumentation and Solar Simulation

Figure 2 is an electrical block diagram that illustrates how PV current-voltage measurements are made. A four-wire (or Kelvin) connection to the device under test allows the voltage across the device to be measured by avoiding voltage drops along the wiring in the current measurement loop. The device under test is illuminated, the load is varied, and the operating point of the device under test changes, which allows the current and voltage points to be captured along the I–V curve [3]. Typically, the entire measurement process is computer controlled.

Two factors greatly influence the design choices of an I–V measurement system: the type of the devices to be tested and the illumination source. For example, it is much easier to design a variable load if all test devices are similar in size and output, such as testing modules at the end of a production line. On the other hand, if a measurement system will be required to handle a wide variety of module sizes and outputs or

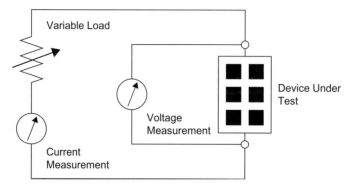

Figure 2 Block diagram of the four-wire Kelvin technique for I−V curve measurements of PV devices.

individual solar cells, then the instrumentation must have the necessary voltage and current ranges. The need to test multijunction devices can greatly complicate the problems that must be solved in a measurement system design.

Any light source used (i.e., a solar simulator or natural sunlight) will have spatial nonuniformities, temporal instabilities, drift of its spectral irradiance, and an illumination time [34]. Outdoor measurements normally have uniform illumination in the test plane (less than 1% spatial variation) and stable irradiance for time periods up to several minutes. A Xe long-arc flash solar simulator can also achieve good spatial uniformities, but these lamps can have measurement times in the 1-ms to 20-ms range. Other flash simulators use a continuously pulsing source and measure one point on the I−V curve at each pulse. Continuous Xe simulators are commercially available for a wide range of sample sizes, from 50-mm square up to as much as 2-m square. Solar simulators with longer measurement times typically have poorer spatial uniformity, from several percent up to as much as 10% or 20%. Two standards have been developed that quantify the performance of solar simulators and can be used as aids in selection and use [35,36].

Using these constraints, the equipment needed for I−V measurements in Figure 2 can be considered [3]. Voltage measurements are easily obtained with commercial voltmeters or high-speed analog-to-digital (A/D) conversion cards. These same instruments can be connected across the sense leads of a four-wire resistor intended for current measurements. In high-current situations, such as source strings in an array, magnetic current probes are very convenient. The variable load can be as simple as a variable resistor, but such loads cannot be used to trace the I–V curve outside of the power quadrant. Computer-programmable DC-power supplies, on the other hand, can be obtained in nearly any voltage and current range needed, and bipolar supplies can operate a PV device in reverse bias or beyond open-circuit voltage (V_{oc}).

In higher-power applications such as larger modules or whole arrays, a popular load for I–V measurements is a capacitor. The capacitor is initially uncharged and connected in series with the PV device to be tested, and any current produced by the device is stored in the capacitor. When the device is illuminated, the voltage across the device is close to zero. As the capacitor charges, the capacitor voltage and the device voltage both increase, thereby sweeping through the I–V curve. The charging stops when the current goes to zero at V_{oc}. Although it cannot operate the device outside of the power quadrant, this technique has the advantage of not dissipating large amounts of power, and it can sweep at high rates. Commercial capacitive sweep systems are available that are portable and can be used to obtain I–V curves of arrays or array segments up to about 100 kW in size.

1.4 Temperature

The I–V curve of a PV device under illumination is a strong function of temperature, which must be accounted for in performance measurements [37]. Typically, I_{sc} has the smallest temperature dependence, which is caused by the semiconductor band gap shifting to longer wavelengths with higher temperatures. V_{oc} and P_{max}, on the other hand, degrade rapidly with increasing temperature [38,39]. These strong dependencies are the reason a fixed temperature is used for SRC.

Temperature measurements of PV devices can be difficult. In general, measured temperatures should be those of the actual semiconductor junctions, but usually the only temperature that can be measured is at the rear surface of a cell. For modules, it not possible to contact individual cells, and one is forced to apply temperature sensors to the rear surfaces. This leads to an error because the rear surface will normally be cooler than the cells laminated inside. Reference [40] states that at 1000 W/m^2, this error is an average of 2.5°C for typical crystalline Si modules. It is possible to correct I–V measurements to a reference temperature if temperature coefficients are known [41]. Another problem with modules is that the temperature can vary from cell to cell, requiring some sort of average if a single value is needed. The same problem is exacerbated in an array, where larger variations can be expected [42]. Temperature coefficients can also be difficult to apply correctly because they can be functions of irradiance and other factors [43].

1.5 Multijunction Devices

A multijunction PV device consists of several individual semiconductor junctions stacked together (also called *subcells*) and connected in series to obtain higher performance. With two subcells, a multijunction solar cell is commonly referred to as a *tandem cell*. In such devices, however, both the I_{sc} and the fill factor (FF) are functions of the incident spectral irradiance, greatly complicating the determination of device performance at SRC [1]. Because all subcells are in series, the same current must flow through each, and the output current is usually limited by the junction generating the smallest current (called the *limiting subcell*). A special case occurs when all subcells generate the same amount of current, which is termed *current matched*. The reason the FF can be a function of the spectral irradiance is that the current generated by a limiting subcell can cause a different subcell to operate in reverse bias and result in a stepped I−V curve in forward bias [1,44−46].

The only way to avoid measuring the wrong I_{sc} and FF is to ensure the currents generated in each subcell under the test light source are equal to those that would be generated under the desired reference spectral irradiance. In general, this can only be accomplished if the spectral response for each subcell is known and if the spectral irradiance of the test light source can be adjusted [1,44,45]. Such a simulator suitable for standard multijunction performance is called a *spectrally adjustable solar simulator*. There are a number of ways an adjustable simulator can be realized, including multiple independently adjustable light sources and selective filtering [44]. A number of procedures can be used to perform spectral matching in adjustable simulators. One procedure is an extension of the reference cell method with spectral mismatch corrections. An iterative process is used in which the spectral mismatch M for each subcell in the test is obtained following a measurement of the simulator spectral irradiance. Based on these results, the simulator spectral irradiance is adjusted and the process is repeated. When all values of M are within an acceptable tolerance of one (usually 2−3%), the process is stopped and the I−V measurements are made [44,45,47]. This procedure will be standardized in the near future [48]. Another procedure relies on a multiple source simulator in which each source can be adjusted independently without changing its spectral distribution. A system of linear equations is then used to obtain the current outputs of one or more reference cells that indicate when the simulator is correctly adjusted [49].

It should be noted that the spectral adjustment of the simulator does not reproduce the reference spectral irradiance; instead, it sets the simulator spectral irradiance so that the test device is operating as if it were illuminated by the reference spectrum.

1.6 Other Performance Ratings

Although performance at SRC is an important parameter that allows measurement comparisons between different laboratories and different PV devices, it is not a realistic indication of the performance that can be expected outdoors, where devices see a range of irradiance conditions and almost never operate at 25°C when the irradiance is greater than 900 W/m^2. One of these ratings is called *performance test conditions* (PTC), the conditions for which are listed in Table 2 [50,51]. This rating is usually used to produce a rating for a system over a period of time (typically one month) when conditions are close to the rating conditions. Data filtering and a regression fit to a linear equation are then used to obtain the performance at PTC. These ratings can be used to monitor the performance of a system over time, and it can be used for DC as well as AC ratings if the system is grid connected. Another system rating uses translations to SRC based on a number of parameters measured outdoors on individual modules [52]. This method employs an empirical air mass versus irradiance correction factor that was called a *measure* of "the systematic influence of solar spectrum" [53], but it should be noted that this factor is not a spectral correction to the reference spectral irradiance and is site specific [54].

There has also been interest in rating modules based on energy production rather than output power at a single fixed condition [55]. A proposed energy rating establishes five reference days of hourly weather and irradiance data: hot-sunny, cold-sunny, hot-cloudy, cold-cloudy, and nice. Hourly module power outputs for each of the five days are obtained and then integrated to obtain the energy output for each condition [56].

Table 2 Performance test conditions

Application	Irradiance (W/m^2)	Ambient temp. (°C)	Wind speed (m/s)
Flat-plate fixed tilt	1000 (global)	20	1
Concentrators	850 (direct normal)	20	1

Module energy ratings have been obtained from laboratory measurements as well as outdoor performance data [57—60].

1.7 Potential Problems and Measurement Uncertainty

Measurement uncertainty analysis is a formal process of identifying and quantifying possible errors and combining the results to obtain an estimate of the total uncertainty of a measurement [61]. An inherent part of this process is understanding the potential problems that can affect the results of any given PV performance measurement. Although an exhaustive discussion of such problems cannot be included here, a number of common problems and pitfalls can be identified. Many of these result from the instrumentation and apparatus used, the characteristics of the device to be tested, or both.

The voltage bias rate and direction can have profound effects on measured I—V data [62—64]. Test samples can exhibit hysteresis when swept toward forward bias and toward reverse bias, resulting in two different values of FF. Note that many I—V measurement systems are unable to sweep in both directions, especially capacitive loads, which can completely hide hysteresis problems. Thin-film devices such as $CuInSe_2$ and CdTe have mid—band-gap states that make them especially vulnerable to errors caused by fast sweep rates and prior light and voltage bias conditions [62]. Electronic loads used with flash simulators can greatly improve or degrade measurement results [65]. Mundane subjects such as area measurements, temperature measurements, and contacts and wiring cannot be ignored [66]. Amorphous Si modules exhibit a dependence of the FF on the incident spectral irradiance, which can affect the results of outdoor performance measurements [67].

A number of uncertainty analyses of PV measurements have been published for general I—V measurements [68,69], spectral corrections [70], and reference cell calibrations [29,71]. Reference [70] concluded that the magnitude of uncertainty in spectral corrections is directly proportional to the size of the spectral mismatch factor.

2. DIAGNOSTIC MEASUREMENTS

A number of diagnostic measurements are widely used in photovoltaic research and development. Two of these date to the earliest days of PV devices: dark I–V and spectral response (also called *quantum efficiency*). Dark I–V shows how a device operates as a p–n junction and can be used to obtain series resistance, shunt resistance, and diode quality factor [38,39]. Although complicated by multiple individual cells connected in series, dark I–V can also be useful for module diagnostics [72]. Many of the same parameters can be derived from light I–V measurements as well [73,74], and with careful design the same measurement system illustrated in Figure 1 can be used for both light and dark I–V measurements. Shunt resistance of individual cells is an important factor for the performance of monolithic thin-film modules such as a-Si, and reference [75] describes a technique for measuring individual shunt resistances in modules.

Spectral response is a fundamental property of solar cells, and it can provide information about optical losses such as reflection and give insights into carrier recombination losses [39]. As previously stated, it is an important parameter for performance measurements because it is used as the basis for spectral mismatch corrections [24,25]. Because of this importance, two standards for spectral response measurements have been developed [76,77]. A wide variety of measurement schemes have been used for spectral response measurements [78–81]. Reference [82] discusses several sources of error that are commonly encountered. Note that special light- and voltage-biasing techniques are required to measure the spectral response of series-connected multijunction devices [47,48,83].

A very useful diagnostic for cells and modules is known as *laser-* (or *light-*) *beam induced current* (LBIC), which produces a map of a device's response by rastering a laser spot across the front surface and measuring the resultant current [84,85]. These maps can easily identify locations of reduced output that can greatly simplify diagnostic investigations, such as looking for cracks in polycrystalline Si cells. LBIC can be used to measure photocurrents and shunt resistances of individual cells in crystalline Si modules [84]. When used with monolithic thin-film modules, LBIC will show regions of reduced photocurrent that might be caused by deposition nonuniformities.

Another technique that has become increasingly popular in recent years is thermal mapping using commercial infrared imaging cameras [86]. These cameras easily measure temperature variations across the surfaces of both modules and arrays, and they are especially useful for locating hot spots, which can be caused by a number of conditions. A localized shunt path within a crystalline Si cell can produce small $20°C$ hot spots when a module is operated in reverse bias. In PV arrays, it is not uncommon for individual cells or a single module to be forced into reverse bias (thereby absorbing rather than generating power) if its output current has been reduced for some reason and the array design does not have adequate bypass diodes. These conditions will result in hot spots, 20°C to 40°C higher than the surrounding cells or modules, which are easily detected in an infrared image.

Ultrasonic imaging is a nondestructive test that can identify voids and delamination in encapsulants. It is useful for studying the condition of solder joints in crystalline Si modules [87].

3. COMMERCIAL EQUIPMENT

Many companies worldwide market PV instrumentation, solar simulators, and complete PV measurement systems. Products are available for testing everything from small individual cells to large modules and small-sized arrays for both space and terrestrial applications. These companies include

- Berger Lichttechnik, Baierbrunn, Germany
- Beval S.A. (Pasan), Valangin, Switzerland
- Daystar, Inc., Las Cruces, NM, USA
- EKO Instruments Trading Co., Ltd., Tokyo, Japan
- Energy Equipment Testing Service Ltd., Cardiff, UK
- The Eppley Laboratory, Inc., Newport, RI, USA
- h.a.l.m elektronik GmbH, Frankfurt/Main, Germany
- Kipp & Zonen B.V., Delft, The Netherlands
- K.H. Steuernagel Lichttechnik, Morfelden-Walldorf, Germany
- LI-COR Environmental Research and Analysis, Lincoln, NE, USA
- NewSun, Nepean, ON, Canada
- NPC Inc., Tokyo, Japan
- Optosolar GmbH, Merdingen, Germany
- Photo Emission Tech., Inc., Newbury Park, CA, USA
- PV Measurements, Boulder, CO, USA
- Shanghai Jiaoda GoFly Green Energy Co., Ltd., Shanghai, China
- Spectrolab, Inc., Sylmar, CA, USA
- Spire Corp., Bedford, MA, USA
- Telecom-STV Co., Ltd., Moscow, Russia
- Thermo Oriel, Stratford, CT, USA

- Thermosenorik GmbH, Erlangen, Germany
- Vortek Industries, Ltd., Vancouver BC, Canada
- Wacom Electric Co., Ltd., Tokyo, Japan

4. MODULE RELIABILITY AND QUALIFICATION TESTING

4.1 Purpose and History

The "holy grail" of module reliability that many people ask for is a single test that, if passed, indicates that a certain module design will last x number of years in use (typically 20 or 30 years). No such test exists, nor can it be developed [88]. The reason for this is that every possible failure mechanism has to be known and quantified. This condition is impossible to meet because some failures may not show themselves for many years and because manufacturers continually introduce new module designs and revise old designs.

Instead of a search for a single test, module reliability testing aims to identify unknown failure mechanisms and determine whether modules are susceptible to known failure mechanisms. Accelerated testing is an important facet of reliability testing, but accelerated tests need to be performed in parallel with real-time tests to show that a certain failure is not caused by the acceleration factor and will never appear in actual use. Qualification tests are accelerated tests, usually of short duration, that are known to produce known failure mechanisms, such as delamination. Thus, passing a qualification test is not a guarantee of a certain lifetime, although extending the duration of a qualification can provide added confidence that a module design is robust and durable [89].

The first PV module qualification tests were developed by the Jet Propulsion Laboratory (JPL) as part of the Low-Cost Solar Array program funded by the U.S. Department of Energy [90–93]. Elements of the Block V qualification sequence included

- temperature cycling,
- humidity-freeze cycling,
- cyclic pressure loading,
- ice-ball impact,
- electrical isolation (hi-pot),
- hot-spot endurance, and
- twisted-mounting surface test.

Following the qualification sequences, test modules were compared with baseline electrical performance tests and visual inspection results to determine whether the design passed or failed. These tests have served as the starting point for all the qualification sequences that have been developed since.

The next development in module qualification was the adoption of the European CEC 502 sequence [94], which was significantly different from Block V and added several new tests:
- UV irradiation,
- high-temperature storage,
- high-temperature and high humidity storage, and
- mechanical loading.

Another significant difference was that CEC 502 lacked humidity-freeze cycling. At the same time, Underwriters Laboratories (UL) developed the UL 1703 safety standard, which has become a requirement for all modules in the US [95]. It incorporated the humidity-freeze, thermal cycling, and hi-pot tests from Block V, plus a large number of other safety-related tests. Note that as a safety standard, UL 1703 does not require a module to retain its performance at a certain level; rather, it simply must not become hazardous as a result of the test sequences.

With the development of commercially available a-Si modules, an investigation into the applicability of Block V to these devices resulted in publication of the so-called interim qualification tests (IQT) [96]. The IQT were similar to the JPL tests but added two tests from UL 1703: surface cut susceptibility and ground continuity. A wet-insulation resistance test was included as a check for electrochemical corrosion susceptibility.

Again building on prior work, Technical Committee 82 (TC-82) of the International Electrotechnical Commission (IEC) produced an international standard for qualification of crystalline Si modules, IEC 1215 (the designation was later changed to 61215), that included elements of all the prior tests [97]. It followed the sequences of JPL's Block V but added many of the new elements in CEC 502. Probably the most significant addition was the 1000-h damp-heat test that replaced the high-temperature and high-humidity storage tests in CEC 502. Before IEC 61215 became a standard, a nearly identical draft version was adopted in Europe as CEC 503 [98]. Since its adoption, IEC 61215 has been a major influence on the reputation of excellent reliability that crystalline Si modules currently have. It has proved to be an invaluable tool for discovering poor module designs before they are sold on the open market [99].

In the United States, another significant development was that of IEEE 1262 [100], which was motivated by the lack of a US qualification standard. In 1989, IEC 61215 was only a draft standard that was several years away from adoption. Also, the IEC document was not suitable for a-Si modules because the UV and outdoor exposures resulted in the initial light-induced degradation inherent in these devices, which makes determination of performance losses due to the qualification sequences very difficult. Therefore, the goal was a document that was applicable to both technologies, which was accomplished with the addition of short thermal-annealing steps to remove the light-induced degradation. IEEE 1262 includes a wet hi-pot test that is nearly impossible to pass if a module has insulation holes or flaws. Otherwise, IEEE 1262 is very similar to IEC 61215. TC-82 solved the problem of a-Si degradation by issuing a separate standard specifically for a-Si modules, IEC 61646 [101]. Rather than thermal annealing, however, IEC 61646 uses light soaking to condition the test modules prior to the qualification testing. In practice, light soaking has been shown to be lengthy and expensive for testing, so it is likely that light soaking will be replaced with thermal annealing in a future version.

4.2 Module Qualification Tests

TC-82 has been working on a revision of IEC 61215 for module qualification that is due to proceed to the Committee Draft for Voting (CDV) stage at the time of this writing [102]. This is an important document for the PV industry, so a brief description of the main test sequences is included below.

The new draft includes a wet insulation test, adopted from IEEE 1262, that all test modules must pass prior to and following the test sequences. A module is immersed in a surfactant solution, and the insulation resistance between the solution and the shorted module leads is measured at 500 V. This test reveals insulation flaws that could be safety hazards in use and also exposes possible paths of moisture intrusion that might result in performance degradation.

4.2.1 Thermal Cycling Sequence

One thermal cycle consists of a $-40°C$ freeze for a minimum of 10 min, followed by an excursion to $+85°C$, again with a 10-min hold. The rate of change of module temperature must not exceed $100°C/h$, and a complete cycle should not last longer than 6 h. This sequence is repeated until 200 cycles are completed.

According to reference [89], thermal cycling "is intended to accelerate thermal differential-expansion stress effects so that design weaknesses associated with encapsulant system, cells, interconnects, and bonding materials can be detected as a direct result of the test." An important change from the previous version of IEC 61215 is the added requirement that a forward-bias current equal to the output current at SRC P_{max} be passed through the test modules when the temperature is above 25°C. Current bias simulates the stresses that solder joints experience in actual use, and the biasing has been shown to reveal poor soldering [103].

4.2.2 Damp-Heat Sequence

This sequence begins with the standard damp heat, which is 1000 h of exposure to 85% relative humidity (RH) at 85°C. Damp heat is followed by the 25-mm ice-ball impact and 2400-Pa mechanical loading tests. Operating conditions as severe as 85% RH and 85°C never occur in actual use, but damp heat stresses the encapsulation and can result in delamination. Note that because of these extremes, it is conceivable that damp heat could produce failures that would not be seen in the field.

4.2.3 UV Exposure, Thermal Cycling, and Humidity-Freeze Sequence

Prior to any stress tests, a short UV exposure of 60 MJ/m^2 (15 kWh/m^2) preconditions the modules. The UV exposure is followed by 50 thermal cycles and then 10 humidity-freeze cycles. Module designs that have tapes inside the encapsulation will typically delaminate during the subsequent thermal or humidity-freeze cycles. This sequence concludes with a robustness of termination test.

4.3 Reliability Testing

As noted previously, qualification testing uses accelerated tests, but it should not be considered reliability testing. However, this does not mean the specific tests that are part of a standard qualification sequence cannot be used for reliability testing. An example is extending the duration of the standard thermal cycle test by increasing the total number of cycles or until failure of the module. Another is combining a qualification test with an added stress, such as damp heat with high-voltage bias [103]. If possible, accelerated tests should be combined with real-time tests. Reference [104] is a standard for solar weathering that uses techniques adapted from the materials industry. This standard uses total UV exposure doses to quantify both real-time and accelerated tests.

4.4 Module Certification and Commercial Services

Product certification is a formal process involving accredited testing laboratories and certification agencies that issue licenses to manufacturers indicating their products have been tested and are in conformance [105]. Certification is well known in other industries, especially for product safety, but it is a fairly recent development for photovoltaics. PV module safety certifications have been available for almost 20 years, but efforts have been made within the past 10 years to initiate certifications based on qualification testing. An example of this is the Global Approval Program for Photovoltaics (PV GAP) [106].

Many independent PV laboratories offer testing services, either as part of a formal certification program or upon request. These include
- SU Photovoltaic Testing Laboratory, Mesa, AZ, USA
- European Commission Joint Research Centre, Environment Institute, Renewable Energies Unit, Ispra, Italy
- TÜV Rheinland, Berlin/Brandenburg, Germany
- Underwriters Laboratories, Inc., Northbrook, IL, USA
- VDE Testing and Certification Institute, Offenbach, Germany

5. MODULE DEGRADATION CASE STUDY

To conclude this chapter, a diagnosis of a degraded polycrystalline Si module is presented as an example that uses many of the techniques outlined here. A small 20-W module was subjected to real-time outdoor weathering with a resistive load, and light and dark I–V measurements were made prior to exposure and at several intervals over a period of four years. At the end of this period, P_{max} under STC as measured with a pulsed Xe simulator had declined by 9.8%, as listed in Table 3. Most of this loss has occurred in the FF, which was down by 6.3%. Series-resistance (R_s) values determined from dark I–V measurements did not show large increases, so a gradual degradation of the solder joints or the cell metallization was not responsible. Examination of the light I–V curves, Figure 3,

Table 3 Polycrystalline Si module I–V parameters

	V_{oc} (V)	I_{sc} (A)	FF	P_{max} (W)	R_s (Ω)
Initial	21.1	1.18	0.733	18.2	0.60
Exposed	20.8	1.15	0.687	16.4	0.43

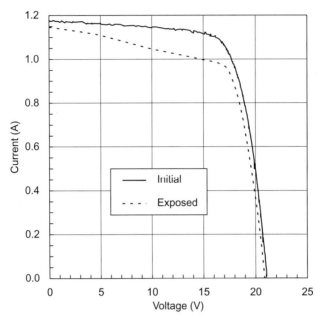

Figure 3 Light I–V curves at SRC of a polycrystalline Si module weathered outdoors for four years.

shows a "stair step," which in a series-connected PV device typically indicates a mismatch in current output between one or more individual cells.

An attempt to locate such a problem was made with an IR camera by placing the module in forward bias (1 A and 22 V) and allowing the cells to heat for several minutes. The resulting IR image, Figure 4, was made of the module rear surface through the polymeric backsheet (shadows from the junction box and the wiring are visible) and shows temperature rises of only about 1°C above the ambient. One or two cells appear to have slight temperature variations from the rest of the module, so this test was inconclusive at best. As it turns out, forward bias is not a very stressful condition for crystalline Si, and it is much easier to see hot spots when a module is shorted while illuminated in sunlight. After just a few minutes with the module in this condition, one cell developed a hot spot greater than 6°C, which was easily visible with an IR camera (Figure 5). This cell is most likely being forced into reverse bias. Looking at the hot cell in Figure 5, it appears one corner is not heating at all, and a visual examination of this cell revealed a crack in this location. Therefore, this cell has an output current about 10% lower than all the others, which caused the loss of fill factor.

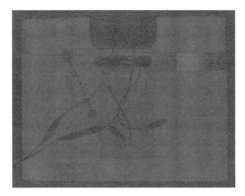

Figure 4 IR thermal image of the rear surface of the module from Figures VI.3 and VI.4. The module was placed at 1 A, 22 V in forward bias and allowed to heat for several minutes.

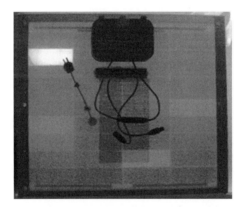

Figure 5 Another IR thermal image of the same module from Figure 3. The module was shorted for several minutes in sunlight before the image was captured.

ACKNOWLEDGMENTS

Preparation of this chapter was supported by the US Department of Energy under contract No. DE-AC36-99-G010337. Steve Rummel, Allan Anderberg, Larry Ottoson, and Tom McMahon performed and assisted with the measurements used in the module degradation case study.

Permissions

This submitted manuscript has been offered by Midwest Research Institute (MRI) employees, a contractor of the US Government Contract No. DE-AC36-99-GO10337. Accordingly, the US Government and MRI retain nonexclusive, royalty-free license to publish or reproduce the published form of this contribution, or allow others to do so, for US Government purposes.

REFERENCES

[1] K.A. Emery, C.R. Osterwald, Efficiency measurements and other performance-rating methods, in: Current Topics in Photovoltaics, vol. 3, Academic Press, 1988.
[2] K. Emery, The rating of photovoltaic performance, IEEE Trans. Electron. Devices 46 (1999) 1928–1931.
[3] K.A. Emery, C.R. Osterwald, Solar cell efficiency measurements, Sol. Cells 17 (1986) 253–274.
[4] ASTM Standard E 490: Standard Solar Constant and Zero Air Mass Solar Spectral Irradiance Tables, in: ASTM Annual Book of Standards, vol. 12.02, ASTM International, West Conshohocken, PA, 2002.
[5] ASTM Standard G 159: Standard Tables for References Solar Spectral Irradiance at Air Mass 1.5: Direct Normal and Hemispherical for a 37° Tilted Surface, in: ASTM Annual Book of Standards, vol. 14.04, ASTM International, West Conshohocken, PA, 2002.
[6] IEC Standard 60904-3: Photovoltaic Devices. Part 3: Measurement Principles for Terrestrial Photovoltaic (PV) Solar Devices with Reference Spectral Irradiance Data. International Electrotechnical Commission, Geneva, 1998.
[7] M. Iqbal, An Introduction to Solar Radiation, Academic Press Canada, Ontario, 1983.
[8] C. Riordan, T. Cannon, D. Myers, R. Bird, Solar irradiance models, data, and instrumentation for PV device performance analyses, in: Proceeding of the 18th IEEE Photovoltaic Specialists Conference, IEEE, 1985, pp. 957–962.
[9] C.R. Osterwald, K.A. Emery, Spectroradiometric sun photometry, J. Atmos. Oceanic Technol. Sept. (2000) 1171–1188.
[10] D. Myers, K. Emery, C. Gueymard, Proposed reference spectral irradiance standards to improve photovoltaic concentrating system design and performance evaluation, in: Proceeding of the 29th IEEE Photovoltaic Specialist Conference, IEEE, 2002.
[11] D. Myers, K. Emery, C. Gueymard, Terrestrial solar spectral modeling tools and applications for photovoltaic devices, in: Proceeding of the 29th IEEE Photovoltaic Specialist Conference, IEEE, 2002.
[12] D.R. Myers, K.A. Emery, D.R. Myers, Uncertainty estimates for global solar irradiance measurements used to evaluate PV device performance, Sol. Cells 27 (1989) 455–464.
[13] D.R. Myers, D.R. Myers, I. Reda, Recent progress in reducing the uncertainty in and improving pyranometer calibrations, J. Sol. Energy Eng. 124 (2002) 44–50.
[14] ASTM Standard G 130: Standard Test Method for Calibration of Narrow- and Broad-Band Ultraviolet Radiometers Using a Spectroradiometer, in: ASTM Annual Book of Standards, vol. 14.04, ASTM International, West Conshohocken, PA, 2002.
[15] ASTM Standard G 138: Standard Test Method for Calibration of a Spectroradiometer Using a Standard Source of Irradiance, in: ASTM Annual Book of Standards, vol. 14.04, ASTM International, West Conshohocken, PA, 2002.
[16] ASTM Standard E 816: Standard Test Method for Calibration of Secondary Reference Pyrheliometers and Pyrheliometers for Field Use, in: ASTM Annual Book of Standards, vol. 14.04, ASTM International, West Conshohocken, PA, 2002.
[17] ASTM Standard E 824: Standard Test Method for Transfer of Calibration from Reference to Field Radiometers, in: ASTM Annual Book of Standards, vol. 14.04, ASTM International, West Conshohocken, PA, 2002.
[18] ASTM Standard E 913: Standard Method for Calibration of Reference Pyranometers with Axis Vertical by the Shading Method, in: ASTM Annual Book of Standards, vol. 14.04, ASTM International, West Conshohocken, PA, 2002.

[19] ASTM Standard E 941: Standard Test Method for Calibration of Reference Pyranometers with Axis Tilted by the Shading Method, in: ASTM Annual Book of Standards, vol. 14.04, ASTM International, West Conshohocken, PA, 2002.
[20] ASTM Standard E 948: Test Method for Electrical Performance of Photovoltaic Cells Using Reference Cells under Simulated Sunlight, in: ASTM Annual Book of Standards, vol. 12.02, ASTM International, West Conshohocken, PA, 2002.
[21] ASTM Standard E 1036: Test Methods for Electrical Performance of Nonconcentrator Terrestrial Photovoltaic Modules and Arrays Using Reference Cells, in: ASTM Annual Book of Standards, vol. 12.02, ASTM International, West Conshohocken, PA, 2002.
[22] IEC Standard 60904-1: Photovoltaic Devices., Part 1: Measurement of Photovoltaic Current-Voltage Characteristics, International Electrotechnical Commission, Geneva, 1987.
[23] C.H. Seaman, Calibration of solar cells by the reference cell method—the spectral mismatch problem, Sol. Energy 29 (1982) 291−298.
[24] ASTM Standard E 973: Test Method for Determination of the Spectral Mismatch Parameter Between a Photovoltaic Device and a Photovoltaic Reference Cell, in: ASTM Annual Book of Standards, vol. 12.02, ASTM International, West Conshohocken, PA, 2002.
[25] IEC Standard 60904-7: Photovoltaic Devices., Part 7: Computation of Spectral Mismatch Error Introduced in the Testing of a Photovoltaic Device, International Electrotechnical Commission, Geneva, 1987.
[26] B. Anspaugh, A verified technique for calibrating space solar cells, in: Proceeding of the 19th IEEE Photovoltaic Specialist Conference, IEEE, 1987, pp. 542−547.
[27] H.W. Brandhorst, Calibration of solar cells using high-altitude aircraft, in: Solar Cells, Gordon and Breach, London, 1971.
[28] K. Bücher, Calibration of solar cells for space applications, Prog. Photovolt. Res. Appl. 5 (1997) 91−107.
[29] C.R. Osterwald, K.A. Emery, D.R. Myers, R.E. Hart, Primary reference cell calibrations at SERI: History and methods, in: Proceeding of the 21st IEEE Photovoltaic Specialists Conference, IEEE, 1990, pp. 1062−1067.
[30] ASTM Standard E 1125: Test Method for Calibration of Primary Non-Concentrator Terrestrial Photovoltaic Reference Cells Using a Tabular Spectrum, in: ASTM Annual Book of Standards, vol. 12.02, ASTM International, West Conshohocken, PA, 2002.
[31] C.R. Osterwald, S. Anevsky, K. Bücher, A.K. Barua, P. Chaudhuri, J. Dubard, et al., The world photovoltaic scale: an international reference cell calibration program, Prog. Photovolt. Res. Appl. 7 (1999) 287−297.
[32] ASTM Standard E 1362: Test Method for Calibration of Non-Concentrator Photovoltaic Secondary Reference Cells, in: ASTM Annual Book of Standards, vol. 12.02, ASTM International, West Conshohocken, PA, 2002.
[33] IEC Standard 60904-6: Photovoltaic Devices, Part 6: Requirements for Reference Solar Modules, International Electrotechnical Commission, Geneva, 1994.
[34] K. Emery, D. Myers, S. Rummel, Solar simulation—problems and solutions, in: Proceeding of the 20th IEEE Photovoltaic Specialists Conference, IEEE, 1988, pp. 1087−1091.
[35] ASTM Standard E 927: Specification for Solar Simulation for Terrestrial Photovoltaic Testing, in: ASTM Annual Book of Standards, vol. 12.02, ASTM International, West Conshohocken, PA, 2002.
[36] IEC Standard 60904-9: Photovoltaic Devices, Part 9: Solar Simulator Performance Requirements, International Electrotechnical Commission, Geneva, 1995.

[37] C.R. Osterwald, T. Glatfelter, J. Burdick, Comparison of the temperature coefficients of the basic I−V parameters for various types of solar cells, in: Proceeding of the 19th IEEE Photovoltaic Specialists Conference, IEEE, 1987, pp. 188−193.
[38] S.M. Sze, Physics of Semiconductor Devices, Wiley, New York, 1981Physics of Semiconductor Devices, Wiley, New York, 1981.
[39] A.L. Fahrenbruch, R.H. Bube, Fundamentals of Solar Cells, Academic Press, New York, 1983.
[40] K. Whitfield, C.R. Osterwald, Procedure for determining the uncertainty of photovoltaic module outdoor electrical performance, Prog. Photovolt. Res. Appl. 9 (2001) 87−102.
[41] C.R. Osterwald, Translation of device performance measurements to reference conditions, Sol. Cells 18 (1986) 269−279.
[42] D.L. King, J.A. Kratochvil, W.E. Boyson, Temperature coefficients for PV modules and arrays: Measurement methods, difficulties, and results, in: Proceeding of the 26th IEEE Photovoltaic Specialists Conference, IEEE, 1987, pp. 1183−1186.
[43] C.M. Whitaker, T.U. Townsend, H.J. Wenger, A. Iliceto, G. Chimento, F. Paletta, Effects of irradiance and other factors on PV temperature coefficients, in: Proceeding of the 22nd IEEE Photovoltaic Specialists Conference, IEEE, 1991, pp. 608−613.
[44] T. Glatfelter, J. Burdick, A method for determining the conversion efficiency of multiple-cell photovoltaic devices, in: Proceeding of the 19th IEEE Photovoltaic Specialists Conference, IEEE, 1987, pp. 1187−1193.
[45] K.A. Emery, C.R. Osterwald, T. Glatfelter, J. Burdick, G. Virshup, A comparison of the errors in determining the conversion efficiency of multijunction solar cells by various methods, Sol. Cells 24 (1988) 371−380.
[46] D.L. King, B.R. Hansen, J.M. Moore, D.J. Aiken, New methods for measuring performance of monolithic multi-junction solar cells, in: Proceeding of the 28th IEEE Photovoltaic Specialists Conference, IEEE, 2000, pp. 1197−1201.
[47] K. Emery, M. Meusel, R. Beckert, F. Dimroth, A. Bett, W. Warta, Procedures for evaluating multijunction concentrators, in: Proceeding of the 28th IEEE Photovoltaic Specialists Conference, IEEE, 2000, pp. 1126−1130.
[48] ASTM Draft Standard: Standard Test Method for Measurement of Electrical Performance and Spectral Response of Nonconcentrator Multijunction Photovoltaic Cells and Modules, ASTM Annual Book of Standards, vol. 12.02, ASTM International, West Conshohocken, PA, Forthcoming.
[49] M. Meusel, R. Adelhelm, F. Dimroth, A.W. Bett, W. Warta, Spectral mismatch correction and spectrometric characterization of monolithic III−V multi-junction solar cells, Prog. Photovolt. Res. Appl. 10 (2002) 243−255.
[50] S. Smith, T. Townsend, C. Whitaker, S. Hester, Photovoltaics for utility-scale applications: Project overview and data analysis, Sol. Cells 27 (1989) 259−266.
[51] C.M. Whitaker, T.U. Townsend, J.D. Newmiller, D.L. King, W.E. Boyson, J.A. Kratochvil, et al., Application and validation of a new PV performance characterization method, in: Proceeding of the 26th IEEE Photovoltaic Specialists Conference, IEEE, 1997, pp. 1253−1256.
[52] D.L. King, J.A. Kratochvil, W.E. Boyson, W. Bower, Field experience with a new performance characterization procedure for photovoltaic arrays, in: Proceeding of the 2nd World Conference and Exhibition on Photovoltaic Solar Energy Conversion, 1998, pp. 1947−1952.
[53] K. Myers, J.A. del Cueto, W. Zaaiman, Spectral corrections based on optical air mass, in: Proceeding of the 29th IEEE Photovoltaic Specialists Conference, IEEE, In press, 2002.
[54] D.L. King, J.A. Kratochvil, W.E. Boyson, Measuring solar spectral and angle-of-incidence effects on photovoltaic modules and solar irradiance sensors, in:

Proceeding of the 26th IEEE Photovoltaic Specialists Conference, IEEE, 1997, pp. 1113—1116.
[55] B. Kroposki, K. Emery, D. Myers, L. Mrig, A comparison of photovoltaic module performance evaluation methodologies for energy ratings, in: Proceeding of the IEEE 1st World Conference on Photovoltaic Energy Conversion, IEEE, 1994, pp. 858—862.
[56] B. Marion, B. Kroposki, K. Emery, J. del Cueto, D. Myers, C. Osterwald, Validation of a Photovoltaic Module Energy Ratings Procedure at NREL. National Renewable Energy Laboratory Technical Report NREL/TP-520-26909. Available from the National Technical Information Service, Springfield, VA, USA, 1999.
[57] D. Chianese, S. Rezzonico, N. Cereghetti, A. Realini, Energy rating of PV modules, in: Proceeding of the 17th European Photovoltaic Solar Energy Conference, 2001, pp. 706—709.
[58] N. Cereghetti, A. Realini, D. Chianese, S. Rezzonico, Power and energy production of PV modules, in: Proceeding of the 17th European Photovoltaic Solar Energy Conference, 2001, pp. 710—713.
[59] D. Anderson, T. Sample, E. Dunlop, Obtaining module energy rating from standard laboratory measurements, in: Proceeding of the 17th European Photovoltaic Solar Energy Conference, 2001, pp. 832—835.
[60] J.A. Del Cueto, Comparison of energy production and performance from flat-plate photovoltaic modules deployed at fixed tilt, in: Proceeding of the 29th IEEE Photovoltaic Specialists Conference, IEEE, In press, 2002.
[61] Guide to the Expression of Uncertainty in Measurement, International Electrotechnical Commission, Geneva, 1995, ISBN 92-67-10199-9.
[62] K.A. Emery, C.R. Osterwald, PV performance measurement algorithms, procedures, and equipment, in: Proceeding of the 21st IEEE Photovoltaic Specialists Conference, IEEE, 1990, pp. 1068—1073.
[63] H.A. Ossenbrink, W. Zaaiman, J. Bishpop, Do multi-flash solar simulators measure the wrong fill factor? in: Proceeding of the 23rd IEEE Photovoltaic Specialists Conference, IEEE, 1993, pp. 1194—1196.
[64] K.A. Emery, H. Field, Artificial enhancements and reductions in the PV efficiency, in: Proceeding of the IEEE 1st World Conference on Photovoltaic Energy Conversion, IEEE, 1994, pp. 1833—1838.
[65] E.G. Mantingh, W. Zaaiman, H.A. Ossenbrink, Ultimate transistor electronic load for electrical performance measurement of photovoltaic devices using pulsed solar simulators, in: Proceeding of the IEEE 1st World Conference on Photovoltaic Energy Conversion, IEEE, 1994, pp. 871—873.
[66] C.R. Osterwald, S. Anevsky, A.K. Barua, J. Dubard, K. Emery, D. King, et al., Results of the PEP'93 intercomparison of reference cell calibrations and newer technology performance measurements, in: Proceeding of the 25th IEEE Photovoltaic Specialists Conference, IEEE, 1996, pp. 1263—1266.
[67] R. Rüther, G. Kleiss, K. Reiche, Spectral effects on amorphous silicon solar module fill factors, Sol. Energy Mat. Sol. Cells 71 (2002) 375—385.
[68] K.A. Emery, C.R. Osterwald, C.V. Wells, Uncertainty analysis of photovoltaic efficiency measurements, in: Proceeding of the 19th IEEE Photovoltaic Specialists Conference, IEEE, 1987, pp. 153—159.
[69] K. Heidler, J. Beier, Uncertainty analysis of PV efficiency measurements with a solar simulator: Spectral mismatch, non-uniformity, and other sources of error, in: Proceeding of the 8th European Photovoltaic Solar Energy Conference, 1988, pp. 554—559.
[70] H. Field, K.A. Emery, An uncertainty analysis of the spectral correction factor, in: Proceeding of the 23rd IEEE Photovoltaic Specialists Conference, IEEE, 1993, pp. 1180—1187.

[71] D.L. King, B.R. Hansen, J.K. Jackson, Sandia/NIST reference cell calibration procedure, in: Proceeding of the 23rd IEEE Photovoltaic Specialists Conference, IEEE, 1993, pp. 1095–1101.
[72] D.L. King, B.R. Hansen, J.A. Kratochvil, M.A. Quintana, Dark current-voltage measurements on photovoltaic modules as a diagnostic or manufacturing tool, in: Proceeding of the 26th IEEE Photovoltaic Specialists Conference, IEEE, 1997, pp. 1125–1128.
[73] M. Chegaar, Z. Ouennoughi, A. Hoffman, A new method for evaluating illuminated solar cell parameters, in: Solid-State Electronics, vol. 45, 2001, pp. 293–296.
[74] J.A. Del Cueto, Method for analyzing series resistance and diode quality factors from field data of photovoltaic modules, Sol. Energy Mat. Sol. Cells 55 (1998) 291–297.
[75] T.J. McMahon, T.S. Basso, S.R. Rummel, Cell shunt resistance and photovoltaic module performance, in: Proceeding of the 25th IEEE Photovoltaic Specialists Conference, IEEE, 1996, pp. 1291–1294.
[76] IEC Standard 60904-8: Photovoltaic Devices, Part 8: Measurement of Spectral Response of a Photovoltaic (PV) Device, International Electrotechnical Commission, Geneva, 1998.
[77] ASTM Standard E 1021: Test Method for Measuring Spectral Response of Photovoltaic Cells, in: ASTM Annual Book of Standards, Vol. 12.02, ASTM International, West Conshohocken, PA, 2002.
[78] J.S. Hartman, M.A. Lind, Spectral response measurements for solar cells, Sol. Cells 7 (1983) 147–157.
[79] R. Van Steenwinkel, Measurements of spectral responsivities of cells and modules, in: Proceeding of the 7th European Photovoltaic Solar Energy Conference, 1987, p. 325.
[80] R. Budde, W. Zaaiman, H.A. Ossenbrink, Spectral response calibration facility for photovoltaic cells, in: Proceeding of the IEEE 1st World Conference on Photovoltaic Energy Conversion, IEEE, 1994, pp. 874–876.
[81] C.R. Osterwald, S. Anevsky, A.K. Barua, K. Bücher, P. Chauduri, J. Dubard, et al., The Results of the PEP'93 Intercomparison of Reference Cell Calibrations and Newer Technology Performance Measurements: Final Report, National Renewable Energy Laboratory Technical Report NREL/TP-520-23477, Available from the National Technical Information Service, Springfield, VA, USA, 1998.
[82] H. Field, Solar cell spectral response measurement errors related to spectral band width and chopped light waveform, in: Proceeding of the 26th IEEE Photovoltaic Specialists Conference, IEEE, 1997, pp. 471–474.
[83] J. Burdick, T. Glatfelter, Spectral response and I–V measurements of tandem amorphous-silicon alloy photovoltaic devices, Sol. Cells 18 (1986) 301–314.
[84] I.L. Eisgruber, J.R. Sites, Extraction of individual-cell photocurrents and shunt resistances in encapsulated modules using large-scale laser scanning, Prog. Photovolt. Res. Appl. 4 (1996) 63–75.
[85] G. Agostinelli, G. Friesen, F. Merli, E.D. Dunlop, M. Acciarri, A. Racz, et al., Large area fast LBIC as a tool for inline PV module and string characterization, in: Proceeding of the 17th European Photovoltaic Solar Energy Conference, 2001, pp. 410–413.
[86] D.L. King, J.A. Kratochvil, M.A. Quintana, T.J. McMahon, Applications for infrared imaging equipment in photovoltaic cell, module, and system testing, in: Proceeding of the 28th IEEE Photovoltaic Specialists Conference, IEEE, 2000, pp. 1487–1490.
[87] D.L. King, M.A. Quintana, J.A. Kratochvil, D.E. Ellibee, B.R. Hansen, Photovoltaic module performance and durability following long-term field exposure, Prog. Photovolt. Res. Appl. 8 (2000) 241–256.

[88] T.J. McMahon, G.J. Jorgensen, R.L. Hulstrom, D.L. King, M.A. Quintana, Module 30 year life: What does it mean and is it predictable/achievable? in: Proceeding of the NCPV Program Review Meeting Available from the National Technical Information Service, Springfield, VA, USA, 2000.

[89] J.H. Wohlgemuth, Reliability testing of PV modules, in: Proceeding of the IEEE 1st World Conference on Photovoltaic Energy Conversion, IEEE, 1994, pp. 889–892.

[90] A.R. Hoffman, R.G. Ross, Environmental qualification testing of terrestrial solar cell modules, in: Proceeding of the 13th IEEE Photovoltaic Specialists Conference, IEEE, 1979, pp. 835–842.

[91] A.R. Hoffman, J.S. Griffith, R.G. Ross, Qualification testing of flat-plate photovoltaic modules, IEEE Trans. Rel. R-31 (1982) 252–257.

[92] Block V Solar Cell Module Design and Test Specification for Intermediate Load Applications, Jet Propulsion Laboratory report 5101-161, Pasadena, CA, 1981.

[93] M.I. Smokler, D.H. Otth, R.G. Ross, The block program approach to photovoltaic module development, in: Proceeding of the 18th IEEE Photovoltaic Specialists Conference, IEEE, 1985, pp. 1150–1158.

[94] Qualification Test Procedures for Photovoltaic Modules, Specification No. 502, Issue 1, Commission of the European Communities, Joint Research Center, Ispra Establishment, 1984.

[95] Standard for Flat-Plate Photovoltaic Modules and Panels, ANSI/UL 1703-1987, American National Standards Institute, New York, 1987, ISBN 1-55989-390-7.

[96] R. DeBlasio, L. Mrig, D. Waddington, Interim Qualification Tests and Procedures for Terrestrial Photovoltaic Thin-Film Flat-Plate Modules. Solar Energy Research Institute technical report SERI/TR-213-3624, Available from the National Technical Information Service, Springfield, VA, USA, 1990.

[97] IEC Standard 61215: Crystalline Silicon Terrestrial Photovoltaic (PV) Modules—Design Qualification and Type Approval, International Electrotechnical Commission, Geneva, 1993.

[98] H. Ossenbrink, E. Rossi, J. Bishop, Specification 503—implementation of PV module qualification tests at ESTI, in: Proceeding of the 10th European Photovoltaic Solar Energy Conference, 1991, pp. 1219–1221.

[99] J. Bishop, H. Ossenbrink, Results of five years of module qualification testing to CEC specification 503, in: Proceeding of the 25th IEEE Photovoltaic Specialists Conference, IEEE, 1996, pp. 1191–1196.

[100] IEEE Standard 1262: IEEE Recommended Practice for Qualification of Photovoltaic (PV) Modules, Institute of Electrical and Electronic Engineers, New York, 1995.

[101] IEC Standard 61646: Thin-Film Terrestrial Photovoltaic (PV) Modules—Design Qualification and Type Approval, International Electrotechnical Commission, Geneva, 1996.

[102] J.H. Wohlgemuth, BP Solar, private communication.

[103] J.H. Wohlgemuth, M. Conway, D.H. Meakin, Reliability and performance testing of photovoltaic modules, in: Proceeding of the 28th IEEE Photovoltaic Specialists Conference, IEEE, 2000, pp. 1483–1486.

[104] ASTM Standard E 1596: Test Methods for Solar Radiation Weathering of Photovoltaic Modules, in: ASTM Annual Book of Standards, Vol. 12.02, ASTM International, West Conshohocken, PA, 2002.

[105] C.R. Osterwald, R. Hammond, G. Zerlaut, R. D'Aiello, Photovoltaic module certification and laboratory accreditation criteria development, in: Proceeding of the IEEE 1st World Conference on Photovoltaic Energy Conversion, IEEE, 1994, pp. 885–888.

[106] C.R. Osterwald, P.F. Varadi, S. Chalmers, M. Fitzgerald, Product certification for PV modules, BOS components, and systems, in: Proceeding of the 17th European Photovoltaic Solar Energy Conference, 2001, pp. 379–384.

CHAPTER III-3

PV System Monitoring

Bruce Cross
Energy Equipment Testing Service Ltd, Cardiff, UK

Contents

1. Introduction — 1071
   1.1 User Feedback — 1072
   1.2 Performance Verification — 1072
   1.3 System Evaluation — 1072
2. Equipment — 1073
   2.1 Displays — 1073
   2.2 Data-Acquisition Systems — 1073
   2.3 Sensors — 1075
3. Calibration and Recalibration — 1075
4. Data Storage and Transmission — 1076
5. Monitoring Regimes — 1076
   5.1 Performance Verification — 1076
   5.2 System Evaluation — 1078
   5.3 Data Gathering, Transmission, and Storage — 1078
   5.4 Data Analysis and Reporting — 1078
References — 1079

1. INTRODUCTION

There are many reasons to monitor a system as expensive and long-term as a PV installation. Careful consideration should be given to the purpose behind the monitoring before developing a specification, and the ethos should be to measure only those variables that are necessary, and the minimum frequency required to give meaningful results, for a period of time over which new information will be produced. These needs for monitoring fall into three main groups:

- User feedback
- Performance verification
- System evaluation

Practical Handbook of Photovoltaics.
© 2012 Elsevier Ltd. All rights reserved.

1.1 User Feedback

This can range from a simple LED on the inverter lid or a user display in a domestic hallway, to a large interactive wall display in the foyer of a corporate building. The common thread to all displays is giving the users an indication that the system is functioning and the benefits that this brings. A PV system appears, to the uninformed user, to do nothing at all—not a good public relations statement! A clear display gives much added value to the system, especially if combined with some graphic or text explaining the concepts (Figure 1).

1.2 Performance Verification

A system may have been financed on the basis of its output, and so the user needs to measure the output and compare to the claims for the system. It may also be the case that the electrical production has to be measured, in order to be sold, or that some 'green certificates' may be evaluated. The complexity and expense of such metering is determined by the number and accuracy of the measurements to be made.

1.3 System Evaluation

For an unusual system it may be worthwhile to measure the detailed operation of the system in order to understand the functioning of its components in detail. This was frequently performed in the 1980s and 1990s when systems were often grant funded, and the main purpose of

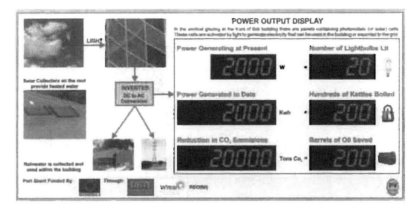

Figure 1 Public display at the William Jefferson Clinton Reading Rooms, Enniskillen, Northern Ireland, showing the PV power generation and also many consequential variables. *(Photo EETS.)*

building the system was to improve knowledge and understanding. The operation of PV systems is now well documented, and the expense of full monitoring, and the evaluation of the data, cannot normally be justified. The European Commission used to require full monitoring of every project they supported [1], but now this level of details seems to have outlived its usefulness.

Of course, these three types of monitoring are not mutually exclusive. A display for user feedback is also necessary in a large system being fully monitored or in a system where the output is being measured [2].

2. EQUIPMENT

2.1 Displays

Displays are the backbone of monitoring. The easiest to fit is a simple indication as part of the inverter. Most PV inverter manufacturers offer an optional display. However, this can place severe constraints on the placing of the inverter, which would normally be in a roof void, electrical switch room, or some other secluded place. If the display is to be effective it must be in a place where it is visible in everyday activities.

Remote displays are easier to site, and may be provided with data from the inverter itself, or by a meter in the cabling from inverter to distribution board. A significant cost to installing this is the routing of the cabling to the display, but there are instruments on the market that avoid this by utilising short-range radio transmission (see Figure 2).

There are many different formats of data that can be displayed: the most popular are the instantaneous power being generated and the total energy to date. However, large displays often include derived values that mean more to the public, such as numbers of lights that are being powered, or the amount of carbon production being offset (see Figure 3). A computer-based monitoring system can often embed that information within a touch-screen-driven information point or have it displayed on the website for the building.

2.2 Data-Acquisition Systems

The main system tends to fall into two types: loggers and computers. The advantage of a logger is its simplicity and robust construction, but its disadvantage is its inflexibility and cost. A computer system, in contrast, may

Figure 2 A domestic PV output display receiving radio-linked data from a sensor next to the inverter. *(Photo EETS.)*

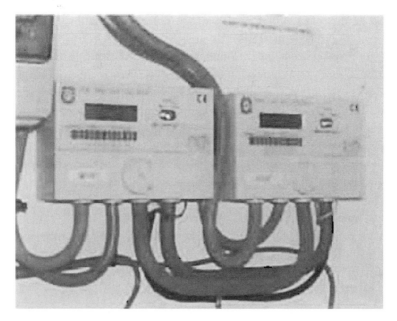

Figure 3 A pair of meters measuring import and export of electricity into a building and outputting pulse data to a monitoring system. *(Photo EETS.)*

Table 1 Typical monitoring variables (see also [3])

Parameter	Sensor	Accuracy	Precision	Comment
Solar radiation	Reference cell	3%	2 W/m^2	Commonly used
	Pyranometer	2%	1 W/m^2	For research only
DC current	Various	1%	0.5%	
DC voltage	Various	1%	0.5%	
Energy	Meter	1%	1 Wh	Alternative to 1 and V above
Ambient temperature	Thermocouple	1°C	1°C	
	PRT	0.2°C	0.2°C	
	Thermistor	1°C	1°C	
Module temperature	As ambient			
Power	AC meter	1%	0.5%	

be slower to set up and commission, but has the advantage of a wider choice of operational modes and custom settings, while the cost may be less for a system based on a desktop PC.

The choice between the types may well be dictated by the type of monitoring strategy for the project: Are the data to be viewed in real time? Are different types of data to be monitored at differing intervals and in differing ways?

2.3 Sensors

There is no limitation to the inputs that may be monitored for a PV systems, but most systems will need to measure the input and output energy, and some environmental and system variables. A list of the more usual variables is given in Table 1.

3. CALIBRATION AND RECALIBRATION

The system should be set up and calibrated preferably in situ. The need for recalibration should be determined whilst considering the length of time for the monitoring, and the accuracy required of the system. The reference cell is particularly critical, but often is the most difficult item to access. If annual recalibration is not practical back in the laboratory, an on-site comparison with a reference device nearby may be adequate. The entire monitoring system can also benefit from a comparative calibration using handheld reference devices (ambient temperature sensors, voltage and current meters, etc.).

4. DATA STORAGE AND TRANSMISSION

The data are generally stored in situ using RAM for a logger, or using a hard drive for a computer system. Loggers often include removable RAM cards, discs, or other magnetic media, as a form of storage and retrieval. PCs may use multiple drives, or daily downloads, as a backup storage.

Having recorded the data, it may be transmitted back to the monitoring organisation by many means. The simplest logging systems may have to be physically carried back to the laboratory and plugged into a special reader device, or a PC serial port. Removable media allow the swapping of the storage medium on site allowing monitoring to continue uninterrupted. Such media are tapes, floppy discs, RAM cards, etc. The only disadvantages are that the new media may not be inserted correctly or that the logger may not be restarted, and the loss of data will not be noticed until the next visit.

Telephonic transmission is often used as it gives the opportunity for frequent downloading of data (reducing the length of any 'lost' periods), and also the chance to 'upload' any changes to the logging schedule. The more sophisticated loggers can initiate a call to a fax or PC to report any faults or out of range signals immediately they are detected.

The advent of the Internet has allowed PCs to connect to a local portal via a local phone line, thus making downloading less expensive anywhere in the world. If a telephone line is not available at a remote site, a cellular phone connection can provide an equivalent facility.

5. MONITORING REGIMES

5.1 Performance Verification

Rapidly changing variables (such as solar radiation) require frequent sampling, although generally only a mean value over a longer period is stored. The accuracy of this measurement may not be high (5% is generally used for irradiance) because of uncertainties along the calibration chain and limited accuracy of absolute reference cell measurements [4,5]. Thus a sample every 10 sec with a mean stored every 10 min is normal. Greater accuracy of solar radiation measurement can be obtained by using a pyranometer, which has a uniform response across the solar spectrum, but without continuous measurement of the sunlight spectrum it will not give an accurate measure of the energy available to the PV cell.

Ambient air temperature is important for correcting the output characteristic of the PV, but an accuracy of 1°C is adequate and readily achieved by most sensor types. Since ambient temperature changes so slowly, samples every 1 min to 10 min are adequate.

Electrical output is readily measured using standard and inexpensive 'watt-hour' meters. Since the integration of instantaneous power to give readings of energy transmitted is a standard activity in all electrical metering, the accuracy required for PV is rarely problematic for standard instruments. The need for measurement in two directions (for the import–export location in a combined generation and load system) is unusual, and two standard meters is often less expensive than a customised single unit. The signal given from these meters is frequently a pulse, which a readily logged by most standard techniques (Figure 4).

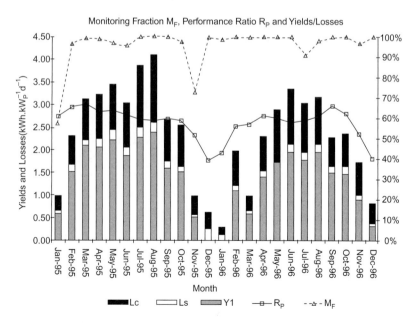

Figure 4 Typical bar graph for reporting. © *TSU. Reprinted with permission.*

5.2 System Evaluation

Whereas performance verification requires only simple measurements of input and output, there is no limit to the variables that may be of interest in a complex system. Since standard systems are well understood, there is little need to monitor except where new technology or systems are being utilised. Interim measurements throughout the system are required to evaluate the component efficiencies and to understand their characteristics.

Many PV inverter manufacturers offer hardware and software to create a monitoring system to display the functioning of a system. These have the advantage of being (usually) well tested and robust. However, they are not easy to modify, and so cannot often meet a very focussed specification such as is required for some national programs.

5.3 Data Gathering, Transmission, and Storage

Electrical measurements of DC and AC power (including phase angle and multiphase measurements) may be required between each component. The standard of care for signal isolation and conditioning is particularly important where high DC voltages and currents are concerned. Keeping these signals free of noise adjacent to high frequency switching devices such as inverters is challenging. The issue of fusing of these signal connections needs careful analysis, as there are a large number of fault conditions possible.

5.4 Data Analysis and Reporting

There is usually a reason for system evaluation, and this will often refer to a particular standard, such as IEC 61724 [6]. This will specify exactly the form and presentation of the output. The general variables used to compare the operation of a PV system are normalised yield values. The array energy output divided by the array nominal power gives the array yield Y_a. Similarly, the final AC energy output of the system gives the final yield Y_f. When this value is divided by the reference yield Y_r, defined as the in-plane irradiation divided by the reference irradiation. The value so obtained is the simplest way to compare the 'quality' of the PV system installed: the performance ratio (PR). PR is normally in the range 50−70%. A detailed discussion of the yield values and other performance indicators can be found in Chapter IIe-3.

Energy values produced at various stages of the plant are useful for verification of particular components, as are the efficiencies at each stage but a detailed knowledge of the system is needed to make these comparisons meaningful, so they are less useful for public information. Similarly, capture losses and system losses are only relevant to detailed analysis of the system [7].

Monthly values of performance ratio, array yield, etc. have become the normal way of disseminating the performance of a PV system [8], and a comparison with existing systems is made easier if this method is continued. Such bar graphs can also be embellished with subcategories of capture losses, system losses, etc. (see Figure 4).

REFERENCES

[1] G. Blaesser, D. Munroe, Guideline for the assessment of photovoltaic plants A: Photovoltaic system monitoring, Report EUR 163 38EN. JRC Ispra, 1995.
[2] Testing, commissioning and monitoring guide for photovoltaic power systems in buildings, ETSU Report S/P2/290, Department of Trade and Industry, UK, 1998.
[3] IEC standard 61194, Characteristic parameters of stand alone photovoltaic systems, 1992.
[4] IEC standard 60904−2, Photovoltaic devices, Part 2: Requirements for reference solar cells, 1988.
[5] IEC standard 60904−6, Part 6: Requirements for reference solar modules, 1994.
[6] IEC standard 61724, Photovoltaic devices, monitoring requirements, 1998.
[7] IEC standard 61829, Crystalline silicon photovoltaic array. On-site measurements of I−V characteristics, 1995.
[8] G. Blaesser, D. Munroe, Guidelines for assessment of photovoltaic plants B: analysis and presentation of monitoring data, Report EUR 163 39 EN, JRC Ispra, 1995.

PART IV

Environment and Health

CHAPTER IV-1

Overview of Potential Hazards

Vasilis M. Fthenakis

National PV EHS Assistance Center, Department of Environmental Sciences, Brookhaven National Laboratory, Upton, New York, USA

Contents

1. Introduction — 1083
2. Overview of Hazards in PV Manufacture — 1084
3. Crystalline Silicon (x-Si) Solar Cells — 1084
   - 3.1 Occupational Health Issues — 1084
   - 3.2 Public Health and Environmental Issues — 1085
4. Amorphous Silicon (a-Si) Solar Cells — 1088
   - 4.1 Occupational Safety Issues — 1088
   - 4.2 Public Health and Environmental Issues — 1089
5. Cadmium Telluride (CdTe) Solar Cells — 1089
   - 5.1 Occupational Health Issues — 1089
   - 5.2 Public Health and Environmental Issues — 1090
6. Copper Indium Diselenide (CIS) Solar Cells — 1090
   - 6.1 Occupational Health and Safety — 1090
   - 6.2 Public Health and Environmental Issues — 1091
7. Gallium Arsenide (GaAs) High-Efficiency Solar Cells — 1092
   - 7.1 Occupational Health and Safety — 1092
8. Operation of PV Modules — 1093
9. Photovoltaic Module Decommissioning — 1094
10. Conclusion — 1095
References — 1095

1. INTRODUCTION

Photovoltaic (PV) technologies have distinct environmental advantages for generating electricity over conventional technologies. The operation of photovoltaic systems does not produce any noise, toxic-gas emissions, or greenhouse gases. Photovoltaic electricity generation, regardless of which technology is used, is a zero-emissions process.

Practical Handbook of Photovoltaics.
© 2012 Elsevier Ltd. All rights reserved.

However, as with any energy source or product, there are environmental, health, and safety (EHS) hazards associated with the manufacture of solar cells. The PV industry uses toxic and flammable substances, although in smaller amounts than many other industries, and use of hazardous chemicals can involve occupational and environmental hazards. Addressing EHS concerns is the focus of numerous studies of the National Photovoltaic EHS Assistance Center at Brookhaven National Laboratory, which operates under the auspices of the US Department of Energy (DOE). More than 150 articles highlighting these studies are posted in the Center's website (www.pv.bnl.gov). This work has been done in cooperation with the US DOE PV Program and the US PV industry, which takes EHS issues very seriously and reacts proactively to concerns. Below is a summary of EHS issues pertaining to the manufacture of crystalline silicon (x-Si), amorphous silicon (a-Si), copper indium diselenide (CIS), copper indium gallium diselenide (CGS), gallium arsenide (GaAs), and cadmium telluride (CdTe), which are currently commercially available.

2. OVERVIEW OF HAZARDS IN PV MANUFACTURE

In manufacturing photovoltaic cells, health may be adversely affected by different classes of chemical and physical hazards. In this chapter, discussion focuses on chemical hazards related to the materials' toxicity, corrosivity, flammability, and explosiveness. These hazards differ for different thin-film technologies and deposition processes. The main hazards associated with specific technologies are shown in Table 1. A listing of hazardous materials used in manufacturing is shown in Table 2.

3. CRYSTALLINE SILICON (x-SI) SOLAR CELLS

3.1 Occupational Health Issues

In the manufacture of wafer-based crystalline silicon solar cells, occupational health issues are related to potential chemical burns and the

Table 1 Major hazards in PV manufacturing

Module type	Types of potential hazards
x-Si	HF acid burns
	SiH_4 fires/explosions
	Pb solder (now being phased out)/module disposal
a-Si	SiH_4 fires/explosions
CdTe	Cd toxicity, carcinogenicity, module disposal
CIS, CGS	H_2Se toxicity, module disposal
GaAs	AsH_3 toxicity, As carcinogenicity, H_2 flammability, module disposal

inhalation of fumes from hydrofluoric acid (HF), nitric acid (e.g., HNO_3) and alkalis (e.g., NaOH) used for wafer cleaning, removing dopant oxides, and reactor cleaning. Dopant gases and vapors (e.g., $POCl_3$, B_2H_3), also are hazardous if inhaled. $POCl_3$ is a liquid, but in a deposition chamber it can generate toxic P_2O_5 and Cl_2 gaseous effluents. Inhalation hazards are controlled with properly designed ventilation systems in the process stations. Other occupational hazards are related to the flammability of silane (SiH_4) and its byproducts used in silicon nitride deposition; these hazards are discussed in the a-Si section below.

3.2 Public Health and Environmental Issues

No public health issues were identified with this technology. The environmental issues are related to the generation of liquid and solid wastes during wafer slicing, cleaning, and etching, and during processing and assembling of solar cells. The x-Si PV industry has embarked upon programs of waste minimization and examines environmentally friendlier alternatives for solders, slurries and solvents. Successful efforts were reported in laboratory and manufacturing scales in reducing the caustic waste generated by etching. Other efforts for waste minimization include recycling stainless-steel cutting wires, recovering the SiC in the slurry, and in-house neutralization of acid and alkali solutions. Finally, the content of Pb in solder in many of today's modules creates concerns about the disposal of modules at the end of their useful life. One x-Si manufacturer has developed and is using Pb-free solders, and has offered the technology know-how to others who are considering such a change [1].

Table 2 Some hazardous materials used in current PV manufacturing

Material	Source	TLV-TWA[a] (ppm)	STEL[b] (ppm)	IDLH[c] (ppm)	ERPG2[d] (ppm)	Critical effects
Arsine	GaAs CVD[e]	0.05	—	3	0.5	Blood, kidney
Arsenic compounds	GaAs	0.01 mg/m^3	—	—	—	Cancer, lung
Cadmium compounds	CdTe and CdS deposition; CdCl$_2$ treatment	0.01 mg/m^3 (dust); 0.002 mg/m^3 (fumes)	—	—	N/A	Cancer, kidney
Carbon tetrachloride	Etchant	5	10	—	100	Cancer, liver, greenhouse gas
Chlorosilanes	a-Si and x-Si deposition	5	—	800	—	Irritant
Copper	CIS deposition	1 mg/m^3 (dust); 0.2 mg/m^3 (fumes)	—	100 mg/m^3	—	
Diborane	a-Si dopant	0.1	—	40	1	CNS[f], pulmonary
Germane	a-Si dopant	0.2	—	—	—	Blood, CNS[f], kidney
Hydrogen	a-Si deposition	—	—	—	—	Fire hazard
Hydrogen fluoride	Etchant	—	C[g] 3	30	20	Irritant, burns, bone, teeth
Hydrogen selenide	CIS sputtering	0.05	—	1	—	Irritant, GI[h], flammable
Hydrogen sulfide	CIS sputtering	10	15	100	30	Irritant. CNS[f]. flammable
Indium compounds	CIS deposition	0.1 mg/m^3	—	—	—	Pulmonary, bone, CI[h]
Lead	Soldering	0.05 mg/m^3	—	—	—	CNS[f], GI[h], blood, kidney, reproductive
Nitric acid	Wafer cleaning	2	4	25	—	Irritant, corrosive

Chemical	Use					Hazards
Phosphine	a-Si dopant	0.3		1	0.5	Irritant, CNS[f], GI[h], flammable
Phosphorous oxychloride	x-Si dopant	0.1				Irritant, kidney
Selenium compounds	CIS deposition	0.2 mg/m^3		1 mg/m^3		Irritant
Sodium hydroxide	Wafer cleaning		C[g]	2 mg/m^3	10 mg/m^3	5 mg/m^3 Irritant
Silane	a-Si deposition	5		—		Irritant, fire and explosion hazard
Silicon tetrafluoride	a-Si deposition	—				
Tellurium compounds	CIS deposition	0.1 mg/m^3				CNS[f], cyanosis, liver

[a] TLV-TWA: Threshold Limit Value, Time Weighted Average is defined by the American Conference of Governmental Industrial Hygienists (ACGIH) as the time-weighted average threshold concentration above which workers must not be exposed during work-shifts (8 h/day, 40 h/week).
[b] STEL: Threshold Limit Value, Short Term Exposure Level is defined by ACGIH as the maximum concentration to which workers can be exposed for a period up to 15 minutes, provided not more than four excursions per day are permitted with at least 60 minutes between exposure periods and provided that daily PEL is not also exceeded.
[c] IDLH: Immediately Dangerous to Life or Health Concentration is defined by the National Institute for Occupational Safety and Health (NIOSH) as the maximum concentration from which one could escape within 30 minutes without any escape-impairing symptoms or any irreversible health effects.
[d] ERPG-2: Emergency Response Planning Guideline-2 is defined by the American Industrial Hygiene Association (AIHA) as the concentration below which nearly al people could be exposed for up to one hour without irreversible or other serious health effects or symptoms that would impair their ability to take protective action.
[e] CVD: chemical vapor deposition.
[f] CNS: central nervous system.
[g] C: Threshold Limit Value-Ceiling is defined by ACGIH as the concentration that should not be exceeded during any part of the working exposure.
[h] GI: gastrointestinal.
—: not available.

4. AMORPHOUS SILICON (a-SI) SOLAR CELLS

Amorphous silicon, cadmium telluride, copper indium selenide and gallium arsenide are thin-film technologies that use about 1/100 of the photovoltaic material used on x-Si.

4.1 Occupational Safety Issues

The main safety hazard of this technology is the use of SiH_4 gas, which is extremely pyrophoric. The lower limit for its spontaneous ignition in air ranges from 2% to 3%, depending on the carrier gas. If mixing is incomplete, a pyrophoric concentration may exist locally, even if the concentration of SiH_4 in the carrier gas is less than 2%. At silane concentrations equal to or greater than 4.5%, the mixtures were found to be metastable and ignited after a certain delay. In an accident, this event could be extremely destructive as protection provided by venting would be ineffective. Silane safety is discussed in detail elsewhere [2–5]. In addition to SiH_4, hydrogen used in a-Si manufacturing, also is flammable and explosive. Most PV manufacturers use sophisticated gas-handling systems with sufficient safety features to minimize the risks of fire and explosion. Some facilities store silane and hydrogen in bulk from tube trailers to avoid frequently changing gas cylinders. A bulk ISO (International Standards Organization) module typically contains eight cylindrical tubes that are mounted onto a trailer suitable for over the road and ocean transport. These modules carry up to 3000 kg of silane. Another option is a single, 450 l cylinder, mounted on a skid, which contains up to 150 kg of silane (mini-bulk). These storage systems are equipped with isolation and flow restricting valves.

Bulk storage decreases the probability of an accident, since trailer changes are infrequent, well-scheduled special events that are treated in a precise well-controlled manner, under the attention of the plant's management, safety officials, the gas supplier, and local fire-department officials. On the other hand, if an accident occurs, the consequences can be much greater than one involving gas cylinders. Currently, silane is used mainly in glow discharge deposition at very low utilization rates (e.g., 10%). To the extent that the material utilization rate increases in the future, the potential worst consequences of an accident will be reduced.

Toxic doping gases (e.g., AsH$_3$, PH$_3$, GeH$_4$) are used in quantities too small to pose any significant hazards to public health or the environment. However, leakage of these gases can cause significant occupational risks, and management must show continuous vigilance to safeguard personnel. Applicable prevention options are discussed elsewhere [5]; many of these are already implemented by the US industry.

4.2 Public Health and Environmental Issues

Silane used in bulk quantities in a-Si facilities may pose hazards to the surrounding community if adequate separation zones do not exist. In the USA, the Compressed Gas Association (CGA) Guidelines specify minimum distances to places of public assembly that range from 80 ft to 450 ft depending on the quantity and pressure of silane in containers in use [6]. The corresponding minimum distances to the plant property lines are 50–300 ft. Prescribed separation distances are considered sufficient to protect the public under worst-condition accidents.

No environmental issues have been identified with this technology.

5. CADMIUM TELLURIDE (CdTe) SOLAR CELLS
5.1 Occupational Health Issues

In CdTe manufacturing, the main concerns are associated with the toxicity of the feedstock materials (e.g., CdTe, CdS, CdCl$_2$). The occupational health hazards presented by Cd and Te compounds in various processing steps vary as a function of the compound specific toxicity, its physical state, and the mode of exposure. No clinical data are available on human health effects associated with exposure to CdTe. Limited animal data comparing the acute toxicity of CdTe, CIS, and CGS, showed that from the three compounds, CdTe has the highest toxicity and CGS the lowest [7]. No comparisons with the parent Cd and Te compounds have been made. Cadmium, one of CdTe precursors, is a highly hazardous material. The acute health effects from inhalation of Cd include pneumonitis, pulmonary edema, and death. However, CdTe is insoluble to water, and as such, it may be less toxic than Cd. This issue needs further investigation.

In production facilities, workers may be exposed to Cd compounds through the air they breathe, as well as by ingestion from hand-to-mouth contact. Inhalation is probably the most important pathway, because of the larger potential for exposure, and higher absorption efficiency of Cd compounds through the lung than through the gastrointestinal tract. The physical state in which the Cd compound is used and/or released to the environment is another determinant of risk. Processes in which Cd compounds are used or produced in the form of fine fumes or particles present larger hazards to health. Similarly, those involving volatile or soluble Cd compounds (e.g., $CdCl_2$) also must be more closely scrutinized. Hazards to workers may arise from feedstock preparation, fume and vapor leaks, etching of excess materials from panels, maintenance operations (e.g., scraping and cleaning), and during waste handling. Caution must be exercised when working with this material, and several layers of control must be implemented to prevent exposure of the employees. In general, the hierarchy of controls includes engineering controls, personal protective equipment, and work practices. Area and personal monitoring would provide information on the type and extent of employees' exposure, assist in identifying potential sources of exposure, and gather data on the effectiveness of the controls. The US industry is vigilant in preventing health risks, and has established proactive programs in industrial hygiene and environmental control. Workers' exposure to cadmium in PV manufacturing facilities is controlled by rigorous industrial hygiene practices and is continuously monitored by medical tests, thus preventing health risks [8].

5.2 Public Health and Environmental Issues

No public health issues have been identified with this technology. Environmental issues are related to the disposal of manufacturing waste and end-of-life modules; these are discussed in Section 9.

6. COPPER INDIUM DISELENIDE (CIS) SOLAR CELLS

6.1 Occupational Health and Safety

The main processes for forming copper indium diselenide solar cells are co-evaporation of Cu, In, and Se and selenization of Cu and In layers in

H$_2$Se atmosphere. The toxicity of Cu, In, and Se is considered mild. Little information exists on the toxicity of CIS. Animal studies have shown that CIS has mild to moderate respiratory track toxicity; in comparing CIS, CGS, and CdTe, CIS was found to be less toxic than CdTe and somewhat more toxic than CGS.

The selenium TLV-TWA of 0.2 mg/m^3 as selenium were set to prevent systemic toxicity, and to minimize the potential of ocular and upper respiratory tract irritation. Interestingly, selenium is an essential element in the human diet and daily intakes of 500−860 μg of selenium are tolerated for long periods [9].

Although elemental selenium has only a mild toxicity associated with it, hydrogen selenide is highly toxic. It has an Immediately Dangerous to Life and Health (IDLH) concentration of only 1 ppm. Hydrogen selenide resembles arsine physiologically; however, its vapor pressure is lower than that of arsine, and it is oxidized to the less toxic selenium on the mucous membranes of the respiratory system. Hydrogen selenide has a TLV-TWA of 0.05 ppm to prevent irritation, and prevent the onset of chronic hydrogen selenide-related disease. To prevent hazards from H$_2$Se, the deposition system should be enclosed under negative pressure and be exhausted through an emergency control scrubber. The same applies to the gas cabinets containing H$_2$Se cylinders in use.

The options for substitution, isolation, work practices, and personnel monitoring discussed for CdTe are applicable to CIS manufacturing as well. In addition, the presence of hydrogen selenide in some CIS fabrication processes requires engineering and administrative controls to safeguard workers and the public against exposure to this highly toxic gas.

6.2 Public Health and Environmental Issues

Potential public health issues are related to the use of hydrogen selenide in facilities that use hydrogen selenide as a major feedstock material. Associated hazards can be minimized by using safer alternatives, limiting inventories, using flow restricting valves and other safety options discussed in detail elsewhere [10]. Emissions of hydrogen selenide from process tools are controlled with either wet or dry scrubbing. Also, scrubbers that can control accidental releases of this gas are in place in some facilities. Environmental issues are related to the disposal of manufacturing waste and end-of-life modules; these are discussed in Section 9.

7. GALLIUM ARSENIDE (GaAs) HIGH-EFFICIENCY SOLAR CELLS

7.1 Occupational Health and Safety

MOCVD is today's most common process for fabricating III−V PV cells; it employs the highly toxic hydride gases, arsine and phosphine, as feedstocks. Similarly to silane and hydrogen selenide handling, the safe use of these hydrides requires several layers of engineering and administrative controls to safeguard workers and the public against accidental exposure. Such requirements pose financial demands and risks that could create difficulties in scaling up the technology to multimegawatt levels. One part of the problem is that today's use of the hydrides in MOCVD is highly ineffective. Only about 2−10% are deposited on the PV panels, as a 10−50 times excess of V to III (As to Ga) is required. Metal-organic compounds are used more effectively, with their material utilization ranging from 20 to 50%. In scaling up to 10-MW/yr production using MOCVD, the current designs of flat-plate III−V modules will require approximately 23 metric tons of AsH_3, 0.7 tons of PH_3, 7 tons of metal organics, and 1500 tons of hydrogen. (These estimates are based on generic data applicable to the largest current MOCVD reactors [e.g., EMCORE Enterprise E400 and Aixtron AIX3000], and carry some unquantified uncertainty. Production for 24 hours a day by one of these reactors could provide about 100 kWp/year, so 100 such reactors will be needed for the 10 MW production-basis we are considering herein. Therefore, larger reactors will be needed at this scale, thereby introducing more uncertainty in our estimates.) These quantities can be effectively delivered only with tube trailers, each carrying 2−3 tons of gases. The potential consequences of a worst-case failure in one of these tube trailers could be catastrophic. On a positive note, however, it is more likely that terrestrial systems will be concentrators, not flat plates, because the former would be less expensive to manufacture. PV cells have been shown to be capable of operation under concentrated sunlight, between 100 and 2000 times the one-sun level (see Parts IId and IIId). A possible practical strength is 500× concentrators; the material requirements for producing such concentrators are 600 times less than those needed for flat plates.

The best way to minimize both the risks associated with certain chemicals and the costs of managing risk is to assess all alternatives during the first steps of developing the technology and designing the facility. These hydrides may be replaced in the future by the use of tertiary butyl arsine

(TBAs) and tertiary butyl phosphine (TBP) [11]; it appears that there are no intrinsic technical barriers to growing PV-quality GaAs with TBAs and GaAsP or GaInP$_2$ with TBP. Until substitutes are tested and implemented, however, it might be prudent to use arsine and phosphine from reduced-pressure containers, which are commercially available. Research efforts are being made in Europe [12] to replace hydrogen by inert nitrogen. Apparently, there is no inherent reason to prohibit such a substitution. However, since molecular hydrogen decomposes to some extent, and atoms participate in the gas-phase chemistry, the PV research community is challenged with learning how to optimize III—V growth conditions with nitrogen.

8. OPERATION OF PV MODULES

The operation of PV systems does not produce any emissions. Although tiny amounts of semiconductor materials are imbedded in the module (e.g., 7 g/m^2 for thin-film technologies), toxic compounds cannot cause any adverse health effects unless they enter the human body in harmful doses. The only pathways by which people might be exposed to PV compounds from a finished module are by accidentally ingesting flakes or dust particles or by inhaling dust and fumes. The photovoltaic material layers are stable and solid, and are encapsulated between thick layers of glass or plastic. Unless the module is ground to a fine dust, dust particles cannot be generated. All the photovoltaic materials examined herein have a zero vapor pressure at ambient conditions. Therefore, it is impossible for any vapors or dust to be generated during normal use of PV modules.

The potential exists for exposure to toxic vapors via inhalation if the modules are consumed in residential fires and people breath the smoke from the fire [13]. However, common US residential fires are not likely to vaporize CdTe and GaAs layers; flame temperatures in roof fires are in the range 800—900°C range. The melting point of CdTe is 1041°C, and evaporation occurs at 1050°C in open air and at about 900°C under nonoxidizing conditions. Sublimation occurs at lower temperatures, but the vapor pressure of CdTe at 800°C is only 2.5 torr (0.003 atm). The melting point of CdS is 1750°C and of GaAs is 1238°C. CIS starts evaporating at 600°C, and a 20% weight loss was measured at 1000°C [14].

The potential for significant photovoltaic material emissions may exist only in large externally fed industrial fires. In any case, the fire itself probably would pose a much greater hazard than any potential emissions of photovoltaic materials.

9. PHOTOVOLTAIC MODULE DECOMMISSIONING

PV modules will have to be decommissioned at the end of their useful life, 20–30 years after their initial installation. In decommissioning these devices, the principal concern will be associated with the presence of Cd in CdTe and CdS solar films and the presence of Pb in x-Si modules if they contain Pb-based solder. If these modules end in a municipal waste incinerator (MWI), the heavy metals will gasify and a fraction of those will be released in the atmosphere. If the MWI is equipped with electrostatic precipitator (ESP) this fraction can be as small as 0.5% with the balance of the heavy metals remaining in the ash. The ash itself will have to be disposed of in controlled landfills.

If the modules end in municipal landfills, then the potential for the heavy metals to leach out in the soil exist. The leachability of metals in landfills, is currently characterized by two elution tests: the US Environmental Protection Agency (EPA) Toxicity Characterization Leachate Profile (TCLP), and the German DEV S4 (Deutsches Einheitsverfahren). In these tests, small pieces (<1 cm^2) of broken modules are suspended and rotated in an elutent for 24 hours. The metals present in the elutent are then measured and compared with limits prescribed by each testing protocol. If the metal concentration exceeds the limits, the modules are demonstrating the metal's leachability and may need to be recycled or disposed of in a hazardous waste landfill: if the metals are not leaching in excessive quantities, the modules can be disposed of in a commercial landfill. Some early CdTe modules have failed the TCLP and the DEV tests [15]. However, the Apollo CdTe modules produced by BP Solar were reportedly passing the TCLP. In exploratory tests with a small number of commercial x-Si modules, some modules failed the TCLP limit for Pb by about 30% [16]. Exploratory tests on CIS modules showed that they pass the test for Se. No tests are reported for GaAs modules. The a-Si modules contain very little hazardous material and easily pass the test. It should be noted that the TCLP test is conservative, as it requires breakage of the whole module to very small pieces, whereas the photovoltaic layer will often be sandwiched between two layers of glass and reasonably isolated from the environment.

The ultimate solution to the PV waste and end-of-life management is recycling of useful materials. Recent studies showed that recycling, based on current collection/recycling infrastructure and on emerging recycling technologies, is technologically and economically feasible [17]. Reclaiming metals from used solar panels in large centralized applications can be done in metal smelting/refining facilities which use the glass as a fluxing agent and recover most of the metals by incorporating them in their product streams. In dispersed operations, small quantities and high transportation costs make this option relatively expensive. Research supported by the US DOE developed technologies for hydro-metallurgical separation that may be used in both small-scale (in-house) and large-scale recycling. These options are being investigated by the photovoltaic industry as part of their proactive long-term environmental strategy to preserve the environmental friendliness of solar cells.

10. CONCLUSION

The manufacture of photovoltaic modules uses some hazardous materials which can present health and safety hazards, if adequate precautions are not taken. Routine conditions in manufacturing facilities should not pose any threats to health and the environment. Hazardous materials could adversely affect occupational health and, in some instances, public health during accidents. Such hazards arise primarily from the toxicity and explosiveness of specific gases. Accidental releases of hazardous gases and vapors can be prevented through choosing safer technologies, processes, and materials; by better use of materials; and by employee training and safety procedures. As the PV industry vigilantly and systematically approaches these issues and mitigation strategies, the risk to the industry, the workers, and the public will be minimized.

REFERENCES

[1] V. Fthenakis, R. Gonsiorawski, Lead-free solder technology from ASE Americas, Workshop Report BNL-67536, 19 October 1999, Brookhaven National Laboratory, Upton, NY 11973, 1999.
[2] V.M. Fthenakis, P.D. Moskowitz, An assessment of silane hazards, Solid State Technol. January (1990) 81–85.

[3] L. Britton, Improve your handling of silane, Semicond. Int. April, 1991.
[4] F. Tamanini, J.L. Chaffee, R.L. Jambar, Reactivity and ignition characteristics of silane/air mixtures, Process Saf. Prog. 17(4) (1998) 243–258.
[5] P.D. Moskowitz, V.M. Fthenakis, A checklist of safe practices for the storage, distribution, use and disposal of toxic and hazardous gases in photovoltaic cell manufacturing, Sol. Cells 31 (1991) 513–525.
[6] CGA P-31–2000, Safe storage and handling of silane and silane mixtures, first edition, Compressed Gas Association, Inc., Arlington, VA, 2000.
[7] V. Fthenakis, S. Morris, P. Moskowitz, D. Morgan, Toxicity of cadmium telluride, copper indium diselenide, and copper gallium diselenide, Prog. Photovoltaics 7 (1999) 489–497.
[8] J. Bohland, K. Smigielski, First Solar's CdTe module manufacturing experience; environmental, health and safety results, Proceeding of the 28th IEEE Photovoltaic Specialists Conference, Anchorage, AK, USA, 2000.
[9] M. Piscator, The essentiality and toxicity of selenium, Proceedings of the 4th International Symposium on Uses of Selenium and Tellurium. in: S.C. Careapella Jr. (Ed.), 1989.
[10] V. Fthenakis, Multi-layer protection analysis for photovoltaic manufacturing, Process Saf. Prog. 20(2) (2001) 1–8.
[11] J. Komeno, Metalorganic vapor phase epitaxy using organic group V precursors, J. Cryst. Growth 145(1–4) (1994) 468–472.
[12] Juelich, MOVPE at Juelich Research Center. Website <http://www.fz-juelich.de/isg/movpe/emovpel.html/>, 2001.
[13] P.D. Moskowitz, V.M. Fthenakis, Toxic materials released from photovoltaic modules during fires: Health Risks, Sol. Cells 29 (1990) 63–71.
[14] H. Steinberger, HSE for CdTe and CIS thin film module operation, IEA expert workshop Environmental aspects of PV power systems, Report No. 97072, E. Niewlaar, E. Alsema (Ed.), Utrecht University, The Netherlands, 1997.
[15] V. Fthenakis, C. Eberspacher, P. Moskowitz, Recycling strategies to enhance the commercial viability of photovoltaics, Prog. Photovoltaics 4 (1996) 447–456.
[16] C. Eberspacher, Disposal and recycling of end-of-life CdTe and Si PV modules, Report prepared for Brookhaven National Laboratory, February 1998.
[17] V.M. Fthenakis, End-of-life management and recycling of PV modules, Energy Policy 28 (2000) 1051–1058.

EDITOR'S NOTE

There should be an awareness of electrical shock hazards, particularly with large PV arrays and when systems for grid-connection are "islanded", i.e., disconnected from the grid. Obviously whenever there is incident light, a PV array presents an electrical potential difference. This applies particularly during installation and dismounting PV modules and arrays, where workers on roofs and structures can be destabilized by otherwise minor shocks. See chapter 11C-3.

CHAPTER IV-2

Energy Payback Time and CO_2 Emissions of PV Systems

Erik Alsema

Department of Science, Technology and Society, Copernicus Institute for Sustainable Development and Innovation, Utrecht University, The Netherlands

Contents

1. Introduction — 1097
2. Energy Analysis Methodology — 1099
3. Energy Requirements of PV Systems — 1100
   3.1 General — 1100
   3.2 Crystalline Silicon Modules — 1100
   3.3 Thin-Film Modules — 1102
   3.4 Other PV System Components — 1104
4. Energy Balance of PV Systems — 1105
5. Outlook for Future PV Systems — 1107
   5.1 Crystalline Silicon Modules — 1107
   5.2 Thin-Film Modules — 1109
   5.3 Other System Components — 1110
   5.4 Energy Payback Time of Future Systems — 1110
6. CO_2 Emissions — 1112
7. Conclusions — 1114
References — 1115

1. INTRODUCTION

Photovoltaic energy conversion is widely considered as one of the more promising renewable energy technologies. It has the potential to contribute significantly to a sustainable energy supplies and it may help to mitigate greenhouse gas emissions. For this reason there is a growing support from governments for photovoltaic R&D programmes and market introduction schemes. In order to fulfil these promises PV technology has to meet two requirements: (1) PV energy generation should have an

Practical Handbook of Photovoltaics.
© 2012 Elsevier Ltd. All rights reserved.

acceptable *cost:performance ratio* and (2) the *net energy yield* for PV systems should be (much) larger than zero. By a positive energy yield we mean that the energy *output* during the lifetime of the PV system must be larger than the energy *input* during the systems life cycle, i.e., for manufacturing of the components and for the installation, maintenance and decommissioning of the PV system. Of course evaluations of the CO_2 mitigation potential of PV technology should be based on expected *net* energy yields. In practice this is seldom done, leading to overly optimistic results for the CO_2 mitigation potential.

In our view, every new energy technology that is promoted as being 'renewable' or 'sustainable' should be subjected to an analysis of its energy balance in order to calculate the net energy yield. It is of great importance that such an energy analysis is not only based on data for present-generation systems but also considers expected improvements in production and energy system technology. Since energy consumption generally has significant environmental implications, the energy analysis may be considered as a first step towards a more comprehensive environmental life-cycle assessment (LCA) [1,2]. Furthermore energy analysis results provide a good indication of the CO_2 mitigation potential of the considered energy technology.

Our objective in this paper is to present estimates of the energy requirements for manufacturing of PV systems and to evaluate the energy balance for a few representative examples of PV system applications. We will also investigate the effects of future enhancements in PV production technology and PV system technology in order to assess the long-term prospects of PV technology as a candidate for a sustainable energy supply and for CO_2 mitigation. We will consider mainly grid-connected PV systems, because these systems have in our view the largest potential in terms of long-term, global energy supply and CO_2 mitigation. With respect to standalone PV systems the energy balance depends very much on the specific application. Therefore, we restrict our discussion to one important type of standalone system, namely 'solar home systems.'

We will begin with a brief discussion on energy analysis methodology and some general assumptions (Section 2). Then we discuss the energy requirements for the components of PV systems (Section 3). In Section 4 we present the energy balance for grid-connected PV systems, followed by an outlook for future PV systems in Section 5. In Section 6, the potential for CO_2 mitigation using PV systems will be assessed. We finish with our conclusions (Section 7).

2. ENERGY ANALYSIS METHODOLOGY

In an energy analysis, a comprehensive account is given of the energy inputs and outputs involved in products or services. The overall energy performance of such products and services is determined by accounting all energy flows in the life cycle (from resource extraction through manufacturing, product use until end-of-life decommissioning). In the case of solar cells, the gross energy requirement E_{in} is determined by adding together the energy input during resource winning, production, installation, operation, and decommissioning[1] of the solar cell panels and the other system components. The energy payback time (EPBT) can now be calculated by dividing the gross energy requirement E_{in} by the annual energy output E_{out}. The energy payback time indicates how long it takes before energy investments are compensated by energy yields. A more comprehensive discussion of the methodological aspects can found in [3] and [4].

Note the following two important points regarding the units we use.
- *Energy units*: Following the conventions within the disciplines of energy analysis and LCA we present all energy data in terms of the *equivalent primary energy*, so, for example, inputs of electricity are calculated back to the extraction of primary energy carriers. The conversion efficiency for the electricity supply system is an often used average value, assumed here to be 0.35. This same efficiency is also used to calculate the primary energy equivalent of the electricity produced by the PV system.
- *Module energy data per m^2*: We present the energy requirements of PV modules in megajoules per square meter module area. The reason is that the energy requirements for module production are generally area dependent and not rated PV peak power dependent. Only after we have established the area-related energy data, we will factor in the power output per square meter module area in order to obtain the module energy requirements per Wp. As a third step, system performance and irradiation assumptions can be taken into account to evaluate the energy balance on a system level. Also some of the balance of system (BOS) components (e.g., supports) will be assessed on m^2 basis.

---

[1] In our case, end-of-life energy flows, e.g., for module recycling, are not counted because of uncertainties and lack of data.

3. ENERGY REQUIREMENTS OF PV SYSTEMS

3.1 General

Over the past decade, a number of detailed studies on the energy requirements for PV modules or systems have been published [5–11]. We have reviewed and compared these and other studies and tried to establish on which data there is more or less consensus and how observed differences may be explained. Based on this review of available data we have established a best estimate of the energy requirement of multicrystalline silicon modules, thin-film modules and other system components [4,12,13]. Here we will present only the main results of these analyses; for a more detailed breakdown of the solar cell production process the reader is referred to the aforementioned publications. Also the reasons for the relatively high uncertainty ($\sim 40\%$) in the estimates can be found there. A recent paper gives an analysis of monocrystalline silicon and CIS modules based on actual measured data at the Siemens plant [14]. The data from this paper have not been included in our earlier review study, but seem to match reasonably with our estimates below.

3.2 Crystalline Silicon Modules

First we discuss crystalline silicon module technology, which is presently the dominant technology in the market. The production process for crystalline silicon modules is discussed in detail in Chapters Ib-1 and Ib-3. Silicon is produced from silica (SiO_2) which is mined as quartz sand. Reduction of silica in large arc furnaces yields *metallurgical-grade silicon*, which has to be purified further to *electronic-grade silicon* before it is suitable for manufacturing of electronic components such as integrated circuits or solar cells. Moreover, both the electronics industry and the PV industry use silicon in a (mono)crystalline form. Therefore, the silicon is melted and subsequently crystallised under carefully controlled conditions so that large blocks or *ingots* of monocrystalline material are obtained. The use by the PV industry of off-specification material which is rejected by the electronics industry, raises questions about a fair allocation of energy consumption between the two end-products [15]. The ingots are sawn into smaller blocks and then into thin *wafers*. Typical wafers are 0.3 mm thick and 100–200 cm^2 in area. After processing the wafers into solar cells they are laid out into a matrix of for example 9 × 4 cells, interconnected, and *encapsulated* between a glass front plate, an EVA lamination foil, and a back-cover foil.

Table 1 Breakdown of the energy requirements for a typical multicrystalline silicon PV module using present-day production technology (in MJ of primary energy per m^2 module area)

Process	Energy requirements (MJ_{prim}/m^2 module)
Silicon winning and purification	2200
Silicon wafer production	1000
Cell/module processing	500
Module encapsulation materials	200
Overhead operations and equipment manufacture	500
Total module without frame)	**4200**
Module frame (aluminium)	400
Total module (framed)	**4600**

In Table 1 we show the energy requirements for a typical module based on multicrystalline silicon. We can see that the silicon winning and purification process is the major energy consumer and responsible for about half of the modules energy requirement. The reason for this is that the purification is performed by a highly energy-intensive process and the purity criteria are laid down by the electronics industry, which consumes about 90% of the electronic-grade silicon. Although solar cell production would also be possible with more relaxed purity standards, dedicated purification of solar-grade silicon has up to now not been commercially feasible because of the relatively small demand for this type of silicon material [16].

The production of the silicon wafer from the electronic-grade silicon also has a significant contribution to the energy consumption, mainly because 60% of the material is lost during the formation of the multicrystalline silicon ingots and the sawing of these ingots into wafers [10]. Altogether the production of the silicon wafers requires about 3200 MJ/m^2 module area, which is about 60% of the total energy requirement for the module.

The energy requirements for cell and module processing are more modest (300 MJ/m^2) as are the energy requirements for the front-cover glass, the lamination foil, and the back-cover foil (200 MJ/m^2). A typical aluminium frame around the module may add an extra 400 MJ/m^2, but frame materials and dimensions can vary considerably. In fact, frameless modules are also available on the market.

Also, note the relatively high energy use for overhead operations (climatisation of production facility, lighting, compressed air, etc.) and for the manufacturing of the production equipment itself (together 500 MJ/m^2). The latter is due to relatively high capital costs of solar cell manufacturing plants [17], costs which also formed the basis for our energy input estimate (cf. [4]).

Based on a conversion efficiency of 13%, we have a power output of 130 Wp/m^2 from the module, so the energy requirements of present-day mc-Si modules can be evaluated as about 32 MJ/Wp without frame and 35 MJ/m^2 with an aluminium frame.

For *mono*crystalline silicon wafers the more elaborate crystallisation process increases the energy requirements, according to our estimate with an additional 1500 MJ/m^2 [12]. However, the recent energy analysis for the Siemens monocrystalline silicon modules [14] differs from our estimates on some points. On one hand, the silicon requirement is lower than we assumed; on the other hand processing energy in crystallisation *and* in cell and module manufacturing is considerably higher than we estimated. For one thing the overhead energy consumption in this US plant seems to be considerable.[2] Overall, the Siemens estimate for a mono-Si module with frame is 6900 MJ/m^2 (5600 kWhe/kWp), which is about 12% higher than our estimate.

3.3 Thin-Film Modules

Regarding thin-film modules, a number of contending cell types and production technologies exists and as yet it is not clear which one of these will be the leading technology in 10–20 years. The production technology for thin-film modules differs significantly from that for crystalline silicon modules. In general, thin-film modules are made by depositing a thin (0.5–10 μm) layer of semiconductor material on a substrate (usually a glass plate). The thin-film deposition can be performed with variety of different techniques, among which are chemical vapour deposition, evaporation, electrolytic deposition, and chemical bath deposition. Depending on the selected deposition technique, the material properties, the material utilisation rate, and the energy consumption for the deposition process will vary. Processes employing elevated temperatures and/or vacuum conditions will generally have a greater energy consumption per m^2

---

[2] Further comparison with, for example, European plants should show whether these data are representative for other plants, too.

Table 2 Breakdown of the energy requirement of an amorphous silicon thin film module using present-day production technology (glass-glass encapsulation; in MJ of primary energy)

Process	Energy requirements (MJ_{prim}/m^2 module)
Cell material	50
Module encapsulation material	350
Cell/module processing	400
Overhead operations and equipment manufacture	400
Total module (frameless)	**1200**
Module frame (aluminium)	400
Total module (framed)	**1600**

processed substrate area. Contact layers are also deposited with chemical vapour deposition (transparent contacts) and evaporation (back contacts). The module is finally encapsulated with a second glass plate or with a polymer film. The module may be left frameless, or it may be equipped with an aluminium or polymer frame.

Today, amorphous silicon (a-Si) is the leading technology in the market, and Table 2 shows the break-down of energy requirements for this type of module, again expressed as MJ_{prim} per m^2 module area [4].

Table 2 shows that the material for the actual solar cell requires relatively little energy. This is, of course, because of the small cell thickness in amorphous silicon and other thin-film technologies. For the rest, the energy requirements are divided about equally between module materials, processing, overhead operations and equipment manufacturing, and, finally, the aluminium frame. Again, frame materials and dimensions may differ greatly between manufacturers. Polyurethane frames for example require much less energy. In this case the module encapsulation was assumed to comprise *two* glass plates because this is most common in present-day thin-film technology. Switching to polymer back covers, like in crystalline silicon technology, may reduce the energy requirement by some 150 MJ/m^2. However, in the case of toxic solar cell materials such as CdTe or $CuInSe_2$ [18] this would be less desirable [19]. Assuming a conversion efficiency of 7% we arrive at an energy requirement for present amorphous silicon modules of 17 MJ/Wp without frame and 23 MJ/Wp including an aluminium frame.

For other thin-film technologies (e.g., CdTe, $CuInSe_2$, organic cell), the energy requirements for processing may be different, depending on the type of deposition processes that are used to lay down the cell materials. For CIS production at Siemens [14] the energy requirement for a module is about 2300 MJ/m^2 without frame and 2870 MJ/m^2 with frame (3070 kWhe/kWp).[3] Here again the overhead energy use seems to be considerable (about 600 MJ/m^2). Regarding CdTe modules, a Japanese study suggests a fairly high processing energy for a specific CdTe deposition process [20].

Furthermore, overhead operations for specific thin films may require more energy, for example, when more extensive waste treatment is required (cadmium recovery). Nonetheless, we expect that the energy requirement of most thin-film modules produced at commercial scale will fall within a range of 1000–2000 MJ/m^2, excluding the frame. Because of the smaller module sizes and the lower efficiency, the energy requirements for the frame will generally be higher than for crystalline silicon modules.

3.4 Other PV System Components

The requirements for the balance of system, that is, all PV system components apart from the modules, will depend largely on the desired application. In grid-connected PV systems, an inverter, cables, and some module support materials will be needed. In autonomous systems a battery for energy storage will be required. The energy requirements for the inverter and cabling are estimated at about 1.6 MJ/Wp, of which about 50% is for the electronic components. The energy for array supports can vary widely. In a recent study of Dutch roof installation systems we found values from 240 MJ/m^2, for a support structure based on lightweight, aluminium profiles on slanted roofs, up to 350 MJ/m^2 for a flat-roof system employing plastic support consoles [21].[4] An analysis by Frankl [6] of PV systems on rooftops and building facades in Italy found a total energy requirement around 700 MJ/m^2 for all BOS components (support, cabling, and inverter). In the same study, the 3.3 MWp ground-mounted system in Serre, Italy, was analysed too, and it was found to require significantly more materials and thus energy, namely, 1850 MJ/m^2.

For autonomous PV systems, the storage battery is a very important component, not only with respect to the costs but also in terms of energy

---

[3] Note that the data was scaled up from 'pre-pilot' scale to commercial scale production.

[4] Notice the difference in units: the energy requirements for converter and cabling are expressed per watt, while material and thus energy requirements for array supports are related to the module area.

requirements. In our LCA study on solar home systems [22], the primary energy requirements for the life cycle of a standard lead-acid starter battery of 100 Ah was estimated at 1100 MJ, *assuming an optimistic 90% recycling rate for scrap batteries.*[5] Unfortunately, the lifetime of starter batteries in solar home systems is quite short: we assumed three years. This means that over a 20-year system lifetime we need 6.7 battery sets, requiring 7300 MJ of primary energy per Ah battery capacity. Other BOS components like cables and charge controllers contribute relatively little to the energy requirement of a solar home system (<10%).

4. ENERGY BALANCE OF PV SYSTEMS

We now look at the energy output of a typical PV system and evaluate the energy payback time. First, we will consider two types of grid-connected PV systems, namely, a rooftop system and a large, ground-mounted system, and different two module technologies. We further assume a system performance ratio of 0.75, which is fairly representative for well-designed PV systems today [23]. As before the conversion efficiency of the conventional electricity supply system is set at 35%, so that 1 kWh of generated PV electricity will save 10.3 MJ of primary energy.

Finally we distinguish three irradiation levels (see Part I):
- high irradiation (2200 kWh/m^2/yr), as found in the southwestern USA and Sahara;
- medium irradiation (1700 kWh/m^2/yr), as found in large parts of the USA and southern Europe; and
- low irradiation (1100 kWh/m^2/yr), as found in central Europe (Germany).

Given these assumptions we evaluate EPBT as the ratio of the total energy input during the system life cycle and the yearly energy generation during system operation (see Figure 1).

Figure 1 shows that the EPBT for present-day grid-connected systems is 2—3 years in a sunny climate and increases to 4—6 years (or more) under less favourable conditions. Also note that the contribution from the BOS[6] and the module frames is significant, especially for the ground-mounted

---

[5] At 0% recycling the energy requirement of the battery increases to 1900 MJ.
[6] We assumed that the inverter life time was half that of the other components, so two inverters are needed over a system life cycle.

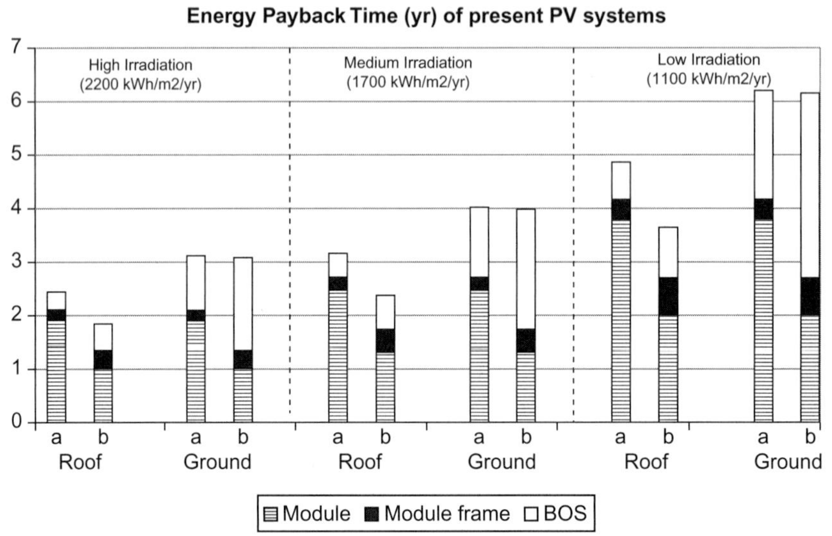

Figure 1 The energy payback time (in years) for present-day PV applications. Two different module technologies: multicrystalline silicon (a) and thin-film amorphous silicon (b) and two types of installation (ground-mounted and roof-integrated) are distinguished.

systems. Regarding thin film technology, we can see that, due to their lower efficiency, the energetic advantages of present a-Si modules are partially cancelled by the higher energy requirements for frames and supports. An energy payback time of 2–6 years may seem rather long, but in view of the expected life time of PV systems of 25–30 years there is still a significant net production of energy.

For solar home systems the concept of energy payback time is more ambiguous and also less interesting, because the SHS is not primarily installed for the energy it produces but rather for the service that it provides (e.g., lighting). The choice of the displaced service supply system (e.g., kerosene lamps) has a huge effect on the resulting EPBT value. Moreover, the quality of the two supply systems can be quite different, for example the light output of the kerosene lamps is much lower than for fluorescent lights and also their health effects should not be ignored. In [22] we have shown that the EPBT of an SHS in comparison with kerosene lamps may vary between 0.8 and 2 years, depending on SHS lay-out and battery recycling rates. However, in comparison with a diesel generator set, the SHS can have an EPBT of 12–20 years. Note that these results are mainly determined by the batteries with their high energy requirement and short lifetime.

A first conclusion from these energy balance considerations can be that grid-connected PV systems already have a significant potential for reducing fossil energy consumption, although it may take a few years of system operation before these savings can actually be cashed in. Autonomous PV systems may save on fossil energy if they replace very inefficient appliances, but they should be valued primarily for the quality of their services and less as an option to save on fossil energy.

Despite the fact that present-day, grid-connected PV systems have a clear energy-saving potential, it would very helpful if their energy balance could be improved further. Otherwise, PV technology may be considered as a less attractive option in comparison with other renewable or energy-saving technologies. Therefore, we investigate in the next section what prospects exist for a further reduction of the energy requirements of PV.

5. OUTLOOK FOR FUTURE PV SYSTEMS

In this section we investigate which improvements in photovoltaic technology may contribute to an improvement of the energy balance of PV systems. The general themes that will be discussed are material efficiency, energy efficiency, new processes, and enhanced module performance. First, we look at crystalline silicon technology. Reduction options specifically for Siemens monocrystalline silicon modules are also discussed in [24].

5.1 Crystalline Silicon Modules

We have seen that in crystalline silicon technology the silicon wafer forms the major contributor to the energy requirements of the module. This offers three avenues for reduction of energy inputs, as given below in decreasing order of probability.

1. Reduced wafer thickness. This possibility is already explored by the PV industry as it offers significant cost advantages. Wafer thickness reductions from the present-day 300–350 μm down to 200 μm or even 150 μm seem possible, so that the silicon requirements can decrease by 30–40% (wafer sawing losses will increase slightly).
2. Other wafer production methods. When blocks of crystalline silicon are sawed into thin wafers about 50% of the material is lost. Furthermore, some 40% is lost in the preceeding process of block

casting, so only 30% of the electronic-grade silicon ends up in the wafer. As there are only limited possibilities for reduction of the sawing losses, novel methods to produce wafers directly from molten silicon or from silicon powder have been investigated since the 1970s. The first approach is now in use in at least one commercial production line (Edge-Defined Film Growth). If such direct wafer production processes become widely employed they may lead to significant reductions in silicon requirements, possibly of the order of 40–60%.

3. Other sources for high-purity silicon feedstock. Table 2 shows us that the high-purity silicon used for solar cells is a very energy-intensive material ($\sim$1100 MJ/kg).[7] Previously, the PV industry relied fully on silicon rejected by the electronics industry because of insufficient quality (off-spec material). With the continued growth of the PV market the supply of off-spec silicon is insufficient so that other feedstock sources have been developed [16]. Because standard electronic-grade silicon is too expensive for PV applications, dedicated silicon purification routes for solar-grade silicon are now available. Not much is known about the energy requirements of such solar-grade silicon processes, but it seems likely that they are lower than that of the standard process produced to the specifications of the electronics industry.

Regarding the actual cell and module production we can expect additional energy reductions from

1. using frameless modules, and
2. scaling up of production plants, resulting in more efficient processing, lower overhead, and less equipment energy.

Summarising, we expect that future multicrystalline silicon production technology may achieve a reduction in energy requirements to around 2600 MJ/m^2, assuming innovations like a dedicated silicon feedstock production for PV applications, improved casting methods and reduced silicon requirements [6,8,11]. This kind of technology will probably become available in the next ten years. If we further assume a future module efficiency of 15% (cf. Table 3), we obtain energy requirements of 17 MJ/Wp for multicrystalline Si technology around 2010. Modules based on *monocrystalline* silicon modules will probably remain somewhat more energy-intensive, at 20 MJ/Wp [6,15,24].

---

[7] Under our assumptions 2 kg of Si feedstock is needed per m^2 module with the present-day technology.

Table 3 Assumptions for module efficiency development for different cell technologies (%)

	2000	2010	2020
Multicrystalline silicon	13	15	17
Thin film	7	10	15

When we look at the situation beyond 2010, then it seems difficult to achieve major energy reductions for wafer-based silicon technology. An energy-efficiency improvement in the production process of 1% per year, as is often found for established production technologies [25], seems therefore a reasonable assumption. Further improvements in the energy requirement per Wp will have to be achieved by improving module efficiency (while not increasing energy consumption). If we assume that in 2020 the efficiency of commercial mc-Si cells has been increased to 20% (the current record for small-area mc-Si cells), then module efficiency would be about 17% and thus the lowest conceivable energy requirement for Si wafer technology might be 13 MJ/Wp.

5.2 Thin-Film Modules

Because the encapsulation materials and the processing are the main contributors to the energy input, the prospects for future reduction of the energy requirement are less clearly identifiable as was the case with c-Si technology. A modest reduction, in the range 10–20%, may be expected in the production of glass and other encapsulation materials. It is doubtful whether displacement of the glass cover by a transparent polymer will lead to a lower energy requirement.

Other, existing trends which may contribute to a lower energy requirement are
1. frameless modules;
2. thinner cells with a reduced processing time and thus a reduction in the processing energy and in the energy for equipment manufacturing; and
3. an increase of production scale, leading to lower processing energy, lower equipment energy, and lower overhead energy.

By these improvements we expect the energy requirement of thin film modules to decrease by some 30%, to 900 MJ/m^2, in the next ten years [5,8,24]. If, concurrently, the module efficiency can be increased to 10%, the energy requirement on a Wp basis may reach the 9-MJ level for frameless modules.

If we try to make projections beyond 2010, we can note that further reductions in the energy requirement below 900 MJ/m^2 do not seem very probable. Like before, we assume a generic 1% per year energy-efficiency improvement in the production process. Only if completely novel module encapsulation techniques are developed, which require much less energy-intensive material, we may obtain a more significant improvement. Furthermore, new methods for cell deposition, which require less processing and less overhead operations, might help to reduce the energy input of thin-film modules. Of course, module efficiency increases will directly improve the energy input per Wp (if energy input per m^2 is constant). In this respect, significant variations may occur between different types of thin films. Moreover, significant efficiency improvements for thin film technology may be achievable. For instance, if we assume a 15% module efficiency for 2020, the energy requirement per Wp may come down to 5–6 MJ.

5.3 Other System Components

We expect that for most balance-of-system components, there is only a limited scope for reduction of the energy requirements. The values that we have given above for roof-integrated and ground-mounted systems are representative of state-of-the art, well-designed installations that cannot be improved much unless revolutionary new concepts are introduced. We will, therefore, only assume a general 1% yearly improvement in the energy efficiency of the production processes, similar to other common materials [25]. Table 4 summarises our expectations for BOS components and for module frames.

5.4 Energy Payback Time of Future Systems

With the estimates given above on the future energy requirements of modules and balance-of-system components and using the assumptions on module efficiency developments, we can determine the expected energy payback time for these future technologies. In Figure 2 these results are given using the same assumptions on system performance ratio and electricity supply efficiency as before, but for only one irradiation level (1700 kWh/m^2/yr). The data for the present situation are also depicted.

We can see that according to our energy input estimates and based on our efficiency assumptions, the EPBT values of future PV technology

Table 4 Energy requirements for balance-of-system components and module frame [6,21]

	Unit	2000	2010	2020
Module frame (Al)	MJ/m^2	400	0	0
Array support — central plant	MJ/m^2	1800	1700	1500
Array support — roof integrated	MJ/m^2	300	270	250
Inverter (2.5 kW)	MJ/W	1.6	1.4	1.2
Battery, lead—acid	MJ/Ah	11	10	9

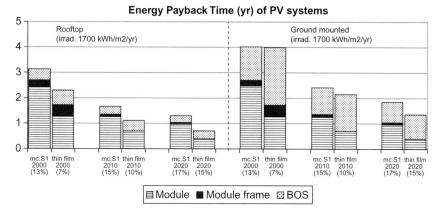

Figure 2 The energy payback time (in years) for two representative PV applications, both for present-day and for future (2010 and 2020) technology. The figures in parentheses denote the assumed module efficiencies for each option. Note that actual payback times will vary with irradiation and system performance.

could improve significantly to less than 1.5 years for rooftop systems and less than 2 years for ground-mounted systems. We also see that thin film modules may gain a significant advantage over c-Si technology when considering rooftop applications, *if the module efficiency can be increased to 10% or higher*. Furthermore, it is quite clear that rooftop installations have a much better potential for low energy payback times than ground-mounted installations.

We conclude that we see good possibilities for future developments in PV technology that result in decreasing energy requirements in component production and thus lead to an increasingly higher potential for fossil energy displacement.

6. CO_2 EMISSIONS

Firstly we note that for PV systems, the energy payback time is also a quite good indicator of the CO_2 mitigation potential because generally more than 90% of the greenhouse gas emissions during the PV system life cycle are caused by energy use [26]. Emissions *not* related to energy use are only found in steel and aluminium production (for the supports and frames) and in silica reduction (for *silicon* solar cells).

To obtain the CO_2 emissions due to the production of a PV system we have to multiply all energy and material inputs with their corresponding CO_2 emission factors. This requires a detailed *life cycle assessment* of greenhouse gas emissions from solar cell manufacturing (and other life cycle stages) and from the life cycle of balance-of-system materials. Such an analysis goes beyond the scope of this chapter. Here we will make a rough estimate of CO_2 emissions by (1) considering only energy-related CO_2 emissions during module production,[8] (2) evaluating all energy inputs in module production as electrical energy,[9] and (3) considering only the aluminium supports employed in roof-integrated systems for the balance-of-system components. We estimate the overall error of these approximations at a maximum of 20%.

We now arrive at a quite important point regarding the CO_2 emissions of PV systems. In comparison to most other energy technologies, the CO_2 emissions of PV occur almost entirely during system *manufacturing* and not during system *operation*. As a consequence, the CO_2 emissions of PV are determined very strongly by the *fuel mix* of the electricity supply system that is employed during the PV system production. In other words, the CO_2 emission of a PV system will depend on the CO_2 emission factor of the local utility system. In this analysis, we have assumed the present-day fuel mix of continental Western Europe (UCPTE region),[10] where about 50% of the electricity is produced by nuclear and hydroelectric plants, as well as 20% by coal, 10% by oil, and 10% by gas-fired plants.

---

[8] Non-energy CO_2 emissions are 0.1 kg/Wp (silica reduction) or less.
[9] Primary energy use for electricity is about 90% of total primary energy consumption for crystalline silicon modules. For thin film modules this share is about 70%, but the remaining 30% is used in glass production where, by chance, the CO_2 emission is similar as our assumption for electricity production (0.055 kg/MJ of used energy).
[10] The Union for the Coordination of Production and Transmission of Electricity. It includes: Austria, Belgium, Bosnia-Herzegovina, Croatia, France, Germany, Greece, Italy, Luxembourg, Macedonia, Montenegro, the Netherlands, Portugal, Serbia, Slovenia, Spain, and Switzerland.

For this utility system, the CO_2 emission factor is presently about 0.57 kg per kWh produced electricity (~ 0.055 kg/MJ_{prim} [27]). We have taken this factor as constant for the considered time period.

Two other parameters, which may not be overlooked when evaluating CO_2 emissions of PV, are, of course, the irradiation at the site of PV system installation (cf. Section 3) and the assumed system lifetime. As before, we have assumed a medium-high irradiation of 1700 kWh/m^2-yr and a system lifetime of 30 years. Note that because of the three parameters mentioned above, CO_2 emission estimates for PV can vary significantly between different studies.

In Figure 3 we have displayed the CO_2 emissions per kWh of supplied electricity for grid-connected rooftop PV systems. For comparison a number of conventional power generation technologies are also depicted. From a recent IEA study [28], we have, furthermore, included estimates for wind turbines and biomass gasification technology as well as their estimate for present-day multicrystalline-silicon PV technology in a rooftop application.[11] The results show that according to our estimates the CO_2 emissions for present PV technology are in the range 40–50 g/kWh which is considerably lower than for fossil-fuel plants but higher than for the two competing renewable energy technologies.[12] Our estimate for mc-Si technology seems somewhat lower than that from the IEA study, but if we consider that the latter probably has assumed a 30% higher CO_2 emission factor for the utility system, the difference becomes negligible (which is to be expected as the respective energy estimates are also comparable). With improving technology, PV-related CO_2 emissions may become significantly lower, around 20–30 g/kWh in the near future, or even 10–20 g/kWh in the longer term. Only in the last case PV systems come into same range as current wind, biomass and nuclear energy.

In our LCA study of solar home systems we found greenhouse gas emissions in the range 11–22 kg/Wp, or 0.6–1.2 kg/kWh. Installation of a solar lighting system could lead to GHG emission reduction around 480 kg/year if it replaces two kerosene lamps [22].

---

[11] Corrected for the different assumptions on irradiation and system lifetime. The assumed fuel mix or the CO_2 emission factor of the electricity supply system is not given exactly ('current German') but IEA study's emission factor is presumably some 30% higher than ours. We did *not* correct for this difference.

[12] For biomass energy it is even argued that the CO_2 emission factor is effectively *zero* because parts of the biomass crop are not harvested and stay in the ground (trunks, etc.), thus forming a CO_2 sink which compensates for the energy consumption in harvesting and transport.

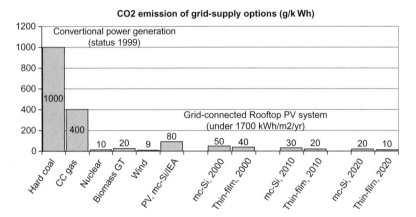

Figure 3 CO_2 emission for grid-connected rooftop PV systems now and expected in the future. For comparison we show emission data for a number of competing energy systems (coal, gas, nuclear data from [27]); wind and biomass energy estimates from a recent IEA study [28]). The PV technology estimate from the same IEA study is also shown. Note that actual CO_2 emissions for PV will vary with irradiation and system performance.

In conclusion, we can say that grid-connected PV systems can supply energy at a considerably lower CO_2 emission rate than current fossil energy technologies, but on the other hand PV is slightly at a disadvantage when compared to biomass and wind energy.

7. CONCLUSIONS

In this chapter we have reviewed the energy viability of photovoltaic energy technology. The question is whether photovoltaic (PV) systems can generate sufficient energy *output* in comparison with the energy *input* required during production of the system components. We evaluated the energy viability mainly in terms of the energy payback time (EPBT). For a grid-connected PV system under a medium-high irradiation level of 1700 kWh/m^2-yr, the EPBT is presently 2.5–3 years for rooftop systems and almost 4 years for large, ground-mounted systems. The share of module frames and BOS in this figure is quite significant. Under other climatic conditions, the EPBT values will be inversely proportional to the irradiation.

Furthermore, we have shown that there are good prospects. In the coming 10 years, the EPBT of rooftop systems will decrease to less than 2 years, if certain improvements, both in production technology and in module performance are realised. For modules based on crystalline silicon, one of the requirements would be a dedicated silicon purification process with substantially lower energy consumption, while for thin-film technology mainly an improved module efficiency would be necessary. Rooftop type systems will maintain their advantage over ground-mounted systems.

For standalone PV systems, like solar home systems, the EPBT is in general a less meaningful indicator. If a very inefficient device like kerosene lamps is replaced by the PV system, the EPBT can be short: 0.8–2.0 years. In this case, almost 500 kg of CO_2 emissions per household per year may be saved.

In case of replacement of other supply systems, like a diesel generator, the energy balance of the PV system is less favourable. This situation will probably not change drastically in the future because the storage battery is the dominating component in these analyses.

CO_2 emissions from rooftop PV systems were calculated assuming that the PV production facility is located in Western Europe. It was found that the specific CO_2 emission at present is about 50–60 g/kWh, which is considerably lower than the emission from existing fossil-fuel electricity plants (400–1000 g/kWh). On the other hand, these CO_2 emission values for PV systems are higher than the estimate for two major competing renewable technologies, wind and biomass (<20 g/kWh). Given the expected technology improvements, the specific CO_2 emission from PV could decrease to 20–30 g/kWh in the next ten years and perhaps even further after 2010.

Finally, we conclude from our analyses that, although the contribution of PV systems to CO_2 mitigation will probably be limited in the next decade [13], PV technology does certainly offer a large potential for CO_2 mitigation when looking beyond 2010.

REFERENCES

[1] E. Nieuwlaar, E.A. Alsema, Environmental Aspects of PV Power Systems—A Report on the IEA PVPS Task 1 Workshop, 25–27 June 1997, Utrecht, The Netherlands, Report 97072, Department of Science, Technology and Society, Utrecht University, 1997.

[2] E. Nieuwlaar, E.A. Alsema, PV power systems and the environment: results from an expert workshop, Prog. Photovolt. 6(2) (1998) 87–90.

[3] IFIAS, International Federation of Institutes for Advanced Study, Workshop Report no, 6: Energy Analysis. Guldsmedhyttan, Sweden, 1974.

[4] E.A. Alsema, Energy requirements of thin-film solar cell modules—a review, Renew. Sust. Energ. Rev. 2(4) (1998) 387–415.
[5] E.A. Alsema, Environmental aspects of solar cell modules, Summary Report, Report 96074, Department of Science, Technology and Society, Utrecht University, 1996.
[6] P. Frankl, A. Masini, M. Gamberale, D. Toccaceli, Simplified life-cycle analysis of PV systems in buildings—present situation and future trends, Prog. Photovolt. 6(2) (1998) 137–146.
[7] G. Hagedorn, E. Hellriegel, Umwelrelevante Masseneinträge bei der Herstellung verschiedener Solarzellentypen—Endbericht—Teil I: Konventionelle Verfahren. Forschungstelle für Energiewirtschaft, München, Germany, 1992.
[8] K. Kato, A. Murata, K. Sakuta, Energy payback time and life-cycle CO_2 emission of residential PV power system with silicon PV module, Prog. Photovolt. 6(2) (1998) 105–115.
[9] G.A Keoleian, G.M. Lewis, Application of life-cycle energy analysis to photovoltaic module design, Prog. Photovolt. 5(1997) 287–300.
[10] J. Nijs, R. Mertens, R. van Overstraeten, J. Szlufcik, D. Hukin, L. Frisson, Energy payback time of crystalline silicon solar modules, in: K.W. Boer (Ed.), Advances in Solar Energy, vol. 11, American Solar Energy Society, Boulder, CO, 1997, pp. 291–327.
[11] G.J.M. Phylipsen, E.A. Alsema, Environmental life-cycle assessment of multicrystalline silicon solar cell modules. Report 95057, Department of Science, Technology and Society, Utrecht University, Utrecht, 1995.
[12] E.A. Alsema, Energy pay-back time and CO_2 emissions of PV systems, Prog. Photovolt., 2000.
[13] E.A. Alsema, E. Nieuwlaar, Energy viability of photovoltaic systems, Energy Policy 28(14) (2000) 999–1010.
[14] K.E. Knapp, T.L. Jester, Empirical investigation of the energy payback time for photovoltaic modules, Sol. Energy 71(3) (2001) 165–172.
[15] E.A. Alsema, P. Frankl, K. Kato, Energy pay-back time of photovoltaic energy systems: present status and prospects, Proceeding of 2nd World Conference on Photovoltaic Solar Energy Conversion, Vienna, 1998.
[16] H.A. Aulich, F.-W. Schulze, Crystalline silicon feedstock for solar cells, Prog. Photovolt.: Res. Appl. 10(2) (2002) 141–147.
[17] T. Bruton, et al., Multimegawatt upscaling of silicon and thin film solar cell and module manufacturing (MUSIC FM), Final Report, BP Solar, Sunbury-on-Thames, 1996.
[18] V. Fthenakis, S. Morris, P. Moskowitz, D. Morgan, Toxicity of cadmium telluride, copper indium diselenide, and copper gallium diselenide, Prog. Photovolt.: Res. Appl. 7(6) (1999) 489–497.
[19] E.A. Alsema, M. Patterson, A. Baumann, R. Hill, Health, safety and environmental issues in thin-film manufacturing, Proceeding of the 14th European Photovoltaic Solar Energy Conference, Barcelona, 1997, pp. 1505–1508.
[20] K. Kato, T. Hibino, K. Komoto, A life-cycle analysis on thin-film CdS/CdTe PV modules, Proceeding of the 11th Photovoltaic Science and Engineering Conference, Hokkaido, Japan, 1999.
[21] E.A. Alsema, E. Nieuwlaar, Life cycle assessment of photovoltaic systems in roof-top installations—an LCA study focused at the contribution of balance-of-system components (in Dutch, with English summary), Report NWS-E-2002–04, Department of Science Technology and Society, Utrecht University, Utrecht, 2002.
[22] E.A. Alsema, Environmental life cycle assessment of solar home systems, Report NWS-E-2000–15, Department of Science, Technology and Society, Utrecht University, Utrecht. <http://www.chem.uu.nl/nws/www/publica/e2000-15.pdf/>, 2000.

[23] T. Erge, U. Hoffman, G. Heilscher, The German 1000-roofs-PV programme—a resume of the 5 years pioneer project for small grid-connected PV systems, Proceeding of the 2nd World Conference on Photovoltaic Solar Energy Conversion, Vienna, 1998, pp. 2648–2651.
[24] K.E. Knapp, T.L. Jester, G.B. Mihalik, Energy balances for photovoltaic modules: status and prospects, Proceeding of the 28th IEEE Photovoltaic Specialists Conference, Anchorage, 2000.
[25] J. de Beer, Potential for industrial energy efficiency improvement in the long term, Thesis, Utrecht, ISBN 90-393-1998-7, 1998.
[26] R. Dones, R. Frischknecht, Life cycle assessment of photovoltaic systems: results of swiss studies on energy chains, Prog. Photovolt. 6(2) (1998) 117–125.
[27] P. Suter, R. Frischknecht, Ökoinventare von Energiesystemen, 3. Auflage, ETHZ, Zürich, 1996.
[28] IEA, Benign Energy? The environmental implications of renewables, International Energy Agency, Paris, 1998.

APPENDICES

APPENDIX A

Constants, Physical Quantities, and Conversion Factors

Tom Markvart, Augustin McEvoy, and Luis Castañer

Name	Symbol	Value
Astronomical unit (mean distance between the Sun and Earth)	R_{SE}	1.496×10^{11} m
Avogadro's number	N_{Av}	6.023×10^{23} molecules/mol
Boltzmann constant	k_B	1.381×10^{-25} J/K
Electron charge	q	1.602×10^{-19} C
Electronvolt	eV	1.602×10^{-19} J
Energy of 1 μm photon		1.240 eV
Free electron mass	m_0	9.109×10^{-31} kg
Permitivity of free space	$\varepsilon_0 = 10^7/4\pi c^2$	8.854×10^{-12} F/m
Permeability of free space	$\mu_0 = 4\pi \; 10^{-7}$	1.257×10^{-6} H/m
Plank's constant	h	6.625×10^{-34} J.s
	$\hbar = h/2\pi$	1.055×10^{-31} J.s
Radius of the Sun	R_s	6.96×10^8 m
Thermal voltage at 300 K	$V_T = k_B T/q$	25.9 mV
Solid angle subtended by the Sun	ω_s	6.85×10^{-5} sterad
	$f_\omega = \omega_s/\pi$	2.18×10^{-5}
Solar constant (mean irradiance outside Earth's atmosphere)	I_σ	1367 W/m^2
Speed of light in vacuum	c	2.998×10^{10} m/s
Stefan-Boltzmann constant	σ	5.670×10^{-8} W/m^2 K^4
Wavelength of 1 eV photon		1240 nm

APPENDIX B

List of Principal Symbols

Tom Markvart, Augustin McEvoy, and Luis Castañer

Quantity		Subscripted quantity Symbol	Name	Usual units
A	Area			m^2, cm^2
B	Beam irradiance (without subscript)			Wm^{-2}
B	Beam irradiation[a]	B_h	Hourly beam irradiation	MJm^{-2}, Whm^{-2} [b]
		B_d	Daily beam irradiation	
		B_m	Monthly mean beam irradiation	
B	Radiative recombination constant			$cm^3\,sec^{-1}$
C	Concentration ratio			—
C	Auger recombination constant	C_{no}	... for electrons	$cm^6\,sec^{-1}$
		C_{no}	... for holes	
C_A	Normalised array size			—
C_n	Battery capacity (in energy units)			Wh
C_S	Number of days of autonomy			days
D	Diffusion constant	D_n	Electron ...	$cm^2\,sec^{-1}$
		D_p	Hole ...	
		D_a	Ambipolar ... $\{ = 2D_n D_p/(D_n + D_p)\}$	
D	Density of localised states (in amorphous semiconductor)	D_{Ct}	... in conduction band	cm^{-3}
		D_{Vt}	... in valence band	
		D_{DB}	... of dangling bonds	
D	Diffuse irradiance (without subscript)			Wm^{-2}

(continued)

Appendix B. (continued)

	Quantity	Subscripted quantity Symbol	Name	Usual units
D	Diffuse irradiation	D_h	Hourly diffuse irradiation	MJ m^{-2} or Whm^{-2} [b]
		D_d	Daily diffuse irradiation	
		D_m	Monthly mean diffuse irradiation	
d	Thickness (of antireflection coating)			μm
E	Energy	E_c	... of the edge of the conduction band	eV
		E_v	... of the edge of the valence band	
		E_g	Energy gap; band gap ($= E_c - E_v$)	
		E_F	Fermi energy	
		E_{Fn}	Electron quasi-Fermi level	
		E_{Fp}	Hole quasi-Fermi level	
		E_a	Activation energy	
		E_A	Energy generated by PV array	Wh or J
		E_{AC}	AC energy	Wh or J
EQE	External quantum efficiency			—
EOT	Equation of time			h or min
ε	Electric field			V m^{-1}
FF	Fill factor	FF_0	Fill factor of ideal solar cell characteristic	—
$F_{1/2}$	$\dfrac{2}{\sqrt{\pi}}\displaystyle\int_0^\infty \dfrac{x^{1/2}}{1+\exp(x-z)}\,dx$			
G	Global irradiance (without subscript)			Wm^{-2}
G	Global irradiation[a]	G_h	Hourly global irradiation	MJm^{-2} or Whm^{-2} [b]
		G_d	Daily global irradiation	
		G_m	Monthly mean global irradiation	
		G_o	Extraterrestrial irradiation (with further suffix indicating time interval)	
G	Carrier generation rate per unit volume			cm^{-3} sec^{-1}
g	Carrier generation function (generation rate per unit distance)			cm^{-1} sec^{-1}

(continued)

Appendix B. (continued)

Quantity		Subscripted quantity Symbol	Name	Usual units
H	Daily irradiation in the plane of the array[a]	H_l	Daily global irradiation in the plane of the array	kWh m^{-2} [b]
		H_d	Daily direct irradiation in the plane of the array	
I	Current	I_{sc}	Short-circuit current	A
		I_{ph}	Photogenerated current	
		I_o	Diode dark saturation current	
		I_m	Current at the maximum power point	
		I_{DC}	Nominal DC current (in PV system)	
IQE	Internal quantum efficiency			—
J	Current density	J_{sc}	Short-circuit current density	Am^{-2}
		J_{ph}	Photogenerated current density	
		J_O	Diode dark saturation current density	
K	Damage constant	K_L	Diffusion length damage constant	—
		K_τ	Lifetime damage constant	cm^2 sec^{-1}
k	Wave vector			cm^{-1}
KT	Clearness index	KT_h	Hourly clearness index ($= G_h/G_{oh}$)	—
		KT_d	Daily clearness index ($= G_d/G_{od}$)	
		KT_{nt}	Monthly mean clearness index ($= (G_d)m/(G_{od})m$)	
L	Diffusion length	L_n	Electron diffusion length	μm
		L_p	Hole diffusion length	
L	Load (daily) energy consumption			Wh, kWh, MWh
L	Losses (system)	L_c	Array capture losses	kWh/(kWp/day)
		L_s	System losses	
L_D	Debye length			μm
ℓ	Drift (collection) length	ℓ_n	Electron drift length	μm
		ℓ_p	Hole drift length	

(continued)

Appendix B. (continued)

Quantity		Subscripted quantity Symbol	Name	Usual units
m	Electron mass	m_o	Free-electron mass	kg
		m_c	Density-of-states effective mass at the bottom of conduction band	
		m_v	Density-of-states effective mass at the top of valence band	
N	Dopant or defect concentration	N_A	Acceptor concentration	cm^{-3}
		N_D	Donor concentration	
		N_B	$= N_A N_D / (N_A + N_D)$	
		N_{eff}	Effective dopant concentration	
		N_ℓ	Concentration of carrier traps or recombination centres	
		N_{dop}	Dopant concentration	
N	Number of modules in a PV array	N_s	... connected in series	—
		N_p	... connected in parallel	
$NOCT$	Nominal operating cell temperature			K
N	Density of states	N_C	... in the conduction band	cm^{-3}
		N_V	... in the valence band	
n	Electron concentration in semiconductor	n_0	... at equilibrium	cm^{-3}
		n_i	Intrinsic carrier concentration	
n_{id}	Diode ideality factor			—
n	Refractive index	n_0	... of material surrounding solar cell	—
		n_{ar}	... of antireflection coating	
		n_{sc}	... of a semiconductor	
P	Power	P_{max}	Power at the maximum power point	W
		P_A	Actual power produced by array	
		P_{eff}	Effective power rating of array (in stand-alone systems)	
		P_o	Nominal power of array	

(continued)

Appendix B. (continued)

Quantity		Subscripted quantity Symbol	Name	Usual units
		P_{AC}	AC power output	
		P_I	Nominal AC power of inverter	
PR	Performance ratio ($= Y_f/Y_r$)			—
PSH	Peak Solar Hours			h
p	Hole concentration in semiconductor	p_0	... at equilibrium	cm^{-3}
QE	Quantum efficiency			—
R	Resistance	R_s	Series resistance	Ω
		R_p	Parallel (shunt) resistance	
r	Fresnel reflection coefficient			—
r	Normalised resistance	r_s	$= R_s I_{sc}/V_{oc}$	—
		r_p	$= R_p I_{sc}/V_{oc}$	
S	Surface recombination velocity			cm sec^{-1}
S	Sunshine duration	S_m	Monthly mean ...	h
SR	Spectral response			AW^{-1}
T	Temperature	T_s	Black body temperature of the Sun	K
		T_c	Cell temperature	
		T_d	Ambient temperature	
		T_n	$= T/300$	
T	Transmission coefficient			—
t	Time			sec. hours
U	Recombination rate per unit volume	U_{rad}	Radiative ...	cm^{-3} sec^{-1}
		U_{Auger}	Auger ...	
		U_{SHR}	SHR ... (at defect)	
V	Voltage	V_{oc}	Open circuit voltage	V
		V_m	Voltage at the maximum power point	
		V_{bi}	Built-in voltage of a p–n junction	
		V_{bat}	Battery voltage	
		V_{DC}	Nominal voltage of PV system	
v	Velocity	v_{sat}	Saturation velocity	cm sec^{-1}
		v_{th}	Thermal velocity	
		v_{oc}	$= qV_{oc}/k_B T$ or $qV_{oc}/n_{id}k_B T$	—

(continued)

Appendix B. (continued)

	Quantity	Subscripted quantity Symbol	Name	Usual units
W	Thickness of a region in solar cell	W_j	Junction width	μm
		W_e	Emitter width	
		W_b	Base width	
		W_f	Width of intrinsic region	
x	Space coordinate			m
x_g			$= E_g/k_B T_s$	—
Y	Yield	Y_A	Array yield	kWh/(kWp/day)
		Y_f	Final yield	
		Y_r	Reference yield	
α	Absorption coefficient			cm^{-1}
α_s	Solar azimuth			degrees or radians
γs	Solar altitude angle			degrees or radians
χ	Electron affinity			eV
δ	solar declination			degrees or radians
ε	Static dielectric constant			—
ε	Irradiance correction to mean solar distance			—
Φ	Photon flux			cm^{-2} sec^{-1}
ϕ	Quasi-Fermi level potential	ϕ_n	Electron ...	V
		ϕ_p	Hole ...	
ø	Potential barrier (of heterojunction)			eV
ϕ	Particle fluence			cm^{-2} sec^{-1}
ϕ	Latitude angle			degrees
η	Efficiency	η_C	Carnot efficiency	—
		η_{CA}	Curzon–Ahlborn efficiency	
		η_L	Landsberg efficiency	
		η_{PT}	Photothermal efficiency	
		η_{Amean}	Mean array efficiency	
		η_I	Inverter efficiency	
		η_{tot}	Overall PV plant efficiency	

(continued)

Appendix B. (continued)

	Quantity	Subscripted quantity Symbol	Name	Usual units
ϑ	Collection efficiency			—
κ	Extinction coefficient			—
λ	Wavelength			μm, nm
μ	Carrier drift mobility	μ_n	Electron mobility	cm^2 V sec
		μ_p	Hole mobility	
ν	Frequency			sec^{-1}, Hz
ρ	Charge density			cm^{-3}
ρ_g	Ground albedo			
σ	Carrier capture cross section			cm^2
τ	Minority carrier lifetime	τ_n	Electron lifetime	sec
		τ_p	Hole lifetime	
		τ_{rad}	Radiative lifetime	
		τ_{Auger}	Auger lifetime	
		τ_{SHR}	SHR lifetime	
		τ_{eff}	Effective	
ψ	electrostatic potential			V
ω	hour angle	ω_S	Sunset or sunrise ...	h
ω	solid angle	ω_S	... subtended by the Sun.	sterad

[a]B and G are the conventional symbols for the direct (beam) and global irradiation, as recommended by the International Solar Energy Society. IEC standard 61724:1998 recommends the symbols H_I and H_d for the daily global and direct irradiation in the plane of the array.
[b]Units $Whm^{-2} h^{-1}$ and $Whm^{-2} day^{-1}$ (or $MJm^{-2} h^{-1}$ and $MJm^{-2} day^{-1}$) are also used for the hourly and daily irradiation.

APPENDIX C

Abbreviations and Acronyms

Tom Markvart, Augustin McEvoy, and Luis Castañer

AC	Alternating current
A/D	Analogue to digital
AFM	Atomic force microscopy
AGM	Absorbent glass mat
AR	Antireflection
ARC	Antireflection coating
a-SI or α-Si	(Hydrogenated) amorphous silicon
AM	Air mass
AOI	Angle of incidence
APCVD	Atmospheric pressure chemical vapour deposition
ASTM	American Society for Testing and Materials
ATOX	Atomic oxygen
BCSC	Buried contact solar cell
BIPV	Building-integrated photovoltaics
BMFT	German Ministry of Education and Research
BOL	Beginning of life
BOS	Balance of system
BR	Bragg reflector
BSF	Back-surface field
BSFR	Back surface field and reflector
BSR	Back surface reflector
BST	British standard time
CAST	Chinese Academy of Space Technology
CBD	Chemical bath deposition
CDV	Committee draft for voting
CEI	Comitato Elettrotecnico Italiano (Italy)
CGA	Compressed Gas Association (USA)
CHP	Combined heat and power, co-generation
CIBSE	Chartered Institution of Building Services Engineers (UK)
CIS	Copper indium di-selenide
CIGS	Copper indium gallium di-selenide
CNES	Centre National d'Etudes Spatiales (France)
CVD	Chemical vapour deposition
CZ	Czochralski

(continued)

Appendix C. (continued)

DC	Direct current
DEV	Deutsches Einheitsverfahren (Germany)
DIN	Deutsches Institut für Normung (Germany)
Dj	Double junction
DOD	Depth of discharge
DOE	Department of Energy (USA)
DOS	Density of states
DR	Directional solidification
DTA	Differential thermal analysis
DTI	Department of Trade and Industry (UK)
DSC or DSSC	Dye sensitised solar cell
EB	Electron beam
EBIC	Electron-beam induced current
EFG	Edge-defined film fed growth
EHL	Environmental Health Laboratories (USA)
EHS	Environmental, health and safety
EMC	Electromagnetic continuous casting
ENS	Einrichtung zur Netzüberwachung mit zugerordnetem allpoligem Schaltorgan/Grid monitoring system with circuit breaker
EOL	End of life
EOT	Equation of time
EP	Electrical performance
EPA	Environmental Protection Agency (USA)
EPBT	Energy pay-back time
EPIA	European Photovoltaic Industry Association
EQE	External quantum efficiency
ESA	European Space Agency
ESD	Electrostatic discharge
ESRA	European Solar Radiation Atlas
ESTI	Eidgenössische Starkstrominspektorat Federal Electric Power Authority (Switzerland)
ETSU	Energy Technology Support Unit for the DTI (UK)
EURECA	European Retrievable Carrier
EVA	Ethylene vinyl acetate
FF	Fill factor
FSF	Front surface field
FZ	Floating zone
GAP	Global Approval Programme (see also PV GAP) (European Union)
GEO	Geostationary orbit
GDP	Gross domestic product
GMT	Greenwich mean time
HIT	Heterojunction with intrinsic thin layer (cell)

(continued)

Appendix C. (continued)

HTM	Hole transporting material
HOMO	Highest occupied molecular orbital
HV	High voltage
IBC	Interdigitated back contact
IEA	International Energy Agency
IEC	International Electrotechnical Commission
IEE	Institution of Electrical Engineers (UK)
IEEE	Institute of Electrical and Electronics Engineers (USA)
INTA	Instituto Nacional de Técnica Aeroespacial (Spain)
IPP	Independent power producers (Portugal)
IQE	Internal quantum efficiency
ISO	International Standards Organisation
ISS	International Space Station
ITO	Indium-tin oxide
JIS	Japan Industry Standard
JPL	Jet Propulsion Laboratory (USA)
JQA	Japan Quality Assurance Agency
LASS	Low angle silicon sheet
LAT	Local apparent time
LBIC	Light-beam induced current
LBSF	Local back-surface Field
LCA	Life cycle assessment
LCR	Load coverage rate
LED	Light emitting diode
LEO	Low earth orbit
LPP	Loss-of-load probability
LMT	Local mean time
LPP	Loss-of-power probability
LPCVD	Low pressure chemical vapour deposition
LPE	Liquid phase epitaxy
LPSP	Loss of power supply probability
LUMO	Lowest unoccupied molecular orbital
LV	Low voltage
MATE	Mars array technology experiment
MBE	Molecular beam epitaxy
mc	Multicrystalline
MDMO-PPV	Poly(2-methoxy-5-(3′,7′-dimethyloctyloxy) 1,4-phenylene vinylene
MEH-PPV	Poly (2-methoxy-5−2′-ethyl-hexyloxy) 1,4-phenylene vinylene)
MEO	Medium-earth orbit
MG	Metallurgical grade
MINP	Metal-insulator np junction (cell)
MIS	Metal-insulator-semiconductor

(continued)

Appendix C. (continued)

MITI	Ministry of Trade and Industry (Japan)
Mj	Multijunction
MOCVD	Metalorganic chemical vapour deposition
MOS	Metal-oxide-semiconductor
MOVPE	Metalorganic vapour phase epitaxy
MPP	Maximum power point
MPPT	Maximum power point tracker
MSD	Mains monitoring units with allocated all-pole switching devices connected in series (also known as ENS)
MV	Medium voltage
NASA	National Aeronautics and Space Administration (USA)
NASDA	National Space Development Agency (Japan)
NEDO	New Energy and Industrial Technology Development Organisation (Japan)
NFPA	National Fire Protection Association (USA)
NIR	Near infrared
NREL	National Renewable Energy Laboratory (USA)
NRL	Naval Research Laboratory (USA)
NRS	Nonreflective
NSRDB	National Solar Radiation Data Base (USA)
O&M	Operation and maintenance
ODC	Ordered defect compound
OECD	Organisation for Economic Cooperation and Development
OECO	Obliquely evaporated contact
OLED	Organic light-emitting diode
OCVD	Open-circuit voltage decay
OMeTAD	2.2′,7,7′-tetrakis(N,N-di-p-methoxyphenylamine)-9,9′-spirobifluorene
ÖVE	Östereichische Verband für Elektrotechnik (Austria)
OVPD	Organic vapour phase deposition
PC	Point contact
PCD	Photoconductive decay
PCBM	1-(3-methoxycarbonyl)-propyl-1-phenyl-(6,6)C_{61}
PECVD	Plasma-enhanced chemical vapour deposition
PEDOT: PSS	Poly(3,4-ethylenedioxythiopene):poly(styrene sulfonate)
PERL	Passivated emitter, rear locally diffused (cell)
PESC	Passivated emitter solar cell
PET	Poly(ethylene terephalate)
PPV	Poly-phenylene vinylene
PR	Performance ratio
PRT	Platinum resistance thermometer

(*continued*)

Appendix C. (continued)

PSH	Peak solar hours
PTC	Performance test conditions
PV	Photovoltaic, photovoltaics
PVD	Physical vapour deposition
PVGAP	PV Global Approval Programme
PVPS	Photovoltaic Power Systems (Programme)
PVUSA	Photovoltaics for Utility Scale Applications
QE	Quantum efficiency
QSSPC	Quasi-steady-state photoconductance (method)
QSSVoc	Quasi-steady state open circuit voltage (method)
R&D	Research and development
RES	Renewable energy sources
RF	Radio frequency
RGS	Ribbon growth on substrate
RH	Relative humidity
ROCOF	Rate of change of frequency
ROW	Rest of the world
RTCVD	Rapid thermal chemical vapour deposition
RTD	Research and technology development
RTP	Rapid thermal process
SA	Solar array
SCA	Solar cell assembly
SCCE	Solar cell calibration experiment
SDL	Surface defect layer
SELF	Solar Electric Light Fund
SEM	Scanning electron microscopy
SG	Specific gravity
SRH	Shockley–Read–Hall
SHS	Solar home system(s)
SIPOS	Semi-insulating polysilicon
Sj	Single junction
SLI	Starting, lighting, and ignition
SMUD	Sacramento Municipal Utility District
SOC	State of charge
SR	String ribbon
SR	Spectral response
SRC	Standard reporting conditions
SSP	Space solar power
STC	Standard test conditions
STAR	Surface texture and enhanced absorption with back reflector
SRV	Surface recombination velocity
SWE	Staebler-Wronski effect

(continued)

Appendix C. (continued)

SWS	Secondary working standard
S-Web	Supporting web
TJ	Tandem junction
TJ	Triple junction
TC	Technical committee
TCO	Transparent conducting oxide
TEC	Thermal expansion coefficient
TEM	Transmission electron microscopy
TPV	Thermophotovoltaics
TTV	Total thickness variation
UCPTE	Union for the Coordination of Production and Transmission of Electricity (USA)
UL	Underwriters Laboratory (USA)
UT	Ultrathin
UV	Ultraviolet
VDE	Verband der Elektrotechnik, Elektronik und Informationstechnik (Germany)
VdEW	Verband der Elektrizitätswirtschaft (Germany)
VHF	Very high frequency
VSE	Verband Schweizerische Elektrizitätswerke (Switzerland)
WCP	Wiring collecting panels
WET	West European time
WMO	World Meteorological Organisation
XRD	X-ray diffraction
x-Si	Crystalline silicon
XTE	X-ray timing explorer
YAG	Yttrium aluminium garnet
ZMR	Zone-melt recrystalization

APPENDIX D

The Photovoltaic Market

Arnulf Jäger-Waldau
European Commission, DG JRC, Institute for Energy, Renewable Energies Unit, 21027 Ispra, Italy*

In 2010, the worldwide photovoltaic market more than doubled, driven by major increases in Europe. For 2010 the market volume of newly installed solar photovoltaic electricity systems varies between 17 GW and 19 GW, depending on the reporting consultancies (Figure 1). This represents mostly the grid-connected photovoltaic market. To what extent the off-grid and consumer product markets are included is not clear, but it is believed that a substantial part of these markets are not accounted for, as it is very difficult to track them. A conservative estimate is that they account for approx. 400 to 800 MW (approx. 1−200 MW off-grid rural, approx. 1−200 MW communication/signals, approx. 100 MW off-grid commercial and approx. 1−200 MW consumer products).

With a cumulative installed capacity of over 29 GW, the European Union is leading in PV installations with a little more than 70% of the total worldwide 39 GW of solar photovoltaic electricity generation capacity at the end of 2010.

1. ASIA AND THE PACIFIC REGION

The Asia and Pacific Region shows an increasing trend in photovoltaic electricity system installations. There are a number of reasons for this development, ranging from declining system prices, heightened awareness, favourable policies and the sustained use of solar power for rural electrification projects. Countries such as Australia, China, India, Indonesia, Japan, Malaysia, South Korea, Taiwan, Thailand, The Philippines, and Vietnam show a very positive upward trend, thanks to increasing governmental commitment towards the promotion of solar energy and the creation of sustainable cities.

The introduction or expansion of feed-in-tariffs is expected to be an additional big stimulant for on-grid solar PV system installations for both

*©Commission of the European Union, with permission

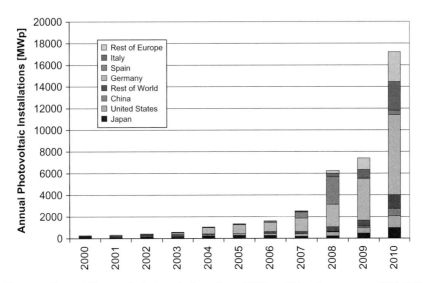

Figure 1 Annual Photovoltaic Installations from 2000 to 2010. *Data source: EPIA [11], Eurobserver [31] and own analysis.*

distributed and centralised solar power plants in countries such as Australia, Japan, Malaysia, Thailand, Taiwan, and South Korea.

The Asian Development Bank (ADB) launched an Asian Solar Energy Initiative (ASEI) in 2010, which should lead to the installation of 3 GW of solar power by 2012 [1]. In their report, ADB states: *Overall, ASEI aims to create a virtuous cycle of solar energy investments in the region, toward achieving grid parity, so that ADB developing member countries optimally benefit from this clean, inexhaustible energy resource.*

Three interlinked components will be used to realise the ASEI target:

Knowledge management. Development of a regional knowledge platform dedicated to solar energy in Asia and the Pacific.

Project development. ADB will provide $2.25 billion[1] (€1.73 billion) to finance the project development, which is expected to leverage an additional $6.75 billion (€5.19 billion) in solar power investments over the period.

Innovative finance instruments. A separate and targeted Asia Accelerated Solar Energy Development Fund is set up to mitigate risks associated with solar energy. The fund will be used for a *buy down* programme to reduce the up-front costs of solar energy for final customers.

---

[1] Exchange rate: 1 € = 1.30 $

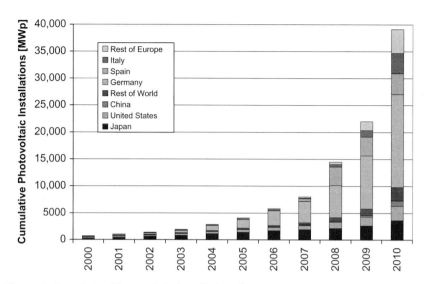

Figure 2 Cumulative Photovoltaic Installations from 2000 to 2009. *Data source: EPIA [11], Eurobserver [31] and own analysis.*

ADB aims to raise $500 million (€385 million) and design innovative financing mechanisms in order to encourage commercial banks and the private sector to invest in solar energy technologies and projects.

Innovative finance instruments. Setting up of a separate and targeted Asia Accelerated Solar Energy Development Fund to mitigate risks associated with solar energy and buy down the up-front costs of solar energy. ADB aims to raise $500 million (€385 million) and design innovative financing mechanisms in order to encourage commercial banks and the private sector to invest in solar energy technologies and projects.

1.1 Australia

In 2010, 383 MW of new solar photovoltaic electricity systems were installed in Australia, bringing the cumulative installed capacity of grid-connected PV systems to 571 MW [2]. The 2010 market was dominated by the increase of grid-connected distributed systems, which increased from 67 MW in 2009 to 378 MW in 2010. The newly installed PV electricity generation capacity in Australia accounted for 20% of the new electricity generation capacity in 2010.

Most installations took advantage of the incentives under the Australian Government's Solar Homes and Communities Plan (SHCP), Renewable

Energy Target (RET) mechanisms and feed-in tariffs in some States or Territories. At the beginning of 2010, eight out of the eleven Australian Federal States and Territories had introduced some kind of feed-in tariff scheme for systems smaller than 10 kWp. All of these schemes have built-in caps which were partly reached that year so that in 2011 only five State schemes are still available for new installations and additional changes are expected in the course of this year.

1.2 India

For 2010, market estimates for solar PV systems vary between 50 MW to 100 MW, but most of these capacities are for off-grid applications. The Indian National Solar Mission was launched in January 2010, and it was hoped that it would give impetus to the grid-connected market, but only a few MW were actually installed in 2010. The majority of the projects announced will come on-line from 2011 onwards.

The National Solar Mission aims to make India a global leader in solar energy and envisages an installed solar generation capacity of 20 GW by 2020, 100 GW by 2030 and 200 GW by 2050. The short-term outlook up until 2013 was improved as well when the original 50 MW grid-connected PV system target in 2012 was changed to 1,000 MW for 2013.

1.3 Japan

In 2010, the Japanese market experienced a high growth, doubling its volume to 990 MW, bringing the cumulative installed PV capacity to 3.6 GW. In 2009 a new investment incentive of ¥70,000 per kW for systems smaller than 10 kW, and a new surplus power purchase scheme, with a purchase price of ¥48 per kWh for systems smaller than 10 kW, was introduced and the start of the discussion about a wider feed-in tariff.

In April 2011, METI (Ministry for Economy, Trade and Industry) announced a change in the feed-in tariffs and increased the tariff for commercial installations from ¥20 to 40 per kWh and decreased the tariff for residential installations to ¥42 per kWh.

As a consequence of the accident at the Fukujima Daiichi Nuclear Power Plant, Prime Minister Naoto Kan announced an overall review of the country's Basic Energy Plan. At the G8 Summit held in Deauville, France, on 26 May 2011, he announced that Japan plans to increase the share of renewable energy in total electricity supply to over 20% in the

2020s. One measure to achieve this goal is to install a PV system on some 10 million houses suitable for it.

1.4 People's Republic of China

The 2010 Chinese PV market estimates are between 530 MW to 690 MW, bringing the cumulative installed capacity to about 1 GW. This is a significant increase from the 160 MW in 2009, but still only 5% to 7% of the total photovoltaic production. This situation will change because of the revision of the PV targets for 2015 and 2020. According to press reports, the National Energy Administration doubled its capacity target for installed photovoltaic electricity systems to 10 GW in 2015 and further up to 50 GW in 2020 [25].

According to the 12th Five-Year Plan, which was adopted on 14 March 2011, China intends to cut its carbon footprint and be more energy efficient. The targets are 17% less carbon dioxide emissions and 16% less energy consumption unit of GDP. The total investment in the power sector under the 12th Five-Year Plan is expected to reach $803 billion (€618 billion), divided into $416 billion (€320 billion), or 52%, for power generation, and $386 billion (€298 billion) to construct new transmission lines and other improvements to China's electrical grid.

Renewable, clean, and nuclear energy are expected to contribute to 52% of the increase and it is planned to increase power generation capacity from non-fossil fuels to 474 GW by 2015.

The investment figures necessary are in-line with a World Bank report stating that China needs an additional investment of $64 billion (€49.2 billion) annually over the next two decades to implement an "energy-smart" growth strategy [33]. However, the reductions in fuel costs through energy savings could largely pay for the additional investment costs according to the report. At a discount rate of 10%, the annual net present value (NPV) of the fuel cost savings from 2010 to 2030 would amount to $145 billion (€111.5 billion), which is about $70 billion (€53.8 billion) more than the annual NPV of the additional investment costs required.

1.5 South Korea

In 2010, about 180 MW of new PV systems were installed in South Korea, about the same as the year before, bringing the cumulative capacity to a total of 705 MW. The Korean PV industry expects a moderate increase for 2011, due to the fact that the feed-in tariff scheme is in its final year and the Korean Government continues its "One Million Green Homes" Project, as well as other energy projects in the provinces. The implementation of the Renewable Portfolio Standard in 2012 has additional consequences, as systems will need to be installed by the end of 2011 to generate electricity.

In January 2009, the Korean Government had announced the Third National Renewable Energy Plan, under which renewable energy sources will steadily increase their share of the energy mix between now and 2030. The Plan covers such areas as investment, infrastructure, technology development and programmes to promote renewable energy. The new Plan calls for a renewable energies share of 4.3% in 2015, 6.1% in 2020, and 11% in 2030.

1.6 Taiwan

In June 2009, the Taiwan Legislative Yuan gave its final approval to the Renewable Energy Development Act, a move that is expected to bolster the development of Taiwan's green energy industry. The new law authorises the Government to enhance incentives for the development of renewable energy via a variety of methods, including the acquisition mechanism, incentives for demonstration projects and the loosening of regulatory restrictions. The goal is to increase Taiwan's renewable energy generation capacity by 6.5 GW to a total of 10 GW within 20 years. In January 2011, the Ministry of Economic Affairs (MOEA) announced the revised feed-in tariffs for 2011. In 2011, the price paid by the state-owned monopoly utility, Taiwan Power, will fall 30% from 11.12 NT\$[2]/kWh (0.264 €/kWh) to 7.33 NT\$/kWh per kWh (0.175 €/kWh) for solar installations, with an exception for rooftop installations which will be eligible for rates of 10.32 NT\$/kWh (0.246 €/kWh).

Despite the favourable feed-in tariff, the total installed capacity at the end of 2010 was only between 19 MW and 20 MW and the annual installation of about 7 to 8 MW was far less than 1% of the 3.2 GW solar cell production in Taiwan that year.

1.7 Thailand

Thailand enacted a 15-year Renewable Energy Development Plan (REDP) in early 2009, setting the target to increase the Renewable Energy

---

[2] Exchange Rate 1 € = 42 NT\$

share to 20% of final energy consumption of the country in 2022. Besides a range of tax incentives, solar photovoltaic electricity systems are eligible for a feed-in premium or "Adder" for a period of 10 years. However, there is a cap of 500 MW eligible for the original 8 THB³/kWh (0.182 €/kWh) "Adder" (facilities in the 3 Southern provinces and those replacing diesel systems are eligible for an additional 1.5 THB/kWh (0.034 €/kWh)), which was reduced to 6.5 THB/kWh (0.148 €/kWh) for those projects not approved before 28 June 2010.

As of October 2010, applications for 1.6 GW, under the *Very Small Power Producer Programme* (VSPP), and 477 MW, under the *Small Power Producer Programme* (SPP), were submitted. In 2010 it is estimated that between 20 MW and 30 MW were actually added, increasing the total cumulative installed capacity to 60 MW to 70 MW.

2. EMERGING MARKETS
2.1 Asia
2.1.1 Bangladesh

In 1997, the Government of Bangladesh established the Infrastructure Development Company Limited (IDCOL) to promote economic development in Bangladesh. In 2003, IDCOL started its Solar Energy Programme to promote the dissemination of solar home systems (SHS) in the remote rural areas of Bangladesh, with the financial support from the World Bank, the Global Environment Facility (GEF), the German Kreditanstalt für Wiederaufbau (KfW), the German Technical Cooperation (GTZ), the Asian Development Bank, and the Islamic Development Bank. Since the start of the programme, more than 950,000 SHS, with an estimated capacity of 39 MW, have been installed in Bangladesh by May 2011.

According to a press report, the Government plans to implement a mega project of setting up 500 MW of PV electrical power generation and the Asian Development Bank (ADB) has, in principal, agreed to provide financial support to Bangladesh for implementing the project within the framework of the Asian Solar Energy Initiative [8,32].

---

³ Exchange Rate 1€ = 44 THB

2.1.2 Indonesia

The development of renewable energy is regulated in the context of the national energy policy by Presidential Regulation No.5/2006 [26]. The decree states that 11% of the national primary energy mix in 2025 should come from renewable energy sources. The target for solar PV is 870 MW by 2024. At the end of 2010 about 20 MW of solar PV systems were installed, mainly for rural electrification purposes.

2.1.3 Malaysia

The Malaysia Building Integrated Photovoltaic (BIPV) Technology Application Project was initiated in 2000 and at the end of 2009 a cumulative capacity of about 1 MW of grid-connected PV systems has been installed.

The Malaysian Government officially launched their GREEN Technology Policy in July 2009 to encourage and promote the use of renewable energy for Malaysia's future sustainable development. By 2015, about 1 GW must come from Renewable Energy Sources according to the Ministry of Energy, Green Technology and Water (KETHHA). The Malaysian Photovoltaic Industry Association (MPIA) proposed a five-year programme to increase the share of electricity generated by photovoltaic systems to 1.5% of the national demand by 2015. This would translate into 200 MW grid-connected and 22 MW of grid systems. In the long-term beyond 2030, MPIA is calling for a 20% PV share. Pusat Tenaga Malaysia (PTM), and its IEA international consultant, estimated that 6,500 MW power can be generated by using 40% of the nation's house rooftops (2.5 million houses) and 5% of commercial buildings alone. To realise such targets, a feed-in tariff is still under discussion, and it is hoped to be under way in the second half of 2011.

First Solar (USA), Q Cells (Germany) and Sunpower (USA) have started to set up manufacturing plants in Malaysia, with a total investment of RM 12 billion and more than 2 GW of production capacities. Once fully operational, these plants will provide 11,000 jobs and Malaysia will be the world's sixth largest producer of solar cells and modules.

2.1.4 The Philippines

The Renewable Energy Law was passed in December 2008 [27]. Under the Law, the Philippines has to double the energy derived from Renewable Energy Sources within 10 years. On 14 June 2011, Energy Secretary, Rene Almendras unveiled the new Renewable Energy Roadmap, which aims to increase the share of renewables to 50% by 2030. The programme will endeavour to boost renewable energy capacity from the current 5.4 GW to 15.4 GW by 2030.

Early 2011, the country's Energy Regulator National Renewable Energy Board (NREB) has recommended a target of 100 MW of solar installations that will be constructed in the country over the next three years. A feed-in tariff of 17 PHP/kWh (0.283 €/kWh)[4] 4 was suggested, to be paid from January 2012 on. The initial period of the programme is scheduled to end on 31 December 2014.

At the end of 2010, about 10 MW of PV systems were installed, mainly off-grid. SunPower has two cell manufacturing plants outside of Manila. Fab. No 1 has a nameplate capacity of 108 MW and Fab. No 2 adds another nameplate capacity of 466 MW.

2.1.5 Vietnam

In December 2007, the National Energy Development Strategy of Vietnam was approved. It gives priority to the development of renewable energy and includes the following targets: increase the share of renewable energies from negligible to about 3% (58.6 GJ) of the total commercial primary energy in 2010, to 5% in 2020, 8% (376.8 GJ) in 2025, and 11% (1.5 TJ) in 2050.

The Indochinese Energy Company (IC Energy) broke ground for the construction of a thin-film solar panel factory with an initial capacity of 30 MW and a final capacity of 120 MW in the central coastal Province of Quang Nam on 14 May 2011.

In March 2011, First Solar broke ground on its four-line photovoltaic module manufacturing plant (250 MW) in the Dong Nam Industrial Park near Ho Chi Minh City.

2.2 Europe and Turkey

Market conditions for photovoltaics differ substantially from country to country. This is due to different energy policies and public support programmes for renewable energies and especially photovoltaics, as well as the varying grades of liberalisation of domestic electricity markets. Within one decade, the solar photovoltaic electricity generation capacity has increased 160 times from 185 MW in 2000 to 29.5 GW in 2010 (Figure 3) [6,11,14,31].

---

[4] Exchange Rate 1 € = 60 PHP

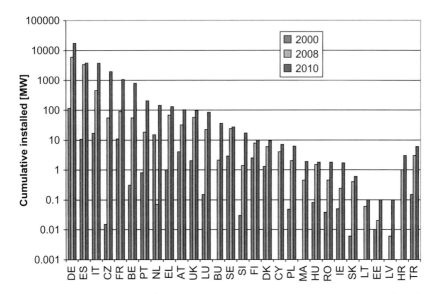

Figure 3 Cumulative installed grid-connected PV capacity in EU + CC. Note that the installed capacities do not correlate with solar resources.

A total of about 58.8 GW of new power capacity was constructed in the EU last year and 2.5 GW were decommissioned, resulting in 56.3 GW of new net capacity (Figure 4) [12,31]. Gas-fired power stations accounted for 28.3 GW, or 48% of the newly installed capacity. According to Platts, about 30 GW of gas-fired power station projects were suspended or cancelled [23]. Solar photovoltaic systems moved to the second place with 13.5 GW (23%), followed by 9.4 GW (16%) wind power; 4.1 GW (7%) MW coal-fired power stations; 570 MW (>1%) biomass; 450 MW (>1%) CSP, 210 MW (>1%) hydro, 230 MW (>1%) peat, and 150 MW (>1%) waste. The net installation capacity for oil-fired and nuclear power plants was negative, with a decrease of 245 MW and 390 MW respectively. The renewable share of new power installations was 40% in 2010.

In the following sub-sections, the market development in some of the EU Member States, as well as Switzerland and Turkey, is described.

2.2.1 Belgium

Belgium showed another strong market performance year in 2010, with new photovoltaic system installations of 420 MW bringing the cumulative installed capacity to 790 MW. However, most of the installations were done

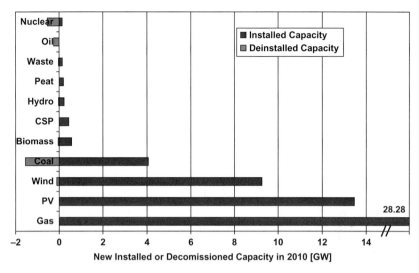

Figure 4 New installed or decommissioned electricity generation capacity in Europe in 2010

in Flanders, where since 1 January 2006 Green Certificates exist with 0.45 €/kWh for 20 years. In Brussels and Wallonia, the Green Certificates have a guaranteed minimum price between 0.15–0.65 €/kWh, depending on the size of the systems and region (Brussels 10 years, Wallonia 15 years).

2.2.2 Czech Republic

In the Czech Republic, photovoltaic systems with about 1.5 GW capacity, were installed in 2010, bringing the cumulative nominal capacity to 1.95 GW exceeding their own target of 1.65 GW set in the National Renewable Action Plan for 2020. The Law on the Promotion of Production of Electricity from Renewable Energy Sources went into effect on 1 August 2005 and guarantees a feed-in tariff for 20 years. The annual prices are set by the Energy Regulator. The electricity producers can choose from two support schemes, either fixed feed-in tariffs or market price + Green Bonus. The 2010 feed-in rate in the Czech Republic was CZK[5] 12.25 per kilowatt hour (0.48 €/kWh).

On 3 February 2010, the Czech transmission system operator, EPS, requested all main distribution system operators (EZ, E-ON, PRE) to stop permitting new renewable energy power plants, due to a virtual risk

---

[5] Exchange rate: 1 € = 25.52 CZK

of instability of the electricity grid caused by intermittent renewable sources, especially photovoltaic and wind. Distribution System Operators (DSO) met the requirement on 16 February 2011. The moratorium still exists and SRES (Czech Association of Regulated Energy Companies) announced that the moratorium will continue until at least September.

A number of legislative changes took place in the second half of the year, which resulted in a lower feed-in tariff for systems larger than 30 kW (5.5 CZK/kWh or 0.216 €/kWh), the phase-out of ground-mounted PV systems from 1 March 2011 onwards and the introduction of a retroactive tax on benefits generated by PV installations.

2.2.3 France

In 2010, 720 MW of PV systems were connected to the grid in France, including about 100 MW which were already installed in 2009. This led to an increase of the cumulative installed capacity to 1.05 GW. However, this positive development came to a sudden stop when the French Prime Minister declared a three-month moratorium on new PV installations above 3 kW and a suspension of projects waiting for grid connection in December 2010.

This rapid growth led to a revision of the feed-in scheme in February 2011, setting a cap of 500 MW for 2011 and 800 MW for 2012 [20]. The new tariff levels only apply to rooftop systems up to 100 kW in size. In addition, those installations are divided into three different categories: residential; education or health; and other buildings with different feed-in tariffs, depending on the size and type of installation. The tariffs for these installations range between 0.2883 €/kWh and 0.46 €/kWh. All other installations up to 12 MW are just eligible for a tariff of 0.12 €/kWh.

2.2.4 Germany

Germany had the biggest market with 7.4 GW [6]. The German market growth is directly correlated to the introduction of the Renewable Energy Sources Act or *"Erneuerbare Energien Gesetz"* (EEG) in 2000 [9]. This Law introduced a guaranteed feed-in tariff for electricity generated from solar photovoltaic systems for 20 years and already had a fixed built in annual decrease, which was adjusted over time to reflect the rapid growth of the market and the corresponding price reductions. Due to the fact that until 2008 only estimates of the installed capacity existed, a plant registrar was introduced from 1 January 2009 on.

The German market showed two installation peaks during 2010. The first one was in June, when more than 2.1 GW were connected to the grid prior to the 13% feed-in cut which took effect on 1 July 2010. The second peak was in December with almost 1.2 GW just before the scheduled tariff reduction of another 13% on 1 January 2011. Compared to 2009, the feed-in tariff has been reduced by 33 to 36% depending on the system size and classification. In June 2011 the Bundesnetzagentur (German Federal Network Agency) announced the results of the PV system installation projection required under the Renewable Energy Sources Act (EEG) in order to determine the degression rates for the feed-in tariffs [7]. According to the Agency approx. 700 MW of PV systems were commissioned between March and May 2011 resulting in a projected annual growth of 2.8 GW, which is below the 3.5 GW threshold set for an additional reduction of the tariffs starting July 2011.

2.2.5 Greece

Greece introduced a new feed-in tariff scheme on 15 January 2009. The tariffs remained unchanged until August 2010 and are guaranteed for 20 years. However, if a grid-connection agreement was signed before that date, the unchanged FIT was applied if the system is finalised within the next 18 months. For small rooftop PV systems, an additional programme was introduced in Greece on 4 June 2009. This programme covers rooftop PV systems up to 10 kWp (both for residential users and small companies). In 2011, the tariffs decreased by 6.8% to 8.5%, depending on the size and location of the installation. In 2010, about 150 MW of new installations were carried out, bringing the total capacity to about 205 MW.

2.2.6 Italy

Italy again took the second place, with respect to new installations and added a capacity of about 2.5 GW, bringing cumulative installed capacity to 3.7 GW at the end of 2010 [14]. At the beginning of July 2011, the total connected PV capacity has surpassed 7 GW [15]. The *Quarto Conto Energia* (Fourth Energy Bill) was approved by the Italian Council of Ministers on 5 May 2011 [13]. The Bill introduced monthly reductions of the tariffs, starting from June 2011 until January 2012 and then another one in July 2012. In addition, the new Bill limits the feed-in tariffs for new systems up until the end of 2016, or until a cap of 23 GW is reached. In addition, separate caps for large systems are set for the second half of 2011 (1.35 GW) and 2012 (1.75 GW).

2.2.7 Spain

Spain is second regarding the total cumulative installed capacity with 3.9 GW. Most of this capacity was installed in 2008 when the country was the biggest market, with close to 2.7 GW in 2008 [11]. This was more than twice the expected capacity and was due to an exceptional race to install systems before the Spanish Government introduced a cap of 500 MW on the yearly installations in the autumn of 2008. A revised Decree (Royal Decree 1758/2008) set considerably lower feed-in tariffs for new systems and limited the annual market to 500 MW, with the provision that two thirds are rooftop mounted and no longer free-field systems. These changes resulted in a new installed capacity of about 100 MW and about 380 MW in 2010.

In 2010, the Spanish Government passed the Royal Decrees 1565/10 [16] and RD-L 14/10 [17]. The first one limits the validity of the feed-in tariffs to 28 years, while the latter reduces the tariffs by 10% and 30% for existing projects until 2014. Both Bills are "retroactive" and the Spanish Solar Industry Association (ASIF) [3] has already announced taking legal actions against them.

2.2.8 United Kingdom

The United Kingdom introduced of a new feed-in tariff scheme in 2010, which led to the installation of approximately 55 MW, bringing the cumulative installed capacity to about 85 MW. However, in March 2011, the UK Government proposed significant reductions of the tariffs, especially for systems larger than 50 kW.

2.2.9 Other European Countries and Turkey

Despite high solar radiation, solar photovoltaic system installation in **Portugal** has only grown very slowly and reached a cumulative capacity of 130 MW at the end of 2010.

The market in **Slovakia** showed an unexpected growth from less than 1 MW installed at the end of 2009 to about 144 MW at the end of 2010. In December 2010, the Slovak Parliament adopted an Amendment to the Renewable Energy Sources (RES) Promotion Act, decreasing the feed-in tariffs and from 1 February 2011 on only solar rooftop facilities or solar facilities on the exterior wall of buildings, with capacity not exceeding 100 kW, are eligible for the feed-in tariff. As a result, larger new solar projects in Slovakia are on hold.

In **Turkey** in March 2010, the Energy Ministry unveiled its 2010–2014 Strategic Energy Plan. One of Government's priorities is to increase the ratio of renewable energy resources to 30% of total energy generation by 2023. At the beginning of 2011, the Turkish Parliament passed a Renewable Energy Legislation which defines new guidelines for feed-in tariffs. The feed-in tariff is 0.133\$/kWh (0.10 €/kWh) for owners commissioning a PV system before the end of 2015. If components *'Made in Turkey'* are used, the tariff will increase by up to \$0.067 (€0.052), depending on the material mix. Feed-in tariffs apply to all types of PV installations, but large PV power plants will receive subsidies up to a maximum size of 600 MWp.

2.3 North America
2.3.1 Canada

In 2010, Canada more than tripled its cumulative installed PV capacity to about 420 MW, with 300 MW new installed systems. This development was driven by the introduction of a feed-in tariff in the Province of Ontario, enabled by the *'Bill 150, Green Energy and Green Economy Act, 2009.'* On the Federal level, only an accelerated capital cost allowance exists under the Income Tax Regulations. On a Province level, nine Canadian Provinces have *Net Metering Rules,* with solar photovoltaic electricity as one of the eligible technologies, *Sales Tax Exemptions* and *Renewable Energy Funds* exist in two Provinces and *Micro Grid Regulations* and *Minimum Purchase Prices* each exist in one Province.

The Ontario feed-in tariffs were set in 2009 and depend on the system size and type, as follows:

- Rooftop or ground-mounted ≤ 10 kW   80.2 ¢/kWh (0.59 €/kWh[6])
- Rooftop > 10 kW ≤ 250 kW   71.3 ¢/kWh (0.53 €/kWh)
- Rooftop > 250 kW ≤ 500 kW   63.5 ¢/kWh (0.47 €/kWh)
- Rooftop > 500 kW   53.9 ¢/kWh (0.40 €/kWh)
- Ground-mounted[7] ★ >10 kW ≤ 10 MW   44.3 ¢/kWh (0.33 €/kWh)

The feed-in tariff scheme has a number of special rules, ranging from eligibility criteria, which limit the installation of ground-mounted PV systems on high-yield agricultural land to domestic content requirements and *additional 'price adders' for Aboriginal and community-based projects.*

---

[6] Exchange Rate 1 € = 1.35 CAD
[7] Eligible for Aboriginal or community adder

Details can be found in the Feed-in Tariff Programme of the Ontario Power Authority [21].

2.3.2 United States of America

With close to 900 MW of new installed PV capacity, the USA reached a cumulative PV capacity of 2.5 GW at the end of 2010. Utility PV installations more than tripled compared to 2009 and reached 242 MW in 2010. The top ten States—California, New Jersey, Nevada, Arizona, Colorado, Pennsylvania, New Mexico, Florida, North Carolina and Texas—accounted for 85% of the US grid-connected PV market [28].

PV projects with Power Purchase Agreements (PPAs), with a total capacity of 6.1 GW, are already under contract and to be completed by 2014 [10]. If one adds those 10.5 GW of projects which are already publicly announced, but PPAs have yet to be signed, this make the total "pipeline" more than 16.6 GW.

Many State and Federal policies and programmes have been adopted to encourage the development of markets for PV and other renewable technologies. These consist of direct legislative mandates (such as renewable content requirements) and financial incentives (such as tax credits). One of the most comprehensive databases about the different support schemes in the US is maintained by the Solar Centre of the State University of North Carolina. The Database of State Incentives for Renewable Energy (DSIRE) is a comprehensive source of information on State, local, utility, and selected Federal incentives that promote renewable energy. All the different support schemes are described therein and it is highly recommended to visit the DSIRE website http://www.dsireusa.org/ and the corresponding interactive tables and maps for details.

REFERENCES

See Appendix E

APPENDIX E

The Photovoltaic Industry

Arnulf Jäger-Waldau
European Commission, DG JRC, Institute for Energy, Renewable Energies Unit, 21027 Ispra, Italy[*]

In 2010, the photovoltaic world market almost doubled in terms of **production** to 23 to 24 GW. The market for installed systems doubled again and values between 16 and 18 GW were reported by various consultancies and institutions. This mainly represents the grid-connected photovoltaic market. To what extent the off-grid and consumer-product markets are included is unclear. The difference of roughly 6 GW to 7 GW has therefore to be explained as a combination of unaccounted off-grid installations (approx. 1—200 MW off-grid rural, approx. 1—200 MW communication/signals, approx. 100 MW off-grid commercial), consumer products (ca. 1—200 MW) and cells/modules in stock.

In addition, the fact that some companies report shipment figures, whereas others report production figures, add to the uncertainty. The difficult economic conditions contributed to the decreased willingness to report confidential company data. Nevertheless, the figures show a significant growth of the production, as well as an increasing silicon supply situation.

The announced production capacities, based on a survey of more than 350 companies worldwide, increased, even with difficult economic conditions. Despite the fact that a number of players announced a scale-back or cancellation of their expansion plans for the time being, the number of new entrants into the field, notably large semiconductor or energy-related companies overcompensated this. At least on paper the expected production capacities are increasing. Only published announcements of the respective companies and no third source info were used. The cutoff date of the info used was June 2011.

It is important to note that production capacities are often announced, taking into account different operation models, such as number of shifts, operating hours per year, etc. In addition, the announcements of the increase in production capacity do not always specify when the capacity will be fully ramped up and operational. This method has of course the setback

[*]©Commission of the European Union, with permission

that (a) not all companies announce their capacity increases in advance and (b) that in times of financial tightening, the announcements of the scale-back of expansion plans are often delayed, in order not to upset financial markets. Therefore, the capacity figures just give a trend, but do not represent final numbers.

If all these ambitious plans can be realised by 2015, China will have about 46.3% of the worldwide production capacity of 102 GW, followed by Taiwan (15.8%), Europe (9.5%) and Japan (6.9%) (Figure 1).

All these ambitious plans to increase production capacities, at such a rapid pace, depend on the expectations that markets will grow accordingly. This, however, is the biggest uncertainty, as the market estimates for 2011 vary between 17 GW and 24 GW, with a consensus value in the 19 GW range. In addition, most markets are still dependent on public support in the form of feed-in tariffs, investment subsidies or tax-breaks.

Already now, electricity production from photovoltaic solar systems has shown that it can be cheaper than peak prices in the electricity exchange. In the second quarter of 2011, the German average price index, for rooftop systems up to 100 kWp, was given with €2,422 per kWp without tax or half the price of five years ago [5]. With such investment costs, the electricity generation costs are already at the level of residential electricity prices in some countries, depending on the actual

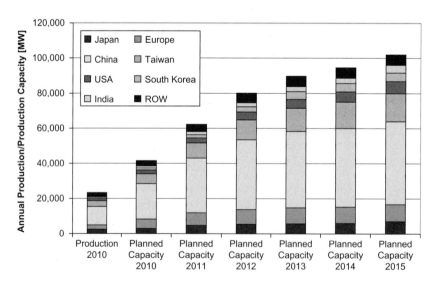

Figure 1 Worldwide PV Production 2010 with future planned production capacity increases

electricity price and the local solar radiation level. But only if markets and competition continue to grow, prices of the photovoltaic systems will continue to decrease and make electricity from PV systems for consumers even cheaper than from conventional sources. In order to achieve the price reductions and reach grid-parity for electricity generated from photovoltaic systems, public support, especially on regulatory measures, will be necessary for the next decade.

1. TECHNOLOGY MIX

Wafer-based silicon solar cells is still the main technology and had around 85% market shares in 2010. Commercial module efficiencies are within a wide range between 12 and 20%, with monocrystalline modules between 14% and 20%, and polycrystalline modules between 12% and 17%. The massive manufacturing capacity increases for both technologies are followed by the necessary capacity expansions for polysilicon raw material.

In 2005, for the first time, production of thin-film solar modules reached more than 100 MW per annum. Since then, the *Compound Annual Growth Rate* (CAGR) of thin-film solar module production was even beyond that of the overall industry, increasing the market share of thin-film products from 6% in 2005 to 10% in 2007 and 16–20% in 2009.

More than 200 companies are involved in thin-film solar cell activities, ranging from basic R&D activities to major manufacturing activities and over 120 of them have announced the start or increase of production. The first 100 MW thin-film factories became operational in 2007, followed by the first 1 GW factory in 2010. If all expansion plans are realised in time, thin-film production capacity could be 17 GW, or 21% of the total 80 GW in 2012, and 27 GW, or 26%, in 2015 of a total of 102 GW (Figure 2).

One should bear in mind that only one third of the over 120 companies, with announced production plans, have produced thin-film modules of 10 MW or more in 2010.

More than 70 companies are silicon-based and use either amorphous silicon or an amorphous/microcrystalline silicon structure. Thirty-six companies announced using $Cu(In,Ga)(Se,S)_2$ as absorber material for

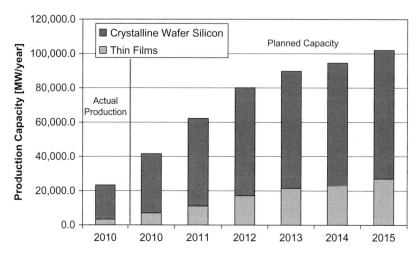

Figure 2 Annual PV Production capacities of thin-film and crystalline silicon based solar modules.

their thin-film solar modules, whereas nine companies use CdTe and eight companies go for dye and other materials.

Concentrating Photovoltaics (CPV) is an emerging technology which is growing at a very high pace, although from a low starting point. About 50 companies are active in the field of CPV development and almost 60% of them were founded in the last five years. Over half of the companies are located either in the United States of America (primarily in California) and Europe (primarily in Spain).

Within CPV there is a differentiation according to the concentration factors[1] and whether the system uses a dish (Dish CPV) or lenses (Lens CPV). The main parts of a CPV system are the cells, the optical elements and the tracking devices. The recent growth in CPV is based on significant improvements in all of these areas, as well as the system integration. However, it should be pointed out that CPV is just at the beginning of an industry learning curve, with a considerable potential for technical and cost improvements. The most challenging task is to become cost-competitive with other PV technologies quickly enough, in order to use the window of opportunities for growth.

---

[1] High concentration >300 suns (HCPV), medium concentration $5 < \text{X} < 300$ suns (MCPV), low concentration <5 suns (LCPV)

With market estimates for 2010 in the 5 to 10 MW range, the market share of CPV is still small, but analysts forecast an increase to more than 1,000 MW globally by 2015. At the moment, the CPV pipeline is dominated by just three system manufacturers: Concentrix Solar, Amonix, and SolFocus.

The existing photovoltaic technology mix is a solid foundation for future growth of the sector as a whole. No single technology can satisfy all the different consumer needs, ranging from mobile and consumer applications, with the need for a few watts to multi MW utility-scale power plants. The variety of technologies is an insurance against a roadblock for the implementation of solar photovoltaic electricity if material limitations or technical obstacles restrict the further growth or development of a single technology pathway.

2. SOLAR CELL PRODUCTION[2] COMPANIES

Worldwide, more than 350 companies produce solar cells. The following chapter gives a short description of the 20 largest companies, in terms of actual production/shipments in 2010. More information about additional solar cell companies and details can be found in various market studies and in the country chapters of this report. The capacity, production, or shipment data are from the annual reports or financial statements of the respective companies or the cited references.

2.1 Suntech Power Co. Ltd. (PRC)

Suntech Power Co. Ltd. (www.suntech-power.com) is located in Wuxi. It was founded in January 2001 by Dr. Zhengrong Shi and went public in December 2005. Suntech specialises in the design, development, manufacturing and sale of photovoltaic cells, modules and systems. For 2010, Suntech reported shipments of 1507 MW, taking the top rank amongst the solar cell manufacturers. The annual production capacity of Suntech Power was increased to 1.8 GW by the end of 2010, and the company plans to expand its capacity to 2.4 GW in 2011.

---

[2] *Solar cell production capacities* means, in the case of wafer silicon based solar cells, only the cells. In the case of thin films, the complete integrated module. Only those companies that actually produce the active circuit (solar cell) are counted. Companies that purchase these circuits for further assembly are not counted.

2.2 JA Solar Holding Co. Ltd. (PRC)

JingAo Solar Co. Ltd. (www.jasolar.com) was established in May 2005 by the Hebei Jinglong Industry and Commerce Group Co. Ltd., the Australia Solar Energy Development Pty. Ltd. and Australia PV Science and Engineering Company. Commercial operation started in April 2006 and the company went public on 7 February 2007. According to the company, the production capacity should increase from 1.9 GW at the end of 2010 to 2.5 GW in 2011. For 2010, shipments of 1,460 MW are reported.

2.3 First Solar LLC. (USA/Germany/Malaysia)

First Solar LLC (www.firstsolar.com) is one of the few companies worldwide to produce CdTe-thin-film modules. The company has currently three manufacturing sites in Perrysburg (USA), Frankfurt/Oder (Germany) and in Kulim (Malaysia), which had a combined capacity of 1.5 GW at the end of 2010. The second Frankfurt/Oder plant, doubling the capacity there to 512 MW, became operational in May 2011 and the expansion in Kulim is on track to increase the production capacity to 2.3 GW at the end of 2011. Further expansions are under way in Meze (AZ), USA, and Dong Nam Industrial Park, Vietnam, to increase the production capacity to 2.9 GW at the end of 2012. The new factory planned in the framework of a joint venture with EdF Nuovelles in France is currently on hold. In 2010, the company produced 1.4 GW and currently sets the production cost benchmark with 0.75 \$/Wp (0.58 €/Wp) in the first quarter of 2011.

2.4 Sharp Corporation (Japan/Italy)

Sharp (www.sharp-world.com) started to develop solar cells in 1959 and commercial production got under way in 1963. Since its products were mounted on "Ume," Japan's first commercial-use artificial satellite, in 1974, Sharp has been the only Japanese maker to produce silicon solar cells for use in space. Another milestone was achieved in 1980, with the release of electronic calculators, equipped with single-crystal solar cells.

In 2010, Sharp had a production capacity of 1,070 MWp/year, and shipments of 1.17 GW were reported [9]. Sharp has two solar cell factories in Japan, Katsuragi, Nara Prefecture, (550 MW c-Si and 160 MW a-Si their triple-junction thin-film solar cell) and Osaka (200 MW c-Si and 160 MW a-Si), one together with Enel Green Power and STMicroelectronics in

Catania, Italy (initial capacity 160 MW at the end of 2011), six module factories and the Toyama factory to recycle and produce silicon. Three of the module factories are outside Japan, one in Memphis, Tennessee, USA with 100 MW capacity; one in Wrexham, UK, with 500 MW capacity; and one in Nakornpathom, Thailand.

2.5 Trina Solar Ltd, PRC (PRC)

Trina Solar (www.trinasolar.com/) was founded in 1997 and went public in December 2006. The company has integrated product lines, from ingots to wafers and modules. In December 2005, a 30-MW monocrystalline silicon wafer product line went into operation. According to the company, the production capacity was 750 MW for ingots and wafers and 1.2 GW for cells and modules at the end of 2010. For 2011, it is planned to expand the capacities to 1.2 GW for ingots and wafers and to 1.9 GW for cells and modules. For 2010, shipments of 1.06 GW were reported.

In January 2010, the company was selected by the Chinese Ministry of Science and Technology to establish a State Key Laboratory to develop PV technologies within the Changzhou Trina PV Industrial Park. The laboratory is established as a national platform for driving PV technologies in China. Its mandate includes research into PV-related materials, cell and module technologies and system-level performance. It will also serve as a platform to bring together technical capabilities from the company's strategic partners, including customers and key PV component suppliers, as well as universities and research institutions.

2.6 Yingli Green Energy Holding Company Ltd. (PRC)

Yingli Green Energy (www.yinglisolar.com/) went public on 8 June 2007. The main operating subsidiary, Baoding Tianwei Yingli New Energy Resources Co. Ltd., is located in the Baoding National High-New Tech Industrial Development Zone. The company deals with the whole set, from solar wafers, cell manufacturing and module production. According to the company, production capacity reached 1 GW in July 2010. A further expansion project to 1.7 GW is ongoing and should be operational at the end of 2011. The financial statement for 2010 gave shipments of 1.06 GW.

In January 2009, Yingli acquired Cyber Power Group Limited, a development stage enterprise designed to produce polysilicon. Through its principle operating subsidiary, Fine Silicon, the company started trial production of solar-grade polysilicon in late 2009 and is expected to reach its full production capacity of 3,000 tons per year by the end of 2011.

In January 2010, the Ministry of Science and Technology of China approved the application to establish a national-level key laboratory in the field of PV technology development, the State Key Laboratory of PV Technology at Yingli Green Energy's manufacturing base in Baoding.

2.7 Q-Cells AG (Germany/Malaysia)

Q-Cells SE (www.qcells.de) was founded at the end of 1999 and is based in Thalheim, Sachsen-Anhalt, Germany. Solar cell production started in mid 2001, with a 12 MWp production line. In the 2010 Annual Report, the company stated that the nominal capacity was 1.1 GW by the end of 2010, 500 MW in Germany, and 600 MW in Malaysia. In 2010, production was 936 MW, 479 MW in Germany, and 457 MW in Malaysia.

In the first half of the last decade, Q-Cells broadened and diversified its product portfolio by investing in various other companies, or forming joint ventures. Since the first half of 2009, Q-Cells has sold most of these holdings and now has one fully owned solar cell manufacturing subsidiary, Solibro (CIGS) with a 2010 production of 75 MW.

2.8 Motech Solar (Taiwan/PRC)

Motech Solar (www.motech.com.tw) is a wholly owned subsidiary of Motech Industries Inc., located in the Tainan Science Industrial Park. The company started its mass production of polycrystalline solar cells at the end of 2000, with an annual production capacity of 3.5 MW. The production increased from 3.5 MW in 2001 to 850 MW in 2010. In 2009, Motech started the construction of a factory in China which should reach its nameplate capacity of 500 MW in 2011. Production capacity at the end of 2010 was given as 1.2 GW (860 MW in Taiwan and 340 MW in China).

In 2007, Motech Solar's Research and Development Department was upgraded to Research and Development Centre (R&D Centre), with the aim not only to improve the present production processes for wafer and cell production, but to develop next generation solar cell technologies.

At the end of 2009, the company announced that it acquired the module manufacturing facilities of GE in Delaware, USA.

2.9 Gintech Energy Corporation (Taiwan)

Gintech (www.gintech.com.tw/) was established in August 2005 and went public in December 2006. Production at Factory Site A, Hsinchu Science Park, began in 2007 with an initial production capacity of 260 MW and increased to 930 MW at the end of 2010. The company plans to expand capacity to 1.5 GW in 2011. In 2010, the company had a production of 827 MW [24].

2.10 Kyocera Corporation (Japan)

In 1975, Kyocera (http://global.kyocera.com/prdct/solar/) began with research on solar cells. The Shiga Yohkaichi Factory was established in 1980 and R&D and manufacturing of solar cells and products started with mass production of multicrystalline silicon solar cells in 1982. In 1993, Kyocera started as the first Japanese company to sell home PV generation systems.

Besides the solar cell manufacturing plants in Japan, Kyocera has module manufacturing plants in China (joint venture with the Tianjin Yiqing Group (10% share) in Tianjin since 2003); Tijuana, Mexico (since 2004); and in Kadan, Czech Republic (since 2005).

In 2010, Kyocera had a production of 650 MW and is also marketing systems that both generate electricity through solar cells and exploit heat from the sun for other purposes, such as heating water. The Sakura Factory, Chiba Prefecture, is involved in everything from R&D and system planning to construction and servicing and the Shiga Factory, Shiga Prefecture, is active in R&D, as well as the manufacturing of solar cells, modules, equipment parts, and devices, which exploit heat. Like solar companies, Kyocera is planning to increase its current capacity of 650 MW in 2010 to 800 MW in 2011 and 1 GW in 2012.

2.11 SunPower Corporation (USA/Philippines/Malaysia)

SunPower (http://us.sunpowercorp.com/) was founded in 1988 by Richard Swanson and Robert Lorenzini to commercialise proprietary high-efficiency silicon solar cell technology. The company went public in November 2005. SunPower designs and manufactures high-performance silicon solar cells, based on an inter-digitated rear-contact design for commercial use. The initial products, introduced in 1992, were high-concentration solar cells with an efficiency of 26%. SunPower also manufactures a 22% efficient solar cell, called Pegasus, that is designed for non-concentrating applications.

SunPower conducts its main R&D activity in Sunnyvale, California, and has its cell manufacturing plant outside of Manila in the Philippines, with 590 MW capacity (Fab. No 1 and No 2). Fab. No. 3, a joint venture with AU Optronics Corporation (AUO), with a planned capacity of 1.4 GW, is currently under construction in Malaysia. Production in 2010 was reported at 584 MW.

2.12 Canadian Solar Inc. (PRC)

Canadian Solar Inc. was founded in Canada in 2001 and was listed on NASDAQ in November 2006. CSI has established six wholly owned manufacturing subsidiaries in China, manufacturing ingot/wafer, solar cells and solar modules. According to the company, it had 200 MW of ingot and wafer capacity, 800 MW cell capacity and 1.3 GW module manufacturing capacity in 2010. The company reports that it is on track to expand their solar cell capacity to 1.3 GW and the module manufacturing capacity to 2 GW, including 200 MW in Ontario, Canada, in 2011. For 2010, the company reported production of 522 MW solar cells and sales of 803 MW of modules.

2.13 Hanwah Solar One (PRC/South Korea)

Hanwah Solar One (www.hanwha-solarone.com) was established in 2004 as Solarfun Power Holdings, by the electricity meter manufacturer, Lingyang Electronics, the largest Chinese manufacturer of electric power meters. In 2010, the Korean company, Hanwha Chemical, acquired 49.99% of the shares and a name change was performed in January 2011. The company produces silicon ingots, wafers, solar cells, and solar modules. The first production line was completed at the end of 2004 and commercial production started in November 2005. The company went public in December 2006 and reported the completion of its production capacity expansion to 360 MW in the second quarter of 2008.

As of 30 April 2011, the company reported the following capacities: 1 GW PV module production capacity, 700 MW of cell production capacity, 415 MW of ingot production capacity and 500 MW of wire sawing capacity. It is planned to expand the module production capacity to 1.5 GW, cell production capacity to 1.3 GW, and ingot and wafer production capacity to 1 GW by the end of 2011.

The 2010 annual production was reported with 360 MW ingots, 387 MW wafers, 502 MW solar cells, and 759 modules.

2.14 Neo Solar Power Corporation (Taiwan)

Neo Solar Power (www.neosolarpower.com/) was founded in 2005 by PowerChip Semiconductor, Taiwan's largest DRAM company, and went public in October 2007. The company manufactures mono- and multicrystalline silicon solar cells and offers their SUPERCELL multicrystalline solar-cell brand with 16.8% efficiency. Production capacity of silicon solar cells at the end of 2010 was 820 MW, and the expansion to more than 1.3 GW is planned for 2011. In 2010, the company had shipments of about 500 MW.

2.15 Renewable Energy Corporation AS (Norway/Singapore)

REC's (www.recgroup.com/) vision is to become the most cost-efficient solar energy company in the world, with a presence throughout the whole value chain. REC is presently pursuing an aggressive strategy to this end. Through its various group companies, REC is already involved in all major aspects of the PV value chain. The company located in Høvik, Norway, has five business activities, ranging from silicon feedstock to solar system installations.

REC ScanCell is located in Narvik, producing solar cells. From the start-up in 2003, the factory has been continuously expanding. In 2010, production of solar cells was 452 MW, with a capacity at year end of 180 MWp in Norway and 550 MW in Singapore.

2.16 Solar World AG (German/USA)

Since its founding in 1998, Solar World (www.solarworld.de/) has changed from a solar system and components dealer to a company covering the whole PV value chain, from wafer production to system installations. The company now has manufacturing operations for silicon wafers, cells, and modules in Freiberg, Germany, and Hillsboro (OR), USA. Additional solar module production facilities exist in Camarillo (CA), USA, and since 2008 with a joint venture between Solarworld and SolarPark Engineering Co. Ltd. in Jeonju, South Korea.

For 2010, solar cell production capacities in Germany were reported at 250 MW and 500 MW in the USA. Total cell production in 2010 was 451 MW, with 200 MW coming from Germany and 251 MW from the USA.

In 2003, the Solar World Group was the first company worldwide to implement silicon solar cell recycling. The Solar World subsidiary, Deutsche Solar AG, commissioned a pilot plant for the reprocessing of crystalline cells and modules.

2.17 Sun Earth Solar Power Co. Ltd. (PRC)

Sun Earth Solar Power (www.nbsolar.com/), or NbSolar, has been part of China's PuTian Group since 2003. The company has four main facilities for silicon production, ingot manufacturing, system integration, and solar system production. According to company information, Sun Earth has imported solar cell and module producing and assembling lines from America and Japan.

In 2007, Sun Earth Solar Power relocated to the Ningbo high-tech zone, with the global headquarters of Sun Earth Solar Power. There the company produces wafers, solar cells, and solar modules. The second phase of production capacity expansion to 350 MW was completed in 2009. Further expansion is planned from 450 MW in 2010, 700 MW in 2011, and 1 GW in 2012. For 2010, shipments of 421 MW were reported [24].

2.18 E-TON Solartech Co. Ltd. (Taiwan)

E-Ton Solartech (www.e-tonsolar.com) was founded in 2001 by the E-Ton Group; a multinational conglomerate dedicated to producing sustainable technology and energy solutions and was listed on the Taiwan OTC stock exchange in 2006.

At the end of 2010, the production capacity was 560 MW per annum and a capacity increase to 820 MW is foreseen for 2011. Shipments of solar cells were reported at 420 MW for 2010.

2.19 SANYO Electric Company (Japan)

Sanyo (http://sanyo.com/solar/) commenced R&D for a-Si solar cells in 1975. 1980 marked the beginning of Sanyo's a-Si solar cell mass productions for consumer applications. Ten years later in 1990, research on the HIT (Heterojunction with Intrinsic Thin Layer) structure was started. In 1992, Dr. Kuwano (former president of SANYO) installed the first residential PV system at his private home. Amorphous Silicon modules for power use became available from SANYO in 1993 and in 1997 the mass production of HIT solar cells started. In 2010, Sanyo produced 405 MW solar cells [24]. The company announced increasing its 2009 production capacity of 500 MW HIT cells to 650 MW by 2011.

At the end of 2002, Sanyo announced the start of module production outside Japan. The company now has a HIT PV module production at SANYO Energy S.A. de C.V.'s Monterrey, Mexico, and it joined Sharp and Kyocera to set up module manufacturing plants in Europe. In 2005, it opened its module manufacturing plant in Dorog, Hungary.

Sanyo has set a world record for the efficiency of the HIT solar cell, with 23% under laboratory conditions [Tag 2009]. The HIT structure offers the possibility to produce double-sided solar cells, which has the advantage of collecting scattered light on the rear side of the solar cell and can therefore increase the performance by up to 30%, compared to one-sided HIT modules in the case of vertical installation.

2.20 China Sunergy

China Sunergy was established as CEEG Nanjing PV-Tech Co. (NJPV), a joint venture between the Chinese Electrical Equipment Group in Jiangsu and the Australian Photovoltaic Research Centre in 2004. China Sunergy went public in May 2007. At the end of 2008, the Company had five selective emitter (SE) cell lines, four HP lines, three capable of using multicrystalline and monocrystalline wafers, and one normal P-type line for multicrystalline cells, with a total nameplate capacity of 320 MW. At the end of 2010, the company had a cell capacity of 400 MW and a module capacity of 480 MW. For 2011, a capacity increase to 750 MW cells and 1.2 GW of modules is foreseen. For 2010, a production of 347 MW was reported.

3. POLYSILICON SUPPLY

The rapid growth of the PV industry since 2000 led to the situation where, between 2004 and early 2008, the demand for *polysilicon* outstripped the supply from the semiconductor industry. Prices for purified silicon started to rise sharply in 2007 and in 2008 prices for polysilicon peaked around 500 $/kg and consequently resulted in higher prices for PV modules. This extreme price hike triggered a massive capacity expansion, not only of established companies, but many new entrants as well. In 2009, more than 90% of total polysilicon, for the semiconductor and photovoltaic industry, was supplied by seven companies: Hemlock, Wacker Chemie, REC, Tokuyama, MEMC, Mitsubishi and Sumitomo. However, it is estimated that now about seventy producers are present in the market.

The massive production expansions, as well as the difficult economic situation, led to a price decrease throughout 2009, reaching about

50–55 $/kg at the end of 2009, with a slight upwards tendency throughout 2010 and early 2011.

For 2010, about 140,000 metric tons of solar grade silicon production were reported, sufficient for around 20 GW, under the assumption of an average materials need of 7 g/Wp [18]. China produced about 45,000 metric tons, or 32%, capable of supplying about 75% of the domestic demand [30]. According to the Semi PV Group Roadmap, the Chinese production capacity rose to 85,000 metric tons of polysilicon in 2010.

In January 2011, the Chinese Ministry of Industry and Information Technology tightened the rules for polysilicon factories. New factories must be able to produce more than 3,000 metric tons of polysilicon a year and meet certain efficiency, environmental, and financing standards. The maximum electricity use is 80 kWh/kg of polysilicon produced a year, and that number will drop to 60 kWh at the end of 2011. Existing plants that consume more than 200 kWh/kg of polysilicon produced at the end of 2011 will be shut down.

Projected silicon production capacities available for solar in 2012 vary between 250,000 metric tons [4] and 410,665 metric tons [9]. The possible solar cell production will in addition depend on the material use per Wp. Material consumption could decrease from the current 7 to 8 g/Wp down to 5 to 6 g/Wp, but this might not be achieved by all manufacturers.

3.1 Silicon Production Processes

The high growth rates of the photovoltaic industry and the market dynamics forced the high-purity silicon companies to explore process improvements, mainly for two chemical vapour deposition (CVD) approaches—an established production approach known as the Siemens process, and a manufacturing scheme based on fluidised bed (FB) reactors. Improved versions of these two types of processes will very probably be the workhorses of the polysilicon production industry for the near future.

3.1.1 Siemens Process

In the late 1950s, the Siemens reactor was developed and has been the dominant production route ever since. About 80% of total polysilicon manufactured worldwide was made with a Siemens-type process in 2009. The Siemens process involves deposition of silicon from a mixture of purified silane or trichlorosilane gas, with an excess of hydrogen onto high-purity polysilicon filaments. The silicon growth then occurs inside an insulated reaction chamber or "bell jar," which contains the gases. The filaments are assembled as electric circuits in series and are heated to the vapour deposition temperature by an external direct current. The silicon filaments are heated to very high temperatures between 1,100–1,175°C at which trichlorsilane, with the help of the hydrogen, decomposes to elemental silicon and deposits as a thin-layer film onto the filaments. Hydrogen Chloride (HCl) is formed as a by-product.

The most critical process parameter is temperature control. The temperature of the gas and filaments must be high enough for the silicon from the gas to deposit onto the solid surface of the filament, but well below the melting point of 1,414°C, that the filaments do not start to melt. Second, the deposition rate must be well controlled and not too fast, because otherwise the silicon will not deposit in a uniform, polycrystalline manner, making the material unsuitable for semiconductor and solar applications.

3.1.2 Fluidised Bed Process

A number of companies develop polysilicon production processes based on fluidised bed (FB) reactors. The motivation to use the FB approach is the potentially lower energy consumption and a continuous production, compared to the Siemens batch process. In this process, tetrahydrosilane or trichlorosilane and hydrogen gases are continuously introduced onto the bottom of the FB reactor at moderately elevated temperatures and pressures. At a continuous rate, high-purity silicon seeds are inserted from the top and are suspended by the upward flow of gases. At the operating temperatures of 750°C, the silane gas is reduced to elemental silicon and deposits on the surface of the silicon seeds. The growing seed crystals fall to the bottom of the reactor where they are continuously removed.

MEMC Electronic Materials, a silicon wafer manufacturer, has been producing granular silicon from silane feedstock, using a fluidised bed approach for over a decade. Several new facilities will also feature variations of the FB. Several major players in the polysilicon industry, including Wacker Chemie and Hemlock, are developing FB processes, while at the same time continuing to produce silicon using the Siemens process as well.

Upgraded metallurgical grade (UMG) silicon was seen as one option to produce cheaper solar grade silicon with 5- or 6-nines purity, but the support for this technology is waning in an environment where higher-purity methods are cost-competitive. A number of companies delayed or suspended their UMG-silicon operations as a result of low prices and lack of demand for UMG material for solar cells.

4. POLYSILICON MANUFACTURERS

World-wide more than 100 companies produce or start up polysilicon production. The following section gives a short description of the ten largest companies in terms of production capacity in 2010. More information about additional polysilicon companies and details can be found in various market studies and the country chapters of this report.

4.1 Hemlock Semiconductor Corporation (USA)

Hemlock Semiconductor Corporation (www.hscpoly.com) is based in Hemlock, Michigan. The corporation is a joint venture of Dow Corning Corporation (63.25%) and two Japanese firms, Shin-Etsu Handotai Company, Ltd. (24.5%) and Mitsubishi Materials Corporation (12.25%). The company is the leading provider of polycrystalline silicon and other silicon-based products used in the semiconductor and solar industry.

In 2007, the company had an annual production capacity of 10,000 tons of polycrystalline silicon and production at the expanded Hemlock site (19,000 tons) started in June 2008. A further expansion at the Hemlock site, as well as a new factory in Clarksville, Tennessee, was started in 2008 and brought total production capacity to 36,000 tons in 2010. A further expansion to 40,000 tons in 2011 and 50,000 tons in 2012 is planned [9].

4.2 Wacker Polysilicon (Germany)

Wacker Polysilicon AG (www.wacker.com), is one of the world's leading manufacturers of hyper-pure polysilicon for the semiconductor and photovoltaic industry, chlorosilanes and fumed silica. In 2010, Wacker increased its capacity to over 30,000 tons and produced 30,500 tons of polysilicon. The next 10,000 tons expansion in Nünchritz (Saxony), Germany, started production in 2011. In 2010, the company decided to build a polysilicon plant in Tennessee with 15,000 tons capacity. The groundbreaking of the new factory was in April 2011, and the construction should be finished at the end of 2013.

4.3 OCI Company (South Korea)

OCI Company Ltd. (formerly DC Chemical) (www.oci.co.kr/) is a global chemical company with a product portfolio spanning the fields of inorganic chemicals, petro and coal chemicals, fine chemicals, and renewable energy materials. In 2006, the company started its polysilicon business and successfully completed its 6,500 metric ton P1 plant in December 2007. The 10,500 metric ton P2 expansion was completed in July 2009 and P3 with another 10,000 metric tons brought the total capacity to 27,000 metric tons at the end of 2010. The debottlenecking of P3, foreseen in 2011, should then increase the capacity to 42,000 tons at the end of the year. Further capacity expansions P4 (20,000 tons by 2012) and P5 (24,000 tons by 2013) have already started (P4) or will commence in the second half of this year (P5).

4.4 GCL-Poly Energy Holdings Limited (PRC)

GCL-Poly (www.gcl-poly.com.hk) was founded in March 2006 and started the construction of their Xuzhou polysilicon plant (Jiangsu Zhongneng Polysilicon Technology Development Co. Ltd.) in July 2006. Phase I has a designated annual production capacity of 1,500 tons, and the first shipments were made in October 2007. Full capacity was reached in March 2008. At the end of 2010, polysilicon production capacity had reached 21,000 tons and further expansions to 46,000 tons in 2011 and 65,000 tons in 2012 are underway. For 2010, the company reported a production 17,850 metric tons of polysilicon.

In August 2008, a joint-venture, Taixing Zhongneng (Far East) Silicon Co. Ltd., started pilot production of trichlorsilane. Phase I will be 20,000 tons, to be expanded to 60,000 tons in the future.

4.5 MEMC Electronic Materials Inc. (USA)

MEMC Electronic Materials Inc. (www.memc.com/) has its headquarters in St. Peters, Missouri. It started operations in 1959 and the company's products are semiconductor-grade wafers, granular polysilicon, ultra-high purity silane, trichlorosilane (TCS), silicon tetraflouride (SiF4), sodium aluminium tetraflouride (SAF). MEMC's production capacity in 2008 was increased to 8,000 tons and to 9,000 tons in 2009 [9].

4.6 Renewable Energy Corporation AS (Norway)

REC's (www.recgroup.com/) vision is to become the most cost-efficient solar energy company in the world, with a presence throughout the whole value chain. REC is presently pursuing an aggressive strategy to this end. Through its various group companies, REC is already involved in all major aspects of the PV value chain. The company located in Høvik, Norway, has five business activities, ranging from silicon feedstock to solar system installations.

In 2005, Renewable Energy Corporation AS ("REC") took over Komatsu's US subsidiary, Advanced Silicon Materials LLC ("ASiMI"), and announced the formation of its silicon division business area, "REC Silicon Division," comprising the operations of REC Advanced Silicon Materials LLC (ASiMI) and REC Solar Grade Silicon LLC (SGS). Production capacity at the end of 2010 was around 17,000 tons [9] and according to the company, 11,460 tons electronic-grade silicon was produced in 2010.

4.7 LDK Solar Co. Ltd. (PRC)

LDK (www.ldksolar.com/) was set up by the Liouxin Group, a company which manufactures personal protective equipment, power tools, and elevators. With the formation of LDK Solar, the company is diversifying into solar energy products. LDK Solar went public in May 2007. In 2008, the company announced the completion of the construction and the start of polysilicon production in its 1,000 metric tons polysilicon plant. According to the company, the total capacity was 12,000 metric tons at the end of 2010, which will be increased to 25,000 tons in 2011. In 2010, polysilicon production was reported at 5,050 tons.

4.8 Tokuyama Corporation (Japan)

Tokuyama (www.tokuyama.co.jp/) is a chemical company involved in the manufacturing of solar-grade silicon, the base material for solar cells. The company is one of the world's leading polysilicon manufacturers and produces roughly 16% of the global supply of electronics and solar grade silicon. According to the company, Tokuyama had an annual production capacity of 5,200 tons in 2008 and has expanded this to 9,200 tons in 2010. In February 2011, the company broke ground for a new 20,000 ton facility in Malaysia. The first phase with 6,200 tons should be finished in 2013.

A verification plant for the vapour to liquid-deposition process (VLD method) of polycrystalline silicon for solar cells has been completed in December 2005. According to the company, steady progress has been made with the verification tests of this process, which allows a more effective manufacturing of polycrystalline silicon for solar cells.

Tokuyama has decided to form a joint venture with Mitsui Chemicals, a leading supplier of silane gas. The reason for this is the increased demand for silane gas, due to the rapid expansion of amorphous/microcrystalline thin-film solar cell manufacturing capacities.

4.9 Kumgang Korea Chemical Company (South Korea)

Kumgang Korea Chemical Company (KCC) was established by a merger of Kumgang and the Korea Chemical Co. in 2000. In February 2008, KCC announced its investment in the polysilicon industry and began to manufacture high-purity polysilicon with its own technology at the pilot plant of the Daejuk factory in July of the same year. In February 2010, KCC started to mass-produce polysilicon, with an annual capacity of 6,000 tons.

4.10 Mitsubishi Materials Corporation (Japan)

Mitsubishi Materials (www.mmc.co.jp) was created through the merger Mitsubishi Metal and Mitsubishi Mining & Cement in 1990. Polysilicon production is one of the activities in their Electronic Materials & Components business unit. The company has two production sites for polysilicon, one in Japan and one in the USA (Mitsubishi Polycrystalline Silicon America Corporation) and is a shareholder (12.25%) in Hemlock Semiconductor Corporation. With the expansion of the Yokkachi, Mie, Japan, polysilicon plant, by 1,000 tons in 2010, total production capacity was increased to 4,300 tons.

REFERENCES TO APPENDIXES D AND E

[1] Asian Development Bank, Asia Solar Energy Initiative: A Primer, ISBN 978-92-9092-314-5, (April 2011).
[2] M. Watt, R. Passey and W. Johnston, PV in Australia 2010—Australian PV Survey Report 2010, Australian PV Association May 2011.
[3] Asociación de la Industria Fotovoltaica (ASIF), <http://www.asif.org/principal.php?idseccion = 565/>.
[4] Johannes Bernreuther and Frank Haugwitz, The Who's Who of Silicon Production, (2010).
[5] Bundesverband Solarwirtschaft, Statistische Zahlen der deutschen Solarwirtschaft, (June 2011).
[6] German Federal Network Agency (Bundesnetzagentur), Press Release 21 March 2011.
[7] German Federal Network Agency (Bundesnetzagentur), Press Release 16 June 2011.

[8] The Daily Star, Target 500 MW solar project, (15 May 2011), <http://www.thedailystar.net/newDesign/news-details.php?nid = 185717/>.
[9] Gesetz über den Vorrang Erneuerbaren Energien (Erneuerbare-Energien-Gesetz—EEG), Bundesgestzblatt Jahrgang 2000 Teil I, Nr. 13, p.305 (29.03.2000).
[10] The US PV Market in 2011—Whitepaper, 2011, Greentech Media Inc., Enfinity America Corporation.
[11] European Photovoltaic Industry Association, 2011, Global Market Outlook for Photovoltaics until 2015.
[12] European Wind Energy Association, Wind in power—2010 European Statistics, (February 2011).
[13] Gazzetta Ufficiale, n. 109, 12 maggio 2011, Ministero dello sviluppo economico, D. M. 5-5-2011; Incentivazione della produzione di energia elettrica da impianti solari fotovoltaici.
[14] Gestore Servici Energetici, Press Release, 15 February 2011.
[15] Gestore Servici Energetici, Aggiornamento: 1 July 2011.
[16] Royal Decree 1565/10, published on 23 November 2010 <http://www.boe.es/boe/dias/2010/11/23/pdfs/BOE-A-2010-17976.pdf/>.
[17] Royal Decree RD-L 14/10, published on 24 December 2010 <http://www.boe.es/boe/dias/2010/12/24/pdfs/BOE-A-2010-19757.pdf/>.
[18] ICIS news, Asia polysilicon prices to firm in 2011 on solar demand,13 January 2011.
[19] Osamu Ikki, PV Activities in Japan, 17(5) (May 2011).
[20] Ministère de l'économie, de l'industrie et de l'emploi, Press Release, 24 February 2011.
[21] Ontario Power Authotity, Feed-In Tariff Programme, 30 September 2009 <http://fit.powerauthority.on.ca/Storage/97/10759_FIT-Program-Overview_v1.1.pdf/>.
[22] Photon International, March 2011.
[23] Platts, *Power in Europe*, January 2011.
[24] PV News, May 2011, published by Greentech Media, ISSN 0739-4829.
[25] Reuters, 06 May 2011, China doubles solar power target to 10 GW by 2015, <http://www.reuters.com/article/2011/05/06/china-solar-idUKL3E7G554620110506/>.
[26] Presidential Regulation 5/2006, *National Energy Policy*, published 25 January 2006.
[27] Republic of the Philippines, Congress of the Philippines, Republic Act No. 9513 December 16, 2008, AN ACT PROMOTING THE DEVELOPMENT, UTILIZATION AND COMMERCIALIZATION OF RENEWABLE ENERGY RESOURCES AND FOR OTHER PURPOSES.
[28] Solar Energy Industry Association (SEIA), U.S. Solar Market Insight, US Solar Industry Year in Review 2010.
[29] Solar Energy Industry Association (SEIA), U.S. Solar Market Insight, 1st Quarter 2011.
[30] Semi PV Group, Semi China Advisory Committee and China PV Industry Alliance (CPIA), China's Solar Future—A Recommended China PV Policy Roadmap 2.0, April 2011.
[31] Systèmes Solaires, le journal du photovoltaique no 5—2011, Photovoltaic Energy Barometer, April 2011, ISSN 0295-5873.
[32] UNB connect, ADB assures fund for 500 MW solar system, 4 June 2011, http://www.unbconnect.com/component/news/task-show/id-49440.
[33] The World Bank, May 2010, Winds of Change—East Asia's Sustainable Energy Future.

APPENDIX F

Useful Web Sites and Journals

Tom Markvart, Augustin McEvoy, and Luis Castañer

There is a large and growing number of Web sites, journals, and newsletters with renewable energy content. This appendix contains a small selection of those that contain useful information relating to photovoltaics and solar cells.

WEB SITES

www.pvpower.com/
Effectively a PV catalogue, components, systems, etc., with a "Learn About" section.

www.solarbuzz.com
PV industry information newsletter with recent regional reports, etc.

www.solaraccess.com/
USA site for information on solar commercial organisations, etc.

www.solarenergy.org/
Solar Energy International: renewable energy education and training.

www.iea-pvps.org/
Photovoltaic Power Systems Programme of the International Energy Agency (IEA). A wealth of information and IEA reports, many of which can be downloaded from the site.

www.pvresources.com
German and European newsletter with useful information.

www.censolar.es
Spanish Solar Energy education and training (in Spanish).

www.ises.org/ises.nsf

International Solar Energy Society: newsletter, conferences, and other information.

www.nrel.gov/pv

U.S. National Renewable Energy Laboratory, Colorado. For resource data on all types of renewable energy techniques, go to www.nrel.gov/rredc/

http://ec.europa.eu/research/energy/eu/research/photovoltaics/index_en.htm

European Union research activities in PV.

www.epia.org/

European PV Industries Association.

www.seia.org/

Solar Energy Industries Association (SEIA), Washington, D.C., USA

www.oerlikon.com/solar/

Specialist in thin-film silicon PV devices.

www.dyesol.com

Australian site for dye-sensitised PV.

www.globalsolar.com/

Specialist in CIS cells.

www.firstsolar.com/en/index.php

Specialist in CdTe thin film cells.

www.ise.fraunhofer.de/

German Institute for Solar Energy Systems, particularly III−V devices (in German).

www.pv.unsw.edu.au

Centre for Photovoltaic Engineering, UNSW. Information about its activities and courses. Specialist in high-efficiency silicon devices.

www.eurosolar.org

Web site of Eurosolar European Association for Renewable Energy.

http://energy.sandia.gov/?page_id = 2727

Sandia National Laboratories, USA: PV activities.

www.pvtaiwan.com/

Taiwan PV information.

www.pvinsider.com

PV newsletter site.

www.soda-is.com

SoDa Project. Detailed solar radiation data for applications including PV, energy in buildings, vegetation, oceanography, and health.

www.satel-light.com

Solar radiation data for Europe; includes a blog on solar-related issues.

www.pv.bnl.gov

Photovoltaic Environmental, Research Assistance Center, Brookhaven National Laboratory: specifically, health and safety issues.

www.ioffe.rssi.ru/SVA/NSM/Semicond/

Russian site, managed by Ioffe Institute, specifically on semiconductor materials and properties.

www.semiconductors.co.uk/

British-based site on semiconductor information; references to conferences and meetings, commercial resources, etc.

www.solarweb.net

Spanish-language renewable energy Web site; includes installation regulations for grid connection in Spain.

JOURNALS

Solar Energy Materials and Solar Cells

published by Elsevier: www.elsevier.com/wps/find/journaldescription.cws_home/505675/description#description

Progress in Photovoltaics

Published by Wiley: http://onlinelibrary.wiley.com/journal/10.1002/%28ISSN%291099-159X

Photon International

Solar Verlag GmbH, Wilhelmstrasse 34, D-52070 Aachen. Germany: www.photon-magazine.com/

Solar Energy

Published by Elsevier; official journal of the International Solar Energy Society: www.elsevier.com/wps/find/journaldescription.cws_home/329/description#description

APPENDIX G

International Standards with Relevance to Photovoltaics

Tom Markvart, Augustin McEvoy, and Luis Castañer

Standards published by the International Electrotechnical Commission. IEC Central Office, 3, rue de Varembé, P.O. Box 131, CH-1211 GENEVA 20, Switzerland. <www.iec.ch/>.

For documents of TC82 "Solar Photovoltaic Energy Systems," the technical committee responsible for PV, see Web site, <www.iec.ch/dyn/www/f?p = 103:23:0: : : : FSP_ORG_ID,FSP_LANG_ID: 1276,25>.

An additional TC47 semiconductor devices subcommittee deals with, among other activities, PV cells from the component aspect and maintains a liaison between relevant committees. For a reference resource on standards, go to <www.pvresources.com/en/standards.php>.

IEC STANDARDS

IEC 60364-7-712: 2002. Electrical installations of buildings. Part 7–712: Requirements for special installations or locations—Solar photovoltaic (PV) power supply systems

IEC 60891: 1987. Procedures for temperature and irradiance corrections to measured I–V characteristics of crystalline silicon photovoltaic devices.

IEC 60904-1: 1987. Photovoltaic devices. Part 1: Measurement of photovoltaic current–voltage characteristics.

IEC 60904-2: 1989. Photovoltaic devices. Part 2: Requirements for reference solar cells.

IEC 60904-3: 1998. Photovoltaic devices. Part 3: Measurement principles for terrestrial photovoltaic (PV) solar devices with reference spectral irradiance data.

IEC 60904-5: 1996. Photovoltaic devices. Part 5: Determination of the equivalent cell temperature (ECT) of photovoltaic (PV) devices by the open-circuit voltage method.

IEC 60904-6: 1994. Photovoltaic devices. Part 6: Requirements for reference solar modules.
IEC 60904-7: 1987. Photovoltaic devices. Part 7: Computation of spectral mismatch error introduced in the testing of a photovoltaic device.
IEC 60904-8: 1998. Photovoltaic devices. Part 8: Measurement of spectral response of a photovoltaic (PV) device.
IEC 60904-9: 1995. Photovoltaic devices. Part 9: Solar simulator performance requirements.
IEC 60904-10: 1998. Photovoltaic devices. Part 10: Methods of linearity measurement.
IEC 61173: 1992. Overvoltage protection for photovoltaic (PV) power generating systems. Guide.
IEC 61194: 1992. Characteristic parameters of stand-alone photovoltaic (PV) systems.
IEC 61215: 1993. Crystalline silicon terrestrial photovoltaic (PV) modules—design qualification and type approval.
IEC 61277: 1995. Terrestrial photovoltaic (PV) power generating systems. General and guide.
IEC 61345: 1998. UV test for photovoltaic (PV) modules.
IEC 61427: 1999. Secondary cells and batteries for solar photovoltaic energy systems. General requirements and methods of testing.
IEC 61646: 1996. Thin-film terrestrial photovoltaic (PV) modules—design qualification and type approval.
IEC 61683: 1999. Photovoltaic systems. Power conditioners. Procedure for measuring efficiency.
IEC 61701: 1995. Salt-mist corrosion testing of photovoltaic (PV) modules.
IEC 61702: 1995. Rating of direct coupled photovoltaic (PV) pumping systems.
IEC 61721: 1995. Susceptibility of a photovoltaic (PV) module to accidental impact damage (resistance to impact test).
IEC 61724: 1998. Photovoltaic system performance monitoring. Guidelines for measurement, data exchange, and analysis.
IEC 61725: 1997. Analytical expression for daily solar profiles.
IEC 61727: 1995. Photovoltaic (PV) systems. Characteristics of the utility interface.
IEC 61829: 1995. Crystalline silicon photovoltaic (PV) array. On-site measurement of I–V characteristics.

IEC/TR261836: 1997. Solar photovoltaic energy systems—Terms and symbols.

IEC/PAS 62111: 1999. Specifications for the use of renewable energies in rural decentralised electrification.

DRAFT IEC STANDARDS

IEC 61215 Ed. 2.0. Crystalline silicon terrestrial photovoltaic (PV) modules—Design qualification and type approval.

IEC 62116. Testing procedure of islanding prevention measures for grid connected photovoltaic power generating systems.

IEC 61727 Ed. 2.0. Characteristics of the utility interface for photovoltaic (PV) systems.

IEC 61730-1 Ed. 1.0. Photovoltaic module safety qualification—Part 1: Requirements for construction.

IEC 61730-2 Ed. 1.0. Photovoltaic module safety qualification—Part 2: Requirements for testing.

IEC 61836 TR Ed. 2.0. Solar photovoltaic energy systems—Terms and symbols.

IEC 61853 Ed. 1.0. Performance testing and energy rating of terrestrial photovoltaic (PV) modules.

IEC 62093 Ed. 1.0. Balance-of-system components for photovoltaic systems—Design qualification natural environments.

IEC 62108 Ed. 1.0. Concentrator photovoltaic (PV) receivers and modules—Design qualification and type approval.

IEC 62109 Ed. 1.0. Electrical safety of static inverters and charge controllers for use in photovoltaic (PV) power systems.

IEC 62116 Ed. 1.0. Testing procedure—Islanding prevention measures for power conditioners used in grid connected photovoltaic (PV) power generation systems.

IEC 62124 Ed. 1.0. Photovoltaic (PV) stand alone systems—Design verification.

IEC 62145 Ed. 1.0. Crystalline silicon PV modules—Blank detail specification.

IEC 62234 Ed. 1.0. Safety guidelines for grid connected photovoltaic (PV) systems mounted on buildings.

IEC 62253 Ed. 1.0. Direct coupled photovoltaic pumping systems—Design qualification and type approval.

IEC 62257 Ed. 1.0. Recommendations for the use of renewable energies in rural decentralised electrification.
IEC 62257-1 TS Ed. 1.0. Recommendations for small renewable energy and hybrid systems for rural electrification—Part 1: General introduction to rural electrification.
IEC 62257-2 TS Ed. 1.0. Recommendations for small renewable energy and hybrid systems for rural electrification—Part 2: From requirements to a range of electrification systems.
PNW 82-304 Ed. 1.0. Photovoltaic module safety qualification—Part 1: Requirements for construction.
PNW 82-306 Ed. 1.0. Photovoltaic module safety qualification—Part 2: Requirements for testing.
PNW 82-314 Ed. 1.0. Procedures for establishing the traceability of the calibration of photovoltaic reference devices.
PWI 82-1 Ed. 1.0. Photovoltaic electricity storage systems.

ASTM STANDARDS

Standards published by ASTM International (formerly American Society for Testing and Materials), 100 Barr Harbor Drive, PO Box C700, West Conshohocken, Pennsylvania, USA 19428–2959; www.astm.org

E490–00a. Standard solar constant and zero air mass solar spectral irradiance tables.
E816–95. Standard test method for calibration of secondary reference pyrheliometers and pyrheliometers for field use.
E824–94. Standard test method for transfer of calibration from reference to field radiometers.
E913–82 (1999). Standard method for calibration of reference pyranometers with axis vertical by the shading method.
E941–83 (1999). Standard test method for calibration of reference pyranometers with axis tilted by the shading method.
E1362–99. Standard test method for calibration of non-concentrator photovoltaic secondary reference cells.
E1125–99. Standard test method for calibration of primary non-concentrator terrestrial photovoltaic reference cells using a tabular spectrum.

E948−95 (2001). Standard test method for electrical performance of photovoltaic cells using reference cells under simulated sunlight.

E973−02. Standard test method for determination of the spectral mismatch parameter between a photovoltaic device and a photovoltaic reference cell.

E973M-96. Standard test method for determination of the spectral mismatch parameter between a photovoltaic device and a photovoltaic reference cell [metric].

E1040−98. Standard specification for physical characteristics of nonconcentrator terrestrial photovoltaic reference cells.

E1036M-96e2. Standard test methods for electrical performance of nonconcentrator terrestrial photovoltaic modules and arrays using reference cells.

E1039−99. Standard test method for calibration of silicon nonconcentrator photovoltaic primary reference cells under global irradiation.

E1830−01. Standard test methods for determining mechanical integrity of photovoltaic modules.

E1799−02. Standard practice for visual inspections of photovoltaic modules.

E1596−99. Standard test methods for solar radiation weathering of photovoltaic modules.

E1524−98. Standard test method for saltwater immersion and corrosion testing of photovoltaic modules for marine environments.

E1171−01. Standard test method for photovoltaic modules in cyclic temperature and humidity environments.

E1597−99. Standard test method for saltwater pressure immersion and temperature testing of photovoltaic modules for marine environments.

E1143−99. Standard test method for determining the linearity of a photovoltaic device parameter with respect to a test parameter.

E1021−95(2001). Standard test methods for measuring spectral response of photovoltaic cells.

E1802−01. Standard test methods for wet insulation integrity testing of photovoltaic modules.

E1462−00. Standard test methods for insulation integrity and ground path continuity of photovoltaic modules.

E1038−98. Standard test method for determining resistance of photovoltaic modules to hail by impact with propelled ice balls.

E2047−99. Standard test method for wet insulation integrity testing of photovoltaic arrays.
E1328−99. Standard terminology relating to photovoltaic solar energy conversion.
E927−91 (1997). Standard specification for solar simulation for terrestrial photovoltaic testing.
E2236−02. Standard test methods for measurement of electrical performance and spectral response of nonconcentrator multijunction photovoltaic cells and modules.
E1036−02. Standard test methods for electrical performance of nonconcentrator terrestrial photovoltaic modules and arrays using reference cells.
E782−95 (2001). Standard practice for exposure of cover materials for solar collectors to natural weathering under conditions simulating operational mode.
E881−92 (1996). Standard practice for exposure of solar collector cover materials to natural weathering under conditions simulating stagnation mode.
E822−92 (1996). Standard practice for determining resistance of solar collector covers to hail by impact with propelled ice balls.
G113−01. Standard terminology relating to natural and artificial weathering tests of nonmetallic materials.
G130−95. Standard test method for calibration of narrow- and broad-band ultraviolet radiometers using a spectroradiometer.
G138−96. Standard test method for calibration of a spectroradiometer using a standard source of irradiance.
G159−98. Standard tables for references solar spectral irradiance at air mass 1.5: direct normal and hemispherical for a 37° tilted surface.
Work item WK558. Reference solar spectral irradiances: Direct normal and hemispherical on 37 tilted surface.

OTHER STANDARDS AND GUIDELINES

Qualification test procedures for photovoltaic modules. **Specification No. 503.** Commission of the European Communities, Joint Research Centre, Ispra, Italy, 2011 <www.pvresources.com/en/standards.php>

ANSI/UL 1703. Standard for flat-plate photovoltaic modules and panels. American National Standards Institute, New York, USA, 1987. <www.ansi.org/>

UL 1741. Inverters, converters, and controllers for use in independent power systems. Underwriters Laboratories, Inc. <www.ul.com/global/eng/pages/>

NFPA 70. National Electrical Code (NEC). 1999. National Fire Protection Association, Quincy, MA 02269, USA. <www.nfpa.org/index.asp?>

ISO/DIS 15387: 2002. Space systems. Space solar cells. Requirements, measurements and calibration procedures. International Organization for Standardization, Geneva, Switzerland, 2002. <www.iso.ch/>

APPENDIX H

Books About Solar Cells, Photovoltaic Systems, and Applications

Tom Markvart, Augustin McEvoy, and Luis Castañer

Belén Cristóbal López, Ana, Antonio Martí Vega and Antonio L. Luque López, *Next Generation of Photovoltaics: New Concepts* (Springer, Heidelberg, Germany: 2011).
Brabec, Christoph, Ullrich Scherf and Vladimir Dyakonov, *Organic Photovoltaics: Materials, Device Physics, and Manufacturing Technologies* (Wiley-VCH, Weinheim, Germany: 2008).
Brendel, R. Thin-Film Crystalline Silicon Solar Cells (Wiley, Weinheim, Germany: 2003).
Bube, Richard H., *Photovoltaic Materials* (Imperial College Press, London: 1998).
Bubenzer, A. and J. Luther, *Photovoltaics Guidebook for Decision Makers* (Springer, Heidelberg, Germany: 2003).
Castañer, L. and S. Silvestre, *Modelling Photovoltaic Systems Using Pspice* (Wiley, Chichester, UK: 2002).
Fonash, Stephen, *Solar Cell Device Physics* (2nd ed.) (Academic Press/Elsevier, Burlington, MA, USA, and Kidlington, England: 2010).
Fraas, Lewis M. and Larry D. Partain, *Solar Cells and Their Applications* (Wiley, Hoboken, NJ, USA: 2010).
Goetzberger, A., J. Knobloch, and B. Voss, *Crystalline Silicon Solar Cells*, (Wiley, Chichester, UK: 1998).
Green, Martin A. *Third Generation Photovoltaics: Advanced Solar Energy Conversion* (Springer Verlag, Berlin, Heidelberg, Germany: 2003, 2006).
Hamakawa Yoshihiro, Thin-Film Solar Cells: Next Generation Photovoltaics and Its Applications (Springer Verlag, Berlin, Heidelberg, Germany: 2010).
Jha, A.R., *Solar Cell Technology and Applications* (Taylor and Francis, Boca Raton, FL, USA: 2010)

Komp, R. J. *Practical Photovoltaics: Electricity from Solar Cells* (Aatec Publishers, Ann Arbor, MI, USA: 2002).

Krebs, Frederik C., *Polymer Photovoltaics: A Practical Approach* (SPIE, Bellingham, WA, USA: 2008).

Kuppuswamy, Kalyanasundaram, *Dye-Sensitized Solar Cells* (EPFL Press, Lausanne: 2010).

Luque, Antonio and Steven Hegedus, *Handbook of Photovoltaic Science and Engineering* (Wiley, Chichester, UK: 2003, 2011).

Luque López, Antonio L. and Viacheslav M. Andreev, *Concentrator Photovoltaics* (Springer, Heidelberg, Germany: 2010).

Markvart, Tom and Castaner Luis, *Solar Cells: Materials, Manufacture and Operation* (Elsevier, Kidlington, UK: 2005).

Markvart, T. (Ed.), *Solar Electricity* (2nd ed.) (John Wiley & Sons, Chichester, UK: 2001).

Messenger, R. A. and J. G. Ventre, *Photovoltaic Systems Engineering* (CRC Press, Boca Raton, FL, USA: 1999).

Nelson, Jenny, *The Physics of Solar Cells* (Imperial College Press, London: 2003, 2004).

Perlin, J. *From Space to Earth: The Story of Solar Electricity* (Harvard University Press, Cambridge, MA, USA: 2002).

Poortmans, Jef, Thin Film Solar Cells: Fabrication, Characterization and Applications (Wiley, Chichester, UK: 2006).

Randall, T. (Ed.), *Photovoltaics and Architecture* (Routledge—Spon Press, London: 2001).

Rau, Uwe, Abou-Ras Daniel and Kirchartz Thomas, *Advanced Characterization Techniques for Thin Film Solar Cells* (Wiley-VCH, Weinheim, Germany: 2011).

Ross, M. and Royer, J. *Photovoltaics in Cold Climates* (James & James, London: 1998).

Scheer, H. *A Solar Manifesto* (James & James, London: 2001).

Smestad, G. P. *Optoelectronics of Solar Cells* (SPIE, Bellingham, WA, USA: 2002).

Levinstein, M., S. Rumyantsev, and M. Shur (Eds.), *Handbook Series on Semiconductor Parameters,* Vols. 1 and 2 (World Scientific, London: 1999).

Palik, E. D. (Ed.), *Handbook of Optical Constants of Solids II* (Academic Press, San Diego, CA, USA: 1991).

Pearsall, T. P. (Ed.), *Properties, Processing and Applications of Indium Phosphide* (IEE/INSPEC, The Institution of Electrical Engineers, London: 2000).

Scharmer, K., and J. Grief (Coordinators), *European Solar Radiation Atlas* (4th ed.) (Les Presses de l' École des Mines de Paris, Paris: 2000).

Scheer, Roland and Schock Hans-Werner, *Chalcogenide Photovoltaics: Physics, Technologies, and Thin Film Devices* (Wiley-VCH, Weinheim, Germany: 2011).

Sze, S. M. *Physics of Semiconductor Devices* (3rd ed.) (Wiley, Hoboken, NJ, USA: 2007).

Wenham, Stuart R., A. Green Martin, Muriel E. Watt, and Richard Corkish (Eds.), *Applied Photovoltaics* (Earthscan, London: 2007).

Würfel, Peter, *Physics of Solar Cells: From Basic Principles to Advanced Concepts* (Wiley-VCH, Weinheim, Germany: 2009).

INDEX

A

Absorbed photon conversion efficiency (APCE), dye-sensitized cells, 511
Absorbent glass mat lead-acid batteries, 745
Absorber preparation, copper-indium-gallium diselenide thin-film solar cells, 333–338
  co-evaporation processes, 334–335, 335f
  epitaxial techniques, 337
  postdeposition air annealing, 337–338
  selinisation processes, 335–337, 336f
Absorption coefficients
  hydrogenated microcrystalline silicon layers, 230, 231f
  semiconductor free-carrier absorption, 47
  semiconductor optical absorption, 44–45, 46f
  thin-film silicon solar cells, p-i-n/n-i-p structures, light trapping, 249–252, 250f
Accumulation layer, cadmium-telluride photovoltaic module, back contact, 301
Acid density, lead-acid batteries, discharge rate and, 749–750
Acid freezing point, lead-acid batteries, 752–753, 753f
Acid stratification, lead-acid batteries, 750–752, 751f, 752f
Activation energy, $Cu(In,Ga)Se_2$ thin-film solar cells, open-circuit voltage, 351
Active technologies, grid-connected photovoltaic systems, 788–789
  frequency variation, 788
  impedance measurement, 788
  power/reactive power variations, 789
Aesthetics, photovoltaic technology, 926–933
  colour, 931, 932f
  point of view, 933, 935f
  shape, 929–930, 931f
  size, 930
  texture and patterns, 931–932, 933f
  translucency, 932, 934f
Ageing, hydrogenated microcrystalline silicon layers, 233–234
Air annealing, $Cu(In,Ga)Se_2$ thin-film solar cells, postdeposition process, 337–338
Air conditioning, solar-powered systems, 996–997, 996f
Air fresheners, solar-powered products, 992
Air mass zero (AM0) efficiencies
  photovoltaic modules and solar cell performance measurement, 1048
  space applications of solar cells, 872–874
Air mass zero spectrum, space applications of solar cells, 865–867, 866f
Air temperature, photovoltaic system monitoring, 1077
All sky irradiation estimation, solar-radiation climatology, 632–637
  direct beam irradiation, inclined planes, 633–634
  ground-reflected irradiation, 636–637
  hourly diffuse irradiation, 634–636
  slope irradiation components, 632–633
  sun-facing surfaces, 636
Alternating current (AC)
  DC to AC power conversion (inversion)
    grid-connected photovoltaic systems, 780–782
    monitoring of, 1078
    photovoltaic system electronics, 703–708
    isolation, 708
    single-phase inverters, 703–707, 704f, 705f, 706f
    three-phase inverters, 707, 707f, 708f

1189

Aluminium-gallium arsenide (Al,Ga)As
solar cells, space applications.
See also Gallium arsenide (GaAs)
solar cells
research background, 400
single-junction III-V cell structure,
403−404, 405f
Aluminium treatment
gettering technique, crystalline silicon
solar cell substrates, 142−143
silicon space cell, 101
Amorphous silicon. *See also* Hydrogenated
amorphous silicon
solar cells, hazard analysis, photovoltaic
systems, 1088−1089
occupational safety issues,
1088−1089
public health and environmental
issues, 1089
thin films
energy requirements, 1103
field experience with, 272
hydrogenated amorphous silicon,
215−224
conductivity, 223−224
doping, 224
gap states, 220−223, 220f, 222f
research background, 210−211
structure, 215−219, 216f
research background, 210−211
tandem solar cells, 254−255, 255f
Amorphous silicon-carbon alloys, 219
Amorphous silicon-germanium alloys
thin-film silicon solar cells, 219
triple-junction solar cells,
255−257, 256f
Angle of incidence
photovoltaic array, annual energy
production, optical losses at high
angles, 646
solar-radiation climatology, 625−627
Angström regression formula, sunshine
data, radiation estimation, 603−604
Angular movements computation, solar-
radiation climatology, 618
altitude, azimuth angles, and astronomical
day length, 624−625, 625f

angle of incidence and vertical/
horizontal angle calculations,
625−627
trigonometric determination, 621−625,
622f
Annealing, organic solar cells, 558
Antireflection coating
crystalline silicon solar cells, low-cost
industrial technologies, 137−138
industrial solar cell technologies, 146
solar cell, 16−17, 18f
Aquatic solar-powered products, 994, 995f
Archaeology and cultural heritage, solar
parks and farms development, 957
Architectural integration, solar cells
aesthetics, 926−933
colour, 931, 932f
point of view, 933, 935f
shape, 929−930, 931f
size, 930
texture and patterns, 931−932, 933f
translucency, 932, 934f
building-integrated photovoltaics,
923−926, 924f, 925f, 926f
modular multifunctionality,
923−924
mounting techniques, 924−926
case studies, 934−940
energy system architecture, 919−921,
920f
British Pavilion, Expo '92 case study,
936
building energy availability, 923
Mont Cenis Conference Centre case
study, 940
Pompeu Fabra Public Library case
study, 938
photovoltaic technology overview,
919−923
research background, 918−919
surface availability in buildings,
921−923, 921f
Array cables, electrical systems,
photovoltaic structures
fault current, 826
production by design, 832−833
protection, 827

Index

Array design, solar parks and farms
  array mountings, 951–952, 952f
  fixed arrays, 949
  solar arrays, 948–952, 949f
  tracking arrays, 950–951, 950f, 951f
Array fault protection, electrical systems, photovoltaic structures, 824–830
  additional fault current sources, 830
  array cable-fault current, 826
  array cable protection, 827
  blocking diodes, 829
  earth faults, 824–825
  fault analysis, 825–826
  line-line faults, 825
  module reverse-current protection, 827–828
  photovoltaic module-reverse current, 826
  string cable-fault current, 825–826, 826f
  string cable protection, 827
  string overcurrent protection, 828–829, 829f
Array voltage, electrical systems, photovoltaic structures, 821–823
ASTM standards, 1184–1186
Astronomical day length, solar-radiation climatology, 624–625, 626f
Atmospheric pressure chemical vapour deposition (APCVD)
  antireflection coating, crystalline silicon solar cells, 137–138
  thin silicon solar cells, 185
    titanium dioxide deposition, 265–266
Atomic oxygen (ATOX) testing, space solar cells, 893–894
Auger effect
  carrier recombination, silicon wafers, 1014–1015
  interdigitated back-contact (IBC) silicon solar cells, 451–452
  semiconductor efficiency limitations, 72
Azimuth angles, solar-radiation climatology, 624–625

B

Back contact solar cells
  cadmium telluride thin-film photovoltaic modules, 299–301, 302f
  interdigitated back-contact silicon
    cell thickness reduction, 471
    concentrator applications, 450–452, 453f, 454f, 455f
    design criteria, 453–458
    efficiency improvement techniques, 468–472
    emitter saturation current density reduction, 470, 470f
    front-surface-field, tandem-junction, and point-contact cells, 456–458, 456f
    light trapping, 471
    low-contact resistance, 470–471
    manufacturing process, 464, 465f
    modelling of, 458–462
    perimeter and edge recombination, 462–464
    point-contact vs. incident power density, 473f
    research background, 449–450
    series resistance, 472
    shrink geometries, 471
    stability, 464–467
    structure, 452f
    target performance, 472, 473f
    thirty percent efficiency goal, 467–468
Back side passivation, crystalline silicon solar cells, low-cost industrial technologies, 138–139, 138f
Back-surface-field (BSF)
  crystalline silicon solar cells, low-cost industrial technologies, 138–139, 138f
  silicon space cell development, 101
  single-junction III-V solar cells, 404, 405f
Band diagram
  Cu(In,Ga)Se$_2$ thin-film solar cells, 346–347, 346f
  heterojunction solar cell, 25f

Band diagram (*Continued*)
  organic semiconductor operation, 548, 549f
  single-junction III-V solar cells, 403, 405f
Band gap. *See also* Indirect band gap
  amorphous silicon thin films, 220–223, 220f
  cadmium telluride intermixing and interdiffusion, back contact, 299–300, 300f, 301f
  Cu(In,Ga)Se$_2$ thin-film solar cells
    energies, 326, 326f, 331f
    wide-gap chalcopyrites, 355–357, 356f
  hydrogenated microcrystalline silicon layers, 228–231
  mechanically stacked multijunction (cascade) cells, space applications, 407, 407f
  multijunction III-V solar cells
    greater than 3 junctions, 436–439, 436f
    lattice constant *vs.*, 423–424
    selection, 421–422, 422f
    upright metamorphic growth, 430–432
  multijunction solar cells, basic properties, 417–419
  semiconductor doping, 56
  semiconductor materials, 34–38
  solar cell efficiencies, 63
Band-to-band Auger recombination
  minority-carrier lifetime, 52
  semiconductor classification, 50
  silicon wafers, 1014–1015
Band-to-band radiative recombination, semiconductor classification, 49
Batteries in photovoltaic systems
  basic properties, 723–724
  black box function of, 727–737
    cost options, 736–737
    lifetime limit, 735–736
    maximum charge rate, 736
    simple black box calculations, 733–737
    voltage specifications, 736

capacity calculations, 768–770
charge control, stand-alone system electronics, 708–711, 709f, 710f, 711f
commissioning, 773
depth of discharge
  maximum depth definition, 768–770
  maximum depth selection, 767–768
design considerations, 770–772
  charge control, 770–771
  internal heating, 771
  overcharge, 770–771
  water loss, 771
disposal, 773–774
duty cycle, 726–727, 728f
  charging energy availability, 726
  normal use, rechargeable batteries, 726
electrochemical properties, 737–739
examples of, 740–741
final specifications, 769–770
functions, 724–725
lead-acid batteries, 741–754
  absorbent glass mat batteries, 745
  acid density, 749–750, 749f
  acid stratification, 750–752, 751f, 752f
  basic principles, 741–742
  capacity variation, 746–749, 747f, 748f
  discharge rate, 746–747, 749–750
  freezing, 752–753, 753f, 754f
  gel-type batteries, 745
  grid alloy construction, 743–744
  grid thickness, 744
  low temperatures and capacity variations, 747–748
  mass-produced and industrial batteries, 745–746
  open *vs.* mass-produced construction, 742–743
  plate-type construction, 728f, 743
  sealed batteries, 744–745
  sulfation and deep discharge, 753–754
lifetime factors for, 756–765
  cycle life, 758

Index   1193

field data *vs.* predicted lifetimes, 762–765
float lifetime, 758–759
predicted lifetimes, 760–762, 761f
sealed batteries, 759–760
maintenance guidelines, 770–774
nickel-cadmium batteries, 754–756
performance definitions, 729–733
   capacity, efficiency and overcharge, 729–730
   charge/discharge rate, 730
   cycle life, 732–733
   depth of discharge/state of discharge, 731, 732–733
   flexible capacity, 730–731
   maximum lifetime, 733
   self-discharge rate, 732
replacement, 773–774
sealed batteries, lifetimes, 759–760
selection guidelines, 765–767
size calculations, 767–770
   seasonal energy balance, stand-alone photovoltaic systems, 670, 671f
temperature control, 772
ventilation environment, 771–772
voltage
   energy balance, stand-alone photovoltaic systems, 660–662, 662f, 668–669
   seasonal energy balance, 669f
   selection guidelines, 767
Beam splitting, solar radiation, stochastic estimation, 616
Bifacial epigrowth, multijunction III-V solar cells, 434
Bilayer architecture, organic solar cells, 550–551, 551f
Black box rules, batteries in photovoltaic systems, 727–737
cost options, 736–737
lifetime limit, 735–736
maximum charge rate, 736
simple black box calculations, 733–737
voltage specifications, 736

Blend additives, organic solar cells, 558–559
Blend composition, organic solar cells, 557–558
Blocking/bypass diodes
   electrical systems, photovoltaic structures, 829
   photovoltaic array, energy production, 655–658, 655f, 656f, 657f
Block scanners, minority-carrier crystalline silicon wafers, 84
Bond angle and length, hydrogenated amorphous silicon, 215, 216f
Boost converter, DC to DC power conversion, photovoltaic systems, 701–702, 701f
Boron-doped Czochralski crystalline silicon wafer, lifetime instabilities and measurement artifacts, boron-doped CZ silicon, 1028–1029
Bourges formula, solar-radiation climatology
   accurate noon declination, 627
   slope irradiation estimation, 633
Bragg reflector (BR), single-junction III-V solar cells, 404–406, 405f
Bring-through passivating oxide, industrial solar cell technologies, 146
British Pavilion, Expo '92 case study, architectural integration of solar cells, 934–936, 936f
Buck-boost converter, DC to DC power conversion, photovoltaic systems, 702, 702f
Buck converter, DC to DC power conversion, photovoltaic systems, 698–701, 699f, 700f
Buffer layer deposition
   cadmium telluride photovoltaic module, back contact, 301
   $Cu(In,Ga)Se_2$ thin-film solar cells, 339–340
   module fabrication, 343
Building-integrated photovoltaics (BIPV), architectural integration, solar cells, 923–926, 924f, 925f, 926f
   modular multifunctionality, 923–924
   mounting techniques, 924–926

Bulk heterojunction architecture, organic solar cells, 551–552, 552f
   microstructure optimisation, 557
Bulk recombination, thin-film silicon solar cells, p-i-n/n-i-p structures, 241–244
Buried-contact solar cell (BCSC), 116f, 118–119
   industrial solar cell technologies, 147–148

C
Cadmium sulfide
   cadmium telluride intermixing and interdiffusion, 295–298, 296f, 297f, 298f
   Cu(In,Ga)Se$_2$ thin-film solar cells, buffer layer deposition, 339–340
   film deposition, 288
   thin-film photovoltaic modules, 288
Cadmium telluride (CdTe)
   solar cells, hazard analysis, 1089–1090
      occupational health issues, 1089–1090
      public health and environmental issues, 1090
   thin-film photovoltaic modules
      applications, 316–319, 317f
         product qualification, 316–317
      basic properties, 284–285, 284f
      critical region improvement, 290–301, 290f
         activation, 299
         back contact, 299–301, 301f, 302f
         best cell performance, 302–303
         charge-carrier lifetime increase, 298–299, 299f
         grain growth-recrystallization, 292–295, 292f, 293f, 294f, 295f
         interdiffusion-intermixing, 295–298, 296f, 297f, 298f
         p-n heterojunction activation, 291–299, 291f
         stability issues, 302
      environmental and health aspects, 314–315
      material resources, 315–316
      film deposition, 285–289
         chemical spraying, 287
         electrodeposition, 287
         screen printing, 287
         vacuum deposition; sublimation and condensation, 286–287
      future research issues, 320
      industrial production, 307–309
      installation examples, 318–319, 318f, 319f
      module integration, 303 306
         cell interconnectivity, 303–304, 303f, 304f
         contacting, 304, 305f
         lamination, 305–306
         production sequence, 306–307
Cadmium-tin-oxides
   film deposition, 289
   thin-film photovoltaic modules, 289
Calculators, solar-powered, 990
Calibration
   photovoltaic modules and solar cells, 1046–1054
      certification and commercial services, 1061
      commercial equipment, 1056–1057
      degradation case study, 1061–1062, 1063f
      diagnostic measurements, 1054–1056
      energy production measurements, 1053–1054
      instrumentation and solar simulation, 1049–1051
      multijunction devices, 1052–1053
      performance test conditions, 1053
      potential problems and measurement uncertainty, 1054
      qualification testing, 1057–1061
      radiometry, 1046–1049, 1047f
      reliability testing, 1057–1061
      temperature, 1051–1052
   photovoltaic system monitoring, 1075–1076
   space solar cells, 883–886
      extraterrestrial methods, 883–884

secondary working standards, 886
synthetic methods, 884–886
Capacitance characterisation, space solar cells, electrical performance testing, 890–891
capacitively-coupled plasma-deposition system, thin silicon solar cells, 260f
layer deposition principles, 223
Capacity, batteries in photovoltaic systems
defined, 729–730
flexibility of, 730–731
lead-acid batteries, variations in, 746–749
size calculations and, 768–770
Carbon dioxide emissions, photovoltaic energy balance, 1112–1114
Carnot efficiency, formula for, 64
Carrier concentrations
interdigitated back-contact (IBC) silicon solar cells, 461
silicon wafer lifetimes, low-excess carrier densities, 1029–1030
Carrier mobility, semiconductors, 42–44
Carrier recombination
semiconductors, 38–40
silicon wafers, 1012–1015, 1013f
lifetime measurements, 1015–1030
Catalytic chemical vapour deposition, thin silicon solar cells, 262
Caughey-Thomas dependency, semiconductor mobility, 42
Cell efficiency, concentrator photovoltaic systems, 840–841, 841f
Cell interconnectivity
cadmium telluride thin-film photovoltaic modules, 303–304, 303f, 304f
silicon thin film solar cells, laser scribing and, 267–269, 267f, 268f
Cell thickness, interdigitated back-contact (IBC) silicon solar cells, reduction of, 471
Certification services, photovoltaic system metrics, 1061
Chalcopyrite structures

$Cu(In,Ga)Se_2$ thin-film solar cells, 325–326, 325f, 354–361, 356f
current research directions, 375–379
industrial production, 379–390
cell production companies, 382–390
technology transfer, 381–382
photovoltaic research background, 374–375
wide-gap chalcopyrites
$CuGaSe_2$ thin-film solar cells, 357–358
$Cu(In,Al)Se_2$ thin-film solar cells, 358
$Cu(In,Ga)S_2$ thin-film solar cells, 358–359
$Cu(In,Ga)Se_2$ thin-film solar cells, 354–361, 356f
basic properties, 354–357
$Cu(In,Ga)(Se,S)_2$ thin-film solar cells, 359
$CuInS_2$ thin-film solar cells, 358–359
graded gap devices, 359–361, 360f
Charge-carrier lifetime, cadmium telluride thin-film photovoltaic modules, 298–299, 299f
Charge-collection efficiency, dye-sensitized cells, small-modulation electron transport and recombination, 517
Charge controllers
batteries in photovoltaic systems, 770–772
photovoltaic system electronics, 715–717
Charge preparation, multicrystalline silicone preparation, 90
Charge rate, batteries in photovoltaic systems, defined, 730
Chargers, solar-powered, 991
Charge separation, organic semiconductor operation, 548–549
Chemical bath deposition, cadmium-sulfide, 288
Chemically textured nonreflecting "black" cell, silicon space cell development, 102f, 103
Chemical spraying, cadmium telluride thin-film photovoltaic modules, 287

Chemical vapour deposition (CVD), silicon thin film solar cells
  atmospheric pressure CVD, 185
  catalytic chemical vapour deposition, 262
  high-temperature silicon-deposition methods, 182–186, 183f
  low-pressure CVD, 186
  low-temparature chemical vapour deposition, 186–187
  rapid thermal chemical vapour deposition, 186
Cladding systems, solar-powered products, 1002
Cleaning process, crystalline silicon solar cells, 134–135
Clearness index, defined, 578
Climatic cycle, energy balance, stand-alone photovoltaic systems, 660, 660f, 661f, 662f
Cloudless sky global radiation
  latitude variations, 579–586, 580f, 581f, 582f, 583f
  solar radiation climatology, monthly mean daily profile estimation, 630
Co-evaporation processes, Cu(In,Ga)Se$_2$ thin-film solar cells
  absorber preparation, 334–335, 335f
  module fabrication, 342–343, 342f
Collection efficiency
  dye-sensitized cells, steady-state measurement, 519–520
  silicon thin film solar cells, p-i-n/n-i-p structures, 241–244
Colloidal semiconductors, semiconductor-electrolyte junction, 490
Colour, photovoltaic technology aesthetics, 931, 932f
Commercial photovoltaic systems. *See also* Industrial solar technologies
  cost of modules, 152–154
  market prices, 130
  nonconsumer solar-powered products, 1003
  performance metrics, 1056–1057
    module certification, 1061

thin film crystalline silicon solar cells, industrial production, 151
Commissioning, batteries in photovoltaic systems, 773
Common equivalent circuits, silicon thin film solar cells, p-i-n/n-i-p structures, shunts, 244–246, 245f
Community engagement and legacy issues, solar parks and farms development, 957–958
Concentrator photovoltaic (CPV) systems
  applications, 854, 855f
  classification, 845–850, 847f, 849f, 850f
  components and operation, 843–845, 845f, 846f
  energy production, 859–860
  future research issues, 860–861
  interdigitated back-contact (IBC) silicon solar cells, 450–452, 453f, 454f, 455f
  objectives of, 837–839, 838f
  physical principles, 839–843, 841f, 842f, 843f, 844f
  production capacity, 1158
  rating and specifications, 854–859
    IEEE qualification procedure, 858–859
    module ratings, 856–858
  tracking-control strategies, 851–853, 852f
Condensation, cadmium telluride thin-film photovoltaic modules, vacuum deposition, 286–287
Conduction band tail
  amorphous silicon thin fillms, gap states, 221–222
  photoelectrochemical cells, 486–487
Conductive parts, earthing of exposed parts, 831–832
Conductivity
  amorphous silicon thin films, 223–224
  microcrystalline silicon layers, 232
Connection approvals, solar parks and farms project development, 958
Construction, photovoltaic systems in
  cooling systems, 814–815
  curtain wall, 812–813, 813f
  facades, 811–816
  fixing systems, 810

ground-mounted systems, 817
loading, 810
maintenance, 815
mechanical strength, 809–810, 814, 815f
new build vs. retrofit, 807, 808, 814
rain screen, 811, 812f
roofs, 806–811, 806f, 807f, 808f
site testing, 816, 816f
standards, 811, 812f
sublayer membranes, 810
substructure, 808, 809f, 813–814
traditional roof interfaces, 810
weatherproofing, 810, 814
Consumer stand-alone solar-powered products, 989–1001
  accent lighting, 991–992
  air fresheners, 992
  aquatic products, 994, 995f
  calculators, 990
  chargers, 991
  deterrent products, 994–995
  garden products, 992–995, 993f
  indoor products, 990–992
  lighting and markers, 992–994, 993f
  watches and clocks, 991, 991f
Contactless sensing methods, silicon wafers, minority-carrier lifetime, 1019
Contact mechanisms
  cadmium telluride thin-film photovoltaic modules, 304, 305f
  silicon wafers, formation monitoring, 1037–1038
Continuity equations, semiconductor transport, 41
Continuous current mode, DC to DC power conversion, photovoltaic systems, buck converter, 699, 700f
Continuous light sources, space solar cells, 888
Cooling systems, construction, photovoltaic systems, 814–815
Copper indium-aluminium diselenide thin-film solar cells, wide gap properties, 358

Copper indium diselenide (CIS) solar cells
  hazard analysis, photovoltaic systems, 1090–1091
  occupational health and safety, 1090–1091
  public health and environmental issues, 1091
  research background, 378–379
  wide-gap chalcopyrite structure, 358–359
Copper indium-gallium diselenide thin-film solar cells
  absorber preparation techniques, 333–338
  band diagram, 346–347, 346f
  band gap energies, 326, 326f, 331f
  chalcopyrite lattice, 325–326
  co-evaporation processes, 334–335, 335f
  defect physics, 328–330
  electrodeposition, 337
  electronic metastabilities, 353–354
  epitaxial absorption deposition, 337
  fill factor, 352–353
  heterojunction solar cell, 331, 332f
    buffer layer deposition, 339–340
    free surface properties, 338–339
    window layer deposition, 340
  high-efficiency solar cells, 331–332
  module production and commercialisation, 341–345
    fabrication technologies, 342–343, 342f
    monolithic interconnections, 341–342, 341f
    radiation hardness, 345
    space applications, 345
    stability, 344–345
    upscaling processes, 343
  open-circuit voltage, 349–352, 350f
  particle deposition, 337
  phase energies, 326–328, 327f
  postdeposition air annealing, 337–338
  research background, 324–325
  screen printing, 337
  selinisation process, 335–337, 336f
  short-circuit current, 348–349, 348f
  wide-gap chalcopyrites, 354–361, 356f

Copper indium-gallium selenium-sulfide thin-film solar cells, wide-gap chalcopyrite structure, 359
Cost analysis
  batteries in photovoltaic systems, 736–737
  commercial photovoltaic modules, 152–154
  dye-sensitized cells, 481
Cost:performance ratio, photovoltaic energy yields, 1097–1098
Coulomb-enhanced Auger recombination, silicon wafers, 1014–1015
Counterelectrodes, dye-sensitized cells
  material properties, 532–533
  redox mediator and, 505–506
CPS Navstars satellites, space solar cell monitoring, orbital anomalies, 905
Critical region improvement, cadmium-telluride thin-film photovoltaic modules, 290–301, 290f
  activation, 299
  back contact, 299–301, 300f, 301f, 302f
  best cell performance, 302–303
  charge-carrier lifetime increase, 298–299, 299f
  grain growth-recrystallization, 292–295, 292f, 293f, 294f, 295f
  interdiffusion-intermixing, 295–298, 296f, 297f, 298f
  p-n heterojunction activation, 291–299, 291f
  stability issues, 302
Crucibles, multicrystalline silicone preparation, 90
Crystalline silicon
  energy requirements, 1100–1102
  modules, future energy potential, 1107–1109
  solar cells
    hazard analysis, 1084–1087
      occupational health issues, 1084–1085
      public health and environmental issues, 1085–1087
    low-cost industrial technologies

back-surface-field and back side passivation, 138–139, 138f
buried contact solar cells, 147–148
cell processing, 131–145
cleaning, 134–135
commercial photovoltaic modules, 152–154, 152f
commercial thin film crystalline cells, 151, 152f
efficiency goals, 130–131
EFG silicon sheets, 150–151
etching, texturing and optical confinement, 132–134, 133f, 134f
fast processing, 143–145
front contact formation, 139–141, 140f
front surface passivation and antireflection coating, 137–138
junction formation, 136–137, 136f
metal-insulator-semiconductor inversion layer cells, 148–149, 150f
screen printing, 145–147, 147f
substrate gettering
  aluminium treatment, 142–143
  phosphorus diffusion, 141–142
substrates, 131–132
substrates, material improvement, 141–143
wafers
  block scanners, 84
  electromagnetic continuous casting, 90–91, 91f
  energy requirements, 1100
  feedstock silicon, 86–87
  float-zone silicon, 87–94
  future energy enhancements, 1107–1109
  microwave photoconductance decay method, 82–83
  minority-carrier lifetimes, 82–86
  nonwafer technologies, 92–94
  photovoltaic manufacturing, 80–86
    geometrical specifications, 80–81
    physical specifications, 81–82
  preparation methods, 87–94

Czochrahki silicon, 87—88, 88f
multicrystalline silicone, 88—90, 89f
production and market share, 79—80
shaping, 94
wafering, 94—96
wire sawing, 95, 95f
Crystal-preparation methods, crystalline silicon wafers, 87—94
Czochrahki silicon, 87—88, 88f
multicrystalline silicone, 88—90, 89f
Cumulative solar irradiance, photovoltaic array, annual energy production, 645—646
Current maxima, electrical systems, photovoltaic structures, 821—823
Current multiplication factor, mono- and multicrystalline silicon modules, 87—88
Current sinks, silicon thin film solar cells, p-i-n/n-i-p structures, 244—246, 245f
Current source inverters, DC to AC power conversion, photovoltaic system electronics, 705—706, 706f
Curtain wall facade, construction, photovoltaic systems, 812—813, 813f
Curzon-Ahlborn efficiency, formula for, 64
Cycle life, batteries in photovoltaic systems
defined, 732—733
lifetime estimations and, 758
Czochralski (CZ) crystalline silicon wafer
buried-contact solar cell development, 117
crystal-preparation methods, 87—88, 88f
lifetime instabilities and measurement artifacts, boron-doped CZ silicon, 1028—1029
production and market share, 79—80
screen-printed high-efficiency silicon solar cells, 112

D

Daily clearness index, solar-radiation climatology, 628—629
Daily cycle
battery lifetime predictions, 760—762, 761f
energy balance, stand-alone photovoltaic systems, 659—660, 660f, 662—663
Daily deficit assessment, seasonal energy balance, stand-alone photovoltaic systems, 670—671, 671f
Daily diffuse irradiation, solar radiation climatology, daily global irradiation and, 631
Daily energy-balance dynamics, electricity consumption, stand-alone photovoltaic systems, 666—668, 667f
Damage constant, semiconductor radiation damage, 54, 54f
Damp-heat sequence, photovoltaic system reliability and qualification testing, 1060
Dangling bonds
amorphous silicon cells, 242
hydrogenated amorphous silicon, 215—217, 217f
silicon thin film solar cells, p-i-n/n-i-p structures, 241—244, 242f
Dark conductivity
amorphous silicon thin films, 223—224, 225f
silicon thin film solar cells, voltage enhancements, 174—175
Dark current curve, dye-sensitized cells, efficiency measurements, 507, 508f
Dark I-V measurements, photovoltaic performance metrics, 1054—1055
Dark saturation current density, Shockley solar cell equation, 68—69
Data-acquisition and analysis, photovoltaic systems, monitoring of, 1073—1075, 1078—1079
Data storage and transmission, photovoltaic system monitoring, 1076, 1078
Debye length, p-n junction, solar cell, 21
Declination estimation, solar-radiation climatology, algorithms for, 623—624
Decommissioning of photovoltaic modules, hazard analysis, 1094—1095
Defect physics, $Cu(In,Ga)Se_2$ thin-film solar cells, 328—330
Degenerate semiconductors, carrier statistics, 39—40

Degradation mechanisms
  interdigitated back-contact (IBC) silicon solar cell manufacturing, 464–467
  organic solar cell production, 560–561
  photovoltaic modules performance metrics, 1061–1062, 1063f
Deject layer model, Cu(In,Ga)Se$_2$ thin-film solar cells, heterojunction formation, 339
Dendritic web technology, crystalline silicon wafers, 92–93
Density of states
  amorphous silicon thin films, gap states, 220–223, 220f
  hydrogenated amorphous silicon semiconductors, 57, 58f
Depletion approximation, p-n junction, solar cell, 22
Depletion layer, photoelectrochemical cells
  light-induced charge separation, 491–492
  semiconductor-electrolyte junction, 488–489
Depletion region, p-n junction, solar cell, 19
Deposition processes, organic solar cells, 562, 563f
Depth of discharge (DOD)
  batteries in photovoltaic systems
    cycle life, 732–733
    defined, 731
    maximum DOD, defined, 768–770
    size selection and, 767–768
  lead-acid batteries, photovoltaic systems, sulphation and, 753–754
Desert-type climates, solar-radiation climatology, 597–598, 598f
Detailed balanced approach, multijunction III-V solar cells, band gap selection, 421–422
Deterrent solar-powered products, 994–995
  vehicle systems, 998–999
Development rationale and site location, solar parks and farms development, 954

Device architectures, organic solar cells, 549–550, 550f
Device ratings and component selection, electrical systems, photovoltaic structures, 823–824
Diagnostic measurements, photovoltaic performance metrics, 1054–1056
Differential spectral response, space solar cells, 886
Diffused emitter, p-n junction, solar cell, 23–25
Diffuse splitting, solar radiation, stochastic estimation, 616
Diffusion length, carrier recombination, silicon wafers, 1012–1013
Diode analysis, silicon wafers, Suns-V_{oc} curves, 1036–1037, 1037f
Diode law, Cu(In,Ga)Se$_2$ thin-film solar cells, open-circuit voltage, 350
Direct beam irradiation, solar-radiation climatology, inclined planes, 633–634
Direct current (DC)
  DC to AC power conversion (inversion)
    grid-connected photovoltaic systems, 780–782
    monitoring of, 1078
    photovoltaic system electronics, 703–708
    isolation, 708
    single-phase inverters, 703–707, 704f, 705f, 706f
    three-phase inverters, 707, 707f, 708f
  DC to DC power conversion, photovoltaic system electronics, 698–703
    advanced topologies, 703, 703f
    boost converter, 701–702, 701f
    buck-boost or flyback converter, 702, 702f
    buck converter, 698–701, 699f, 700f
    stand-alone systems, maximum power point tracker, 711–712
  electrical systems, photovoltaic structures, ratings, 821–823

grid-connected photovoltaic systems, injection, 784
photovoltaic performance metrics, DC installation defects, 971–972
Direct gap semiconductors, 38
Directly driven motor loads, photovoltaic systems, batteries in, 725
Direct sunlight calibration, space solar cells, 885
Discharge rate, batteries in photovoltaic systems
　acid density in lead-acid batteries and, 749–750
　defined, 730
　lead-acid batteries, capacity variation and, 746–747, 747f
Discontinuous current mode, DC to DC power conversion, photovoltaic systems, buck converter, 699, 700f
Display equipment, photovoltaic systems, monitoring of, 1073–1075, 1074f
Disposal procedures, batteries in photovoltaic systems, 773–774
Dopant concentration dependence, semiconductor carrier mobility, 44f
Doping effects
　amorphous silicon thin films, 224
　cadmium telluride intermixing and interdiffusion, back contact, 299–300
　hydrogenated microcrystalline silicon layers, 233
　semiconductor solar modelling, 55–57
Double-insulated cable, electrical systems, photovoltaic structures, 832
Double printing, screen-printed high-efficiency silicon solar cells, 114–115, 114f
D-π-A dyes, dye-sensitized cells, 530
Dual-junction principle
　multijunction cell research and, 427–429
　tandem and multijunction solar cells, 252, 252f
Duty cycle, batteries in photovoltaic systems, 726–727, 728f
　charging energy availability, 726
　normal use, rechargeable batteries, 726

Dye sensitization process
　dye-sensitized cells, 529–531
　photoelectrochemical cells, 493–495
Dye-sensitized cell (DSC)
　current status and operational principles, 497–499, 498f, 499f
　device characterization, 506–527, 507f
　efficiency measurements, 507–509, 508f
　electron concentration measurements, 520–521
　external/internal quantum efficiencies, 509–512, 510f
　mesoporous electrode, internal potential, 521–522
　photo-induced absorption spectroscopy, 522–524, 523f, 525f
　photovoltage/photocurrent light intensity and, 514–515, 514f
　small-modulation electron transport and recombination, 515–519, 516f
　Stark effect, 525–527, 526f
　steady-state quantum efficiency measurements, 519–520
　toolbox, 512–524
electron transfer, 496–534, 497f
　current status and operational principles, 497–499, 498f, 499f
　electron injection and excited state decay, 501–502
　electron recombination; oxidized dyes or electrolyte species, 504–505
　electron-transfer processes, 499–506, 500f, 501f
　mesoporous oxide film, electron transport, 503–504
　oxidized dye regeneration, 502–503
　redox mediator and counterelectrode reaction, 505–506
future research issues, 535–536
material components and devices, development of, 527–534
　counterelectrodes, 532–533
　dyes, 529–531
　electrolytes, 531–532

Dye-sensitized cell (DSC) (*Continued*)
  mesoporous oxide working electrodes, 528–529
  module development, 533–534, 533f, 534f
  photoelectrochemical cells, 483–496
    dye sensitization process, 493–495, 494f
    electrolyte, 487–488, 487f
    energy and potential levels, 484
    light-induced charge separation, 491–492
    operational principles, 495–496, 498f
    semiconductor-electrolyte junction, 488–491, 489f
    semiconductors, 485–487, 485f
  research background, 480–482

E

Earth faults, electrical systems, photovoltaic structures, 824–825
Earthing arrangements, electrical systems, photovoltaic structures, 830–832
ECN, Buildings 31 and 42 case study, architectural integration of solar cells, 936–937, 937f
Ecology and nature conservation, solar parks and farms development, 956–957
Edge-defined film fed growth (EFG)
  crystalline silicon solar cell substrates, 131–132
  crystalline silicon wafers, 92–93
  silicon sheet solar cell production, 150–151
Edge recombination, interdigitated back-contact (IBC) silicon solar cells, 462–464
Effective lifetime, silicon wafers, surface recombination velocity, 1021
Efficiency metrics
  batteries in photovoltaic systems, defined, 729–730
  commercially produced crystalline silicon solar cells, 130–131
  dye-sensitized cells, 507–509, 508f

interdigitated back-contact (IBC) silicon solar cells
  improvement strategies, 468–472
  thirty-percent improvement strategy, 467–468
semiconductors
  lifetime properties, 86f
  limitations, 72–73
  Shockley solar cell equation, 67–72
  thermodynamic efficiencies, 64–65
Einstein relations, semiconductor transport, 41
Electrical appliances, electricity consumption, stand-alone photovoltaic systems, 664–668
Electrical systems, photovoltaic structures
  available products, 715–718
    charge controllers, 715–717
    inverters, 717–718
  DC to AC power conversion (inversion), 703–708
    isolation, 708
    single-phase inverters, 703–707, 704f, 705f, 706f
    three-phase inverters, 707, 707f, 708f
  DC to DC power conversion, 698–703
    advanced topologies, 703f, 703
    boost converter, 701–702, 701f
    buck-boost or flyback converter, 702, 702f
    buck converter, 698–701, 699f, 700f
  electromagnetic compatibility, 718–719, 719f
  installation guidelines
    array fault protection, 824–830
    codes and regulations, 820–821
    DC ratings, 821–823
    device ratings and component selection, 823–824
    earthing arrangements, 830–832
    labelling requirements, 834
    maximum current, 822–823
    maximum voltage, 821
    production by design, 832–833, 832f, 833f
  local electricity utility-photovoltaic systems connected to, 714–715

Index 1203

monitoring of electrical output, 1077, 1077f
properties and limitations, 697–698
solar-powered products design, 1005, 1005f
stand-alone photovoltaic systems, 708–714
  battery charge control, 708–711, 709f, 710f, 711f
  maximum power point tracking, 711–713, 712f
  power inversion, 713–714
Electrical testing, space solar cells
  basic principles, 882–883
  capacitance characterisation, 890–891
  performance testing, 887–889
  relative spectral response, 889–890, 890f
  reverse characterisation, 890, 891f
Electricity consumption, load characteristics, stand-alone systems, 664–668
Electroabsorption, dye-sensitized cell characterization, 525–527, 526f
Electrochemical impedance spectroscopy (EIS), dye-sensitized cell analysis, 513–514
Electrochemical properties, batteries in photovoltaic systems, 737–739
Electrochromism, dye-sensitized cell characterization, 525–527, 526f
Electrodeposition
  cadmium telluride thin-film photovoltaic modules, 287
  $Cu(In,Ga)Se_2$ thin-film solar cells, 337
Electrodes
  batteries in photovoltaic systems, 737–739
  organic solar cells
    degradation, 561–562, 561f
    voltage generation, 555–556
Electrolytes
  batteries in photovoltaic systems, 737–739
  dye-sensitized cells
    electron transport, 504–505
    photo-induced absorption spectroscopy, 522–524
    material properties, 531–532

photoelectrochemical cells, 487–488, 487f
Electromagnetic compatibility, photovoltaic system electronics, 718–719, 719f
Electromagnetic continuous casting (EMC), crystalline silicon wafers, 90–91, 91f
Electron concentration measurements, dye-sensitized cells, 520–521
Electron current density, semiconductor transport equations, 40–41
Electron diffusion length
  dye-sensitized cells, small-modulation electron transport and recombination, 518
  dye-sensitized cells, steady-state measurement, 519–520
Electronic materials, organic solar cells, 544–548, 545f
Electronic metastabilities, $Cu(In,Ga)Se_2$ thin-film solar cells, 353–354
Electron injection, dye-sensitized cell electron transfer, 501–502
Electron transfer, dye-sensitized cells, 496–534, 497f
  current status and operational principles, 497–499, 498f, 499f
  electron injection and excited state decay, 501–502
  electron recombination; oxidized dyes or electrolyte species, 504–505
  electron-transfer processes, 499–506, 500f, 501f
  mesoporous oxide film, electron transport, 503–504
  oxidized dye regeneration, 502–503
  redox mediator and counterelectrode reaction, 505–506
  small-modulation electron transport and recombination, 515–519
Electrostatic discharge testing, space solar cells, 895
Emitter formation, silicon wafer lifetimes, 1034–1035

Emitter saturation current density
 interdigitated back-contact (IBC) silicon solar cells, reduction strategy, 470, 470f
 silicon wafers, lifetime measurements, 1024–1027, 1026f
Emitter wrap-through (EWT) cells, rear-contacted cell development, 121–122, 123, 123f
Encapsulation
 Cu(In,Ga)Se$_2$ thin-film solar cells, module fabrication, 343
 organic solar cells, 562
 silicon film solar modules, 269
End voltage, lead-acid batteries, capacity dependence on, 748–749, 748f
Energy balance, stand-alone systems
 climatic cycle, 660–662, 661f, 662f
 energy payback time calculations, 1105–1107, 1106f
 load characteristics, 663–668, 664f
  daily energy-balance dynamics, 666–668, 667f
  electricity consumption, lighting and electrical appliances, 664–668
 research background, 659–663
 schematic, 660f
 seasonal balance, 668–671, 669f, 670f, 671f
Energy dependence
 effiencies in terms of, 65–66
 photoelectrochemical cells, 484
 semiconductor radiation damage, 54, 55f
Energy gap. See Band gap
Energy-harvesting efficiency, multijunction III-V solar cells, 436–439, 436f
Energy payback time (EPBT)
 carbon dioxide emissions, 1112–1114
 energy balance in photovoltaic systems, 1105–1107, 1106f
 future systems perspectives, 1110–1111
 photovoltaic energy analysis, 1099
Energy units, photovoltaic energy analysis, 1099
Energy-vs.-distance diagram, photoelectrochemical cells, semiconductor-electrolyte junction, 488–489, 489f
Energy yields
 architecture-solar cell integration, 919–921, 920f
 British Pavilion, Expo '92 case study, 936
 building energy availability, 923
 Mont Cenis Conference Centre case study, 940
 Pompeu Fabra Public Library case study, 938
 solar energy availability in buildings, 923
 concentrator photovoltaic systems, 859–860
 photovoltaic array
  annual energy production, 645–646
  mismatch losses and blocking/bypass diodes, 655–658, 655f, 656f, 657f
  module orientation, 651–652
   fixed tilt arrays, 651–652, 653f
   tracking arrays, 652
  normal array power, 648–649
  peak solar hours, 646–648, 647f
  statistical analysis, 653–655, 654f
  temperature dependence, array power output, 650–651, 650f
 photovoltaic performance metrics, 964–965
  analysis methodology, 1099
  annual yields, 967–969
  energy requirements of system, 1100–1105
  low yield analysis, 969–972
  overview, 1097–1098
  shading effects, 970–971
Environmental issues
 cadmium telluride thin-film photovoltaic modules, 314–315
 carbon dioxide emissions, photovoltaic energy balance, 1112–1114
 hazard analysis, photovoltaic systems amorphous silicon solar cellls, 1088–1089

public health and environmental
  issues, 1089
cadmium telluride solar cells,
  1089—1090
  public health and environmental
    issues, 1090
copper indium diselenide cells,
  1090—1091
  public health and environmental
    issues, 1091
crystalline silicon solar cells,
  1084—1087
  public health and environmental
    issues, 1085—1087
future research issues, 1095
gallium arsenide high-efficiency solar
  cells, 1092—1093
manufacturing hazards, 1084
overview, 1083—1084
photovoltaic modules
  decommissioning, 1094—1095
  operation, 1093—1094
solar parks and farms development,
  956—957
  landscaping and visual change, 955,
    955f
  light reflection and aviation, 955—956
  noise, vibration, traffic, and air quality,
    956
  regulatory issues, 959
  surface water drainage, flooding, and
    ground conditions, 956
space solar cells, 891—895, 893f
  atomic oxygen, 893—894
  electrostatic discharge, 895
  humidity, 895
  micrometeoroids, 894
  radiation testing, 892—893
  thermal cycling, 894
  ultraviolet radiation, 893
  vacuum, 894
Epitaxial compound films, Cu(In,Ga)Se$_2$
  thin-film solar cells, 337
Equivalent circuits, ideal solar cell, 10—12,
  11f
Equivalent primary energy, photovoltaic
  energy analysis, 1099

Etching, crystalline silicon solar cells,
  132—134
Ethyl-vinyl-acetate (EVA) sealant,
  cadmium telluride thin-film
  photovoltaic modules, 305—306
EUCLIDES cells, I-V curves, concentrator
  photovoltaic systems, 841, 842f,
  843f, 847f
European Communication Satellite (ECS),
  space solar cell monitoring, orbital
  anomalies, 901—905
European Retrievable Carrier (EURECA),
  postflight monitoring, 907
European Solar Radiation Atlas (ESRA)
  cloudless sky global radiation, 579, 580f
  simplified radiation climatology
    components, 587—588, 589f
  solar irradiation, 575—576
    data sequences, 586—587, 587f
Excess electron concentration, silicon
  wafers, minority-carrier lifetime,
  1017
Excited state decay, dye-sensitized cell
  electron transfer, 501—502
Exciton diffusion length, organic
  semiconductors, 547
Exciton utilisation
  organic semiconductors, 545—548
  semiconductor efficiencies, 73
Extended spectral-response analysis, silicon
  thin film solar cells, light trapping,
  173, 174f
External optical elements, silicon thin film
  solar cells, light trapping, 169—170,
  170f
External quantum efficiency (EQE) curve
  dye-sensitized cells, 509—512
  multijunction III-V solar cells
    band gap selection, 421—422, 422f
    series interconnectivity, 425
  quantum well solar cells, 430, 431f
  silicon thin film solar
    cells, p-i-n/n-i-p structures,
    recombination, 242—243, 243f,
    244f
Extraterrestrial calibration
  solar-radiation climatology

Extraterrestrial calibration (*Continued*)
    horizontal angles and hourly global radiation values, 627–632
    mean daily profiles, 629
    space solar cells, 883–884
        high-altitude aircraft calibration, 884
        high-altitude balloon calibration, 883–884
        lost twin method, 884
        photovoltaic engineering testbed, 884
        space shuttle methods, 884

F

Facades
    architectural integration, solar cells, 924f, 926, 928f
    construction, photovoltaic systems, 811–816
        curtain wall, 812–813, 813f
        rain screen, 811–813, 812f
    solar-powered products, 1002, 1002f
Failure analysis, photovoltaic system performance metrics, 981–982
Fast processing techniques, crystalline silicon solar cell development, 143–145
Fault analysis, electrical systems, photovoltaic structures, 824–830
    additional fault current sources, 830
    array cable-fault current, 826
    array cable protection, 827
    blocking diodes, 829
    earth faults, 824–825
    fault analysis, 825–826
    line-line faults, 825
    module reverse-current protection, 827–828
    photovoltaic module-reverse current, 826
    string cable-fault current, 825–826, 826f
    string cable protection, 827
    string overcurrent protection, 828–829, 829f
Feed-in tariffs, solar park and farm development, 946
    incentive schemes and, 960

Feedstock silicon, crystalline silicon wafers, 86–87
    high-purity sources, 1108
Fermi levels
    Cu(In,Ga)Se$_2$ thin-film solar cells, heterojunction formation, 339
    hydrogenated microcrystalline silicon layers
        conductivities, 232
        doping, 233
    photoelectrochemical cells, 484
        electrolytes, 487–488
        operational principles, 487–488
        semiconductor-electrolyte junction, 488–491
        semiconductors, 485
    p-n junction, solar cell, 19
    semiconductor carrier statistics, 38–40
Field effect transistors (FETs), photovoltaic system electronics, 716–717
Field experience, silicon thin film solar cells, 272
Fill factor
    Cu(In,Ga)Se$_2$ thin-film solar cells, 352–353
        electronic metastabilities, 353–354
    dye-sensitized cell efficiency measurements, 507–508
    ideal solar cell, 11–12
    Shockley solar cell equation, 69–71
    silicon thin film solar cells, p-i-n/n-i-p structures, recombination, 242–243
    silicon wafers, 1038–1040
    tandem and multijunction solar cells, 253
        performance measurement, 1052–1053
Film deposition, cadmium-telluride thin-film photovoltaic modules, 285–289
    chemical spraying, 287
    electrodeposition, 287
    screen printing, 287
    vacuum deposition; sublimation and condensation, 286–287

Financing issues, solar parks and farms development, 958–959
  cadmium telluride thin-film photovoltaic module production, 309f
Five-junction III-V solar cells, development of, 439
Fixation systems, construction, photovoltaic systems, 810
Fixed arrays
  solar parks and farms, 949
  tilt arrays, module orientation, 651–652, 653f
Flat roof construction, architectural integration, solar cells, 921, 921f, 925, 927f
Flight experiments, space solar cell monitoring, 895–901
Floating junction structure, crystalline silicon solar cells, 136–137
Float service life, batteries in photovoltaic systems, 733
  lifetime estimations and, 758–759
Float-zone silicon
  crystalline silicon wafers, 91–92, 93f
  interdigitated back-contact (IBC) silicon solar cells, 455
  wafering process, carrier recombination, 1013, 1013f
Fluidised bed process, silicon production, 1169–1170
Fly-back converter, DC to DC power conversion, photovoltaic systems, 702, 702f
Four-junction III-V solar cells, development of, 436–439, 436f, 438f
Free-carrier absorption, semiconductors, 47f, 48–50
Free surface properties, Cu(In,Ga)Se$_2$ thin-film solar cells, heterojunction formation, 338–339
Freezing, lead-acid batteries, photovoltaic systems, 752–753, 753f, 754f
Frequency detection, grid-connected photovoltaic systems, passive technologies, 787–788
Frequency variation, grid-connected photovoltaic systems, active technologies, 788
Front contact formation, crystalline silicon solar cells, 139–141, 140f
"Frontier zones," thin-film silicon solar cells, p-i-n/n-i-p structures, intrinsic layer, internal electric field formation in, 238–240, 239f
Front-surface-field (FSF) solars, interdigitated back-contact design, 456–458, 456f
Front surface passivation, crystalline silicon solar cells, low-cost industrial technologies, 137–138
Fuel mix, carbon dioxide emissions, photovoltaic energy balance, 1112–1113

G

Gallium arsenide (GaAs) solar cells
  doping effects, 56–57
  hazard analysis, photovoltaic systems, 1092–1093
    occupational health and safety, 1092–1093
  multijunction III-V solar cells
    bifacial epigrowth, 434
    tunnel diodes, 424–425, 424f
  semiconductor carrier mobility, 43
  space applications
    germanium substrates, 401
    research background, 400
    single-junction III-V cell structure, 403–404, 405f
  surface recombination, 51
Gap states. *See also* Band gap
  amorphous silicon thin films, 220–223
  hydrogenated microcrystalline silicon layers, 228–231
Garden solar-powered products, 992–995, 993f
Gas consumption, silicon thin fiilm solar cells, layer deposition principles, 260
Gate turn-off thyristors (GTOs), DC to AC power conversion, photovoltaic system electronics, three-phase inverters, 707

Gel type lead-acid batteries, photovoltaic systems, 745
Geographic location, solar parks and farms, 947–948
Geometrical specifications
  crystalline silicon wafers, 80–81
  thin silicon solar cells, light trapping, 168–169, 169f
Gerischer diagram, photoelectrochemical cells, dye sensitization, 494–495, 494f
German 1000-Roofs-PV-Programme trends, long-term performance and reliability, 973–974, 974f
Germanium
  AlGaAs-GaAs solar cells, space applications, 401
  gallium arsenide solar cells, space applications, substrates, 401
  lattice-matched triple-junction solar cells on, 429–430, 429f
    inverted metamorphic growth, 432–434, 433f
    upright metamorphic growth, 430–432, 432f
Gettering, crystalline silicon solar cells
  aluminium treatment, 142–143
  phosphorus diffusion, 141–142
  substrate material improvement, 141–143
Glass substrates
  cadmium telluride thin-film photovoltaic modules, 289
  Cu(In,Ga)Se$_2$ thin-film solar cells, 332
  silicon thin film solar cells, p-i-n/n-i-p structures, 236–237, 237f
Global irradiance profiling, solar radiation, stochastic estimation, 615
  diffuse and beam splitting, 616
Global short-wave radiation, defined, 577–578
Global sunlight calibration, space solar cells, 884–885
Graded gap devices, wide-gap chalcopyrites, 359–361, 360f
Grain boundaries and growth

cadmium telluride thin-film photovoltaic modules, recrystallisation, 292–295, 293f, 294f, 295f
  silicon thin film solar cells, minority-carrier recombination, 176–177, 176f, 177f
Grid alloys, lead-acid batteries, photovoltaic systems, 743–744
Grid-connected photovoltaic systems. *See also* Stand-alone photovoltaic systems
  active methods, 788–789
    frequency variation, 788
    impedance measurement, 788
    power/reactive power variations, 789
  architecture for, 681–684, 682f
  batteries in, 725
  cumulative array output, 683f
  energy balance, 1105–1107
    carbon dioxide emissions, 1113
  energy requirements, 1104
  integration issues, 783–784
    DC injection, 784
    power quality, 783–784
    radio frequency suppression, 784
    safety, 783
  inverters, 717, 718
    structure and operating principles, 784–785, 785f
  islanding, 785–789
  long-term performance metrics, 972–977
  passive methods, 787–788
    ROCOF technique/phase jump detection, 788
    voltage and frequency detection, 787–788
  performance metrics, performance indicators, 966
  regulatory issues, 789–803
    draft IEC standard 62776, 790–798
    parallel connection agreement, 798
    post- or pre-notification, 803
    Tariff Agreement, 798–802
    technical connection guidelines, 789–790
    type testing, 790–798, 798f

Index 1209

research background, 780—783, 781f, 782f
solar-powered products, 1001—1002
 cladding and facade products, 1002, 1002f
 roofing products, 1001, 1002f
Grid optimisation, multijunction III-V solar cells, 426, 427f
Grid parity, solar parks and farms development
 future projections, 961—962, 961f
 incentive schemes and, 959—960
Grid thickness, lead-acid batteries, photovoltaic systems, 744
Ground, grounded (USA). *See* earth, earthed (Europe)
Ground-mounted photovoltaic systems, construction, 817
Ground-reflected irradiation, solar-radiation climatology, 636—637

H

Hazard analysis, photovoltaic systems
 amorphous silicon solar cellls, 1088—1089
  occupational safety issues, 1088—1089
  public health and environmental issues, 1089
 cadmium telluride solar cells, 1089—1090
  occupational health issues, 1089—1090
  public health and environmental issues, 1090
 copper indium diselenide cells, 1090—1091
  occupational health and safety, 1090—1091
  public health and environmental issues, 1091
 crystalline silicon solar cells, 1084—1087
  occupational health issues, 1084—1085
  public health and environmental issues, 1085—1087
 future research issues, 1095

gallium arsenide high-efficiency solar cells, 1092—1093
 occupational health and safety, 1092—1093
manufacturing hazards, 1084
overview, 1083—1084
photovoltaic modules
 decommissioning, 1094—1095
 operation, 1093—1094
Health issues
 cadmium telluride thin-film photovoltaic modules, 314—315
 hazard analysis, photovoltaic systems
  amorphous silicon solar cellls, 1088—1089
   occupational safety issues, 1088—1089
   public health and environmental issues, 1089
  cadmium telluride solar cells, 1089—1090
   occupational health issues, 1089—1090
   public health and environmental issues, 1090
  copper indium diselenide cells, 1090—1091
   occupational health and safety, 1090—1091
   public health and environmental issues, 1091
  crystalline silicon solar cells, 1084—1087
   occupational health issues, 1084—1085
   public health and environmental issues, 1085—1087
  future research issues, 1095
  gallium arsenide high-efficiency solar cells, 1092—1093
   occupational health and safety, 1092—1093
  manufacturing hazards, 1084
  overview, 1083—1084
  photovoltaic modules
   decommissioning, 1094—1095
   operation, 1093—1094

Heat sink, concentrator photovoltaic systems, 841, 843f
Helmholtz layer, photoelectrochemical cells, semiconductor-electrolyte junction, 489
Henry's construction, photovoltaic energy loss, 71, 71f
Heteroface solar cell, 27
Heterojunction solar cells, 25–27, 25f
  bulk heterojunction architecture, organic solar cells, 551–552, 552f
  cadmium telluride thin-film photovoltaic modules, p-n heterojunction improvement, 291–299, 291f
  conduction discontinuity, 26f
  Cu(In,Ga)Se$_2$ thin-film solar cells, 331, 332f
    buffer layer deposition, 339–340
    free surface properties, 338–339
    window layer deposition, 340
High-altitude aircraft calibration, space solar cells, 884
High-altitude balloon calibration, space solar cells, 883–884
High-efficiency solar cells
  Cu(In,Ga)Se$_2$ thin-film solar cells, 331–332
  interdigitated back-contact silicon
    cell thickness reduction, 471
    concentrator applications, 450–452, 453f, 454f, 455f
    design criteria, 453–458
    efficiency improvement techniques, 468–472
    emitter saturation current density reduction, 470, 470f
    front-surface-field, tandem-junction, and point-contact cells, 456–458, 456f
    light trapping, 471
    low-contact resistance, 470–471
    manufacturing process, 464, 465f
    modelling of, 458–462
    perimeter and edge recombination, 462–464

    point-contact vs. incident power density, 473f
    research background, 449–450
    series resistance, 472
    shrink geometries, 471
    stability, 464–467
    structure, 452f
    target performance, 472, 473f
    thirty percent efficiency goal, 467–468
  multijunction III-V solar cells
    application fields and reference conditions, 419–421
    band-gap selection, 421–422, 422f
    band gaps vs. lattice constant, 424f
    bifacial epigrowth, 434
    characterisation, 425
    concentrator solar cell design, 425–426, 426f, 427f
    inverted metamorphic growth, 432–434, 433f
    lattice-matched triple-junction cells, germanium deposition, 429–430
    more than triple junctions, 436–439, 438f
    nanowire cells, 440
    optical beam splitting, 439–440
    quantum well solar cells, 430
    research background, 417–419, 418f, 427–440, 428f, 429f
    schematic structure, 418f
    silicon deposition, 434–436, 435f
    tunnel diodes, 424–425, 424f
    upright metamorphic growth, germanium deposition, 430–432, 432f
  photovoltaic system performance metrics, 977
  silicon solar cells
    future research issues, 124–125
    HIT cell, 120–121, 120f
    laboratory development, 100–111, 101f
      PERL cell design, 109f, 110–111
      rear passivated cells, 108–109, 108f
      space cells, 100–104, 102f, 105f

terrestrial cells, 104–108, 106f, 107f
laser-processed cells, 116–120
  buried-contact cells, 116–118, 116f
  laser-doped, selective-emitter solar cells, 119–120, 119f
  semiconductor finger solar cell, 118–119, 118f
rear-contacted cells, 121–124
  emitter wrap-through cells, 123, 123f
  metal wrap-through cells, 124
  rear-junction solar cells, 122–123, 122f
research background, 100
screen-printed cells, 111–116
  hot-melt and stencil printing, 115
  limitations, 115–116
  paste improvements, 114
  performance, 113
  plated seed layers, 115
  selective emitter and double printing, 114–115, 114f
  structure, 111–113, 112f
Highest occupied molecular orbital (HOMO)
  organic semiconductors
    exciton utilisation, 545–546
    voltage enhancements, 554
  photoelectrochemical cells, dye sensitization, 493–494
High-level injection (HLI), interdigitated back-contact (IBC) silicon solar cells, 457, 459
High-temperature silicon-deposition methods, silicon thin film solar cells, 181–187
  chemical vapour deposition, 182–186, 183f
High-voltage range, grid-connected photovoltaic systems, 780
HIT cell, development of, 120–121, 120f
"H" metallization pattern, screen-printed high-efficiency silicon solar cells, 112, 112f
Hole current density, semiconductor transport equations, 40–41

Horizontal angle calculations, solar-radiation climatology, 625–627
  extraterrestrial estimation, 627–628
  hourly diffuse radiation estimation, 630–632
Hot-melt printing, screen-printed high-efficiency silicon solar cells, 115
Hot spot formation, photovoltaic array, energy production, mismatch losses, 655
Hot wire deposition, silicon thin film solar cells, 262
Hourly diffuse irradiation estimation, solar-radiation climatology, inclined planes, hourly horizontal irradiation and, 634–636
Hourly global radiation values, solar radiation climatology, 627–632
  diffuse radiation, horizontal surfaces, 630–632
  estimation of, 630–632
  extraterrestrial estimation, 627–628
  stochastic estimation, 614–616
Hourly horizontal irradiation, solar-radiation climatology, inclined planes, hourly diffuse irradiation estimation, 634–636
Household solar-powered products, 996–997
Hubble Space Telescope Solar Array 1, postflight monitoring, 906–907
Humidity testing
  photovoltaic system reliability, 1060
  space solar cells, 895
Humid tropics, solar-radiation climatology
  diffuse radiation, 597
  examples, 594–596, 595f, 596f
  simplified radiation conditions, 590–597, 591f
Hybrid inorganic-organic devices, organic solar cells, 559
Hydrogenated amorphous silicon. *See also* Amorphous silicon
  semiconductor properties, 57–58
  silicon thin film solar cells, 215–224
    conductivity, 223–224
    doping, 224

Hydrogenated amorphous silicon
(*Continued*)
   gap states, 220–223, 220f, 222f
   research background, 210–211
   structure, 215–219, 216f
Hydrogenated microcrystalline silicon.
    See also Microcrystalline materials
   thin film solar cells, 225–234
     ageing, 233–234
     conductivities, 225f, 232, 232f
     doping, 233
     impurities, 233, 234f
     optical absorption, gap states, and defects, 228–231, 230f, 231f
     research background, 213
     structure, 216f, 217f, 218f, 225–228, 226f, 227f
Hydrogen passivation, silicon thin film solar cells, 178, 179f

I

Ideality factor, two-diode solar cell, 13–15, 14f
Ideal solar cell, electrical characteristics, 10–12
IEC Standards, 1181–1183
   draft 62776, grid-connected photovoltaic systems, 790–798
IEEE specifications, concentrator photovoltaic systems, 858–859
Impact ionisation
   semiconductor efficiencies, 72–73
   semiconductor recombination, 48, 49f
Impedance measurement, grid-connected photovoltaic systems, active technologies, 788, 789f
Impurities, hydrogenated microcrystalline silicon layers, 233, 234f
Impurity photovoltaic effect, semiconductor efficiencies, 73
Incentive schemes, solar parks and farms development, 959–960
Incident photon to current conversion efficiency (IPCE)
   dye-sensitized cell measurements, steady-state quantum efficiency measurements, 519–520

dye-sensitized cells, 509, 510f
Inclined planes, solar-radiation climatology
   all sky irradiation estimation, direct beam irradiation, 633–634
   direct beam irradiation, 633–634
   photovoltaic systems, 601–602, 602f, 606
   quantitative processing, 606
Indirect band gap
   hydrogenated microcrystalline silicon layers, 230
   semiconductors, 38
Indium cells
   copper indium-gallium diselenide thin-film solar cells
     absorber preparation, 334
     defect physics, 330
     space applications, 401
Indium phosphide semiconductor, majority-carrier mobility, 44, 45f
Indium tin oxide (ITO)
   film deposition, 289
   silicon film solar cells, substrate configuration, 266–267
   thin-film photovoltaic modules, 289
Indoor solar-powered products, 990–992, 996f
Industrial solar technologies, 145–151
   cadmium telluride thin-film photovoltaic modules, 307–309
   dye-sensitized cells, 482
   lead-acid batteries, 745–746
   nonconsumer solar-powered products, 1003–1004
     commercial applications, 1003
     industrial applications, 1003–1004
     transportation, 1003, 1004f
   production capacities, 1155, 1158f
   technology mix, 1157–1159
Ingot process, silicon solar cells, lifetime measurements, 1027–1028
Injection efficiency, dye-sensitized cells, steady-state measurement, 519–520
Installation guidelines, photovoltaic systems construction
   cooling systems, 814–815
   curtain wall, 812–813, 813f
   facades, 811–816

fixing systems, 810
ground-mounted systems, 817
loading, 810
maintenance, 815
mechanical strength, 809–810, 814, 815f
new build *vs.* retrofit, 807, 808, 814
rain screen, 811, 812f
roofs, 806–811, 806f, 807f, 808f
site testing, 816, 816f
standards, 811, 812f
sublayer membranes, 810
substructure, 808, 809f, 813–814
traditional roof interfaces, 810
weatherproofing, 810, 814
electrical systems
array fault protection, 824–830
codes and regulations, 820–821
DC ratings, 821–823
device ratings and component selection, 823–824
earthing arrangements, 830–832
labelling requirements, 834
maximum current, 822–823
maximum voltage, 821
production by design, 832–833, 832f, 833f
Integral roof construction, photovoltaic systems, 806, 806f
Integrated modules
cadmium telluride thin-film photovoltaic modules, 303–306
cell interconnectivity, 303–304, 303f, 304f
contacting, 304, 305f
lamination, 305–306
solar-powered products, photovoltaic integration in, 988–989
Integration issues, grid-connected photovoltaic systems, 783–784
DC injection, 784
power quality, 783–784
radio frequency suppression, 784
safety, 783
Interannual variability, solar-radiation climatology, 602–604

sunshine data, radiation estimation, 603–604
Interdiffusion, cadmium-telluride thin-film photovoltaic modules, 295–298
Interdigitated back-contact (IBC) silicon cell thickness reduction, 471
concentrator applications, 450–452, 453f, 454f, 455f
design criteria, 453–458
efficiency improvement techniques, 468–472
emitter saturation current density reduction, 470, 470f
front-surface-field, tandem-junction, and point-contact cells, 456–458, 456f
light trapping, 471
low-contact resistance, 470–471
manufacturing process, 464, 465f
modelling of, 458–462
perimeter and edge recombination, 462–464
point-contact *vs.* incident power density, 473f
research background, 449–450
series resistance, 472
shrink geometries, 471
stability, 464–467
structure, 452f
target performance, 472, 473f
thirty percent efficiency goal, 467–468
Interface states, $Cu(In,Ga)Se_2$ thin-film solar cells, heterojunction formation, 339
Intermediate reflector, micromorph tandem cells, 258
Intermixing, cadmium telluride thin-film photovoltaic modules, 295–298, 296f, 297f, 298f
Internal electric field, silicon solar thin film cells, p-i-n/n-i-p structures, 234–241, 236f
intrinsic layer, formation in, 238–240, 239f
intrinsic layer, reduction and deformation, 241

Internal heating, batteries in photovoltaic systems, 771
Internal potential measurement, dye-sensitized cells, mesoporous electrode, 521–522
Internal quantum efficiencies, dye-sensitized cells, 509–512
Intrinsic layers (i layers)
  hydrogenated microcrystalline silicon layers, impurities, 233
  silicon thin film solar cells, p-i-n/n-i-p structures, internal electric field, 236–237
    formation, 238–240, 239f
    reduction and deformation, 241
Inverted metamorphic (IMM) growth, lattice-matched triple-junction solar cells on, 432–434, 433f
Inverters
  grid-connected photovoltaic systems, structure and operating principles, 784–785, 785f
  photovoltaic system electronics, 717–718
  performance metrics, 972
Ionisation energies, $Cu(In,Ga)Se_2$ thin-film solar cells, 328–329
Iron-contaminated silicon, lifetime instabilities and measurement artifacts, boron-doped CZ silicon, 1028–1029
Irradiance/irradiation
  carbon dioxide emissions, photovoltaic energy balance, 1113
  concentrator photovoltaic systems, 839–840
  defined, 577–578
Islanding, grid-connected photovoltaic systems, 785–789
Isolated converters
  DC to AC power conversion, photovoltaic system electronics, 708
  photovoltaic systems, DC to DC power conversion, 703
Isotropic surface texturing, crystalline silicon solar cells, 134

Italian photovoltaic demonstration projects
  long-term performance and reliability, 976, 977f
  user experience with, 980, 981f

J
Junction formation, crystalline silicon solar cells, 136–137, 136f

K
Kerf loss, crystalline silicon solar cell substrates, 131–132
KT algorithm, daily global radiation, time series estimation, 612–613

L
Labelling, electrical systems, photovoltaic structures, 834
Lamination, cadmium telluride thin-film photovoltaic modules, 305–306
Landsberg efficiency, formula for, 64
Landscaping and visual change, solar parks and farms development, 955, 955f
Land usage, solar parks and farms, 947–948
Large-area point-focus parabolic concentrator, 849f
Large power systems, space applications of solar cells, 877–878, 878f
Laser-beam induced current (LBIC), photovoltaic performance metrics, 1055
Laser-doped, selective-emitter (LSDE) solar cells, development of, 119–120, 119f
Laser flash photolysis, dye-sensitized cells, 522
Laser grooved buried contact metallization, crystalline silicon solar cells, front contact formation, 139
Laser-processed high-efficiency silicon solar cells, 116–120
  buried-contact cells, 116–118, 116f
  laser-doped, selective-emitter solar cells, 119–120, 119f
  semiconductor finger solar cell, 118–119, 118f

Laser scribing, silicon thin film solar cells, 267–269, 267f
Lattice constants
   mechanically stacked multijunction (cascade) cells, space applications, 407, 407f
   multijunction III-V solar cells
      band gap vs., 423–424
      silicon deposition, 434–436, 435f
      triple-junction solar cells, germanium deposition, 429–430
Layer deposition, silicon thin film solar cells, 259–262, 260f
Lead-acid batteries
   photovoltaic systems, 741–754
      absorbent glass mat batteries, 745
      acid density, 749–750, 749f
      acid stratification, 750–752, 751f, 752f
      basic principles, 741–742
      capacity variation, 746–749, 747f, 748f
      discharge rate, 746–747, 749–750
      freezing, 752–753, 753f, 754f
      gel-type batteries, 745
      grid alloy construction, 743–744
      grid thickness, 744
      low temperatures and capacity variations, 747–748
      mass-produced and industrial batteries, 745–746
      plate-type construction, 728f, 743
      sealed batteries, 744–745
      sealed vs. mass-produced construction, 742–743
      sulfation and deep discharge, 753–754
   stand-alone photovoltaic system electronics, 709, 709f
Levelised cost of solar electricity (LCSE), interdigitated back-contact (IBC) silicon solar cells, 452
Life cycle analysis, carbon dioxide emissions, photovoltaic energy balance, 1112
Lifetime properties

batteries in photovoltaic systems, 756–765
   blackbox limit calculations, 735–736
   cycle life, 758
   field data vs. predicted lifetimes, 762–765
   float lifetime, 758–759
   maximum lifetime, 733
   predicted lifetimes, 760–762, 761f
   sealed batteries, 759–760
boron-doped CZ silicon and iron-contaminated silicon, 1028–1029
cadmium telluride thin-film photovoltaic modules, charge-carrier lifetime, 298–299
carbon dioxide emissions, photovoltaic energy balance, 1113
dye-sensitized cells, small-modulation electron transport and recombination, 517
interdigitated back-contact (IBC) silicon solar cells, 455
minority-carrier semiconductors, 51–53
   crystalline silicon wafers, 82–86, 85f
silicon ingots, 1027–1028
silicon thin film solar cells, minority-carrier recombination, 176–178, 176f, 177f, 178f, 179f
silicon wafers
   bare wafer measurements, 1033–1034
   carrier recombination, 1012–1015, 1013f
   device voltage and, 1030–1031
   emitter formation, 1034–1035
   emitter saturation current density, 1024–1027
   instabilities and measurement artifacts, 1028–1030
   low excess carrier densities, 1029–1030
   minority-carrier lifetime, 1015–1030
   monitoring applications, 1032–1035
   surface recombination velocity, 1021–1024
   voltage specification and, 1030–1031
Light-harvesting efficiency (LHE)
   dye-sensitized cells, 509–512

Light-harvesting efficiency (LHE) (*Continued*)
  organic solar cells, 553, 554f
Light-induced charge separation, photoelectrochemical cells, 491–492
Light-induced degradation effect, hydrogenated amorphous silicon, 217–219
Lighting systems
  electricity consumption, stand-alone photovoltaic systems, 664–668
  solar-powered products
    accent lighting, 991–992
    garden products, 992–995, 993f
    household products, 997
Light intensity, dye-sensitized cell analysis, photovoltage/photocurrent as function of, 514–515, 514f
Light reflection and aviation issues, solar parks and farms development, 955–956
Light trapping
  interdigitated back-contact (IBC) silicon solar cells, 471
  organic solar cells, 554
  silicon thin film solar cells, 165–173, 167f
    effects assessment, 170–173
    extended spectral-response analysis, 173, 174f
    external optical elements, 169–170, 170f
    geometrical or regular structuring, 168–169, 169f
    implementation methods, 167–170
    p-i-n/n-i-p structures, 249–252, 250f, 251f
    random texturing, 168
    short-circuit current analysis, 170–172, 171f
    sub-band-gap reflection analysis, 172, 173f
  solar cell, 17–19
Line-line faults, electrical systems, photovoltaic structures, 825
Liquid-phase epitaxy (LPE) silicon thin film solar cells, silicon deposition and crystal growth, 187, 188f
  single-junction III-V solar cells, 404, 405f
Lithium-ion batteries, photovoltaic systems, 740, 741
Lithography technology, interdigitated back-contact (IBC) silicon solar cell manufacturing, 464, 465f
Live conductors, earthing of, 830–831
Load characteristics
  construction, photovoltaic systems, 810
  energy balance, stand-alone systems, 663–668, 664f
    daily energy-balance dynamics, 666–668, 667f
    electricity consumption, lighting and electrical appliances, 664–668
  solar parks and farms, load factor optimisation, 946–947
Local apparent time, solar-radiation climatology estimation, 620–621
Local electric utility
  photovoltaic system connection with, 714–715
  solar parks and farms, local electricity network and usage, 947
Local mean time (LMT) conversion, solar-radiation climatology estimation, 620–621
Long-range order, hydrogenated amorphous silicon, 215
Long-term performance metrics, photovoltaic systems, 972–977
  German 1000-Roofs-PV-Programme trends, 973–974, 974f
  Italian PV programme trends, 976
  Switzerland demonstration projects, 974–976, 975f
  system performance efficiencies, 977, 978f
Loss of load parameter, seasonal energy balance, stand-alone photovoltaic systems, 669–670, 670f, 671f

Lost twin extraterrestrial calibration, space solar cells, 884
Low-angle silicon sheet (LASS) technology, crystalline silicon wafers, 93–94
Low-contact resistance, interdigitated back-contact (IBC) silicon solar cells, 470–471
Low-cost industrial technologies, crystalline silicon solar cells
  back-surface-field and back side passivation, 138–139, 138f
  buried contact solar cells, 147–148
  cell processing, 131–145
  cleaning, 134–135
  commercial photovoltaic modules, 152–154, 152f
  commercial thin film crystalline cells, 151, 152f
  efficiency goals, 130–131
  EFG silicon sheets, 150–151
  etching, texturing and optical confinement, 132–134, 133f, 134f
  fast processing, 143–145
  front contact formation, 139–141, 140f
  front surface passivation and antireflection coating, 137–138
  junction formation, 136–137, 136f
  metal-insulator-semiconductor inversion layer cells, 148–149, 150f
  screen printing, 145–147, 147f
  substrate gettering
    aluminium treatment, 142–143
    phosphorus diffusion, 141–142
  substrates, 131–132
  substrates, material improvement, 141–143
Lowest unoccupied molecular orbital (LUMO)
  organic semiconductors
    exciton utilisation, 545–546
    voltage enhancements, 554
  photoelectrochemical cells, dye sensitization, 493–494
Low-level injection, interdigitated back-contact (IBC) silicon solar cells, 459

Low-pressure chemical vapour deposition (LPCVD), silicon thin film solar cells, 186
Low-temperature chemical vapour deposition (LTCVD), silicon thin film solar cells, 186–187
  substrate configuration, 266–267
Low-voltage range, grid-connected photovoltaic systems, 780
Low-work metals, high-efficiency terrestrial solar cells, 106–107
Low yield analysis, photovoltaic performance metrics, 969–972

M

Maintenance, construction, photovoltaic systems, 815
Majority-carrier mobility, semiconductors, 42
Manufacturing hazards, photovoltaic systems, 1084
Maritime European Communication Satellite (MARECS), space solar cell monitoring, orbital anomalies, 901–905
Market conditions, photovoltaic technology, 1137, 1138f
  Asia and Pacific region, 1137–1145
  Australia, 1139–1140
  Bangladesh, 1143
  Belgium, 1146–1147
  Canada, 1151–1152
  commercial photovoltaic market prices, 138
  crystalline silicon wafers, production and market share, 79–80
  Czech Republic, 1147–1148
  Czochralski crystalline silicon wafer, production and market share, 79–80
  emerging markets, 1143–1152
  Europe, 1145–1151, 1146f, 1147f
  France, 1148
  Germany, 1148–1149
  Greece, 1149
  India, 1140
  Indonesia, 1144
  Italy, 1149

Market conditions, photovoltaic
technology (*Continued*)
Japan, 1140–1141
Malaysia, 1144
North America, 1151–1152
organic solar cells, 544
People's Republic of China, 1141
Philippines, 1144–1145
South Korea, 1141–1142
Spain, 1150
Taiwan, 1142
Thailand, 1142–1143
Turkey, 1145–1151
United Kingdom, 1150
United States, 1152
Vietnam, 1145
Markov transition matrices (MTM), daily global radiation, time series estimation, 612
Mass action law, semiconductor carrier statistics, 38–39
Mass-produced lead-acid batteries, 745–746
Material resources, cadmium-telluride thin-film photovoltaic modules, 315–316
Matlab software, photovoltaic system design, 688
Maximum charge rate, batteries in photovoltaic systems, calculations for, 736
Maximum current, electrical systems, photovoltaic structures, 822–823
Maximum lifetime, batteries in photovoltaic systems, 733
Maximum power point tracking (MPPT)
charge controllers, photovoltaic systems, 717
photovoltaic array energy production mismatch loss measurement, 657–658
nominal array power, 648
photovoltaic modules and solar cells, 1046
photovoltaic performance meetrics, 972
stand-alone photovoltaic system electronics, 711–713, 712f

Maximum voltage
electrical systems, photovoltaic structures, 821
power point voltage dependence, irradiance level, array annual energy production, 646
Mean daily profiles, solar-radiation climatology, 629–630, 632
Measurement uncertainty analysis, photovoltaic performance metrics, 1054
Mechanically stacked multijunction (cascade) cells, space applications, 407–409, 407f, 408f
Mechanical strength standards
construction guidelines, photovoltaic systems, 809–810, 814, 815f
solar-powered products design, 1006
Mechanical testing, space solar cells, 895
Mechanical texturing, crystalline silicon solar cells, 132–134, 134f
Melt-growth techniques, silicon thin film solar cells, 181, 182f
Mesoporous oxide films, dye-sensitized cells
electron transport, 503–504
internal potential determination, 521–522
working electrodes, 528–529
Metal-insulator-NP (MINP) junction cell, high-efficiency terrestrial solar cells, 105–106, 106f
Metal-insulator-semiconductor inversion layer (MIS-IL) solar cell, industrial production, 148–149
Metallurgical-grade silicon, 1100, 1169–1170
Metal organic chemical vapour deposition (MOCVD)
AlGaAs-GaAs solar cells, space applications, 400–401
gallium arsenide solar cells, space applications, germanium substrates, 401
single-junction III-V solar cells, 403
Metal pastes
industrial solar cell technologies, 145

screen-printed high-efficiency silicon solar cells
  costs and efficiency limitations, 112
  improved compounds, 114
Metal wrap-through (MWT) cells, rear-contacted cell development, 121–122, 123f, 124
Metastability, Cu(In,Ga)Se$_2$ thin-film solar cells, 353–354
Microcrystalline materials. *See also* Hydrogenated microcrystalline silicon
  silicon thin film solar cells, 163
    hydrogenated microcrystalline silicon layers, 225–234
      ageing, 233–234
      conductivities, 225f, 232, 232f
      doping, 233
      impurities, 233, 234f
      optical absorption, gap states, and defects, 228–231, 230f, 231f
      structure, 216f, 217f, 218f, 225–228, 226f, 227f
    research background, 212
    silicon thin film solar cells, tandem solar cells, 257–259, 257f
Micrometeoroids, space solar cells, testing for, 894
Micromorph tandem cells, 257–259, 257f
Microstructure optimisation, organic solar cells, bulk heterojunction architecture, 557
Microwave photoconductance decay method, crystalline silicon wafers, 82–83
Midgap defects
  amorphous silicon thin films, 221
  hydrogenated microcrystalline silicon layers, 231
  silicon thin film solar cells, 210–211
Minority-carrier systems
  lifetime properties
    cadmium telluride thin-film photovoltaic modules, 299–301, 299f
    silicon wafers, 1015–1030

recombination, silicon thin film solar cells, voltage enhancements, 176–178, 176f, 177f, 178f, 179f
semiconductors
  lifetime, semiconductor solar modelling, 51–53
  mobility, 42
  physical specifications, 82–86
  radiation damage, 53–54
MIR Solar Array, postflight monitoring, 908
MIS inversion layer solar cell technology, junction formation, 137
Mismatch losses, photovoltaic array, energy production, 655–658, 655f, 656f, 657f
Mobility gap, amorphous silicon thin films, 222
Modified equivalent circuits, silicon thin film solar cells, p-i-n/n-i-p structures, shunts, 244–246, 245f
Modified square-wave inverters, DC to AC power conversion, photovoltaic system electronics, 704, 705f
Modular solar cell systems
  building-integrated photovoltaics, 923–924
  cadmium telluride thin-film photovoltaic modules
    applications, 316–319, 317f
    product qualification, 316–317
    basic properties, 284–285, 284f
    critical region improvement, 290–301, 290f
      activation, 299
      back contact, 299–301, 300f, 301f, 302f
      best cell performance, 302–303
      charge-carrier lifetime increase, 298–299, 299f
      grain growth-recrystallization, 292–295, 292f, 293f, 294f, 295f
      interdiffusion-intermixing, 295–298, 296f, 297f, 298f
      p-n heterojunction activation, 291–299, 291f
      stability issues, 302

Modular solar cell systems (*Continued*)
  environmental and health aspects, 314–315
    material resources, 315–316
  film deposition, 285–289
    chemical spraying, 287
    electrodeposition, 287
    screen printing, 287
    vacuum deposition; sublimation and condensation, 286–287
  future research issues, 320
  industrial production, 307–309
  installation examples, 318–319, 318f, 319f
  module integration, 303–306
    cell interconnectivity, 303–304, 303f, 304f
    contacting, 304, 305f
    lamination, 305–306
    production sequence, 306–307
    ten-MW production, 310–313
  copper indium-gallium diselenide thin-film solar cells, 341–345
    fabrication technologies, 342–343, 342f
    monolithic interconnections, 341–342, 341f
    radiation hardness, 345
    space applications, 345
    stability, 344–345
    upscaling processes, 343
  crystalline silicon, future energy potential, 1107–1109
  dye-sensitized cells, 533–534, 533f
  encapsulation
    electrical systems, photovoltaic structures, 90
    silicon thin film solar cells, 269
  energy data, 1099
  energy requirements, 1101
  hazard analysis
    decommissioning, 1094–1095
    operation, 1093–1094
  orientation
    electrical systems, photovoltaic structures, module-reverse current, 826, 827–828
    photovoltaic array, energy production, 651–652
      fixed tilt arrays, 651–652, 653f
      tracking arrays, 652
    photovoltaic specifications and performance metrics, deviation from, 970
    power rating, photovoltaic array, annual energy production, 646
  silicon thin film solar cells
    encapsulation, 269
    field experience, 272
    laser scribing and cell interconnection, 267–269, 267f, 268f
    performance factors, 270–272, 271f
    substrate materials and transparent contracts, 263–267, 263f, 265f
    thin-film deposition, 259–262, 260f
  thin film modules, 1109–1110
Molybdenum film, $Cu(In,Ga)Se_2$ thin-film solar cells, absorber preparation, 333
Monitoring
  photovoltaic systems
    calibration and recalibration, 1075–1076
    data analysis and reporting, 1078–1079
    data storage and transmission, 1076, 1078
    equipment, 1073–1075, 1074f
    performance verification, 1072, 1076–1077, 1077f
    system evaluation, 1072–1073, 1078
    user feedback, 1072, 1072f
  silicon wafers, 1031–1040
    contact formation, 1037–1038
  space solar cells, 895–908
    flight experiments, 895–901
    orbital anomalies, solar array, 901–906
    returned solar arrays, postflight investigation, 906–908
    solar array performance, 901
Monocrystalline silicon modules
  current multiplication factor, 87–88

energy requirements, 1102
voltage multiplication factor, 822
Monolithically interconnect module (MIM) solar cells, space applications, 876–877, 876f
Monolithic interconnections
   Cu(In,Ga)Se$_2$ thin-film solar cells, 341–342, 341f
   dye-sensitized cells, 533–534, 533f, 534f
   solar-powered products, 989, 990f
     calculators, 990
     design issues, 1004–1006, 1005f
Monolithic multijunction (cascade) cells, space applications, 409–412, 410f
Mont Cenis Conference Centre case study, architectural integration of solar cells, 938–940, 939f
Monthly average values, solar radiation climatology, monthly average daily diffuse irradiation, 630–631
Monthly means estimation, solar-radiation climatology, daily mean profiles from, 629–630
Morphology, organic solar cell production, 561
Mott-Schottky capacitance analysis, photoelectrochemical cells, semiconductors, 485–486
Mott-Wannier excitons, organic semiconductors, 545–546, 546f
Mountainous areas, solar-radiation climatology, 598–601, 599f
Mounting techniques
   building-integrated photovoltaics, 924–926
   solar parks and farms, array mountings, 948–952, 952f, 953f
Multicrystalline silicon
   crystal-preparation methods, 88–90, 89f
     charge preparation, 90
     crucibles, 90
   current multiplication factor, 87–88
   energy requirements, 1101
     future developments, 1108
   voltage multiplication factor, 822

Multifunctional photovoltaic modules, building-integrated photovoltaics, 923–924
Multijunction solar cells. *See also* Tandem and multijunction solar cells
   high-efficiency III-V solar cells
     application fields and reference conditions, 419–421
     band-gap selection, 421–422, 422f
     band gaps *vs.* lattice constant, 424f
     bifacial epigrowth, 434
     characterisation, 425
     concentrator solar cell design, 425–426, 426f, 427f
     inverted metamorphic growth, 432–434, 433f
     lattice-matched triple-junction cells, germanium deposition, 429–430
     more than triple junctions, 436–439, 438f
     nanowire cells, 440
     optical beam splitting, 439–440
     quantum well solar cells, 430
     research background, 417–419, 418f, 427–440, 428f, 429f
     schematic structure, 418f
     silicon deposition, 434–436, 435f
     tunnel diodes, 424–425, 424f
     upright metamorphic growth, germanium deposition, 430–432, 432f
   performance measurement, 1052–1053
   space applications, 407–412
     mechanically stacked cells, 407–409, 407f, 408f
     monolithic cells, 409–412, 410f
     research background, 402
Multiplication factors, electrical systems, photovoltaic structures, 87–94
Multiquantum wells, multijunction (cascade) structures, space applications, 411–412
Muneer algorithms, solar-radiation climatology, estimation process, 634, 635
Müser photo-thermal efficiency, formula for, 64

N

Nanocrystalline materials
  semiconductor-electrolyte junction, 490
  silicon thin film solar cells, 163
    research background, 213–214
    space applications of solar cells, 875–877
Nanowire solar cells, multijunction III-V solar cells, 440
Nernst equation, photoelectrochemical cells, electrolytes, 487–488
Net energy yield, photovoltaic systems, 1097–1098
Neural networks, maximum power point tracking, stand-along photovoltaic systems, 713
Neutral environment, space applications of solar cells, 869–870
New build vs. retrofit installation, photovoltaic systems, 807, 808, 814
Nickel-cadmium batteries, photovoltaic systems
  ban on, 740
  basic properties, 754–756
Nickel-iron batteries, photovoltaic systems, 740
Nickel-metal hydride batteries, photovoltaic systems, 740
n-i-p solar cells, silicon thin film, 234–252
  i layer, internal electric field formation, 238–240, 239f
  i layer, internal electric field reduction and deformation, 241
  internal electric field, 234–241, 236f, 237f
  light trapping, 249–252, 250f, 251f
  recombination and collection, 241–244, 242f, 243f, 244f
Nitride passivation, high-efficiency terrestrial solar cells, 105f
Noise, vibration, traffic, and air quality, solar parks and farms development, 956
Nominal array power, photovoltaic array energy production, 648–649, 648f, 649f
Nominal efficiency, Shockley solar cell equation, 68
Nominal operating cell temperature (NOCT)
  peak solar hours, 647–648
  photovoltaic array power output, temperature dependence, 650–651
Nonconsumer solar-powered products, 1003–1004
  commercial applications, 1003
  industrial applications, 1003–1004
  transportation, 1003, 1004f
Nondegenerate semiconductors, carrier statistics, 39–40
Nonwafer technologies, crystalline silicon wafers, 92–94
Noon declination, solar-radiation climatology estimation, 627
Normal hydrogen electrode (NHE), photoelectrochemical cells, 484
Normalised losses, photovoltaic performance metrics, 965
n-type silicon wafer, HIT cell development, 120–121, 120f

O

Off-grid inverters, photovoltaic system electronics, 717
I-V curve
  photovoltaic modules and solar cell performance measurement, instrumentation and solar simulation, 1049–1050, 1062f
  semiconductors
    cadmium telluride thin-film photovoltaic modules, p-n heterojunction improvement, 291–299, 291f
  solar cell, 12f
    dark characteristics, 15f
    practical characteristics, 13–15, 14f
    series and parallel resistance, 14f
    superposition principle, 12, 13f
  silicon wafers, device-level characterization, 1035–1040
  space solar cells, electrical performance testing, 888–889

One-sun AM0 efficiency curve
  mechanically stacked multijunction (cascade) cells, 407, 408f
  minority-carrier lifetime, silicon wafers, 1016–1017
Open-circuit voltage
  Cu(In,Ga)Se$_2$ thin-film solar cells, 349–352, 350f
    wide-gap chalcopyrites, 355, 356f
  dye-sensitized cell analysis, 514
  dye-sensitized cells, small-modulation electron transport and recombination, 517
  high-efficiency terrestrial solar cells, 105–106
  interdigitated back-contact (IBC) silicon solar cells, 461
  organic solar cells, 554–555, 555f
  screen-printed high-efficiency silicon solar cells, 113
  Shockley solar cell equation, 68, 70f
  silicon wafers, lifetime properties and, 1030–1031
Operating temperature, photovoltaic array, annual energy production, 646
Optical absorption
  hydrogenated microcrystalline silicon layers, 228–231, 230f
  semiconductor carrier generation, 44–47
    band-to-band transitions, 44–45, 45f
    free-carrier absorption, 48–50
    hydrogenated amorphous silicon semiconductors, 58, 59f
Optical beam splitting, multijunction III-V solar cells, 439–440
Optical confinement
  crystalline silicon solar cells, 132–134
  solar cell technology, 17–19
Optical gap, amorphous silicon thin films, 222–223, 222f
Optical losses
  Cu(In,Ga)Se$_2$ thin-film solar cells, 348–349, 348f
  photovoltaic array, annual energy production, angle of incidence increase, 646

Orbital anomalies, space solar cell monitoring, 901–906
Ordered Defect Compound (ODC), Cu(In,Ga)Se$_2$ thin-film solar cells, heterojunction formation, 338
Organic solar cells
  alternative architectures, 559
  annealing, 558
  bilayer architectures, 550–551, 551f
  blend additives, 558–559
  blend composition, 557–558
  bulk heterojunction architecture, 551–552, 552f
    microstructure optimisation, 557
  deposition procedures, 562, 563f
  device architectures, 549–550, 550f
  electronic materials, 544–548, 545f
  encapsulation, 562
  excitons, organic semiconductors, 545–548, 546f
  future research issues, 563
  market advances, 544
  operating principles, 548–552, 549f
  performance optimisation, 552–559
    electrode optimisation, 555–556
    light harvesting, 553, 554f
    optical trapping, 554
    voltage generation, 554–556, 555f
  production issues, 559–562
    degradation morphology, 561
    electrode degradation, 561–562, 561f
    materials degradation, 560–561
    stability, 560
    substrates, 560
  research background, 544
  tandem cells, 556–557, 556f
Overcharge, batteries in photovoltaic systems, 770–771
  defined, 729–730
Over roof construction, photovoltaic systems, 807, 807f
Oxide passivation, high-efficiency terrestrial solar cells, 105f
Oxidized dye regeneration, dye-sensitized cell electron transfer, 502–503
  photo-induced absorption spectroscopy, 522–524

Oxidized dye regeneration, dye-sensitized cell electron transfer (*Continued*)
  semiconductor-oxidized dye regeneration, 504–505

P

Parallel Connection Agreement, grid-connected photovoltaic systems, 798
Partially clouded conditions, solar-radiation climatology, 584–586, 585f
  global and diffuse irradiance estimates, 593–594
Particle deposition, Cu(In,Ga)Se$_2$ thin-film solar cells, 337
Particulate environment, space applications of solar cells, 870
Passivated emitter, rear locally (PERL) diffused cell
  design features, 110–111
  rear passivated cells, 109, 109f
Passivated emitter solar cell (PESC) structure, high-efficiency terrestrial solar cells, 107, 107f
Passive technologies, grid-connected photovoltaic systems, 787–788
  ROCOF technique/phase jump detection, 788
  voltage and frequency detection, 787–788
Patterning, photovoltaic technology aesthetics, 931–932, 933f
Peak solar hours
  concept, definition, and illustration, 646–648, 647f
  statistical analysis using, 653–654, 654f
Penetration depth, silicon thin film solar cells, p-i-n/n-i-p structures, light trapping, 249–252, 250f
Percentage of unused energy, seasonal energy balance, stand-alone photovoltaic systems, 669–670
Performance metrics. *See also* Testing
  batteries in photovoltaic systems, 756–765
    blackbox limit calculations, 735–736
    cycle life, 758

  field data *vs.* predicted lifetimes, 762–765
  float lifetime, 758–759
  maximum lifetime, 733
  predicted lifetimes, 760–762, 761f
  sealed batteries, 759–760
cadmium telluride thin-film photovoltaic modules, 302–303
  long-term performance and reliability, 972–977
    German 1000-Roofs-PV-Programme trends, 973–974, 974f
    Italian PV programme trends, 976
    Switzerland demonstration projects, 974–976, 975f
    system performance efficiencies, 977, 978f
  operational performance results, 964–972
    DC installations, defects from, 971–972
    inverter problems, 972
    low yield analysis, 969–972
    performance indicators, 964–967, 966f, 967f
    PV module specifications, deviation from, 970
    results summary, 967–969, 969f
    shading, 970–971
organic solar cells, 552–559
photovoltaic modules and solar cells, 1046–1054
  certification and commercial services, 1061
  commercial equipment, 1056–1057
  degradation case study, 1061–1062, 1063f
  diagnostic measurements, 1054–1056
  energy production measurements, 1053–1054
  instrumentation and solar simulation, 1049–1051
  multijunction devices, 1052–1053
  performance test conditions, 1053
  potential problems and measurement uncertainty, 1054
  qualification testing, 1057–1061

radiometry, 1046–1049, 1047f
reliability testing, 1057–1061
temperature, 1051–1052
silicon thin film modules, 270–272, 271f
space solar cell monitoring, 901
user experience, 978–982
  failure reports, 981–982
  Italian PV power plant, 980, 981f
  PV sound barrier, Switzerland, 979, 980f
  small residential systems, 978–979, 979f
Performance ratio
  photovoltaic performance metrics, 965
  annual results, 968, 969f
  photovoltaic system monitoring, 1079
  seasonal energy balance, stand-alone photovoltaic systems, 669–670
Performance test conditions (PTC), performance metrics using, 1053–1054
Perimeter recombination, interdigitated back-contact (IBC) silicon solar cells, 462–464
Perturb and observe (P&O) algorithm, stand-alone systems, maximum power point tracker, 712–713
Phase diagram, Cu(In,Ga)Se$_2$ thin-film solar cells, 326–328, 327f
Phase jump detection, grid-connected photovoltaic systems, 788
Phosphorus diffusion, gettering technique, crystalline silicon solar cell substrates, 141–142
Photoconductance decay (PCD), silicon wafers, minority-carrier lifetime, 1015–1020
Photoconductivity, amorphous silicon thin films, 223
Photoelectrochemical cells (PEC), 483–496
  dye sensitization process, 493–495, 494f
  electrolyte, 487–488, 487f
  energy and potential levels, 484
  light-induced charge separation, 491–492
  operational principles, 495–496, 495f
  research background, 483
  semiconductor-electrolyte junction, 488–491, 489f
  semiconductors, 485–487, 485f
Photogenerated current, AM1.5 spectrum, 68, 68f
Photo-induced absorption spectroscopy (PIA), dye-sensitized cells, 522–524, 523f, 525f
Photovoltage/photocurrent, dye-sensitized cell analysis
  light intensity, 514–515, 514f
  small-modulation electron transport and recombination, 515–519, 516f
Photovoltage rise method, dye-sensitized cells, small-modulation electron transport and recombination, 518–519
Photovoltaic crystalline silicon wafers
  geometric specifications, 80–81
  physical specifications, 81–82
Photovoltaic engineering testbed, space solar cells, 884
Photovoltaic systems. See also Grid-connected photovoltaic systems; Stand-alone photovoltaic systems
  design tools, 684–691
    simulation tools, 687–691
    sizing tools, 684–687
  electronics
    available products, 715–718
      charge controllers, 715–717
      inverters, 717–718
    DC to AC power conversion (inversion), 703–708
      isolation, 708
      single-phase inverters, 703–707, 704f, 705f, 706f
      three-phase inverters, 707, 707f, 708f
    DC to DC power conversion, 698–703
      advanced topologies, 703, 703f

Photovoltaic systems (*Continued*)
    boost converter, 701−702, 701f
    buck-boost or flyback converter, 702, 702f
    buck converter, 698−701, 699f, 700f
    electromagnetic compatibility, 718−719, 719f
    local electricity utility-photovoltaic systems connected to, 714−715
    properties and limitations, 697−698
    stand-alone photovoltaic systems, 708−714
        battery charge control, 708−711, 709f, 710f, 711f
        maximum power point tracking, 711−713, 712f
        power inversion, 713−714
Physical specifications
    crystalline silicon wafers, 81−82
    minority-carrier semiconductors, 82−86
    space solar cells, 895
Physical vapour deposition (PVD), Cu(In, Ga)Se$_2$ thin-film solar cells, 332
p-i-n solar cells, 27−28, 28f
    interdigitated back-contact (IBC) silicon solar cells, 470
        emitter saturation current density, 470f
    series resistance, 29f
    silicon thin film, 234−252
        internal electric field, 234−241, 236f, 237f
        intrinsic layer
            formation, 238−240, 239f
            reduction and deformation, 241
        light trapping, 238−240, 241, 249−252, 250f, 251f
        minority-carrier recombination, 178, 178f
        recombination and collection, 241−244, 242f, 243f, 244f
Pioneer Venus Orbiter SA, space solar cell monitoring, orbital anomalies, 905−906
Planning consent

solar parks and farms, 960
solar parks and farms development, 954−958
Plasma enhanced chemical vapour deposition (PECVD)
    amorphous silicon thin films, doping effects, 224
    antireflection coating, crystalline silicon solar cells, 137−138
    edge-defined film fed growth silicon sheet production, 150−151
    hydrogenated microcrystalline silicon layers, 225−228, 226f
    silicon thin film solar cells, 181, 183f
        layer deposition principles, 260
    thin-film silicon solar cells, 210
Plated seed layers, screen-printed high-efficiency silicon solar cells, 115
Plate systems, lead-acid batteries, photovoltaic systems, 728f, 743
p-n junction solar cell, 19−25
    cadmium-telluride thin-film photovoltaic modules, heterojunction, 291−299, 291f
    diffused emitter, 23−25
    junction parameters, 19−23, 20f
    metal-insulator-semiconductor inversion layer combined with, 149
    open circuit, 21f
    thin silicon solar cells, minority-carrier recombination, 178, 178f
    uniform emitter and base, 23
Point-contact (PC) solar cell, interdigitated back-contact (IBC) silicon solar cells, 453f
    concentrator applications, 452
    design, 456−458, 456f
Point of view, photovoltaic technology aesthetics, 933, 935f
Poisson equation, semiconductor transport, 41
Polysilicon technology
    emitters, interdigitated back-contact (IBC) silicon solar cells, 470

high-efficiency terrestrial solar cells, 105–106
manufacturers, 1170–1173
production capacity and supply, 1167–1170
Pompeu Fabra Public Library case study, architectural integration of solar cells, 937–938, 938f
Portable solar-powered products, 997–1001
vehicle systems, 999–1001, 1000f
Postflight investigation, space solar cell monitoring, returned solar arrays, 906–908
Post-notification requirements, grid-connected photovoltaic systems, 803
Power density optimisation, solar parks and farms, 947
Power inversion, stand-alone photovoltaic system electronics, 713–714
Power quality, grid-connected photovoltaic systems, 783–784
Power-to-mass ratio, multijunction III-V solar cells, 419–420
Power variation, grid-connected photovoltaic systems, active technologies, 789
Predicted battery lifetimes, field data comparisons, 762–765
Pre-notification requirements, grid-connected photovoltaic systems, 803
Primary cells, photovoltaic modules and solar cell performance measurement, 1049
Priority despatch legislation, solar parks and farms development, 959
Prismatic cover slips, silicon thin film solar cells, light trapping, 169–170, 170f
Production issues
 cadmium telluride thin-film photovoltaic modules, 306–307
 industrial solar production capacity, 1155, 1158f

organic solar cells, 559–562
 degradation morphology, 561
 electrode degradation, 561–562, 561f
 materials degradation, 560–561
 stability, 560
 substrates, 560
polysilicon manufacturing, 1170–1173
solar cell production companies, 1159–1167
Product qualification, cadmium-telluride thin-film photovoltaic modules, 316–317, 317f
Programming languages, photovoltaic system design, 688–691
Project development issues, solar parks and farms, 953–959, 957f
 archaeology and cultural heritage, 957
 community engagement and legacy issues, 957–958
 connectiion approvals, 958
 development rationale and site location, 954
 ecology and nature conservation, 956–957
 financing issues, 958–959
 landscaping and visual change, 955, 955f
 light reflection and aviation, 955–956
 noise, vibration, traffic, and air quality, 956
 planning consent, 954–958
 surface water drainage, flooding, and ground conditions, 956
Protection by design, electrical systems, photovoltaic structures, 832–833, 832f, 833f
Pspice simulation program, photovoltaic system design, 688
Public health. *See* Health issues
Pulsed light sources, space solar cells, 888
Pulse width modulation (PWM), DC to DC power conversion, photovoltaic systems
 boost converter, 702
 buck converter, 700–701

Q

Qualification testing, photovoltaic systems, 1057–1061
Quantitative processing, solar-radiation climatology, photovoltaic design, 605–609
 inclined planes, 606
Quantum efficiency
 dye-sensitized cell measurements, 519–520
 hydrogenated microcrystalline silicon layers, 233, 234f
 photovoltaic performance metrics, 1054–1055
 solar cell, 15–16
Quantum well solar cells, multijunction III-V technology, 430, 431f
Quarter-wavelength rule, antireflection coating, solar cell optics, 17
Quasi-Fermi levels, dye-sensitized cells, mesoporous electrode, internal potential measurement, 521–522
Quasi-steady-state photoconductance (QSSPC), minority-carrier lifetime, silicon wafers, 1016

R

Radiation damage
 Cu(In,Ga)Se$_2$ thin-film solar cells, 345
 semiconductor solar modelling, 53–55
 space applications of solar cells, 873–874, 874f
Radiation testing, space solar cells, 892–893, 893f
Radiative band-to-band recombination, silicon wafers, 1014–1015
Radio frequency suppression, grid-connected photovoltaic systems, 784
Radiometry, photovoltaic modules and solar cell performance measurement, 1046–1049
Rain screen facade, construction, photovoltaic systems, 811–813, 812f
Raman crystallinity, hydrogenated microcrystalline silicon layers, 228
 conductivities, 232, 232f
Raman spectroscopy, hydrogenated microcrystalline silicon layers, 228
Random texturing, silicon thin film solar cells, light trapping, 168
Rapid thermal chemical vapour deposition, silicon thin film solar cells, 186
Rapid thermal processing (RTP)
 Cu(In,Ga)Se$_2$ thin-film solar cells, selinisation processes, 336
 crystalline silicon solar cells
  fast processing applications, 143, 144f
  junction formation, 136–137
Ratings, concentrator photovoltaic systems, 854–859
RCA cleaning, crystalline silicon solar cells, 134–135
Reactive ion etching, crystalline silicon solar cells, 132–134
Reactive power, grid-connected photovoltaic systems, 789
Rear-contacted cells, high-efficiency silicon solar cell development, 121–124
 emitter wrap-through cells, 123, 123f
 metal wrap-through cells, 124
 rear-junction solar cells, 122–123, 122f
Rear-junction solar cells, high-efficiency silicon solar cell development, 122–123, 122f
Rear passivated cells, high-efficiency solar cells, laboratory development, 108–109, 108f
Recalibration, photovoltaic system monitoring, 1075–1076
Rechargeable batteries, photovoltaic systems, 726
Recombination
 Cu(In,Ga)Se$_2$ thin-film solar cells
  defect physics, 328–330
  open-circuit voltage, 350, 350f
  short-circuit current losses, 348–349
 dye-sensitized cells, electron transport photo-induced absorption spectroscopy, 522–524

semiconductor-oxidized dye or electrolyte species, 504–505
small-modulation electron transport and recombination, 515–519
interdigitated back-contact (IBC) silicon solar cells
  high efficiency requirements, 455
  modelling parameters, 459
  perimeter and edge recombination, 462–464
  velocity requirements, 455
semiconductor classification, 48–53
  bulk recombination, 48–50
  surface recombination, 50–51
silicon thin film solar cells
  minority-carrier recombination, 176–178, 176f, 177f, 178f, 179f
  p-i-n/n-i-p structures, 241–244
silicon wafers, carrier recombination, 1012–1015
tandem and multijunction solar cells, junction principle, 253, 254f
Recrystallisation
  cadmium telluride thin-film photovoltaic modules, grain boundaries and growth, 293f, 294f, 295f
  silicon thin film solar cells, 181–182, 183f, 184f, 185f
Redox potential
  dye-sensitized cell electron transfer, counterelectrode reaction, 505–506
  photoelectrochemical cells, 484
    electrolytes, 487, 487f
Reference cells, space solar cells, electrical performance testing, 888
Reflection coefficient, solar cell, 18f
  antireflection coating, 16, 18f
Refractive index, antireflection coating, solar cell optics, 16–17
Regeneration kinetics
  dye-sensitized cells, electron transport, photo-induced absorption spectroscopy, 522–524
  oxidized dye regeneration, dye-sensitized cell electron transfer, 502–503

Regulatory issues
  electrical systems, photovoltaic structures, 820–821
  grid-connected photovoltaic systems, 789–803
    draft IEC standard 62776, 790–798
    parallel connection agreement, 798
    post- or pre-notification, 803
    Tariff Agreement, 798–802
    technical connection guidelines, 789–790
    type testing, 790–798, 798f
  solar parks and farms, 959–960
    consent issues, 960
    energy and environment policy, 959
    incentive schemes, 959–960
Relative optical path length, photovoltaic modules and solar cell performance measurement, 1048
Relative performance metrics, silicon thin film modules, 272
Relative spectral response, space solar cells, electrical performance testing, 889–890, 890f
Reliability, photovoltaic systems, 972–977
  German 1000-Roofs-PV-Programme trends, 973–974, 974f
  Italian PV programme trends, 976
  module reliability and qualification testing, 1057–1061
  Switzerland demonstration projects, 974–976, 975f
  system performance efficiencies, 977, 978f
Reorganization energy, photoelectrochemical cells, electrolytes, 487–488
Resistivity measurements, silicon wafers, 1031–1032
Resonant converters, DC to DC power conversion, photovoltaic systems, 703
Retrofit vs. new build installation, photovoltaic systems, 807, 808, 814
Reverse characterisation, space solar cells, electrical performance testing, 890, 891f

Reverse current, electrical systems, photovoltaic structures, module-reverse current, 826
Ribbon-growth on substrate (RGS) technology, crystalline silicon wafers, 92–93
Ribbon technologies
 crystalline silicon solar cell substrates, kerf loss, 131–132
 crystalline silicon wafers, 92–94
ROCOF technique, grid-connected photovoltaic systems, 788
Roof construction, photovoltaic systems, 806–811
 architectural integration, solar cells, 921–923, 921f
 integral roofs, 806, 806f
 over roof, 807, 807f
 solar-powered products, 936, 1002f
 tiles, 807, 808f
 traditional roof interfaces, 810, 811f
Ruthenium complex (N3)
 dye-sensitized solar cells, 496–497, 497f
 heteroleptic complexes, 530

S

Safety issues. *See also* Health issues
 grid-connected photovoltaic systems, 783
 hazard analysis, photovoltaic systems
  amorphous silicon solar cellls, 1088–1089
   occupational safety issues, 1088–1089
  cadmium telluride solar cells, 1089–1090
   occupational health issues, 1089–1090
  copper indium diselenide cells, 1090–1091
   occupational health and safety, 1090–1091
  crystalline silicon solar cells, 1084–1087
   occupational health issues, 1084–1085
  future research issues, 1095

gallium arsenide high-efficiency solar cells, 1092–1093
 occupational health and safety, 1092–1093
 manufacturing hazards, 1084
 overview, 1083–1084
 photovoltaic modules
  decommissioning, 1094–1095
  operation, 1093–1094
Saturation current density
 interdigitated back-contact (IBC) silicon solar cells, 470
 silicon wafers, lifetime measurements, 1024–1027
Schottky barrier, cadmium-telluride thin-film photovoltaic modules, back contact, 300
Screen-printed cells
 cadmium telluride thin-film photovoltaic modules, 287
 Cu(In,Ga)Se$_2$ thin-film solar cells, 337
 crystalline silicon solar cells, front contact formation, 140
 high-efficiency silicon solar cell development, 111–116
  hot-melt and stencil printing, 115
  limitations, 115–116
  paste improvements, 114
  performance, 113
  plated seed layers, 115
  selective emitter and double printing, 114–115, 114f
  structure, 111–113, 112f
 industrial solar cell technologies, 145–147
Sealed construction, lead-acid batteries
 lifetime estimations, 759–760
 photovoltaic systems, 742–745
  absorbent glass mat, 745
  gel type, 745
Seasonal energy balance, stand-alone photovoltaic systems, 668–671, 669f, 670f, 671f
Secondary cells, photovoltaic modules and solar cell performance measurement, 1049

Secondary standards, space solar cell calibration, 886
Selective-emitter printing
  industrial solar cell technologies, 146
  screen-printed high-efficiency silicon solar cells, 114, 114f
Selenisation processes, copper indium-gallium diselenide thin-film solar cells
  absorber preparation, 335–337, 336f
  module fabrication, 342–343
Selenium cells. See Copper indium-gallium diselenide solar cells
Selenium vacancy, thin-film solar cells
  defect physics, 326
  postdeposition air annealing, 337–338
Self-discharge rate, batteries in photovoltaic systems, defined, 732
Semiconductor-electrolyte junction, photoelectrochemical cells, 488–491
Semiconductor finger solar cell, laser processing, 118–119, 118f
Semiconductors
  dye-sensitized cells, electron transport, semiconductor-oxidized dye or electrolyte species, 504–505
  organic materials, exciton properties, 545–548
  photoelectrochemical cells, 485–487, 485f
  solar cell modelling
    band structure, 34–38
    carrier generataion, optical absorption, 44–47
    carrier mobility, 42–44
    carrier statistics, 38–40
    heavy doping effects, 55–57
    hydrogenated amorphous silicon properties, 57–58
    overview, 33–34
    radiation damage, 53–55
    recombination, 48–53
    transport equations, 40–41
Semi-insulation polysilicon (SIPOS) contact passivation

high-efficiency terrestrial solar cells, 105–106
  interdigitated back-contact (IBC) silicon solar cells, 470
SEMI™ M61000 standard, photovoltaic crystalline silicon wafer requirements, 80–81
Sensors, photovoltaic system monitoring, 1075
Series resistance
  interdigitated back-contact (IBC) silicon solar cells, reduction of, 472
  silicon thin film solar cells, p-i-n/n-i-p structures, 248–249
  silicon wafers, 1038–1040
  solar cells, 29, 29f
  stand-alone photovoltaic system electronics, 709–710, 710f, 711f
Shading effects, photovoltaic performance metrics, 970–971
Shallow junction "violet" cell, silicon space cell development, 101–103, 102f
Shape, photovoltaic technology aesthetics, 929–930, 931f
Shaping process, crystalline silicon wafers, 94
Sheet silicon technologies, crystalline silicon solar cell substrates, kerf loss, 131–132
Shockley-Anderson model, heterojunction solar cell, 25–26, 26f
Shockley-Queisser estimates
  nominal efficiency, 67f
  ultimate efficiency, 66, 67f
Shockley-Read-Hall model
  hydrogenated amorphous silicon semiconductors, 58
  interdigitated back-contact (IBC) silicon solar cells, 451–452
  p-i-n junction solar cell, 27
  p-n junction, solar cell, 22–23
  semiconductor recombination rate, 50
  silicon wafers
    carrier recombination, 1014
    surface recombination velocity, 1021
Shockley solar cell equation
  efficiency metrics and, 67–72

Shockley solar cell equation (*Continued*)
  ideal solar cell, 10–11
Short-circuit current
  Cu(In,Ga)Se$_2$ thin-film solar cells, 348–349, 348f
  dye-sensitized cell analysis, 514
  organic solar cells, performance optimisation, 552–553
  silicon thin film solar cells, light trapping, 170–172, 171f
  silicon wafers, minority-carrier lifetime, 1018–1019
Short-range order, hydrogenated amorphous silicon, 215
Shrink geometries, interdigitated back-contact (IBC) silicon solar cells, 471
Shunts
  silicon thin film solar cells, p-i-n/n-i-p structures, 244–247, 245f, 246f, 247f
  stand-alone photovoltaic system electronics, 709–710, 710f, 711f
Siemens Process, silicon production, 1168–1169
Silicon deposition and crystal growth. *See also* Crystalline silicon; Polysilicon technology
  multijunction III-V solar cells, 434–436
  thin film silicon solar cells, 179–187, 180f
    atmospheric pressure chemical vapour deposition, 185
    high-temperature chemical vapour deposition, 182–186, 183f
    high-temperature deposition methods, 181–187
    liquid-phase epitaxy, 187, 188f
    low-pressure chemical vapour deposition, 186
    low-temperature chemical vapour deposition, 186–187
    melt-growth techniques, 181, 182f
    rapid thermal chemical vapour deposition, 186
    recrystallisation, 181–182, 183f, 184f, 185f
    substrate properties, 179–180

Silicon-Film™ technology, commercial thin film crystalline silicon solar cells, 151, 152f
Silicon ingots
  carrier lifetime measurements, 1033
  lifetime measurements in, 1027–1028
Silicon solar cell. *See also* Crystalline silicon
  concentrator photovoltaic systems, 843, 844f
  electronic-grade silicon, 1100
  maximum power point tracking, 713
  production processes, 1168–1170
    fluidised bed process, 1169–1170
    Siemens Process, 1168–1169
  space applications
    laboratory development, 100–104, 102f, 105f
    III-V solar cell technology, historical review, 399–402
  structure, 10f
Silicon wafering process
  carrier lifetime and voltage characteristics, 1030–1031
  carrier recombination, 1012–1015
  lifetime instabilities and measurement artifacts, 1028–1030
  minority-carrier lifetimes, 1015–1030
  process monitoring and control, 1031–1040
  research background, 375
Simulation tools
  photovoltaic modules and solar cell performance measurement, 1049–1051
  photovoltaic system design, 687–688
Sine wave inverters, photovoltaic system electronics, 717–718
Single-axis tracking linear concentrator, 849f, 852f
Single-junction III-V cell structure
  multijunction cell research and, 427–429
  space applications, 403–407
    AlGaAs-GaAs structures, 403–404, 405f

Index

GaAs-based cells, Ge substrate, 406—407
internal Bragg reflector, 404—406
Single-phase inverters, DC to AC power conversion, photovoltaic system electronics, 703—707, 704f, 705f, 706f
Site selection, solar parks and farms
design issues, 946—948
development process and, 954
Site testing, construction, photovoltaic systems, 816, 816f
Size calculations
batteries in photovoltaic systems, 767—770
seasonal energy balance, stand-alone photovoltaic systems, 670, 671f
photovoltaic technology aesthetics, 930
solar-powered products design, 1005—1006
tools for, photovoltaic system design, 684—687
Sloped surfaces, architectural integration, solar cells, 922, 922f, 925—926, 929f
Slope irradiation analysis
all sky irradiation estimation, 632—633
inclined surfaces, hourly horizontal diffuse irradiation, 634—635
quantitative processing, solar radiation climatology, London case study, 606—608, 607f, 608f
Small power systems
performance metrics and user experience, 978—982
space applications of solar cells, 875—877, 875f, 876f
SoDa project, solar radiation, stochastic estimation, 613—614
Sodium, $Cu(In,Ga)Se_2$ thin-film solar cells, absorber preparation, 333
Soiling, photovoltaic array, annual energy production, 646
Solar altitude calculation, solar-radiation climatology, 624—625
Solar cell design. See also Organic solar cells

actual and planned production capacities, 382—390, 382f
efficiency rankings, 73
electrical characteristics
ideal solar cell, 10—12
I-V characteristic, 13—15, 14f, 15f
quantum efficiency and spectral response, 15—16
series and parallel resistance, 14f
global production, 374—375, 374f, 375f
industrial cell production companies, listing of, 382—390
optical properties, 16—19
antireflection coating, 16—17
light trapping, 17—19, 18f
series resistance, 29
structure, 10f
heterojunction cells, 25—27, 25f, 26f
p-i-n junction, 27—28, 28f
p-n junction cell, 19—25, 29f
Solar flares, space applications of solar cells, 868—869
Solar geometry computation, solar-radiation climatology, 618—627
angle of incidence and vertical and horizontal shadow angles, 625—627, 626f
angular movement, 618
local mean time/local apparent time conversion, 620—621
noon declination accuracy, 627
time systems, 618—620
trigonometric determination, 621—625
Solar parks and farms
defined, 944—946
design issues, 946—953
array mountings, 951—952, 952f
fixed arrays, 949
load factor optimisation, 946—947
local electricity network and usage, 947
location and land usage, 947—948
photovoltaic technology selection, 948, 949f
power conditioning, 953
power density optimisation, 947
site selection, 946—948

Solar parks and farms (*Continued*)
  solar arrays, 948–952, 949f
    tracking arrays, 950–951, 950f, 951f
  future expectations, 962
  grid parity, 961–962, 961f
  historical background, 944–946, 945f, 946f
  project development issues, 953–959, 957f
    archaeology and cultural heritage, 957
    community engagement and legacy issues, 957–958
    connectiion approvals, 958
    development rationale and site location, 954
    ecology and nature conservation, 956–957
    financing issues, 958–959
    landscaping and visual change, 955, 955f
    light reflection and aviation, 955–956
    noise, vibration, traffic, and air quality, 956
    planning consent, 954–958
    surface water drainage, flooding, and ground conditions, 956
  regulatory issues, 959–960
    consent issues, 960
    energy and environment policy, 959
    incentive schemes, 959–960
Solar-powered products
  design issues, 1004–1006
    electrical characteristics, 1005
    mechanical and structural properties, 1006
    size, 1005–1006
  future trends in, 1006–1007
  genesis of, 988–989
  grid-connected products, 1001–1002
    cladding and facade products, 1002, 1002f
    roofing products, 1001, 1002f
  household products, 996–997
    lighting products, 997
    ventilation and air conditioning, 996–997, 996f
  nonconsumer products, 1003–1004
    commercial applications, 1003
    industrial applications, 1003–1004
    transportation, 1003, 1004f
  photovoltaic integration in, 988–989
  portable and transportable products, 997–1001
    portable electronics, 999–1001, 1000f
    vehicle products, 997–999, 999f
  stand-alone consumer products, 989–1001
    accent lighting, 991–992
    air fresheners, 992
    aquatic products, 994, 995f
    calculators, 990
    chargers, 991
    deterrent products, 994–995
    garden products, 992–995, 993f
    indoor products, 990–992
    lighting and markers, 992–994, 993f
    watches and clocks, 991, 991f
  thin-film cells, 989, 990f
Solar-radiation climatology
  all sky irradiation estimation, 632–637
    direct beam irradiation, inclined planes, 633–634
    ground-reflected irradiation, 636–637
    hourly diffuse irradiation, 634–636
    slope irradiation components, 632–633
    sun-facing surfaces, 636
  cloudless sky global radiation, latitude variations, 579–586, 580f, 581f, 582f, 583f
  daily clearness index, 628–629
  data sequences, 586–587, 587f
  definitions and associated units, 577–604
  desert-type climates, 597–598, 598f
  future research issues, 604, 637–638
  hourly global and diffuse horizontal irradiation, 627–632
    estimation of, 630–632
  humid tropics
    diffuse radiation, 597
    examples, 594–596, 595f, 596f
    simplified radiation conditions, 590–597, 591f

interannual variability, 602—604
  radiation estimation, sunshine data, 603—604
mean daily profiles, monthly solar irradiation estimation, 629—630
mountainous areas, 598—601, 599f
overcast sky conditions, 586
partially clouded conditions, 584—586, 585f
  global and diffuse irradiance estimates, 593—594
photovoltaic systems
  design criteria, 609
  inclined planes systems, 601—602, 602f, 606
  quantitative processing, 605—609
  research background, 575—577
  slope irradiation, London case study, 606—608, 607f, 608f
  variability in, design implications, 578—579
simplified radiation conditions, 587—601
  European examples, 588—590, 589f
  humid tropics, 590—597, 591f
solar geometry computation, 618—627
  angle of incidence and vertical and horizontal shadow angles, 625—627, 626f
  angular movement, 618
  local mean time/local apparent time conversion, 620—621
  time systems, 618—620
  trigonometric determination, 621—625
stochastic generation, 610—617
  daily values, hourly values generation, 614—616
  data generalisation probolem, 611
  general principles, 610—611
  global climate variations, limitations of, 611
  global radiation, diffuse and beam splitting, 616
  progress assessment, 616—617
  SoDa project improvements in, 613—614
  time series estimation, KT-based methods, 612—613
Solar simulator calibration, space solar cells, 885—886
Solar spectrum variation, photovoltaic array, annual energy production, 646
Space applications of solar cells
  calibration, 883—886
    extraterrestrial methods, 883—884
    secondary working standards, 886
    synthetic methods, 884—886
  Cu(In,Ga)Se$_2$ thin-film solar cells, 345
  monitoring, 895—908
    flight experiments, 895—901
    orbital anomallies, solar array, 901—906
    returned solar arrays, postflight investigation, 906—908
    solar array performance, 901
  multijunction solar cells, 407—412, 419—420
    mechanically stacked cells, 407—409, 407f, 408f
    monolithic cells, 409—412, 410f
  operating procedures
    air mass zero spectrum, 866f
    AM0 efficiencies, 872—874
    large power systems, 877—878, 878f
    neutral environment, 869—870
    particulate environment, 870
    radiation damage, 873—874, 874f
    research background, 863—865, 864f
    small power systems, 875—877, 875f, 876f
    solar flares, 868—869
    space missions, 865—871
    thermal environment, 870—871
    trapped radiation environment, 867, 868f
  research background, 882—883
  silicon space cell
    III-V solar cell technology, historical review, 399—402
    laboratory development, 100—104, 102f, 105f

Space applications of solar cells (Continued)
  shallow junction "violet" cell, 101–103, 102f
  single-junction III-V cell structure, 403–407
    AlGaAs-GaAs structures, 403–404, 405f
    GaAs-based cells, Ge substrate, 406–407
    internal Bragg reflector, 404–406
  testing, 886–895
    electrical performance, 887–891, 890f, 891f
    environmental testing, 891–895, 893f
    physical characteristics and mechanical tests, 895
Space shuttle calibration, space solar cells, 884
Spectral irradiances, photovoltaic modules and solar cell performance measurement, 1047f, 1048
Spectrally adjustable solar simulator, tandem and multijunction solar cells, performance measurement, 1052–1053
Spectral mismatch error, photovoltaic modules and solar cell performance measurement, 1049
Spectral response
  hydrogenated microcrystalline silicon layers, impurities, 233, 234f
  photovoltaic performance metrics, 1055
  silicon thin film solar cells, p-i-n/n-i-p structures, 244
  solar cell, 15–16
Spectrum testing, space solar cells, electrical performance, 888
Square-wave switching, DC to AC power conversion, photovoltaic system electronics, 704, 705f
Stability
  cadmium telluride thin-film photovoltaic modules, 302
  Cu(In,Ga)Se$_2$ thin-film solar cells, module fabrication, 344–345
  interdigitated back-contact (IBC) silicon solar cells, 464–467

organic solar cell production, 560
space solar cells, electrical performance, 888
Staebler-Wronski effect (SWE)
  amorphous silicon cells, 242
  field experience with, 272
  silicon thin film solar cells
    gap states, 221
    hydrogenated amorphous silicon, 215–217, 218f
    micromorph tandem cells, 257–258
    research background, 210–211
Stand-alone photovoltaic systems. See also Grid-connected photovoltaic systems
  architectural integration, 923, 925f, 926, 929f
  electronics, 708–714
    battery charge control, 708–711, 709f, 710f, 711f
    maximum power point tracking, 711–713, 712f
    power inversion, 713–714
  energy balance in
    climatic cycle, 660–662, 661f, 662f
    load characteristics, 663–668, 664f
    daily energy-balance dynamics, 666–668, 667f
    electricity consumption, lighting and electrical appliances, 664–668
    research background, 659–663
    schematic, 660f
    seasonal balance, 668–671, 669f, 670f, 671f
  energy requirements, 1104–1105
  inverters for, 717
  modelling research, 673–674
  performance indicators, 966–967
  sizing, 674–680
    energy balance and, 674–676, 676f
    solar pumping systems, 680
    supply reliability, 676–680, 678f, 679f
  solar-powered consumer products, 989–1001
    accent lighting, 991–992

Index 1237

air fresheners, 992
aquatic products, 994, 995f
calculators, 990
chargers, 991
deterrent products, 994–995
garden products, 992–995, 993f
indoor products, 990–992
lighting and markers, 992–994, 993f
watches and clocks, 991, 991f
Standard reporting conditions, photovoltaic modules and solar cell performance measurement, 1046–1047
Standards
concentrator photovoltaic systems, 854–859
construction, photovoltaic systems, 811, 812f
international standards for photovoltaics, 1181
photovoltaic performance metrics, 964–972
space solar cell calibration, secondary standards, 886
Stark effect, dye-sensitized cell characterization, 525–527, 526f
State of charge (SOC), batteries in photovoltaic systems, defined, 731
Static concentrator, 850f
Statistical analysis, photovoltaic array, energy production, 653–655, 654f
Steady-state (quantum efficiency) measurements
dye-sensitized cells, 519–520
minority-carrier lifetime, silicon wafers, 1016
Stencil printing, screen-printed high-efficiency silicon solar cells, 115
Stochastic generation, solar-radiation climate data, 610–617
daily values, hourly values generation, 614–616
data generalisation probolem, 611
general principles, 610–611
global climate variations, limitations of, 611
global radiation, diffuse and beam splitting, 616

progress assessment, 616–617
SoDa project improvements in, 613–614
time series estimation, KT-based methods, 612–613
String cable-fault current, electrical systems, photovoltaic structures, 825–826, 826f
String cable protection, electrical systems, photovoltaic structures, 827
String overcurrent protection, electrical systems, photovoltaic structures, 828–829
String ribbon technology, crystalline silicon wafers, 92–93
Structural properties, solar-powered products design, 1006
Sub-band-gap reflection analysis, silicon thin film solar cells, light trapping, 172, 173f
Sublayer membranes, photovoltaic system construction, 810
Sublimation, cadmium telluride thin-film photovoltaic modules, vacuum deposition, 286–287
Sublinear cell structure, interdigitated back-contact (IBC) silicon solar cells, 457
Substrates
cadmium telluride thin-film photovoltaic modules, 289
crystalline silicon solar cells, 131–132
material improvement, 141–143
film deposition, 289
organic solar cell production, 560
silicon thin film solar cells
materials and transparent contacts, 263–267, 263f, 265f
silicon deposition and crystal growth, 179–180
thinning, 188–190, 189f, 190f
Substructure installation, photovoltaic systems, 808, 809f, 813–814
Sulfation, lead-acid batteries, photovoltaic systems, 753–754
Sun-facing surfaces, solar-radiation climatology, estimation process, 636

Sun-path diagram, solar-radiation climatology, 625f
Suns-V_{oc} curves, silicon wafers, diode analysis, 1036–1037, 1037f, 1039f
Superlinear cell structure, interdigitated back-contact (IBC) silicon solar cells, 457
Superposition principle, solar cell I-V characteristic, 12, 13f
Superstrate configuration, silicon thin film solar cells, 263–264, 263f
Supperlattice structures, multijunction (cascade) structures, space applications, 411–412
Supporting web (S-Web) technology, crystalline silicon wafers, 93–94
Surface availability in buildings, architectural integration, solar cells, 921–923, 921f
Surface photovoltage response, silicon wafers, minority-carrier lifetime, 1018–1019
Surface recombination
  semiconductor classification, 50–51
  silicon wafers, lifetime measurements, velocity measurements, 1021–1027
Surface texture
  Cu(In,Ga)Se$_2$ thin-film solar cells, heterojunction formation, 338–339
  solar cells, 18f
Surface water drainage, flooding, and ground conditions, solar parks and farms development, 956
Swiss photovoltaic demonstration projects
  long-term performance and reliability, 974–976, 975f
  sound barrier system, 979, 980f
Switching frequency, DC to DC power conversion, photovoltaic systems, buck converter, 701
Synthetic calibration, space solar cells, 884–886
  differential spectral response, 886
  direct sunlight, 885
  global sunlight, 884–885
  solar simulator, 885–886

T

Tandem and multijunction solar cells
  interdigitated back-contact design, 456–458, 456f
  organic solar cells, 556–557, 556f
  performance measurement, 1052–1053
  silicon thin film, 214–215
  general principles, 252–254, 252f, 254f
  hydrogenated silicon tandems, 254–255, 255f
  microcrystalline amorphous tandems, 257–259, 257f
  triple-junction amorphous cells, silicon-germanium alloys, 255–257, 256f
  space applications
    mechanically stacked multijunction (cascade) cells, 408–409
    monolithic multijunction (cascade) cells, 409–412
Target performance, interdigitated back-contact (IBC) silicon solar cells, 472, 473f
Tariff Agreement, grid-connected photovoltaic systems, 798–802
Technical connection guidelines, grid-connected photovoltaic systems, 789–790
Telemetry systems, nonconsumer solar-powered products, 1004
Temperature controls
  batteries in photovoltaic systems, 772
  space solar cells, electrical performance testing, 889
Temperature dependence
  lead-acid batteries, capacity reduction and, 747–748, 748f
  photovoltaic array power output, 650–651, 650f
  photovoltaic modules and solar cell performance measurement, 1051–1052
  semiconductor carrier mobility, 43f
Temps and Coulson algorithm, solar-radiation climatology, estimation process, 636

Ternary compounds, chalcopyrite solar cell research, 376
Terrestrial solar cells, high-efficiency development, 104–108, 105f
Testing. *See also* Performance metrics
  photovoltaic modules and solar cells, 1046–1054
    certification and commercial services, 1061
    commercial equipment, 1056–1057
    degradation case study, 1061–1062, 1063f
    diagnostic measurements, 1054–1056
    energy production measurements, 1053–1054
    instrumentation and solar simulation, 1049–1051
    multijunction devices, 1052–1053
    performance test conditions, 1053
    potential problems and measurement uncertainty, 1054
    qualification testing, 1057–1061
    radiometry, 1046–1049, 1047f
    reliability testing, 1057–1061
    temperature, 1051–1052
  space solar cells, 886–895
    electrical performance, 887–891, 890f, 891f
    environmental testing, 891–895, 893f
    physical characteristics and mechanical tests, 895
Texturing
  crystalline silicon solar cells, 132–134, 133f
  photovoltaic technology aesthetics, 931–932, 933f
Thermal cycling sequence, photovoltaic system reliability and qualification testing, 1059–1060
Thermal environment
  space solar cells, 870–871
    thermal cycling testing, 894
Thermal mapping, photovoltaic performance metrics, 1055–1056
Thermodynamic efficiencies, formulas for, 64–65

Thermophotovoltaics (TPV), multijunction III-V solar cells, 420
Thick silicon layers, silicon thin film solar cells, 163–164
Thin films
  cadmium telluride thin-film photovoltaic modules
    applications, 316–319, 317f
    product qualification, 316–317
    basic properties, 284–285, 284f
    critical region improvement, 290–301, 290f
    activation, 299
    back contact, 299–301, 300f, 301f, 302f
    best cell performance, 302–303
    charge-carrier lifetime increase, 298–299, 299f
    grain growth-recrystallization, 292–295, 292f, 293f, 294f, 295f
    interdiffusion-intermixing, 295–298, 296f, 297f, 298f
    p-n heterojunction activation, 291–299, 291f
    stability issues, 302
    environmental and health aspects, 314–315
    material resources, 315–316
  film deposition, 285–289
    chemical spraying, 287
    electrodeposition, 287
    screen printing, 287
    vacuum deposition; sublimation and condensation, 286–287
  future research issues, 320
  industrial production, 307–309
  installation examples, 318–319, 318f, 319f
  module integration, 303–306
    cell interconnectivity, 303–304, 303f, 304f
    contacting, 304, 305f
    lamination, 305–306
    production sequence, 306–307
  copper gallium diselenide solar cells, 357–358
  future research issues, 361

Thin films (*Continued*)
  copper indium-aluminium diselenide solar cells, 358
  copper indium-gallium diselenide solar cells, 358–359
    absorber preparation techniques, 333–338
    band diagram, 346–347, 346f
    band gap energies, 326, 326f, 331f
    chalcopyrite lattice, 325–326
    co-evaporation processes, 334–335, 335f
    defect physics, 328–330
    electrodeposition, 337
    electronic metastabilities, 353–354
    epitaxial absorption deposition, 337
    fill factor, 352–353
    heterojunction solar cell, 331, 332f
      buffer layer deposition, 339–340
      free surface properties, 338–339
      window layer deposition, 340
    high-efficiency solar cells, 331–332
    module production and commercialisation, 341–345
      fabrication technologies, 342–343, 342f
      monolithic interconnections, 341–342, 341f
      radiation hardness, 345
      space applications, 345
      stability, 344–345
      upscaling processes, 343
    open-circuit voltage, 349–352, 350f
    particle deposition, 337
    phase energies, 326–328, 327f
    postdeposition air annealing, 337–338
    research background, 324–325
    screen printing, 337
    selinisation process, 335–337, 336f
    short-circuit current, 348–349, 348f
    wide-gap chalcopyrites, 354–361, 356f
  copper indium-gallium selenium sulfur solar cells, 359
  copper indium sulfide solar cells, 358–359
  energy requirements, 1102–1104
  graded-gap devices, 359–361, 360f
  modules, 1109–1110
  silicon cells
    basic properties, 210–215
    defined, 162–163
    future research issues, 273–274
    high-temperature growth methods, 190–196
    hydrogenated amorphous silicon layers, 215–224
      conductivity, 223–224
      doping, 224
      gap states, 220–223, 220f, 222f
      structure, 215–219
    hydrogenated microcrystalline silicon layers, 225–234
      ageing, 233–234
      conductivities, 225f, 232, 232f
      doping, 233
      impurities, 233, 234f
      optical absorption, gap states, and defects, 228–231, 230f, 231f
      structure, 216f, 217f, 218f, 225–228, 226f, 227f
    light trapping, 165–173, 167f
      effects assessment, 170–173
      extended spectral-response analysis, 173, 174f
      external optical elements, 169–170, 170f
      geometrical or regular structuring, 168–169, 169f
      implementation methods, 167–170
      random texturing, 168
      short-circuit current analysis, 170–172, 171f
      sub-band-gap reflection analysis, 172, 173f
    module production and performance
      encapsulation, 269
      field experience, 272
      laser scribing and cell interconnection, 267–269, 267f, 268f
      performance factors, 270–272, 271f

substrate materials and transparent contracts, 263–267, 263f, 265f
thin-film deposition, 259–262, 260f
p-i-n and n-i-p functioning, 234–252
  i layer, internal electric field formation, 238–240, 239f
  i layer, internal electric field reduction and deformation, 241
  internal electric field, 234–241, 236f, 237f
  light trapping, 249–252, 250f, 251f
  recombination and collection, 241–244, 242f, 243f, 244f
  series resistance problems, 248–249
  shunts, 244–247, 245f, 246f, 247f
research background, 162–164
results summary, 190–196, 191f
silicon deposition and crystal growth, 179–187, 180f
  atmospheric pressure chemical vapour deposition, 185
  high-temperature chemical vapour deposition, 182–186
  high-temperature deposition methods, 181–187
  liquid-phase epitaxy, 187, 188f
  low-pressure chemical vapour deposition, 186
  low-temperature chemical vapour deposition, 186–187
  melt-growth techniques, 181, 182f
  rapid thermal chemical vapour deposition, 186
  recrystallisation, 181–182, 183f, 184f, 185f
  substrate properties, 179–180
substrate thinning, 188–190, 189f, 189f, 190f
tandem and multijunction solar cells, 214–215
  general principles, 252–254, 252f, 254f
  hydrogenated silicon tandems, 254–255, 255f

microcrystalline amorphous tandems, 257–259, 257f
triple-junction amorphous cells, silicon-germanium alloys, 255–257, 256f
voltage enhancements, 174–178
minority-carrier recombination, 176–178, 176f, 177f, 178f, 179f
solar-powered products, 989
indoor products, 990–992, 996f
Thirty-percent efficiency strategy, interdigitated back-contact (IBC) silicon solar cells, 467–468
III-V semiconductors
  high-efficiency multijunction solar cells
    application fields and reference conditions, 419–421
    band-gap selection, 421–422, 422f
    band gaps vs. lattice constant, 424f
    bifacial epigrowth, 434
    characterisation, 425
    concentrator solar cell design, 425–426, 426f, 427f
    inverted metamorphic growth, 432–434, 433f
    lattice-matched triple-junction cells, germanium deposition, 429–430
    more than triple junctions, 436–439, 438f
    nanowire cells, 440
    optical beam splitting, 439–440
    quantum well solar cells, 430
    research background, 417–419, 418f, 427–440, 428f, 429f
    schematic structure, 418f
    silicon deposition, 434–436, 435f
    tunnel diodes, 424–425, 424f
    upright metamorphic growth, germanium deposition, 430–432, 432f
  single-junction III-V cell structure, 403–407
    AlGaAs-GaAs structures, 403–404, 405f

III-V semiconductors (*Continued*)
  GaAs-based cells, Ge substrate, 406–407
  internal Bragg reflector, 404–406
Three-phase inverters, DC to AC power conversion, photovoltaic system electronics, 707, 707f, 708f
Three-stage co-evaporation process, Cu(In, Ga)Se$_2$ thin-film solar cells, 334–335
Thyristors, DC to AC power conversion, photovoltaic system electronics, three-phase inverters, 707
Tiled roof construction, photovoltaic systems, 807, 808f
"Tiler's pattern," passivated emitter, rear locally (PERL) diffused cell design, 110
Time series estimation, daily global radiation, KT-based methods, 612–613
Time systems, solar-radiation climatology, geometric computation, 618–620
Tin dioxide
  thin-film photovoltaic modules, 288–289
  transparent conducting oxide films, 288–289
Titanium dioxide
  dye-sensitized cells, 497–499, 499f
    mesoporous working electrodes, 528–529
    operating principles, 499f
    small-modulation electron transport and recombination, 516
    Stark effect, 525–527, 526f
  thin film solar cells, materials and transparent contacts, 265–266
TMEC-cleaning, crystalline silicon solar cells, 135
Tracking arrays. *See also* Maximum power point tracking (MPPT)
  concentrator photovoltaic systems, 851–853, 852f
  module orientation, 652
  solar parks and farms, 950–951, 950f, 951f

Transfer techniques, silicon thin film solar cells, 164
Transient absorption spectroscopy (TAS), dye-sensitized cells, 522
Transient decay method, silicon wafers
  minority-carrier lifetime, 1015–1016
  surface recombination velocity, 1021
Translucency, photovoltaic technology aesthetics, 932, 934f
Transparent conducting oxide (TCO) layers
  cadmium telluride thin-film photovoltaic modules, 288–289
  film deposition, 288–289
  HIT cell development, 120–121
  silicon thin film solar cells, p-i-n/n-i-p structures
    internal electric field, 237–238
    light trapping, 250, 251f
    series resistance, 248
Transparent contacts, silicon thin film solar cells, 263–267
Transportable solar-powered products, 997–1001
Transportation systems, nonconsumer solar-powered products, 1003, 1004f
Transport equations, semiconductors, 40–41
Trapped radiation environment, space applications of solar cells, 867, 868f
Trigonometric determination, solar geometry, solar-radiation climatology estimation, 621–625, 622f
Triple-junction solar cells
  amorphous silicon-germanium alloys, 255–257, 256f
  lattice-matched III-V, germanium deposition, 429–430 429f
  micromorph tandem cells, 259
  space applications, monolithic multijunction (cascade) cells, 410–411
Tunnel diodes, multijunction III-V solar cells, 424–425, 424f

Tunnel junction, tandem and multijunction solar cells, 253, 254f
Type testing guidelines, grid-connected photovoltaic systems, 790−798, 798f

U

Ultimate efficiency, Shockley-Queisser estimates, 66
Ultraviolet radiation
  photovoltaic system reliability and qualification testing, 1060
  space solar cells, environmental testing, 893
Uniform emitter and base, p-n junction, solar cell, 23
Uniformity testing, space solar cells, electrical performance, 888
Upscaling, Cu(In,Ga)Se$_2$ thin-film solar cells, module fabrication, 343
User feedback, photovoltaic systems
  performance metrics, 978−982
    failure reports, 981−982
    Italian PV power plant, 980, 981f
    PV sound barrier, Switzerland, 979, 980f
    small residential systems, 978−979, 979f
    system monitoring, 1072, 1072f

V

Vacuum deposition, cadmium-telluride thin-film photovoltaic modules, sublimation and condensation, 286−287
Vacuum testing, space solar cells, 894
Valence band tail
  amorphous silicon thin films, gap states, 221
  hydrogenated microcrystalline silicon layers, 231
Validation, solar radiation, stochastic estimation, daily irradiation generation models, 613−614
Variable intensity method (VIM) analysis, silicon thin film solar cells, p-i-n/n-i-p structures, shunt problems, 246
Vehicle systems, solar-powered products, 997−999, 998f, 999f
Ventilation
  batteries in photovoltaic systems, 771−772
  solar-powered systems, 996−997, 996f
Vertical angle calculations, solar-radiation climatology, 625−627
Voltage enhancements
  organic solar cells, 554−555, 555f
  electrode optimisation, 555−556
  silicon thin film solar cells, 174−178
  minority-carrier recombination, 176−178, 176f, 177f, 178f, 179f
Voltage source inverters, DC to AC power conversion, photovoltaic system electronics, 705−706
Voltage specifications
  batteries in photovoltaic systems, 736
    lead-acid batteries, capacity dependence on, 748−749, 748f
    size selection and, 767
  electrical systems, photovoltaic structures
    maximum voltage, 821
    variations, designing for, 822
    voltage multiplication factor, 822
  grid-connected photovoltaic systems, passive technologies, 787−788
  silicon wafers, carrier lifetime and, 1030−1031

W

Wafering process
  crystalline silicon wafers, 94−96
  multijunction III-V solar cells, silicon deposition, 435−436
  silicon solar cells
    carrier lifetime and voltage characteristics, 1030−1031
    carrier recombination, 1012−1015
    lifetime instabilities and measurement artifacts, 1028−1030
    minority-carrier lifetimes, 1015−1030
    process monitoring and control, 1031−1040

Wafering process (*Continued*)
  research background, 375
Watches and clocks, solar-powered, 991, 991f
Watering conditions, crystalline silicon wafers, 95
Water loss, batteries in photovoltaic systems, 771
Water-pumped photovoltaic systems, batteries in, 725
Weatherproofing, photovoltaic system construction, 810, 814
Wide-gap chalcopyrites
  thin-film solar cells
    copper gallium diselenide cells, 357–358
    copper indium-aluminium deselenide cells, 358
    copper indium-gallium diselenide cells, 354–361, 356f
    copper indium-gallium selenium-sulphide cells, 359
    copper indium-gallium sulphide cells, 358–359
    copper indium sulphide cells, 358–359
    graded gap devices, 359–361, 360f
Window layer deposition, Cu(In,Ga)Se$_2$ thin-film solar cells, 340
Wireless technologies, nonconsumer solar-powered products, 1003
Wire sawing technology, crystalline silicon wafers, 94–95, 95f
Working electrodes, mesoporous oxide films, dye-sensitized cells, 528–529

X

X-ray Timing Explorer, space solar cell monitoring, orbital anomalies, 905
XRI concentrator, 850f

Z

Zinc blende structure, chalcopyrite lattice, 325–326, 325f
Zinc oxide
  copper indium-gallium diselenide thin-film solar cells
    module fabrication, 343
    short-circuit current losses, 349
    window layer deposition, 340
  silicon thin film solar cells, titanium dioxide deposition, 266
Zinc oxide-aluminium
  film deposition, 289
  thin-film photovoltaic modules, 289
Zone-melting recrystallization, silicon thin film solar cells, 181–182, 184f
Z parameter, silicon thin film solar cells, light trapping, 166, 167f

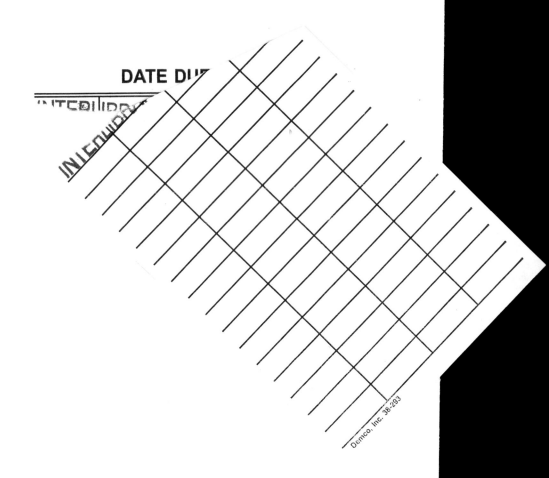